AF616408

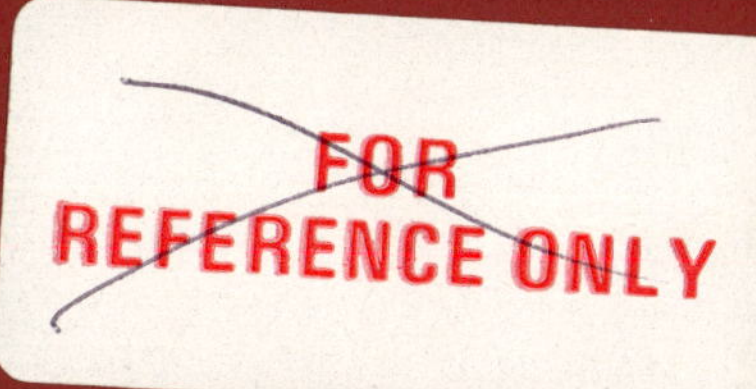
FOR
REFERENCE ONLY

CONCISE ENCYCLOPEDIA OF
MAGNETIC & SUPERCONDUCTING MATERIALS

ADVANCES IN MATERIALS SCIENCE AND ENGINEERING

This is a new series of Pergamon scientific reference works, each volume providing comprehensive, self-contained and up-to-date coverage of a selected area in the field of materials science and engineering. The series is being developed primarily from the highly acclaimed *Encyclopedia of Materials Science and Engineering*, published in 1986. Other titles in the series are listed below.

BEVER (ed.)
Concise Encyclopedia of Materials Economics, Policy & Management

BROOK (ed.)
Concise Encyclopedia of Advanced Ceramic Materials

CAHN & LIFSHIN (eds.)
Concise Encyclopedia of Materials Characterization

CARR & HERZ (eds.)
Concise Encyclopedia of Mineral Resources

CORISH (ed.)
Concise Encyclopedia of Polymer Processing & Applications

KELLY (ed.)
Concise Encyclopedia of Composite Materials

MAHAJAN & KIMERLING (eds.)
Concise Encyclopedia of Semiconducting Materials & Related Technologies

MOAVENZADEH (ed.)
Concise Encyclopedia of Building & Construction Materials

SCHNIEWIND (ed.)
Concise Encyclopedia of Wood & Wood-Based Materials

WILLIAMS (ed.)
Concise Encyclopedia of Medical & Dental Materials

NOTICE TO READERS

Dear Reader
If your library is not already a standing order/continuation order customer to the series **Advances in Materials Science and Engineering**, may we recommend that you place a standing/continuation order to receive immediately upon publication all new volumes. Should you find that these volumes no longer serve your needs, your order can be cancelled at any time without notice.

CONCISE ENCYCLOPEDIA OF

MAGNETIC & SUPERCONDUCTING MATERIALS

Editor

JAN EVETTS

University of Cambridge, UK

Executive Editor

ROBERT W CAHN FRS

University of Cambridge, UK

Senior Advisory Editor

MICHAEL B BEVER

MIT, Cambridge, MA, USA

PERGAMON PRESS

OXFORD • NEW YORK • SEOUL • TOKYO

UK — Pergamon Press Ltd, Headington Hill Hall, Oxford OX3 0BW, England

USA — Pergamon Press, Inc., 660 White Plains Road, Tarrytown, New York 10591-5153, USA

KOREA — Pergamon Press Korea, KPO Box 315, Seoul 110-603, Korea

JAPAN — Pergamon Press Japan, Tsunashima Building Annex, 3-20-12 Yushima, Bunkyo-ku, Tokyo 113, Japan

---

First edition 1992

Library of Congress Cataloging in Publication Data
Concise encyclopedia of magnetic & superconducting
materials / editor, J. E. Evetts. — 1st ed.
p. cm. — (Advances in materials science and engineering)
Includes index.
1. Magnetic materials—Encyclopedias. 2.
Superconductors—Encyclopedias. I. Evetts, J. II. Title:
Concise encyclopedia of magnetic and superconducting
materials. III. Series.
QC764.52.C66 1992
620.1'1297—dc20 91–37830

British Library Cataloguing in Publication Data
Evetts, J. (Jan)
Concise encyclopedia of magnetic and superconducting
materials. — (Advances in materials science and
engineering)
I. Title II. Series
537.622

ISBN 0–08–034722–3

(∞)™ The paper used in this publication meets the minimum requirements of the American National Standard or Information Sciences—Permanence of Paper for Printed Library Materials, ANSI Z39.48–1984.

Printed and bound in Great Britain by Butler & Tanner Ltd, Frome, Somerset

CONTENTS

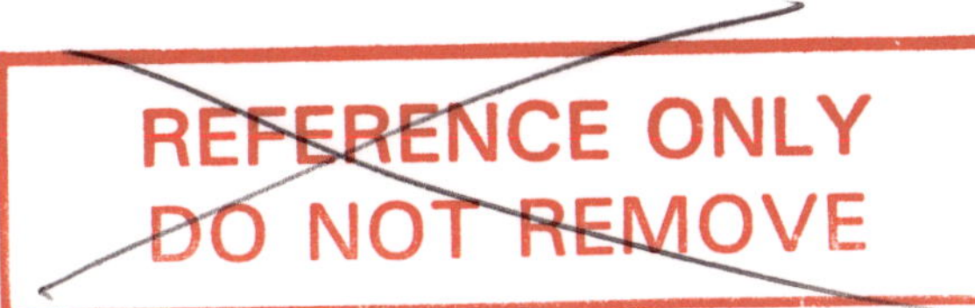

HONORARY EDITORIAL ADVISORY BOARD

FOREWORD

In the time since its publication, the *Encyclopedia of Materials Science and Engineering* has been accepted throughout the world as the standard reference about all aspects of materials. This is a well-deserved tribute to the scholarship and dedication of the Editor-in-Chief, Professor Michael Bever, the Subject Editors and the numerous contributors.

During its preparation, it soon became clear that change in some areas is so rapid that publication would have to be a continuing activity if the Encyclopedia were to retain its position as an authoritative and up-to-date systematic compilation of our knowledge and understanding of materials in all their diversity and complexity. Thus, the need for some form of supplementary publication was recognized at the outset. The Publisher has met this challenge most handsomely: both a continuing series of Supplementary Volumes to the main work and a number of smaller encyclopedias, each covering a selected area of materials science and engineering, are in the process of publication.

Professor Robert Cahn, the Executive Editor, was previously the editor of an important subject area of the main work and many other people associated with the Encyclopedia have contributed or will contribute to its Supplementary Volumes and derived Concise Encyclopedias. Thus, continuity of style and respect for the high standards set by the *Encyclopedia of Materials Science and Engineering* are assured. They have been joined by some new editors and contributors with knowledge and experience of important subject areas of particular interest at the present time. Thus, the Advisory Board is confident that the new publications will significantly add to the understanding of emerging topics wherever they may appear in the vast tapestry of knowledge about materials.

The appearance of the Supplementary Volumes and the series *Advances in Materials Science and Engineering* is an event which will be welcomed by scientists and engineers throughout the world. We are sure that it will add still more luster to a most important enterprise.

Walter S Owen
Chairman
Honorary Editorial Advisory Board

EXECUTIVE EDITOR'S PREFACE

As the publication of the *Encyclopedia of Materials Science and Engineering* approached, Pergamon Press resolved to build upon the immense volume of work which had gone into its creation by embarking on a follow-up project. This project had two components. The first was the creation of a series of Supplementary Volumes to the Encyclopedia itself. The second component of the new project was the creation of a series of Concise Encyclopedias on individual subject areas included in the Main Encyclopedia to be called *Advances in Materials Science and Engineering*.

These Concise Encyclopedias are intended, as their name implies, to be compact and relatively inexpensive volumes (typically 300–600 pages in length) based on the relevant articles in the Encyclopedia (revised where need be) together with some newly commissioned articles, including appropriate ones from the Supplementary Volumes. Some Concise Encyclopedias will offer combined treatments of two subject fields which were the responsibility of separate Subject Editors during the preparation of the parent Encyclopedia (e.g., dental and medical materials).

Eleven Concise Encyclopedias have been contracted. These and their editors are listed below.

Concise Encyclopedia of Advanced Ceramic Materials	Prof. Richard J Brook
Concise Encyclopedia of Building & Construction Materials	Prof. Fred Moavenzadeh
Concise Encyclopedia of Composite Materials	Prof. Anthony Kelly CBE, FRS
Concise Encyclopedia of Magnetic & Superconducting Materials	Dr Jan Evetts
Concise Encyclopedia of Materials Characterization	Prof. Robert W Cahn FRS & Dr Eric Lifshin
Concise Encyclopedia of Materials Economics, Policy & Management	Prof. Michael B Bever
Concise Encyclopedia of Medical & Dental Materials	Prof. David Williams
Concise Encyclopedia of Mineral Resources	Dr Donald D Carr & Prof. Norman Herz
Concise Encyclopedia of Polymer Processing & Applications	Mr Patrick J Corish
Concise Encyclopedia of Semiconducting Materials & Related Technologies	Prof. Subhash Mahajan & Prof. Lionel C Kimerling
Concise Encyclopedia of Wood & Wood-Based Materials	Prof. Arno P Schniewind

Most of the new or substantially revised articles in the Concise Encyclopedias have been or will be published in one of the three Supplementary Volumes, which are designed to be used in conjunction with the Main Encyclopedia. The Concise Encyclopedias themselves, however, are "free-standing" and are designed to be used without reference to the parent Encyclopedia.

The Executive Editor is personally responsible for the selection of topics and authors of articles for the Supplementary Volumes. In this task, he has the benefit of the advice of the Senior Advisory Editor and of other members of the Honorary Editorial Advisory Board. The Executive Editor is responsible for appointing the Editors of the various Concise Encyclopedias and for supervising the progress of these volumes.

Robert W Cahn FRS
Executive Editor

EDITOR'S PREFACE

The *Concise Encyclopedia of Magnetic & Superconducting Materials* draws together in a single volume articles that cover all major aspects of the science and technology of magnetic and superconducting materials. When I was asked by Professor Robert W. Cahn, the Executive Editor of the series, to edit this volume, I was immediately attracted by the challenge of producing a unified and balanced work that would enable readers to draw parallels and experience contrasts in the conceptual treatment and practical technology of these materials. Having myself worked with magnetic and superconducting materials since the 1960s, I have become acutely aware that these two important classes of material have much in common, both in the phenomenology that underlies their behavior and in many of the issues that relate to their fabrication and optimization as bulk materials or thin films for application in technology. I believe that the treatment of these distinct materials, each at their own characteristic stage of development, in a single volume brings a conceptual freshness to our appreciation of both classes of material that helps to highlight deficiencies in our overall knowledge base and corresponding opportunities for new advances. Although the emphasis is on materials science and technology, the subject matter extends from physics to engineering design and applications and it is expected that the *Concise Encyclopedia of Magnetic & Superconducting Materials* will be useful to engineers, teachers, managers and scientists from a wide range of disciplines and many areas of manufacturing industry.

Because of the remarkable advances of recent years, this Concise Encyclopedia is based mainly on newly commissioned articles; about a third of the entries are drawn from the *Encyclopedia of Materials Science and Engineering* which was published in 1986, although many authors of articles in rapidly developing fields have rewritten or extensively updated their articles for this volume. In the area of magnetic materials there are 30 updated articles and 32 articles that are essentially new, while in the area of superconductivity three articles have been updated and 52 articles have been newly commissioned. The articles have been selected to give comprehensive and balanced coverage to each of three broad subject areas: first, there are articles devoted to particular classes of material; second, there are those that describe and elucidate material properties and phenomena that are common to different materials; and, third, there are articles that treat the general area of applications dealing with issues of materials selection and fabrication as well as the design and principles of operation of devices for particular applications. Extensive coverage has been given to articles concerned with the phenomenology of technical magnetism and superconductivity since the effective characterization and optimization of magnetic and superconducting materials depends critically on an understanding of the mechanisms underlying their various properties.

I am grateful to all the authors who responded to my invitation to write for this Concise Encyclopedia; they are all recognized authorities in their fields and I respect the special efforts they made with their contributions to accommodate the guidelines on length, style and content that were necessary to achieve a balanced and comprehensive coverage of the subject. Although the work is not intended as a source book for data on materials, an effort has been made to include, at every point, sufficient data to equip the reader with a reasonably quantitative understanding of the subject. An important part of each article is the list of references; these are selective rather than comprehensive and provide the reader with avenues for further study. At a time when much public attention has been focused on developments in the field of high-temperature superconductivity, considerable effort has been made to retain objectivity in the selection of topics and coverage. The aim has been to produce a work that will have lasting value and relevance over many years. In practice, this has not been too difficult as there is no area of high-temperature super-conductivity that cannot be seen, on pragmatic inspection, to have roots that go back over many years and, sometimes, over decades. In such an extensive work with so many contributing authors there is, inevitably, the occasional overlap; where this occurs articles are cross-referenced to each other. Considerable attention has been given to harmonizing notation, equations and preferred units, although I have chosen not to force total uniformity where the common practice differs from one area of the subject to another. I have, however, insisted on the use of SI units as the primary unit throughout, although other units are sometimes given in parentheses.

I would like to acknowledge the encouragement and guidance of Professor Robert Cahn, the Executive Editor, and to thank Mr Michael Mabe and Mr Peter Frank of Pergamon Press for their support and advice. I am also particularly indebted to Mr Paul Aslin for his assistance in maintaining day-to-day contact with authors and in coordinating and assembling the manuscript copy. Finally, it is with sadness and regret that I acknowledge the valued contributions of two authors who are now deceased.

Professor John Bardeen, who inspired so many through his contributions to the theory of superconductivity over 35 years, completed a balanced and carefully worked article *Superconducting Materials: BCS and Phenomenological Theories* in January 1991, a few days before he died. Dr Gilbert Y. Chin, who recently retired as director of the Passive Components Research Laboratory at AT&T Bell Laboratories, passed away in May 1991. Dr Chin contributed the key introductory article *Magnetic Materials: An Overview* and coauthored with his friend and colleague Dr Jack Wernick a further ten articles on specialist magnetic materials.

Jan Evetts
Editor

GUIDE TO USE OF THE ENCYCLOPEDIA

This Concise Encyclopedia is a comprehensive reference work covering all aspects of magnetic and superconducting materials. Information is presented in a series of alphabetically arranged articles which deal concisely with individual topics in a self-contained manner. This guide outlines the main features and organization of the Encyclopedia, and is intended to help the reader to locate the maximum amount of information on a given topic.

Accessibility of material is of vital importance in a reference work of this kind and article titles have therefore been selected not only on the basis of article content but also with the most probable needs of the reader in mind. An alphabetical list of all the articles contained in this Encyclopedia is to be found on pp. xv and xvi.

Articles are linked by an extensive cross-referencing system. Cross-references to other articles in the Encyclopedia are of two types: in text and end of text. Those in the body of the text are designed to refer the reader to articles that present in greater detail material on the specific topic under discussion at that point. They generally take one of the following forms:

> ...which is fully described in the article *Superconducting Materials: Basic Theoretical Concepts.*
>
> ...other applications of magnetic materials (see *Magnetic Storage Media*).

The cross-references listed at the end of an article serve to identify broad background reading and to direct the reader to articles that cover different aspects of the same topic.

The nature of an encyclopedia demands a higher degree of uniformity in terminology and notation than many other scientific works. The widespread use of the SI system of units has determined that such units be used in this Encyclopedia. It has been recognized, however, that in some fields Imperial units are more generally used. Where this is the case, Imperial units are given with their SI equivalent quantity and unit following in parentheses. Where possible the symbols defined in *Quantities, Units, and Symbols*, published by the Royal Society of London, have been used.

All articles in the Encyclopedia include a bibliography giving sources of further information. Each bibliography consists of general items for further reading and/or references which cover specific aspects of the text. Where appropriate, authors are cited in the text using a name/date system as follows:

> ...as was recently reported (Smith 1988).
>
> Jones (1984) describes...

The contributor's name and the organization to which they are affiliated appear at the end of each article. All contributors can be found in the alphabetical List of Contributors, along with their full postal address and the titles of the articles of which they are authors or coauthors.

The article *An Introduction to Magnetic and Superconducting Materials* provides an overview of the subject area and discusses in brief the issues covered in detail by the articles in the body of the work.

The most important information source for locating a particular topic in the Encyclopedia is the multilevel Subject Index, which has been made as complete and fully self-consistent as possible.

ALPHABETICAL LIST OF ARTICLES

AN INTRODUCTION TO MAGNETIC AND SUPERCONDUCTING MATERIALS

by Jan Evetts

The essential characteristics of magnetic order or superconducting order are only seen on cooling below a critical ordering temperature that differs from material to material. In both magnetic and superconducting materials the electronic structure becomes remarkably organized with the appearance of long-range coherence and there is an accompanying change in the specific heat. Although there are other classes of ordered material, magnetic and superconducting materials are of special significance because they relate so directly to the root and branch of electromagnetism, magnetic induction and electric current. It is for this reason that they pervade every avenue of the technological world from microelectronics and mass-data storage to medicine and heavy engineering. This Concise Encyclopedia draws together for the first time in a single volume the remarkable advances seen in recent years in the discovery of new materials and the improvement of existing materials, in the elucidation of a new and deeper understanding of their phenomenology and in the expansion and diversification of their application in technology.

Magnetic and superconducting materials have much in common. They have both experienced, during the 1980s, a revolution that has revitalized and galvanized their respective fields. In each case the impetus has come from the discovery of new materials—the high-critical-temperature oxide superconductors and the ultrahard and ultrasoft ferromagnetic materials based on rare-earth intermetallics and transition metal–metalloid metallic glasses, respectively. These discoveries have, in turn, stimulated a reevaluation of a wide range of basic mechanisms and phenomena. Effects that were of marginal significance have assumed major importance for applications and have also led to a deeper understanding of basic science. Examples are giant magnetoresistance in magnetic multilayers and flux creep in high-temperature superconductors. Not infrequently, a new discovery for one class of material is familiar ground for the other. Thus, current interest in the possibility of quantum mechanical tunnelling of flux lines in superconductors is a problem that was investigated in the case of domain walls in magnetic rare earths and rare-earth compounds in the early 1970s.

The strongest unifying aspect of magnetic and superconducting materials, one that sets them apart from other classes of material, is the complex magnetic substructure that they both exhibit. Most of the static and dynamic properties of bulk materials and many of the properties of thin films and devices depend on the response of the flux structure to generalized forces. This is the basis of a rich phenomenology where the parallels between the two classes of materials are striking and often lead to new insights.

The articles in the Encyclopedia can be viewed as falling broadly into three categories: articles on material properties and phenomena, articles on classes of material and articles on devices and applications.

1. Material Properties and Phenomena

The properties of a material may be roughly divided into *intrinsic* or primary properties characteristic of the hypothetical "ideal" material and *extrinsic* properties that depend on the microstructure or defect structure of the material. Examples of intrinsic properties are the critical ordering temperature T_c, the saturation magnetic moment of a magnetic material M_s and the upper critical field for a superconductor H_{c2}.

Extrinsic properties may vary by many orders of magnitude as the defect structure changes. Examples are the coercive field H_c in a magnetic material and the critical current density J_c in a superconductor which can both vary by up to 7–8 orders of magnitude. As always, such a division, on closer inspection, is not so straightforward. Thus, in certain materials, generally those with very high anisotropy, there can be intrinsic contributions to properties commonly regarded as extrinsic. For example, H_c and J_c in principle go to zero in an ideal defect-free magnetic material or type II superconductor; in practice, in rare-earth intermetallic materials, the narrow domain walls may be pinned at the atomic planes and, in anisotropic layer-structure superconductors, the flux lines can be pinned at the individual layers. More ambiguous still is the treatment of heterogeneous or inhomogeneous materials. In certain limits strongly heterogeneous materials may be treated as homogeneous "averaged" structures; examples are nanocrystalline soft magnetic materials and one class of granular superconductors. Extreme examples of materials that derive novel bulk or averaged properties by virtue of their nanoscale heterogeneity are magnetic or superconducting multilayer thin films or heterostructures, currently an active area of research.

The articles that broadly deal with material properties and phenomena are listed in Table 1. The classification is loose and includes articles on intrinsic properties and extrinsic properties as well as articles on measurements, important phenomena and theoretical background. Very often it is appropriate to have corresponding articles in the areas of magnetism and superconductivity; thus, material anisotropy is important for each class of material. A few years ago anisotropy in superconductors was a specialist topic of little technological relevance; however, the properties of the new class of oxide layer-structure superconductors depend crucially on their anisotropy. Other topics that are common are, for example, optical properties, high-frequency properties, irradiation effects and articles on magnetic substructure. On the other hand, there are topics which are only of sufficient importance to merit an article in the case of one class of material. Thus, aspects of magnetoelastic phenomena and magnetostriction pervade magnetism, while, although the volume and elastic moduli of a material change when it becomes superconducting, the effects are small and their relevance is limited to the pinning of flux lines by lattice defects and dislocations in metallic superconductors.

Different aspects of the domain structure in magnetic materials and the magnetic flux structure in superconductors are the subject of a number of articles. Domain walls and flux lines are distinct entities that can move in response to generalized forces and form the basis of a rich phenomenology that is the key to the understanding of a broad range of static and dynamic properties of magnetic and superconducting materials. Historically in the development of the subject, the flux line has been seen as the simpler and the domain wall the more complex entity; this has determined the generality of models and mechanisms developed. The flux lines in a type II superconductor are thought of as a collection of identical current vortices, each containing one flux quantum, that repel each other and, in the absence of microstructural defects, usually form a stable hexagonal lattice. In contrast, there can be domain walls of several different types within one material and they may have a complex internal structure and a variable defect structure that changes both the domain wall self-energy and its dynamic characteristics. Recently, increasing interest in strongly anisotropic layer-structure superconductors, often with weak coupling between the layers, has altered this view; in a certain sense, the parallels between magnetic materials and superconductors have become stronger with the increasing complexity of the flux line and the variability of its energy and structure within one material.

The most powerful unifying concept is that of "hardenability"; both domain walls and flux lines can be pinned by microstructural defects leading directly to "hard" magnetic materials and hard superconductors with high H_c and J_c, respectively. When the pinning forces are low, magnetic materials are said to be "soft" while superconductors are "reversible." Although the immediate analogy is lost, thermodynamic reversibility corresponds to a lack of hysteresis, an essential feature of a soft material. Concomitant with the idea of pinning is that of a pinning well and, at temperatures above absolute zero, the idea of thermally activated depinning, variously termed creep, reptation, thermal relaxation or after-effects. An issue of current interest and controversy for flux line pinning and the critical current in high-temperature oxide superconductors is the

Table 1
Articles on material properties and phenomena

Magnetic materials	Superconducting materials
Anisotropy in Magnetic Materials	AC Losses in Superconducting Materials
Barkhausen Effect	Anisotropy in Superconducting Materials
Coercivity and Domain Wall Pinning	Coherence Length, Proximity Effect and Fluctuations
Demagnetizing Factors	Critical Current Density
Domains and Domain Walls	Critical State Model
High-Frequency Properties of Magnetic Materials	Flux Creep Phenomena
Magnetic Materials: Basic Concepts	Flux Line Lattice
Magnetization Imaging	Flux Pinning and the Summation of Pinning Forces
Magnetocaloric Effect	High-Frequency Impedance of Superconducting Materials
Magnetoelastic Phenomena	Josephson Effects in Weak Links
Magnetooptics	Magnetic Properties of Superconductors
Magnetotransport Phenomena	Measurements in Superconducting Materials
Measurements in Magnetic Materials	Nonequilibrium Superconductivity
Micromagnetic Calculations	Optical Properties of Superconductors
Power Loss in Magnetic Materials	Radiation Effects on Superconducting Materials
Radiation Effects on Magnetic Materials	Resistive Transition and Flux Flow in Superconducting Materials
	Superconducting Materials: Basic Theoretical Concepts
	Superconducting Materials: BCS and Phenomenological Theories
	Superconducting Materials: Physical Properties
	Superconducting Materials Under Pressure
	Transport Currents and Force-Free Current Flow
	Tunnelling Spectroscopy

mechanism for the occurrence of an "irreversibility line" $H_{irr}(T)$ that lies in the magnetic field–temperature plane, such that for sufficiently high applied magnetic fields or temperatures a material becomes completely reversible, with zero critical current density. The argument turns on whether the irreversibility line is determined solely by thermally activated depinning or depends also on temperature- and field-dependent modifications to the magnetic flux line structure (e.g., melting or thermally activated flux line cutting and crossjoining). The controversy is by no means academic since its resolution will determine the extent to which the irreversibility line is an intrinsic feature of a material or whether it can be shifted to higher magnetic fields and temperatures by optimizing the defect structure of the material. Although thermally activated reversibility is a novel concept for superconducting materials, it was the subject of detailed experimental and theoretical study for domain walls in shallow pinning wells in the early 1970s. At the same time, the issue of quantum mechanical tunnelling of narrow domain walls was treated as a natural response to the observation that creep did not seem to reduce with temperature quite as expected on a simple picture of thermal activation.

A second area of rich correspondence relates to the motion or flow of domain walls and flux lines. Major contributions to the important concepts of the motion of domain walls in conservative and nonconservative pinning fields were made in the 1950s by C. P. Bean at the General Electric Research Laboratory. Although formulated for domain walls, these ideas are just as sound and relevant for flux lines moving in a pinning field and also relate directly to phase slip in a weak superconducting link. Within a few months of the discovery of type II superconductivity in the early 1960s, this experience with domain wall motion and pinning was translated and iterated into the now famous Bean critical state model. The rigorous treatment of the motion of domain walls and flux lines can be expressed phenomenologically using the formalism of irreversible thermodynamics and Onsager reciprocity. The analogies of crosseffects such as the Nernst effect and Peltier effect arise naturally. For flux lines the concept of a thermodynamic flux line driving force also emerges from the formalism, together with an interesting conceptual division of the total electric current into a transport current and a magnetization current (a division that B. D. Josephson has shown meaningful even for conduction electrons in a simple normal metal such as copper).

The main emphasis of the Encyclopedia is on issues of materials science and engineering; however, articles outlining the basic concepts of magnetism and superconductivity are included as a necessary background. A full microscopic theoretical treatment of most of the complex phenomena involving domain walls or flux lines is not possible. There are, however, powerful phenomenological theories that stem from the fact that both the ferromagnetic and superconducting states are ordered states that derive from an order–disorder transition at a critical ordering temperature. The powerful general theory introduced by L. D. Landau and V. L. Ginzburg, treating ordering in terms of an order parameter (which, for superconductors, is complex with both amplitude and phase) that varies with temperature and also spatially, was applied with extraordinary success to both superconducting and magnetic materials and is known as the Ginzburg–Landau theory for superconductors and the Landau–Ginzburg theory when applied to magnetic ordering phenomena.

A more microscopic approach is possible when treating some of the intrinsic properties of magnetic and superconducting materials. Examples are magnetotransport properties in magnetic materials, magnetic exchange coupling in magnetic multilayers, and nonequilibrium superconductivity and tunnelling spectroscopy in superconductors. These are active fields with particular relevance and application for multilayer structures and devices. This is also an area where magnetism and superconductivity occasionally make real contact rather than contact by analogy as, for instance, with the study of spin-polarized tunnelling into a superconductor or the properties of superconductor–insulator–superconductor (SIS) tunnel junctions with magnetic tunnel barriers.

2. Magnetic and Superconducting Materials

Since the late 1970s, a series of remarkable discoveries has transformed the way scientists and engineers view magnetic and superconducting materials. This is illustrated in Fig. 1 where trends in the critical temperature T_c for superconducting materials and the maximum energy product $(BH)_{max}$ for magnetic materials are displayed. It is clear that these properties have changed both more extensively and in a shorter period of time than anyone was able to predict; however, much more is apparent. First, the changes are seen to have occurred in new classes of material, thus bringing new challenges in material processing, new issues in engineering design and even problems in the sourcing of appropriate raw materials. Second, the field of materials has broadened remarkably; both magnetic and superconducting materials now include within their ranks metals, alloys, intermetallics, metallic glasses, oxides, nitrides and carbides and, most recently, organic materials, with unexpected advances in T_c for the fullerenes and in the first room-temperature value for $(BH)_{max}$ in an organic ferromagnet. Apart from T_c and $(BH)_{max}$, many other properties show similarly dramatic changes and are treated in turn in different articles in this volume; notable examples are seen in improvements in soft magnetic materials and in the critical current density of superconductors, in the almost twofold increase in Curie temperature T_C

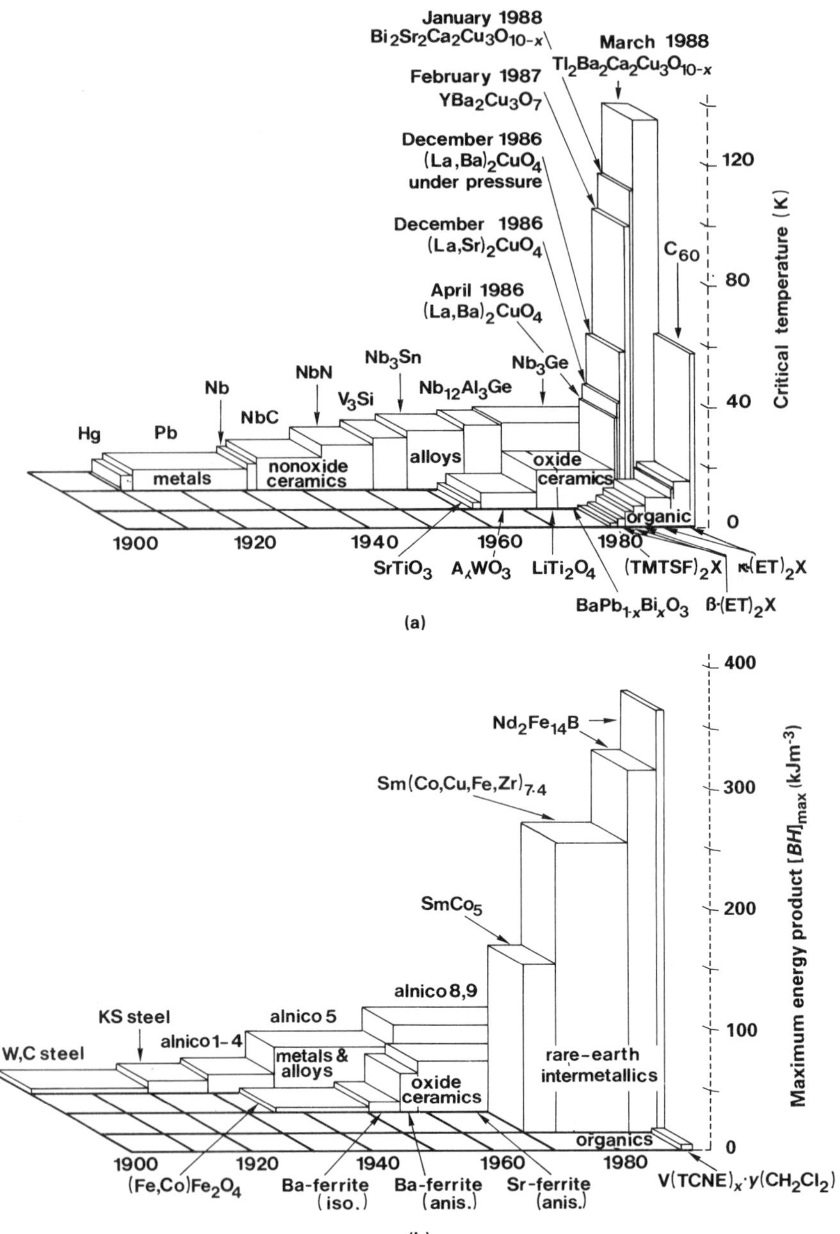

Figure 1
(a) Trends in the critical ordering temperature T_c for different classes of superconductor; (b) trends in the maximum energy product $(BH)_{max}$ at 300 K for different classes of hard magnetic material

in interstitial nitrides and carbides based on certain rare-earth intermetallics and in new high-performance magnetocaloric materials for magnetic refrigeration.

Articles on specific classes of magnetic or superconducting material are listed in Table 2. Such a list cannot be exhaustive; the selection has been made subjectively with emphasis on materials with intrinsic interest, novelty or particular relevance for applications. Materials are sometimes more conveniently defined by function rather than by structural class, such as magnetostrictive materials or magnetooptic

Table 2
Articles on classes of materials

Magnetic materials	Superconducting materials
Alnicos	A15 Compound Superconductors
Cobalt–Platinum Alloys	Amorphous Superconductors
Cobalt–Rare-Earth Compounds	Chevrel Phases
Ferrites, Hard	Granular Superconductors
Ferrites, Soft	Heavy Fermion Metals
Iron–Aluminum and Iron–Aluminum–Silicon Alloys	Niobium–Titanium Alloys
Iron and Low-Carbon Steels for Magnets	Nitride and Carbonitride Superconductors
Iron–Chromium–Cobalt Alloys	Organic Superconductors: Anisotropic Molecular Metals
Iron–Cobalt Alloys	Organic Superconductors: Doped C_{60} (Fullerene)
Magnetic Alloys: Intermediate Coercivity Materials	Oxide Superconductors: Ceramic Processing
Magnetic Materials: An Overview	Oxide Superconductors: Electron Microscopy
Magnetooptic Materials	Oxide Superconductors: Physical Properties
Magnetostrictive Materials	Oxide Superconductors: Solid-State Chemistry
Manganese–Aluminum–Carbon Alloys	Oxide Superconductors: Thin-Film Deposition
Metallic Glasses	Superconducting Materials: An Overview
Nanocrystalline Soft Magnetic Materials	Superconducting Materials: Elements
Neodymium–Iron–Boron Alloys	
Nickel–Iron Alloys	
Organic and Molecular Ferromagnets	
Oxide Glasses as Magnetic Materials	
Paramagnetic Materials	
Rare-Earth Intermetallics	
Rare-Earth Metals and Alloys	
Silicon Steels	
Steels for Magnets	

materials. This, inevitably, leads to overlap and some repetition. Materials definition by function touches on more applied subjects, coming closer to the areas of materials processing and raising issues of engineering design for specific applications. Indeed, much of the progress over the years has been through optimization of existing materials rather than discovery of new materials. In the excitement generated by a new discovery, the important fact is often lost that the act of discovery is only the first step towards an engineering material. The stages of materials assessment, development and optimization to a particular specification for a chosen market are equally challenging and rewarding activities undertaken by materials scientists and engineers.

3. *Devices and Applications*

Articles on particular devices and applications are listed in Table 3. Although both magnetic and superconducting materials find diverse application and have remarkable future potential, the stage of technological development and market penetration for the two classes of material is totally different. The magnetics industry is mature, effectively in its third generation and serves a multibillion dollar market worldwide that has seen strong growth decade by decade. There is no doubt that remarkable changes have taken place since the early 1970s but this has mainly been by the improvement of material already in the marketplace or by the substitution of one material by another in an existing market. Against this background the many novel applications for magnetic materials find easy and unobtrusive acceptance. The superconductivity industry is, by contrast, in its infancy; its markets are uncertain and it has yet to find its way into the vernacular of the engineering world. It depends for its application on the cryogenics industry which is itself developing rapidly, with new and improved materials changing the cost, efficiency and complexity of cryogenic systems. The likely impact of the new high-T_c superconductors is difficult to judge so soon after their discovery in 1986. As with most advanced materials, the speed with which they can be translated into bulk products or devices in the marketplace depends, largely, on innovation and the development of skills in materials processing. There are already niche applications for thin-film materials in microwave components and sensors and bulk materials are at the demonstrator stage (e.g., small motors and solenoids). The intensity and quality of the research effort in this area makes it likely that, in the near future, materials will be able to be fabricated reproducibly to a sufficiently attractive specification for major applications to become feasible. However,

Table 3
Articles on devices and applications

Magnetic devices	Superconducting devices
Bubble Domain Materials	AC Applications of Superconducting Materials
Design with Magnetic Materials	Biomagnetic Measurements
Design with Magnetic Materials: Computer Simulation	Magnetic Energy Storage
Electromagnetic–Acoustic Transducers	Magnetic Separation
Electromagnetic Forming	Microwave Applications of Superconducting Materials
Elongated-Single-Domain Magnets	Millimeter-Wave Detection
Fluids, Magnetic	Multifilamentary Superconducting Composites
Inks, Magnetic	Nuclear Magnetic Resonance Imaging
Magnet Wire	SQUID Design, Fabrication and Application
Magnetic Storage Media	Superconducting Cavities
Magnetic Units and Material Specification	Superconducting Device Fabrication
Magnetooptic Storage Media	Superconducting Devices and Electronics
Microwave Materials and Devices, Magnetic	Superconducting Magnets
Nondestructive Evaluation Using Magnetism	Superconducting Materials: Large-Scale Applications
Powder Cores, Magnetic	Superconducting Materials: Specification
Recording Head Technology	Thin Films, Multilayers and Devices, Superconducting
Reed Contact Materials	Voltage-Standard Devices
Sensors and Devices, Magnetic	
Shields, Magnetic	
Temperature Compensation Materials	
Thin Films and Multilayers, Magnetic	

there are other factors that determine the scale of application as is seen in the case of A15 compound superconducting multifilamentary wires that were developed to impressive maturity in the mid-1980s but have found only a limited market.

Generally, the gap between the discovery of materials and their application can be seen to be narrowing. In part, this must be because of the increasing collective experience of dealing with advanced materials of every type, whether metallic, ceramic, organic or composite. A further factor accelerating the application of magnetic and superconducting materials is the current trend to support focused research by initiatives or programs on a national or international scale. Examples are seen in the Concerted European Action on Magnets (CEAM) program, the National Banner Program to coordinate superconducting power system development in the USA and the Research and Development Association for Future Electron Devices (FED) in Japan.

The prediction of future advances in the discovery and application of materials has always been hazardous. It is indeed sobering to reflect that in the mid-1960s "metallic glass melt-spun ribbon" would have been the source of wonder and consternation—wonder at the control and reproducibility achievable in a totally novel fabrication process and at the remarkable combination of extremes of mechanical hardness and magnetic softness, and consternation because it was confidently considered at that time that the amorphous state could not be a source of ferromagnetism. At about the same time, much energy was also being expended on arguments as to whether the limiting maximum temperature for superconductivity was 25 K or 35 K. Whatever the eventual prospects for applications, it is certain that magnetic and superconducting materials have a fascinating quality and charm beyond due measure that will always engage the enthusiastic attention of scientist, engineer and layperson alike.

A15 Compound Superconductors

Before the discovery of the high-T_c oxide superconductors by Bednorz and Müller (1986), only compounds forming in the A15 type structure were known to exhibit T_c values above 20 K. Compounds of this type are characterized by the formula A_3B, where A is always a transition metal, while B can be either a transition or a nontransition metal. The interest in A15 compounds began with the discovery of V_3Si ($T_c = 17$ K) and Nb_3Sn ($T_c = 18$ K) by Hardy and Hulm (1953), the highest reported value in these compounds being obtained by Gavaler et al. (1974), who observed the onset of superconducting transitions above 23 K in sputtered Nb_3Ge films. By 1991, 70 binary A15 compounds had been found; some of them are metastable and must be produced under nonequilibrium conditions (with respect to room temperature), that is by sputtering, chemical vapor deposition (CVD), high pressure or fast quenching. In some cases, for example $Nb_3Al_{0.8}Ge_{0.2}$, A15 compounds are stabilized by ternary additions (pseudobinaries), but no case of a real ternary A15 compound has ever been reported.

Interest in A15 compounds resides not only in their unusual physical properties, but also in their technological importance; these type II superconductors have very high upper critical fields ($H_{c2}(0) = 28$ T for Nb_3Sn) and have a high current carrying capacity—two necessary conditions for the production of high magnetic fields. This interest is not limited to the widely used laboratory magnets with fields up to 20.7 T, but is also directed towards future large-scale applications, as for example fusion magnets. It should be mentioned here that these important technological achievements could not have been realized without considerable knowledge of the physical metallurgy of these complex systems.

The T_c values of the superconducting A15 compounds are listed in Table 1. It is apparent that the A15 crystal structure alone does not lead to the unusual electronic properties necessary for a high critical temperature. Favorable conditions for superconductivity are expected in narrow band systems; such a situation is encountered in niobium-based and to a lesser degree in vanadium-based A15 compounds, provided that the atom B in these cases is a nontransition element, such as aluminum, gallium, silicon, germanium or tin. A notable feature of A15 type compounds is the coincidence of optimum superconducting properties and structural instability, experimentally observed in the compounds Nb_3Sn and V_3Si. It has been shown theoretically that the martensitic cubic-tetragonal phase transformation at low temperatures in both compounds has its origins in the particular electronic structure.

1. The Crystal Structure

The ordered phase of the Cr_3Si type, called A15 within the nomenclature of the "Strukturbericht," belongs to the space group *Pm*3*n*. At the stoichiometric composition, the A and B atoms are located on the two lattice sites 6*c* and 2*a*, respectively, of the unit cell represented in Fig. 1. The B atoms form a body-centered-cubic (bcc) sublattice, while the A atoms are located

Table 1
Superconducting transition temperatures of binary A15 phases

Nontransition elements	T_c (K)	Transition elements	T_c (K)
Ti_3Sb	6.5	Ti_3Ir	4.2
		Ti_3Pt	0.5
$Zr_{80}Sn_{20}$[a]	0.92		
Zr–Pb	0.76	Zr_3Au	0.9
$Zr_{\sim 3}Bi$[b]	3.4		
		$V_{29}Re_{71}$	8.4
V–Al[c]	14	$V_{50}Os_{50}$	5.7
V_3Ga	15.9	$V_{65}Rh_{35}$	$\simeq 1$
V_3Si	17.0	$V_{63}Ir_{37}$	1.7
$V_{\sim 3}Ge$	6	$V_{\sim 3}Pd$	0.08
$V_{\sim 3}Ge$[c]	11	V_3Pt	3.7
$V_{\sim 79}Sn_{\sim 21}$	3.8	$V_{76}Au_{24}$	3
$V_{77}As_{23}$	0.2	$Nb_{75}Os_{25}$	1.0
$V_{76}Sb_{24}$	0.8	$Nb_{75}Rh_{25}$	2.6
		$Nb_{72}Ir_{28}$	3.2
Nb_3Al	19.1		
		Nb_3Pt	11
Nb_3Ga	20.7		
		$Nb_{\sim 3}Au$	11.5
$Nb_{\sim 3}In$[b]	9.2		
$Nb_{82}Si_{18}$[a]	4.4	$Ta_{85}Pt_{15}$	0.4
		$Ta_{\sim 80}Au_{20}$	0.55
Nb–Si[c]	11–17		
Nb–Ge[a]	17	$Cr_{72}Ru_{28}$	3.4
Nb–Ge[c]	23	$Cr_{73}Os_{27}$	4.7
Nb_3Sn	18	$Cr_{78}Rh_{22}$	0.07
Nb–Sb	2	$Cr_{82}Ir_{18}$	0.75
$Nb_{\sim 3}Bi$[b]	3		
		$Mo_{40}Tc_{60}$	13.4
$Ta_{\sim 3}Ge$[c]	8	$Mo_{\sim 65}Re_{\sim 35}$[c]	$\simeq 15$
$Ta_{\sim 3}Sn$	8.3	$Mo_{75}Os_{25}$	13.1
$Ta_{\sim 3}Sb$	0.7	$Mo_{78}Ir_{22}$	8.5
		$Mo_{82}Pt_{18}$	4.6
Mo_3Al	0.58		
Mo_3Ga	0.76	$W_{\sim 60}Re_{\sim 40}$[c]	11
$Mo_{77}Si_{23}$	1.7		
$Mo_{77}Ge_{23}$	1.8		

a Rapid quenching b High-pressure synthesis c Film deposition techniques

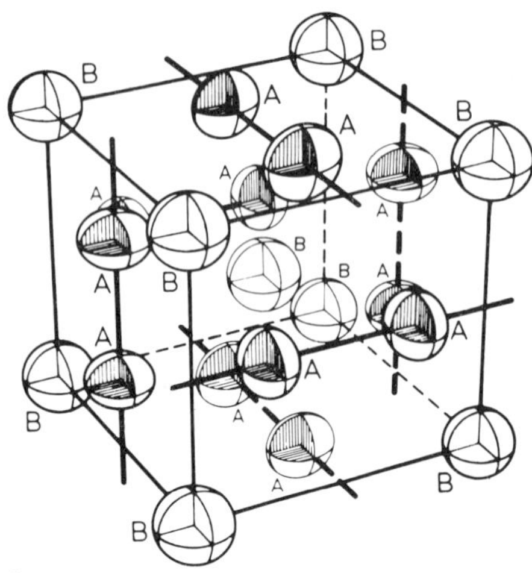

Figure 1
The A15 type structure A_3B, space group *Pm3n*

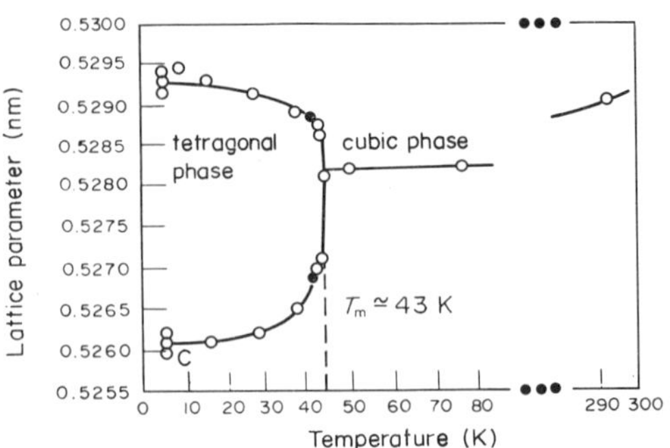

Figure 2
Structural instability in Nb_3Sn

on the faces and form three families of orthogonal chains in the ⟨100⟩ directions throughout the crystal. Each A atom has two nearest neighbors at the distance $AA_1 = a/2$ (a is the lattice parameter), and is coordinated by 4 B atoms at $AB = \sqrt{5}a/4$ and by 8 A atoms separated by $AA_2 = \sqrt{6}a/4$. The coordination number is thus 2, 6 or 14, depending on the number of considered neighbors. The B atoms have 12 nearest neighbors at the distance $AB = \sqrt{5}a/4$, the coordination number thus being 12. Along the chains, the nearest neighbor distance AA_1 is unusually short: as an example, the distance between niobium atoms in Nb_3Ge is 0.2575 nm, which is 9.9% shorter than the distance between niobium atoms in the elementary niobium unit cell, 0.2858 nm. The resulting anisotropy is schematically depicted in Fig. 1. The lattice parameter of an A15 compound A_3B with fixed element A (or B) increases linearly with the Goldschmidt radius of the element B (or A). This correlation is represented in the so-called Nevitt plot.

2. The Low-Temperature Structural Instability of V_3Si and Nb_3Sn

A structural instability at low temperatures has so far only been observed in the compounds V_3Si and Nb_3Sn. The fact that this transition was not observed in other compounds with high electronic density of states (e.g., V_3Ga, Nb_3Al, etc.) apparently restricts the validity of a correlation between the occurrence of structural instability and a particular electronic structure. However, it was proposed by Flükiger et al. (1981) that the occurrence of the low-temperature cubic-tetragonal phase transition in A15 type compounds is strongly correlated to the particular conditions of stability at high temperature: V_3Si and Nb_3Sn are the only A15 type compounds simultaneously fulfilling the three following requirements:

(a) high electronic density of states,

(b) stability at the precise stoichiometric composition, and

(c) the occurrence of perfect long-range atomic ordering.

Due to the low energy involved in the low-temperature phase transition, it is seen that even a small deviation from these three conditions can lead to the suppression of this phenomenon, while other physical properties (e.g., phonon softening, temperature variation of the magnetic susceptibility, etc) will not be fundamentally affected. All other compounds with T_c values above 15 K show small, but significant deviations from these conditions, thus preventing the phase transition.

The variation of the lattice parameter at $T_M = 43$ K, the temperature of the cubic-tetragonal martensitic transition in Nb_3Sn, is represented in Fig. 2. Thus, The observed distortion essentially corresponds to a variation of the distance between niobium atoms along the ⟨100⟩ chains. In the framework of the Landau theory, a sudden deformation is predicted at $T = T_M$, corresponding to a first-order transition. This is experimentally confirmed for Nb_3Sn, while the transition in V_3Si cannot be described by first-order criteria. Precursors of the tetragonal distortion are observed in a softening of the acoustic phonons at temperatures above T_M. The considerable decrease of the elastic constants with decreasing temperature in both systems, Nb_3Sn and V_3Si, is in contrast to the usual thermal behavior of metals.

3. Theoretical Descriptions

3.1 The Eliashberg Theory

Conventional superconductors can be described by means of the Eliashberg theory, which reduces to the BCS theory if the electron–phonon interaction is

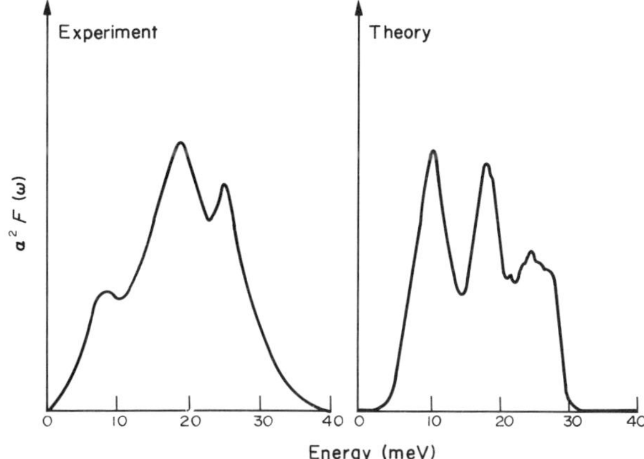

Figure 3
Measured and calculated Eliashberg function for Nb_3Sn

weak (see *Superconducting Materials: BCS and Phenomenological Theories*). The value of T_c is estimated using two parameters: μ^*, the coulomb pseudopotential and λ, the electron–phonon interaction parameter. Experimentally, μ^* can be determined by means of tunnel spectroscopy, while λ is calculated from specific heat data (Junod et al. 1983). Theoretically, the so called Eliashberg function $\alpha^2(\omega)F(\omega)$ is of particular importance. This function is determined by means of the phonon dispersion curves, the line width of the latter being influenced by the interaction with the electrons (see *Tunnelling Spectroscopy*). An example of the difference between calculated and experimentally determined Eliashberg functions is given by Fig. 3, which illustrates the case of Nb_3Sn. Based on the Eliashberg theory, a high value of T_c for a given compound is expected for a high electronic density of states $N(E_F)$ at the Fermi level E_F, and a low average phonon frequency. In particular, the last property explains the phonon softening usually observed in A15 compounds at temperatures above T_c.

3.2 The Linear (or One-Dimensional) Model

A very simple model explaining both the occurrence of high-T_c values and of structural instabilities in A15 type compounds is the linear model of Labbé and Friedel (1966a, b). This model assumes that the A atoms on the orthogonal chains have a decisive influence on the superconducting properties. It only takes into account the d electrons, the role of the s electrons being restricted to crystal cohesion rather than to superconductivity. The bands corresponding to d electrons are calculated by a tight binding approximation, an overlapping occurring only between neighboring atoms on a chain, which reflects their small mutual distance with respect to that between atoms belonging to neighboring chains. In this one-dimensional approximation, all interactions between chains or between A and B atoms are neglected. For the d bands, the energy dispersion

$$E(k) = E_a + E_m \cos(ka)$$

is assumed, where E_m is the overlap integral for the orbitals of neighbor atoms, which has different values for $d_{3z^2-r^2}$, for d_{xz} and d_{yz} or for $d_{x^2-y^2}$. The corresponding density of states is

$$N(E) \simeq (E_m{}^2 - (E - E_a)^2)^{-1/2}$$

This function has the characteristic singularities of a one-dimensional system at $k=0$, $E = E_a + E_m$ and $k = \pi/a$, $E_a - E_m$. Since the three chains [100], [010] and [001] lead to an identical density of states, the system has a threefold degeneracy. A further assumption in this model is the proximity of the Fermi level to a singularity of the density of states. This favors a high value of T_c, but gives, in addition, the possibility of lowering the energy of the electronic system by a tetragonal distortion. Under the assumption of the invariance of the volume at the tetragonal transition, the lattice parameters will undergo the variation

$$a \rightarrow a(1 + \varepsilon)$$

and

$$b = c \rightarrow a(1 - \tfrac{1}{2}\varepsilon)$$

This means that the [100] chain is no longer degenerate with the [010] and the [001] chains. Now the overlap integrals will take the values

$$E_m \rightarrow E_1 = E_m \exp(-\tfrac{1}{2}qa\varepsilon)$$

and

$$E_m \rightarrow E_2 = E_m \exp(\tfrac{1}{4}qa\varepsilon)$$

where q is the Slater parameter describing the decay of the radial component of the wave function. The corresponding density of states can now be represented in Fig. 4, with a twofold degenerate [010], [001] and a simple ([100]) component. Such a distortion, caused by the suppression of the degeneracy of the

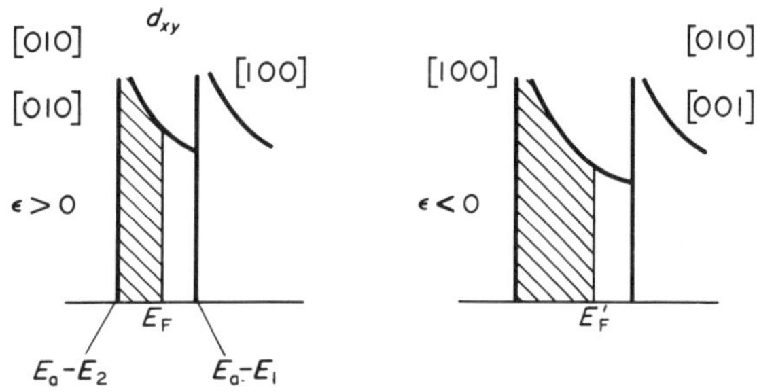

Figure 4
Electronic density of states in A15 type compounds calculated with the linear model

electronic d band structure is an example of a Jahn–Teller transformation.

The energy change ΔE in the ground state (at $T=0$) as a consequence of the tetragonal distortion can be estimated within the linear model. For $E_F > E_a - E_1$,

$$E = 3\varepsilon^2[(1/16)a^2q^2\int_{E_m}^{E_F} E_n(E)dE - E_F^2 N(E_F)] + 3/2A'$$

It follows that the tetragonal structure is only stable for $\Delta E < 0$, that is, $\sin(k_F\ a) < a^2q^2E_m/2A'$. This means that the Fermi energy E_F must be in the vicinity of the singularity of the density of states, E_m. For V_3Si, the characteristic parameters should be $a = 0.47$ nm, $q = 10\ nm^{-1}$, $E_m = 1$ eV and $A' = 20$ eV per atom, in order that the difference between E_F and E_m should be 1.25% of the band width, $2E_m$. From similar calculations it can also be shown that, for this system, the cubic structure ($a = b = c$) cannot be stable at $T=0$.

For V_3Si, the temperature variation of various experimentally observed quantities, such as the paramagnetic susceptibility, the elastic constants or the sound attenuation, has been qualitatively described by means of the linear model. The strong temperature variation of these properties can be essentially explained by the fact that the Fermi energy is situated at the vicinity of a singularity of the electronic density of states. The kind of singularity itself is of secondary importance; the strong variation of $N(E)$ close to E_F is the relevant parameter, as follows from the fact that the approximation of $N(E)$ by a step function or by a parabolic function is also sufficient for the qualitative description of the observed phenomena.

The observed agreement with some physical properties shows the importance of independent, orthogonal chains for the electronic properties of the A15 structure. However, the analysis of other A15 type compounds with lower electronic density of states than V_3Si suggests that account must be taken of an interaction between atoms belonging to different orthogonal chains as well as between atoms on the chain sites and on the body-centered sites.

3.3 Interaction Between Chains

On the basis of x-ray measurements, Staudenmann (1977) determined the distribution of the valence electron density in V_3Si. This procedure consists of an accurate measurement of the single crystal x-ray amplitudes, followed by a Fourier transformation. After subtracting the contributions of the core electrons, the "valence density" is obtained, while the subtraction of the atom contribution leads to the "deformation density." The valence electron density of V_3Si is represented in Fig. 5. The main features are

(a) the bond between adjacent vanadium atoms is by far the strongest covalent interaction in the crystal;

(b) a weaker but significant bond is directed from the silicon atoms to the middle of two adjacent vanadium atoms, perpendicular to the chain; and

(c) there is no interaction between vanadium atoms belonging to different chains.

These observations indicate that in spite of its ability to explain some unusual features of V_3Si, the linear model remains a first approximation, neglecting all interactions other than those between adjacent vanadium atoms. It is interesting to compare these results with those of a more recent band structure calculation for Nb_3Sn by Weber (1984), who used the augmented plane wave (APW) method within the framework of the tight binding approximation. A marked peak in the density of states is found in the vicinity of the Fermi energy, thus confirming the results of the linear model. In contrast to the latter, however, the density of states does not diverge, partly as a consequence of the coupling between the chains, but also between atoms belonging to cubic and to chain sites.

The sharp peak of $N(E)$ is due to the presence of two relatively flat bands near E_F, which originate from a doubly degenerate Γ_{12} state at the point Γ, $\boldsymbol{k} = (0,0,0)$, that is, at the center of the Brillouin zone. These bands are the result of a "mixture" of the various atomic orbitals, not considered in the linear model. An important contribution to the Γ_{12} bands arises from the d electrons, which are oriented in the z direction, corresponding to the orbitals $d_{3z^2-r^2}$, d_{xz} and d_{yz}. A remarkable point is the twofold degeneracy of the Γ_{12} bands at the Γ point (reflecting the cubic symmetry), in contrast to the threefold degeneracy in the linear model (arising from the three independent orthogonal chains). The degeneracy at the Γ point is also raised in the model of Weber (1984), thus confirming the Jahn–Teller character of the distortion.

The effect of interactions other than between adjacent chain atoms can be visualized by studying the x-ray investigation by Staudenmann (1977) of the compound Cr_3Si, which does not exhibit a superconducting transition down to 0.001 K. It appears that the chromium–chromium bond is much less pronounced than the corresponding vanadium–vanadium bond in Fig. 5. For this compound, the electronic density of state is very low and the magnetic susceptibility does not vary as a function of temperature; the behavior exhibited by this compound is far removed from that of V_3Si or Nb_3Sn. A similar example to Cr_3Si is represented by the low-T_c compound Nb_3Sb ($T_c = 2$ K), where E_F is situated far from a peak in $N(E)$; for compounds of this kind, the interactions between chains and also between atoms on chain and cubic sites are considerable and the conditions of the linear model are far from being fulfilled.

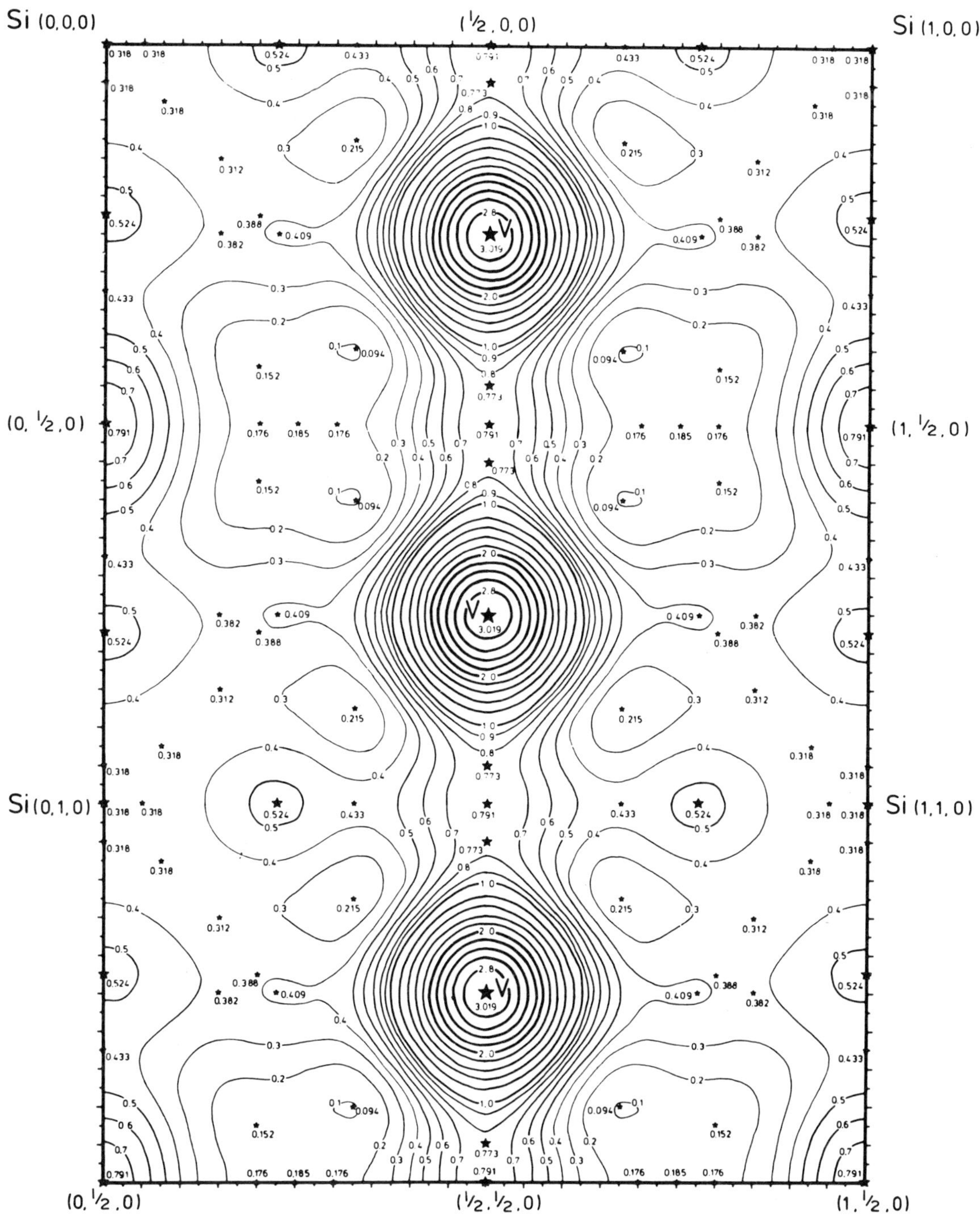

Figure 5
Valence electron density (e Å^{-3}) in V_3Si (after Staudenmann 1977)

4. A15 Phase Fields and Superconductivity

At equilibrium, the formation temperatures of the most interesting A15 compounds are situated well above 1500 °C and, for several compounds, the maximum value of T_c is measured for atomic compositions corresponding to an extreme limit of the phase field, stable only at high temperatures. It is thus of fundamental interest to know the high temperature relationships of this portion of the phase diagram accurately. The precise knowledge of the formation conditions of a given A15 phase is a necessary condition for the preparation of homogeneous, well-characterized alloys suitable for low-temperature measurements. It is an unexpected result of research in the field of superconductivity that the search for higher T_c values contributed to a more precise determination of phase diagrams at temperatures above 2000 °C. A detailed review of the known A15 phase fields, containing a large number of references, has been published by Flükiger (1981).

4.1 Vanadium-Based A15 Type Compounds with Nontransition B Elements

The A15 phase fields of the systems V_3Si, V_3Ga, V_3Ge and V_3Al, together with the known variation of T_c as a function of temperature are shown in Fig. 6.

(*a*) *V_3Si*. According to Jorda et al. (1982) the A15 phase in the vanadium–silicon system forms congruently and is stable between 19 at.% and 25 at.% silicon. Large single crystals can be grown in the whole range either by zone melting or by recrystallization at 1870 °C. There is a linear increase in T_c from 19 at.% to 25 at.% silicon, where a maximum of 17.1 K is achieved. At stoichiometry, a decrease of 0.2 K with respect to the highest hypothetical value is due to the cubic–tetragonal phase transformation at 21 K.

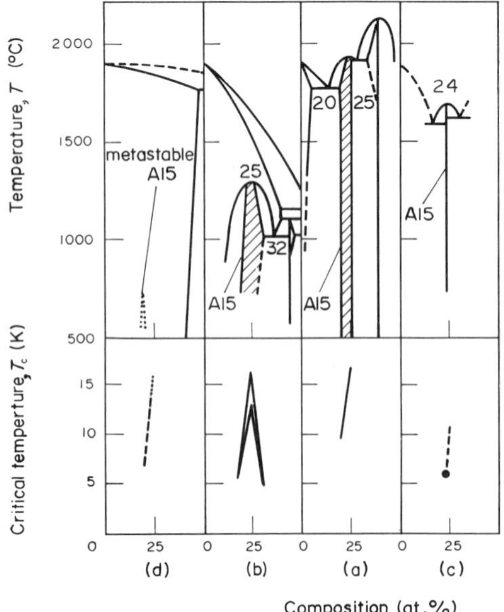

Figure 6
A15 phase fields and superconductivity in (a) vanadium–silicon, (b) vanadium–gallium, (c) vanadium–germanium and (d) vanadium–aluminum

(*b*) *V_3Ga*. The A15 phase V_3Ga forms by a congruent reaction from the bcc solid solution at 1295 °C and is stable from 20 at.% to 32 at.% gallium. The T_c value shows a marked maximum of 15.9 K at stoichiometry and decreases almost linearly on both sides. Such behavior has so far been encountered only in two other A15 type compounds, V_3Pt and Nb_3Pt. The value of T_c in this system is considerably influenced by changes in the degree of atomic ordering, the largest effect being observed for the stoichiometric composition. The compound V_3Ga has the highest electronic density of states of all known A15 type compounds, including V_3Si: $\gamma = 24\ \mathrm{mJ\ K^{-2}\ g^{-1}\ at^{-1}}$. In contrast to the latter, however, it does not exhibit a low-temperature phase transformation, which is explained by the fact that perfect ordering in this compound cannot be reached, the highest order parameter being $S = 0.98$ (Flükiger et al. 1987).

(*c*) *V_3Ge*. As shown in Fig. 6, the A15 phase in the vanadium–germanium system forms congruently from the melt. The homogeneity range of V_3Ge at equilibrium is very narrow and is centered around 23.5 ± 0.5 at.% germanium. Although the stoichiometric composition is not included in the equilibrium phase field, it can be approached by forming the pseudobinary systems $V_3(Ge_{1-x}Ga_x)$ and $V_3\,(Ge_{1-x}Al_x)$. By substituting germanium by aluminum or gallium, the T_c value of V_3Ge (6.2 K) increases at two distinctly different rates. The higher initial rate reflects the shift of the low vanadium limit of the pseudobinary A15 phase towards stoichiometry, the lower one indicates that for $x > 0.2$ the stoichiometric composition is contained in the pseudobinary A15 phase field. The A15 phase is stable up to $x = 1$ for gallium, whereas $x = 0.4$ is the aluminum solubility limit. For V_3Ge, the extrapolated T_c would be about 10 K, which was confirmed by Somekh and Evetts (1977) on thin films produced by sputtering.

(*d*) *V_3Al*. The A15 phase in the vanadium–aluminum system is not stable at equilibrium. From the pseudobinary system $V_3(Al_{1-x}Ga_x)$, it follows that the formation temperature of V_3Al would be of the order of 600–700 °C; this appears to be just below the temperature where diffusion is high enough to permit the formation of an A15 compound, the lowest known formation temperature for this structure being 900 °C for V_3Ni. The V_3Al phase is thus on the verge of stability and can so far only be produced by nonequilibrium methods. The hypothetical value for stoichiometric V_3Al can be extrapolated to 17 K from the pseudobinary series $V_3(X_{1-x}Al_x)$, where X = Ge, Ga, As, Sb, Sn, and Si.

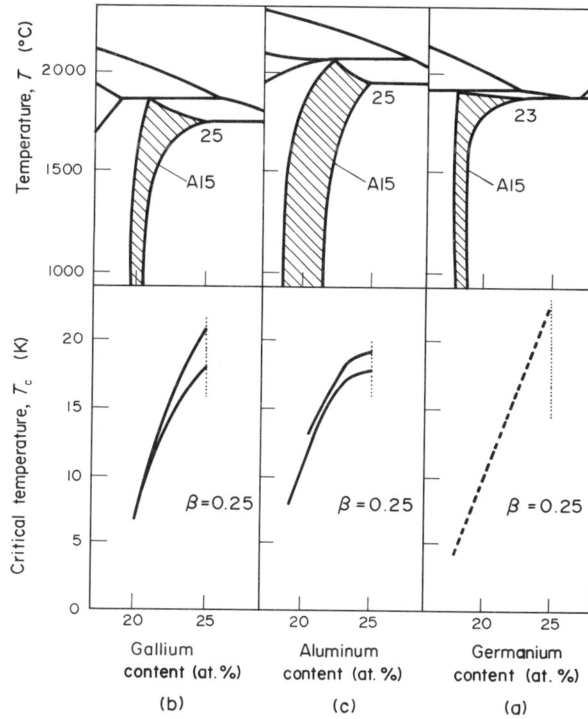

Figure 7
A15 phase field and superconductivity in (a) the niobium–germanium, (b) the niobium–gallium and (c) the niobium–aluminum systems

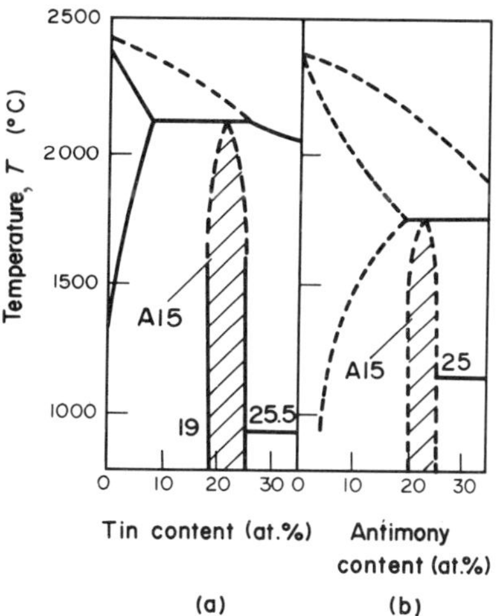

Figure 8
The A15 phase fields in (a) the niobium–tin and (b) the niobium antimony systems

4.2 Niobium-Based A15 Type Compounds with Nontransition B Elements

The three A15 compounds Nb_3Ge, Nb_3Ga and Nb_3Al exhibit the highest values of T_c known in metallic superconductors: 23.0 K, 20.7 K and 19.1 K, respectively. The A15 phase field in these three systems (see Fig. 7) has common features, which can be characterized as follows:

(i) peritectic formation,

(ii) tetragonal and bcc neighboring phases,

(iii) strongly temperature-dependent low niobium phase limit, and

(iv) the stoichiometric composition is, if at all, only stable at high temperatures, at the temperatures of the next lower peritectic or eutectic line.

Another type of A15 phase field is observed in Nb_3Sn and Nb_3Sb, as shown in Fig. 8. Both systems form peritectically, but the boundary A15/A15 + L reaches unusually low temperatures, 1050 °C and 1150 °C, respectively. The highest solubility of tin and antimony, respectively, occurs at these temperatures.

(*a*) *Nb_3Ge*. With $T_c = 23$ K, the compound Nb_3Ge exhibits the highest T_c value, not only among the A15 type compounds, but also among all known metallic superconductors. The most striking feature of the Nb_3Ge phase field is that the stoichiometric composition is not included at equilibrium, the highest solubility of germanium being 23 at.%, as shown in Fig. 7. At 1000 °C, the homogeneity range of the A15 phase is very narrow and is centered at 18 at.% germanium with a total width of 1 at.%. There is a difference between the kinetics of segregation in Nb_3Ge and that of Nb_3Ga or Nb_3Al. It is particularly difficult to retain compositions above 19 at.% germanium, even by using very fast quenching rates. The highest T_c onset value for a bulk sample, 17 K, was obtained after splat cooling, the final composition being approximately 23 at.% germanium. The variation of T_c as a function of the germanium content is almost linear: $T_c = 4.2$, 6.0 and 17 K for 17.5, 18.5 and 23 at.% germanium, respectively. The highest value of 23 K was measured on sputtered films prepared by Gavaler et al. (1974), the effective composition of which is very close to 25 at.% germanium.

(*b*) *Nb_3Ga*. The A15 phase field determined by Jorda et al. (1977), together with the variation of T_c as a function of the gallium content is represented in Fig. 7. The highest value of T_c, 20.7 K, was obtained by combined heat treatments improving both stoichiometry and atomic ordering. It is possible to separate the effects of composition and atomic ordering; once the stoichiometric composition has been reached by argon jet quenching from 1740 °C, a further increase of T_c from 18 K to 20.7 K can be obtained after prolonged heat treatments at 650 °C.

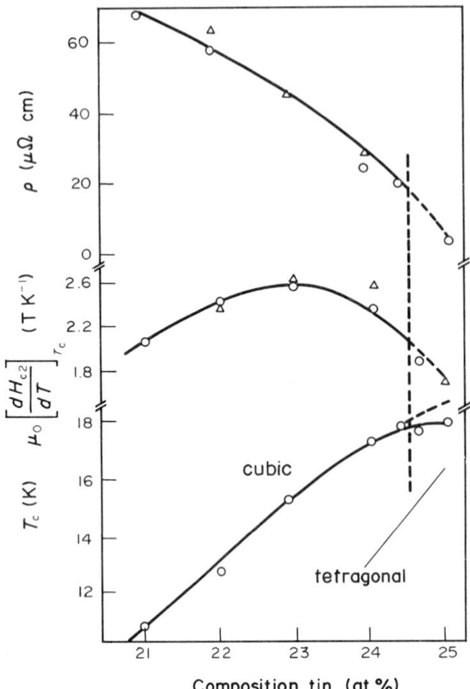

Figure 9
T_c, initial slope of $H_{c2}(T)$ and ρ_0 in the A15 phase of the niobium–tin system as a function of the tin content

(c) Nb_3Al. The niobium–aluminum phase diagram determined by Jorda et al. (1980) shows a peritectic formation of the A15 phase (Fig. 7). This figure also shows the strongly temperature-dependent aluminum-rich limit of this phase, similar to the Nb_3Ge and Nb_3Ga systems; the stoichiometric composition is metastable at room temperature and is only stable at 1940 °C. Stoichiometric Nb_3Al with a lattice parameter $a = 0.5180$ nm was prepared by argon jet quenching from this temperature (the second peritectic line in the niobium–aluminum phase diagram).

The value of T_c as a function of the aluminum content shows a slower rate of increase for compositions approaching 25 at.% aluminum. This "saturation" of T_c is unusual for A15 type compounds, for which an almost linear variation of T_c with composition is in general observed (another exception, the compound Nb_3Sn, will be discussed in the following). A possible explanation for this behavior could be given by peculiarities in the shape of the electronic density of states curve in the vicinity of the Fermi energy.

Flükiger et al. (1981) have shown that the long-range atomic order parameter of Nb_3Al is quite high, ranging from $S_a = 0.95$ to 0.97. In the same work, an attempt was made to detect whether a decrease of S_a for compositions close to stoichiometry could be responsible for the saturation of T_c. However, the observed decrease of S_a from the compositions $\beta = 0.231$ to 0.245, $\Delta S_a = 0.02$, was of the same order as the measuring error, and the question still remains open.

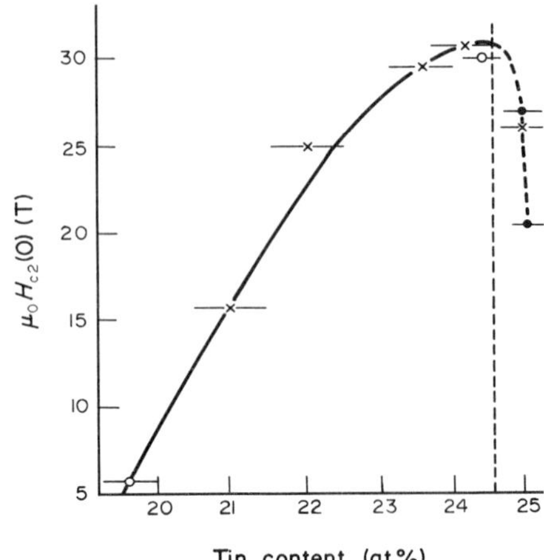

Figure 10
Variation of $H_{c2}(0)$ as a function of the tin content in Nb_3Sn

(d) Niobium–tin. The phase diagram of the niobium–tin system has been determined by Charlesworth et al. (1970) and is reproduced in Fig. 8. In contrast to the preceeding systems, the stoichiometric composition in the A15 phase Nb_3Sn is stable at equilibrium. The variation of T_c as a function of the tin content reveals a linear variation of T_c with the content, as shown in Fig. 9. The slight saturation effect of T_c close to the stoichiometric composition is in this case caused by the martensitic phase transformation occurring at 43 K for compositions above 24.5 at.% tin. This transformation also has a small effect on the electronic density of states, as determined from specific heat measurements.

There is another physical quantity undergoing a much stronger variation than T_c in the narrow composition range between 24.5 at.% tin and 25 at.% tin: ρ_0, the electrical resistivity in the normal state at temperatures just above T_c. Indeed, Fig. 9 shows that the electrical resistivity changes from 20 μΩ cm at 24.4 at.% tin to less than 4 μΩ cm at 25 at.% tin. This sharp change is a direct consequence of the perfectly ordered state in Nb_3Sn. This has consequences for the upper critical magnetic field B_{c2}, which is related to the electrical resistivity ρ_0 by the formula $B_{c2} = T_c\gamma\rho_0$ where γ is the electronic specific heat. This is the reason for the peculiar variation of B_{c2} as a function of the composition in the niobium–tin system (Fig. 10), showing that the maximum of B_{c2} does not occur at the stoichiometric composition, but somewhere around 24 at.% tin, as a combination of the opposing variation of both T_c and γ on the one hand and of ρ_0 on the other. The variation of B_{c2} is of considerable technical importance, since it influences the behavior of the critical current density J_c at high magnetic fields in Nb_3Sn multifilamentary wires (see Sect. 6). In particular, it explains why the addition of ternary and quaternary elements to Nb_3Sn leads to a considerable increase of J_c at high fields ($B > 14$ T); This is due to a strong enhancement of ρ_0.

5. *Atomic Ordering in A15 Type Compounds*

5.1 The Bragg–Williams Long-Range Order Parameter

Partial disorder in the A15 structure (Fig. 1) is introduced by a random site exchange between A and B atoms on their respective lattice sites, 6*c* and 2*a*. The degree of atomic ordering in an A15 type compound of the general formula $A_{1-\beta}B_{\beta}$ can be described in different ways:

(a) the occupation factors r_a and r_b where r_a is defined by the fraction of 6*c* sites occupied by A atoms and r_b by the fraction of 2*a* sites occupied by B atoms, or

(b) by the Bragg–Williams order parameters S_a and S_b for each lattice site

$$S_a = \frac{r_a - (1-\beta)}{1-(1-\beta)} = \frac{r_a - (1-\beta)}{\beta}$$

and

$$S_b = \frac{r - \beta}{1 - \beta}$$

The Bragg–Williams order parameter is defined in such a way that $S = 1$ for the perfectly ordered lattice and $S = 0$ for complete disordering (or random distribution). These two states correspond to $r_a = 1$ (or $r_b = 1$) and $r_a = 0.75$ ($r_b = 0.75$) at the composition β. At the stoichiometric composition $\beta = 0.25$, the degree of ordering can be described by one parameter: $S = S_a = S_b$.

The known order parameters in niobium- and vanadium-based A15 type compounds are represented in Fig. 11, where the value of S (or S_a for nonstoichiometric compounds) and its variation after various heat treatments are plotted as a function of the valence electron number of the B element (column in the periodic system). It is seen that there are two regions where S (or S_a) deviates significantly from perfect ordering: for the transition elements B = osmium and rhenium and for the nontransition elements B = gold, aluminum and gallium. These regions correspond to the limits of the series of stable

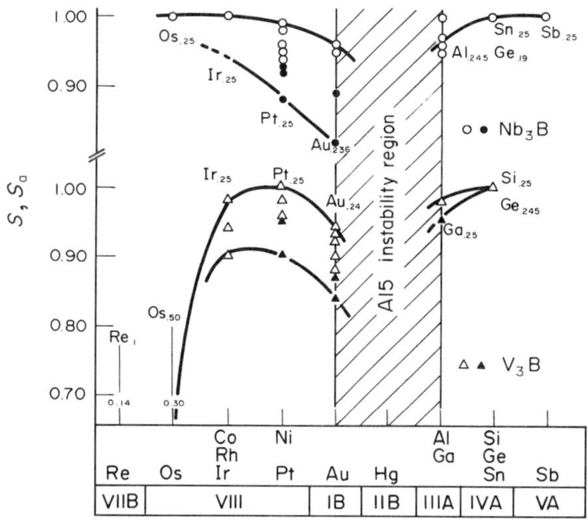

Figure 11
Atomic order parameters in niobium- and vanadium-based A15 type compounds

A15 type compounds, thus suggesting a correlation between the degree of atomic ordering and the stability of the A15 phase.

5.2 Vacancy Diffusion in A15 Type Compounds: The "Virtual Site"

In the tightly packed A15 structure, each atom is closely surrounded by its neighbors (see Fig. 1). The atoms would need to be considerably compressed before any two could squeeze past one another and interchange their positions as required for a change of the order parameter. Among various possible diffusion mechanisms, vacancy diffusion is the most probable.

The vacancy diffusion mechanism is based on the fact that at thermal equilibrium each solid at a temperature above zero contains a certain number of vacant lattice sites. An atom will now jump into a neighboring vacancy, thus creating a new vacant site. Atoms and vacancies undergo a series of position exchanges, in order that a very small number of vacant lattice sites is sufficient to induce a substantial diffusion. In the A15 type structure, the situation is more complex than for example in the bcc or fcc structures; indeed, the 6*c* and the 2*a* sites are not equivalent from the point of view of electronic bonding, interactions between atoms on 6*c* or 2*a* sites being characterized by metallic bonding, while the intrachain bonding between two nearest A neighbors is of the covalent type, as shown by Staudenmann (1977).

The question arises whether the covalent bonding between two neighboring A atoms on the chain sites still resides if one of them is next to a 6*c* vacancy.

Welch et al. (1984) have shown by means of pair potential calculations that such an individual vacancy of an A atom on a 6*c* site is unstable. The state of lower energy corresponds to a configuration where one of the two A atoms adjacent to the single 6*c* vacancy is shifted towards a new site which is equidistant from the next two A neighbors. This particular type of vacancy, also called "split vacancy," is correlated to the occupation of a nonequilibrium or "virtual" lattice site. This site is located at the region of overlap between two neighboring A atoms on a chain. The vacancy diffusion mechanism in the A15 structure thus comprises at least two steps, in contrast to the one-step mechanism acting in simple bcc or fcc structures. The concept of the virtual site has been useful as a basis for understanding the solid-state diffusion mechanism in Nb_3Sn wires (Welch et al. 1984) and the occurrence of homogeneous disordering in A15 type compounds submitted to irradiation with high-energy particles (Flükiger 1989).

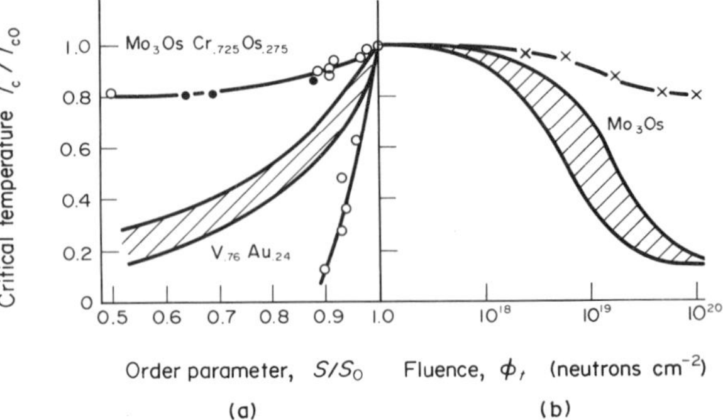

Figure 12
Variation of T_c/T_{c0} as a function of (a) S/S_0 and (b) the fluence (φ_t) for various A15 type compounds—hatched area: Nb_3Ge, Nb_3Sn, Nb_3Pt, Nb_3Au, Nb_3Al, V_3Ga and V_3Si

5.3 Variation of the Low-Temperature Properties with Atomic Ordering.

The variation of T_c for a series of A15 type compounds as a function of the order parameter has been plotted in Fig. 12a. All the order parameter values in this figure have been determined by refining diffractometric data obtained from irradiated or quenched samples. In order to compare the data, a normalized representation was chosen: T_c/T_{c0} vs S/S_0 (or S_a/S_{a0} for nonstoichiometric systems). The variation of T_c with the degree of atomic ordering is smallest for Mo_3Os and $Cr_{0.725}Os_{0.275}$ and is largest for $V_{0.76}Au_{0.24}$, while the intermediate hatched area is common to the Nb_3Ge, Nb_3Sn, Nb_3Pt, $Nb_{0.766}Au_{0.234}$, V_3Ga, V_3Si, V_3Pt and Nb_3Al systems.

In order to illustrate the analogy between thermally induced and radiation-induced disordering, the representation T_c/T_{c0} vs the neutron fluence φ_t has been plotted in Fig. 12b.

There have been attempts to ascribe a universal behavior of T_c/T_{c0} vs φ_t in A15 type compounds. However, it has to be taken into account that this behavior is connected to the individual density of states in each compound, which must fulfill the requirement of being sufficiently high. From this point of view, $V_{0.76}Au_{0.24}$ is an exception which may possibly arise from the strongly enhanced spin susceptibility in this compound. There is thus no physical reason which could be advanced for explaining a "universal" behavior of T_c/T_{c0} vs the fluence φ_t: It is certainly more correct to say that there is a similarity in the dependence of T_c on the degree of atomic ordering. This is due to the correlated change in $N(E_F)$ with irradiation, under the condition that $N(E_F)$ is sufficiently high in order to dominate over the simultaneous effect of phonon hardening with disordering.

There is a contrast between the behavior of high- and low-T_c A15 type compounds with disordering:

(a) high-T_c compounds show a lowering of T_c and $N(E_F)$ and an enhancement of the Debye temperature Θ_D; and

(b) low-T_c compounds show an enhancement of T_c and $N(E_F)$ and a decrease of Θ_D.

The origin of the very different behavior in these two classes of materials may be connected to a substantial difference in their pair potentials, in particular in the interaction between the A and the B elements, as follows from the calculations of Chudinov et al. (1984) for Nb_3Sn and Mo_3Si. This is thought to be the basis for understanding the differences between the compounds based on vanadium and niobium on the one hand and chromium and molybdenum on the other.

5.4 Variation of the Electronic Specific Heat with the Order Parameter

The variation of the electronic specific heat γ as a function of the order parameter has been plotted in Fig. 13 for several A15 type compounds. The contrast between the variation of γ/γ_0 and T_c/T_{c0} is immediately apparent when comparing Fig. 12 with Fig. 13; all specific heat curves are practically situated on the same curve, including $V_{0.76}Au_{0.24}$. This demonstrates that the different behavior of T_c with ordering for various A15 type compounds cannot be explained by the variation of $N(E_F)$ alone. From Fig. 13 it follows that the initial decrease of γ/γ_0 is strongest for the most ordered state. For example, the decrease of S by 5% causes a decrease of γ/γ_0 by 20%, while a further decrease of S by 5% lowers the heat capacity by 10% only (i.e., half the preceding value). Below $S = 0.7$, saturation effects start to occur; as shown for Mo_3Si, both modifications, the highly disordered A15 structure and the amorphous structure exhibit very similar values of the electronic density of states.

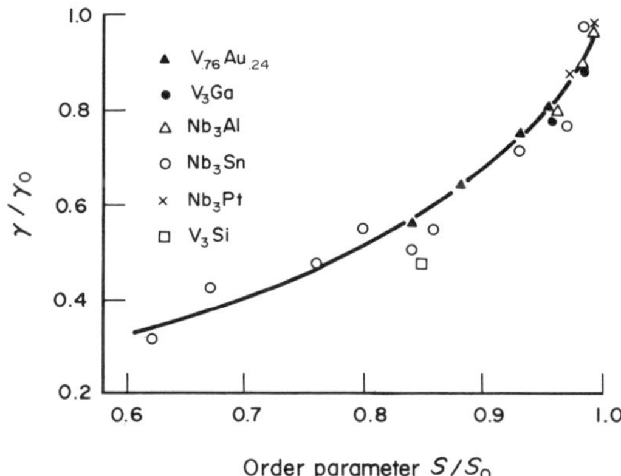

Figure 13
γ/γ_0 vs S/S_0 for V_3Ga, V_3Si, V_3Au, Nb_3Pt, Nb_3Al and Nb_3Sn

5.5 Effect of Composition and Atomic Ordering on the Electrical Resistivity

The variation of ρ_0 in V_3Si has been studied in detail as a function of composition on a series of single crystals. The results are shown in Fig. 14, where the RRR value has been plotted rather than ρ_0, due to the irregular shape of the crystals (where RRR is defined as the ratio between the electrical resistivities at 300 K and at 20 K). A striking feature of Fig. 14 is the sharp variation of the RRR quotient when approaching the stoichiometric composition. Between 22 at.% silicon and 24.5 at.% silicon, RRR increases from 1 to 6, while the further addition of only 0.5 at.% silicon causes an increase up to RRR = 80! The nearly symmetrical behavior with respect to the stoichiometric composition indicates that ρ_0 is mainly dependent on the perturbation due to wrongly occupied sites, the identity of the excess atom being of secondary importance. The behavior shown in Fig. 14 is certainly the most expressive way to demonstrate that this system is perfectly ordered; at the vicinity of the stoichiometric composition, very small perturbations of the system (e.g. inhomogeneities, small deviations from stoichiometry, impurities, small amounts of ternary additives or deviations from perfect ordering) may cause a considerable increase of ρ_0.

6. Critical Currents in A15 Type Compounds

Over the years, there has been a steadily growing requirement for superconducting high-field magnets (see *Superconducting Magnets*). At present, the highest magnetic field produced by a superconducting magnet (without a normal conducting hybrid insert) is 20.7 T at 1.8 K in a bore of 20 mm diameter. This magnet consists of three concentric superconducting magnets: an external NbTi magnet, an intermediate

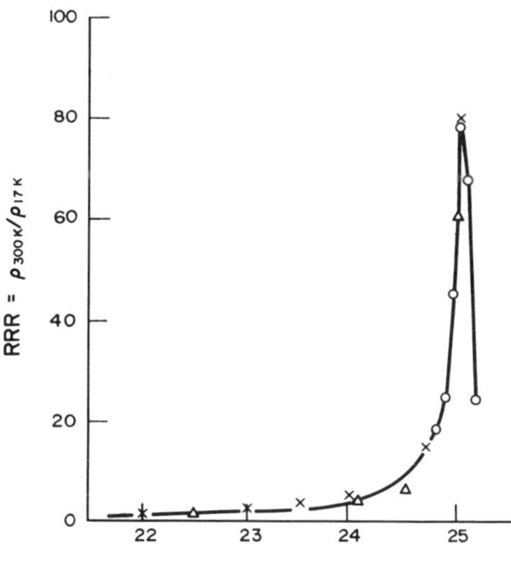

Figure 14
Variation of the RRR value in V_3Si as a function of the silicon content

Nb_3Sn magnet and the central magnet, consisting of titanium-alloyed Nb_3Sn wires.

In the near future, superconducting magnets with fields above 10 T are still expected to be wound with wires based on A15 type superconductors. Wire based on compounds crystallizing in other phases with even higher upper critical fields, for example tapes of Bi(2223) (see *Oxide Superconductors: Solid-State Chemistry*) or Chevrel phases ($PbMo_6S_8$) (see *Chevrel Phases*), show without any doubt promising current carrying properties, but still need considerable development before they can be applied commercially in high-field magnets.

The numerous methods reported so far for producing multifilamentary wires based on A15 type compounds all have one thing in common: during mechanical deformation to the final wire diameter, all the constituents have to be ductile, the inherently brittle A15 phase being formed at the end of the deformation process by a reaction at temperatures well below 1000 °C. These ductile components are ordinarily combinations of materials such as niobium, copper, copper–tin and tantalum.

It is unfortunate that just for the two binary A15 type compounds exhibiting the highest T_c values, Nb_3Ge ($T_c = 23$ K) and Nb_3Ga ($T_c = 20.7$ K), no method has been developed to produce the required multifilamentary configuration. This may appear relatively surprising in view of the relative ease with which both compounds can be formed as thin films or tapes, either by sputtering, coevaporation or CVD. This illustrates that the occurrence of a high-T_c value in a compound does not necessarily imply its suitability

Figure 15
Cross section of a multifilamentary Nb_3Sn wire, prepared by means of the internal tin diffusion technique (diameter 0.8 mm, 7260 filaments—after Xu et al. 1987)

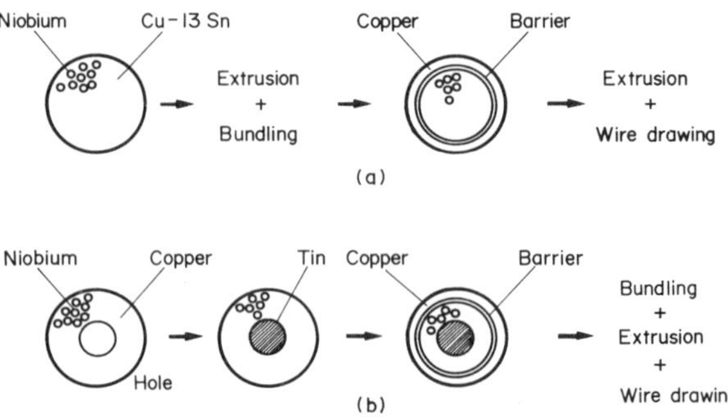

Figure 16
Schematic representation of (a) the bronze route (with intermediate anneals) and (b) the internal tin diffusion process (without anneals) for producing multifilamentary Nb_3Sn wires

for practical superconducting wires fulfilling the necessary current carrying and stability requirements at high fields.

A typical Nb_3Sn superconducting wire of 1 mm diameter consists of several thousand well separated filaments, 3–5 μm diameter, which are embedded in a matrix of copper–tin bronze. This characteristic configuration with a large number of fine filaments (which are in addition twisted around the wire axis) satisfies the various stability requirements (electrical, thermal and mechanical stability—see *Multifilamentary Superconducting Composites*). An example is shown in Fig. 15; it is seen how a tantalum barrier separates the filament/matrix region from the external copper sheath. Since the function of the copper sheath is to control the thermal stability, a low specific resistivity is required. This can only be obtained if the tantalum barrier prevents the diffusion of tin into copper.

Nb_3Sn is a very brittle phase. The fact that it is possible to produce such wires as represented in Fig. 15, to wind them to a coil and to operate them under full load without degradation is due to the fortunate coincidence that all stability criteria lead to one and the same requirement—the filaments must be very fine ($\leqslant 5\,\mu m$).

A further favorable condition allows the wires to withstand the Lorentz forces in a coil in a reversible manner for axial strains up to several tenths of a percent. As a consequence of the different thermal expansion coefficients of copper–tin bronze and Nb_3Sn filaments, the latter are under compression (also called precompression state), which enhances the fracture strain from 0.1% to 0.8%, thus constituting a mechanical reserve.

All known methods for producing Nb_3Sn multifilamentary wires are based on the so-called bronze diffusion process, where tin atoms diffuse out of a copper–tin bronze to react with the niobium to produce Nb_3Sn, the reaction temperature being of the order of 700 °C. The most competitive methods for producing industrial Nb_3Sn wires are the bronze route and the internal tin diffusion processes. The main difference between the processes is the necessity of intermediate anneals during the bronze route wire drawing process, in contrast to the internal tin diffusion process, where the whole wire deformation procedure can be performed without intermediate anneals. This factor is of considerable economical importance in favor of the latter. In addition, the amount of tin in the copper matrix, which for the bronze route is limited to 13 wt% (due to the phase diagram limitations), can be raised to 20 wt%, which has a positive effect on the critical current density. Both processes are schematically represented in Fig. 16.

The reaction proceeds by the diffusion of tin, the more mobile species, through the Nb_3Sn layer that has already formed to react with niobium at the Nb_3Sn/Nb core interface; the diffusion is primarily grain boundary or short circuit diffusion, particularly at lower reaction temperatures (Glowacki and Evetts 1989).

After reaction, the Nb_3Sn layer reaches a thickness of several micrometers and only a small part at the center of the core of each filament consists of unreacted niobium. The grain size of the A15 phase at the Nb_3Sn/Cu–Sn interface is approximately 100–140 nm, while at the Nb_3Sn/Nb interface, it can reach up to 1 μm (columnar growth). As will be shown in Sect. 6.1, there are different gradient fields in Nb_3Sn wires which need to be controlled in order to optimize the critical current density.

6.1 Optimization of the Critical Current Density in Nb_3Sn Wires

An Nb_3Sn multifilamentary wire is an extremely complex construction, not only because of its microscopic configuration, but also because of the fact that the value of J_c results from the superposition of three different gradient fields (Flükiger 1989).

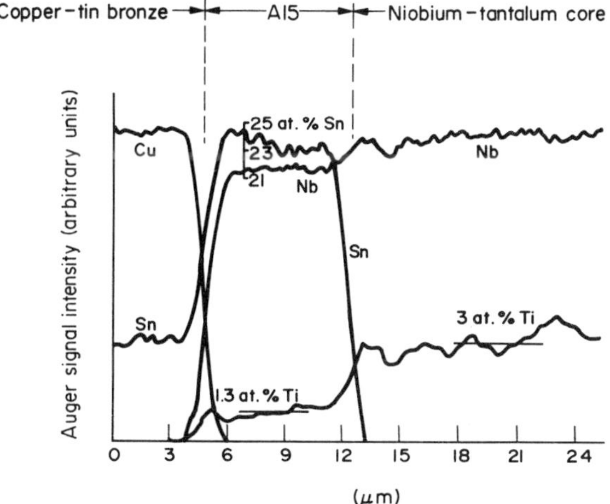

Figure 17
Auger scanning analysis of the cross section of an Nb–3Ti/Cu–7.5Sn (at. %) wire after reacting for 64 h at 750 °C (after Drost et al. 1984)

(*a*) *Concentration gradient.* By means of Auger spectroscopy it has been shown that the tin content in the Nb_3Sn layer varies between 25 at.% and about 22 at.% from the interface with the bronze to that with the unreacted niobium (see Fig. 17). The center of gravity lies between 23 at.% and 24.5 at.% tin after reaction at 700 °C. This means that T_c as well as B_{c2} decrease strongly towards the center of the filament. Thus, the superconducting properties of the A15 layer are far from being homogeneous; at high fields, only a tubular region of the filament will contribute to the current carrying mechanism, the thickness of the tube tending to zero for $B \rightarrow B_{c2}$.

(*b*) *Gradient in the grain size.* The critical current density J_c depends on the pinning density at the grain boundaries and thus on the grain size. Various grain structures have been observed so far: at the niobium interface large, columnar grains (up to 1 μm long) with a strong texturing perpendicular to the wire axis, followed by very fine grains with statistical orientations (80–200 nm after reaction of 3 d at 700 °C). It is obvious that the fine grains, which also have a higher tin content, will be essentially responsible for the current transport at low fields, but in particular at high fields. It has often been overlooked in the literature that the effect of the gradient fields (*a*) and (*b*) superpose, thus seriously influencing the effective current carrying volume. This "tubular volume" is maximum at low fields and decreases gradually with increasing field. The value of J_c is thus considerably greater than the value simply obtained by dividing the critical current I_c through the measured A15 cross section. This may also be the reason why it is so difficult to establish correct scaling laws for Nb_3Sn wires (see *Flux Pinning and the Summation of Pinning Forces*). It is even doubtful if the mentioned gradient fields allow the application of any scaling law at all, the conditions being different at each applied field.

(*c*) *Mechanical stress gradient.* It was shown by neutron diffraction experiments that the state of precompression in Nb_3Sn multifilamentary wires leads to an elastic tetragonal distortion of the A15 lattice. The distortion $(1-c/a)$ depends on the orientation of the A15 grains with respect to the axis. For the (100) parallel to the axis, a maximum value of $(1-c/a) = 0.004$ is observed. If the (100) direction forms an angle of 45° with the wire axis, the grains remain cubic (i.e., $1-c/a=0$). The stress distribution in the wire can thus be represented by the change of $(1-c/a)$ as a function of the grain orientation with respect to the wire axis. In addition to precompression arising from differential thermal contraction of the bronze matrix and the filaments, there is also a complex stress distribution within each reacted Nb_3Sn layer that arises from the 37% increase in volume of the niobium as tin diffuses through the existing Nb_3Sn layer to react to form further Nb_3Sn. The large stresses induced control the grain structure and morphology of the growing layer (Wallach and Evetts 1986). The stress distribution in an Nb_3Sn wire can thus be represented by a superposition of a hydrostatic and a nonhydrostatic component. Both stress components have an influence on J_c that becomes more important as the field increases.

6.2 Enhancement of J_c by Alloying

The observed enhancement of J_c in alloyed Nb_3Sn multifilamentary wires is a consequence of the perfectly ordered state in Nb_3Sn. The critical current density J_c of Nb_3Sn multifilamentary wires at high fields (i.e., above 11 T) has been considerably enhanced in recent years by the addition of third and fourth elements. Tantalum or titanium additives to multifilamentary Nb_3Sn wires have now reached the stage of commercial application (see *Critical Current Density*).

Based on the Ginzburg–Landau–Abrikosov–Gor'kov (GLAG) formalism for dirty type II superconductors, it is expected that small additions to Nb_3Sn will cause an increase of ρ_0 without a drastic depression of T_c and γ, thus leading to an enhancement of the upper critical field B_{c2} (see *Magnetic Properties of Superconductors*). Indeed, for bulk Nb_3Sn and also for samples prepared by coevaporation, there is experimental evidence that, for a few atomic percent of additives, the residual resistivity ρ_0 increases sufficiently to give an enhancement in B_{c2} while the specific heat coefficient γ decreases and the critical temperature T_c even rises slightly.

The most effective method to enhance J_c at high fields consists of alloying with small quantities ($\leqslant 3$ at.%) of tantalum or titanium to Nb_3Sn. As shown in Fig. 18, the enhancement of J_c is mainly observed at

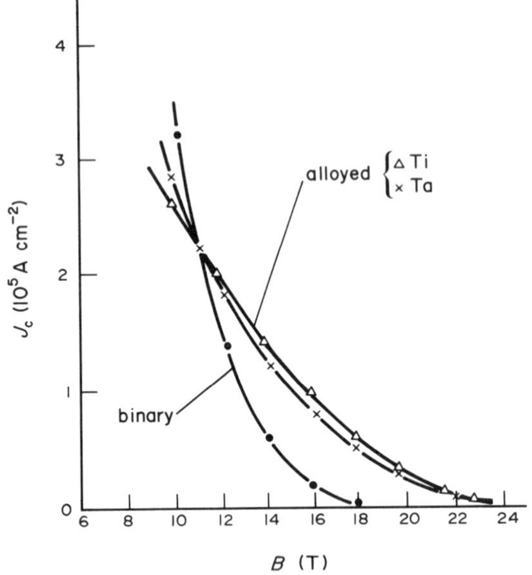

Figure 18
Critical current density in the A15 layer of Nb_3Sn wires with titanium, tantalum and nickel plus zinc additions (after Drost et al. 1984)

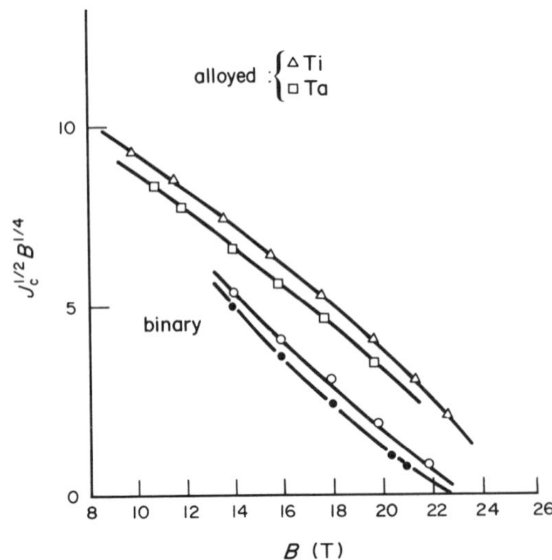

Figure 19
Kramer plots for 19 core Nb_3Sn wires with tantalum, titanium and nickel plus zinc and without additions, yielding H^*_{c2} values of around 25 T and 22 T, respectively

fields exceeding 11 T. The very considerable enhancement of J_c at high fields in alloyed Nb_3Sn wires is a direct consequence of the perfect ordering in this A15 type compound. Indeed, the addition of 3 at.% tantalum enhances the low original value of ρ_0 up to 35 μΩ cm due to the perturbation of the A15 lattice. The enhancement of the electrical resistivity is correlated to a rise of B_{c2} by 2–3 T, which explains the effect on J_c at higher fields.

The usual way to estimate the upper critical field of multifilamentary Nb_3Sn wires consists of establishing a Kramer plot, $J_c^{1/2}B^{1/4}$ vs B, based on J_c measurements and to extrapolate linearly to zero. The intersection occurs at a field called B^*_{c2} which reflects the upper critical field at 4.2 K, the temperature at which J_c was measured. The Kramer plots should not be overinterpreted since they cannot be applied to all superconducting structures (see *Flux Pinning and the Summation of Pinning Forces*). Nevertheless, these plots give an approximate value for B^*_{c2} in Nb_3Sn wires, which is useful in determining relative changes of the upper critical field after different reaction conditions.

The Kramer plot for the wires characterized in Fig. 18 is shown in Fig. 19. It is seen that alloyed Nb_3Sn wires have appreciably higher B^*_{c2} values compared with those of binary (unalloyed) wires.

See also: Critical Current Density; Multifilamentary Superconducting Composites; Superconducting Materials: An Overview; Superconducting Materials: Physical Properties

Bibliography

Batterman B W, Barrett C S 1964 Crystal structure of superconducting V_3Si. *Phys. Rev. Lett.* 13: 390–2

Bednorz J G, Müller K A 1986 Possible high-T_c superconductivity in the Ba–La–Cu–O system. *Z. Phys. B* 64: 189–96

Charlesworth J P, MacPhail I, Madsen P E Jr 1970 The phase diagram of the system Nb–Sn. *J. Mater. Sci.* 5: 580–94

Chudinov V G, Moseev N V, Goschitskii B N, Protasov V I 1984 Possible amorphization and phase separation of intermetallic Mo_3Si compound under radiation. *Phys. Stat. Sol. A* 85: 435–40

Drost E, Specking W, Flükiger R 1984 Superconducting and metallurgical properties of bronze processed Nb_3Sn wires with Ni and Zn additions and comparison with Ta and Ti additions. *Cryogenics* 24: 622–8

Flükiger R 1981 Phase diagrams of superconducting materials. In: Foner S, Schwartz B B (eds.) 1981 *Superconductors Materials Science.* Plenum, New York, pp. 511–604

Flükiger R 1989 Characterization of low T_c superconductors and the mechanism of critical current transport. In: Levy F, Martinoli P (eds.) 1989 *Proc. 31ème Cours. Assoc. Vaudoise Cherch. Physique.* Case postale 101, Lausanne, Switzerland, pp. 149–98

Flükiger R, Jorda J L, Junod A, Fischer P 1981 Superconductivity, atomic ordering and stoichiometry in the A15 type phase Nb_3Al. *Appl. Phys. Commun.* 1: 9–30

Flükiger R, Küpfer H, Jorda J L, Muller J 1987 Effect of atomic ordering and composition changes on the electrical resistivity of Nb_3Al, Nb_3Sn, Nb_3Ge, Nb_3In, V_3Si and V_3Ga. *IEEE Trans. Magn.* 23: 980–3

Gavaler J R 1973 Superconductivity in Nb–Ge films above 22 K. *Appl. Phys. Lett.* 23: 480–2

Gavaler J R, Janocko M A, Jones C K 1974 Superconductivity and metastability in alloys of the Mo–Re system. *Proc. LT13*, Vol. 3. Plenum. New York, p. 558–62

Glowacki B A, Evetts J E 1989 Influence of cyclic pumped diffusion on the morphology and microstructure of A15 layers in bronze processed multifilamentary wire. *IEEE Trans. Magn.* 25: 2200–3

Hardy G F, Hulm J D 1953 Superconducting silicides and germanides. *Phys. Rev.* 89: 884–5

Jorda J L, Flükiger R, Muller J 1977 On the formation and stability of the A15 phase in the Nb–Ga system. *J. Less Common Met.* 55: 249–64

Jorda J L, Flükiger R, Muller J 1978 The phase diagram of the Nb–Ge system. *J. Less Common Met.* 62: 25–37

Jorda J L, Flükiger R, Muller J 1980 A new metallurgical investigation of the Nb–Al system. *J. Less Common Met.* 75: 227–39

Jorda J L, Muller J 1982 The V rich part of the V–Si phase diagram. *J. Less Common Met.* 84: 39–45

Junod A, Jarlborg T, Muller J 1983 Heat-capacity analysis of a large number of A15-type compounds. *Phys. Rev. B* 27: 1568–85

Junod A, Muller J, Rietschel H, Schneider E 1978 Chaleur spécifique et transformation martensitique dans le système $Nb_{1-x}Sn_x$. *J. Phys. Chem. Solids* 39: 317–28

Kurmaev E Z, Werfel F, Brummer O, Flükiger R 1977 X ray photoemission spectra of valence electrons in V_3X and Nb_3X compounds. *Solid State Commun.* 21: 239–43

Labbé J, Friedel J 1966a Instabilité électronique et changement de phase cristalline de composés du type V_3Si à basse température. *J. Phys. (Paris)* 27: 153–65

Labbé J, Friedel J 1966b Effet de la température sur l'instabilité électronique et le changement de phase cristalline des composés du type V_3Si à basse température. *J. Phys. (Paris)* 27: 303–8

Machlin E S, Wang S H 1976 On the stability of the A15 structure. *J. Phys. Chem. Solids* 37: 555–65

Machlin E S, Wang S H 1978 A15—a structure stabilized by crystal field distortion of "exotic" elements at chain sites. *J. Phys. Chem. Solids* 39: 607–15

Mirmelshtein A V, Karkin A E, Arkhipov V Y, Voronin V I 1984 Superconductivity and phonon-spectrum softening in Mo_3Ge compounds after fast neutron irradiation. *Fiz. Metall. Metalloved.* 58: 1008–11

Nölscher C, Saemann-Ischenko 1985 Superconductivity and crystal and electronic structures in hydrogenated and disordered Nb_3Ge and Nb_3Sn layers with A15 structure. *Phys. Rev. B* 32: 1519–31

Somekh R E, Evetts J E 1977 The sputtering of high T_c A15 Nb_3Si and Nb_3Ge. *Solid State Commun.* 24: 733–7

Staudenmann J L 1977 The electron distribution in superconducting alloys. *Solid State Commun.* 23: 121–5

Staudenmann J L 1978 The electron charge distribution in V_3–Si. *Solid State Commun.* 26: 461–8

Testardi L R 1975 Structural instability and superconductivity in A15 compounds. *Rev. Mod. Phys.* 47: 637–48

Wallach E R, Evetts J E 1986 The development of microstructures in multifilamentary bronze route A15 composites. *Adv. Cryog. Eng.* 32: 911–23

Weber W 1984 Electronic structure of complex systems. In: Phariseau P, Temmerman W (eds.) 1984 *Electronic Structure of Complex Systems*, NATO-ASI Series. Plenum, New York, p. 422–68

Welch D O, Dienes G J, Lazareth O W Jr, Hatcher R D 1984 Defects and diffusion mechanisms in Nb_3Sn. *J. Phys. Chem. Solids* 45: 1225–42

Xu J Q, Specking W, Obst B, Seibt E, Flükiger R 1987 Superconducting and metallurgical properties of Nb_3Sn wires processed by internal Sn route including hydrostatic extrusion. *Cryogenics* 29: 87–95

R. Flükiger
[Université de Genève, Geneva, Switzerland]

AC Applications of Superconducting Materials

With the development in the 1960s and 1970s of high-field, high-current-density superconductors operating at liquid helium temperatures many applications in power equipment were envisaged (Foner and Schwartz 1974). Initially, such applications as dc generator field windings met with some success. However, ac applications were hindered because of the magnitude of the ac losses. Type II superconductors are lossy in time-varying magnetic fields (see *AC Losses in Superconducting Materials*). The way around this problem during the 1970s was to develop filamentary superconducting wire with smaller and smaller diameter filaments. By the early 1980s, wire fabrication technology for fine filamentary superconductors was sufficiently developed to make the prospect of incorporating superconductors in ac power equipment design attractive, the individual filament diameter in these wires having been reduced to below the micrometer level (Dubots et al. 1984, Hlasnik 1984). Continuing research and development during the 1980s has overcome such problems as proximity coupling between closely spaced filaments (see *Coherence Length, Proximity Effect and Fluctuations* for an explanation of this effect). It is now possible to commercially produce long lengths of very-low-loss ultrafine multifilamentary wire intended for use in power equipment (see *Multifilamentary Superconducting Composites*). The filament size in these wires is around 100 nm and the ac losses are typically an order of magnitude lower than those of copper in a similar situation (Cave et al. 1989). Some possible applications are fault current limiters, lightweight 50–60 Hz power transformers, power generators with both stator and rotor made from superconducting wire, and quick response power regulators that can inject or absorb both active and reactive power using an energy storage magnet (Février 1987, Verhaege et al. 1989; *Superconducting Materials: Large-Scale Applications*).

1. Multifilamentary Wires for 50–60 Hz AC Applications

Use of fine filamentary wires in superconducting electrical machines is advantageous because of high current densities (several 10^9 A m^{-2}) and low ac losses. These properties lead to lighter and more efficient machines. Furthermore, a high electrical resistance beyond the conductor critical current leads to an autocurrent limiting effect during fault conditions. This results in only a small increase in magnetic forces and surge currents and permits a simplification of the mechanical design. A further property necessary for use of multifilamentary wire with ac currents and fields is thermoelectromagnetic stability (i.e, the wire must not quench suddenly to the normal state) (Guéraud et al. 1989). Stability is especially important in low applied fields where the critical current density is high (see *Critical Current Density*). The filament diameter and twist pitch, matrix material, positioning of the filamentary region and the wire diameter all influence the stability (the smaller the diameter the more stable the wire) and this has led to the design of different wire configurations for different applications depending on the intensity of the magnetic field (Lacaze et al. 1991).

1.1 Basic Wire Design Considerations

Recalling that 1 W dissipated at helium temperatures requires 500–1000 W of refrigerator power at ambient temperature reinforces the fact that ac losses in these materials must be kept to a minimum. Accurate prediction of ac losses is complex, involving both hysteretic and eddy current contributions (see *AC Losses in Superconducting Materials*; Février 1983). However, under certain conditions, approximate formulae can be applied. For filaments that are completely penetrated by the changing magnetic field (which is usually the case for ultrafine filaments subjected to moderate field variations) the overall hysteretic losses in W m^{-3} are given approximately by

$$\alpha d_f\, J_{cov} \left[1 + \left(\frac{J_{ov}}{J_{cov}}\right)^2\right] \left|\frac{dB_t}{dt}\right| \tag{1a}$$

where α is a wire-dependent constant (of order unity), d_f is the effective filament diameter (whose lower limit is the filament diameter and which increases gradually as successive layers of filaments become saturated with current, see *AC Losses in Superconducting Materials*), J_{cov} is the overall composite critical current density (which depends on both field and temperature), J_{ov} is the overall current density and $|dB_t/dt|$ is the instantaneous variation of the transverse magnetic induction. The eddy current contribution also in W m^{-3} can be estimated from

$$\beta \sigma p^2 \left|\frac{dB_t}{dt}\right|^2 \tag{1b}$$

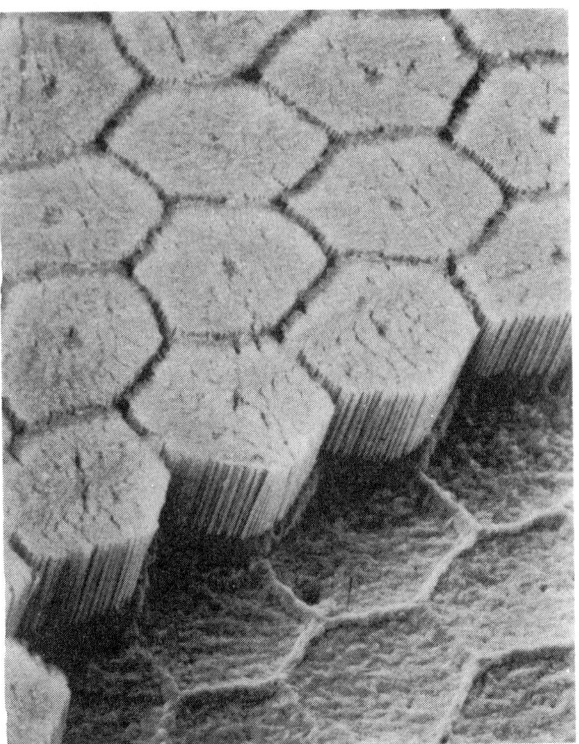

Figure 1
View of an ultrafine multifilamentary wire, near the center of the wire; each filamentary packet is about 8 μm in diameter and contains 1038 Nb–Ti filaments separated from each other by a highly resistive Cu–Ni matrix (AISA)

where β is a wire-dependent constant (of order unity), σ is the matrix conductivity (in the filamentary region) and p is the filament twist pitch. Thus, from Eqn. (la), it can be seen that while keeping the overall critical current density high, a lowering of the hysteretic losses is achieved through a reduction in filament diameter. Similarly, Eqn. (lb) shows that eddy current losses can be diminished by decreasing the interfilament matrix conductivity and by shortening the filament twist pitch.

Niobium–titanium is used for the most part in industrial applications due to its ductility (see *Niobium–Titanium Alloys*). Figure 1 shows a micrograph of the filamentary structure near the center of a high-performance ultrafine multifilamentary wire made by AISA (Alsthom Intermagnetic S.A.). This wire contains 956 340 Nb–46.5%Ti filaments embedded in a Cu–30%Ni matrix. Such a wire is produced by a multiple hot extrusion, drawing and compaction process similar to that described in the article *Multifilamentary Superconducting Composites*. After typically three such stages filament sizes can be reduced to about 100 nm. For the wire depicted in Fig. 1 at a

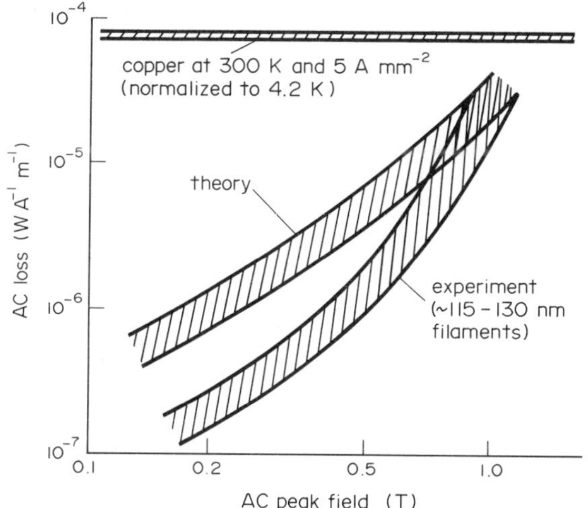

Figure 2
Schematic ac losses for ultrafine filamentary wires measured in $W A^{-1} m^{-1}$; the comparison to copper carrying $5 \times 10^6 A m^{-2}$ has been normalized to 4.2 K using a refrigerator efficiency of 500 (the true copper losses at 300 K are about $4 \times 10^{-2} W A^{-1} m^{-1}$)

final wire diameter of 0.15 mm the filaments are 131 nm in diameter with a twist pitch of 1.2 mm and are separated by 126 nm of resistive copper–nickel matrix (with a resistivity of about 4×10^{-7} Ω m at liquid helium temperatures, several thousand times more resistive than copper at the same temperature). Nb_3Sn fine multifilamentary wires, although more fragile, have a higher critical temperature (18 K instead of 9 K) and upper critical field (25–30 T instead of 11 T) and are also under development (Kubota et al. 1986).

1.2 Special Features of AC Loss in Ultrafine Niobium–Titanium Multifilamentary Wires

Measurements of ac loss for superconductors are described in the article *Measurements in Superconducting Materials.* A convenient test method is to wind the wire into a small test coil which is then subjected to a sinusoidal ac current (usually at 50–60 Hz). In this configuration the field and current vary together. Figure 2 shows ac losses for wires with different size filaments compared to calculated losses. The loss expressed in $W A^{-1} m^{-1}$ is obtained by dividing the total loss in $W m^{-3}$ by the J_{cov} and is used to obtain a more meaningful comparison between wires with different J_{cov} (see Eqn. (la)).

A first feature shown by these results is that the losses are about an order of magnitude lower than those of copper (carrying $5 \times 10^6 A m^{-2}$, a typical copper current density) confirming the suitability of such conductors in ac applications. A second feature is that the losses become lower as the ac field is reduced. The main reason for this can be seen from Eqns. (1) which show that the hysteretic losses and eddy current losses should vary proportionally to B_{max} and B^2_{max}, respectively.

However, a third feature of these results is that the theoretical calculations do not always accurately predict the experimental loss and thus there are also other phenomena at play (Cave 1989). The origin of this discrepancy is due to two effects. First, proximity coupling may occur when the filaments become too closely spaced (see *Coherence Length, Proximity Effect and Fluctuations*). This effect increases the effective diameter of the filaments and thus the losses. It may be controlled by simply increasing the filament spacing or by reducing the normal metal coherence length; the latter can be achieved by either increasing the matrix resistivity or by adding magnetic impurities (typically 0.5% manganese) into the matrix between the filaments (Goldfarb et al. 1989). Second, for non-proximity-coupled ultrafine filaments subjected to small ac amplitudes, the ac losses can be substantially lower than those calculated by the Bean critical state model (see *AC Losses in Superconducting Materials*; Takacs and Campbell 1988). This is because the full critical state is not attained when the flux vortex motion is less than the vortex spacing (see *Critical State Model*). Wires with the highest performance benefit from these effects when the ac field amplitudes are substantially lower than 1 T.

2. AC Applications at Industrial Frequencies

Many applications of superconductivity at industrial frequencies are covered in the article *Superconducting Materials: Large-Scale Applications.* The areas of interest for ac power uses fall under the major headings of power conversion (generators, motors and transformers), power transmission (high-power transmission cables) and power control, especially under fault conditions (fault current limiters, low-temperature power electronics and superconducting power regulators). By combining these elements, entirely superconducting power generation and control systems can be envisaged. However, the first likely applications will be where superconducting equipment can be incorporated into power systems to increase overall reliability and efficiency and where lightweight and compact powertrains are advantageous. Figure 3 shows a laboratory scale transformer and fault current limiter; the latter device is able to limit the buildup of the surge current due to its quick (<100 μs) response time. Figure 4 shows the stabilizing effect of a power regulator on the frequency variations that can occur after a short circuit fault. If these oscillations become too large the power generators become desynchronized. The power regulator can absorb energy from, or inject energy into, the line at just the right times in order to dampen the oscillations.

Figure 3
(a) A laboratory scale 50 Hz monophase 200 kVA superconducting transformer; (b) a 25 kV fault current limiter (courtesy Alsthom-Alcatel Recherche)

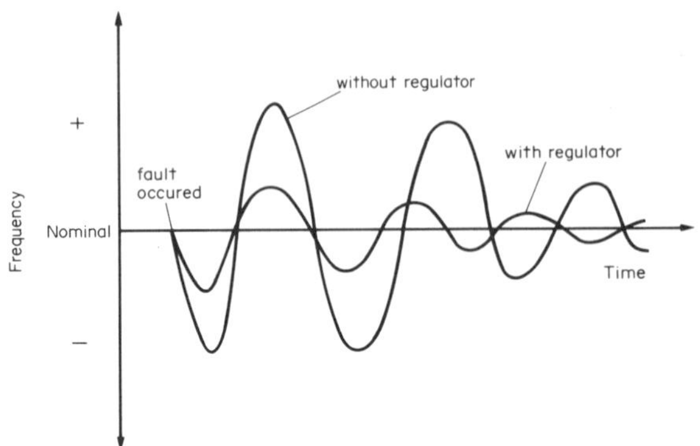

Figure 4
Schematic of the dampening effect on frequency oscillations during a fault condition using rapid, controlled power exchange from a power regulator

3. Impact of High-Temperature Superconductivity

With the advent of ceramic high-temperature superconductors many exciting new possibilities are emerging. These are tied to lower refrigeration costs as liquid helium (4.2 K) can be replaced by liquid nitrogen (77 K, at which temperature 1 W dissipated requires only a 10–20 W ambient refrigerator power) and a very high normal state resistance (around $10^{-5}\,\Omega\,\mathrm{m}$); this is particularly encouraging for fault current limiter designs. However, these ceramic materials are inherently brittle and much work needs to be done before they take their place in ac power equipment (see *Critical Current Density*; *Oxide Superconductors: Ceramic Processing*). It should be noted that low- and high-temperature superconductors have properties that differ from each other as well as from normal metals such as copper. Therefore, when incorporating superconductors into power equipment their special properties and requirements should be considered from the outset of machine design. Such an approach may often lead to new ways of realizing the desired function of the machine, instead of designs

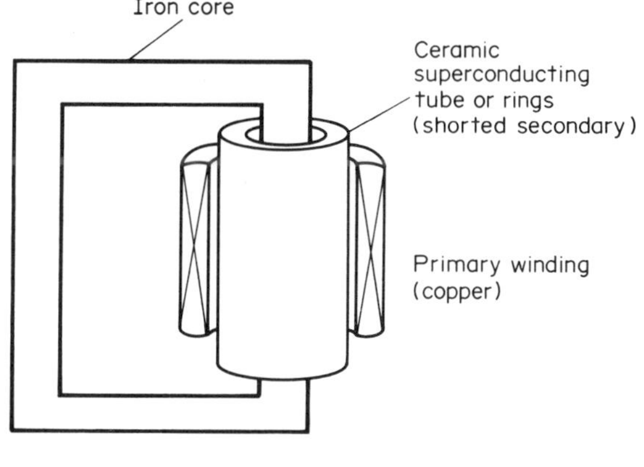

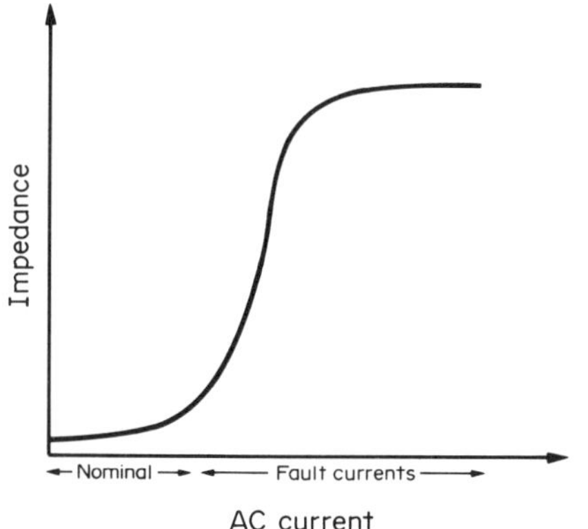

Figure 5
Principle of a high-temperature superconductor iron core fault current limiter

where copper conductors are simply replaced with superconductors. Such an example is a fault current limiter (Bashkirov et al. 1991) (see Fig. 5) that uses an ambient temperature iron core that passes through a set of closely spaced high-temperature ceramic superconducting rings. The current to be limited is passed through a winding exterior to the rings. During normal conditions the induced currents in the superconducting rings do not allow the magnetic field created by the winding to penetrate to the iron core. However, during a fault condition this field penetrates through the superconducting rings resulting in an increased impedance which in turn limits the fault current.

As the development of these new materials, and also low-temperature superconductors, progresses it is expected that they will play a more and more important role in ac power applications.

See also: AC Losses in Superconducting Materials; Critical Current Density; Critical State Model; Superconducting Materials: Large-Scale Applications

Bibliography

Bashkirov Yu A, Fleishman L S, Patsayeva T Yu, Sobolev A N, Vdovin A B 1991 Current-limiting reactor based on high-T_c superconductors. *IEEE Trans. Magn.* 27(2): 1089–92

Cave J R 1989 Electromagnetic properties of ultra-fine filamentary superconductors. *Cryogenics* 29: 304–11

Cave J R, Février A, Verhaege T, Lacaze A, Laumond Y 1989 Reduction of AC losses in ultra-fine multifilamentary NbTi wires. *IEEE Trans. Magn.* 25(2): 1945–8

Dubots P, Février A, Renard J C, Goyer J, Hoang Gia Ky 1984 Behavior of multifilamentary Nb–Ti conductors with very fine filaments under AC magnetic fields. *J. Phys. (Paris), Colloq.* 45: 467–70

Février A 1983 Losses in a twisted multifilamentary superconducting composite submitted to any space and time variations of the electromagnetic surrounding. *Cryogenics* 23(4): 185–200

Février A 1987 Latest news about superconducting AC machines. *Proc. 10th Int. Conf. Magnet Technology*. Boston, MA

Foner S, Schwartz B (eds.) 1974 *Superconducting Machines and Devices: Large Systems Applications*. Plenum, New York

Goldfarb R B, Ried D L, Kreilick T S, Gregory E 1989 Magnetic evaluation of Cu–Mn matrix material for fine-filament NbTi superconductors. *IEEE Trans. Magn.* 25(2): 1953–5

Guéraud A, Tavergnier J P, Février A, Laumond Y, Lacaze A, Dalle B, Ansart A 1989 Thermo-electromagnetic stability of ultra-fine multifilamentary superconducting cables for industrial frequency use. *Proc. 11th Int. Conf. Magnet Technology*. Tsukuba, Japan

Hlasnik I 1984 Prospects of multifilamentary superconductor AC 50 Hz applications. *J. Phys. (Paris), Colloq.* 45: 459–66

Kubota Y, Okon H, Ogasawara T 1986 Development of a Nb_3Sn multifilamentary composite conductor for AC use. *Cryogenics* 26(12): 654–9

Lacaze A, Laumond Y, Tavergnier J P, Février A, Verhaege T, Dalle B, Ansart A 1991 Coils performances of superconducting cables for 50/60 Hz applications. *IEEE Trans. Magn.* 27(2): 2178–81

Takacs S, Campbell A M 1988 Hysteresis losses in superconductors with very fine filaments. *Supercond. Sci. Technol.* 1(1): 53–6

Verhaege T, Tavergnier J P, Février A, Laumond Y, Bekaled M, Collet M, Pham V D 1989 25 kV superconducting fault current limiter. *Proc. 11th Int. Conf. Magnet Technology*. Tsukuba, Japan

J.R. Cave
[Hydro-Québec, Varennes, Canada]

A. Février
[Alsthom-Alcatel Recherche, Marcoussis, France]

AC Losses in Superconducting Materials

Superconductors only have zero resistance under dc conditions; a number of mechanisms give losses when the magnetic field is changed. At high frequencies, the dominant loss comes from the normal electrons and excitations across the superconducting gap (see *High-Frequency Impedance of Superconducting Materials*). At low frequencies, the dominant effects are due to the movement of flux vortices. The forces on the flux lattice may be pinning forces or the viscous drag due to the flux flow resistivity. This applies equally to flux tubes in type I superconductors and flux lines or flux vortices in type II superconductors (see *Magnetic Properties of Superconductors*). However, a quantitative theory only exists for the latter and even then only in certain limits (see *Resistive Transition and Flux Flow in Superconducting Materials*).

The local dissipation is $\boldsymbol{J}.\boldsymbol{E}$ per unit volume, where $\boldsymbol{J}$ is the transport current density and $\boldsymbol{E} = \boldsymbol{B} \times \boldsymbol{v}$ is the electric field with $\boldsymbol{v}$ equal to the flux line velocity. This is consistent with the picture of flux lines being forced past pinning centers because $\boldsymbol{J}.\boldsymbol{E} = \boldsymbol{J}.(\boldsymbol{B} \times \boldsymbol{v}) = (\boldsymbol{J} \times \boldsymbol{B}).\boldsymbol{v}$. Since $\boldsymbol{J} \times \boldsymbol{B}$ is the force exerted on the vortices, this can be interpreted as the rate at which the driving force does work. Except at very high flux line velocities, it can be assumed that $J = J_c$, the critical current density.

A second way of calculating the loss is from the hysteresis in the magnetization since the work dissipated per unit volume during a minor loop excursion is $\int B\,dM$. Usually it is possible to approximate to either the low penetration or the complete penetration regime (see *Critical State Model*). The two limits are illustrated on typical magnetization curves in Fig.1.

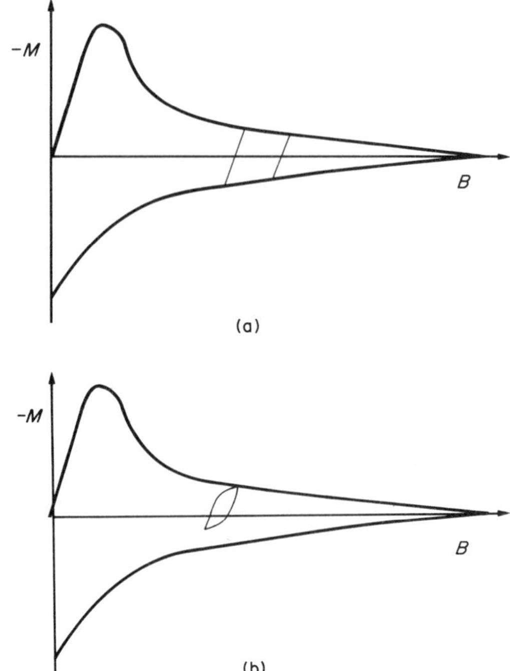

Figure 1
Typical magnetization curves: (a) large penetration, $B \gg \mu_0 J_c d$; (b) small penetration, $B \ll \mu_0 J_c d$, where d is the sample width

1. Order of Magnitude

First, an order of magnitude calculation will be given which will be justified in more detail later. If a superconductor is in an oscillating magnetic field of amplitude B, there will be no loss for both very low and very high critical current densities since, in each case, the magnetic hysteresis must tend to zero. Just as for the skin effect in the calculation of eddy current losses, the maximum loss is when the field just penetrates to the center of the sample, when the magnetization will be of the order of B/μ_0. Hence, the loss per cycle will be of the order of B^2/μ_0 per unit volume. In general, this is an upper limit for the loss. Losses at lower amplitudes will be reduced by the ratio of the penetration to the sample width $B/\mu_0 J_c d$, where d is the width. At higher amplitudes the losses scale as the reciprocal of the same factor $\mu_0 J_c d/B$. Thus, at the crossover, the loss changes from one proportional to the cube of the amplitude (B^3) to one proportional to the amplitude (B).

Hysteresis loss is normally quoted as a loss per cycle and since the volume of superconductor is normally fixed by the field required, it is usual to quote the loss per cycle per unit volume of superconductor. Losses are minimized by having either very thick material or very thin material. Thick wires are not possible with low-temperature superconductors (LTSs) because they are unstable against flux jumps but, at liquid nitrogen temperatures, thick monolithic high-temperature superconductor (HTS) material may prove more practical for some applications.

2. Bean Model Losses for Constant Critical Current Density

Initially, the commonest situation of an oscillating magnetic field, first with no transport current and then with a transport current added, will be treated. Constant J_c is assumed, which is extremely accurate if the external field is much larger than the field due to currents in the superconductor, so these results can be used to calculate the losses in the windings of magnets.

If the critical current density does not vary significantly with the flux density B, it is relatively easy to calculate the losses from the Bean model in simple geometries (see *Critical State Model*). The first calculation was by Kamper (1962). He used what is now known as the Bean model, although the model had not been published when the paper was submitted and had been derived independently by London (1963).

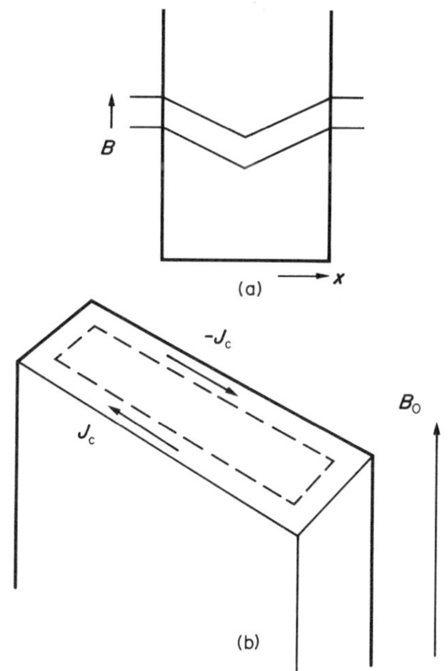

Figure 2
Slab geometry: (a) field distribution, (b) the current paths; the current is parallel to the electric field and $\boldsymbol{B}_0$ is the applied magnetic field

2.1 Large-Amplitude Magnetic Hysteresis

The loss for a slab geometry with an oscillating magnetic field parallel to the face of the slab (see Fig. 2) will be calculated, first, from the induced electric field and, second, from the magnetization.

If the magnetic flux has penetrated to the center of the slab, the rate of change of flux density dB/dt is everywhere the same as that at the surface. Hence, from Faraday's law of induction, the electric field at a distance x from the center of the slab is $E = x\,dB/dt$ (see *Critical State Model*). The local dissipation per unit volume is therefore $xJ_c\,dB/dt$ so that the dissipation per unit volume integrated over a slab of half width d is $\frac{1}{2}dJ_c\,dB/dt$. If the total field change is ΔB, then loss is $\frac{1}{2}dJ_c\Delta B$. For a complete cycle in the large amplitude limit ($B \gg \mu_0 J_c d$), it is permissible to ignore the small region just after the peak fields where the slab is not fully penetrated by the critical state and assume this expression for the whole cycle. The hysteresis loss for a zero to peak amplitude B is $2dJ_cB$ J m^{-3} per cycle.

The same result can be obtained from the area of the hysteresis loop in the magnetization curve. The magnetization is $\pm\frac{1}{2}J_cd$ so that the area of a complete loop of amplitude $\pm B$ is $2dJ_cB$ as was found before.

The following conclusions can be drawn in this regime. First, the loss is proportional to the size of the superconductor. Second, it is proportional to the critical current density. (However, to provide a given magnetic field, the volume of material required is inversely proportional to the critical current density so no advantage is gained by reducing J_c.) Third, the loss per cycle is independent of frequency. Fourth, the overall shape introduces a numerical factor but surface roughness in unimportant. The loss for superconductors of various shapes can be found from the magnetization (see *Critical State Model*). For a sphere of radius a the dissipation is $(3\pi/32)aJ_c\,dB/dt$, for a long cylinder with diameter $2a$ parallel to the field it is $\frac{1}{3}aJ_c\,dB/dt$ and for a transverse cylinder it is $(2/3\pi)$ $aJ_c\,dB/dt$.

2.2 Small-Amplitude Magnetic Hysteresis

The second easily calculable regime is for small penetrations. The Bean model is used to calculate the flux distribution through the cycle and the electric field is calculated as before. Integration gives the loss. The result for a slab is $\frac{2}{3}B^3/(\mu^2_0J_c)$ J m^{-2} per cycle. The loss is now proportional to the surface area of the sample so the loss per unit volume is inversely proportional to the sample size. It is $\frac{2}{3}B^3/(\mu^2_0J_cd)$ which can be written $(\frac{2}{3}B^2/\mu_0)(B/\mu_0J_cd)$ and is consistent with the order of magnitude calculated in Sect.1.

In slab geometry, this expression is exact until the flux reaches the center, when more complicated expressions can be calculated for the complete cycle. For more complicated shapes, a reasonable approximation is obtained if this expression is used as the local loss per unit area of surface, where B is the local flux density at the surface assuming perfect diamagnetism.

2.3 Rotating Fields

The effect of a uniform rotating field can be found in the large penetration regime by considering the work done by pulling the flux lines through the frictional force exerted by the pinning centers. At every point the dissipation is BJ_cv where v is the local velocity perpendicular to the flux lines. The result for a round wire transverse to the field is $(8/3)aJ_c$ J m^{-3} per revolution which is π times that for the oscillating field (Harrowell 1971). For small penetrations the critical state must be solved, but a similar factor occurs and, in general, a linear approximation, which gives a factor of 2, will overestimate the loss (Pang et al. 1981).

2.4 Magnetic Hysteresis and Transport Currents

If the external field is constant, the loss due to an alternating transport current in a slab is the same as for an oscillating field whose amplitude is that produced at the surface by the oscillating current. For a round wire the calculation was first carried out by London (1963). However, these calculations are normally only applicable to zero external field, when the assumption of constant J_c is a poor one, so the calculation has not found much practical use.

The commonest case is when the main effect is the oscillating external field, with the current oscillation producing a much smaller field. For large penetrations, the addition of a transport current shifts the electric center from the geometric center of the sample to the outside edge, as the current reaches the critical value. At this point the flux must travel twice as far through the sample as when no current is flowing, so the loss is approximately doubled. In general, a good approximation to the loss is to multiply it by a factor $1 + kj^2$, where j is the ratio of the transport current to the critical value and k is a numerical constant less than or equal to one (Campbell 1982). The value of k depends on the shape, and whether the current is constant or oscillating with the field. For rectangular conductors with constant current, $k = 1$ and, if the current varies in proportion to the field, $k = \frac{1}{3}$.

At the critical current there is a simple and general expression for the loss. For any conductor symmetric about the y axis in a large applied field in the y direction, the loss is $aJ_c\, dB/dt$ where a is the half width perpendicular to the field.

For small penetrations the effect of the current can be taken into account by adding the field of the current to the surface field. Thus, a dc transport current will not affect the losses in this regime since only the variable component of the surface field determines the loss.

2.5 Dynamic Resistance

It is found that when an oscillating magnetic field is applied to a wire carrying a dc current, a dc voltage appears across the end of the wire. This is called the dynamic resistance and a full account has been given by Ogasawara et al. (1976). A simple explanation of the effect will be given.

First, it must be realized that calculating the voltage measured at the ends of a superconducting wire is not straightforward, since the electric field is varying over the cross section and the end voltage is not the average electric field multiplied by the length (see *Critical State Model*). The voltage is not all dissipative since part of the work goes into changing the magnetic energy and part into dissipation. To get at the loss it is normal to integrate round a complete cycle.

Figure 3 shows the flux distribution in a slab carrying a steady current in an oscillating external field for an amplitude large enough to penetrate the sample. If the electric field is calculated as described previously and the voltage integrated over a whole cycle, the average is not zero. This can be understood by seeing that in Fig. 3a the flux in the central region is moving to the right and in the other half of the cycle (Fig. 3c) it continues to move in the same direction. There is therefore a net transfer of flux across the sample each cycle, caused by the magnetic field oscillation, even though the current through the wire is constant. The induced voltage appears as the dynamic resistance.

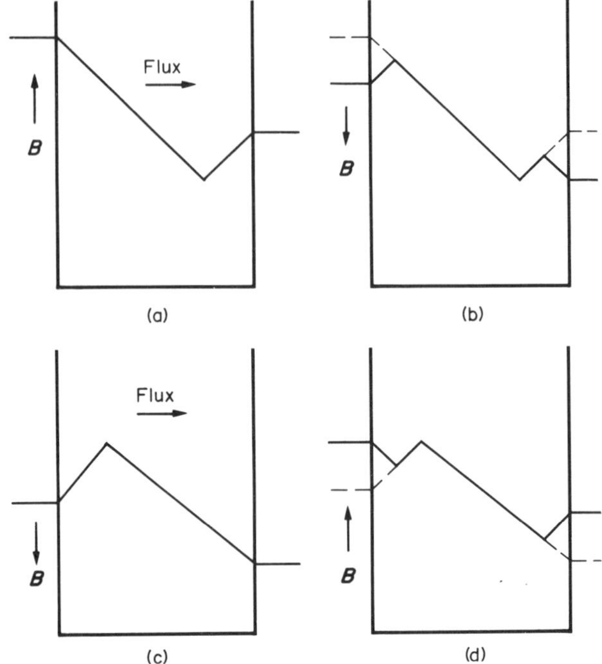

Figure 3
Flux distribution in a wire carrying a dc current in an oscillating field: (a) and (c) show the main part of the cycle, (b) and (d) the distribution near the turning points

This is not an extra loss to that calculated in Sect. 2.4. That result remains valid, but the appearance of the dynamic resistance shows that, instead of all the work being done by the source of the applied field, the magnetic work is reduced by an amount which is made up by the current source. When the critical current is reached the magnetization becomes completely reversible (it is then the average of increasing and decreasing magnetization curves) and all the energy is supplied by the current source.

3. Extensions of the Bean Model

The Bean model is very successful in most situations but a number of problems can arise, particularly at low fields and for low oscillation amplitudes. This section deals with losses in circumstances in which the Bean model does not work well.

3.1 Surface Barriers

In reversible samples or very fine filaments, the dominant hysteresis may come from the surface. This is characterized by a surface barrier to the entry and exit of flux B_s which can be treated by applying the Bean model to a very narrow surface region of very high J_c. The loss is found from the area of the hysteresis loop and for an amplitude B is $4BB_s/\mu_0$ J m^{-3} per cycle. More detailed calculations are given by Clszek et al. (1989)

3.2 Low Amplitudes

Low amplitudes refers to amplitudes which are not large enough to build up a critical flux gradient in the superconductor. This is not well defined but, in general, the vortices must move several vortex spacings if the Bean model is to be applied. Simple models predict that the movement will be completely reversible for displacements less than $a_0/2\pi$, where a_0 is the vortex spacing (Campbell 1978). For a magnetic field ripple of amplitude b applied with a steady background field B, the vortex displacement $y < bd/B$ where d is the sample half thickness or penetration distance, whichever is the smaller. If $y > a_0/2\pi$ is required, then $b > (B\Phi_0)^{1/2}/2\pi d$ where Φ_0 is the flux-quantum. For a field of 1 T in a sample 6 μm thick, this is an amplitude of about 3 mT. The loss in this regime can be found by assuming a shape for the force–displacement curve of the flux lines and that its curvature is due to unpinning, so that the hysteresis loop is a Rayleigh loop (Campbell 1971). In the general case this will involve calculating the flux profile numerically and the problem becomes complex. However, in practice, this situation usually occurs in very thin wires or small grains which are much smaller than the pinning penetration depth λ_{pin} (see *Critical State Model*). In this case, it is possible to assume that the flux density and, hence, the vortex displacements are, to a first approximation, the same as if there were no pinning forces. One can then find an analytic expression for the loss, which is much less than that predicted by the Bean model (Takacs and Campbell 1988). The amplitude is determined by a parameter c which is the ratio of the vortex displacement at the surface to the maximum reversible vortex movement, typically about 30 nm, or the vortex spacing divided by 2π.

The Bean model expression is then multiplied by a factor of

$$1 - (6/c^2)(1 - \exp - c) - (2/c)(2 + \exp - c)$$

This reduction in loss to well below the Bean model value has been observed experimentally in fine multifilaments (see *AC Applications of Superconducting Materials*).

3.3 Low Fields and Variable J_c

In principle, the arguments given previously can be used at low applied fields. However, there are a number of problems in using the theory in this regime. They arise from the difficulty of obtaining accurate measurements of J_c at low fields, its rapid variation with B so that the flux profile must be found by integration and the need to include the effects of the reversible magnetization, particularly if part of the cycle is below the lower critical field H_{c1} (see *Magnetic Properties of Superconductors*). A reasonable approximation to the critical current density in this regime is $J_c = k/(B + B_0)$, where k and B_0 are constants. Losses using this expression have been published, but the algebra is rather complex (e.g., Müller 1989). This regime is normally of interest for practical purposes for surface fields below H_{c1} when the effects are dominated by surface roughness; however, the results are also important where ultrafine filaments are used at 50 Hz.

3.4 Fields Below H_{c1}

Losses reduce dramatically in fields below H_{c1}. There are two mechanisms in this regime. One is the entry of flux into surface asperities where field concentrations make the local field exceed H_{c1}. This is difficult to quantify although Melville (1971) has calculated losses in this regime with a number of adjustable parameters which depend on surface topography. The second mechanism is the movement of flux trapped in the sample. This emerges from the surface where it can be moved by the external field leading to a loss. This is also difficult to quantify. Trapped flux will arise due to the earth's field and also due to any overload which has raised fields above H_{c1} at any time in the past. It is necessary to raise a cable above T_c after an overload to remove trapped flux and restore losses to the original value.

4. Composite Superconducting Wires

It was originally thought that flux jumps would limit the usefulness of type II superconductors in magnets. The problem was solved by drawing the material into very fine filaments (see *Multifilamentary Superconducting Composites*). This required a conducting matrix both to carry away heat and to transfer current in the event of a filament breaking. However, solving the stability problem introduced a second problem, that of eddy current losses in the matrix. If straight superconducting filaments are embedded in copper the flux must either cross the filaments when the magnetic field is raised, giving large hysteresis loss, or diffuse from the ends through the copper matrix, giving a large eddy current loss. (For a kilometer of copper at 4 K, the time constant for the decay of eddy currents is 200 years.) To reduce this problem the filaments are twisted, so reducing inductive coupling to the external field and bringing the time constant down to a value characteristic of the twist pitch rather than the wire length. The result was the very successful multifilamentary twisted superconductor developed at the Rutherford Laboratory (Wilson et al. 1970). The process of reducing losses by reducing the filament diameter has been continued with remarkable success so that it is now possible to buy material with a filament diameter of less than 100 nm that can be used with low losses at 50 Hz.

In superconducting composites the eddy current loss (often called coupling current loss) is a major loss mechanism which must be added to the hysteresis loss of the superconductor. A third loss component is the loss in surface filaments which carry the coupling currents back to the other side of the wire. The relative sizes of the three components depend on the frequency and amplitude.

There is a very large amount of literature on this subject and only a summary can be given. For a more complete list of references see *AC Applications of Superconducting Materials*; Campbell (1982). The first calculation treated the filaments as a loop and calculated the induced voltage across the wire using Faraday's law of induction (Morgan 1970). This circuit approach has proved very useful for applications to power frequencies. A second approach is to average the properties of the wire and treat it as an anisotropic continuum (Carr 1974). Carr solved the equations for the potential and this has probably been the most used approach. A development of this is to solve for the electric field rather than for the potential (Campbell 1982). The equations are simpler and more geometries can be treated. This arises because interest is normally centered on long conductors in a large uniform transverse magnetic field. If the curve is in the z direction and one puts $\partial/\partial z = 0$ in $\nabla \times \boldsymbol{E} = -(0, \dot{\boldsymbol{B}}_y, 0)$ then $E_z = x\dot{\boldsymbol{B}}_y$. The electric field in the z direction is independent of the superconducting filaments and the rate at which flux penetrates to the axis is the same as in a solid copper conductor. The local electric field must be perpendicular to the filaments and is associated with the fact that the filaments force the flux to travel at least one twist pitch down the length of the wire to reach the wire axis. It is this transverse field which generates the major part of the coupling loss. The longer the pitch the faster the flux must move to reach the axis in the same time, since it cannot cross the filaments, and the larger the loss. To this eddy current loss must be added the hysteresis loss of the filaments which has been treated previously.

4.1 Round Multifilamentary Wires at Low Amplitude and Frequency

A simple physical argument which gives the field distribution and loss is as follows. Consider a conductor rotating about its axis in a uniform field $\boldsymbol{B}$. If the twist pitch of the superconducting filaments is p the flux is screwed along the wire like a nut on a bolt because it cannot cross the filaments. The longitudinal velocity is $\omega p/2\pi$ where ω is the angular velocity of the wire. Since $\boldsymbol{E} = \boldsymbol{B} \times \boldsymbol{v}$, this flux movement generates a uniform electric field perpendicular to both the axis of the wire and the applied magnetic field. Its value is $B\,\omega p/2\pi$ and the current density is $\sigma B\omega p/2\pi$, where σ is the transverse conductivity of the composite. The current flows uniformly across the wire and is picked up at the surface by superconducting filaments. They carry the current back to the opposite side, half a pitch twist down the wire so current is conserved (see Fig. 4.). This is an exact self-consistent solution for low amplitudes and frequencies. The loss is $\sigma(B\omega p/2\pi)^2$ W m^{-3}.

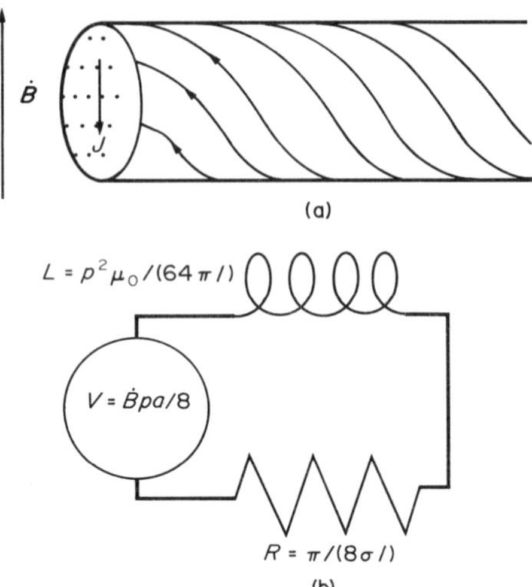

Figure 4
Round multifilamentary wire: (a) the field and current directions, (b) the equivalent circuit

The system consists of diamagnetic filaments in a copper matrix which is a linear material, so the rotating field can be decomposed into two independent oscillating fields at right angles, each with its corresponding oscillating current and each contributing half the loss. The loss for an oscillating field of zero to peak amplitude B is therefore $\frac{1}{8}\sigma(B\omega p/\pi)^2$ W m^{-3}. This is the same as that for a copper sheet whose thickness is $p\sqrt{3}/\pi$ so that the effect of twisting is to create decoupled copper sections with a thickness of the same order as the twist pitch.

In this context, low frequency means that the magnetic field generated by eddy currents is small compared with the external field (i.e., the twist pitch is larger than the skin depth of copper so that the internal field is equal to the applied field. Low amplitude means that the field is small compared with the field required to penetrate the filaments and also that the returning currents do not saturate the outer layer of filaments.

4.2 High-Frequency Effects

Consideration will be given first to the effect of increasing the frequency. This is best expressed as a function of the time constant of the system (Kwasnitza 1977, Campbell 1982). The eddy currents produce a back field proportional to the value of B in the material so one can write $B = B_0 - \tau\dot{B}$ where B_0 is the applied field and τ is a constant defined as the time constant. The system is essentially a voltage generator applied to an inductance and resistance. The time constant is similar to the effective time constant of a slab of thickness, equal to the twist pitch. The loss for an external amplitude B_0 is $2\pi B_0^2\omega\tau/\mu_0(1+\omega^2\tau^2)$ J m^{-3} per cycle. As the period decreases below the time constant, eddy currents in the copper exclude the field from the material so that losses go through a maximum of $\pi B_0^2/\mu_0$ per unit volume of conductor when the skin depth is of the order of the twist pitch ($\omega\tau = 1$) and decrease to zero at very high frequencies. The field exclusion also reduces the hysteresis loss from the filaments since the internal field is $B_0/(1+\omega^2\tau^2)^{1/2}$

The wire can be represented by the equivalent circuit of Fig. 4b. The values are for a wire of length l and radius a and geometrical constants are chosen so that the current in the circuit i is the same as the current across a diameter of the wire. The component values are

$$V = \dot{B}pa/8$$
$$i = \sigma\dot{B}pal/\pi$$
$$R = \pi/(8\sigma l)$$
$$\tau = \mu_0\sigma p^2/(8\pi^2)$$
$$L = \mu_0 p^2/(64\pi l)$$

4.3 High-Amplitude Effects

If the frequency is low the loss mechanisms are coupling loss and filament hysteresis loss. The coupling loss is proportional to B^2 while the filament loss is proportional to B^3 at low amplitudes and to B after full penetration of the filaments. Therefore, the filament loss is dominant in an intermediate range of amplitudes around the filament penetration field.

4.4 Saturation and Other Effects

For high frequency and high amplitude the flux pushes straight past both superconductor and copper and the twist pitch is irrelevant. The wire behaves as a solid cylinder with average properties and the filaments are said to be saturated. The condition for the filaments beginning to saturate is $dB/dt > \omega\tau\mu_0 aJ_{cov}$ where a is the wire radius and J_{cov} the overall critical current density of the wire. Below full saturation the saturation of the outer filaments leads to Bean-type losses with the wire diameter, rather than the filament diameter, as the relevant size.

The various regimes are illustrated in Fig.5. The dark areas indicate saturated areas of superconductor and the map is divided into regions of different dominant loss mechanism. The center is eddy current loss in the copper. At low frequencies, filament loss is dominant whereas at intermediate amplitudes and at high frequencies saturation of the outer filaments is dominant. The axes are in terms of the dimensionless quantities $\omega\tau$ and B_{pw}, where B_{pw} is the penetration field of the complete wire, $\mu_0 aJ_{cov}$. A second dimensionless parameter determines the relative positions

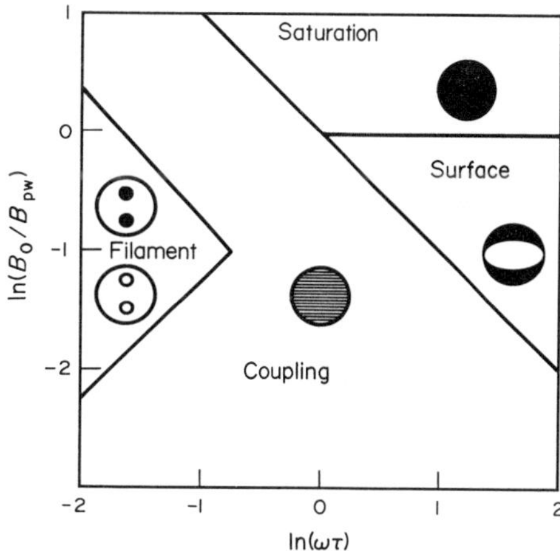

Figure 5
Map showing the dominant loss in various regimes

of the filament loss and saturation loss. This is fd/a where f is the volume fraction of superconductor, d is the filament radius and a is the wire radius.

Further corrections can be added to take account of the effect of transport currents and the proportion of material occupied by superconductor. Details and references can be found in Pang et al. (1981). The twisting does not decouple filaments as far as transport current is concerned and in zero field the effective size is that of the full wire. However, in an applied field a subtle effect caused by the need to conserve flux between the moving electric centers of the filaments leads to an almost uniform distribution of transport current in the composite.

4.5 Cables

More complications arise when the composite is wound into a cable, with contact between the wires determined by the solder holding it together. The circuit model now has generators V_f corresponding to the field across the cable as a whole. V_f is determined

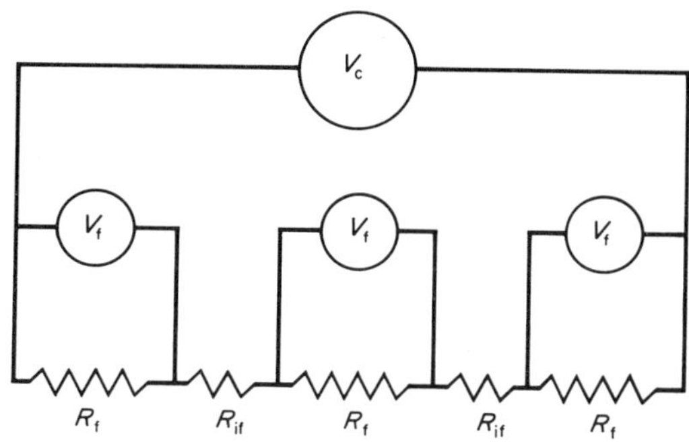

Figure 6
The equivalent circuit for a three-strand cable

by the filament twist pitch and V_c is determined by the pitch of the cable. Figure 6 shows the low-frequency circuit. The resistance between each wire R_{if} may be very low and, if the cable voltage is not equal to the sum of the wire voltages, a finite voltage across the interwire contacts can give large losses and a greatly increased time constant (Campbell 1982). Ideally the cable should have the same pitch and sense as the filaments, although this may not be possible in practice. The combination of eddy currents in composite wires and cables with flux creep in both transport currents and magnetization currents in filaments means that the design of large magnet systems is a very complex matter which can still give some unpleasant surprises.

See also: AC Applications of Superconducting Materials; Critical State Model; Multifilamentary Superconducting Composites

Bibliography

Campbell A M 1971 The interaction distance between flux lines and pinning centres. *J. Phys. C* 4: 3186–98

Campbell A M 1978 Pinning parameters in the PbTl system. *Philos. Mag. B* 37: 149–74

Campbell A M 1982 A general treatment of losses in multifilamentary superconductors. *Cryogenics* 22: 3–16

Carr W J 1974 AC loss in a twisted superconducting wire. *J. Appl. Phys.* 45: 929–38

Clszek M, Teklel P, Koslowski G 1989 *Supercond. Sci. Technol.* 1: 360

Harrowell R V 1971 AC losses in rotating and reciprocating superconductors and normal conductors. *J. Phys. D* 4: 1769–75

Kamper R A 1962 AC loss in superconducting lead bismuth. *Phys. Lett.* 2: 290–4

Kwasnitza K 1977 Scaling laws for the AC loss of multifilamentary superconductors. *Cryogenics* 17: 616–19

London H 1963 Alternating current losses in superconductors of the second kind. *Phys. Lett.* 6: 162–5

Melville P H 1971 Theory of AC loss in type II superconductors in the Meissner state. *J. Phys. C* 4: 2833–48

Morgan G W 1970 Theoretical behaviour of twisted multicore superconducting wire in a time varying magnetic field. *J. Appl. Phys.* 41: 3673–9

Müller K-H 1989 Susceptibility of high temperature superconductors in a critical state model. *Physica C* 159: 717–25

Ogasawara T, Yasukochi K, Nose S, Sekizawa H 1976 Effective resistance of current carrying superconducting wire in oscillating magnetic fields, 1: single core composite cable. *Cryogenics* 16: 33–8

Pang C Y, Campbell A M, McLaren P G 1981 Losses in Nb/Ti multifilamentary composites when exposed to transverse alternating and rotating fields. *IEEE Trans. Magn.* 17: 134–7

Sumiyoshi F, Matsuyama M, Noda M, Matshushita T, Funaki K, Iwakuma M, Yamafuli K 1986 *Jpn. J. Appl. Phys.* 25: 148–50

Takacs S, Campbell A M 1988 Hysteresis losses in superconductors with fine filaments. *Supercond. Sci Technol.* 1: 53–6

Wilson M N, Waters C R, Lewin J D, Smith P F, Spurway A H 1970 Experimental studies of filamentary superconducting composites. *J. Phys. D* 3: 1517–46

A. M. Campbell
[University of Cambridge,
Cambridge, UK]

Alnicos

The alnicos are a very important group of permanent-magnet alloys which are used in a wide range of electrical and electronic devices and machines. They are essentially heat-treated Fe–Co–Ni–Al–Cu alloys which can be divided into two main subgroups: (a) the isotropic alnicos 1–4 which have a cobalt content in the range 0–20 wt%; and (b) the anisotropic alnicos 5–9 which have a cobalt content in the range 22–40 wt% and a titanium content of 5–8.5 wt%. The anisotropy in the latter alloys is produced by controlled cooling or isothermal heat treatment in a saturating magnetic field. The best permanent-magnet properties in both the isotropic and anisotropic alnicos are produced by controlled cooling followed by tempering for several hours at about 600 °C.

1. Isotropic Alnicos 1–4

The first member of the alnicos (numbered, rather strangely, as alnico 3) has the approximate composition Fe_2NiAl. After cooling at about 30 K s^{-1} from the single-phase body-centered-cubic (bcc) state at 1100 °C followed by tempering for 4 h at 650 °C, a flux coercivity $_BH_c = 41$ kA m^{-1} and a maximum energy product of $(BH)_{max} = 11$ kJ m^{-3} can be obtained. The latter can be increased to 12 kJ m^{-3} by the addition of about 3% copper, but the best properties are obtained by the addition of 1% titanium and the application of a tensile stress giving $(BH)_{max} = 17$ kJ m^{-3}, $_BH_c = 63$ kA m^{-1} and $B_r = 0.67$ T. Significant improvements in the energy products of Fe–Ni–Al–Cu alloys can also be obtained by adding up to 25% cobalt. The cobalt-containing isotropic alloys are known as alnicos 1, 2 and 4. Typical *B–H* demagnetization curves for alnicos 1–4 are shown in Fig. 1. The extra-high coercivity of alnico 2 results from the relatively high cobalt and titanium contents. The composition and magnetic properties of the isotropic alnicos 1–4 and those of the anisotropic alnicos 5–9 are given in Table 1, with fuller details given by Heck (1974 pp. 686–701).

During cooling of the alnicos 1–4 from about 1250 °C the bcc α phase undergoes a spinodal decomposition into two bcc phases, α_1 and α_2. The α_1 phase is iron rich or iron–cobalt rich and is strongly ferromagnetic, whereas the α_2 phase is nickel–aluminum rich and is weakly ferromagnetic. The α_1 phase forms a complex three-dimensional interlocking system of

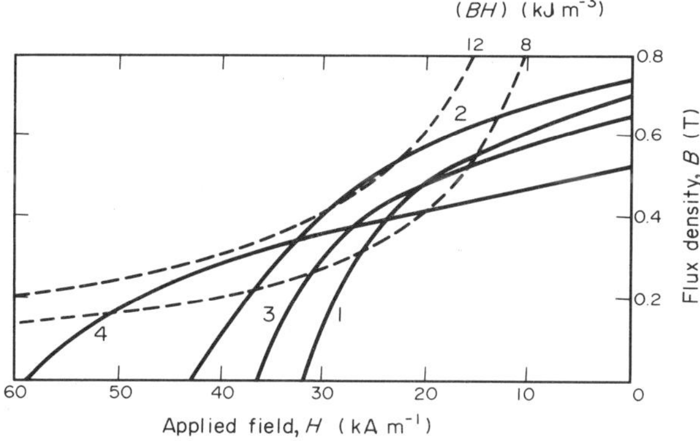

Figure 1
Typical B–H demagnetization curves for isotropic alnicos 1–4

"regularly" spaced elongated particles (not well defined) in which the axes of elongation are approximately parallel to the ⟨100⟩ directions. Such a microstructure is characteristic of phase separation by spinodal decomposition. (When phase separation occurs by nucleation and growth the particles are "randomly" distributed.) The coercivity of the alloys can therefore be attributed principally to the shape anisotropy of the iron- or iron–cobalt-rich α_1 phase particles. If it can be assumed for simplicity that the particles are very approximately prolate ellipsoids of revolution (i.e., elongated ellipsoids or rods) then, assuming that magnetization reversal occurs by coherent rotation, the intrinsic coercivity H_c of a single particle is given by

$$H_c \simeq (D_z - D_x)M_s \tag{1}$$

where D_z and D_x are the respective demagnetization factors (see *Demagnetizing Factors*) perpendicular and parallel to the axis of elongation of the particle and M_s is its saturation magnetization. Strictly speaking, M_s should be replaced by ΔM_s, the difference in the saturation magnetization of the α_1 phase particles and the surrounding nickel–aluminum-rich matrix (α_2 phase). The observed coercivities are usually the flux coercivities $_BH_c$ (see Table 1) rather than the intrinsic coercivities H_c, but for the alnicos, H_c is usually not more than 10% greater than $_BH_c$. Since the microstructures of the isotropic alnicos 1–4 are very complex, it is not possible to make a quantitative comparison between the observed and theoretical coercivities, but there is little doubt that the coercivity is due principally to the shape anisotropy. This conclusion has been confirmed by simulating the "magnetic" microstructure of Fe_2NiAl by embedding ferromagnetic rods parallel to the three mutually orthogonal axes [100], [010] and [001] in a nonferromagnetic matrix. Comparison of the magnetic torque curves obtained in a (100) plane of a single crystal of Fe_2NiAl with the torque curves of the simulated specimen shows that the magnetic behavior is essentially the same.

2. Anisotropic Alnico 5

The most important advance in the development of the alnicos was made when it was found that the energy product, coercivity and remanence could be increased significantly by the controlled cooling of the cobalt-containing alloys ~20–25% cobalt) from 1200 °C in a saturating magnetic field. This thermomagnetic treatment produces anisotropic magnets in which the preferred direction of magnetization is parallel to the direction of the field applied during cooling. The magnetic properties in the preferred direction are further improved by subsequent tempering for several hours at about 600 °C. The application of a field during the latter treatment, however, has no effect on the final magnetic properties. The effect of annealing at around 600 °C is to increase the difference in the saturation magnetic polarizations of the α_1 and α_2 phases, which in turn increases the coercivity, remanence and energy product.

The principal reason for the improvement in the magnetic properties after the thermomagnetic (and tempering) treatments is the increase in the "fullness" or "squareness" factor of the B–H demagnetization curve and the remanence; the increase in the coercivity has relatively little effect on the energy product. A typical B–H demagnetization curve for alnico 5 is shown in Fig. 2.

3. Grain-Oriented Alnico 5

Further improvements in the magnetic properties of alnico 5 can be achieved by grain orienting the alloy before the thermomagnetic and tempering treatments. This is done by casting the alloys in heated or exothermic molds onto steel or copper slabs (preferably water cooled). For iron-base alloys such as alnico 5, it has been found that when they are allowed to solidify on a cold surface the grains tend to grow with their axes parallel to the ⟨100⟩ directions and preferentially perpendicular to the cold surface. This results in a "semicolumnar" alloy in which the columnar axis (of the elongated grains) is parallel to one of the ⟨100⟩ directions, for example [001].Grain-oriented alnico 5, often referred to as alnico 5DG (DG means directed grain), has significantly better properties than alnico 5, but the optimum magnetic properties are of course obtained for single crystals. The B–H demagnetization curves for alnico 5DG and single-crystal alnico 5 are shown in Fig. 2. The highest energy products yet reported for these magnets are about 58 kJ m^{-3} and 88 kJ m^{-3} respectively. Unfortunately, single-crystal alnico 5 magnets are difficult and

Table 1
Compositions and magnetic properties of isotropic alnicos 1–4 and anisotropic alnicos 5–9

	Composition (wt%)[a]						Magnetic properties		
Characteristics	Alnico	Ni	Al	Co	Cu	Others	B_r (T)	${}_BH_c$ (kA m^{-1})	$(BH)_{max}$ (kJ m^{-3})
Isotropic, Co-free	3	24–30	12–14	0	0–3	0–1 Ti	0.5–0.6	54–40	10
Isotropic, low Co	1,4	21–28	11–13	3–5	2–4	0–1 Ti	0.55–0.75	56–36	11–12
Isotropic, medium Co	2	16–20	9–11	12–14	3–6	0–1 Ti	0.65–0.8	50–40	13–14
Isotropic, extra-high ${}_BH_c$	2	18–21	8–10	17–20	2–4	4–8 Ti 0–1 Nb	0.6–0.7	72–60	14–16
Field-treated random grain	5	12–15	7.8–8.5	23–25	2–4	0–0.05 Ti 0–1 Nb	1.2–1.3	52–46	40–44
Field-treated random grain, higher ${}_BH_c$	6	13–16	7.8–8.5	23–25	2–4	1–1.5 Ti 1–2 Nb	1.05–1.15	62–56	30–36
Field-treated random grain, high ${}_BH_c$	7	18	8	24	3	5	0.74	85	24
Field-treated random grain, high ${}_BH_c$	8	14–16	7–8	32–36	4	4–6 Ti 0–1 Nb	0.8–0.9	140–110	40–45
Field-treated random grain, extra-high ${}_BH_c$	8	14–15	7–8	37–40	3	7–8 Ti	0.74–0.78	170–150	44–48
Field-treated directed-grain form of alnico 5	5DG	13–15	7.8–8.5	24–25	2–4	0–1 Nb	1.3–1.4	62–56	56–64
Field-treated columnar-grain form of alnico 8	9	14–16	7–8	32–36	4	0–1 Nb 0.3 S	1.0–1.1	140–110	60–75

a Balance Fe

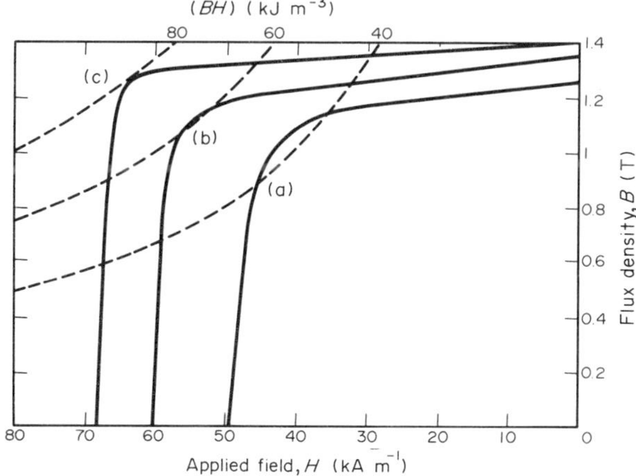

Figure 2
Typical *B–H* demagnetization curves for (a) alnico 5, (b) alnico 5DG and (c) single-crystal alnico 5

expensive to produce on a commercial scale. Alnico 5 and alnico 5DG or their equivalents are among the most widely used permanent-magnet alloys.

High-coercivity forms of polycrystalline alnico 5, known as alnico 6 and alnico 7, can be produced by adding small quantities of titanium and niobium, but the higher coercivities are obtained at the expense of the remanence and the energy product (see Table 1).

4. *Microstructure of Alnico 5*

Electron microscopic and magnetic investigations of alnico 5 have shown that the axes of elongation of the iron–cobalt-rich α_1 phase particles are usually very close to the direction of the thermomagnetic field but, due to their elastic and interfacial energies, there is also a metallurgical preference for them to form with their axes parallel to the $\langle 100 \rangle$ direction which is closest to the field direction. The best properties are obtained when the "thermomagnetic" field is applied parallel to one of the $\langle 100 \rangle$ directions. This condition can be satisfied only for single crystals of known orientation, but it can be partially satisfied for columnar crystallized alnico 5DG. In the electron micrograph of alnico 5 shown in Fig. 3, it can be seen that the iron–cobalt-rich particles (dark phase) are rodlike, about 30 nm in diameter and about 120 nm long. Particles of this size are single domains, so magnetization reversal cannot occur by nucleation and movement of Bloch–Lifshitz domain walls (see *Coercivity and Domain Wall Pinning*; *Domains and Domain Walls*), which for iron are of the order of 100 nm wide and are therefore much too broad to be nucleated in such small particles. Since the degree of elongation (the length–diameter ratio of the particles, l/d) is around 4–6, the difference in the demagnetization

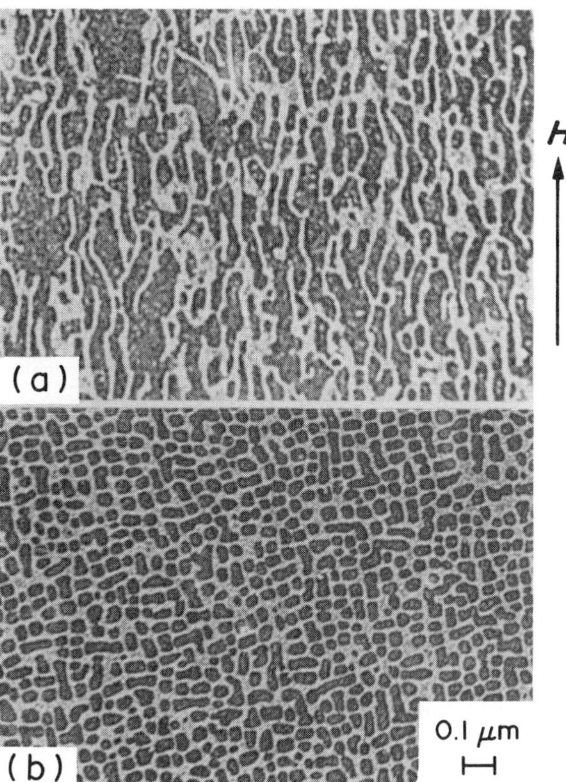

Figure 3
Microstructure of alnico 5: (a) in a plane showing the elongation of the iron–cobalt-rich particles parallel to the applied field during thermomagnetic treatment and (b) in a plane perpendicular to the direction of the field (note the regularity of the particle spacing characteristic of phase separation by spinodal decomposition)

factors perpendicular and parallel to the axis of elongation (D_z–D_x) is about 0.4 (McCurrie 1982), compared with the maximum value of 0.5 for infinitely elongated prolate ellipsoids of revolution. Assuming that the iron–cobalt-rich particles have a saturation magnetization M_s of about 1.7 MA m^{-1}, substitution into Eqn. (1) gives a theoretical intrinsic coercivity H_c of around 700 kA m^{-1}. Since the observed coercivities, even in single crystals ($_BH_c \simeq 60$ kA m^{-1}), are very much less than this, it is clear that magnetization reversal cannot occur by coherent rotation, as is assumed in the derivation of Eqn. (1). There is considerable evidence to show that magnetization reversal occurs by the complex incoherent mechanism known as "curling" (McCurrie 1982), which occurs in lower fields and, hence, accounts for the lower observed coercivities, although because of the complexity of the magnetic behavior, full quantitative agreement between the experimental and theoretical values is lacking.

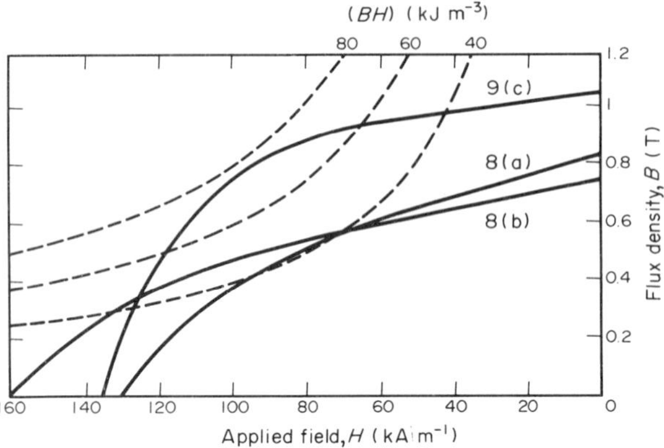

Figure 4
Typical *B–H* demagnetization curves for (a) alnico 8, (b) extra-high-coercivity alnico 8 and (c) columnar alnico 9

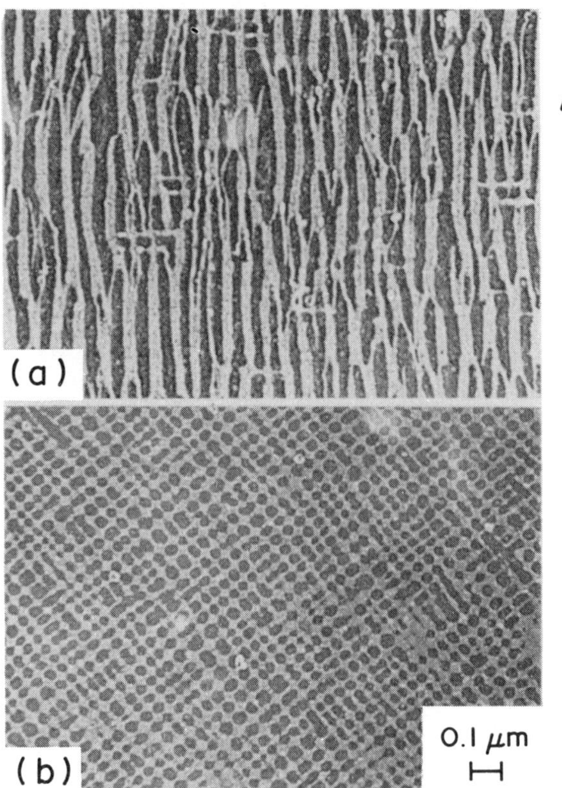

Figure 5
Microstructure of alnico 8: (a) in a plane showing the elongation of the iron–cobalt-rich particles parallel to the applied field during thermomagnetic treatment and (b) in a plane perpendicular to the applied field direction (note the regularity of the particle spacing, as in Fig. 3)

5. Anisotropic Alnico 8

The high-energy products of alnico 5DG are achieved principally as a result of the improvement in the squareness of the *B–H* demagnetization curve and the increase in the remanence in the thermomagnetic field direction which is parallel to the columnar axis. The coercivity, however, remains almost the same as that in alnico 5 in which the grains are randomly oriented. For many applications where greater stability of the magnetic properties is required, it is necessary to use permanent magnets with higher coercivities and it is for these applications that the high-coercivity alnico 8 alloys were developed. These usually contain 32–36 wt% cobalt and 4–5 wt% titanium—the elements which are responsible for the increase in the coercivity. Typical coercivities are in the range 90–120 kA m^{-1} with a $(BH)_{max}$ of around 40–49 kJ m^{-3}. Unfortunately, the high coercivities are obtained at the expense of the remanences ($B_r \simeq 0.8$ T), which are significantly lower than those for alnicos 5 and 5DG. A typical *B–H* demagnetization curve for alnico 8 is shown in Fig. 4. Even higher coercivities, up to a maximum of 190 kA m^{-1}, can be obtained by increasing the cobalt and titanium contents to about 38% and about 8.5%, respectively (Fig. 4). Sometimes, 1–2% of the titanium content is replaced by niobium. Typical compositions and magnetic properties of alnico 8 alloys are given in Table 1.

Alnico 8 alloys are homogenized at 1250 °C to produce the high-temperature bcc phase and then cooled at a controlled rate to about 820 °C (i.e., at a temperature just below the Curie temperature of the iron–cobalt-rich α_1 phase particles), at which temperature they are given an isothermal heat treatment in a saturating magnetic field (~300 kA m^{-1}) for a few minutes (~5–10 min). After this treatment they are tempered for about 6 h at 650 °C, followed by about 24 h at 550 °C. The isothermal heat treatment in the magnetic field is necessary because the presence of the relatively large Ti^{2+} ions lowers the rates of diffusion in the alloys, so sufficient time at 820 °C must be allowed for the preferred growth of the particles (in the field direction) to occur. This thermomagnetic heat treatment, combined with the higher cobalt content and the titanium content, increases the elongation and saturation magnetization of the particles and, hence, their coercivity.

Electron micrographs of alnico 8 (Fig. 5a, b) clearly show the rodlike shape of the particles and their regular spacing, which is characteristic of phase separation by spinodal decomposition. Since the average length and diameter of the particles are about 30 nm and about 600 nm, respectively, it may be assumed that they are single domains with elongation $l/d \simeq 20$, which corresponds to $D_z–D_x \simeq 0.5$. With $M_s \simeq 1.7 \times 10^6$ A m^{-1}, Eqn. (1) gives a theoretical intrinsic coercivity H_c of around 800 kA m^{-1}. Since the latter is very much larger than the maximum observed coercivity of around 190 kA m^{-1}, it is clear that, in spite of the high elongation of the particles, magnetization reversal does not occur by coherent rotation but rather by curling.

6. Columnar Alnico 9

The magnetic properties of polycrystalline alnico 8 can be significantly improved by columnar crystallization, producing the alloy known as alnico 9. Unfortunately, the high titanium content of alnico 8 decreases its grain size and makes columnar crystallization difficult. However, this problem can be overcome by the addition of small quantities of sulfur, selenium or tellurium which increase the grain size. The fully columnar grain structure is obtained by casting the alloy in a heated or exothermic mold with a base plate consisting of a chilled steel or copper slab (preferably water cooled). This results in the preferential growth of grains in which one of the ⟨100⟩ axes grows perpendicular to the plane of the chill plate. The alloy is then homogenized at about 1250 °C to produce the bcc α phase, cooled at a carefully controlled rate to about 820 °C and held at this temperature for about 5–10 min in a saturating magnetic field (~ 300 kA m^{-1}) parallel to the [001] columnar axis. Following this, the alloy is given the same two-stage tempering treatment as alnico 8. Although the coercivity of alnico 9 is about the same as that of alnico 8, there is a significant increase in the remanence parallel to the [001] columnar axis (see the demagnetization curve in Fig. 4). The maximum energy products for alnico 9 alloys are typically in the range 60–75 kJ m^{-3}. These improved properties parallel to the columnar axis are obtained at the expense of those perpendicular to and at other angles to the columnar axis. However, in many applications this is not a serious disadvantage. The best magnetic properties are obtained with single crystals which are magnetized in the ⟨100⟩ direction which was parallel to the field during the thermomagnetic treatment. For such a single crystal an energy product of 107 kJ m^{-3} with a coercivity $_BH_c$ of 122 kA m^{-1} can be obtained. This energy product is the highest yet reported for any alnico. Typical compositions and magnetic properties of alnico 9 alloys are given in Table 1.

Since alnico 9 alloys are produced by columnar crystallization of alnico 8, it seems reasonable to assume that the microstructure, particle dimensions and volume fractions are almost identical to those of alnico 8.

7. Sintered Alnicos

All the alnicos can also be made by sintering the powdered alloys at 1250–1350 °C. After sintering, they are given a heat treatment appropriate to their composition. The resulting alloys have magnetic properties inferior to those of the cast alloys, but they have the important advantage over the latter that small magnets with intricate shapes can be manufactured which would be very difficult or very expensive to produce by the usual casting and grinding processes. Typical magnetic properties for sintered alnicos 2,5,6 and 8 are shown in Fig. 6.

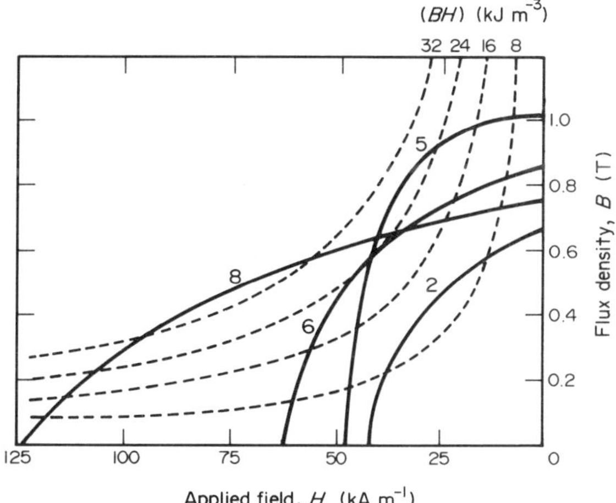

Figure 6
Typical B–H demagnetization curves for sintered alnicos 2, 5, 6 and 8

8. Bonded or Molded Alnico Magnets

Very small magnets can be produced by bonding alnico powders in a thermoplastic binder so that they can be shaped as required. Since the volume fraction of the magnetic material is <1, these magnets have magnetic properties inferior to those of sintered alnicos.

9. Effects and Theory of Thermomagnetic Treatment of Alnicos 5–9

Numerous experiments have shown that the principal effect of heat treatment (either cooling or isothermal) in a magnetic field is to elongate the iron–cobalt-rich particles in a direction which is parallel to or close to the applied field direction. The best magnetic properties are obtained when the field is parallel to a ⟨100⟩ direction, namely, the [001] columnar axis. The elongation of the iron–cobalt-rich particles parallel to the field direction can be explained as follows. When the homogeneous alloy is cooled from the high-temperature single-phase (bcc α) state to a critical temperature T_s of around 850 °C, it undergoes a phase separation by spinodal decomposition to form the α_1 and α_2 phases. The resulting iron–cobalt-rich particles (α_1 phase) are approximately spherical and regularly spaced. As they cool below their Curie temperature T_C (T_s–$T_C \simeq 20$–30 °C) in the presence of a magnetic field (preferably of saturating strength), they grow preferentially in the field direction. This elongation occurs because the magnetic free energy of the particles is lower when the axis with the lowest demagnetization factor (i.e., the longest axis) is in the direction of the applied field. Although the elongation increases the surface area of the particles and, hence, the interface area between them and the matrix, the resulting increase in the interfacial energy is small compared with the reduction in the magnetic energy which accompanies the elongation.

It has been established that the addition of cobalt to the alnicos reduces the interfacial energy. This accounts for the fact that thermomagnetic treatment is most effective in the alloys of high cobalt content, such as alnicos 5–9. The particle elongation in the alloys is also favored by the addition of cobalt because it increases the difference in the magnetic polarizations of the iron–cobalt-rich particles and the matrix and, hence, further reduces the magnetic energy when the particles are elongated in the field direction.

The thermomagnetic treatment of alnico 5 alloys produces an induced overall shape-anistropy energy constant K_u of around 150 kJ m^{-3} with the "easy" or preferred axis of magnetization parallel to the field direction during heat treatment (see *Anisotropy in Superconducting Materials*). In alnicos 8 and 9, $K_u \simeq 230$ kJ m^{-3}. The higher coercivities of alnicos 8 and 9 compared with that of alnicos 5, 6 and 7 can therefore be attributed to the increase in the shape anisotropy resulting from (a) the increase in the saturation magnetization of the iron–cobalt-rich particles, due to the increased cobalt content, and (b) the increased elongation of the particles due to both the cobalt and the titanium contents.

10. Applications of Alnico Magnets

Alnico permanent magnets, particularly alnico 5 and alnico 5DG, are very widely used. The most important applications are in loudspeakers (although they are rapidly being replaced by $BaFe_{12}O_{19}$ and $SrFe_{12}O_{19}$, which are very much cheaper since they do not contain either cobalt or nickel (see *Ferrites, Hard*)), watt-hour meters, moving-coil ammeters and voltmeters, electric motors, generators and alternators. The high Curie temperature of the alnicos and their magnetic and structural stability up to about 550 °C make them virtually indispensable in applications where they are exposed to high ambient temperatures. They can also be used for short times at around 600 °C without serious deterioration in their magnetic properties. The alnicos are also used in many other devices (e.g., see Bradley 1971 pp. 159–209, Heck 1974, McCaig 1977 pp. 220–308).

See also: Anisotropy in Magnetic Materials; Cobalt–Rare-Earth Alloys; Coercivity and Domain Wall Pinning; Ferrites, Hard; Measurements in Magnetic Materials; Neodymium–Iron–Boron Alloys

Bibliography

Bradley F N 1971 *Materials for Magnetic Functions.* Hayden, New York

Cullity B D 1972 *Introduction to Magnetic Materials.* Addison-Wesley, Reading, MA

de Vos K J 1969 Alnico permanent magnet alloys. In: Berkowitz A E, Kneller E (eds.) 1969 *Magnetism and Metallurgy*, Vol. 1. Academic Press, New York, Chap. 9, pp. 473–512

Heck C 1974 *Magnetic Materials and their Applications.* Butterworth, London

McCaig M 1977 *Permanent Magnets in Theory and Practice.* Pentech, London

McCurrie R A 1982 The structure and properties of alnico permanent magnet alloys. In: Wohlfarth E P (ed.) 1982 *Ferromagnetic Materials. A Handbook on the Properties of Magnetically Ordered Substances*, Vol. 3. North-Holland, Amsterdam, Chap. 3, pp. 107–88

Parker R J 1990 *Advances in Permanent Magnetism.* Wiley, New York

Tebble R S, Craik D J 1969 *Magnetic Materials.* Wiley, New York

R. A. McCurrie
[University of Bradford, Bradford, UK]

Amorphous Superconductors

Amorphous superconductors are a class of superconducting materials that lack long-range, translational ordering in their atomic arrangement. Since the discovery of superconductivity in amorphous bismuth and gallium (a-Bi, a-Ga) by Bückel and Hilsch in the 1950s, noncrystalline superconductors have been the subject of numerous experimental and theoretical studies. Recently, interest in this field has been focused on amorphous transition-metal (TM) superconductors as opposed to the amorphous non-transition-metal (non-TM) superconductors investigated earlier.

The techniques used to synthesize amorphous superconducting materials include quenching from the vapor (obtained by evaporation or sputtering), particle irradiation, ion mixing or ion implantation, and liquid quenching. The technique of quenching from the melt is capable of producing large quantities of metallic glasses in the form of ribbons, flakes or powder suitable for large-scale engineering applications. Vapor deposition and particle bombardment

Table 1
The T_c of some amorphous (or highly disordered) superconductors and their crystalline counterparts

Metals	Amorphous or highly disordered (K)	Crystalline (K)
Nontransition metals		
Pb	7.16	7.2
$Pb_{50}Bi_{50}$	6.99	8.1
Ga	8.4	6 (β-Ga) 1.08 (α-Ga)
Bi	6.1	<0.05
Be[a]	10	0.026
Al[a]	6	1.18
Transition metals		
Nb	5.7	9.2
Mo	8–9	0.9
Nb_3Ge	3.6	23
Nb_3Si	3.9	18
Nb_3Sn	3	18
V_3Si	1.5	17
Mo_3Si	7.5	1.4

a Extrapolated values

methods are used to prepare thin films amenable to small-scale electronic device applications.

In general, the Cooper pairs in superconductors are formed from time-reversed electron states, no matter how complicated these states are in real space as a result of scattering from various disorders. Hence, the basic mechanism of superconductivity is not expected to be different in disordered or crystalline materials. There are, however, several interesting differences between amorphous superconducting materials and their crystalline counterparts; these differences will be discussed.

1. Systematics of the Transition Temperature

1.1 Nontransition Metals

Amorphous non-TM superconductors usually show an enhancement in transition temperature T_c over that of corresponding crystalline phases. For example, a-Bi thin films become superconducting at about 6 K, whereas the crystalline phase, a semimetal, remains normal down to 0.05 K. The value of T_c for a-Ga is found to be 8.4 K, significantly higher than the T_c of the crystalline phases ($T_c = 1.08$ K for α-Ga and 6 K for β-Ga). A more complete listing of the T_c values of amorphous superconductors and their crystalline counterparts is given in Table 1. It is noted that T_c for amorphous non-TM superconductors tends to increase with valence and with atomic number for a given valence (Johnson 1981).

1.2 Transition Metals

A systematic study by Collver and Hammond (1973) on amorphous 4*d* and 5*d* TM elements and alloys thereof indicates that the well-known Matthias rule for crystalline superconductors (see *Superconducting Materials: An Overview*; *Superconducting Materials: Elements*) is significantly altered for the amorphous state. The variation of T_c as a function of e/a (the average valence electrons–atom ratio) shows a sharp peak at $e/a = 6.4$, corresponding to a half-filled *d* shell (see Fig. 1). More recent work has revealed that T_c as a function of e/a is more symmetric than the original Collver–Hammond curve and similar systematics are also observed in some liquid-quenched TM-metalloid metallic glasses. The T_c of amorphous TM alloys made from elements widely separated in the periodic table is always significantly lower than that of the original Collver–Hammond alloys composed of nearest 4*d* or 5*d* elements. The T_c of alloys mixing the 4*d* and 5*d* elements shows an even more intriguing behavior. For example, whereas the values of T_c for a-MoRe alloys are slightly higher than those for the a-MoRu alloys, lower T_c values have been found for all the a-MoRh alloys in the same range of e/a. The T_c values for a-NbIr, a-NbRh, a-TaRh and a-TaIr alloys are considerably lower than those for corresponding amorphous near-neighbor TM alloys and show essentially no variation of e/a. These data along with others are plotted in Fig. 1. (Recent T_c data on amorphous alloy systems containing 3*d* elements and noble metals, for example a-ZrCu, a-ZrFe and a-ZrNi, are not included in this plot and should be treated separately.) It is clear from Fig. 1 that there is no universal curve of T_c as a function of e/a for all amorphous TM alloys. It has been suggested that the effects of spin fluctuations and charge transfer may be responsible for this.

At present, there is no satisfactory interpretation for all the systematics of T_c previously described. The data shown in Fig. 1 do suggest, however, that there is an upper limit of T_c for a given e/a in amorphous 4*d* and 5*d* TM alloys as roughly outlined by the modified Collver–Hammond curve. The fact that the peak of this upper-bound curve coincides with half-filled *d* shells suggests that there is an atomic parameter characterizing the e/a dependence of T_c. This point can be qualitatively discussed with the aid of the well-known McMillan formula:

$$T_c(\theta_D/1.45)\exp\left[\frac{-1.04(1+\lambda)}{\lambda-\mu^*(1+0.62\lambda)}\right] \quad (1)$$

where λ is the dimensionless electron–phonon coupling parameter, μ^* is the effective electron–electron Coulomb pseudopotential and θ_D is the Debye temperature. The parameter λ can be expressed as

$$\lambda = N(E_F)\langle I^2\rangle/M\langle\omega^2\rangle \quad (2)$$

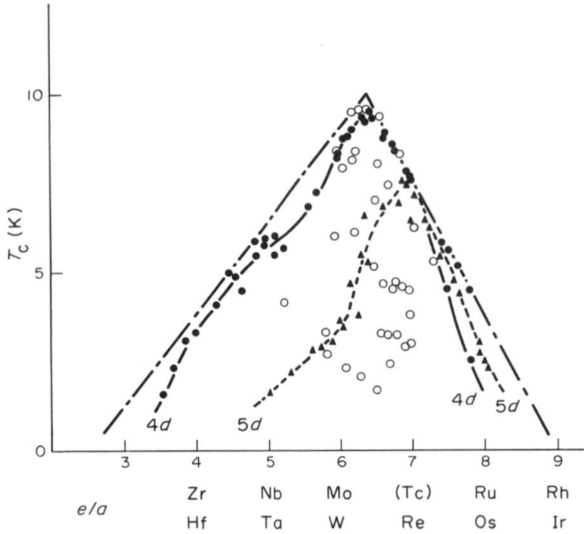

Figure 1
The T_c of amorphous TM elements and alloys as a function of e/a. The original Collver–Hammond curves for amorphous TM superconductors are shown for (●) 4*d* and (▲) 5*d* metals; also shown are (○) the T_c values of various amorphous TM alloys made from elements widely separated in the periodic table and from mixing the 4*d* and 5*d* metals. For a given e/a, there is an upper limit of T_c as outlined by the triangular peak (—·—·)

where $N(E_F)$ is the bare electronic density of states at the Fermi level, $\langle\omega^2\rangle$ is an average square phonon frequency and $\langle I^2\rangle$ is the Fermi-surface average of the electron–phonon interaction strength.

In terms of tight-binding models, it has been suggested that the Collver–Hammond curve essentially reflects the e/a dependence of $N(E_F)$ and that the ratio $\langle I^2\rangle/\langle\omega^2\rangle$ is a constant throughout the 4*d* series. A systematic study of the low-temperature specific heat of some of the more stable Collver–Hammond alloy films is crucial in determining $N(E_F)$ quantitatively as a function of e/a. It should be pointed out that the empirical and calculated values of $\langle I^2\rangle$ for the 4*d* transition metals (Butler 1977) are insensitive to structural change and their e/a variation closely mimics the T_c upper limit shown in Fig. 1.

2. Strength of the Electron–Phonon Interaction

The coupling between electrons and phonons is fundamentally important to the occurrence of superconductivity and is of particular interest in the case of amorphous superconductors in view of possible enhancement of the coupling strength through structural disorder as suggested by several theories. The strength of the electron–phonon interaction can be estimated from (a) the ratio of Bardeen–Cooper–Schrieffer (BCS) energy gap to transition temperature, $2\,\Delta(0)/k_BT_c$, and (b) the parameter λ (see *Superconducting Materials: BCS and Phenomenological Theories*). The BCS gap can be measured directly by superconductive tunnelling experiments or estimated indirectly from calorimetric measurements. The parameter λ can be determined using either the McMillan formula for T_c or the technique of superconductive tunnelling spectroscopy (see *Tunnelling Spectroscopy*) through

$$\lambda = 2\int_0^{\infty} d\omega\, \omega^{-1}\alpha^2(\omega)F(\omega) \tag{3}$$

where $\alpha^2(\omega)F(\omega)$ is the Eliashberg electron–phonon spectral function.

The results accumulated so far clearly lead to three conclusions.

First, amorphous non-TM superconductors are all characterized by a very strong electron–phonon coupling (i.e., $2\,\Delta(0)/k_BT_c \sim$ 4–5 and $\lambda > 2$), as are their crystalline counterparts.

Second, amorphous TM or TM-based superconductors are found to be in the BCS weak-coupling limit. In particular, the electron–phonon coupling strength of several strong-coupled A15 superconductors is found to be reduced to the BCS limit in the amorphous state (see *A15 Compound Superconductors*). These experimental findings suggest that structural disorder alone does not necessarily give rise to a significantly enhanced electron–phonon interaction.

Third, a detailed analysis of the tunnelling data for various amorphous non-TM superconductors has revealed that the Eliashberg function is enhanced and varies linearly with ω in the low-frequency region as predicted by various theories. There is no significant low-frequency enhancement found in $\alpha^2F(\omega)$ for amorphous nitrogen-stabilized niobium and molybdenum films. The function $\alpha^2F(\omega)$, in fact, follows the phonon density of states $F(\omega)$ based on a structural model of dense random packing of hard spheres. From this finding, it is concluded that the observed low-frequency enhancement in $\alpha^2F(\omega)$ for amorphous non-TM superconductors is essentially a manifestation of an increase in the electron–phonon interaction caused by structural disorder. The lack of such enhancement in amorphous TM superconductors is attributed to the "nearly localized" nature of *d* electrons, which are relatively insensitive to structural disorder.

In a more general perspective, this result can also be understood by noting the difference between the TM and non-TM amorphous superconductors in the effectiveness of the phonon modes in contributing to the electron–phonon coupling. Unlike amorphous TM superconductors, the phonon density of states in non-TM superconductors (crystalline and amorphous) is distributed over a range from 0 to approximately 10 meV and is most efficient in achieving a strong electron–phonon interaction (Tsuei 1981).

3. Critical Fields

Amorphous superconductors are all characterized by relatively large upper critical fields H_{c2} in view of their generally low T_c values. This is mainly a manifestation of a short electron mean free path ($l \lesssim 1$ nm) leading to a short zero-temperature coherence length $\xi(0)$ (see *Magnetic Properties of Superconductors*). The upper critical field at 0 K is related to $\xi(0)$ by

$$H_{c2}(0) = \Phi_0/2\pi\xi^2(0) \tag{4}$$

where $\Phi_0 = h/2e$ is the flux quantum. Typical values of $\mu_0 H_{c2}(0)$ are about 2–5 T (20–50 kG) for amorphous non-TM superconductors and are in the range 10–40 T (100–400 kG) for the noncrystalline TM or TM-based superconductors. The lower critical field $H_{c1}(T)$ for several metallic glasses has been measured; $\mu_0 H_{c1}(0)$ values are typically of the order of 1–10 mT (10–100 G). Both $H_{c2}(T)$ and $H_{c1}(T)$ are found to vary linearly with temperature over a wide range of temperatures below $T_c(\sim 0.5T_c$ to $T_c)$. Examples of dH_{c2}/dT near T_c are about $-2\,\mathrm{T\,K^{-1}}$ to $-3\,\mathrm{T\,K^{-1}}$ for amorphous TM superconductors and about $-0.5\,\mathrm{T\,K^{-1}}$ for amorphous simple metals. The values of dH_{c1}/dT are about $1\,\mathrm{mT\,K^{-1}}$.

From critical-field measurements, values for $\xi(0)$ and the Ginzburg–Landau parameter κ can be estimated; the values are of the order of 5 nm and 60, respectively. Such large values of κ represent a true indication of the extreme type II, or "dirty limit," behavior expected for amorphous superconductors. Detailed studies of the temperature dependence of $H_{c2}(T)$ for several amorphous TM alloys indicate that H_{c2} is enhanced at low temperatures by the effect of weak localization as predicated by the FEM theory (Fukuyama et al. 1984). The amount of the H_{c2} enhancement depends on the alloy system and can be as large as 20% (Poon 1985a, b). It should be mentioned that the quantum correction effects of weak localization and interaction also manifest themselves in suppressing T_c (Fukuyama et al. 1984) and reducing the temperature coefficient of the normal-state resistivity.

4. Flux Pinning Phenomena

The pinning of the flux line lattice (FLL) in superconductors (see *Flux Line Lattice*) can be studied by measuring the critical current density $J_c(T, H)$ as a function of temperature and applied field. Results of critical-current measurements on amorphous superconductors show that J_c at $H = 0$ is in general relatively low ($\sim 10^7$–$10^8\,\mathrm{A\,m^{-2}}$) and that it decreases rapidly with increasing field. Such a weak FLL pinning can probably be attributed to the homogeneity and isotropy, on a scale of $\xi(0)$, intrinsic in all amorphous materials. It is exactly this characteristic property which makes direct use of amorphous superconductors in high-field applications less attractive. A possible solution to this problem is to incorporate, by appropriate heat treatments, some second-phase precipitates in an amorphous superconducting matrix (Johnson 1981). Homogeneous amorphous superconductors, however, are ideal for studying flux pinning phenomena in terms of existing theory (see *Flux Pinning and the Summation of Pinning Forces*). The fact that amorphous TM or TM-based superconductors are weak coupled and in the extreme type II limit simplifies the analysis of the experimental data considerably, because theoretical expressions in the dirty limit are highly developed. Furthermore, pinning behavior in these materials (in the form of carefully prepared thin films) is expected to be insensitive to complications arising from pinning by surface irregularities, edge dislocations and buckling of the flux lines.

A recent fundamental study of flux pinning in amorphous films of Nb_3Ge and Nb_3Si (Kes and Tsuei 1981) led to the following results.

(a) Owing to the large κ and low J_c, the FLL in amorphous thin-film superconductors is essentially two dimensional.

(b) The pinning force density $F_p(=J_cB$, where B is flux density) as a function of perpendicular field and temperature is found to be in good agreement with the theory of collective pinning (see *Flux Pinning and the Summation of Pinning Forces*).

(c) A sharp peak in F_p is observed near H_{c2}. A pinning-mediated renormalization of the shear modulus of the FLL is probably responsible for this peak effect.

(d) The F_p for samples with relatively large J_c shows a domelike behavior characteristic of strong pinning. By annealing, J_c decreases by up to two orders of magnitude and both two-dimensional collective pinning and the peak effect are recovered, supporting the view that weak pinning is an essential requirement to observe the collective behavior. It has also been demonstrated that the flux pinning can serve as a sensitive probe to study structural relaxation in amorphous metals.

The concept of collective flux pinning along with the consideration of the thermal fluctuation effects has been used to understand the flux-creep-related phenomena in high-T_c cuprate superconductors (Feigel'man and Vinokur 1990). It is interesting to note that both amorphous superconductors and high-T_c copper oxides are characterized by short coherence length and relatively high normal-state resistivity except that there is a large difference in T_c and temperature scale.

5. *Potential Applications*

Most amorphous non-TM superconductors are structurally stable only at very low temperatures. For example, a-Bi films are found to crystallize at 20 K and the amorphous phase of gallium films transforms into a crystalline state (β-Ga, $T_c = 6.4$ K) at about 15 K. Further annealing at 60 K results in the room-temperature stable phase α-Ga with a T_c of 1.08 K. In sharp contrast, many TM or TM-based superconducting alloys can resist crystallization at temperatures up to more than 1000 K. In addition to this phase stability, these materials are ductile, flexible and possess extremely high mechanical strength, comparable with that of single crystals. Potential areas for using amorphous superconductive materials include high-field magnets, especially for use in an environment of high-level radiation, and possibly some specialized vortex memory devices for computer applications.

See also: Metallic Glasses; Superconducting Materials: An Overview; Superconducting Materials: Elements

Bibliography

Bergmann G 1976 Amorphous metals and their superconductivity. *Phys. Rep.* 27: 159–85

Butler W H 1977 Electron–phonon coupling in the transition metals. *Phys. Rev. B* 15: 5267–82

Collver M M, Hammond R H 1973 Superconductivity in "amorphous" transition-metal alloy films. *Phys. Rev. Lett.* 30: 92–95

Feigel'man M V, Vinokur V M 1990 Thermal fluctuations of vortex lines, pinning, and creep in high-T_c superconductors. *Phys. Rev. B* 41: 8986–90

Fukuyama H, Ebisawa H, Maekawa S 1984 Bulk superconductivity in weakly localized regime. *J. Phys. Soc. Jpn.* 53: 3560–7

Johnson W L 1981 Superconductivity in metallic glasses. In: Guntherodt H, Beck H (eds.) 1981 *Glassy Metals 1*. Springer, Berlin, Chap. 9, pp. 191–223

Kes P H, Tsuei C C 1981 Collective flux pinning phenomena in amorphous superconductors. *Phys. Rev. Lett.* 47: 1930–4

Poon S J 1985a Amorphous and highly disordered metallic superconductors. *Physica B* 135: 259–66

Poon S J 1985b Localization effects on superconductivity in homogeneous metallic glasses. *Phys. Rev. B* 31: 7442–5

Tsuei C C 1981 Amorphous superconductors. In: Foner S, Schwartz B B (eds.) 1981 *Superconductor Materials Science*. Plenum, New York, Chap. 12, pp. 735–56

C. C. Tsuei
[IBM, Yorktown Heights, New York, USA]

Anisotropy in Magnetic Materials

Experimental measurements have shown that for each magnetic crystal there are directions in which the magnetization is attained in the lowest field strength (in general, they coincide with the principal crystal symmetry directions). Such directions are called the axes or directions of easy magnetization (or simply the easy axes or directions). For example, in pure iron these are the ⟨100⟩ directions and in nickel they are the ⟨111⟩ directions at any temperature below the Curie point. It has also been shown that in crystals there are directions of hard magnetization (simply the hard directions), in which much larger fields are needed to magnetize the crystal than in others. Hence, it follows that the free energy of the crystal contains a component that depends only on the direction of the spontaneous magnetization and which is minimum (or maximum) when this is parallel to an easy (hard) direction. This component is called the magnetocrystalline anisotropy. Generally, the magnetocrystalline anisotropy is intrinsic to the material and reflects the symmetry of the crystal.

Besides the magnetocrystalline anisotropy several other kinds of magnetic anisotropy exist in magnetic materials:

(a) surface anisotropy,
(b) exchange anisotropy,
(c) shape anisotropy,
(d) stress anisotropy,
(e) induced anisotropy (by field or stress annealing), and
(f) growth-induced anisotropy.

The term magnetic anisotropy simply means that the magnetic properties depend on the direction in which they are measured. The magnetic anisotropy plays an important role in a variety of properties of magnetic materials. It is responsible for the shape and size of magnetic domains, the shape of the magnetization curve and hysteresis loop, the magnitude of magnetic permeability, coercive force, ferromagnetic resonance field and many other properties.

1. *Magnetocrystalline Anisotropy*

Phenomenologically, the magnetocrystalline anisotropy can be expressed uniformly for all magnetic crystals of corresponding symmetry independently of the microscopic mechanism responsible for this effect.

For a saturated crystal, the magnetocrystalline anisotropy energy density E_{ma} may be formally expressed as an expansion series:

$$E_{ma} = K_0 + K_{ij}\alpha_i\alpha_j + K_{ijkl}\alpha_i\alpha_j\alpha_k\alpha_l + \dots \quad (1)$$

Where the α_i are the direction cosines of the saturation magnetization M_s relative to the rectangular cartesian system of coordinated axes and the K_i are the anisotropy constants of the material.

For cubic crystals Eqn. (1) reduces to

$$E_{ma} = K_0 + K_1(\alpha_1^2\alpha_2^2 + \alpha_1^2\alpha_3^2 + \alpha_2^2\alpha_3^2) + K_2\alpha_1^2\alpha_2^2\alpha_3^2 + K_3(\alpha_1^2\alpha_2^2 + \alpha_1^2\alpha_3^2 + \alpha_2^2\alpha_3^2)^2 + \cdots \tag{2}$$

where the α_i refer to the $\langle 100\rangle$ cubic axes.

The polynomial in Eqn. (1) represents a very simple form of expressing the angular dependence of the magnetocrystalline anisotropy but it is frequently not the most suitable form. For some applications the most convenient series expansion for E_{ma} is in terms of the spherical harmonics Y_l^m:

$$E_{ma} = \sum_{l=0}^{\infty} \sum_{m=-l}^{l} K_l^m \; Y_l^m(\theta, \varphi) \tag{3}$$

where θ and φ are the polar and azimuthal angles of magnetization, respectively.

Since the spherical harmonics form a complete orthonormal system, the expansion of Eqn. (3) is especially suitable when the anisotropy has to include terms of higher order. Several expressions for E_{ma} for crystals of various symmetry can be developed:

(a) hexagonal symmetry

$$E_{ma} = K_0 + K_1 \sin^2\theta + K_2 \sin^4\theta + K_3 \sin^6\theta + K_4 \sin^6\theta \cos 6\varphi \tag{4}$$

(b) trigonal symmetry

$$E_{ma} = K_0 + K_1 \sin^2\theta + K_2^1 \sin^3\theta \cos\theta \cos 3\varphi + K_3 \sin^6\theta + K_3^1 \sin^6\theta \cos 6\varphi + K_3^2 \sin^3\theta \cos^3\theta \cos 3\varphi \tag{5}$$

(c) tetragonal symmetry

$$E_{ma} = K_0 + K_1 \sin^2\theta + K_2 \sin^4\theta + K_3 \sin^4\theta \cos^2\varphi \sin^2\varphi \tag{6}$$

(d) orthorhombic symmetry

$$E_{ma} = K_0 + \sin^2\theta\,(K_1\cos^2\varphi + K_2\sin^2\varphi) + \sin^4\theta(K_3\cos^4\varphi + K_4\sin^2\varphi\cos^2\varphi + K_5\sin^4\varphi) + \sin^2\theta\cos^2\theta(K_6\cos^2\varphi + K_7\sin^2\varphi) \tag{7}$$

1.1 Microscopic Origin

In general, magnetocrystalline anisotropy is determined by the spin state of the magnetic ions and by the symmetry of their atomic arrangement in the lattice. Microscopic theory of the magnetocrystalline anisotropy for a real material presents rather formidable problems since detailed knowledge of the electronic structure is required for such calculations. Moreover, very often strong spin–lattice coupling should be taken into account. As a result, detailed calculations of magnetocrystalline anisotropy constants have only been carried through for a few magnetic materials.

One source of magnetocrystalline anisotropy is the classical interatomic magnetic dipole–dipole interaction. Since the potential in this case may be accurately evaluated, the contribution to the magnetocrystalline anisotropy constants from this mechanism may be obtained by summing over all the atomic sites. Apart from this pure magnetic interaction, the other sources of two-ion magnetocrystalline anisotropy are electrostatic in nature, involving anisotropic exchange between the spins of two or more atoms. Examples of such interactions are pseudodipolar or Dzyaloshinski–Moriya exchange interactions. In the two-ion models the anisotropic exchange affects the spin state via the anisotropic electrostatic interactions.

In the single-ion mechanism of magnetocrystalline anisotropy the spins of magnetic ions (mostly localized) situated in the crystalline electrostatic field arising from the neighboring ions will have preferred directions with respect to the symmetry axes of that field. The spins are coupled to the lattice through the spin–orbit coupling.

The single- and two-ion mechanisms apply to systems with localized moments and therefore would not be expected to be valid for itinerant systems. The theory of magnetocrystalline anisotropy for transition metal alloys and compounds is based on detailed knowledge of the band structure and its dependence on the direction of the magnetization.

(*a*) *The single-ion model.* This model is based on the assumption that the magnetocrystalline anisotropy energy is determined by the quantum states of a single ion only. It means that this model permits the calculation of the macroscopic properties from the microscopic data by superposition. The Hamiltonian $\mathscr{H}$ used to calculate the energy levels of the individual ions consists of two parts:

$$\mathscr{H} = \mathscr{H}_c + \mathscr{H}_0 \tag{8}$$

$\mathscr{H}_c$ is the crystal field Hamiltonian. Its form can be deduced from group theoretical considerations and therefore involves only a few parameters (e.g., only one parameter for a $3d$ ion and two for a $4f$ ion in a site of cubic symmetry). The parameters describing the strength of a crystal field, called crystal field parameters, are usually determined from experimental (mostly optical) data.

The other term $\mathscr{H}_0$ of the Hamiltonian in Eqn. (8) describes the exchange interaction between magnetic ions, which is assumed to be isotropic:

$$\mathscr{H}_0 = g\beta \boldsymbol{H}_m \boldsymbol{S} \tag{9}$$

where g is the g factor of the magnetic ion, β is the Bohr magneton and $\boldsymbol{H}_m$ is the effective field (in the molecular field approximation).

Diagonalization of $\mathscr{H}$ for a specific direction of magnetization gives the energy levels E_i of the individual magnetic ions and permits the calculation of the total free energy F of the system (per unit volume):

Table 1
Anisotropy constants of some ferrimagnetic spinels (at room temperature)

Crystal	K_1 (10^3 J m^{-3})	K_2 (10^3 J m^{-3})
Fe_3O_4	−11	−28
$MnFe_2O_4$	−3.4	
$NiFe_2O_4$	−6.5	
$CoFe_2O_4$	270	300
$MgFe_2O_4$	−3.9	
$LiFe_5O_8$	−8.5	

Source: Krupicka and Zaveta (1975)

$$F = -k_B T \sum_{\text{ions}} \ln \sum_i \exp(-E_i/k_B T) \tag{10}$$

The evaluation of F for several crystallographic directions and comparison with Eqn. (1) gives directly the values of anisotropy constants. The magnitude of the magnetocrystalline anisotropy constants is controlled by the strength of the crystal field, by the electronic configuration, by the strength of the spin–orbit coupling and by the effective field H_m. For ions with an orbitally nondegenerate ground state, the spin–orbit contributes only in higher-order terms of perturbation theory and thus the magnetocrystalline anisotropy constants are small, as in the case of *S*-state ions (Fe^{3+}, Mn^{2+}, Gd^{3+}). Usually for orbitally degenerate ions, the spin–orbit coupling contributes to the first-order terms, giving rise to large anisotropic effects.

Magnetocrystalline anisotropy in oxides with cubic structure is usually determined by the single-ion model. The values of the anisotropy constants of some ferrimagnetic spinels are given in Table 1.

Large anisotropic effects are expected to exist in rare-earth ions (with the exception of S-state ions Gd^{3+} and Eu^{2+}). In this case the spin–orbit interaction is sufficiently strong to lock together spin ($\boldsymbol{S}$) and orbital ($\boldsymbol{L}$) momentum making J ($\boldsymbol{J} = \boldsymbol{L} + \boldsymbol{S}$) a good quantum number. The anisotropy is then determined solely by the crystal field leading in some cases to giant single-ion magnetocrystalline anisotropy. It is interesting that the main contribution to the magnetic anisotropy in permanent magnet materials like intermetallic rare earth–iron–boron or rare earth–cobalt is due to the single-ion crystal-field-induced anisotropy of the magnetic rare-earth ions (see Table 2).

In Eqns. (1–7) the magnetic anisotropy at a given temperature has been expressed formally in terms of the anisotropy constants. The magnetocrystalline anisotropy is expected to depend on temperature since, at any temperature T, the thermal excitations cause local deviations of the spin direction from the average direction of magnetization. The distribution of spins about the direction of magnetization determines the value of the reduced spontaneous magnetization $u(T) = M(T)/M(0)$ at temperature T. This distribution also determines the magnetocrystalline anisotropy at temperature T, since the total anisotropy energy may be obtained from a suitable average of the anisotropy energy at absolute zero, taken over the spread of individual spin directions appropriate to temperature T. For the single-ion case, Callen and Callen (1963) have shown that the temperature dependences of all magnetocrystalline anisotropy constants are given by

$$K_l^m(T) = K_l^m(0)\frac{I_{l+1/2}(\mathscr{L}^{-1}(u))}{I_{1/2}(\mathscr{L}^{-1}(u))} \tag{11}$$

where I is the hyperbolic Bessel function and $\mathscr{L}^{-1}$ is the inverse of the Langevin function.

It has been shown (Callen and Shtrikman 1965) that this dependence (Eqn. (11)) is valid at least for a wide class of statistical models, including those based on the molecular field and spin wave theory.

The low-temperature ($u \simeq 1$) expansion of the Bessel functions leads to the famous $l(l+1)/2$ power law:

$$K_l^m(T)/K_l^m(0) = u^{l(l+1)/2} \tag{12}$$

At high temperatures $K_l^m(T)$ varies as u^l.

(*b*) *Two-ion mechanism.* The single-ion anisotropy energy vanishes identically for an effective spin of one-half and is very small for ions with orbital singlet ground state (e.g., for S-state ions such as Mn^{2+} and Fe^{3+}). In such cases, the two-ion mechanisms play a decisive role in magnetocrystalline anisotropy. The most important two-ion mechanism is based on pseudodipolar interactions (or magnetic dipole–dipole interactions) described by the following spin Hamiltonian:

$$\mathscr{H}_{ij} = D(r_{ij})\,[S_i^z S_j^z - \boldsymbol{S}_i\boldsymbol{S}_j/3] \tag{13}$$

Table 2
Anisotropy constants of $R_2Fe_{14}B$ compounds (at room temperature)

Element R	K_1 (10^6 J m^{-3})	K_2 (10^6 J m^{-3})
Yttrium	1.1	
Cerium	1.44	
Praseodymium	5.5	
Neodymium	4.5	0.66
Samarium	−12.0	0.29
Gadolinium	0.9	
Terbium	5.9	
Dysprosium	4.0	
Holmium	4.8	
Erbium	−0.03	
Thulium	−0.03	

Source: Cook and Rossiter (1989)

where the z axis is along the line joining the atoms, r_{ij} is the interatomic distance and $\boldsymbol{S}_i$ is the spin operator of i-type atoms.

For pure magnetic dipole–dipole interaction, constant D has the form:

$$D = -\frac{3g^2\beta^2}{r_{ij}^3} \tag{14}$$

It should be noted that the pseudodipolar interaction does not contribute in the first order to the magnetocrystalline anisotropy constants in crystals of cubic symmetry. If, however, the crystal has a lower symmetry than the cubic crystal, as in a hexagonal crystal, the pseudodipolar interaction may contribute to the magnetic anisotropy and it is believed that in some hexagonal ferrites this contribution may be substantial.

The pseudodipolar interaction may contribute to the cubic anisotropy but only in the higher-order perturbation terms. These are, however, usually rather small and can be neglected in most practical cases.

For completeness, the Dzyaloshinski–Moriya anisotropic exchange interaction term should be mentioned:

$$\mathscr{H}_{\text{D-M}} = \boldsymbol{D}(\boldsymbol{S}_1 \times \boldsymbol{S}_2) \tag{15}$$

It appears because of the combined effect of spin–orbit and exchange interactions. The Dzyaloshinski–Moriya interaction appears in the second-order processes of the perturbation which give rise to the term

$$D \simeq J_{\text{ex}}(\Delta g/g) \tag{16}$$

where J_{ex} is the exchange constant and $\Delta g = g - 2$ is the deviation from the free-electron value $g \simeq 2$.

The Dzyaloshinski–Moriya interaction is responsible for 'weak ferromagnetism' observed in some magnetic oxides (e.g., in orthoferrites and orthochromites).

The calculation of the temperature dependence of the anisotropy constants in the framework of two-ion models is more complicated than in the case of the single-ion model. In the single-ion model it is assumed that there is no correlation between neighboring spins, each spin being orientated independently by the molecular field. However, at low temperatures, the two ion interacting spins are highly correlated and the two-spin Hamiltonian (see Eqn. (13)) transforms as twofold spherical harmonics leading to a u^3 temperature variation of the anisotropy constants. In the uncorrelated limit the single-spin functions of the two-spin Hamiltonian transform separately as singlefold harmonics and therefore the crossover to the u^2 temperature dependence is expected at temperatures approaching the Curie point.

The two-ion mechanism (dipole–dipole interaction) may contribute to the magnetocrystalline anisotropy of noncubic crystals and it is believed that in some hexagonal ferrites this contribution may be substantial. Representative data on magnetic anisotropy of hexagonal ferrites are given in Table 3.

Table 3
Anisotropy constants of some hexagonal ferrites (at room temperature)

Crystal	K_1 (10^5 J m^{-3})
$BaFe_{12}O_{19}$	3.3
$BaFe_{18}O_{27}$	3.0
$BaFe_{17}CoO_{27}$	0.2
$BaCo_2Fe_{16}O_{27}$	− 1.9
$BaFe_{17}ZnO_{27}$	2.4

Source: Krupicka and Zaveta (1975)

(*c*) *Itinerant electron model.* The magnetic properties of 3*d* metals and alloys are well described by the itinerant electron model. In this model, the main contribution to the magnetic anisotropy is due to *d* electrons. The role of *s–p* electrons is to provide a nonspherical background in which the degeneracy of the atomic 3*d* levels is lowered due to the crystal field. Magnetic anisotropy results from the shifts of the energy band and from the redistribution of the occupied and empty states near the Fermi level which are induced by the spin–orbit coupling if the magnetization changes its direction. Both contributions are of the same order of magnitude but they are small and sometimes (e.g., for nickel) have opposite signs.

It should be mentioned that our present understanding of magnetocrystalline anisotropy in itinerant magnets is limited to a few well-studied cases (e.g., nickel, iron and cobalt) whose electronic structure is known in detail. Table 4 gives the anisotropy constants of some cubic metals and intermetallic compounds (at room temperature).

2. Surface Anisotropy

Magnetism at surfaces and at interfaces is modified due to reduced symmetry, the lower coordination number and the presence of highly localized surface

Table 4
Anisotropy constants of some cubic metals and intermetallic compounds (at room temperature)

Crystal	Fe	Ni	$ErFe_2$	$DyFe_2$	$TbFe_2$	$HoFe_2$
K_1(10^3 J m^{-3})	45	− 5	− 330	2100	− 7600	580

and interface states. One of the most striking properties of surface and interface magnetism is surface magnetic anisotropy.

The concept of surface anisotropy was introduced by Néel. In his theory a phenomenological, nearest neighbor pseudodipolar interaction is used to predict the variation of magnetic anisotropy with exposed surface:

$$E_{surf} = \left(-\frac{3g^2\beta^2}{r^3} + l\right)\left(\cos^2\varphi - \frac{1}{3}\right) + q\left(\cos^4\varphi - \frac{30}{35}\cos^2\varphi + \frac{3}{35}\right) \quad (17)$$

where φ is the angle between magnetization and the vector $\boldsymbol{r}$. Coefficients l and q describe the pseudodipolar and pseudoquadrupolar coupling, respectively. The magnitude of the pseudodipolar coupling constant l may be estimated from the magnetostriction constants.

In addition to the two-ion contribution, considered by Néel, in many cases (particularly for nonmetallic magnetic materials) the single-ion mechanism should be taken into account.

The origin of surface anisotropy of metals and alloys lies in the symmetry restrictions imposed on the allowed electron states by the presence of the surface (or two surfaces in the case of the ultrathin films and multilayers). In itinerant models the changes in magnetic anisotropy due to the surface are associated with electronic states extending into the interior of the magnetic materials in contrast with the behavior for localized models (single-ion or two-ion). Within the localized model the surface anisotropy is associated with effects occuring only at the surface planes.

According to the group theoretical approach, the surface symmetry defines the shape of the surface anisotropy. For example, the surface anisotropy of cubic materials for the (001) plane has tetragonal symmetry and the corresponding anisotropy energy can be written in the form

$$E_{surf} = K_1^{\perp}(\alpha_x^4 + \alpha_y^4) + K_1^{\parallel}\alpha_z^4 + K_u\alpha_z^2 \quad (18)$$

where α_x, α_y and α_z are directional cosines with respect to the [100], [010] and [001] crystallographic axes.

Surface anisotropy is observed experimentally in ultrathin films and in nanoscale multilayers. For some systems, surface anisotropy is large enough to change the overall magnetic anisotropy considerably. The magnitude of K_u is of the order of 10^{-4}–10^{-3} J m^{-2}.

3. Exchange Anisotropy

Exchange anisotropy is an interfacial effect between two magnetic materials (usually ferromagnetic or ferrimagnetic and antiferromagnetic) described by the formula

$$E_{exch} = -K\cos\theta \quad (19)$$

where θ is the angle between M_s and some direction (usually it is identified with the easy direction in antiferromagnetic subsystems).

The exchange anisotropy has its origins in peculiar properties of antiferromagnetic and ferromagnetic subsystems exchange coupled across the interface. To induce the exchange anisotropy, an external magnetic field should be applied at temperature T_a such that $T_C > T_a > T_N$ (T_C is the Curie temperature and T_N is the Néel temperature). When the system is cooled, still in the field, to far below T_N, the antiferromagnetic ordering that occurs will be dictated by exchange interactions at the interface. Subsequent reversal of the magnetic field will cause the magnetization of the ferromagnetic subsystem to rotate while the magnetic moments in the antiferromagnetic subsystem, having a high anisotropy, are unaffected far from the interface. If the reverse field is removed, the magnetic moments in the antiferromagnetic subsystem force the magnetic moments in the ferromagnetic subsystem to turn up, restoring their original direction. Thus, the materials with exchange anisotropy have two unusual characteristics: a hysteresis loop displaced along the field axis and high-field rotational hysteresis. However, it should be mentioned that these two features of exchange anisotropy may or may not be present at the same temperature.

4. Shape Anisotropy

The expression for the magnetostatic energy (self-energy) of a permanently magnetized body in the shape of a prolate spheroid (with semimajor axis c and semiminor axes of equal length a) can be written as

$$E_{me} = [(M_s\cos\theta)^2 N_c + (M_s\sin\theta)^2 N_a]/2 \quad (20)$$

where θ is the angle between magnetization vector $\boldsymbol{M}_s$ and the c axis and N_c and N_a are demagnetizing coefficients along the c and a axes, respectively (see *Demagnetizing Factors*).

Equation (20) has an angle-dependent term of exactly the same form as uniaxial magnetocrystalline anisotropy energy (see Eqn. (4)):

$$E_{me} = [(N_a - N_c)M_s^2\sin^2\theta]/2 \quad (21)$$

Thus, shape alone can be a source of magnetic anisotropy (called the shape anisotropy).

The shape anisotropy constant K_s is given by:

$$K_s = [(N_a - N_c)M_s^2]/2 \quad (22)$$

Since $N_c < N_a$, the c axis plays the role of easy axis.

The shape anisotropy is responsible for perpendicular easy axis magnetic anisotropy observed in some amorphous thin films. Macroscopic magnetic anisotropy in amorphous films is associated with anisotropically shaped heterogeneities that have their largest dimensions mutually aligned. Such inhomogeneities may be present in the form of voids or regions differing in composition from their surroundings.

5. Stress Anisotropy

The effect of mechanical stress σ on a magnetic body of cubic symmetry is described by the magnetoelastic energy E_{ml}:

$$E_{ml} = -\frac{3}{2}\lambda_{100}\sigma(\alpha_x^2\gamma_x^2 + \alpha_y^2\gamma_y^2 + \alpha_z^2\gamma_z^2) - 3\lambda_{111}\sigma(\alpha_x\alpha_y\gamma_x\gamma_y + \alpha_x\alpha_z\gamma_x\gamma_z + \alpha_y\alpha_z\gamma_y\gamma_z) \quad (23)$$

where α_x, α_y and α_z are the directional cosines of $\boldsymbol{M}_s$; γ_x, γ_y and γ_z are the directional cosines of σ; and λ_{100} and λ_{111} are the magnetostriction constants (see *Magnetoelastic Phenomena*).

When the magnetostriction is isotropic, so that $\lambda_{100} = \lambda_{111} = \lambda_s$, Eqn. (23) reduces to

$$E_{ml} = -\frac{3}{2}\lambda_s\sigma\cos^2\theta \quad (24)$$

where θ is the angle between $\boldsymbol{M}_s$ and σ.

The relation in Eqn. (24) is of exactly the same form as Eqn. (4) for uniaxial magnetocrystalline anisotropy. It means that stress can create uniaxial anisotropy (stress anisotropy) with anisotropy constant K_s:

$$K_s = \frac{3}{2}\lambda_s\sigma \quad (25)$$

The axis of stress is an easy axis if $\lambda_s\sigma > 0$.

Stress anisotropy can make a significant contribution to magnetic anisotropy in magnetic thin films. In this case, stresses are produced by the film deposition process or by thermal expansion mismatch between a magnetic film and its substrate.

6. Induced Anisotropy

In some cases, when a magnetic material is heated (at a certain $T < T_C$) in the presence of a magnetic field or applied stress, a preferential orientation of individual atoms or of nearest neighboring like-atom pairs may develop, which leads to a directional ordering and consequently to an additional magnetic anisotropy. This additional anisotropy is referred to as induced anisotropy. It is important to realize that directional ordering is not due to the magnetic field applied during annealing, but to the local magnetization. All that the applied field does is to saturate the magnetic material, thereby making the direction of magnetization uniform throughout.

The microscopic origin of the induced anisotropy is the same as in the case of magnetocrystalline anisotropy but, since the induced anisotropy is related to the local anisotropic arrangements in the lattice, it usually exhibits lower symmetry than that of the crystal as a whole. As for magnetocrystalline anisotropy, a formal expansion of the induced anisotropy energy E_{ind} can be derived from symmetry considerations. For a cubic crystal, the lower terms are usually written as:

$$E_{ind} = -F(\alpha_x^2\beta_x^2 + \alpha_y^2\beta_y^2 + \alpha_z^2\beta_z^2) - G(\alpha_x\alpha_y\beta_x\beta_y + \alpha_x\alpha_z\beta_x\beta_z + \alpha_y\alpha_z\beta_y\beta_z) \quad (26)$$

where the actual direction of magnetization at the measured temperature T_m is described by the direction cosines α_x, α_y and α_z while its direction at the annealing temperature T_{an} is described by the direction cosines β_x, β_y and β_z.

For pure isotropic magnetic materials (e.g., for amorphous alloys):

$$G = 2F$$

and

$$E_{ind} = -K_u\cos^2(\boldsymbol{\alpha},\boldsymbol{\beta}) \qquad K_u = F \quad (27)$$

Because inducing anisotropy always means lowering the free energy, the coefficients F, G and K_u are expected to be positive. The induced anisotropy constant K_u depends on T_m, T_{an} and the chemical composition if the annealing has been long enough to reach the thermodynamical equilibrium at T_{an} and may depend on the history of the sample if this is not the case. The process of approaching thermal equilibrium may in crystalline alloys be described by a simple relaxation equation with a single relaxation time τ:

$$\frac{dE_{ind}}{dt} = \frac{E_{ind} - E_{ind}^{\infty}}{\tau} \quad (28)$$

where E_{ind}^{∞} describes the equilibrium value of E_{ind} towards which the system is relaxing.

The solution of Eqn. (28) has the form:

$$E_{ind}(t) = E_{ind}(t=0)\exp(-t/\tau) + E_{ind}^{\infty}[1 - \exp(-t/\tau)] \quad (29)$$

It should be mentioned that the relaxation times for F and G terms need not be identical.

Controlling the induced anisotropy is extremely important for any technological application. This problem seems to be of particular interest for amorphous magnetic materials. The fundamental differences that exist between amorphous alloys and their crystalline counterparts are related to the fact that the atomic structure of glasses, unlike the crystalline structure, is highly degenerate and nonunique. Such a structure has many internal degrees of freedom, which are described in terms of atomic short-range order. The atomic short-range order can be easily changed by heat treatment or mechanical deformation. Relaxation in amorphous alloys depends on a spectrum of relaxation times (Kronmüller 1984, Hodson and Evetts 1986). Among the properties that arise due to the atomic rearrangement is, first of all, magnetic anisotropy induced by magnetic (or stress) annealing. Annealing the amorphous alloys in a magnetic field (at elevated temperatures, but below the Curie point) induces an anisotropy in the direction of the applied field. The order of magnitude of the field-induced anisotropy in amorphous alloys is comparable with that observed in crystalline materials. Moreover, the field-induced anisotropy in metallic glasses is due to directional atomic pair ordering, just as in the crystalline alloys.

In contrast to the field-induced anisotropy, the stress-induced anisotropy may appear even after annealing above the Curie temperature; the magnitude of this kind of anisotropy can be much larger than that induced by field annealing. Since the stress-induced anisotropy occurs also in zero-magnetostrictive alloys, the anisotropy is not only due to internal stresses, although such stresses, if they are present, may give some contribution to the resulting induced anisotropy. The magnitude of K_u in metallic glasses is of the order of 10^2–10^3 J m^{-3}.

Stress annealing induces the uniaxial anisotropy K_u that is composed of two contributions: the anelastic K_{an} (recoverable) and plastic K_{pl} (permanent). Both contributions are proportional to the stress σ applied during the stress annealing. For most alloys studied so far K_{an} is negative (i.e., the hard axis is parallel to the direction of the tensile stress applied during annealing) and K_{pl} is positive or zero. The microscopic origin of K_u is the same as in the case of the field-induced anisotropy (i.e., the local atomic rearrangement).

7. Growth-Induced Anisotropy

When certain mixed magnetic oxides (e.g., garnets or spinels) are grown it is observed that additional anisotropy is induced under particular growth faces. This anisotropy originates from the growth process, which causes an ordering of the magnetic ions (or sometimes nonmagnetic ions or even oxygen vacancies) along certain growth directions. The ordering of the magnetic ions owing to the growth conditions induces a reduction in symmetry. With respect to the origin of this ordering, it is expected that the equilibrium distribution of magnetic ions can be restored by annealing treatments implying the disappearance of the growth-induced anisotropy. Two models are usually used to account for the growth-induced anisotropy:

(a) the site preference model, and

(b) the pair-ordering model.

The site preference model is based on the assumption that within a crystal lattice there exists the preferential occupation of a certain type of the crystallographic positions which are different with respect to the coordination of the neighboring magnetic ions. In mixed oxides, a fraction of one kind of magnetic ion may enter preferentially a particular type of position, depending on the growth direction. These positions although crystallographically equivalent then become magnetically inequivalent.

The pair-ordering model assumes a short-range pair ordering between two kinds of magnetic ions and, therefore, this model additionally introduces the pair preference parameter (the angle between the pair-bond direction and the growth facet normal).

Usually growth-induced anisotropy is treated in a phenomenological manner by considering the symmetry restriction. For example, the general form of the growth-induced anisotropy energy of lowest order in the direction cosines of magnetization α_i has the following form for cubic crystals:

$$E_{\text{growth}} = A(\alpha_x^2\beta_x^2 + \alpha_y^2\beta_y^2 + \alpha_z^2\beta_z^2) + B(\alpha_x\alpha_y\beta_x\beta_y + \alpha_x\alpha_z\beta_x\beta_z + \alpha_y\alpha_z\beta_y\beta_z) \quad (30)$$

where β_x, β_y and β_z denote the direction cosines of the normal to the growth face.

The growth-induced anisotropy in garnet thin films is crucial to the existence of bubble domains and, consequently, is the basis for magnetic bubble technology (see *Bubble Domain Materials*). The magnitude of the growth-induced anisotropy in garnet thin films is of the order of 10^2–10^4 J m^{-3}.

8. Experimental Methods

There are three main methods of measuring magnetic anisotropy:

(a) magnetization measurements along different directions.

(b) torque measurements, and

(c) ferromagnetic resonance.

The first method can generally be applied to single crystals as well as to polycrystalline materials. Anisotropy constants of single crystals may be determined in two ways:

(a) by fitting a calculated magnetization curve to the observed one, or

(b) by measuring the area included between the magnetization curves for two different crystal directions.

The analysis of the second derivative d^2M/dH^2 as a function of H gives information on the magnetic anisotropy of polycrystalline materials.

The second method is based on the measurement of the torque that is exerted on a sample being suspended by a string in a homogeneous magnetic field.

This mechanical torque is measured as a function of magnetic field orientation and from the analysis of the torque curves the anisotropy constants can be evaluated.

The third method utilizes the dependence of the resonance frequency on the magnetic anisotropy. In the resonance method the ratio of the anisotropy constant to the saturation magnetization is obtained, in contrast to the torque method, in which the anisotropy constants are measured directly.

See also: Magnetic Materials: Basic Concepts; Measurements in Magnetic Materials

Bibliography

Aubert G 1980 Magnetic anisotropy of ferro- and ferrimagnets. *J. Magn. & Magn. Mater.* 19: 396–408

Bayreuther G 1989 Magnetic surfaces. *Hyperfine Interact.* 47: 237–49

Birss R R 1964 *Symmetry and Magnetism.* North-Holland, Amsterdam.

Callen E R, Callen H B 1963 Static magnetoelastic coupling in cubic crystals. *Phys. Rev.* 129: 578–93

Callen H B, Shtrikman S 1965 A probability density common to molecular field and collective excitation theories of ferromagnetism. *Solid State Commun.* 3: 5–8

Chikazumi S 1964 *Physics of Magnetism.* Wiley, New York

Cook J S, Rossiter P L 1989 Rare-earth iron boron supermagnets. *CRC Crit. Rev. Solid State Mater. Sci.* 15: 509–50

Cullity B O 1972 *Introduction to Magnetic Materials.* Addison-Wesley, Reading, MA

Darby M I, Isaac E D 1974 Magnetocrystalline anisotropy of ferro- and ferrimagnets. *IEEE Trans. Magn.* 10: 259–304

du Tremolet de Lacheisserie E 1987 Pair model of anisotropy and magnetostriction in soft amorphous magnets. *J. Magn. & Magn. Mater.* 67: 102–14

Hansen P 1978 Magnetic anisotropy and magnetostriction in garnets. *Proc. Int. Sch. Phys. Enrico Fermi* 70: 56–133

Hodson P D, Evetts J E 1986 Polarization "crossover" of field induced magnetic anisotropy in ferromagnetic amorphous alloys. *J. Magn. & Magn. Mater.* 59: 81–8

Kronmüller H 1984 Magnetic aftereffects, induced anisotropies, and the role of two-level systems in amorphous alloys. *J. Magn. & Magn. Mater.* 41: 366–75

Krupicka S, Zaveta K 1975 Anisotropy, induced anistropy and related phenomena. In: Craik D J (ed.) 1975 *Magnetic oxides.* Wiley, London, p. 235–87

Szymczak H, Lachowicz H K 1988 Stress-induced magnetic phenomena in metallic glasses. *IEEE Trans. Magn.* 24: 1747–51.

H. Szymczak
[Polskiej Akademii Nauk, Warsaw, Poland]

Anisotropy in Superconducting Materials

Anisotropic superconductors differ from isotropic materials in many of their properties. The differences are pronounced in layered or intercalated compounds (e.g., $NbSe_2$), in organic superconductors (e.g., κ-$(BEDT\text{-}TTF)_2Cu(NCS)_2$) and in the high-temperature superconductors (HTSs), all of which are strongly anisotropic. In this article, only the major effects of anisotropy on magnetic behavior will be outlined. In addition, this review will be restricted to effects for which details on the atomic length scale are unimportant; that is, to describe the properties of interest, materials are considered as anisotropic but uniform.

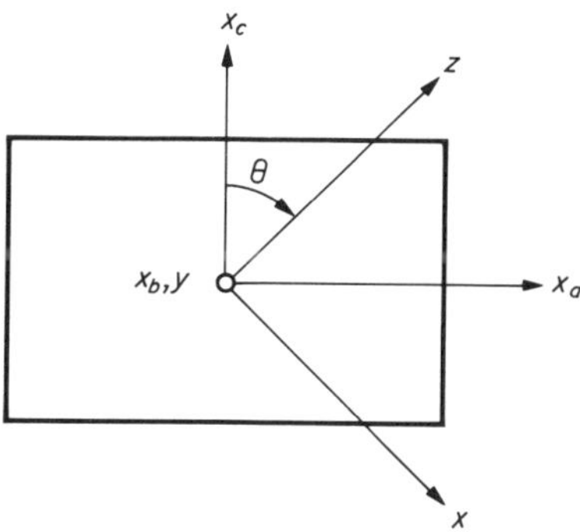

Figure 1
The *ac* plane of a crystal; axes *b* and *y* are normal to the figure plane. The coordinate frame *x*, *y*, *z* is often chosen with the axis *z* along the direction of vortices ***B***

1. Coherence Length, Penetration Depth and Mass Tensor

The properties of isotropic superconductors are described in terms of two characteristic lengths: the penetration depth λ and the coherence length ξ (see *Magnetic Properties of Superconductors*; de Gennes 1966, Tinkham 1975). The length λ describes the decay length for changes of the magnetic field (at a constant density n_s of superconducting pairs); $\lambda \propto M^{1/2}$, where M is the mass of the current carriers. The length ξ determines the length scale over which n_s changes; $\xi \propto M^{-1/2}$. The simplest way to account for the major effects of anisotropy is to introduce the "mass tensor" M_{ik} in place of the isotropic M. This amounts to replacing λ^2 and ξ^2 with the tensors $\lambda^2 M_{ik}/M$ and $\xi^2 M_{ik}^{-1} M$, respectively, where $i, k = x, y, z$ and $M = (M_a M_b M_c)^{1/3}$, which is the geometric mean of the masses in the three principal crystal directions a, b and c. It is convenient to define the dimensionless masses $m_{ik} = M_{ik}/M$. The tensor m_{ik} is diagonal in a coordinate system chosen to have x, y and z axes along the three principal directions a, b and c; these coordinates are called $\{x_a, x_b, x_c\}$ (see Fig. 1). In this system the diagonal components of m_{ik} are m_a, m_b and m_c; they have the normalization property $m_a m_b m_c = 1$.

Hence, within the mass tensor approach, an anisotropic superconductor is characterized by two (average) lengths λ and ξ, and two mass ratios, for example m_a/m_c and m_b/m_c (the third mass being determined by the above normalization). In the uniaxial case, $m_a = m_b$ and only one mass ratio remains.

In some special situations, not all of these four quantities are necessary to describe the anisotropic behavior of the material; in general, however, they are all needed. To choose the relevant values of the penetration depth, for example, it should be remembered that the masses enter the kinetic part of the energy associated with the persistent current. If the current flows everywhere parallel to one of the principal directions, say along the a axis, then the length over which the field and current decay is $\lambda_a = \lambda m_a^{1/2}$. If the current direction, for example x, is tilted relative to all principal axes, the penetration depth is given by $\lambda m_{xx}^{1/2}$; to find m_{xx}, m_{ik} must be transformed from $\{x_a, x_b, x_c\}$ to the frame where x is one of the coordinate axes. In most cases, however, the current direction within a sample changes from point to point. The current distribution cannot then be described in terms of any one particular penetration depth; that is, the tensor character of the (squared) penetration depth should be taken into account explicitly.

2. Dense Flux Line Lattices

The magnetic field penetrates type II superconductors in the form of magnetic flux lines or vortices (see *Flux Line Lattice*; *Magnetic Properties of Superconductors*), each containing the flux quantum $\Phi_0 = h/2e = 2.07 \times 10^{-15}$ Wb. The penetration depth λ gives roughly the linear dimension of the vortex cross section. The density n_s, which vanishes at the vortex axis, reaches its full zero-field value (at a given temperature) at approximately a distance ξ from the axis. For all known HTSs, $\lambda \gg \xi$. Physically, this means that the fraction of the volume occupied by vortex cores (where n_s changes) is small, provided the intervortex distance a_0 is large with respect to the core size ξ. In other words, if the magnetic induction $B \ll B_{c2}$ (at the upper critical field B_{c2} the cores overlap), n_s can be considered constant. In this domain, the magnetic properties are well described by an anisotropic version of the London theory that assumes that the magnetic part F of the energy is given (in SI units) by

$$2\mu_0 F = \int \boldsymbol{b}^2 + \lambda^2 m_{1k}\, \nabla_i \times \boldsymbol{b}\, \nabla_k \times \boldsymbol{b}\, dV \qquad (1)$$

where $\boldsymbol{b}$ is the microscopic magnetic induction (to distinguish from $\boldsymbol{B}$, the macroscopic average of $\boldsymbol{b}$); summation over repeated indices is implied. Due to the repulsive interaction, the vortices (which are parallel to $\boldsymbol{B}$) arrange in periodic flux line lattices (FLLs) in the plane normal to $\boldsymbol{B}$ (at a given density B/Φ_0).

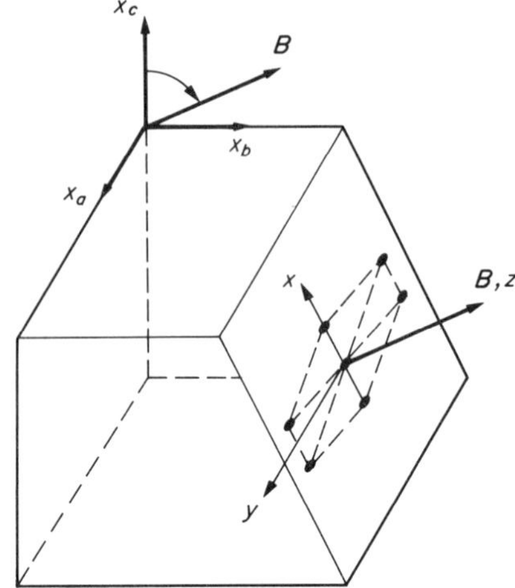

Figure 2
The flux line lattice in a uniaxial superconductor is shown schematically in the plane perpendicular to $\boldsymbol{B}$

Equation (1) makes it possible to compare the energies of different possible FLLs. For a uniaxial ($m_a = m_b$) crystal in high fields ($B_{c1} \ll B < B_{c2}$), the FLL with minimum energy is made of isosceles triangles with a side-to-base ratio of $(1 + 3m_c/m_{zz})^{1/2}/2$, where $\boldsymbol{z} = \boldsymbol{B}/B$ is the direction of the magnetic induction and $m_{zz} = m_a \sin^2\theta + m_c \cos^2\theta$, with θ being the angle between $\boldsymbol{B}$ and the c crystal direction. The FLL in equilibrium is locked upon the crystal for a given orientation of $\boldsymbol{B}$: the triangle base is directed along the projection of $\boldsymbol{c}$ onto the plane normal to $\boldsymbol{B}$ and the height of the triangle is situated in the ab crystal plane (see Fig. 2) (Balatskii et al. 1986, Campbell et al. 1988). Any rotation of the FLL as a whole around $\boldsymbol{B}$ causes an energy increase, a feature absent in isotropic superconductors.

3. Free Energy and Magnetization

Knowing the equilibrium FLL structure, it is possible to evaluate the free energy given in Eqn. (1) in terms of $\boldsymbol{B}$ at intermediate fields ($\Phi_0/\lambda^2 \ll B \ll \Phi_0/\xi^2$) (Campbell et al. 1988):

$$2\mu_0 F = B^2 + (\Phi_0/4\pi\lambda^2)B^* \ln(\Phi_0\beta/2\pi\xi^2 B^*) \qquad (2)$$

where the constant β is of the order of unity and $B^* = (m_{ik}B_iB_k)^{1/2}$ is an invariant quantity independent of the choice of coordinates. In the crystal frame, $B^{*2} = m_aB_a^2 + m_bB_b^2 + m_cB_c^2$; the same quantity can be written in the system with $\hat{z}$ directed along $\boldsymbol{B}$ as $B^* = Bm_{zz}^{1/2}$. Having $F(\boldsymbol{B})$, it is possible to find the field $\boldsymbol{H} = \partial F/\partial \boldsymbol{B}$ (see *Transport Currents and Force-Free Current Flow*) and the magnetization $\boldsymbol{M} = \boldsymbol{B}/\mu_0 - \boldsymbol{H}$:

$$\left.\begin{aligned} M_i &= -m_{ik}B_k\varphi(B^*) \\ \varphi(B^*) &= \frac{\Phi_0}{8\pi\lambda^2 B^*\mu_0}\ln\frac{\Phi_0\beta'}{2\pi\xi^2 B^*} \end{aligned}\right\} \quad (3)$$

where $\beta' = \beta/e$. Thus, the magnetization $\boldsymbol{M}$ is not parallel to $\boldsymbol{B}$ unless the field is directed along one of the principal crystal directions; that is, the magnetic response of anisotropic superconductors is not purely diamagnetic.

To extract the longitudinal (diamagnetic) part $\boldsymbol{M}_{lg}$ and the transverse $\boldsymbol{M}_{tr}$ (which are parallel and perpendicular, respectively, to $\boldsymbol{B}$), the coordinates can be used with the z axis parallel to $\boldsymbol{B}$. In this frame, $m_{ik}B_k = m_{iz}B_z = m_{iz}B$; therefore,

$$\left.\begin{aligned} M_{lg} &= M_z = m_{zz}B\varphi(B^*) \\ (M_{tr})_n &= -m_{nz}B\varphi(B^*) \end{aligned}\right\} \quad (4)$$

where the subscript n stands for either x or y. In general, the transverse magnetization is of the same order of magnitude as the longitudinal one. In strongly anisotropic materials, M_{tr} may exceed M_{lg}; for example, in $Bi_2Sr_2Ca_1Cu_2O_8$, M_{tr} is larger than M_{lg} by an order of magnitude when the $\boldsymbol{B}$ direction is a few degrees away from the *ab* plane (Tuominen et al. 1990). Note that the ratio $|\boldsymbol{M}_{tr}/\boldsymbol{M}_{lg}| = (m_{xz}^2 + m_{yz}^2)^{1/2}/m_{zz}$ is expressed exclusively in terms of the anisotropy parameters; this result also holds near B_{c2} (Kogan and Clem 1981).

Most of the high-T_c superconductors are uniaxial or almost uniaxial materials. In this case, the general formulae given in Eqn. (4) can be specified. The coordinate frame with z along the vortex axes ($\boldsymbol{B}$) can always be chosen so that the axis $\boldsymbol{y}$ is situated in the *ab* crystal plane as in Fig.1. The tensor m_{ik} is diagonal in the crystal frame with $m_a = m_b < m_c$ (the usual situation for layered materials); being transformed to the new frame it reads

$$\left.\begin{aligned} m_{xx} &= m_a\cos^2\theta + m_c\sin^2\theta \\ m_{yy} &= m_a, \quad m_{xy} = m_{yz} = 0 \\ m_{zz} &= m_a\sin^2\theta + m_c\cos^2\theta \\ m_{xz} &= (m_a - m_c)\sin\theta\cos\theta \end{aligned}\right\} \quad (5)$$

where θ is the angle between the c axis (the symmetry axis of the crystal) and z (or $\boldsymbol{B}$).

One of the consequences of the nonzero transverse magnetization is the mechanical torque $\tau = \boldsymbol{M} \times \boldsymbol{B}$ with the magnitude $|\boldsymbol{M}_{tr}|B$ (per unit volume) that acts on the FLL when $\boldsymbol{B}$ is not along any of the principal crystal directions. This torque can be measured, offering an accurate method to extract the anisotropy parameters (Kogan 1988, Farrell et al. 1989, 1990a). Physically, the nonzero transverse magnetization and the torque occur due to the tendency of persistent vortex currents to flow as close as energetically possible to the *ab* plane, where the masses are minimum.

This reduces the kinetic contribution to the free energy (see the second term in the free energy given in Eqn. (1)).

4. Vortex Interaction at Low Densities

The domain of small fields, where B is of the order of the lower critical field B_{c1}, is more difficult to treat than the region of intermediate fields, where the relatively simple expression given in Eqn. (2) for the free energy is available. It is necessary to investigate the interaction between vortices separated by distances of the order of λ. In the isotropic case, the interaction energy per unit length of two parallel vortices separated by r is proportional to the the field b of the first vortex at the core location of the second: $\varepsilon = \Phi_0 b(r)/\mu_0$. The field $\boldsymbol{b}$ has only one component, parallel to the axis of the vortex it belongs to. When r is large with respect to the core size ξ, London theory applies and yields a monotonically decreasing $\varepsilon(r) \propto K_0(r/\lambda)$ with K_0 being the modified Bessel function (i.e., the interaction force is everywhere repulsive). In the anisotropic situation, the direction of the field $\boldsymbol{b}$ of an isolated vortex is no longer parallel to the vortex axis (Kogan 1981, Balatskii et al. 1986). Moreover, the direction of $\boldsymbol{b}$ changes in space in a complicated way. It has been shown, however, that the interaction ε is still given by the isotropic expression, in which the total field $b(r)$ is replaced with the projection $b_z(r)$ along the direction of the vortex axes (Balatskii et al. 1986, Kogan 1990). Thus, the problem of interaction is reduced to that of the spatial distribution $b_z(x, y)$ of an isolated vortex. This problem has been solved numerically (Grishin et al. 1990, Kogan et al. 1990). An example is shown in Fig.3 of $b_z(x, y)$ for a vortex tilted at $\theta = 60°$ to the c axis of a strongly anisotropic uniaxial material with $(m_c/m_a)^{1/2} = 55$ (corresponding to $Bi_2Sr_2Ca_1Cu_2O_8$). A shallow minimum in the domain of *negative* $b_z(x, y)$ is clearly seen in the perspective plot. Another vortex situated in the vicinity of the one at the origin feels this potential well; the interaction energy is minimum at $x = 0.8\lambda$, $y = 0$. Due to this unusual interaction, the vortices have a tendency to form chains (Buzdin and Simonov 1990, Grishin et al. 1990), an effect which was recently seen in decoration experiments (Bolle et al. 1991).

5. Vortex Core and Upper Critical Field

The London approach fails in considering phenomena where the structure of vortex cores is relevant. The Ginzburg–Landau theory is then used, which allows for variation in the density of superconducting carriers n_s (see *Flux Line Lattice*; *Superconducting Materials: Physical Properties*). A few results obtained within this framework will be outlined. In the isotropic case, the upper critical field, at which the

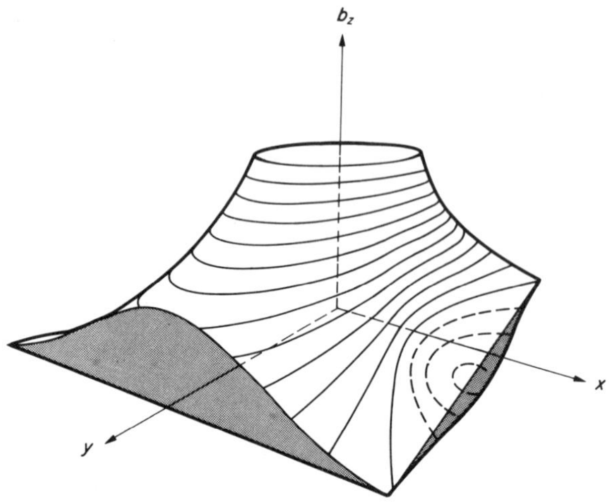

Figure 3
Perspective plot of the field component $b_z(x, y)$ parallel to the vortex axis for a vortex at 60° to the c axis (not shown) of a uniaxial crystal; $(m_c/m_a)^{1/2} = 55$. The y axis is parallel to the ab planes of the crystal. The solid curves are contours of constant energy of a vortex situated at a point (x, y) and parallel to the central vortex at $x = y = 0$. Dashed contours correspond to negative energies

bulk of the superconductor goes normal, is given by $B_{c2} = \Phi_0/2\pi\xi^2$. Assuming that at B_{c2} the normal vortex cores fill all of the sample volume, the area of the core cross section is obtained as $S = 2\pi\xi^2$. Looking for B_{c2} in the anisotropic case, it is noted that the upper critical field in a given direction is a scalar thermodynamic quantity (i.e., it is invariant with respect to the choice of coordinates). The simplest scalar that can be formed out of m_{ik} and the field $\boldsymbol{B}$ is $m_{ik}B_iB_k$. It can be guessed (by analogy with the isotropic result) that, in the anisotropic case, the phase transition happens if $(m_{ik}B_iB_k)^{1/2} = \Phi_0/2\pi\xi^2$. Introducing the unit vector $\boldsymbol{e}$ in the direction of $\boldsymbol{B}$ gives

$$B_{c2} = \frac{\Phi_0}{2\pi\xi^2(m_{ik}e_ie_k)^{1/2}} = \frac{\Phi_0}{2\pi\xi^2 m_{zz}{}^{1/2}} \tag{6}$$

where the first formula is written in the frame-independent form and the second one is written in a frame for which the z axis is directed along the field. This result has been obtained by Tilley (1965) for a uniaxial material ($m_a = m_b$), and by Klemm and Clem (1980) for a general case of three different masses. In terms of the angles α, β, and γ between $\boldsymbol{H}$ and the principal directions a, b and c, the quantity $m_{ik}e_ie_k = m_a\cos^2\alpha + m_b\cos^2\beta + m_c\cos^2\gamma$.

Equation (6) suggests that in an anisotropic material the core cross section has an area

$$S = 2\pi\xi^2(m_{ik}e_ie_k)^{1/2} = 2\pi\xi^2\, m_{zz}{}^{1/2} \tag{7}$$

In the uniaxial case with $m_c > m_a$, the semimajor axis of the elliptical core equals $\xi(2m_am_c)^{1/2}$ and lies in the ab crystal plane; the semiminor axis is $\xi(2m_am_{zz})^{1/2}$ with m_{zz} given in Eqn. (5) ($\cos\theta = \boldsymbol{B.c}/B$). Thus, when the vortex orientation changes from the c axis ($\theta = 0$) towards the ab plane ($\theta = \pi/2$), the core deforms from a circle of radius $\xi(2/m_a)^{1/2}$ via a sequence of ellipses with a constant semimajor axis $\xi(2/m_a)^{1/2}$ (situated in the ab plane) and a shrinking semiminor axis; the latter reaches a minimum of $\xi(2/m_c)^{1/2}$ at $\theta = \pi/2$.

6. Other Properties of Anisotropic Superconductors

Although the question of anisotropy is of special relevance for HTSs, all of which are strongly anisotropic, the subject has a long history. Most of the "conventional" low-T_c materials are only weakly anisotropic. Nevertheless, the anisotropy of such quantities as B_{c2} and of the energy gap Δ, a microscopic characteristic of superconductors, has been studied extensively (Weber 1977). It is worth noting that the effects of anisotropy are richer than those outlined previously within the mass tensor approach. For example, in cubic materials, for which the m_{ik} approach predicts no anisotropy at all, the B_{c2} of pure vanadium and niobium experimentally shows anisotropy on the level of several percent at low temperatures (Weber 1977). The anisotropy of the energy gap (and, therefore, of the tunnelling characteristics) is of particular importance in studies of the microscopic mechanisms of superconductivity in such materials as "heavy fermions.", organic and other "exotic" superconductors (see *Superconducting Materials: An Overview*). Artificially prepared multilayered superconductors, in which superconducting layers (S) alternate with insulating (I) or normal (N) ones, are of course, also anisotropic. The properties of "thick" S–N multilayers (the layer thickness $d \gg \xi$) can be described in terms of the proximity effect (see *Coherence Length, Proximity Effect and Fluctuations*, Dobrosavljevic and Raffy 1974). When the coherence length ξ is comparable with the spacing between S layers, the behavior is governed by the Josephson coupling of the layers (Lawrence and Doniach 1970). In materials with $\lambda \gg \xi$, the effect of the Josephson interlayer coupling on the field and current distributions (governed by λ) is weak. However, at distances of the order ξ, the Josephson coupling substantially changes predictions of the Ginzburg–Landau or London theories; in particular, effects associated with the vortex core structure acquire many peculiar features (reviewed by Bulaevskii (1990)).

Only some of the magnetic properties of anisotropic superconductors in thermodynamic equilibrium have been touched upon in this article. Many non-equilibrium properties, such as critical currents, dissipation and fluctuation phenomena in anisotropic

Table 1
Anisotropy parameter for uniaxial materials

Material	γ	Source
$NbSe_2$	3.3	Bulaevskii (1990)
$YBa_2Cu_3O_7$[a]	7.7	Farrell et al. (1990a)
$Bi_2Sr_2Ca_1Cu_2O_8$	55	Farrell et al. (1989)
κ-(BEDT-TTF)$_2$Cu(NCS)$_2$	200	Farrell et al. (1990)b)

a This value is obtained from torque data on a twin-free crystal at $T > 75$ K. Decoration experiments at $T = 4.2$ K show deviations from tetragonal (uniaxial) symmetry and yield for this orthorhombic material $m_a/m_b \simeq 1.2$ and $m_c/m_a \simeq 5.5$ (Vinnikov et al. 1988, Dolan et al. 1989)

superconductors, also show strong anisotropy. A description of these effects, however, would take the reader well beyond the main goal of this article, which has been to give a brief introduction to this rapidly expanding subject. In conclusion, Table 1 lists the anisotropy parameter $\gamma = (m_c/m_a)^{1/2}$ (the ratio of the B_{c2} parallel to the a axis to that along the c direction, which enters the expressions for many other anisotropic quantities) for several uniaxial materials; it shows the strength of anisotropy effects in different classes of superconductors.

See also: Flux Line Lattice; Magnetic Properties of Superconductors; Oxide Superconductors: Physical Properties; Superconducting Materials: Physical Properties

Bibliography

Balatskii A V, Burlachkov L I, Gor'kov L P 1986 Magnetic properties of anisotropic type-II superconductors. *Sov. Phys. JETP (Engl. Transl.)* 63: 866–71

Bolle C A, Gammel P L, Grier D G, Murray C A, Bishop D J, Mitzi D B, Kapitulnik A 1991 Observation of a commensurate array of flux chains in tilted flux lattices in Bi–Sr–Ca–Cu–O single-crystals. *Phys. Rev. Lett.* 66: 112–15

Bulaevskii L N 1990 Macroscopic theory of layered superconductors. *Int. J. Mod. Phys.* 4: 1849–77

Buzdin A I, Simonov A Yu 1990 Penetration of inclined vortices into layered superconductors. *JETP Lett. (Engl. Transl.)* 51: 191–5

Campbell L J, Doria M M, Kogan V G 1988 Vortex lattice structures in uniaxial superconductors. *Phys. Rev. B* 38: 2439–43

de Gennes P G 1966 *Superconductivity of Metals and Alloys.* Benjamin, New York

Dobrosavljevic L, Raffy H 1974 Flux line lattice in anisotropic type II superconductors. *Phys. Status Solidi B* 64: 229

Dolan G J, Holtzberg F, Feild C, Dinger T R 1989 Anisotropic vortex structure in $Y_1Ba_2cu_3O_7$. *Phys. Rev. Lett.* 62: 2184–7

Farrell D E, Allen C J, Haddon R C, Chichester S V 1990b Superconducting effective-mass anisotropy in κ-(BEDT-TTF)$_2$Cu(NCS)$_2$ [where BEDT-TTF is bis (ethylenedithio)tetrathiafulvalene]. *Phys. Rev. B* 42: 8694–7

Farrell D E, Bonham S, Foster J, Chang Y C, Jiang P Z, Vandervoort K G, Lam D J, Kogan V G 1989 Giant superconducting anisotropy in $Bi_2Sr_2Ca_1Cu_2O_8 + \delta$. *Phys. Rev. Lett.* 63: 782–5

Farrell D E, Rice J P, Ginzberg D M, Liu J Z 1990a Experimental evidence of a dimensional crossover in $Y_1Ba_2Cu_3O_7$-δ. *Phys. Rev. Lett.* 64: 1573–6

Grishin A M, Martynovich A Yu, Yampolskii S V 1990 Magnetic-field inversion and vortex chains in anisotropic superconductors. *Zh. Eksp. Teor. Fiz.* 97: 1930–46

Klemm R A, Clem J R 1980 Lower critical field of anisotropic type-II superconductors. *Phys. Rev. B* 21: 1868–75

Kogan V G 1981 London approach to anisotropic type-II superconductors. *Phys. Rev. B* 24: 1572–5

Kogan V G 1988 Uniaxial superconducting particle in intermediate magnetic fields. *Phys. Rev. B* 38: 7049–50

Kogan V G 1990 Forces upon vortices in anisotropic superconductors. *Phys Rev. Lett.* 64: 2192–4

Kogan V G, Clem J R 1981 Uniaxial type–II superconductors near the upper critical field. *Phys. Rev. B.* 24: 2497–505

Kogan V G, Nakagawa N, Thiemann S 1990 Interaction of vortices in uniaxial superconductors. *Phys. Rev. B* 42: 2631–4

Lawrence W E, Doniach S 1970 In: Kanda E (ed.) 1970 *Proc. 12th Int. Conf. Low Temperature Physics.* Keigaku, Tokyo, p. 361

Morris R C, Coleman R V, Bhandari R 1972 Superconductivity and magnetoresistance in $NbSe_2$. *Phys. Rev. B* 5: 895–901

Tilley D R 1965 Critical fields and flux penetration pattern in anisotropic type-II superconductors. *Proc. Phys. Soc. London* 85: 1177–84

Tinkham M 1975 *Introduction to Superconductivity.* McGraw–Hill, New York

Tuominen M, Goldman A M, Chang Y Z, Jiang P Z 1990 Magnetic-anisotropy of high-T_c superconductors. *Phys. Rev. B* 42: 412–9

Vinnikov L Ya, Gurevich L A, Emelchenko G A, Osipyan Y A 1988 Direct observation of the Abrikosov vortices in the single-crystal of a high-temperature superconductor $YBa_2Cu_3O_x$. *JETP Lett. (Engl. Transl.)* 47: 131–4

Weber H W (ed.) 1977 *Anisotropy Effects in Superconductors.* Plenum, New York

V. G. Kogan and J. R. Clem
[Iowa State University, Ames, Iowa, USA]

B

Barkhausen Effect

The term Barkhausen effect refers to rapid localized changes (Barkhausen jumps) in the magnetization of a ferromagnetic body. As a jump occurs, the associated magnetic-induction field varies, and a time-varying voltage which may be displayed on an oscilloscope will be induced in an appropriately oriented inductive search coil in proximity with the body. If the aggregate magnetization of a polycrystalline body is cyclically varied by an alternating applied field, numerous jumps occur in random sequence, inducing in the search coil a corresponding train of voltage pulses which, if amplified and fed to a loudspeaker, produce noiselike sound (Barkhausen noise). The voltage signal itself is also called Barkhausen noise. Figure 1 illustrates the basic arrangement for observing Barkhausen noise.

The Barkhausen effect occurs in ferromagnetic materials such as iron and ferritic steels, cobalt, nickel, numerous alloys and ferrimagnetic compounds (ferrites). Several applications of Barkhausen-noise analysis for nondestructive evaluation have been investigated, notably the assessment of mechanical stress and metallurgical grain size.

This article reviews briefly the physical origin of the Barkhausen effect and the manner in which it is influenced by material properties, describes the basic features of instrumentation and concludes with remarks on the intrinsic limitations and future possibilities of Barkhausen noise analysis.

1. *Magnetic Domains and Barkhausen Jumps*

The Barkhausen effect is explained by ferromagnetic-domain theory, which is described in the article *Domains and Domain Walls*.

The aggregate magnetization of a body is reduced by subdivision into magnetic domains within each of which the magnetic-dipole-moment density $\boldsymbol{M}$ is uniform, but with $\boldsymbol{M}$ directed differently in adjacent domains. Separating adjacent domains is a relatively thin domain wall within which $\boldsymbol{M}$ (or a component of $\boldsymbol{M}$) rotates about an axis normal to the wall. Domain walls increase exchange and anisotropy energies; this is offset by a reduction in net magnetostatic energy. Through an appropriate arrangement of domains and intervening walls, the aggregate magnetization of a finite body can be reduced to zero (or nearly so). In polycrystalline bodies, grain boundaries and other material inhomogeneities may result in very complex domain structures.

When an external magnetic field is applied to an unsaturated ferromagnetic body, magnetostatic forces tend to align $\boldsymbol{M}$ along the applied field, precipitating a new domain arrangement of minimum total energy. Domains in which $\boldsymbol{M}$ is favorably aligned with respect to the applied field enlarge in volume at the expense of adjacent domains less favorably oriented. Also, entirely new domains (with more favorable orientation of $\boldsymbol{M}$) may be created, old ones being destroyed. The movement of domain walls may occur smoothly, but imperfections such as voids and inclusions may pin a wall until there is sufficient energy to overcome the local barrier, at which point the wall jumps to a new position (see *Coercivity and Domain Wall Pinning*). Large-scale rearrangements of domain structure also occur abruptly. Such rapid, localized jumps in $\boldsymbol{M}$ are Barkhausen jumps.

Barkhausen jumps may be induced by applied mechanical stress via the magnetoelastic effect (see *Magnetoelastic Phenomena*). Conversely, when a jump occurs, the local lattice strain varies rapidly, creating a stress wave which can be detected by acoustic-emission techniques. Barkhausen jumps may

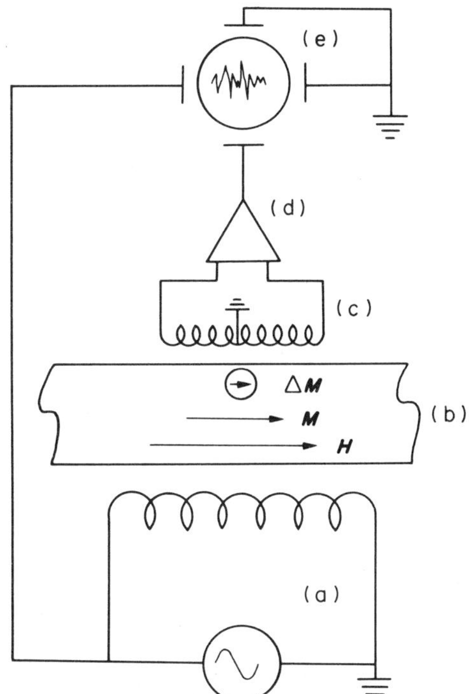

Figure 1
Basic arrangement for observing the Barkhausen effect by the inductive method: (a) applied field source; (b) ferromagnetic specimen; (c) detection coil; (d) amplifier; and (e) oscilloscope

also be induced by thermal expansion and contraction.

Barkhausen jumps are accompanied by rapid changes in magnetic induction; if the surrounding material is electrically conductive (as is iron), eddy currents are induced which dissipate energy and attenuate the rate of change of the net induction field. Consequently, Barkhausen jumps can be inductively detected only as they occur near the surface of most specimens.

Although the Barkhausen signal train has the appearance of noise, some autocorrelation exists, reflecting the sequential triggering of one or more secondary jumps by primary jumps. In single crystals, some domain structure recurs during successive magnetization cycles; this presumably also occurs in polycrystalline specimens, but to what extent is not known.

2. *Stress Measurement*

If additional strain is imposed on the spontaneous lattice strain of a ferromagnetic body, an energetically preferred orientation of $\boldsymbol{M}$ with respect to the principal axes of imposed strain exists. In iron, the effect of moderate strain (up to the yield point) is to render one of the cube axes energetically preferred (see *Anisotropy in Magnetic Materials*). For uniaxial tension, the preferred axis is the one most nearly parallel to the stress axis; for uniaxial compression, the preferred axis is the one most nearly perpendicular to the stress axis. Multiaxial fields have more complicated effects. If a slowly varying magnetic field is applied, the details of the resulting Barkhausen jumps are affected by lattice strain, with a concomitant effect on the inductively detected Barkhausen noise; such effects have been empirically correlated with the stress state.

Figure 2 illustrates a simple instrument developed for stress measurement. The low-frequency applied field is supplied by a C-shaped electromagnet. To reduce the effect of specimen geometry, a Hall magnetometer probe senses the low-frequency magnetic-field component parallel to the surface; a feedback circuit ensures that the field follows a selected wave form. A small multiturn inductive coil with its axis parallel to the applied field and located as near the specimen surface as practicable senses Barkhausen jumps in its immediate vicinity. The Barkhausen noise signal is then processed in some manner, the simplest being a measurement of the peak power produced during a single magnetization reversal, or the integrated power over such a reversal. Reproducibility is improved by averaging over several reversals. Other signal-processing procedures such as pulse-height analysis and spectrum analysis have also been studied.

Figure 2 also shows the general type of calibration curve obtained for uniaxial stress, the applied field, the stress axis and the sensing coil being aligned. In general, satisfactory correlation is obtained between measurements of uniaxial surface stress by Barkhausen-noise analysis and by x-ray diffraction, although anomalies exist. Less satisfactory results have been obtained for biaxial surface stress. For most structural steels the instrumentation described above has a calibration curve that "saturates" at moderate stress (well below yield). This is in part a reflection of the fact that the magnetoelastic effect is itself nonlinear and saturates at moderate stress. More sophisticated instrumentation and signal processing may extend the useful range.

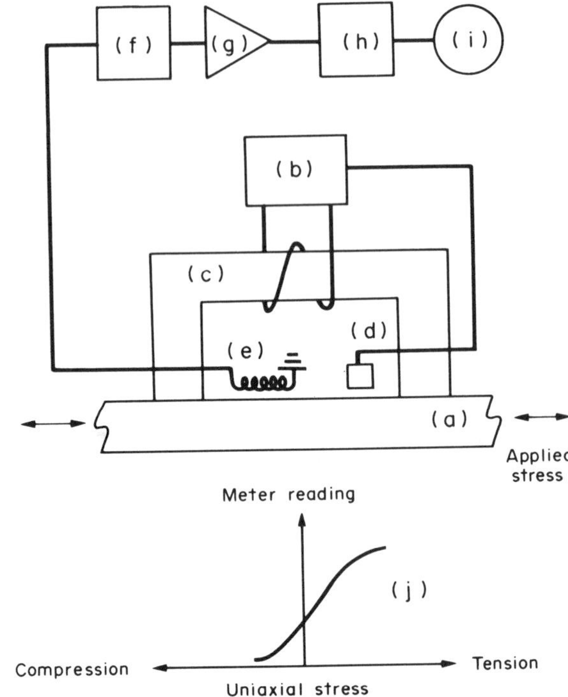

Figure 2
Instrumentation for stress measurement: (a) ferromagnetic specimen; (b) current supply; (c) electromagnet; (d) Hall probe; (e) detection coil; (f) high-pass filter; (g) amplifier; (h) signal processor; (i) output display; and (j) general shape of calibration curve

3. *Grain Size Measurement*

In polycrystalline materials, magnetic domains and single grains do not, in general, coincide. However, fine-grain materials with random grain orientation (untextured) have, on average, smaller domains than do large-grain materials. Consequently, some correlation between grain size and Barkhausen-noise characteristics is expected. Instrumentation, similar in most respects to that illustrated in Fig. 2, has been empirically calibrated against grain size for selected materials.

4. Conclusion

The Barkhausen effect is an extremely complex phenomenon which is strongly influenced by: (a) material properties including chemical composition, phase constitution, grain size and orientation, dislocations, inclusions, voids, grain boundaries and temperature; (b) the magnitude and distribution of both microscopic stress (associated with x-ray line broadening) and macroscopic stress (associated with x-ray line shifts), whether applied or residual; and (c) the direction, uniformity and rate of change of the applied magnetic field. The detected Barkhausen noise is further influenced by the location, orientation and electrical characteristics of the detection coil, and by the bandwidth and noise figure of the amplifier and associated signal-processing circuitry. To date, no satisfactory means has been found for distinguishing unambiguously among these factors; necessary systematic studies are lacking. Thus, if one factor is to be assessed, all others must be closely controlled. Absolute measurements are possible only by empirical calibration using reference specimens for which absolute factor values have been determined by other methods.

The usefulness of Barkhausen-noise analysis for nondestructive evaluation may in future be extended by improvements in instrumentation and in signal processing. Detailed studies of possible correlations between various material properties and the stochastic properties of the detected signal may reduce the ambiguities and thereby greatly enhance the usefulness of the method.

See also: Nondestructive Evaluation Using Magnetism

Bibliography

Gardner C G, Matzkanin G A, Davidson D L 1971 The influence of mechanical stress on magnetization processes and Barkhausen jumps in ferromagnetic materials. *Int. J. Nondestr. Test.* 3: 131–69

Kittel C, Galt J K 1956 Ferromagnetic domain theory. *Solid State Phys.* 3: 439–564

Matzkanin G A, Beissner R E, Teller C M 1979 *The Barkhausen Effect and Its Application to Nondestructive Evaluation*, Report NTIAC-79-2. Nondestructive Testing Information Analysis Center, Southwest Research Institute, San Antonio, TX

Stierstadt K 1966 Der magnetische Barkhausen-Effekt. In: Höhler G (ed.) 1966 *Springer Tracts in Modern Physics*, Vol. 40. Springer, Berlin, pp. 2–106

Sundstroem O, Poerroenen K 1979 The use of Barkhausen noise analysis in nondestructive evaluation. *Mater. Eval.* 37: 51–6

C. G. Gardner
[University of Houston, Houston, Texas, USA]

Biomagnetic Measurements

Biomagnetism studies magnetic fields originating in biological tissues, with special consideration to those generated inside the human body. These fields may be produced by

(a) magnetic contaminants accumulated in some tissues such as the lungs of people exposed to occupational pollution, or intentionally administered to specific patients to be used as magnetic markers;

(b) iron accumulation in the liver of patients affected by several diseases such as thalassemia major and hemochromatosis; or

(c) ionic, namely bioelectric, currents inside and outside the membrane of excitable cells, such as Purkinje fibers and neurons.

The most studied biomagnetic fields are, indeed, those linked to bioelectric activity in the brain (magnetoencephalogram) and in the heart (magnetocardiogram). It is possible to record these fields by the use of suitable instrumentation and, from the field distribution over the investigated area of the body such as for example a portion of the scalp, it is possible to localize the source of epileptic activity, or that responsible for a specific cerebral response to a sensory stimulus. In heart measurements it is possible to follow the cardiac muscle activation, to detect and, possibly, localize an abnormal conduction pathway. In this framework, biomagnetic measurements provide information similar to that obtainable by the well-known electric potential measurements (electroencephalogram, electrocardiogram). In addition to the intrinsic difference between the two techniques, the main advantage of the biomagnetic method is, however, its ability to very precisely identify the location of the field source (if any). On the other hand, as will be explained, electric measurements are influenced by variations in the conductivity of the tissues, due to specific anatomical features, so that an accurate source localization would require a complete and exact knowledge of all the conductivities and shapes of the surrounding medium. The biomagnetic method should therefore be regarded as a new, powerful tool to perform a functional investigation of the human body and, possibly, to provide a functional imaging of various organs.

1. Biomagnetic Fields and Measuring Devices

The intensity of biomagnetic fields may be deduced from Maxwell's equations. Nevertheless, for all applications the quasistatic approximation is well satisfied, thus allowing the use of the Biot–Savart law. The term magnetic field will be used henceforth, although its proper definition would be magnetic induction, since the two terms are strictly proportional. Figure 1 summarizes field strengths and frequency ranges of the most representative biomagnetic fields. The field strength units currently used in biomagnetism are the femtotesla (fT), with 1 fT = 10^{-15} T and the picotesla (pT) with 1 p = 10^{-12} T. By comparison, the earth's magnetic field is 50 μT.

A twofold problem must be faced in the detection of biomagnetic fields: the extreme weakness of the signals and the strength of other "disturbing" fields, commonly regarded as "noise." Figure 1 shows that most of the biomagnetic activity features an intensity below 50 pT, which is a million times weaker than the earth's magnetic field, and a frequency range confined below 1 kHz. The weakness and the frequency content of these signals force the experimenter to use superconducting instrumentation, since ordinary room-temperature equipment would anyway be limited by Johnson noise (Romani et al. 1982a). The device used is the SQUID (superconducting quantum interference device) (see *SQUID Design, Fabrication and Application*) which is basically an ultra-low-noise magnetic flux–voltage converter. Early biomagnetic systems utilized rf SQUIDS, whereas today microfabricated dc SQUIDs are commonly used. Figure 2 shows a typical single-channel biomagnetic system. A "sensing" coil is located inside the tail of a cryogenic Dewar, in order to be placed as close as possible to the subject, for reasons which will become evident. This coil is connected to an "input" coil, tightly coupled to the SQUID by means of a superconducting loop, and the SQUID is located at a certain distance above the coil, especially if it is contained inside a superconducting shield that might modify the field pattern in the detecting area. The room-temperature amplifier is usually located at the top of the cryogenic insert, whereas the SQUID control unit is positioned far away, to avoid magnetic interference. Subsequently, the output of the SQUID electronics may be filtered and analog-to-digital (AD) converted to be stored onto magnetic disk just as with regular electric data.

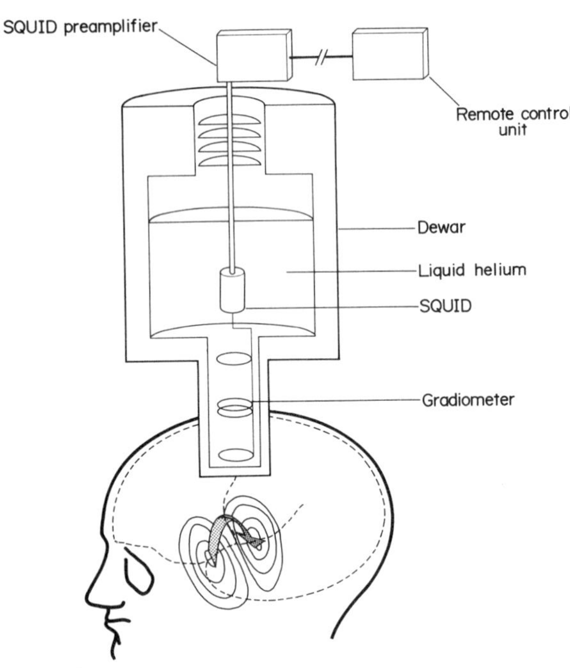

Figure 2
Schematic of a single-channel biomagnetic system: the pickup coil is a second-order axial gradiometer for use in an unshielded environment; the SQUID may be either rf or dc; the SQUID remote control unit is generally placed near the experimenter to check SQUID functioning

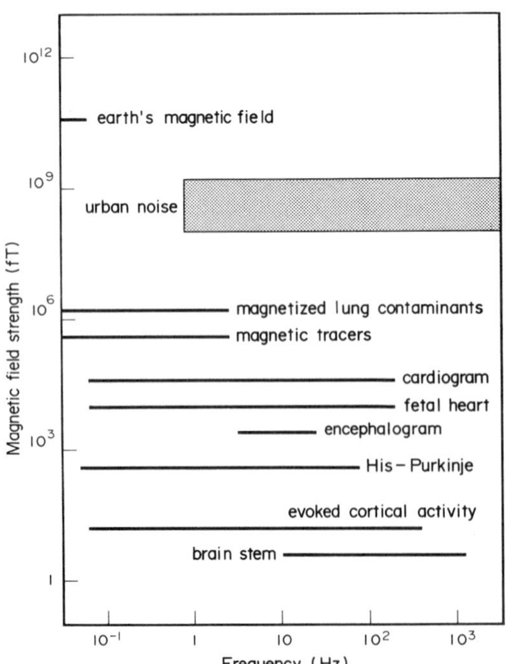

Figure 1
Field strength and frequency range characterizing typical biomagnetic fields and major noise sources

Biomagnetic fields are much smaller than other existing magnetic fields, which constitute background noise for the experimenter. Therefore, even if the instrumentation were noiseless, successful detection of biomagnetic signals would be possible only by adequate discrimination between this field and other noise signals, in order to remove the latter. There are two solutions for this problem: the adoption of pick-up coils with a suitable geometry or the use of magnetically shielded rooms. The simplest detection coil to be coupled to the SQUID is a single coil, namely a magnetometer. However, it is possible to build a detection coil insensitive to constant fields by using two identical, oppositely wound, coaxial coils, since each coil senses an equal field but with opposite polarity. Furthermore, if the two coils are separated by a distance D along their axis, the sensor is referred to as a first-order gradiometer and distinguishes between near ($< D$) and far ($> D$) sources, in the sense that sensitivity is maintained only for close sources. The most widespread detection coil in unshielded environments is the second-order gradiometer. It is formed by three coaxial coils (with the same area and $+N$, $-2N$, $+N$ turns, see Fig. 2), that is, by two first-order gradiometers arranged back to back. This geometry is also insensitive to constant gradients. Other sensors, suitable for microfabrication, may be achieved by displacing the coils in a transverse direction, so that they lie on a plane (planar gradiometers). Indeed, these latter sensors will be used more and more in the near future, as they can be integrated directly into the SQUID loop, thus increasing the instrument sensitivity.

The rejection of signals coming from distant sources provided by the detection coil may not be adequate in noisy environments such as hospitals and laboratories. In these cases the use of a magnetically shielded room is preferable. These rooms are made by alternating layers of high permeability materials (e.g., μ-metal) for dc and low-frequency shielding, and of high-conductivity materials (e.g., pure aluminum) for high-frequency shielding. A shielding factor of about 1000 is obtained for dc fields, whereas much better shielding is achieved at the frequency of the power line and its harmonics. The best choice seems to be a magnetically shielded room to be used in connection with first-order gradiometers. A typical noise figure featured by commercial systems inside a magnetically shielded room is about $10\,\text{fT}\,\text{Hz}^{-1/2}$, whereas laboratory prototypes have shown a noise figure down to $2\text{–}3\,\text{fT}\,\text{Hz}^{-1/2}$ in similar conditions (see *SQUID Design, Fabrication and Application*).

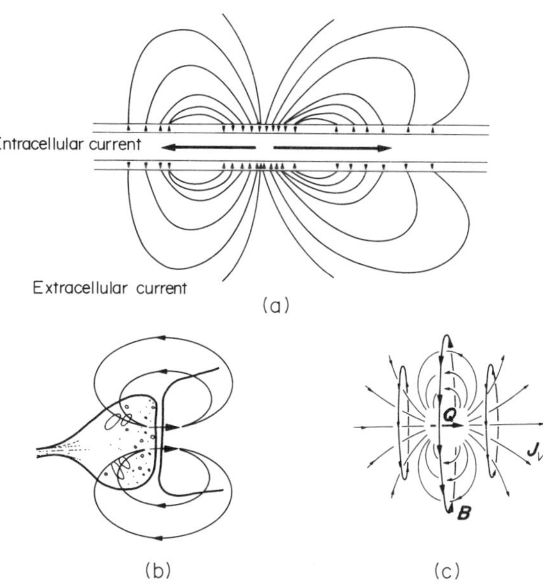

Figure 3
Schematic diagram of bioelectric activity in the nervous system: (a) action potential travelling along an axon modelled with two opposite current dipoles (complete analysis must include intracellular and extracellular current); (b) postsynaptic potential and single current dipole model for interpreting cortical activity; and (c) current dipole model with related return currents

2. *Modelling and Source Localization*

In order to successfully identify the sources of the measured magnetic activity, it is necessary to model the sources themselves in a simple mathematical form. The simplest model sources are the magnetic dipole and the current dipole (Williamson and Kaufman 1981).

The magnetic dipole permits modelling fields originating from a magnetic particle inside living tissues. It is characterized by a magnetic moment $\boldsymbol{m}$. It is worth stressing that the field strength features an inverse cube dependence on the distance from the dipole.

A more complex situation is encountered in modelling ionic currents (see Fig. 3). The magnetic field recorded over the human brain as a consequence of a sensory stimulus is believed to be generated by postsynaptic potentials. The current pattern associated with this cerebral event may be modelled by a current dipole (i.e., a short segment of current I and length L) the properties of which are fully described by the dipole moment $\boldsymbol{Q} = I\boldsymbol{L}$. The magnetic field generated by this source has an orientation given by the right hand rule, and a strength, expressed in $\text{A}\,\text{m}^{-1}$, that features an inverse square dependence on the distance from the source. In the case of a current pulse travelling along an axon (action potential), two current dipoles are used, one for the depolarization front and the other for repolarization. It is worth noting that the current dipole represents only the intracellular current, which is not explaining all the electrical activity. Since the current dipole is immersed in a conducting medium, the extracellular or return current must also be considered in the evaluation of the magnetic field. This current is fixed by the shape of the medium where these dipoles are located. In the case of brain measurements, it is convenient to model, at least in a first approximation, the head as a homogeneous sphere. In this case, the magnetic field normal to the sphere is affected only by intracellular currents, thus giving direct information about neural functioning.

With these arguments it is possible, by measuring

the magnetic field outside the human head, to localize the source inside it. To obtain a successful localization, it is convenient to record simultaneously the magnetic field all over the scalp, or at least over a large portion of it. Presently, the available systems feature 20–30 channels, thus allowing the simultaneous detection of the magnetic field in 20–30 points over the head on a surface of about 200 cm^2. This is not enough to achieve a complete topographic mapping of the brain, so the efforts are now devoted to the development of larger multichannel instruments able to obtain a single-shot magnetic field map of the whole head.

3. *Magnetic Contaminants*

The magnetic field generated by magnetic contaminants inside biological tissue may be due either to an accumulation of ferromagnetic materials, or to an accumulation of paramagnetic substances in an external magnetic field (e.g., the earth's magnetic field).

Ferromagnetic materials may be stored in the lungs of arc welding workers or coal miners. By recording the magnetic field it is possible to keep track of the amount of magnetic dust present in the lungs, in order to evaluate the risk of occupational diseases. Another application consists in using small magnetized steel spheres (e.g., 1 mm diameter) to be used as magnetic markers. For the evaluation of the gastrointestinal transit time, such a sphere is ingested by the subject. By measuring the magnetic field generated by the sphere it is possible to localize the marker inside the abdomen every few minutes, thus following in real time the gastrointestinal activity (Di Luzio et al. 1989).

Paramagnetic substances inside an external magnetic field also generate a detectable field. Such fields may be measured near the liver of patients affected by hemochromatosis and thalassemia, where an excess concentration of paramagnetic iron is found. The strength of such fields is proportional to the external applied field. The earth's field may be used, but preferably the subjects are placed inside a stronger field to enhance the signal-to-noise ratio. In clinical practice, iron overload in the liver is estimated by recurrent biopsy, with evident psychological stress for the patients. The results obtained with the magnetic approach are in quite good agreement with those provided by biopsy, without any stress for the patient (Brittenham et al. 1982).

4. *Ionic Currents Generated by Active Cells*

The possibility of obtaining direct information on fast occurring displacement of ions inside living tissues is the most outstanding feature of the biomagnetic method, particularly when studying currents flowing inside the human body, where other invasive methods cannot be used.

Leading studies on isolated axons were carried out by Wikswo and co-workers, showing that the action potential moving along a giant axon can be modelled by two opposite current dipoles. Furthermore, they developed a complete theory of electric and magnetic signals generating inside axons immersed in a conducting medium (Roth and Wikswo 1985). A possible application for this study is the detection of injuries in the nerves, where the conduction properties of the nerve are reduced or even missing.

The study of human brain functions is the area where the biomagnetic method had the major impact. Usually, brain studies are separated into two parts: evoked fields and spontaneous activity. The measurement of magnetic fields elicited by an external stimulus gives detailed information about the underlying processes. Many sensory inputs may be stimulated in order to record a brain response. The major working areas include the auditory system, the somatosensory system and the visual system. The neuromagnetic studies on the structural organization of the neural sources generating auditory evoked fields showed that there is a tonotopic organization of the auditory cortex (Romani et al. 1982b). This tonotopic organization was already known from animal studies, but only the neuromagnetic method provided the needed accuracy in source localization to prove it in humans. A great deal of work has been carried out studying the evoked field from the somatosensory cortex, since the single current dipole model seems to hold well in this context. Localizations of sources elicited by somatosensory stimuli showed consistent agreement with the somatosensory homunculus. The investigation of spontaneous brain activity, particularly epileptic activity, was the first major success of the biomagnetic method. With magnetic measurements it is possible to localize, within a few millimeters, focal epileptic activity (Barth et al. 1982). Interesting studies are also being carried out on generalized activity. However, in these measurements the demand for an instrument able to simultaneously record the magnetic field over the whole head is mandatory, as the brain activity is not stationary and, therefore, a time reference, like the stimulus trigger, is not available. An example of neuromagnetic localization of an epileptic focus is illustrated in Fig. 4.

Significant results in clinical studies have also been achieved in the detection of magnetic signals generated by the heart electrical activity. By recording magnetic fields during the cardiac cycle it is possible to follow the heart activation in order to look for abnormal pathways of the heart conduction system (Wolf–Parkinson–White syndrome). Other significant results include the evaluation of the risk for sudden death in arrythmogenic diseases and the follow up of implanted hearts to predict rejection phenomena.

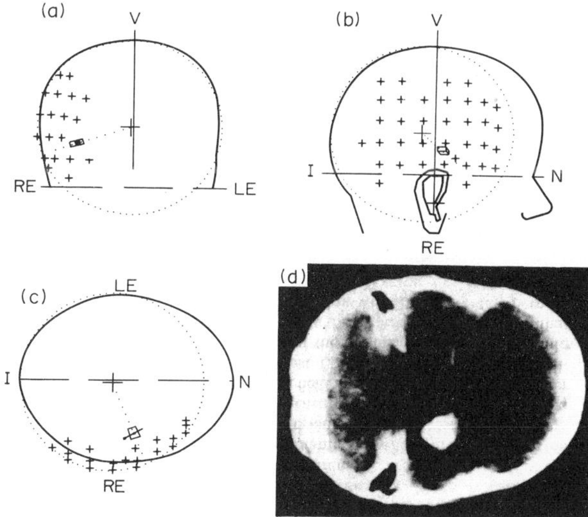

Figure 4
Example of neuromagnetic localization of an epileptic focus: (a)–(c) neuromagnetic localization and (d) CT scan showing a clear-cut calcification in the right uncus region, in good agreement with (a)–(c) (after Ricci et al. 1987)

5. *Future Developments*

The measurement of the magnetic field originating in biological tissues may provide useful clinical information on heart and brain functioning in a completely noninvasive way. In brain measurements, epileptic foci may be accurately localized without the need for depth electrodes. In heart measurements, the heart activation system may be examined without the need for intracardial catheterization. Additionally, the adoption of magnetic markers may replace radiopaque markers, for instance in the study of gastrointestinal activity, removing the need for using x rays. New multichannel instruments are now being developed and are expected to enhance the reliability of magnetic source localization and to improve the possibility of automatic data analysis in clinical practice.

See also: Nuclear Magnetic Resonance Imaging; SQUID Design, Fabrication and Application; Superconducting Device Fabrication

Bibliography

Atsumi K, Katila T, Kotani M, Williamson S J, Ueno S (eds.) 1988 *Biomagnetism '87*. Tokio Denky University Press, Tokyo

Barth D S, Sutherling W, Engel J, Beatty J 1982 Neuromagnetic localization of epileptiform spike activity in the human brain. *Science* 218: 891–4

Brittenham G M, Farrell D E, Harris J W, Feldman E S, Danish E H, Muir W A, Tripp J H, Bellon E M 1982 Magnetic-susceptibility measurement of human iron stores. *N. Engl. J. Med.* 307: 1671–5

Di Luzio S, Comani S, Romani G L, Basile M, Del Gratta C, Pizzella V 1989 A biomagnetic method for studying gastro-intestinal activity. *Il Nuovo Cimento Note Brevi D* 11: 1853–9

Ricci G B, Romani G L, Salustri C, Pizzella V, Torrioli G, Buonomo S, Peresson M, Modena I 1987 Study of focal epilepsy by multichannel neuromagnetic measurements. *Electroencephalogr. Clin. Neurophysiol.* 66: 358–68

Romani G L, Williamson S J, Kaufman L 1982a Biomagnetic instrumentation. *Rev. Sci. Instrum.* 53: 1815–45

Romani G L, Williamson S J, Kaufman L 1982b Tonotopic organization of the human auditory cortex. *Science* 212: 1339–40

Roth B J, Wikswo J P Jr 1985 The magnetic field of a single axon. A comparison of theory and experiment. *Biophys. J.* 48: 93–109

Williamson S J, Hoke M, Stroink G, Kotani K (eds.) 1989 *Advances in Biomagnetism*. Plenum, New York

Williamson S J, Kaufman L 1981 Biomagnetism. *J. Magn. & Magn. Mater.* 22: 129–201

Williamson S J, Romani G L, Kaufman L, Modena I (eds.) 1983 *Biomagnetism: An Interdisciplinary Approach*. Plenum, New York

V. Pizzella
[Istituto di Elettronica dello Stato Solido, Rome, Italy]

G. L. Romani
[Università G. D'Annunzio, Chieti, Italy]

Bubble Domain Materials

Solid solutions based on rare-earth iron garnets have emerged as an important class of soft magnetic materials for use in bubble domain memories first reported by Bobeck (1967). A magnetic bubble is a small stable cylindrical domain (typically 1–4 μm in diameter) of reverse magnetization perpendicular to the surface of a thin magnetic plate or film. Stable bubbles can be generated, moved under the influence of a field gradient and annihilated so as to perform memory and logic functions.

1. *Rare-Earth Iron Garnets*

The rare-earth iron garnets are the basis for practical bubble domain materials. The general formula is

$$\begin{aligned} R_3Fe_5O_{12} &= R_3^{3+}\,Fe_2^{3+}\,[(Fe^{3+}\,O_4)^{5-}]_3 \\ &= R_3^{3+}\,Fe_2^{3+}\,Fe_3^{3+}\,O_{12}^{2-} \end{aligned} \quad (1)$$

where R is a rare-earth or yttrium ion. The structure is cubic, with space group I*a*3*d*. There are eight formula units per cell. The 24 R ions occupy the 24*c* positions, the 40 iron ions occupy the 16 *a* and 24 *d* positions and the oxygen ions occupy the 96 *h* positions, which

Table 1
Room-temperature magnetostriction constants, λ_{100} and λ_{111}, of rare-earth iron garnets

R	λ_{100} ($\times 10^6$)	λ_{111} ($\times 10^6$)
Sm	21.0	−8.5
Eu	21.0	1.8
Gd	0	−3.1
Tb	−3.3	12.0
Dy	−12.5	−5.9
Ho	−3.4	−4.0
Er	2.0	−4.9
Tm	1.4	−5.2
Yb	1.4	−4.5
Lu	−1.4	−2.4
Y	−1.4	−2.4

Source: Iida (1967)

require three parameters to specify the oxygen coordinates. The 24 Fe_3^{3+} ions occupy the tetrahedral (*d*) sites (each coordinated to four oxygen ions), and the 16 Fe_2^{3+} ions occupy the octahedral sites (each coordinated to six oxygen ions). Each R ion is coordinated to eight oxygen ions in a so-called dodecahedral site.

The R–iron garnets are ferrimagnetic. The moments of the Fe^{3+} ions ($3d^5$ configuration) on the tetrahedral sites, although parallel to one another, are antiparallel to the moments of the Fe^{3+} ions on the octahedral sites. Since there are three tetrahedral ions to every two octahedral ions, there is a net imbalance of five unpaired electrons. This results in relatively low net magnetic moments and saturation magnetizations near room temperature. The moment of the R-ion sublattice contributes somewhat to the net moment near room temperature. The moments of gadolinium and heavier R ions couple antiferromagnetically to the net moment of the iron sublattice and, since the magnetization of the R-ion sublattice becomes increasingly dominant at low temperatures, compensation temperatures occur below room temperature (see *Magnetic Materials: Basic Concepts*). The moments of the light rare-earth ions couple ferromagnetically to the net moment of the iron sublattices and no compensation points are observed. $Y_3Fe_5O_{12}$ exhibits a room-temperature saturation magnetization $\mu_0 M_s \cong 0.175$ T. The saturation magnetizations of the light R–iron garnets are slightly higher. They are much lower for the heavy R–iron garnets. The magnetocrystalline anisotropy of the pure iron garnets is of the order of 500 J m^{-3} (Nielsen 1979). The magnetostriction constants are given in Table 1.

2. *Practical Garnet Bubble Domain Materials*

Stable bubble domains are formed when an external bias field is applied normal to the plane of a single-crystal film; this film must be nearly free of imperfections detrimental to domain-wall motion. For a given material to be suitable for economical mass-memory devices, a number of requirements must be met. A uniaxial anisotropy K_u sufficient to sustain stable bubbles must exist normal to the film. The saturation moment must be low so that the bubbles are large enough to handle, yet it must allow for storage densities to be greater than 200 kbits cm^{-2}; coercivities must also be low (< 150 A m^{-1}). For a data rate of Mbit s^{-1}, mobilities of a few hundred cm^2 s^{-1} A^{-1} are required. Extremely small bubbles (< 1 μm) also lead to higher bit rates for a given domain-wall velocity. Since the bubbles move about four or five bubble diameters per step, generation and detection problems arise which are attributable in part to the present state of the technology of generating thin-film circuits. The anisotropy field, $H_k = 2K_u/\mu_0 M$, must be greater than M_s, but for high mobility it must not be too large. In terms of bulk magnetic properties, the equilibrium bubble diameter d_p is of order

$$d_p \sim (AK_u)^{1/2}/2\pi M_s^2 \tag{2}$$

where A is the exchange constant. Film thickness must be of the order of the bubble diameter. The minimum field required to move a domain (wall-motion coercivity) should be about 60–150 A m^{-1}. The uniaxial anisotropy K_u, as well as M_s, must be relatively independent of temperature near room temperature.

Materials parameters that need to be considered in designing bubble materials for field-access device technology have been discussed by Blank et al. (1979). A material quality factor Q is given by

$$Q = (H_k/M_s) = 2K_u/\mu_0 M_s^2 \tag{3}$$

For bubble domains to be stable Q must be greater than unity; however, the wall mobility is inversely proportional to Q so, to attain high bubble mobility for high data rates, Q must not be too large. A general range recommended is $1 < Q < 10$.

A material length parameter l, which is related to the optimum film or platelet thickness, is given by

$$l = 4(AK_u)^{1/2}/\mu_0 M_s^2 \tag{4}$$

The material parameters l and Q are related through the domain wall width δ_w:

$$\delta_w = \pi l/2Q \tag{5}$$

Combining Eqns. (3) and (4):

$$K_u = 4Q^2 A/l^2 \tag{6}$$

and

$$M_s = (8AQ/\mu_0)^{1/2}/l \tag{7}$$

In general, when the thickness of the magnetic layer is approximately $4l$, the bubble diameter can be approximated by the relation

Table 2
Site preferences for nonmagnetic ions in rare-earth iron garnets

Ion	Approximate fraction in site		
	Dodecahedral	Octahedral	Tetrahedral
Ca^{2+}	1.0		
Sr^{2+}	1.0		
Al^{3+}		0.15	0.85
Ga^{3+}		0.1	0.90
In^{3+}		1.0	
La^{3+}	1.0		
Lu^{3+}	0.95	0.05	
Sc^{3+}		1.0	
Y^{3+}	1.0		
Ge^{4+}		0.02	0.98
Si^{4+}			1.0
Ti^{4+}		1.0	
Zr^{4+}		1.0	

$$d_p \simeq 8l \simeq 32(AK_u)^{1/2}/\mu_0 M_s^2 \quad (8)$$

Defect (e.g., dislocations, precipitates) densities must be low ($\leqslant 3\,cm^{-2}$) for good yields of mass-memory chips.

The crystal chemistry of the garnets offers a wide latitude for the control (and adjustment) of M_s, temperature coefficient of magnetization and uniaxial anisotropy through the substitution of other cations (magnetic and nonmagnetic). Site preferences for nonmagnetic ions are summarized in Table 2. A uniaxial anisotropy of the single-crystal solid-solution garnet films presently being used for devices is growth-induced. It has its genesis in the fortunate discovery (Bobeck et al. 1970) that a uniaxial anisotropy existed under certain growth faces in some flux-grown solid-solution garnets. Although the microscopic origin of the growth-induced anisotropy is not completely understood, it is clear that it is due to a nonrandom distribution of substituted ions, particularly substitutions on the dodecahedral (R ion) sites. The relative size of R ions plays an important role.

The garnet solid-solution compositions now being employed by most of the industry are based on the (Y, Lu, Sm, Ca)$_3$(Fe, Ge)$_5$O$_{12}$ system (Blank et al. 1979, Blank and Licht 1980). In this system, Ge^{4+} on tetrahedral sites dilutes the Fe^{3+} lattice for M_s control, Ca^{2+} is used to compensate for the Ge^{4+} charge, and the combination of rare-earth ions gives the required H_k and lattice constant. Several such materials and their properties are listed in Table 3. The lattice-matched single-crystal films are grown onto (111) nonmagnetic gadolinium–gallium garnet substrates, which are essentially defect free, via liquid-phase epitaxy from molten-oxide fluxes, such as $PbO–B_2O_3$, contained in platinum crucibles.

3. Other Magnetic Oxides

Historically, as suggested by Sherwood, the rare-earth orthoferrites $RFeO_3$ were the first materials to be considered for bubble devices (Bobeck 1967), and early device studies were made on orthoferrite materials. These materials are canted antiferromagnets with small values of M_s and, therefore, large bubbles ($> 10\,\mu m$). They were eventually ruled out as suitable bubble domain materials because they did not meet the required material properties already discussed. Hexagonal ferrites, such as $PbFe_{12}O_{19}$, although exhibiting $\mu_0 M_s \cong 0.4\,T$, which can be lowered by substituting appropriate ions for iron, were eliminated from consideration because of low domain velocities.

4. Amorphous Metallic Films

Amorphous metallic films based on rare-earth $3d$ transition metal alloys, such as Gd_xCo_y and $(Gd_{1-x}Co_x)_{1-y}Mo_y$ (Chaudhari et al. 1973), have been considered as materials for bubble domain devices. Although these materials offer several advantages, mainly economic, over garnet films, they are not being used for practical devices because of several disadvantages (e.g., high electrical conductivity, chemical nonuniformity of films). Laboratory (i.e., information storage) devices have, however, been made and characterized (Eschenfelder 1980a, b, c). Amorphous $3d$ transition metal–rare earth alloys, as well as garnet films, are now being considered for use as erasable magnetooptic recording media (see *Magnetooptic Storage Media*; Hansen and Heitmann 1989).

Table 3
Properties of several bubble domain garnet compositions

Property	Composition			
	A[a]	B[b]	C[c]	D[d]
Bubble diameter (μm)	3	1.7	1.3	1.1
Lattice parameter (nm)	1.2383	1.2383	1.2383	1.2383
$\mu_0 M_s$ (T)	0.03	0.051	0.0511	0.0784
K_u (mJ cm^{-3})	2.3000	3.6000	2.6645	7.0600
A (pJ cm^{-1})	0.025	0.027	0.2	~0.029
L (μm)	0.35	0.19	0.14	0.12
Curie temperature (K)	473			
H_k (MA m^{-1})	139	140.9	104.2	180.7
Q	5.3	3.5	2.6	2.9
Mobility (cm^2 s^{-1} A^{-1})	260	~300	500	~300
$10^{-6}\lambda_{111}$	−1.9	−2.2	−1.7	−2.65

Source: Blank et al. (1979)
a $Y_{1.2}Lu_{0.5}Sm_{0.4}Ca_{0.9}Ge_{0.9}Fe_{4.1}O_{12}$ b $Y_{1.0}Lu_{0.7}Sm_{0.5}Ca_{0.8}Ge_{0.8}Fe_{4.2}O_{12}$
c $La_{0.6}Lu_{2.1}Sm_{0.3}Ga_{0.9}Fe_{4.1}O_{12}$ d $La_{0.2}Lu_{1.6}Sm_{0.6}Ca_{0.6}Ge_{0.6}Fe_{4.4}O_{12}$

5. *New Materials*

A "hard bubble" possesses a complex domain wall structure and moves in the "wrong" direction and with extreme reluctance in a field gradient. It was first observed in epitaxial films of high crystalline perfection. Several techniques have been suggested for suppressing hard bubbles; use of a multilayered film, ion implantation of such ions as H^+, He^+ and Na^+ (Wolfe and North 1972) and use of very thin (~30 nm) permalloy films deposited directly on a garnet film (Lin and Keefe 1973, Takahashi et al. 1973). Ion implantation has become an important part of magnetic bubble technology (North et al. 1978).

The growth-induced anisotropy of garnet bubble domain materials is due to preferential occupation of magnetic rare-earth ions on the dodecahedral sites. It has been shown (Luther et al. 1980) that the substitution of Ir^{4+} ions on the octahedral iron sites, with Mg^{2+} ions for charge compensation, yields a large growth-induced anisotropy in garnets even without magnetic rare earths. For example, such a substitution in yttrium–iron garnet (YIG) results in a K_u as large as $+4.5\ kJ\ m^{-3}$ for a (111) film. This is larger than the previously reported high of $+3.0\ kJ\ m^{-3}$ for $Sm_{1.2}\ Lu_{1.8}Fe_5O_{12}$ (Bullock et al. 1974).

Stable bubbles and large in-plane anisotropy have been obtained for (110) films by use of growth-induced anisotropy in $(EuLaCa)_3(FeGaAl)_5O_{12}$ compositions (Wolfe et al. 1976, LeCraw et al. 1977) and stress-induced anisotropy in garnets containing Mn^{3+} ions (Breed et al. 1980).

Dynamic conversion in high-mobility (~1000 $cm^2\ s^{-1}\ A^{-1}$) bubble garnets refers to the limit in attainable bubble velocities. This is a result of the conversion of normal bubble domains into a complex domain state, similar to a hard bubble, during rapid displacement. The limiting velocity is directly proportional to the effective gyromagnetic ratio g_{eff} (Slonczewski 1973, Thiele 1973). To increase this velocity, dynamic conversion must be suppressed. LeCraw et al. (1975) have shown that dynamic conversion can be suppressed by formulating garnets containing europium ($g_{eff} \gg 2$). For a film of $Eu_{1.45}Y_{0.45}Ca_{1.1}Fe_{3.9}Si_{0.6}Ge_{0.5}O_{12}$, a g factor greater than 30 was obtained, increasing the usable domain wall velocity before onset of dynamic conversion by more than a factor of two over comparable films with $g = 2$.

Single-crystal garnet films in the system $(Bi, Y, Lu)_3(FeGa)_5O_{12}$ have also been reported as being useful for $\leqslant 1$ µm bubbles (Robertson et al. 1981). Lattice-matched films can be grown by liquid-phase epitaxy onto gadolinium–gallium garnet substrates. Films yielding (a) 1 µm and (b) 0.8 µm bubbles possessed the following properties:

(a) $Q = 3.6$, $\mu_0 M_s = 0.06$ T, ferromagnetic resonance linewidth ΔH at 10 GHz = 222.8 $A\ m^{-1}$; and

(b) $Q = 2.7$, $\mu_0 M_s = 0.093$ T and $\Delta H = 318.3\ A\ m^{-1}$.

The dynamic coercive fields for these films were less than 80 $A\ m^{-1}$. An improved BiYIG composition for 1 µm bubbles has been reported by Luther et al. (1980). Calcium and silicon were substituted for the gallium in the composition researched by Robertson et al. (1981). The approximate composition of the improved material is $Y_{1.9}Bi_{0.5}Ca_{0.6}Fe_{4.4}Si_{0.6}\ O_{12}$. This material exhibits larger room-temperature M_s and Q, a high Curie temperature and a reduced temperature dependence of the collapse field. These materials are most promising for extended-temperature-range or high-speed applications.

6. *Other Applications for Garnets*

The high cost per bit of bubble memories, compared with semiconductor dynamic random access memories (DRAMS), resulted in diminished excitement and utility for bubble memories. Present applications appear to be limited to the military (Scott and de Aranjo 1989). Companies known to manufacture bubble memories are Memtech and Magnesys in the USA and Sagem in France. Bismuth-substituted garnets $(YbTbBi)Fe_5O_{12}$ have been proposed as Faraday rotators; Faraday rotation coefficients of $-2400°\ cm^{-1}$ and $-1600°\ cm^{-1}$ at wavelengths of 1.31 µm and 1.55 µm, respectively, have been exhibited by these materials (see *Magnetooptic Materials*; Nakajima et al. 1988). Films of the garnets in the system $(GdPrTmBi)_3(FeGa)_5O_{12}$ have been proposed for use as magnetooptic read heads (Ferrand et al. 1988).

See also: Magnetic Storage Media; Rare-Earth Intermetallics

Bibliography

Blank S L, Licht S J 1980 The simultaneous multiple dipping of magnetic bubble garnets. *IEEE Trans. Magn.* 16: 604–9

Blank S L, Nielsen J W, Biolsi W A 1976 Preparation and properties of magnetic garnet films containing divalent and trivalent ions. *J. Electrochem. Soc.* 123: 856–63

Blank S L, Wolfe R, Luther L C, LeCraw R C, Nelson T J, Biolsi W A 1979 Design and development of single-layer, ion-implantable small bubble materials for magnetic bubble devices. *J. Appl. Phys.* 50: 2155–60

Bobeck A H 1967 Properties and device applications of magnetic domains in orthoferrites. *Bell Syst. Tech. J.* 46: 1901–25

Bobeck A H, Della Torre E 1979 *Magnetic Bubbles*. North-Holland, Amsterdam

Bobeck A H, Spencer E G, Van Uitert L G, Abrahams S C, Barns R L, Grodkiewicz W H, Sherwood R C, Schmidt P H, Smith D H, Walters E M 1970 Uniaxial magnetic garnets for domain wall "bubble" devices. *Appl. Phys. Lett.* 17: 131–6

Breed D J, de Geus W, Enz U 1980 Fast wall motion garnet films with orthorhombic anisotropy. *J. Appl. Phys.* 51: 2780–3

Bullock D C, Carlo J J, Mueller D W, Brewer T L 1974 Material and submicron bubble device properties of $(LuSm)_3 Fe_5 O_{12}$. *AIP Conf. Proc.* 24: 647–8
Chaudhari P, Cuomo J J, Gambino R J 1973 Amorphous metallic films for bubble domain applications. *IBM J. Res Dev.* 17: 66–8
Eschenfelder A H 1980a *Magnetic Bubble Technology*. Springer, New York
Eschenfelder A H 1980b Amorphous films for bubbles. In: Wohlfarth 1980
Eschenfelder A H 1980c Crystalline films for bubbles. In: Wohlfarth 1980
Ferrand B, Chamboz B, Gay J C, Olivier M 1988 Liquid phase epitaxy of garnet double layers for magnetooptic read heads. *IEEE Trans. Magn.* 24: 2563–4
Geller S 1960 Garnets: Magnetic interactions and distribution ions in the garnets. *J. Appl. Phys.* 31: 30S–37S
Geller S, Williams H J, Espinosa G P, Sherwood R C 1964 Importance of intrasublattice magnetic interactions and of substitutional ion type in the behavior of substituted yttrium iron garnets. *Bell Syst. Tech. J.* 43: 565
Gilleo M A 1980 Ferromagnetic insulators: Garnets. In: Wohlfarth 1980
Hansen P, Heitmann H 1989 Media for erasable magneto-optic recording. *IEEE Trans. Magn.* 25: 4390–6
Iida S 1967 Magnetostriction constants of rare earth iron garnets. *J. Phys. Soc. Jpn.* 22: 1201–9
LeCraw R C, Blank S L, Villa-Coliero G P, Pierce R D 1975 High speed bubble garnets based on large gyromagnetic ratios (high *g*). *AIP Conf. Proc.* 29: 91–5
Le Craw R C, Pierce R D, Blank S L, Wolfe R 1977 Effect of lanthanum on growth-induced anisotropy in LPE bubble garnet films. *IEEE Trans. Magn.* 13: 1092–4
Lin Y S, Keefe G E 1973 Suppression of hard bubbles by a thin Permalloy layer. *Appl. Phys. Lett.* 22: 603–4
Luther L C, LeCraw R C, Gyorgy E M, Blank S L 1980 Large growth-induced anisotropy to preferential occupation of the iron sites in garnet. *Appl. Phys. Lett.* 36: 779–81
Nakajima K, Nomi Y, Ishikama H, Machida K 1988 A new improved $(YbTbBi)_3 Fe_5 O_{12}$ epitaxial thick film. *IEEE Trans. Magn.* 24: 2565–7
Nielsen J W 1979 Magnetic bubble materials. *Annu. Rev. Mater. Sci* 9: 87–121
North J C, Wolfe R, Nelson T J 1978 Applications of ion implantation to magnetic bubble devices. *J. Vac. Sci. Technol.* 15: 1675–84
Robertson J M, Algra H A, Breed D J 1981 Garnet compositions for submicron bubbles with low clamping constants. *J. Appl. Phys.* 52: 2338–40.
Scott J F, de Aranjo C A P 1989 Ferroelectric memories. *Science* 246: 1400–5
Slonczewski J C 1973 Theory of domain-wall motion in magnetic films and platelets. *J. Appl. Phys.* 44: 1759–70
Takahashi M, Nishida H, Kobayoshi T, Sugita Y 1973 Hard-bubble-free garnet epitaxial films: The garnet–permalloy composite structure. *J. Phys. Soc. Jpn.* 34: 1416
Thiele A A 1973 Excitation spectrum of magnetic domain walls. *Phys. Rev. B* 7: 391–7
Wohlfarth E P 1980 Ferromagnetic Materials, Vol. 2. North-Holland, Amsterdam
Wolfe R, LeCraw R C, Blank S L, Pierce R D 1976 (110) bubble garnet films with growth-induced orthorhombic anisotropy. *Appl. Phys. Lett.* 29: 815–17
Wolfe R, North J C 1972 Suppression of hard bubbles in magnetic garnet films by ion implantation. *Bell Syst. Tech. J.* 51: 1436–44

J. H. Wernick
[Bellcore, Red Bank, New Jersey, USA]

G. Y. Chin[†]
[AT&T Bell Laboratories, Murray Hill, New Jersey, USA]

C

Chevrel Phases

The Chevrel phases $M_xMo_6X_8$ (where M is a metal and X is a chalcogen) are the first family of ternary materials in which superconductivity has been observed. The structure was established by Chevrel et al. (1971) and superconductivity was discovered soon after by Matthias et al. (1972). The maximum value of T_c reaches 15 K for $PbMo_6S_8$ and the report by Fischer et al. (1973) of the high values of the critical field B_{c2} in several of these compounds immediately focused attention onto these new materials. Later, the discovery of a coexistence between magnetism and superconductivity in rare-earth molybdenum chalcogenides also contributed to the understanding of these materials. More recently, interest in Chevrel phases has increased since some of them present similarities with the high-temperature superconductors (short coherence length). For instance, it is important to investigate the possible correlation between intrinsic pinning mechanisms and the coherence length for low-T_c materials that also have short coherence lengths. On the other hand, $PbMo_6S_8$ is also one of the potential materials for the third generation of superconductors for ultrahigh-field applications.

1. *Structure of Chevrel Phases*

All the ternary molybdenum chalcogenide Chevrel phases $M_xMo_6X_8$ are built up from the basic Mo_6X_8 unit depicted in Fig. 1a. This unit appears as a pseudocube with one chalcogen at each corner and one molybdenum located approximately in the center of each face. The basic structure of the Chevrel phases is made from the stacking of these Mo_6X_8 units in an almost cubic unit cell as shown in Fig. 1b. Along the cube axis there is a fairly large free space around either the origin of the unit cell or between the adjacent planes of the Mo_6X_8 unit. Large (ionic radii) M elements are located close to the origin whereas the positions of small M elements are found to be strongly delocalized around the origin, allowing them to have more than one M element per unit cell. Therefore, depending on the ionic radii of the M element, the Chevrel compounds can be classified into two groups, referred to hereafter as the stoichiometric and the nonstoichiometric compounds. Both groups have a rhombohedral structure ($R\bar{3}$) at high temperature; however, at low temperature, many of the nonstoichiometric compounds make a structural transition into a triclinic structure ($P\bar{1}$).

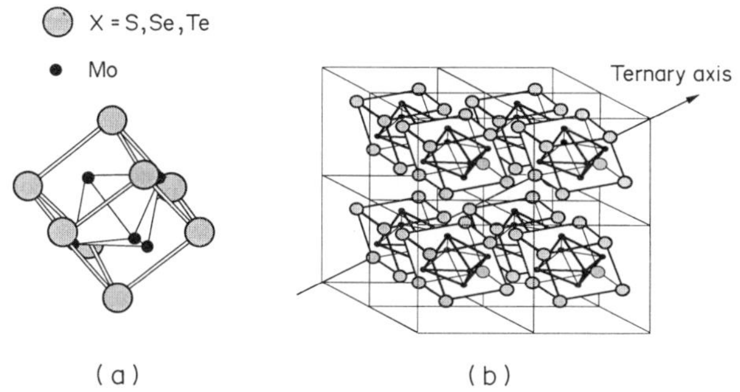

Figure 1
(a) Basic building block Mo_6S_8 of the Chevrel phase and (b) arrangement of the rhombohedral crystal structure

1.1 Stoichiometric Compounds MMo_6X_8

On the list of stoichiometric compounds given in Table 1, the telluride-based Chevrel phases are not

Table 1
Properties of the stoichiometric Chevrel phases

		Sulfide					Selenide				
Compound	Valence	T_c (K)	B_{c2} (T)	M[a]	S[b]	T_s[c] (K)	T_c (K)	B_{c2} (T)	M	S	T_s (K)
Mo_6X_8		1.8	—[d]		R[e]/R	nst[f]	6.3	8–18		R/R	nst
$NaMo_6X_8$	Na^{+1}	4.2	—		R	—	—	—		T[g]	—
$CaMo_6X_8$	Ca^{+2}	ns[h]	ns		R/T	~100	ns	ns		R	—
$SrMo_6X_8$	Sr^{+2}	ns	ns		R/T	~140	ns	ns		R	—
$BaMo_6X_8$	Ba^{+2}	ns	ns		R/T	~170	ns	ns			
$ScMo_6X_8$	Sc^{+3}	3.6	—		R						
YMo_6X_8	Y^{+3}	2.3	0.5		R	—	6.2	—		R	—
$LaMo_6X_8$	La^{+3}	7.0	~6	nm	R	—	11.2	45–55	nm	R	—
$CeMo_6X_8$	Ce^{+3-4}	ns	ns	AF[i]?	R	—	ns	ns	?	R	—
$PrMo_6X_8$	Pr^{+3}	4.0	—	?	R	—	9.2	21	—	R	—
$NdMo_6X_8$	Nd^{+3}	3.5	—	AF	R	—	8.0	—	—	R	—
$SmMo_6X_8$	Sm^{+3}	2.4	—	AF	R	—	6.6	—	—	R	—
$EuMo_6X_8$	Eu^{+2}	ns	ns	AF	R/T	109	ns	ns	—	R	—
$GdMo_6X_8$	Gd^{+3}	1.4	0.12	AF	R	—	5.4	—	AF	R	—
$TbMo_6X_8$	Tb^{+3}	1.7	0.20	AF	R	—	5.7	—	?	R	—
$DyMo_6X_8$	Dy^{+3}	2.1	0.13	AF	R	—	5.8	—	—	R	—
$HoMo_6X_8$	Ho^{+3}	2.2	ns	F[j]	R	—	6.1	—	AF	R	—
$ErMo_6X_8$	Er^{+3}	1.9	0.12	AF	R	—	6.2	—	AF	R	—
$TmMo_6X_8$	Tm^{+3}	2.1	—	AF	R	—	6.3	—	—	R	—
$YbMo_6X_8$	Yb^{+2}	9.1	—	?	R	—	5.8	—	—	R	—
$LuMo_6X_8$	Lu^{+3}	2.0	—	nm	R	—	6.2	—	nm	R	—
$AgMo_6X_8$	Ag^{+1}	8.5			R/R	nst	5.8	—		R	—
$InMo_6X_8$	In^{+3}	ns	ns		R/T	?	8.1	—		R	—
$PbMo_6X_8$	Pb^{+2}	10–15	45–60		R/R	nst	3.6	~4		R	—
$SnMo_6X_8$	Sn^{+2}	10–14	30–40		R/R	nst	4.8	—		R	—
$ThMo_6X_8$		ns	ns		R	—	ns	ns		R	—
UMo_6X_8		ns	ns		R	—	ns	ns		R	—
$NpMo_6X_8$		—	—		R	—	5.6?	—		R	—
$PuMo_6X_8$							—	—		R	—
$AmMo_6X_8$							—	—		R	—

a Magnetic structure b Room-temperature/low-temperature structure c Structural transition temperature d No information available e Rhombohedral f No structural transition g Triclinic h Nonsuperconducting i Antiferromagnetic j Ferromagnetic

Table 2
Properties of the nonstoichiometric Chevrel phases

	Sulfide					Selenide				
Compound	X	T_c (K)	B_{c2} (T)	S[a]	T_s[b] (K)	X	T_c (K)	B_{c2} (T)	S	T_s (K)
$Li_xMo_6X_8$	1	5.5	—[c]	R[d]	—	1	3.9	—	R	—
	3	ns[e]	ns	R	—	3	ns	ns	R	—
$Mg_xMo_6X_8$	1	3.5	—	R	—					
$Ti_xMo_6X_8$						1.2	ns	ns	T[f]	—
$V_xMo_6X_8$						1.2	ns	ns	T	—
$Cr_xMo_6X_8$	2	ns	ns	T	—	1.2	ns	ns	T	
$Mn_xMo_6X_8$	1	ns	ns	R	—	1.2	ns	ns	T	
$Fe_xMo_6X_8$	1	ns	ns	R/T	100	1.2	ns	ns	T	—
	2	ns	ns	R/T	400					
$Co_xMo_6X_8$	2	ns	ns	R/T	220	1	ns	ns	R	—
$Ni_xMo_6X_8$	1	ns	ns	R/R	nst[g]	0.66	ns	ns	T	—
	2	ns	ns	R/T	—	1.6	ns	ns	R	—
$Pd_xMo_6X_8$	1.6	ns	ns	—	—					
$Cu_xMo_6X_8$	1.2	5.6	—	R/?	?	1.4	5.8	—	R/T	165
	1.8	10.8	14–16	R/T	269	2.6	4.6	—	R	—
	3.2	6.4	5–8	R/?	?					
	4	ns	ns	R/R	nst					
$Zn_xMo_6X_8$	1.2	3.6	—	R/T	320	1.2	ns	ns	R/T	130
$Cd_xMo_6X_8$	1	1.5	—	R						

a Room-temperature/low-temperature structure b Structural transition temperature c No information available d Rhombohedral e Nonsuperconducting f Triclinic g No structural transition

reported since these ternary compounds are rarely formed. The majority of these stoichiometric compounds keep their rhombohedral structure down to low temperature and they are characterized by a rhombohedral angle slightly smaller than 90°, with the exception of intermediate-size M elements such as silver and indium which have angles larger than 90°. This is the manifestation of a delocalization of the M element which statistically can occupy six positions around the origin. For large M atoms, these positions are so close to each other that they appear as a single site. With indium, this delocalization becomes important enough so that at low temperatures the indium atom freezes in one position, thus breaking the rhombohedral symmetry (Yvon 1979). This type of structural transition is frequently observed in the nonstoichiometric compounds. To a certain extent, silver and indium establish the link between the two groups of Chevrel phases. Another type of phase transition is observed in several compounds having divalent M elements such as europium, calcium, strontium and barium (Flükiger and Baillif 1982). For these compounds, the low-temperature structure is also triclinic but it involves a very small displacement of the molybdenum positions within the cluster instead of the position of the M element. This phase transition comes from electronic instabilities at the Fermi surface and strongly affects the physical properties of these compounds.

1.2 Nonstoichiometric Compounds $M_xMo_6X_8$ ($1 < x < 4$)

The nonstoichiometric compounds have M elements with small ionic radii and a list of them is given in Table 2. The M element statistically occupies 12 crystallographic positions shown in Fig. 2b. The six inner sites are located around the origin and the six outer sites are located along the rhombohedral axis roughly on the surface of the pseudocube formed with the eight chalcogens surrounding the origin. Since these positions are well separated, up to four M atoms can be inserted into this structure. At low temperature, the majority of these compounds undergo a structural transition into a triclinic structure ($P\bar{1}$). However, this structural transition affects the Mo_6 cluster very little and therefore does not necessarily prevent the superconducting state, as with the structural transition observed in the stoichiometric compounds with divalent M elements.

2. Synthesis of Chevrel Phases

2.1 Solid-State Diffusion Process

The Chevrel-phase compounds can be synthesized by a solid-state diffusion technology carried out in an inert and sealed quartz crucible. The final product is obtained either from direct reaction of the appropriate amount of the constituents at temperatures ranging from 900 °C to 1200 °C or from the reaction of precursors: metal–chalcogenide (e.g., PbS, PbSe), molybdenum–chalcogenide (e.g., MoS_2, $MoSe_2$) and molybdenum. To achieve "pure" samples, this first reaction is generally followed by one or two annealings at a temperature ranging between 1000 °C and

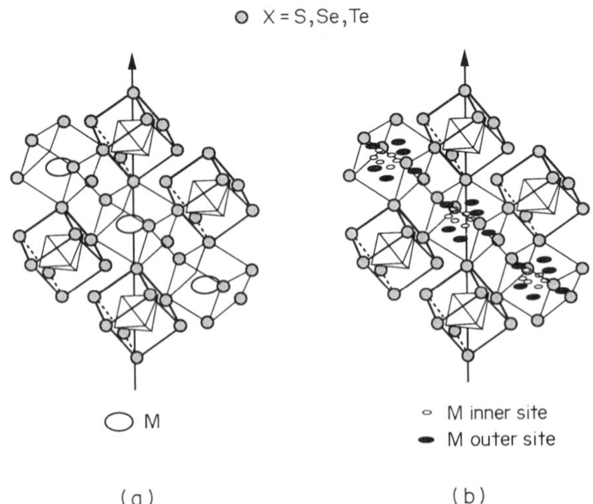

Figure 2
Rhombohedral crystal structure of (a) stoichiometric (MMo_6X_8) Chevrel phase and (b) nonstoichiometric ($M_xMo_6X_8$) Chevrel phase

1200 °C. However, compounds produced by this route are generally contaminated by oxygen released by the quartz tube. This contamination considerably affects the superconducting properties, especially the value of T_c (Hinks et al. 1983).

2.2 Melting Process

The relatively high vapor pressure of the chalcogen and sometimes of the M element makes it difficult to melt these phases in an open crucible without important losses. However, these losses can be made negligible if a large pressure (2 kbar) of an inert gas is applied around the sample. The best results have been obtained with $Cu_xMo_6S_8$ and $EuMo_6S_8$ which melt congruently. Grains of the order of several cubic millimeters have been produced by this route (Flükiger and Baillif 1982). Attempts to melt $PbMo_6S_8$ hardly achieved pure compounds since this phase melts peritectically. The liquid-encapsulating technique has also been successfully used for $Cu_xMo_6Se_8$. However, the risk of contamination of the Chevrel phases by the encapsulating liquid appears non-negligible.

2.3 Crystal Growth

Major efforts have been made to grow single crystals of many Chevrel phases. Good crystals were grown by gas transport using halogens as transport agent. Good single crystals were also obtained from the melt of the phase by a modified Bridgman–Stockbarger technique. However, the common process is now to use an off-stoichiometry starting product and hold it at a temperature close to 1600 °C in a sealed molybdenum crucible (Horyn et al. 1989). Crystals of the order of $1 \times 1 \times 1\ mm^3$ can be easily produced for almost all the Chevrel phases.

2.4 Thin Films

Thin films of MMo_6S_8 (M = Pb, Sn, Ag or Cu) were produced by sputtering of a target of MMo_6S_8 or the appropriate amount of MoS_2, molybdenum and M elements (Woollam et al. 1982). For $Cu_xMo_6S_8$ the coevaporation of molybdenum and copper in the presence of a hot sulfur vapor (H_2S or sulfur) gives interesting results. The best results are obtained with sapphire substrates, since the sapphire thermal expansion is close to that for Chevrel phases. Thin films of silver and copper molybdenum sulfide can be obtained *in situ* if the substrate temperature is held at a temperature in the range 750–900 °C, the lower limit corresponding to the temperature of the phase formation. *In situ* preparation of lead Chevrel-phase thin films has always failed since the partial vapor pressure of lead above 800 °C is already too large and leads to the loss of the metal. Nevertheless, good $PbMo_6S_8$ films have been produced by annealing the films under lead atmosphere. These films can be produced by directly reacting a molybdenum film with an appropriate amount of lead and/or MoS_2.

3. Physical Properties of Chevrel Phases

3.1 Critical Temperature

Within the framework of the Bardeen–Cooper–Schrieffer (BCS) theory, the critical temperature is expected to depend essentially on the density of states at the Fermi level, on the phonon spectrum and on the electron–phonon interaction. The isotope effect on T_c has shown that phonon modes of the Mo_6S_8 unit influence predominantly T_c (Pobell et al. 1982). This conclusion is also supported by the high value of T_c found in the pseudobinary $Mo_6S_6I_2$ ($T_c = 14$ K) which does not contain any M element and therefore has no phonon modes associated with it. This suggests that T_c, in first approximation, is strongly dependent on the charge transfer to the Mo_6 cluster, behavior which is experimentally observed for each chalgogenide family (Yvon 1979). In the sulfide compounds $T_{c,\,max}$ is found with divalent M elements, whereas for the selenide $T_{c,\,max}$ occurs with trivalent M elements.

The absence of superconductivity in the majority of compounds with a divalent M element is very intriguing since, for instance, the highest critical temperatures are found for sulfur compounds with lead and tin. These "high" values of T_c could reflect some lattice instabilities since all the nonsuperconducting divalent sulfide Chevrel phases undergo a structural transition from the rhombohedral structure into a triclinic one. The origin of this instability is not clear, but is certainly electronic. For $EuMo_6S_8$, the absence of superconductivity is correlated with the structural transition and not with the magnetic nature of the europium ion (Decroux et al. 1988). Recently, similar conclusions have been drawn for other nonsuperconducting divalent Chevrel phases, where pressure also prevents the structural transition and simultaneously restores the superconducting state.

3.2 Critical Field

Among the superconducting properties of the Chevrel phases, the most interesting one is undoubtedly the capacity of some compounds to sustain very high magnetic fields in the superconducting state. Tables 1 and 2 report the critical fields at zero temperature for some Chevrel phases. Most of the superconducting Chevrel phases have rather large critical fields B_{c2}, with $PbMo_6S_8$ having the highest B_{c2} with 60 T. These high values of B_{c2} come from the particular structure of these phases where the conduction electrons are rather localized on the Mo_6 clusters because of the large distance between the clusters (Decroux and Fischer 1982). This localization of the electron orbital motion explains the large value of the first limitation referred to as the orbital critical field B^*_{c2}. The second limit to B_{c2} is the paramagnetic critical field B_p due to the spin polarization of the conduction electrons.

The paramagnetic limitation is the mechanism that causes the anomalous temperature behavior of B_{c2} of compounds containing europium ions, where the

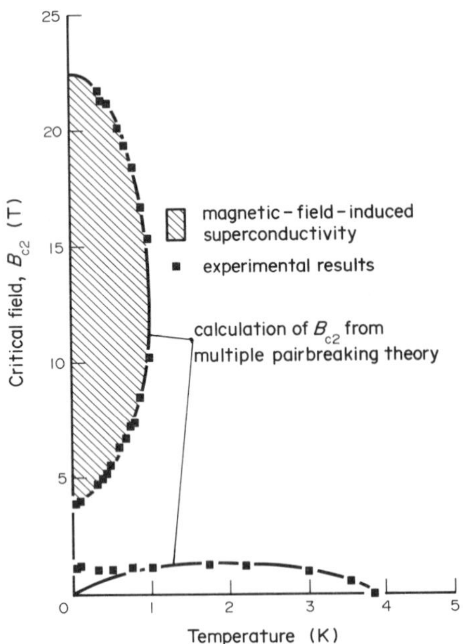

Figure 3
Critical field vs temperature for $Eu_{0.75}Sn_{0.25}Mo_6S_{7.2}Se_{0.8}$ showing the magnetic-field-induced superconductivity: this particular behavior is based on the Jaccarino–Peter effect

magnetic interaction between the magnetic ions and the conduction electrons can be described by a pseudonegative internal field. Figure 3 shows the so-called magnetic-field-induced superconducting state, a state predicted in 1962 by Jaccarino and Peter. At low field, superconductivity is first destroyed by the spin alignment. Once the spin alignment is completed, a further increase of the external field decreases the spin polarization allowing the superconducting state to be restored. Finally, at high fields a combination of orbital and paramagnetic effects destroys the superconducting state.

3.3 Magnetism and Superconductivity

(*a*) *Chevrel phase with 3d magnetic ions.* As has already been stated, compounds containing 3*d* magnetic ions are not superconducting. These compounds belong to the nonstoichiometric group and some of the magnetic ions statistically occupy the six delocalized positions close to the Mo–Mo intercluster link. The magnetic interaction with the conduction electrons is expected to be strong enough to destroy the superconductivity. The strong influence of these magnetic ions is better displayed in pseudoternary compounds where the superconductivity of nonmagnetic compounds is destroyed by the insertion or substitution of less than 1 at.% 3*d* magnetic ions.

(*b*) *Rare-earth molybdenum chalcogenides.* Most of the rare-earth (RE) molybdenum chalcogenides, which belong to the stoichiometric group, are superconducting although the concentration of magnetic ions reaches 7 at.%. This results directly from the existence of the two metallic sublattices (i.e., the Mo_6 cluster and the M element sublattice) which are isolated from each other by the chalcogen lattice. Furthermore, as the distances between RE ions are rather large, the direct exchange interaction between RE is weak, leading to very low ordering temperatures. Most of these magnetic phases stay in the paramagnetic state below the temperature where superconductivity appears. At very low temperature, most of the RE chalcogenides are antiferromagnetic. $HoMo_6S_8$ is the unique compound having a ferromagnetic transition whereas $TmMo_6S_8$ stays in the paramagnetic state down to 70 mK. Europium and cerium chalcogenides are the four nonsuperconducting RE chalcogenides. The absence of superconductivity in the former is attributed to the structural transition whereas the latter is explained on the basis of the mixed valence effect of the magnetic ions. This important class of compounds has provided a unique tool to investigate the interplay between magnetism and superconductivity (Pena and Sergent 1989, Fischer 1990).

(*c*) *Antiferromagnetic compounds.* Coexistence of antiferromagnetism and superconductivity is observed in many of the RE Chevrel phases. Among them, $HoMo_6Se_8$ is particularly interesting since it shows a sinusoidally modulated magnetic structure with a wavelength of 10.1 nm which slightly increases as the temperature decreases, indicating a tendency towards a ferromagnetic state. This particular sinusoidal structure is induced by the superconducting state which prevents the ferromagnetic state occurring by forcing the magnetization to oscillate.

(*d*) *Ferromagnetic compounds.* $HoMo_6S_8$ becomes superconducting at $T_{c1} = 1.2$ K and at a lower temperature $T_{c2} = 0.69$ K it reenters into the normal state. Neutron diffraction studies have unambiguously demonstrated the correlation between the occurrence of a uniform ferromagnetic state and the disappearance of superconductivity (Thomlinson et al. 1982). In a small interval above T_{c2} (0.06 K) it has been observed, using neutrons, that a superconducting state forces the ferromagnet into an oscillatory state with a modulation period of 25 nm. If the superconductivity could be avoided, the ferromagnetic state would appear at the temperature of the occurrence of the oscillatory state.

4. Critical Current Density

For practical applications the critical current density J_c is one of the most important physical parameters.

In contrast to the critical temperature and the critical field, the critical current density is not an intrinsic property of the superconductor. At magnetic fields typically less than 0.4 of the critical field B_{c2}, the J_c is determined by pinning of magnetic flux lines. Pinning is based on all kinds of defects such as voids, dislocations, grain boundaries and precipitates and is, therefore, strongly dependent on the metallurgy (extrinsic property). At higher magnetic fields the interaction between flux lines becomes important; the pinning depends on the elastic properties of the flux line lattice which has a more intrinsic character (see *Flux Pinning and the Summation of Pinning Forces*). Although a complete understanding of the mechanisms that determine the J_c of Chevrel phases is not available, the present knowledge will be reviewed in Sect. 5.

In many superconductors there is only one acting pinning mechanism for various magnetic fields and temperatures. In very inhomogeneous superconductors this can be different and more than one pinning mechanism may be observed. In the case of Chevrel phases where the mechanism for pinning the flux lines normally does not change as a function of the magnetic field and temperature, the critical current density can be described by a so-called scaling law:

$$J_cB = gB_{c2}^m(T)b^p(1-b)^q \tag{1}$$

with the reduced magnetic field $b = B/B_{c2}$ where B is the applied magnetic field. The prefactor g is given by the microstructure, $B_{c2}^m(T)$ describes the temperature dependence and $b^p(1-b)^q$ determines the field dependence of the critical current density J_c. The exponents m, p and q are determined experimentally (see *Critical Current Density*). When J_cB is normalized, a unique curve for different temperatures is obtained. In this case, only one measurement of J_c at one temperature as a function of field is necessary in order to calculate J_c for other temperatures or magnetic fields. The characteristic behavior of the normalized J_cB as a function of the reduced field is shown in Fig. 4. Taking Eqn. (1), the m exponent may be estimated to be $m = 2.4 \pm 0.2$ and the field dependence can be approximated by $b^{0.5}(1-b)^2$. The latter is responsible for the maximum of J_cB at the reduced field $b \simeq 0.2$. In comparison with the technical superconductor Nb_3Sn, this is a signature that the J_c of $PbMo_6S_8$ at magnetic fields lower than 10 T is dominated by grain boundaries. Studying the prefactor g at $(J_cB)_{max}$ as a function of the mean grain size also gives indications that this may be the case. However, it can be shown that the efficiency of grain boundaries in $PbMo_6S_8$ samples is quite modest. For instance, an individual flux line need move only about 1 nm from the grain boundary to get unpinned in the aforementioned field region. It is obvious that by appropriate grain-boundary engineering the critical current density may be optimized. An additional argument for nonoptimized grain boundaries in $PbMo_6S_8$ may be derived by a careful analysis of measurements of the ac susceptibility of high-quality bulk samples (see *Measurements in Superconducting Materials*). A distinction must be made between two different critical current densities. During the cool down, and by applying a small ac magnetic field, microscopic shielding currents in individual grains will be induced flowing with a critical current density J_c^G. When the temperature is further decreased, macroscopic shielding currents start to flow between grains. The critical current density related to these currents is affected by the grain boundaries and is called the transport critical current density J_c^T. There are measurements available that enable a lower-bound estimate to be made of J_c^G of $PbMo_6S_8$ in the range of $5 \times 10^9\,A\,m^{-2}$ at 9 T and 4.2 K (Cattani et al. 1990). This value differs with respect to the macroscopic critical current density J_c^T, which is about 20 times and ten times smaller in bulk and wire samples, respectively. The aforementioned lower-bound estimate of $5 \times 10^9\,A\,m^{-2}$ at 9 T and 4.2 K has been measured in $PbMo_6S_8$ grains up to 20 μm in size and is far from optimal. Work is going on to dope $PbMo_6S_8$ in order to improve the grain boundaries as well as to shift $(J_c^TB)_{max}$ to higher reduced fields. This is of particular interest to the high-field value of the critical current density.

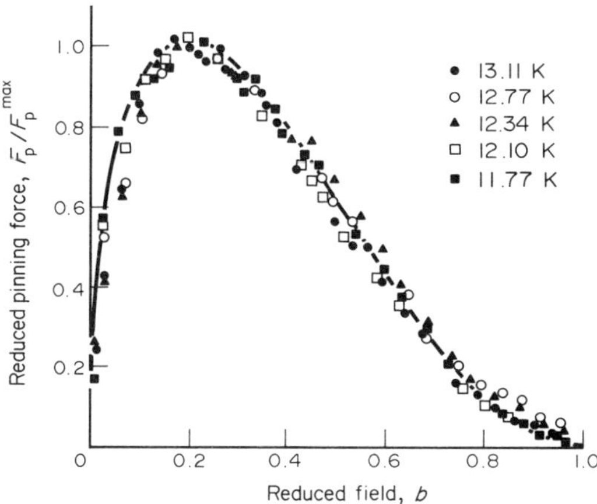

Figure 4
Reduced pinning force vs reduced field $b = B/B_{c2}$ of a $PbMo_6S_8$ bulk sample at different temperatures: the fitted curve has been calculated by $F_P/F_P^{max} \propto b^{0.5}(1-b)^2$

Less information is available concerning the elastic or plastic behavior of flux lines, which determines the J_c at higher reduced fields. Depending on the sample preparation and treatment, it seems that more than one flux line system is acting; the extremely short coherence length in these materials may lead to an intrinsic weakness for flux cutting (see *Transport Currents and Force-Free Current Flow*). Another conclusion coming out of flux line elasticity experiments is that the critical current density at higher magnetic fields is determined by a quite dense but not very efficient defect structure.

5. *Wires for High-Field Applications*

Due to the exceptionally high critical fields of some of the Chevrel phases, it was recognized very early that these materials could be employed for the generation of extremely high steady-state magnetic fields. For this purpose, the Chevrel phase must be fabricated in the form of a wire. Account must be taken of the fact that the most interesting compound $PbMo_6S_8$ must be synthesized by a solid-state diffusion process (powder metallurgy process), so the manufacture of wires is a quite difficult task.

Two types of Chevrel-phase powders for wires are used, the so-called prereacted powder and the precursor powder. In the case of a prereacted powder, the Chevrel phase is already present at the beginning of the wire fabrication process. After deformation a recovery annealing to reconnect the grains, typically at about 700 °C, is necessary. The precursor powder is a mixture of different powders which contain the elements to form the Chevrel phase. For instance, the precursor powder for $PbMo_6S_8$ may be a mixture of lead or PbS, MoS_2 or Mo_2S_3 and molybdenum. In this case, after the drawing process the wire needs a thermal treatment in the range of 1000 °C for the formation of the superconducting phase. A typical cross section of a Chevrel-phase monofilamentary wire is shown in Fig. 5. In the center is the Chevrel-phase superconductor, surrounded by a barrier and an outer matrix material. The latter is necessary for thermal stress compensation (see later). In choosing the barrier material, account must be taken of the fact that it should be nonreactive with the Chevrel phase at elevated temperatures of up to 1000 °C. The ideal barrier is molybdenum with the disadvantage that deformation is only possible at high temperatures, typically between 500 °C and 1000 °C. Other barriers used are niobium or tantalum which may be deformed at room temperature but show chemical reactions with the Chevrel phase above about 800 °C. A common problem for all these barriers is their rather low thermal expansion with respect to the Chevrel phase. This means that the Chevrel phase will come under tensile strain when cooled to the temperature of liquid helium (4.2 K), leading to the formation of cracks. For this reason, an outer matrix is essential to adjust the thermal expansion of the barrier. The outer matrix material may also be designed for mechanical and/or electrical stabilization of the wire.

Typical steps for the fabrication of Chevrel-phase wires are as follows. After cold or hot pressing of the powder the resulting billet is machined and passed into a cylinder which serves as the barrier. After

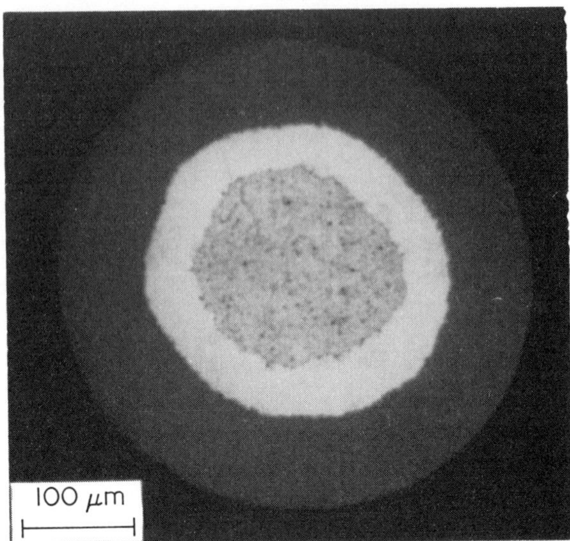

Figure 5
Cross section of a monofilamentary $PbMo_6S_8$ wire with a stainless steel matrix and a molybdenum barrier (courtesy of Metallwerk Plansee, Reutte)

sealing under vacuum the billet is put into the outer matrix cylinder. The latter is also sealed and the whole billet is reduced in diameter by extrusion followed by drawing. As already mentioned, if using molybdenum as a barrier the whole deformation process must be at elevated temperatures. Monofilamentary Chevrel-phase wires have been fabricated up to about 1 km in length. The feasibility of the fabrication of multifilamentary wires has also been shown.

The critical current densities of $PbMo_6S_8$ wires without diffusion barrier and matrix are in the range up to 5×10^8 A m^{-2} at 10 T and 2×10^8 A m^{-2} at 20 T, both at 4.2 K (see *Critical Current Density*; Seeber et al. 1989). Unfortunately, at the present time these values cannot be maintained over the whole length of the wire. The main reasons for this are microcracks and chemical inhomogeneities (in the case of precursor powders). There are preliminary results available concerning the critical current density under uniaxial strain in $PbMo_6S_8$ and $SnMo_6S_8$ wires (Goldacker et al. 1989). It is interesting to note that $PbMo_6S_8$ may be strained up to 0.65% without irreversibility in J_c (Nb_3Sn up to 0.5%). This value may be increased by compressive prestrain of the superconductor by choosing another matrix–barrier fraction. In addition, no magnetic field dependence of the J_c as a function of strain has been observed up to 24 T (Ekin et al. 1985).

See also: A15 Compound Superconductors; Multifilamentary Superconducting Composites; Oxide Superconductors: Ceramic Processing; Rare-Earth Intermetallics

Bibliography

Cattani D, Cors J, Decroux M, Fischer Ø 1990 Intrinsic critical current density of $PbMo_6S_8$. *Physica B* 165–166: 1409–10

Chevrel R, Sergent M, Prigent J 1971 On the new sulfur phase ternaries of molybdenum. *J. Solid State Chem.* 3: 515–19

Decroux M, Fischer Ø 1982 Critical fields of ternary molybdenum chalcogenides. In: Fischer and Maple 1982, Vol. 34, pp. 57–98

Decroux M, Lambert S E, Maple M P, Guertin R P 1988 Normal and superconducting properties of $EuMo_6S_8$ in high magnetic fields. *J. Low Temp. Phys.* 73: 283–303

Ekin J E, Yamashita T, Hamasaki K 1985 Effect of uniaxial strain on the critical current and critical field of Chevrel phase $PbMo_6S_8$ superconductors. *IEEE Trans. Magn.* 21: 474–7

Fischer Ø 1990 Magnetic superconductors. In: Buschow K H J, Wohlfarth E P (eds.) 1990 *Ferromagnetic Materials*, Vol. 5. Elsevier, Amsterdam, pp. 466–550

Fischer Ø, Maple M B (eds.) 1982 *Topics in Current Physics*. Springer, Berlin

Fischer Ø, Odermatt R, Bongi G, Jones H, Chevrel R, Sergent M 1973 On the superconductivity in the ternary molybdenum sulfides. *Phys. Lett. A* 45: 87–8

Flükiger R, Baillif R 1982 Metallurgy and structural transition in ternary molybdenum chalcogenide, In: Fischer and Maple 1982, Vol. 32, pp. 113–41

Goldacker W, Specking W, Weiss F, Rimikis G, Flükiger R 1989 Influence of transverse compressive and axial tensile stress on the superconductivity of $PbMo_6S_8$ and $SnMo_6S_8$. *Cryogenics* 10: 955–60

Hinks D G, Jorgensen J D, Li H C 1983 Structure of the oxygen point defects in $SnMo_6S_8$ and $PbMo_6S_8$. *Phys. Rev. Lett.* 51: 1911–14

Horyn R, Pena O, Geantet C, Sergent M 1989 Kinetics of destruction of Mo–S binary phases and crystal growth of rare earth molybdenum chalcogenide. *Supercond. Sci. Technol.* 2: 71–90

Jaccarino V, Peter M 1962 Ultra-high-field superconductivity. *Phys. Rev. Lett.* 9: 290–2

Matthias B T, Marezio M, Barz H E, Corenzwit E, Cooper A S 1972 High temperature superconductors, the first ternary system. *Science* 175: 1465–6

Pena O, Sergent M 1989 Rare earth based Chevrel phases $REMo_6X_8$: crystal growth, physical and superconducting properties. *Prog. Solid State Chem.* 19: 165–281

Pobell F, Rainer D, Wühl H 1982 Electron–phonon interaction in Chevrel-phase compounds. In: Fischer and Maple 1982, Vol. 32, pp. 251–77

Seeber B, Decroux M, Fischer Ø 1989 Status and prospects of superconducting Chevrel phase wires for high magnetic field applications. *Physica B* 155: 129–35

Thomlinson W, Shirane G, Lynn J W, Moncton D E J 1982 Neutron scattering studies of magnetic ordering in ternary superconductors. In: Fischer and Maple 1982, Vol. 34, pp. 229–48

Woollam J A, Alterovitz S A, Luo H-L 1982 Thin film ternary superconductors. In: Fischer and Maple 1982, Vol. 32, pp. 143–64

Yvon K 1979 Bonding and relationship between structure and physical properties in Chevrel phases compounds $M_xMo_6X_8$ (M = metal X = S,Se,Te). In: Kaldis E (ed.) 1979 *Current Topics in Materials Science*, Vol. 3. North-Holland, Amsterdam, pp. 55–129

M. Decroux and B. Seeber
[Université de Genève, Geneva, Switzerland]

Cobalt–Platinum Alloys

Cobalt–platinum permanent magnet alloys were discovered in 1936 by Jellinghaus. The equiatomic alloy (77 wt% platinum) eventually gave the best value of the coercivity and maximum energy product ($H_{cJ} = 400\,kA\,m^{-1}$ and $(BH)_{max} = 72\,kJ\,m^{-3}$) of all permanent magnets until the cobalt–rare-earth magnets were developed (see *Cobalt–Rare-Earth Compounds*).

The alloy is ductile upon quenching from above 1000 °C. In this state the material retains the high-temperature face-centered-cubic structure. After aging at about 600 °C, this phase partially transforms to an ordered tetragonal phase with reduced saturation magnetization but extremely large crystal anisotropy, resulting in a steep rise of coercivity. The remanence is about 0.65 T.

The primary applications of cobalt–platinum alloys have been in watches, hearing aids and travelling-wave tubes. The market now, however, is practically filled by the lower-cost cobalt–rare-earth alloys.

Bibliography

Gaunt P 1966 Magnetic properties of platinum–cobalt near the equiatomic composition. II. Mechanism of magnetic hardening. *Philos. Mag.* 13: 579–88

Heck C 1974 *Magnetic Materials and Their Applications*. Crane-Russak, New York

McCaig M 1977 *Permanent Magnets in Theory and Practice*. Wiley, New York

McCurrie R A, Gaunt P 1966 Magnetic properties of platinum–cobalt near the equiatomic composition. I. Experimental data. *Philos. Mag.* 13: 567–77

G. Y. Chin[†]
[AT&T Bell Laboratories, Murray Hill, New Jersey, USA]

J. H. Wernick
[Bellcore, Red Bank, New Jersey, USA]

Cobalt–Rare-Earth Compounds

Cobalt–rare-earth alloys are a group of hard magnetic materials comprising cobalt and rare-earth elements with or without minor constituents of other transition metal elements such as iron, copper and zirconium. They are based on intermetallic phases of the type

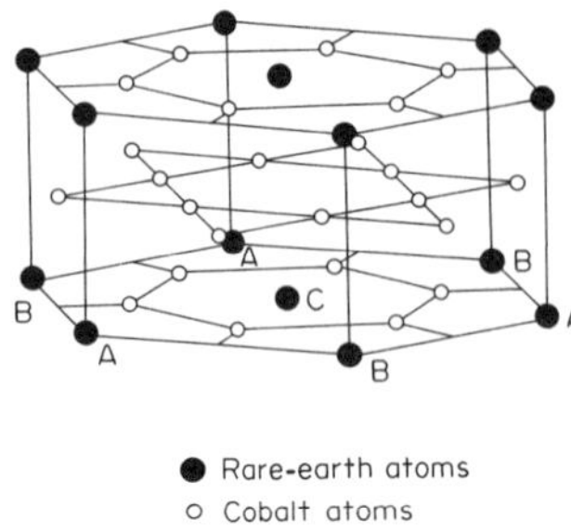

Figure 1
The Co_5R structure; in the basal plane, rare-earth atoms form a triangular plane sublattice, which can be divided into three equivalent triangular sublattices A, B and C

Co_5R and/or $Co_{17}R_2$ whose primary magnetic properties, saturation magnetization and magnetocrystalline anisotropy lead to excellent permanent-magnet properties. Among the rare earths, samarium is the most commonly used because it provides the best permanent-magnet properties. Cerium and cerium misch metal are also used for economy. The other rare-earth elements are sometimes employed in combination with samarium to meet special requirements; for example, praseodymium is added for higher flux density and gadolinium for lower temperature dependence of flux density.

Most of the permanent magnets based on cobalt–rare-earth alloys exhibit several times higher magnetic energy than conventional magnets such as alnicos and hard ferrites. They also exhibit very high stability of magnetic flux density even at elevated temperatures, a property unattainable with any other existing permanent-magnet materials. However, they are expensive, mainly because of their high raw material costs, and their applications are therefore restricted to devices in which high performance, especially under heavy-duty conditions, is of primary importance.

1. Crystallographic and Magnetic Properties

The crystal structure of Co_5R is illustrated in Fig. 1. This structure consists of two different layers of atoms. One layer is composed of two kinds of atoms in the proportion of one rare-earth atom to two cobalt atoms with the rare-earth atoms arranged so as to form a triangular plane array with the cobalt atoms at the center of each triangle. This layer alternates with another layer consisting of only cobalt atoms.

The crystal structure of $Co_{17}R_2$ (see Fig. 2.) is closely related to the Co_5R structure. One rare-earth atom is substituted by a pair of cobalt atoms at one-third of the rare-earth sites in the Co_5R structure. The triangular R sublattice is divided into three equivalent triangular sublattices A, B and C as indicated in Fig. 1. The position of the substitution is assigned arbitrarily to one of these sublattices, for example, A. Then the position of the substitution in the neighbouring similar layer can be either B or C. Thus the order of the substitution can be either (a) ABCABC or (b) ABABAB along the *c* axis, giving rise to either (a) a rhombohedral or (b) a hexagonal structure. Crystallographic data for Co_5R and $Co_{17}R_2$ are given in Table 1.

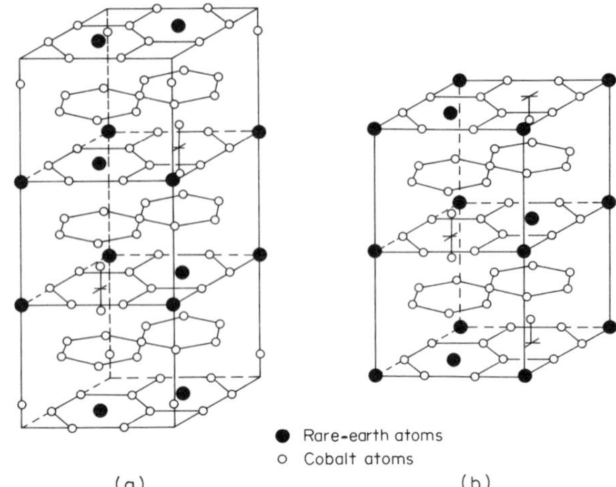

Figure 2
The $Co_{17}R_2$ structure: (a) rhombohedral form and (b) hexagonal form

Cobalt–rare-earth alloys are suitable for making hard magnetic materials owing to their primary magnetic properties: saturation magnetization M_s, Curie temperature T_C and magnetocrystalline anisotropy constant K_1. Table 2 lists experimental values of M_s and T_C of Co_5R and $Co_{17}R_2$ compounds. The mag-

Table 1
Crystallographic data for Co_5R and $Co_{17}R_2$ compounds

	Co_5R		$Co_{17}R_2$			
			Rhombohedral		Hexagonal	
R	*a* (nm)	*c* (nm)	*a* (nm)	*c* (nm)	*a* (nm)	*c* (nm)
La	0.5105	0.3966				
Ce	0.4926	0.4020	0.8370	1.2193	0.8371	0.8316
Pr	0.5024	0.3988	0.8427	1.2265		
Nd	0.5012	0.3978	0.8407	1.2257		
Sm	0.4989	0.3981	0.8379	1.2212	0.8360	0.8515
Gd	0.4976	0.3973	0.8365	1.2184	0.8351	0.8125
Tb	0.4946	0.3980	0.8344	1.2190	0.8348	0.8125
Dy	0.4933	0.3983	0.8310	1.2070	0.8328	0.8125
Ho	0.4911	0.3993	0.8332	1.2201	0.8320	0.8113
Er	0.4883	0.4007			0.8310	0.8113
Tm	0.4863	0.4017			0.8285	0.8095
Yb					0.8271	0.8089
Lu					0.8247	0.8093
Y	0.4937	0.3978	0.8344	1.2190	0.8341	0.8125

Table 2
Saturation magnetization M_s and Curie temperature T_C of Co_5R and $Co_{17}R_2$ compounds

R	Co_5R M_s (T)	Co_5R T_C (°C)	$Co_{17}R_2$ M_s (T)	$Co_{17}R_2$ T_C (°C)
La	0.909	567		
Ce	0.870	464	1.16	810
Pr	1.203	639	1.38	898
Nd	1.228	637	1.39	877
Sm	1.07	747	1.20	917
Gd	0.363	741	0.75	936
Tb	0.236	707	0.65	907
Dy	0.437	687	0.68	879
Ho	0.606	760	0.83	900
Er	0.727	780	0.91	913
Tm	0.750	747	1.15	909
Yb				911
Lu				923
Y	1.061	714	1.25	894

netic moments of the cobalt atoms align parallel to each other. A subgroup of Co_5R characterized by R having no magnetic moment (yttrium, lanthanum, cerium and lutetium) thus results in ferromagnetism due to the cobalt atoms. The other rare-earth elements having incomplete $4f$ shells also contribute to magnetism. According to a general rule governing alloys of transition metals and rare-earth elements, the spins of rare-earth atoms couple antiparallel to those of cobalt atoms. However, the total angular momentum J for a $4f$ shell that is less than half filled is given by $L - S$ and is antiparallel to the spin angular momentum. As a result, magnetic moments of light rare-earth atoms couple parallel to those of cobalt atoms. On the contrary, $4f$ shells that are more than half filled, where the rule $J = L + S$ is valid, result in antiparallel coupling of rare-earth and cobalt moments. The Curie temperature is primarily determined by the exchange interaction between cobalt atoms. Exchange interactions between rare-earth and cobalt atoms are much weaker and contribute to the Curie temperature to a lesser extent. As a consequence, the Curie temperature is higher as the concentration of cobalt atoms increases. Thus, the $Co_{17}R_2$ phase has a higher Curie temperature than the Co_5R phase for a given R.

The magnetocrystalline anisotropy constants K_1 of Co_5R and $Co_{17}R_2$ compounds are listed in Table 3. The subgroup of Co_5R with R having zero spin angular momentum exhibits roughly the same value of K_1 ($\sim 5\ \mathrm{MJ\,m^{-3}}$). Gadolinium, with a half-filled $4f$ shell and a spherically symmetric wave function, does not primarily contribute to the anisotropy and therefore can be included in the same category. Thus, an approximately invariant contribution to K_1 of $3d$ electrons of the cobalt sublattice in the Co_5R phases can be assumed. Hence, the contribution of rare-earth atoms can be evaluated approximately. It is seen from Table 3 that samarium contributes positively to K_1 and neodymium negatively. A corresponding consideration of the $Co_{17}R_2$ phases suggests that the anisotropy constants due to the cobalt sublattice are negative and one order of magnitude smaller than those in the Co_5R phases. Therefore, the contribution of the rare-earth atoms to K_1 is more important than in the Co_5R phases. In practice, only samarium can give rise to significantly large positive values of K_1. Thus, the $Co_{17}R_2$ phases are of practical interest only when R consists totally, or at least mainly, of samarium.

2. Phase Diagrams

The occurrence of various phases in binary cobalt–rare-earth alloys is very much alike for all the light rare-earth elements (lanthanum, cerium, praseodymium, neodymium and samarium) with few exceptions. The cobalt–samarium system is described in the following discussion not only because it is typical of the group, but also because it is particularly important for permanent-magnet applications. The phase diagram of cobalt–samarium is shown in Fig. 3. $Co_{17}Sm_2$ is formed congruently from the liquid phase, whereas Co_5Sm is formed by a peritectic reaction between $Co_{17}Sm_2$ and the liquid phase. Co_7Sm_2 is also formed peritectically. Co_5Sm is stable above 800 °C but decomposes eutectoidally into $Co_{17}Sm_2$ and Co_7Sm_2 below that temperature. However, the Co_5Sm phase

Table 3
The uniaxial magnetocrystalline anisotropy constant K_1 of Co_5R and $Co_{17}R_2$ compounds

R	Co_5R K_1 ($\mathrm{MJ\,m^{-3}}$)	$Co_{17}R_2$ K_1 ($\mathrm{MJ\,m^{-3}}$)
La	5.9	
Ce	5.3	− 0.6
Pr	8.1	− 0.6
Nd	0.7	− 1.1
Sm	17.2	3.3
Gd	4.6	− 0.5
Tb	NC[a]	− 3.3
Dy	NC[a]	− 2.6
Ho	3.6	− 1.0
Er	3.8	0.72
Tm		0.43
Yb		1.04
Lu		− 0.2
Y	5.2	− 0.34

a NC, noncollinear spin structure

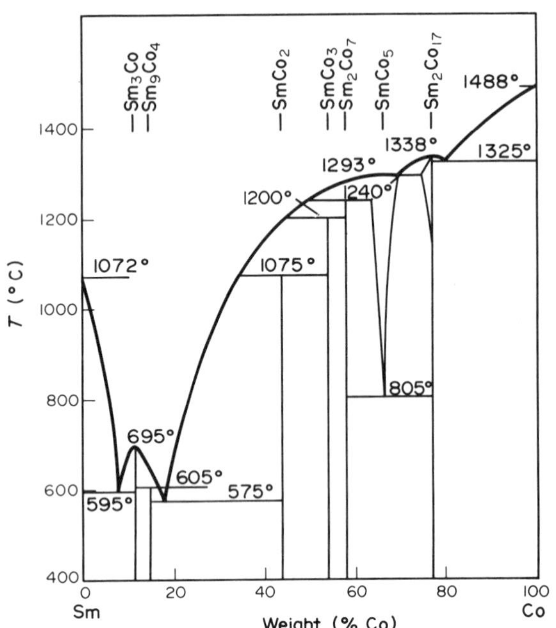

Figure 3
Phase diagram for the samarium–cobalt system

normally exists in a metastable state at room temperature as the decomposition takes place slowly. Co_5Sm has a homogeneity region at elevated temperatures that extends towards higher cobalt concentrations. In contrast, $Co_{17}Sm_2$ has a homogeneity region extending towards lower cobalt concentrations. Thus, Co_5Sm and $Co_{17}Sm_2$ are mutually soluble to a considerable degree. The similarity of the crystal structures as described in Sect. 1 may account for the existence of these homogeneity regions. A partial disorder of the substituted rare-earth sites should be allowed to accommodate nonstoichiometric compositions in the basic lattice.

3. *Coercivity Mechanism*

Assuming coherent rotation of magnetization, the intrinsic coercive force ${}_IH_c$ of a single domain particle is given by the anisotropy field $H_k = 2K_1/M_s$. The ${}_IH_c$ values obtained in practice are much smaller than the anisotropy field. In real crystals, the magnetization reversal takes place through nucleation of reverse domains, followed by displacement of domain walls. Therefore, one of two mechanisms, either hard nucleation of reverse domains or pinning of domain walls, has to account for the coercivity (see *Coercivity and Domain Wall Pinning*). In fact, magnetization behavior corresponding to these two mechanisms is found in actual materials as schematically illustrated in Fig. 4. In the case of a nucleation-controlled coercive force (Fig. 4a), initial magnetization increases steeply with magnetizing field, indicating the existence of easily movable domain walls. Coercive force is developed, depending on the maximum applied magnetizing field, showing that once reverse domains are eliminated in a grain, nucleation of reverse domains cannot occur easily. In the case of a pinning-controlled coercive force (Fig. 4b), the magnetization remains nearly zero up to a value H_p, the pinning field, before increasing steeply to saturation. The coercive force is independent of the applied magnetizing field greater than H_p.

The nucleation-controlled coercive force is observed in fine-grained cobalt–rare-earth materials with high magnetocrystalline anisotropy, typically in Co_5Sm and $T_{17}Sm_2$, in both particulate and sintered forms. The pinning-controlled coercive force is observed widely in T_zR ($5 < z < 9$) materials, where T stands for a combination of cobalt and one or more of the other transition metals, copper being the indispensable component. Pinning sites are assigned to epitaxial boundaries between Co_5R and $Co_{17}R_2$ phases, the difference in domain wall energies in both phases being responsible for setting up potential barriers for wall displacement.

4. *Permanent Magnets*

As already mentioned, light rare-earth elements (except neodymium) form Co_5R compounds poten-

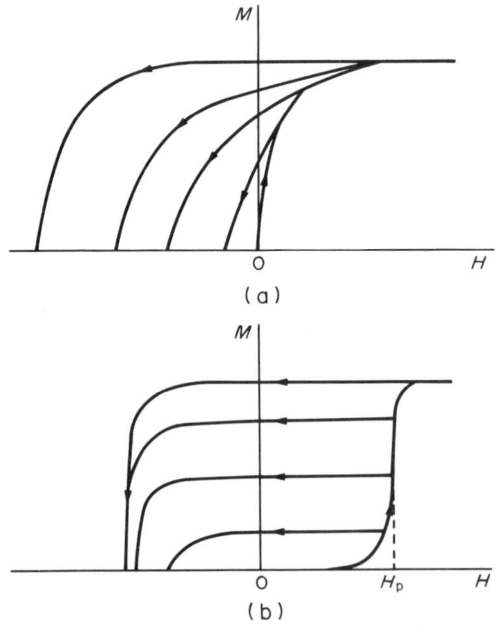

Figure 4
Hysteresis M–H loops corresponding to (a) a nucleation-controlled coercive force and (b) a pinning-controlled coercive force

Table 4
Magnetic properties of various permanent magnets based on cobalt–rare-earth alloys

Composition	B_r (T)	$_bH_c$ (kA m^{-1})	$_iH_c$ (kA m^{-1})	$(BH)_{max}$ (kJ m^{-3})	α[d] (% °C^{-1})
Co_5Sm[a]	1.06	792	1360	224	
Co_5Sm[b]	0.91	716	1430	159	− 0.04
Co_5Sm[c]	0.86	660	1430	143	
$Co_5MM_{0.5}Sm_{0.5}$	0.90	605		159	
$Co_5Sm_{0.5}Pr_{0.3}Nd_{0.2}$[a]	1.05	770	1150	210	
$Co_5Sm_{0.58}Gd_{0.42}$	0.64	454	3020	74	− 0.003
$(Co_{0.72}Fe_{0.14}Cu_{0.14})_5Ce$	0.72	398	420	99	− 0.09
$(Co_{0.73}Fe_{0.2}Cu_{0.05}Zr_{0.02})_{7.5}Ce_{0.5}Sm_{0.5}$	1.06	648	744	210	− 0.04
$(Co_{0.72}Fe_{0.2}Cu_{0.055}Zr_{0.025})Sm_{0.56}Gd_{0.44}$	0.85	640	950	136	− 0.004
$(Co_{0.715}Fe_{0.2}Cu_{0.06}Zr_{0.025})_{7.5}Sm$	1.07	750	1700	216	
$(Co_{0.73}Fe_{0.2}Cu_{0.05}Zr_{0.02})_{7.5}Sm$	1.08	760	1000	224	− 0.03
$(Co_{0.69}Fe_{0.2}Cu_{0.1}Zr_{0.01})_{7.4}Sm$	1.12		533	240	
$(Co_{0.65}Fe_{0.28}Cu_{0.05}Zr_{0.02})_{7.8}Sm$[a]	1.20		1110	263	

a Laboratory best value b Commercial high grade c Commercial average d Reversible temperature coefficient of flux density

tially suitable for use in permanent magnets (see Table 4). Among these, Co_5Sm has exceptionally high magnetocrystalline anisotropy and has been successfully made into permanent magnets having the highest intrinsic coercive force. The best published value of the maximum energy product is 224 kJ m^{-3}, whereas commercial grade magnets have an energy of 140–160 kJ m^{-3}. Partial substitution of praseodymium for samarium results in an even higher maximum energy product. However, no one has succeeded in obtaining superior energy values with pure Co_5Pr magnets. Misch metal, a natural mixture of light rare-earth metals, is the cheapest and the most abundant among rare-earth metals. Therefore, misch-metal-based magnets are preferred for large-scale applications. However, to keep the coercive force at a useful level, samarium must constitute 15–20% of the total rare-earth additions. Partial substitution of samarium by gadolinium in Co_5Sm yields magnets with a reduced temperature dependence of magnetic induction at a loss to its absolute value. The heavy rare-earth elements, such as erbium and holmium, have similar effects.

There is a group of cobalt rare-earth magnets whose coercivity is generated by the pinning of domain walls due to fine precipitates of a secondary magnetic phase. The precipitates can be either of a $T_{17}R_2$ phase in T_5R host alloys or of a T_5R phase in $T_{17}R_2$ host alloys. These phenomena are widely found in these T_zR alloys. The copper, sometimes with a secondary additive such as zirconium, has the effect of increasing mutual solubility between T_5R and $T_{17}R_2$ phases and thus realizing a single phase (T_5R, $T_{17}R_2$ or a disordered phase of $TbCu_7$ type) at high temperatures. This phase decomposes into T_5R and $T_{17}R_2$ phases at lower temperatures resulting in the required fine coherent two-phase structure. The compositions $(Co,Fe,Cu)_5Sm$ and $(Co,Fe,Cu)_5Ce$ were first found to exhibit precipitation-hardened coercivity. The coercive force is controllable within a wide range by controlling the amount of copper. Iron is added to improve the saturation magnetization. The $(Co,Fe,Cu)_5Ce$ alloys yield magnets having a relatively low maximum energy product of 70–100 kJ m^{-3}. They are successfully employed in large-scale applications because they include cerium, the cheapest and most abundant elemental rare-earth metal. The other typical composition is $(Co,Fe,Cu,Zr)_{7.5}Sm$, which yields magnets having a maximum energy product of 200–260 kJ m^{-3}, better than that of the Co_5Sm sintered magnets. The mixture of cerium and samarium is used in a variety of practical magnets to achieve a compromise between cost and performance. A z-value between 5 and 7.5 is selected, depending on the ratio of cerium to samarium, so as to achieve the best possible magnetic properties for the respective ratio. The $(Co,Fe,Cu,Zr)_z(Sm, Ce)$ magnets have become increasingly popular because they save expensive metals, samarium and cobalt, compared with Co_5Sm magnets.

The combination of samarium and a heavy rare earth such as gadolinium applied to the T_zR alloys yields magnets having extremely low temperature dependence of magnetic flux density. This is because the magnetic moment of the heavy rare earth, which is coupled antiparallel to that of cobalt, decreases more rapidly with temperature and compensates the decrease of the magnetic moment of cobalt.

5. *Preparation of Cobalt–Rare-Earth Magnets*

Cobalt–rare-earth starting alloys are prepared by two alternative methods, that is, by melting the ingredient metals or by calciothermic reduction of rare-earth oxides in the presence of cobalt metal or cobalt oxide powder. The latter process is cheaper because rare-earth reduction and alloying are simultaneously completed; it is often adopted for processing binary cobalt–samarium alloys. However, this process is less popular for multicomponent alloys because of the difficulty encountered in controlling the amounts of additional transition metals other than cobalt. These alloys are pulverized into fine particles preferably of about 3 μm average size. A nitrogen jet stream and a cyclone are usefully employed for milling, because a narrow particle-size distribution is obtained and deleterious contaminants such as carbon are minimized. The powder is compacted in a die to 50–65% of bulk density with a pressure of 50–100 MPa, usually with a magnetic field of about 800 kA m^{-1} applied to achieve grain orientation. The compacted bodies are sintered usually between 1100 °C and 1200 °C depending on composition. The sintered bodies are further heat treated at various temperatures depending on composition. In the case of Co_5Sm magnets, the sintered bodies are aged at 900 °C to improve the coercive force by removing physical as well as chemical inhomogeneities which give rise to nucleation sites. The aged bodies are rapidly cooled to room temperature to prevent the incorporation of unwanted magnetic phases due to the eutectic decomposition (into Co_7Sm_2 and $Co_{17}Sm_2$). Sintered bodies of copper-containing T_zR alloys are isothermally aged at a temperature of 400–900 °C. The objective of this aging process is to introduce a fine two-phase microstructure to provide pinning sites rather than to eliminate nucleation sites. Sintered bodies are sometimes heated at successively lower temperatures in the given temperature range. This process is called step aging.

An alternative process to step aging is continuously controlled cooling in the given temperature range. Step aging is effective in developing a high coercive force, especially for T_zR alloys containing both copper and zirconium.

See also: Magnetic Materials: An Overview; Neodymium–Iron–Boron Alloys; Rare-Earth Intermetallics; Rare-Earth Metals and Alloys

Bibliography

Kumar K 1988 Re Tm_5 and Re_2 Tm_{17} permanent magnet development. *J. Appl. Phys.* 63(6): R13–57

Strnat K J 1988 Rare-earth–cobalt permanent magnets. In: Wolfarth E P, Buschow K H J (eds.) 1988 *Ferromagnetic Materials*, Vol. 4. Elsevier, Amsterdam, pp. 131–209

Y. Tawara
[Shin-Etsu Chemical Company,
Kawasaki City, Japan]

Coercivity and Domain Wall Pinning

Since the early 1970s the development of high-quality magnetically hard and soft materials has progressed dramatically due to the discovery of new alloys and structurally tailored materials. In amorphous non-magnetostrictive alloys, minimum coercive fields H_C of 0.1–1 A m^{-1} have been realized (Warlimont 1988) and in hard magnetic materials based on rare earth–transition metal intermetallic compounds, maximum coercive fields of 5 MA m^{-1} have been obtained (Kronmüller et al. 1988a, b). This variation of H_C by eight orders of magnitude is comparable with the variations observed for the electrical conductivity in insulators and semiconductors. The origin of the wide variation of H_C is related to both the variation of intrinsic material parameters and to the influence of different types of microstructures on the magnetization processes.

The main intrinsic parameters governing the coercive field are the spontaneous magnetization M_S, the magnetocrystalline anisotropy constant K_1 and the saturation magnetostriction λ_S. Microstructures affecting H_C are all types of defects leading to inhomogeneities of the intrinsic magnetic material parameters: point defects, dislocations, antiphases and grain boundaries, surface inhomogeneities and precipitates. In amorphous alloys, structural fluctuations of intrinsic parameters as well as double-well systems and quasidislocations determine the coercive field (Evetts et al. 1978, Kronmüller 1981a, b).

Magnetic hardening in general is based on two types of magnetization process:

(a) reversal of magnetization by a rotation process or a nucleation mechanism, and
(b) reversal of magnetization by domain wall displacements.

The first one of these two processes is connected with the existence of small isolated particles or grains which are magnetically saturated and consist of one single domain (e.g., in sintered magnets). The second case is realized in as-cast bulk materials optimized by special pretreatments. Recently, hard magnetic materials have also been produced by rapid quenching and mechanical alloying (Schultz 1988) where the hardening mechanism may consist of a combined mechanism—suppression of the expansion of a reversed nucleus by a pinning process of the domain wall boundaries of the nuclei.

The absolute upper bound of H_C is given by the so-called nucleation field for homogeneous rotation (Brown 1963):

$$\mu_0 H_N = \frac{2K_1}{M_s} - N_{\mathrm{eff}}\mu_0 M_S \qquad (1)$$

where the first term represents the magnetocrystalline field and the second term takes care of demagnetizing

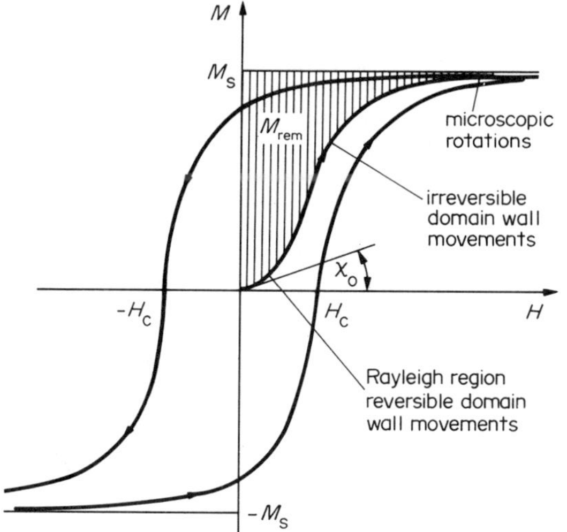

Figure 1
Definition of characteristic parameters of the hysteresis loop

fields characterized by an effective demagnetization factor N_{eff} (see *Demagnetizing Factors*). Figure 1 gives a definition of H_C by the hysteresis loop and Fig. 2 gives a comparison of the experimentally achieved coercive fields with the theoretical nucleation field H_N for some prominent magnetic materials. According to Fig. 2, only 20–30% of the maximum achievable H_C values are realized with somewhat larger values in the case of laboratory magnets. In hard magnetic materials the coercive field can be satisfactorily described by a modification of Eqn. (1) (Kronmüller 1987, Kronmüller et al. 1988a, b, Sagawa and Hirosawa 1988):

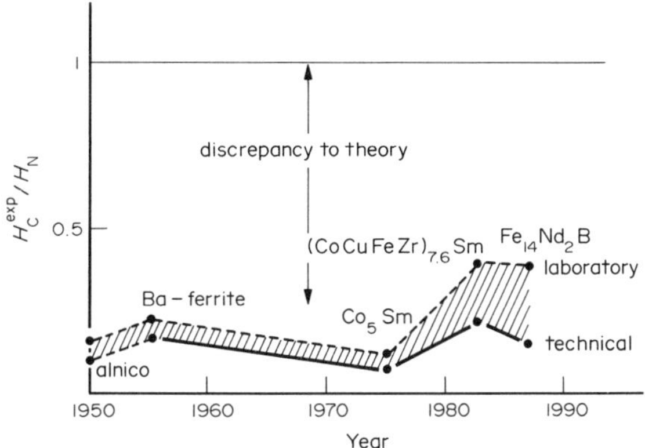

Figure 2
Ratio H_C^{exp}/H_N of prominent permanent magnets

$$\mu_0 H_C = \frac{2K_1}{M_S}\alpha - N_{\text{eff}}\mu_0 M_S \tag{2}$$

where α takes care of the effects of the microstructure on H_C. In the case of nucleation-hardened ferromagnets, α results from regions where the magnetocrystalline anisotropy is reduced. For pinning hardened magnets, α describes the strength of the domain wall pinning centers.

It is the aim of this article to summarize the results derived for different types of pinning center as obtained on the basis of the theory of micromagnetism (see *Micromagnetic Calculations*). Four different groups of pinning phenomena will be considered: intrinsic pinning by the atomic structure, pinning by individual planar defects, pinning by statistical fluctuations of a multitude of defects and pinning by the fluctuations of intrinsic material parameters in amorphous alloys.

1. Intrinsic Coercive Field of Narrow Domain Walls

In the conventional theory of micromagnetism, the spin arrangement within a domain wall is described by the vector $\boldsymbol{M}_S(r)$ of the spontaneous magnetization. For planar domain walls in uniaxial crystals, the orientation of $\boldsymbol{M}_S$ is defined by the angle φ between $\boldsymbol{M}_S$ and the c axis, where φ obeys the micromagnetic equation (Kronmüller 1978):

$$2A\frac{d^2\varphi}{dz^2} - K_1\sin^2\varphi = 0 \tag{3}$$

(A = exchange constant) with the well-known solution

$$\sin^2\varphi = \frac{1}{\text{ch}(\pi z/\delta_B)} \tag{4}$$

and with the wall width δ_B and the specific wall energy γ_B, where

$$\delta_B = \pi(A/K_1)^{1/2}, \qquad \gamma_B = 4(AK_1)^{1/2} \tag{5}$$

The results given by Eqns. (4) and (5) are valid for domain walls with $\delta_B \gg a$ where a corresponds to the lattice parameter. In this case, the domain wall averages over the discrete atomic structure and, therefore, the coercive field of an ideal crystal vanishes.

The situation is different in the case of narrow domain walls where δ_B is only a few lattice constants. In this case, the wall energy depends on the position of the domain wall with respect to the atomic lattice. Two different types of configurations of 180° walls are shown in Fig. 3 for Co_5Sm. In configuration A the center of the domain wall lies on a lattice plane, whereas in configuration B the center of the domain wall lies just in the middle between two neighboring

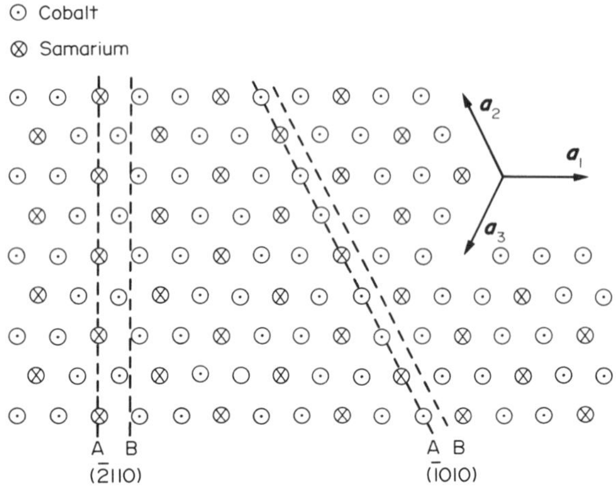

Figure 3
Atomic structure of the basal plane of Co_5Sm: the two configurations A and B are shown for the $(\bar{2}110)$ and the $(\bar{1}010)$ 180° wall

lattice planes. The corresponding spin structures for these two configurations are shown schematically in Fig. 4 for the $\{\bar{2}110\}$ and the $\{\bar{1}010\}$ domain walls. According to Fig. 4, the spin rotation by 180° takes place within a few lattice planes and as a consequence the micromagnetic continuum theory fails. Applying the concepts of lattice theory the specific wall energies γ_A and γ_B of Co_5Sm have been determined previously (Hilzinger and Kronmüller 1972) within the framework of the Heisenberg model for the symmetric positions A or B. Intermediate positions are only stable for applied magnetic fields and are characterized by the z coordinate parallel to the domain wall normal. In first approximation, $\gamma(z)$ may be written as

$$\gamma(z) = \gamma_B + \Delta\gamma \sin^2(\pi z/D) \tag{6}$$

where D denotes the periodicity distance between crystallographically equivalent planes ($D = 3a/2$ for the $(\bar{2}110)$ domain wall and $D = a(3/2)^{1/2}$ for the $(\bar{1}010)$ domain wall). The amplitude, $\Delta\gamma = \gamma_A - \gamma_B$,

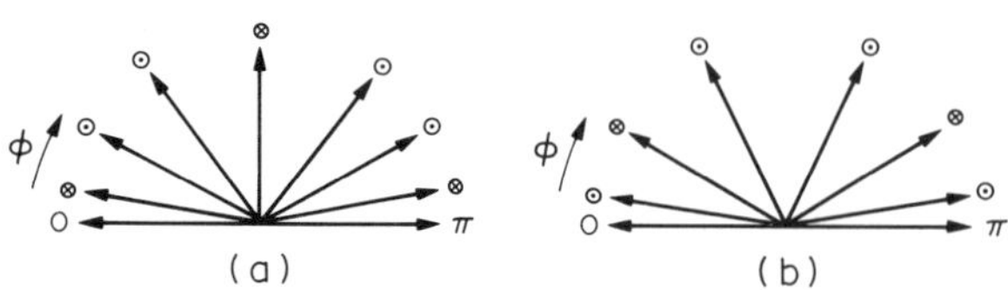

Figure 4
Schematic representation of spin orientation in atomic layers parallel to the domain wall: (a) configuration A and (b) configuration B

can be written as

$$\Delta\gamma = c\frac{2A}{a}\exp(-\pi\delta_B/D) \tag{7}$$

where A denotes the exchange constant and c, a numerical factor, is given by

$$c = 10\left|\frac{S^R J^{RCo}}{S^{Co} J^{CoCo}} - 1 + \frac{K^R}{K^{Co}} - 1\right| \tag{8}$$

Here, S^{Co} and S^R denote the spin quantum numbers of cobalt and rare-earth atoms, respectively, J^{CoCo} and J^{RCo} denote the exchange integrals between cobalt and rare-earth atoms and cobalt atoms, respectively, and K^R and K^{Co} correspond to the local anisotropy constants of rare-earth and cobalt atoms, respectively.

The critical field required to overcome the so-called Peierls potential $\gamma(z)$ follows from the point of inflection of the Peierls potential, giving

$$\mu_0 H_C = \frac{\pi\Delta\gamma}{2DM_S} \tag{9}$$

The intrinsic coercive fields derived from Eqns. (7–9) are only of the order of 10^4 A m^{-1} for Co_5Sm, dysprosium and some rare-earth alloys (Kütterer et al. 1977, Arnaudas et al. 1986). However, H_C may be neglected for materials such as cobalt, iron or nickel. Furthermore, narrow domain walls have a tendency to move by thermal activation of planar kink loops (Egami 1973), as shown in Fig. 5. If, locally, the domain wall forms a displacement by one atomic distance like a pancake, this pancake expands spontaneously above a critical diameter r_c, thus reducing the intrinsic coercive field drastically above 0 K.

In conclusion, the magnetic Peierls potential of domain walls is an interesting feature of hard magnetic materials; however, it is not sufficient to produce a strong permanent magnet.

2. *Pinning Domain Walls by Planar Defects*

2.1 Pinning by Thin Planar Defects

In real materials, planar defects such as grain boundaries, stacking faults, antiphase boundaries (APB) and phase boundaries (PB) between precipitates (e.g., Co_5Sm in a $Co_{17}Sm_2$ matrix) may act as rather effective pinning centers. Pinning domain walls by planar defects may be treated by a micromagnetic continuum approach (Kronmüller 1973, Friedberg and Paul 1975) as well as by the discrete Heisenberg model (Hilzinger and Kronmüller 1975).

Consider a planar defect with N atomic layers of spacing d. The domain wall is assumed to be parallel to the planar defect of width $D = Nd$. For a calculation of the wall energy, the domain wall is divided into

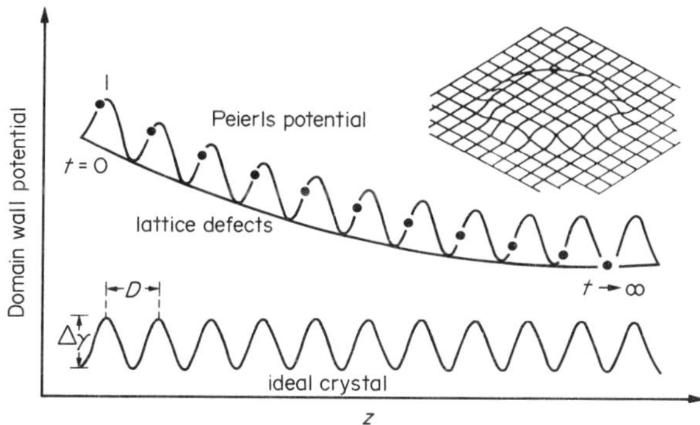

Figure 5
Model of the domain wall potential of narrow domain walls due to the superposition of the Peierls potential and the long-range potential due to defects: points show successive positions of the domain wall under the action of a demagnetizing field; the upper right corner shows a model for the planar domain wall kink

three regions as shown in Fig. 6. The external regions I and III are treated by a continuum approach. The internal region II includes the defect where the ith layer is characterized by an exchange constant $A^{i,i+1}$ and an anisotropy constant K^i_1. The unperturbed regions I and III are characterized by the exchange constant A and the anisotropy constant K_1. The total domain wall energy per unit area is given by

$$\gamma = 2(AK)^{1/2}(1 - \cos\varphi_1 + 1 + \cos\varphi_N)$$

$$+ (K_1 \sin^2\varphi_1 + 2\sum_{i=2}^{N-1} K^i_1 \sin^2\varphi_i + K^N_1 \sin^2\varphi_N)\frac{d}{2}$$

$$+ \sum_{i=1}^{N-1} \frac{2A^{i,i+1}}{d}[1 - \cos(\varphi_{i+1} - \varphi_i)] \quad (10)$$

where φ_i denotes the rotation angle of $\boldsymbol{M}_S$ with respect to the c axis. In order to avoid overlapping effects, only half of the crystalline energy of the transition regions for $\varphi = \varphi_1$ and φ_N have been taken into account. The wall energy given by Eqn. (10) has to be minimized with respect to the angles φ_i, and the coercive field then follows from the maximum force $d\gamma/dz$ exerted by the planar defect on the domain wall:

$$\mu_0 H_C = \frac{\pi}{3\sqrt{3}}\frac{2K_1}{M_S}\frac{d}{\delta_B}\left|\sum_{i=1}^{N-1}\left(\frac{A}{A^{i,i+1}} - \frac{K^i_1}{K_1}\right)\right| - N_{eff}\mu_0 M_S \quad (11)$$

If it is assumed that all perturbed n atomic layers have the same properties, $A^{i,i+1} = A'$ and $K^i_1 = K'_1$, Eqn. (11) may be written as

$$\mu_0 H_C = \frac{\pi}{3\sqrt{3}}\frac{2K_1}{M_S}\frac{nd}{\delta_B}\left|\frac{A}{A'} - \frac{K'_1}{K_1}\right| - N_{eff}\mu_0 M_S \quad (12)$$

where $W = nd$ corresponds to the width of the perturbed layer. Comparing Eqns. (11) and (12) with Eqn. (2) it is evident that the α parameter is given by

$$\alpha^{pin} = \frac{\pi}{3\sqrt{3}}\frac{d}{\delta_B}\left|\sum_{i=1}^{N-1}\left(\frac{A}{A^{i,i+1}} - \frac{K^i_1}{K_1}\right)\right| \quad (13)$$

In the case of equivalent perturbed layers, for $K^i_1 = 0$, $A/A' = 2$ and $nd = 0.25\,\delta_B$, a value $\alpha^{pin} = \pi\sqrt{3}/18 \simeq 0.3$ is obtained, thus showing that the ideal coercive field is reduced by 70%. If a more refined planar defect is considered with a linear decrease of $A^{i,i+1}$ from 1.0 A to 0.5 A and of 1.0 K_1 to 0.1 K_1 within the first five planes, and a similar linear increase up to the tenth plane, $\alpha^{pin} = \pi(\sqrt{3}/9)W/\delta_B$ is obtained, giving $\alpha^{pin} = 0.151$. Obviously, a smooth transition between the matrix and the defect region leads to considerably smaller coercive fields.

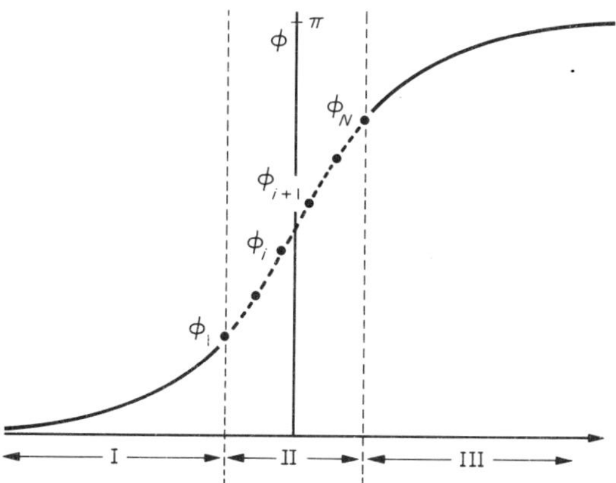

Figure 6
Definition of regions I, II and III for a domain wall at a planar defect

2.2 Pinning By Extended Planar Defects

If the planar defect is wider than the wall width, the coercive field in a first approximation is given by

$$\mu_0 H_C = \frac{1}{2M_S}\frac{d\gamma(z)}{dz}\bigg|_{max} - N_{eff}\mu_0 M_S \quad (14)$$

where $d\gamma(z)/dz$ denotes the maximum slope of the wall energy being given by $\gamma(z) = 4[AK_1(z)]^{1/2}$. In the following it is assumed that only K_1 shows an appreciable inhomogeneity which may be described (Kronmüller et al. 1988a, b) by

$$K_1(z) = K_1(\infty) - \frac{\Delta K}{ch^2(z/r_0)} \tag{15}$$

with $K_1(\infty)$ denoting the K_1 value in the perfect material and ΔK the decrease of K_1 within the inhomogeneity of half-width r_0. Inserting Eqn. (15) into Eqn. (14) gives

$$\mu_0 H_C = \frac{2K_1(\infty)}{M_S}\frac{2\delta_B}{3\pi r_0} - N_{eff}\mu_0 M_S$$

$$= \frac{\gamma_B}{3M_S r_0} - N_{eff}\mu_0 M_S \tag{16}$$

which corresponds to a parameter $\alpha^{pin} = 2\delta_B/3\pi r_0$. Inserting $\delta_B/r_0 = 0.5$ gives $\alpha^{pin} = 0.1$. Figure 7 summarizes the results for narrow and wide planar defects with dependences linear in r_0 for $r_0 \ll \delta_B$, and proportional to $1/r_0$ for $r_0 > \delta_B$. There exists an intermediate region where both approximations fail. In Fig. 7 it is indicated by the broken curve, showing how the transition between narrow and wide domains may take place. In particular, Fig. 7 shows that the maximum α^{pin} parameter is obtained for $r_0/\delta_B \sim 0.6$ giving a value of $\alpha^{pin} = 0.3$. Accordingly, the largest coercive field which may be achieved by domain wall pinning is at most 30% of the ideal coercive field.

A permanent magnet that is based on the pinning of domain walls by planar defects is the multiphase CoSm magnet of composition $Sm(Co_{0.68}Cu_{0.08}Fe_{0.22}Zr_{0.02})_{7.6}$. After an annealing treatment at 1160 °C followed by a 24 h isothermal annealing at 840 °C and a subsequent slow cooling down to 400 °C in 3 h, this alloy consists of a pyramidal cell structure where the interior of the cells consists of an iron-rich 17:2 phase and the cell walls of a copper-rich 5:1 phase, as shown in Fig. 8 (Livingston and Martin 1977, Lee

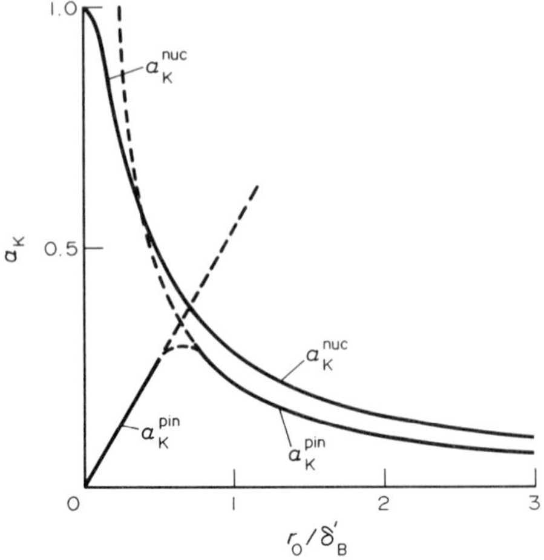

Figure 7
The α parameters for pinning and nucleation in a perturbed region of halfwidth r_0

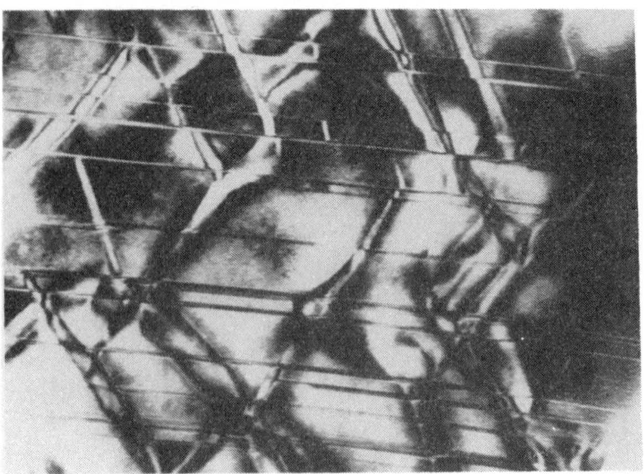

Figure 8
Cellular microstructure of processed $Sm(Co,Fe,Cu,Zr)_{7.6}$ showing rhombic cells and coherent cell boundaries

1981, Ray 1984). Domain walls are pinned within cell walls which correspond to regions of reduced magnetocrystalline energy (Durst et al. 1988). An analysis of the intrinsic material parameters of this two-phase material gives, for the 17:2 phase, $\mu_0 M_S = 1.320$ T and $K_1 = 5 \times 10^6$ J m^{-3} and, for the 5:1 phase, $\mu_0 M_S = 0.800$ T and $K_1 = 1.9 \times 10^6$ J m^{-3}. With these values the ideal coercive field of the 17:2 phase would be about 10 MA m^{-1}. Experimentally, a value of H_C around 3.5 MA m^{-1} has been obtained, showing that the material has been ideally optimized (i.e., the conditions $r_0/\delta_B \sim 0.6$ and $K_1' < K_1$ are realized by appropriate pretreatments).

According to these results the width of the phase boundary between the 17:2 and the 5:1 phase is $r_0 = 3$ nm ($\delta_B \simeq 5$ nm), that is, it is much smaller than the average cell diameters of 100–150 nm.

The existence of strong pinning centers which act as trapping centers for domain walls is also demonstrated by the properties of minor hysteresis loops as shown in Fig. 9. After applying a certain maximum field the hysteresis branches in the first quadrant behave nearly reversibly for increasing and decreasing fields because the magnetization takes place by pushing domain walls into the cell walls where they remain trapped (Durst et al. 1988). In agreement with the properties of a pinning hardened magnet saturation is just achieved at a field of $H = 3.5$ MA m^{-1}, which corresponds to the coercive field.

3. Statistical Pinning Theory of Domain Walls

3.1 Basic Results

In Sect. 2 the case of one domain wall and one defect interacting with each other was considered. In many

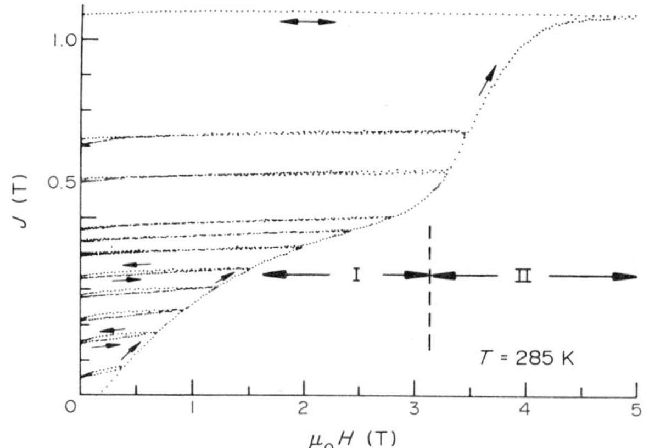

Figure 9
Minor hysteresis branches of processed Sm(Co,Fe, Cu,Zr)$_{7.6}$ as a function of the maximum applied field

systems, however, the domain walls interact with a multitude of defects (e.g., in domains of volume $10^{-4}\,cm^3$ the number of dislocations in a plastically deformed material is of the order of 10^8).

A similar situation holds for point defects or small precipitates. For a general treatment of the statistical interaction force, consider the individual force $v(z-z_j)$ exerted by a defect at position z_j on a domain wall at position z. The coordinate z is taken parallel to the domain wall normal. The total force acting on a domain wall is the sum

$$V(z)=\sum_j v(z-z_j) \qquad (17)$$

over all defects interacting with the domain wall. The statistical field of force $V(z)$, shown schematically in Fig. 10, can be characterized by the following parameters:

(a) the average wavelength $2L_0$ defined as twice the average distance between adjacent zeros, $V(z)=0$;

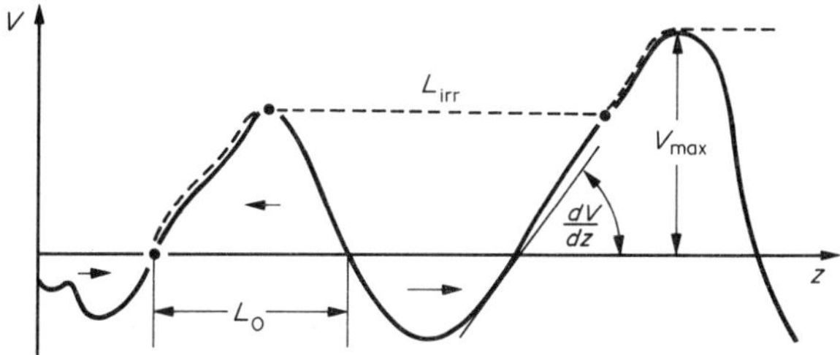

Figure 10
Definition of characteristic parameters of the statistical field of force acting on a domain wall

(b) the average value $\bar{V}_{max}$ of the maxima of $V(z)$; and

(c) the average value $\overline{1/R}=\overline{1/(dV/dz)}$ of the reciprocal slopes of $V(z)$ at $V(z)=0$.

The parameters (a)–(c) previously have been derived by means of correlation functions. Extensive treatments have been given (Träuble 1966, Pfeffer 1967, Kronmüller 1970, 1972). All investigations are based on a normal distribution of the force V. The probability $f(V)$ to find an interaction force between V and $V+dV$ is given by a normal distribution

$$f(V)=\frac{1}{(2\pi B_0)^{1/2}}\exp(-V^2/2B_0) \qquad (18)$$

Similarly, the slopes $R=dV/dz$ obey a normal distribution

$$g(R)=\frac{1}{(2\pi B_1)^{1/2}}\exp(-R^2/2B_1) \qquad (19)$$

The correlation functions B_0 and B_1 are related to the individual interaction force v of a defect as follows:

$$B_0=\frac{F_B N}{L_3}\int_{-L_{3/2}}^{L_{3/2}}\left[v^2(z)-\langle v(z)\rangle^2\right]dz \qquad (20)$$

$$B_1=\frac{F_B N}{L_3}\int_{-L_{3/2}}^{L_{3/2}}\left\{\left[\frac{dv(z)}{dz}\right]^2-\left\langle\frac{dv(z)}{dz}\right\rangle^2\right\}dz \qquad (21)$$

Where F_B is the domain wall area, L_3 is the domain wall distance and N is the defect density.

The aforementioned parameters are related to the correlation functions B_0 and B_1:

$$\left.\begin{aligned}2L_0&=2\pi(B_0/B_1)^{1/2}\\ \overline{1/R}&=(\pi/2B_1)^{1/2}\\ \bar{V}_{max}&=(B_0/2\pi)^{1/2}\end{aligned}\right\} \qquad (22)$$

The characteristic properties of the virgin magnetization curve and the hysteresis loop, that is, the Rayleigh region $M=\chi_0 H+\alpha_R\mu_0 H^2$, with the initial susceptibility χ_0 and the Rayleigh constant α_R, and the coercive field are given by

$$\left.\begin{aligned}\mu_0 H_C&=\frac{\pi^{1/2}}{M_S F_B\,|\cos\varphi_0|}\bar{V}_{max}\left(\ln\frac{L_3}{2L_0}\right)^{1/2}\\ \chi_0&=\frac{\mu_0(2M_S\cos\varphi_0)^2 F_B}{L_3}\left(\overline{\frac{1}{R}}\right)\\ \alpha_R&=\frac{\mu_0|2M_S\cos\varphi_0|^3F_B^2}{(2\pi)^2}\frac{L_0}{L_3}\frac{1}{(\bar{V}_{max})^2}\end{aligned}\right\} \qquad (23)$$

where φ_0 is the angle between the applied field and $\boldsymbol{M}_S$ within the domains. If we make use of the fact that the average wavelength is approximately given by $2L_0 \simeq 4\delta_B$ the following useful relations between H_C, χ_0 and α_R may be derived:

$$\left.\begin{aligned}\chi_0 H_C &= M_S \cos\varphi_0 \frac{\pi^{1/2}}{2}\frac{L_0}{L_3}\left(\ln\frac{L_3}{2L_0}\right)^{1/2}\\ \alpha_R \frac{\mu_0 H_C}{\chi_0} &= \frac{8}{3\pi^{1/2}}\left(\ln\frac{L_3}{2L_0}\right)^{1/2}\\ \frac{\chi_0^2}{\alpha_R} &= \frac{3\pi}{16}\frac{L_0}{L_3}\mu_0 M_S\cos\varphi_0\end{aligned}\right\}\quad(24)$$

Equation (23) also predicts the dependence on the defect density:

$$H_C \propto N^{1/2},\qquad \chi_0 \propto 1/N^{1/2},\qquad \alpha_R \propto 1/N \quad (25)$$

3.2 Application to Different Defect Structures

(a) Dislocations. In general, these have a magnetostrictive interaction with domain walls due to the elastic interaction energy of the magnetostrictive elastic stresses of the domain wall and the long-range elastic stresses of the dislocations. The interaction force is most simply obtained from the Peach–Köhler formula (Peach and Köhler 1950):

$$\boldsymbol{v}(z) = \int d\boldsymbol{l} \times (\boldsymbol{\sigma}.\boldsymbol{b}) \quad (26)$$

where $d\boldsymbol{l}$ corresponds $\boldsymbol{l}$ to the line element of the dislocation, $\boldsymbol{b}$ to its Burgers vector and $\boldsymbol{\sigma}$ to the magnetostrictive elastic stresses of the domain wall. The elastic stresses of a domain wall have been determined previously by Rieder (1959) for cubic and hexagonal crystals. In order to avoid the complexities resulting from the elastic and the magnetostrictive anisotropies, only the case of isotropy will be considered in both cases. Choosing a coordinate system with the x_1 axis parallel to the easy direction, the x_3 (z) axis parallel to the domain wall normal and the x_2 axis perpendicular to x_1 and x_3, the magnetostrictive stresses of a 180° domain wall for an isotropic crystal are given by

$$\begin{aligned}\sigma_{11} &= 3G\lambda_S \sin^2\varphi\\ \sigma_{22} &= -3G\lambda_S\sin^2\varphi\\ \sigma_{12} &= -\frac{3}{2}G\lambda_S\sin 2\varphi\end{aligned}\quad(27)$$

where G is the shear modulus and λ_S is the magnetostriction. The interaction force exerted parallel to the domain wall normal by a general dislocation lying parallel to the domain wall plane is written as

$$v_3(z) = l_1 b_2 \sigma_{22} - l_2 b_1 \sigma_{11} + (l_1 b_1 - l_2 b_2)\sigma_{12} \quad (28)$$

where l_i and b_i correspond to the components of the dislocation line and of the Burgers vector, respectively. With l_1, $b_2 = 0$, this gives $v_3(z) = 3b_1 l_2 G\lambda_S \sin^2\varphi$. Inserting this result into Eqn. (20) and using Eqns. (21–23) $(b_1 = b, l_2 = l)$, gives

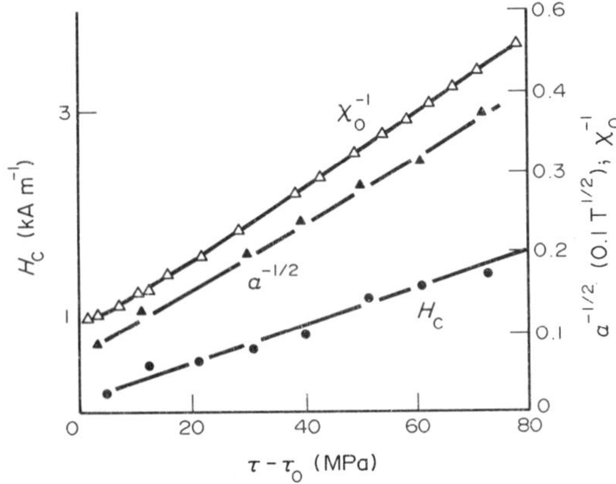

Figure 11
The dependence of H_0, χ_0^{-1} and $\alpha_R^{-1/2}$ on plastic deformation of a nickel single crystal

$$2L_0 = \delta_B\sqrt{5}$$

$$\mu_0 H_C = \frac{Gbl\lambda_S}{\pi^{1/2}M_S F_B\cos\varphi_0}\left(6NF_B\delta_B\right)^{1/2} \times\left(\ln\frac{L_3}{2L_0}\right)^{1/2} \quad (29)$$

Equation (29) contains all important intrinsic and microstructural parameters of a plastically deformed material. Of special importance is the dependence of H_C on the dislocation density N. In the case of plastically deformed fcc crystals, N is related to the applied flow stress τ according to Kronmüller (1967)

$$\tau - \tau_0 = 0.3\, bG(Nl)^{1/2} \quad (30)$$

Measuring the characteristic parameters of the hysteresis loop and using Eqns. (25) and (30), linear behavior of $\tau - \tau_0$ is expected for the quantities H_C, $1/\chi_0$ and $1/\alpha^{1/2}$. Figure 11 represents these quantities as a funtion of $\tau - \tau_0$ for a plastically deformed nickel single crystal showing the expected linear relation (Köster 1967, Kronmüller 1970). A further interesting result of Eqn. (29) is the dependence of H_C on the magnetic intrinsic properties, given by

$$\mu_0 H_C \propto \frac{\lambda_S}{M_S}\left(\frac{A}{K_1}\right)^{1/4} \propto \delta_B^{1/2} \quad (31)$$

With increasing anisotropy constant K_1 or decreasing wall width δ_B the pinning effect of dislocations decreases as $\delta_B^{1/2}$ (Träuble 1966). Similarly, the initial susceptibility can be determined from Eqns. (21–23), giving

$$\chi_0 = \mu_0 \left(\frac{5}{6}\right)^{1/2} [M_S \cos \varphi_0]^2$$

$$\times \left(\frac{F_B}{L_3}\right)^{1/2} \frac{\delta_B{}^{1/2}}{|blG\lambda_S|} \frac{1}{N^{1/2}} \qquad (32)$$

According to this result, the dependence of χ_0 on δ_B is the same as for H_C; however, with increasing magnetostriction and dislocation density, χ_0 decreases whereas H_C increases.

(*b*) *Dislocation dipoles*. In many cases, the dislocation structure of plastically deformed materials is composed of dislocation dipoles which may be formed by cyclic deformation or by agglomerations of vacancies or interstitial atoms. For a determination of the interaction forces $v(z)$, the Peach–Köhler formula given in Eqn. (26) can be rearranged as follows (Kronmüller 1981a):

$$v_3(z) = \int_f d\mathbf{Q}.\nabla_3 \boldsymbol{\sigma} \qquad (33)$$

where the dyadic product

$$dQ = df.\boldsymbol{b} \qquad (34)$$

corresponds to the differential displacement tensor of the dipole with df corresponding to the surface element of the area spread out by the dislocation loop and $\boldsymbol{b}$ denoting its Burgers vector. In the special case of rectangular dislocation loops of width D extending parallel to the x_2 axis and with a loop area inclined by an angle ε with respect to the x_1 axis, the interaction force is given by

$$v_3(z) = -3\pi D l_3 G\lambda_S \sin\varepsilon\, \delta_B{}^{-1} \sin\varphi\, (b_1 \sin 2\varphi + b_2 \cos 2\varphi) \qquad (35)$$

Inserting Eqn. (35) into Eqns. (20)—(23) gives, for $\varepsilon = \pi/2$,

$$\mu_0 H_C = \frac{12\sqrt{\pi}}{|M_S \cos\varphi_0|} \frac{1}{(30F_B)^{1/2}} |G\lambda_S \Delta V|$$

$$\times \left(\frac{N_{\rm dip}}{\delta_B}\right)^{1/2} \left(\ln \frac{L_3}{2L_0}\right)^{1/2} \qquad (36)$$

where $\Delta V = |Dl_2 b_1|$ corresponds to the volume expansion or contraction due to the dislocation dipole and $N_{\rm dip}$ denotes the volume density of dislocation dipoles. The main temperature dependence of H_C derived from Eqn. (36) can be written as

$$\mu_0 H_C(T) \propto \frac{\lambda_S}{M_S} \delta_B{}^{-1/2} \qquad (37)$$

In contrast to the case of individual dislocations, dislocation dipoles lead to an increase of H_C with decreasing wall width. It is of interest to note that the initial susceptibility decreases according to a $\delta_B{}^{3/2}$ law with decreasing wall width.

(*c*) *Point defects*. The results of the preceding section can be easily applied to point defects by interpreting the dipole tensor $\mathbf{Q}$ as that of a point defect, thus giving

$$v_3(z) = \mathbf{Q}.\nabla_3 \sigma \qquad (38)$$

If an isotropic point defect is considered with $\mathbf{Q} = \mathbf{I}Q_0$ ($\mathbf{I}$ = unit tensor), the force acting on a domain wall is found to vanish. For an anisotropic defect of orthorhombic symmetry $Q_{11} \neq Q_{22} \neq Q_{33}$ the interaction force $v_3(z) = 3\pi G\lambda_S \delta_B{}^{-1}(Q_{22} - Q_{11}) \sin\varphi \sin 2\varphi$ leads to a coercive field of

$$\mu_0 H_C = \frac{12\pi^{1/2}}{|M_S \cos\varphi_0|} \frac{G\lambda_S |Q_{22} - Q_{11}|}{(30F_B)^{1/2}}$$

$$\times \left(\frac{N_{\rm dip}}{\delta_B}\right)^{1/2} \left(\ln \frac{L_3}{2L_0}\right)^{1/2} \qquad (39)$$

Equation (39) is equivalent to the result for dislocation dipoles, the volume dilatation ΔV just being replaced by the dipole tensor.

3.3 *Amorphous Materials*

(*a*) *Intrinsic fluctuations*. In amorphous ferromagnetic materials neither dislocations nor grain boundaries as in crystalline alloys are present. Therefore, amorphous alloys are soft magnetic materials with minimum coercive field (Hilzinger et al. 1978, Kronmüller 1981b, Luborsky 1983). In spite of the lack of conventional defects in amorphous alloys, the technically realized coercive fields are still comparable with those of permalloy. Obviously magnetic inhomogeneities exist in amorphous alloys leading to a moderate intrinsic pinning of domain walls. Four types of intrinsic magnetic inhomogeneities may be of importance: fluctuations of the exchange energy δA, fluctuations of the local magnetic anisotropy δK, magnetoelastic coupling energy of elastic dipoles and clusters of atomic short-range order. Assuming that δA and δK may be described by a locally uniaxial anisotropy, the magnetic interaction energies with the domain wall may be written as

$$\delta\xi_{\rm exch} = \delta A \left(\frac{d\varphi}{dx_3}\right)^2 \langle\Omega\rangle$$

$$\delta\xi_{\rm an} = \delta K \langle\Omega\rangle \cos^2\theta \qquad (40)$$

where Ω is the average atomic volume of magnetic ions, φ is the rotation angle of M_S within the domain wall and $\cos\theta = \sin\theta_0 \cos(\varphi - \varphi_0)$ with θ_0 and φ_0 denoting the polar and azimuthal angles, respectively, of the local anisotropy axis with respect to the coordinate system of the domain wall. With Eqns. (40) and

(23), the coercive field of these intrinsic fluctuations is given by

$$\mu_0 H_C = \frac{\pi^{1/2}}{M_S \cos\varphi_0} \frac{\langle\Omega\rangle^{1/2}}{(2F_B\delta_B)^{1/2}} \times \left(\frac{8}{15}\langle\delta K^2\rangle + \frac{11}{15}\langle\delta A^2\rangle \frac{\pi^4}{\delta_B{}^4}\right)^{1/2} \left(\ln\frac{L_3}{2L_0}\right)^{1/2} \quad (41)$$

The fluctuations of the magnetic anisotropy energy are only of importance in hard magnetic materials such as iron–terbium alloys. The fluctuations of the exchange energy may be calculated on the basis of a nearest neighbor model. In the case of $Fe_{40}Ni_{40}(PB)_{20}$ alloys, the nickel atoms may be considered as nonmagnetic and the total exchange interaction of a certain iron atom is proportional to the number of nearest neighbor iron atoms. In the case of a dense random packing model with 12 nearest neighbors, the probability w_n of finding n nearest neighbor atoms is given by

$$w_n = \binom{12}{n} c_{Fe}^n (1 - c_{Fe})^{12-n} \quad (42)$$

where c_{Fe} corresponds to the atomic density of iron atoms. Using Eqn. (42), $\langle\delta A^2\rangle$ may be determined by assuming that the exchange energy of n Fe–Fe pairs is $(n/r_0)J_{FeFe}/3$ where r_0 corresponds to the nearest neighbor distance and J_{FeFe} denotes the exchange integral between iron atoms. A detailed calculation gives (Kronmüller 1981a)

$$\langle\delta A^2\rangle = \frac{1}{9r_0{}^2} \sum_{n=0}^{12} w_n n^2 J_{FeFe}^2 = \frac{13}{3r_0{}^2} J_{FeFe}^2 \quad (43)$$

$$\mu_0 H_C^{min} = \frac{\pi^{5/2}}{3M_S} \frac{\Omega_{Fe}{}^{1/2}}{(2F_B)^{1/2}} \frac{J_{FeFe}}{r_0 \delta_B{}^{5/2}} \times \left(\frac{143}{5}\right)^{1/2} \left(\ln\frac{L_3}{2L_0}\right)^{1/2} \quad (44)$$

Equation (44) corresponds to the absolute minimum of H_C achievable in a magnetically soft amorphous alloy. Inserting realistic values for the material properties into Eqn. (44), an intrinsic coercive field of $H_C = 3 \times 10^{-3}\,A\,m^{-1}$ is derived ($F_B = 6 \times 10^{-6}\,m^2$, $\delta_B = 3 \times 10^{-7}\,m$, $\mu_0 M_S = 1.29\,T$, $\Omega_{Fe} = 2.2 \times 10^{-27}\,m^3$, $J_{FeFe} = kT_C/8$, $T_C = 1000\,K$).

(*b*) *Elastic dipoles*. The minimum coercive field H_C^{min} is expected to hold in nonmagnetostrictive amorphous alloys with vanishing local anisotropy. In magnetostrictive materials, elastic stress sources must be assumed to govern the coercive field. In a number of papers it has been shown that elastic dipoles of dimensions 1–10 nm are present in amorphous alloys in a high density of $10^{18}\,m^{-3}$ (Kronmüller et al. 1979, Kronmüller 1981a, Grimm and Kronmüller 1983). Accordingly, H_C should be described by Eqn. (39) predicting a temperature dependence of the type $H_C(T) \propto (\lambda_S/M_S)K_1{}^{1/4}$. This relation has been tested in Fig. 12 for various nickel–iron-based amorphous alloys, showing, in fact, the predicted temperature dependence of H_C (Kronmüller et al. 1979, Gröger and Kronmüller 1981). The different points in Fig. 12 refer to measurements in the temperature range 100–310 K. Figure 13 represents a test of the self-consistency conditions given in Eqn. (24), for iron–nickel-based alloys. It should be noted that the nonmagnetostrictive (cobalt, iron, nickel)-based alloys do not fit the self-consistency relations because surface irregularities become quite important in these cases.

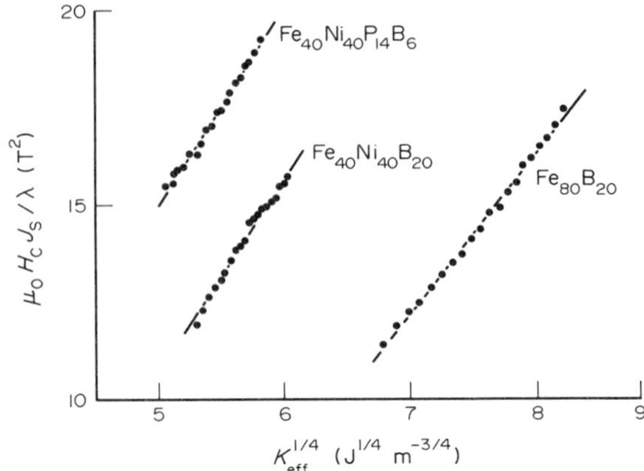

Figure 12
Test of the temperature dependence of H_C for magnetostrictive iron–nickel-based alloys (K_{eff} denotes the effective uniaxial anisotropy of the amorphous alloys)

3.4 Surface Irregularities

In low-coercivity materials, the natural surface roughness may have a significant effect on H_C (e.g., rapidly quenched amorphous ribbons have a natural surface roughness due to the uneven roller as well as due to inhomogeneous solidification processes). In general, the contact surface reveals a large surface roughness $\Delta T_r/T_r \simeq 0.05$–$0.1$, whereas the free side of the ribbon is rather smooth with $\Delta T_r/T_r \simeq 0.01$ where T_r is the ribbon thickness (Becker 1981, Wang and Kronmüller 1982). The coercive field resulting from the fluctuations of the surface roughness is based on the varying local domain wall energy which, for a domain wall area $L_2\ T_r(z)$, is written as

$$\xi_\gamma(z) = \gamma_B L_2 T_r(z) \quad (45)$$

where L_2 denotes the lateral extension. Assuming that $T_r(z)$ varies with a wavelength λ, the surface roughness leads to the following coercive field (Kronmüller 1981):

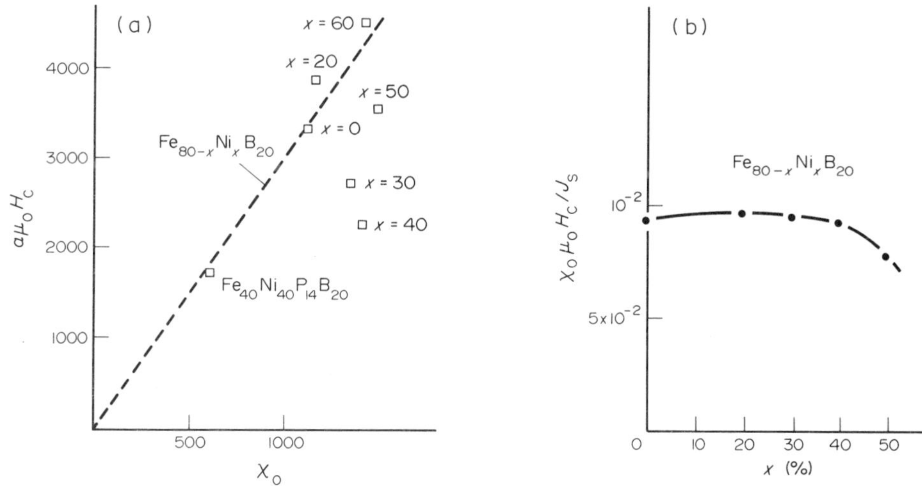

Figure 13
Test of Eqn. (24) for iron–nickel-based alloys: (a) $\alpha_R\mu_0H_C \propto \chi_0$; and (b) $\chi_0\mu_0H_C/J_S$ = constant where $J_S = \mu_0M_S$

$$\mu_0H_C^{\text{surf}} = \frac{\pi}{M_S}\frac{\Delta T_\tau}{\langle T\rangle\,\lambda}\left(\frac{L_2}{2\lambda}\right)^{1/2}\gamma_B\left(\ln\frac{L_3}{2L_0}\right)^{1/2} \tag{46}$$

Equation (46) reveals a number of interesting features which may be tested by changing the topological parameters of the surfaces by special treatments. Figure 14 shows the influence of different thinning procedures on the coercive field of an $Fe_{40}Ni_{40}P_{14}B_6$ ribbon. In all cases, H_C^{surf} depends linearly on $1/\langle T_r\rangle$ as predicted by Eqn. (46). The increase is largest for grinding with glass-paper and smallest for electropolished samples. Equation (46) gives reasonable values for H_C^{surf} for microstructural parameters corresponding to the real situation. For the parameters of Ni–Fe–PB alloys, $\mu_0M_S = 1.3$ T, $\Delta T_r/\lambda = 0.1$, $\Delta T_r/\langle T_r\rangle = 0.05$, $\gamma_B = 10^{-4}\,\text{J m}^{-2}$, $\lambda \simeq T = 50\,\mu\text{m}$, $L_2 = L_3 = 100\,\mu\text{m}$, $H_C^{\text{surf}} = 0.5\,\text{A m}^{-1}$.

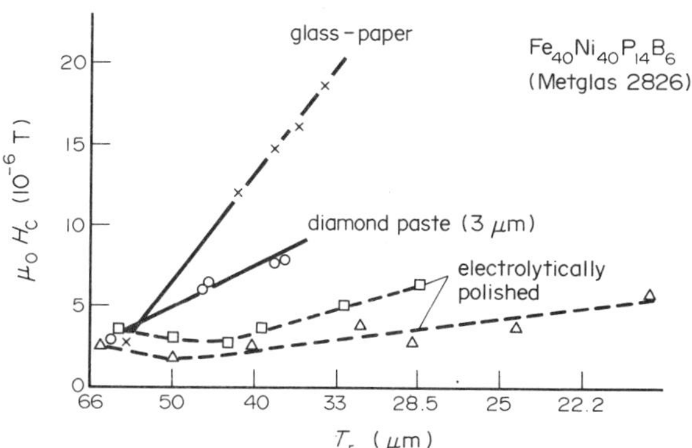

Figure 14
The coercive field of $Fe_{40}Ni_{40}P_{14}B_6$ as a function of the ribbon thickness for different types of surface roughness

See also: Domains and Domain Walls; Flux Pinning and the Summation of Pinning Forces; Magnetization Imaging; Micromagnetic Calculations

Bibliography

Arnaudas J I, Del Moral A, Abell J S 1986 Intrinsic coercive field in pseudobinary cubic intermetallic compounds. I: $Dy_xY_{1-x}Al_2$ and $Dy_{1-\delta}Al_2$. *J. Magn. & Magn. Mater.* 61: 370–80

Becker J J 1981 Effects of surface treatment and thinning on magnetic properties of rapidly quenched amorphous alloy ribbons. *J. Appl. Phys.* 52: 1905–7

Brown W F Jr 1963 *Micromagnetics.* Interscience, New York

Durst K-D, Kronmüller H, Ervens W 1988 Investigation of the magnetic properties and the demagnetization processes of an extremely high coercive $Sm(Co,Fe,Cu,Zr)_{7.6}$ permanent magnet. *Phys. Status Solidi A* 108: 403–16; 705–19

Egami T 1973 Theory of Bloch wall tunnelling. *Phys. Status Solidi B* 57: 211–24

Evetts J E, Howarth W, Gibbs M R J 1978 The micromagnetic properties of an amorphous alloy. *Proc. 3rd Int. Conf. Rapidly Quenched Metals.* Metals Society, London, pp. 127–32

Friedberg R, Paul D I 1975 New theory of coercive force of ferromagnetic materials. *Phys. Rev. Lett.* 34: 1234–7

Grimm H, Kronmüller H 1983 Investigation of structural defects in the amorphous ferromagnetic alloy $Fe_{40}Ni_{40}P_{14}B_6$. *Phys. Status Solidi B* 117: 663–74

Gröger B, Kronmüller H 1981 The effect of internal stresses on the magnetization curve of amorphous ferromagnetic FeNi-alloys. *Appl. Phys.* 24: 287–97

Hilzinger H R, Kronmüller H 1972 Spin configuration and intrinsic coercive field of narrow domain walls in Co_5R-compounds. *Phys. Status Solidi B* 54: 593–604

Hilzinger H R, Kronmüller H 1975 Investigation of Bloch

wall pinning by antiphase boundaries in RCo_5 compounds. *Phys. Lett. A* 51: 59–60
Hilzinger H R, Mager A, Warlimont H 1978 Amorphe ferromagnetische Werkstoffe—Magnetische Grundlagen, Eigenschaften und Anwendungen. *J. Magn. & Magn. Mater.* 9: 191–9
Köster E 1967 Verformungs- und Temperaturabhängigkeit der Magnetisierungsprozesse in Nickeleinkristallen. *Phys. Status Solidi* 19: 655–71
Kronmüller H 1970 Statistical theory of Rayleigh's law. *Z. Angew. Phys.* 30: 9–13
Kronmüller H 1972 Magnetic techniques for the study of dislocations in ferromagnetic materials. *Int. J. Nondestr. Test.* 3: 315–50
Kronmüller H 1973 Microstructure and micromagnetism. *AIP Conf. Proc.*, Vol. 10. American Institute of Physics, New York, pp. 1006–25
Kronmüller H 1981a Domain walls and the coercive field of amorphous ferromagnets. *J. Phys. (Paris)* 42: 1285–92
Kronmüller H 1981b Theory of the coercive field of amorphous ferromagnetic alloys. *J. Magn. & Magn. Mater.* 24: 159–67
Kronmüller H 1987 Theory of nucleation fields in inhomogeneous ferromagnets. *Phys. Status Solidi B* 144: 385–96
Kronmüller H, Durst K-D, Hock S, Martinek G 1988a Micromagnetic analysis of the magnetic hardening mechanisms in RE-Fe-B magnets. *J. Phys. (Paris), Colloq.* 49(8): 623–28
Kronmüller H, Durst K-D, Sagawa M 1988b Analysis of the magnetic hardening mechanism in RE-Fe-B permanent magnets. *J. Magn. & Magn. Mater.* 74: 291–302
Kronmüller H, Fähnle M, Domann M, Grimm H, Grimm R, Gröger B 1979 Magnetic properties of amorphous ferromagnetic alloys. *J. Magn. & Magn. Mater.* 13: 53–70
Kütterer R, Hilzinger H R, Kronmüller H 1977 The temperature dependence of the coercive field of Co_5Sm magnets. *J. Magn. & Magn. Mater.* 4: 1–7
Lee R W 1981 The future of rare earth–transition metal magnets of type RE_2TM_{17}. *J. Appl. Phys.* 52: 2549–53
Livingston J D, Martin D L 1977 Microstructure of aged (Co,Cu,Fe)Sm magnets. *J. Appl. Phys.* 48: 1350–4
Luborsky F E 1983 Application-oriented magnetic properties. In: Luborsky F E (ed.) 1983 *Amorphous Metallic Alloys*. Butterworths, London, pp. 360–80
Peach M, Koehler J S 1950 The forces exerted on dislocations and the stress fields produced by them. *Phys. Rev.* 80: 436–9
Pfeffer K-H 1967 Zur Theorie der Koerzitivfeldstärke und Anfangssuszeptibilität. *Phys. Status Solidi* 21: 857–72
Ray A E 1984 Metallurgical behaviour of $Sm(Co,Fe,Cu,Zr)_7$ alloys. *J. Appl. Phys.* 55: 2094–6
Rieder G 1959 Eigenspannungen in unendlichen geschichteten und elastisch anisotropen Medien, insbesondere in Weisschen Bezirken und in geschichteten Platten. *Abh. Braunschw. Wiss. Ges.* 11: 20–61
Sagawa M, Hirosawa S 1988 Coercivity and microstructure of R–Fe–B sintered permanent magnets. *J. Phys. (Paris), Colloq.* 49 (8): 617–22
Schultz L 1988 Formation of amorphous metals by mechanical alloying. *Mater. Sci. Eng.* 97: 15–23
Träuble H 1966 Magnetisierungkurve und magnetische Hysterese ferromagnetischer Einkristalle. In: Seeger A (ed.) 1966 *Moderne Probleme der Metallphysik*, Vol. 2. Springer, Berlin, pp. 157–475
Wang Y J, Kronmüller H 1982 The influence of surface conditions on the magnetic properties in amorphous alloys $Fe_{40}Ni_{40}B_{20}$ and $Co_{58}Ni_{10}Fe_5Si_{11}B_{16}$. *Phys. Status Solidi A* 70: 415–21
Warlimont H 1988 The impact of amorphous metals on the field of soft magnetic materials. *Mater. Sci. Eng.* 99: 1–10

H. Kronmüller
[Max-Planck-Institut für Metallforschung, Stuttgart, Germany]

Coherence Length, Proximity Effect and Fluctuations

The phenomenological London theory for superconductivity was able to explain the Meissner effect in terms of a rigid wave function under the application of an external magnetic field (see *Superconducting Materials: BCS and Phenomenological Theories*; *Superconducting Materials: Physical Properties*). Under this assumption, F. and H. London calculated the value of the London penetration depth λ_L:

$$\lambda_L^2 = \frac{m}{\mu_0 e^2 n_s} \tag{1}$$

where m is the effective mass of the electron and n_s is the superfluid density, which at $T=0$ is equal to the electron density in the normal state. Since the ratio m/n does not vary by more than one order of magnitude among common metals, λ_L does not vary much from one superconductor to another: roughly by a factor of 4 from say lead (39 nm) to the high-T_c oxides (about 140 nm in $YBa_2Cu_3O_7$). No explicit dependence of λ_L on T_c is predicted.

The main modification to the London theory proposed by Ginzburg and Landau (GL theory) was the introduction of a finite correlation length ξ, somewhat unfortunately called the coherence length. This is the minimum length scale over which the superfluid density may vary with only a moderate increase of the free-energy density. Ginzburg and Landau showed that under small applied fields, the spatial variation of the superfluid density remains negligible and the London result (Eqn. (1)) is recovered, whatever the respective values of λ and ξ. The effect of a finite coherence length is only apparent in strong applied fields and when the GL parameter $\kappa = \lambda/\xi$ is larger than $1/\sqrt{2}$ (see *Magnetic Properties of Superconductors*). As shown by Pippard, when $\xi > \lambda$ nonlocal corrections lead to an effective value of λ somewhat larger than that given by the London theory.

1. Coherence Length

1.1 Coherence Length in the GL Theory

Ginzburg and Landau introduced the coherence length in the framework of their treatment of the superconducting phase transition as a second-order transition, occurring at the critical temperature T_c. The existence of a coherence length that diverges at the transition is a general property of all second-order phase transitions:

$$\xi(t) = a\xi_0(1-t)^{-\nu} \qquad (2)$$

where a is a numerical coefficient of order unity, ξ_0 is the zero temperature coherence length, t is the reduced temperature (T/T_c) and ν is the critical exponent for the coherence length. In the mean-field approximation $\xi(t)$ can be obtained from the GL free-energy functional (see *Superconducting Materials: Physical Properties*):

$$\mathbb{F}_s = \mathbb{F}_n + A|\Delta|^2 + \frac{B}{2}|\Delta|^4 + C|\nabla\Delta|^2 \qquad (3)$$

where the coefficients $A = N(0)\ (t-1)$, $B = 0.098N(0)/(k_B T_c)^2$, $C = 0.55\,\xi_0^2\ N(0)\ \chi(\alpha)$, $N(0)$ is the normal state density of states and $\chi(\alpha)$ is the Gorkov impurity function, as defined in the article *Flux Pinning and the Summation of Pinning Forces*. Here Δ is the pair potential, equal to the superconducting energy gap when it is uniform across the sample.

The pair potential Δ is related to the order parameter Ψ by $|\Delta|^2 = \hbar^2|\Psi|^2/(4mC)$. In some cases the free-energy functional is written in terms of the *reduced*-order parameter $\psi = \Psi/\Psi_\infty$, where Ψ_∞ is the equilibrium order parameter at the temperature under consideration. Sometimes it is convenient to use the pair *amplitude* $F = \Delta/V$, where V is the BCS interaction parameter (e.g., in a normal metal with $V = 0$ in contact with a superconductor, the pair potential is zero but the pair amplitude is finite, as will be discussed in Sect. 3). In simplified terms $(FV)^2$ and $(F/\pi N(0))^2$ are proportional to n_s in the two regions.

In the notation used here

$$\xi^2 = \frac{C}{|A|} \qquad (4)$$

Since $A \propto 1-t$, $\nu = \frac{1}{2}$ and λ is found in this approximation to diverge with the same exponent. In the clean limit, therefore, $\xi(t) = 0.74\,\xi_0\,(1-t)^{-1/2}$ while in the dirty limit $\xi(t) = 0.85\,(\xi_0 l)^{1/2}\,(1-t)^{-1/2}$ where l is the normal state electron mean free path. The applicability of the mean field approximation to the superconducting phase transition will be returned to later.

The GL theory is phenomenological and does not predict what the value of ξ_0 should be. However, it does offer a simple method for measuring $\xi(t)$ through the expression for the nucleation field:

$$\mu_0 H_{c2}(t) = \frac{\Phi'_0}{2\pi\xi^2(t)} \qquad (5)$$

where $\Phi_0 = 2.07 \times 10^{-15}$ T m^2 is the flux quantum. If the superconductor is type II, H_{c2} as given by Eqn. (5) is the upper critical field. If it is type I, H_{c2} is the supercooling field. Coherence lengths in the clean and dirty limits, obtained from H_{c2} measurements, are given in Tables 1 and 2, respectively. Measurements of the supercooling field are usually affected by the Saint James–de Gennes surface nucleation phenomenon which occurs at a field $H_{c3} = 1.69\ H_{c2}$ in regions of the sample where the surface is parallel to the applied field (Saint James and de Gennes 1963, Deutscher 1967). Once the superconducting state has nucleated in some region, it spreads all over the sample. Hence H_{c3} rather than H_{c2} is usually the measured nucleation field and ξ must be calculated accordingly.

When $H_{c3} > H_c$, a stable superconducting surface sheath of thickness ξ exists below the nucleation field. Its thickness has been measured directly in lead and was found to be about twice the coherence length (Deutscher 1967).

Table 1
Coherence lengths in the clean limit

	T_c (K)	ξ_0 (μm)	$\xi_0 T_c$ (μm K)
Aluminum	1.19	1.20	1.4
Indium	3.40	0.33	1.1
Tin	3.72	0.26	0.97
Gallium	5.90	0.16	0.94
Lead	7.20	0.08	0.58
Niobium	9.25	0.035	0.32
$YBa_2Cu_3O_7$	93	0.0015	0.14

1.2 Coherence Length in the BCS Theory

The microscopic origin of ξ became clear only with the advent of the BCS theory: it is the average radius of a Cooper pair. The value of ξ_0 can be simply estimated from the uncertainty principle:

$$\xi_0 \simeq \frac{hv_F}{\pi\Delta} \qquad (6)$$

where v_F is the Fermi velocity and 2Δ is the energy necessary to break a Cooper pair. In the weak coupling limit of the BCS theory, $2\Delta = 3.5\,k_B T_c$. Since v_F

Table 2
Coherence lengths in the dirty limit (practical superconductors)

	T_c (K)	$\xi(T=0)$ (μm)
Nb–56 at.%Ti	10	0.0050
Nb_3Sn	17	0.0040
$(Nb, Ta)_3Sn$	18	0.0035

does not vary much from one metal to another, the coherence length is essentially inversely proportional to the critical temperature (Table 1). As a rule of thumb, ξ_0 (μm) $\simeq T_c^{-1}$ (K^{-1}). Since λ_L is basically independent from T_c, low-T_c superconductors should be type I and high-T_c superconductors should be type II. This is well borne out experimentally. Superconductors with a T_c lower than that of lead (7.2 K) are type I, superconductors with a higher T_c are type II (niobium, the A15 compounds, the Chevrel phases and the high-T_c oxides).

The effect of impurities on λ and ξ has been calculated by Gorkov. In the limit where the electron mean free path l is much shorter than ξ_0 (the so-called dirty limit):

$$\lambda = 0.64\,\lambda_L(\xi_0/l)^{1/2} \tag{7a}$$

$$\xi = 0.85(\xi_0 l)^{1/2} \tag{7b}$$

As first observed by Shubnikov, the introduction of impurities can turn a type I superconductor into a type II material. Nonmagnetic impurities do not affect the critical temperature (Anderson theorem). However, magnetic impurities strongly depress T_c; for a low concentration x of paramagnetic impurities with total angular momentum J this is given by the BCS Gorkov theory

$$\frac{dT_c}{dx} \simeq \frac{-\pi^2 N(0) f^2 (g-1)^2 J(J+1)}{2k_B}$$

where f is the exchange coupling constant and g is the gyromagnetic ratio. Eqns. (5) and (7b) show that H_{c2} is inversely proportional to the mean free path and, hence, proportional to the normal state resistivity, that is, to the concentration of impurities (see *Magnetic Properties of Superconductors*). This result is in excellent agreement with experiment (Serin 1969) and is put into practice in the commercial Nb–Ti wires, which have an upper critical field of about 10 T, roughly 30 times larger than that of pure niobium.

A short coherence length, achieved either in a clean high-T_c superconductor or in a dirty low-T_c superconductor, is therefore the key to practical applications of superconductivity in strong applied fields. However, in general, ξ remains much larger than interatomic distances. The high-T_c oxides are an exception to that rule: along the CuO planes, ξ is about 2 nm in $YBa_2Cu_3O_7$; in the perpendicular direction it is only 0.3 nm. Practical superconductors such as NbTi and Nb_3Sn have coherence lengths of the order of 5 nm.

2. *Thermal Fluctuations of the Superconducting Order Parameter*

2.1 Quasi-Mean-Field Regime

Thermodynamic fluctuations of the superconducting order parameter can be estimated from the GL free-energy functional. The lowest free energy is achieved for $\Delta = \Delta_0$, $\Delta_0 = (-A/B)^{1/2}$ below T_c and $\Delta_0 = 0$ above T_c. Any other configuration $\Delta = \Delta_0 + \delta\Delta(r)$ can be decomposed into its Fourier components, each of them being characterized by an amplitude and a wavelength. The Fourier components with the largest amplitude have a wavelength of order ξ. This is because, for larger wavelengths, the energy cost is high because of the large volume involved and, for shorter wave lengths, the gradient term in Eqn.(3) is prohibitive. Hence, the volume of a typical fluctuation is ξ^3. Its amplitude can be estimated by taking its energy to be of the order of $k_B T$. Developing $\mathbb{F}_s$ to second order in $\delta\Delta$, one obtains below T_c

$$-2A\langle|\delta\Delta|^2\rangle\xi^3 \simeq k_B T \tag{8}$$

Apart from numerical coefficients, the relative amplitude of the fluctuations is given by:

$$\frac{\langle|\delta\Delta|^2\rangle}{\Delta_0^{\,2}} = \frac{k_B T}{\Delta\mathbb{F}\xi^3} \tag{9}$$

where $\Delta\mathbb{F}$ is the condensation energy per unit volume at equilibrium, $\Delta\mathbb{F} = A^2/2B = \mu_0 H_c^2/2$, where H_c is the thermodynamic critical field.

The relative amplitude of the fluctuations is thus essentially given by the ratio of the thermal energy $k_B T$ to the value of the condensation energy per coherence volume. Eqn.(9) can be immediately generalized to the case of a general dimensionality d by replacing ξ^3 by ξ^d. The relative amplitude of the fluctuations diverges as

$$\frac{\langle|\delta\Delta|^2\rangle}{\Delta_0^{\,2}} \propto (1-t)^{-(4-d)/2} \tag{10}$$

where $d < 3$ is achieved when one or more of the dimensions of the sample is smaller than the coherence length $\xi(t)$.

In the preceding analysis the fluctuations have been treated as independent of each other (quasi-mean-field approximation). Equations (9) and (10) are thus only valid in the limit where the relative amplitude of the fluctuations is much smaller than unity. This approximation is in general excellent for conventional low-T_c superconductors (LTS), but not for the high-T_c oxides (HTS).

The effects of fluctuations are very small in bulk LTS (e.g., their effect on the heat capacity is negligible; this is not the case for the high-T_c oxides). They are larger for low dimensionalities (thin films, wires, small grains) because the relative amplitude of the fluctuations then diverges with a stronger exponent (see Eqn.(10)). The most easily detected fluctuation effect is an excess conductivity $\Delta\sigma$ above T_c. In two dimensions within the quasi-mean-field approximation (Azlamazoff and Larkin 1968)

$$\Delta\sigma = \frac{e^2}{16\hbar}(t-1)^{-1} \tag{11}$$

where the conductivity is expressed in $(\Omega/\text{square})^{-1}$. In a d-dimensional sample, the conductivity diverges as $(t-1)^{-(4-d)/2}$, that is, with the same exponent as that of the relative amplitude of the fluctuations given by Eqn.(10). Additional terms to the excess conductivity considered by Maki and Thompson lead to a stronger divergence.

The effect of fluctuations on the conductivity is most easily detected when the background normal state conductivity is small, that is, in dirty films (Glover 1971) or wires. Another case where the excess conductivity is easily observed is that of granular superconductors (Deutscher and Dodds 1977) (see *Granular Superconductors*).

On the contrary, the effect of fluctuations on the diamagnetism (which diverges with the same exponent as that of the conductivity) is most easily observed in clean samples, with a large coherence length. Fluctuation diamagnetism has been studied in low-T_c intercalated compounds.

2.2 Critical Fluctuation Region

The mean-field approximation is valid as long as the relative amplitude of the fluctuations as given by Eqn.(10) is much smaller than unity. However, since this amplitude diverges at T_c, there exists a temperature range near T_c where this approximation fails. This range is called the critical region (for a recent review see Ginzburg 1988). It is customary to measure it on the reduced temperature scale $\varepsilon = (T - T_c)/T_c$. If ε_c is the reduced temperature at which the relative amplitude of the fluctuations is of order unity, it follows from Eqn.(9) that

$$\varepsilon_c = \alpha \left[\frac{k_B T_c}{\Delta \mathbb{F}(0)\xi_0^3} \right]^2 \qquad (12)$$

where α is a numerical factor (actually important, $\alpha = 2^{-11}\pi^{-2}$) and $\Delta\mathbb{F}(0)$ is the condensation energy at $T = 0$. According to the BCS theory:

$$\Delta\mathbb{F}(0) = \frac{1}{2}N(0)\Delta(0)^2 \qquad (13)$$

where $N(0)$ is the normal state density of states. In the weak coupling limit, $\Delta(0)$ is proportional to T_c and since as seen before $\xi_0 \propto T_c^{-1}$, it follows that $\varepsilon_c \propto T_c^4$. ε_c is extremely small for LTS, typically $\varepsilon_c < 10^{-12}$. This fully justifies the use of the GL mean-field approximation. However, for the HTS ε_c is much larger ($\varepsilon_c \simeq 10^{-3}$) and the critical region can be studied experimentally (Salamon et al. 1990).

As seen from Eqn.(12), the width of the critical region is determined by the value of the condensation energy in a coherence volume at $T = 0$, compared with $k_B T_c$. However, $U_0 = \Delta\mathbb{F}\xi^3$ is also the typical energy scale for core pinning of vortices (see *Flux Pinning and the Summation of Pinning Forces*). Through the coherence length, there exists therefore an intimate relationship between the critical temperature, the width of the critical region and the ability of the superconductor to pin vortices. Notice that as T_c goes up, the upper limit for U_0 (clean limit) goes down ($U_0 \propto T_c^{-1}$). In LTS, ε_c is small and vortices are easily pinned even close to T_c but, in HTS, ε_c becomes significant and vortex pinning cannot be achieved when T is of the order of T_c (Deutscher 1990). Full use of their extremely high upper critical field (several hundred tesla) can only be made at temperatures significantly lower than T_c. Furthermore, the quasi-mean-field approximation appears to be inadequate for the description of the vortex state near $H_{c2}(T)$ in the bismuth- and thallium-based oxide superconductors (Kes et al. 1992).

3. Proximity Effect

When a normal metal (N) is in good electrical contact with a superconductor (S), Cooper pairs can leak from S to N. If the electron–electron interaction in N is zero ($V_N = 0$), and in the absence of magnetic impurities and applied magnetic fields, Cooper pairs in N are only destroyed by the thermal energy $k_B T$. Their decay length is then given by (Deutscher and de Gennes 1969)

$$\xi_N = K^{-1} = \frac{\hbar v_{FN}}{2\pi k_B T} \qquad (14)$$

where v_{FN} is the Fermi velocity in N. When pair breaking is also induced by magnetic impurities and applied magnetic fields, a multipair breaking model in which the various depairing factors simply add can be used. The value of K is then expressed in terms of a depairing variable α (see Yang and Finnemore 1984 and *Granular Superconductors*).

The probability amplitude for finding a Cooper pair in N at a distance $|x|$ from the SN interface is given by

$$F(x) = \varphi(x)\exp(-K|x|) \qquad (15)$$

where $\varphi(x)$ is a slowly varying function of x. The probability amplitude F is defined as the ratio of the pair potential Δ to the BCS interaction parameter V. If $V_N = 0$, $\Delta_N = 0$, but F is finite (see later for the boundary conditions for F at the SN interface).

Eqn. (14) is valid when the mean free path in N, l_N, is large compared to the decay length as given by Eqn. (14) (clean limit). In the dirty limit

$$K^{-2} = \left(\frac{\hbar D_N}{2\pi k_B T} \right) \qquad (16)$$

where D_N is the normal state coefficient of diffusion in N, $D_N = v_{FN} l_N/3$.

If N is a superconductor at some lower temperature $T_{cN} < T$, K^{-1} diverges as $T \to T_{cN}$. Properties of the SN contact near T_{cN} can be treated with the GL formalism. Very close to T_{cN}, $K^{-1} \propto (T - T_{cN})^{-1/2}$, symmetrical with the temperature dependence of the

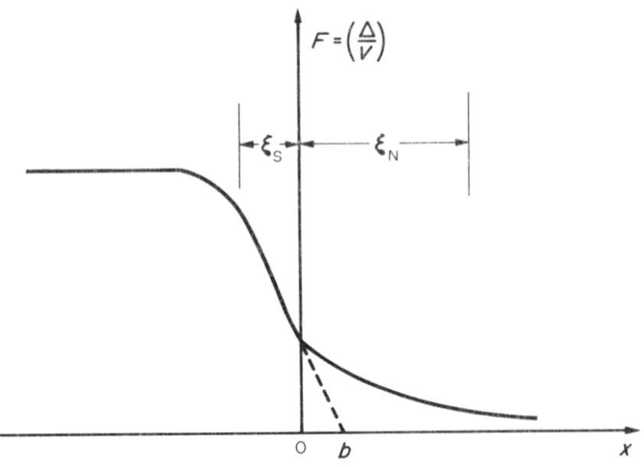

Figure 1
Behavior of the pair amplitude F near an SN interface. For simplicity, it is assumed that the normal state density of states is the same in N and in S (otherwise, F is discontinuous at the interface) and that the boundary resistance is negligible (which also leads to a discontinuity of F). A case is illustrated where the coefficient of diffusion in S is shorter than in N. In this case, the extrapolation length b is shorter than $K^{-1}(=\xi_N)$

coherence length in the superconducting state below T_{cN}. At $T = T_{cN}$, the exponential decay law (Eqn. (15)) must be replaced by a power law dependence $F(x) = F(0)\ [x_0/(x + x_0)]$ (Deutscher and de Gennes 1969). At temperatures sightly above T_{cN}, this power law persists near the interface. It crosses over to the exponential decay far from the interface where the pair amplitude is sufficiently small so that the nonlinear term in the GL functional becomes negligible.

At low temperatures, K^{-1} is a macroscopic length scale. For instance, in copper at 1 K, K^{-1} is of the order of 1 μm. This allows the easy observation of the proximity effect in superconductivity. This is in contrast to other ordering phenomena, such as magnetism, where the basic length scale is much shorter.

In the absence of any significant boundary resistance (which may arise either from imperfections of the boundary, such as a thin insulating layer, or from a mismatch of the normal state properties, that is, different Fermi velocities and electron effective masses), the boundary conditions at the SN interface are (de Gennes 1964)

$$F/N(0) \text{ continuous} \tag{17a}$$

$$D(dF/dx) \text{ continuous} \tag{17b}$$

The pair amplitude is depressed on the S side over a length $\xi(T)$ and has a tail on the N side over a length $K^{-1}(T)$ (see Fig.1). For this reason K^{-1} is alternatively expressed as the normal state coherence length.

The consequences of the proximity effect are (Deutscher and de Gennes 1969) (a) in zero field, a depressed critical temperature of SN bilayers, an excitation spectrum showing states within the gap if measured from the S side and gapless superconductivity if measured from the N side; (b) in strong applied fields, a reduction of the upper critical field of thin films and suppression of the surface sheath above H_{c2} in bulk superconductors; and (c) in weak applied fields, existence of a Meissner effect on the N side of the contact.

A simplified discussion of the proximity effect on the transition temperature and on the upper critical field can be made in terms of a length b defined by

$$\left|\frac{dF}{dx}\right| = \frac{F}{b} \tag{18}$$

where the value of F and its derivative are taken at the interface on the S side. According to Eqn. (18), b is equal to the distance from the interface to the intercept of the linear extrapolation of $F(x)$ (into N) with the x axis, which is why it is called the extrapolation length (Fig. 1).

In the dirty limit, and in the absence of any boundary resistance, b is given by:

$$b = (\rho_N/\rho_S)K^{-1}\coth(Kd_N) \tag{19}$$

where ρ_N and ρ_S are the normal state resistivities and d_N and d_S are the thicknesses of N and S, respectively. The critical temperature T_{cNS} of a bilayer is given by

$$T_{cNS} = T_{cS} - \frac{\pi^2}{8}\xi_S^2(T=0)\frac{1}{(d_S + b)^2} \tag{20}$$

This expression is only valid in the GL approximation, that is, when T_{cNS} is not too different from T_{cS}.

In the same approximation, the parallel upper critical field of a bilayer is given by (Orsay Group 1966)

$$B_{c2}^2 = \frac{1}{\beta}\left(\frac{2\Phi_0}{\pi d_S{}^2}\right)^2 \frac{d_S{}^2}{\xi^2(0)}(1 - t_{cNS}) \tag{21}$$

where β is a function of (b/d_S).

One of the most fundamental manifestations of the proximity effect is the existence of a Meissner effect on the N side of SN contacts. This induced Meissner effect was revealed by the study of the susceptibility of SN contacts in weak fields, which showed that a field applied parallel to the interface is screened out at a distance ρ_{scr} from the interface (Deutscher et al. 1969). This induced diamagnetism is easily identified because it is destroyed in fields that are much smaller than the characteristic fields of the S side. The screening length ρ_{scr} was found to be typically of the order of 100 nm in the conventional liquid (helium) range. If N is itself a superconductor at a temperature T_{cN}, ρ_{scr} diverges when T_{cN} is approached. This effect is felt well above T_{cN}, $\rho_{scr} = 1$ μm in zinc ($T_{cN} = 0.92$ K) at 1.6 K. If N is a "truly" normal metal, ρ_{scr} diverges as $T \to 0$. In the limit of very small applied fields and large N thickness ($d_N \gg \rho_{scr}$)

$$\rho_{scr} = K^{-1}\{\ln[K^{-1}/\lambda(0)] - 0.116\} \tag{22}$$

where $\lambda(0)$ is the value of the local penetration depth at the SN interface. As soon as the temperature is sufficiently low that $[K^{-1}/\lambda(0)] > 1$, ρ_{scr} is typically of the order of K^{-1}.

At low temperatures, as ρ_{scr} becomes larger, it is also more sensitive to the applied field:

$$\sigma_{scr}(H) = K^{-1} \ln(2H_0/H) \qquad (23)$$

where H_0 is a field characteristic of the N side, $H_0 = [\Phi_0 K/2\pi\lambda(0)]$. Notice that $H_0 \rightarrow 0$ as $T \rightarrow 0$.

These relations apply for a semi-infinite N layer. The main effect of a finite N thickness is that for $d_N \simeq K^{-1}$ and $[K^{-1}/\lambda(0)] > 1$, a first-order transition is observed at a breakdown field H_b at which the pair amplitude in N collapses and ρ_{scr} varies essentially from d_N to zero (Orsay Group 1967). A good approximation for H_b is

$$H_b = 3.8 H_0 \exp(-Kd_N) \qquad (24)$$

There are at least two important applications where the proximity effect is important. One is in the fabrication of practical Josephson devices. For LTS electronics, niobium, being a refractory metal, is a preferred choice of material for stable, cyclable devices. Unfortunately, it does not have a good insulating oxide. The solution is then to use a niobium film with a thin aluminum coating and to oxidize the surface of the aluminum. The operation of the device is based on the proximity-induced superconductivity in the aluminum coating (see *Superconducting Device Fabrication*). Another example is in the design of ultrathin filamentary wires, composed of very fine Nb–Ti filaments in a copper matrix, for ac applications. In these wires, the distance between the filaments is so small that the copper matrix is effectively superconducting; to minimize ac losses the proximity coupling has to be reduced either by separating the filaments or by suppressing the proximity coupling by alloying the copper with magnetic impurities such as manganese (see *AC Applications of Superconducting Materials*).

Special mention should be made of proximity effects where the S side is a high-T_c oxide. In this case ξ_S is very small, typically of the order of 1 nm. This makes the fabrication of high-T_c devices quite difficult, since electron tunnelling occurs into a surface region of thickness ξ_S, whose quality is hard to control. One solution consists of using a high-T_c–low-T_c (or normal metal) contact, taking care, however, to use a *high-resistivity* normal side. This is essential because of the high normal state resistivity of the oxides: as can be seen from Eqn. (19), the extrapolation length b is very small when $\rho_N \ll \rho_S$, and if b is very small the pair amplitude will be destroyed right at the interface (Fig. 1).

Proximity-coupled weak link structures and tunnel junctions have been made using thin proximity-coupled layers of silver or gold deposited onto $YBa_2Cu_3O_7$ and other high-T_c oxides. Various device heterostructures have been investigated including SNS, SNS′ and SNIS′ junctions, where S′ is usually lead or niobium (Akoh et al. 1990, Gijs et al. 1990). Properties of these junctions have been calculated within a BCS framework by Kupriyanov and Likharev (1991). Recently, proximity effects have been measured using a semiconducting layer of perovskite-type oxide (e.g., $La_{1.5}Ba_{1.5}Cu_3O_{7-y}$). It appears that the proximity decay length K^{-1} can be as large as 65 nm in this system (Tarutani et al. 1991).

The small leakage of Cooper pairs into an insulator is sufficient to weaken the pair amplitude at the interface of an HTS with an insulator (Deutscher and Muller 1987). The extrapolation length b at an SI interface is of the order of ξ_0^2/a where a is an interatomic distance: since in the oxides ξ_0 is only a few interatomic distances, b is not much larger than ξ_0 and as a rule the pair amplitude will be reduced at all surfaces and interfaces. This explains small critical currents and granular behavior in polycrystalline oxides; a disordered region at grain boundaries is sufficient to depress the pair amplitude at the surfaces of the adjacent grains. This depression results in an anomalous temperature variation of the critical current of grain boundary junctions, $I_c \propto (T - T_c)^2$ instead of the linear dependence of usual Josephson junctions (Gross et al. 1990). It also explains why I_c scales with the normal state resistance R_N of the junction as R_N^{-2} instead of the usual dependence R_N^{-1} (Deutscher and Chaudhari 1991). Both effects contribute towards reducing the critical current density in randomly oriented polycrystalline samples.

Bibliography

Akoh H, Camerlingo C, Takada S 1990 Anisotropic Josephson junctions of YBaCuO/Au/Nb film sandwiches. *Appl. Phys. Lett.* 56(15): 1487–9

Azlamazoff L G, Larkin A L 1968 Effect of fluctuations on the properties of a superconductor above the critical temperature. *Sov. Phys. Solid State* 10: 875–80

de Gennes P G 1964 Boundary effects in superconductors. *Rev. Mod. Phys.* 36: 225–37

Deutscher G 1967 Contribution a l'étude expérimentale de la supraconductivité de surface. *J. Phys. Chem. Solids* 28: 741–82

Deutscher G 1990 The potential for large critical currents in the high T_c oxides. Transport properties of superconductors. *Prog. High Temp. Supercond.* 25: 2–10

Deutscher G, Chaudhari P 1991 Scaling behavior of the critical current of grain boundary junctions. *Phys. Rev. B* 44: 4664–5

Deutscher G, de Gennes P G 1969 Proximity effects. In: Parks R D (ed.) 1969 *Superconductivity*. Dekker, New York, pp. 1005–34

Deutscher G, Dodds S 1977 Critical field anisotropy and fluctuation superconductivity in granular aluminium films. *Phys. Rev. B* 16: 3936

Deutscher G, Hurault J P, van Dalen P A 1969 Electrodynamic properties of superconducting contacts. *J. Phys. Chem. Solids* 30: 509–20

Deutscher G, Muller K A 1987 Origin of extrinsic critical currents and glass behavior in the high-T_c oxides. *Phys. Rev. Lett.* 59: 1745–7

Gijs M A M, Sholten D, van Rooy Th, Gerrits A M 1990 $YBa_2Cu_3O_{7-x}$/Ag–Al/Al_2O_3/Pb tunnel junctions based on the superconducting proximity effect. *Appl. Phys. Lett.* 67: 2600–2

Ginzburg V L 1988 Macroscopic theory for superconductors with small coherence length. *Phys. C* 153: 1617–21

Glover R E III 1971 Superconducting fluctuation effects above the transition temperature. In: Chilton F (ed.) 1971 *Superconductivity*. North-Holland, Amsterdam, pp. 3–31

Gross R, Chaudhari P, Dimos D, Gupta A, Koren G 1990 Thermally activated phase slippage in high-T_c grain boundary Josephson junctions. *Phys. Rev. Lett.* 64(2): 228–31

Kes P H, van der Beek C J, Maley M P, McHenry M E, Huse D A, Menken M J V, Menovsky A A 1992 Field suppression of the phase transition in $Bi_2Sr_2CaCu_2O_8$. *Phys. Rev. Lett.* (in press)

Kupriyanov M Yu, Likharev K K 1991 Towards the quantitative theory of the high-T_c Josephson junctions. *IEEE Trans. Magn.* 27: 2460–3

Orsay Group 1966 Strong field effects at the surface of a superconductor. In: Brewer D (ed.) 1966 *Quantum Fluids*. North-Holland, Amsterdam, pp. 26–67

Orsay Group 1967 Field dependence of the superconducting properties induced by the proximity effect: breakdown fields. *Phys. Kondens. Mater.* 6: 307–24

Saint James D, de Gennes P G 1963 Onset of superconductivity in decreasing fields. *Phys. Lett.* 7: 306–9

Salamon M B, Inderhees S E, Rice J P, Ginsberg D M 1990 Heat capacity of untwinned $YBa_2Cu_3O_7$ in magnetic fields: dimensional cross-over near T_c. *Phys. A* 168: 283–90

Serin B 1969 Type II superconductors: experiments. In: Parks R D (ed.) 1969 *Superconductivity*. Dekker, New York, pp. 925–76

Tarutani Y, Fukazawa T, Kabasawa U, Tsukamoto A, Hiratani M, Takagi K 1991 Superconducting characteristics of a planar-type HoBaCuO–LaBaCuO–HoBaCuO junction. *Appl. Phys. Lett.* 58: 2707–9

Yang H C, Finnemore D K 1984 Pair-breaking mechanisms in superconductor–normal metal–superconductor junctions. *Phys. Rev. B* 30(3): 1260–5

G. Deutscher
[Tel Aviv University, Tel Aviv, Israel]

Critical Current Density

Soon after the discovery of superconductivity by Kamerlingh Onnes in 1911 attempts were made to construct superconducting coils for the generation of high magnetic fields without dissipation. It was found, however, that when the current density in a superconductor exceeded a certain critical value the material made a transition to a resistive state. In a type I superconductor the material becomes resistive when the sum of the applied field and the self-field of the current exceeds the thermodynamic critical field (Silsbee rule). In type II superconductors the critical current density J_c is determined by the interaction of the flux line lattice with sample heterogeneity (see *Flux Line Lattice*; *Flux Pinning and the Summation of Pinning Forces*). In general, the material microstructure needs to be precisely controlled to achieve a high J_c. Much of the thrust of applied research over the years has been directed towards understanding and optimizing J_c. Initially there were fundamental doubts about the prospects for large values of J_c in high-temperature superconductors (HTSs); however, very high J_c values have now been demonstrated under certain conditions and major efforts are being made to develop processing routes for high-J_c wires and tapes (see *Oxide Superconductors: Ceramic Processing*).

Although the critical temperature T_c sets an upper limiting temperature for the possible exploitation of a superconducting material, it is the value of J_c that determines the practical prospect for real engineering application. Typically, in commercial wires, a J_c is required in the range 10^4–10^7 A cm^{-2}. Since J_c decreases with increasing temperature and applied field, any conductor will be of value for application only up to certain fields at a given temperature. Commercial materials have not found application above about two-thirds of the critical temperature and upper critical field $B_{c2}(T)$ and, in the case of niobium–titanium wire, practical design considerations sometimes favor operating at temperatures as low as 1.8 K, a reduced temperature of $T/T_c \approx 0.2$. For HTSs the pressure to operate in liquid nitrogen at 77 K places particularly stringent demands on the critical current (at 77 K, $T/T_c \approx 0.8$ for $YBa_2Cu_3O_7$) and it is likely that initial applications will be at a lower value of the reduced temperature.

The critical current density is not uniquely defined but depends on the criterion used for its measurement (Evetts 1983). Figure 1 shows a schematic of surfaces of constant electric field $E(B, T, J)$ that define J_c as a function of field and temperature. The shape of these curves is strongly materials dependent. Inhomogeneous materials show a gradual approach to zero J_c at high fields and temperatures and it is usual to define $B^*_{c2}(T)$ and T^*_c by a power law extrapolation for a given voltage criterion. The extrapolation will depend on the appropriate pinning model, for NbTi a linear extrapolation such as $1 - B/B_{c2}$ is used, while for Nb_3Sn the extrapolation is usually as $(1 - B/B_{c2})^2$, known as a Kramer plot (Kramer 1973). Because most superconducting materials exhibit a curved resistive transition, it is usual to specify a commercial material in terms of a J_c defined at a particular voltage criterion, for instance 10^{-5} V m^{-1}, together with an "n value" obtained by fitting a power law to the resistive transition (see *Resistive Transition and Flux Flow in Superconducting Materials*). Alternatively, resistive criteria of $10^{-12}\,\Omega$ m and $10^{-14}\,\Omega$ m are sometimes used for conductors or cables as they relate directly to the overall dissipation in an engineering specification. For commercial conductors, the n value is generally in the range 20–100 and varies with applied field and temperature. The practical consequence of a low n

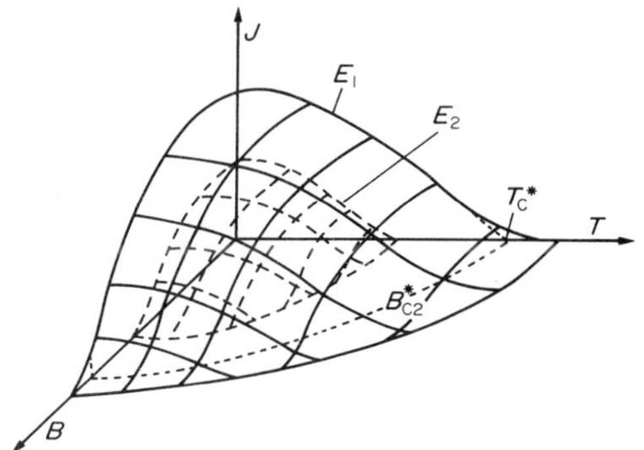

Figure 1
A schematic diagram showing surfaces of constant electric field (or voltage) in (B–T–J) space for two different fields E_1 and E_2; the dotted curve is the extrapolation to B^*_{c2} for the criterion $E_1 = E(B, T, J_c)$

value is that a commercial conductor has to be used at a maximum current density well below the J_c value quoted at $10^{-5}\,V\,m^{-1}$. Typically, for $n \sim 30$ the working current density in a magnet designed for persistent-mode operation may be restricted to less than one-half of the J_c specification at $10^{-5}\,V\,m^{-1}$.

Granular materials and HTSs can show curved resistive transitions with n values as low as two or three. Furthermore, J_c becomes negligibly small for temperatures and applied magnetic fields which place the material above a line termed the irreversibility line. Below this line the material can sustain a critical current and exhibit hysteresis and irreversibility in the motion of the flux vortex lattice (see *Critical State Model*); above the irreversibility line the material behaves reversibly and the critical current density is zero. The irreversibility line can be expressed as a phase boundary $B_{irr}(T)$ in the B–T plane. Empirically it is observed that $B_{irr}(T) \sim [1-(T/T_c)]^n$ where $n \simeq 1.5$ for $YBa_2Cu_3O_7$ and $n \simeq 3$–4 for bismuth-based oxide superconductors. The origin of the irreversibility line is still the subject of some controversy. The ability to sustain a critical flux gradient by flux pinning is certainly reduced by giant thermally activated flux creep at high temperatures and fields (see *Flux Creep Phenomena*) and the critical current density will be correspondingly reduced by an amount dependent on the chosen critical current criterion. It has also been proposed that flux lattice "melting transitions" and other possible phase changes in the flux lattice (see *Flux Line Lattice*) may have a strong influence on J_c. Experience with low-temperature superconductors (LTSs) has shown that neither flux lattice disorder nor a reduction of the shear modulus leads to an abrupt reduction of J_c in isotropic three-dimensional materials. However, Suenaga et al. (1991) have shown that both NbTi and NbSn exhibit an irreversibility line close to T_c which is probably related to flux line cutting. Since HTSs exhibit both strong anisotropy and many of the features of weakly coupled superconducting multilayers, it is possible that thermally activated vortex cutting could lead to an additional melting transition relating to the decoupling of the layers of the structure, thus drastically increasing flux mobility and reducing J_c. This constitutes an alternative mechanism for an irreversibility line; in practice, because there is no generally accepted criterion for reversibility, it is not easy to discriminate quantitatively between different models. Since HTSs cannot find application above the irreversibility line, a major issue is whether or not $B_{irr}(T)$ can be moved to higher fields and temperatures by optimizing the microstructure. This question is returned to in Sect. 4.

1. Critical Currents and Microstructure

Although the maximum theoretical supercurrent that a superconductor may carry, termed the depairing current density, is very large, most materials have J_c values that are orders of magnitude below the depairing limit. For instance, at 4 K the highest J_c values in commercial niobium–titanium approach 3% of the depairing limit, which corresponds roughly to the supercurrent density $J_c = H_c/\lambda$ for which the kinetic energy of a Cooper pair approaches the superconducting gap energy 2Δ. Here H_c is the thermodynamic critical field and λ the magnetic penetration depth. There is a parallel with the ideal strength of solids and the maximum coercive force of a magnetic material; in each case, the ideal "hardness" is limited by the nucleation and motion of some entity which interrupts the long-range order of the ideal material. For a mechanical solid this is a dislocation, for a magnetic material it is the domain wall and for a type II superconductor it is the flux line. In a hard material the microstructure is optimized so as to prevent the motion of the disruptive entity. In type II superconductors J_c depends on the pinning of flux lines or flux vortices and a material with a high J_c is termed a hard superconductor. Although there is a sound quantitative understanding of flux pinning and the summation of pinning forces in simple microstructures, the treatment of hard superconductors with very high critical currents is more qualitative (see *Flux Pinning and the Summation of Pinning Forces*). Flux pinning models fall into three distinct classes according to the state of perfection of the flux vortex lattice:

(a) class I—lightly distorted lattice (no defects),
(b) class II—distorted lattice (with lattice defects), and
(c) class III—grossly distorted lattice (polycrystalline or amorphous lattice).

A classification according to the pin strength is not adequate; the pin distribution and geometry are also crucial factors. As the applied field or temperature changes, a material can pass from one regime to another. Only samples with a dilute random distribution of weak point pins satisfy criteria for the collective pinning models of class I. A major characteristic of class II behavior is the observation of a history effect in the voltage–current characteristic that is related to a variable flux lattice defect density (Evetts and Mullock 1985). Class II behavior is associated with stronger random point pins and line pins as the lattice softens elastically with increasing applied field. The essential feature of class III behavior is the stabilization of gross lattice distortion by the pinning field. Gross flux lattice defects such as grain boundaries, normally unstable in a perfect lattice, are stabilized by the material defect structure. Samples with extended planar defects such as grain boundaries, cell walls and extended phase boundaries or precipitates show class III behavior over the entire field range. There is evidence that at applied fields sufficiently close to B_{c2} most samples make a transition to class III behavior.

In a grossly distorted flux lattice the critical current density is determined by one of two critical failure modes: pin breaking or pin avoidance. In the former case, the critical current is exceeded when the flux vortex driving force is sufficient to pull vortices from their pinning centers. The latter process depends on the little understood mechanisms of flux shear and flux cutting (see *Transport Currents and Force-Free Current Flow*). Within a bulk three-dimensional sample flux lattice shear alone cannot lead to effective pin avoidance; flux cutting or other phase-slip-related processes are also necessary and defect structures that involve double kinks or dislocation loops may be involved.

Since elementary pinning forces vary approximately linearly as $1-b$ at high fields where $b=B/B_{c2}$, a strong pinning material whose J_c depends on pin breaking will tend to exhibit a global pinning force density J_cB that also approaches B_{c2} linearly as $1-b$ (see *Flux Pinning and the Summation of Pinning Forces*). Commercial NbTi composites clearly fall into this category. On the other hand, pin avoidance by a flux shear process depends on the flux lattice shear modulus c_{66} which approaches B_{c2} as $(1-b)^2$ (Kramer 1973). Such a dependence is characteristic of commercial polycrystalline A15 bronze route filaments.

The quantitative analysis of strong pinning materials with critical currents that have been optimized empirically is difficult because the maximum critical current is generally achieved as a compromise between the extreme microstructural disorder necessary for strong pinning and the inevitable degradation of the primary parameters T_c and B_{c2} that occurs in heavily defective microstructures. In most high-critical-current materials the critical current density is spatially nonuniform and can be described by a Gaussian or normal distribution with a mean critical current and a standard deviation. For practical critical current criteria the measured J_c is several standard deviations smaller than the mean critical current density and the standard deviation of the critical current density is closely related to the n value for the resistive transition (see *Resistive Transition and Flux Flow in Superconducting Materials*).

Figure 2
Schematic of flux distortion at $k=k_{BZ}$: (a) unstable tilt distortion and (b) stable tilt distortion

A critical current density that varies as $(1-b)^2$ results in a poor high-field performance. It is of importance therefore to attempt to avoid the microstructural conditions that favor the flux cutting and shear processes leading to pin avoidance. All strong pinning potentials lead to severe flux lattice distortion; however, certain distortions are stable against shear and cutting while others promote it. The flux lattice constants exhibit dispersion $c_{ij}=c_{ij}(k)$ and soften strongly when subject to short-range periodic disturbances (see *Flux Line Lattice*). Short-range periodic distortion is of particular relevance in commercial NbTi and Nb_3Sn because their pinning fields tend to induce periodic lattice distortion. Thus, distortion near the Brillouin radius $k_{BZ}\sim\pi/a$ of the lattice can be a low-energy precursor to flux shear and flux cutting. Figure 2 shows schematically examples of flux lattice tilt distortion near k_{BZ}. The tilt distortion of Fig. 2a is characteristic of that produced by a fine grain Nb_3Sn microstructure and favors flux shear and cutting, while the stable tilt distortion of Fig. 2b is characteristic of the pinning field in commercial NbTi which consists of highly elongated subcells decorated with precipitates with the long axis of the cells orientated perpendicular to the Lorentz driving force (see *Niobium–Titanium Alloys*).

2. Critical Currents in Model Systems

Although it is not yet possible to relate microstructure to critical current quantitatively in materials with the highest critical currents, much information for the optimization of pinning microstructures can be obtained by studying simpler systems. There are a number of examples of model systems where a quantitative link between microstructure and critical current density has been clearly established (see *Flux Pinning and the Summation of Pinning Forces*). The clearest picture emerges when the density or strength of pins

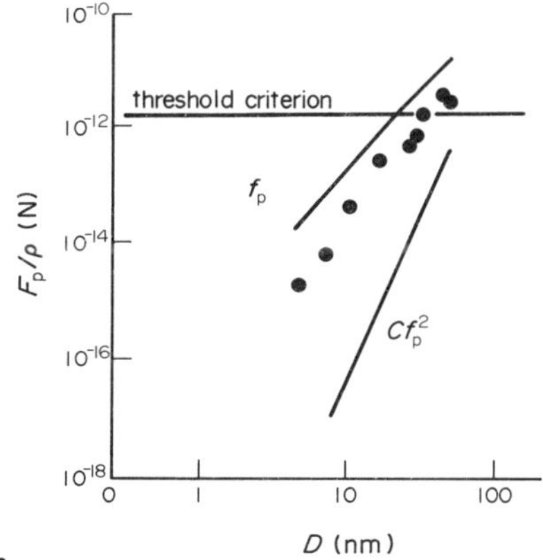

Figure 3
The specific pinning force $Q = F_p/\rho$ vs void diameter in niobium at $B/B_{c2}=0.55$, $T=4.2$ K; lines representing the direct summation $Q=f_p$ and the statistical theory $Q = Cf_p^2$ are indicated for comparison (after H. C. Freyhardt in Kramer 1978)

can be varied in a material whose B_{c2} and Ginzburg–Landau parameter κ remain unchanged. The global pinning force density $F_p = BJ_c$ can then be related to pin density and strength as a function of applied field and temperature enabling a quantitative test of pinning theories and the prediction of critical current trends in more complex materials. An example of the variation of the global pinning force density expressed as a specific pinning force $Q = F_p/\rho$ for a density ρ of voids of diameter D is shown in Fig. 3 for a wide range of void diameters. Also shown are lines representing a direct summation of the elementary pinning forces f_p and the Labusch statistical summation theory (Labusch 1969). The threshold criterion is the point at which it is predicted that the weak pinning centers cease to be effective on a single vortex picture (see *Flux Pinning and the Summation of Pinning Forces*).

The global pinning force frequently exhibits scaling behavior with reduced field $b=B/B_{c2}$, B_{c2} and κ:

$$F_p = BJ_c = S\frac{B_{c2}^n}{\kappa^m}b^p(1-b)^q$$

where the constant S is a geometrical function of the microstructure. This simple formulation fails at low flux densities and at phase lines where there is a change in the basic pinning mechanism due, for instance, to matching effects, dimensional crossover or the onset of thermally activated processes.

Extensive scaling behavior is seen when the microstructure consists of a random distribution of second-phase precipitates that are large compared with the flux vortex spacing. The pinning has been shown to be due effectively to the surface pinning of flux vortices at the sharp change associated with the extended planar surface of a precipitate. In the case of normal precipitates in an otherwise defect-free material, J_c has been shown to be proportional to the total surface area of precipitate per unit volume of material S_v (Campbell and Evetts 1972):

$$BJ_c = 0.6\ S_v B_{c2}^2 b^{1/2}(1-b)/\mu_0\kappa^3 \qquad (1)$$

Figure 4a shows the close agreement of the experimental field dependence and accurate scaling at different temperatures; Fig. 4b confirms the variation as the square of B_{c2} (Table 1 presents the data for these graphs). Investigation of a wide range of different types and sizes of precipitate indicates an inverse dependence of BJ_c on the precipitate diameter for particles with diameter greater than twice the vortex spacing. There is also a volume fraction term; for a fixed value of S_v the pinning decreases as $1 - 1.5f$ where f is the volume fraction of precipitate. This sets a limit of about 30% on the volume fraction of normal precipitate that can be incorporated into a superconducting matrix to increase J_c. Such information is, for example, important for pinning by normal Y_2BaCuO_5 precipitates in the high-T_c oxide superconductor $YBa_2Cu_3O_7$ (see *Flux Pinning and the Summation of Pinning Forces*; *Oxide Superconductors: Ceramic Processing*).

3. High Critical Current Densities in LTSs

The highest critical current densities at high fields are seen in superconductors that combine a high T_c and B_{c2}, and BJ_c frequently varies as B_{c2}^n with n between 5/2 and 3/2. Since $B_{c2} = 2\pi\Phi_0/\xi_0 l$ in the dirty limit, commercial materials are frequently ternary alloys with a high normal state resistivity and a correspondingly short electron mean free path l in the normal state. The coherence length $\xi_0 \sim hv_F/k_BT_c$ also tends to be short for the higher-T_c materials, particularly those involving transition metals where the Fermi velocity v_F is usually low. For commercial (Nb, X)$_3$Sn ternary superconductors, B_{c2} is maximum for a normal state resistivity of about 35 μΩ cm; at the maximum the Ginzburg–Landau coherence length $\xi = (\xi_0 l)^{1/2}$ is of the order of 6 nm and B_{c2} is in excess of 25 T. One consequence of a short ξ is a reduced diameter for the normal core of a vortex; as a result the core displaces a smaller condensation energy per unit length and the pinning energy for interaction with defects U_0 tends to be small. Unless the phenomenological pinning energy, or activation energy, is some 40–50 times the thermal energy k_BT, flux creep becomes a major factor reducing J_c determined on a voltage criterion. The phenomenological activation energy which may be the energy barrier associated with a "flux bundle" rather than a single vortex is denoted by U or sometimes U_b to distinguish it from the pinning energy U_0 for a single vortex interacting with an isolated defect (see *Flux Creep Phenomena*). For Nb–50 wt%Ti, $U_b/k_BT \sim 480$ at 4 K falling to 180 at 7.5 K, the latter

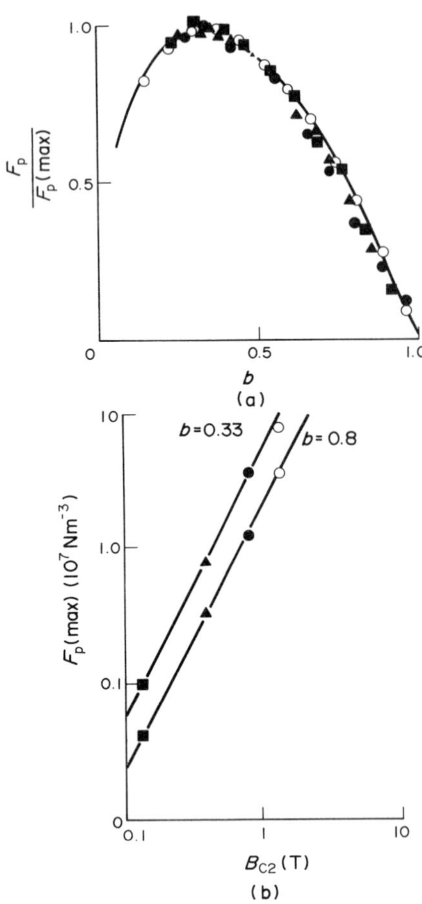

Figure 4
Scaling of the global pinning force density F_p for flux pinning by large normal precipitates: (a) variation of pinning force with reduced field for a lead–bismuth eutectic for the conditions given in Table 1, the solid line is $b^{1/2}(1-b)$; (b) variation of F_p with B_{c2} at different temperatures, the upper curve shows values of F_p at the peak, $b = 0.33$; the lower curve is for $b = 0.8$; the solid lines correspond to slope $n = 2$

Table 1
Data for lead–bismuth eutectic (see Fig. 4)

Symbol	T (K)	$t = T/T_c$	B_{c2} (T)	F_p (10^7 Nm^{-3}) $b = 0.33$	$b = 0.8$
○	4.45	0.53	1.350	7.66	3.52
●	6.19	0.72	0.815	3.40	1.22
▲	7.45	0.89	0.384	0.78	0.328
■	8.10	0.96	0.127	0.099	0.0415

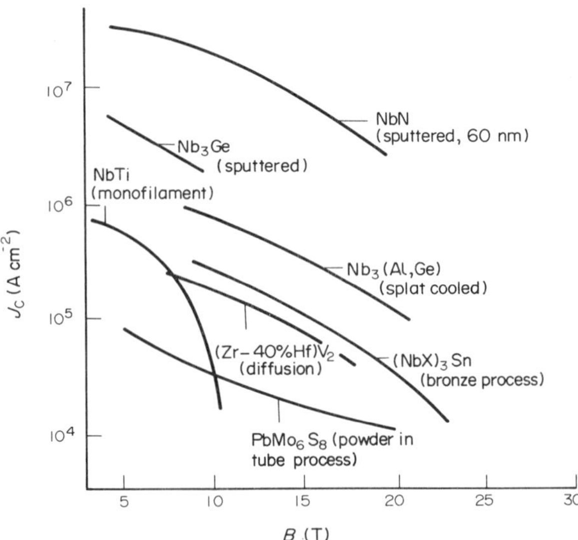

Figure 5
Variation of $J_c(B)$ in high-J_c LTS materials produced by different methods (after Flükiger 1984)

temperature corresponds to a reduced temperature for niobium–titanium that is approximately equivalent to 77 K for the HTS oxide $YBa_2Cu_3O_7$. In LTSs flux creep can be detected at 4 K in large persistent-mode magnets and becomes rapid at temperatures near T_c, but is not the major problem that is seen for HTSs.

Figure 5 shows critical current densities at 4 K for a range of high-J_c LTS materials with potential for application (Flükiger 1984). In practical commercial conductors the overall critical current density will invariably be lower because the conductor cross section has to include support material, insulation and any normal conductor required for stabilization (see *Multifilamentary Superconducting Composites*). Figure 6 shows J_c data expressed as a current density with respect to the overall conductor cross section in advanced commercial or prototype LTS multifilamentary wires. The overall J_c value alone is not a sufficient specification for commercial conductors; the n value of the resistive transition is an equally important parameter.

4. *High-Temperature Superconductors*

The prospects for early application of HTSs depend primarily on achieving a high critical current density (10^4–10^6 A cm^{-2}) in high magnetic fields. Although the oxide superconductors have a high T_c and very large values of B_{c2}, initial HTS materials were polycrystalline ceramics that exhibited granular behavior with low critical currents that were suppressed

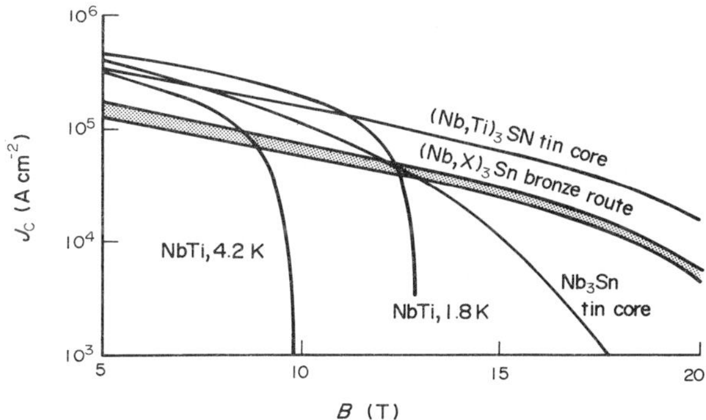

Figure 6
Comparison of *overall* $J_c(B)$ variation for different advanced commercial or prototype LTS multifilamentary conductors

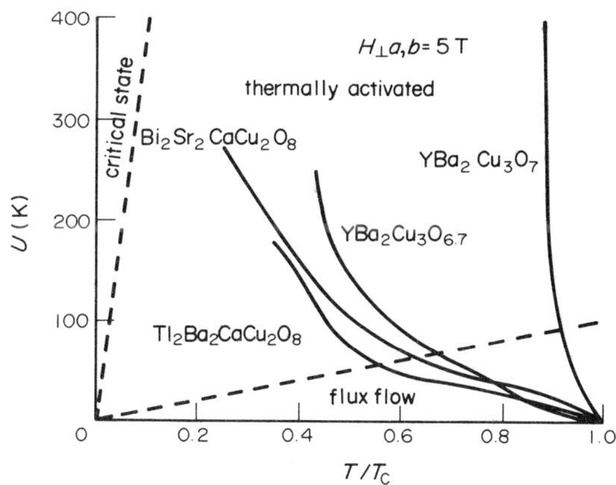

Figure 7
The activation energy U determined from the resistivity at 5 T as a function of reduced temperature for different HTS materials; the different regions for flux flow (TAFF) $U < k_BT$, flux creep $1 < U/k_BT < 40$ and the critical state $U > 40k_BT$ are indicated for $T_c \sim 100$ K (after Palastra et al. 1991)

strongly by fields of a few millitesla. However, it is now clear that most HTS single-crystal materials and epitaxial thin films can be prepared with strong flux pinning and, at low temperatures, can support very high critical current densities comparable or greater than most LTSs. The problem lies in the strong reduction of J_c with applied field as the temperature is raised, with J_c becoming negligibly small for temperatures and fields that place the material above the irreversibility line.

The variation of the irreversibility line $B_{irr}(T)$ depends on the material, the measurement technique and the critical current criterion; however, it clearly lies at lower temperatures and fields for the more anisotropic HTS materials. There is some controversy as to whether or not the irreversibility line can be moved to higher fields and temperatures by changing the defect structure of a material. If the irreversibility line depends primarily on the low value of U/k_BT, where U is here the phenomenological energy barrier for flux creep, then increasing U will shift the irreversibility line. For $U = 50$ meV, the flux creep rate at 4 K ($U/k_BT = 150$) is some 10^{60} times slower than that at 77 K ($U/k_BT = 8$). Because of the exponential dependence on U, even small increases in U or decreases in T can enormously reduce the creep rate. In principle, U may be increased without changing the J_c value calculated in the absence of thermal activation by rearranging the distribution of pinning sites so that they are grouped on vortex lines to give a larger effective U. If flux creep determines the irreversibility line, the introduction of extended planar pinning centers should be very effective in shifting $B_{irr}(T)$. If, on the other hand, the irreversibility line is primarily determined by a melting transition or by breakdown of the vortex structure by a phase slip or cutting process between weakly coupled layers in highly anisotropic materials, it is likely that changes in the microstructure or defect structure will have little effect on J_c at high temperatures and applied fields. A number of results give a clear indication that $B_{irr}(T)$ can be shifted by altering the pinning by irradiation or otherwise. In particular, it has been shown that the linear damage tracks produced by heavy ion irradiation lead to a significant shift of the irreversibility line to higher fields in $YBa_2Cu_3O_7$ (Civale et al. 1991). In the case of bismuth and titanium compounds, $B_{irr}(T)$ appears to be somewhat less dependent on pinning. Figure 7 shows the temperature dependence of U determined from resistivity data for four different compounds (Palastra et al. 1991). The boundary between the critical state and flux creep is defined arbitrarily at $U/k_BT = 40$; the boundary between flux creep and thermally activated flux flow (TAFF) is indicated for $U/k_BT = 1$ (see *Flux Creep Phenomena*).

The initially disappointing values of J_c in HTSs have improved rapidly since 1988 and significant improvements continue to be reported at regular intervals. The situation for J_c is shown at 4 K in Fig. 8 and at 77 K in Fig. 9. The figures represent an encapsulation of typical $J_c(B)$ data as of October 1991 in a fast moving field. The voltage criteria and the conditions and methods of measurement differ from material to material and the available lengths of "wire" and "tape" material vary from a few centimeters to tens of meters. It is particularly significant that epitaxial thin-film materials have such high J_c values sometimes exceeding 10% of the depairing limit. From a practical viewpoint, any isolated report of a new high for J_c whether in a thin-film or bulk material sets a new benchmark on what can, in due course, be expected of properly designed technological conductors. In polycrystalline materials there are two requirements in addition to the requirement for strong flux line pinning (see *Oxide Superconductors:*

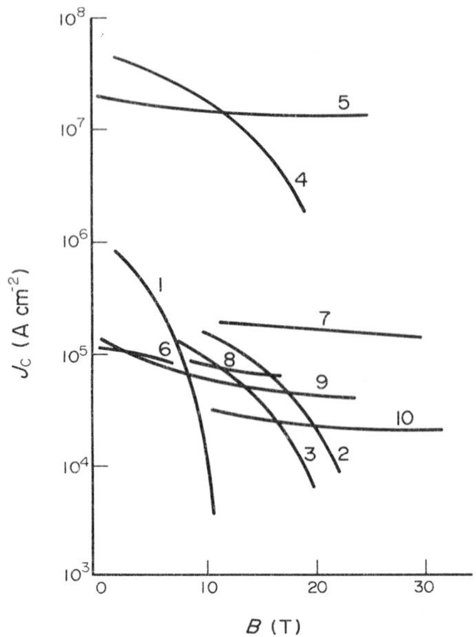

Figure 8
Typical critical current densities as functions of field for various LTS and HTS materials at 4 K: (1) NbTi; (2) $(Nb,Ti)_3Sn$; (3) Nb_3Sn; (4) NbN sputtered thin film; (5) YBaCuO sputtered thin film; (6) YBaCuO tape, melt textured; (7) $Bi_2Sr_2Ca_1Cu_2O_x$ tape cast, melt processed; (8) $Bi_2Sr_2Ca_1Cu_2O_x$, dip coated; (9) $Bi_2Sr_2Ca_1O_x$ rolled tape; and (10) $Bi_2Sr_2Ca_1Cu_2O_x$ round wire

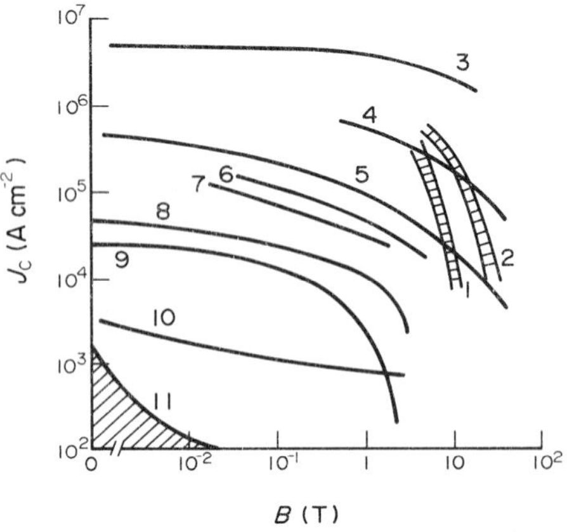

Figure 9
Comparison of current densities of various HTS at 77 K and LTS at 4 K (note the logarithmic scale for B): (1) Nb–Ti (4 K); (2) $(Nb,X)_3Sn$ (4 K); (3) sputtered YBaCuO thin film; (4) CVD-deposited YBaCuO film, $B_\|ab$ planes; (5) CVD-deposited YBaCuO film, $B_\|$ c axis; (6) powder melt processed (PMP) rod YBaCuO; (7) melt textured growth (MTG) rod YBaCuO; (8) mechanically processed tape $(BiPb)_2Sr_2Ca_2Cu_3O_x$; (9) melt processed tape $Bi_2Sr_2Ca_1Cu_2O_x$; (10) YBaCuO wire; (11) polycrystalline granular samples

Ceramic Processing). First, the individual grains have to be aligned or textured so that the layers of the crystal structure lie in the direction of the current and are misoriented by less than 2–3° across a grain boundary. Second, the grains must have a morphology that is elongated in the direction of the current flow so that the need for current to cross grain boundaries is minimized. Although many grain boundaries, particularly high-angle grain boundaries, support very low current densities or even act as low-conductance tunnel barriers, the origin of these effects in terms of boundary atomic structure or impurity segregation is not understood. It has been shown that some high-angle boundaries in $YBa_2Cu_3O_7$ can carry quite large critical currents (Babcock et al. 1990) and, for reasons that are still not clear, the $Bi_2Ca_2Sr_1Cu_2O_x$ compound appears not to suffer a high-angle grain boundary problem at low temperatures.

See also: Flux Pinning and the Summation of Pinning Forces; Multifilamentary Superconducting Composites; Resistive Transition and Flux Flow in Superconducting Materials; Transport Currents and Force-Free Current Flow

Bibliography

Babcock S E, Cai X Y, Kanser D L, Larbalestier D C 1990 Weak-link-free behaviour of high-angle $YBa_2Cu_3O_{7-x}$ grain boundaries in high magnetic fields. *Nature* 347: 167–9

Campbell A M, Evetts J E 1972 Flux vortices and transport currents in type II superconductors. *Adv. Phys.* 21: 199–428

Civale L, Marwick A D, Worthington T K, Kirk M A, Thompson J R, Krusin-Elbaum L, Sun Y, Clem J R, Holtzberg F 1991 Vortex confinement by columnar defects in $YBa_2Cu_3O_7$ crystals: enhanced pinning at high fields and temperatures. *Phys. Rev. Lett.* 67: 648–51

Evetts J E 1983 The characterization of superconducting materials—conflicts and correlations. *IEEE Trans. Magn.* 19: 1109–19

Evetts J E, and Mullock S J 1985 Collective flux pinning and history effects in a defective flux vortex lattice. In: Matsushita T, Yamafuji K, Irie F (eds.) 1985 *Proc. Int. Symp. Flux Pinning and Electrodynamic Properties in Superconductors.* Matsukuma, Fukuoka, Japan, pp. 94–7

Flükiger R 1984 Advanced superconducting materials. *J. Phys. (Paris), Colloq.* 45: 365–72

Kramer E J 1973 Scaling laws for flux pinning in hard superconductors. *J. Appl. Phys.* 44: 1360–70

Kramer E J 1978 Fluxoid–defect interactions in irradiated superconductors. *J. Nucl. Mater.* 72: 5–33

Labusch R 1969 Calculation of the critical field gradient in type II superconductors. *Cryst. Lattice Defects* 1: 1–16

Palastra T T M, Batlogg B, Schneemeyer L F, Waszczak J V 1991 The role of anisotropy in the dissipative behaviour of high-temperature superconductors. *Phys. Rev. B* 43: 3756–9

Suenaga M, Ghosh A K, Youwen Xu, Welch D O 1991

Irreversibility temperature of Nb_3Sn and NbTi. *Phys. Rev. Lett.* 66: 1777

J. E. Evetts
[University of Cambridge, Cambridge, UK]

P. H. Kes
[Kamerlingh Onnes Laboratory, Leiden, The Netherlands]

Critical State Model

Many of the properties of type II superconductors may be explained in terms of the critical state model. This model describes how the spatial distribution of magnetic flux within a sample changes when the applied fields or currents change. The model therefore describes the magnetization variation and, since the flux distribution is related uniquely to the transport current density, it also describes the internal patterns of current flow (see *Transport Currents and Force-Free Current Flow*). The model was introduced independently by Bean (1962) and London (1963) and is generally referred to as the Bean model or Bean–London model. Although originally the model seemed to arise from the "sponge" or filamentary model for type II superconductors and is now commonly viewed as a consequence of the mixed state or vortex state, it is essentially a result only of the voltage–current (*VI*) characteristic and is not dependent on any microscopic picture. The critical state phenomena seen in type II superconductors are an early example of the now topical subject of self-organized criticality (Bak and Chen 1991).

According to the Bean model any change in external magnetic field induces screening currents at the surface of an arbitrarily shaped sample (the qualitative pattern of currents is similar to high-frequency screening currents induced in copper). As soon as the current locally exceeds the critical current density the material enters the resistive state or flux flow state (see *Resistive Transition and Flux Flow in Superconducting Materials*) and magnetic flux penetrates into the sample at a rate determined by the *VI* characteristic. The resulting disturbance of the flux at a greater distance from the surface induces a current and the currents continue to spread into the sample to a sufficient depth to screen the surface changes from the interior or, if this is not possible, until the entire sample supports the critical current density and is said to be in the critical state. Although the current spreads to fill the entire sample cross section the change in magnetic flux density may be small particularly if the sample is in a large applied field. If the *VI* characteristic is very sharp the critical current density J_c will be well defined; more commonly the characteristic is curved, in which case the critical currents induced by changes in applied fields will gradually decay by flux creep (see *Flux Creep Phenomena*).

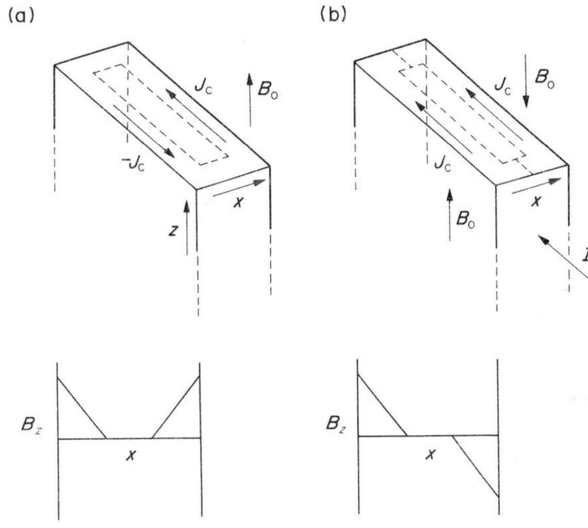

Figure 1
Schematic of the flux profile and current flow for (a) a slab in an applied field B_0 and (b) a sample with similar dimensions carrying a total current I sufficient to generate a field B_0 at the surface of the sample

Once a sample has been put into the critical state reversing the field change leads to a second set of negative currents penetrating from the surface. Thus, all regions of the sample are in one of three states. There may be zero current, $+J_c$ or $-J_c$. Application of a current from external sources instead of a change in the applied field is treated by considering the magnetic field due to the current as an additional applied field (see Fig. 1).

In terms of flux lines the model describes the way an increase in external field will push more flux into the surface of the sample. The flux is prevented from moving into the interior by pinning centers and a flux gradient builds up until the magnetic pressure overcomes the pinning forces. The critical state is a static force balance between the driving force $\boldsymbol{B} \times \boldsymbol{J}$ and the pinning forces in the material, where $\boldsymbol{J} = \nabla \times \boldsymbol{H}$ is the transport current and $\boldsymbol{H}$ is the thermodynamic field (see *Transport Currents and Force-Free Current Flow*). In many cases J_c can be assumed constant, but the model applies equally to situations in which J_c is a function of B. The force balance equation

$$|(\boldsymbol{B} \times (\nabla \times \boldsymbol{H}))| = BJ_c(B) \qquad (1)$$

is known as the *critical state equation* and its solutions determine the distributions of flux, current and magnetization in a sample. The solutions depend on sample shape and orientation with respect to the applied field and on the previous history of current and field changes. In addition J_c may vary spatially depending on the local flux density and any sample inhomogeneity; in anisotropic materials J_c will also depend on the direction of current flow. It is usual to treat the sample as a nonmagnetic medium so that $B = \mu_0 H$ and assume that current flow is perpendicular to the flux lines. The validity of this is discussed in more detail in Sect. 2.

The critical state equation is a powerful general expression of the critical state model that is entirely adequate for most situations (Campbell and Evetts 1972). Since the introduction of the model in 1962 there have been many attempts to modify or extend the critical state model. Most of these treatments essentially introduce an analytic form for $J_c(B)$, for example $J_c(B) = \alpha(B_0 + B)^{-1}$. These developments of the critical state model have no generality and should be regarded as fitting routines. The variation of $J_c(B)$ depends on the pinning microstructure and its functional form can change in different field and temperature regimes (see *Flux Pinning and the Summation of Pinning Forces*); $J_c(B)$ has no fundamental relevance to the critical state model which may be applied equally to low- and high-pinning materials, to inhomogeneous materials and to granular materials.

There are limits in which the critical state model is difficult to apply and also fundamental limitations to the model for small displacements of flux and small samples. These limits will be discussed in Sect. 2.

1. Critical State in Slabs, Cylinders and Spheres

1.1 Critical Currents and Magnetization

The magnetization of a type II superconductor is one of the commonest measured quantities and is often used to derive J_c when electrical contacts cannot conveniently be made to a sample (see *Measurements in Superconducting Materials*). The critical state equation is readily solved for samples in the form of cylinders or rectangular bars. At high applied fields the assumption of a constant J_c is a good approximation for homogeneous isotropic samples and simple expressions are given in Sect. 1.2 relating J_c to the magnetization per unit volume $\boldsymbol{M}$.

The total moment of a sample of volume V is

$$\boldsymbol{m} = \boldsymbol{M}V = \frac{1}{2}\int (\boldsymbol{r} \times \boldsymbol{j})\, dV \qquad (2)$$

where $\boldsymbol{j} = (1/\mu_0)\nabla \times \boldsymbol{B}$ is the local average current density including magnetization currents (see *Transport Currents and Force-Free Current Flow*). The magnetization can be viewed as a sum over the moments of elemental current loops, $\delta \boldsymbol{m}_i = \boldsymbol{I}_i \times \boldsymbol{S}_i$ where $\boldsymbol{I}_i$ is the current flowing round loops $\boldsymbol{S}_i$, thus

$$\boldsymbol{m} = \boldsymbol{M}V = \sum_i \boldsymbol{I}_i \times \boldsymbol{S}_i \qquad (3)$$

In long samples of arbitrary cross section with zero demagnetization $\mu_0 \boldsymbol{M}$ is equal to the difference between the applied induction and the average internal induction.

In slab geometry with the applied field in the direction of the z axis and the plane of the slab in the yz plane the critical flux density gradient for $B = \mu_0 H$ is $dB/dx = \pm \mu_0 J_c(B)$. For constant J_c this gives a linear flux profile. Any solution can have a constant B added to it and remain valid, provided the correct value of $J_c(B)$ is used. At high applied fields the applied field is much greater than the field due to the currents and a Bean model calculation with a constant J_c is usually very accurate; Fig. 2 shows a number of examples of the critical state flux profile for different current-applied field trajectories. The solutions of the critical state equation for constant J_c in long samples with arbitrary cross section can be modelled simply and visualized in a very practical way by use of a "sand model" (Campbell and Evetts 1972); examples for elliptical and square cross sections are illustrated in Fig. 3. It can be seen that if a long sample does not have axial symmetry the critical state flux front will in general form a cusp when it has penetrated a distance equal to the minimum radius of curvature of the sample surface. Until the critical state reaches this point it forms smooth contours of constant spacing, which are also flow lines for the current, with a flux gradient that is everywhere constant. After forming a cusp the critical state will exhibit a line of discontinuity extending inwards from the cusp at which the current changes its direction sharply. If the critical current is isotropic the line of discontinuity will exactly bisect the angle formed by each flow line. When the critical current is anisotropic the spacing between flow lines will no longer be constant and cusps with corresponding lines of discontinuity will form even in a cylindrical sample (see Sect. 1.3).

1.2 Isotropic Samples

Table 1 lists expressions for the magnetization of common sample shapes assuming that the material is in the full critical state and that J_c is constant. The magnetization is in A m^{-1}, the critical current density is in A m^{-2} and the sample dimensions are in meters. Results in the literature are often expressed in mixed units and there is potential confusion depending on whether or not the magnetization is taken as M[emu cm^{-3}] or $4\pi M$[G]. If the expressions in Table 1 are multiplied by a factor $4\pi/10$ and J_c is in A cm^{-2} and sample dimensions are in centimeters then the resultant magnetization is $4\pi M$[G].

In practice the magnetization in Table 1 is added to, or subtracted from, the equilibrium reversible magnetization M_{rev} that would be exhibited by the superconductor in the absence of flux pinning (see *Magnetic Properties of Superconductors*). Figure 4 indicates the magnetization loop and flux profiles including M_{rev}. A magnetization loop may be used to determine critical current densities and estimate the ac

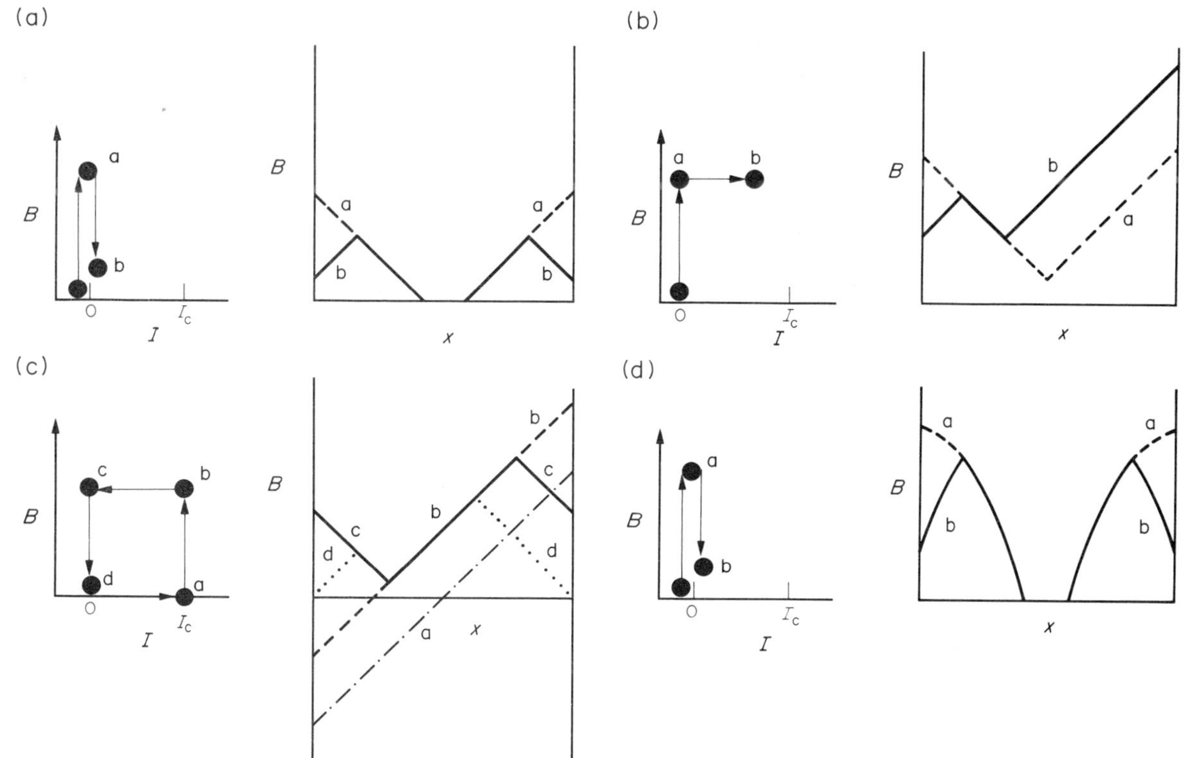

Figure 2
Schematic of the critical state flux profile; the different current–applied field trajectories are indicated in the *IB* diagrams: (a–c) Bean model with J_c constant; (d) as (a) but $J_c(B)$ decreasing strongly with increasing B

losses under different conditions (see *AC Losses in Superconducting Materials*). If $J_c(B)$ is a slowly varying function of B, the use in the expression of Table 1 of the arithmetic mean of the values of the critical state $M(B)$ for increasing and decreasing applied fields gives an excellent approximation for $J_c(B)$. For normal particle pinning with $J_c \simeq b^{-1/2}(1-b)$ where $b = B/B_{c2}$ the approximation is accurate to better than 1% for $b > 0.4$ (Coote et al. 1972).

When $J_c(B)$ is a strong function of B, as is often the case at low fields, more complex curve fitting and iteration is necessary to extract $J_c(B)$ from magnetization measurements; in such cases the measurement and analysis of various major and minor loop magnetization traces provides additional diagnostic information (Cave et al. 1991). The starting point is usually the selection of an appropriate parameterized expression for $J_c(B)$. Empirical expressions that are commonly used are $J_c(B) = \alpha(B_0 + B)^{-1}$, sometimes referred to as the Kim critical state model (Kim et al. 1964, Hulbert 1965, Chen and Goldfarb 1989) and also $J_c(B) = \alpha B^{-n}$ (Campbell and Evetts 1972, Yeshurun et al. 1990). Analytic expressions may be derived for different prominent features of the magnetization curve in terms of α and n. For example, for a cylindrical sample in an axial field, the slope of the magnetization curve $M(B)$ at the point of inflection in increasing applied field is simply $n/2$ while the extremum of the curve occurs for $M = t\mu_0 B$ where

$$t = \frac{(n+1)}{(2n+3)}\left[1 + \left(\frac{n}{n+2}\right)^{1/2}\right] \qquad (4)$$

For $n = 1$, $t = 0.632$ while for $n = 2$, $t = 0.731$, independent of both the size of the sample and the strength of the pinning. These results are in the approximation $M \gg M_{rev}$ and are generally valid for all but the thinnest or most reversible samples.

When the applied field is parallel to the short axis of a sample, demagnetization corrections (self-field effects) become important at low applied fields; the critical state for disk-shaped samples has been computed by Däumling and Larbalestier (1989). Demagnetization effects are also important in the case of transverse cylinders or spheres. For small flux penetrations the surface field is similar to that in the diamagnetic limit and the corresponding flux penetration can be approximated by that for a slab in the same applied field. For larger penetrations numerical solutions are necessary.

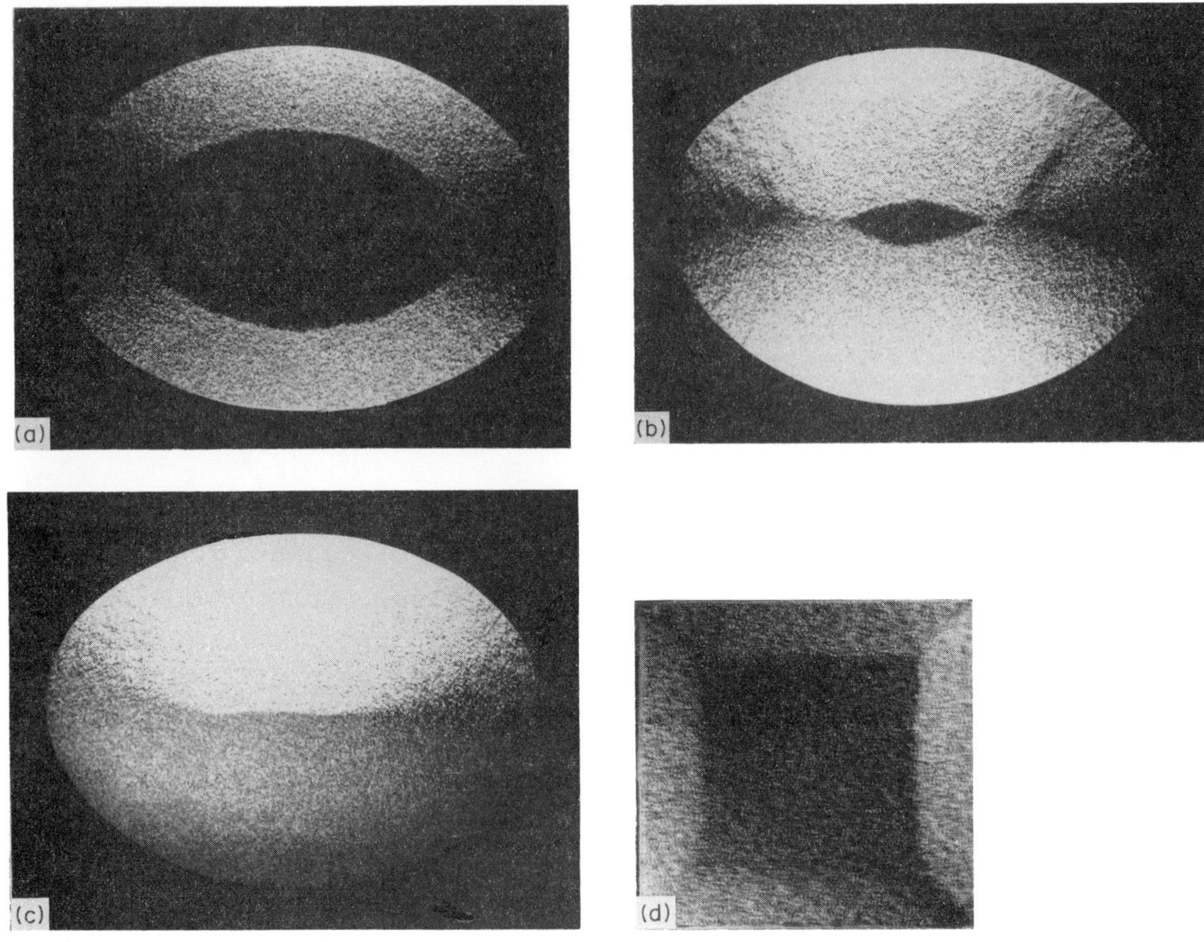

Figure 3
Sand analogue of the critical state model for long samples obtained by filling a container with the required cross section, flattening the top and then lowering the walls: (a–c) three stages for an elliptical cross section, the cusps at the centers of curvature can be clearly seen; (d) square cross section

When the flux distribution becomes curved the full driving force expression $\boldsymbol{B} \times (\nabla \times \boldsymbol{H})$ in Eqn.(1) involves terms arising from the line tension of flux lines. An example is a round wire carrying a current I in zero applied field. For constant J_c and $I_c = \pi a^2 J_c$ the azimuthal flux density as a function of the radial distance r from the center of the wire is

$$B_\theta = (\mu_0/2)\, J_c r[1 - (a/r)^2(1 - I/I_c)] \qquad (5)$$

This expression is of limited practical application since the low values of B mean that J_c is very dependent on B and the effects of M_{rev} should also be taken into account. However, it can be used accurately in large applied fields to calculate the θ component of B and to calculate losses (see *AC Losses in Superconducting Materials*). The flux vortices are also curved in the case of rectangular slabs or films carrying transport currents in a perpendicular applied field. The critical state current distribution in wires and tapes of different shapes has been determined by Norris (1971) using conformal transformations. For an infinite slab the flux gradient is zero and it is solely the curvature of the flux lines that provides the driving force. In a thick slab the critical state spreads from the top and bottom surfaces as the current is increased (Campbell and Evetts 1972). If the slab is initially cooled in a perpendicular applied field the flux lines at the center of the slab remain straight until the critical state penetrates to the center plane when the current reaches the critical current. Thin films, however, are usually thinner than the pinning penetration depth λ_{pin}

Table 1
The magnetization for a full critical state established in the samples indicated for different applied field directions

Sample shape and field orientation	M (A m^{-1})
Cylinder diameter $2a$	
$B\parallel$ axis	$J_c a/3$
$B\perp$ axis	$4J_c a/3\pi$
Infinite slab, thickness d	
$B\parallel$ face	$J_c d/2$
Square section bar ($d\times d$)	
$B\parallel$ axis	$J_c d/6$
$B\perp$ face	$J_c d/4$
Sphere radius a	$3\pi J_c a/32$
Disk, $2a$, thickness d	
$B\perp$ face	$J_c a/3$
Rectangular section bar ($b\times d$ with $b>d$)	
$B\parallel$ axis	$\frac{3b-d}{12b}J_c d$

(see *Measurements in Superconducting Materials*); the critical state model begins to break down in this limit and the flux profile can no longer be viewed as spreading inwards from the surface of the sample. There can be no significant changes in B so the current must be provided by a local distortion of each vortex; the nature of the critical state in this geometry is considered further in Sect. 2.

In rotating machines and torque magnetometers the change in applied field consists of a rotation in the field direction. The Bean model can also be used for changes in field direction. The change in applied induction $\delta\boldsymbol{B}$ which is to be screened is perpendicular to $\boldsymbol{B}$, so the induced magnetic moment is also perpendicular to the instantaneous applied field direction.

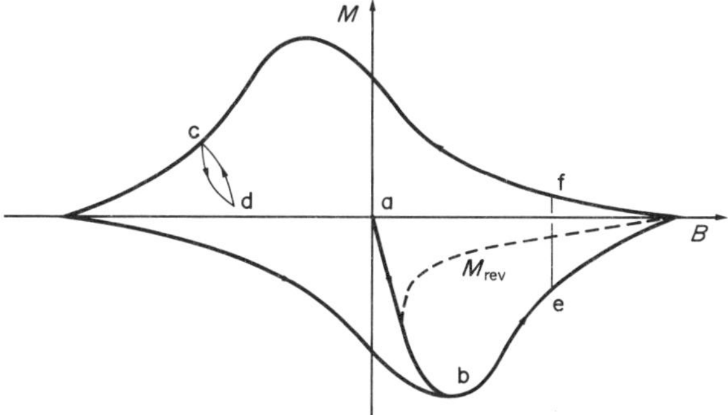

Figure 4
A typical hysteretic magnetization loop including the reversible magnetization M_{rev}. The figure also indicates the initial curve from zero induction (ab), a minor loop excursion typically experienced by a superconductor under low amplitude ac conditions (cdc) and a pair of magnetization values used to extract $J_c(B)$ information as described in the text (ef)

1.3 Anisotropic Samples

The critical current density is anisotropic in any material with an anisotropic distribution of pinning centers; most fabrication techniques impart anisotropy to J_c particularly those involving rolling or wire drawing. The high-T_c oxide superconductors exhibit strong J_c anisotropy that has its origin in anisotropy of the intrinsic superconducting parameters (see *Anisotropy in Superconducting Materials*). The treatment of the critical state is simplest in rectangular or square section samples when J_c is anisotropic (Gyorgy et al. 1989, Sauerzopf et al. 1990). The solution shows lines of discontinuity in current flow somewhat similar to the isotropic case, however, the current flow contours are no longer equally spaced and the tangent of the angle of the line of discontinuity is the ratio of the critical current densities for current flow along each of the sides of the rectangle. Figure 5 gives three characteristic examples for the full critical state for the samples with different J_c anisotropies. The magnetization for the full critical state is given in each case. In other cases the full angular variation of J_c is required for a quantitative solution; a cylindrical rod, for instance, will display cusps and a line of discontinuity in current flow similar to that found for a sample with elliptical cross section in the isotropic limit (see Fig. 3). An example of the characteristic pattern of lines of discontinuity for an angular variation of J_c that has a minimum somewhere between the x and y directions is given in Fig. 5c. When the critical current is anisotropic the best procedure is to make magnetization measurements on samples of different shapes and sizes to extract scaling information.

2. *Deviations from the Critical State Model*

In certain regimes the critical state model becomes difficult to apply; in other regimes the basic concepts are seen to break down. An example of the former is when the applied field is low, of order H_{c1}, the lower critical field. In this regime the reversible magnetization is large and $\mu_0\boldsymbol{B}=\boldsymbol{H}+\boldsymbol{M}_{rev}$ defines the thermodynamic field $\boldsymbol{H}$. The transport current density $\nabla\times\boldsymbol{H}$ is very different from the total current density and the flux gradient is not simply related to the pinning forces. A second regime where the critical state model is not obviously applicable is when there is a "longitudinal" current component parallel to $\boldsymbol{B}$. For large samples the flux structure can approach a *force-free configuration* in which case $\boldsymbol{B}\times\boldsymbol{J}=0$; suggestions have been made for extensions to the critical state

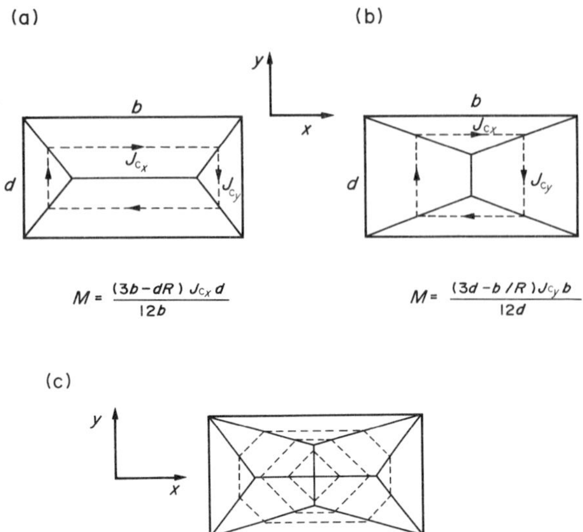

Figure 5
The critical state for a long rectangular sample with anisotropic J_c: the pattern of characteristic lines of current discontinuity depend on whether $R(d/b)$ is greater or less than unity where $R = J_{cx}/J_{cy}$ is the ratio of critical current density in the x and y directions: (a) $R < 1$, $R(d/b) < 1$; (b) $R > 1$, $R(d/b) > 1$ (the expression for the full critical state magnetization is given in each case); and (c) general angular variation of the critical current with a maximum J_{cx}, a lower J_{cy} and a minimum J_c at some intermediate angle (the dashed lines are lines of constant B)

model to include flux changes when the current is not perpendicular to B (see *Transport Currents and Force-Free Current Flow*). Longitudinal currents occur not only when currents are applied but also in simple magnetization experiments. If a cylinder or thin film is at an angle to an applied field, currents are generated with a component parallel to the field. In a magnetization experiment the sample has to be aligned to an accuracy of the thickness divided by the length of the sample in order to detect the true parallel magnetization. One way round this is to cut a film into narrow strips.

A second source of deviation from the critical state model is thermally activated flux creep (see *Flux Creep Phenomena*). When the effective activation energy for flux creep is less than about $40k_BT$ deviations from the low-temperature critical state model begin to occur. The field and current profiles within a sample become time dependent, however, if the flux creep can be expressed in terms of a modified *VI* characteristic the essential framework of the critical model remains.

A major limitation to the application of the critical state model occurs for small-amplitude displacements of the flux vortices or for small samples. The model assumes that as soon as a vortex moves it experiences the full pinning force. In practice, for small displacements the restoring force is linear and reversible and the critical state is not achieved until the flux lines move a characteristic interaction distance d. Typical values of the interaction distance are in the region of one-quarter of a vortex spacing. The effect is to add an exponential tail to the Bean model flux profile. At the leading edge of the critical state profile the tail extends a distance λ_{pin} known as the *pinning penetration depth* (see *Measurements in Superconducting Materials*). Similar deviations from the critical state occur at any sharp discontinuities in the critical state profile where the current reverses or changes direction although the rounding is not always as great as λ_{pin}. Figure 6 shows the modified flux profiles that occur. They are found by solving the flux conservation equation $dy/dx = -b/B$ and the force balance equation $B\,db/dx = -\mu_0 F(y)$. Here $F(y) \simeq BJ_c(1-\exp(-y/d))$ is the restoring force per unit volume for a flux displacement y. The applied field is $B+b$ where b is the incremental field whose change is tending to establish a critical flux gradient $\mu_0 J_c = dB/dx$. Unless $y > d$ the critical state will not be fully established. Setting b constant throughout the cross section gives an upper bound estimate for the surface flux vortex displacement $y = -ba/B$ for a sample with half width a. Thus for the case of 10 μm particles or filaments in a 1 T field with an applied field ripple of 10 mT the surface flux motion is 1 nm. Since the vortex spacing is 50 nm at this field, it is clearly impossible to create a critical state in samples of this size with an applied ripple amplitude of only 10 mT.

The distortion of the critical state is particularly severe for films of thickness approaching λ_{pin} in perpendicular applied fields. Figure 7 illustrates the flux profile for a film for different values of t/λ_{pin} where $2t$ is the film thickness.

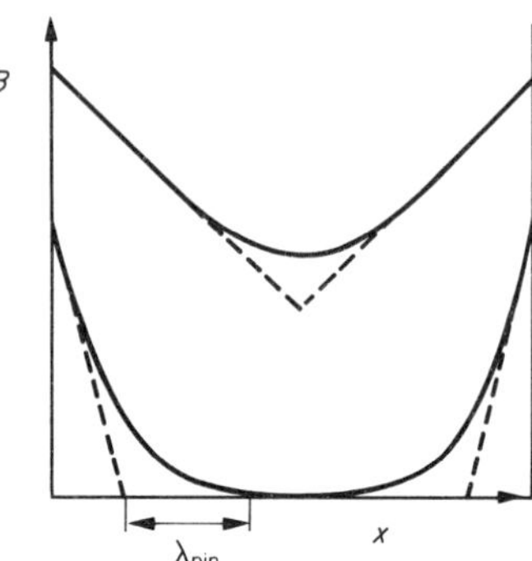

Figure 6
Deviations from the critical state model for low applied fields (lower curves) and for high applied field (upper curve); the deviation extends over distances of up to λ_{pin}

3. Surface Barriers, Inhomogeneous Samples and Granular Materials

The surface of a sample may pin flux lines and its treatment can be regarded as a particular example of an inhomogeneous material (Campbell and Evetts 1972). The material is regarded as being covered with a layer of stronger pinning material with an increased flux gradient over the bulk of the material. In fact, in some wire samples a layer of increased pinning is actually introduced during processing. The composite material then exhibits two different critical state regions with different critical flux profiles and the analysis proceeds as before. When the inhomogeneity is more general the flux profile may be calculated using the relevant critical state equation for each region with appropriate boundary conditions.

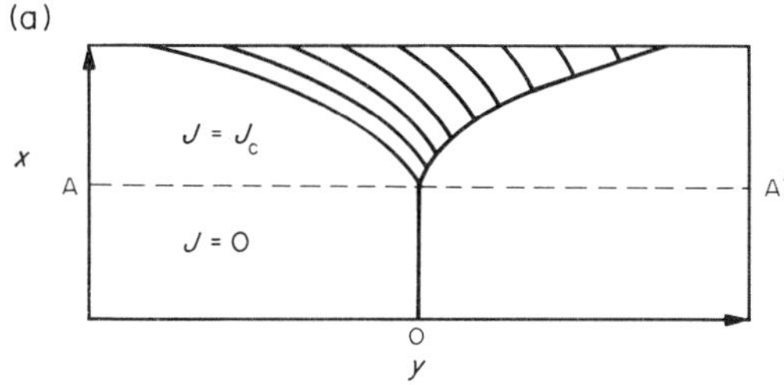

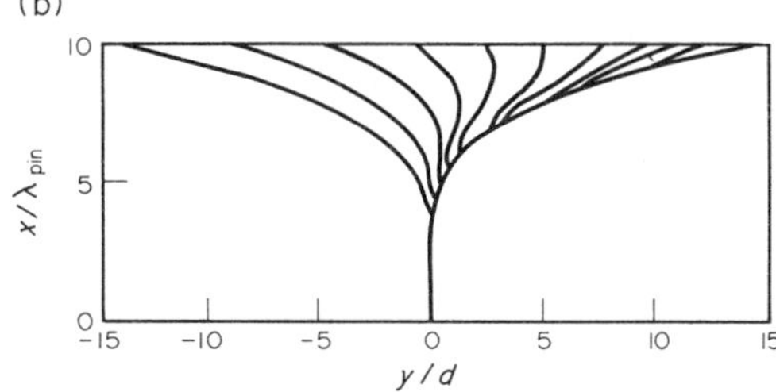

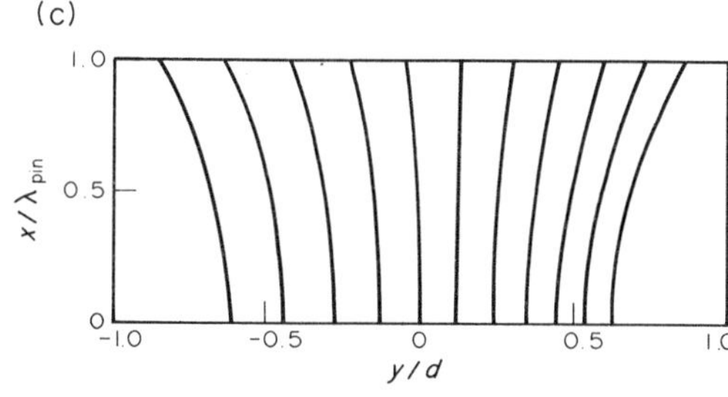

Figure 7
The critical state indicated by the field direction in samples of different thickness expressed as a ratio t/λ_{pin} where $2t$ is the film thickness and d is the interaction distance. The applied field is applied in the x direction, perpendicular to the film and the current is changed in $0.1\,I_c$ steps from $+I_c/2$ to $-I_c/2$: (a) profiles according to the critical state model, the current flow has penetrated to AA′; (b) profiles for $t = 10\lambda_{pin}$, slight departure from the critical state model and (c) $t = 2\lambda_{pin}$, flow almost uniform, breakdown of the critical state model

It can be shown that a shunted Josephson junction (see *Josephson Effects in Weak Links*) obeys the same equations as a single pinning center. A two- or three-dimensional array of weak links behaves as an array of pinning centers and can be treated using the critical state model. When this is applied to granular materials (see *Granular Superconductors*) the treatment depends on the size and superconducting characteristics of the grains making up the body. For very small grains the critical state cannot be established within the grains and the current distribution depends only on the weak links between grains. For larger grained granular materials the magnetic properties of the grains become important. At low applied fields they are perfectly diamagnetic which leads to an effectively reversible magnetic component. Since it is linear we can redefine H by $\mu\mu_0 H = B$ where as usual B is the average flux density in the material. Here μ is the effective permeability which is something like the volume fraction of normal material. However, it should be defined by measurement of the magnetic properties rather than calculated from an assumed microstructure. If J_c is the critical current density of the weak links averaged over many grains, then in slab geometry for constant J_c the critical state solution is $H = J_c x$ and $B = \mu\mu_0 J_c x$. This treatment depends on the grains being perfectly diamagnetic. Once significant penetration of the grains takes place the situation becomes much more complex with a coupled secondary critical state within the grains leading to trapped flux and a wide range of history effects (Evetts and Glowacki 1988).

Bibliography

Bak P, Chen K 1991 Self-organized criticality. *Sci. Am.* 264: 46–53

Bean C P 1962 Magnetization of hard superconductors. *Phys. Rev. Lett.* 8: 250–3

Campbell A M, Evetts J E 1972 Flux vortices and transport currents in type II superconductors. *Adv. Phys.* 21: 199–421

Cave J R, Critchlow P R, Lambert P, Champagne B 1991 Calculation of magnetic flux profiles and deduction of critical current densities for type II superconductors. *IEEE Trans. Magn.* 27(2): 1379–82

Chen D-X, Goldfarb R B 1989 Kim model for magnetisation of type II superconductors. *J.Appl. Phys.* 66: 2489–500

Coote R I, Evetts J E, Campbell A M 1972 Flux line pinning by large normal particles in type II superconductors. *Can. J. Phys.* 50: 421–7

Däumling M, Larbalestier D C 1989 Critical state in disc shaped superconductors. *Phys. Rev. B* 40: 9350–3

Evetts J E, Glowacki B A 1988 Relation of critical current irreversibility to trapped flux and microstructure in polycrystalline $YBa_2Cu_3O_7$. *Cryogenics* 28: 641–9

Gyorgy E H, van Dover R B, Jackson K A, Schneemeyer L F, Waszczak J V 1989 Anisotropic critical currents in $Ba_2YCu_3O_7$ analyzed using an extended Bean model. *Appl. Phys. Lett.* 55: 283–7

Hulbert J A 1965 Magnetisation of a cylindrical hard superconductor. *J. Phys. D* 16: 1657–66

Kim Y B, Hempstead C F, Strnad A R 1964 *Rev. Mod. Phys.* 36: 43

London H 1963 Alternating current losses in superconductors of the second kind. *Phys. Lett.* 6: 162–5

Norris W T 1971 Calculation of hysteresis loss in hard superconductors: polygonal section conductors. *J. Phys. D* 4: 1358–64

Sauerzopf F M, Wiesinger H P, Weber H W 1990 Anisotropic current flow and demagnetisation corrections in the Bean model. *Cryogenics* 30: 650–5

Yeshurun Y, McElfresh M W, Malozemoff A P, Hagerhorst-Trewhella J, Mannhart J, Holtzberg F, Chandrashekhar G V 1990 Effect of a geometrical length scale on remanent magnetization and critical currents in YBaCuO and BiSrCaCuO crystals. *Phys. Rev. B* 42: 6322–31

J. E. Evetts and A. M. Campbell
[University of Cambridge,
Cambridge, UK]

D

Demagnetizing Factors

Demagnetizing (or demagnetization) factors are used to correct data on particular magnetic samples to give results intrinsic to the magnetic material. The demagnetizing correction adjusts the applied field H_a by the demagnetizing field H_d to yield the internal magnetic field strength

$$H = H_a + H_d$$

Magnetic susceptibility, permeability and the shape of the magnetization curve as a function of field are affected by the demagnetizing correction. The demagnetizing field H_d is related to the magnetization M of the sample by the geometric demagnetizing factor N:

$$H_d = -NM$$

Sometimes the symbol D is used for N (see *Measurements in Magnetic Materials*). In SI units, $0 \leqslant N \leqslant 1$; in cgs units, $0 \leqslant N \leqslant 4\pi$. The demagnetizing correction is important in materials with large magnetic susceptibilities (or permeabilities) or in samples whose relative dimension parallel to the applied field is small.

This article examines theoretical demagnetizing factors for ellipsoids of revolution and right circular cylinders. The applied field is assumed to be uniform and along a principal axis. Empirically, an "effective" demagnetizing factor of a sample of arbitrary geometry at a given state of magnetization may be deduced from the measured external susceptibility $\chi_{ext} = dM/dH_a$ provided that the internal susceptibility $\chi = dM/dH$ for the material is known:

$$N = \chi_{ext}^{-1} - \chi^{-1}$$

Demagnetizing factors should not be confused with the process of reducing the remanent magnetization of samples, similarly known as demagnetization.

1. *Ellipsoids*

The source of the demagnetizing field is usually taken to be fictitious magnetic poles (sometimes termed free poles) on the surface of a magnetized specimen. In ellipsoids, the poles are distributed in such a way that the fields H_a, H_d and H, and the magnetization M, are uniform. For ellipsoids of revolution (spheroids), N_z for the axis of symmetry (longitudinal or polar axis) is a function of the aspect ratio γ of the ellipsoid and independent of susceptibility χ (Osborn 1945, Stoner 1945):

$$N_z = (1-\gamma^2)^{-1}[1-\gamma(1-\gamma^2)^{-1/2}\cos^{-1}\gamma] \qquad (\gamma<1,\ \text{oblate})$$

$$N_z = \tfrac{1}{3} \qquad (\gamma = 1,\ \text{sphere})$$

$$N_z = (\gamma^2-1)^{-1}[\gamma(\gamma^2-1)^{-1/2}\cosh^{-1}\gamma - 1] \qquad (\gamma>1,\ \text{prolate})$$

Here, γ is the ratio of the longitudinal (polar) axis to the transverse (equatorial) axis. These formulae occasionally appear in different, but algebraically equivalent, forms in terms of the ratio of the long axis to the short axis, equal to γ^{-1} for oblate spheroids or in terms of the eccentricity, equal to $(1-\gamma^2)^{1/2}$ for oblate spheroids and $(\gamma^2-1)^{1/2}/\gamma$ for prolate spheroids.

For the general ellipsoid,

$$N_x + N_y + N_z = 1$$

where the subscripts x, y and z refer to the three semiaxes. For ellipsoids of revolution, with the z axis equal to the axis of symmetry,

$$N_x = N_y = \tfrac{1}{2}(1 - N_z)$$

Representative values of N_z and N_x are listed in Table 1. In the limit of small γ an oblate spheroid approximates a circular film with $N_x = N_y \simeq \frac{1}{4}\pi\gamma - \gamma^2$ and $N_z \simeq 1 - \frac{1}{2}\pi\gamma + 2\gamma^2$.

2. *Cylinders*

There are two types of demagnetizing factors commonly used for cylinders. The magnetometric demagnetizing factor N_m is the ratio $-\langle H_d\rangle_v/\langle M\rangle_v$, where $\langle\ \rangle_v$ indicates an average over the volume of the cylinder. The fluxmetric (or ballistic) demagnetizing factor N_f is the ratio $-\langle H_d\rangle_s/\langle M\rangle_s$, where $\langle\ \rangle_s$ indicates an average over the center plane of the cylinder. Magnetometric factors are used with mag-

Table 1
Longitudinal and transverse demagnetizing factors N_z and N_x as functions of γ for ellipsoids of revolution

γ	N_z	N_x
0	1	0
0.25	0.7036	0.1482
0.5	0.5272	0.2364
1	0.3333	0.3333
2	0.1736	0.4132
4	0.07541	0.4623
∞	0	0.5

netometers that sense the entire sample volume, such as vibrating-sample magnetometers, SQUID magnetometers and ac susceptometers. Fluxmetric factors are used when magnetization is measured ballistically, with a short search coil closely wrapped around the center of a long sample.

The demagnetizing factors N_m and N_f depend on γ and χ. Susceptibility χ is assumed to be constant, independent of both H and position in the sample. For nonlinear or hysteretic materials, N may be taken to be a function of H, with the corresponding susceptibility approximated by $\chi(H) = dM/dH$. With H_a along the cylinder axis, M and H_d are both nonuniform except in two cases. When $\chi = 0$, M is uniform. The approximation $\chi = 0$ is used for saturated ferromagnets, diamagnets and paramagnets. When $\chi \to \infty$, H_d is uniform and equal to $-H_a$. The condition $\chi \to \infty$ applies to soft ferromagnetic materials. When $\chi = -1$, $(M + H_d)$ is uniform and equal to $-H_a$. That is, flux density $B = \mu_0(H_a + H_d + M) = 0$, as for superconductors in the shielding state.

2.1 Zero Susceptibility

The magnetometric demagnetizing factor for a uniformly magnetized cylinder ($\chi = 0$) with ratio of length to diameter γ is (Joseph 1966)

$$N_{mz} = 1 - [4/(3\pi\gamma)]\{(1+\gamma^2)^{1/2}[\gamma^2 F(\kappa) + (1-\gamma^2)E(\kappa)] - 1\}$$

where $F(\kappa)$ and $E(\kappa)$ are the complete elliptic integrals of the first and second kind of modulus κ, which is defined by

$$\kappa = (1+\gamma^2)^{-1/2}$$

As is true for ellipsoids (Brown 1962), $N_{mx} + N_{my} + N_{mz} = 1$ and $N_{mx} = N_{my} = \frac{1}{2}(1 - N_{mz})$. Several values of $N_{mz}(0)$ and $N_{mx}(0)$ are listed in Table 2. The fluxmetric (ballistic) demagnetizing factor is (Joseph 1966)

$$N_{fz} = 1 - (2/\pi)(\gamma/k)[F(k) - E(k)]$$

where the modulus k is defined by

$$k = (1 + \tfrac{1}{4}\gamma^2)^{-1/2}$$

Table 2
Axial and radial magnetometric demagnetizing factors N_{mz} and N_{mx} as functions of γ for cylinders with $\chi = 0$, ∞ and -1

γ	$N_{mz}(0)$	$N_{mx}(0)$	$N_{mz}(\infty)$	$N_{mx}(\infty)$	$N_{mz}(-1)$	$N_{mx}(-1)$
0	1	0	1	0	1	0
0.25	0.6346	0.1827	0.5712	0.1618	0.6764	0.2136
0.5	0.4745	0.2628	0.4111	0.2371	0.5258	0.2928
1	0.3116	0.3442	0.2590	0.3154	0.3692	0.3669
2	0.1819	0.4091	0.1409	0.3829	0.2341	0.4237
4	0.09835	0.4508	0.06635	0.4319	0.1361	0.4596
∞	0	0.5	0	0.5	0	0.5

Table 3
Axial fluxmetric (ballistic) demagnetizing factors N_{fz} as functions of γ for cylinders with $\chi = 0$ and ∞

γ	$N_{fz}(0)$	$N_{fx}(\infty)$
0	1	1
0.5	0.4221	0.3769
1	0.2322	0.2275
2.5	0.06544	0.09260
5	0.01889	0.03988
10	0.004927	0.01530
25	0.0007981	0.003766
50	0.0001999	0.001211
∞	0	0

For the fluxmetric demagnetizing factors, $N_{fx} + N_{fy} + N_{fz} < 1$. Although they are seldom used, several values of $N_{fz}(0)$ are listed in Table 3.

2.2 Nonzero Susceptibilities

Axial and radial magnetometric demagnetizing factors N_{mz} and N_{mx} for $\chi \to \infty$ and $\chi = -1$ may be deduced from published values of electric and magnetic polarizability of perfectly conducting cylinders as described by Chen et al. (1991). They are listed in Table 2. For $\chi > 0$, $N_{mx} + N_{my} + N_{mz} < 1$, and for $\chi < 0$, $N_{mx} + N_{my} + N_{mz} > 1$. Fluxmetric demagnetizing factors N_{fz} for $\chi \to \infty$ have been evaluated precisely by Templeton and Arrott (1987) (Table 3). The radial fluxmetric factors N_{fx} are rarely needed. A complete treatment of N_{mz} and N_{fz} for cylinders, as functions of γ and χ, is given by Chen et al. (1991).

See also: Design with Magnetic Materials; Measurements in Magnetic Materials; Measurements in Superconducting Materials; Shields, Magnetic

Bibliography

Brown W F Jr 1962 *Magnetostatic Principles in Ferromagnetism.* North-Holland, Amsterdam, pp. 49–53; 83–7; 187–92

Chen D-X, Brug J A, Goldfarb R B 1991 Demagnetizing factors for cylinders. *IEEE Trans. Magn.* 27: 3601–19

Joseph R I 1966 Ballistic demagnetizing factor in uniformly magnetized cylinders. *J. Appl. Phys.* 37: 4639–43

Osborn J A 1945 Demagnetizing factors of the general ellipsoid. *Phys. Rev.* 67: 351–7

Stoner E C 1945 The demagnetizing factor for ellipsoids. *Philos. Mag.* (7th Series) 36: 803–21

Templeton T L, Arrott A S 1987 Magnetostatics of rods and bars of ideally soft ferromagnetic materials. *IEEE Trans. Magn.* 23: 2650–2

R. B. Goldfarb
[National Institute of Standards and Technology, Boulder, Colorado, USA]

Design with Magnetic Materials

A typical magnetic design may pass through several phases involving innovation, analysis and modelling, and experimental verification. From the perspective of electrical characterization, magnetic devices may be graphically represented using scales of volt seconds vs current (λ vs i), hence they are described as generalized inductors. The same unifying format (λ vs i) is used in this article to describe actual inductors, transformers without output loading and electromechanical transducers in which a mechanical output is obtained at the changing airgap. The λ vs i electrical characteristic curve may be obtained by systematically translating the B vs H magnetic characteristic which represents the design choice of the selected magnetic material. The translation of the horizontal and vertical axes is by core dimensions and turns according to Ampère's law and Faraday's law, respectively. The dominant properties of the magnetic material may be approximated by a piecewise-linear model. The analysis and modelling phase of the design process is given in detail for inductors and transformers. Basic physical principles are discussed for rotating machines.

1. General Considerations

1.1 The Nature of Design

It is convenient to define four main phases of the design process:

(a) a creative or innovative phase in which new ideas are formulated into a novel approach to be explored;

(b) a phase of analysis and modelling in which the implications of the novel approach are made quantitative and specific (carrying out analysis and modelling is the heart of the design process, as reviewed in this article);

(c) a verification phase in which the new device is built and evaluated experimentally in order to prove the concept and/or to confirm the specifications; and

(d) a possible product engineering phase in which the device may undergo redesign from perspectives such as cost reduction, ease of manufacture or ease of repair.

In reviewing these four steps, it is clear that a firm understanding of fundamental principles is a useful background for an ability to innovate (step (a)) and so this article emphasizes basic and fundamental concepts. The analysis and modelling phase of working out the design details is an area of design activity that is undergoing major change. It is becoming common practice to use computer programs to assist with the selection of core geometry and materials, specification of wire sizes and windings placement, and determination of losses and of temperature rise. The experimental step of proving the concept is particularly relevant for designs that are centered on new ideas. Alternatively, some design projects that involve the routine implementation of well-established principles may require validation of specifications rather than proof of concept. For instance, large power transformers are usually custom designed to meet specific requirements and are produced in quantities of one. Often a great quantity of data exists from a history of similar designs. The design process of such products is evolutionary rather than revolutionary so that steps (b) and (d) are merged.

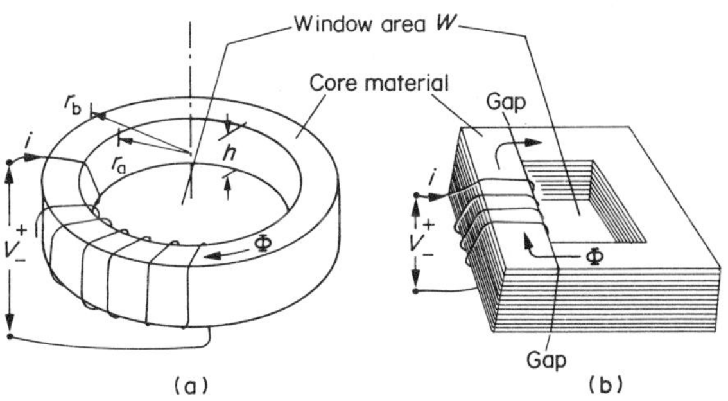

Figure 1
Examples of magnetic cores wound with n turns of wire (windings are shown schematically): (a) toroidal core without airgap and (b) core of C–I laminations with airgaps

1.2 Electrical Properties of Magnetic Devices

Before considering procedures and criteria for design with magnetic materials, it is appropriate to consider the nature of the device being designed.

First, it is important to recognize that the end product of design with magnetic materials is described electrically as a generalized inductance. That is, the electrical description has units of volt seconds vs current, whereas a simple resistor is described by voltage vs current. In contrast to the general class of resistive devices, which may be either linear or nonlinear, the generalized inductance of magnetic materials invariably has only a limited range of linearity. Clearly of overriding significance, this perspective underlies the following discussion of Figs. 1 and 2.

A closely related concept is the view of the design process as a translation of the magnetic core properties of B vs H into the proportional electrical properties of λ vs i. Two example magnetic cores are shown in Fig. 1. Figure 2 shows core properties and illustrates the general mechanics of the translation concept. The translation is carried out according to Eqns. (1) and (3) and the accompanying text in Sect. 2.1. Section 2.2 models the dominant properties of B vs H

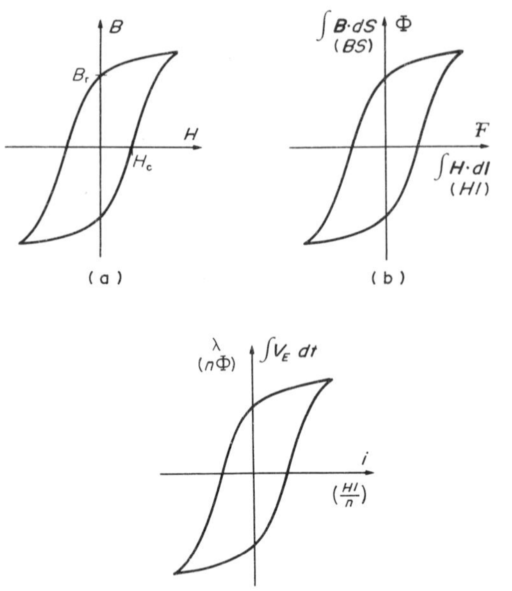

Figure 2
Illustration of design mechanics for ideal core geometry: (a) bulk property of material, (b) specific core shape and (c) turns of wire

in order to give a perspective for the creative process as well as for modelling and analysis. The translation concept is applied in Sect. 3 to the design of an inductor.

The previous paragraph considers the general electrical property of λ vs i of a single set of n turns of wire on a magnetic core. In extending our perspective to transducer design, we may now add a second set of turns and redirect our interest to the conversion of energy between two systems. The transducer is defined as a transformer in Sect. 3 where the core and windings are immobile and input energy and output energy are defined electrically at the respective windings. Alternatively, a transducer with moveable components may convert electrical energy into mechanical energy or vice versa. In either case the input energy parameters are no longer a simple characterization of the device itself, but are strongly affected by whether energy is supplied by the device output. Thus, a relatively small amount of energy is required to characterize an unloaded transducer of high efficiency.

A transducer of the transformer type transforms electrical energy by means of the magnetic system to which both input and output windings are coupled. An electromechanical transducer converts energy between three systems: electrical, magnetic and mechanical. A special case of electromechanical transducer develops its magnetic field by using a permanent magnet rather than an electromagnet. There is considerable interest in the potential of battery-driven actuators and motors for a great variety of applications in automatic portable equipment.

2. Basic Principles

To introduce the elementary principles of design and to define terms, an example of a core with windings is considered. Figure 1 shows two magnetic cores wound with current-carrying wires, illustrative of a wide range of magnetic products. If electrical signals V and i are applied with the polarity shown, a magnetic flux Φ will increase in the direction indicated according to the right-hand rule. The flux path of the toroidal core has no airgap, which is a necessary condition for the practical realization of specific magnetic properties such as extremely high permeability. Toroidal cores are available in several materials: ferrites, powdered iron or powdered permalloy, and tape-wound cores built up of many wraps of one continuous strip of magnetic metal alloy.

The C core of Fig. 1b illustrates a built-up stack of laminations punched from sheet core material, usually an iron alloy. The C–I configuration has a cost advantage in that the coil may be wound separately before the core is assembled. The flux path crosses two airgaps that tend to dilute the effect of the core itself, so that the volt second vs current properties at the winding terminals are also affected by the gap dimensions. Generalizations of C-core geometry include the E core, which is essentially a composite of two C cores with the winding usually on the center leg, and the pot core, which is like an E core of revolution.

The design of static magnetic components involves four sets of decisions in the process of meeting the design requirements and of working within the fundamental limitations of magnetics:

(a) choice of core material;
(b) specification of core geometry, configuration, size and airgap (if any);
(c) selection of winding configuration, number of turns n, wire size, insulation, wire placement details and shielding (if any); and
(d) choice of final packaging, terminals, mounting and potting (if any).

The sequential decisions (a), (b) and (c) also correspond directly to Figs. 2a–c.

Only cores with single windings, as in Fig. 1, are considered here in developing the basic ideas of magnetic design. In treating inductor design concepts, Sect. 3 also includes considerations of geometrical design and temperature rise that are also applicable to transformers. A transformer results from adding a second winding to either core of Fig. 1. The electrical properties measured at the transformer primary winding reflect not only the core properties but also the properties of the electrical load that may be connected to the secondary winding of the transformer.

2.1 Fundamental Field Equations

Magnetic design is based on field theory, circuit theory and the physics of materials. Basic field equations for design with magnetic cores include Ampère's law:

$$\oint \boldsymbol{H}.d\boldsymbol{l} = I_{\text{enclosed}} = ni = \mathscr{F} \tag{1}$$

Where the closed-line integral path links n turns of wire, each carrying current i, and where $\boldsymbol{H}$ is the magnetic field strength on the path. Faraday's law states that

$$V_E = -\oint \boldsymbol{E}.d\boldsymbol{l} = n\frac{d\Phi}{dt} = \frac{d\lambda}{dt} \tag{2}$$

where E is the electric field strength on the path, whose integral is the induced voltage V_E, and $d\Phi/dt$ is the time derivative of the magnetic flux enclosed by the path. Equation (2) defines, for example, the induced voltage at the terminals of the coil in Fig. 1. It is convenient to define $\Delta\lambda = n\Delta\Phi$ as the total change of flux linkage at the winding, then to integrate Eqn. (2) to obtain

$$\Delta\lambda = n\Delta\Phi = \int V_E dt \tag{3}$$

It now remains to combine the foregoing two analytical steps by considering the constitutive relation of how magnetic flux density B depends on H and therefore how λ depends upon I. For a wire-wound coil of cross section S and a flux path mean length l, then $B = \Phi/S$ and $H = ni/l$. Both B and H are average quantities that, in principle, can also be defined by spatial derivatives.

Figure 2 shows a B–H hysteresis curve representing a specific material and illustrates the concept that similar curves exist for flux vs potential and for volt seconds vs current. This concept is valid for gapless toroidal cores as in Fig. 1a. The three graphs of Fig. 2 correspond to the sequential design steps of choosing a material, choosing the core geometry and deciding the number of turns. The horizontal and vertical axes of the graphs are scaled in accordance with the toroidal core dimensions and the number of turns, as implied by Eqns. (1) and (3), respectively.

The total induction B consists of a free-space component and a material component:

$$B = \mu_0 H + J \tag{4}$$

where the magnetic constant of free space is $\mu_0 = 4\pi \times 10^{-7}\,\text{H m}^{-1}$ and the magnetic material property is indicated as an induction J, sometimes called magnetic polarization. Magnetization curves for permanent-magnet materials are often given as J vs H instead of B vs H as in Fig. 2a.

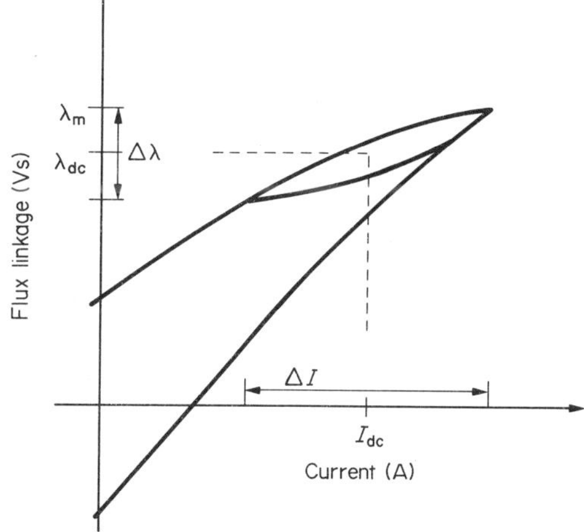

Figure 3
Magnetization curve for hysteretic inductor with unipolar current

There are many core configurations for which the foregoing treatment is incomplete because the core has an airgap in the flux path, as in the C core of Fig. 1. The effect of a short airgap of length l_g may be shown by a tilting or shearing of the characteristic (see Fig. 2b or 2c) by the magnetic potential Bl_g/μ_0.

The increase in magnetic potential $\mathscr{F}$ may also be written as ΦR_g, where $R_g = l_g/\mu_0 S$ is the reluctance of the airgap. The airgap reluctance R_g is added to the reluctance of the core material R_m, thereby decreasing the total core permeance $P = 1/(R_m + R_g)$. Permeance is defined as $\Phi/\mathscr{F}$, which has the units of slope in Fig. 2b. The shearing effect for an ideal inductor core is shown in Fig. 3.

The underlying equations for such magnetostatic calculations may be found by substituting Eqn. (4) into Maxwell's equation $\nabla.\boldsymbol{B} = 0$ to give

$$\mu_0 \nabla.\boldsymbol{H} = -\nabla.\boldsymbol{J} = \rho_m \tag{5}$$

where ρ_m is a volume density of bound magnetic poles. The field may be calculated using the magnetostatic Coulomb's law:

$$d\boldsymbol{H}(r_2) = \hat{\boldsymbol{a}}_R \frac{dQ_m}{4\pi\mu_0 R^2} \tag{6}$$

where $\hat{\boldsymbol{a}}_R$ is the unit vector of $\boldsymbol{R}$, the vector distance from the source point r_1 to the field point r_2. The point source dQ_m may be defined as either a surface pole or a volume pole:

$$dQ_m = \boldsymbol{J}.d\boldsymbol{s} \quad \text{or} \quad dQ_m = \rho_m dV \tag{7}$$

Equation (6) is also a basic starting point for various magnetostatic calculations, which in this article includes demagnetizing factors and models for permanent magnets. The basic physical phenomenon

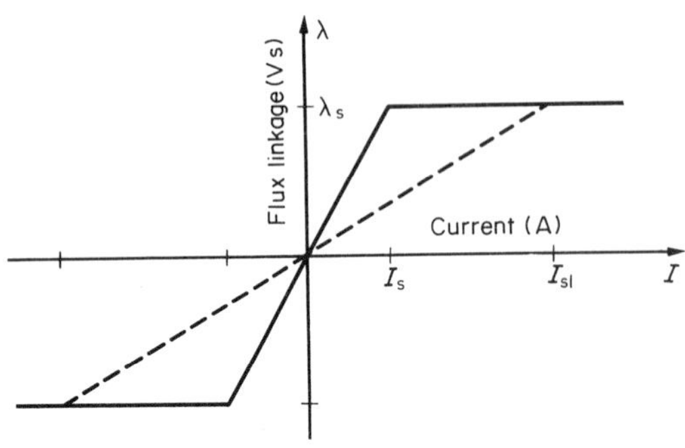

Figure 4
Magnetization curve for an ideal inductor; dashed line shows effect of a larger airgap in trading L for I_{max}

of a magnetic pole distribution at a material–free space interface gives rise to different models that depend on the relative dimensions of the material and free space.

In Fig. 1b the free space is a "short airgap" in a C core across which a magnetic potential is developed. A short airgap is usually modelled as a magnetic circuit element, reluctance, through which flux Φ flows to develop a magnetic potential $\mathscr{F}$. The effect of adding airgap reluctance is to decrease circuit permeance P and thereby to decrease the inductance L. This effect is shown graphically in Fig. 4 and is discussed further in Sect. 3.1. The change of permeance may generate a mechanical force in a transducer and, finally, the airgap reluctance may define the operating point for a permanent magnet.

The concept of demagnetizing factors (see Section 2.3) is most applicable to simple shapes of magnetic specimens that are isolated in free space (for other cases see *Design with Magnetic Materials: Computer Simulation*).

Some fundamental equations for transducers describe their conversion of energy between the three systems: electrical, magnetic and mechanical. Turning again to the generalized inductance curve λ vs i of Fig. 2c, it may be observed that electric energy

$$E = \int i \, d\lambda \tag{8}$$

is defined as an area with dimensions given by the λi product. Similarly, magnetic energy $E = \int \mathscr{F} \, d\Phi$ is defined in Fig. 2b as an area with units of the $\Phi\mathscr{F}$ product. Furthermore, mechanical energy $E = \int F dx$ has units of force times displacement Fx or of torque times angular displacement $T\theta$ for rotating systems.

In Figs. 2b and 2c, the areas enclosed by the hysteresis loops represent energy loss in joules per cycle of excitation, which if multiplied by frequency in hertz yields power loss in watts (see *Power Loss in Magnetic Materials*). The lossy curves of Fig. 2 may be contrasted with the ideal inductance curve of Fig. 4, in which lossless energy storage is

$$E = \tfrac{1}{2}\lambda_s I_s = \tfrac{1}{2} L I_s^2 \tag{9}$$

This expression defines an area in the first quadrant as the area left of the magnetization curve. A moveable airgap, as in Fig. 4, thus defines a change of energy that may be stored in the magnetic field. A governing equation for energy conversion results from the definition of force as the gradient of energy

$$F = -\nabla E \tag{10}$$

Applying Eqn. (10) to Eqn. (9) may define a force as proportional to dL/dx.

A current flowing in a magnetic field experiences a force per unit length

$$dF = I \, dl \times B \text{ or } F = BIl \tag{11}$$

where force, field and current are all mutually perpendicular. Equations (10) and (11) are applied to electromechanical transducers in Sect. 4.

2.2 Piecewise-Linear Model

Four properties of a general magnetization curve are the saturation limitation of induction B, the slope of the characteristic, its double-valued static width and the common tendency of the B–H loop to become wider as the frequency of measurement is increased. One approach to magnetic design uses a simple piecewise-linear approximation to each of these four effects, defined as parameters J_{sat}, permeability μ, coercivity H_c and dynamic width parameter g_e, respectively. Such definitions may lead to analytical relations between Eqns. (1) and (3), rather than only empirical relations.

A square-hysteresis-loop material may be modelled approximately as $H = H_c \times$ (sign of dB/dt), where the parenthesis multiplier of ± 1 provides counterclockwise traversal around the hysteresis loop. Linear material may be modelled ideally as intrinsic induction proportional to applied field strength through the magnetic susceptibility χ:

$$J = \mu_0 \chi H \tag{12}$$

by which Eqn. (4) reduces to

$$B = \mu_0 H(1 + \chi) = \mu_0 \mu_r H = \mu H \tag{13}$$

where μ_r is the relative permeability. For an alloy core, the dynamic increase in hysteresis-loop width due to eddy currents may be modelled very approximately as $\Delta H = g_e dB/dt$. The condition for nonsaturation $|J| \leqslant J_{sat}$ is required for these models. The condition of core saturation implies that its flux can no longer increase, which from Eqn. (2) suggests that V_E is zero. This concept may be represented by a parallel switch that shorts out the core model. Electrical components that represent the ideal magnet parameters (current source, inductor, resistor and saturation switch) may be derived using the concepts of Fig. 2.

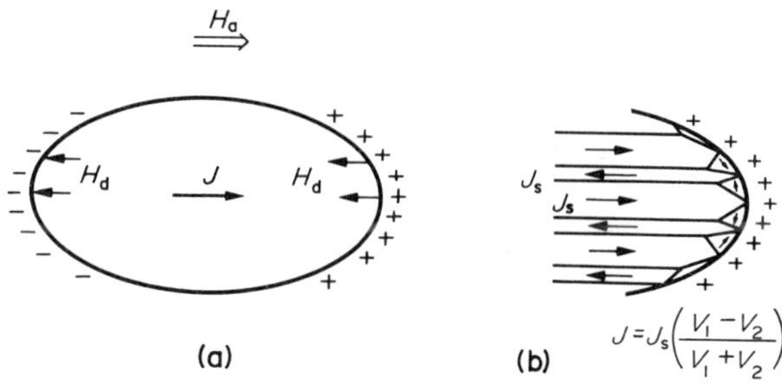

Figure 5
Response of an isolated permeable specimen to applied field H_a: (a) continuum model for J and (b) definition of J as average over discrete domains

2.3 Demagnetizing Factors

The hysteresis curves of Fig. 2 and the piecewise-linear model of hysteresis loop properties in Sect. 2.2 both assumed that the cores provide a closed magnetic flux path without airgap. The existence of even a short airgap (see Fig. 3) may be enough to dominate or obscure the intrinsic magnetic properties. This section describes the other extreme of magnetic design, in which the magnetic specimens of interest may be separated from each other by large airgaps or may be single isolated particles not in a core configuration.

When an H field is applied to magnetize an isolated permeable specimen of finite dimensions, its induction J must become zero at each end. The fact that J becomes zero in the direction of the applied field means that its spatial derivative has a value in Eqn. (5), which is a mathematical model for the existence of magnetic poles at the extremities of the sample. Such a condition is shown in Fig. 5, in which the magnetic poles at the ends give rise to a demagnetizing field H_d inside the sample. The specimen shown in Fig. 5 is drawn as an ellipsoid because this shape possesses significant magnetostatic properties of field uniformity. An ellipsoidal material of intrinsic susceptibility χ, if immersed into a previously uniform applied field H_a, will become uniformly magnetized throughout to a value

$$J = \frac{\mu_0 \chi H_a}{1 + N_d \chi} \tag{14}$$

and will also develop a uniform demagnetizing field

$$H_d = -N_d M = -N_d J/\mu_0 \tag{15}$$

where N_d is called the demagnetizing factor (see *Demagnetizing Factors*). We note that in Eqn. (14) the effective χ is reduced from the intrinsic value by the denominator factor $1 + N_d\chi$. In other words, the applied field H_a must be larger by H_d than the net internal field required to magnetize a toroidal sample of the same material. Any ellipsoid possesses three mutually orthogonal demagnetizing factors which sum to unity as $N_{dx} + N_{dy} + N_{dz} = 1$. Equation (14) may be derived by combining Eqn. (15) with

$$J = \mu_0 \chi H_{in} \qquad \text{where} \qquad H_{in} = H_a + H_d \tag{16}$$

The demagnetizing factor is determined from the geometry of the ellipsoid; its minimum value lies in the direction of the longest axis and thereby defines the easiest direction for the sample to be magnetized, which is the direction of minimum magnetostatic energy.

Figure 5b illustrates the domain nature of ferromagnetism in which the net induction J, represented mathematically in Eqn. (12) as a continuum variable, actually occurs physically as a volume average over discrete domains that are each magnetized to the saturation value J_s in alternate directions. Evidently the poles distribute as a smooth function although they are derived from a discrete domain configuration (see *Domains and Domain Walls*).

3. Inductors and Transformers

Inductors and transformers are similar in that both devices use low-coercivity magnetic cores with turns of wire. An inductor may be used as a low-loss, current-limiting impedance that isolates a voltage source from its load, as a frequency-dependent impedance $Z = j\omega L$ in a wave filter or as an energy storage element $E = \frac{1}{2}LI^2$ (where L is inductance). Except for autotransformers that have only one winding with taps, transformers have at least two sets of windings that are linked by a common flux. The magnetic coupling between windings is the cause of the basic transformer function which may be used to provide voltage sources with no common (ground) terminal or to provide a multiplication (division) of voltage (current) and hence of impedance level. The characteristic of λ vs I, as in Fig. 2c, should be fairly linear for inductors, a special condition that is generally not required of transformers. For either device, there exist three design limitations.

(a) Core saturation limits the area under the waveform of terminal voltage according to Eqn. (3); for the case of inductors, the volt second limitation may appear as a limiting LI product.

(b) The current rating may be limited by the allowable current in the windings, usually specified as a maximum value of current density $j = I/a$, where a is the area of a single wire.

(c) The temperature rise of the device due to losses in the core and windings may impose a restriction.

3.1 Inductors

Analytical design criteria given here for inductors start with a piecewise-linear approximation for the core material and include geometrical considerations for the core configuration and winding.

An ideal material for inductors might have constant permeability, zero coercivity and zero dynamic width. For such a fictitious material, Figs. 2a–c would have constant slopes designated, respectively, by permeability μ, permeance $P = \mu S/l$ and inductance $L = n^2P$. It is convenient to assume abrupt saturation of the material at $J_s \simeq B_s$ and of the resulting inductor at $\lambda_s = nB_sS$, as shown in Fig. 4. Maximum current I_m for linear operation is implied by

$$\lambda_m = LI_m \tag{17}$$

where it is understood that the maximum practical flux linkage is less than the saturation value ($\lambda_m < \lambda_s$). It must also be understood that the permeability of real materials varies with the measurement method; production measurements of inductance should be made with the same type of signal as in the intended use of the device.

(*a*) *Effect of airgap*. The permeance of a core without airgap is approximately

$$P = \mu S/l = 1/R_m \quad \text{Henry (turn)}^{-2} \tag{18}$$

whereas an airgap of length l_g will cause l in Eqn. (18) to be replaced by $(l_m + \mu_r l_g)$, where l_m is the mean length of the flux path in the material and μ_r is the relative permeability of the material. The inductance thus decreases with increasing airgap and also becomes less sensitive to variations in the core material. In power supply filter applications it is common to specify a fixed airgap, thereby trading off inductance for current according to Eqn. (17). The modified Eqn. (18) decreases in accuracy for larger airgaps because of flux fringing and may require a correction factor. A deleterious effect of flux fringing is increased heat generated in the portion of the conductor windings that are located in the fringing flux. Metal powder cores possess essentially a distributed airgap; they may be treated in design using Eqn. (18), using certain commercially available values of permeability.

(*b*) *Geometrical factors*. In contrast to the design of very-high-Q inductors in which low losses may be the prime consideration, design factors are now described that are intended for choosing cores of minimum size and weight. So far, the core cross section S, the core window area W, the number of turns n and the wire area a have been defined. It is useful now to define two further terms, K and $\bar{l}$, the winding fill factor and length of the mean turn, respectively. The fill factor K is the fraction of the window area that is filled by conductor, $K = na/W$, where K is less than unity because of the natural space between the wires and the space of wire insulation, for example. When the window area of a core is filled to its maximum practicable K, the winding configuration and thus $\bar{l}$ are defined by the dimensions of the core. It follows that core geometry can be selected using tabulated values of area product WS or of geometrical constant $K_G = KS^2W/\bar{l}$, described as follows.

(*c*) *Core selection*. In power supply design it often happens that the required L and I_m are both specified and may be used as a basis for either of two methods of core selection. The expression for twice the stored energy may be written in several ways:

$$LI_m^2 = \lambda_m I_m = (nB_mS)ja = B_mSjKW \tag{19}$$

where the right-hand term can be solved for WS and used to select a core, independent of the number of turns, provided that B_m and j are chosen in a prudent manner. A rule-of-thumb value for j may be some multiple of $200\,\mathrm{A\,cm^{-2}}$. For the second method, current density j is proportional to winding resistance R:

$$R = j\rho n\bar{l}/I = j\rho L\bar{l}/(B_mS) \tag{20}$$

where ρ is the resistivity of the winding at the temperature of operation. Combining Eqns. (19) and (20) gives

$$L^2I_m^2 = B_m^2RK_G/\rho \tag{21}$$

which can be solved for K_G and used to choose a core to meet specified values of I, L and R for the desired inductor. It is assumed that core catalogs will include listings of WS and K_G values for each core, in addition to the basic core dimensions. With either method, after choosing the core, the number of turns and the wire size may be specified to complete the analytical part of the design.

The design of inductors that carry a direct component of current as well as a fluctuating component

$$I_m = I_{dc} + \tfrac{1}{2}\Delta I \tag{22}$$

must take into account the distinctions (Fig. 3) between maximum, dc and ac values for current, flux linkage and flux density. In Eqns. (19–21), I, B and λ may be taken as maximum values whereas, in the core loss Eqn. (23), B_m is defined as the maximum of the fluctuating component $\tfrac{1}{2}\Delta B$ of Fig. 3.

(*d*) *Losses and temperature rise*. The historical method for determining whether device losses cause an excessive temperature rise is experimental: measurement of the device temperature for actual or simulated operating conditions. Experimental verification is usually preceded by estimates of power dissipation in the windings and core as part of the design process. The winding I^2R loss is straightforward to calculate, since I_m and winding resistance R are both known. Graphs of experimental data for core loss density are available from core material manufacturers, often given in $\mathrm{W\,kg^{-1}}$ and showing the dependence on B_m and frequency.

Some suppliers of core materials also offer empirical approximations to the graphical data in the form of values for A, α and β in the equation for power loss density:

$$p = Af^{\alpha}B_m^{\beta} \tag{23}$$

The temperature rise above ambient may be found by multiplying the power dissipation by an effective thermal resistance in KW^{-1}. This thermal parameter includes three heat flow processes (conduction, radiation and convection) which all depend upon the packaging of the device. Computer-aided design tools are now available to assist with the thermal design including the identification of local hot spots.

3.2 Transformers

The analytical criteria for transformer design will be preceded by a review of voltage and current transformation equations. A transformer consists of a core with two or more sets of windings, usually with one winding over the other on the same spool. Assuming the same value of flux linkage for each winding, by Faraday's law the voltage developed across each winding is proportional to the number of turns:

$$\frac{V_1}{n_1} = \frac{V_2}{n_2} = \frac{d\varphi}{dt} \tag{24}$$

The equation for current transformation according to Ampère's law is

$$n_1 I_1 - n_2 I_2 = Hl \tag{25}$$

which becomes

$$n_1 I_1 = n_2 I_2 \tag{26}$$

by neglecting the magnetic potential drop Hl around the core. A load resistance $R_L = V_2/I_2$ connected across the secondary winding has an apparent resistance at the primary winding of $V_1/I_1 = (n_1/n_2)^2 R_L$, the impedance transformation relationship.

The assumption of equal flux linking each winding is only approximately true. A coefficient of coupling K may be defined for each winding as the ratio of coupled flux to total flux produced by the ampere turns of that winding. The leakage coefficient $1-K$ may be used to define leakage inductances (see Fig. 6).

(*a*) *Volt second limitation.* Equation (24) concerns only the ratio of primary to secondary turns, whereas the actual number of turns may be defined by Eqn. (3). The key concept is that the area under one half-cycle of a periodic voltage cannot exceed $2\lambda_s$ if core saturation is to be avoided. For a periodic input voltage of frequency f and of half-cycle average amplitude V_1, Eqn. (3) leads to

$$V_1 = 4fn_1 B_m S \tag{27}$$

or, for a sinusoidal input voltage of V_{rms} to

$$V_{rms} = \sqrt{2}\pi f n_1 B_m S \tag{28}$$

(*b*) *Current limitation.* For a transformer with a single primary winding and a single secondary winding of n_1 and n_2 turns, respectively, there is a space limitation imposed by the core window area and the geometrical fill factor defined in Sect 3.1(b):

$$KW = n_1 a_1 + n_2 a_2 = (n_1 I_1 + n_2 I_2)/j \tag{29}$$

assuming that the current density j is the same for each winding. For transformers with two secondary windings used alternately, it may be desirable to use a larger instantaneous current density, for example by a factor of $\sqrt{2}$, since the winding is used with 50% duty cycle. The spatial requirements then may be determined from

$$\begin{aligned} KW &= n_1 a_1 + n_2 a_2 + n_2 a_2 \\ &= (1/j)(n_1 I_1 + \sqrt{2} n_2 I_2) \end{aligned} \tag{30}$$

Equation (29) or (30) may be combined with other analysis and used as a basis for core size selection. There may sometimes be a need to specify current density j, on which the winding resistance depends. The allowable internal voltage drop IR at full-load current may be specified as an allowed effective winding resistance R for a given load resistance. For equal power loss in the primary and secondary windings (Fig. 6),

$$R = r_2 + r_1(n_2/n_1)^2 \simeq 2r_2 \tag{31}$$

where $r_2 = j\rho l/I_2 = j\rho n_2 \bar{l}/I_2$. Since load resistance $R_L = V_2/I_2$ it follows that

$$r_2/R_L = j\rho n_2 \bar{l}/V_2 = j\rho \bar{l}/4fB_m S \tag{32}$$

(*c*) *Core selection.* If Eqn. (27) is solved for n_1 and again for n_2 then substituted into Eqn. (29) the result may be rearranged to give the same area product basis as Eqn. (19) for choosing the core size:

$$WS = \frac{V_1 I_1 + V_2 I_2}{4fB_m jK} \tag{33}$$

Alternatively, if solved for current density and substituted in, Eqn. (32) becomes

$$r_2/R_L = P\rho/8f^2 B_m^2 K_G \tag{34}$$

where $P = V_1 I_1 \simeq V_2 I_2$ and where $K_G = KS^2W/\bar{l}$ is the geometrical constant previously described in Sect. 3.1(*b*) for selecting cores for inductors.

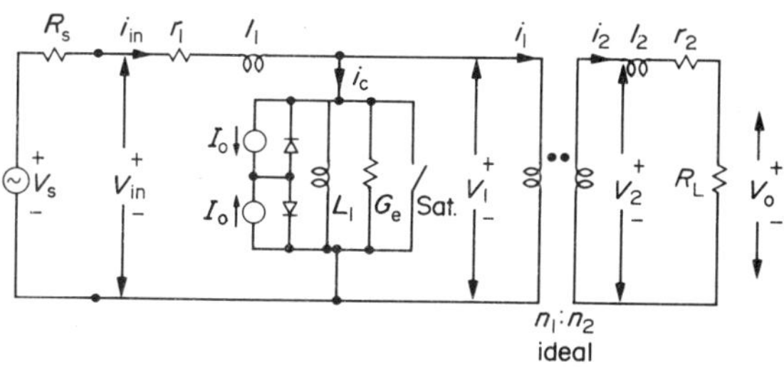

Figure 6
Circuit model of transformer including source resistance R_s, load resistance R_L, winding resistances r_1 and r_2, leakage inductances l_1 and l_2 and a piecewise-linear model of the core, referred to the primary

(*d*) *Transformer losses and temperature rise.* In a manufacturing environment where specific transformer designs consist of implementations of established design principles, prior experience may include loss and temperature information from which extrapolations may be made to the new design. In the absence of such an empirical data base, analytical estimates may be made of transformer losses and of temperature rise using the same considerations as for inductors (Eqn. (23)).

(*e*) *Packaging.* This article previously mentioned packaging as one of the four sets of decisions made by magnetics designers in addition to specifying the magnetic material, the core geometry and the windings. In fact, the packaging considerations overlap with the other three classes of design decisions.

Core mounting considerations may include the need to hold the core together, to secure it in position and to conduct away the dissipated heat. If the core material is stress sensitive, the core may require its own separate package for isolation from the stress of the windings or from the stress of encapsulation.

The basic configuration of the windings depends greatly on the style of core geometry, for example an uncut toroid cannot use a bobbin. Second-order effects, such as the trade-off of leakage inductance vs winding capacitance for a pulse transformer, may determine a specific core configuration as well as the placement of the windings. A high voltage rating for the windings, or a low capacitance requirement, may be implemented by extra insulation tape between the layers of windings or, in extreme cases, by separating the windings. Encapsulation of the windings to provide mechanical support and heat removal may also serve to increase the voltage rating by reducing corona breakdown, but encapsulation can also increase winding capacitance unless measures are taken.

The packaging function involves mechanical, electrical and thermal considerations for the core, for the windings and for the device as a whole. Cost and repairability are also considerations for this largely empirical aspect of magnetic design.

4. *Electromechanical Transducers*

Space constraints allow descriptions of only a restricted set of transducers: those involving translational motion of a coil or of the magnetic flux path and rotary motion of motors. The basic force equation (Eqn. (11)) for the moving coil transducer is given in Sect. 2.1. A loudspeaker is an example of a moving coil system using a permanent magnet to provide the magnetic field. An acoustic cone is affixed to a coil in a magnetic field which is driven back and forth by the current it carries as electric energy is converted to acoustic energy. The same principle is used in a variety of small motors, often switching current into different fixed coils according to their position relative to a rotating array of magnets. A translational class of movable core transducers may be described electrically by a generalized inductance graph of λ vs i, as illustrated in Fig. 7, which shows the stored energy for two different airgaps. The crosshatched area left of the line defines the stored energy according to Eqn. (8).

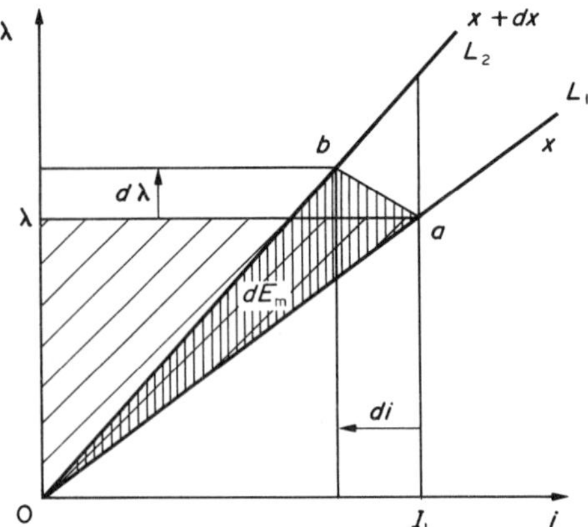

Figure 7
Differential mechanical energy dE_m associated with armature movement dx

A simple relay is one example of such a machine. At zero current a spring holds the gap to its largest amount, corresponding to inductance L_1. As the current is increased, a magnetic force of attraction is developed which tends to close the gap and at some current I_1 the magnetic force exceeds the threshold force of the spring. The relay snaps shut, increasing the inductance to L_2. The release or dropout current is typically less than the pickup current. The magnetic force is given by the derivative of the energy, as

$$F = \frac{dE}{dX} = \frac{d}{dx}(\tfrac{1}{2}LI^2) = \tfrac{1}{2}I^2\frac{dL}{dx} \quad (35)$$

A somewhat similar relationship may be used to describe a reluctance motor, in which the inductance varies with angle.

Motors may be considered as transducing electrical input energy to supply mechanical output. Most motors may also function in the inverse mode as generators that transduce mechanical input energy to supply electrical output, but that function is not described here. Torque is developed in typical motors as an interaction between two magnetic circuits: a stator which is fixed in space and a rotor which turns. Many innovative arrangements are possible, exploiting semiconductor switches and magnetic field sensors to advantage. In applying Eqn. (10), the stator may consist of an array of simple coils whose inductance is caused to vary by the rotation of a passive asymmetrical rotor. This is called a reluctance motor. Alternatively, the rotor may possess rotational symmetry and may use current to develop its field interaction with the stator. Such currents may be supplied to selected coils of wire in the rotor by use of brush contacts on the motor shaft or currents may be induced into the rotor structure by transformerlike magnetic excitation from the stator windings. Motors may be designed to operate from direct current or from alternating current as from the power utility. Most ac motors of fractional horsepower ratings use only single-phase excitation and thus require special provisions for starting. Multiple horsepower motors usually use three-phase excitation of the stator to achieve a spatially rotating magnetic field.

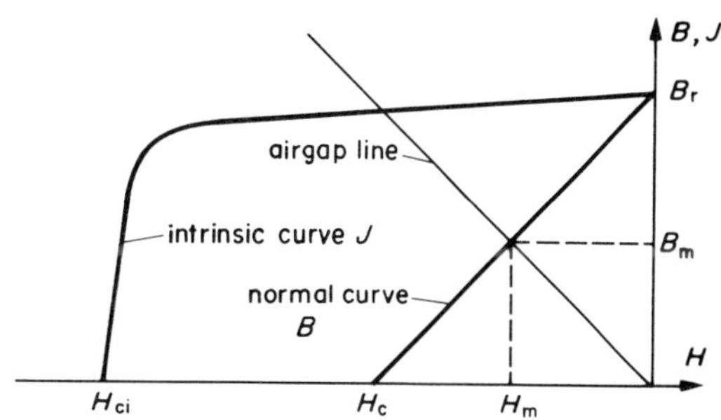

Figure 8
Demagnetizing curves B vs H and $B-\mu H-\mu_0 H=J$ vs H for Nd–Fe–B permanent magnet materials

5. *Permanent Magnets*

Magnetic materials have two general properties B_r and H_c, defined in Fig. 2a, which for permanent magnets are both large. A residual induction B_r indicates that the material stays partially magnetized after having been driven to saturation by a large value of H. The ratio B_r/B_{sat} is a measure of permanent-magnet quality and may ideally approach unity. A large value of coercivity H_c is necessary for the magnet to remain magnetized as an isolated specimen, that is when it is not in the geometry of a closed flux path as in Fig. 1. To apply Fig. 5 to a permanent magnet, note that induction J may exist for a permanent magnet in the absence of applied field H_a, provided the magnet coercivity exceeds the demagnetizing field ($H_c > H_d$).

A large value of H_c is the key property of a permanent magnet. In comparison with the most highly permeable (low-H_c) materials for inductors and transformers, coercivity H_c for a modern permanent magnet material may be greater by a factor of 10^6 or more. A breakthrough in permanent-magnet development occurred in the 1960s with the development of cobalt–rare-earth permanent-magnet materials of extremely high coercivity, $\mu_0 H_c/B_r > 1$, a criterion that was formerly regarded as theoretically impossible to achieve (see *Cobalt–Rare-Earth Compounds*). Many engineers expected revolutionary changes to take place in the design of rotating machines using permanent magnets but the high cost of cobalt, and its limited world supply, mitigated against major product changes. A second breakthrough took place in the 1980s with the discovery of Nd–Fe–B alloys that are comparable with cobalt–rare-earth materials (see *Neodymium–Iron–Boron Alloys*). Interestingly, the more recent component materials are relatively inexpensive and abundant and an eventual impact on machine design is anticipated.

Design with permanent magnets makes use of the second-quadrant properties of the experimental B–H hysteresis curve, since the magnet supplies energy to the system. The energy is in the form of a magnetic field in an airgap. A typical criterion for static field designs may require that the permeance of the airgap should be matched to the magnet. This achieves the maximum energy in the airgap for a minimum magnet volume and hence minimum cost. Figure 8 shows a graphical construction of an airgap line on the second quadrant of a B–H curve, the so-called demagnetization curve, for data that are representative of the new Nd–Fe–B alloys. Two curves for the magnet are shown. The "normal" curve of B vs H is diagonal with a slope of approximately μ_0, which may be explained by Eqn. (4) as the incremental slope when material induction J is constant at its saturation value. The "intrinsic" curve of J vs H reveals the material properties more clearly, showing that J remains fairly constant at saturation until H approaches the negative value of intrinsic H_c. The prominent knee in the second quadrant curve of J vs H would be evident in the third quadrant of B vs H, which is not shown in Fig. 8. The optimum airgap load line intersects the normal curve at $B_m \simeq \frac{1}{2}B_r$.

See also: Design with Magnetic Materials: Computer Simulation; Magnetic Materials: An Overview; Magnetic Units and Material Specification; Shields, Magnetic

Bibliography

Adams C H 1970 Power rating of magnetic cores. *IEEE Trans. Magn.* 6(1): 65

Bradley F N 1971 *Materials for Magnetic Functions*. Hayden, New York

Heck C 1974 *Magnetic Materials and Their Applications*. Crane–Russak, New York

Huijer E, Watson J K, Dove D B 1980 Magnetostatic effects in I Bars. *IEEE Trans. Magn.* 16 (1): 120–6.

Massachusetts Institute of Technology 1943 *Magnetic Circuits and Transformers*. Wiley, New York.

Slemon G R, Straughen A 1982 *Electric Machines*. Addison-Wesley, Reading, MA

Wahlgren W W 1958 How epoxies are used to improve large and small transformer designs. *Insulation* 4(6): 2–5

Watson J K 1985 *Applications of Magnetism*. J. K. Watson, University of Florida, Gainesville, FL

J. K. Watson
[University of Florida, Gainesville, Florida, USA]

Design with Magnetic Materials: Computer Simulation

Design is the most creative part of the manufacturing process and, although it is often regarded as difficult to organize and plan in a consistent and predictable way, the benefits of intelligent simulation are enormous. The central problem in design is to be able to determine all the significant parameters (material properties, geometry, etc.) to meet the performance goals (magnetic field shape, forces, access, etc.) subject to practical constraints (cost, environment, time scales, etc.). It is therefore a problem of optimization over many disciplines. The most important component of a design software system is the solver, a computer program for evaluating the magnetic field produced by a configuration defined by the designer. The normal procedure of using a field computer code is when the designer iterates towards the optimal solution by a more-or-less heuristic method (cut and try) supplemented by personal creativity and experience. To carry out this process efficiently, the software system has three stages: a preprocessor to enter the data, a solver to compute the numerical solution of the field equations and a postprocessor to extract the significant engineering parameters.

1. *Field Equations*

The appropriate field equations for static and low-frequency magnets are Faraday's law, $\nabla \times \boldsymbol{E} = -\frac{\partial \boldsymbol{B}}{\partial t}$, and Ampère's law, $\nabla \times \boldsymbol{H} = \boldsymbol{J}$, where $\boldsymbol{B}$, $\boldsymbol{E}$ and $\boldsymbol{H}$ are the field vectors and $\boldsymbol{J}$ is the current density. These equations are normally sufficient when radiation effects can be neglected. The field vectors are not independent since they are further related by the material constitutive properties: $\boldsymbol{B} = \mu \boldsymbol{H}$ and $\boldsymbol{J} = \sigma \boldsymbol{E}$ (Ohm's law), where μ and σ are the material permeability and conductivity, respectively. In practice, μ and σ may often be field-dependent quantities and, furthermore, some materials will exhibit both anisotropic and hysteretic effects. It cannot be overstated that accurate knowlege of the material properties is the single most important factor in obtaining reliable simulations.

Since the field vector $\boldsymbol{B}$ satisfies a zero divergence condition ($\nabla . \boldsymbol{B} = 0$), it can be expressed in terms of a magnetic vector potential $\boldsymbol{A}$ (i.e., $\boldsymbol{B} = \nabla \times \boldsymbol{A}$) and it follows, from Faraday's law, that $\boldsymbol{E} = -(\frac{\partial}{\partial t}\boldsymbol{A} + \nabla V)$, where V is the electric scalar potential. Neither $\boldsymbol{A}$ nor V are completely defined since the gradient of an arbitrary scalar function can be added to $\boldsymbol{A}$ and the time derivative of the same function can be subtracted from V without affecting the physical quantities $\boldsymbol{E}$ and $\boldsymbol{B}$. These changes to $\boldsymbol{A}$ and V are the gauge transformations, and uniqueness is usually ensured by specifying the divergence (gauge) of $\boldsymbol{A}$ and sufficient boundary conditions. If $\nabla . \boldsymbol{A} = -\mu\sigma V$ (Lorentz gauge) is selected, then the field equations in terms of $\boldsymbol{A}$ and V are

$$\nabla \times \frac{1}{\mu}\nabla \times \boldsymbol{A} = -\sigma\frac{\partial \boldsymbol{A}}{\partial t} + \sigma\nabla\left(\frac{1}{\mu\sigma}\nabla . \boldsymbol{A}\right) \tag{1}$$

$$-\mu\sigma^2\frac{\partial V}{\partial t} + \nabla . \sigma\nabla V = 0 \tag{2}$$

where σ has been assumed piecewise constant. This approach decouples the vector potential from the scalar potential. In the global region where $\sigma = 0$ and $\nabla \times \boldsymbol{H} = \boldsymbol{J}_s$, these equations reduce to

$$\nabla . \mu\nabla\varphi = 0 \tag{3}$$

where φ is known as the reduced magnetic scalar potential with $\boldsymbol{H} = \boldsymbol{H}_s - \nabla\varphi$ for a source field $\boldsymbol{H}_s$.

The main reason for using potentials rather than field components directly is that the difficulties arising from field discontinuities at material interfaces can be avoided. For the important class of two-dimensional problems there will be only one component of $\boldsymbol{A}$, A_z parallel to J_z the excitation current density, with field components given by $B_x = \frac{\partial A_z}{\partial y}$ and $B_y = -\frac{\partial A_z}{\partial x}$. In this case, the field equations (Eqns. (1,2)) reduce to

$$\frac{\partial(1/\mu\frac{\partial A_z}{\partial x})}{\partial x} + \frac{\partial(1/\mu\frac{\partial A_z}{\partial y})}{\partial y} = \sigma\frac{\partial A_z}{\partial t} + J_z \tag{4}$$

In the statics limit (dc), the time-dependent term in Eqn. (4) vanishes and the field is described by the Poisson equation. For fields involving time at least two types can be distinguished: the time-harmonic (ac) case, in which the fields are periodic at a given frequency ω (i.e., $A_z = A_o \exp(j\omega t)$), and the transient case, in which the time dependence is arbitrary.

2. *Numerical Methods: Finite Elements*

Numerical solutions for the field equations are now routine for a very large number of problems encountered in magnet design; these include, for example, two-dimensional models of a particular configuration taking into account nonlinear, anisotropic and even hysteretic effects.

The situation for complete three-dimensional models is not so encouraging because of the escalating computer resources needed as the problem size increases. Nevertheless, solutions for nonlinear statics devices are regularly obtained in industry and time-dependent solutions will become more cost effective as parallel computer hardware architectures develop.

The widespread use of computer-based solutions has come about largely because of the generality of the finite-element method (FEM). In this method, the problem space is subdivided (discretized) into finite regions (elements) over which the solution is assumed to follow a simple local approximating trial function (shape functions) (see Fig. 1a). In the simplest situation, a particular element could be a plane triangle defined by its three vertex or nodes and a component of the solution of the statics case of Eqn. (4) may be approximated by

$$A_x \simeq U = \alpha_1 + \alpha_2 x + \alpha_3 y = \sum N_i U_i \qquad (5)$$

In other words, it is assumed that, within the element, the solution can be approximated by a linear law and since the triangles have three nodes it is natural to select an equation that is linear in x and y with three parameters. The functions N_i are called the local shape functions and the parameters U_i are the solution values at the nodes ($N_1 = 1$ at node 1 and $N_1 = 0$ at nodes 2 and 3; in general, $N_i(x_k, y_k) = \delta_{ik}$).

The finite elements can be integrated into a numerical model for the whole problem space either by (a) the variational method, in which the total energy of the system is expressed in terms of the finite-element trial functions and then minimized to determine the best solution, or (b) the weighted residual method, in which the formal error (residual), arising by substituting the trial functions into the defining equation, is weighted by a suitably chosen function and then integrated over the problem domain. The best fit for the trial function parameters can then be obtained by equating the integral to zero. Both methods lead to a set of algebraic equations and are equivalent if the weighting functions are chosen to be the trial functions (Galerkin's method). At the element level, the residual R_i is given by

$$R_i = \left[\int_{\text{elem}} \nabla N_i \; \kappa \nabla N_j \, d\Omega \right] u_j + \int_{\text{elem}} N_i \, Q \, d\Omega \qquad (6)$$

where κ and $Q(J_z)$ denote the material reluctivity ($\frac{1}{\mu}$) and current sources, respectively. If the integrals in Eqn. (6) are expressed as k_j and f_i the expression becomes

$$k_{ij} = \int_{\text{elem}} \kappa \left[\frac{\partial N_i}{\partial x}\frac{\partial N_j}{\partial x} + \frac{\partial N_i}{\partial y}\frac{\partial N_j}{\partial y} \right] d\Omega \qquad (7)$$

$$f_i = \int_{\text{elem}} N_i Q \, d\Omega \qquad (8)$$

The integrals given in Eqns. (7) and (8) can be readily evaluated. These equations can be assembled for the whole problem by superposition taking account of the boundary conditions and removing the redundancy at shared nodes. At interelement boundaries, the solution is forced to be continuous but the normal flux (i.e, $\kappa \partial u/\partial n$) will only be continuous in a *weak* sense, that is, the discontinuity will depend upon the mesh refinement.

The finite-element method provides a systematic technique for replacing the continuum field equations with a set of discrete algebraic equations which can be solved by standard procedures. In Fig. 2, a typical field map is shown for a permanent magnet machine modelled by a computer simulator that can take into account nonlinearity and permanently magnetized materials. Although hysteresis effects can be included, the computational resources required can be prohibitive because of the vector nature of magnetization. The magnetic material must be characterized by a large number of measurements to take account of the minor loops and from these the convolution integrals necessary to obtain the constitutive relationships can be evaluated. These characteristics must then be followed through time; this can be implemented by solving at a discrete set of time points given the initial conditions in the material.

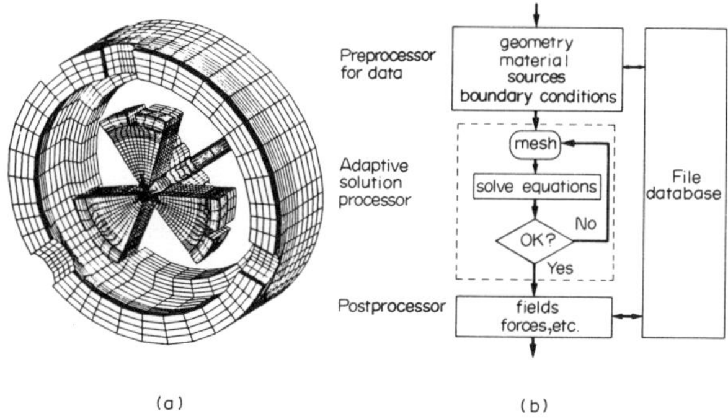

Figure 1
CAD system for magnet simulation: (a) cyclotron finite-element mesh and (b) flow diagram

3. Integral Methods

An alternative procedure is to solve the field equations in their integral form; for example, in magnetostatics the magnetization vector $\boldsymbol{M}$ given by $\boldsymbol{M} = (\mu - 1)\,\boldsymbol{H}$ can be used instead of $\boldsymbol{H}$ to derive an integral equation over all ferromagnetic domains of the problem:

$$\boldsymbol{M}(\boldsymbol{r}) = (\mu - 1)\left[\boldsymbol{H}_s(\boldsymbol{r}') - \frac{1}{4\pi}\nabla \int_{\Omega} \boldsymbol{M}(\boldsymbol{r}').\nabla\left(\frac{1}{R}\right) d\Omega \right] \qquad (9)$$

where R is the distance between the source and field point. For problems with linear materials, Eqn. (9)

reduces to integrations over the bounding surfaces of materials in terms of the magnetic scalar potential:

$$4\pi\varphi = -\int_{\Gamma} \frac{1}{R}\frac{\partial\varphi}{\partial n} - \varphi\frac{\partial(1/R)}{\partial n}\, d\Gamma \qquad (10)$$

Equation (10) can be solved numerically by the boundary-element method (BEM) in which the active surfaces are discretized into elements. The advantages of integral formulations compared with the standard differential approach using finite elements are (a) only active regions need to be discretized, (b) the far field boundary condition is automatically taken into account and (c) the fields recovered from the solution are usually very smooth. Unfortunately, the computational costs rapidly escalate as the problem size increases due to the complexity of the system coefficients and the fact that the resulting matrix is full, whereas in the differential approach the coefficients are simple and the matrix is sparse.

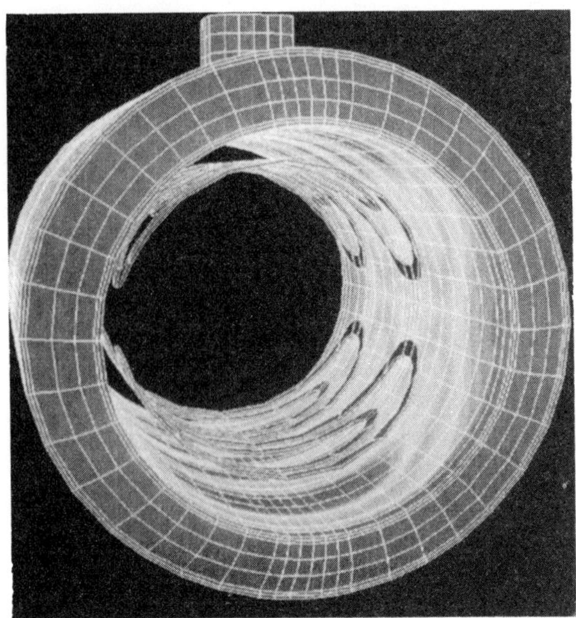

Figure 3
Eddy currents in an MRI body scanner

4. Modern Design Environment

The most common system used in software packages is one in which the preprocessor includes data input, model building and mesh (element) generation. Although fully automated meshing is now a practical possibility, it needs to be combined with error estimation in order to allow the generation of optimal meshes. This approach is becoming quite common for two-dimensional systems and can be expected in three-dimensional systems before long. Figure 1 illustrates such a field simulation environment and Fig. 1b is a flow diagram of a magnetic field simulator in which the solution processor includes an adaptive mesh generator controlled by *a posteriori* error estimation. This avoids the costly and essentially heuristic task of mesh generation which, in the past, had to be performed by the designer. Figure 3 shows a solution for magnetic resonance imaging (MRI) body scanner gradient coils with the current density induced on the inner surface of the cryostat (see *Nuclear Magnetic Resonance Imaging*).

Figure 2
Finite-element solution of a four-pole synchronous motor

See also: Design with Magnetic Materials

Bibliography

Miya K (ed.) 1990 Compumag Tokyo proceedings. *IEEE Trans. Magn.* 26(2)

Silvester P P, Ferrari R L 1990 *Finite Elements for Electrical Engineers*, 2nd edn. Cambridge University Press, Cambridge

Zienkiewicz O C 1990 *The Finite Element Method*, 3rd edn. McGraw-Hill, New York

C. W. Trowbridge
[Vector Fields, Kidlington, UK]

Domains and Domain Walls

Magnetic materials such as ferromagnets and ferrimagnets, which can maintain a resultant magnetic moment even in the absence of a magnetizing field, are characterized by an internal micromagnetic domain structure. This subdivision of the material into small volumes, which are themselves uniformly magnetized to a value equal to the maximum or saturation magnetization of the body M_s, enables the material to accommodate a continuous range of magnetic states. These range from complete demagnetization to full saturation as the applied field is increased from zero (see Fig. 1a). On reduction of the field from saturation, the magnetization follows a hysteresis loop and a remanent magnetization is retained at zero applied field.

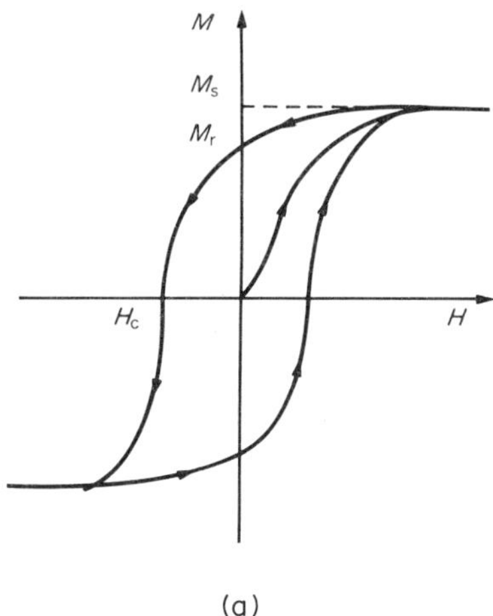

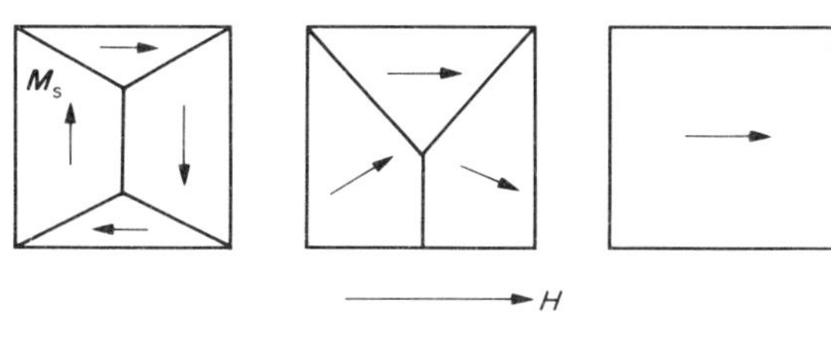

Figure 1
(a) The magnetization–applied field hysteresis loop of a typical ferromagnetic specimen showing the saturation magnetization M_s, the remanent magnetization M_r and the coercivity or coercive field H_c; and (b) a schematic diagram of domain rearrangement to saturation in an applied field H (the arrows mark the local direction of M_s)

In magnetically soft materials, the hysteresis loop can be cycled in very small fields, typically $< 400\ \mathrm{A\,m^{-1}}$. This may be compared with the internal field of $10^9\ \mathrm{A\,m^{-1}}$ which aligns the atomic moments in each domain (Weiss 1907).

The rearrangement of the domain structure, by the growth of suitably magnetized domains and the shrinking of others and by the rotation of the magnetization at higher fields, is illustrated schematically in Fig. 1b. The domain concept explains the hysteretic behavior, the possibility of a remanent magnetization and saturation in fields that are minute in comparison with the internal field. The domain structure is determined by the magnetic, crystallographic and microstructural properties of the material.

Single crystals and polycrystalline materials have different domain structures. The surface of the body has a distinct effect on its domain structure and, at small dimensions, so does its size and shape. The form of any domain structure is such as to reduce the total magnetic energy of the material to a minimum. The most important individual contributions to this energy are magnetic anisotropy, exchange and magnetostatic energies (see *Anisotropy in Magnetic Materials*; *Magnetic Materials: Basic Concepts*; *Micromagnetic Calculations*).

1. Domain Concepts and Typical Structures

Figure 2 illustrates domain structures that might be observed in specimens of iron, iron–silicon and cobalt. They are typical of structures in materials with cubic and uniaxial, hexagonal crystal structures.

The domain structure in the ideal iron single crystal (Fig. 2a) reflects the condition that the "easy" directions of magnetization in body-centered-cubic (bcc) iron are the $\langle 100\rangle$ crystal directions (i.e., magnetization along $\langle 100\rangle$ involves the minimum magnetocrystalline anisotropy). It is also clear that the magnetization vector is always parallel to the surface of the crystal. This ensures that $\nabla . \boldsymbol{M}$ at the surface is everywhere zero and that no magnetostatic energy is involved in the structure. The magnetostatic energy reflects a limiting tendency for the body to be fully demagnetized in zero applied field. For a uniformly magnetized body, the magnetostatic energy density $E_m = \frac{1}{2}DM_s^2$ per unit volume, where D is a constant known as the demagnetizing factor. D is a function of the dimensions of the body and, except for ellipsoids of revolution, it is generally a complex function (see *Demagnetizing Factors*). The triangular, prismatic domains which avoid magnetostatic energy in the structure are often called flux closure domains or simply closure domains.

In the grain-oriented, polycrystalline iron–silicon alloy (Fig. 2b) the domain structure again follows these crystallographic and magnetostatic "rules," with the domain magnetization parallel either to

⟨100⟩ or the next easiest directions ⟨110⟩ and also parallel to the surface of the specimen. In this specimen, the crystal size is larger than the equilibrium domain size and the domain pattern is determined by the relative orientations of the crystals. In polycrystalline materials with very small grains, such as many thin films, the domain width is much greater than the crystal size and the domain structure is not determined by the relative orientations of the grains.

The cobalt crystal (Fig. 2c) has only one easy axis of magnetization, the [0001] c axis. A result of this is that closure domains are not energetically favorable near the top and bottom surfaces as this would involve considerable anisotropy energy. However, a structure of narrow slablike or blocklike domains serves to reduce magnetostatic energy at the top and bottom surfaces of the crystal by subdividing the free magnetic pole distribution. So-called "reverse spike" domains and undulating walls also help in this process.

Kittel (1946) showed that, for the simple structure in Fig. 2d, the domain spacing or width d is given by

$$d = (E_w L/1.7\ M_s^2)^{1/2} \quad (1)$$

where E_w is an energy associated with the boundary between the domains, L is the thickness of the crystal

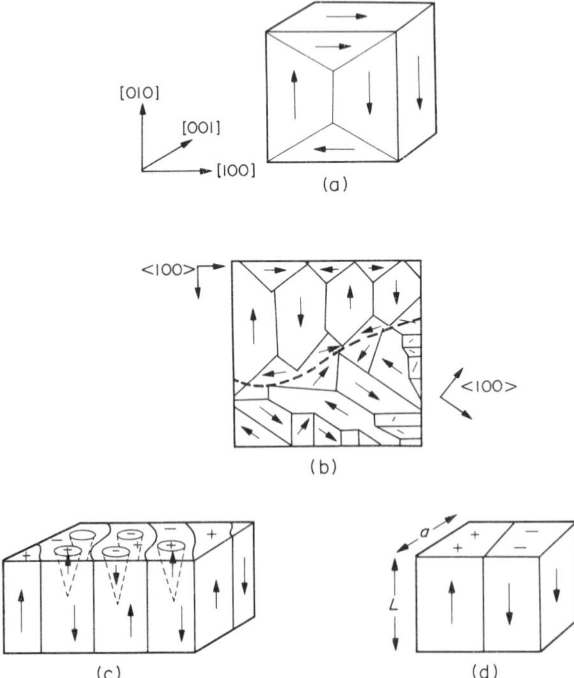

Figure 2
Types of domain structure that might be found in (a) an iron single-crystal cube, (b) two crystals taken from a polycrystalline Fe–3%Si alloy sheet (the structure at the grain boundary will be confused), (c) a cobalt single crystal and (d) a two-domain uniaxial crystal.

and M_s is the saturation magnetization. As will be shown, E_w for cobalt is about $10^{-2}\,\mathrm{J\,m^{-2}}$ and, for a 1 cm thick crystal of cobalt with $M_s = 1400\,\mathrm{kA\,m^{-1}}$, d is obtained as about 5.35 μm.

If a ferromagnetic or ferrimagnetic body becomes very small it will not contain domain walls and the particle is said to be a single-domain particle (Frenkel and Dorfman 1930, Kittel 1946). This occurs when the domain boundary energy becomes greater than any other magnetic energy contributing to the structure. If the structure in Fig. 2d containing one domain boundary is considered, the size at which the body will become a single domain is determined by a balance between the boundary region energy over the area aL of the wall given by $E_w aL$ and the magnetostatic energy of a uniformly magnetized crystal given by $\frac{1}{2} DM_s^2 a^2 L$. The critical size is then given by the condition

$$\tfrac{1}{2} DM_s^2 a^2 L < E_w aL$$

$$a < 2E_w / DM_s^2 \quad (2)$$

If it is further assumed that a is approximately equal to L and that the particle approximates to a sphere, for which $D = \frac{1}{3}$, then a for a cobalt particle is obtained as approximately 30 nm. For elongated particles, such as prolate ellipsoids or needles, D is reduced and the critical single-domain size is greater (see *Elongated-Single-Domain Magnets*). Very small particles behave as large paramagnetic moments and exhibit superparamagnetism which is characterized by an absence of hysteresis and, of course, by zero remanence.

A related modification to the domain structure occurs in the case of thin-film materials. Depending on the relative magnitudes of anisotropy and magnetostatic energies, thin films can be magnetized in a direction parallel or perpendicular to the film plane. In general, a need to reduce magnetostatic energy forces the magnetization vector into the plane of the film but, if the easy axis is perpendicular to the film plane, uniaxial materials with strong magnetocrystalline anisotropy, such as cobalt, hexagonal ferrites and garnets, will be magnetized normal to the film. At small thicknesses, when essentially a single-domain state is theoretically favorable, thin films are magnetized normally or in the plane. For film thicknesses on a monolayer or nanometer scale, an additional surface-induced anisotropy (Néel 1954, Bruno 1988) can play a part in determining the equilibrium orientation of the magnetization.

The boundary region between each domain is called the domain wall. It is a two-dimensional sheet of finite width and it is the region in which the atomic moment vectors rotate to accommodate the change in magnetization orientation between adjacent domains. As such, it is determined principally by a balance between magnetic exchange and anisotropy energies, although other magnetic energies can be involved as will be seen. Other micromagnetic structures of lower dimension (i.e., lines and points) exist and also represent local changes of magnetization direction.

In some elements, particularly some of the rare-earth metals, complex long-range spin structures occur. These often change with temperature and result in unusual magnetic phase behavior. They do not fit with the treatment in this article and are considered in *Rare-Earth Metals and Alloys*.

2. Simple Domain Walls

To illustrate a simple type of domain boundary, an early model due to Landau and Lifshitz (1935) will be used, which improved and rationalized an idea by Bloch (1932). Figure 3a follows Kittel and Galt (1956) and shows a wall N moments or spins wide. The exchange energy density across the wall can be written as

$$E_{\mathrm{ex}} = \frac{A\theta^2}{N} = A\left(\frac{d\theta}{dx}\right)^2 \tag{3}$$

where θ is the total rotation angle across the wall and A is the exchange constant given by $A = 2JS^2/a$, where J is the exchange integral, S is the spin quantum number and a is the interatomic distance. The anisotropic energy associated with the spin moments pointing away from the easy axis of magnetization (i.e., the direction held by the domain magnetization) is a function $K(\theta)$ in terms of the magnetocrystalline anisotropy constants K_n and the angle θ (see *Anisotropy in Magnetic Materials*). The total wall energy per unit area is given by

$$E_{\mathrm{w}} = \int_{-\infty}^{\infty} \left[A\left(\frac{d\theta}{dx}\right)^2 + K(\theta)\right] dx \tag{4}$$

The equilibrium condition for θ is obtained when these two terms are equal:

$$A\left(\frac{d\theta}{dx}\right)^2 = K(\theta) \tag{5}$$

For a magnetically uniaxial material, $K(\theta)$ can be expressed to a sufficient accuracy by $K(\theta) = K_1 \sin^2\theta$, where K_1 is the first anisotropy constant. Substituting into Eqn. (4) gives

$$E_{\mathrm{w}} = 2(AK_1)^{1/2} \int_0^{\pi} \sin\theta \, d\theta = 4(AK_1)^{1/2} \tag{6}$$

The wall can be characterized by a width δ given by the rate of rotation of the spin moments from Eqn. (5):

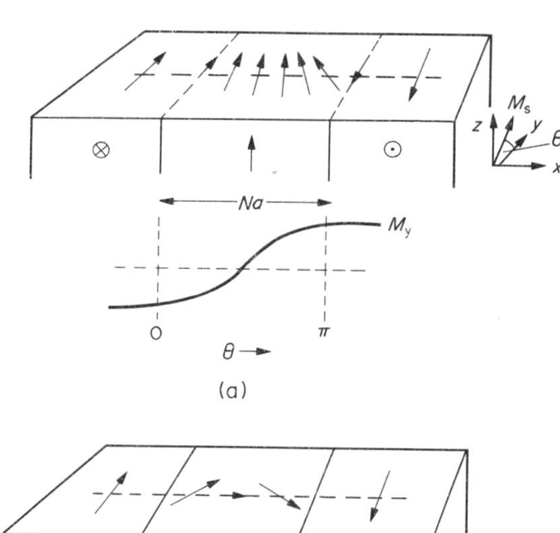

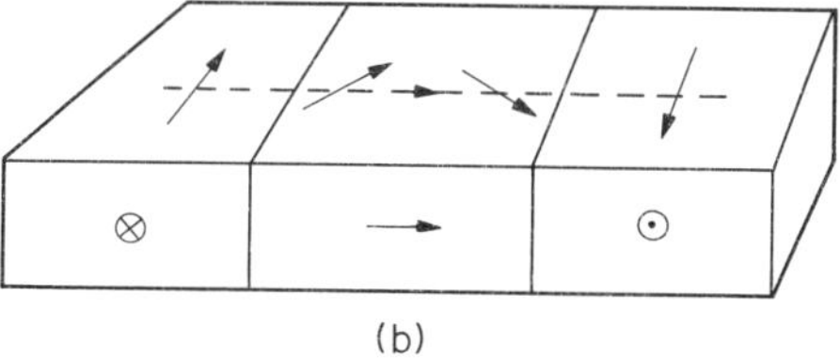

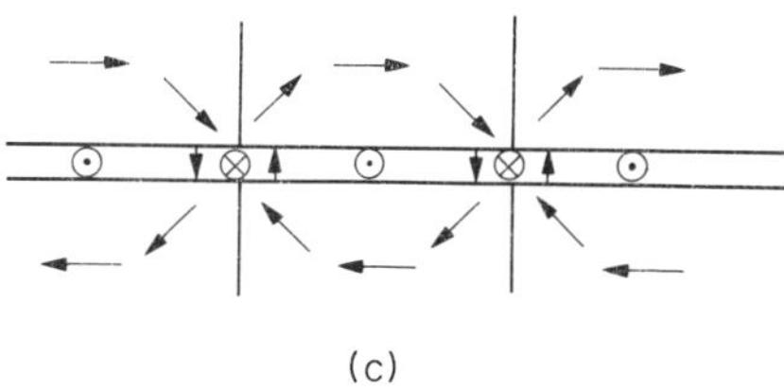

Figure 3
Notional spin structures in (a) a Bloch wall (Na is the width of the wall, N being the number of atoms or spins across the wall and a the interatomic distance, and θ is the angle between M_s and the y axis), (b) a Néel wall and (c) a cross-tie wall. The diagram is a view of the surface of the film

$$A\left(\frac{d\theta}{dx}\right)^2 = K_1 \sin^2\theta \tag{7}$$

and

$$\delta = \pi\frac{dx}{d\theta} = \pi\left(\frac{A}{K_1}\right)^{1/2} = \pi\delta_0 \tag{8}$$

For cobalt, $A = 1.3 \times 10^{-11}\,\mathrm{J\,m^{-1}}$ and $K_1 = 5.3 \times 10^5\,\mathrm{J\,m^{-3}}$ giving a value of $E_{\mathrm{w}} = 10^{-2}\,\mathrm{J\,m^{-2}}$ and $\delta = 1.55 \times 10^{-8}\,\mathrm{m}$ or 15.5 nm. The parameter δ_0 is a reduced width and in many calculations δ/δ_0 is different from π. It is clear that the width and energy of such a Bloch wall are narrower and greater for strongly magnetic and anisotropic materials.

The width of domain walls is of some importance in the consideration of magnetic coercivity mechanisms in materials. In soft materials with low anisotropy, the domain walls often appear to arrange themselves without regard to the easy and hard directions but, in materials with high anisotropy, the orientation of domain walls is strongly related to the crystal symmetry of the material. The previous example considers a 180° wall with the magnetization on each side parallel and antiparallel to the [0001] *c* axis. In cubic crystals of iron, as indicated in Fig. 2a, 90° walls also occur and in nickel with ⟨111⟩ easy directions 180°, 109° and 71° walls can be observed. Table 1 gives some estimates of domain wall energies and widths in a variety of materials.

A factor ignored in the simple calculation given is the magnetostatic energy associated with the component of magnetization out of the surface at the intersection of the domain boundary with the surface of the specimen (Néel 1955). This contribution becomes significant in the case of thin films where these intersections at the top and bottom surfaces are close together. In the wall in Fig. 3b the spins rotate about an axis parallel to the wall and normal to the film plane. There are no free poles on the surface of the film for this Néel wall but there may be a contribution on the faces of the wall. A further modification to the structure which gives a periodic reduction in this contribution along the wall is the cross-tie wall, shown in Fig. 3c (Huber et al. 1958). A simple calculation (Jones and Middleton 1974) of the relative energies of these walls for Permalloy (nickel–iron alloy) thin films shows that cross-tie walls should be the favored structure at film thicknesses below about 80 nm; in reality it is the Néel-type wall which is observed at very small (<20 nm) film thicknesses.

3. Walls, Lines and Points

A recognition of the importance of magnetostatic energy in micromagnetics (see *Micromagnetic Calculations*) and in domain wall calculations has produced models with involved spin structures. Some of the principal refinements are the introduction of possible asymmetric walls and vortex spin structures. The simple walls discussed previously are one-dimensional and estimates of their energies and widths have been made by analytical, Ritz and numerical solutions (e.g., Jones and Middleton 1974). Lower energy values, and probably more realistic structures, can be obtained in models where the magnetization direction $\boldsymbol{M}_s$ varies through the specimen thickness and flux closure and stray field free arrangements occur. Figures 4a and 4b show examples of vector plots of the magnetization distribution in a two-dimensional Bloch wall and in a two-dimensional Néel wall, respectively, in a thin film. The diagrams show cross sections through the film and the walls. The length of the arrows represents the calculated component of magnetization in the plane of the figure. The magnetization in the domains at each side of the walls may be directed either into or out of the plane of the figure, as may $\boldsymbol{M}_s$ in the center of the Bloch wall in Fig. 4a. These early schemes clearly show asymmetry and more recent numerical modelling (Hubert 1974, Aharoni and Jakubovics 1990) has produced improved suggestions.

Table 1
Domain wall energies and widths in a variety of materials

Material	K ($\mathrm{J\,m^{-3}}$)	δ (nm)	E_w ($\mathrm{J\,m^{-2}}$)
Cobalt (180°)	5.3×10^5	15.5	10^{-2}
Iron (90°)	4.7×10^4	50	10^{-3}
Nickel (71°)	5.1×10^3	110	1.8×10^{-4}
Nickel–iron (Permalloy)	10^2	1000	10^{-4}
$Nd_2Fe_{14}B$	5×10^6	4	10^{-2}

Some information from Lilley (1950)

The cross-tie wall of Fig. 3c is an example of a complex structure that avoids magnetostatic energy by subdividing a Néel wall with Bloch lines. These essentially one-dimensional structures extend through the thickness of the specimen and, in low-anisotropy materials, provide flux closure for the Néel segments. The spin rotation close to these is Bloch in character, with a component of $\boldsymbol{M}_s$ perpendicular to the film surface, hence the name Bloch lines. Such linear features also occur in subdivided walls in bulk material, where they connect segments in a Bloch wall in which the normal component of $\boldsymbol{M}_s$ points alternately normal to the top and bottom surfaces of the specimen. As they connect Bloch segments, their spin structure contains some Néel character.

The structure and behavior of domain walls are central to the working of many magnetic devices such as thin-film sensors and magnetic recording heads and information storage media. The development of one particular device—the bubble domain memory—in which small cylindrical domains are a memory element drew much attention to the structure of domain walls (Malozemoff and Slonczewski 1979). Bubble domain materials are covered in the article *Bubble Domain Materials*, but the typical structure of stripe and cylindrical bubble domains is illustrated in Fig. 5a. Figures 5b, c give a more detailed view of the walls round a bubble showing vertical Bloch lines within the wall.

The thin crystal or film shown in Fig. 5a has an easy axis normal to its plane and when partly magnetized in an applied field or at remanence it can contain elongated stripe domains and/or cylindrical bubble domains. The cylindrical domains are called bubbles because when observed in plan view (Fig. 5c) they appear circular and give the impression of the projection of a sphere. The bubble in Fig. 5b contains two Bloch lines extending from the top surface of the film

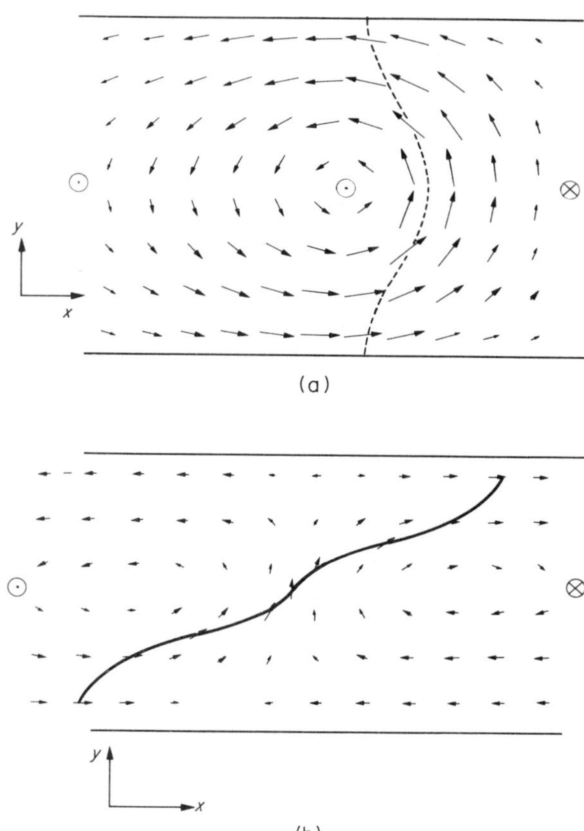

Figure 4
Spin structures in thin films: (a) two-dimensional Bloch wall distribution (after Labonte 1969) (along the dashed line the spin moment is in the plane of the diagram) and (b) two-dimensional asymmetric Néel-type distribution (after Hubert 1969) (along the full line the moment is in the plane of the diagram)

to the bottom and Fig. 5c contains four lines. It can be seen in Fig. 5c that the spin distribution across the lines can rotate in a clockwise or an anticlockwise sense.

The dynamic behavior of bubble domains in epitaxial garnet films can only be explained if some bubbles at rest contain complex spin structures of the type shown in Fig. 5b, c. In motion, bubble domain walls can develop a more complex internal structure. Vertical Bloch lines (which really have Néel character) were first experimentally observed in bubbles by electron microscopy (see *Magnetization Imaging*; Grundy et al. 1972). The presence of horizontal Bloch lines and structures known as Bloch points is also necessary for a full understanding of bubble dynamics. The motion of vertical Bloch lines along a stripe domain wall is the basis of a more advanced memory concept—the Bloch line memory.

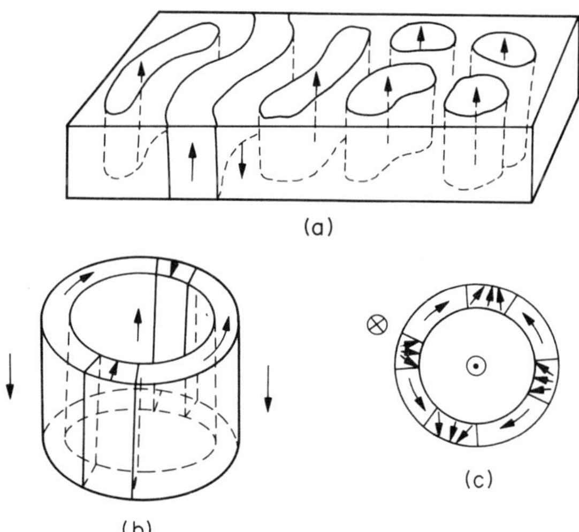

Figure 5
(a) Stripe and bubble domains in a garnet thin film, (b) Bloch lines in a bubble domain wall and (c) plan view of a bubble domain wall with a more detailed representation of the spin distribution (note that several lines may exist in a wall and that M_s can rotate in either sense)

See also: Anisotropy in Magnetic Materials; Barkhausen Effect; Coercivity and Domain Wall Pinning; Magnetic Materials: An Overview; Magnetic Materials: Basic Concepts; Magnetization Imaging; Micromagnetic Calculations; Power Loss in Magnetic Materials

Bibliography

Aharoni A, Jakubovics J P 1990 Structure and energy of 90° domain walls in thin ferromagnetic films. *IEEE Trans. Magn.* 26: 2810–12

Bloch F 1932 Theory of the exchange problem and of residual ferromagnetism. *Z. Phys.* 74: 295–335

Bruno P 1988 Magnetic surface anisotropy of cobalt and surface roughness effects within Néel model. *J. Phys. F* 18: 1291–8

Frenkel J, Dorfman J 1930 Spontaneous and induced magnetisation in ferromagnetic crystals. *Nature* 126: 274–5

Grundy P J, Hothersall D, Jones G A, Middleton B K, Tebble R S 1972 The formation and structure of cylindrical magnetic domains in thin cobalt crystals. *Phys. Status Solidi* A 9: 79–88

Huber E E, Smith D O, Goodenough J B 1958 Domain wall structure in Permalloy films. *J. Appl. Phys.* 29: 294–5

Hubert A 1969 Stray field free magnetisation configurations. *Phys. Status Solidi* 32: 519–34

Hubert A 1974 *Theorie der Domanenwande in Geordneten Medien*. Springer, Berlin

Jones G A, Middleton B K 1974 A review of domain walls in thin films. *Int. J. Magn.* 6: 1–18

Kittel C 1946 Theory of the structure of ferromagnetic domains in thin films and small particles. *Phys. Rev.* 70: 965–71

Kittel C, Galt J K 1956 *Ferromagnetic Domain Theory*, Solid State Physics, Vol. 3. Academic Press, New York, p. 437
LaBonte A E 1969 Two dimensional Bloch-type domain walls in ferromagnetic films. *J. Appl. Phys.* 40: 2450–8
Landau L, Lifshitz B 1935 On the theory of the dispersion of magnetic permeability in ferromagnetic bodies. *Phys. Z. Sowjetunion* 8: 153–69
Lilley B A 1950 Energies and widths of domain boundaries in ferromagnets. *Phil. Mag.* 41: 792–813
Malozemoff A, Sloncewski J 1979 *Magnetic Domain Walls in Bubble Materials*, Applied Solid State Sciences, Advances in Materials and Device Research, Suppl. 1. Academic Press, New York
Néel L 1954 Anisotropie magnetique superficiel et surstructure d'orientation. *J. Phys. Rad.* 15: 225–39
Néel L 1955 Energie des parois de Bloch dans les couches minces. *C. R. Acad. Sci. Paris* 241: 533–7
Weiss P 1907 L'hypothese du champ moleculaire et la propriete ferromagnetique. *J. Phys. (Paris)* 6: 661–90

P. J. Grundy
[University of Salford, Salford, UK]

E

Electromagnetic–Acoustic Transducers

The ultrasonic nondestructive evaluation (NDE) of materials, either for the detection and evaluation of flaws or for the measurement of material properties, depends critically on the performance of transducers which convert electrical energy into ultrasonic waves. Piezoelectric materials are usually used for transducers, and these must be coupled to the part via a solid or liquid bond. This bond can degrade the quality of inspection by introducing irreproducibility, wasted time, operating temperature limits and restrictions on the types of waves that can be excited. The electromagnetic–acoustic transducer (EMAT) is a device that overcomes these limitations by directly coupling to ultrasonic waves in a metal without the need for a couplant.

1. Physical Principles

In nonmagnetic metals, the physical principles of the EMAT are quite simple. Consider first a single wire, as shown in Fig. 1a. When this is driven by a dynamic current I_ω and placed next to a conducting body, eddy currents J_ω will be induced in the surface of that body. If, in addition, there is a magnetic field H_0 present, Lorentz forces will be exerted on these eddy currents. The generation of ultrasonic waves occurs as a result of the reaction of the material to these forces. The same structure can receive, as well as generate, ultrasonic waves by means of reciprocal processes.

In ferromagnetic materials, such as steel, additional mechanisms are found. Considering again the generation case, dynamic magnetic fields H_ω penetrate the material as well as the eddy currents. These dynamic fields modulate (M_ω) the magnetic state M_0 established by the static bias and, via magnetostrictive coupling, launch ultrasonic waves. In addition, direct interactions can occur between the magnetization of the material and the applied fields. Again, reciprocal processes exist whereby waves can be detected.

An important distinction between the Lorentz force and the ferromagnetic mechanisms is their dependence on the strength of the bias field. Whereas the Lorentz forces increase linearly with the bias field, the magnetostrictive and direct magnetic forces have a complex, material-dependent field variation. These can be used to advantage to obtain maximum efficiency at relatively low bias fields, but must be used with caution since piece-to-piece changes in magnetic properties occasionally cause unexpected efficiency variations.

2. Probe Configurations

A practical EMAT probe is constructed from elements such as shown in Fig. 1a. The wire is wound into a coil which induces an eddy-current distribution $\boldsymbol{J}$ into the near-surface volume of the metal. In the nonmagnetic case, this interacts with the static magnetic bias B_0 to produce a driving body force $f = \boldsymbol{J} \times \boldsymbol{B}_0$. Probe design is accomplished by shaping the coil and the magnets so that the driving force distribution matches as closely as possible the desired profile of the mode to be launched. In many cases the electromagnetic skin depth is sufficiently small with respect to the elastic wavelength that this force distribution can be approximated by a surface stress.

In the ferromagnetic case, the expressions for the driving force distributions are considerably more complex, involving the field-dependent magnetization

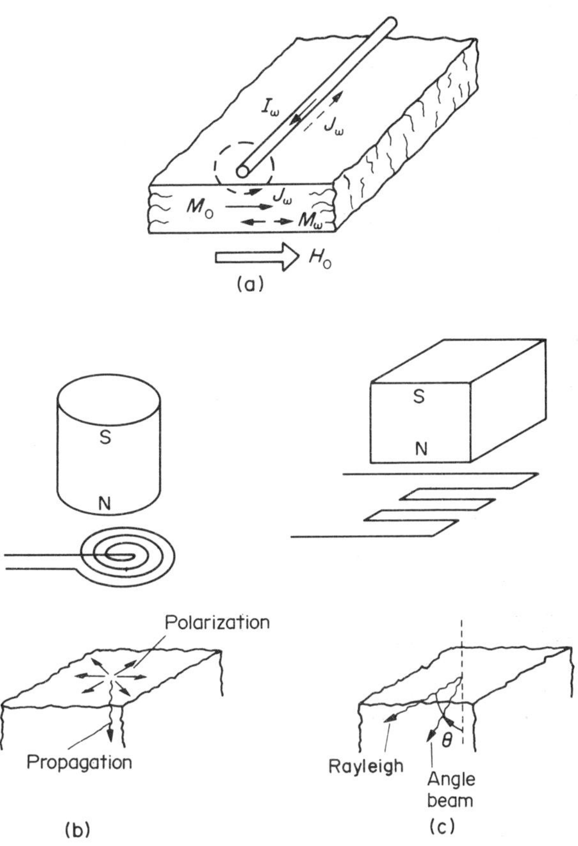

Figure 1
EMAT configurations: (a) principles, (b) spiral coil and (c) meander coil

and magnetostriction of the material. These force distributions may not have the same directions as the Lorentz forces. Moreover, the notion of an effective surface stress is not always appropriate, since higher-order moments of the depth-dependent driving forces may make important contributions to the generation process.

Figures 1b and 1c illustrate two of the simplest EMAT configurations that have found industrial application. The spiral coil configuration is used to generate shear waves, propagating normal to a part surface. The meander coil configuration excites Rayleigh waves (waves confined near surfaces), Lamb waves (guided modes of plates) and angle beams (plane longitudinal or shear waves inclined at an angle θ to the surface normal).

The polarization of ultrasonic waves is an important parameter which can be controlled more readily by EMATs than by other transducer types due to the fact that the driving force direction can be changed by altering the direction of B_0. In the spiral EMAT, the shear waves are radially polarized. In flaw detection applications, the resulting shaped mode is insensitive to on-axis flaws, but this can be overcome by using a two-probe system in which the probes are offset so that their radiating lobes overlap. Alternatively, the probe design can be made more complex to produce a plane-polarized wave.

The meander coil EMAT (Fig. 1c) differs from the spiral coils in that it is designed to excite a spatially periodic stress. If the period of this stress is matched to the wavelength of a Rayleigh (surface wave on a halfspace) or Lamb (plate) mode, then a constructive interference of the wavelets launched by the individual elements of the transducer occurs and maximum efficiency is obtained. This happens when the probe is driven at a frequency $f = V/D$ where V is the phase velocity of the mode and D is the coil period. At frequencies above that required to excite Rayleigh waves on halfspaces, the constructive interference occurs for plane waves propagating at an angle into the body. In this case, the angle with respect to the surface normal is given by $\theta = \sin^{-1}(V/fD)$. One important practical consequence of this is the fact that the angle can be rapidly scanned by electronically changing the frequency. In nonmagnetic materials with the bias configuration shown in Fig. 1c, the meander coil excites a vertically polarized wave whose displacement is in the plane of propagation. The orthogonally polarized, horizontal shear waves can be excited by the same meander coil in ferromagnetic materials when the magnetic bias is oriented parallel to the long elements of the coil. Various EMAT configurations are available for special purposes. This ability to selectively excite specific polarizations and radiation patterns is an important advantage of EMATs.

3. *Operational Characteristics*

The primary operational characteristics of any transducer are efficiency, polarization, bandwidth and beam pattern. The characteristics of EMATs are best discussed by comparing them with the more commonly used piezoelectric transducers.

Because of ohmic losses, EMATs have an efficiency of conversion of electrical to mechanical energy that is substantially less than that of piezoelectric probes. Whereas the latter have conversion losses that are typically 0–10 dB, EMAT losses in nonmagnetic materials are of the order of 40–60 dB. Ferromagnetic mechanisms can be used to reduce these losses in certain materials. Although these losses are large and rule out certain applications, they can often be overcome. On transmission, it is possible to use higher electrical driving powers than can be used with piezoelectrics, which suffer dielectric breakdown at high voltages. On reception, signal averaging techniques can be used to suppress noise. By these and other techniques, satisfactory signal-to-noise ratios can often be achieved despite the large conversion losses.

An advantage of the EMAT is the ability to operate without couplant. An important operational parameter is the variation in signal strength as a function of distance between the probe and the part. Typically, the efficiency decreases exponentially with distance. For spatially periodic probes, the rate is much greater than for uniformly polarized probes since the fields are much more tightly confined. A typical value for a surface wave probe is 20 dB $\mathrm{mm}^{-1}\,\mathrm{MHz}^{-1}$. The frequency dependence arises because of the smaller spatial period, and hence more rapid self-cancellation, of the higher-frequency probes. For the nonperiodic probes such as the spiral coil, the lift-off dependence is not as severe, and does not directly depend on frequency. A value of a few decibels per millimeter might be expected for a typical probe.

Another important characteristic of EMATs is their ability to operate at extremes of temperature. Operation as high as 700 °C has been reported. High-temperature operation can, of course, considerably modify or eliminate ferromagnetic contributions to the transduction.

Because EMATs have no mechanically resonant elements, they can exhibit wide bandwidths. The bandwidth of the bulk-wave EMATs is limited primarily by the electrical bandwidth of the coil and receiver circuitry and by radiation conditions. The spiral coil can be quite broadband, but Rayleigh-wave generation by a meander coil is narrowband since the condition for constructive interference is only exactly satisfied at one frequency.

Beam patterns of EMATs are similar, to first order, to those of piezoelectric probes since the radiated waves are governed by the same laws of diffraction. They only differ to the extent that the driving stresses differ.

4. Applications

Electromagnetic–acoustic transducers are finding a rapidly increasing number of field applications in which their high-temperature, high-speed operation or their ability to excite unusual wave types overcome their lower efficiency. These applications can be based on either the detection of signals reflected from material discontinuities or on the precise measurement of time delays. In the latter case, the absence of spurious signal phase shifts associated with couplant thickness variation offers an additional advantage of noncontact operation. Recent flaw detection applications include control of welding processes, inspection of austenitic cladding and weldments, monitoring of tubing and plate integrity and detection of railroad wheel and rail cracks during operation. Precise time delay measurements lead to thickness gauging and measurement of stress and texture (preferred grain orientation). As the role of on-line control expands in materials processing, EMATs should find an increasing range of applications.

5. Summary

Electromagnetic–acoustic transducers provide a means for inspecting metallic or magnetic materials without a couplant. By exciting a wide variety of wave types at elevated temperatures and high speeds, they broaden the range of problems that can be successfully addressed with NDE and open the way for improved automation. The direct magnetic and magnetostrictive response of ferromagnetic materials can be utilized to enhance the efficiency of these devices and excite different wave polarizations.

See also: Nondestructive Evaluation Using Magnetism; SQUID Design, Fabrication and Application

Bibliography

Alers G A, Huebschen G, Maxfield B, Repplinger W, Salzburger H J, Thompson R B, Wilbrand A 1991 Electromagnetic acoustic transducers. *Nondestructive Testing Handbook, Ultrasonics*, Sect. 10, Pt. 3. American Society for Nondestructive Testing, Columbus, OH, pp. 334–48

Dobbs E R 1973 Electromagnetic generation of ultrasonic waves. In: Mason W P, Thurston R N (eds.) 1973 *Physical Acoustics*, Vol. 10. Academic Press, New York, pp. 127–91

Frost H M 1979 Electromagnetic-ultrasound transducers: principles, practice, and applications. In: Mason W P, Thurston R N (eds.) 1979 *Physical Acoustics*, Vol. 14. Academic Press, New York, pp. 179–275

Thompson R B 1990 The physical principles of measurements with EMAT transducers. In: Thurston R N, Pierce A D (eds.) 1990 *Ultrasonic Measurement Methods, Physical Acoustics*, Vol. 19. Academic Press, New York, pp. 157–200

Wilbrand A 1987 Quantitative modeling and experimental analysis of the physical properties of electromagnetic-ultrasonic transducers. In: Thompson D O, Chimenti D E (eds.) 1987 *Review of Progress in Quantitative Nondestructive Evaluation*, Vol 7A. Plenum, New York, pp. 671–80

R. B. Thompson
[Iowa State University, Ames, Iowa, USA]

Electromagnetic Forming

Electromagnetic forming makes use of high-strength, pulsed magnetic fields to form metallic workpieces. The basic operating principle of this forming process, first developed around 1958, is as follows. A high-voltage capacitor is discharged through a work coil enclosing a tubular workpiece (see Fig. 1). The discharging current sets up an alternating magnetic field in an axial direction between the work coil and the tube and induces a circumferential eddy current in the tube. The skin effect restricts the eddy current to the outside surface of the tube; there is no magnetic field beneath this surface. The magnetic field applied to the tube is identical to a magnetic pressure radially directed towards the center of the tube. If strong enough, this magnetic field will change the shape of the tube by compression.

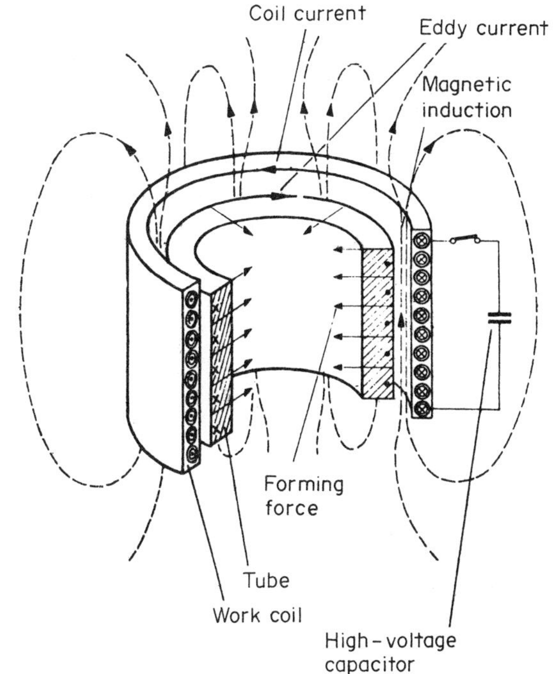

Figure 1
Operating principle of the electromagnetic forming process

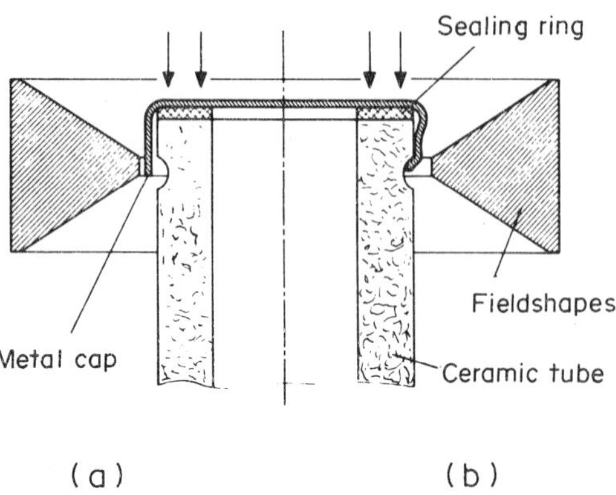

Figure 2
Pressing on the end cap of a high-voltage fuse: (a) before forming; (b) after forming

Conventional electromagnetic forming machines consist of a capacitor bank, the charging and discharging circuit and monitoring and protective devices. The capacitors are charged to a preselectable voltage (normally 8 kV maximum), so that between 2 kJ and 84 kJ of electrical energy is available according to the size of the capacitor bank. The discharging circuit, together with the inductance and ohmic resistance of the work coil, acts as a damped electrical oscillating circuit. The first current maximum attains values from 50 kA to 150 kA at a discharging frequency of around 10 kHz. Short-time high-strength magnetic inductions are generated which exert a pressure several times 100 MN mm^{-2} on the workpiece. Owing to the strong damping of the discharging current, only the first half-wave of the discharge deforms the workpiece. The work coil itself is exposed to strong mechanical alternating forces throughout the forming process. Conventional work coils are mechanically stabilized, electrically isolated, often water-cooled to dissipate the resistance losses and have a service life of around 10^6 discharges.

Fieldshapers are generally used to concentrate the magnetic pressure on those parts of the workpiece that are to be deformed. The fieldshaper has a disklike or cylindrical geometry with an internal hole to accept the workpiece; it is basically a secondary coil with one winding. With the aid of adapted fieldshapers, the work coil can be used to form workpieces of different diameters. As an example, Fig. 2 shows a sectional view of a fieldshaper used in the manufacture of high-voltage fuses. The special form and arrangement have the effect of concentrating the magnetic pressure on the edge of the fuse cap. The sealing pressure applied mechanically to the top of the cap prior to electromagnetic forming is subsequently retained by the bent edge. Correct selection of capacitor energy makes it a simple matter to bend the edge of the cap without damaging the ceramic tube.

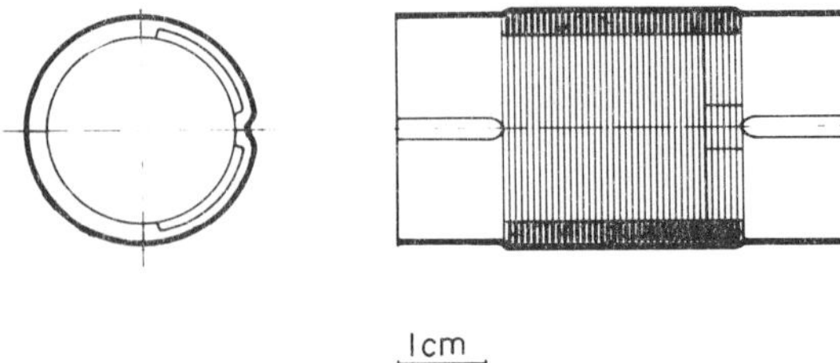

Figure 3
Contactless assembly of a dc motor stator

Table 1
Advantages and limitations of electromagnetic forming

Advantages
Low costs per forming operation
Contactless forming (also through protective coverings)
Forming pressure acts equally on all sides
A controlled pressure can be directed to definite zones of the workpiece
High forming rate and extremely short forming times
Easy adaptation to different workpieces by suitably formed fieldshapers
No moving parts in the electromagnetic forming machine
Production can be easily automated
Assemblies and equipment can be joined without close tolerances being required at the joints
Pipes can be compressed or widened (sheet metal can be deep-drawn, but this process is rarely used in practice)
Limitations
The materials to be formed must have a relatively high electrical conductivity (copper, aluminum, and their alloys)
The materials should be highly ductile
Wall thicknesses are limited to a few millimeters
The forming distances must be only a few millimeters

Figure 3 shows another application of electromagnetic forming. The stator laminations of a dc motor are held by an assembly mandrel. An aluminum tube is positioned over them and compressed over its whole length. The laminations are held together in this way. Seats for the two endshields are also produced by compressing the tube ends to the size of the mandrel.

Further applications of electromagnetic forming include the production of electrical connections and the assembly of maintenance-free ball joints and shock absorbers. The advantages and limitations of electromagnetic forming are summarized in Table 1.

Bibliography

Brower D F 1963 Metalforming with pulsed magnetic fields. *IEEE Trans. Ind. Electron. Control Instrum.* 10: 77–82

Dietz H, Lippmann H J, Schenk H 1969 Metallumformung mit hohen gepulsten Magnetfeldern. *Schweiz. Tech. Z.* 66: 500–10
Jones F W 1965 *An Investigation of the Magnetic-Pulse Forming Process.* AEC Research and Development Report Y-1483. Union Carbide, Oak Ridge, TN
Noland M C, Gadberry H M, Loser J B, Sneegas E C 1967 *High-Velocity Metalworking: A Survey*, NASA SP-5062. National Aeronautics and Space Administration, Washington, DC, Chap. 2

H. Schenk
[Siemens, Erlangen, Germany]

Elongated-Single-Domain Magnets

Elongated-single-domain (ESD) magnets are composed of ESD particles such as iron–cobalt embedded in a nonmagnetic matrix such as lead. Early magnetic theory, developed in the late 1940s, indicated that high coercivity is expected in fine magnetic particles which contain no domain walls and thus reverse their magnetization by rotation. The General Electric Company succeeded in commercially developing a family of such ESD magnets by electrodepositing elongated iron–cobalt particles into a mercury cathode and, after suitable extraction, compacting them into a lead matrix (Falk 1966). They are commercially known as Lodex. The magnetic properties, however, never met the expectations of theory.

There were three grades of Lodex available commercially. One grade, Lodex 40, 41, 42 and 43, was isotropic and prepared by pressing without field. A second grade, Lodex 30, 31, 32 and 33, was anisotropic and prepared by pressing with field applied perpendicular to the field direction. The third grade, Lodex 36, 37 and 38, was anisotropic and prepared by extrusion. Table 1 lists four grade numbers used by the Magnetic Materials Producers Association (1987).

The salient advantages of ESD magnets were dimensional precision and magnetic uniformity. Hence, they were often used in timing motors, meters, thermostats and relays. Owing to the intricate processing facilities required, however, production was limited to a single plant. This plant was closed at the end of 1988 and commercial production of ESD magnets has stopped.

Table 1
ESD magnets

Magnet (wt%)[a]	B_r (T)	B_r (kG)	H_c (kA m^{-1})	H_c (Oe)	$(BH)_{max}$ (kJ m^{-3})	$(BH)_{max}$ (MG Oe)
ESD 31 (20.7 Fe, 11.6 Co)	0.50	5.0	80	1000	18.3	2.3
ESD 32 (18.3 Fe, 10.3 Co)	0.68	6.8	76	960	23.9	3.0
ESD 41 (20.7 Fe, 11.6 Co)	0.36	3.6	77	970	8.7	1.1
ESD 42 (18.3 Fe, 10.3 Co)	0.48	4.8	66	830	9.9	1.25

a Balance Pb

See also: Iron–Cobalt Alloys

Bibliography

Falk R B 1966 A current review of Lodex permanent magnet technology. *J. Appl. Phys.* 37: 1108–12
Heck C 1974 *Magnetic Materials and Their Applications.* Crane-Russak, New York
McCaig M 1977 *Permanent Magnets in Theory and Practice.* Wiley, New York
Magnetic Materials Producers Association 1987 *Standard Specifications for Permanent Magnet Materials*, MMPA Standard No. 0100–87. MMPA, Evanston, IL

G. Y. Chin[†]
[AT&T Bell Laboratories, Murray Hill, New Jersey, USA]

J. H. Wernick
[Bellcore, Red Bank, New Jersey, USA]

Ferrites, Hard

Ferrites with high coercivity are called hard ferrites. The most important hard ferrites are the hexaferrites or M-type ferrites, characterized by the compound $MeFe_{12}O_{19}$ (Me = Ba, Sr, Pb) with the hexagonal magnetoplumbite or M structure. Their hard magnetic nature is due to the large magnetocrystalline anisotropy, with the hexagonal c axis as the preferred direction. M-ferrites are widely used as permanent-magnet materials, well known under various trade names, such as Ferroxdure, Koerox, Oxid and Ferrimag (Stäblein 1982).

This article concentrates on M-ferrites for permanent-magnet applications, but other applications of M-ferrites and other hard ferrites are discussed briefly. Unless specifically noted, the term ferrite is to be understood to mean M-ferrite.

1. General Characteristics of Ferrite Magnets

Ferrite magnets (Van den Broek and Stuijts 1977) were first introduced commercially by Philips under the trade name Ferroxdure, the isotropic form in 1952 and the anisotropic (crystal-oriented) form in 1954. Since then, ferrites have gradually acquired a dominant position in the permanent-magnet market (see Table 1), mainly by replacing the established alnico magnets (see *Alnicos*) in many applications.

The dominance in market value has become less in recent years due to the growing importance of the new rare-earth-based metallic magnet material $Nd_2Fe_{14}B$ (see *Neodymium–Iron–Boron Alloys*). This material has largely extended the separate market of low-volume, high-price and high-performance magnets for miniaturized systems, which was only represented before by the very expensive SmCo-based magnets (see *Cobalt–Rare-Earth Compounds*). The most important feature behind the great economic success of ferrite magnets is the low price per unit of available magnetic energy (see Table 1). This is particularly due to the relatively low cost and wide availability of the raw materials.

Compared with its direct competitor material (Al–Ni–Co), ferrite also has some technical advantages: the low B/μ_0H value at $(BH)_{max}$, the high chemical stability and the high electrical resistivity. The first has led to applications where large demagnetizing fields are present such as flat loudspeakers and compact dc motors. The last has allowed new applications at high frequencies where eddy currents are a major problem for metallic magnets.

On the other hand, ferrite magnets also have some disadvantages. The low value of $(BH)_{max}$ requires relatively big magnets. In addition, being a ceramic, it is brittle. Finally, the temperature coefficients of remanence $\alpha(B_r)$ and coercivity $\alpha(H_c)$ are relatively high.

Table 1
Comparison of the main permanent-magnet materials on the world market (magnetic values refer to most current grades)

	Ferrite	NdFeB	SmCo	Al–Ni–Co
B_r (mT)	370	1100	890	1200, 700
$_jH_c$ (kA m^{-1})	255	> 1000	1200	50, 150
$(BH)_{max}$ (kJ m^{-3})	25.5	215	150	40
B/μ_0H at $(BH)_{max}$	1.0	1.0	1.0	20, 5
Density (kg m^{-3})	4650	7400	8300	7300
$\rho(\Omega$ m)	10^4	10^{-6}	5×10^{-7}	5×10^{-7}
T_C (K)	750	585	995	860
$\alpha(B_r)$ (% K^{-1})	− 0.2	− 0.13	− 0.05	− 0.02
$\alpha(H_c)$ (% K^{-1})	+ 0.4	− 0.6	− 0.3	− 0.03
World production 1990				
(kt)	240	1.8	1.1	6.5
(million US$)	1100	310	310	230
Price ratio per unit[a]				
weight	1[b]	37	75	7.5
magnetic energy	1[b]	7.0	23	7.5

a Price per unit magnetic energy = price per unit weight × density/$(BH)_{max}$ b Ferrite is taken as reference

2. Crystal Structure

The magnetoplumbite crystal structure (Adelsköld 1938) is characterized by the space group $P6_3/mmc$. There is a close packing of oxygen and barium ions with iron ions at interstitial sites. The structure is built up from cubic blocks with the spinel structure and hexagonal blocks containing the barium ions. The iron ions are located at five different crystallographic positions. The dimensions of the unit cell, which contains two formula units, and the theoretical densities are given in Table 2 for the three main ferrites (BaM, SrM and PbM).

The M structure allows various substitutions. The most important ones are listed in Table 3. If the substituting ion does not have the nominal charge, it must be compensated. Some examples are $BaFe_{12-2x}Ti_x^{4+}Co_x^{2+}O_{19}$, $BaFe_{12-x}Fe_x^{2+}O_{19-x}F_x^-$, $Ba_{1-x}La_x^{3+}Fe_{12-x}Fe_x^{2+}O_{19}$ and $Ba_{1-2x}Na_x^+La_x^{3+}Fe_{12}O_{19}$.

3. Intrinsic Magnetic Properties

The magnetism of $BaFe_{12}O_{19}$ is due to the Fe^{3+} ions, each carrying a spin moment of five Bohr magnetons (μ_B). These are aligned by exchange interaction, either

Table 2
Crystallographic data and primary magnetic properties (at room temperature) of M-ferrites

Property	BaM	SrM	PbM
Crystallographic data			
Cell constants[a]			
a (nm)	0.588	0.586	0.588
c (nm)	2.317	2.303	2.302
Molecular weight	1111	1062	1181
Theoretical density (kg m^{-3})	5320	5130	5690
Magnetic properties[b]			
Saturation polarization, J_s (mT)	478	478	402
Specific saturation polarization, σ_s (μT m^3 kg^{-1})	90.5	93.4	70.7
Anisotropy constant, K_1 (kJ m^{-3})	325	357	220
Curie temperature, T_C (K)	740	750	725

a Adelsköld (1938) b Shirk and Buessem (1969) for BaM and SrM, Pauthenet and Rimet (1959) for PbM

parallel or antiparallel. Ions of the same crystallographic position are aligned parallel, thus constituting a ferromagnetic sublattice. The interaction between neighboring ions of different sublattices is due to superexchange by way of oxygen. The theory predicts that the ionic moments are inclined to be antiparallel, the more so the nearer the angle Fe–O–Fe approaches 180° and the distance Fe–O–Fe becomes smaller. The resulting most-probable spin configuration leads to a net magnetic moment per cell of eight Fe^{3+} ionic moments, that is, $40\mu_B$. This corresponds to a saturation polarization of 668 mT at 0 K for BaM, which agrees with experimental values (Smit and Wÿn, 1959, Kojima 1982).

The energy associated with the (anti)parallel coupling of the ionic moments is characterized by the exchange constant A, which is difficult to calculate exactly for superexchange in complex structures. Experimentally, it has been found that $A \simeq 5.1 \times 10^{-12}$ J m^{-1}, as derived from domain-width observations (on PbM). The temperature dependence of A, as shown in Fig. 1, is very small at room temperature (Gemperle et al. 1963).

The temperature dependence of the saturation polarization J_s for BaM is shown in Fig. 1. The J_s–T curve is much less curved than the usual Brillouin function (see *Magnetic Materials: Basic Concepts*), resulting in a relatively low J_s value at room temperature (478 mT) and a relatively high temperature coefficient (-0.2% K^{-1}).

The magnetization is strongly bound to the hexagonal c axis, owing to a high magnetocrystalline anisotropy. This anisotropy originates from spin–orbit coupling of the iron ions and is characterized by the anisotropy constant K_1. Higher-order constants (K_2, K_3) are negligibly small. The temperature dependence of K_1 is shown in Fig. 1 for BaM. Room-temperature values of K_1 for BaM, SrM and PbM are given in Table 2, together with other primary magnetic properties.

Based on the values for A, J_s and K_1 (Fig. 1), a number of secondary properties may be derived which govern the magnetic state (domain structure) or the mechanism of magnetization reversal. For instance, for BaM at room temperature, the anisotropy field $H_k = 2K_1J_s^{-1} = 1360$ kA m^{-1}, the (Kittel) critical particle diameter for single domain behavior of spherical particles $D_c = 72\mu_0(AK_1)^{0.5}J_s^{-2} = 0.5$ μm,

Table 3
Reported substitutions in $BaFe_{12}O_{19}$

$Ba^{2+}_{1-x}A^{k+}_{x}Fe^{3+}_{12-y}B^{l+}_{y}O^{2-}_{19-z}C^{m-}_{z}$ $[(k-2)x + (l-3)y = (m-2)z]$

A			B				C
$k=1$	2	3	$l=2$	3	4	5	$m=1$
Na	Sr	La	Mg	Cr	Ti	Sb	F
K	Pb	Pr	Mn	Mn	Ir	As	
Rb	Ca	Nd	Fe	Co	Ge	V	
Ag		Sm	Co	Al	Sn	Ta	
Tl		Eu	Ni	Ga	Zr	Nb	
		Bi	Zn	In			
			Cu	Sb			
				Sc			
				Ru			

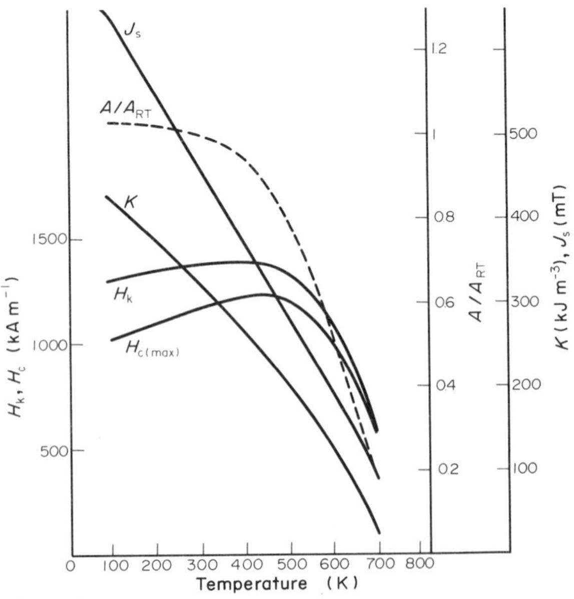

Figure 1
Temperature dependence of J_s, K_1, H_k and $H_{c(max)}$ for BaM (Smith and Wijn 1959) and of A/A_{RT} for PbM (Gemperle et al. 1963)

and $H_{c(max)} = H_k - NJ_s\mu_0^{-1} = 1132\,kA\,m^{-1}$, where $\mu_0 = 4\pi \times 10^{-7}\,T\,mA^{-1}$ is the permeability of free space and N is the self-demagnetizing factor parallel to the c axis, being about 0.6 for the usual platelet-shaped ferrite particles (see *Demagnetizing Factors*). $H_{c(max)}$ is the field required for magnetization reversal by coherent rotation of a single domain particle or an assembly of aligned, noninteracting particles. Thus, $H_{c(max)}$ represents an upper limit to the coercivity. However, magnetization reversal in ferrites generally occurs by nucleation and propagation of domain walls and the values of coercivity fall considerably below $H_{c(max)}$ (see also Sect. 4).

The temperature dependence of H_k and $H_{c(max)}$ is also shown in Fig. 1. It appears that $H_{c(max)}$ increases with increasing temperature up to about 300 °C. This agrees with the temperature dependence of experimental H_c values.

The intrinsic properties may be modified by substitution (see Table 3). Barium can be fully replaced by strontium or lead, and partly by calcium (<40 mol.%). CaM, stabilized with 0.03 mol.% La_2O_3, is also possible. The intrinsic properties of these M-ferrites vary somewhat (see Table 2) and other factors, such as sintering behavior and price of raw materials, often dictate their commercial viability. Although magnets based on CaM or PbM are feasible, large-scale production is concentrated on BaM and SrM. High-quality magnets are generally based on SrM and somewhat lower-priced magnets are based on BaM.

Substitution for Fe^{3+} has a drastic effect on the intrinsic magnetic properties. Partial substitution by aluminum or chromium decreases J_s without affecting K_1 seriously, resulting in larger H_k and H_c values. Substitution by Ti^{4+} and Co^{2+} causes a considerable decrease in K_1; the uniaxial anisotropy ($K_1 > 0$) may even change into planar anisotropy ($K_1 < 0$). Intermediate magnetic structures are also possible; for example, preferred directions on a conical surface around the c axis are observed for In^{3+} substitution. There are a few substitutions by which the K_1 value is increased while the J_s value is hardly affected, such as substitution of Fe^{3+} by Ru^{3+} or by Fe^{2+} compensated by La^{3+} at barium sites. Finally, there are some substitutions which may reduce the temperature coefficient of the remanence: $As^{3+/5+}$ compensated by Fe^{2+} and Cu^{2+} compensated by Me^{4+} or Me^{5+} (Kojima et al. 1982).

4. Microstructural Aspects

Apart from the intrinsic properties, microstructural factors are also important for the realization of the desired magnetic properties. The most important microstructural factors are particle size and shape, volume fraction of ferrite phase and degree of alignment of the c axes (Stäblein 1982).

The remanence B_r is a product of saturation polarization J_s, fraction of ferrite phase s and degree of alignment f. Typical values of s are 0.9 for sintered materials and 0.6 for plastic-bonded materials, the low value for the latter being due to the presence of the binder. The alignment factor f ranges from a value of 0.5 for isotropic magnets to about 0.9 for anisotropic magnets.

The coercivity H_c strongly depends on the microstructure (see *Coercivity and Domain Wall Pinning*). In most cases detailed relationships are not known. We concentrate here on fired or annealed materials where crystal imperfections do not play a significant role.

Typical ball-milled powder particles are somewhat larger than D_c and magnetization reversal proceeds by nucleation and growth of reversed domains. Nucleation is the limiting factor when the powder is sufficiently fine, say 0.5–1.5 μm (Schippan and Hempel 1985). In that case, the expression for the coercivity of a single crystallite becomes $H_c = H_n - NJ_s\mu_0^{-1}$, where H_n is the internal field needed for nucleation, approximated as $H_n = k_1 H_k$ ($k_1 \leqslant 1$). There are theoretical and experimental indications that k_1 increases with decreasing crystallite thickness (Holz 1970).

For an assembly of crystallites, f and s are also involved (Kojima 1982, Stäblein 1982). In the case of noninteracting single domain particles, random orientation causes a decrease of the polarization coercivity $_JH_c$, in theory by a factor of 0.48. The $_JH_c$ of specially prepared fine powders may indeed be close to this theoretical value. On the other hand, owing to the different reversal mechanism, the $_JH_c$ of typical ball-milled powders increases when f decreases. Mutual particle interactions give rise to an increase of the term $NJ_s\mu_0^{-1}$. In theory, this increase is proportional to s, but in practice this is only found after special dispersing techniques (Schippan and Hempel 1985).

With respect to typical polycrystalline materials (grain size ~ 1 μm) some complications arise:

(a) the H_n values of the individual crystallites may be lowered by exchange interaction across the grain boundary; and

(b) magnetization reversal on a macroscale may be nonuniform, proceeding by the initiation and growth of large multicrystal reversed regions.

Only a few studies are available, which mainly refer to grain-oriented materials. The influence of exchange is usually neglected (i.e., the H_n of the individual crystallites remains important). Starting from a distribution in H_n and a uniform distribution of reversed regions, a quite realistic demagnetization curve can be calculated (Heinecke 1966). On the other hand, the existence of nonuniform reversal has been demonstrated and, on this basis, an approximate expression for $_JH_c$ has been derived:

$${}_JH_c = k_1 H_k - N(B_r + J_s)\mu_0^{-1}$$

where k_1 and N refer to the average values (Kools 1985, 1988).

This expression shows the influence of various microstructural factors: grain size (on k_1), grain shape (on N), s and f (on B_r). Typical values both for isotropic and anisotropic materials are $k_1 \cong 0.5$ and $N \cong 0.9$.

The *quality* of a magnet is characterized by a combination of B_r and H_c. For static applications at high $B/\mu_0 H$, the $(BH)_{max}$ value can be used. For dynamic applications, however, $(BH)_{max}$ is less indicative, since it is hardly sensitive or is even insensitive to improved H_c values. In that case the product of B_r and ${}_JH_c$ is more suited (Van Heffen 1980). The factor K (see Table 4) is a measure of the processing quality (costs). It is in principle independent of the aluminum or chromium content with which the B_r/H_c ratio can be varied while the processing remains equivalent.

5. Commercial Ferrite Magnet Materials

The ferrite crystallites are bonded to solid bodies either by sintering or by plastic bonding and may be randomly oriented (isotropic) or aligned with the c axis in one direction (anisotropic). Depending on the binder used, plastic-bonded ferrites may be flexible or rigid (Stäblein 1982). The anisotropic sintered form is by far the most important one. These groups represent different application fields, as shown in Table 4.

Table 4 summarizes the most important characteristics of different material groups (F_1–F_9) and Fig.2 shows typical demagnetization curves. Where more than one grade exists within one group, the extreme B_r values are given in Table 4. Compared with plastic-bonded ferrites, sintered materials have higher magnetic quality and closer dimensional control (when machined). Plastic-bonded ferrites may be preferred because of their lower price or their special mechanical properties and shaping capabilities. Anisotropic materials are used when high magnetic quality is essential, whereas the cheaper isotropic material is preferred when low magnetic quality is acceptable or when the properties have to be isotropic.

Figure 3 provides a survey of typical processing routes employed in the production of ferrite. The manufacture of high-grade anisotropic SrM (Van den Broek and Stuijts 1977) is taken as an example. The raw materials are dry powders of $SrCO_3$, Fe_2O_3 and additives such as SiO_2 and B_2O_3. The desired quantities (Fe_2O_3–SrO molar ratio $\simeq 5.5$) are weighed and dry mixed. The mixture is granulated in a disk pelletizer to granules of about 5 mm and subsequently prefired at about 1250 °C in air. During prefiring, the raw materials react to form the desired compound ($SrFe_{12}O_{19}$). The hard prefired granules are wet-milled in steel ball mills to a fine powder. The resulting thick suspension (slurry) is processed to a crystallographically aligned pressed product by pressure filtration (wet pressing) in a magnetic field. After drying, the compacts are sintered at about 1250 °C in air. During sintering, anisotropic shrinkage occurs, about 15% perpendicular and about 25% parallel to the preferred direction. Since accurate dimensional control is not possible, the pole faces are ground afterwards.

The microstructural requirements impose demands on the sintering process and preceding operations. During sintering, considerable densification has to occur without significant grain growth. These more or less contradictory demands can be achieved satisfactorily by having small particles in the slurry and by using sinter additives such as SiO_2. For a high degree of alignment in the pressed product, the particles in the slurry should be single crystals and mobile. This in turn requires a small grain size of the prefired granule. Control of grain growth during prefiring can be attained by using excess SrO in combination with additives such as B_2O_3 and SiO_2. In addition, conversion to SrM must be sufficient, otherwise a high B_r cannot be obtained.

6. Applications

Owing to its low price, ferrite has replaced other magnet materials in existing systems, either with or without modifications to the system (Van den Broek and Stuijts 1977). This is particularly the case with static applications where small demagnetizing fields are involved. A typical example is the application of flat ferrite rings instead of high-metallic center core magnets in loudspeaker systems. The high H_c has stimulated the development of new systems, especially in dynamic applications where periodically high demagnetizing fields are present. A typical example is the electric motor with its strong armature reaction fields (Van Heffen 1980). New electric motors are being developed which require very high ${}_JH_c$ values lying far beyond the range of Al–Ni–Co materials, for example the starter motor, requiring ${}_JH_c > 320\ \mathrm{kA\,m^{-1}}$.

The main ferrite products are anisotropic segments for motors in cars (F_3–F_5), anisotropic rings for loudspeakers (F_2) and large anisotropic blocks for ore separators (F_3). These products together represent about 70% of the total production. It appears that large-scale production is concentrated on systems requiring relatively large magnets. In this case the greatest advantage is taken of the low material price. The remainder of production is distributed over a wide range of applications, involving mostly small magnets, such as shavers, mixers and coffee grinders.

Applications of ferrites other than permanent magnets are in the field of microwaves, magnetic bubble memories, magnetic recording and magnetooptics. M-ferrites are of interest for resonance-type microwave devices (e.g., isolators, filters, circulators)

Table 4
Commercial barium and strontium ferrite materials (typical values)

		Me[a]	B_r (mT)	$_BH_c$ (kA m^{-1})	$_JH_c$ (kA m^{-1})	K^b (mT)	$(BH)_{max}$ (kJ m^{-3})	μ^c_{rec}	H_s (kA m^{-1})	$\alpha(B_r)^d$ (% K^{-1})	$\alpha(H_c)^d$ (% K^{-1})	ρ (Ω m)	Density (kg m^{-3})	Typical application	Market share (%)
Sintered															
Isotropic	F_1	Ba	215	145	>240	>311	7.4	1.2		−0.2	0.4	10^4		rotors for cycle dynamos	~10
Anisotropic low $_JH_c$	F_2	Ba	400	160	165	466	29.5	1.1	560	−0.2	0.48	10^4	4900	flat rings for loudspeakers	
medium $_JH_c$	F_3	Sr	370	245	255	472	25.5	1.1	875	−0.2	0.37	10^4	4650	segments for windscreen wiper motors	
			425	250	260	529	33.6		955		0.37	10^4	4900		~70
high $_JH_c$	F_4	Sr	380	280	320	508	26.8	1.1	1115	−0.2	0.30	10^4	4700	segments for blower motors	
			400	295	330	532	30.5		1100		0.29	10^4	4850		
very high $_JH_c$	F_5		385	285	360	529	27.6	1.1	1200	−0.2	0.26	10^4	4850	segments for starter motors	
Plastic-bonded															
Flexible isotropic	F_6	Ba	125	88	190	201	2.8	1.15	800	−0.2	0.4	10^7	3100	doorcatches (e.g. for refrigerators), small dc motors (e.g., for fuel pumps)	~10
			145	96	190	221	3.6	1.15	800			10^6	3700		
anisotropic	F_7		200	140	200	280	8.0	1.05		−0.2	0.4				
			250	176	240	346	12.0	1.05							
Rigid isotropic	F_8	Ba	80	58	190	156	0.9	1.15	800	−0.2	0.4	10^8	2500	correction magnets for TV	
			155	104	190	231	4.4	1.15	800	−0.2	0.4	10^4	3900		~10
anisotropic	F_9	Sr	245	100	260	349	12.0	1.05	800	−0.2	0.4	10^5	3500	small dc motors	
			270	196	260	374	14.0		800			10^5	3900		

Source: McCaig and Clegg (1987); *Philips Components Data Handbook* (1991)
a Normally applied M-ferrite b $K(\text{mT}) = B_r(\text{mT}) + 0.4\,_JH_c(\text{kA m}^{-1})$, quality factor independent of the aluminum content c Recoil permeability d α = temperature coefficient (0.8 kA m^{-1} K^{-1} for BaM and 0.95 kA m^{-1} K^{-1} for SrM)

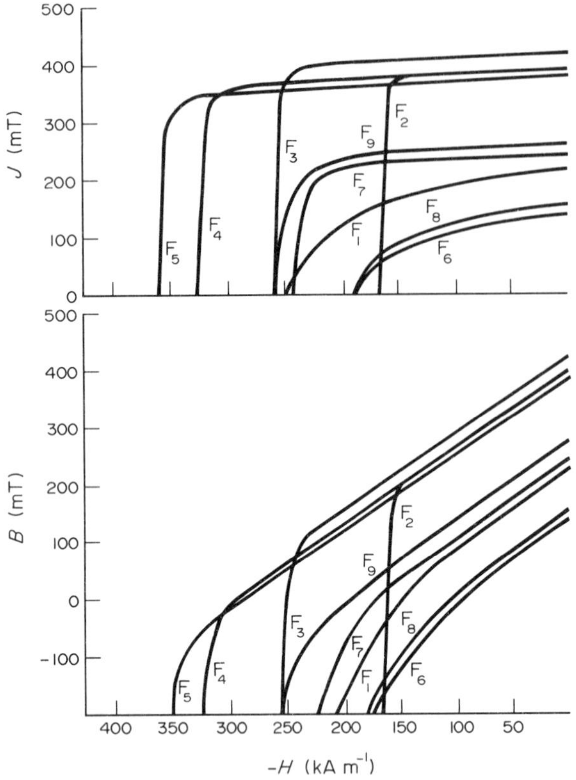

Figure 2
Typical demagnetization curves for the material groups of Table 3 (highest-quality values) (after McCaig and Clegg 1987, Philips Components Data Handbook 1991)

(Winkler and Dötsch 1979). Below 20 GHz such devices normally employ garnet or spinel ferrites in combination with a bias magnet. At high frequencies the required bias field becomes impracticably high ($>570\,kA\,m^{-1}$). M-ferrites are then preferred because of their large anisotropy field, which acts as built-in bias field and provides a resonance frequency of about 50 GHz in a small tuning field. For broader frequency coverage, various substitutions can be applied by which H_k is altered. For instance, ZnTi substitution is suitable for the 10–50 GHz range and aluminium substitution for the 50–100 GHz range. The efficiency of the resonance device is improved when the width of the resonance absorption line is reduced. For this reason sintered M-ferrites must at least be in the anisotropic form. Much better results, however, can be obtained with single-crystal materials.

Magnetic bubbles are cylindrical reversed domains in a thin crystal with its dominant preferred axis normal to its wide surface (see *Bubble Domain Materials*). They were first observed in a thin BaM crystal. The mobility of M-ferrite bubbles, however, has proved to be rather slow. The dominant bubble materials today are the garnets.

Interest in barium ferrite as a base material for particulate recording media has increased considerably since the invention of perpendicular recording in 1977 (see *Magnetic Storage Media*). This recording system is more suitable for high-density recording than the classical longitudinal recording system. Apart from thin metal films, particulate media can be used for perpendicular recording. The barium ferrite

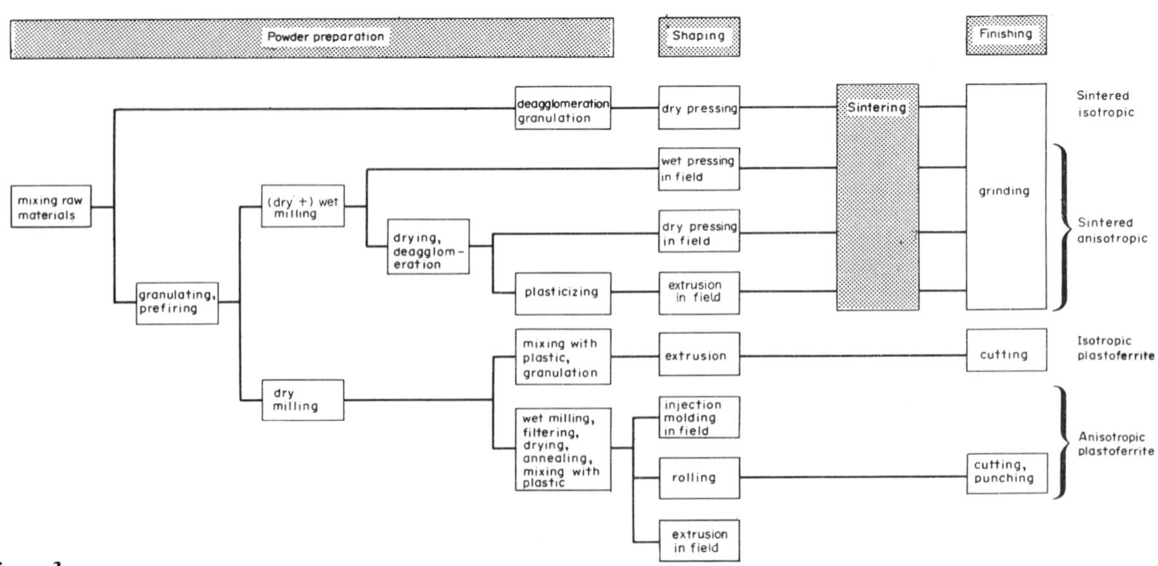

Figure 3
Typical processing routes for M-ferrite magnets (after Van den Broek and Stuijts 1977)

medium is particularly suitable since the platelet particle shape with the preferred axis normal to the platelet plane allows perpendicular alignment during the coating on the substrate film. The barium ferrite powder has a small particle size (~50 nm diameter) and a low coercivity, in the range 30–150 kA m^{-1}, which is adjusted by the extent of substitution (e.g., with cobalt plus titanium). To control the particle shape and size, special powder preparation techniques are applied such as crystallization from a B_2O_3 glass or hydrothermal synthesis.

Barium ferrite media allow high-density recording even in the low-coercivity range where ferrite heads are applicable, and have high output at short wavelengths. Moreover, and perhaps most characteristic, they allow high-speed contact duplication. For these reasons, barium ferrite media are expected to be suitable for high-density recording on tapes and disks for advanced analog and digital recording systems. Disks are already on the market (e.g., a 4 MByte flexible disk).

Despite promising characteristics (Sharrock 1989, Noda et al. 1990) barium ferrite media have not yet found their way to the mass production market. A variety of reasons for this includes outstanding technical problems, the still relatively high price and the fact that competing classical media have improved considerably in recent years.

7. *Other Hard Ferrites*

The definition of hard ferrites as iron-base oxide compounds possessing significant intrinsic magnetic anisotropy means that most other hexagonal ferrites as well as cobalt ferrite are to be included among the hard ferrites. The first commercial ceramic magnets, such as vectolite, were based on Fe_3O_4–$CoFe_2O_4$ mixed crytals. Minimum magnet specifications were $B_r = 0.16$ T and ${}_JH_c = 80$ kA m^{-1}. The magnetic hardness stems from an induced anisotropy, introduced by slowly cooling from 300 °C in a magnetic field, thus effecting ordering of the highly anisotropic cobalt ions. With the rise of M-ferrite magnets, those based on cobalt ferrite have gradually been ousted. Today, cobalt ferrite, in the form of γ-Fe_2O_3–$CoFe_2O_4$ mixed crystals, is sometimes used in tape recording.

In addition to M-ferrites, other hexagonal ferrites, such as $BaZn_2Fe_{12}O_{27}$ (Zn_2W), $BaZnFe_{14}O_{23}$ (ZnX) and $Ba_2ZnFe_{18}O_{30}$ (ZnU), also possess uniaxial anisotropies comparable with the M-ferrites, while values of J_s can be even higher. Thus, these compounds should in principle make comparable or even superior magnets. Commercial development, however, has been lacking, due to more difficult processing technology.

See also: Ferrites, Soft; Magnetic Materials: An Overview; Magnetic Storage Media

Bibliography

Adelsköld V 1938 X-ray studies on magneto-plumbite PbO. $6Fe_2O_3$, and other substances resembling "beta-alumina", Na_2O. Al_2O_3. *Ark. Kemi Mineral. Geol. A* 12(29): 1–9

Gemperle R, Shtolts E V, Zeleny M 1963 The temperature dependence of the domain structure of magnetoplumbite. *Phys. Status Solidi* 3: 2015–28

Heinecke U 1966 Uber ein Modell zur Berechnung der Entmagnetisierungskurven von Hartferriten. *Phys. Status Solidi* 18: 569–79

Holz A 1970 Formation of reversed domains in plate-shaped ferrite particles. *J. Appl. Phys.* 41: 1095–6

Kojima H 1982 Fundamental properties of hexagonal ferrites. In: Wohlfarth 1982, Chap. 5, pp. 305–91

Kools F 1985 Factors governing the coercivity of sintered anisotropic M type ferrites. *Proc. M.M.A.*, Grenoble, France

Kools F 1988 The effect of Al_2O_3 substitution on the microstructure and coercivity of polycrystalline $SrFe_{12}O_{19}$. *Proc. ICF$_5$*, Bombay, India

McCaig M, Clegg A G 1987 *Permanent Magnets in Theory and Practice*. Pentech, London

Noda M, Okazaki Y, Hara K, Ogisu K 1990 Characteristics of barium ferrite tape for magnetic contact duplication. *IEEE Trans. Magn.* 26: 81–6

Pauthenet R, Rimet G 1959 Variations thermiques des constantes d'anisotropie et de l'aimantation spontanée de la magnétoplombite. *C. R. Hebd. Seances Acad. Sci.* 249: 1875–7

Philips Components Data Handbook 1991 MA02—Permanent magnets. Philips, Eindhoven, The Netherlands

Schippan R, Hempel K A 1985 The influence of microstructural properties on the coercivity of presintered strontium hexaferrite. *Proc. ICF$_4$*, San Francisco, CA

Sharrock M P 1989 Particulate recording media: a review. *IEEE Trans. Magn.* 25: 4374–89

Shirk B T, Buessem W R 1969 Temperature dependence of M_s and K_1 of $BaFe_{12}O_{19}$ and $SrFe_{12}O_{19}$ single crystals. *J. Appl. Phys.* 40: 1294–6

Smit J, Wijn H P 1959 *Ferrites*. Philips Technical Library, Eindhoven, The Netherlands

Stäblein H 1982 Hard ferrites and plastoferrites. In: Wohlfarth 1982, Chap. 7, pp. 441–602

Van den Broek C A M, Stuijts A L 1977 Ferroxdure. *Philips Tech. Rev.* 37: 157–75

Van Heffen H J H 1980 Ceramic permanent magnets for d.c. motors. *Electron. Compon. Appl.* 3: 22–30; 120–5

Winkler G, Dötsch H 1979 Hexagonal ferrites at millimeter wavelengths. *Proc. 9th European Microwave Conf.* Microwave Exhibitions and Publishers, Sevenoaks, UK, pp. 13–24

Wohlfarth E P (ed.) 1982 *Ferromagnetic Materials. A Handbook on the Properties of Magnetically Ordered Substances*, Vol. 3. North-Holland, Amsterdam

F. Kools
[Philips Composants, Evreux, France]

Ferrites, Soft

The term ferrite, in the most general sense, applies to a wide variety of commercially important magnetic oxide ceramics. The utility of these materials is based on a unique combination of three characteristics: high electrical resistivity (particularly at frequencies greater than 10 kHz), a wide range of magnetic properties and ease of fabrication using standard ceramics techniques. "Soft" ferrites have low coercive forces ($< 80\ \mathrm{A\,m^{-1}}$). These materials are generally used when the dynamic behavior of the magnetic properties plays an important role in the application. Examples include high-frequency inductors and transformers, switching cores for memory applications and magnetic elements in microwave components.

Table 1
Magnetic moments, saturation flux densities and Curie temperatures of selected ferrites

Ferrite	$n\mu_B^a$	$n\mu_B^b$	B_s (298 K) (mT)	T_C (K)
$NiFe_2O_4$	2.0	2.3	328	858
$MnFe_2O_4$	5.0	4.6	535	573
$Mn_{0.5}Zn_{0.5}Fe_2O_4$	7.5	7.0	388	413
$Y_3Fe_5O_{12}$	5.0	5.0	175	553
$Y_3Fe_4AlO_{12}$	0	1.7	51	430
$Gd_3Fe_5O_{12}$	16.0	15.2	0.6	564

a Theoretical values b Experimental values

1. Crystal Chemistry

In most contexts the term ferrite refers specifically to the spinel ferrites with the general formula $MO.Fe_2O_3$. In the simple spinel compound, M refers to a divalent cation, typically Mg^{2+}, Mn^{2+}, Ni^{2+}, Zn^{2+}, Fe^{2+} or sometimes Co^{2+} or Cu^{2+}. The spinel crystal structure is cubic, with a face-centered-cubic oxygen anion lattice providing the framework and cations occupying one tetrahedral (A) and two octahedral (B) sites—AB_2O_4.

Magnetic iron garnets, in their simplest form, have a general formula $\{R_3^{3+}\}[Fe_2^{3+}](Fe_3^{3+})O_{12}$, where R^{3+} is Y^{3+}, one of the heavier magnetic rare-earth ions such as Sm^{3+}, Gd^{3+}, Eu^{3+} or Tb^{3+}, or Ca^{2+}. These ions occupy the large ($r \sim 0.1$ nm) dodecahedral $\{c\}$ sites in the garnet lattice. The Fe^{3+} on the three tetrahedral (d) sites can be replaced by Al^{3+}, Ga^{3+} or V^{5+} (to provide charge balance for any Ca^{2+} substitution), while In^{3+} or Cr^{3+} can replace Fe^{3+} ions on the somewhat larger octahedral $[a]$ sites. In the case of the a and d sites there can be some ambiguity in site occupancy.

The complex garnet unit cell, with 64 cations and 96 oxygen ions, tolerates very little variation in stoichiometry. This behavior has several important implications. Sintering and ion migration effects are slower in garnets, while the electrical resistivity can be considerably higher than in the case of spinels.

2. Magnetic Properties

2.1 Static Properties

One characteristic common to the magnetic properties of all ferrites is ferrimagnetism. In most magnetic metals and alloys, the strong exchange interaction between neighboring magnetic atoms results in a parallel alignment of their magnetic moments. In contrast, the magnetic interactions between neighboring magnetic cations in ferrites are weaker, as they must occur indirectly via the oxygen ion intermediaries. Moreover, the resultant interactions give rise to antiparallel alignment of neighboring magnetic moments. In the spinel and garnet structures, two or more cation sites with unequal occupancies exist. Thus, in general, antiparallel alignment between neighbors leaves a resultant magnetic moment. Magnetic moments, expressed as numbers of Bohr magnetons ($n\mu_B$), for several ferrites are given in Table 1. Differences between measured and theoretical values are due to nonideal site distributions and to deviations from the nominal cation moments. The resultant saturation flux densities at room temperature B_s (298 K) are also listed in Table 1.

The Curie temperature T_C in ferrites is generally lower than in metals, as the exchange interactions in ferrites are weaker. Curie temperatures of typical ferrites are given in Table 1.

In garnets with three magnetic sublattices, such as $Gd_3Fe_5O_{12}$ (GdIG), the interesting phenomenon of compensation occurs. At low temperatures the large net moment on the Gd^{3+} lattice dominates, while at higher temperatures the stronger Fe^{3+}–Fe^{3+} exchange dominates. The contributions cancel out exactly at the compensation temperature. This phenomenon can be put to practical use when a magnetic material with a near-zero M_s–T behavior is required. In GdIG, the M_s–T curve is flat between 200 °C and 250 °C.

Two other intrinsic magnetic properties of ferrites, magnetocrystalline anisotropy and magnetostriction, also have an important influence on the behavior of ferrites. The magnetic anisotropy energy causes $\mathbf{M}_s$ to prefer alignment with a specific crystal axis. In most ferrites the magnetocrystalline anisotropy constant K_1 is small and negative, leading to an alignment with the $\langle 111 \rangle$ axis. This behavior is associated with a negative K_1 contribution of Fe^{3+} ions on octahedral sites. Additions of Fe^{2+} or Co^{2+} ions to the lattice provide positive contributions to K_1 (and a tendency toward $\langle 100 \rangle$ alignment). An important consequence of this behavior is the possibility of achieving a near-zero K_1, at least over a limited temperature range. Low values of K_1 are critical to the performance of soft ferrites.

In soft ferrites it is also important to reduce magnetostriction—magnetic anisotropy induced by applied stress. Ions with quenched orbital moments, such as

Fe^{2+} and Co^{2+}, have magnetic moments which are tightly coupled to the lattice and consequently introduce appreciable magnetostriction. Hence, doping with these ions to reduce K_1 may well increase the total magnetic anisotropy of the constrained crystallites within a ferrite body.

2.2 Dynamic Properties

The dynamic response of a soft ferrite to an applied field is determined by the two magnetization processes—domain wall motion and magnetization rotation within the domains. Domain wall motion is particularly sensitive to extrinsic material characteristics such as grain size and grain-boundary structure, the presence of inclusions or pores within the grains, impurity levels and stresses.

The permeability μ is most commonly used to characterize the magnetic response of a ferrite to an applied field: $B = \mu H$. For small changes in field ($<50\ \mathrm{A\,m^{-1}}$) the response is essentially linear and the initial permeability μ_i characterizes the response. However, outside the linear regime soft ferrites exhibit hysteresis. An example of a B–H or hysteresis loop for a nickel ferrite is shown in Fig. 1. Definitions of μ commonly used to describe the nonlinear behavior include amplitude permeability μ_A, incremental permeability μ_Δ and maximum permeability μ_m (see Fig. 1).

Two other parameters used to characterize the hysteresis loop are the remanence B_r and the coercivity H_c. Both are sensitive to microstructure and anisotropy effects. The remanence of a randomly oriented, polycrystalline ferrite, for example, will achieve a value approaching $0.87B_s$ when magnetocrystalline anisotropy dominates and M_s is aligned with the $\langle 111 \rangle$ axis (i.e., $K_1 < 0$).

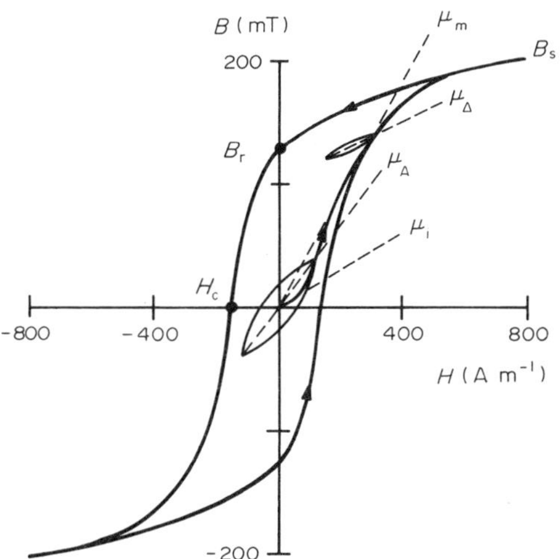

Figure 1
Hysteresis loop for a polycrystalline ferrite: the virgin curve defines μ_i and minor loops define μ_A, μ_m and μ_Δ

2.3 Losses

Loss for linear ferrites can be considered in terms of a generalized permeability:

$$\mu(\omega) = \mu' - j\mu'' \tag{1}$$

The loss factor is

$$\tan\delta = \mu''/\mu' = Q^{-1} \tag{2}$$

where Q is the quality factor.

At low signal levels, losses in soft ferrites are divided into three parts: hysteresis losses, eddy current losses and residual losses. Hysteresis losses are due to irreversible domain wall motion and are determined by the area of the hysteresis loop. The small signal loop in Fig. 1, which can be represented by a pair of parabolic expressions, is termed a Rayleigh loop. Under these conditions the loss tangent is given by

$$\tan\delta_H = 4vH_M/3\pi\mu_A \tag{3}$$

where v is the hysteresis coefficient and H_M is the maximum field. The magnitude of the hysteresis loss is, in general, related to the magnitude of the coercivity and is thus determined by intrinsic factors such as anisotropy and extrinsic factors such as inclusions, pores and grain boundaries.

Eddy current losses are a result of electrical currents induced in the ferrite body by the ac magnetic field. Thus, for a given microstructure, the eddy current contribution is directly related to the dc electrical resistivity. The dominant mechanism for conduction in ferrites is the easy transfer of electrons between Fe^{2+} and Fe^{3+} ions on the octahedral sites. Thus, the resistivity is determined by the ferrous ion content. Resistivities of MnZn ferrites range from $2\ \Omega\,\mathrm{cm}$ to $1000\ \Omega\,\mathrm{cm}$, while NiZn ferrites exhibit higher resistivities—as high as $10^9\ \Omega\,\mathrm{cm}$.

Measurements of dielectric properties of ferrites as a function of frequency show a behavior consistent with that of a composite material: the individual grains with a relatively low resistivity are surrounded by a thin ($\gtrsim 1$ nm) high-resistivity layer at the grain boundaries. This effect has been shown to be associated with segregation of CaO and SiO_2 at the boundaries (Akashi 1961). Therefore, low eddy current losses are best achieved in a fine-grained material in which the amount and distribution of additives is carefully controlled. The loss tangent of a ferrite core due to eddy currents is found to be proportional to frequency f and independent of field:

$$\tan\delta_E = (\pi\mu_0/\beta)\mu\sigma f \tag{4}$$

where β is a core geometry parameter, σ the bulk conductivity and μ_0 the permeability of free space.

Measurements of loss as a function of H_M and f provide a measure of $\tan \delta_H$ and $\tan \delta_E$. The residual loss is the remaining component after subtracting the hysteresis and eddy current contributions from the total loss.

The frequency dependence of μ_i for MnZn ferrites exhibits a dispersion behavior with a relaxation time $\tau = 1/\omega_0$, characteristic of Fe^{2+}–Fe^{3+} electron hopping. At room temperature ω_0 lies below 2 MHz.

High-resistivity materials such as NiZn ferrite are relatively free of Fe^{2+} and do not exhibit low-frequency relaxation processes. These materials exhibit a resonance behavior at ω_0. This resonance is ascribed to (a) domain wall resonance at frequencies below 100 MHz or (b) resonance of the rotational permeability in the anisotropy field at frequencies about 100 MHz.

2.4 Disaccommodation

Some relaxation processes observed in soft ferrites have relaxation times from several seconds to several years. The mechanisms associated with these processes involve migration of cation vacancies in the spinel lattice. This stabilization process inevitably results in the stabilization of a specific domain pattern, which lowers permeability. The decay of permeability with time is called disaccommodation. A strong perturbation, such as a mechanical or magnetic shock or a rise in temperature above the Curie temperature, results in recovery of a high permeability.

The disaccommodation factor (DAF) is defined as the change in permeability over a decade in time, say 10–100 min, after some perturbation or demagnetization treatment:

$$\mathrm{DAF} = (1/\mu^2)(\mu_2 - \mu_1)/[\log_{10}(t_2/t_1)] \qquad (5)$$

Disaccommodation factors can vary from a few parts per million in carefully prepared materials to several percent for ferrites with high cation vacancy concentrations.

3. *Processing Technology*

The methods used to fabricate ceramic ferrite shapes are simple, in principle. Raw material (i.e., Fe_2O_3, $MnCO_3$, ZnO, $NiCO_3$) powders are first blended and reacted (calcined) at high temperature. The partially converted mixture is then ball milled to form a fine (particle size ~ 1 μm) homogeneous powder. A binder is added and the dried powder is pressed into the desired shape. The binder is then burned off and the parts are sintered to form a strong, high-density core. The cores are machined to final dimensions, if necessary.

The major use of ferrite cores is in the "entertainment" market (yokes for television tubes and ferrite antenna rods). As the requirements for these applications are modest, the described process will produce adequate material with only moderate attention to critical steps in the process. More demanding applications, including low-loss ferrites for telecommunications filters, memory cores, high-permeability ferrites and microwave ferrites, require constant attention to every process step to achieve high quality consistently. The more critical aspects of the process include (a) raw material purity, (b) particle size distribution after milling, (c) pressing characteristics of powder, (d) binder behavior during burnoff, and (e) control of sintering temperature and atmosphere.

4. *Applications*

The wide range of applications for soft ferrites requires a corresponding range of special properties. In every case, however, it is the combination of high resistivity and low cost which is the compelling reason for the choice of ferrite over alternative materials.

4.1 Low-Loss Ferrites

Ferrite-core inductors are widely used in the telecommunications field to provide precise values of inductance in LC filters for frequency-division multiplexers. Parameters of most importance include the initial permeability μ_i and its stability with respect to temperature (TF) and time (DAF). Of equal importance is the inductor loss, which is determined by the μQ product of the ferrite core.

The most critical frequency-control applications require inductance stability corresponding to permeability changes of less than 1% over the temperature range and lifetime of the component. As most technical ferrites require Fe^{2+} to control permeability and TF, reduction in DAF is accomplished through the minimization of vacancy content.

The highest-permeability materials exhibit an undesirable temperature behavior (see Fig. 2). The sharp peak, associated with $K_1 \rightarrow 0$, is well defined in these large-grained, stress-free, homogeneous materials. A flatter temperature response is achieved by (a) adjusting the position of the $K_1 = 0$ peak so that it lies below the minimum operating temperature and (b) broadening the peak, through the introduction of inhomogeneity in the material.

At frequencies below 2 MHz, MnZn ferrites generally exhibit higher μQ values, as permeabilities are higher. Above 2 MHz, however, eddy current losses associated with excess Fe^{2+} ions limit application. The more nearly stoichiometric NiZn ferrites find application into the GHz range (see *High-Frequency Properties of Magnetic Materials*). NiZn ferrites achieve their optimum loss properties when cobalt additions are present. These materials are annealed at a low temperature (~ 500 °C) before use,

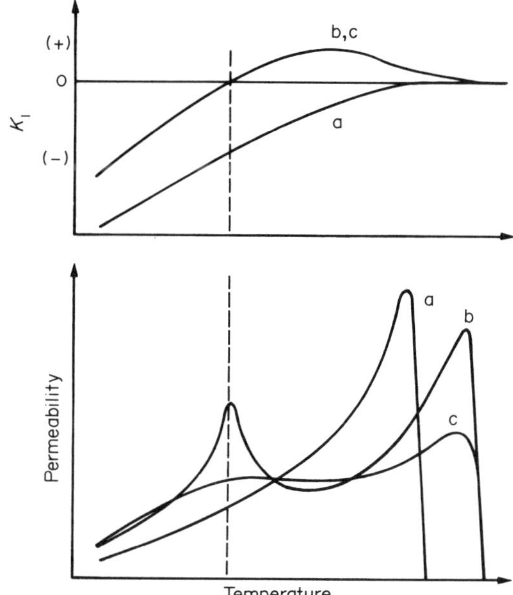

Figure 2
Temperature dependence of permeability and K_1 for MnZn ferrites; increasing Fe^{2+} content shifts $K_1 = 0$ to lower temperature: (a) $Mn_{0.61}Zn_{0.31}Fe_{2.08}O_4$; (b) $Mn_{0.56}Zn_{0.31}Fe_{2.13}O_4$ (Ohta 1963); (c) addition of 2.5 mol.% TiO_2 reduces μ and TF (after Stijntjes et al. 1970)

to produce high values of μQ and improved time stability.

4.2 High-Permeability Ferrites

High-permeability ferrites are used as core material for transformers whenever a wide frequency response is desired (i.e., wideband and pulse transformers). Below 50 kHz, permalloy alloys, in laminated tape or dust form, are generally used in this application, as permalloy has a higher μ and B_s. However, at higher frequencies eddy current effects degrade the permeability to levels below that of ferrites. MnZn ferrites are used up to about 1 MHz. From 1 MHz to 300 MHz, NiZn ferrites find use. Above this frequency range, ferrite permeabilities are not significantly greater than that of air.

To achieve permeabilities greater than 15 000 reproducibly, many conditions must be met. The choice of the bulk composition must be consistent with the simultaneous conditions $\lambda_s \to 0$ (where λ_s is the magnetostrictive constant, see *Magnetoelastic Phenomena*) and $K_1 \to 0$ in the vicinity of the operating temperature. High-purity raw materials, careful material preparation and control of sintering conditions are necessary to achieve the large grain size (>20 µm) needed to attain high domain wall mobility. Intragranular porosity must be avoided, as must residual stresses due to chemical inhomogeneity or uneven pressing conditions.

For a given transformer design, the bandwidth (or pulse response) is determined by the effective permeability μ_0 of the core structure, which is often assembled from two ferrite parts. The effective permeability is strongly perturbed by the presence of any air gaps in the structure. Thus, to utilize ferrite permeabilities above 20 000 in standard structures, gaps due to rough or uneven mating surfaces or to glue thicknesses must be reduced to less than 1 µm. The ability to lap and polish MnZn ferrite parts to provide flat, smooth mating surfaces is clearly equally important to the achievement of high permeability.

4.3 Ferrites for Power Applications

The use of ferrite cores in power transformers has increased dramatically as operating frequencies of power supplies have continued to rise. As in other applications, the low eddy current loss at high frequency is the major attraction of ferrites for these applications.

The ferrite core in a power transformer serves as the energy storage medium for a switched-mode (ac → dc) or converter (dc → dc) power supply. Desired characteristics include a high B_s, which determines the storage capacity of the core, and a low power loss, which determines the transfer efficiency.

High power densities, together with loss in the core and winding, result in heating of the core. In many designs it is the maximum permissible operating temperature of the core which limits the transformer performance. For this reason reduction in power loss is critical. In addition, it is important that core characteristics be "tuned" to the operating temperature, which may be from 60 °C to as high as 120 °C.

The desire for high B_s indicates the use of MnZn ferrites for this application. Materials with $B_s > 400$ mT at 100 °C are available. As with low-loss materials (see Sect. 4.1), losses are reduced by (a) minimizing hysteresis losses and magnetostriction at the operating temperature, (b) adding calcium and impurities to provide high-resistivity grain boundaries and (c) controlling microstructure. In addition, it is important to reduce porosity to achieve near-theoretical values of B_s. Power ferrites now available exhibit power losses at 200 kHz and 100 mT of less than 50 mW g^{-1} when measured at 100 °C.

4.4 Square-Loop Ferrites

Applications for square-loop materials include switches, shift registers and multiaperture logic and storage devices. The square hysteresis loop clearly defines two equally stable remanent states, $+B_r$ and $-B_r$, corresponding to the "0" and "1" digits.

Characteristics important to the performance of a high-speed ferrite core memory include (a) high remanence ratio B_r/B_s, (b) fast switching time τ_s, (c) a small but well-controlled coercive force H_c, (d) high B_s and (e) low temperature coefficients.

A square hysteresis loop in a ferrite is achieved when B_r/B_s approaches the theoretical value of 0.87. To meet this requirement K_1 should be negative and magnetostriction λ_s must be small. Compositions which meet requirements on K_1 and λ_s are generally mixed manganese ferrites with additions of magnesium, lithium, copper or zinc. The requirements on microstructure include low porosity and small grain size.

Switching time τ_s is dependent on the applied field H_a:

$$\tau_s = S_w/(H_a - H_0) \qquad (6)$$

where H_0 is a threshold field and S_w is the switching constant. Most square-loop materials exhibit S_w values around $100\,\mathrm{A\,m^{-1}\,\mu s^{-1}}$. Empirical evidence suggests that S_w increases when Fe^{2+} ions are present.

It is of vital importance to manufacture cores with a high degree of uniformity of properties. Parameters which must be controlled carefully include chemistry, mechanical dimensions, grain size, porosity and residual stresses. Even with the tightest controls, 100% testing is required to assure the performance of square-loop cores.

4.5 Microwave Ferrites

Microwave ferrite devices make use of the nonreciprocal interaction between an electromagnetic field and a ferromagnetic medium (see *High-Frequency Properties of Magnetic Materials*; *Microwave Materials and Devices, Magnetic*). These devices operate at frequencies from 300 MHz up to 50 GHz.

The source of the nonreciprocal interaction is the gyromagnetic response of a ferrite saturated by a dc field H_0 and excited by a small perpendicular ac signal $\boldsymbol{h}$ (on xOy plane). The equation of motion describing this response is similar to that for a gyroscope:

$$d\boldsymbol{M}/dt = -\gamma(\boldsymbol{M} \times \boldsymbol{H}) + \text{damping} \qquad (7)$$

where γ is the gyromagnetic ratio ($\gamma = 35\,\mathrm{MHz\,(kA\,m^{-1})^{-1}}$). Damping is associated with coupling between spin waves excited by the ac signal and the lattice. For materials magnetically isotropic on the xOy plane the solution of this equation gives a tensory permeability:

$$\begin{bmatrix} b_x \\ b_y \\ b_z \end{bmatrix} = \begin{bmatrix} \mu & -\mathrm{j}\kappa & 0 \\ \mathrm{j}\kappa & \mu & 0 \\ 0 & 0 & \mu_0 \end{bmatrix} \begin{bmatrix} h_x \\ h_y \\ h_z \end{bmatrix} \qquad (8)$$

where $\mu = 1 + \chi_{xx} = 1 + \chi_{yy}$ and $\mathrm{j}\kappa = -\chi_{xy} = \chi_{yx}$. The terms μ and κ are complex when damping is present.

It is the off-diagonal terms $\pm\mathrm{j}\kappa$ which give rise to the nonreciprocal effects. For example, if the ac signal is propagating in a direction parallel to the dc field H_0, the scalar permeabilities depend on the polarization of the field. Figure 3 shows the real and imaginary components of $\mu_+ = \mu - \kappa$ and $\mu_- = \mu + \kappa$, the permeabilities for a positively and negatively polarized signal, respectively, with respect to H_0. The positively polarized signal experiences a resonance absorption (ferromagnetic resonance) at $H_0 = \gamma\mu_0/\omega$, while the negatively polarized signal response is relatively insensitive to changes in H_0.

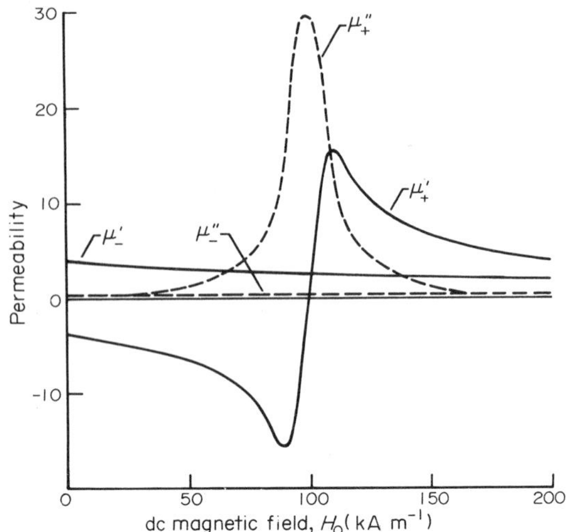

Figure 3
Effective permeabilities for circular polarization in an infinite ferrite medium; computed for a frequency of 3.5 GHz, $B_s = 300$ mT and $\Delta H = 20\,\mathrm{kA\,m^{-1}}$

Microwave ferrite devices exploit differences in μ_+ and μ_- in several ways. Resonance isolators use the highly efficient absorption at H_0 to attenuate selectively signals reflected backward down a waveguide, while presenting a very low impedance to forward signals. Other devices, such as phase shifters and field-displacement isolators, utilize differences in μ_+ and μ_-, both above and below the resonance peak. Three-port circulators transmit signals selectively from one port to another ($1 \rightarrow 2$, $2 \rightarrow 3$ or $3 \rightarrow 1$) while strongly attenuating other signal paths.

Both spinel and garnet ferrites are used in microwave devices. Material characteristics of concern to the user include (a) dielectric losses (usually associated with the presence of Fe^{2+} ions), (b) magnetic losses (primarily associated with such contributions to ferromagnetic resonance linewidth ΔH as porosity, spin–orbit coupling and magnetocrystalline anisotropy) and (c) M_s and its temperature coefficient.

Garnets find major application at lower frequencies (<5 GHz), where lower values of M_s are desired to minimize natural resonance effects. The low dielectric losses ($\tan\delta \sim 10^{-4}$) and narrow ΔH values (2–4 $\mathrm{kA\,m^{-1}}$) are among the desirable properties.

Above about 5 GHz, spinel ferrites based on MnMg, NiAl and LiZn substitutions are commonly used. The higher M_s values can be effectively used at higher frequencies. However, linewidths are large (10–

40 kA m^{-1}) as a result of anisotropy broadening, while spin wave linewidths are small. Dielectric losses also tend to be higher ($\tan\delta \sim 5 \times 10^{-4}$) because of finite Fe^{2+} levels.

See also: Ferrites, Hard; Magnetic Materials: An Overview; Magnetic Materials: Basic Concepts; Magnetostrictive Materials; Microwave Materials and Devices, Magnetic

Bibliography

Akashi T 1961 Effect of the addition of CaO and SiO_2 on the magnetic characteristics and microstructures of manganese–zinc ferrites ($Mn_{0.68}Zn_{0.21}Fe_{2.11}O_{4+\delta}$). *Trans. Jpn. Inst. Met.* 2: 171–6

Craik D J (ed.) 1975 *Magnetic Oxides.* Wiley, London

Gyorgy E M 1975 The imperfect solid-magnetic properties. In: Hannay N B (ed.) 1975 *Defects in Solids, Treatise on Solid State Chemistry*, Vol. 2. Plenum, New York, Chap. 6, pp. 395–467

Ohta K 1963 Magnetocrystalline anisotropy and magnetic permeability of Mn–Zn–Fe ferrites. *J. Phys. Soc. Jpn.* 18: 685–90

Smit J, Wijn H P J 1959 *Ferrites: Physical Properties of Ferrimagnetic Oxides in Relation to their Technical Applications.* Wiley, New York

Snelling E C 1988 *Soft Ferrites: Properties and Applications.* Butterworth, London

Stijntjes T G W, Klerk J, van Groenou A B 1970 Permeability and conductivity of Ti-substituted MnZn ferrites. *Philips Res. Rep.* 25: 95–107

Wohlfarth E P (ed.) 1980/82 *Ferromagnetic Materials. A Handbook on the Properties of Magnetically Ordered Substances*, Vol. 2/3. North-Holland, New York

R. C. Sundahl Jr.
[Signal Research Laboratory, Des Plaines, Illinois, USA]

Y. S. Kim
[Bell Laboratories, Allentown, Pennsylvania, USA]

Fluids, Magnetic

It is unlikely that any liquid will be found that is intrinsically ferromagnetic. Nickel, cobalt and iron, for example, are only ferromagnetic in the solid state. It is fluid media composed of solid magnetic particles of subdomain size (see *Magnetic Materials: Basic Concepts*) colloidally dispersed in a liquid carrier that are the basis for the highly stable, strongly magnetizable liquids known as magnetic fluids or ferrofluids.

1. Phenomena of Magnetic Fluids

Magnetic fluids exhibit fascinating, unique phenomena. Attraction of ferrofluid against gravity to the poles of a magnet illustrates the most characteristic response (see Fig. 1). Less obviously, the fluid surrounding a vertical current-carrying wire rises in a conical meniscus around the wire. Also, the field of a centrally disposed wire prevents capillary disintegration of a fluid cylinder into droplets. Numerous other phenomena relating to fluid configurational stability are recognized, such as the onset of peaks containing the liquid in a repeating geometric pattern on the surface of a ferrofluid subjected to a uniform, perpendicular magnetic field. A related phenomenon occuring with fluid confined between flat plates produces a labyrinthine pattern (see Fig. 2); liquid magnetic bubbles may be formed, analogous to solid-state magnetic-bubble domains, and application to optical displays of information has been developed.

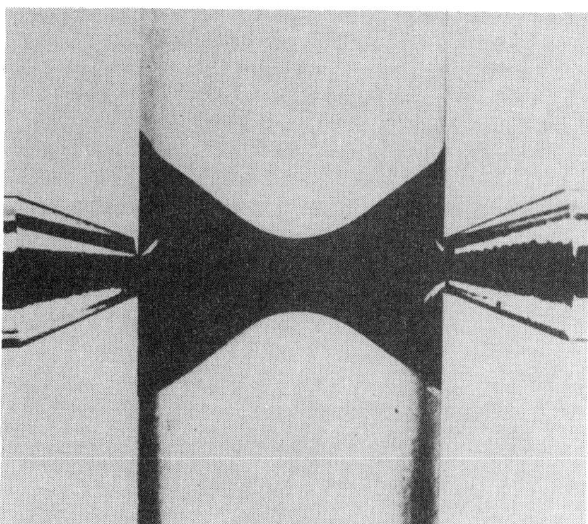

Figure 1
Photograph of the opaque, colloidal magnetic fluid positioned and held against the force of gravity by the gradient magnetic field of two pole pieces (after Rosensweig et al. 1968)

The familiar buoyancy relationships of Archimedes are subject to revision with magnetic fluids. Thus, when immersed in a magnetic fluid, dense nonmagnetic objects may be floated with application of a magnetic-field gradient. An object may be levitated in the interior of the fluid space near a point of minimum field; extraordinary also is the discovery that a permanent magnet levitates itself within a container of ferrofluid.

While ordinary fluids are free of body couples, it is found that a rotating magnetic field induces circulating motion in ferrofluid. The motion is directly related to the generation of free surface couple accompanied by the appearance of antisymmetric stress.

None of these phenomena depends on electric current flow in the ferrofluid as there is none; the forces are due to magnetic polarization.

Among other effects, optical birefringence, spin-glass behavior, magnetoelectric effect and modification of ultrasound propagation are reported in magnetic fluids.

Figure 2
Liquid magnetic labyrinth resulting from application of uniform magnetic field to a mixture of magnetic fluid and a clear immiscible fluid confined between closely spaced (400 μm) glass plates: the width of a path in the labyrinth is about 1 mm and applied field is perpendicular to the plates (after Romankiw et al. 1975)

2. Colloidal Stability of the Ferrofluid

The particles of a ferrofluid are subdomain, so each is magnetized uniformly to saturation and the maximum energy to separate a particle pair is given as $\mu_0 M^2 V/12$, where μ_0 is the permeability of free space ($4\pi \times 10^{-7}\,\mathrm{H\,m^{-1}}$), M is the magnetization ($\mathrm{A\,m^{-1}}$) and V is the volume ($\mathrm{m^3}$). Thermal energy kT tends to disrupt agglomerates, so the ratio of these energy terms defines a coupling coefficient and yields a criterion for the requisite monodisperse particle size:

$$d \leqslant (72kT/\pi\mu_0 M^2)^{1/3} \qquad (1)$$

For magnetite at 298 K, Eqn. (1) yields a diameter d of 7.8 nm (78 Å). Thus, ferrofluids have little in common with the relatively coarse 0.5–40 μm suspensions that solidify in a field in magnetic clutches.

Van der Waals forces also tend to agglomerate the particles of a ferrofluid. However, the particles may be kept apart with a surface-adsorbed layer of a long-chain molecule having one or more polar heads anchoring the molecule to the particle surface. This furnishes an elastic protective layer with the mechanism known as steric repulsion.

The energetics of these processes are sufficiently understood that semiquantitative calculations can be done. Computer Monte Carlo simulation of clustering in a system of oversize particles is illustrated in Fig. 3. In the absence of an external field the particles cluster as shown as open chains having no particular orientation, while in a strong external field the chains are straight and parallel.

A ferrofluid with hydrocarbon carrier fluid remains in existence (i.e., is stable) after 25 years of shelf life.

3. Preparation of Ferrofluids

There are two broad methods of preparing magnetic ferrofluids—size reduction and precipitation. It is remarkable that size reduction by grinding, discovered in the early 1960s, can reduce bulk material to sizes of the order of tens of nanometers. Grinding is done in the presence of a dispersing agent and a solvent for time durations of the order of 1000 h. Oleic acid in kerosene carrier furnishes the classical composition. A proportion of dispersant to solid is utilized corresponding to a monolayer coating in the finished product.

In chemical precipitation, iron (II) and iron (III) ions in aqueous solution are coprecipitated in the molar ratio of about 2:1 using ammonium hydroxide. To maintain the magnetite product in the small colloidal size range a peptization step is included, in which particles are transferred to a heated organic phase containing the dispersing agent.

Certain ferrofluids produced in aqueous carrier may be dispersed in alternate carriers by exchanging the surface-adsorbed dispersant species. Alternate carriers have included paraffinic and aromatic hydrocarbons, esters and diesters, perfluorocarbons and polyphenyl ether.

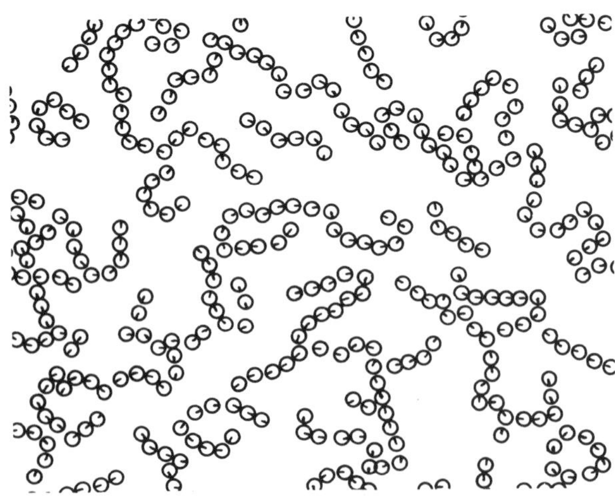

Figure 3
Microscopic view of magnetic particle clusters in magnetic fluids having a relatively large value of the coupling coefficient. Each circle represents a cobalt particle of about 15 nm diameter. The bar inside any circle indicates the direction of magnetization (after Chantrell et al. 1980)

Certain paramagnetic salt solutions (holmium nitrate, manganese chloride) are appreciably magnetizable in a high field and offer additional features such as transparency.

4. Properties of Magnetic Fluids

4.1 Magnetization

Each magnetic particle in a colloidal ferrofluid possesses an inherent magnetic moment m. At equilibrium the tendency for the moment to align with an applied field is counteracted by the disorienting influence of thermal agitation. The relaxation mechanism has two origins: reversal of magnetic moment within the grain (Néel) and rotation of particles in the liquid (Brownian mechanism). The resulting superparamagnetic relationship for mean component $\bar{m}$ in the direction of the field is

$$\bar{m}/m = \coth \xi - \xi^{-1} = L(\xi) \quad (2)$$

where the argument of the Langevin function is

$$\xi = \mu_0 mH/k_B T \quad (3)$$

The saturation magnetization of ferrofluids ranges up to about $6 \times 10^4\,\mathrm{A\,m^{-1}}$ (800 G), with initial susceptibility in the range 1–5 in the most concentrated preparations. Mössbauer spectra indicate that the bonding of organic molecules pins the moments of surface atoms, yielding an apparent dead surface layer.

4.2 Viscosity

Ferrofluid retains its flowability in the presence of a magnetic field even when magnetized to saturation. Nonetheless, the rheology is affected by the presence of the field, since the particles are inhibited from rotating freely around any axis except that established by the field direction. As a result, the rate of viscous dissipation of energy increases during shear flow. A magnetoviscous expression assuming the domain magnetization is locked with respect to the particle crystal axis has been derived for dilute suspensions:

$$\frac{\eta_s}{\eta_0} = 1 + \frac{5}{2}\varphi + \frac{3}{2}\varphi \left(\frac{\xi - \tanh \xi}{\xi + \tanh \xi}\right) \sin^2\beta \quad (4)$$

where η_s is the mixture viscosity; η_0 is the carrier viscosity; φ is the solids volume fraction; the term $\frac{5}{2}\varphi$ is the Einstein contribution due to pure strain perturbed by presence of a sphere; ξ is again the argument of the Langevin function (see Eqn. (3)); and β is the angle between the magnetic field and the local fluid vorticity vector. The maximum increment due to magnetoviscous influence is $\frac{3}{2}\varphi$, a prediction that is supported in experiments in dilute suspensions. Experiments with concentrated colloids exhibit a ratio of colloid viscosity with and without field up to nearly 4, with the value correlated by the group $\Gamma = \gamma\eta_0/\mu_0 MH$, where γ is the shear rate. For $\Gamma > 10^{-4}$ the viscosity ratio is less than 1.5 and approaches unity as Γ increases.

4.3 Stress and Force Density

In slowly changing applied fields the magnetization vector of a ferrofluid is collinear with the local magnetic field vector and the stress tensor **T** takes the form

$$\mathbf{T} = \mu_0 \left[\int_0^H \left(\frac{\partial Mv}{\partial v}\right)_{H,T} dH + \frac{H^2}{2}\right]\mathbf{I} + \boldsymbol{HB} \quad (5)$$

where v is specific volume. An arbitrary surface within the fluid having unit normal $\boldsymbol{n}$ experiences a traction force $\boldsymbol{n}.\mathbf{T}$. Introducing this force expression into the momentum balance for the fluid yields the continuum description of ferrohydrodynamics that successfully describes quasistatic magnetic flows; the appearance of a surface force density directed normal to an interface often plays a key role in the flows, particularly as the changing shape of a fluid interface changes the local magnetic field and the forces acting on the fluid.

When a field changes orientation rapidly, the finite relaxation rate of the magnetization produces flow fields with unsymmetric stresses due to the presence of a magnetic body couple. **T** from Eqn. (5) remains valid and contributes terms to the momentum equation arising from the noncollinearity of $\boldsymbol{H}$ and $\boldsymbol{B}$.

5. Applications

Nonmagnetic particles in a thin layer of magnetized fluid comprise relatively diamagnetic holes that repel each other and self-organize into a periodic lattice; the effect furnishes a macroscopic model of two-dimensional molecular melting.

The unique phenomena of magnetic fluids have led to such commercialization as zero-leakage rotary-shaft seals, inertial dampers and fluid-cooled loudspeakers. Sink/float separation of particles according to mass density is employed for sorting materials. Many additional applications have been demonstrated in optics, sensors, printing, transducers, surface grinding, medicine and other fields.

See also: Magnetic Materials: An Overview; Magnetic Materials: Basic Concepts

Bibliography

Berkovsky B (ed.) 1978 *Thermomechanics of Magnetic Fluids, Theory and Applications*. Hemisphere, Washington, DC

Chantrell R W, Bradbury A, Popplewell J, Charles S W 1980 Particle cluster configuration in magnetic fluids. *J. Phys. D* 13: L119–22

Kamiyama S, Rosensweig R E 1987 Magnetic fluids bibliography. *J. Magn. & Magn. Mater.* 65: 401–39

Papell S S 1965 Low viscosity magnetic fluid obtained by the colloidal suspension of magnetic particles. US Patent No. 3,215,572
Romankiw L T, Slusarczuk M G, Thompson D A 1975 Liquid magnetic bubble. *IEEE Trans. Magn.* 11: 25–8
Rosensweig R E 1975 Ferrofluid compositions and method of making the same. US Patent No. 3,917,538
Rosensweig R E 1985 *Ferrohydrodynamics*. Cambridge University Press, New York
Rosensweig R E, Ezekiel F D, Miskolczy G 1968 Magnetic fluid shaft seal. *Mach. Des.* 40: 145–50
Skjeltorp A T 1983 One- and two-dimensional crystallization of magnetic holes. *Phys. Rev. Lett.* 51(25): 2306–9

R. E. Rosensweig
[Exxon Research and Engineering Company, Annandale, New Jersey, USA]

Flux Creep Phenomena

In many applications of superconductors, large transport current densities J in high magnetic fields are required. As all high-field superconductors are type II superconductors, large critical current densities J_c can only be reached by pinning the vortices to defects of the superconducting material. The interaction in the flux line lattice with defects is treated in the article *Flux Pinning and the Summation of Pinning Forces*. The strength of this interaction can successfully be increased in conventional superconductors by appropriate thermal and mechanical treatment during wire production. This results in high-field superconducting coils with slow changes of the trapped field in the persistent mode. Strictly speaking, at the absolute zero, the trapped flux should remain constant whenever $J < J_c$. Dissipation should only occur when $J > J_c$ (i.e., when the Lorentz force is larger than the pinning force). In this regime, which is referred to as the flux flow regime, the flux lines move rather steadily (see *Resistive Transition and Flux Flow in Superconducting Materials*). In the flow regime, a superconductor behaves as an ohmic resistor with a flow resistivity ρ_f comparable to ρ_n, the normal state resistivity. At nonzero temperatures, Anderson and Kim (1964) proposed that flux lines could move from one pinning region to an adjacent one through thermally activated hopping. The resulting flux creep is generally weak in conventional superconductors and can only be observed at current densities very close to J_c. Furthermore, it depends logarithmically on time, which implies that the same amount of creep occurs between 1 s and 1000 s as between 1 year and 1000 years.

The interest in flux creep increased strongly after Müller et al. (1987) reported (a) large and nonexponential relaxation of the magnetic moment of ceramic $(La, Ba)_2CuO_{4-y}$ samples and (b) the existence of nonergodicity with respect to field versus nonfield cooled susceptibility measurements. The transition temperature T_{irr} above which the superconductor behaves reversibly was found to vary with magnetic field as $T_c - T_{irr} \propto H^{2/3}$. As nonexponential relaxation and a line separating ergodic from nonergodic regions (the so-called de Almeida–Thouless line) are well-known characteristic features of spin glasses, Müller et al. claimed that the high-T_c superconductors behave as superconducting glasses. Superconducting glass behavior is theoretically found for a randomly dilute lattice of Josephson junctions and for randomly coupled superconducting clusters in magnetic fields. Yeshurun and Malozemoff (1988) and many other workers pointed out, however, that the effects reported by Müller et al. could also be qualitatively understood by means of thermally activated flux motion (TAFM) in high-T_c superconductors. The purpose of this article is to review the theory of TAFM and check some of its predictions concerning critical currents and magnetization relaxation and in magnetic fields.

1. Flux Motion in Type II Superconductors

Although pinning forces and driving forces act on individual flux lines, a macroscopic description of flux motion will be adopted here by considering the induction $\boldsymbol{B}(\boldsymbol{r}, t)$ defined as an average of the microscopic field over several penetration depths λ. In the presence of a macroscopic transport current $\boldsymbol{J}(\boldsymbol{r}, t)$ the force density $\boldsymbol{F}$ on the vortices is given by

$$\boldsymbol{F}_L = \boldsymbol{J} \times \boldsymbol{B} = -\boldsymbol{B} \times \nabla \times \boldsymbol{H} \tag{1}$$

where $\boldsymbol{H}$ is the field generated by externally imposed currents (see *Transport Currents and Force-Free Current Flow*). In type II superconductors with high κ values, $\boldsymbol{B} \cong \mu_0 \boldsymbol{H}$ as the microscopic screening currents associated with the equilibrium magnetization are much smaller than the generated transport current. In this limit, the transport current density $\boldsymbol{J} = \nabla \times \boldsymbol{H}$ may conveniently be approximated by the mean microscopic current density $\boldsymbol{j} = (1/\mu_0)\nabla \times \boldsymbol{B}$. The Lorentz force $\boldsymbol{F}_L$ induces a displacement of flux lines with an average velocity $\boldsymbol{v}$. The electric field associated with this displacement is $\boldsymbol{E} = \boldsymbol{B} \times \boldsymbol{v}$ and from Maxwell's equation it follows that

$$\frac{\partial \boldsymbol{B}(\boldsymbol{r},t)}{\partial t} = -\nabla \times (\boldsymbol{B} \times \boldsymbol{v}) \tag{2}$$

Neglecting the small Hall angle in type II superconductors (i.e., assuming that flux lines are moving perpendicularly to both $\boldsymbol{j}$ and $\boldsymbol{B}$) gives

$$\frac{\partial \boldsymbol{B}}{\partial t} = -\nabla \times \left[\frac{\boldsymbol{B} \times \boldsymbol{j} \times \boldsymbol{B}}{B} f(j)\right] \tag{3}$$

In many situations, the current $\boldsymbol{j}$ flows perpendicularly to $\boldsymbol{B}$ and Eqn. (3) takes the simpler form

$$\frac{\partial \boldsymbol{B}}{\partial t} = -\nabla \times [B f(j)\boldsymbol{j}] \tag{4}$$

where $B \equiv |\boldsymbol{B}|$ is the magnitude of the induction. Since $\nabla \times \boldsymbol{B} = \mu_0 \boldsymbol{j}$, Eqns. (3) and (4) are thus partial differential equations for $\boldsymbol{B}$. The function $f(j)$ depends on the flux motion regime. In the flux flow regime, according to the Stephen–Bardeen theory, $f(j) = \rho_n / B_{c2}$, where ρ_n is the normal state resistivity and B_{c2} is the upper critical field (see *Magnetic Properties of Superconductors*).

In the TAFM regime

$$f(j) = 2x_{\text{hop}}\omega_0 \frac{\sinh(\alpha j)}{j} \exp(-\varepsilon) \tag{5}$$

with $\varepsilon = U_0/k_B T$ and $\alpha = \varepsilon / j_c$, where j_c is the critical current, k_B is Boltzmann's constant, T is the temperature and U_0 is the energy barrier for thermally activated flux line hopping in the absence of current. The sinh term is due to the fact that in the presence of a transport current j the activation energy $U^+(j)$ for forward hopping in the direction of the Lorentz force differs from $U^-(j)$ for backward hopping. The simplest model assumes a sawtooth pinning potential with periodicity $2x_{\text{hop}}$ for which $U^{\pm}(j) = U_0\,(1 \mp j/j_c)$ and the effective jump rate is

$$\omega \cong \tfrac{1}{2}\omega_0 \exp[-\varepsilon(1 - j/j_c)] - \tfrac{1}{2}\omega_0 \exp[-\varepsilon(1 + j/j_c)] \tag{6}$$

where ω_0 is the attempt frequency. As a result of the interaction between flux lines, hopping occurs in the form of flux bundles in a correlated volume V_c over a distance $2x_{\text{hop}}$. The force acting on a bundle is $V_c \boldsymbol{j} \times \boldsymbol{B}$ and the change in energy for a hopping length $2x_{\text{hop}}$ perpendicular to both current and field is $2V_c jBx_{\text{hop}}$ and $U^{\pm} = U_0\ (1 \mp V_c jBx_{\text{hop}})$. Consequently, $\alpha = V_c B x_{\text{hop}}/kT$ and $\varepsilon = j_c V_c B x_{\text{hop}}/kT$. For an isolated flux line, $V_c B = \Phi_0 L_c$ and it is expected with $x_{\text{hop}} \simeq \xi$, the coherence length, that $U_0 \simeq j_c \Phi_0 L_c \xi$, where the pinning length L_c depends on the disorder in the flux line lattice. With $\xi < L_c < a_0$, where a_0 is the periodicity of the flux lattice, $j_c \Phi_0 \xi^2 \lesssim U_0 \lesssim j_c\, \Phi_0\, a_0 \xi$ is evaluated, which leads to $0.02 \lesssim U_0 \lesssim 0.2\,\text{eV}$ with $j_c = 10^{12}\,\text{A m}^{-2}$, $\Phi_0 = 2.05 \times 10^{-15}\,\text{T m}^2$, $\xi = 10^{-9}\,\text{m}$ and $a_0 = 10^{-8}\,\text{m}$. These values are typical for high-T_c superconductors. For conventional superconductors, typically $U_0 \approx 2\,\text{eV}$ since $\xi \approx 10^{-8}\,\text{m}$. The short coherence length in high-T_c superconductors is thus the origin of the large flux creep relaxation in these materials (see *Coherence Length, Proximity Effect and Fluctuations*).

2. *Thermally Activated Flux Flow*

At low current densities (i.e., when $\alpha j < 1$), the function $f(j)$ is independent of current and equal to $= \alpha x_{\text{hop}}\, \omega_0 \exp(-\varepsilon)$, where TAFF stands for thermally activated flux flow. The time evolution of the induction is then given by

$$\mu_0 \frac{\partial \boldsymbol{B}}{\partial t} = f_{\text{TAFF}} \nabla \times [B \nabla \times \boldsymbol{B}]$$

$$= f_{\text{TAFF}}(B\nabla^2 \boldsymbol{B} - (\nabla B) \times \nabla \times \boldsymbol{B}) \tag{7}$$

In sufficiently high magnetic fields, when $\boldsymbol{B}(\boldsymbol{r}, t) \cong \boldsymbol{B}_0$, where $\boldsymbol{B}_0$ is the induction averaged over the sample, the second term may be neglected. Eqn. (7) then reduces to a diffusion equation:

$$\frac{\partial \boldsymbol{B}}{\partial t} = D\nabla^2 \boldsymbol{B} \tag{8}$$

with a constant diffusion coefficient $D = \mu_0^{-1} \alpha x_{\text{hop}} \times B_0 \omega_0 \exp(-\varepsilon)$. Solutions of this diffusion equation have been determined for many geometries and initial conditions. For a slab geometry (i.e., an infinite lamella limited in the x direction by the planes $x = \pm R$) in a constant external magnetic field H_0 along the z axis, in the long-time limit, $B(x, t) \cong B_0 - B_1 \exp(-t/\tau) \cos(\pi x / 2R)$ with $\tau = 4R^2/\pi^2 D$ and $B_0 = \mu_0 H_0$. Inserting this solution into Eqn. (7) shows that the second term is $B_1 \exp(-t/\tau)/B_0$ times smaller than the first one.

However, the second term in Eqn. (7) is not always negligible. For example, if the external field is slowly increasing linearly with time so that $H(t) = ht$, which means that the induction $B_s(t)$ at the surfaces of the slab varies as $B_s(t) = bt$, the solution of Eqn. (7) with ε and α constant is $B(x, t) = \pm (b\mu_0/f_{\text{TAFF}})^{1/2}\,(x - R) + bt$ for $|x| > a$ and $B(x,t) = 0$ for $|x| < a$ as long as $a(t) = R - (bf_{\text{TAFF}}\mu_0)^{1/2}\, t$ is positive (i.e., as long as the flux which penetrates the slab from each side has not reached the center of the slab). This exact solution is obviously not a solution of the approximate diffusion equation (Eqn. (8)) as $d^2B/dx^2 = 0$ for a linear flux profile.

To remain within the TAFF regime, it is required that $dB/dx < \mu_0/\alpha$ which is satisfied whenever

$$\frac{dB_s}{dt} = b < \frac{\mu_0 x_{\text{hop}} \omega_0}{\alpha} \exp(-\varepsilon) \tag{9}$$

With typical values for high-T_c superconductors (i.e., $\alpha = 10^{-10}\,\text{m}^2\,\text{A}^{-1}$, $\varepsilon = 25$, $x_{\text{hop}}\omega_0 = 10\,\text{m s}^{-1}$, Eqn. (9) implies that $b < 10^{-6}\,\text{T s}^{-1}$. At a sweep rate $b = 10^{-6}\,\text{T s}^{-1}$, it would take approximately $10^{10}\,\text{s}$ before the flux reaches the center of a slab 2 μm thick. This example shows that usual ac measurements are certainly not in the TAFF regime except near T_c where ε has significantly decreased (Kes et al. 1989).

Finally, it is interesting to point out that although TAFF is completely different from flux creep, the solution given has the same functional dependence as for the penetration of flux into a sample in the Bean critical state model (see *Critical State Model*).

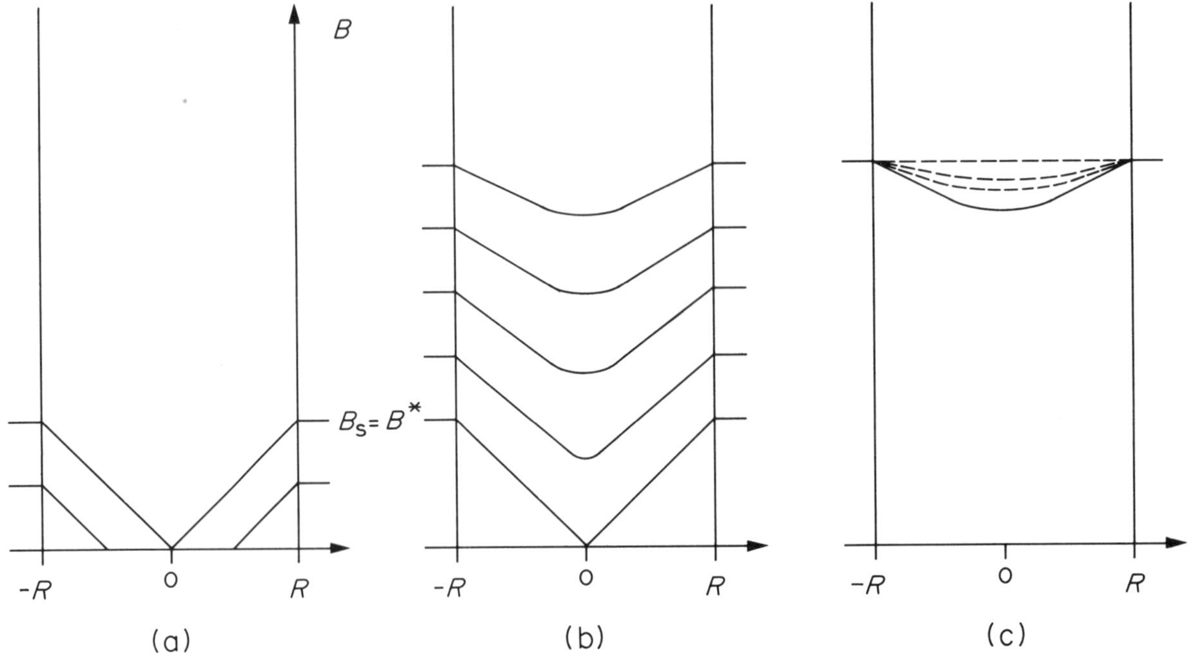

Figure 1
Time evolution of the flux profile in a superconducting slab during sweeping up of the external magnetic field in the (a) partial penetration and (b) full flux penetration regimes; (c) magnetization relaxation after the field is stabilized at a given value. The critical current $j_c = \varepsilon/\alpha$ is assumed to be constant. The flattening of the flux profile in (a) is due to relaxation effects

3. *Flux Creep*

In the flux creep regime (i.e., when the current density satisfies the condition $3 < \alpha j \leqslant \varepsilon$) the function $f(j)$ is strongly nonlinear in j and the differential equation for $\boldsymbol{B}(\boldsymbol{r}, t)$ in the flux creep regime is

$$\frac{\partial \boldsymbol{B}}{\partial t} = -\nabla \times \left[\frac{\boldsymbol{j}}{j} B x_{\text{hop}} \omega_0 \exp(-\varepsilon + \alpha j)\right] \quad (10)$$

where $j = |\boldsymbol{j}|$ and $\nabla \times \boldsymbol{B} = \mu_0 \boldsymbol{j}$. For a discussion of the behavior of $\boldsymbol{B}(\boldsymbol{r}, t)$ predicted by this equation, an infinite slab is again considered with planes at $x = \pm R$ and a magnetic field parallel to the z axis; that is, $\boldsymbol{B}(x, t) = (0, 0, B_z(x, t))$. From Eqn. (10) it follows that

$$\frac{\partial B_z}{\partial t} = x_{\text{hop}} \omega_0 \exp(-\varepsilon) \frac{\partial}{\partial x}\left[B_z \exp\left(\frac{\alpha}{\mu_0}\left|\frac{\partial B_z}{\partial x}\right|\right)\right] \quad (11)$$

which can be interpreted as an equation of continuity (Beasley et al. 1969) for the density of flux lines $n^{\square} = B_z/\Phi_0$.

3.1 *Partial Penetration During Field Sweep*

The time evolution of the flux profile in a superconducting slab for a linear sweep of the external magnetic field with the flux not completely penetrating the slab is given in Fig. 1a. A general solution of Eqn. (11) is not known except for the case where α is field independent and $B_s \equiv B_z(\pm R, t) = bt$ with, at time zero, $B_z(x, t = 0) = 0$. As for the corresponding problem in the TAFF regime, $B_z(x, t) = \pm \mu_0 j_0 (x - R) + bt$ for $|x| > a$ and $B_z(x, t) = 0$ for $|x| < a$ as long as $a(t) = R - bt/\mu_0 j_0$ is positive. The current j_0 flowing in the sheet $a < x \leqslant R$ is given by the implicit relation

$$b = \frac{dB_s}{dt} = \mu_0 j_0 \omega_0 x_{\text{hop}} \exp(-\varepsilon + \alpha j_0) \quad (12)$$

and is thus a function of the sweep rate.

In the conventional superconductors with $T_c < 25$ K and activation energies of the order of 1 eV, ε is larger than 500 and consequently from Eqn. (12) it follows that $j_0 = \varepsilon/\alpha \equiv j_c$ to a very good approximation. Any increase of the field will thus result in setting up a critical state with $j = j_c = U_0/(\Phi_0 x_{\text{hop}} L_c)$. In high-$T_c$ superconductors, U_0 is typically 0.05 eV and $T_c \simeq$ 100 K. Taking $\varepsilon \simeq 20$, $j_c = \varepsilon/\alpha = 10^{10}\,\text{A m}^{-2}$, $\mu_0 x_{\text{hop}} \omega_0 = 10^{-4}\,\text{T m}^2\,\text{A}^{-1}\,\text{s}^{-1}$ and $dB_s/dt = 10^{-2}\,\text{T s}^{-1}$, it is found that $j_0 \cong 1.7 \times 10^9\,\text{A m}^{-2}$, which is approximately six times smaller than the maximum current j_c compatible with flux creep. With $dB_s/dt = 1\,\text{T s}^{-1}$, $j_0 \cong 3.6 \times 10^9\,\text{A m}^{-2}$.

These examples illustrate that the current j_0 induced by sweeping the field is significantly lower than j_c even for relatively high sweep rates (see *Critical Current Density*). The notion of critical state is thus ill defined in systems with $\varepsilon \lesssim 20$ such as the high-T_c superconductors.

3.2 Total Penetration During Field Sweep

The time evolution of the flux profile in a slab for a linear sweep in the complete penetration regime is given in Fig. 1b. When the left and right flux fronts have met at the center of the slab (i.e., when the field B_s at the surface of the slab is larger than $B^* = \mu_0 j_0 R$) the field profile does not remain exactly linear as this would lead to a pileup of flux lines at the midplane of the slab. The spatial variation of $\partial B_z/\partial x$ (or equivalently of $j(x, t)$) is, however, weak since otherwise $\exp[-\varepsilon + (\alpha/\mu_0)\,|\partial B_z/\partial x|]$ would lead to wildly varying values of $\partial B_z/\partial t$. To obtain an approximate solution for $B_s(t) = bt > B^*$, Eqn. (11) is integrated over the slab to obtain

$$\mu_0 \frac{dM}{dt} = -2x_{\text{hop}}\omega_0 B_s(t)\exp[-\varepsilon + \alpha j_s(t)] + 2R\frac{dB_s}{dt} \quad (13)$$

and the magnetic moment is approximated by $M(t) = R^2 j_s(t)$, where $j_s(t)$ is the current density at the surface of the slab. The solution of this differential equation is, for $bt \gtrsim \mu_0 Rt/\alpha$,

$$j_s(t) = j_c\left\{1 - \frac{1}{\varepsilon}\ln\left[\frac{B_s(t)x_{\text{hop}}\omega_0}{R(dB_s/dt)}\right]\right\} \quad (14)$$

As $M(t) \propto j_s(t)$, this relation shows explicitly that the magnetic moment measured by sweeping continuously the external magnetic field depends on the sweep rate. It also depends on the field B_s. This is a remarkable result since in these calculations it has been assumed that both ε and α are constant which implies that the critical current j_c (which is not accessible to experiment) and the activation energy for flux line hopping are *field independent*. It is, however, often a weak effect, which is masked by the intrinsic field dependence of the activation energy and the critical current in real materials.

On the other hand, Eqn. (14) predicts a sizeable temperature dependence of j_s. For example, with $B_s = 1\,\text{T}$, $x_{\text{hop}}\omega_0 = 10\,\text{m s}^{-1}$, $R = 10^{-3}\,\text{m}$ and $b = 10^{-2}\,\text{T s}^{-1}$, $j_s = j_c(1 - 13.8\,k_BT/U_0)$, which implies that at constant j_c and U_0 the hysteretic shielding current j_s should virtually vanish at 72 K for an activation energy of 1000 K ($\sim$ 86 meV). Above this temperature, the system is not in the flux creep regime but in the TAFF regime.

3.3 Magnetization Relaxation

Figure 1c shows the time evolution of the flux profile in a slab for relaxation in a fixed external magnetic field. After having swept the field to a certain value B_s, it is kept constant and the relaxation of the superconducting current governed by Eqn.(13) is considered with $dB_s/dt = 0$, which in terms of current can be written

$$\frac{dj(t)}{dt} = -\frac{2x_{\text{hop}}\omega_0 B_s}{\mu_0 R^2}\exp[-\varepsilon + \alpha j(t)] \quad (15)$$

The origin of time is chosen as the beginning of the relaxation and thus $j(0) = j_s$, where j_s is the current just before the sweep is stopped (Eqn.(14)). The solution of Eqn.(15) is

$$j(t) = j_s - \frac{1}{\alpha}\ln\left[1 + \exp(-\varepsilon + \alpha j_s)\frac{t}{\tau}\right] \quad (16)$$

where

$$\tau = \left(\frac{\mu_0 R^2}{2x_{\text{hop}}\omega_0 B_s}\right)\frac{j_c}{\varepsilon} \quad (17)$$

The relaxation time τ, which depends both on the dimension of the sample and the magnetic field, is typically 10^{-6} s when $B_s = 1\,\text{T}$, $j_c = 10^{10}\,\text{A m}^{-2}$, $\varepsilon = 25$, $R = 10^{-4}\,\text{m}$ and $x_{\text{hop}}\omega_0 = 10\,\text{m s}^{-1}$. In the logarithmic decay regime (i.e., when $\tau\exp(\varepsilon - \alpha j_s) \ll t < \tau\exp\varepsilon$),

$$j(t) \simeq j_c\left[1 - \frac{1}{\varepsilon}\ln\left(\frac{t}{\tau}\right)\right] \quad (18)$$

$$\frac{dj}{d\ln t} = -\frac{1}{\alpha} \quad (19)$$

and for the normalized derivative

$$S \equiv -\frac{dj}{j(t)\,d\ln t} = \frac{l}{\varepsilon - \ln(t/\tau)} \quad (20)$$

For the values chosen and $t = 10^3$ s, $S \cong (\varepsilon - 22)^{-1}$. This shows that S may be quite different from $\varepsilon^{-1} = k_BT/U_0$ as often assumed. A schematic illustration of the relaxation of a flux line density gradient is shown in Fig. 2.

An interesting relation is

$$j_s - j(t) = \frac{1}{\alpha}\ln\left[\frac{2\alpha t(dB_s/dt)}{\mu_0 R}\right] \quad (21)$$

as all parameters are accessible to experiment (see Eqn. (19)). This equation might thus be used to test the flux creep model or, in inhomogeneous samples, to determine an effective grain size R.

3.4 Sinusoidal Pinning Potential

All the relations given so far are valid for a sawtooth potential for which the forward and backward activation energies $U^+(j)$ and $U^-(j)$ depend linearly on the current density. For continuously differentiable pinning potentials (e.g., a sinusoidal potential), it is found that $U^+(j) \cong U_0(1 - j/j_c)^{3/2}$. The presence of the

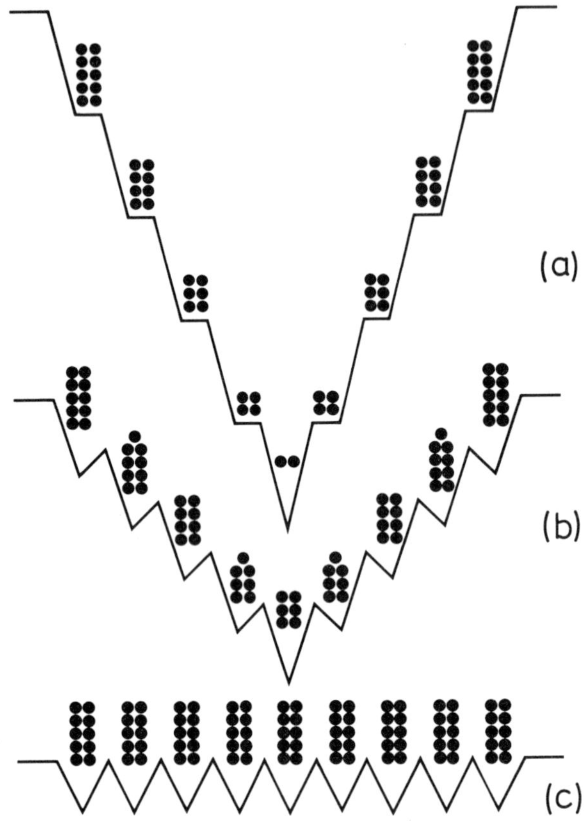

Figure 2
Schematic representation of the flux creep relaxation of a flux line gradient in a slab in a constant external field B_s as a function of time. (a) At $t = 0$ the system is assumed to be in the critical state (Bean model) and $U^+(j_c) = 0$. The number of flux lines in each pinning region is equal to the number of dots. At the surface it is equal to B_s/Φ_0. (b) For $t < \tau \exp \varepsilon$, the magnetization and current (which are proportional to the gradient in the density of flux lines) decrease as ln t. (c) In the long-time limit (i.e., when $t > \tau \exp(\varepsilon)$, the relaxation is in the TAFF regime and is well described by a diffusion equation

3/2 power leads to the following relations (Geshkenbein and Larkin 1989, Griessen 1991):

$$j_s(t) \cong j_c\left(1 - \left(\frac{1}{\varepsilon}\right)^{2/3}\left\{\ln\left[\frac{B_s(t)x_{hop}\omega_0}{R(dB_s/dt)}\right]\right\}^{2/3}\right) \quad (14a)$$

$$j(t) \cong j_c\left\{1 - \left(\frac{1}{\varepsilon}\right)^{2/3}\left[\ln\left(\frac{t}{\tau_{3/2}}\right)\right]^{2/3}\right\} \quad (18a)$$

$$\frac{dj}{d\ln t} \cong -\frac{2j_c}{3\varepsilon^{2/3}[\ln(t/\tau_{3/2})]^{1/3}} \quad (19a)$$

and

$$S \cong \frac{2}{3\{\varepsilon^{2/3}[\ln(t/\tau_{3/2})]^{1/3} - \ln(t/\tau_{3/2})\}} \quad (20a)$$

with

$$\tau_{3/2} = \frac{\mu_0 R^2}{2x_{hop}\omega_0 B_s}\frac{2j_c}{3\varepsilon^{2/3}} \quad (17a)$$

The derivatives $dj/d\ln t$ and S are weakly time dependent. Within the time windows accessible to most experiments the upward curvature is, however, barely noticeable.

3.5 Collective Pinning and Current Dependence of Activation Energy

In the preceding sections, it has always been assumed that the size of the bundles of flux lines which hop from one pinning region to another and the hopping distance are independent of the current densities. In fact, for the case of single flux lines, it is expected, in analogy with domain wall theory, that the hopping distance x_{hop} is related to the length L of a hopping flux line segment, so that $x_{hop} \propto L^{\zeta}$. An estimate of the pinning energy $U^+(j)$ can then be obtained as follows. The energy gained due to the Lorentz force is $\Phi_0 jL\, x_{hop} \propto j\, L^{\zeta+1}$. The elastic deformation energy of the vortex lattice is proportional to $(x_{hop}/L)^2 L$ (i.e., the strain squared times the length of the hopping flux line). As both energies should be of comparable magnitude, it is found that

$$U^+(j) = G\, j^{(2\zeta - 1)/(\zeta - 2)} \quad (22)$$

where $G = G(B, T)$ depends in general on field and temperature in a complicated manner as discussed by Feigelman and Vinokur (1990) (see also Feigelman et al. 1989). Since typically $\zeta \approx 0.6$, the activation energy diverges in the limit of small transport current as $U^+(j) = G\, j^{-\nu}$ with $\nu = (2\zeta - 1)/(2 - \zeta) > 0$ and the corresponding flux creep equation is

$$\frac{dj(t)}{dt} = -\frac{2x_{hop}\omega_0 B_s}{\mu_0 R^2}\exp(-Gj^{-\nu}/k_B T) \quad (23)$$

in which the current dependence of $x_{hop} \propto j^{\zeta/(\zeta-2)}$ is neglected since the j dependence in the exponential is much stronger. At sufficiently long times (i.e., $t \gg \tau^*$) the solution of Eqn. (23) is

$$j(t) \cong \frac{G^{1/\nu}}{\left[k_B T \ln\left(\frac{t}{\tau^*}\right)\right]^{1/\nu}} \quad (24)$$

where

$$\tau^* = \frac{\mu_0 R^2}{2x_{hop}\omega_0 B_s}\frac{k_B T}{\nu G}j_s^{\nu+1} \quad (25)$$

Eqn. (24) leads to the temperature dependences for $t \gg \tau$ of $j(t) \propto T^{-1/\nu}$, $dj/d\ln t \propto T^{-1/\nu}$ and $S = -d\ln j/$

$d \ln t = $ constant. In the case of strong currents and collective pinning of individual flux lines (where the hopping bundle is reduced to a segment of flux line of length intermediate to the coherence length and the flux line spacing), $\zeta = 0.6$ and $\nu = 1/7$. At smaller currents, bundles of flux lines are hopping and the effective exponent ν in the relation $U^{+}(j) = Gj^{-\nu}$ decreases to 3/2 when the dimension of the bundle perpendicular to B is comparable to the penetration depth λ. At even smaller currents, $\nu = 7/9$.

4. Experimental Results

Although many flux creep experiments have been carried out, only a limited number of investigations have proved useful for a comparison with theory. The discussion here will be restricted to data for single crystals, grain-aligned samples and epitaxial films gathered in magnetic fields sufficiently high so that full flux penetration has occurred (Griessen et al. 1990).

In all cases, except near the irreversibility line (see *Critical Current Density*), a $\ln t$ dependence of $j(t)$ or $M(t)$ is found. This indicates that TAFM is important in high-T_c superconductors. The limited number of decades (less than four) in most experimental studies is, however, too small to discriminate between the standard flux creep (Eqn. (18)), the sinusoidal pinning potential model (Eqn. (18a)) and the collective pinning theory (Eqn. (24)).

Experimental data for the normalized decay rate $S = -d \ln M / d \ln t = -d \ln j / d \ln t$ are shown in Fig. 3 for various $YBa_2Cu_3O_{7-\delta}$ samples. Although there is a wide spread in the results at higher temperatures, it is remarkable to see that below 15 K all samples (single crystals, epitaxial films, grain-aligned samples) behave in the same manner: (a) S increases with T and (b) the values are similar to each other, $S = 0.015 \pm 0.01$ at 4.2 K. The increase of S with T is consistent with the standard flux creep model and $S = 0.015$ implies that $U_0 \cong 31$ meV with $\ln t/\tau \cong 20$. The results for neutron-irradiated grain-aligned samples in high fields (4 T and 8 T) for which S continues to increase with T up to approximately 50 K can also be understood in terms of flux creep with $U_0 \cong$ 150 meV.

For some samples, a plateau is found in S for $20\,\mathrm{K} \lesssim T \lesssim 50\,\mathrm{K}$. This seems to support the collective pinning model for which $S \cong [\nu \ln(t/\tau^*)]^{-1}$ is essentially independent of temperature. In this regime, $S \cong 0.025$ implies with $\ln(t/\tau^*) \cong 20$ that $\nu = 2$, a value which is outside the range of the theoretical values previously mentioned. Another explanation of this occurrence of a plateau in $S = S(T)$ might be that a single activation energy does not adequately describe high-T_c superconductors. For a model with a distribution of activation energies, it can be shown that the effective activation energy is always an increasing function of T since more and more relaxation processes have fully relaxed before the first-data point is taken (Hagen and Griessen 1989, Griessen 1990, Niel and Evetts 1991, 1992). In such a model, ε remains approximately constant and $S \cong$ constant. Models incorporating distributions of activation energies can also reproduce data with a peak in $S(T)$ such as observed for grain-aligned samples (● and ▲ in Fig. 3) and small single crystals (▼). The behavior of $S(T)$ for these samples is certainly not in agreement with the collective pinning model as it would require rather large values of the exponent ν.

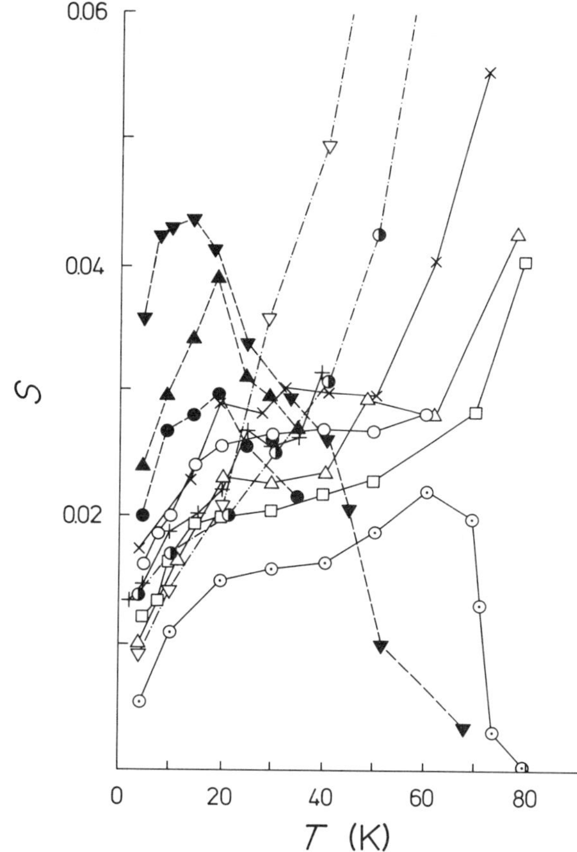

Figure 3
Normalized relaxation rate $S \equiv -d \ln M/d \ln t = d \ln j/d \ln t$ for various $YBa_2Cu_3O_{7-\delta}$ samples in a magnetic field along the c axis: ▲ and ●, grain-aligned samples ($B = 1$ T and 2 T, respectively); △, ◑ and ▽, grain-aligned samples after neutron irradiation ($B = 1$ T, 4 T and 8 T, respectively); ▼ and ○, single crystal ($B = 1.5$ T and 1 T, respectively); □, proton-irradiated single crystal ($B = 1$ T); + and ⊙, electron beam evaporated films on $SrTiO_3$ substrates ($B = 1$ T for both); ×, sputtered films on MgO substrate ($B = 0.9$ T)

Much more systematic work on relaxation phenomena and dissipation effects which influence the resistive transition in magnetic fields will be needed to identify the correct theoretical description. In all the models presented so far it has been assumed that the Lorentz force on a flux line bundle is the force driving flux creep relaxation. The large broadening of resistive transitions of high-T_c superconductors in magnetic fields seems, however, to be only weakly dependent on the angle between $\boldsymbol{j}$ and $\boldsymbol{B}$. This indicates that other dissipative processes are playing a role. Another problem is that $S(T)$ has been found to remain sizeable down to 1.2 K in epitaxial films and down to 0.1 K in $YBa_2Cu_3O_7$ single crystals (Fruchter et al. 1991). Furthermore, it seems that $\lim S(T)_{T\to 0} \neq 0$. This suggests that tunnelling of flux lines is effective at very low temperatures.

See also: Flux Line Lattice; Magnetic Properties of Superconductors; Resistive Transition and Flux Flow in Superconducting Materials

Bibliography

Anderson P W, Kim Y B 1964 Hard superconductivity: theory of the motion of Abrikosov flux-lines. *Rev. Mod. Phys.* 36: 39–45

Beasley M R, Labusch R, Webb W W 1969 Flux-creep in type-II superconductors. *Phys. Rev.* 181: 682–700

Campbell A M, Evetts J E 1972 Flux vortices and transport currents in type-II superconductors. *Adv. Phys.* 21: 199–428.

Feigelman M V, Geshkenbein V B, Larkin A I, Vinokur V M 1989 Theory of collective flux creep. *Phys. Rev. Lett.* 63: 2303–6

Feigelman M V, Vinokur V M, 1990 Thermal fluctuations of vortex lines, pinning and creep in high-T_c superconductors. *Phys. Rev. B* 41: 8986–90

Fruchter L, Malozemoff A P, Campbell I A, Sanchez J, Konczykowski M, Griessen R, Holzberg F 1991 Low-temperature magnetic relaxation in $YBa_2Cu_3O_{7-\delta}$: evidence for quantum tunneling of vortices. *Phys. Rev. B* 43: 8709–12.

Geshkenbein V B, Larkin A I, 1989 Time dependence of the magnetic moment of high-temperature superconductors. *Sov. Phys. JETP (Engl. Transl.)* 68: 639–41

Griessen R 1990 Resistive behavior of high-T_c superconductors—influence of a distribution of activation energies. *Phys. Rev. Lett.* 64: 1674–7.

Griessen R 1991 Thermally activated flux motion near the absolute zero. *Physica C* 172: 441–9

Griessen R, Lensink J G, Schröder T A M, Dam B 1990 Flux-creep and critical currents in epitaxial high T_c films. *Cryogenics* 30: 563–8

Hagen C W, Griessen R 1989 Distribution of activation energies for thermally activated flux-motion in high T_c superconductors: an inversion scheme. *Phys. Rev. Lett.* 72: 2857–60

Huebener R P 1979 *Magnetic Flux Structures in Superconductors*, Solid State Sciences, Vol. 6. Springer, Berlin

Kes P H, Aarts J, van den Berg J, van der Beek C J, Mydosh J A 1989 Thermally assisted flux-flow at small driving forces. *Supercond. Sci. Technol.* 1: 242–8

Müller K A, Takashige M, Bednorz J G 1987 Flux trapping and superconductive glass state in La_2CuO_{4-y}: Ba. *Phys. Rev. Lett.* 58: 1143–6

Niel L, Evetts J E 1991 Flux creep in superconductors in terms of an activation energy spectrum of randomly distributed pinning centres. *Europhys. Lett.* 15: 453–8

Niel L, Evetts J E 1992 Thermally activated hopping of vortices in high temperature superconductors with a distribution of pinning energies. *Supercond. Sci. Technol.* 5: 347–50

Yeshurun Y, Malozemoff A P 1988 Giant flux-creep and irreversibility in an Y–Ba–Cu–O crystal: an alternative to the superconducting-glass model. *Phys. Rev. Lett.* 60: 2202–6.

R. Griessen
[Vrije Universiteit, Amsterdam,
The Netherlands]

Flux Line Lattice

When a type II superconductor is put into a magnetic field H then at small fields ($H < H_{c1}$) the magnetic field is completely expelled from its interior (Meissner state). In large fields ($H > H_{c2}$) the superconductor becomes normal conducting and the magnetic field penetrates completely. The lower and upper critical fields of the material, H_{c1} and H_{c2}, depend on the temperature T and are approximately proportional to $1 - T^2/T_c^2$ where T_c is the superconducting transition temperature. In the Shubnikov or "mixed" state ($H_{c1} < H < H_{c2}$), magnetic flux can penetrate the superconductor in the form of flux lines or flux vortices (see also *Magnetic Properties of Superconductors*). Flux lines are tiny current vortices that arrange to a more or less regular lattice, the flux line lattice (FLL). The existence of an FLL was predicted in 1957 by A. A. Abrikosov when he found that the phenomenological theory of superconductivity, formulated in 1950 by V. L. Ginzburg and L. D. Landau (the GL theory), had periodic solutions $\psi(x, y)$ (the complex GL or order parameter) and $B(x, y)$ (the magnetic induction) where x, y and z are Cartesian coordinates with z along the applied field H. Abrikosov interpreted this solution as an FLL, with the zeros of ψ and maxima of B defining the flux line centers. Each flux line carries one quantum of magnetic flux, $\Phi_0 = h/2e = 2.07 \times 10^{-15}\,\mathrm{T\,m^2}$ (h = Planck's constant, e = charge of the electron). An FLL with triangular symmetry and spacing a_0 thus has an average induction

$$\bar{B} = \frac{2\Phi_0}{\sqrt{3}a_0^2} \tag{1}$$

For example, in niobium at $T = 4.2\,\mathrm{K}$, $\bar{B} \leqslant \mu_0 H_{c2} = 0.39\,\mathrm{T}$ which yields $a_0 \geqslant 0.08\,\mu\mathrm{m}$.

The existence of the FLL was confirmed first by Cribier and his co-workers in 1964 who observed

Bragg reflections from the FLL in small-angle neutron scattering experiments. In 1967, Träuble and Essmann observed the FLL in an electron microscope after decorating the points where flux lines emerge from the surface of the superconductor with tiny ferromagnetic particles suspended in a helium atmosphere (see Fig. 1a). For a review of this method and of some properties of the FLL see Brandt and Essmann (1987). The FLL in the ceramic high-temperature superconductor $YBa_2Cu_3O_7$ was observed by the decoration method first by Gammel et al. (1987) (Fig. 1b).

Although a full theoretical description of the static and dynamic FLL has only been given for essentially homogeneous materials in particular limiting cases, the concept of the flux line and FLL has proved extremely durable and it has been possible to build an elaborate self-consistent phenomenological framework to describe the behavior of type II superconductors under very diverse conditions. The FLL may be distorted and generally curved in three dimensions, it can be elaborately defective (see Fig. 2), and can even be made amorphous, and individual flux lines may cut and cross-join with each other in a process that involves phase slip and dissipation. The additional concepts of a flux line driving force, flux pinning and flux flow enable the properties of inhomogeneous materials and the dynamic response of the FLL in the presence of transport currents and heat flow to be treated. (see *Resistive Transition and Flux Flow in Superconducting Materials*; *Transport Currents and Force-Free Current Flow*).

The underlying strength of the flux line concept lies in the fact that the flux line is an extended quantum unit locked into a quantum mechanical phase continuum. Investigation of the extremes and limits for the application of powerful concepts has always been of interest and the flux line has been no exception. Over the years there has been study of the overall static and dynamic stability and the effects of thermal energy on flux lines in thin films (Fisher 1980), weakly coupled multilayers (Lawrence and Doniach 1971), anisotropic superconductors (Tilley 1965) and inhomogeneous superconductors (Gubser et al. 1980). With the discovery of high-temperature superconductors (HTS) there has been renewed interest and a resurgence of debate on the special features of the FLL at high temperatures, particularly in anisotropic materials and multilayers.

As the superconducting critical temperature increases the effects of thermal disorder on the FLL becomes more important. Thermal fluctuations can in certain limits lead to melting of the FLL and, for weakly coupled layered systems, this eventually results in the complete breakdown of the FLL concept. In the initial sections on the equilibrium FLL structure and its elastic properties the effect of thermal fluctuations will be neglected.

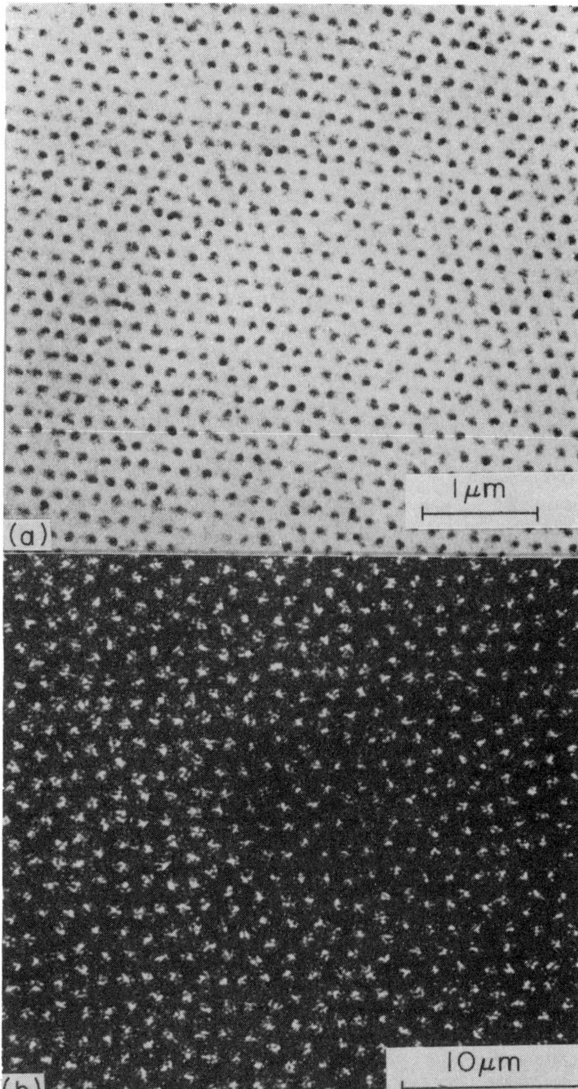

Figure 1
FLL made visible by a decoration method in the electron microscope: (a) niobium (flux lines dark). $T = 1.2$ K, $\mu_0 H = 98.5$ mT, $\bar{B} = 82$ mT and $a_0 = 0.17$ μm (courtesy of Uwe Essmann, Max-Planck-Institut) and (b) $YBa_2Cu_3O_7$ (flux lines light), $T = 15$ K, $\mu_0 H \simeq \bar{B} \simeq 2$ mT and $a_0 = 1.2$ μm. At this low induction the FLL is strongly disordered due to pinning by material inhomogeneities and possibly by thermal fluctuations (courtesy of David Bishop AT&T Bell Laboratories)

1. Equilibrium FLL

Many of the properties of type II superconductors may be investigated through the phenomenological

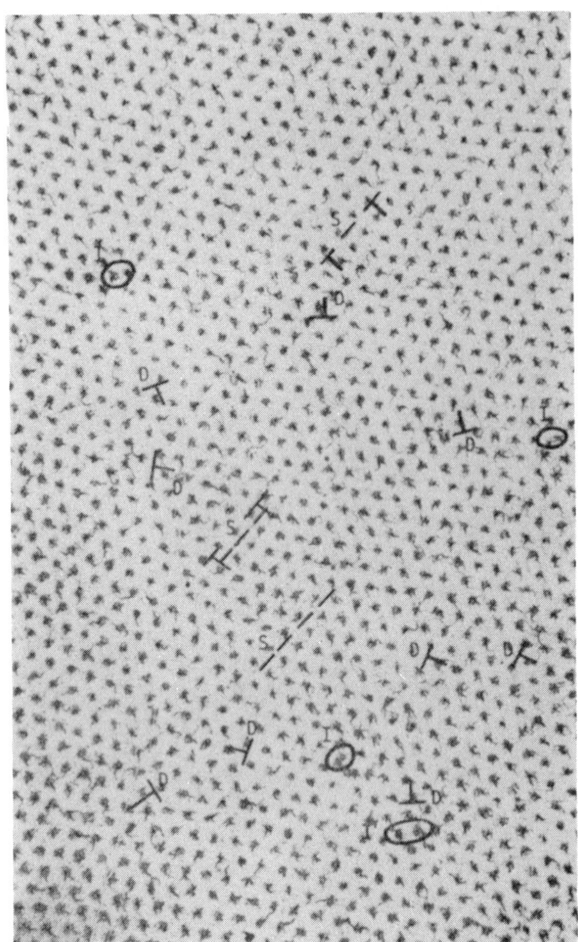

Figure 2
Defective flux lattice in Pb–6 at.%In, edge dislocations (D), point defects (I) and stacking faults (S) are clearly visible: screw dislocations cannot be imaged but are associated with lattice shear and current transport parallel to the mean flux direction in the FLL; the vortex lattice spacing $a_0 = 0.74$ μm (after Dew-Hughes 1971)

GL theory and its development in the time-dependent GL theory (TDGL). The GL theory has been rigorously confirmed and extended from the microscopic (BCS) theory by Gor'kov in the Ginzburg–Landau–Abrikosov–Gor'kov (GLAG) theory. For most purposes, the GL theory is adequate and, in particular, it yields transparently simple solutions in the limit of high temperatures and both low and high inductions. Treatment of more complex problems seen in anisotropic materials, multilayers and inhomogeneous materials can be attempted using a modified GL approach.

1.1 Ginzburg–Landau Theory for Isotropic Superconductors

The essential properties of the FLL may be obtained from the GL theory (see *Superconducting Materials: Physical Properties*). Although this phenomenological theory was derived for temperatures T close to T_c, most of the GL results remain qualitatively correct at all $T < T_c$. For an exception see Sect. 1.3. GL introduces two characteristic lengths which both diverge as $(T_c - T)^{-1/2}$ when $T \rightarrow T_c$. The magnetic penetration depth λ is the length over which the magnetic flux density $\boldsymbol{B} = \nabla \times \boldsymbol{A}$ varies ($\boldsymbol{A}$ is the vector potential). The coherence length ξ is the length over which the complex order parameter ψ can vary. $|\psi|^2$ is proportional to the density of the superconducting electrons or Cooper pairs. $\psi(\boldsymbol{r}) = \Delta(\boldsymbol{r})/\Delta_{BCS}$ where $\Delta(\boldsymbol{r})$ is the gap function of the miscroscopic theory of superconductivity of Bardeen, Cooper and Schrieffer (BCS) and Δ_{BCS} is the maximum value $|\Delta|$ can attain at that temperature. Thus $|\psi| \leqslant 1$.

The nearly temperature-independent ratio $\lambda/\xi = \kappa$, the GL parameter, completely characterizes the type of solutions and thus the behavior of the superconductor. In particular, for $\kappa < 1/\sqrt{2}$ the superconductor is of type I and for $\kappa > 1/\sqrt{2}$ it is of type II. Lead, tin and most superconducting metals are type I superconductors. Niobium, vanadium, most alloys and the ceramic superconductors are of type II and can thus contain an FLL.

The GL equations for ψ and $\boldsymbol{A}$ and the appropriate boundary conditions of a given problem follow from a variational principle by minimizing the functional $F\{\psi, \boldsymbol{A}\}$ which is a free energy. In reduced units with length unit λ, field unit $\sqrt{2}H_c$ (H_c is the thermodynamic critical field) and energy density unit $\mu_0 H_c^2$, the free energy of a superconductor, referred to its normal conducting state at $B = 0$, is

$$F = \int \left[-|\psi|^2 + \frac{1}{2}|\psi|^4 + \left| \left(\frac{\nabla}{i\kappa} - \boldsymbol{A}' \right) \psi \right|^2 + B'^2 \right] d^3r \tag{2}$$

where $\boldsymbol{A}'$ and $\boldsymbol{B}'$ are in reduced units of $\mu_0\sqrt{2}\,H_c$. Note that the only material parameter in Eqn. (2) is κ. The first two terms are the superconducting condensation energy, which is a minimum when $|\psi| = 1$. The third term is the kinetic energy of the supercurrent (of density $(\boldsymbol{A}' - \nabla\varphi/\kappa)|\psi|^2$ where φ is the phase of ψ) plus a term $(\nabla|\psi|/\kappa)^2$ which shows that $|\psi|$ tends to be a smooth function. The last term is the energy of the magnetic field.

The applied field H, which does not appear in F, plays the role of an external pressure that forces the superconductor into some equilibrium state with average induction $\bar{B}$. The field H that establishes this state is obtained as $H = \partial F/\partial \bar{B}$. This law follows from minimization of the Gibbs free energy $G = F - \bar{B}H$. Special solutions forced by the applied field are the Meissner state ($|\psi| \equiv 1$, $B \equiv 0$, $H < H_c$ for type I and

$H < H_{c1}$ for type II superconductors), the mixed or Shubnikov state with an FLL ($|\psi|^2$ and B periodic, $0 < \bar{B} < \mu_0 H_{c2}$, $H_{c1} < H < H_{c2}$) and the normal conducting state ($\psi \equiv 0$, $B \equiv \mu_0 H$, $H > H_{c1}$ for type I and $H > H_{c2}$ for type II superconductors):

$$\left.\begin{aligned} \mu_0 H_{c1} &= \frac{\Phi_0(\ln \kappa + \varepsilon)}{4\pi\lambda^2} \\ \mu_0 H_c &= \frac{\Phi_0}{2\sqrt{2}\pi\lambda\xi} \\ \mu_0 H_{c2} &= \frac{\Phi_0}{2\pi\xi^2} \end{aligned}\right\} \quad (3)$$

with $\varepsilon \ll 1$ for $\kappa \gg 1$ and $\varepsilon = 1 + \ln\sqrt{2}$ for $\kappa = 1/\sqrt{2}$. For $\kappa = 1/\sqrt{2}$, $H_{c1} = H_c = H_{c2}$ and, in general, $H_{c1} \simeq H_c/\sqrt{2}\kappa$ and $H_{c2} = H_c\sqrt{2}\kappa$.

Type I superconductors formally exhibit $H_{c1} > H_{c2}$, thus an FLL cannot exist in them. However, in type I superconductors with a demagnetization factor $N > 0$ ($N = 0$ for long cylinders in parallel fields, $N = \frac{1}{3}$ for spheres and $N = 1$ for thin disks in perpendicular fields (see also *Demagnetizing Factors*)) there may be an intermediate state that consists of alternating superconducting and normal conducting layers in which $B \equiv 0$ and $B \equiv \mu_0 H_c$, respectively. This state occurs in the field range $(1 - N)H_c < H < H_c$. It is stable since the wall energy between these lamellae is positive; that is, the wall length tends to be as short as possible (see *Magnetic Properties of Superconductors*).

In contrast to type I superconductors, the wall energy in type II superconductors is negative; the walls will thus curl spontaneously until the longest possible wall (of thickness $\simeq \xi$) or finest subdivision is achieved. The result of this tendency is just the FLL: small normal vortex cores of radius $\simeq \xi$ are surrounded by a superconducting phase, each vortex carrying the smallest possible flux $\Phi = \Phi_0$. A lattice of flux lines that carry more flux, $\Phi = n\Phi_0$ with $n > 1$ (n is an integer), is a solution that has higher energy than a lattice of single flux quanta. Any distorted lattice of flux lines, straight or curved, has higher energy than the perfect FLL in a homogeneous medium. In inhomogeneous media with spatially varying material parameters λ and ξ, spontaneous distortion of the FLL will occur (see Sect. 1.3). An interesting feature of the GL equations is that, for the special case $\kappa = 1/\sqrt{2}$, any solution has the same Gibbs free energy G and is thus a possible solution for the ground state: perfect or distorted FLLs with one or more flux quanta, flux tubes (these are formally flux lines with $n \gg 1$) or lamellae, and the homogeneous states $\psi = 1$, $B = 0$ and $\psi = 0$, $B = \mu_0 H_c$.

1.2 FLL at Low and High Inductions

At low inductions $\bar{B} < 0.25\,\mu_0 H_{c2}$ and not too small values of $\kappa > 1.4$, the FLL may be considered as consisting of individual vortices with well-separated cores. The fields of parallel flux lines, centered at positions $\boldsymbol{r}_\nu$, then superimpose and yield

$$B(r) \simeq \sum_\nu \frac{\Phi_0}{2\pi\lambda^2} K_0\{[(\boldsymbol{r} - \boldsymbol{r}_\nu)^2 + \boldsymbol{r}_c^2]^{1/2}/\lambda\} \quad (4)$$

where $K_0(x)$ is a modified Bessel function with limits $K_0 \simeq -\ln x$ for $x \ll 1$ and $K_0 \simeq (\pi/2x)^{1/2} \exp(-x)$ for $x \gg 1$. The isolated vortex fields in Eqn. (4) have a range λ and a core radius $r_c \simeq \xi$. In the limit $r_c \to 0$, the London limit, Eqn. (4) coincides with the result of a theory established in 1935 by F. and H. London. This London theory coincides with the GL theory if one puts $|\psi| = 1$. It yields "London vortices" with zero core diameter and a logarithmic divergence of B at the vortex cores. This divergence is removed by the GL theory which provides a finite r_c. The vortex core is defined as the region where the order parameter decreases from its nearly constant value between the flux line cores:

$$|\psi(\boldsymbol{r})|^2 \simeq \prod_\nu \{1 - \exp[-(\boldsymbol{r} - \boldsymbol{r}_\nu)^2/r_c^2]\} \quad (5)$$

At low flux densities the flux lines interact with each other by a pair potential, which within the GL and London theories is

$$V(r) = \frac{\Phi_0^2}{2\pi\lambda^2\mu_0} K_0(r/\lambda) \quad (6)$$

This is a monotonically decreasing and thus repulsive potential. The energy of the FLL is then the sum of all pair interactions, plus the self-energy J of the N vortices ($J = \Phi_0 H_{c1}$ for $B \ll \mu_0 H_{c2}$):

$$F = \tfrac{1}{2}\sum_\mu \sum_\nu V(|\boldsymbol{r}_\mu - \boldsymbol{r}_\nu|) + NJ \quad (7)$$

Eqns. (4–7) apply to arbitrary arrangements of parallel flux lines with separations larger than $\simeq 4\xi$ and for $\kappa > 1.4$. More general approximate expressions for all separations, arbitrary κ values, and straight or curved flux lines are given by Brandt and Essmann (1987).

For a regular FLL, $B(\boldsymbol{r})$ is periodic and may be written as a two-dimensional Fourier series:

$$B(\boldsymbol{r}) = \sum_K B_K \cos \boldsymbol{Kr} \quad (8)$$

where $\boldsymbol{K} = \boldsymbol{K}_{mn}$ (m, $n = 0, \pm 1, \pm 2, \ldots$) are the reciprocal lattice vectors of the FLL. For the triangular lattice $K_{mn}^2 = (16\pi^2/3a_0^2)(m^2 + mn + n^2)$. At low inductions, the Fourier coefficients are

$$B_K \simeq \bar{B}/(1 + K^2\lambda^2 + K^4 r_c^4) \quad (9)$$

The term $K^4 r_c^4$ provides a cutoff at $K \simeq r_c^{-1}$ and smears the logarithmic singularity of $B(r)$ at the vortex centers over a range of radius $\simeq r_c$ (cf. Eqn. (4)).

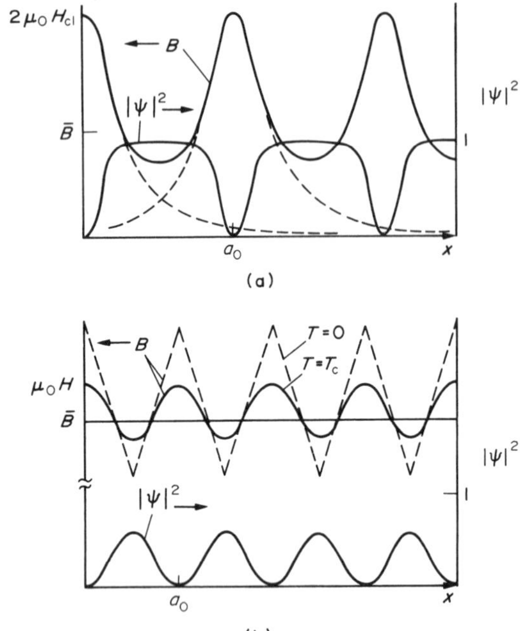

Figure 3
Profiles of magnetic field B and order parameter $|\psi|^2$ along a line connecting neighboring flux lines in pure niobium: (a) low induction $\bar{B} \simeq 0.1\mu_0 H_{c2}$ (the dashed lines give the profiles of isolated flux lines) and (b) high induction $\bar{B} \simeq 0.7\mu_0 H_{c2}$ (the dashed line is the exotic field profile at zero temperature)

At higher inductions the vortex cores overlap and, therefore, the vortex fields no longer superimpose linearly. In this case, the shapes of $\psi(\mathbf{r})$ and $B(\mathbf{r}) - \bar{B}$ may be obtained by solving the linearized GL equations; these apply when H is close to H_{c2} since then $|\psi|^2 \ll 1$ holds. The amplitude of this solution is then obtained from the nonlinear terms. The resulting $|\psi|^2$ and B are smooth periodic functions with Fourier coefficients decreasing rapidly with increasing K (see Fig. 3):

$$|\psi(\mathbf{r})|^2 = \frac{1 - \bar{B}/\mu_0 H_{c2}}{1.16 - 0.16/2\kappa^2} \sum_K a_K \cos \mathbf{Kr} \qquad (10)$$

$$B_K = \mu_0 M a_K \quad (K \neq 0) \qquad (11)$$

where, for the triangular lattice,

$$a_K = (-1)^{m+mn+n} \exp[-(\pi/\sqrt{3})(m^2 + mn + n^2)] \quad (12)$$

The magnetization $M = \bar{B}/\mu_0 - H$ is given by

$$M = -\frac{(H_{c2} - H)}{1.16(2\kappa^2 - 1)} \qquad (13)$$

Numerical computations show that the approximate analytic solutions Eqns. (10–13) apply in the large field range $0.5 < \bar{B}/\mu_0 H_{c2} < 1$.

1.3 FLL at Low Temperatures

At lower temperatures ($T < T_c$), the GL results are still approximately valid for sufficiently impure superconductors; that is, for alloys with electron mean free path l smaller than the BCS coherence length ξ_0. For clean type II superconductors ($l \gg \xi_0$) at $\bar{B} \simeq \mu_0 H_{c2}$, $\psi(\mathbf{r})$ retains its GL form (Eqn. (10)) but $B(\mathbf{r})$ changes (Delrieu 1972). At $T = 0.625\,T_c$ the minima and saddle points of $B(\mathbf{r})$ interchange their positions. At $T = 0$, the field profile along the lines connecting neighboring flux line centers is zigzag shaped, the maximum and the three minima (per unit cell) attain conical shape, the field in the vortex centers exceeds the applied field ($B_{max} = \mu_0(H + 1.8|M|)$) and the saddle points exhibit a three-fold rotational axis (see Fig. 4). This exotic behavior of the microscopic magnetic field at $T = 0$ is due to the slow decrease of the Fourier coefficients $B_K \sim K^{-3}$:

$$B_K \simeq 0.363\mu_0 M(-1)^{m+mn+n}(m^2 + mn + n^2)^{-3/2} \quad (K \neq 0) \qquad (14)$$

The B_K for arbitrary lattice symmetry and temperatures $0 \leqslant T < T_c$ are given by Brandt (1988).

A further modification which extends the GL picture to low temperatures is the introduction of three GL parameters $\kappa_1(T)$, $\kappa_2(T)$ and $\kappa_3(T)$ which coincide

Figure 4
Lines of equal $B(x, y)$ for a type II superconductor at $\bar{B} \simeq 0.2\mu_0 H_{c2}$ (the circles are centered at the flux line positions where B is at a maximum)

with κ at $T=T_c$ but increase slightly when T is lowered. κ_1 appears in the relation $H_{c2}=\sqrt{2}\kappa_1 H_c$, κ_2 in the slope of $M(H)$ (Eqn. (13)) and κ_3 in $H_{c1}=H_c(\ln\kappa_3+\varepsilon)/\sqrt{2}\kappa_3$. Here, again, the correction to the GL result is largest for pure superconductors. For pure niobium, which has $\kappa=0.72$, this correction has interesting consequences. At $T\simeq T_c$ its magnetization curve looks almost like that of a type I superconductor, but with decreasing T, H_{c1} decreases (in reduced units with $\sqrt{2}H_c=1$), H_{c2} increases, the slope of $-M(H)$ decreases and the area under the magnetization curve H_c^2/μ_0 stays constant. As a consequence, the magnetization curve near H_{c1} takes an S shape which, by a Maxwell construction, yields a finite jump of M for H slightly smaller than the "overheated" H_{c1}. This abrupt jump from $\bar{B}=0$ to $\bar{B}=B_0$ may be explained by an attractive interaction between the flux lines at large distances which, together with the usual repulsion at short distances, yields a finite equilibrium distance $a_0=(1.155\,\Phi_0/B_0)^{1/2}$. As $T\rightarrow T_c$ the GL behavior with mere repulsive interaction (Eqn. (6)) and with $B_0=0$, results in $a_0=\infty$. Pure niobium at $T<T_c$ thus combines features of type I and type II superconductors: in the range $(1-N)H_{c1}<H<H_{c1}$ it exhibits lamellae or tubes containing an FLL with $\bar{B}=B_0$, embedded in a Meissner phase with $B=0$, or Meissner tubes embedded in an FLL.

The results of this section going beyond the GL theory were obtained from the BCS theory extended by Gor'kov to a spatially varying gap parameter $\Delta(\boldsymbol{r})$.

1.4 Anisotropic Superconductors

Decoration experiments on single-crystal samples indicate that the FLL is anisotropic for most materials, however the anisotropy is usually very small. Table 1 gives the morphology of the FLL for different LTS. Figure 5 shows an example of a square lattice observed in a Pb–1.6%Tl sample. Anisotropy in many HTS oxides and in layer compounds such as $NbSe_2$ is much stronger. Apart from the increase in analytic complexity, the theoretical treatment of the equilibrium FLL for anisotropic superconductors is very similar to that for isotropic materials. Anisotropic formulations of the London and GL theories are of value in the different limits discussed in the previous sections. It is notable that, because of the very high B_{c2} in many HTS, the London theory gives an accurate treatment of the FLL over a wide field range. It only becomes necessary to use the anisotropic GL theory when the flux line cores begin to overlap and for the other problems that involve variation in the density of superconducting pairs (e.g., pinning or flux cutting calculations). The GL theory can be generalized for noncubic materials by introducing a phenomenological anisotropic effective mass tensor $\mathbf{m}_{ik}$ for the superconducting pairs which is normalized and dimensionless so that the diagonal components m_a, m_b and m_c in the three principal crystal directions have the property $m_a m_b m_c=1$ (see *Anisotropy in Superconducting Materials*). This is equivalent to replacing λ^2 with the tensor $\lambda^2\mathbf{m}_{ik}$ and ξ^2 with $\xi^2\mathbf{m}_{ik}^{-1}$, so that in the direction of the principal axes $\lambda_a=\lambda m_a^{1/2}$, $\lambda_b=\lambda m_b^{1/2}$, $\lambda_c=\lambda m_c^{1/2}$, $\xi_a=\xi m_a^{-1/2}$, $\xi_b=\xi m_b^{-1/2}$ and $\xi_c=\xi m_c^{-1/2}$.

For an FLL of individual well-separated vortex cores the superimposed fields given by Eqn. (4) for an isotropic material becomes for flux lines parallel to the a axis (in the London limit with $r_c\rightarrow 0$),

$$\boldsymbol{B}(\boldsymbol{r})=\sum_i\frac{\Phi_0}{2\pi\lambda_b\lambda_c}K_0\left\{\left[\left(\frac{y-y_i}{\lambda_c}\right)^2+\left(\frac{z-z_i}{\lambda_b}\right)^2\right]^{1/2}\right\} \tag{15}$$

$\boldsymbol{B}(\boldsymbol{r})$ is a solution of the anisotropic London equation

$$\boldsymbol{B}+\nabla\times(\boldsymbol{\Lambda}.\nabla\times\boldsymbol{B})=\Phi_0\sum_i\delta(\boldsymbol{r}-\boldsymbol{r}_i) \tag{16}$$

where $\boldsymbol{\Lambda}$ is the tensor $\lambda^2\mathbf{m}_{ik}$ and $\delta(\boldsymbol{r}-\boldsymbol{r}_i)$ is a two-dimensional delta function that defines the position of vortices and accounts for flux quantization. For an isolated vortex centered on the x axis (a axis), the magnetic field in the London limit is

$$B_x(y,z)=\frac{\Phi_0}{2\pi\lambda_b\lambda_c}K_0\cdot\left\{\left[\left(\frac{y}{\lambda_c}\right)^2+\left(\frac{z}{\lambda_b}\right)^2\right]^{1/2}\right\} \tag{17}$$

The decay distance along the y axis (b axis) is approximately exponential with decay length λ_c. The penetration depth is λ_c and not λ_b because the screening currents on the y axis are in the direction of the z axis (c axis); they therefore involve an electron density associated with the c direction and if this density is low the screening is correspondingly reduced and the screening length increased. The contours of constant B_x are ellipses $(y/\lambda_c)^2+(z/\lambda_b)^2=\text{constant}$; these are also the local streamlines. Figure 6 shows a schematic cross section for a vortex drawn with the ratios of the effective masses approximately equal to that in $YBa_2Cu_3O_7$. The anisotropic GL theory provides a finite core (set to zero in Eqn. (15)) defined by the ellipse

$$\left(\frac{y}{\lambda_c}\right)^2+\left(\frac{z}{\lambda_b}\right)^2=\left(\frac{1}{\kappa_a}\right)^2 \tag{18}$$

where $\kappa_a=\kappa m_a^{-1/2}=\kappa m_b^{1/2}m_c^{1/2}=\lambda_b/\xi_c=\lambda_c/\xi_b$ is an effective κ for a vortex aligned along the a axis. The semimajor axes of the core are $\xi_c=\xi m_c^{-1/2}$ along the z axis (c axis) and $\xi_b=\xi m_b^{-1/2}$ along the y axis (b axis). The corresponding upper critical field for B parallel to the a axis is $B_{c2a}=\Phi_0/2\xi_b\xi_c$.

The FLL in an anisotropic material no longer has triangular symmetry for a general field direction, and the average induction B is no longer parallel to the magnetic field H (Kogan 1988). The FLL free energy

Table 1
Morphology of the vortex lattice in single-crystalline type-II superconductors for different orientations of the applied magnetic field

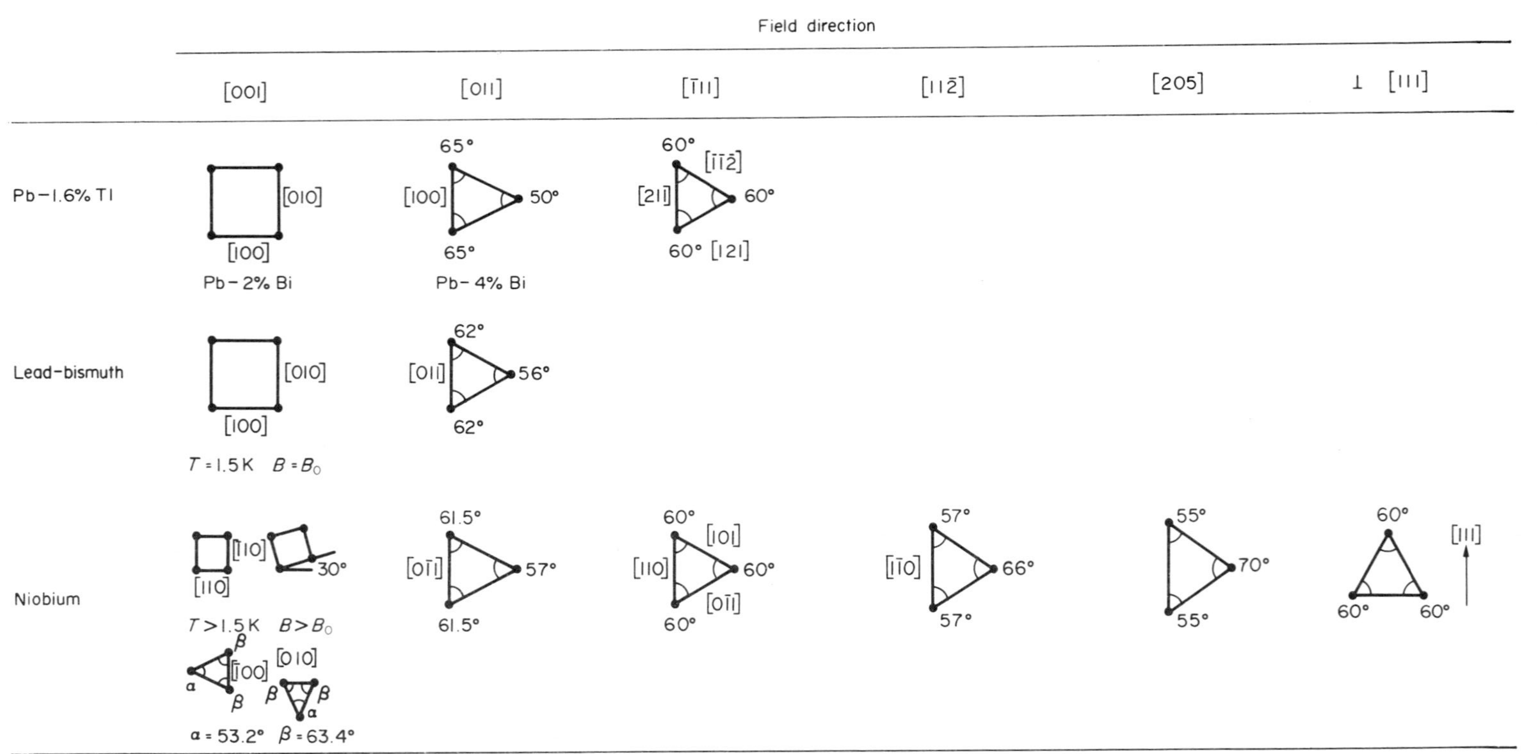

Source: Shelton (1977)

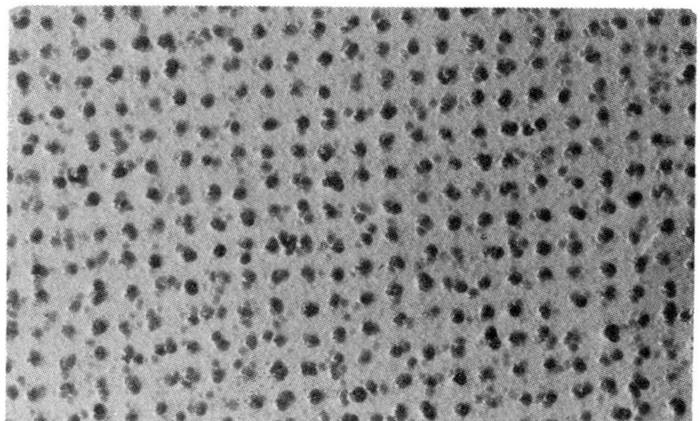

Figure 5
Square vortex lattice in Pb–1.6 wt%Tl single crystal with the applied field in [001] direction: the vortex lattice is oriented with close packed directions parallel to [010] and [100], $\kappa \sim 0.72$, $T = 1.2$ K; $B = 36.5$ mT, the vortex lattice spacing $a_0 = 0.26$ μm (after Obst 1971)

varies as the lattice is rotated about an axis parallel to B; the FLL in equilibrium is therefore locked to the crystal. At low inductions for some directions of B the interaction between parallel flux lines may at certain distances become attractive. This is a consequence of a nonmonotonic variation in $B(\boldsymbol{r})$ which develops an oscillating tail with regions of reversed field. This interaction can lead to the formation of vortex chains at low inductions in strongly anisotropic superconductors (Bolle et al. 1991).

So far the treatment has been restricted to materials that are anisotropic but uniform. If the coherence length becomes comparable to the lattice parameter of the crystal the discreteness of the structure has to be taken into account. This is important for HTS where ξ_c is less than the c axis lattice repeat at low temperatures. Since $\xi \sim (T_c - T)^{-1/2}$ the superconducting coherence length diverges near T_c and the three-dimensional anisotropic GL approach will be valid. At lower temperatures there will be a crossover to a more complicated behavior that can be better approximated by treating the material as a stack of weakly coupled two-dimensional superconducting layers; this has important consequences for the FLL.

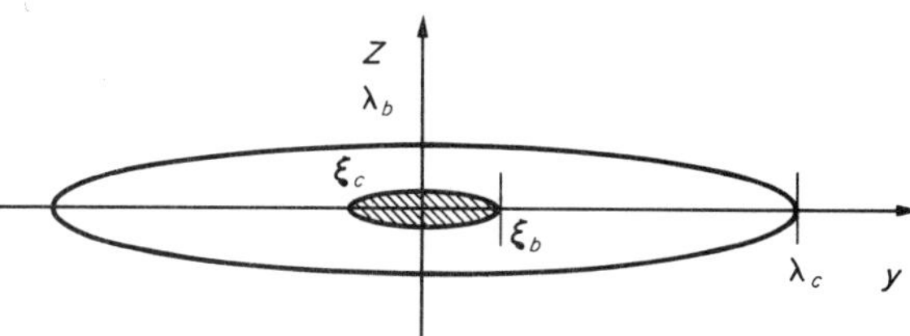

Figure 6
Schematic cross section of an isolated vortex lying along the x-axis (a axis) in an anisotropic type II superconductor: the outer ellipse is a contour of constant B_x and also the streamline for current flow; the shaded region is the vortex core and the coherence lengths ξ_b and ξ_c are about 100 times smaller than the penetration depths λ_c and λ_b

1.5 Layered Superconductors

Layered superconductors have been the subject of research since the 1960s. As a class of superconducting materials they combine some of the features of anisotropic three-dimensional superconductors, of two-dimensional superconductors and of granular superconductors. The characteristic behavior of the layered superconductor depends on the strength of the coupling and can change with temperature and applied field leading to crossover phenomena sometimes referred to as phase boundary effects.

Examples of layered superconductors are transition metal dichalcogenides such as TaS_2, intercalation compounds based on them (e.g., TaS_2 (pyridine)$_{1/2}$), the HTS layer structure oxides and artificial SIS or SNS multilayers (see *Thin Films, Multilayers and Devices, Superconducting*). The coupling between layers can vary from purely magnetically coupled layers whose properties depend on screening currents induced in one layer by the magnetic structure of a neighboring layer, to Josephson coupled tunnel junction heterostructures where supercurrents tunnel from layer to layer through an insulating barrier layer. The strongest coupling corresponds to proximity coupled SNS multilayers which become anisotropic three-dimensional superconductors when the normal state coherence length ξ_N is greater than the interlayer spacing.

A powerful phenomenological theory of great generality was proposed by Lawrence and Doniach in 1971. This modified GL theory follows a tight-binding approximation by replacing the continuous order parameter $\psi(\boldsymbol{r})$ with a sum over a series of $\psi_n(x, y)$ for layers positioned at $z_n = ns$. The free energy functional $F\{\psi, \boldsymbol{A}\}$ corresponding to Eqn. (2) for the isotropic case sums over the individual ψ_n and contains an additional coupling term,

$$\sum_n \frac{\xi_c^2}{s} |\psi_{n+1} - \psi_n \exp(iI_n)|^2 \qquad (19)$$

where the same reduced units are used as in Eqn. (2) and

$$I_n(\boldsymbol{r}) = \frac{2\pi}{\Phi_0} \int_{ns}^{ns+s} A_z(\boldsymbol{r})\, dz$$

For $\xi_c > s$, $F\{\psi, \boldsymbol{A}\}$ reduces to that for the anisotropic GL theory of the previous section. In the opposite limit $\xi_c \ll s$, the weakly coupled Josephson limit is obtained.

The equilibrium FLL in layered superconductors is poorly understood and is likely to be an area of major activity for some years. In the limit of very weakly coupled superconducting layers the FLL breaks down into a magnetic structure that is better viewed as almost independent two-dimensional arrays of flux spots (or point vortices) localized on the superconducting layers. The weak correlations between flux spots from layer to layer compete with the disordering effect of thermal energy. If a stack of point vortices remains coupled for a sufficient length of time, either because the temperature is low or the coupling strong, then an FLL can be defined. The energy required to cause flux slippage or phase slippage can be identified with the flux line cutting energy. The characteristics of the FLL in layered structures that distinguishes it most clearly from the FLL in less anisotropic materials is its low flux cutting energy. Flux cutting and cross-joining phenomena observed in LTS mainly during flux flow under force-free conditions, therefore, are expected to occur more generally in layered superconductors.

The structure of the FLL in layered superconductors depends very much on the direction of the applied field; for B perpendicular to the layers the FLL consists of stacks of point vortices coupled magnetically and also by proximity currents in the region between superconducting layers (Feigel'man et al. 1990). For B parallel to the layers the structure of flux lines is more complex and, in artificial SS'S multilayers, may involve crossover at critical fields and temperatures when the core of a vortex changes its equilibrium position from one layer to another (Tachiki and Takahashi 1991). In the case of a Josephson coupled layer the vortex core will be entirely in the nonsuperconducting region and will cease to display a line singularity in the order parameter. Such a vortex is termed a Josephson vortex. The order parameter in the superconducting layers remains high; the superconducting phase changes according to the Josephson relation (see *Josephson Effects in Weak Links*) with sharp changes between the layers where the superelectron density is low. The Josephson core lies at the center of the Josephson current that flows through the Josephson medium defined by the multilayer system in much the same way as for vortices in a granular superconductor or a weak link network of tunnel junctions.

The nature of the vortex structure when the vortices are neither perpendicular nor parallel to the layers is still the subject of research. The magnetic structure can be viewed as stacks of point vortices connected by Josephson strings. It is likely that the FLL in such systems will exhibit a complex defect structure of kinks and a mosaic substructure to accommodate the incommensuracy of the vortex spacing and layer repeat; its static and dynamic properties will be strongly affected by thermal disorder.

The magnetic properties of both HTS and LTS granular materials (see *Granular Superconductors*) have been modelled in a similar way by considering the GL free energy functional for superconducting grains coupled by Josephson intergranular currents. When the intragranular condensation energy E_g is small compared with the Josephson coupling energy E_J, the theory reduces exactly to the GL theory in the dirty limit with modified characteristic lengths ξ_J and λ_J which are both much larger than the grain sizes (Clem 1988). In the opposite limit $E_g \gg E_J$, the intergranular currents are too small to suppress ψ within the grains. The behavior of the granular material is then dominated by the critical currents of the weak links between grains. Abrikosov vortices can exist within the grains and motion of the flux structure is accompanied by phase slip in the weak links. At high temperatures when $k_BT > E_J$ thermally activated phase slippage occurs and a time-averaged voltage occurs across any weak link that carries a current.

2. *Elasticity of the FLL*

The equilibrium FLL is only found in single-crystal samples in a uniform applied field when the transport current is zero. Otherwise the FLL is invariably distorted and defective. Although any stationary magnetic structure is in local equilibrium it has proved valuable to treat the distorted FLL as a distorted elastic medium in metastable equilibrium under the influence of local and long-range forces arising from the internal FLL defect structure, interaction with crystal lattice inhomogeneity and the effect of external electromagnetic constraints.

On a scale greater than $\lambda' = \lambda/(1-b)^{1/2}$ the FLL in an isotropic superconductor may be treated as a uniaxial elastic continuum in terms of elastic moduli c_{ij}. Deformation on a shorter scale than λ' as shown has to take account of both the discreteness of the FLL and elastic nonlocality, arising because the range of the vortex interaction is larger than the vortex spacing. The elastic constants then depend on the wavelength of the distortion and the FLL becomes soft for short-range tilt and compression distortions.

2.1 Uniform Distortion in Isotropic Superconductors

When the FLL is subjected to a uniform deformation its increase in energy, the elastic energy, may be expressed in terms of three elastic moduli: for uniaxial compression (c_{11}), shear (c_{66}) and tilt (c_{44}). The modulus for isotropic compression (bulk modulus) is $c_{11}-c_{66}$. From thermodynamic considerations it is found that

$$c_{11} - c_{66} = \bar{B}^2 \frac{dH}{d\bar{B}} \qquad (20)$$

$$c_{44} = \bar{B}H \tag{21}$$

The shear modulus $c_{66} \ll c_{11} \simeq c_{44} \simeq \bar{B}^2/\mu_0$ cannot be obtained in this way since the magnetization curves, or $\bar{B}(H)$, are practically the same for triangular and square FLLs. At very low inductions ($\bar{B} \ll \mu_0 H_{c2}$) where the self energy J of the flux lines (defined by Eqn. (7)) does not depend on $\bar{B}$, it is possible to derive a formula that relates the shear modulus

$$c_{66} = \frac{\bar{B}}{16\Phi_0}\sum_\nu [R_\nu^2 V''(R_\nu) + 3R_\nu V'(R_\nu)] \tag{22}$$

(where V is the pair potential and R_ν are the flux line positions in an ideal triangular lattice) to $F(\bar{B})$ (Eqn. (7)) and thus to the measurable applied field $H(\bar{B}) = \partial F/\partial \bar{B}$. This formula, however, is of little practical use since at $\bar{B} \ll \mu_0 H_{c2}$, or $H \simeq H_{c1}$, the measured magnetization curves are strongly influenced by pinning. From the GL equations it is found for the triangular FLL that

$$c_{66} \simeq \mu_0 H_{c2}^2 \frac{b(1-b)^2}{8\kappa^2}\left(1 - \frac{1}{2\kappa^2}\right)(1 - 0.29b) \tag{23}$$

where $b = \bar{B}/\mu_0 H_{c2}$.

For deformations varying little over an effective penetration depth $\lambda' = \lambda/(1-b)^{1/2}$ the elastic energy of the FLL is given by the "local" expression

$$U_{\text{elast}} = \frac{1}{2}\int \{c_{11}(s_{x/x} + s_{y/y})^2 + c_{66}[(s_{x/y} + s_{y/z})^2 + (s_{x/x} - s_{y/y})^2] + c_{44}(s_{x/z}^2 + s_{y/z}^2)^2\}\, d^3r \tag{24}$$

where $\mathbf{s}(\mathbf{r}) = (s_x;\, s_y;\, 0)$ is the flux line displacement field, and $s_{x/x} = \partial s_x/\partial x$, $s_{x/y} = \partial s_x/\partial y$ and so on are the strain components.

2.2 General Deformation, Nonlocality

When $\mathbf{s}(\mathbf{r})$ varies considerably over λ' then the "nonlocal" theory of elasticity applies. In terms of the Fourier transformed displacement field $\tilde{\mathbf{s}}(\mathbf{k}) = (\tilde{s}_x;\, \tilde{s}_y;\, 0)$, where $\mathbf{k} = (k_x;\, k_y;\, k_z)$ is the wave vector, the general expression for the elastic energy (Brandt 1986) is

$$U_{\text{elast}} = \frac{1}{2}\int_{\text{BZ}} \frac{d^3k}{8\pi^3} \sum_{\alpha,\beta} \tilde{s}_\alpha(\mathbf{k})\tilde{s}_\beta^*(\mathbf{k})\Phi_{\alpha\beta}(\mathbf{k}) \tag{25}$$

Here the integration is over the first Brillouin zone (BZ) of the FLL and over $-\infty < \mathbf{k}_z < \infty$, $\Phi_{\alpha\beta}(\mathbf{k})$ is the elastic matrix, and α and β denote Cartesian components x and y, respectively. In general, $\tilde{\mathbf{s}}(\mathbf{k})$ and $\Phi(\mathbf{k})$ are periodic in $\mathbf{k}$ space. A useful approximation to Φ, valid for small $k \ll k_{\text{BZ}}$ (πk_{BZ}^2 is the area of the BZ, $k_{\text{BZ}}^2 = 2b/\xi^2$) is

$$\Phi_{\alpha\beta}(\mathbf{k}) = k_\alpha k_\beta[c_{11}(k) - c_{66}] + \delta_{\alpha\beta}[(k_x^2 + k_y^2)c_{66} + k_z^2 c_{44}(k)] \tag{26}$$

For Eqn. (20),

$$c_{11}(k) - c_{66} = \mu_0 H_{c2}^2 \frac{b^2}{1 + k^2/k_h^2}\frac{1 - 1/2\kappa^2}{1 + k^2/k_\psi^2} \tag{27}$$

and for Eqn. (21),

$$c_{44}(k) = \mu_0 H_{c2}^2\left[\frac{b^2}{1 + k^2/k_h^2} + \frac{b(1-b)}{\kappa^2}\right] \tag{28}$$

where $k^2 = k_x^2 + k_y^2 + k_z^2$, $k_h^2 = (1-b)/\lambda^2$ and $k_\psi^2 = 2\kappa^2 k_h^2$. Eqns. (27) and (28) apply if either $b\kappa^2 \gg 1$ holds or $b > 0.3$; in these cases the vortex fields strongly overlap. The two terms in $c_{44}(k)$ are equal when $k^2 \simeq k^2_{\text{BZ}}$. The compression and tilt moduli are dispersive; that is they depend on the wavelength $2\pi/k$ of the strain field. The FLL is softer with respect to short-wavelength deformations than it is with respect to uniform deformations. The ratios $c_{44}(0)/c_{44}(k) \simeq b\kappa^2/(1-b)$ and $c_{11}(0)/c_{11}(k) \simeq 2b^2\kappa^2/(1-b)^2$ are very large when $\kappa \gg 1$ ($\kappa \simeq 200$ for $YBa_2Cu_3O_7$) or when $H \simeq H_{c2}$ (see Fig. 7).

The nonlocality of the elastic response of the FLL has several important consequences.

(a) The flux line displacements caused by small pins become larger as compared with the result of the usual local theory in Eqn. (24). This fact makes the summation of random pinning forces more effective and yields larger critical current densities.

(b) The core structure of defects in the FLL (vacancies, interstitials and edge and screw dislocations) is modified. This facilitates the generation and reaction of such defects, and the mutual cutting of flux lines.

(c) The thermal fluctuation of flux line positions is greatly enhanced.

As an example the displacement $s(0)$ of the FLL at the position of a point pinning force f and the mean square thermal displacement $\langle s^2 \rangle$ of the flux line positions shall be considered. Both quantities are given by the same integral (in the continuum approximation):

$$\frac{2s(0)}{f} = \frac{\langle s^2 \rangle}{k_B T} \simeq \int_{\text{BZ}} \frac{d^3k}{8\pi^3}[c_{66}(k_x^2 + k_y^2) + c_{44}(k)k_z^2]^{-1} \simeq \frac{1}{\mu_0 H_{c2}^2}\frac{\kappa/\pi}{b(1-b)}\left[1 + \left(\frac{b\kappa^2/2}{1-b}\right)^{1/2}\right] \tag{29}$$

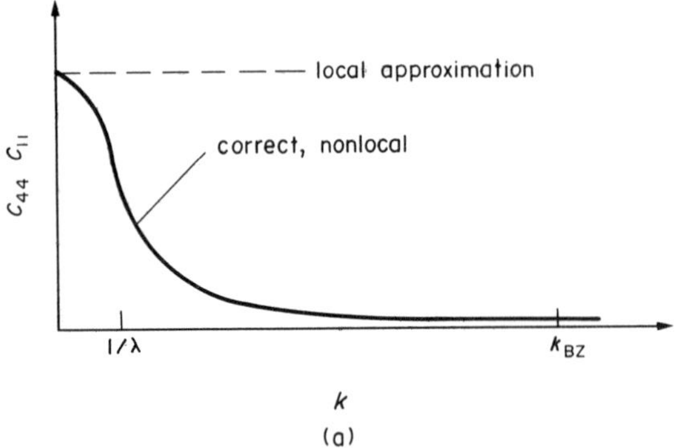

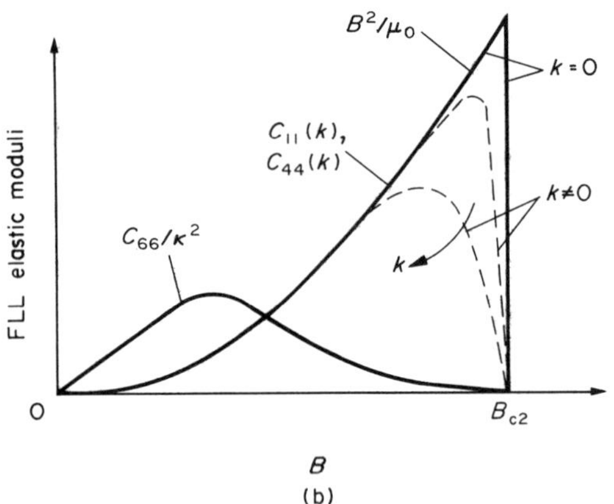

Figure 7
(a) Nonlocal tilt and compressional moduli c_{44} and c_{11} of the FLL versus the strain wave vector $\boldsymbol{k}$. (b) The dispersive moduli $c_{11} \simeq c_{44}$ and the shear modulus c_{66} of the FLL vs induction B for various strain wave vectors (κ is the GL parameter)

The first term in the square brackets is the local result, which is obtained by replacing $c_{44}(k)$ by $c_{44}(0)$ in the integral. The second term is the nonlocal "correction" which is typically very large and even diverges at H_{c2}.

2.3 Anisotropic Elasticity

For uniform deformation the local elastic moduli c_{11} and c_{44} in an anisotropic material are essentially the same as in isotropic materials. This follows directly from the gauge invariance of the complex order parameter $\psi = |\psi| \exp(i\varphi)$ which ensures that the flux density B is determined by the positions of the vortex cores at lines of phase singularity. Since the vortex core can be identified in a Josephson-coupled layered superconductor the moduli are also almost unchanged. There is a small fractional change of order $\mu_0 M/B \ll 1$. The behavior of the shear modulus c_{66} is more complex; for B parallel to the c axis the FLL is almost isotropic in the ab plane for uniaxial materials with $\lambda_a = \lambda_b = \lambda_{ab}$. However, for B perpendicular to c the anisotropy of c_{66} is extremely large (Kogan and Campbell 1989). For shear parallel to the ab plane c_{66} (hard) is a factor $(\lambda_c/\lambda_{ab})^4$ larger than c_{66} (soft) when shear is along the c axis. The isotropic London value is related to anisotropic values,

$$c_{66}\,(\text{isotropic}) = (c_{66}\,(\text{hard})\; c_{66}\,(\text{soft}))^{1/2} = B\Phi_0/16\pi\mu_0\lambda^2$$

For $YBa_2Cu_3O_7$, $\lambda_c/\lambda_{ab} \sim 5$ so that c_{66} (hard)/c_{66} (soft) ~ 600. For nonuniform deformation, the nonlocal (dispersive) moduli $c_{11}(k)$ and $c_{44}(k)$ are anisotropic. The nonlocal elastic moduli have been calculated for B parallel to the c axis using the anisotropic GL theory (Houghton et al. 1989). In particular, the tilt modulus becomes

$$c_{44}(k) = \frac{B^2}{\mu_0}\left(\frac{(1-b)}{1-b+k_{\|}^2\lambda_{ab}^2+k_{\perp}^2\lambda_c^2} + \frac{(1-b)}{k_{BZ}^2\lambda_c^2}\right) \quad (30)$$

where $k\|$ and $k\perp$ are the components of k along and perpendicular to the c axis, respectively. The softening of the FLL is additionally enhanced by anisotropy for short-wavelength distortion; the modulus $c_{44}(k)$ is reduced by $(\lambda_{ab}/\lambda_c)^2$ as compared with the isotropic value.

The complete elastic theory including three different nonlocal tilt moduli and three compressional moduli (for B and the displacements parallel and perpendicular to the c axis) have been derived from the London theory by Sudbø and Brandt (1991a,b). The correction term to c_{44}(k), the last term in Eqn. (30), was shown to exhibit logarithmic dispersion at short tilt wavelengths and to be related to the line tension of the isolated vortex. In anisotropic superconductors this line tension does *not* coincide with the vortex self energy as in isotropic superconductors. Line tension and tilt modulus have a sharp peak for B perpendicular to the c axis, and are minimum for B parallel to the c axis. In contrast, the self energy is a smooth function of the tilt angle, with a maximum for B parallel to and a sharp minimum (or cusp) for B perpendicular to the c axis.

3. *Curved and Disordered FLL*

So far, the treatment of the equilibrium structure and elastic response of the FLL has been conceptually straightforward although the detailed handling of the equations for anisotropic materials and layer structures can be complex. In most systems of interest, pinning inhomogeneities and thermal fluctuations lead to general curvature and disorder of the FLL. This is currently an area of great interest and some considerable controversy. The argument turns on the significance of the phase diagram for different vortex states in the BT plane and, in particular, the relevance of the FLL melting transition for systems with flux pinning and its relation to the irreversibility line $B_{irr}(T)$. These issues will be discussed further in Sect. 4; however, essential ingredients for the discussion are flux line cutting or phase slip in a system of curved flux lines and the magnitude of mean square fluctuations of the vortex positions caused by thermal fluctuations or by a random pinning field.

3.1 Curved Flux Lines and Flux Cutting

The isotropic GL and London theories can be extended to describe a quite arbitrary arrangement of vortex lines by expressing the nonlinear elastic free energy in terms of a repeated sum over products of line integrals along pairs of interacting vortex lines (Brandt 1986). Furthermore, the anisotropic London limit equations apply at all temperatures and over a significant field range, particularly for high-κ materials and HTS. The resulting equations apply to arbitrarily curved, entangled and even crossed and looped vortices and enable quantitative calculations to be made of the energy of FLL defects and distortions of the FLL in the neighborhood of defects. In addition, the energy of interaction of vortices and their spontaneous curvature can be estimated as they approach closely prior to cutting and cross-joining. The interaction between two straight vortices tilted against each other by an angle α and with separation d is, in the isotropic case,

$$U = \frac{\Phi_0^2 \cot \alpha}{2\mu_0 \lambda} \exp(-d/\lambda)$$

A finite core size and spontaneous bending of the vortices will reduce the barrier for cutting (see *Transport Currents and Force-Free Current Flow*). A very strong reduction of the cutting barrier has indeed been obtained recently by Sudbø and Brandt in numerical computations allowing for vortex curving and taking account of the anisotropy of the superconductor. Relatively weak twisting of two vortices was found to lead to spontaneous cutting.

3.2 Thermal Fluctuations, Frequency Spectrum

Although at low temperatures thermal fluctuations cause little disruption to the static FLL structures described, the frequency spectrum for thermal fluctuations in the FLL provides the only fully quantitative approach to describing the attempt frequency in flux creep phenomena and flux cutting.

In the case of HTS, the FLL is soft (because of large λ) and the temperature can be high. Therefore, thermal fluctuations of the vortex position, measured by the mean square $\langle |s(r)|^2 \rangle$ can be large. Ascribing to each elastic mode of energy $\tilde{s}_\alpha(k)\tilde{s}_\beta(k)\ \Phi_{\alpha\beta}(k)$ (cf. Eqn. (25)) an average energy $k_B T/2$, $\langle s^2 \rangle$ is obtained as the integral in k space over $k_B T[\Phi_{xx}^{-1}(k) + \Phi_{yy}^{-1}(k)]$. Explicitly, this gives, generalizing Eqn. (29) to the anisotropic case (Houghton et al. 1989),

$$\langle s^2 \rangle \cong k_B T \left(\frac{\mu_0}{4\pi B \Phi_0 c_{66}} \right)^{1/2} \left(\frac{b\kappa^2/2}{1-b} \right)^{1/2} \frac{\lambda_c}{\lambda_{ab}} \quad (31)$$

or, at $B \ll B_{c2}$

$$\langle s^2 \rangle \cong k_B T \mu_0 \left(\frac{4\pi}{B\Phi_0^3} \right)^{1/2} \lambda_{ab} \lambda_c \quad (32)$$

where $\lambda_a = \lambda_b = \lambda_{ab}$. The three factors in Eqn. (31) have the following origin: first there is the local term, then a correction factor from nonlocal elasticity $\simeq (B/B_{c1})^{1/2} \gg 1$ and, finally, a correction factor from the anisotropy $\lambda_c/\lambda_{ab} \geqslant 5$. Both correction factors increase the thermal fluctuation $\langle s^2 \rangle$. Further enhancement is expected when going beyond the continuum model and by including anharmonicity (nonlinear elasticity). There is some recent evidence that the effect of fluctuations generally near T_c in HTS is strong enough to seriously affect the FLL phase transition at B_{c2} (see *Coherence Length, Proximity Effect and Fluctuations*).

The time scale and frequency spectrum of characteristic thermal fluctuations of the FLL are determined by elastic and viscous forces in the fluctuating FLL (see *Resistive Transition and Flux Flow in Superconducting Materials*). The characteristic fluctuation frequency $\Gamma/2\pi$ is closely linked to the relaxation rate of an excited elastic mode, $\exp(-\Gamma t)$. For a compressional mode

$$\Gamma_{comp} \simeq \frac{2\pi B(1-b)\rho_n}{\Phi_0 \mu_0 \kappa^2}$$

where ρ_n is the normal state resistivity. Shear modes have a lower fluctuation frequency and slower relaxation rate

$$\Gamma_{shear} \simeq \left(\frac{k_z}{k} \right)^2 \Gamma_{comp}$$

For $YBa_2Cu_3O_7$ at $T = 77$ K, $B = 1$ T and with $\lambda_{ab} = 200$ nm, $\lambda_c = 1000$ nm and the mean square thermal displacement $(\langle s^2 \rangle)^{1/2} \simeq 3.2$ nm; this is comparable to the coherence length ξ_{ab} and only $0.065\, a_0$ where $a_0 = 49$ nm is the FLL spacing. At the same field and temperature the fluctuation frequency $\Gamma/2\pi \simeq 20$ GHz for a normal state resistivity $\rho_n = 5 \times 10^{-7}/\Omega$ m and $\kappa = 100$. For NbTi at 4 K and $B = 1$ T with $\rho_n = 7 \times 10^{-7}\ \Omega$m and $\kappa = 30$, $\Gamma/2\pi \simeq 200$ GHz, while for high-purity low κ niobium at $B = 0{\cdot}1$ T, $\Gamma/2\pi \simeq 5$ GHz.

3.3 Fluctuations due to Random Pinning Forces

The mean square fluctuations of the vortex positions caused by random pinning forces, $\langle s_p{}^2 \rangle$, were first calculated by Larkin and Ovchinnikov (1979). At low temperatures the distortions of the FLL caused by pinning always greatly exceed the thermal fluctuations $\langle s_{th}{}^2\rangle$. For point pinning forces, thermal-induced disorder and pinning-induced disorder are related approximately in the continuum approximation (Eqn. (29)) by $\langle s_p^2\rangle/\langle s_{th}^2\rangle \sim U_0/k_BT$ where U_0 is the pinning energy. For LTS this ratio is usually large (e.g., for Nb–50 wt%Ti: $U_0/k_BT \sim 480$ at 4 K falling to 180 at 7.5 K). For strong pinning the vortex distortion should be greater than about $a_0/2\pi$, otherwise the pinning drops below the pinning threshold for single pinning centers and the weaker collective pinning models apply with much smaller levels of vortex distortion (see *Critical Current Density*). Extreme levels of pinning induced disorder lead to the formation first of a defective FLL and ultimately an amorphous arrangement of vortices. This state should not be confused with the "vortex glass" state (Fisher 1989) where the FLL becomes rigidly pinned or "frozen" against thermal activation by virtue of the large size of the correlated volume that interacts with the pinning array.

4. Concluding Remarks

The quantitative understanding of the magnetic structure of a type II superconductor is a prerequisite to the interpretation of many static and dynamic phenomena including flux pinning and the critical current density. There is by now a general consensus on all the main features of the FLL in different materials although a full quantitative treatment is lacking in certain areas. However, the problem of combining the effect of a pinning field and thermal fluctuations has so far proved intractable in all but the simplest cases. This constitutes an area of major current effort and controversy since it relates directly to the irreversibility line $B_{irr}(T)$ and the issue as to whether it is essentially an intrinsic property of the equilibrium FLL or depends also on the strength and geometry of the pinning field (Brandt 1991). Although the issue has been highlighted by the high-temperature oxide superconductors because of its importance for critical currents at high fields and temperature, all aspects of the problem are now known to be present in conventional LTS, granular materials or layered superconductors.

The problem is commonly expressed in terms of whether or not there is an FLL "melting transition" and the effect this may have on the critical current. However the melting transition is often not defined closely and in practice, takes different forms in different materials, in particular the behavior of two-dimensional and three-dimensional materials is different and also the limits of weak collective pinning and strong pinning require quite different treatments. Since there can be transitions from quasi-two-dimensional to three-dimensional behavior and from strong to weak pinning in a given HTS material as the temperature or field is changed, both the meaning and relevance of FLL melting has to be assessed in each regime separately.

Much of the discussion turns on the reduction of the shear modulus at high fields and temperatures and a corresponding "melting" with respect to FLL shear induced by thermal fluctuations. This melting transition is of fundamental importance in two-dimensional systems but largely irrelevant in three-dimensional materials. In the former case it is associated with the Berezinsky – Kosterlitz – Thouless transition (Gubser et al. 1980) and, in general, the melting behavior of thin two-dimensional films, both with and without pinning, is well understood and in broad agreement with theory. The melting temperature coincides with the irreversibility line in this limit, both for weak collective pinning and strong point pins; in the latter case, flux flow occurs by pin avoidance (see *Flux Pinning and the Summation of Pinning Forces*), the unpinned vortices moving freely between the pinned vortices.

In the case of three-dimensional materials, an FLL melting transition leading to easy shear of the FLL does not by itself reduce the critical current in strong pinning systems. Flux shear has to be accompanied by flux cutting and cross-joining for steady-state flux flow to occur. In strong pinning three-dimensional systems, the irreversibility line is either determined by thermally activated depinning (pin breaking) or by thermally activated flux cutting (pin avoidance) which will certainly depend on the characteristic FLL distortions induced by the pinning field (see *Critical Current Density*). In systems with weak collective pinning, there are other possible melting transitions related to the vortex glass state although their relevance for the irreversibility line is still uncertain. Many HTS materials fall into the category of quasi-two-dimensional materials with easy sites for flux cutting or phase slip. The issue of the dependence of the irreversibility line on the pinning field is still unresolved in these intermediate materials.

There is accumulating experimental evidence that $B_{irr}(T)$ can be moved to significantly higher fields and temperatures by strong pinning in $YBa_2Cu_3O_7$ which is a good three-dimensional material over a wide temperature and field range; this is strong support for a $B_{irr}(T)$ determined by the abrupt onset of thermal depinning. In the case of the more anisotropic bismuth and thallium compounds, $B_{irr}(T)$ appears to be somewhat less dependent on pinning. The development of a full understanding of flux pinning, flux flow and irreversibility in these materials will depend on careful interpretation of the widest possible range of experiments on carefully characterized materials.

See also: Magnetic Properties of Superconductors; Resistive Transition and Flux Flow in Superconducting Materials; Superconducting Materials: Physical Properties; Transport Currents and Force-Free Current Flow

Bibliography

Bolle C A, Gammel P L, Grier D G, Murray C A, Bishop D J, Mitzi D B, Kapitulnik A 1991. Observation of a commensurate array of flux chains in tilted flux lattices in Bi–Sr–Ca–Cu–O single crystals. *Phys. Rev. Lett.* 66: 112–15

Brandt E H 1986 Elastic and plastic properties of the flux line lattice in type II superconductors. *Phys. Rev. B* 34: 6514–17

Brandt E H 1988 Magnetic field density of perfect and imperfect flux line lattices in type II superconductors. *J.Low Temp. Phys.* 73: 355–89

Brandt E H 1991 Thermal depinning and "melting" of the flux-line lattice in high T_c superconductors. *Int J. Mod. Phys. B* 5: 751–95

Brandt E H, Essmann U 1987 The flux line lattice in type II superconductors. *Phys. Status Solidi B* 144: 13–38

Campbell A M, Evetts J E 1972 Flux vortices and transport currents in type II superconductors. *Adv. Phys.* 21: 199–428

Clem J R 1988 Granular and superconducting-glass properties of the high-temperature superconductors. *Physica* 153–5: 50–5

De Gennes P G 1966 *Superconductivity of Metals and Alloys*. Benjamin, New York

Delrieu J M 1972 Singularities in the magnetic field map of type II pure superconductors. *J. Low Temp. Phys.* 6: 197–219

Dew-Hughes D 1971 The metallurgical enhancement of type II superconductors. *Rep. Prog. Phys.* 34: 821–73

Essman U, Träuble H 1971 The magnetic structure of superconductors. *Sci. Am.* 224: 75–84

Feigel'man M V, Geshkenbein V B, Larkin A I 1990 Pinning and creep in layered superconductors. *Physica* 167: 177–87

Fisher D S 1980 Flux lattice melting in thin film superconductors *Phys. Rev. B* 22: 1190–9

Fisher M P A 1989 Vortex glass superconductivity – a possible new phase in bulk high T_c oxides. *Phys. Rev. Lett.* 62: 1415–18

Gammel P L, Bishop D J, Dolan G J, Kwo J R, Murray C A, Schneemeyer L F, Waszczak J V 1987 Observation of hexagonally correlated flux quanta in $YBa_2Cu_3O_7$ *Phys. Rev. Lett.* 60: 2592–5

Gubser D U, Francavilla T L, Wolf S A, Leibowitz J R (eds) 1980 *Inhomogeneous Superconductors – 1979*, AIP Conference Proceedings Number 58. American Institute of Physics, New York

Houghton A, Pelcovits R A, Sudbø A 1989 Flux lattice melting in high-T_c superconductors. *Phys. Rev. B* 40: 6763–70

Huebener R P 1979 *Magnetic Flux Structures in Superconductors*. Springer, Berlin

Kogan V G 1988 Uniaxial superconducting particle in intermediate magnetic fields. *Phys. Rev. B* 38: 7049–50

Kogan V G, Campbell L J 1989 Elastic properties of flux lattices in anisotropic high-T_c superconductors. *Phys. Rev. Lett.* 62: 1552–5

Larkin A I, Ovchinnikov Yu N 1979 Pinning in type II superconductors. *J. Low Temp. Phys.* 34: 409–28

Lawrence W E, Doniach S 1971 Theory of layer structure superconductors. In: Kanda E (ed.) 1971 *Proc. 12th Int. Conf. Low Temperature Physics, LT12*. Academic Press, Kyoto, Japan, pp. 361–2

Saint James D, Sarma G, Thomas E J 1969 *Type II Superconductivity*. Pergamon, Oxford

Sudbø A, Brandt E H 1991a Nonlocal elastic properties of flux-lines in anisotropic superconductors in an arbitrarily oriented field. *Phys. Rev. B* 43: 10482–8

Sudbø A, Brandt E H 1991b Flux line tilt moduli in anisotropic superconductors. *Phys. Rev. Lett.* 66: 1781–4

Tachiki M, Takahashi S 1991 Theory of superconductivity in multilayer systems. *Physica B* 169: 121–9

Tilley D R 1965 Critical fields and flux penetration pattern in anisotropic type II superconductors. *Proc. Phys. Soc. London* 85:1177–84

Tinkham M 1975 *Introduction to Superconductivity*. McGraw-Hill, New York

Weber H W (ed) 1977 *Anisotropy Effects in Superconductors*. Plenum, New York

E. H. Brandt
[Max-Planck-Institut für Metallforschung, Stuttgart, Germany]

J. E. Evetts
[University of Cambridge, Cambridge, UK]

Flux Pinning and the Summation of Pinning Forces

Large-scale applications of superconductors require large critical current densities J_c, high critical temperatures T_c and strong critical fields B_{c2}. Large critical currents have been achieved by appropriate treatment during wire fabrication, yielding an optimal defect morphology in the material. Defects cause variations in the free energy of the flux line lattice (FLL), by which the flux lines (or vortices) are trapped. Thus, defects act as pinning centers preventing the motion of flux lines and enable a superconductor to carry dissipation-free transport currents with densities of at least 10^4–10^6 A cm^{-2} in fields as strong as 20 T (see *Critical Current Density*).

Although the concepts of flux pinning have been known for many years, a fully quantitative theoretical analysis is still a complicated problem. Only for simple defect systems and geometries have some successes recently been obtained. The problem actually consists of three issues: (a) the elementary pinning interactions between the FLL and various defects must be computed; (b) the elastic and plastic properties of the FLL should be known; and (c) the contributions of the individual pinning centers must be summed over the distribution of pinning centers. This finally results in a prediction of the pinning force density F_p which is related to J_c by the expression

$$\boldsymbol{F}_p = \boldsymbol{J}_c \times \boldsymbol{B} \quad (1)$$

The elastic properties are treated in a separate article (see *Flux Line Lattice*) and the discussion in this article will therefore be concentrated on the elementary pinning interaction and the summation problem.

1. Defect–Flux Line Interaction

Crystal defects locally alter the material properties and consequently the superconducting parameters in their environment. These local changes may couple to the periodic variations of both the order parameter and the local field that are characteristic of the mixed state. A first classification of the elementary interactions discriminates between these coupling mechanisms: core interaction and magnetic interaction, respectively.

1.1 Magnetic Interaction

Examples of magnetic interaction are the effect of surfaces parallel to the applied magnetic field (this might be the external surface as well as some large precipitate interface) and thickness variations of thin films for fields normal to the film. In the first case, the current distribution around a vortex core near the surface is forced to change in such a way that the normal component at the boundary vanishes. Theoretically, this is achieved by assuming an antivortex image at the other side of the boundary which results in an attractive surface–vortex interaction. This interaction is opposed by the repulsion due to the superconducting screening currents generated by the external field. The net effect is a potential barrier for flux entry and flux exit that is most effective just above the lower critical field H_{c1}. The barrier decreases with increasing field (De Gennes 1966 pp. 76–80).

In the case of thin films with thickness variations, the vortices are trapped at the sites of smallest thickness where the line energy of the vortex is minimum. Evidently, the typical length scale related to the magnetic interaction is the penetration depth λ. In materials with a large Ginzburg–Landau parameter κ, this kind of interaction is therefore small. It generally disappears with increasing magnetic field.

1.2 Core Interaction

The coupling to the variation of the order parameter $|\Psi|^2$ (the density of Cooper pairs) is the origin of flux pinning for almost all defects, including dislocations, point defects, voids, grain boundaries and precipitates. Defects deviate from the surrounding material by having a different density, elasticity, electron–phonon coupling or electron mean free path. The first three properties give rise to a local change in T_c, whereas the last predominantly leads to a variation in κ. Consequently, a distinction can be made between δT_c pinning and $\delta\kappa$ pinning.

The core pin mechanism follows from the Ginzburg–Landau free energy as derived from the microscopic theory (De Gennes 1966 pp. 171):

$$F = \int d^3r \left(A|\Delta|^2 + \frac{B}{2}|\Delta|^4 + C|\partial\Delta|^2 + \frac{h^2}{2\mu_0} \right) \quad (2)$$

where $A = N(0)(1-t)$, $B = 0.098\,N(0)/(k_BT_c)^2$, $C = 0.55\,\xi_0^2 N(0)\chi(\alpha)$, $t = T/T_c$ and $\partial = -i\nabla - (2e/\hbar)\boldsymbol{A}$. $\boldsymbol{A}$ is the vector potential related to the local field $\boldsymbol{h}$; $\Delta(\boldsymbol{r})$ is the pair potential related to the order parameter via $|\Delta|^2 = \hbar^2|\Psi|^2/(4mC)$ where m is the electron mass; $\alpha = 0.882\,\xi_0/l_{tr}$ the impurity parameter (i.e., the ratio of the Bardeen–Cooper–Schrieffer (BCS) coherence length ξ_0 and the mean free path l_{tr}); $\chi(\alpha)$ is the Gorkov impurity function; and $N(0)$ is the density of states at the Fermi surface (see *Superconducting Materials: Physical Properties*). The defects perturb the coefficients A, B and C, as well as the vector and pair potentials. Because F is supposed to be minimized with respect to Δ and $\boldsymbol{A}$, the pin energy in first order can be obtained by only taking into account the variations of the coefficients (e.g., Campbell and Evetts 1972, Thuneberg 1989). However, this is only allowed if the relevant size of the defect is small compared with the distance over which Δ varies.

It is clear from Eqn. (2) that the first two terms are involved in the case of T_c deviations, whereas l_{tr} fluctuations only lead to a perturbation of the third term. In the older literature, however, Eqn. (2) is usually written in reduced units which then eliminates the contribution of the third term and mixes $\delta\kappa$ and δT_c effects into the first and second terms. Thus, the use of reduced units incorrectly destroys the clear distinction between both pin mechanisms. It is important to retain this distinction, because it gives rise to a different field and temperature dependence for the various elementary interactions which will help to determine the nature of the predominant pinning mechanism. In addition to temperature and field effects, size and spatial distributions of the pinning centers are also important. For extended defects the crystal anisotropy and the defect orientation with respect to the field direction are involved.

(*a*) *Electron scattering*. The theory has been worked out and reviewed by Thuneberg (1989). The concept is illustrated by deriving from Eqn. (2) the elementary pinning potential

$$\Omega_p(r) = \mu_0 H_c^2 \xi^2 g(\alpha)\xi_0 \int d^3r'\, |\partial\psi(\boldsymbol{r}')|^2 \delta(1/l_{tr}(\boldsymbol{r}-\boldsymbol{r}')) \quad (3)$$

where $\xi^2 = 0.55\xi_0^2\chi(\alpha)(1-t)^{-1}$, $g(\alpha) = 0.882\,d\ln(\chi)/d\alpha$ ($g(\alpha) = 0.85$ for $\alpha = 0$ and $g(\alpha) = l_{tr}/\xi_0$ for $\alpha \gg 1$), H_c is the thermodynamic critical field and $\delta(l_{tr}^{-1})$ is the extra electron-scattering by the defect. A reduced-order parameter has also been introduced: $|\psi|^2 = \hbar^2|\Psi|^2/(4m\xi^2\mu_0H_c^2)$ where ξ is here the Ginzburg–Landau coherence length (see *Superconducting Materials: Physical Properties*). The elementary pinning force f_p is obtained by computing the maximum value of the gradient of $\Omega_p(r)$. The extra scattering of the electrons at the defect decreases locally ξ, which means that this pin mechanism always results in an attractive interaction.

In the case of a small void (size $D < (\xi_0^{-1} + l_{tr}^{-1})^{-1}$) the factor $\delta(l_{tr}^{-1})$ can be replaced by a δ function times the scattering cross section $\pi D^2/4$, thereby illustrating that the pin energy is proportional to the condensation energy multiplied by the defect volume enhanced by a factor ξ_0/D or l_{tr}/D for the clean or dirty limit, respectively. For extended defects, such as grain boundaries or large, flat precipitates, parallel to the flux lines, the situation is more complicated because the integral over the entire defect should be evaluated. In addition, the scattering probability of the defect should be determined from other experiments. A special case in which the crystal anisotropy also has to be considered is that of the grain boundary in a bicrystal.

Regarding field and temperature dependence, two field regimes should be distinguished (see *Flux Line Lattice*). In small fields ($B < 0.2\,B_{c2}$) the vortices are isolated and the order parameter changes over a distance ξ. In large fields the vortices overlap, which yields (near B_{c2})

$$|\psi|^2 \simeq 0.86\,(1-b)\left\{1-\frac{1}{3}\left[\cos\left(x-\frac{y}{\sqrt{3}}\right)k_0 + \cos\left(\frac{2y}{\sqrt{3}}\right)k_0 + \cos\left(x+\frac{y}{\sqrt{3}}\right)k_0\right]\right\} \quad (4)$$

where $k_0 = 2\pi/a_0$, $a_0 = 1.075\,(\Phi_0/B)^{1/2}$, the FLL parameter, and $b = B/B_{c2}$. For small fields the expression for f_p contains a factor ξ^{-3} ($\sim (1-t)^{3/2}$), while for large fields a factor $(2\pi/a_0)^{-3}$ appears ($\sim B^{3/2}$). Since $a_0/\xi = 2.69\,b^{-1/2}$, the difference is $12.7\,b^{3/2}$, which is of the order of unity at $b = 0.18$. Accordingly, field and temperature dependences in both regimes are distinctly different; that is, f_p for isolated vortices should be independent of B but it is proportional to $b^{3/2}(1-b)$ near B_{c2}. Typically, for a void ($D = 1$ nm) and an isolated vortex, $f_p \simeq 0.3 \times 10^{-14}$ N, where $\mu_0 H_c = 0.1$ T and $\xi_0 = 10$ nm has been used.

(*b*) *T_c Variations.* Pinning by various kinds of dislocations has been extensively studied in the past (see Campbell and Evetts 1972). The formalism used differs somewhat from that already described. Two effects are distinguished, both related to the normal character of the vortex core: a larger density (ΔV effect) and larger elastic constants (ΔE effect). These effects, though very small, give rise to a periodic stress and elasticity field linked with the variation of $|\Psi|^2$. The pin potential arises via the coupling to the strain field around the defects and is linear in the strain for the ΔV effect and quadratic for the ΔE effect. Accordingly, these effects are referred to as first and second-order interactions. Typical values of $\Delta V/V$ and $\Delta E/E$ are 10^{-7} and 3×10^{-5}, respectively. Some typical values of f_p or f_{pl} (f_p per unit length) of an isolated vortex taking again $\mu_0 H_c = 0.1$ T and $\xi_0 = 10$ nm are: edge dislocation perpendicular to a flux line $f_p \simeq 10^{-13}$ N; edge or screw dislocation parallel to a flux line $f_{pl} \simeq 5 \times 10^{-10}\,\mathrm{N\,m^{-1}}$ and $10^{-5}\,\mathrm{N\,m^{-1}}$, respectively; dislocation loop ($D = 10$ nm) perpendicular to the field direction $f_p \simeq 3 \times 10^{-14}$ N. Moreover, $f_p \propto D^2$ (see also Kramer 1978).

Precipitates may be considered as δT_c defects as well. The pin potential can be estimated to be at most $\mu_0 H_c^2 V/2$ for dielectric inclusions of volume V. For conducting precipitates the proximity effect should be taken into account giving rise to a much lower value. No better theory exists at present. As to the field and temperature dependence for δT_c pinning, f_p now contains a factor ξ^{-1} or $(2\pi/a_0)^{-1}$ in small and large fields, respectively. This kind of mechanism may lead to attractive or repulsive interactions depending on the defect characteristics.

1.3 Concluding Remarks

From the previous discussion it follows that a general formula for f_p cannot be given. For a specific material the predominant defect structure should first be determined and then the pin interaction estimated. Defect morphology, pin mechanism, characteristic length scales and field orientation and regime must be taken into account. Reasonable estimates can now be made for the most important pinning centers, although the situation for precipitates is not yet firmly settled.

So far, only single-defect interactions have been considered. Whether this simplification is allowed depends on the concentration of pinning centers and the range of the pin interaction. The effectiveness of the pinning force will considerably decrease when the pins strongly overlap, that is, if the distance between the pins is much less than the interaction range. A good example is an amorphous superconductor. Although the defect concentration is very large, the pinning is small, since on the relevant length scale ξ the material is homogeneous. Only density or stress modulations with wavelengths comparable to ξ will be effective.

2. *Summing the Pinning Forces*

The pinning centers in real materials have a random site and size distribution. In order to compute the total force on the FLL, it is usually assumed that all defects have the same size, so that only the positional distribution has to be considered. Consequently, there will be a distribution of pinning forces determined by the mutual positions of the pinning centers and the flux lines. The latter are defined by the zeros of the order parameter. In addition, it is assumed that the pins are small ($D \ll \xi$) in at least one direction perpendicular to the field.

For a perfectly uniform FLL the effect of a random force distribution will average to zero. It can therefore be concluded that the deviations from perfect periodicity are essential for a net effect. This has been pointed out by Labusch (1969) in a first attempt to solve the statistical summation problem taking into account the elastic deformations of the FLL in the presence of pinning centers. Labusch suggested that only pins with f_p larger than a certain threshold value f_{th} could be effective because they are able to produce plastic defects in the FLL and disrupt the long-range order (see also Campbell and Evetts 1972). However, this idea leads to a paradox, since large critical current densities were observed in materials with pinning centers for which $f_p \ll f_{th}$, as shown by Kramer (1978). It turns out that the threshold paradox disappears when one accounts for the collective effects that occur in dense pinning systems.

The elastic response of the FLL to the force exerted by a specific pinning center is mainly determined by the shear and tilt moduli c_{66} and c_{44}, respectively (see *Flux Line Lattice*). The effect of the other pins is to keep the FLL fixed. Such a mean-field approach is only reasonable if the range of the FLL interaction is small compared with the distance between the pins. However, because of the electromagnetic origin, the interaction between the vortices has a range λ that can be very large, especially in high-κ materials. Moreover, real pin systems are often very dense ($n_p^{1/3} \ll \lambda$, where n_p is the concentration of pinning centers), which suggests that a collective treatment of the pinning forces is more appropriate. In the particular case of weak pins, for which the individual defect–flux line interactions only lead to elastic deformations with strains very much less than unity, the summation problem can be solved. This was shown for the first time by Larkin and Ovchinnikov (1979) in their theory of collective pinning.

2.1 Concepts of Collective Pinning

The local displacements due to weak pins will be very small ($\ll a_0$), so that around each flux line a hexagonal FLL can be well defined. Over longer distances, however, the displacements may accumulate as in a random-walk process causing a gradual breakdown of the order in the FLL. This is expressed by the monotonic growth of the displacement correlation function

$$g(\boldsymbol{r}) \equiv \langle [\boldsymbol{u}(\boldsymbol{r}) - \boldsymbol{u}(0)]^2 \rangle \tag{5}$$

with distance $|\boldsymbol{r}|$. Here $\boldsymbol{u}$ is the displacement in the *xy*-plane (perpendicular to the field) and the origin may be located at the core of any flux line. When $g(\boldsymbol{r})$ is of order a_0^2, the positional long-range order is destroyed. However, for the summation of pinning forces, it is more relevant to know the criterion for which the pin interactions become uncorrelated. This happens when $g(\boldsymbol{r}) = r_f^2$, where r_f is the distance in which $f(x, y)$ changes from $-f_p$ to $+f_p$, that is, $r_f \simeq \xi$ for $b < 0.2$ and $r_f \simeq a_0/2$ for $b > 0.2$.

The function $g(r)$ was first derived by Larkin and Ovchinnikov for an infinitely large system in which the order breaks down in the directions both parallel and perpendicular to the field. This is called three-dimensional disorder (see the review papers by Kes (1987) and Brandt and Essmann (1987) for references). The corresponding longitudinal and perpendicular correlation lengths L_c and R_c are defined by

$$g_3(0, L_c) = g_3(R_c, 0) = r_f^2 \tag{6}$$

where L_c and R_c are functions of c_{66} and c_{44} and of the pin correlation function (pin "strength") which for small point pins is given by

$$W = n_p \langle f^2 \rangle \simeq 0.5\, n_p f_p^2 \tag{7}$$

The average is taken for a uniform distribution of pinning centers over a unit cell of the FLL. The relation between L_c and R_c depends on the elastic moduli in a quite natural manner

$$L_c = (c_{44}/c_{66})^{1/2} R_c \tag{8}$$

It should be noted that all these quantities depend in a specific way on the temperature and field, as well as on the defect morphology.

The physical meaning of the correlation lengths is illustrated by deriving qualitative expressions for L_c and R_c. Instead of a continuous displacement field, correlated regions with volume $V_c \simeq R_c^2 L_c$ may be considered, which are weakly coupled to each other so that they can move independently over a distance of the order of r_f. Inside V_c the FLL distortions may be ignored. The size of the correlated volume is determined by minimizing the free energy which consists of two terms: the elastic energy related to the mutual displacements of the correlated regions and the work done by the pinning centers in V_c to effect this displacement. This model readily provides the pinning force density as well, since in order to unpin the entire FLL each correlated volume has to be unpinned independently. A statistical argument states that the fluctuation (per unit volume) in the net effect of $n_p V_c$ pinning centers is

$$F_p = J_c B \simeq (W/V_c)^{1/2} \tag{9}$$

The situation becomes very simple for a thin-film amorphous superconductor in a perpendicular field. Since $L_c \gg R_c$ and W is small, L_c can be easily made larger than the thickness d of the film. In this case the FLL deformations in the field direction may be ignored, so that only two-dimensional disorder in the *xy*-plane remains. This gives $V_c \simeq dR_{c,2D}^2$ and

$$R_{c,2D} \simeq \frac{r_f c_{66}}{(W/d)^{1/2}} \tag{10}$$

$$F_{p,2D} \simeq \frac{W}{(c_{66} r_f d)} \tag{11}$$

The only unknown parameter is W, thereby offering a means to obtain information about f_p.

In the Larkin–Ovchinnikov theory it is assumed that elastic deformations solely determine the correlation lengths. The effect of dislocations *in the FLL* is not considered, although their existence is known from decoration experiments. In fact, almost equiva-

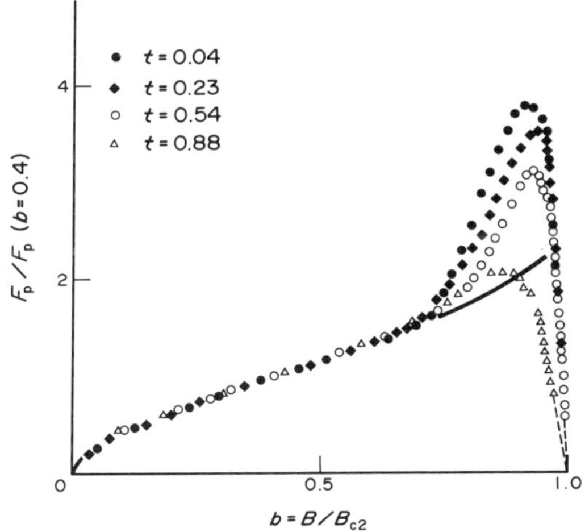

Figure 1
Typical pinning force density vs field for two-dimensional collective pinning in a thin-film amorphous superconductor: experimental data for amorphous Nb_3Ge film ($d = 1.24$ μm) at several temperatures ($T_c = 3.81$ K); the solid line represents the ideal theoretical behavior

lent expressions can be derived by modelling the disorder in two dimensions by a square array of edge dislocations. This illustrates that the concept of computable correlation lengths is less settled. The statistical argument that leads to Eqn. (9), has a more general validity and may be used to obtain V_c from experiment, if W is known.

2.2 Two-Dimensional Collective Pinning

In thin films of amorphous Nb_3Ge two-dimensional collective pinning has been observed (see Fig. 1 and Wördenweber and Kes (1989)). The full line represents the theory fitting W at $b = 0.4$. The field and temperature dependence of W show that the pinning is caused by quasidislocation loops. The $R_c/a_0(b)$ values computed from these data display a dome-shaped behavior with a maximum at $b = 0.3$, typically of the order of 40, although values as large as 200 have also been determined for more homogeneous films. Both at small fields and close to H_{c2}, R_c decreases steeply demonstrating the increase of disorder related to the softening of the FLL (i.e., decrease of c_{66} to zero). For values of R_c/a_0 between 15 and 20 the data in Fig. 1 start to deviate from the theoretical curve. It can be demonstrated that this peak effect is caused by the creation of edge dislocations in the FLL when locally the shear strength is surpassed. At the onset of the peak effect, history effects are frequently observed. It has been shown very recently that the steep decrease in F_p is a precursor of the melting of the two-dimensional FLL. It is not yet clear whether the field of the maximum F_p or of the zero F_p is the melting field B_m. However, it can be convincingly shown that B_{c2} is larger than previously assumed. The horizontal scale in Fig. 1 should thus rather be B/B_m. Nevertheless, it turns out that this analysis remains valid, except for numerical values that may change by at most a factor of two.

2.3 Three-Dimensional Collective Pinning

From Eqn. (8), L_c/d can be calculated. For thick enough films it is expected that a transition to three-dimensional disorder can be observed at $L_c/d = 0.5$, which in fact has been observed (Fig. 2). At the transition field the pinning force increases by almost an order of magnitude. Further experiments reveal a different thickness dependence below and above the transition, confirming the crossover in dimensionality. Below the transition the two-dimensional collective pinning theory describes the results very well and even predicts the transition field correctly. Above the transition, however, the three-dimensional collective pinning expressions result in a much larger correlated volume than is determined from the data using Eqn. (9). The far-too-small value of V_c suggests that the disorder is determined by dislocations (Mullock and Evetts 1985). In particular, screw dislocations created at the film surface may move freely along the flux lines since they have no Peierls potential. This picture is justified by the observed IV curves (see Wördenweber and Kes (1989)).

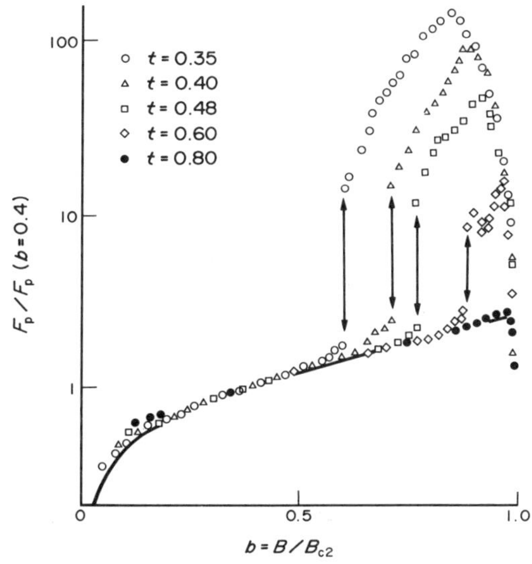

Figure 2
Typical pinning force density vs field for a thick, amorphous superconducting film showing the transition from two-dimensional disorder at low fields to three-dimensional disorder at high fields: experimental data for amorphous Nb_3Ge film ($d = 17.95$ μm) at several temperatures ($T_c = 4.3$ K); the transitions are indicated by the arrows; the solid line represents the two-dimensional theory (note the log scale for F_p)

The fact that these features are demonstrated in materials with weak pinning centers, does suggest that the "pure" three-dimensional collective pinning situation is not stable and therefore not observable. Instead, a highly defective FLL seems to be a better representation, certainly in the case of stronger pinning centers. For the ultimate limit of disorder (i.e., an amorphous FLL), the collective-pinning theory offers an estimation for the correlation length referred to as the single flux-line approximation:

$$L_c \simeq [B^4 a_0^6 r_f^2 (1-b)^2 / (\mu_0^2 \lambda^4 W)]^{1/3} \tag{12}$$

This description is expected to apply close to B_{c2} where c_{66} becomes very small. In some cases, using $F_p = (W/a_0^2 L_c)^{1/2}$, good agreement is found over a much larger field regime. Since the field dependence of W is given by $b^p(1-b)^2$ with $p = 1$ for δT_c pinning and $p = 3$ for $\delta\kappa$ pinning, it follows that F_p scales with field according to $b^{(4p+3)/6}(1-b)$. A maximum occurs at $b = 5/7$ for $p = 3$ and at $b = 7/13$ for $p = 1$.

2.4 Real Pinning Systems

Features observed in strong pinning materials are dome-shaped $F_p(B)$ curves which, close to B_{c2}, behave either like $(1-b)$ or $(1-b)^2$ (see *Critical Current Density*). Scaling is frequently seen (i.e., $F_p\ (B, T) \propto f(T)\ b^p(1-b)^q$) as also is "saturation" (e.g., in NbTi by Matsushita and Küpfer (1988)). (Note that these authors use a different definition of saturation.) Saturation is here taken to mean that the $F_p(B)$ curves change in shape and increase up to a certain maximum envelope curve when the pinning strength is enhanced. The envelope curve is dome shaped and represents the optimum pinning force achievable for the defect morphology under consideration. The nonsaturated curves typically show a much sharper peak effect closer to B_{c2}. History effects are also often reported in these cases. For further reading see Yamafuji (1985).

In a pioneering paper, Kramer (1973) pointed out that pin breaking gives a $(1-b)$ behavior, since it is characterized by a uniform motion of the FLL and because the elementary pinning forces decrease proportional to $(1-b)$ when B_{c2} is approached. On the other hand, a very inhomogeneous pinning-force distribution would lead to a shear flow of less-strongly pinned regions of the FLL along strongly pinned (bundles of) flux lines. Being proportional to c_{66}, the shear flow mechanism yields a $(1-b)^2$ behavior. Although "Kramer-plots" have been very successful in describing $F_p(B)$ curves, a more profound comprehension of the mechanism for bulk materials is still needed. Recent experiments in thin films which simulated the shear flow behavior proved that the FLL for all fields behaves like an incompressible fluid with a small but finite shear modulus (see Pruijmboom et al. 1989). It was shown that the shear modulus is considerably reduced by the effect of edge dislocations in the FLL. These results support the view that pin breaking occurs in the case of extended defects perpendicular to the flux flow direction, whereas flux line shear would occur for smaller defects with a more inhomogeneous distribution.

In general, scaling is observed in materials that also show saturation, whereas history effects coincide with nonsaturated $F_p(B)$ curves without scaling. In terms of the collective pinning model one might speculate that for intermediate pinning strengths, the FLL contains many edge and screw dislocations, dislocation loops, dislocation dipoles and other defects. Obviously, dislocations can be created or annihilated by flux line flow giving rise to new flux line arrangements and history effects (Evetts and Mullock 1985). In the thin-film geometry with straight flux lines, it is easy to visualize how dislocation glide can make the FLL softer or more disordered, thereby giving it the opportunity to adapt itself more effectively to the distribution of pinning centers. For three-dimensional disorder, flux line cutting may need to be allowed in order to model the FLL rearrangements. On the other hand, it should be realized that screw dislocations and screw-dislocation loops (causing a twist of the FLL around the field direction) can only occur by virtue of the presence of pinning centers and, in addition, that they easily glide along the flux lines. Therefore, it might be easier to annihilate and create dislocations in an FLL than in a crystal lattice.

3. High-Temperature Superconductors

3.1 Special Properties

The recently discovered oxide superconductors (see *Oxide Superconductors: Ceramic Processing*; *Oxide Superconductors: Electron Microscopy*; *Oxide Superconductors: Physical Properties*; *Oxide Superconductors: Solid-State Chemistry*; *Oxide Superconductors: Thin-Film Deposition*) deserve special attention because of their layered structure. The (super)conductivity is concentrated in the CuO_2 double or triple planes along the *ab* direction. In the case of $YBa_2Cu_3O_7$ (YBCO) the electronic coupling between the double (triple) layers is established via the conducting CuO chain plane, whereas in the case of $Bi_2Sr_2CaCu_2O_8$ (Bi:2212) and Bi:2223 coupling is established via the nonmetallic BiO layers. This is the reason for the very anisotropic properties of YBCO and the extreme anisotropy of Bi:2212. It is believed that the thallium compounds behave like the bismuth compounds. The flux lines in such anisotropic superconductors (see *Anisotropy in Superconducting Materials*) also have a very anisotropic nature, leading in the extreme case to segmentation of the flux line into weakly coupled "pancake" vortices located in the superconducting double layers stacked on top of each other. The following concentrates on the YBCO and Bi:2212 compounds and the two field orientations $H_\|$

and $H_\perp$ referring to parallel or perpendicular to the *a–b* planes, respectively.

With the field in the *c* direction ($H_\perp$), the FLL consisting of pancake vortices retains its hexagonal symmetry (see *Flux Line Lattice*). The weak interlayer coupling, however, leads to quasi-two-dimensional behavior—as if the material had been cleaved into thin films of thickness *s*—above a crossover field $B_{2D} = \Phi_0/\gamma s^2$, where γ is the anisotropy parameter and *s* is the distance between the double layers. The respective values of γ for YBCO and Bi:2212 are 29 and 3000 (Farrell et al. 1989) and of *s* are 1.2 nm and 1.5 nm. The tilt modulus c_{44} is essentially reduced by a factor γ which favors phenomena like FLL melting and cutting; for example, in Bi:2212 one estimates a melting temperature $T_m \simeq 30$ K for $B > B_{2D} \simeq 1$ T. Since thermal depinning and FLL melting are intimately related, the effect of pinning is restricted to the low-temperature part of the *H–T* diagram. In addition, H_{c2} is very large at low *T*, which means that in most experiments the isolated vortex limit is valid.

With the field along the planes ($H_\parallel$) the hexagonal symmetry is replaced by an isosceles lattice structure with lattice parameters $a_\parallel \simeq a_0\gamma^{1/4}$ and $a_\perp \simeq a_0\gamma^{-1/4}$. In the extreme case, the normal vortex core is replaced by a uniform, very-small-order parameter with a screening current pattern like that of a Josephson vortex. The pinning force against motion along the planes will be very small. However, for motion in the *c* direction the pinning will be extremely large because of the intrinsic modulation of the order parameter in the layered superconductor. Theoretically, the intrinsic pinning yields critical current densities near the depairing limit, as was shown by Ivlev and Kopnin (1990).

Another remarkable property of the high-temperature superconductors (HTS) is the very short coherence length. This property makes them very sensitive to material disorder, especially in YBCO where disorder leads to oxygen deficiency and a local decrease of T_c. If the disorder is located at grain boundaries and if the defective layer is thicker than ξ_0, superconducting coherence between the grains is destroyed. Such grain boundaries act rather as weak links than as the strong pinning centers in Nb_3Sn. On the other hand, as long as superconducting coherence is preserved, a small ξ yields large elementary pin interactions. The optimal defect size for point pinning will be of order $D \simeq \xi$. Because of the quasi-two-dimensional nature of the FLL, the most effective pinning centers will be situated in the CuO_2 planes.

3.2 Pin Mechanisms of HTS

Shortly after the discovery of YBCO it was generally assumed that the twin planes in this material might be effective pinning centers. In (quasi)single crystals the twin planes have orientations parallel to the (110) and ($1\bar{1}0$) lattice planes. Therefore, they only contribute to the pinning if the field is parallel and the Lorentz force is perpendicular to the planes. Nevertheless, it has been shown by Kwok et al. (1990) that the pinning is enhanced with both the driving force and the field along the planes. A possible explanation is that the twin planes may act as traps for point defects which then provide the pinning force component along the planes. It should be noted that the vortices are attracted to the twin planes which, in addition to their regular shape, explains why an enhancement is found.

Other effective pinning centers might be the frequently observed stacking faults in YBCO. They consist of intergrowths of other phases with more than one CuO chain plane, such as $YBa_2Cu_{3.5}O_7$ and $YBa_2Cu_4O_8$. In the latter compound T_c is very sensitive to pressure which helps to increase the pin interaction because the ΔV effect (see Sect. 1.2 (*b*)) is proportional to dT_c/dP. For the bismuth compounds, intergrowth of minority phases like Bi:2201 or Bi:2223 might possibly act as pinning defects.

Recently, it has been observed by scanning tunnelling microscopy that *c*-axis oriented YBCO films can contain a high density of screw dislocations oriented along the growth direction surrounded by spiral-staircase terraces (see Gerber et al. 1991, Hawley et al. 1991). In this case the main source of pinning is possibly not the screw dislocation itself but rather the surface roughness with the vortices preferentially sitting in the valleys. The observation that F_p is inversely proportional to the film thickness may support this view, although another explanation in terms of misfit dislocations at the substrate interface would also be possible.

A well-known property of perovskites is their large number of point defects due to oxygen vacancies, substituted atoms, impurity atoms and so on. In particular, defects in the CuO_2 planes might be effective pinning centers, since they couple to the variation of the order parameter, which is the largest in the CuO_2 layers. It is also very likely that the electron scattering by oxygen vacancies is more effective than by impurity atoms. Therefore, oxygen vacancies in the CuO_2 planes seem to be realistic candidates for pinning centers in all HTS.

Apart from the intrinsic defects considered so far, fundamental studies of flux-pin mechanisms often rely on artificial pinning centers created, for instance, by means of neutron or ion irradiation. For a recent review the reader is referred to Weber and Crabtree (1991). Other ways of creating pinning centers exploit special heat treatments of off-stoichiometric compositions, for example, the melt-process melt-quench (MPMQ) procedure introduced by Murakami et al. (1989) for YBCO. It is believed that the pinning in this case is enhanced by small Y_2BaCuO_5 precipitates.

3.3 Elementary Interaction of Twin Planes and Oxygen Vacancies: Optimum Pinning Force

In order to verify which is the predominant pinning mechanism, it is necessary to know the elementary pinning force f_p or the force per unit length f_{pl} for the defect under consideration. For twin planes and oxygen vacancies this problem has been solved assuming that the formalism of Thuneberg for $\delta\kappa$ pinning by enhanced electron scattering is appropriate. In Sect. 1.2 the high-field situation (overlapping vortices) was discussed, which is the most important case for conventional superconductors. For HTS the low-field case is more relevant. It turns out that for isolated vortices the gradient energy arising from the factor $|\partial\psi|^2$ in Eqn. (3) is less important than the contribution from the vector potential describing the kinetic energy of the screening currents. Using Clem's (1975) solution for $\psi(\mathbf{r})$ and $\mathbf{A}(\mathbf{r})$ and a conjecture for $\delta(l_{tr}^{-1})$, Kes and Van den Berg (1990) derived the elementary interaction f_{pl} between a twin plane and a parallel flux line. The conjecture consists of a simple exponential decay of the scattering probability

$$\delta\left(\frac{l}{l_{tr}}\right) = \frac{p_{tr}}{x_0} \exp\left(-\frac{x}{x_0}\right) \qquad (13)$$

where x is the distance from the twin plane and $x_0^{-1} = \xi_0^{-1} + l_{tr}^{-l}$. This approach leaves the scattering probability p_{tr} as an unknown parameter (see Thuneberg (1989) for a more detailed treatment). This parameter is probably not very large for pure twin planes, since the resistance in YBCO does not seem to depend on the twinning. If the twin planes accumulate a high density of point defects, the scattering may be enhanced giving rise to more effective pinning. To obtain the volume pinning force F_p directed normal to an array of twin planes, a direct summation approach can be used leading to

$$F_p \simeq f_{pl}/La_0 \qquad (14)$$

where L denotes the average distance between the twin planes. However, the observation that F_p hardly changes between twinned and untwinned single crystals of YBCO indicates that other defects, such as oxygen vacancies, play a more dominant role.

To obtain an expression for the interaction between an oxygen vacancy and an isolated flux line in the pure limit, one can use the fact that the pin potential for a point pin with unitary scattering, derived numerically by Thuneberg et al. 1984, is very well described by a negative Gaussian profile: $\exp(-r^2/r_0^2)$, where $r_0 = 0.81\xi(T)$ and r is the distance from the defect in the plane normal to the vortex. It is assumed that the size of the defect $D \ll r_0$. The actual interaction is given by

$$f(r) \simeq 2.3\, f_p \frac{r}{r_0} \exp\left(-\frac{r^2}{r_0^2}\right) \qquad (15)$$

and the maximum by

$$f_p \simeq \frac{10.2}{(1+t)^4} \frac{B_c^2(T)}{\mu_0 \xi(T)} \sigma_{tr} \xi_0 \qquad (16)$$

where the factor $(1+t)^4$ accounts for the temperature dependence of the pin potential below $t \simeq 0.6$ and $\sigma_{tr} = \pi D^2/4$. An estimate for Bi:2212 inserting $\xi(0) = 0.74\,\xi_0 \simeq 2.5$ nm, $B_c(0) \simeq 0.3$ T and $D = 0.28$ nm (the diameter of O^{2-}), yields $f_p \simeq 6.5 \times 10^{-14}$ N for an oxygen vacancy at $T = 0$. Note that this force does not depend on $\xi(0)$ because of the enhancement factor ξ_0/D, and that f_p is comparatively large by virtue of a large value of $B_c(0)$.

In the case of oxygen vacancies it is necessary to use the collective pinning concept to derive F_p and J_c. For isolated vortices it must be taken into account that the defects only give rise to a pinning force when they are located in the vortex core, say $r < 2r_0$. The procedure will now be illustrated for Bi:2212 in the quasi-two-dimensional limit $B > 1$ T assuming the density of vacancies per double layer is $n_\square = 3.5 \times 10^{17}\,\mathrm{m}^{-2}$. There are $n_\square \pi \xi^2$ (~ 8) vacancies per vortex core, which shows that the fluctuation of the site distribution gives rise to a net pinning force:

$$F_c \simeq (n_\square A_{core} \langle f^2(r) \rangle_{cell})^{1/2} = \{0.89\, n_\square \pi \xi^2 f_p^2\}^{1/2} \qquad (17)$$

A_{core} is the effective core area that follows from averaging $f^2(r)$ over the lattice cell yielding the final result in Eqn. (17). The macroscopic pinning force is obtained by dividing F_c by the volume of the pancake vortex sa_0^2. Estimates for Bi:2212 yield $F_c = 1.7 \times 10^{-13}$ N, $F_p = 5 \times 10^{10}\,B\,\mathrm{N\,m^{-3}}$ (with B in tesla), and $J_c = 5 \times 10^{10}\,\mathrm{A\,m^{-2}}$. The pin energy of each pancake vortex is $U_c \simeq F_c\xi \simeq 34$ K.

An idea of the optimum pinning force for a random distribution of pins can now also be obtained. It should be realized that the oxygen vacancies are very effective pins and that every vortex is pinned independently. The question is where to gain pin energy. Large precipitates and their interfaces can never be as effective. Metallic substitutions will also be less effective because of a smaller enhancement factor. The optimum therefore follows from $F_c \propto n_\square^{1/2} D^2$ and the requirement that $n_\square D^2 \ll 1$, since otherwise T_c would be largely suppressed or the enhancement factor would be reduced by multiple scattering of quasiparticles. A very optimistic estimate would be $D \simeq \xi(0)$ and $n_\square \simeq (3\xi(0))^{-2}$ giving rise to $n_\square^{1/2} D^2 \simeq 8 \times 10^{-10}$ m, compared with 5×10^{-11} m above. So, the most optimistic $J_c(0) \simeq 8 \times 10^{11}\,\mathrm{A\,m^{-2}}$, which is close to the depairing current $J_0 \simeq 2 \times 10^{12}\,\mathrm{A\,m^{-2}}$. An enhancement of a factor 5 to 10 leading to a $J_c < 5 \times 10^{11}\,\mathrm{A\,m^{-2}}$ and a pin energy $U_c(0)$ between 170 K and 340 K seems more realistic.

3.4 Concluding Remarks

From the estimates in the preceding section an impression may be given that the practical critical current densities may become very large. However, the small pin energies must also be considered. An energy of 34 K leads to a very large flux creep effect (see *Flux Creep Phenomena*). The actual current densities are therefore much smaller than J_c (see

Critical Current Density). The creep effect at a temperature far below the melting or depinning temperatures will eventually decay logarithmically and will be practically unobservable after a few decades. The current density J_∞ attained after such a decay is the relevant value for practical applications. This J_∞ may be one or two orders of magnitude smaller than the estimated $J_c(0)$. From the final result of Sect. 3.3 it follows that progress can still be made, especially at low temperatures, by optimization of the pinning energy.

See also: Critical Current Density; Flux Line Lattice; Resistive Transition and Flux Flow in Superconducting Materials; Superconducting Materials: Physical Properties

Bibliography

Brandt E H, Essmann U 1987 The flux-line lattice in type II superconductors. *Phys. Status Solidi B* 144: 13–38

Campbell A M 1971 The interaction distance between flux lines and pinning centers. *J. Phys. C* 4: 3186–98

Campbell A M, Evetts J E 1972 Flux vortices and transport currents in type II superconductors. *Adv. Phys.* 21: 199–428

Clem J R 1975 Simple model for the vortex core in a type II superconductor. *J. Low Temp. Phys.* 18: 427–34

De Gennes P G 1966 *Superconductivity of Metals and Alloys.* Benjamin, New York

Evetts J E, Mullock S J 1985 Collective flux pinning and history effects in a defective flux vortex lattice. In: Matsushita T, Yamafuji K, Irie F (eds.) 1985 *Proc. Int. Symp. Flux Pinning and Electromagnetic Properties in Superconductors.* Matsukuma Press, Fukuoka, Japan, pp. 94–7

Farrell D E, Bonham S, Foster J, Chang Y C, Jiang P Z, Vandervoort K G, Lam D J, Kogan V 1989 Giant superconducting anisotropy in $Bi_2Sr_2CaCu_2O_{8+\delta}$. *Phys. Rev. Lett.* 63: 782–5

Gerber C, Anselmetti D, Bednorz J G, Mannhart J, Schlom D G 1991 Screw dislocations in high-T_c films. *Nature* 350: 279–80

Hawley M, Raistrick J D, Beery J G, Houlton R J 1991 Growth mechanism of sputtered films of $YBa_2Cu_3O_7$ studied by scanning tunneling microscopy. *Science* 251: 1587–9

Ivlev B J, Kopnin N B 1990 Flux creep and flux pinning in layered high temperature superconductors. *Phys. Rev. Lett.* 64: 1828–31

Kes P H 1987 Recent developments in flux pinning. *IEEE Trans. Magn.* 23: 1160–7

Kes P H, Van den Berg J 1990 Flux pinning and thermally activated depinning in single crystals of high temperature superconductors. In: Narlikar A V (ed.) 1990 *Studies of High Temperature Superconductors*, Vol. 5. Nova Science Publisher, New York, pp. 83–117

Kramer E J 1973 Scaling laws for flux pinning in hard superconductors. *J. Appl. Phys.* 44: 1360–70

Kramer E J 1978 Fluxoid-defect interactions in irradiated superconductors. *J. Nucl. Mater.* 72: 5–33

Kwok W K, Welp U, Crabtree G W, Vandervoort K G, Hulscher R, Lui J Z 1990 Direct observation of dissipative flux motion and pinning by twin boundaries in $YBa_2Cu_3O_{7-\delta}$ Single Crystals. *Phys. Rev. Lett.* 64: 966–9

Labusch R 1969 Calculation of the critical field gradient in type II superconductors. *Cryst. Lattice Defects* 1: 1–16

Larkin A I, Ovchinnikov Yu N 1979 Pinning in type II superconductors. *J. Low Temp. Phys.* 34: 409–28

Matsushita T, Küpfer H 1988 Enhancement of the superconducting critical current from saturation in NbTi wire. *J. Appl. Phys.* 63: 5048–59

Mullock S J, Evetts J E 1985 Collective flux pinning in a defective flux vortex lattice. *J. Appl. Phys.* 57: 2588–92

Murakami M, Morita M, Doi K, Miyamoto K 1989 A new process with the promise of high J_c in oxide superconductors. *Jpn. J. Appl. Phys.* 28: 1189–94

Pruijmboom A, Kes P H, Van Der Drift E, Radelaar S 1989 Flux-line shear studied in artificially structured superconducting double layers. *Cryogenics* 29: 232–5

Thuneberg E V 1989 Elementary pinning potentials in superconductors with anisotropic Fermi surface. *Cryogenics* 29: 236–44

Thuneberg E V, Kurkijarvi J, Rainer D 1984 Elementary flux-pinning potential in type II superconductors. *Phys. Rev. B* 29: 3913–23

Weber H W, Crabtree G W 1991 Neutron irradiation effects in high-T_c single crystals In: Narlikar A V (ed.) 1991 *Studies of High Temperature Superconductors*, Vol. 9. Nova Science Publisher, New York

Wördenweber R, Kes P H 1989 Peak effects in two- and three-dimensional collective pinning. *Cryogenics* 29: 321–7

Yamafuji K 1985 On the theory of flux pinning. In: Matsushita T, Yamafuji K, Irie F (eds.) 1985 *Proc. Int. Symp. Flux Pinning and Electromagnetic Properties in Superconductors.* Matsukuma Press, Fukuoka, Japan, pp. 18–25

P.H. Kes
[Rijksuniversiteit Leiden, Leiden,
The Netherlands]

G

Granular Superconductors

Granular superconductors are a special class of composites in which an array of superconducting particles are separated from one another by a nonsuperconducting medium. The occurrence of granularity is widespread. In some materials the composite nature of the granular superconductor is clear, as in the case of fine multifilamentary composites or in evaporated granular aluminum. In other materials the granularity seems to be intrinsic to the superconductor, examples being the oxide superconductors, particularly when polycrystalline, and Chevrel-phase materials. The very short superconducting coherence length in these materials predisposes them to exhibit nonsuperconducting regions or regions of weakened superconductivity at microstructural features such as grain boundaries or stacking faults. The high-critical-temperature layer structure oxide superconductors show an additional strong anisotropy (see *Anisotropy in Superconducting Materials*). In certain regimes of temperature and applied field the individual layers decouple, the material then behaving as a stack of two-dimensional superconducting layers separated by nonsuperconducting regions. Layered superconductors exhibit a complex magnetic structure (see *Flux Line Lattice*) and a description of their behavior draws on concepts that are common to granular materials. This is not surprising since a layered SIS or SNS superconductor is essentially a one-dimensional granular superconductor.

One limit of the granular materials is the artificial weak link network (WLN) formed by interconnecting an array of superconducting tunnel junctions in two or three dimensions (see *Thin Films, Multilayers and Devices, Superconducting*).

Granular superconductors provide an important class of materials for practical devices because they are composites in which the strength of the supercurrent coupling and phase locking between grains can be controlled by changing the processing variables. In the strong coupled regime, where the grain-to-grain supercurrents are high, these composites can be used for conductor material for magnet windings. In the weaker coupling regime, they can be used to control flux motion in electronic devices.

1. Granular Materials

The classical example of a granular material is granular aluminum (g-Al) which is prepared by evaporating aluminum in an oxygen atmosphere to produce an array of relatively pure aluminum particles separated by Al_2O_3 barriers (Ekin 1975). Within each aluminum particle, the superconductivity is very strong in that the pair potential Δ is large and the phase locking is strong between the ground state electrons of the aluminum particles. By contrast, the Al_2O_3 barriers between the particles behave as Josephson junctions or weak links where the pairing is diminished. This permits a phase difference or phase slip between the ground state electrons in one particle relative to its neighboring particles. In a composite material of this kind, the superconducting transition takes place in stages (Patton et al. 1980). The electrical resistance of g-Al, for example, can be made to drop abruptly at the superconducting transition T_c of the aluminum grains but does not go all the way to zero until some lower temperature T_s where the grain-to-grain phase coupling is strong enough to support a transport current as a supercurrent. Expressing these ideas in terms of energies, it can be shown that there are two broad classes of granular material depending on the magnitude of the Josephson coupling energy E_J relative to the superconducting condensation energy for a grain E_g. The former depends on the maximum Josephson current between grains I_c, thus $E_J = \hbar I_c/2e$ (see *Josephson Effects in Weak Links*), and the latter depends on the volume V_g of a grain; $E_g = \frac{1}{2}\mu_0 H_{cg}{}^2 V_g$, where H_{cg} is the thermodynamic critical field for the grain. If the energy of the system is expressed in terms of a Ginzburg–Landau-like free energy functional involving E_J and E_g the important dimensionless parameter is $\varepsilon = E_J/2E_g$ (Clem et al. 1987). When $\varepsilon \gg 1$, current-induced suppression of the superconducting order parameter becomes important and the theory reduces exactly to the Ginzburg–Landau (GL) theory in the dirty limit with the Josephson coupling energy playing the role of the supercurrent kinetic energy (see *Superconducting Materials: Physical Properties*). The characteristic lengths appearing in the GL theory, λ_J the magnetic penetration depth and ξ_J the coherence length, are both larger than the grain size. In the opposite limit, $\varepsilon \ll 1$, any intergranular currents are too small to suppress the order parameter. Granular oxide superconductors are usually in this limit; the individual grains are often sufficiently large to contain a number of flux vortices at high fields. In this case, the penetration and pinning of flux within each grain has to be considered and magnetic phenomena depend on two sets of characteristic lengths: λ_J and ξ_J for the granular medium and λ_g and ξ_g for the grains (Clem 1988).

In the thin-film field, there are many examples of granular composites that have an irregular distribution of barriers connecting the superconducting islands. Thin-film g-NbN (see *Nitride and Carbonitride Superconductors*) for example, is a material commonly used in high-temperature superconducting devices such as high-T_c superconducting quantum interference devices (SQUIDS). Here, an irregular void structure along the grain boundaries of NbN provides the weak link barriers (Gubser et al. 1980) for samples specifically processed to produce them. Similarly, thin-film granular In–In_2O_3 composites have proved to be a very fruitful prototype material for the study of new physical phenomena such as vortex lattice melting and the transition from insulating to superconducting behavior (Gammel et al. 1988). To understand materials such as g-Al, g-NbN and g-In/In_2O_3, it is helpful to recall that there is a random character to the strength of the coupling between particles just by the nature of the preparation technique. The physical properties, however, are often regular because the grain size is small compared with λ_J and ξ_J. In this regime, the electromagnetic response of the material averages over many grains and the material can be made to behave like a uniform superconductor with adjustable superconducting behavior.

It also is possible to prepare thin films with a regular array of barriers using photolithographic techniques to prepare micrometer-size islands of superconductor separated by normal metal barriers (see Resnick et al. 1981). A typical example might be an array of circular lead islands 3 μm in diameter spaced in a triangular array with 4 μm between centers on a silver substrate. This type of superconductor–normal (SN) metal composite has provided a particularly good material for the study of vortex dynamics, the spontaneous nucleation of vortex–antivortex pairs and the broad field of Kosterlitz–Thouless transitions.

An excellent review of the state of this field in 1979 is given by Gubser et al. (1980).

2. *Order Parameter and Critical Current*

To understand the physical properties and design features of granular composites, it is very helpful to discuss a model system to illustrate the spatial variation of the order parameter near the boundary and the critical current of a barrier separating two superconducting reservoirs (see *Coherence Length, Proximity Effect and Fluctuations*). For insulating barriers, the classical g-Al case of Al–Al_2O_3–Al barriers is chosen. For normal barriers, the case of Pb–Ag–Pb is chosen.

For most g-Al, a Cooper pair incident on the barrier usually has a less than 1% chance of being transmitted through the barrier. The barrier is of the order of 1 eV high and internal reflection is a very local phenomenon within a few atomic spacings of the boundary. This means that the gap parameter $\Delta(x)$ is a constant equal to 0.35 meV up to the surface and the critical current density, as shown in Fig. 1a. The maximum Josephson current is given by the Ambegaokar–Baratoff expression (Ambegaokar and Baratoff 1963):

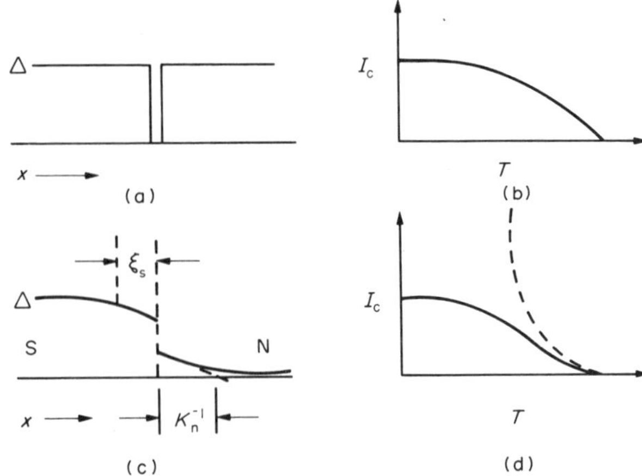

Figure 1
Properties of SNS and SIS barriers: (a) pair potential as a function of position for SIS, (b) I_c vs T for SIS, (c) pair potential as a function of position for SNS and (d) I_c vs T for SNS for both nonmagnetic impurities (dash) and magnetic impurities (solid)

$$I_c = \frac{\pi\Delta(T)}{2eR_n}\tanh\left(\frac{\Delta}{2k_BT}\right)$$

This is a function that rises linearly near T_c and saturates below $T/T_c = 0.6$, as shown in Fig. 1b.

This is to be contrasted with a thin-film Pb–Ag–Pb SNS junction where Cooper pairs move a long way into the N layer and the normal electrons from the N layer move about a coherence distance ξ_s into the S layer before they form pairs (de Gennes 1964), as shown in Fig. 1c. For the lead superconducting reservoirs, the spatial variation of Δ is governed by

$$\xi_{Pb} = \frac{0.2\,\hbar v_F}{k_BT_c}$$

which is about 80 nm (v_F is the Fermi velocity). Similarly, magnetic fields penetrate a distance

$$\lambda_{Pb}^2 = \frac{m}{\mu_0 e^2 n_s}$$

which is about 180 nm in lead. When Cooper pairs in the lead are incident on the SN boundary, they have about a 70% chance of transmission into the N layer and they continue to move into the normal metal until a phonon breaks the pair (Toplicar and Finnemore 1977). In the clean limit, this decay length is

$$K_n^{-1} = \frac{\hbar v_F}{2\pi k_BT}$$

which is about 100 nm at 4.2 K, a large value. This is

illustrated by Fig. 1c. As the silver becomes dirty, this formula becomes

$$K_n^{-1} = \left(\frac{\hbar v_F l}{6\pi k_B T}\right)^{1/2}$$

where l is the normal state mean free path. For $l = 1$ nm, this leads to a K_n^{-1} of 5.7 nm.

The value of Δ is uniform deep in the interior of a superconducting lead film; it drops a bit within a distance ξ of the edge (80 nm for lead) due to the injection of normal electrons from the silver. At the SN interface, the de Gennes boundary conditions have been shown to apply (Toplicar and Finnemore 1977) so the ground state wave function is continuous ($\Psi_s = \Psi_n$) which means that

$$\left(\frac{\Delta}{N(0)V}\right)_s = \left(\frac{\Delta}{N(0)V}\right)_n$$

where $N(0)V$ is the BCS interaction constant. This means that Δ drops by about a factor of five at the interface as shown in Fig. 1c. The critical current density J_c is governed by the minimum value of Δ in the center of the junction and is given by

$$J_c = J_0(1 - T/T_c)^2 \exp(-K_n d_n)$$

in zero magnetic field (Hsiang and Finnemore 1980). The $(1 - T/T_c)^2$ term reflects the temperature dependence of Δ in the S layer. The $\exp(-K_n d_n)$ term reflects the rate of decay of Δ in the N layer. As T goes to zero, K_n^{-1} approaches infinity and $J_c \simeq 10^7$ A cm^{-2}, the value governed by Δ in the N layer at the interface. For clean N-layer barriers such as cadmium between two lead reservoirs, $J_c = 10^4$ A cm^{-2} at 6 K, a temperature close to T_c for lead (Hsiang and Finnemore 1980). It is surprising how much supercurrent these clean N-layer barriers can carry.

3. Pair Scattering and the Depairing Variable

In the design of high-performance materials, it is critical to control supercurrent transmission at the boundaries. For the SNS case there are four variables that can be used to systematically vary this coupling: temperature, magnetic field, nonmagnetic impurity scattering and magnetic impurity scattering. A multi-pair breaking model in which the various depairing factors simply add describes the data well (Hsiang and Finnemore 1980, Tinkham 1980). For a bulk superconductor, the value of T_c is governed by

$$\ln\left(\frac{T_c}{T_{c0}}\right) = \psi\left(\frac{1}{2}\right) - \psi\left(\frac{1}{2} + \frac{\alpha}{2\pi k_B T_c}\right)$$

where ψ is the digamma function, T_{c0} is the pure metal transition temperature, T_c is the transition temperature of the alloy and α is the depairing variable (in units of energy). In addition, it is found that

$$K_n = \left(\frac{\hbar v_F l}{6\pi k_B T}\right)^{1/2} \left[6\left(\frac{1}{2} + \frac{\alpha}{\pi k_B T}\right)\right]^{1/2}$$

where

$$\alpha = \alpha_B + \alpha_s$$

where

$$\alpha_B = DeB$$

and

$$\alpha_s = \hbar/\tau_s$$

where τ_s is the spin scattering rate and $D = \frac{1}{3}v_F l$ is the electron diffusivity. Various contributions to α simply add.

It is useful to see how different variables affect the pair breaking rate and J_c in the barrier. Here, α is just the pair breaking rate expressed in energy units. The decay length of Δ in the barrier is

$$K_n^{-1} = \xi_n \left[6\left(\frac{1}{2} + \frac{\alpha}{\pi k_B T}\right)\right]^{-1/2}$$

where ξ_n is the value of K_n^{-1} when there is only phonon depairing. Critical currents are then exponential in K_n, $J_c \simeq \exp(K_n d_n)$. In the clean limit with $H = 0$ and no magnetic impurities,

$$K_n^{-1} = \xi_n = \frac{\hbar v_F}{2\pi k_B T}$$

As nonmagnetic impurities are added, a diffusive motion of the electrons sets in and

$$K_n^{-1} = \left(\frac{\hbar v_F l}{6\pi k_B T}\right)^{1/2}$$

If a magnetic field is applied,

$$\alpha = DeB = \frac{v_F leB}{3}$$

This means that the depairing rate increases with B but decreases with increasing nonmagnetic impurity scattering (Yang and Finnemore 1984). This seems strange at first. With nonmagnetic scattering, the field must be higher than in the clean limit to get the same pair breaking rate. Physically this arises because pairs are broken when they have precessed by 180° out of phase in the field. With no scattering, this precession proceeds with no inhibition. With scattering, the precession is frequently reversed with the diffusive motion and it takes longer to precess to 180°. As magnetic impurities are added, each spin scattering breaks a pair and α_s is independent of B and T. Typically, it might take 0.5% manganese in copper to quench the supercurrent coupling between NbTi filaments which are 100 nm apart (see *AC Applications of Superconducting Materials*). Magnetic impurities are special in that they give a J_c vs T curve that saturates as T goes to zero just as for insulating barriers. In the design of a magnet wire, where filament-to-filament coupling must be controlled, all of these factors need to be remembered.

To appreciate the quantitative aspects of conductor design, all of these depairing energies are to be compared with $\pi k_B T = 2.6$ meV for $T = 10$ K. For nonmagnetic impurity scattering, $\alpha_s = 0.52\ lB$ in meV where l and B are in nanometers and tesla, respectively. For $B = 1$ T and $l = 20$ nm, $\alpha = 0.19$ meV and $\alpha/\pi k_B T = 0.07$. For many magnetic impurity ions, l_s tends to be about 100 times longer than the nonmagnetic scattering mean free path l. Hence, for magnetic impurity scattering where $\alpha_s = 990/l$ in meV where l is in nanometers, an $l_s = 1000$ nm gives $\alpha \simeq 1$ meV and $\alpha/\pi k_B T \simeq 0.4$. A little magnetic scattering suppresses J_c very significantly. The case of nonmagnetic impurity scattering is shown by the dashed line of Fig. 1d and the magnetic impurity case is shown by the solid line.

Control of the critical currents for granular materials with insulating barriers (SIS junctions) is more difficult to predict. For this case, the coupling frequently has a tunnelling origin with high barriers, of the order of 1 eV, and very short distances, of the order of angstroms. Usually, these interfaces are only a few atomic layers thick and it is difficult to write analytic expressions for the critical currents. Rather, it is necessary to plot out the performance characteristics such as J_c as a function of the processing variables. Hence, there is no formalism comparable to the SNS case.

4. High-T_c Oxide Superconductors

Very early in the study of high-T_c materials, it became clear that this class of materials was extremely prone to granular behavior. Flux appeared to leak into the sample along grain boundaries, along twin planes and, in some cases, along other types of unspecified weak links within single crystals. Decoration studies gave direct confirmation of these effects and the temperature dependence of the slope of the magnetization curve in the Meissner regime dM/dH confirmed these effects quantitatively (see *Measurements in Superconducting Materials*). If the grain size is comparable to the penetration depth, dM/dH data can be analyzed to give R/λ (Finnemore et al. 1987), where R is the effective radius of the volume in which the screening currents flow. Values of R were consistently in the micrometer range, much smaller than the sample size. In modest magnetic fields of about 0.1 T, there were weak links for flux penetration distributed on a scale comparable to the penetration depth. Granular oxide superconductors generally lie in the limit $\varepsilon \ll 1$. Characterization of their properties is complex because, in addition to the properties of the intergrain material characterized by λ_J and ξ_J, the individual grains can trap flux and have their own characteristic lengths λ_g and ξ_g (Clem 1988).

There are fundamental reasons why granular and weak link behaviour is so pervasive. In the high-T_c materials, ξ is very small, of the order of 1.5 nm in the *ab* plane or about four unit cells. Along the *c* axis it is even shorter with $\xi = 0.8$ nm or less than one unit cell. In classical superconductors, by contrast, $\xi \simeq 100$ nm so that in the volume of any one Cooper pair there are millions of other Cooper pairs and the pair–pair correlations are very strong. In the high-T_c materials, the pairing is much more local and there are only about a dozen Cooper pairs overlapping any one Cooper pair. Hence, the phase locking is much more fragile and far more susceptible to the effects of local disturbances such as oxygen deficiencies, cation site exchange, interstitials, or stress from dislocations. Because $\xi = 0.2\hbar v_F / k_B T_c$, the high values of T_c and the relatively low values of v_F lead to low coherence volumes. A calculation of the energy associated with the core of the vortex is $E = (\frac{1}{2}\mu_0 H_c^2)\ (\pi\xi_{ab}^2)$ of the order of 20 meV nm^{-1}. Here ξ_{ab} is the coherence distance in the *ab* plane. If the defect causing the pinning is about one unit cell long in the *c*-axis direction, the pinning energy will be about $2kBT$ at 77 K. Hence, any slight disturbance disrupts the phase locking and leads to flux motion (see *Critical Current Density*; *Flux Line Lattice*; *Resistive Transition and Flux Flow in Superconducting Materials*).

Acknowledgements

Ames Laboratory is operated for the US Department of Energy by Iowa State University under Contract No. W-7405-ENG-82. This work was supported by the Director of Energy Research and the Office of Basic Energy Sciences.

See also: Coherence Length, Proximity Effect and Fluctuations; Josephson Effects in Weak Links; Superconducting Materials: Physical Properties

Bibliography

Ambegaokar V, Baratoff A 1963 Tunneling between superconductors. *Phys. Rev. Lett.* 10: 486–9

Clem J R 1988 Granular and superconducting-glass properties of high-temperature superconductors. *Physica C* 153: 50–5

Clem J R, Bumble R, Raider S I, Gallagher W J, Shih Y C 1987 Ambegaokar–Baratoff–Ginzburg–Landau crossover effects on the critical current density of granular superconductors. *Phys. Rev. B* 35: 6637–42

de Gennes P G 1964 Boundary effects in superconductors. *Rev. Mod. Phys.* 36: 225–37

Ekin J W 1975 Critical currents in granular superconductors. *Phys. Rev. B* 12: 2676–81

Finnemore D K, Shelton R N, Clem J R, McCallum R W, Ku H C, McCarley R E, Chen S C, Kogan V 1987

Magnetization of superconducting lanthanum copper oxides. *Phys. Rev. B* 35: 5319–22

Gammel P L, Hebard A F, Bishop D J 1988 Flux lattice melting in amorphous composite In–InO_x two-dimensional superconductors. *Phys. Rev. Lett.* 60: 144–7

Gubser D A, Wolf S A, Francavilla T L, Feldman J L 1980 In: Gubser D U, Francavilla T L, Liebowitz J R, Wolf S A 1980 *Inhomogeneous Superconductors*, American Institute of Physics Conference Proceedings, Vol. 58. AIP, New York, p. 159

Hsiang T Y, Finnemore D K 1980 Superconducting critical currents for thick, clean superconductor–normal metal–superconductor junctions. *Phys. Rev. B* 22: 154–63

Patton B R, Lamb W, Stroud D 1980 In: Gubser D U, Francavilla T L, Liebowitz J R, Wolf S A 1980 *Inhomogeneous Superconductors*, American Institute of Physics Conference Proceedings, Vol. 58. AIP, New York, p. 13

Resnick D J, Garland J C, Boyd J T, Shoemaker S, Newrock R S 1981 Kosterlitz–Thouless transition in proximity coupled superconducting arrays. *Phys. Rev. Lett.* 47: 1542–5

Tinkham M 1980 *Introduction to Superconductivity*. Krieger, Malabar, FL, pp. 265–73

Toplicar J R, Finnemore D K 1977 Electron tunneling study of the superconducting proximity effect in Pb–Cd. *Phys. Rev. B* 16: 2072–80

Yang H C, Finnemore D K 1984 Pair-breaking mechanisms in superconductor normal-metal superconductor junction. *Phys. Rev. B* 30: 1260–5

D. K. Finnemore
[Iowa State University, Ames,
Iowa, USA]

H

Heavy Fermion Metals

Heavy fermion metals are a subclass of the valence fluctuation metals, a class of intermetallic compounds with rare-earth or actinide component, which show certain drastic anomalies in their thermal, magnetic and transport properties. In particular, at the Fermi energy of these metals, the effective mass m^* of the conduction electrons is from a few tens to a few thousand times larger than the free electron mass m_e. The heavy fermion subclass is defined at the upper end of this range, at $m^* > 400m_e$.

The large effective masses of the valence fluctuation metals are due to the incipient delocalization of the f electrons of the rare-earth or actinide component. These f electrons are valence electrons, like the outer (sd) electrons, but spatially they are closer to the nucleus. Therefore they usually remain localized, even in metals, and then the conduction electrons derive from the outer (sd) electrons only. In the heavy fermion metals, the f electrons begin to delocalize noticeably for the first time.

Heavy fermion metals were discovered in the mid-1970s. Several years later a few were found to be superconductors below 1 K. All previous experience had shown that conventional Bardeen–Cooper–Schrieffer (BCS) superconductivity does not survive the interaction with even minute concentrations of delocalizing f electrons. This prospect of a new mechanism for superconductivity has triggered intensive experimental and theoretical work on heavy fermion metals.

1. *Effective Mass and Bandwidth*

The effective mass m^* is a measure of the ease with which a conduction electron moves from atom to atom through the crystal. The conduction electrons of the solid derive from the valence electrons of the constituent atoms. In atoms and solids alike, the absolute speed of the valence electrons is of the order of $v = h/m_e d$, where h is the Planck constant and d is a distance of the order of the atomic radius. The corresponding kinetic energy $W = h^2/2m_e d^2$ is of the order of a few electron volts (a few 10^4 K). If an electron moves at this speed in a straight line through the solid (e.g., as in copper) its translational (effective) mass is close to the free electron value m_e. If, however, it circles at this speed $N \gg 1$ times around a given atom before moving on to the next, its translational (effective) speed and its translational kinetic energy (the "bandwidth") are reduced to $v^* = v/N$ and to $W^* = W/N$, respectively, and its translational (effective) mass is increased to $m^* = Nm_e$. For electrons that stay forever at a given atom (e.g., those electrons in the completed inner atomic shells) $N \to \infty$, the effective mass is infinite and the bandwidth is zero. The electrons are then termed localized. A heavy fermion metal with $m^* = 10^3 m_e$ has a nearly localized electron with a bandwidth W^* of only a few tens of kelvin. Such metals can be studied experimentally at thermal energies larger than the bandwidth; that is, far into the nondegenerate (classical) regime of the Fermi gas, in contrast to the usual metals with $m^* \simeq m_e$, which are accessible to experiment only in the degenerate (quantum) regime, at $T \ll W/k_B \simeq 10^4$ K, where k_B is Boltzmann's constant. The bandwidths W^* of the heavy fermion metals also lie well below the energy of the average phonon; that is, in these metals the effective speed v^* of the heavy electrons is smaller than that of the nuclei, rather than the other way around as is usual. Therefore, the elementary excitations of such a solid can no longer in general be divided into separate electronic and phononic excitations.

2. *Valence Fluctuations, Fractional Valence and Lifetime Widths*

The delocalization of f electrons in metals is viewed more concretely in terms of a fluctuation of the rare-earth or actinide atom between two states with adjacent valences q and $q+1$ in the configurations $f^{n+1}(sd)^q$ and $f^n(sd)^{q+1}$. In both configurations the atom is actually electrically neutral, as it should be in a metal. In $f^{n+1}(sd)^q$, an electron is in the inner f shell (i.e., localized), while in $f^{n+1}(sd)^{q+1}$ it has moved to the outer (sd) shell, where it is then one of the conduction electrons. Averaging over time, this fluctuation implies a nonintegral (fractional) occupation of both the f and the (sd) shells, namely $0 < v < 1$ for $f^n(sd)^{q+1}$ and $1 - v$ for $f^{n+1}(sd)^q$. The fractional valence $q + v$ and its change with temperature and pressure can be detected by several techniques, such as measurements of the lattice constant, of the Curie constant, of the Mössbauer isomer shift and of the structure of the L_{III} x-ray absorption edge. Fractional valence has been detected in all valence fluctuation metals, including the heavy fermion metals. The observed values of $q + v$ tend to cluster near those values that would be expected from the degeneracies of the ground states of the f^n and f^{n+1} shells at equipartition. The equipartition values are 3.14 for cerium, 2.89 for ytterbium and 3.53 for uranium. In other words, in most valence fluctuation metals with cerium and ytterbium, the fractional valence is close to three (see Table 1). Usually the fractional valences are weak functions of temperature but strong functions of pressure.

Table 1
Magnetic ordering temperatures T_M, superconducting transition temperatures T_c, linear specific heat coefficients γ_0, fractional occupations of the higher valence state v, and magnetic bandwidths W^* at 300 K and near $T=0$ for some representative metals with cerium, ytterbium and uranium with delocalizing f shells

System	T_M (K)	T_c (K)	γ_0 ($mJ\ mol^{-1}K^2$)	v	$W^*(300)$ (K)	$W^*(0)$ (K)
CeB_6	3.15			0.03	40	
$CeAl_2$	3.8			0.04	60	
$CeAl_3$			1700	0.12	100	10
$CeCu_6$			1500	0.12	120	8
$CeCu_2Si_2$			600–900	0.06	$\simeq 100$	10
$CeCu_{2.2}Si_2$		0.9	700–1100	0.11	$\simeq 100$	130
$CeBe_{13}$			130	0.12	170	280
$CePd_3$			36	0.29	200	300
$CeSn_3$			40	0.02	250	400
αCe			11	0.20	2300	2300
YbPd	0.5			0.8	130	
YbCuAl			260	0.8	50	30
$YbCu_2Si_2$			125	0.82	50	40
$YbAl_2$			12	0.4	700	700
URu_2Si_2	17	1.5	40		300	15
UBe_{13}		0.9	700	$\simeq 0.5$	70	100
UPt_3		0.45	415	$\simeq 0.5$	70	100
UAl_2			90	$\simeq 0.5$	300	250
USn_3			170			
$UAuPt_4$			720			

With the onset of delocalization of the f shell, the configurations $f^{n+1}(sd)^q$ and $f^n(sd)^{q+1}$ acquire finite lifetimes τ_{n+1} and τ_n, which cause finite lifetime widths $\Delta\varepsilon_{n+1}=h/\tau_{n+1}$ and $\Delta\varepsilon_n=h/\tau_n$ for all states of the f^{n+1} and f^n configurations. The longer lifetime belongs to the configuration with q, if $v<v_e$, and to that with $q+1$, if $v_e<v$, where v_e is the equipartition value of v. Since in the configuration $f^{n+1}(sd)^q$ the fluctuating electron circles the rare-earth or actinide atom as a local f electron, while in the configuration $f^n(sd)^{q+1}$ it moves as an ordinary conduction electron with $m^*\simeq m_e$, the lifetime widths may be identified with with the bandwidths W^*:

$$\left.\begin{aligned}\Delta\varepsilon_{n+1}&=h/\tau_{n+1}=W^*_{n+1}\\ \Delta\varepsilon_n&=h/\tau_n=W^*_n\end{aligned}\right\}\qquad(1)$$

These lifetime widths can be detected directly by inelastic magnetic neutron scattering and indirectly in the temperature dependence of the static susceptibility and of the resistivity.

3. Thermal Properties

Experimentally, the effective mass m^* at the Fermi level is extracted from the coefficient γ_0 of the linear term of the specific heat $C(T)$ measured at helium temperatures, via

$$\left.\begin{aligned}\Delta(C/T)\equiv C(T)/T-\beta T^2&=\gamma_0\\ &=gk_B/W^*\\ &=gk_Bm^*/m_eW\quad(T/W^*\ll 1)\end{aligned}\right\}\qquad(2)$$

where βT^3 is the low-temperature contribution of the phonons to the specific heat, W^* is the f-bandwidth (K) and g is the degeneracy of the groundstate of the local f shell. Table 1 exhibits the full range of γ_0 values observed in valence fluctuation metals. In the classical regime ($T/W^*>1$), $\Delta(C/T)$ goes to zero rapidly. The integral over $\Delta(C/T)$ from $T/W^*\ll 1$ to $T/W^*\gg 1$ gives the entropy $\Delta S=k_B\ln g$ of the narrow band. One finds, for example, $g=2$ for some cerium heavy fermion compounds such as $CeCu_6$, which have sufficiently large splittings of the $J=5/2$ Hund's rule multiplet of the local f^1 shell caused by the crystalline electric fields of the solid. The linear specific heat coefficient γ_0 decreases rapidly in a sufficiently strong magnetic field H, namely when the Zeeman splitting μH becomes larger than $W^*=gk_B/\gamma_0$, where μ is the magnetic moment of the crystal field ground state of the local f^1 shell. Then a Schottky anomaly develops instead, which corresponds to a two-state system with energy splitting $2\mu H$. This proves that the large γ_0 is indeed due to a very narrow doubly degenerate band of conduction electrons that is intersected by the Fermi level and that derives from the magnetic-doublet crystal-field groundstate of the local f electrons. (For the exceptional behavior of the heavy fermion superconductors see Sect. 6.)

4. *Magnetic Properties*

The magnetic properties of heavy fermion metals are dominated by the f shell of the rare-earth or actinide atoms. They differ significantly from compounds with a strictly local f shell only at low temperatures, at $T \leqslant W^*$. At integral valence, the magnetic properties of rare-earth and actinide metals are well understood (see *Rare-Earth Intermetallics*; *Rare-Earth Metals and Alloys*). First, the magnetic moments seen in the high-temperature Curie constants are almost identical with those of the Hund's rule ground state of the stable f^n shell in isolated atoms, except for a small permanent change of the moment caused by a local magnetic polarization of the (*sd*) conduction electrons. Second, at temperatures of the order of 100 K and below, the susceptibility reflects the splitting of the Hund's rule ground state into crystal-field multiplets by the electric field of the environment. Finally, the interaction between the f^n magnetic moments on neighboring atoms via the polarization of the (*sd*) conduction electrons induced by the f electrons leads to magnetic order below a critical temperature T_M, which ranges from a few hundred kelvin to a few tenths of a kelvin, depending on n and on the structure of the environment of f^n.

The main effect of the onset of delocalization of the f electron on the magnetic properties is a strong tendency towards a nonmagnetic groundstate of the metal; this is caused by the now finite lifetime width of all its magnetic Zeeman levels. While a lifetime width does not lift any Zeeman degeneracies and does not prevent the usual Zeeman splitting in an external magnetic field or in the exchange field of the neighbor, it does reduce the resulting magnetic polarization (i.e., the susceptibility). Schematically, the temperature dependence of the susceptibility of a heavy fermion metal with cerium in low fields may be written as

$$\chi(T) = \frac{(1-v)\mu_1^2}{3k_B(T+T_M+W_1^*(T))}$$
$$\chi_0 \rightarrow \frac{(1-v)\mu_1^2}{3k_B W_1^*} \quad (T, T_M \ll 1) \qquad (3)$$

Here, μ_1 is the magnetic moment of the delocalizing f^1 shell and W_1^* is its lifetime width (μ_0 of f^0 happens to be zero). At high temperatures, in the classical regime, $\chi(T)$ is close to the Curie–Weiss behavior of the local f^1 shell but with a reduced effective moment and with a Curie–Weiss temperature which is increased by W_1^* with respect to the T_M (in Eqn. (3), the usually observed antiferromagnetism is assumed). Both $1-v$ and W_1^* are therefore measurable by $\chi(T)$. At low temperatures, χ becomes a constant, similar to the Pauli–Sommerfeld theory of ordinary metals, provided $T_M \ll W^*$. The magnetic order is then suppressed, and such a metal is said to have a nonmagnetic groundstate. χ_0 then gives another measurement of the heavy fermion bandwidth W^*. Indeed, the ratio $(\chi_0/\gamma_0)(\mu_i^2/3k_B^2 g_i)$ is found to be close to unity for most valence fluctuation metals, including the nonsuperconducting heavy fermion metals. This tendency can, for example, be read from the constancy of the product $\gamma_0 W^*(0)$. (In Table 1 $W^*(0)$ is taken either from the static susceptibility via Eqn. (3) or from the quasielastic linewidth of the magnetic neutron spectra, or from both.) However, for the heavy fermion superconductors, this ratio is closer to 0.2 (see Sect. 6).

It can happen that the magnetic ordering temperature T_M is not small compared with the bandwidth W^*; that is, T_M and W^* are comparable, or even cross each other as functions of pressure or concentration. (For a rare-earth valence fluctuator, T_M can be estimated by de Gennes scaling from isostructural reference metals with other rare-earth atoms with strictly localized f electrons.) This can lead to very complex behavior and is more likely to occur in the heavy fermion metals than in the valence fluctuation metals with smaller m^* (larger W^*). Some such cases are listed in Table 1. The tendency for magnetic order is seen to decrease from uranium to cerium to ytterbium. This is consistent with the decreasing size of the f electron in this order (i.e., with the decreasing size of the exchange integral between the local f electrons and the light (*sd*) conduction electrons). If there is magnetic order, the specific heat no longer varies linearly with temperature near $T=0$. No value can then be given for γ_0. This does not mean that there is then no narrow f band, but merely that $W^* < T_M$ (note that CeB_6 and $CeAl_2$ have slightly fractional v; that is, the f shells are delocalizing, although only weakly, with $W^* > 0$ nevertheless).

W^* can also be measured directly by inelastic magnetic neutron scattering, in terms of the energy width of the Zeeman levels. It was by this powerful technique (which also measures the crystal field splittings) that the key feature was found which distinguishes heavy fermion metals from valence fluctuation metals with similar fractional valence but smaller effective mass: in both types of system W^* is of similar magnitude (a few ten to a few hundred kelvin) at temperatures of order 100 K and above (see W^* (300) in Table 1) and is then roughly independent of temperature, but when the temperature crosses the crystal field splittings on its way to zero, W^* decreases in the heavy fermion metals, by a factor of up to ten, while it stays roughly constant in the others (compare

$W^*(300)$ with $W^*(0)$ in Table 1). Apparently, the fluctuation rate from and to the other valence state depends on the f crystal field level; in the heavy fermion metals, this rate happens to be much smaller in the crystal field groundstate than in the excited crystal field states, but not in the other valence fluctuation metals.

5. Transport Properties

For $T \to 0$, the resistivity $\rho(T)$ of valence fluctuation compounds approaches zero, as it should in a metal (if the lattice is sufficiently free of defects). However, when the temperature approaches W^* from below, $\rho(T)$ increases much more rapidly than the usual phonon contribution $\rho_p(T)$ observed in isostructural reference compounds with stable (strictly local) f electrons. The so-called resistivity anomaly $\Delta\rho(T) = \rho(T) - \rho_p(T)$ rises proportional to T^2 up to $T \simeq W^*$, then goes through a point of inflection, and then usually saturates above $T \simeq 2W^*$ at values of order 100 μΩ cm. Thereafter it remains constant up to the melting point. In some cases, especially in the heavy fermion metals, $\Delta\rho(T)$ goes through a maximum at $W^* \leqslant T \leqslant 2W^*$, before saturating at temperatures above about 100 K. This behavior can be understood in the simple picture of conduction electron scattering in a dynamic alloy: the valence fluctuation metal with the f-carrying atoms in two configurations $f^n(sd)^{q+1}$ and $f^{n+1}(sd)^q$ with concentrations v and $(1-v)$, respectively, may be regarded in first approximation as a metallic alloy (a solid solution) of two different types of atoms, whose valence differs by one. In such alloys, the resistivity anomaly follows the relation

$$\Delta\rho(v) = Av(1-v) \tag{4}$$

where A is a constant, which increases with the difference in the charge density of the outer valence electrons (here with the difference between $(sd)^{q+1}$ and $(sd)^q$). With v from the fractional valence, as determined by the techniques mentioned in Sect. 2, $A \simeq 1$ mΩ cm is found from a large set of data. The high-temperature saturation of $\Delta\rho(T)$ is simply a consequence of the well-established temperature independence of $v(T)$ at high temperatures, and the occasional increase at $\Delta\rho(T)$ with decreasing T in the case of the maxima of some heavy fermion metals is consistent with a temperature dependence of $v(T)$, as found experimentally in the same region of temperature. This shows that at temperatures above W^*, the valence fluctuation compound exhibits the resistivity of an alloy, on top of the normal phonon contribution. In a static alloy, the resistivity anomaly would remain high and constant for $T \to 0$. The actually observed steep drop of $\Delta\rho(T)$ towards zero at $T < W^*$(e.g., by more than 100 μΩ cm within the last few kelvin in the case of $CeCu_2Si_2$) is a consequence of the temporal valence fluctuation; that is, the fact that the alloy is actually dynamic, on a time scale of order $\tau \simeq h/W^*$. The charge density on the atoms performing valence fluctuations on the time scale τ is identical on all atoms if averaged over a time $t \gg \tau$. In a pure crystal, the conduction electrons occupy a given Bloch state with momentum $\hbar k$ for a time which has a lower limit of $t_T > h/(k_B T)$. Therefore, in the region $T < W^*$, the conduction electrons exist in Bloch states for times that are longer than the valence fluctuation time τ, the more so the lower the temperature. As the temperature falls the charge density of the valence fluctuating atoms becomes ever closer to the mean charge density, until at $T = 0$ a perfect crystal is obtained with the same mean charge on each f-carrying atom. There is no conduction electron scattering in a perfect crystal (i.e., $\Delta\rho(T) \to 0$ for $T \to 0$).

There are also large anomalies in the thermopower, the Lorentz ratio, the Hall effect and the magnetoresistance (which all behave differently at $T < W^*$ and at $T > W^*$) but these are too numerous and too complicated to enumerate here.

6. Superconducting Properties

Four heavy fermion metals are superconductors, with critical temperatures T_c of order 1 K: $CeCu_{2.2}Si_2$, UBe_{13}, UPt_3 and URu_2Si_2 (see Table 1). URu_2Si_2 also exhibits antiferromagnetic order at $T_m = 17$ K and does not really qualify as a heavy fermion metal, judging from γ_0. However, as mentioned in Sect. 4, in the case of sufficiently strong magnetic order (at $W^* \leqslant T_M$) there is still a bandwidth due to the delocalizing f electron, and this width is found to be about 15 K just above T_M by neutron scattering. In this sense URu_2Si_2 qualifies. At T_c the specific heat shows jumps of order $\gamma_0 T_c$ (see *Superconducting Materials: Physical Properties*). This proves not only that the superconductivity is a bulk property, but also that it is a property of the heavy electrons, and not just of the ordinary (sd) conduction electrons with their much lower mass. On the other hand, in a magnetic field the static susceptibility with the magnetic formfactor of the f electrons does not decrease below T_c, as it should if the superconducting electrons were of f character. A smaller but finite linear specific heat coefficient remains near $T = 0$; that is, a superconducting gap seems to develop over only part of the Fermi surface. The superconductivity is of extreme type II, with a very large ratio $\kappa^2 \simeq H_{c2}/H_{c1}$ of the upper and lower critical fields (see *Magnetic Properties of Superconductors*). In UBe_{13} the derivative dH_{c2}/dT is nearly infinite at T_c.

Recently, the case for unconventional (non-BCS) superconductivity has been strengthened by the evidence for the existence of more than one superconducting phase in UPt_3 and possibly also $(UTh)Be_{13}$. There is evidence for phase multiplicity in the double discontinuities in the specific heat and also in the unusual temperature dependence of the linear thermal expansion and velocity of sound; for a recent review see Taillefer et al. (1991).

The first three heavy fermion superconductors have very large γ_0 but their W^* does not increase significantly between room and helium temperature, in contrast to the nonsuperconducting heavy fermion metals (see Table 1); the product $\gamma_0 W^*$ is nearly an order of magnitude smaller than in the nonsuperconducting heavy fermion metals. Also, for the superconductors $CeCu_{2.2}Si_2$ and UBe_{13}, $\Delta S \leqslant 0.5 k_B \ln 2$ (i.e., $g < 2$). Thirdly, contrary to the nonsuperconducting heavy fermion metals, the specific heat is insensitive to large magnetic fields for the superconductors (see Sect. 3). The distinction between nonsuperconducting and superconducting heavy fermion metals is clearest in $CeCu_{2+\varepsilon}Si_2$, which exists in both versions (superconducting at $\varepsilon = 0.2$ and nonsuperconducting at $\varepsilon = 0$). For $\varepsilon = 0.2$, $W^*(T)$ behaves as in UBe_{13} and UPt_3, but also as in $CeBe_{13}$ and in $CePd_3$, while for $\varepsilon = 0$ it behaves as in $CeCu_6$ and in $CeAl_3$. It is worth noting that the superconducting transition temperature of $CeCu_{2.2}Si_2$ jumps to about 2 K at a pressure near 4 GPa, while at the same time W^* increases strongly, as seen in the temperature dependence of the resistivity. All this suggests that the large linear specific heat coefficients γ_0 of the heavy fermion superconductors do not originate from narrow f-bands; the magnetically determined bandwidth of the f levels is nearly an order of magnitude larger than suggested by Eqn. (1). The origin of their large γ_0 (or the large effective mass of the fermions) remains uncertain, and so does the mechanism of the heavy fermion superconductivity, in spite of the intense experimental and theoretical activity of recent years.

See also: Rare-Earth Intermetallics; Rare-Earth Metals and Alloys; Superconducting Materials: Physical Properties

Bibliography

Assmus W, Fulde P, Lüthi B, Steglich F (eds.) 1988 *Crystal Field Effects and Heavy Fermion Physics* North-Holland, Amsterdam

Gupta L C, Malik S K (eds.) 1987) *Theoretical and Experimental Aspects of Valence Functions.* Plenum, New York

Taillefer L, Flouquet J, Lonzarich G G 1991 Normal and superconducting phases of heavy fermions. *Physica B* 169: 257–70

D. K. Wohlleben
[Universität zu Köln, Cologne, Germany]

High-Frequency Impedance of Superconducting Materials

It was realized early in the development of the subject that measurements of the high-frequency response of both normal metals and superconductors, such as the microwave surface impedance by cavity perturbation techniques (Pippard 1953) and the infrared transmission or reflection factors (Tinkham 1961), could provide important information about the electron dynamics and details of the electronic states. This article will discuss the high-frequency response of superconductors in zero magnetic field (excluding any discussion of the high-frequency properties of Josephson weak links and mobile magnetic flux within superconductors). High-field effects and practical details are discussed in the article *Microwave Applications of Superconducting Materials.*

1. Complex Conductivity and Surface Impedance

When a high-frequency electromagnetic field falls on a good conductor, the field is usually confined to a shallow skin depth. When the surface is flat on the scale of the skin depth, the surface impedance is defined as

$$Z_s = E_\| / H_\| = R_s + iX_s \tag{1}$$

where $E_\|$ and $H_\|$ are the tangential components of the electric and magnetic fields at the surface. For unit surface current density, the surface resistance R_s is a measure of the rate of energy dissipation in the surface and the surface reactance X_s is a measure of the peak inductive energy stored in the surface. When in the local limit (where the electronic mean free path is much less than the skin depth), the superconductor may be described in the usual way by a complex, frequency-dependent conductivity

$$\sigma_s(\omega,T) = \sigma_1(\omega,T) - i\sigma_2(\omega,T) \tag{2}$$

whose real part σ_1 is associated with the lossy response of the normal excitations (electrons) and whose imaginary part σ_2 is associated with the response of the superelectrons (Cooper pairs). The surface impedance and the complex skin depth δ are then given by the classical skin effect formulae

$$Z_s = (i\omega\mu_0/\sigma_s)^{1/2}, \qquad \delta = Z_s/i\omega\mu_0 \tag{3}$$

All theories agree in showing that σ_2 is proportional to $1/\omega$ at low frequencies. At very low frequencies σ_1 may be ignored and it follows that

$$Z_s = i\omega\mu_0\lambda, \qquad \delta = \lambda \tag{4}$$

where λ is the frequency-independent magnetic penetration depth, equal to $(\omega\mu_0\sigma_2)^{-1/2}$. At low frequencies or low temperatures, where σ_1 is finite but much smaller than σ_2, it follows that

$$Z_s = \frac{1}{2}\mu_0^2\omega^2\lambda^3\sigma_1 + i\omega\mu_0\lambda \tag{5}$$

to first order in σ_1. Evidently, the superconductor is only truly lossless at zero frequency or if σ_1 falls to zero, which it will do at $T = 0$ provided there is a gap in the excitation density of states, as in the case of conventional superconductors.

2. The London Theory

According to the phenomenological theory of F. and H. London, the effect of the electric field within the skin depth is to accelerate the superelectrons. In this theory, the rate of change of supercurrent is given by $\dot{\boldsymbol{J}}_s = (n_s e^2/m_e)\boldsymbol{E}$, where n_s is the effective number density of superelectrons. This gives an imaginary conductivity of

$$\sigma_2 = n_s e^2/\omega m_e \tag{6}$$

with the supercurrent lagging the electric field by $\pi/2$. The penetration depth is now $(m_e/\mu_0 n_s e^2)^{1/2}$; n_s rises with falling temperature below T_c, reaching n, the conduction electron density, at $T = 0$.

3. The Mattis–Bardeen Theory

So far only superelectrons that have a local relation between the electric field and supercurrent have been considered. This occurs so long as the BCS coherence length ξ and the electronic mean free path l are both much smaller than $|\delta|$. It should be noted that the London theory is only valid for a local superconductor which is also in the clean limit, $l \gg \xi$. A much more general calculation of Z_s based on the BCS theory for a homogeneous, isotropic superconductor in weak ac magnetic fields was provided by the Mattis–Bardeen (MB) theory in 1958 (Waldram 1964). For arbitrary values of l, ξ and δ the MB calculation of Z_s is a complicated matter, but in certain limits simple results emerge.

(*a*) *Dirty local limit* ($l \ll \xi, |\delta|$). In this limit, the MB theory finds that the ratio

$$\left(\frac{\sigma_s}{\sigma_n}\right)_{\text{dirty}} = \left(\frac{\sigma_1}{\sigma_n}\right)_{\text{MB}} - \mathrm{i}\left(\frac{\sigma_2}{\sigma_n}\right)_{\text{MB}} \tag{7}$$

is a function of frequency and reduced temperature only, independent of ξ and l. Here, σ_n is the corresponding (real) conductivity of the normal metal. This limit applies to many alloy superconductors. (When the MB conductivity ratios are quoted it is usually these dirty local limit ratios that are referred to. Note carefully that they do not apply in all situations.)

(*b*) *London limit* ($\xi\, l \ll |\delta|$). In this limit, the effective conductivity ratio is

$$\left(\frac{\sigma_s}{\sigma_n}\right)_{\text{London}} = \left(\frac{\sigma_1}{\sigma_n}\right)_{\text{MB}} - \mathrm{i}\frac{\xi}{l}\left(\frac{\sigma_2}{\sigma_n}\right)_{\text{MB}} \tag{8}$$

In this limit, since σ_n is proportional to l, the effective value of σ_2 is independent of l; it is, in fact, the conductivity of the London theory, given by Eqn. (6). Most high-T_c superconductors appear to be in this limit.

(*c*) *Pippard limit* ($\xi, l \gg |\delta|$). In this limit, the classical skin effect theory breaks down since this case is outside the range of the local limit. It turns out, however, that

$$\frac{Z_s}{Z_n} = \left(\frac{\sigma_s}{\sigma_n}\right)_{\text{MB}}^{-1/3} \tag{9}$$

where Z_n is the surface impedance of the normal metal in the extreme anomalous limit ($l \gg |\delta|$), which is proportional to $(\mathrm{i}\omega\mu_0)^{2/3}(\sigma_n/l)^{-1/3}$, the constant of proportionality depending on how the electrons are scattered at the surface. A number of pure superconductors with low transition temperatures, such as aluminum, are near this limit.

4. Details of the Mattis–Bardeen Conductivity Ratios

The behavior of $\sigma_s(\omega,T)$ as predicted by the MB theory can be qualitatively understood in terms of the energy gap Δ in the excitation density of states. At $T = 0$ and for frequencies below the gap frequency ω_g (defined as $2\Delta/\hbar$ and typically of the order of 1 THz for conventional superconductors), $\sigma_1 = 0$ since there are no excitations present above the gap. However, for $\omega > \omega_g$, σ_1 rises rapidly corresponding to a loss mechanism where the infrared photons create excitations in pairs out of the ground state. At low frequencies, σ_2 has a $1/\omega$ dependence, as in the London theory.

At finite temperatures, σ_2 is reduced in magnitude, again as in the London theory. Thermal excitations are now present, so that $\sigma_1 > 0$ below ω_g. Figure 1 shows the MB prediction for $\sigma_1(T)$ for selected frequencies above and below ω_g. Note that σ_1 decreases like exp $(-\Delta/kT)$ for $T \ll T_c$. The corresponding surface resistance is approximately proportional to $\omega^\alpha|\exp(-\Delta/k_B T)$, where the exponent α varies from 1.5 in the Pippard limit to 2 in the local limit. Just below T_c, σ_1 exhibits a broad maximum. This enhancement of σ_1 is larger at lower frequencies and is due to two characteristic features of the BCS theory: the enhanced density of states at the gap edge and the coherence factors for electromagnetic absorption.

The MB predictions for the frequency and temperature dependence of σ_S have been successfully verified by experiments at infrared and microwave frequencies for conventional superconductors under a wide range of conditions.

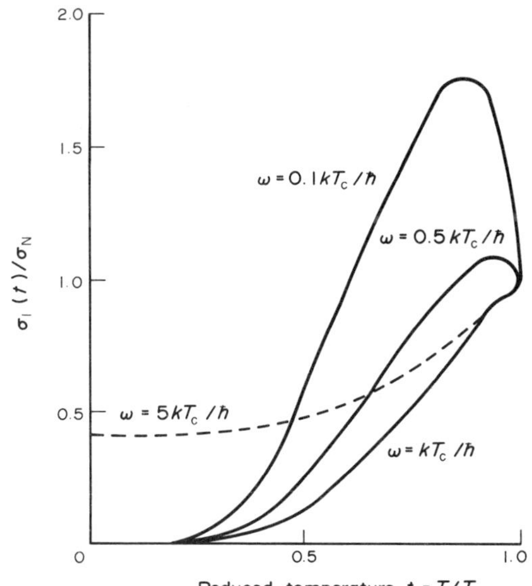

Figure 1
The temperature dependence of σ_1/σ_n at selected frequencies for a BCS superconductor ($2\Delta(0) = 3.52\,kT_c$)

5. High-Frequency Response of High-T_c Superconductors

Measurements of the infrared response of single crystals of high-T_c superconductors are currently controversial (see *Optical Properties of Superconductors*). It is clear that there is a very large anisotropy and evidence of a large gap has been found. In some cases, however, this "gap" persists above T_c and may not be associated with the superconductivity.

The large body of microwave measurements of high-T_c materials has been reviewed by Piel and Müller (1991). The anisotropy is very large, with low conductivity in the c-axis direction. In all high-T_c materials, for currents flowing in the ab plane, the penetration depth is around 140 nm, the coherence length is around 2.5 nm and the typical electron mean free path appears to be a few tens of nanometers, so the materials should be in the local London limit.

Figure 2 shows $R_S(\omega)$ for $YBa_2Cu_3O_{7-\delta}$ (YBCO) at 4.2 K at low microwave fields compared to some conventional superconductors at the same reduced temperature, the bands representing varying sample quality (note that pure copper is in the extreme anomalous limit at 4.2 K). The surface resistance of Nb_3Sn ($T_c \simeq 18$ K) is much lower than that of niobium ($T_c = 9.2$ K) at 4.2 K since it is at a much lower reduced temperature. In bulk and thick film polycrystals of YBCO, the results are variable and the losses are larger than in single-crystal material, probably due to the large number of intergrain contacts.

Even in single crystals and epitaxial thin films the results are not always reproducible. What is certain is that the unusually short coherence length results in fluctuation effects visible near T_c and there is some evidence that the characteristic BCS enhancement of σ_1 below T_c is absent. At low temperatures there is evidence for a term in $\lambda(T)$ proportional to T^2 (which is also seen in low-frequency susceptibility measurements). Figure 3 shows the typical surface resistance of polycrystalline and thin-film YBCO as a function of temperature at 10 GHz, compared with pure, single-crystal niobium and Nb_3Sn. Note that at low temperatures, the surface resistance of even thin-film YBCO is at least three orders of magnitude higher than that of conventional superconductors; it reaches a limiting residual resistance as the temperature decreases, rather than continually decreasing (as in the case of niobium) as the number of excitations decreases. The origin of these residual losses is controversial. If they prove to be intrinsic, then there could be evidence that some of the excitations have zero gap, perhaps because of an unconventional pairing mechanism. However, since at least part of the loss changes with doping levels, it seems more probable that these losses are due to localized effects.

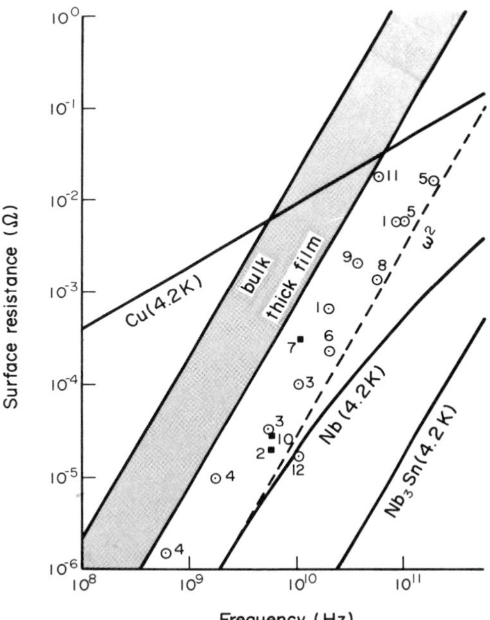

Figure 2
The frequency dependence of R_S in YBCO at 4.2 K: ⊙, epitaxial thin film results; ■, single-crystal results—1, Wuppertal Uni.; 2, Cornell Uni.; 3, DSRC/NJ; 4, MIT Lincoln Labs.; 5, UCLA; 6, Los Alamos; 7, Boston NE Uni; 8, Rockwell ISC; 9, Washington NRL; 10, Cambridge Uni., UK; 11, GEC–Marconi, UK; 12, Hewlett–Packard. The band of data for polycrystalline YBCO represents varying sample quality (adapted from Piel and Müller 1991)

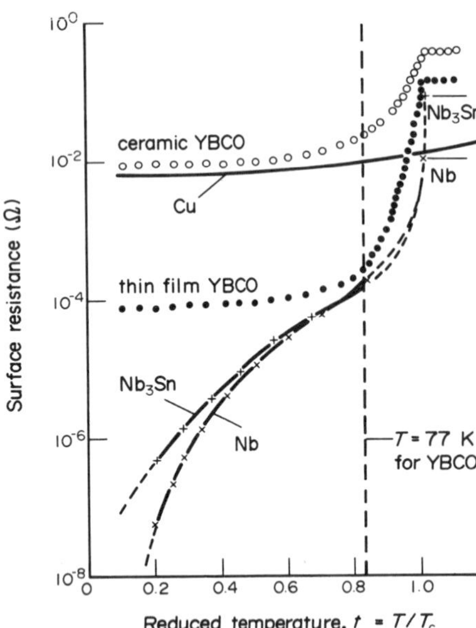

Figure 3
Typical results for the temperature dependence of R_S, quadratically scaled to 10 GHz. The results shown are for polycrystalline YBCO (Birmingham Uni./ICI), epitaxial thin-film YBCO on an $LaAlO_3$ substrate (UCLA) and crystalline niobium and Nb_3Sn (Wuppertal Uni.); the results for copper are plotted against the same reduced temperature as for the YBCO samples ($T_c = 93$ K); the dotted vertical line corresponds to 77 K for YBCO on this scale

See also: AC Losses in Superconducting Materials; Microwave Applications of Superconducting Materials; Optical Properties of Superconductors; Superconducting Cavities; Superconducting Devices and Electronics

Bibliography

Piel H, Müller G 1991 The microwave surface impedance of high-T_c superconductors. *IEEE Trans. Magn.* 27(2): 854–7

Pippard A B 1953 An experimental and theoretical study of the relation between magnetic field and current in a superconductor. *Proc. R. Soc. London, Ser. A* 216:547–68

Tinkham M 1961 Superconductivity. In: de Witt C, Dreyfus B, de Gennes P G (eds.) 1961 *Low Temperature Physics.* Gordon and Breach, New York, pp. 167–76

Waldram J R 1964 The surface impedance of superconductors. *Adv. Phys.* 13: 1–88

A. Porch
[University of Birmingham, Birmingham, UK]

J.R. Waldram
[University of Cambridge, Cambridge, UK]

High-Frequency Properties of Magnetic Materials

Materials in which a spontaneous magnetization arises due to an antiparallel coupling of unequal atomic moments are called ferrimagnets. Several ferrimagnetic oxides with spinel, garnet and hexagonal crystal structures, often called ferrites, are the materials of choice for signal processing devices at microwave and millimeter-wave frequencies due to relatively small dielectric and magnetic losses (see *Ferrites, Soft*). Since ferrites are insulators, the microwaves can penetrate the material. In the presence of a static magnetic field, the interaction between the microwave magnetic field and the static magnetization leads to a variety of effects such as ferromagnetic resonance, spin-wave instabilities, and autooscillations and chaos. The purpose of this article is to provide a basic overview of such high-frequency phenomena in microwave ferrites.

Ferromagnetic resonance (FMR) occurs when a ferrite is subjected simultaneously to a static field and a transverse microwave field. At resonance, the frequency of the microwave field is equal to the natural precession frequency of the net magnetization about the static field. The line width or the width of the resonance line at half the maximum power absorbed is a measure of magnetic losses in the material (Sparks 1964). In FMR experiments at high microwave power levels, two important effects are observed: (a) premature saturation of the main resonance and (b) subsidiary absorption. These transverse-pump spin-wave instability processes result in a loss of microwave power due to the excitation of spin waves at the microwave signal frequency and at half the signal frequency, respectively (Lax and Button 1962). Such nonlinear losses at large signal levels are also observed for static fields parallel to the microwave field (Sparks 1964). For microwave powers above the threshold for transverse and parallel pump instabilities, low-frequency oscillations are observed in the power absorbed by ferrimagnets such as yttrium iron garnet. Such autooscillations become nonperiodic or chaotic at very high power levels (Wigen et al. 1990). The basic phenomenon of FMR and the related high-power effects form the basis for a variety of high-frequency magnetic devices such as isolators, tunable filters, power limiters, frequency doublers and magnetostatic amplifiers (see *Microwave Materials and Devices, Magnetic*; Helszajn 1969). In recent years, devices such as delay lines, signal-to-noise enhancers and filters based on long-wavelength spin waves or magnetostatic waves in planar magnetic samples have been proposed (Webb 1988) and light scattering techniques have been used for the direct detection and characterization of magnetic excitations in such device structures (Srinivasan et al. 1987). In the following sections, these high-frequency phenomena in ferrites are briefly discussed.

1. Ferromagnetic Resonance

The response of a ferrite sample to a static field H_0 along the z axis and a superimposed transverse microwave magnetic field of frequency ω can be described in terms of an effective susceptibility tensor whose components show a singularity when $\omega = \omega_0 = \gamma H_0$, where γ is the gyromagnetic ratio. This is defined as the condition for FMR for an infinite medium and the resonance is observed by varying either H_0 or ω. In samples of finite dimension, the static field seen by atomic moments is influenced by the sample shape, magnetic anisotropy and porosity in polycrystals. The resonance condition, taking into consideration the sample shape, is given by Kittel's equation (Lax and Button 1962)

$$\omega_0^2 = [\gamma H_0 + (N_x - N_z)\omega_m][\gamma H_0 + (N_y - N_z)\omega_m] \quad (1)$$

where N_x, N_y and N_z are the demagnetizing factors and ω_m is equal to the product of γ and the saturation induction M_s for the sample. In FMR experiments on an actual ferromagnetic medium, the singularity predicted for the susceptibility tensor components is not observed. Losses associated with the motion of the precessing magnetization vector result in a finite width for the resonance absorption, and the full width ΔH at the half-maximum is a measure of such losses. For most ferrites and garnets the saturation induction varies from 0.05 T to 0.5 T, γ is about 35 MHz(kA m^{-1})$^{-1}$, the anisotropy field is usually small, and ΔH at x-band frequencies for polycrystalline samples ranges from 4 kA m^{-1} to 40 kA m^{-1} (von Aulock 1965). Single crystals of yttrium iron garnet (YIG) show line widths as small as 80 A m^{-1} and are useful as high-Q resonators in tunable filters. Ferrites are also used in resonance isolators (Helszajn 1969). Hexagonal ferrites with planar and uniaxial anisotropy are suitable for devices operating at millimeter-wave and higher frequencies (Smit and Wijn 1959). The anisotropy field which can be as high as 1.5×10^6 A m^{-1} in hexagonal ferrites adds to the external field in Eqn. (1) for static fields along the uniaxial direction and the corresponding resonance frequency is about 50 GHz for a saturated sample in zero external field (see *Microwave Materials and Devices, Magnetic*).

The overall magnetic losses associated with the motion of the magnetization vector can be described by the line width ΔH or by a relaxation time T. The energy stored in the uniformly precessing spins or uniform precession magnons is usually lost to the lattice through the excitation of phonons and appears as heat. This energy transfer process is called the spin–lattice relaxation. Energy would also be transformed from the uniform mode to nonuniform modes or spin waves due to the interaction between spin dipoles which in turn lose the energy to the lattice. Such a process is called spin–spin relaxation. The various sources of line broadening in ferrites are slow- and fast-relaxing impurities, valence exchange, eddy current process, two-magnon process, Ksuya–LeCraw process, anisotropy, porosity and surface pits. In the presence of slow- or fast-relaxing impurities, such as rare-earth ions in YIG, energy is directly coupled from the uniform mode to the lattice and the line broadening is due to the spin–lattice relaxation process. In ferrites containing both the divalent and trivalent iron ions, the valence-exchange or charge-transfer relaxation mechanism is a source of line broadening. Other sources of losses in ferrites are anisotropy, porosity in polycrystals and surface pits. These processes promote the excitation of nonuniform modes and long-wavelength spin waves or magnetostatic waves (Sparks 1964). Such excitations, which are considered next, are also the basis for several signal processing devices.

2. Spin Waves and Magnetostatic Modes

The ground state of a ferromagnet corresponds to a state in which the atomic moments are aligned parallel to each other. In the excited state, waves of spin reversal called spin waves are present in the material that result in a disturbance of the completely ordered state (see *Magnetic Materials: Basic Concepts*). The quanta of spin waves are magnons. Such magnetic excitations have energies that correspond to photon energies in the microwave region of the electromagnetic spectrum so that spin waves can be excited with an external microwave signal in ferrites.

The frequencies of spin waves are determined mainly by the strength of exchange interactions in a magnetic material. The interaction which results in a parallel alignment of atomic moments in a ferromagnet also strongly opposes any local fluctuation in the direction of the atomic moments. Thus, as the wavelength λ of spin waves decreases, the spin reversal occurs over a short range and the exchange energy associated with the magnon increases. For an ellipsoidal sample, the spin-wave frequency is given by the expression (Sparks 1964)

$$\omega_k = [(\gamma H_0 - N_z\omega_m + \gamma D k^2)(\gamma H_0 - N_z\omega_m + \gamma D k^2 + \gamma M_S \sin^2\theta_k)]^{1/2} \quad (2)$$

In this spin-wave dispersion relation, the constant D is proportional to the strength of exchange interactions, $k = 2\pi/\lambda$ is the magnon wave number and θ_k is the angle between the applied field and the spin-wave propagation direction. The frequency ω_k is minimum for propagation parallel to the static field and is maximum for perpendicular propagation. Thus, the uniform precession magnons excited under FMR can be degenerate with a large number of spin waves with different k and θ_k values provided $\omega_k = \omega_0$ and the major contribution to the FMR line width arises from the excitation of such degenerated spin waves. The dispersive properties of spin waves can be utilized to fabricate delay lines (Helszajn 1969).

In thin metal films such as cobalt and Permalloy, it is possible to excite standing spin waves with wave number $k = n\pi/\delta$, where n is an integer and δ is the film thickness. The exchange parameter D can be determined from data on the static field or frequency separation between modes corresponding to different n values (Lax and Button 1962). The spin-wave resonance experiments are useful for the investigation of surface anisotropy and surface inhomogeneities in thin films (Wigen 1978).

The dispersion relation in Eqn. (2) is valid for spin waves with wavelengths much smaller than the sample size. For long-wavelength spin waves the boundary conditions at the sample surface become important. Such excitations are called magnetostatic modes and can be excited by placing the sample in a nonuniform microwave field. Walker classified the magnetostatic modes using a scheme based on the symmetry of the microwave field (Lax and Button 1962). Damon and Eshbach (1961) solved for magnetostatic mode spectra in planar samples and predicted the existence of surface waves and volume waves. Such magnetostatic waves (MSW) can be excited in a YIG film by placing the sample in contact with a stripline transducer. For suitable static fields applied parallel to the film plane and stripline direction, surface waves are excited in the YIG film by the stripline power. In order to excite backward volume waves, it is necessary to apply the static field parallel to the film plane and perpendicular to the stripline. Forward volume waves are excited for static fields perpendicular to the sample plane. The terms "backward" and "forward" derive from the group velocity of the particular wave of interest. Magnetostatic wave propagation in YIG films provides the physical basis for a promising array of devices such as filters, delay lines, resonators and signal-to-noise enhancers (Adam and Stitzer 1980, Webb 1988). The MSW excitations in device structures can be detected and characterized with the technique of Brillouin light scattering (Srinivasan et al. 1987).

3. High-Power Effects

The response of magnetized ferrites to small transverse microwave fields was discussed in Sect. 1. In high-power FMR experiments an additional absorption peak at static fields near the main resonance and a substantial broadening and lowering of the main resonance are observed. These effects are termed, respectively, subsidiary absorption and premature saturation of the main resonance. At high powers, energy can be coupled from the uniform mode to spin waves through the nonlinear instability of the magnetization. A similar nonlinear effect is also observed for static fields parallel to the microwave field and is termed parallel pumping (Sparks 1964).

Suhl investigated theoretically the effects of high microwave powers under the transverse pump configuration. The subsidiary resonance arises due to coupling between the first order of the amplitude of the uniform precession mode and spin waves at half the excitation frequency and is known, therefore, as a first-order nonlinear process. The instability is expected to occur at an excitation frequency ω only if spin-wave states with frequency $\omega_k = \omega/2$ are available. The energy is transferred from the uniform mode to half-frequency spin waves due to demagnetizing fields and exchange fields, and at a critical power level the power transfer becomes catastrophic resulting in an exponential growth in the spin-wave population. According to theory, the threshold power is minimum for $\theta_k = \pi/4$ spin waves. In particular, the critical power is relatively small when the main resonance overlaps the subsidiary absorption. Under such conditions, the subsidiary absorption is observed as a reduction in the susceptibility at the main resonance (Lax and Button 1962).

The mechanism for the premature saturation of the main resonance is the transfer of energy from the uniform mode to spin waves at the excitation frequency and the nonlinear process is known as a second-order spin-wave instability. The process involves annihilation of two uniform precession magnons and creation of a pair of magnons with wave vectors $\boldsymbol{k}$ and $-\boldsymbol{k}$. The threshold power for this instability is minimum for spin waves propagating parallel to the static field (Helszajn 1969).

In parallel pumping, the power absorbed by a ferrite is very small for microwave fields below a certain critical value and the power absorbed increases dramatically above the critical field. The absorption is due to coupling of power from the microwave field to half-frequency spin-wave pairs with equal and opposite momentum. The critical field for the instability is lowest for $\theta_k = \pi/2$ magnons. Parallel pumping is a useful tool for measuring the relaxation frequency for individual magnons. Such a measurement is possible from data on static field dependence of the threshold field (Sparks 1964).

Spin-wave instabilities under parallel or perpendicular pump configuration in ferrites can be used for microwave amplification. Several other devices such as frequency doublers and power limiters also depend on the nonlinear high-power response or spin-wave instability processes in ferrites (Helszajn 1969). However, in many devices the nonlinear effects generally lead to an increased insertion loss and device degradation.

A recent development regarding nonlinear excitations in ferrites concerns the detection of magnetic solitons in thin films of YIG. Solitons exist in media that are both nonlinear and dispersive. They propagate with constant velocity and without change in shape. De Gasperis et al. (1987) have recently reported the observation of a significant decrease in the propagation loss associated with magnetic excitations in YIG films due to soliton creation above a certain threshold power. Such magnetic solitons are expected to play a major role in the development of high-Q filters and resonators in the microwave region.

4. Autooscillations and Chaos

Nonlinear effects in ferrites will now be considered for microwave powers well above the threshold for spin-wave instabilities. In general, an increase in the power absorbed by YIG samples is observed as the microwave power exceeds the threshold for the first- and second-order processes. For microwave fields above a second threshold, the power absorbed is found to contain ac components at a frequency ranging from 10 kHz to 2 MHz. Such autooscillations have been found to occur for all three basic resonant conditions: FMR, subsidiary absorption and parallel pumping. The oscillations were initially attributed to the excitation of standing spin waves in the sample. They are in fact due to nonlinear interactions between spin-wave modes. When the input power is increased above the threshold for autooscillations, additional peaks corresponding to various subharmonics are observed in the frequency spectra for the absorbed power. These multiple peaks then evolve into a broad continuous spectrum characteristic of chaos. The specific ways in which the evolutions occur are termed "routes to chaos." The autooscillations and the subsequent subharmonic bifurcations and chaos have been the subject of extensive research in recent years. Details on experimental observations and theory of autooscillations and chaos are discussed in an article by Wigen et al. (1990).

Chaos in spin-wave instability was first observed for FMR saturation in YIG spheres. A period-doubling route to chaos and period windows were observed at high powers in this experiment. A similar period-doubling and odd-period bifurcation route to chaos was reported for parallel pumping in YIG spheres. Subsidiary absorption experiments in YIG films have shown several new routes to chaos (Srinivasan et al. 1988).

Several theoretical models based on interactions between the uniform mode and a degenerate spin-wave mode, uniform mode and an entire manifold of spin waves, and half-frequency spin waves with wave number $\pm k$ have been proposed for the interpretation of autooscillations at high powers. According to these models, nonlinear interactions result in a dynamic exchange of energy between competing modes, leading to oscillations in the magnon occupation number. The frequency of oscillation depends on the interaction parameters, dissipation and pumping intensity. These oscillations show bifurcations and chaotic behavior at high powers. Current theories generally ignore the effects of magnetocrystalline anisotropy on autooscillations. However, a substantial crystallographic anisotropy in autooscillation properties has been reported by Zakharov et al. (1975). Several microwave devices depend on the parametric excitation of spin waves above some critical power for their operation. The onset of autooscillations and chaos affect their performance. A complete understanding of these phenomena is essential for the optimization of present devices and the development of new devices.

See also: Ferrites, Soft; Magnetic Materials: Basic Concepts; Microwave Materials and Devices, Magnetic

Bibliography

Adam J D, Stitzer S N 1980 A magnetostatic wave signal-to-noise enhancer. *Appl. Phys. Lett.* 36: 485–7

Damon R W, Eshbach J R 1961 Magnetostatic modes of a ferromagnetic slab. *J. Phys Chem. Solids* 19: 308–20

De Gasperis P, Marceli R, Miccoli G 1987 Magnetostatic soliton propagation at microwave frequency in magnetic garnet films. *Phys. Rev. Lett.* 59: 481–4

Helszajn J 1969 *Principles of Microwave Ferrite Engineering*. Wiley, London

Lax B, Button K J 1962 *Microwave Ferrites and Ferrimagnetics*. McGraw-Hill, New York

Smit J, Wijn H P J 1959 *Ferrites*. Wiley, New York

Sparks M 1964 *Ferromagnetic Relaxation Theory*. McGraw-Hill, New York

Srinivasan G, Chen M, Patton C E 1988 Observation of auto-oscillations and chaos in subsidiary absorption in yttrium iron garnet. *J. Appl. Phys.* 64: 5480–2

Srinivasan G, Patton C E, Emtage P R 1987 Brillouin light scattering on yttrium iron garnet films in a magnetostatic wave device structure. *J. Appl. Phys.* 61: 2318–26

von Aulock W H 1965 *Handbook of Microwave Ferrite Materials*. Academic Press, New York

Webb D C 1988 Microwave magnetic thin-film devices. *IEEE Trans. Magn.* 24: 2799–804

Wigen P E 1978 Magnetic anisotropy and magnetostriction in garnets. In: Paoletti A (ed.) 1978 *Physics of Magnetic Garnets*. North-Holland, Amsterdam, pp. 196–269

Wigen P E, McMichael R D, Jayaprakash C 1990 Route to chaos in the magnetic garnets. *J. Magn. & Magn. Mater.* 84: 237–46

Zakharov V E, L'vov V S, Starobinets S S 1975 Spin-wave turbulence beyond the parametric excitation threshold. *Sov. Phys. Usp.* 17: 896–912

G. Srinivasan and B. U. M. Rao
[Oakland University, Rochester, Michigan, USA]

I

Inks, Magnetic

Magnetic inks are special cases, from the point of view of composition and properties, of the general type of magnetic fluids originally described by Rosensweig et al. (1965). The reason for the differentiation from general magnetic fluids is that these inks are specifically adapted for printing purposes.

The original preparation of ferrofluid materials started in the USA with the work at NASA, Avco and Ferrofluidics Corporation in the 1960s. Later, simultaneous work in the study and preparation of various types of colloidal magnetic suspensions was carried out at IBM, the US Bureau of Mines, by Khalafalla and Reimers, in the UK by Charles and Popplewell and in Japan at Nittetsu Mining and Matsumoto Yushi-Seiyaku.

Magnetic inks can be made to generate character patterns on paper or other media by either of two general ways: (a) the ink is attracted into a magnetized spot in a tape or drum, which is later transferred to the paper: or (b) the ink is directly applied in discrete amounts to the paper, thereby generating dot patterns. In the first case, the ink can be in aqueous or nonaqueous solution; in the second, the ink is ejected from the nozzle of a jet printing system and is water based. In general, these fluids are colloidal systems made of a solid material dispersed in a liquid carrier. The magnetic material is typically magnetite (Fe_3O_4), subdomain in size, and more or less permanently suspended in a liquid carrier with the aid of dispersing agents, surfactants. The liquid carrier employed is usually a nonaqueous solvent, often an organic solvent of the nonpolar type. Examples of nonaqueous solvents used in the preparation of ferrofluids are aliphatic hydrocarbons, (e.g., heptane, decane, mineral oil and kerosene), halogenated hydrocarbons (e.g., carbon tetrachloride and trichloroethylene), aromatic solvents (e.g., benzene and toluene) and silicone oils. The dispersing agent is included in the ferrofluid formulation to prevent aggregation of magnetic material in the nonaqueous solvent, which could otherwise lead to flocculation and deposition of the particles. The dispersing aid can be one or more surfactants or wetting agents, and is applied to coat the surfaces of the individual magnetic particles. For colloidal magnetic dispersion in nonaqueous solvents, the coating layer is usually made up of aliphatic carboxylic acids with 8–24 carbon atoms.

1. Water-Based Inks for Ink Jets

Figure 1 illustrates printing with a magnetic ink jet, showing the sequences of ink drop generation, deflection and placement on paper. In this technology, the intelligence mechanism is contained in the fluid in the form of an inherent magnetic moment, so that magnetic field gradients can be used to deflect and position ink droplets on paper according to prearranged patterns (Fig. 1). The ink is a dispersion of particles rather than a pure aqueous solution, since even supersaturated solutions of the strongest magnetic metals do not possess enough magnetism to generate a ferrofluid. Therefore, the magnetic jet uses a heterogeneous system—an aqueous magnetic colloid. For this particular kind of colloid, the essential parameters to consider are: magnetic moment, particle size, electrical charge and chemical stability. The use of colloids in which the nuclei are made of materials of high magnetic moment, such as iron, cobalt and nickel, has been considered. However, particles of ferromagnetics when sufficiently small are single domain in character and possess an intrinsic permanent magnetic dipole moment μ, such that

$$\frac{2\mu^2}{a^3} = \frac{I_s^2 a^3 \pi}{18} \tag{1}$$

where μ is the magnetic dipole moment, I_s is the saturation magnetization and a is the particle radius. At certain particle size, the inline dipole–dipole interaction energy becomes greater than thermal energy and the particles tend to agglomerate in linear chains.

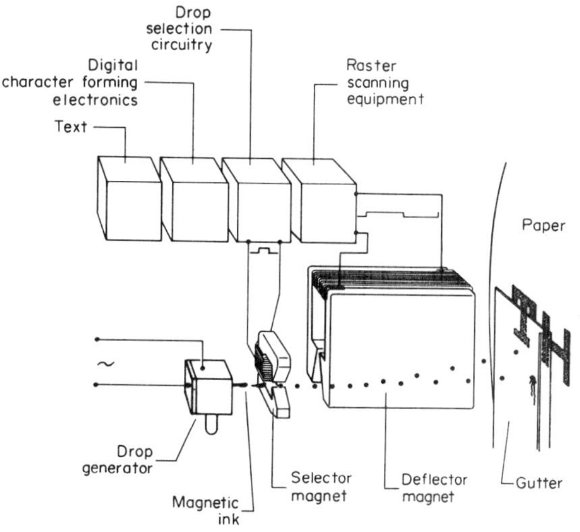

Figure 1
Magnetic ink jet system. Drops of ink are steered by nonuniform magnetic fields: deflection in the horizontal plane selects the drops for printing, vertical deflection by the deflector magnet controls placement of drops on paper

For example, for cobalt, at particle size greater than 5 nm interaction energy overcomes the thermal energy. Furthermore, systems containing iron or nickel as nuclei present problems in chemical stability due to oxidation and hydrolysis in aqueous solutions. The magnetic colloids considered here have been synthetically prepared by precipitation of magnetite (Fe_3O_4) and dispersion in aqueous media. In this type of aqueous colloid, stabilization is mainly due to formation of an electrostatic double layer. The stability of these heterogeneous aqueous systems is governed by the overall energy of interaction E between particles, which can generally be expressed as

$$E = \underbrace{K_1 \exp(-\psi a)}_{\text{electrostatic double layer repulsion}} - [\underbrace{K_2 a^3 x^{-3}}_{\text{magnetic attraction}} + \underbrace{K_3 x^{-6}}_{\text{van der Waals attraction}}] \qquad (2)$$

where K_1, K_2 and K_3 are constants, a is the particle radius. x is the interparticle distance and ψ is the electrokinetic potential which is a function of the charge of the double layer around the particles. The higher the value of $K_1 \exp(-\psi a)$, the more stable the system is. This electrostatic mechanism was established by attaching various charged surfactants to the particles to give the particles different values of Zeta potential ξ, and determining suspension stability by ultracentrifugation.

In addition, referring to Eqn. (2), the stability of the suspended particles is enhanced by forces of repulsion due to (a) steric hindrance or entropic repulsion and (b) hydration or, in general, solvation of adsorbed surfactant molecules. The last two effects (although difficult to treat mathematically) are known to combine in some aqueous systems, creating repulsive forces and an energy barrier which stabilizes the magnetic colloids even in the absence of large electric repulsive fields. For example, stable inks have been made using only uncharged nonionic surfactants and highly hydrated polyoxyethylenated surfactant layers (see *Fluids, Magnetic*).

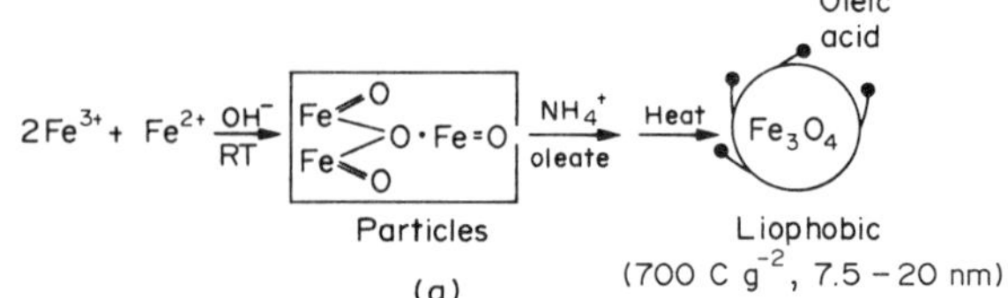

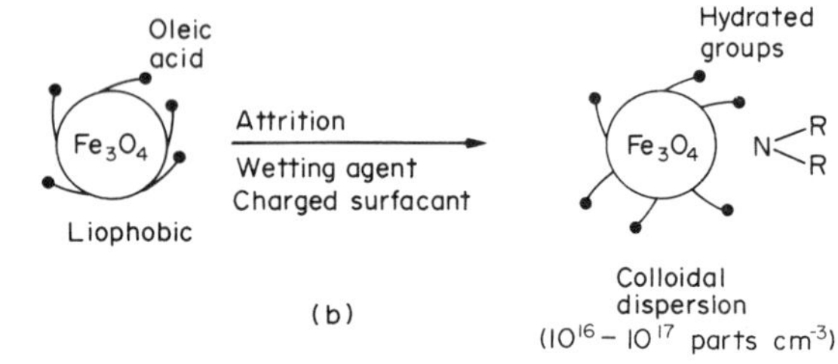

Magnetic fluid + Polyethylene glycol-glycerol-ethylene-monobutyl ether ⟶ Magnetic ink

(c)

Figure 2
The three principal steps in the preparation of magnetic ink or ink jet printing: (a) the solid phase, (b) the magnetic fluid and (c) conversion into ink

2. *Ink Formulation*

The preparation of jet inks involves three principal steps. These are illustrated in Fig. 2.

2.1 Oleate Coverage

As shown in Fig. 2, the initial particle generation is a simple precipitation reaction. In the subsequent coating of the particles by oleate, however, care has to be taken to use the right amount of oleate and to add it slowly and with constant stirring so that an optimum particle coverage is achieved. This step is critical to avoid later agglomeration problems and high erratic viscosity in the final fluid. For calculation of the amount of oleate required, it is first estimated from absorption data that each oleic acid molecule occupies an area site of $\tau = 0.205\,\text{nm}^2$ on the Fe_3O_4 surface. Hence, the total amount of oleate needed per gram of Fe_3O_4 would be $|A/\tau|\,(M_r/N_A)$, where A is the total area available for coverage, M_r is the molecular weight of oleic acid and N_A is the Avogadro number. A is derived as follows:

$$A = 4\pi r^2 N_p \qquad (3)$$

where r is particle radius and N_p is the number of Fe_3O_4 particles per gram of fluid, given by

$$N_p = \frac{3\pi r^3}{4\rho_{Fe_3O_4}} \qquad (4)$$

Here, $\rho_{Fe_3O_4}$, is the specific gravity of the magnetite suspension. In the ink formulations described in this article and for an average particle diameter of 10 nm, the optimum amount of ammonium oleate is 0.13–0.16 g per gram of Fe_3O_4.

2.2 Surfactant Coating

The second major step in the formation of the ink is the dispersion in water of the hydrophobic mass of oleic acid-coated Fe_3O_4. This is done by means of surfactants—one part of the surfactant molecule (the hydrocarbon moiety) attaches to the oleic acid-coated surface, while the hydrophilic moiety of the surfactant

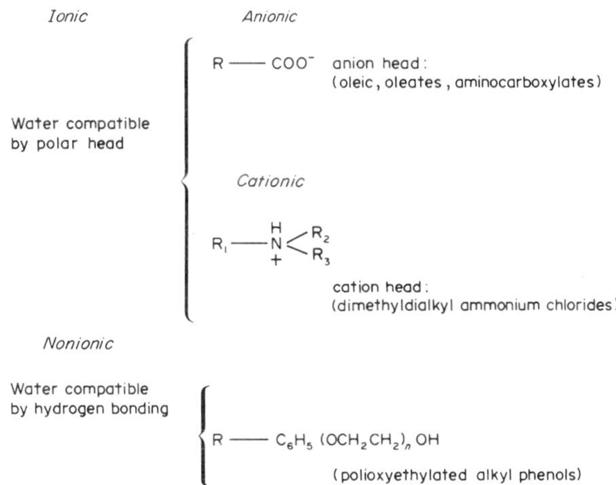

Figure 3
Surfactants used with magnetic fluids

provides hydration, thus rendering the particle compatible with water. This establishes particle stabilization by formation of double or multiple layers of surfactant coating. In effect, absorption studies using surface tension techniques indicate that most water-soluble surfactants do not attach directly to bare Fe_3O_4, but they do bind strongly to the oleic acid-coated particle. For the magnetic fluids described here, three different types of surfactants are used, depending on the nature of the hydrophilic group as shown in Fig. 3.

In the anionic and cationic surfactants, water solubility is provided by the saltlike structure of the (+) or (−) polar heads. In the nonionic surfactants, the ether oxygen of the ethylene oxide groups provides considerable hydrogen bonding with the water molecules. In particular, Triton N-101 (where R is a nonylphenol group and $n = 9$) has a sharp lowering effect of the surface tension and provides good particle wettability. Obviously, anionic and cationic surfactants cannot be present simultaneously in a fluid because their charges neutralize and coagulation occurs. From the point of view of colloidal stability, one of the most successful surfactant combinations is a mixture of Triton N-101 (0.17 g per gram Fe_3O_4) and Arquad (0.05 g per gram Fe_3O_4).

The third step—digestion with a mixture of glycerol and polyethylene glycol—improves colloidal stability by hydration, reduces micellar formation and lowers the viscosity of the ink. Ethylene glycol butyl ether optimizes the drying time of the ink on paper.

A typical magnetic ink for ink jet, prepared as described, has a magnetic moment of 24–26 A m^2 kg^{-1} (240–260 C g^{-2}), a viscosity of 0.010–0.012 Pa s and a pH of 5–7.

See also: Magnetic Materials: An Overview; Magnetic Materials: Basic Concepts

Bibliography

Jacobs I S, Bean C P 1963 Fine particle magnetism. In: Rado G T, Suhl H (eds.) 1963 *Spin Arrangements and Crystal Structure, Domains, and Micromagnetics*, Magnetism, Vol. 3. Academic Press, New York, pp. 271–351

Kaiser R, Miskolczy G 1970 Magnetic properties of stable dispersions of subdomain magnetic particles. *J. Appl. Phys. B* 41: 1064–72

Reimers G W, Khalafalla S E 1972 *Preparing Magnetic Fluids by a Peptizing Method.* US Bureau of Mines Technical Progress Report No. 59. US Bureau of Mines, Washington, DC

Rosensweig R E 1979 Fluid dynamics and science of magnetic liquids. *Adv. Electron. Electron Phys.* 48: 103–99

Rosensweig R E Nestor J W, Timmins R S 1965 *Ferrohydrodynamic Fluids for Direct Conversion of Heat Energy*, AIChE Symposium Series No. 5. American Institute of Chemical Engineers, New York

Tanford C 1961 *Physical Chemistry of Macromolecules.* Wiley, New York

Zahn M, Shenton K E 1980 Magnetic fluids bibliography, *IEEE Trans. Magn.* 16: 387–415

C. J. Sambucetti
[IBM Deutschland, Sindelfingen, Germany]

Z. Kovac
[IBM Thomas J Watson Research Center, Yorktown Heights, New York, USA]

Iron–Aluminum and Iron–Aluminum–Silicon Alloys

Iron–aluminum and iron–aluminum–silicon alloys of commercial importance are characterized by high resistivity, high hardness, high permeability and low hysteresis loss. The commercial binary alloys contain in excess of 12% aluminum and have been given the trade names Alfer (13% aluminum), Alfenol (16% aluminum) and Alperm (17% aluminum); at present they find limited application for magnetostrictive transducers. The iron–aluminum–silicon alloys of commercial importance contain 4–7% aluminum and 7–13% silicon. They are referred to as the Sendust family of alloys. They exhibit high initial and maximum permeabilities, high resistivities and low hysteresis losses. Because of their high resistivity, strain insensitivity and hardness, they are used mainly for recording heads and synchronous motors.

The magnetization of iron–aluminum binary alloys decreases with increasing aluminum content up to 15% aluminum in a manner corresponding to simple dilution. The first anisotropy constant K_1 goes to zero at 12–14% aluminum. The magnetostriction constant λ_{111} (i.e., magnetostriction along the [111] direction) increases and tends towards zero with increasing aluminum content and therefore is small in the range of composition of commercial interest. On the other hand, λ_{100} first rises sharply with aluminum content, reaches a maximum near 10% aluminum and then decreases. A combination of large λ and low K_1 in the 12–14% aluminum range led to the development of Alfer for transducer use.

Iron–aluminum alloys beyond 8% aluminum begin to exhibit effects due to an order–disorder transformation (Fe_3Al, DO_3-type structure). The magnetic properties are therefore sensitive to cooling rate and magnetic field annealing. Initial and maximum permeabilities, coercive force and residual induction are sensitive to cooling rate, particularly in the 14–16% aluminum range. Highest permeabilities are obtained for alloys containing 17% aluminum. This alloy, Alperm, has found applications as a recording head material because of its high hardness and resistivity along with good permeability.

The room-temperature ductility of iron–aluminum alloys below 7% aluminum is very good. Limited room-temperature ductility is obtained at high aluminum contents because of the order–disorder transformation. Hot working is easily accomplished in these alloys.

Aluminum and silicon have the greatest effect on the resistivity of iron and are nearly equal in their effect. Isotropic iron–aluminum alloys possess magnetic properties comparable to iron–silicon alloys of equivalent composition, particularly in the dilute range ($<6\%$ aluminum).

Within the Fe–Al–Si system, initial relative permeabilities μ_i as high as 35 000, maximum relative permeabilities μ_m as high as 110 000, resistivities of approximately $100\ \mu\Omega\,cm$ and hysteresis losses of about 15% of that exhibited by grain-oriented 4% Si–Fe have been achieved (see *Silicon Steels*). Certain compositions possess very low magnetostriction and anisotropy. Optimum magnetic properties are obtained over a narrow composition range centered at 5.4% aluminum and 9.6% silicon (Sendust alloys). These alloys are hard and brittle and thus limited in their use, in cast form, to dc applications and as insulated powder cores for high-frequency inductors. Sputtered, thick ($12\ \mu m$) Sendust films suitable for recording heads have been prepared (Shibaya and Fukuda 1977). The films exhibit saturation flux densities B_m of 1.1 T, comparable to those of cast Sendust, and effective permeabilities of 240 at 20 MHz. More recent work (Minakata 1987) has shown that the magnetic properties of Sendust are suitable for thin-film recording heads. A permalloy film on ferrite with a subsequent thicker Sendust layer on the permalloy has been proposed for video recording heads (Sillen et al. 1988).

Commercially available Sendust cores exhibit $\mu_i = 10\,000–30\,000$, $\mu_m = 80\,000–160\,000$ and losses of $240\ kJ\,m^{-3}$. Core losses for both Sendust and 4% Mo–permalloy at $B_m = 0.6$ T are comparable, but at higher frequencies the loss values become more favorable for Sendust laminated cores. Yamamoto and Utsushikawa (1976) have reported that a 2–4% nickel addition to Sendust, after an appropriate heat treatment, resulted in decreased coercivity and increased maximum permeability. This alloy was referred to as "Super Sendust."

See also: Magnetic Materials: An Overview; Magnetic Materials: Basic Concepts; Magnetostrictive Materials; Recording Head Technology

Bibliography

Chin G Y, Wernick J H 1980 Iron–aluminum alloys. In: Wohlfarth E P (ed.) 1980 *Ferromagnetic Materials*, Vol. 2. North-Holland, Amsterdam, pp. 114–23

Minakata R 1987 Magnetic properties of Fe–Al–Si films prepared by a dc triode sputtering method. *IEEE Trans. Magn.* 23: 3236–8

Shibaya H, Fukuda I 1977 Preparation by sputtering of thick sendust film suitable for recording head core. *IEEE Trans. Magn.* 13: 1029–35

Sillen C W M P, Ruigrok J J M, Vangroenou A B, Enz U 1988 Permalloy/Sendust metal-in-gap head. *IEEE Trans. Magn.* 24: 1802–4

Tebble R S, Craik D J 1969 *Magnetic Materials*. Wiley, New York

Yamamoto T, Utsushikawa Y 1976 Effects of Ni content on magnetic properties of a new high permeability magnetic alloy "super sendust" in the Fe–Si–Al–Ni system. *J. Jpn. Inst. Met.* 40: 975–81

J. H. Wernick
[Bellcore, Red Bank, New Jersey, USA]

G. Y. Chin[†]
[AT&T Bell Laboratories, Murray Hill, New Jersey, USA]

Iron and Low-Carbon Steels for Magnets

Iron and low-carbon steels are generally used as soft magnetic materials because of the ease with which they magnetize and demagnetize under the influence of a small magnetic field. These materials, low-carbon steels in particular, find extensive applications in small ac motors and other electrical machinery. The consumption rate for such steels is estimated to be about 10^6 t per year in the USA, which constitutes approximately 57% of the total electrical steel consumed. Rising energy costs have prompted the motor industry to produce energy-efficient motors by utilizing improved design and better quality steels. Superior properties in steels are obtained with the proper selection of composition and processing. This has been made possible through advances in steelmaking technology and a better understanding of the relationship between the structure and properties of steels.

1. Basic Considerations

For dc applications, a soft magnetic material should possess as high a saturation magnetization as possible and at the same time be able to reach the saturation value with a minimum magnetizing force. The performance of the material is usually described by quoting permeability, $\mu = B/\mu_0 H$, where H is the magnetizing force required to attain a particular induction level B. Iron is generally considered to be a suitable candidate for such purposes. However, in ac applications, low-carbon steels with high permeability and low core loss are required to reduce the total energy loss.

2. Permeability

Permeability improvements at inductions below the "knee" of the B/H (hysteresis) curve are obtained with large grain sizes and elimination of inclusion-forming impurities (e.g., carbon, nitrogen, oxygen and sulfur), because they interfere with the domain wall movements (see *Coercivity and Domain Wall Pinning*; *Domains and Domain Walls*). In the knee region, favorable crystallographic orientations are also required to achieve improved permeability because the magnetization process is controlled by both domain wall motion and rotation. However, similar benefits at high inductions are only realized through a superior texture. Such a texture is generally characterized by the preponderance of the {100} and {110} components at the expense of {111} and {211} orientations. The simple reason for this is the presence of the easiest direction of magnetization ⟨100⟩. Based on this, an empirical texture-related parameter has been developed to determine the quantitative effects of textural changes on the permeability of low-carbon steels (Rastogi 1975). Improved textures in steels are developed with the selection of appropriate alloying elements and/or processing (Rastogi 1977).

3. Core Loss

According to the classical model, the total ac core loss P_T is composed of the hysteresis loss P_H, classical eddy current P_E and anomalous or unaccounted loss P_A and is described as

$$P_A = P_T - (P_H + P_E)$$

The calculation of P_E assumes complete flux penetration and constant permeability, which is not true for ferromagnetic materials because of the variation in permeability with magnetizing force. This apparently results in the underestimation of the eddy current loss and the observed discrepancy is represented by the anomalous loss. The ratio of the true eddy current loss $P_T - P_H$ to the classical eddy current loss P_E is called the anomaly factor and is generally greater than one for low-carbon steels (Rastogi 1976, 1977), except when the domain size is extremely small. The reduction in P_E at a given test frequency and induction is readily accomplished by decreasing the sheet thickness and increasing the steel resistivity with the addition of alloying elements. Larger grain size, lower volume fraction of inclusions and favorable orientations are required to achieve improvements in the hysteresis loss P_H. Therefore, it is evident that, at a particular frequency and induction, P_H and P_E should be kept as low as possible to effect significant improvements in the core loss.

An empirical relationship (Rastogi and Shapiro 1973) has also been used to predict the relative importance of different parameters on the 1.5 T core loss, which is defined as

$$P_T = \frac{11.33 t^{1.35} G^{0.065}}{(B_{30}/B_s)^{3.09} \rho^{0.40}} \quad (\mathrm{W\,kg^{-1}})$$

where t is thickness (m), G is the average number of grain boundary intercepts per millimeter, ρ is resistivity (μΩ cm) and B_{30}/B_s is a texture-related parameter.

4. Effect of Processing and Composition

Magnetic properties of low-carbon steels (also known as lamination steels) are controlled by both processing and composition. Advances in steelmaking technology have led to the development of cleaner products with low levels of carbon and nitrogen. In addition, continuous casting is being heavily favored over ingot casting for producing more uniform products at lower cost. The processing scheme for most lamination steels normally involves hot rolling, pickling, cold rolling, annealing and temper rolling prior to decarburization and testing. Both batch-annealing and continuous-annealing processes are employed in the production of lamination steels. In many instances, continuous annealing is preferred to batch annealing to improve permeability in the final product through the formation of a more favorable texture. In the development of improved steels, the interactive effects of composition and processing are also considered.

Table 1
Magnetic properties of iron

Grade	Maximum permeability	Hysteresis loss at 1.0 T (W kg^{-1})
High-purity iron (99.99%)	100000	~0.08
Ingot iron (99.6–99.8%)	5000	~2.12

Source: *Metals Handbook* (1980)

Commercially pure iron is often used for electromagnets and other dc devices (see Table 1). Ideally, the highest saturation magnetization and permeability in iron is obtained with the removal of impurities (carbon, nitrogen, oxygen, sulfur). This can be achieved by annealing iron at high temperatures (1200–1300 °C) in a pure dry hydrogen atmosphere. The elimination of these impurities also favors grain growth, which is conducive to lower hysteresis loss. Since such a process is commercially difficult and very expensive, anneals for iron sheets are usually performed at approximately 800 °C to achieve a carbon level ⩽0.005 wt%. The detrimental effects arising from nitrogen aging are minimized by controlling its level in steels to below 0.006 wt%. In comparison, the adverse effects resulting from the oxide and sulfide particles are less pronounced because of large particle sizes. Stabilized low-carbon steels are also used as inexpensive substitutes for pure iron in certain applications. Despite respectable permeability for nominally pure iron, it is not considered suitable for ac devices due to the high core loss associated with the low resistivity.

5. Low-Carbon Steels

Relatively low prices and good magnetic properties have led to a greater use of low-carbon steels in motors and small transformers. These are used as sheet products in three different conditions: (a) full-hard, (b) fully processed and (c) semiprocessed. The selection of a steel for a particular application is largely based on its magnetic characteristics. Cold-rolled steel can be supplied in one of three conditions: (a) full-hard (cold-rolled condition), (b) semiprocessed (cold-rolled, annealed and temper-rolled) and (c) fully processed (annealed). Both full-hard and semiprocessed steels are always decarburized to develop optimum magnetic properties through carbon removal and grain growth. This is accomplished by annealing steels at approximately 790 °C for 1.25 h in an atmosphere consisting of 9% H_2 and 91% N_2 with a dew point above 13 °C. A detailed description of this process employed by lamination-steel users is given by Shapiro (1978). Most steel is purchased in the semiprocessed condition and the remainder finds applications as full-hard or fully processed products in less critical devices. The reason for this is the better magnetic properties of the semi-processed products compared with the other two products.

Product quality is usually measured by permeability and core-loss values at 1.5 T. Of these properties, the core loss is used as a basis for selecting a product for a specific application. The following discussion of steels supplied in each of the three conditions is confined to 0.46 mm thick products. Most low-carbon steels for electrical applications generally contain 0.04–0.06 wt% carbon, 0.05–0.15 wt% phosphorus, 0.35–0.80 wt% manganese, 0.006–0.25 wt% aluminum, 0.05–0.25 wt% silicon, and 0.015–0.025 wt% sulfur. Typical properties of these steels are listed in Table 2.

Table 2
Typical properties of 0.46 mm thick low-carbon steels at 1.5 T and 60 Hz

Steel	Condition	Core loss ($W\,kg^{-1}$)	Relative permeability
Standard product (0.05% C, 0.40% Mn, 0.12% P, 0.025% S)	full-hard	8.50	2000
Con-Core[a]	fully processed	9.90	
Locore N[b] (silicon steel)	fully processed	9.90	
Vacuum-degassed extra-low-carbon steel (0.006% C)	fully processed	8.40	
Locore M-47[b] (silicon steel)	fully processed	8.40	
Standard product	semiprocessed	6.60	2200
Improved product (0.04% C, ⩽0.60% Mn, 0.12% P, 0.025% S)	semiprocessed	6.10	2300
New steel[c]	semiprocessed	5.50	2300

a US Steel trade name b Republic Steel trade names c Inland product

5.1 Full-Hard Steels

A full-hard steel is a cold-rolled steel which is decarburized in the form of punched laminations to develop optimum magnet properties. A typical decarburized full-hard steel (Table 2) exhibits a 1.5 T core loss of 8.50 $W\,kg^{-1}$ with a relative permeability of 2000. Recent studies at the Inland Steel Company have shown that lowering initial carbon and sulfur levels from 0.05 wt% to 0.02 wt% and from 0.025 wt% to 0.015 wt%, respectively, results in a 0.75 $W\,kg^{-1}$ decrease in the core loss along with a modest gain in permeability. Improved properties are primarily attributed to the lower volume fraction of nonmetallic inclusions and larger grain size. Additional improvement in the core loss can also be obtained with increased levels of manganese and silicon through higher resistivity. These observations indicate that the potential for producing low-loss full-hard steel has yet to be fully exploited.

5.2 Fully Processed Steels

Fully processed steels were developed as less costly alternatives to the low- and intermediate-level silicon steels. These materials are produced on a continuous-annealing line with the final carbon level as low as economically feasible, this being a function of the initial carbon content and the annealing practice. Annealing temperatures in the range 790–820 °C are usually employed to achieve efficient decarburization and larger grain size. Core-loss values in Table 2 indicate that these steels are comparable to the silicon steels. Fully processed products are also coated for the purpose of good interlaminar resistance and improved punching.

5.3 Semiprocessed Steels

Advances in steel metallurgy have led to significant improvements in the magnetic properties of decarburized semiprocessed steels. In recent years, changes in steel composition and processing have produced steady improvements in the 1.5 T core loss of a decarburized semiprocessed steel, as shown in Table 2. This approach has led to the development by the Inland Steel Company of a lamination product with a core loss of 5.50 $W kg^{-1}$ and relative permeability over 2000. This core loss is about 1.10 $W kg^{-1}$ lower than the typical value of 6.60 $W kg^{-1}$ reported for a standard product. In the development of such products, silicon and aluminum levels are usually kept below 0.25 wt%, whereas the manganese content can be up to 0.80 wt%. The composition restrictions for silicon and aluminum are designed to achieve high permeability in these steels. Further improvement in these properties can be expected with the processing of such a steel containing $\leqslant$0.015 wt% carbon and $\leqslant$0.010 wt% sulfur. Besides, a lower carbon level reduces the anneal time, thereby providing economic benefits to users. It has been recently demonstrated that further refinements in steel composition and processing lead to the development of a 4.40 $W kg^{-1}$ lamination steel with high permeability. In addition to good magnetic properties, the surface roughness of the steel should be between 5 µm and 12 µm to achieve a high stacking factor for improved motor performance.

6. Applications

Low-carbon steels are largely used in small motors. It is recognized that steels with low core loss and high permeability are needed to produce energy-efficient motors. However, the relative significance of these properties in terms of improving motor performance is determined by motor design parameters. It appears that proper design modifications may be necessary to optimize product performance in a particular ac device. The use of improved and new lamination steels in energy-efficient motors should result in substantial energy savings. Because of escalating energy costs, demands for better products are expected to grow rapidly in the future. The current drive for energy conservation has created a greater awareness among steel producers and users of the impact of both material properties and design parameters on motor performance.

See also: Magnetic Materials: An Overview; Steels for Magnets

Bibliography

Metals Handbook, 9th edn., Vol. 3 1980 Magnetically soft materials. American Society for Metals, Metals Park, OH, pp. 598–603

Rastogi P K 1975 Effect of chromium on the magnetic properties and texture of non-oriented steel. *AIP Conf. Proc.* 26: 556–7

Rastogi P K 1976 Magnetic properties and texture of unalloyed non-oriented electrical steel. *AIP Conf. Proc.* 34: 61–2

Rastogi P K 1977 Effect of manganese and sulfur on the texture and magnetic properties of non-oriented steel. *IEEE Trans. Magn.* 13: 1450–80

Rastogi P K, Shapiro J M 1973 A statistical model for the total AC core loss of lamination steels. *IEEE Trans. Magn.* 9: 122–4

Shapiro J M 1978 Optimizing the decarburization of punched electrical steel laminations. *Mechanical Working and Steel Processing XVI.* Iron and Steel Society (AIME), New York, pp. 56–77

P. K. Rastogi
[Inland Steel Research Laboratories,
East Chicago, Indiana, USA]

Iron–Chromium–Cobalt Alloys

The Fe–Cr–Co permanent-magnet alloys were discovered by Kaneko et al. (1971). The metallurgy and magnetic properties of these alloys are remarkably similar to those of the alnicos (see *Alnicos*). However, whereas the alnicos are brittle and must be cast and ground to finished shape or shaped by powder metallurgy techniques, these new alloys are cold formable at room temperature. Hence, normal metal-forming operations such as rolling, wire drawing and stamping can be carried out relatively easily. In addition, equivalent magnetic properties can be attained with much less cobalt.

1. Physical Metallurgy

Permanent-magnet alloys in the Fe–Cr–Co system range from about 2% to 25% cobalt and from about 20% to 35% chromium (all compositions are quoted in weight percent). At temperatures above 1200–1300 °C the alloys exist as a single-phase body-centered-cubic (bcc) α phase. Below this temperature to about 700 °C the α phase tends to coexist with the face-centered-cubic (fcc) γ phase. For high-chromium contents a brittle σ phase may also appear in this temperature range.

Of these three phases, only the α phase is ferromagnetic at room temperature. It can be retained at room temperature if quenched from the high-temperature single-phase regime. Aging at low temperature then decomposes the α phase spinodally into a highly ferromagnetic iron-rich α_1 phase and a less ferromagnetic chromium-rich α_2 phase, both of which maintain the bcc structure. It is this spinodal decomposition that results in the permanent-magnet properties of the Fe–Cr–Co system. The spinodal decomposition temperature is about 550 °C for the Fe–30%Cr binary and increases to about 650 °C for the Fe–20%Co–30%Cr ternary. Hence, cobalt is desirable since it raises the spinodal decomposition. On the other hand, cobalt is a γ former, which is undesirable. In practice, the γ phase in the ternary can be suppressed by the further addition of α formers such as aluminum, niobium, titanium, vanadium and zirconium. This has been achieved for alloys near the Fe–15%Co–28%Cr composition. Another alternative is to stay in the ternary, but at a lower cobalt content. The kinetics of decomposition, however, are then decreased.

Transmission electron microscopy studies have shown that the primary mechanism of coercivity in Fe–Cr–Co alloys is pinning of domain walls, not an unreasonable mechanism since in many cases both α_1 and α_2 phases remain ferromagnetic at room temperature. However, a quantitative theory has not yet been evolved and the situation in which α_2 becomes nonmagnetic is unclear. In any case, experimentally the optimum microstructure for permanent-magnet properties in the Fe–Cr–Co system appears to consist of highly aligned particles of α_1 and α_2 matrix, with the particles size about 30–40 nm in diameter by about 150 nm in length, and with as large a difference in magnetization between the two phases as possible. Aligned elongated structures can be achieved by heat treatment in a magnetic field, as for the alnicos, including cast columnar structures that favor the ⟨001⟩ growth axis. In addition, the ductile nature of the Fe–Cr–Co alloys also permits the development of aligned elongated structures using a deformation aging process (Jin 1979).

In addition to the anisotropic magnets, isotropic magnets with unaligned structures have been prepared. Preparation using powder metallurgy techniques has also been demonstrated. Figure 1 shows typical demagnetization curves of several Fe–Cr–Co magnets.

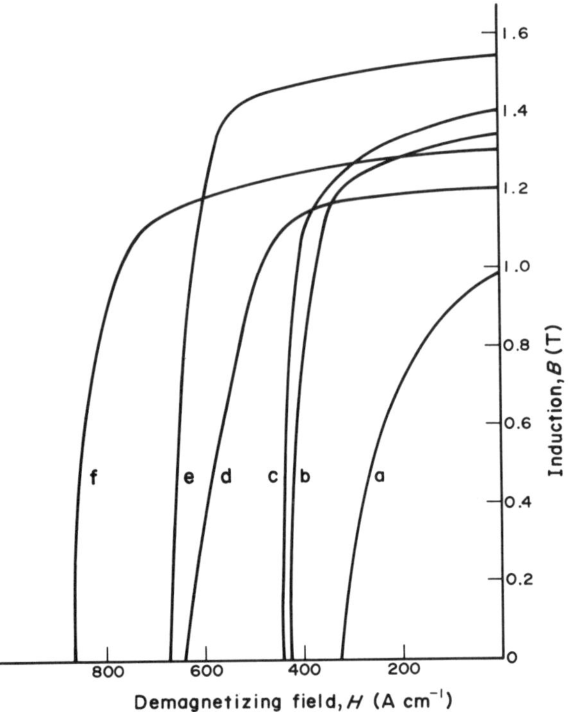

Figure 1
Typical demagnetization curves of Fe–Cr–Co magnets: (a) Fe–10.5%Co–28%Cr (isotropic); (b) Fe–5%Co–30%Cr; (c) Fe–12%Co–25%Cr (sintered); (d) Fe–11.5%Co–33%Cr–2%Cu (deformation-aged); (e) Fe–15%Co–24%Cr–3%Mo (⟨100⟩ columnar); (f) Fe–23%Co–32%Cr–2%Cu (deformation-aged)

2. Isotropic Magnets

Since impurities such as carbon, oxygen and nitrogen are detrimental to the magnetic and mechanical properties of Fe–Cr–Co alloys, vacuum induction melting is preferred. The cast ingots are converted to sheet or rod stock following conventional wrought-mill practice. The finished part, which may be cold worked or annealed, may be aged isothermally below the spinodal temperature, followed by stepped cooling to develop the permanent-magnet properties. However, the optimum aging temperature is quite sensitive to slight composition variations. An improved method which is less sensitive consists of first cooling at a fast rate from above the spinodal temperature to an intermediate holding temperature below the spinodal temperature, followed by a much slower cooling rate. The rationale behind the two-stage treatment is that the initial fast cool establishes the optimum particle size and the second slow cool establishes the optimum composition profile between the two phases. The initial cooling rate is lower, the lower the cobalt content (e.g., 100 °C h^{-1} for a 10.5% cobalt alloy, 200 °C h^{-1} for a 15% cobalt alloy).

Representative magnetic properties of isotropic magnets are shown in Table 1. In general, as the cobalt content increases, the coercivity and energy product increase while the remanence slightly decreases.

Table 1
Typical magnetic properties of Fe–Cr–Co alloys

Alloy (wt%, balance Fe)	B_r (T)	H_c (kA m^{-1})	$(BH)_{max}$ (kJ m^{-3})
Isotropic magnets			
23Co, 31Cr	0.84	50	18
10.5Co, 28Cr	0.98	32	14
7Co, 28Cr	0.97	26	11
Anisotropic magnets by field aging			
25Co, 30Cr, 3Mo	1.15	62	40
15Co, 24Cr, 3Mo	1.54	67	76
⟨100⟩ columnar			
15Co, 22Cr	1.56	51	66
12Co, 25Cr	1.45	50	61
9Co, 27Cr	1.30	46	50
7Co, 28Cr	1.20	41	42
5Co, 30Cr	1.34	42	42
2Co, 33Cr	1.24	38	34
Anisotropic magnets by deformation aging			
23Co, 32Cr, 2Cu	1.30	86	78
16Co, 32Cr, 2Cu	1.29	70	65
11.5Co, 32Cr, 2Cu	1.15	65	50
7Co, 32Cr, 2Cu	1.18	42	34
Sintered magnets			
25Co, 29.5Cr, 3Mo	0.95	69	27
12Co, 25Cr	1.40	44	42
5Co, 31Cr	1.23	40	35

3. *Anisotropic Magnets Prepared by Magnetic Field Aging*

If a magnetic field is applied during the initial aging treatment below the spinodal temperature, the decomposed particles tend to elongate and align in the direction of the field. The result is a squaring up of the hysteresis loop and an increase in the energy product in the direction of the applied field.

For the Fe–Cr–Co alloys, a magnetic field strength exceeding about 500 A cm^{-1} is needed. Also, in the Fe–Cr–Co ternary, there is no strong tendency for the particles to grow in the ⟨100⟩ direction as is common in the alnicos. Consequently, crystallographic texture is relatively unimportant. The addition of molybdenum, however, favors the particle growth in the ⟨100⟩ direction. In this case, a ⟨100⟩ columnar structure such as that obtained by directional solidification further improves the energy product.

Table 1 lists typical magnetic properties of magnetic-field-aged alloys. The highest energy product to date for magnetic-field-aged material, at 76 kJ m^{-3}, is obtained in an Fe–15%Co–24%Cr–3%Mo alloy that has been directionally solidified. The alloy with the highest energy product per unit of cobalt is the Fe–2%Co–33%Cr alloy. However, cooling rates of the order of 0.1 °C h^{-1} are required as the kinetics of decomposition become rather slow.

4. *Anisotropic Magnets Prepared by Deformation Aging*

Deformation aging is a technique of achieving magnetic anisotropy in an elongated precipitate structure through plastic deformation rather than magnetic field treatment. The technique makes use of the ductile quality of the Fe–Cr–Co alloys.

In deformation aging, the alloy is first cooled through the spinodal temperature to develop oversized, nearly spherical particles. An anisotropic deformation such as extrusion, wiredrawing or rolling then elongates and aligns these particles. The material is then reaged to develop an optimum composition profile for magnetics.

Since particle size, shape and alignment are important in developing the magnetic properties, the nature and amount of deformation is important. Wiredrawing results in improved properties over those obtained by rolling and 70% reduction of area is about optimum. A typical operation on an Fe–11.5%Co–33%Cr (Chromindur III) alloy involves cooling a rod-shaped sample from 680 °C at 40 °C h^{-1} to 610 °C, wiredrawing to 67% reduction in area and reheating at 600 °C followed by controlled cooling, ranging from 15 °C h^{-1} initially to 4 °C h^{-1} finally, to 500 °C. Values of B_r = 1.2 T, H_c = 616 A cm^{-1} and $(BH)_{max}$ = 44 kJ m^{-3} are obtained. These values compare well with those of alnico 5.

The substitution of 2% copper for iron in the deformation-aged alloys results in an increase of 4–8 kJ m^{-3} in $(BH)_{max}$ for the same cobalt and chromium content, primarily due to an increase in H_c. An Fe–23%Co–2%Cu–33%Cr alloy reaches a value of $(BH)_{max}$ = 78 kJ m^{-3}, the highest value reported to date in the Fe–Cr–Co alloys.

Table 1 lists typical magnetic properties of deformation-aged alloys.

5. *Sintered Magnets*

Although it is often advantageous or even necessary to prepare magnets of brittle alloys such as the alnicos and the cobalt–rare-earth alloys by powder metallurgy, there is no such limitation in the ductile Fe–Cr–Co alloys. Nevertheless, the prior existence of powder metallurgy facilities or the economical production of a large number of small complex parts may dictate the use of powder techniques.

Sintered magnets can be prepared from a blend of elemental powders or from atomized prealloyed powders. The compacted powders are generally sintered in dry hydrogen at 1300–1400 °C for a few hours and then quenched in water prior to the usual low-temperature aging treatment.

As in all sintered magnets, the magnetic properties are generally inferior to those of solid magnets on account of the porosity. Some typical properties are shown in Table 1.

6. Applications

Since the Fe–Cr–Co alloys are ductile and the magnetic properties of the isotropic Fe–Cr–Co magnets are generally superior to those of the cobalt magnet steels Cunife and Vicalloy, applications include, for example, speedometers, hysteresis and timing motors and antitheft labels. The anisotropic magnets having magnetic properties similar to alnico 5 should find widespread application in motors, generators and loudspeakers.

By far the most extensive application of Fe–Cr–Co permanent-magnet alloys to date has been the use of an isotropic Fe–10.5%Co–28%Cr magnet, called Chromindur II, in a telephone receiver design (Chin et al. 1978, Jin et al. 1980). The cup-shaped magnet requires extensive drawing from a blank. The ductile nature of Chromindur II permits high-speed room-temperature forming, whereas the Remalloy magnet that it replaces had to be hot drawn at about 1250 °C.

See also: Iron–Cobalt Alloys; Magnetic Materials: An Overview

Bibliography

Chin G Y 1980 New magnetic alloys. *Science* 208: 888–94

Chin G Y, Plewes J T, Wonsiewicz B C 1978 New ductile Cr–Co–Fe permanent magnet alloys for telephone receiver applications. *J. Appl. Phys.* 49: 2046–8

Jin S 1979 Deformation-aged Cr–Co–Fe permanent magnet alloys. *IEEE Trans. Magn.* 15: 1748

Jin S, Chin GY 1987 Fe–Cr–Co magnets. *IEEE Trans. Magn.* 23: 3187–92

Jin S, Chin G Y, Wonsiewicz B C 1980 A low cobalt ternary Cr–Co–Fe alloy for telephone receiver magnet use. *IEEE Trans. Magn.* 16: 139–46

Kaneko H, Homma M, Nakamura K 1971 New ductile permanent magnet of Fe–Cr–Co system. *AIP Conf. Proc.* 5: 1088–92

G. Y. Chin[†]
[AT&T Bell Laboratories, Murray Hill, New Jersey, USA]

J. H. Wernick
[Bellcore, Red Bank, New Jersey, USA]

Iron–Cobalt Alloys

The technological importance of iron–cobalt-base alloys is a consequence of their useful intrinsic properties (high saturation inductions and high Curie temperatures) and the ability to obtain high permeabilities and low coercivities through composition and processing control. However, these alloys, particularly the soft magnetic alloys, are relatively costly and they are only used where cost is not a factor. Examples of these applications are aircraft electric generators (where high permeabilities and flux densities allow for weight and size reductions), cores requiring exceptionally high flux densities, high-temperature magnetic components and, to a limited extent, magnetostrictive transducers.

The addition of cobalt to iron increases the saturation magnetization of iron, peaking at around 2.45 T at room temperature for a 35% cobalt alloy. The Curie temperature also rises with cobalt content up to 15% cobalt.

The iron–cobalt phase diagram is characterized by the existence of a complete range of face-centered-cubic (fcc) solid solutions at elevated temperatures, with the transformation of the fcc alloys on cooling to a disordered body-centered-cubic (bcc) solid solution and ordering of this phase at lower temperatures for compositions between about 30% and 70% cobalt. The ordered phase exhibits the CsCl-type (B2) structure. The order–disorder reaction on cooling is exceedingly rapid and plays an important role in determining the magnetic and mechanical properties of these materials.

The first anisotropy constant K_1 of iron is positive and decreases as cobalt is added. It is zero near 40% cobalt and 50% cobalt for the disordered and ordered phases, respectively. A maximum in permeability occurs at around 50% cobalt.

Saturation magnetostriction along the [111] direction, λ_{111}, is approximately zero at 25% cobalt and rises to around 25×10^{-6} at 50% cobalt λ_{100} exceeds λ_{111} at all cobalt contents and reaches a maximum of 150×10^{-6} at 50% cobalt for the disordered alloy. Ordering appears to lower λ_{100} somewhat. The high λ_{100} value qualifies the 50%Co–50%Fe alloy as one of the best magnetostrictive transducer alloys available, but its high cost has prevented large-scale use for transducer applications.

The maximum saturation induction exhibited by Fe–35%Co led Stanley and Yensen (1947) to investigate the effect of chromium additions on magnetic and mechanical properties. An alloy containing 0.5% chromium was shown to be capable of being rolled into thin sheets. After suitable annealing heat treatments, which also reduced the carbon content to below 0.005%, the following magnetic properties were obtained: saturation induction $B_s = 2.4$ T, coercivity $H_c = 50.14$ A m^{-1}, remanence $B_r = 1.15$ T initial relative permeability $\mu_i = 650$, maximum relative permeability $\mu_m = 10000$ and resistivity $\rho = 20$ μΩ cm. Dc losses for 0.043 cm sheet were 6.6 W kg^{-1} and 220 W kg^{-1} at 60 Hz and 600 Hz, respectively. A commercial Fe–Co–Cr alloy is available under the name Hiperco "27" (27% cobalt and 0.6% chromium). A commercial binary alloy, Hyperm Co 35, contains 35% cobalt and is supplied in the form of castings or as parts machined from them. Some mechanical and dc magnetic properties of Hiperco "27" and Hyperm Co 35 are given in Table 1.

Table 1
Mechanical and dc magnetic properties of Hiperco "27" and Hyperm Co 35

Property	Hiperco "27"	Hyperm Co 35
Tensile strength (MPa)	517.50	
Yield strength (MPa)	262.20	
Elongation (%)	10	
Reduction of area (%)	50	
Rockwell hardness	B85/92	
Coercive force ($A m^{-1}$)	198.95	198.95
Permeability (max)	2800	
Flux density at max permeability (T)	1.6	
Saturation induction (T)	2.36	2.4
Resistivity (μΩ cm)	19	40

Table 2
Properties of Permendur (regular) and Supermendur (high-purity Co–Fe–2% V)

Property	Permendur	Supermendur
Electrical resistivity (Ω cm)	25×10^{-6}	25×10^{-6}
Saturation induction (T)	2.4	2.4
Remanent induction (T)	1.5	2.215
Magnetic permeability	4000–8000	92500
Coercive force ($A m^{-1}$)	3.98×10^2	15.92
Magnetostriction coefficient[a]	$+60 \times 10^{-6}$	$+60 \times 10^{-6}$

a At $7.96 \times 10^3 A m^{-1}$

The most important iron–cobalt soft magnetic alloys are Permendur (49% Fe–49% Co–2% V) and Supermendur. The high permeability of Permendur persists ("endures") to high flux densities—hence the name. Preparation of thin sheets ($\sim 2 \times 10^{-6}$ cm) usually starts by hot working of the ingot in the fcc state, followed by a magnetic anneal below the fcc→bcc transformation temperature (below the Curie temperature), but above the order–disorder temperature, and then rapidly cooling to room temperature should further reduction be required. The 2% vanadium improves the ductility and markedly increases (by about four times) the resistivity of the basic 50% Fe–50% Co. Major et al. (1989) have shown that additions as small as 0.25% tantalum or niobium can be used as an alternative to 2% vanadium to improve the ductility of the 50% cobalt alloy, such that it can be cold worked. The importance of this discovery relates to the higher saturation that can be achieved as a result of the ability to substitute the vanadium addition.

By use of the highest-purity commercial vanadium, iron and cobalt obtainable at the time, use of controlled-atmosphere melting, and wet and dry hydrogen annealing treatments and a magnetic anneal, Gould and Wenny (1957) showed that B_r, H_c and μ_m could be markedly improved. They named the material prepared in this fashion Supermendur. A comparison of the properties of Permendur (regular) and Supermendur is shown in Table 2.

It has been shown (Major et al. 1975) that the addition of 4.5% nickel to the basic 49%Fe–49%Co–2%V alloy further enhanced elevated-temperature ductility and magnetic properties. Thus, the alloy can be used at high temperatures, for example, as a rotor material for advanced high-speed generators.

Electroplated thin films of Co–10%Fe exhibit properties comparable with Permalloy and the suggestion has been made that it could be a replacement for Permalloy for use in high-density longitudinal and perpendicular recording (Liao 1987).

See also: Magnetic Alloys: Intermediate Coercivity Materials; Magnetic Materials: An Overview; Magnetic Materials: Basic Concepts

Bibliography

Chin G Y, Wernick J H 1980 Soft magnetic metallic materials. In: Wohlfarth E P (ed.) 1980 *Ferromagnetic Materials*, Vol. 2. North-Holland, Amsterdam, pp. 55–188

Gould H L B, Wenny D H 1957 Supermendur: New rectangular loop magnetic material. *Conf. Magnetism and Magnetic Materials*, AIEE Special Publication T-91. American Institute of Electrical Engineers, New York, pp. 675–81

Liao S H 1987 High-moment CoFe thin films. *IEEE Magn.* 23: 2981–3

Major R V, Martin M C, Branson M W 1975 Development of cobalt–iron soft magnetic alloys with improved mechanical properties for use in electrical generators. *Proc. 2nd European Physical Society Conf. Soft Magnetic Materials.* Wolfson Centre for Magnetics Technology, Cardiff, UK, pp. 53–60

Major R V, Samadian V 1989 Physical metallurgy and properties of a new high saturation Co–Fe alloy. *J. Mater. Eng.* 11: 27–30

Stanley J K, Yensen T D 1947 Hyperco—a magnetic alloy. *Trans. Am. Inst. Electr. Eng.* 66: 714–18

J. H. Wernick
[Bellcore, Red Bank, New Jersey, USA]

G. Y. Chin†
[AT&T Bell Laboratories, Murray Hill, New Jersey, USA]

J

Josephson Effects in Weak Links

In 1962, Josephson predicted that a zero-voltage supercurrent should be able to flow between two superconducting electrodes separated by a thin insulating barrier. He further predicted that if a voltage difference V were maintained across the junction, the current would be an alternating current of frequency $f = 2eV/h$, where e is the charge of the electron and h is Planck's constant. These predicted effects have been fully confirmed and now bear his name. Although his prediction was based on a microscopic theoretical analysis of the quantum mechanical tunnelling of electrons through the barrier layer, it is now clear that the effects are much more general and occur whenever two strongly superconducting electrodes are connected by a "weak link." As illustrated in Fig. 1, the weak link can be an insulating layer as Josephson originally proposed, a normal metal layer made weakly superconductive by the so-called proximity effect (see *Coherence Length, Proximity Effect and Fluctuations*), or simply a short narrow constriction in an otherwise continuous superconducting material.

Each weak link is characterized by a critical current I_c which is the maximum dc supercurrent it can carry and is also the amplitude of the ac supercurrent that flows if a voltage is applied. This critical current is a measure of how strongly the two superconducting electrodes are coupled through the weak link. The coupling free energy is $E_J = (\hbar/2e)I_c$ (where $\hbar = h/2\pi$) and depends on how thin and of what material the barrier is, or, in the case of constriction weak links, on the cross-sectional area and length of the neck. In most applications, I_c lies in the range of a microampere to a few milliamperes. The lower limit stems from the requirement that E_J exceed the energy of thermal fluctuations (i.e., that $I_c > 2ek_BT/\hbar$); otherwise, fluctuations wash out the supercurrent. (Although the numerical factor $2ek_B/\hbar$ is only $\sim 0.04\,\mu A\,K^{-1}$, I_c must be sufficient to overcome an effective noise temperature which may approach room temperature rather than the temperature of the superconductor, unless the electrical leads are very well screened.) At the other extreme are the weak links coupling together the strongly superconducting grains in a granular bulk superconductor such as the ceramic oxide high-temperature superconductors (see *Granular Superconductors*). Here, a critical current of the order of 1 mA between adjacent 1 μm diameter grains would be needed to obtain a useful bulk critical current density of order $10^5\,A\,cm^{-2}$.

1. Relation to Bulk Superconductivity

The relation of the Josephson effects to bulk superconductivity will now be examined, based on the Ginzburg–Landau (GL) theory of 1950 which is commonly used to discuss those superconducting phenomena that do not depend explicitly on the microscopic nature of the superconducting state (e.g., Tinkham 1985). In this theory, the superconducting electrons are described by a complex pseudowavefunction

$$\Psi(\boldsymbol{r}) = |\Psi(\boldsymbol{r})|\exp[i\varphi(\boldsymbol{r})] \tag{1}$$

with the physical interpretation that

$$|\Psi(\boldsymbol{r})|^2 = n_s/2 \tag{2}$$

is the number density of superconducting electrons. The supercurrent density can then be written

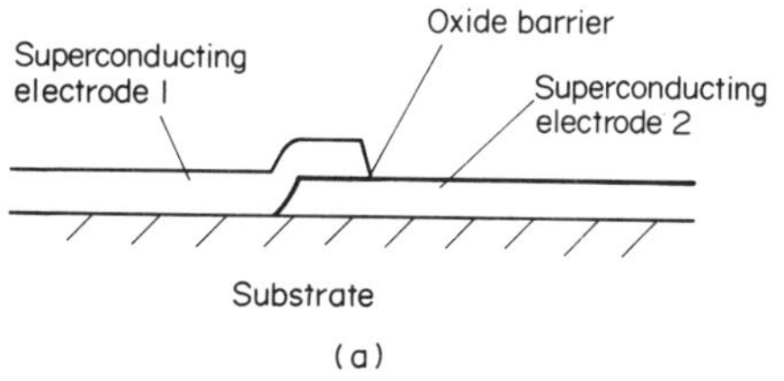

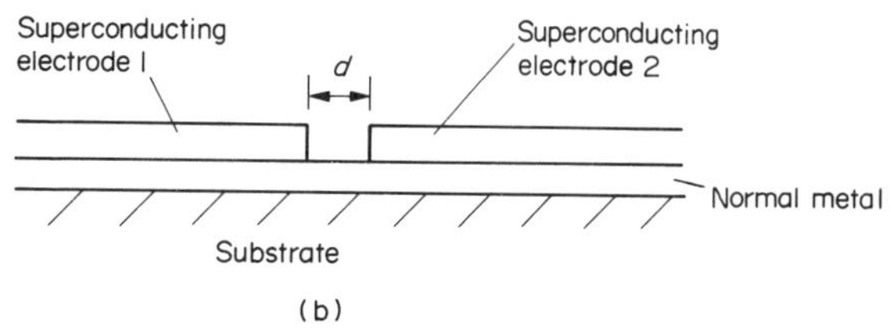

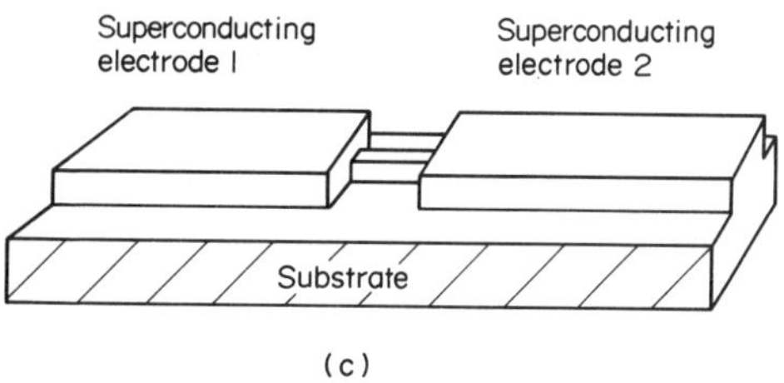

Figure 1
Schematic diagrams showing representative configurations of (a) SIS, (b) SNS and (c) ScS (constriction) weak links

$$J_s = n_s e v_s \tag{3}$$

where the supercurrent velocity is

$$v_s = \frac{\hbar}{2m}\left(\nabla\varphi - \frac{2\pi A}{\Phi_0}\right) \tag{4}$$

where $2m$ is the mass of a superconducting Cooper pair of electrons, A is the magnetic vector potential, and $\Phi_0 = h/2e = 2.07 \times 10^{-15}\,\mathrm{T\,m^2}$, the superconductive flux quantum. This expression for the supercurrent density is derived in the GL theory, but it is plausible simply by analogy with the expression for current in quantum mechanics (see *Magnetic Properties of Superconductors*). Both terms in Eqn. (4) are required to yield a unique value for v_s, since the forms of both $A(r)$ and $\varphi(r)$ depend on the arbitrary choice of gauge for A, but the combination appearing in Eqn. (4) is gauge invariant. If no magnetic field is present, it is usually convenient to set $A = 0$, in which case the phase $\varphi(r)$ is unique (*modulo* 2π) apart from an overall additive constant.

Now consider a one-dimensional current flow through a thin wire of uniform cross section; continuity of J_s in Eqn. (3) requires (for $A = 0$) that $d\varphi/dx \propto v_s \propto 1/n_s(x)$. Thus, if n_s drops near zero in a normal metal or insulating barrier region in the wire, the phase gradient in this region must become very large compared with that in the strongly superconducting regions where $n_s(x)$ is large. The same effect occurs if the material has a very narrow constriction which requires v_s to rise to compensate for a sharp reduction in the current carrying cross section. In both cases, if the link is weak enough, the phase gradient in it is so large that the phase gradient required to drive the current through the strongly superconducting regions can be neglected and the phase can be treated as constant in them. Thus, the current will be determined solely by the total *phase difference* $\Delta\varphi$ between the strong superconductors on either side of the weak link, not by details of the local *phase gradient* $\nabla\varphi$. This is the essential property of a generic Josephson junction.

2. The DC Josephson Effect

It is plausible that the phase gradient everywhere in the weak link should scale with the total $\Delta\varphi$, so that I_s will increase in proportion to $\Delta\varphi$ until some maximum supercurrent I_c is reached and superconductivity breaks down. Such behavior is indeed found in constriction weak links which are longer than the GL coherence length ξ (see *Superconducting Materials: Physical Properties*). However, if the link is shorter than about ξ, the internal phase gradient associated with a $\Delta\varphi$ outside the range from $-\pi$ to $+\pi$ would be energetically prohibitive, and the current is uniquely determined by the overall phase difference in this range *modulo* 2π. The simplest generalization of $I_s \propto \Delta\varphi$ which is bounded and periodic over 2π is $I_s \propto \sin(\Delta\varphi)$, in agreement with Josephson's result.

$$I_s = I_c \sin(\Delta\varphi) \tag{5}$$

(A more general Fourier expansion in $\Delta\varphi$ may be required if the link is not sufficiently short.) Finally, the magnetic field is introduced by defining a gauge-invariant phase difference

$$\gamma \equiv \Delta\varphi - \left(\frac{2\pi}{\Phi_0}\right)\int A.ds \tag{6}$$

where the integration is from one electrode of the weak link to the other, so that the general expression for the supercurrent in an ideal Josephson junction is

$$I_s = I_c \sin\gamma \tag{7}$$

The numerical value of the critical current I_c will depend on the details of the weak link, but it typically rises linearly with $(T_c - T)$ near T_c and saturates at a maximum value as $T \to 0$. In the case of an ideal tunnel junction between two identical superconductors, Ambegaokar and Baratoff showed that

$$I_c(T) = \left[\frac{\pi\Delta(T)}{2eR_n}\right] \tanh\left[\frac{\Delta(T)}{2k_B T}\right] \tag{8}$$

where R_n is the resistance of the junction in the normal state and $\Delta(T)$ is the BCS energy gap. Thus, at $T = 0$, ideally $I_c = \pi\Delta(0)/2eR_n$, where $\Delta(0) = 1.76\,k_B T_c$. Equation (8) also holds, as shown by Aslamozov and Larkin, in the GL approximation near T_c for a sufficiently short and narrow constriction weak link in metal with a short electronic mean free path l. Kulik and Omel'yanchuk have used more sophisticated theories which are valid all the way to $T = 0$ to work out $I_c(T)$ for short metallic constrictions. The results are qualitatively similar to Eqn. (8) but at $T = 0$ reach values which are 1.32 and 2 times greater than given by Eqn. (8) for the dirty and clean limits, respectively. An important property of Eqn. (8) is that the $I_c R_n$ product is an invariant for ideal junctions of a given material and temperature, so that a junction with high resistance in the normal state will have a low critical current in the superconducting state.

If the weak link is an SNS (superconducting–normal metal–superconducting) junction in which the thickness d of the normal barrier is *not* thin compared with the normal metal coherence length ξ_n, then I_c falls with the thickness of the normal layer approximately as $\exp(-d/\xi_n)$ while R_n increases only linearly with d, so the $I_c R_n$ product can fall far below the ideal value given by Eqn. (8). In the clean limit, $\xi_n^0 = \hbar v_F / 2\pi k_B T$, but if the electronic mean free path $l \ll \xi_n^0$, then the appropriate expression for ξ_n is the geometric mean of $l/3$ and ξ_n^0. The temperature dependence of I_c is also quite different from Eqn. (8) if $d > \xi_n(T)$, since I_c then rises exponentially with decreasing temperature as $\exp(-\alpha T)$ or $\exp(-\beta T^{1/2})$, depending on whether ξ_n is in the clean or dirty limit (see *Coherence Length, Proximity Effect and Fluctuations*).

2.1 Effect of Magnetic Field

A generalization of Eqn. (7) which is very important for applications is inclusion of the effect of an applied magnetic field. For simplicity, the effect of the field on the superconducting electrodes themselves is ignored, leaving only its effect on the gauge-invariant phase difference in the Josephson effect via the $\boldsymbol{A}$ term in Eqn. (6). To avoid sensitivity to the arbitrary gauge choice, only the gauge-invariant magnetic flux Φ through a specified contour is used. The basic element of the analysis is the summation of the supercurrent passed by two Josephson contacts (denoted 1, 2) between a pair of strong superconducting electrodes (denoted A, B) (see Fig. 2a). The line integral of $\boldsymbol{A}$ around a contour passing through both links and the interior of the electrodes gives Φ, the enclosed flux, since $\boldsymbol{B} = \nabla \times \boldsymbol{A}$. Because the electrodes are assumed to be strongly superconducting compared with the weak links, $\boldsymbol{v}_s$ in them can be taken to be zero. Then Eqn. (4) implies that $\boldsymbol{A} = (\Phi_0/2\pi)\nabla\varphi$ in the electrodes, so

$$\Phi = \oint \boldsymbol{A}.d\boldsymbol{s} = (\Phi_0/2\pi)\int_{\text{electrodes}} \nabla\varphi.d\boldsymbol{s} + \int_{\text{links}} \boldsymbol{A}.d\boldsymbol{s}$$

Because the phase φ must be single valued, $\int_{\text{electrodes}}\nabla\varphi.d\boldsymbol{s}$ plus the sum of the finite phase differences across the links must be zero, *modulo* 2π. From this, it follows that the sum of the gauge-invariant phase differences (taken in the same direction around the contour), or the *difference* in the phase differences γ_1 and γ_2 (if both are taken from electrode A to electrode B) must be $2\pi\Phi/\Phi_0$. Thus, unless Φ is an integral multiple of Φ_0, γ_1 and γ_2 cannot simultaneously be $\pi/2$ to give the greatest supercurrent, so the maximum supercurrent of the parallel combination must be less than $I_{c1} + I_{c2}$. In fact, if $I_{c1} = I_{c2} = I_c$, then the maximum supercurrent can easily be shown to be $I_{max} = 2I_c|\cos(\pi\Phi/\Phi_0)|$. This relation, plotted in Fig. 2a, is the basis of the dc SQUID magnetometer, which permits Φ to be measured to an extremely small fraction of Φ_0 (see *SQUID Design, Fabrication and Application*).

The same principle can be applied to determine the maximum supercurrent of an extended junction penetrated by flux Φ in the plane of the junction, as sketched in Fig. 2b. When screening by the junction supercurrents can be neglected, $\Phi \simeq HL(d + 2\lambda)$, where H is the ambient magnetic field, L, is the length of the junction perpendicular to the field, d is the thickness of the barrier and λ is the penetration depth of the field into the superconducting electrode on either side of the barrier. If the junction is rectangular, I_{max} is reduced from its zero-field value by a factor $|\sin(\pi\Phi/\Phi_0)/(\pi\Phi/\Phi_0)|$, a form often referred to as a Fraunhofer diffraction pattern by analogy with the case of light passing through a narrow rectangular slit. If the local critical supercurrent density $J_c(x)$ is not uniform, this simple result is replaced by a Fourier transform of $J_c(x)$; this provides a useful diagnostic technique. If the junction area is round instead of rectangular, a similar analysis yields an Airy diffraction pattern. Peterson and Ekin (1989) have argued that the Airy dependence $|J_1(x)|/(x/2)$ is more plausible than the simple $|\sin x|/x$ formula for modelling the reduction in the maximum supercurrent between grains in a granular high-temperature superconductor (HTS). In either case, it is presumed that the random areas of the intergranular junctions wash out the periodic modulation of I_{max}, leaving the envelope dependence $1/(\pi\Phi/\Phi_0) \propto 1/H$ for rectangular contacts, or the $1/H^{3/2}$ dependence for circular ones, to account, at least in part, for the strong reduction of the measured critical currents of granular HTS materials in a magnetic field.

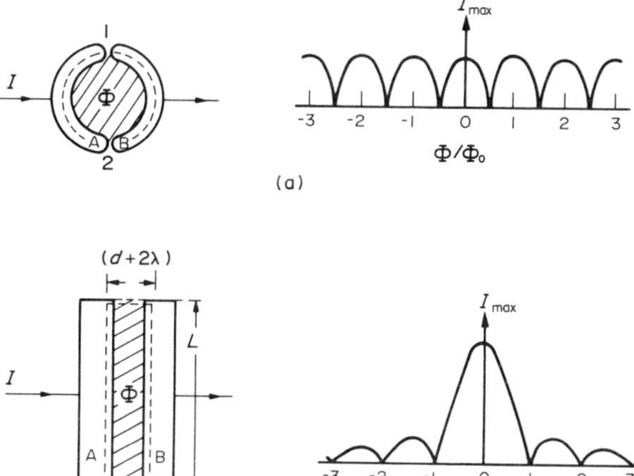

Figure 2
Modulation of I_{max} by magnetic field: (a) two-junction dc SQUID and (b) extended junction, showing Fraunhofer pattern (the graphs are accompanied by schematic sketches showing the areas within which the flux Φ is found)

3. *The AC Josephson Effect*

To discuss the ac Josephson effect, a second Josephson equation is needed to supplement Eqn. (7) by giving the time evolution of γ. If a voltage difference V exists between the two electrodes, transfer of a Cooper pair of charge $2e$ involves an energy change of $2eV$ which, by the Planck–Einstein relation, can be equated to a quantum energy $\hbar\omega$. This motivates the equation

$$\omega = \frac{d\gamma}{dt} = \frac{2eV}{\hbar} \tag{9}$$

which Josephson showed gives the relation between instantaneous frequency and voltage in a Josephson junction.

Although $I_c(T)$ suffices to characterize the zero-voltage dc properties of a weak link, a more complete description is required for the finite-voltage situations involved in the ac Josephson effect. This is usually provided by the resistively and capacitively shunted junction (RCSJ) model, in which the physical Josephson junction is modelled by an ideal one described by Eqn. (7), shunted by a resistance R and a capacitance C. The resistance R builds in dissipation in the finite-voltage regime, without affecting the lossless dc regime, while C reflects primarily the geometric shunting capacitance between the two electrodes. The appropriate magnitude of R will be of order R_n in an SNS junction or in an SIS (superconductor–insulator–superconductor) junction near T_c. At lower temperatures, however, the appropriate R for a high-quality SIS junction rises approximately as $R_n \exp(\Delta/k_B T)$ (for $V < V_g \equiv 2\Delta/e$) because of the freezing out of the so-called normal electrons or *quasiparticles*, which have excitation energies at or above the energy gap Δ. In this connection, it is useful to review the dc I–V curve (see Fig. 3b) which results from quasiparticle tunnelling alone (see *Tunnelling Spectroscopy*), if the Josephson supercurrent is suppressed by a strong magnetic field or by thermal fluctuations. For $V < V_g$, only thermally excited quasiparticles can tunnel, giving a high resistance; for $V > V_g$, the applied voltage can break Cooper pairs, giving a tunnel resistance comparable with R_n. The linear damping resistor R in the RCSJ model should be chosen to represent this highly nonlinear conductance as well as possible in the relevant voltage regime.

Within the RCSJ model, the time dependence of the phase γ in the presence of an externally supplied bias current can be derived by equating the total junction current to the bias current I as follows:

$$I = I_c \sin\gamma + \frac{V}{R} + C\frac{dV}{dt}$$

Eliminating V in favor of γ, the second-order differential equation

$$\frac{d^2\gamma}{d\tau^2} + Q^{-1}\frac{d\gamma}{d\tau} + \sin\gamma = \frac{I}{I_c} \qquad (10)$$

is obtained, in which a dimensionless time variable $\tau = \omega_p t$ is introduced, with $\omega_p = (2eI_c/\hbar C)^{1/2}$ being the so-called plasma frequency of the junction and the quality factor $Q = \omega_p RC$. (This Q is identical with $\beta_c^{1/2}$, where β_c is a frequently used damping parameter introduced by Stewart and McCumber.)

So long as $I < I_c$, a static solution of Eqn. (10) with $\gamma = \sin^{-1}(I/I_c)$ is allowed. However, if $I > I_c$, only time-dependent solutions exist and they determine even the dc I–V curve of the junction. For example, if C is small so that $Q \ll 1$, the dc I–V curve has the simple limiting form $V = R(I^2 - I_c^2)^{1/2}$, which smoothly interpolates between $V = 0$ for $I < I_c$ and Ohm's law $V = IR$ for $I \gg I_c$. It is worth noting, however, that when I only slightly exceeds I_c, the dc voltage is actually the time average of a series of pulses, occurring with frequency $f = 2eV_{dc}/h$, each with $V_{max} \simeq 2I_c R$ and $\int V\,dt = h/2e = \Phi_0$. Note that, with the Josephson frequency relation of Eqn. (9), this pulse area implies a total phase slip by 2π. This phase slip occurs as γ increases from around $\pi/2$, where the supercurrent is maximum (but still $< I$), up through larger values of γ which yield less supercurrent and, hence, require more normal current, requiring in turn higher voltage (and, hence, higher $d\gamma/dt$) to drive it, until $\gamma \simeq 5\pi/2 (= \pi/2$ *modulo* $2\pi)$ and a low value of $d\gamma/dt$ is reached again. This concept of a phase slip by 2π being the elementary resistive process in superconductors is a very important one.

When a thermal noise current is included as an additional drive term in Eqn.(10), Ambegaokar and Halperin (1969) showed that the simple I–V relation for an overdamped junction is fundamentally modified. In particular, as shown in Fig. 3a, they found that there is always a *finite* resistance, even below I_c, because of thermally activated phase-slip processes. Although this resistance increases as I approaches I_c, it has a nonzero limiting value R_0 as $I \to 0$, which is related to the normal state resistance by

$$R_0/R_n = [I_0(u/2)]^{-2} \propto u\exp(-u) \qquad (11)$$

where $u = \hbar I_c/ek_B T$ is the normalized activation energy, I_0 is the modified Bessel function and the exponential dependence holds when $u \gg 1$. This exponentially small linear resistance at low currents is the single-junction analog of the so-called TAFF (thermally activated flux flow) resistance in bulk superconductors (see *Flux Creep Phenomena*; *Resistive Transition and Flux Flow in Superconducting Materials*), in which the activation energy is thought to be the energy to move a fluxon rather than for phase slip in a single junction (Tinkham 1988).

When C is large enough so that $Q > 1$, the I–V curve becomes hysteretic. Neglecting thermally activated processes, upon increasing I from zero, $V = 0$ until I_c, at which point V jumps discontinuously up to a finite voltage V, corresponding to a "running state" in which the phase difference γ increases steadily at the rate $2eV/\hbar$. In the simple RCSJ model this $V \simeq I_c R$, but in an ideal tunnel junction at $T \ll T_c$ the voltage jumps up to near the energy gap voltage $V_g = 2\Delta/e$ (Fig. 3b). If I is now reduced below I_c, V decreases somewhat, but it does not drop back to zero until a "retrapping current" $I_r \simeq 4I_c/\pi Q$ is reached. The physical reason for the $1/Q$ dependence is that the energy stored in the capacitance during the high-voltage (phase slip) part of the cycle must be dissipated in about one cycle of the oscillation of γ in order to prevent another phase slip from occurring and, thus, to allow the system to retrap back into the zero-voltage state.

If an ac drive current of frequency ω_{rf} is added to the dc bias current, constant voltage steps (the Shapiro steps in Fig. 3c) appear on the I–V curve at

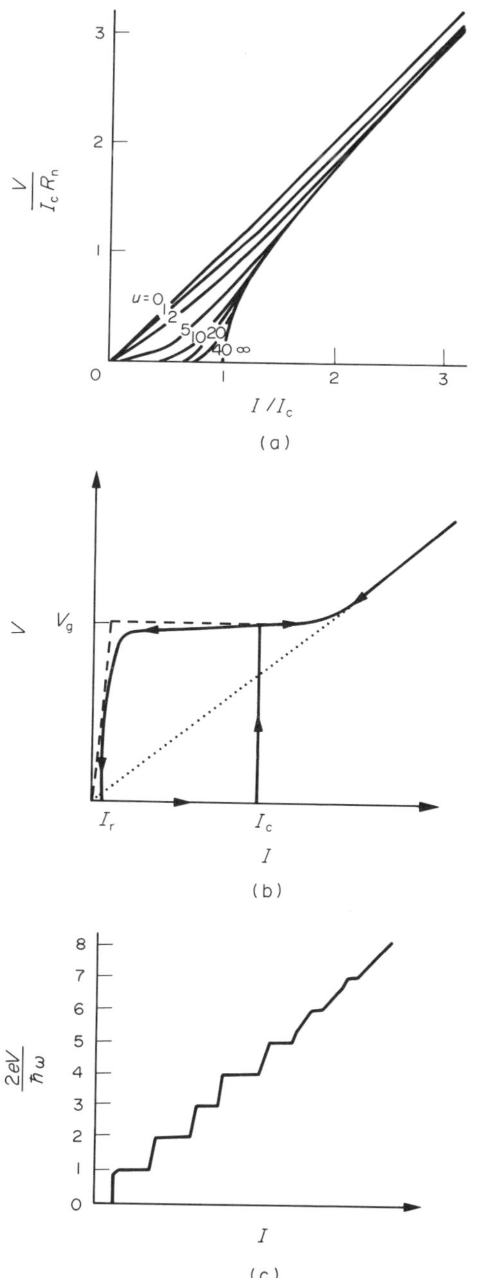

Figure 3
Schematic I–V curves of Josephson weak links: (a) overdamped junction, showing the effect of thermal activation according to Ambegaokar and Halperin, for various values of $u = \hbar I_C/ek_BT$, (b) underdamped (hysteric) junction, also showing the underlying quasiparticle tunnelling characteristic (dashed line) and the normal state I–V curve (dotted line) and (c) junction in presence of rf drive, showing Shapiro steps

voltages $V_n = n(\hbar\omega_{rf}/2e)$ at which the (angular) frequency $2eV/\hbar$ of the Josephson current synchronizes with harmonics of the drive frequency (see *Millimeter-Wave Detection*). (These Shapiro steps should not be confused with the Fiske steps, which occur at voltages for which the Josephson frequency is harmonically related to the frequency of one of the internal electromagnetic resonance modes of the junction.) If the rf drive is effectively voltage biased, the half width of the nth Shapiro step can be shown to be $I_c J_n(2eV_{rf}/\hbar\omega_{rf})$, where J_n is a Bessel function. For rf current bias, numerical solutions are required to determine the step widths; they are found to oscillate with drive amplitude in a manner qualitatively resembling the Bessel function dependence of the voltage-bias case. Since the voltages of these Shapiro steps are determined by only fundamental constants and the drive frequency (which can be known extremely accurately), they have been adopted as the basis for a voltage standard. High-capacitance junctions are typically used for this application but certain parameter combinations must be avoided for which the current-biased case leads to *chaotic* or nonperiodic solutions of Eqn. (10) rather than the phase-locked periodic solutions which lead to ideal constant voltage steps (see *Voltage–Standard Devices*).

See also: Coherence Length, Proximity Effect and Fluctuations; Superconducting Device Fabrication; Superconducting Materials: BCS and Phenomenological Theories; Thin Films, Multilayers and Devices, Superconducting

Bibliography

Ambegaokar V, Halperin B I 1969, Voltage due to thermal noise in the dc Josephson effect. *Phys. Rev. Lett.* 22: 1364–6
Barone A, Paterno G 1982 *Physics and Applications of the Josephson Effect*. Wiley, New York
Josephson B D 1965 Supercurrents through barriers. *Adv. Phys.* 14: 419–51
Likharev K K 1979 Superconducting weak links. *Rev. Mod. Phys.* 51: 101–59
Peterson R L, Ekin J W 1989 Airy pattern, weak-link modeling of critical currents in high-T_c superconductors. *Physica C* 157: 325–33
Ruggiero S T, Rudman D A (eds.) 1990 *Superconducting Devices*. Academic Press, San Diego, CA
Tinkham M 1985 *Introduction to Superconductivity*. Krieger, Malibar, FL
Tinkham M 1988 Resistive transition in high-temperature superconductors. *Phys. Rev. Lett.* 61: 1658–61
Tucker J R, Feldman M J 1985 Quantum detection at millimeter wavelengths. *Rev. Mod. Phys.* 57: 1055–113
Waldram J R 1976 The Josephson effects in weakly coupled superconductors. *Rep. Prog. Phys.* 39: 751–821

M. Tinkham
[Harvard University, Cambridge, Massachusetts, USA]

M

Magnet Wire

Magnet wire consists of a solid electrical conductor, with a flexible coating or covering of solid insulation which, when wound into coil form and energized, produces a magnetic field. Magnet wire finds application in transformers, motors, generators, electromagnets and other electrical equipment.

1. Conductors

Copper has traditionally been used as the conductor for magnet wire because of its good electrical conductivity, its relative abundance and its excellent physical properties. Aluminum conductors were introduced in the 1950s as a lower-cost alternative. The conductivity of aluminum is only 60% of that of copper and thus direct replacement of copper by aluminum in a given device requires an aluminum conductor of much larger cross section. This often means an enlarged design to make space for the larger wire. Both copper and aluminum must be annealed in the magnet as used; this is often accomplished in the wiremaking machine for smaller sizes of wire, but requires preannealing for larger wires.

Where special characteristics are required, the copper is often coated with other metals. For example, tin is used where rapid soldering is desired and nickel is used to protect copper from oxidation at high temperatures.

Copper and aluminum wires are produced in a large range of circular, square and rectangular sizes (e.g., circular, sizes are 0.0124–11.68 mm for copper and 0.404–5.0 mm for aluminum).

2. Polymer Coatings

Magnet-wire coatings are usually applied in a wet state, as a polymer solution. The coating process involves forming a large number of coats, each being heated to a baked state on the wire. The final thickness or build (double thickness) is offered in four recognized categories. In the USA these are referred to as single, heavy, triple and quadruple builds.

There are over 25 different polymer combinations used in magnet wire (Table 1). The polymers are arranged in temperature classes according to ASTM specification D2307 (ASTM 1968). Certain polymers appear alone as "sole-coat" enamels. Others appear as base-coat enamels with top-coat polymers (those shown after the slash marks in Table 1). Various property advantages are gained by use of overcoat enamels. Thus polyamides (e.g., nylon) add toughness, lubricity and windability to base coats. A polyamide–imide overcoat adds toughness and also gives a higher thermal stability to polyester base coats. Poly(vinyl butyral) and polyamides both add mechanical strength and are solderable along with such solderable bases as acrylic and polyurethane. Some of these overcoat polymers such as epoxy, polyester and poly(vinyl butyral) are bondable by direct application of heat or solvent (Fink and Beaty 1978).

The major polymers used in coatings are poly(vinyl acetal) (Formvar), polyurethane, polyamide (nylon), polyester (poly(ethylene terephthalate)), THEIC (tris(2-hydroxyethyl)isocyanurate) polyester, polyester imide, polyamide imide, polyester amide imide and polyimide. THEIC polyester is produced by a cyanuration process which improves thermal endurance. Imidization also improves thermal endurance (see Sect. 5.5).

3. Fiber and Film Coverings

Fiber and film coverings are served or wrapped onto magnet wire in a continuous butt-lap or overlap fashion. The butt-lap mode produces few or no ridges on the surface, promoting a closer packing of turns in a winding. The overlap mode produces better dielectric strength, especially with film-tape wrap (Saums and Pendleton 1973).

The early fiber wraps (e.g., cotton, silk, nylon thread and paper) have largely been superseded by synthetic fibers. An exception is paper, as used in oil for high-voltage windings. The newer fibers are

Table 1
Magnet-wire coatings

Class	Coating
105	acrylic, acrylic/nylon, nylon, Formvar, Formvar/nylon, oleoresinous, polyurethane, Formvar–isocyanate, Formvar/poly(vinyl butyral), cellulose acetate
130	epoxy, epoxy/poly(vinyl butyral), epoxy/polyester, epoxy/epoxy cement, polyurethane/nylon/poly(vinyl butyral)[a]
155	polyethylene terephthalate, THEIC polyester, solderable polyester, polyester/nylon, polyester/polyester bond
180	polytetrafluoroethylene, polyester imide, polyester amide imide
200	THEIC polyester/polyamide imide, polyester imide/polyamide imide, polyester amide imide/polyamide imide
220	polyamide imide, polyimide

a Cement coat over nylon

fiberglass, polyester (Dacron) and aromatic polyamide (e.g., Nomex, a high-temperature polyamide). Nomex, cellulosic paper and asbestos paper are forms of fibers that are readily applied to the larger magnet wires. The films are cellulose acetate, polyester (Mylar) and polyimide (Kapton) materials.

The largest group of coverings for magnet wire is that of fiberglass. It is useful on medium and larger circular sizes as well as with all square and rectangular wires. When bonded with phenolic resin, fiberglass is used as a covering for class 130 applications. With silicone bonding, fiberglass coverings are used for classes 180–200 and with polyamide bonding they are used for class 220 applications.

Nomex-paper-covered magnet wire is useful up to 220 °C; however, Nomex is most frequently used as ground insulation in such applications as dry-type transformers and high-temperature motor slot cells and phase insulation.

Polyimide (Kapton) film wrap is a very highly heat-resistant covering used primarily for square and rectangular wire and large circular wire (Saums and Pendleton 1973).

Asbestos paper treated with a binder to minimize moisture problems has a proven capability for ground insulation in classes 180, 200 and 220. It has not been used as a magnet-wire insulation to any great extent, however.

4. Composite Insulation

Composite insulation for magnet wire takes two forms: dual-fiber coverings and fiber-over-film coatings. Dacron is used with fiberglass as dual-fiber wrap with a bonding resin to increase scuff resistance. It is also possible to heat the dual-fiber wrap sufficiently to partially melt the Dacron to form a bond for the glass fibers. This latter combination may or may not have a binding resin added.

Fiber-over-film coating comprises fiberglass or Dacron–fiberglass as the outer wrapper and a polymer film as the inner coating. This fiber–film combination provides the high dielectric strength of the film and the long-time spacing safety factor of the fiber. Many of the polymer combinations in Table 1 can be used in this insulation structure on magnet wire (Fink and Beaty 1978).

5. Types of Magnet Wire

5.1 Ultrafine Enamelled Magnet Wire

The ultrafine wire category comprises circular sizes from 0.0447 mm down to 0.0124 mm. These magnet wires are used in extremely small coils for watches and electronic applications. They are available in six enamel types: oleoresinous, polyester, polyurethane, poly(vinyl acetal), polyimide and poly(vinyl butyral) as a bond coat over a polyurethane base coat. They require dustfree wire enamelling, handling, storage, winding and treating.

5.2 Self-Lubricated Magnet Wire

Most magnet wires are sold with either a light oil or a fine wax added to the surface to reduce the coefficient of friction between the turns. By adding fillers with a low coefficient of friction to the coating polymer, the coated wire becomes "self-lubricated"; that is, it possesses a low coefficient of friction in the winding operation. As a result, the wire can be wound into a denser coil with smaller dimensions, reduced wire length and greater operating efficiency.

5.3 Self-Bonding Magnet Wire

Top-coat polymers with properties of softening and flowing under heat or solvent action provide a self-bonding action for cementing turns of a coil together. Such action makes coils rigid without the postvarnishing and baking that are normally required. Deep bonding in a large-cross-section coil is attained without resorting to vacuum impregnation. Some polyester bond coats achieve a degree of crosslinking during the melt stage to prevent softening in later use.

5.4 Solderable Magnet Wire

Certain polymers, such as polyamide, polyurethane, acrylic and some polyesters, have the property of complete depolymerization and evaporation at soldering temperatures (350–450 °C). These coatings are very useful in small coils where insulation removal is undesirable economically or mechanically.

5.5 Thermally Stable Magnet Wire

Using the AIEE/IEEE #57 test (American Society for Testing and Materials 1968) as reference, polymer chemists and research engineers have searched since the 1950s for thermally stable magnet-wire enamels. Six methods have evolved.

(a) Fluoridization produces such polymers as polytetrafluoroethylene and polychlorotrifluoroethylene, both lacking in the mechanical properties needed for magnet-wire coatings.

(b) Cyanuration (THEIC) is successful with polyester (poly(ethylene terephthalate)). The THEIC-modified polyester has a 25 °C higher thermal index than the unmodified polyester when used with an amide–imide overcoat.

(c) Imidization, as performed with polyester imide, polyester amide imide and polyamide imide, also raises the thermal index. Even without top coats the first two polymers are 25 °C higher in thermal index than unmodified polyester, whereas polyamide imide is 45 °C higher (i.e., 115 °C higher than that of the unmodified polyamide). Polyimide is the totally imidized polymer in the family and has exhibited thermal indices in the region of 250 °C. Thus, imidization has proved highly successful in providing thermally stable wire enamel coatings.

(d) Change of conductor metal from copper to aluminum produces an increase of 15–20 °C in the thermal index. Some enamels are more sensitive to this change than others. The US National Electrical Manufacturers Association recognizes a 220 °C level for polyamide imide-overcoated THEIC polyester on aluminum and a 200 °C rating for the same coating on copper (National Electrical Manufacturers Association 1984).

(e) Use of overcoating enamels, particularly polyamide imide, has proved to be a compromise step. THEIC polyester with a polyamide imide overcoat produces a 200 °C thermal index on copper while maintaining desirable properties and reasonable costs. The overcoat adds both to the mechanical properties and to the thermal stability.

(f) Silicones were a development of the 1950s. While proving successful for fiberglass bonding, silicones for film coating have been found to require extensive modification to meet specifications for magnet-wire enamels. Blending of the silicone resin with polyesters was carried out to the extent that very little silicone remained. Indeed, this was the industry's early look at polyester (poly(ethylene terephthalate)) as a magnet-wire enamel. In the 1960s these silicones were used unmodified with a glass frit and ceramic powder to make the space-age 600 °C magnet-wire coating (Pendleton and De Vries 1962, Pendleton 1963, 1966).

5.6 *Multiple Magnet Wire*

Twinned, triplet and Litz wire (many strands of magnet wire interwoven or transposed to minimize the skin effect at high frequencies) are available. The bifilar and trifilar types are usually bonded in parallel form.

5.7 *Magnet Strip*

Long metal sheets or strip, bare or insulated, are wound into spiral coils for the low-voltage legs of power transformers. If bare, the strip is interleaved with cellulosic paper, Nomex paper, asbestos paper, mica paper, or one of several polymer films. Strip-wound coils have excellent heat-dissipation properties.

5.8 *Hollow-Conductor Magnet Wire*

Hollow conductors with magnet-wire insulation are used with a flowing liquid coolant in the center where a severe heating problem exists (Saums 1965).

See also: Multifilamentary Superconducting Composites

Bibliography

American National Standards Institute 1968 *Magnet Wire*, ANSI Standard No. C-9. 100. ANSI, New York

American Society for Testing and Materials 1968 *Relative Thermal Endurance of Film-Insulated Round Magnet Wire*, ASTM Specification D2307. ASTM, Philadelphia, PA

Brancato E L, Johnson L M, Walker H P 1959 Functional evaluation of motorette insulation systems. *Electr. Manuf.* 93(3): 146–53

Clark F M 1962 *Insulating Materials for Design and Engineering Practice*. Wiley, New York, pp. 766–87

Department of the Navy 1970 *Magnet Wire, Enamelled Round Copper*, MIL-W-583-C. Department of the Navy, Philadelphia, PA

Devins J C 1984 *Annual Report, Conf. Electrical Insulation and Dielectric Phenomena*. IEEE, New York

Fink D G, Beaty H W 1978 *Standard Handbook for Electrical Engineers*, 11th edn. McGraw-Hill. New York

Goba F A 1969 Bibliography on thermal aging of electrical insulation. *IEEE Trans. Electr. Insul.* 4(2): 31–58

Institute of Electrical and Electronics Engineers 1960 *Tentative Method of Test for Thermal Stability of Film Insulated Round Magnet Wire*, IEEE Standard No. 57. IEEE, New York

Institute of Electrical and Electronics Engineers 1983 *Proc. 16th Electrical/Electronics Insulation Conf.* IEEE, New York

National Electrical Manufacturers Association 1984 *Magnet Wire, Enamelled Round Copper*, MW-1000. NEMA, Washington, DC

Paloniermi P, Kadotani K, Kako Y, Matsunobu K, Aki F 1983 Voltage–endurance curves for combined thermal and electrical aging of coil insulation. *IEEE Trans. Electr. Insul.* 18(5): 553–5

Pendleton W W 1963 Advanced magnet-wire systems. *Electro-Technol.* 72(4): 95–102

Pendleton W W 1966 Electrical conductors for high temperature operation. *Wire Prod.* 41: 908–12; 953–5

Pendleton W W, De Vries D H 1962 New high temperature magnet wires and their application in rotating equipment. *Proc. 5th Electrical/Electronics Insulation Conf. AIEE.* American Institute of Electrical Engineers, New York, pp. 36–9

Saums H L 1965 Magnet wire, strip, hollow conductors, and super conductors. In: Samuelson L (ed.) 1965 *Insulation, Directory/Encyclopedia*. Lake, Chicago, IL

Saums H L, Pendleton W W 1957 Enameled magnet wire. *Electr. Manuf.* 60(4): 129–48

Saums H L, Pendleton W W 1958 Evaluation and applications of fiber-insulated magnet wire. *Electr. Manuf.* 61(2): 93–103

Saums H L, Pendleton W W 1973 *Materials for Electrical Insulating and Dielectric Functions*. Hayden, New York

Saums H L, Pendleton W W, Olson E H 1956 Compatibility tests of enameled wires in closed systems. *Proc. Conf. AIEE.* American Institute of Electrical Engineers, New York, Paper 560

W. W. Pendleton
[Muskegon, Michigan, USA]

Magnetic Alloys: Intermediate Coercivity Materials

Permanent magnet alloys of Co–Fe–Mo, Co–Fe–V, Co–Ni–Fe and Cu–Ni–Co with coercivities in the

10–50 kA m^{-1} range can be considered intermediate coercivity alloys. These lie between the semihard magnets with coercivities between 1 kA m^{-1} and 10 kA m^{-1} and the higher coercivity magnets such as alnicos, ferrites and rare earths. Generally, the magnetic hardening in these intermediate coercivity alloys is due to a solid-state phase decomposition, either through precipitation of a second phase in the case of Co–Fe–Mo and Co–Fe–V, or through a spinodal decomposition in the cases of Cu–Ni–Fe and Cu–Ni–Co.

1. Cobalt–Iron–Molybdenum

The Co–Fe–Mo permanent magnet alloys are significant from both a historical and a commercial standpoint. Historically, these alloys were the first synthetic permanent magnets free of carbon. They were developed in the early 1930s, foreshadowing the development of the much more important alinicos. Commercially, these alloys, known as Remalloy, were quickly developed for full-scale usage in telephone receivers right up to 1980. During the height of the usage over 450 000 kg of 20 Remalloy (20%Mo–12%Co–68%Fe) were produced annually.

The Remalloy family generally contains 12% cobalt and molybdenum in the 15–20% range. Increasing the molybdenum content lowers the remanence but increases the coercivity. Magnetic hardening is developed by oil quenching from 1250 °C followed by aging at 675 °C. Being hot workable, the material can be formed into the desired shape above 1200 °C. This was the method used in the production of cup-shaped telephone receiver magnets. Values for the 15% and 20% molybdenum Remalloys are shown in Table 1.

Commercial production of Remalloy essentially stopped by mid-1980 as a result of its replacement by the more economical and ductile Fe–Cr–Co magnets, Chromindur (see *Iron–Chromium–Cobalt Alloys*).

2. Cobalt–Iron–Vanadium

The Co–Fe–V alloy family containing about 50% cobalt ranges from magnetically soft (2% vanadium, called 2V-Permendur and Supermendur) to semihard (6–16% vanadium, called Vicalloy). In addition to modifying the magnetic hardness, vanadium renders ductility to the generally brittle cobalt–iron alloys in this composition range.

Table 1
Magnetic properties of Co–Fe–Mo permanent magnet alloys

Alloy (wt%, balance Fe)	B_r (T)	H_c (kA m^{-1})	$(BH)_{max}$ (kJ m^{-3})
15 Mo, 12 Co	0.97	20	8.0
20 Mo, 12 Co	0.90	28	10.4

Table 2
Magnetic properties of Co–Fe–V permament magnet alloys

Alloy (wt%, balance Fe)	B_r (T)	H_c (kA m^{-1})	$(BH)_{max}$ (kJ m^{-3})
10 V, 52 Co (Vicalloy I)	0.75	20	64
14 V, 52 Co (Vicalloy II wire)	1.00	42	28
8 V, 53 Co, 4 Cr (Koerflex 300 wire)	1.15	30	25
8 V, 53 Co, 4 Cr (Koerflex 300 wire)	1.05	24	15
5 V, 53 Co, 4 Cr (Koerflex 200)	1.30	10	8.8

The Vicalloys were discovered in 1938 by Kelsall and Nesbitt. The most widely used commercial composition today, called Vicalloy I, contains 10% vanadium, 52% cobalt and 38% iron. The material is usually hot worked to suitable size, quenched from 1200 °C and cold worked to thin sheets or fine wires if desired, followed by aging at 600 °C to develop coercivity. Typical magnetic properties are shown in Table 2.

Metallurgically, Vicalloy I exists in the face-centered-cubic γ phase at high temperature. Quenching converts this phase, to the body-centred-cubic α phase. Aging at 600 °C precipitates some γ phase to produce an $\alpha + \gamma$ mixture. If the vanadium content is raised to 12–14%, the high-temperature γ phase is retained on cooling and the material is only weakly magnetic. Cold deformation, however, converts the γ phase to α. Aging of the cold-worked alloy in the 600 °C range again produces an $\alpha + \gamma$ mixture with concomitant magnetic hardening. These phase relations are actually very complex. The reader is referred to studies by Pinnel and Bennett (1973) and Mahajan et al. (1974).

The sequence of quenching, cold working and aging is known as the Vicalloy II treatment, in contrast to Vicalloy I where cold working is unnecessary. Nowadays, Vicalloy I and II are labels applied to the ~10% and ~12–14% vanadium compositions, respectively, even though the 10% vanadium alloy is generally cold worked prior to aging. Properties of a 14% vanadium Vicalloy II are shown in Table 2.

In Germany, Fahlenbrach and his co-workers have extended earlier work to include partial substitution of vanadium by chromium. These are known as Koerflex alloys. Properties of several Koerflex alloys are shown in Table 2.

Because Vicalloy is ductile, it has been made into thin sheets and fine wires for such uses as armatures for telephone receivers and antitheft labels.

Table 3
Magnetic properties of Cu–Ni–Fe and Cu–Ni–Co permanent magnet alloys

Alloy (wt%)	B_r (T)	H_c ($kA\,m^{-1}$)	$(BH)_{max}$ ($kJ\,m^{-3}$)
60 Cu, 20 Ni, 20 Fe	0.55	42	11.0
50 Cu, 21 Ni, 29 Co	0.34	53	6.4
35 Cu, 24 Ni, 41 Co	0.53	36	8.0

3. Copper–Nickel–Iron and Copper–Nickel–Cobalt

Cu–Ni–Fe and Cu–Ni–Co permanent magnet alloys are among the few alloys that are ductile, that is, they can be extensively worked. The Cu–Ni–Fe alloys, in particular, can be formed even after magnetic hardening.

The most widely used commercial Cu–Ni–Fe permanent magnet alloys is called Cunife, with a nominal composition of 20% iron, 20% nickel and 60% copper. The permanent magnet properties are developed from quenching from 1040 °C, tempering at 650 °C, followed by cold drawing or rolling and aging at 650 °C (McCaig 1977 p. 151). A large amount of deformation prior to aging is necessary to develop the best permanent magnet properties. Typical properties are listed in Table 3.

The permanent magnet state of Cunife is attributed to spinodal decomposition of the high-temperature γ (fcc) phase. The cold formability permits its use in speedometers and timing motors where parts can be precision stamped. On the other hand, the cast ingot has severe segregation problems and cannot be hot worked. For this reason, the ingot size is limited to diameters of about 3 cm.

The Cu–Ni–Co alloys are similar to the Cu–Ni–Fe alloys. The magnetic properties of two typical alloys, Cunico I (29%Co–21%Ni–50%Cu) and Cunico II (41%Co–24%Ni–35%Cu) are shown in Table 3. These are no longer produced commercially.

See also: Alnicos; Iron–Chromium–Cobalt Alloys; Magnetic Materials: An Overview; Magnetic Materials: Basic Concepts

Bibliography

Bozorth R M 1951 *Ferromagnetism*. Van Nostrand, New York
Fahlenbrach H 1970 Fundamental studies on Vicalloy alloys. *Cobalt* 49: 196–205
Heck C 1974 *Magnetic Materials and Their Applications*. Crane-Russak, New York
McCaig M 1977 *Permanent Magnets in Theory and Practice*. Wiley, New York
Mahajan S, Pinnel M R, Bennett J E 1974 Influence of heat treatments on microstructures in an Fe–Co–V alloy. *Metall. Trans.* 5: 1263–72
Pinnel M R, Bennett J E 1973 The metallurgy of Remendur: effects of processing variations. *Bell Syst. Tech. J.* 52: 1325–40

G. Y. Chin[†]
[AT&T Bell Laboratories, Murray Hill, New Jersey, USA]

J. H. Wernick
[Bellcore, Red Bank, New Jersey, USA]

Magnetic Energy Storage

One of the potentially most beneficial applications of superconductivity for the energy producing and consuming world is superconducting magnetic energy storage (SMES). The basic purpose of SMES is to optimize the use of electrical energy by providing an extremely efficient means of storing reserve energy for retrieval during periods of greater demand. In this application, the zero resistance feature of superconductivity provides the means of preserving energy in the form of a dc electric current, or an equivalent magnetic field, without dissipation or loss of that energy. The principal benefit of SMES from an electrical utility standpoint is that additional and relatively inexpensive energy produced at night by baseline generators can be stored and retrieved during daytime peak demand periods in place of generating additional energy through the costly means of peaking generators. Thus, SMES provides the means for diurnal load levelling.

1. Research Efforts

Production and operating cost considerations reveal that SMES is most attractive for storage capacities of 1000 MW h or greater. However, at these capacities the size of the magnet becomes significant. For example, a 5000 MW h solenoid requires a radius of 500 m and a height of 30 m. Designs for such large magnets have been addressed by a number of extensive research and development programs, the first and longest running having begun at the University of Wisconsin–Madison in 1970 (Boom and Peterson, 1972, Boom et al. 1988). Other programs have been carried out by the Los Alamos Scientific Laboratory during the late 1970s and early 1980s (Rogers 1981), by Bechtel beginning in the early 1980s (Loyd et al. 1986) and in Japan since the mid-1970s (Masuda and Shintomi 1987). Present activities in the USA center around a program funded by the US Department of Defense to develop a scaled down engineering test model (ETM) version of SMES (Verga 1990). This 20 MW h SMES unit is expected to be constructed by

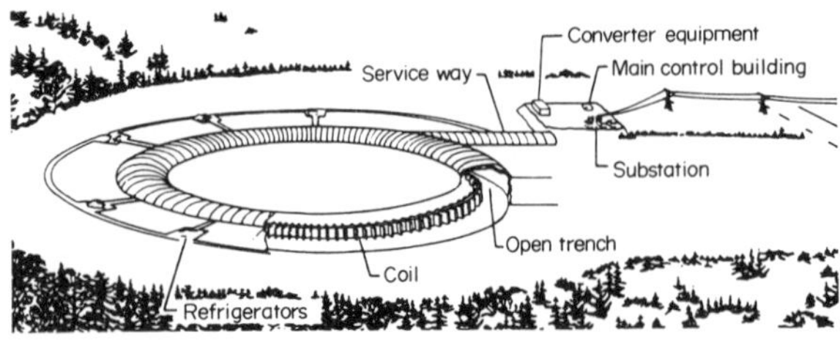

Figure 1
Artist's sketch of test SMES system (after Verga 1990)

1995 and will have the additional feature of availability for use as a large, quick power source for the ground-based systems of the US Strategic Defense Initiative.

2 SMES Design Issues

A number of issues must be addressed in the design of an SMES system, for example, the overall magnet style, the conversion of electricity between ac and dc circuits, the mechanical structure, the materials and design for the conductor, including the superconducting strand and the stabilizer, and the use of liquid helium for conductor stability. The manner in which each of these issues influences the overall SMES system design is detailed in the following sections. An artist's sketch of a probable test SMES unit is shown in Fig. 1.

2.1 Magnet Style

Through the use of the virial theorem it has been shown that any type of superconducting energy storage magnet supported solely by cold structure will be economically unfeasible. Therefore, the first condition for the "best" shape of a magnet is one which makes optimal use of support from inexpensive warm structure such as rock or soil.

Detailed optimizations of various superconducting energy storage magnets reveal that low-aspect-ratio solenoids are the preferred style. Further, the shape of the solenoid can be optimized by examining the expressions for the surface area A, radial force F_r and ampere meters IS:

$$A = Q_A(\beta)E^{2/3}/B_M{}^{4/3} \quad (1)$$

$$F_r = Q_{F_r}(\beta)E^{2/3}B_M{}^{2/3} \quad (2)$$

$$IS = Q_{IS}(\beta)E^{2/3}/B_M{}^{1/3} \quad (3)$$

These expressions are derived from the fundamental expressions for the energy E, inductance L and mid-plane field B_M:

$$E = \tfrac{1}{2}LI^2 \quad (4)$$

$$L = 2\pi\mu_0 n^2 R^2 lK(\beta) \quad (5)$$

$$B_M = \mu_0 nIK'(\beta) \quad (6)$$

In these equations β is the aspect ratio given by the solenoid length l divided by the radius R, n is the number of turns per unit length and $K(\beta)$, $K'(\beta)$, $Q_A(\beta)$, $Q_{F_r}(\beta)$ and $Q_{IS}(\beta)$ are all geometric factors dependent on β. Various costs such as materials and construction of dewar walls and excavation are proportional to surface area A. The costs of radial struts communicating the Lorentz forces to the surrounding soil are proportional to F_r and the conductor costs are proportional to required ampere meters IS. Eqns. (1–3) show that each of these costs is proportional to $E^{2/3}$ and to its respective geometric function $Q(\beta)$. Closer examination of the individual functions $Q(\beta)$ reveals an optimum β range of less than 0.1. Other cost considerations influence the magnet design as well, but this abbreviated explanation demonstrates that low–aspect–ratio solenoids are the preferred magnet type for SMES.

2.2 Electrical Conversion Bridge

One of the keys to the overall efficiency of the SMES system is the conversion of ac to dc power and back again. As proposed by Boom and Peterson (1972), the Graetz bridge circuit accomplishes this task and provides round trip efficiencies of better than 95%. It should be noted that in the SMES system electrical energy is stored as electrical energy, while in other energy storage systems, such as pumped hydro, batteries or flywheels which have efficiencies near 70%, the round trip process requires a conversion in the type of energy. The large efficiency of SMES in turn provides an economical means for storage of large quantities of energy. A second benefit realized by the electrical conversion system is that the change from storing to retrieving of energy can be made in less than 35 ms, thus giving the SMES system a great degree of dynamic response. This feature allows for excellent load following and for immediate resource availability.

2.3 Mechanical Structure

Primary considerations governing the type of structure for low-β solenoids are: (a) sufficient mass for axial compression loading, (b) a means of accommodating differential thermal contraction on cool down and (c) sufficient communication of the radial Lorentz forces to the surrounding soil or rock. The first of these requirements is typically met by using high-strength aluminum alloys. The aluminum also serves the additional purpose of providing the heat-absorbing mass during an emergency shutdown. If a growing resistive zone or some other emergency fault is detected in an SMES, protection of all components

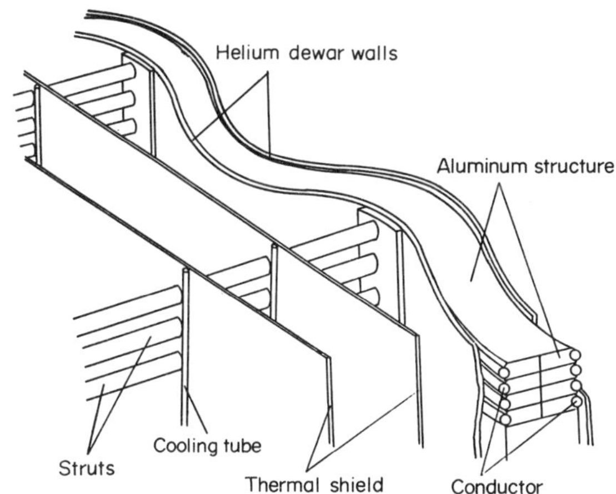

Figure 2
Structural features of SMES system including structural aluminum for axial compression, radial ripple for cooldown and struts for radial forces

is best achieved by quickly "dumping" all the helium and converting the electrical energy to evenly distributed thermal energy. In this event the aluminum prevents excessive temperatures from being realized in the coil. The second requirement is satisfied through the use of a radially rippled design; the ripple amplitude decreases as the structure contracts on cooldown to liquid helium temperatures. The final structural requirement is met through the use of glass-fiber-reinforced epoxy struts. Such struts have large compressive strength and low thermal conductivity to minimize heat leak from the warm soil to the cold structures. These design issues are collectively displayed in Fig. 2.

2.4 Conductor Design

The heart of the SMES design is the composite conductor. Although a wide variety of conductor designs has been proposed, the choice of materials routinely begins with NbTi superconductor filaments in a copper matrix (see *Niobium–Titanium Alloys*). Most SMES designs require optimization of superconductor current densities at fields in the range 3 T to 5 T, with present J_c values ranging from 3000 A mm^{-2} to 3300 A mm^{-2} at 5 T and 4.2 K. In most SMES conductor designs, the remaining and overwhelming fraction of conductor cross section is composed of high-purity aluminum in order to minimize joule heat production during any transient resistive or normal zones.

Residual resistivity ratio (RRR) values greater than 1000 are typical for high-purity aluminum and should be compared with maximum RRR values of 200 for copper. Cyclic strain, as would be realized during the daily operation of an SMES system, will decrease the RRR value of the high-purity aluminum. However RRR values of between 300 and 600 can be maintained when strains are kept below 0.1%.

2.5 Helium Cooling

One further design issue that influences conductor design and stability and which is crucial for the reliability of the entire SMES system is that of the use of liquid helium for cooling. In this design issue, subcooled superfluid helium at 1.8 K is found to be superior to all other cooling methods. For example, a saturated bath of 4.2 K helium results in unacceptably low allowable conductor current densities, while a system incorporating forced flow helium must deal with unmanageably large pressure drops or numerous parallel flow paths due to the large path lengths inherent to an SMES magnet. The clear advantage of a 1.8 K system is the increased allowable current densities due to both the lower temperatures and the enhanced cooling properties of superfluid helium. Heat flow through He II (the phase of liquid helium exhibiting superfluid properties) obeys the nonlinear Gorter–Mellink equation $q=[f^1(T)\nabla T]^{1/3}$ from which it is not possible to gain a value for thermal conductivity in the usual way. For purposes of comparison, however, heat flux values of order 1 W cm^{-2} (typical for engineering applications) result in an effective thermal conductivity $q/\nabla T$ of the order of 10^3 W cm^{-1} K^{-1}. This can be compared with values of the order of 1 W cm^{-1} K^{-1} for copper. The choice of the superfluid cooling still allows for a great deal of variety in conductor design. For example, one proposal uses a cable-in-conduit design with He II filling the interstices between the conductor filaments in the conduit. Another design uses a monolithic conductor cooled by an adjacent open annulus of He II. The first of these designs greatly simplifies dewar design, but is an inherently less stable design. The second design must deal with a complicated dewar and vacuum vessel; however, the open bath best utilizes the excellent heat transfer properties of He II and therefore allows for a high degree of conductor stability. Future refinements of SMES designs will follow improvement in conductor design.

See also: Multifilamentary Superconducting Composites; Superconducting Magnets; Superconducting Materials: Large-Scale Applications

Bibliography

Boom R W, Eyssa Y M, Abdelsalam M K, Huang X 1988 Superconductive magnetic energy storage (SMES) system studies for electrical utility usage at Wisconsin. In: Kwok H S, Shaw D T (eds.) 1988 *Superconductivity and Applications.* Elsevier, Amsterdam, p. 417

Boom R W, Peterson H A 1972 Superconductive energy storage for power systems. *IEEE Trans. Magn.* 8(2): 701–3

Loyd R J, Nakamura T, Schoenung S M, Lieurance D W, Hilal M A, Rogers J D, Purcell J R, Hassenzahl W V 1986

Design and cost of a utility scale superconducting magnetic energy storage plant. In: Fast R W (ed.) 1986 *Advances in Cryogenic Engineering*, Vol. 31. Plenum, New York, pp. 103–11

Masuda M, Shintomi T 1987 The conceptual design of a utility-scale SMES. *IEEE Trans. Magn.* 23(2): 549–52

Rogers J D 1981 Superconducting magnetic energy storage (SMES) program. Report No. LA-8777-PR. Los Alamos Scientific Laboratory, Los Alamos, NM

Verga R L 1990 Superconducting magnetic energy storage and other large scale SDI cryogenic applications programs. In: Fast R W (ed.) 1990 *Advances in Cryogenic Engineering*, Vol. 35A. Plenum, New York, pp. 555–64

J. M. Pfotenhauer
[University of Wisconsin–Madison,
Madison, Wisconsin, USA]

Magnetic Materials: An Overview

Magnetic materials play a vital role in modern society. They are found in common home appliances such as telephones, washing machines, radio and television sets and audio and video recorders. They are key components of motors, generators and transformers that provide the electrical power for manufacturing and transportation. Computer and communications technologies require mass storage of data in the form of magnetic recording tapes and disks. The annual world magnetic materials market in 1988 was over US\$5 × 10^9, with the added value of components made with these materials perhaps at 10–20 or more times. For example, while the market for magnetic particles used as recording media is about US\$5 × 10^8, the market for the finished tapes and disks is US\$$10^{10}$.

In this article, the various magnetic materials of technological importance will be surveyed, with an emphasis on recent developments. Much progress has been made since the early 1980s, particularly in grain-oriented silicon steel, amorphous alloys, rare-earth permanent magnets and recording media. A detailed assessment of the situation in 1985 is given by the National Materials Advisory Board (1985).

1. Classification of Magnetic Materials

Magnetic materials are classified as soft, with values of coercivity $H_c \lesssim 10^3\,\mathrm{A\,m^{-1}}$ or hard, with $H_c \gtrsim 10^4\,\mathrm{A\,m^{-1}}$. Between these values the term semihard is often used. Figure 1 displays the magnetic properties of the most prominent commercial materials in the form of saturation polarization versus coercivity. The intrinsic coercivity varies by seven orders of magnitude, ranging from a value of around $0.1\,\mathrm{A\,m^{-1}}$ for the softest nickel–iron alloys to that of around $10^6\,\mathrm{A\,m^{-1}}$ for the hardest rare-earth cobalt magnets. The cobalt–iron alloys have the highest values of saturation polarization (2.43 T for Fe–35%Co) while the ferrites have the lowest values (0.2–0.4 T).

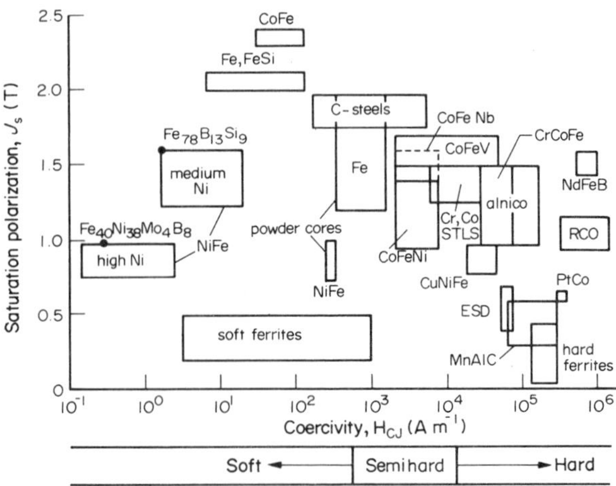

Figure 1
Commercial magnetic materials

2. Soft Magnetic Materials

The major classes of soft magnetic materials in use today are iron and low-carbon steels, iron–silicon alloys, nickel–iron alloys, iron–cobalt alloys and soft ferrites. Table 1 lists selected properties of typical commercial materials; representative dc magnetization curves are shown in Fig. 2.

Commercially pure iron is used in dc applications such as pole pieces of electromagnets. For ac applications such as small motors, the higher-resistivity low-carbon steels and nonoriented silicon steels are used (see *Iron and Low-Carbon Steels for Magnets*). These two materials constitute the largest magnetic materials usage by weight, about two and one million tonnes per year, respectively. For power and distribution transformers whose weight and efficiency are important the lower core loss oriented silicon steels are employed (see *Silicon Steels*).

For electronic transformer and inductor applications in the 1–100 kHz range, nickel–iron alloys known as permalloys are used (see *Nickel–Iron Alloys*). These have desirable combinations of high initial permeability μ_i and good ductility for forming into very thin strips. The most common in this category are 48% nickel with μ_i around 11 000 and 4%Mo–80%Ni with μ_i around 80 000. The cobalt–iron alloys, usually containing 2%V–49%Co, tend to be used in high-temperature applications because of their high Curie temperature (~980 °C), or in airborne devices where weight savings from the high

Table 1
Soft magnetic materials

Name	Composition	Permeability		Coercivity, H_c ($A\,m^{-1}$)	Retentivity, B_η (T)	B_{max} (T)	Resistivity ($\mu\Omega$ m)
		Initial	Maximum				
Ingot iron	99.8%Fe	150	5000	80	0.77	2.14	0.10
Low-carbon steel	99.5%Fe	200	4000	100		2.14	1.12
Silicon iron, unoriented	Fe–3%Si	270	8000	60		2.01	0.47
Silicon iron, grain oriented	Fe–3%Si	1400	50000	7	1.20	2.01	0.50
4750 alloy	Fe–48%Ni	11000	80000	2		1.55	0.48
4-79 permalloy	Fe–4%Mo–79%Ni	40000	200000	1		0.80	0.58
Superalloy	Fe–5%Mo–80%Ni	80000	450000	0.4		0.78	0.65
2V-Permendur	Fe–2%V–49%Co	800	450000	0.4		0.78	0.65
Supermendur	Fe–2%V–49%Co		100000	16	2.00	2.30	0.40
Metglas[a] 2605SC	$Fe_{81}B_{13.5}Si_{3.5}C_2$		300000	3	1.46	161	1.35
Metglas[a] 2605S-2	$Be_{78}B_{13}S_9$		600000	2	1.35	1.56	1.37
MnZn ferrite	H5C2[b]	10000		7	0.09	0.40	1.5×10^5
MnZn ferrite	H5E[b]	18000		3	0.12	0.44	5×10^4
NiZn ferrite	K5[b]	290		80	0.25	0.33	2×10^{12}

a Allied Corporation trademark b TDK ferrite code

saturation polarization are important (see *Iron–Cobalt Alloys*).

Iron–boron-based amorphous alloys, prepared by rapid quenching from the melt, have the potential for replacing crystalline materials in several areas (see *Metallic Glasses*). Some have properties similar to the best grades of the permalloys while others exhibit core losses substantially below those of the oriented silicon steels.

Because of their exceptionally high electrical resistivity, ferrites are used at high frequencies, beyond about 100 kHz (see *Ferrites, Soft*). The primary applications are electronic transformers and inductors,

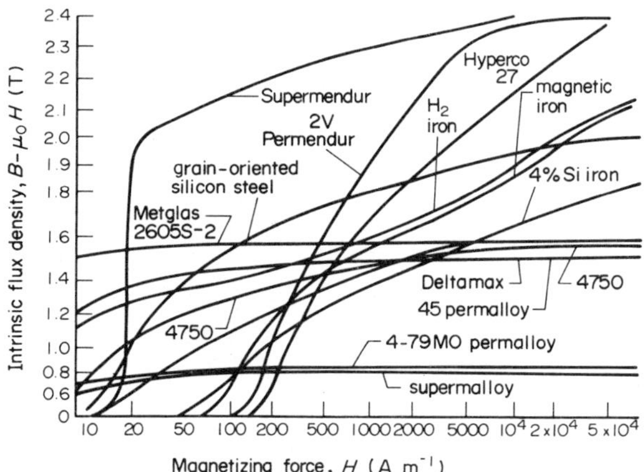

Figure 2
Magnetization curves (dc) of some commercial soft magnetic materials

deflecting yokes and flyback transformers in television sets, and in switched-mode power supplies. Ferrites have a spinel crystal structure with the general formula $MeFe_2O_4$. In manganese zinc ferrite Me is some combination of Mn, Zn and Fe, while in nickel zinc ferrite, Mn is replaced by Ni. MnZn ferrites have higher saturation but lower resistivity and are used up to about 1 MHz, beyond which the higher resistivity NiZn ferrites are generally preferred. Ferrites are also used in microwave devices such as isolators and circulators which make use of the nonreciprocal propagation characteristics close to or at a gyromagnetic resonance frequency between about 1 GHz and 100 GHz (see *High-Frequency Properties of Magnetic Materials*; *Microwave Materials and Devices, Magnetic*).

2.1 Grain-Oriented Silicon Steels

Grain-oriented silicon steels, containing about 3% silicon and oriented with the {110}⟨001⟩ texture, are primarily used in power and distribution transformers in the electrical utility industry. Here, core loss at 50 Hz or 60 Hz is the single most important factor. The core loss is manifested as an operating cost over the life of the transformer. This cost in turn comes from supplying the energy to produce the electricity lost and the capital investment in necessary equipment for energy production, and enters directly in transformer design cost as loss evaluation. Since the oil crisis of 1973–1974, core loss evaluation has assumed particular importance, providing materials suppliers with an economic incentive to cut core loss.

Such incentive, together with potential competition from amorphous metals, has resulted in substantial core loss improvements in recent years. The major improvements have come from the development of high-induction grades of superior {110}⟨001⟩ texture, surface treatment to refine the 180° domains, and thin-gauge material with sharp texture. Lower core loss also comes from design for lower-induction operation.

The high-induction, grain-oriented material was first introduced by Nippon Steel in 1968. Since then, similar grades have been developed by Kawasaki Steel, Allegheny Ludlum and Terni/Ceentro Sperimentale Metallurgico. As compared with conventional grain-oriented grades, the high-induction material shows an average misorientation of 3° instead of 7°, resulting in an induction of 1.97 T (vs 1.83 T) at $H = 800\ \mathrm{A\ m^{-1}}$, and a 20% lower core loss at high induction (1.7 T). In the Nippon Steel process, AlN is used as a strong grain growth inhibitor which then permits the use of a single, large-reduction, cold-rolling step prior to final recrystallization to a sharp {110}⟨001⟩ texture.

Detailed studies of the domain structure revealed that low loss is associated with a well-defined pattern of very fine 180° domains. These studies led to the development of surface treatments such as mechanical scribing and laser irradiation, further reducing the loss by as much as 10%. Since the primary effect of these treatments is domain refinement by an induced stress, they are effective for stacked cores of power transformers that do not require a stress-relief anneal. For wound cores in distribution transformers, the associated stress-relief anneal would nullify the benefits of the surface treatment. To overcome this difficulty, antimony has been deposited in the laser-irradiated regions. As the material is stress-relief annealed, the antimony diffuses below the surface, creating a region of weak magnetization. The result is a permanent domain refinement and consequent improvement in core loss. Alternatively, controlled local indentation by a gear-type roll produces an optimum strain which, upon stress-relief annealing, produces a locally recrystallized structure that again refines the domain structure and lowers the core loss.

In addition, since eddy current loss is reduced with decreased thickness, a thinner-gauge material should lead to reduced loss. Both conventional and high-induction grades have now been reduced to 0.23 mm in commercial practice, with an attendant loss improvement of over 10% as compared with a 0.30 mm material. The challenge here was to retain the sharp texture during processing to the thinner gauge. Also at the thinner gauge, surface smoothness becomes important. In laboratory studies, highly grain-oriented 2.93% silicon steel 0.15 mm thick, chemically polished to produce a smooth surface and followed with domain refinement by laser irradiation and tensile stressed to $1.0\ \mathrm{kg\ mm^{-2}}$, exhibits a core loss close to that of amorphous alloys. In fact, by cold rolling grain-oriented silicon steel to thicknesses comparable to those of amorphous alloys (0.03–0.04 mm) and then annealing the sheet in vacuum above 1200 °C to cause tertiary recrystallization, even lower losses could be obtained after suitable domain refining techniques were applied (Arai et al. 1989). It remains to be seen whether such measures can be translated into economic large-scale production. The core losses of this and some other materials discussed are shown in Table 2.

Table 2
Trends in grain-oriented silicon steels

Material	Thickness (mm)	Core loss 50 Hz, 1.7 T ($\mathrm{W\ kg^{-1}}$)	Core loss 50 Hz, 1.3 T ($\mathrm{W\ kg^{-1}}$)
Conventional	0.27	1.27	0.71
High induction	0.27	1.03	0.57
High induction	0.23	0.90	0.48
High induction	0.15	0.66	0.37
Laser scribed	0.15	0.53	0.30
Polish plus laser	0.15	0.40	0.20
Tertiary recrystallized	0.07	0.35	0.17
Metglas[a] 2605SC	0.04		0.16
Tertiary recrystallized	0.03	0.21	0.13

a Allied Corporation trademark

2.2 Amorphous Alloys

Transition metal–metalloids with compositions near $(\mathrm{Fe, Co, Ni})_{80}(\mathrm{B, C, Si})_{20}$ exhibit an amorphous structure when quenched from the melt. These alloys represent a radically new family of soft magnetic alloys that has emerged since the late 1970s. The first commercial alloys in the useful form of continuous ribbons was produced by Allied Corporation in 1973. Since then many new compositions have been developed. The current Allied Metglas products catalog contains no less than nine commercially available alloys. These may be divided into three major classes, with representative compositions and properties shown in Table 3.

(a) *Iron-based* alloys of saturation induction in the 1.6–1.8 T range: these were originally developed as low-loss substitutes for the grain-oriented silicon steels for power distribution applications, but with suitable composition modifications and heat treatment have also been found to possess good high-frequency (up to ~100 kHz) characteristics equal to or better than those of the crystalline nickel–iron alloys.

(b) *Iron–nickel-based* alloys of saturation induction in the 0.90 T range which mimic the crystalline nickel–iron alloys: these have lower magnetostriction λ_s and better corrosion resistance than the iron-based alloys, but are more expensive.

Table 3
Properties of some amorphous magnetic alloys

Alloy	B_s (T)	H_c ($A\,m^{-1}$)	λ_s (10^{-6})	ρ ($\mu\Omega\,m$)	T_C (°C)	Core loss 60 Hz, 1.4 T ($W\,kg^{-1}$)	Core loss 20 kHz, 0.2 T ($kW\,m^{-3}$)
Iron-based							
$Fe_{81}B_{13.5}Si_{3.5}C_2$	1.61	3.2	30	1.30	370	0.3	300
$Fe_{78}B_{13}Si_9$	1.56	2.4	27	1.30	415	0.23	
$Fe_{67}Co_{18}B_{14}Si_1$	1.80	4.0	35	1.30	415	0.55	
$Fe_{79}B_{16}Si_5$	1.58	8.0	27	1.25	405	1.2	58
Iron–nickel based							
$Fe_{40}Ni_{38}Mo_4B_{18}$	0.88	1.2	12	1.60	353		200
Cobalt based							
$Co_{67}Ni_3Fe_4Mo_2B_{12}Si_{12}$	0.72	0.4	0.5	1.35	340		43

Source: Smith (1982)

(c) *Cobalt-based* alloys with near-zero magnetostriction properties: these have the highest permeabilities and lowest losses coupled with stress insensitivity, but are also the most expensive.

Historically, starting with $Fe_{40}Ni_{40}P_{14}B_6$ (Metglas 2826) in 1973, the high-induction route led to the development of $Fe_{81}B_{13.5}Si_{3.5}C_2$ (Metglas 2605SC) in 1979, $Fe_{78}B_{13}Si_9$ (Metglas 2605S-2) in 1980 and $Fe_{67}Co_{18}B_{14}Si_1$ (Metglas 2605CO) in 1984. The last alloy exhibits the highest value of saturation induction (1.8 T) of commercially available amorphous alloys.

The high-frequency route first led to the iron–nickel-based $Fe_{40}Ni_{38}Mo_4B_{18}$ (Metglas 2826MB) in 1978, with properties equivalent to the crystalline nickel–iron alloys. Subsequently, it was discovered that the iron-based alloys could be developed to exhibit good high-frequency properties as well. These include $Fe_{79}B_{16}Si_5$ (Metglas 2605S-3) and $Fe_{75}Ni_4Mo_3B_{16}Si_2$ (Metglas 2605SM), developed in 1981, and $Fe_{76.5}B_{16}Si_5Cr_2C_{0.5}$ (Metglas 2605S-3A), developed in 1984, as well as the aforementioned $Fe_{67}Co_{18}B_{14}Si_1$ alloy. They are annealed at a temperature to precipitate a small fraction (1–5 vol.%) of submicrometer size bcc iron particles, which contribute to domain refinement and lower eddy current loss at high frequencies.

Another high-frequency route led to the cobalt-based near-zero magnetostriction compositions, with the development of $Co_{67.4}Fe_{4.1}Ni_3Mo_{1.5}B_{12.5}Si_{11.5}$ (Metglas 2705M) and $Co_{76}Fe_2Mn_4B_{12}Si_6$ (Metglas 2705MN) in 1982.

By far the greatest potential economic payoff for amorphous alloys is in power distribution transformers (up to 500 kW), which are being installed at a rate of about one million a year in the US. These transformers account for over 30×10^6 kW h, or about US\$$10^9$ of energy loss per year. Since the core loss of amorphous alloys such as Metglas 2605S-2 is about one-quarter that of grain-oriented silicon steel, the potential savings amount to several hundred million dollars a year. In 1983, the Electric Power Research Institute funded a program for General Electric to build a thousand 25 kW transformers using commercial production techniques. The successful completion of that program led General Electric, Westinghouse, Osaka and others to offer amorphous metal distribution transformers in early 1986. To prepare for potential large-scale commercialization, Allied Corporation has built a 60 000 t per year production plant in South Carolina. The material price, originally at US\$300 kg^{-1} in 1978, is now down to about US\$3.00 kg^{-1}, or less than twice that of oriented silicon steel. How much of the silicon steel market will be replaced by the amorphous alloys will depend on pricing trends, overall design considerations, continued improvements of both materials and trends in loss evaluations which in turn depend on such unpredictable factors as inflation rate and oil price.

3. *Hard Magnetic Materials*

The most important commercial hard magnetic materials in use today are ferrites, rare-earth magnets, and alnicos. Worldwide sales in 1987 were estimated at about US\$$1.46 \times 10^9$, with ferrites at 65%, rare-earth magnets 22% (18% RE–Cobalt, 4% NdFeB), alnicos 11%, and others at 2%. Motors and loudspeakers are the largest consumers of hard magnets. Figure 3 shows the historical progress in energy products of these three major classes of materials. The rise of the rare-earth magnets is clearly spectacular. Some magnetic properties of selected hard magnetic materials are shown in Table 4. Representative demagnetization curves are shown in Fig. 4.

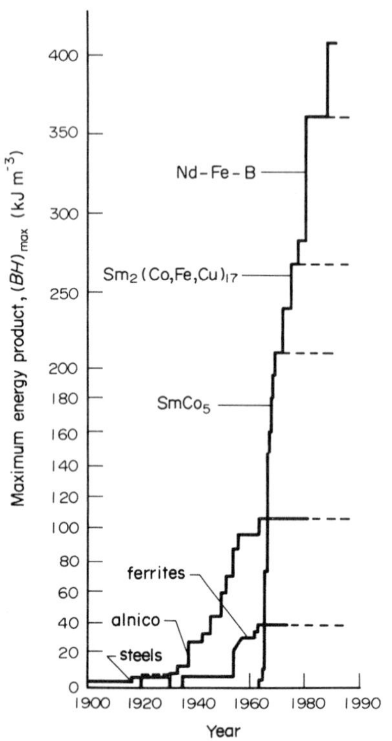

Figure 3
Progress in energy product for hard magnetic materials

3.1 Alnicos

The alnico alloys have moderately high energy product ($(BH)_{max}$ = 40–70 kJ m^{-3}), high remanent induction (B_r = 0.7–1.35 T) and moderate coercivity (H_c = 40–160 kA m^{-1}). They are a family of iron-based alloys containing aluminum, nickel and cobalt plus about 3% copper. Alnicos 1–4 are isotropic, whereas Alnico 5 and upwards are anisotropic via heat treatment in a magnetic field. Alnico 5 is the most widely used of this family. The alnicos are brittle and therefore produced by casting or by powder metallurgy.

The alnicos had been widely used in loudspeakers previously. In 1972, for example, their share of the permanent magnet market was 40%. However, since 1978 when a temporary shortage of cobalt, a key ingredient in the alnicos, forced the cobalt price to increase sevenfold, the consequent massive conversion to redesign using ferrites has reduced the alnico market share to 11% in 1987.

3.2 Ferrites

Hard ferrites have the general formula $MFe_{12}O_{19}$ where M is Ba or Sr. In contrast to the alnicos, the ferrites exhibit large coercivity (H_{cJ} = 200–300 kA m^{-1}) but low B_r (0.42 T). While barium ferrites are still produced in large quantity, the more recently developed strontium ferrites have greater coercivity and are increasingly produced. Ferrites are produced by ceramic techniques and are often bonded in plastic for low-cost, large-volume production. As a result of the very low cost combined with adequate magnetic properties, ferrites are the dominant hard magnetic materials in use.

3.3 Rare-Earth Magnets

The rare-earth magnets were developed in the 1960s and became commercially available in about 1970. The first alloys developed were based on the intermetallic compounds, RCo_5 and R_2Co_{17}, particularly

Table 4
Properties of some hard magnets

Material	B_r (T)	H_c (kA m^{-1})	$(BH)_{max}$ (kJ m^{-1})
17% cobalt steel (Fe–18.5%Co–3.75%Cr–5%W–0.75%C)	1.07	13	6
Alnico 2 (Fe–10%Al–19%Ni–13%Co–3%Cu)	0.75	45	14
Alnico 5 (Fe–8%Al–14%Ni–24%Co–3%Cu)	1.28	51	44
Alnico 9 (Fe–7%Al–15%Ni–35%Co–4%Cu–5%Ti)	1.05	120	72
Barium ferrite (BrO. $6Fe_2O_3$)	0.43	170	36
Strontium ferrite (SrO. $6Fe_2O_3$)	0.42	250	36
$SmCo_5$	0.87	640	144
$Sm(Co_{0.68}Cu_{0.10}Fe_{0.21}Zr_{0.01})_{7.4}$	1.10	510	240
$Nd_2Fe_{14}B$	1.23	880	290
Lodex 32 (19.2%Fe–10.8%Co–63%Pb–76%Sb)	0.74	75	28
Cunife (Fe–60%Cu–20%Ni)	0.54	44	12
Vicalloy (Fe–10%V–52%Co)	0.84	19	7
Chromindur II (Fe–10.5%Co–28%Cr)	0.98	30	16
Fe–23%Cr–15%Co–3%V–2%Ti	1.35	44	44
77%Pt–23%Co	0.65	360	73
70%Mn–29.5%Al–0.5%C	0.56	180	44

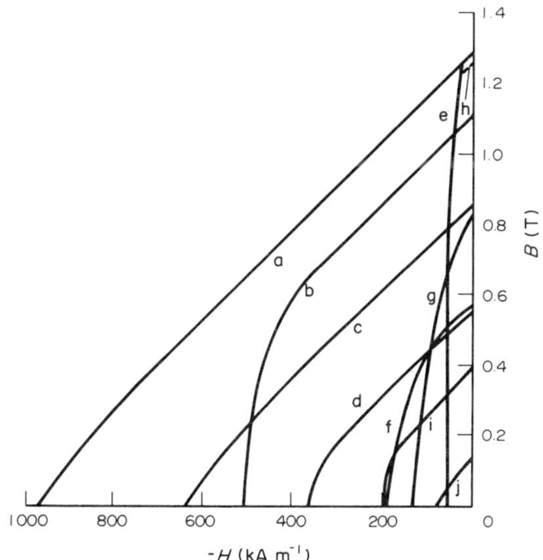

Figure 4
Demagnetization curves of representative hard magnetic materials: (a) $Nd_2Fe_{14}B$, (b) $Sm(Co, Cu, Fe, Zr)_{7.4}$, (c) $SmCo_5$, (d) bonded $SmCo_5$, (e) alnico 5, (f) Mn–Al–C, (g) alnico 8, (h) Cr–Co–Fe, (i) ferrite and (j) bonded ferrite

the Sm series (see *Cobalt–Rare-Earth Compounds*). $SmCo_5$ has the $CaCu_5$ hexagonal structure with a moderate value of saturation induction (0.97 T) and an extremely high value of magnetocrystalline anisotropy ($K_1 = 17.2$ MJ m^{-3}). Sm_2Co_{17} also has a hexagonal crystal structure with a higher value of saturation induction (1.20 T) but a lower value of $K_1 = 3.3$ MJ m^{-3}.

The magnets based on $SmCo_5$ are single-phase magnets. Laboratory magnets have attained values of $B_r = 1$ T, $H_{cJ} = 3200$ kA m^{-1} and $(BH)_{max} = 200$ kJ m^{-3}, although typical commercial products are in the $(BH)_{max} = 130$–160 kJ m^{-3} range. The mechanism of coercivity is based on nucleation of domains and wall pinning at grain boundaries. Large values of coercivity can be achieved only in fine particle (1–10 μm) form or in fine-grained sintered magnets prepared from such powders.

The magnets based on Sm_2Co_{17} are of the precipitation hardened multiphase type, via partial substitution of cobalt by copper. Magnetic hardening arises from domain wall pinning at the precipitates. The substitution was first discovered in the $SmCo_5$ and $CeCo_5$ compounds, and resulted in a dramatic increase in H_{cJ} to nearly 2400 kA m^{-1} in bulk alloys, but the saturation induction became too low. The addition of iron helped to recover B_s partially, but it was still inadequate. Further increases in B_s were made by moving the samarium: cobalt ratio from 1:5 toward 2:17 at the sacrifice of coercivity. Later on a small amount of zirconium was added to enable a further increase in B_s with an additional decrease in the Sm:(Co, Fe, Cu) ratio toward 2:17, and an increase in iron at the expense of cobalt and copper without the loss of coercivity. A commercial alloy $Sm(Co_{0.68}$-$Cu_{0.10}Fe_{0.21}Zr_{0.01})_{7.4}$ has values of $B_r = 1.10$ T, $H_{cJ} = 520$ kA m^{-1} and $(BH)_{max} = 240$ kJ m^{-3}. Laboratory studies have achieved values of $(BH)_{max} = 264$ kJ m^{-3}. Large values of H_{cJ} to 2100 kA m^{-1} have also been obtained by raising the zirconium content. For a more detailed account of these developments and a reference to the original literature, see Chin (1985).

The newest rare-earth magnet family, based on $Nd_2Fe_{14}B$, was developed in 1983 by Sumitomo Special Metals and General Motors (see *Neodymium–Iron–Boron Alloys*). This compound has a tetragonal crystal structure with a magnetocrystalline anisotropy of $K_1 = 4.5$ MJ m^{-3}, $J_s = 1.61$ T and $T_C = 312$ °C. The combination of high K_1 and J_s enables the attainment of record values of energy product, as high as 405 kJ m^{-3} in laboratory samples. Commercial production units average in the 160–300 kJ m^{-3} range.

Except for the low Curie temperature which translates to a large temperature dependence of the magnetic induction, the Nd–Fe–B magnets are superior to all other magnet materials. Thus it is little wonder that the market share of Nd–Fe–B had gone up to 15% in 1989, despite becoming available only six years earlier, surely one of the most outstanding success stories in the history of technology transfer.

It should be noted that while the Nd–Fe–B magnets are prepared by powder metallurgy techniques, the approaches taken by Sumitomo and General Motors are quite different. Sumitomo uses the conventional approach of pulverizing a cast ingot while General Motors uses the rapid-solidification, melt-spinning process of making ribbons followed by pulverization. While dense magnets can be prepared from powders made by either method, powder made by the melt-spun process has exhibited greater stability and, hence, practically all plastic-bonded Nd–Fe–B magnets use this type of powder.

3.4 Other Hard Magnet Materials

(a) Fe–Cr–Co magnet alloys. These alloys were first reported in 1971. They are analogous to the alnicos in metallurgical structure and magnetic properties. However, they exhibit better ductility and achieve equivalent magnetic properties with less cobalt. The early work concentrated on the high-cobalt regime typified by the 23%Co–28%Cr–1%Si–Fe alloy with values of $(BH)_{max} = 42$ kJ m^{-3}. Attention was later shifted to a medium-cobalt content of around 15%, often including a few percent niobium, aluminum, vanadium, titanium or zirconium. Subsequently, lower cobalt in the 5–12% range was developed without the need for additives.

Highest-energy products in the Fe–Cr–Co family include the magnetic-field-aged 22%Cr–18.5%Co–3%Mo single crystal at 91 kJ m^{-3} and 24%Cr–15%Co–3%Mo columnar grained alloy at 76 kJ m^{-3}, and the deformation-aged 23%Cr–23%Co–2%Cu at 78 kJ m^{-3}. For details, see Jin and Chin (1987).

The Fe–Cr–Co alloys can be almost one-to-one substitutes for the alnicos. The good ductility and formability also qualify them as substitutes for Cunife and Vicalloy in stamped parts, thin sheets and wires. The largest usage to date has been an isotropic alloy Fe–28%Cr–10.5%Co called Chromindur II, which was developed to replace Remalloy (12%Co–20%Mo–Fe) as the bias magnet for a telephone receiver. At the height of production, more than 10^7 units were made annually. Subsequently, the good formability led to a new design of the receiver utilizing a miniaturized Chromindur II cup magnet with one-tenth the weight, resulting in significant cost savings in materials and assembly. For a summary of Fe–Cr–Co alloys, see *Iron–Chromium–Cobalt Alloys* and Jin and Chin (1987).

(*b*) *Cu–Ni–Fe alloys*. These alloys, commercially known as Cunife with a composition of 60%Cu–20%Ni–20%Fe, are ductile and can be stamped even in the hard magnetic state (see *Magnetic Alloys: Intermediate Coercivity Materials*). However, ingot sizes are limited to around 30 mm in diameter since they cannot be hot worked. Although the magnetic properties are rather modest ($(BH)_{max}$ ~ 12 kJ m^{-3}), Cunife magnets are used in automobile speedometers and timing motors where parts are precision stamped at high speed.

(*c*) *V–Co–Fe alloys*. These alloys, which are also ductile, are commercially known as Vicalloy, with a nominal composition of 10%V–52%Co–38%Fe. Values of $(BH)_{max}$ are modest, at around 6 kJ m^{-3}. Vicalloy is used widely in antitheft labels and thin-foil magnets due to its formability into thin sheets of about 0.05 mm thickness.

(*d*) *Elongated-single-domain* (*ESD*) *magnets*. These magnets, under the trade name Lodex, are made by electrodeposition of elongated iron–cobalt particles into a mercury cathode and subsequent compaction into a lead matrix (see *Elongated-Single-Domain Magnets*). The magnets have values of $(BH)_{max}$ up to 28 kJ m^{-3} and H_c up to 90 kA m^{-1}. Since the materials can be pressed to dimensional precision and magnetic uniformity, they are often used in timing motors, meters, thermostats and relays. Owing to the intricate processing facilities required, however, production was limited to a single plant which was closed at the end of 1988.

(*e*) *Platinum–cobalt magnets*. These magnets, near the equiatomic composition, exhibit excellent combinations of coercivity (H_{cJ} = 400 kA m^{-1}) and energy product ($(BH)_{max}$ = 72 kJ m^{-3}). The principal applications have been in hearing aids, watches and travelling wave tubes, but nearly all of them have been switched to the less expensive rare-earth magnets.

(*f*) *Mn–Al–C magnets*. These magnets exhibit energy products in the alnico 5 range (~44 kJ m^{-3}) but with much larger coercivity ($\leqslant$240 kA m^{-1}). The properties are optimized, however, by a relatively expensive hot extrusion process to produce a crystalline texture (see *Manganese–Aluminum–Carbon Alloys*). Currently, production is limited to a single plant in Japan for internal consumption of one manufacturer only.

(*g*) *Magnet steels*. These were probably the first synthetic material (~1930) fabricated specifically for permanent magnets, and with modifications are still used to a small extent in some hysteresis motors. These are carbon steels (~1% carbon) containing cobalt, tungsten and chromium. The hard magnetic properties are developed by quenching from around 1000 °C to develop a martensitic structure. The energy product of about 6 kJ m^{-3} is rather low.

4. Magnetic Recording Materials

Vast amounts of audio and video information and digital data from computers are stored in magnetic tapes and disks. Increasingly, the trend is toward higher-density recording to satisfy the growing demands for greater storage capacity, more portable products and digital rather than analog format. This trend has led to the development of new recording materials and new recording system technology (see *Magnetic Storage Media*; *Magnetooptic Storage Media*). For example, in the 1980s the audio compact disk burst onto the scene. Optical disks for data storage came on board in 1985, and magnetooptic recording systems are in commercial production. Also in development are digital audiotape and videotape technologies. It is no wonder that magnetic recording materials represent the fastest growth segment of the magnetics industry.

Magnetic recording materials function as hard magnet materials with modest coercivity in the 25–100 kA m^{-1} range. Several factors, however, set them apart from the traditional hard magnet market. First, instead of being used in the traditional bulk form, these materials are dispersed in an organic medium or deposited as a film. Second, the methods of preparation of the particles are primarily chemical rather than pyrometallurgical. Third, the magnetic property requirements are but one of many physical, chemical and mechanical property requirements of a complex recording system. Hence, the history of the progress in the development of recording materials has been driven by the progress in system design and vice versa.

Currently, the most widely used technique of preparing recording media is the dispersion of an organic

Table 5
Magnetic recording materials

	γ-Fe_2O_3	Fe_2O_3–Fe_3O_4	Co–γ-Fe_2O_3	CrO_2	Barium ferrite	Iron
B_r (T)	0.11	0.15	0.15	0.15	0.12	0.30
H_c (kA m^{-1})	26	37	52	45	64	120

slurry of particulates, either on a plastic substrate (tape or flexible disk) or on a rigid disk (usually an aluminum–magnesium alloy). More recently, continuous films are deposited using sputtering or evaporation, particularly on the smaller rigid disks.

There are three major classes of magnetic recording particles: γ-Fe_2O_3 and its modifications, CrO_2 and iron. Recently, particulates of barium ferrite have started commercial production. Some of the characteristics are shown in Table 5. Because of low cost and continuing improvements, γ-Fe_2O_3 dominates the usage today. Acicular particles of α-(FeO)OH (goethite) are first precipitated from $FeSO_4$ by NaOH, dehydrated to α-Fe_2O_3 (hematite), reduced to Fe_3O_4 (magnetite), and finally reoxidized to γ-Fe_2O_3 (maghemite). Coercivity is in the 25 kA m^{-1} range.

By stopping the Fe_3O_4 to γ-Fe_2O_3 oxidation process part way, coercivity can be increased to about 40 kA m^{-1}. The increase is thought to be due to stress between the mixed compounds.

Finally, by coating the γ-Fe_2O_3 particles with a layer of cobalt, for example by precipitation of $Co(OH)_2$ on the surface, coercivity is increased to the 50 kA m^{-1} range. This modification is the most common in production today.

Particles of CrO_2 also have high values of coercivity (45 kA m^{-1}). They are made by thermal decomposition of CrO_3 under pressure and in the presence of H_2O.

Barium ferrite has long been known for its bulk permanent magnetic properties and very fine particles of this material have demonstrated exciting potential for high-density recording. Problems with temperature-dependent coercivity are being addressed with additions of substitutive elements such as cobalt and titanium.

Metal particles have the highest value of coercivity and remanant induction, more than twice that of Co–γ-Fe_2O_3. They are generally prepared by reduction of the oxide in a reducing atmosphere. Metal particles are increasingly used in high-density recordings such as 8 mm videotapes and in the emerging digital audiotape and videotape technologies.

Continuous thin films have advantages over particulates in terms of higher density and thinner coatings. Video disks for instant replays are made by electroplating a thin film of cobalt–phosphorus or CoNi–P onto an aluminum substrate. Audiotapes have been made by vacuum deposition of FeCo alloys. The biggest inroad since the late 1980s, however, has been the large-scale production of rigid disks made of sputtered CoNi films on aluminum substrates. They now account for over half of the rigid disk market, and are almost exclusively used for disks up to 5.25 in (130 mm) diameter.

The discussion thus far pertains to the standard format of longitudinal recording, where the magnetization is parallel to the recording surface. By contrast, Iwasaki in 1975 proposed the perpendicular recording format, with the promise of higher-density recording. Highly oriented cobalt–chromium alloy films near the $Co_{80}Cr_{20}$ composition with a columnar grain structure can be sputter deposited to yield a strong perpendicular anisotropy. This has been, and continues to be, a topic of extensive investigations ranging from materials chemistry, microstructure and stability to head design and recording theory (Ouchi and Iwasaki 1987).

Another recording technique, which is in early production stage, is magnetooptic recording. Amorphous thin films of rare-earth transition metals such as GdTbFe and CoTbFe exhibit perpendicular anisotropy and a Curie temperature in the 100–200 °C regime where the temperature dependence of coercivity and magnetization is steep. A laser beam impinging on the film will heat it locally so that a weak magnetic field can reverse the direction of magnetization there. Upon removal of the beam and return to room temperature, the local magnetization remains reversed. The recorded bit of information can be read using the magnetooptic effect of the film. Erasure is accomplished by irradiating with a laser beam together with a bias field in the opposite direction.

The salient advantages of magnetooptic recording are the potential high recording density ($\sim$120 Mbit cm^{-2}) and noncontact between write/read head and the recording media. A large effort worldwide has been underway for some time and many materials problems remain. These include film uniformity and reproducibility, defect minimization, and corrosion protection. By 1989, several companies had started offering magnetooptic recording systems in both the 3.5 in (90 mm) and 5.25 in (130 mm) formats with memory capacities up to the gigabyte range.

5. Future Prospects

As indicated throughout this article, activities in magnetic materials remain vibrant. In the soft magnetic materials area, large-scale commercialization of the amorphous metals appears imminent. Along with the continued concern for energy, the amorphous metals have stimulated dramatic improvements in the traditional oriented silicon steels. Similarly, in the hard magnetic materials area, the impressive entry of the rare-earth magnets, first samarium–cobalt and now Nd–Fe–B, will spur a continued search for still higher-energy materials. The prediction is that, by the end of the 1990s, rare-earth magnets will have passed ferrites as the market leader. In magnetic recording, there is a full platter of various particles and films to satisfy both longitudinal and perpendicular recording modes as well as the emergence of very-high-density magnetooptic recording (see *Magnetooptic Materials*). The insatiable demands for ever higher storage densities and faster retrieval times will guarantee continued innovations in this area.

In the future new developments may be looked for from processing innovations. The melt-quenching technique which led to the development of amorphous metals has also been applied to crystalline high-silicon steels in the 4–6% silicon range. Such high-silicon alloys normally have excellent soft magnetic properties but poor ductility. The latter is vastly improved through melt quenching. Similarly, melt quenching has led to the General Motors process for preparing the new Nd–Fe–B permanent magnets. Thus, melt-quenching techniques should open the way to other materials yet to be discovered, as well as improved properties of current materials (see *Nanocrystalline Soft Magnetic Materials*).

Production of magnetic multilayers and heterostructures with novel properties or special combinations of properties is another active and promising area. Multilayer films can be produced by sputtering, molecular beam epitaxy (MBE) or organometallic chemical vapor deposition (OMCVD). These techniques permit the precision fabrication of synthetic superlattices where structure and properties can be tailored. The first of the new synthetic magnetic superlattices was gadolinium–yttrium prepared by MBE. Here, the ferromagnetic gadolinium layers are separated by nonmagnetic yttrium layers, and the gadolinium layers are coupled magnetically across these yttrium layers. The coupling can be ferromagnetic (parallel alignment) or antiferromagnetic (antiparallel) depending on the yttrium layer thickness. More recently, synthetic dysprosium–yttrium superlattices have been created. While bulk dysprosium undergoes a transition from an antiferromagnetic basal plane spiral structure to a ferromagnetic structure at 90 K, the transition is absent in films less than 7.6 nm thick. Furthermore, in the superlattice the spiral wavevector of the antiferromagnetic structure "locks in" at low temperature, in contrast to the continuous change with temperature for bulk dysprosium (Kwo et al. 1988). Recent developments are cobalt–platinum multilayers for high-density magnetooptic recording and superlattices such as iron–chromium and cobalt–copper with giant magnetoresistive properties (see *Thin Films and Multilayers, Magnetic*).

In a similar vein, there is theoretical and experimental evidence for enhanced two-dimensional magnetism at surfaces and interfaces (Freeman and Fu 1987). The magnetic moment of iron, nickel and cobalt surfaces are enhanced by 10–30% over the bulk values, while the chromium surface becomes ferromagnetic with a strong moment. In addition, monolayers and bilayers of the ferromagnetic metals do not show significant loss of magnetism when sandwiched between nonmagnetic layers.

Thus, the synthetic superlattices and sandwiches are radically new materials with new magnetic behavior and continued research in this direction holds the prospect of producing technologically useful magnetic materials.

Bibliography

Arai K I, Ishiyama K, Mogi H 1989 Iron loss of tertiary recrystallized silicon iron. *IEEE Trans. Magn.* 25: 3949–54

Boll R 1979 *Soft Magnetic Materials*. Heyden, London

Chin G Y 1985 Magnetic materials. In: Meyers M, Inal O T (eds.) 1985 *Frontiers in Materials Technology*. Elsevier, Amsterdam, p. 415

Chin G Y, Wernick J H 1981 Magnetic materials, bulk. *Encyclopedia of Chemical Technology*, 3rd edn., Vol. 14, Wiley, New York, pp. 646–86

Croat J J 1989 Status of rapidly solidified Nd–Fe–B permanent magnets. *IEEE Trans. Magn.* 25: 3550–4

Freeman A J, Fu C L 1987 Strongly enhanced 2D magnetism at surfaces and interfaces. *J. Appl. Phys.* 61: 3356–61

Higuchi A, Hirosawa S 1989 Sintered Nd–Fe–B permanent magnets, *IEEE Trans. Magn.* 25: 3555–60

Jacobs I S 1979 Magnetic materials and applications: a quarter-century overview. *J. Appl. Phys.* 50: 7294–306

Jin S, Chin G Y 1987 Fe–Cr–Co magnets. *IEEE Trans. Magn.* 23: 3187–92

Kwo J, Hong M, McWhan D B, Yafet Y, Fleming R M, Disalvo F J, Waszczak J V, Majkrzak C F, Gibbs D, Goldmann A I 1988 magnetic superlattices. *J. Phys. (Paris), Colloq.* 49(8): 1651–5

McCaig M 1977 *Permanent Magnets in Theory and Practice*. Wiley, New York

National Materials Advisory Board 1985 *Magnetic Materials*, Publication NMAB-426. National Academy Press, Washington, DC

Ouchi K, Iwasaki S 1987 Recent subjects and progresses in research on Co–Cr perpendicular magnetic recording media. *IEEE Trans. Magn.* 23: 2443–8

Parker R 1990 *Advances in Permanent Magnetism*. Wiley, New York

Salsgiver J A 1986 Future prospects for high quality grain-oriented silicon steels. *J. Met.* 38: 27–31

Smith C H 1982 Magnetic shielding to multi-gigawatt magnetic switches—ten years of amorphous magnetic applications. *IEEE Trans. Magn.* 18: 1376–81
Wernick J H, Chin G Y 1981 Magnetic materials, thin film. *Encyclopedia of Chemical Technology*, 3rd edn., Vol. 14. Wiley, New York, pp. 686–707
Wohlfarth E P (ed.) 1989–1992 *Ferromagnetic Materials*, Vols. 1–3. North-Holland, New York

G. Y. Chin[†]
[AT&T Bell Laboratories, Murray Hill, New Jersey, USA]

Magnetic Materials: Basic Concepts

Of the vast variety of materials available, only a small fraction are magnetic. This is especially true if "magnetic" is defined as meaning ferromagnetic in the sense of a magnetic-flux-carrying capability even in the absence of an applied magnetic field (which in essence is the practical definition used in most other articles concerned with magnetic materials in this encyclopedia). The observation is still true if the definition is broadened to encompass all materials known to have some type of ordered magnetic state, including the much larger family (particularly of nonmetals) of antiferromagnetic materials. Thus, even adhering to the broader definition, which is appropriate for any general discussion of the subject, the question still remains of why only certain materials are magnetic while most are not. The question is both pertinent and convenient because in the attempt to answer it in progressive stages a conceptual framework can be developed to account for some of the basic features of magnetic materials.

1. Magnetic Atoms in Materials

Most aspects of magnetism in materials ultimately derive from the electronic states of the constituent atoms. It is noteworthy in this respect that, in addition to their orbital motion, electrons have a dynamic property, entirely quantum mechanical in origin, known as spin. Crudely described, every electron acts like an electrically charged body that is spinning about a fixed axis, thus creating (according to Ampère's law) a magnetic moment along the axis of spin. The magnetic moment associated with an electron spin has a universal constant value called the Bohr magneton μ_B. By way of contrast, the magnetic moment of an atomic nucleus, whose protons and neutrons also have spin, is generally several orders of magnitude smaller than μ_B and can usually be neglected.

Thus, for most practical purposes, the electrons of a material constitute its elementary magnets, and if nature were such that all their moments tended to align parallel to each other the world could be a fantastically magnetic place. This interesting possibility is, however, denied categorically by the Pauli exclusion principle, which prescribes that any pair of electrons in the same orbital state must have their spins (and associated moments) antiparallel, that is, pointing in opposite directions ("up" and "down"). Since every electron shell surrounding an atomic nucleus consists of some integral number of orbital states, each of which can be occupied by at most two electrons, it follows that for any fully occupied electron shell there is complete vectorial cancellation of the spin moments. Moreover, for a full shell the orbital moments also cancel out (in zero applied field). This situation prevails—with every electron shell completely occupied—in the atoms of all but certain elements. The exceptions are the various transition groups (3*d*, 4*d*, 5*d*), the lanthanides (4*f*) and the actinides (5*f*). The labels in parentheses refer to the electron shell that is only partly filled; the other (inner-core) electron shells of these atoms have lower energies and are therefore fully occupied. In these exceptional cases, an atom can have a nonzero magnetic moment since some of the electrons in the partly filled shell can lie in singly occupied orbital states with parallel spin. Indeed, electrostatic forces within an atom favor the separation of the electrons into different orbitals, tending towards a maximum inequality between the number of electrons of opposing spin. This is known as Hund's rule and applies even to atoms in a solid as long as the electrostatic fields imposed by the crystalline environment (the so-called crystal fields) have a negligible effect on the relative orbital energies.

Granted that as free atoms the transition group, lanthanide (rare-earth) and actinide atoms are uniquely capable of having nonzero magnetic moments, the question of particular interest here is whether they can continue to do so when they are located in a crystal. In the case of an ionic insulator, the answer is fairly straightforward and depends on the crystal-field effects mentioned previously. For example, the ferrous (Fe^{2+} ion has six electrons in its 3*d* shell, which can accommodate up to ten electrons, that is, a pair of electrons of opposing spin in each of five orbitals. According to Hund's rule, Fe^{2+} will have five 3*d* electrons with spin up and one with spin down, resulting in a net spin moment of $4\mu_B$, and this situation exists in many simple ferrous salts (e.g., $FeCl_2$), where the sole effect of the crystal fields is to virtually eliminate or "quench" any orbital contribution to the ionic moment. However, in certain complex compounds in which the local symmetry about the Fe^{2+} ion is low, two of the 3*d* orbitals are raised in energy relative to the other three; consequently, the latter are each occupied by an electron pair of opposing spin, resulting in a net spin moment of zero! This example also serves to illustrate that the magnetic moment of a transition group ion in an insulator always corresponds to an integral number of Bohr magnetons. The same is true for rare-earth ions in insulators, although in this case the orbital moment is not quenched and combines additively or subtractively with the spin moment.

The formation of magnetic moments in a metal is a more complex process, except for the case of the rare earths (see *Rare-Earth Metals and Alloys*). The relative simplicity of magnetic moment formation in a rare-earth atom in a metal stems from the fact that its $4f$ electrons are tightly bound to the nucleus (as they are in an insulator) and do not participate in the interatomic binding, which is accomplished entirely by the itinerant conduction electrons. Hence, the partly filled $4f$ electron shell again produces a magnetic moment that is integral in μ_B per atom. Since the conduction electrons are nearly free, the energies of their orbital states are spread out over a broad continuous band (the conduction band), in contrast to the sharp $4f$-electron energy distribution, and their contribution to the atomic moment is usually well under one μ_B (band polarizations not being restricted to integers in μ_B per atom—as will be described below).

Magnetic moment formation has no analogous simplicity in transition group metals and alloys, whose potential sources of magnetism—the d electrons—are nearly as free as the conduction electrons and enter significantly into the interatomic binding. Since the partially filled d electron shell of a transition group atom in a metal interacts strongly with the local environment, its orbital-state energies form a continuous band that is broad but generally narrower than (and overlapped by) the conduction band. This situation is pictured schematically in Fig. 1a, where the density of states (the number of states per atom per increment of energy) is plotted as a function of the energy for both the d electrons and the conduction (s) electrons. The integrated area under each curve corresponds to the total number of available states per atom (ten for the d electron band and two for the conduction electron band). The states in these two overlapping bands are occupied by electrons in accordance with Fermi–Dirac statistics, which are appropriate for itinerant electrons. This means that at low temperatures all the states are fully occupied up to a maximum energy (the Fermi energy E_F), such that the areas under the occupied portions of both bands (shown shaded in the figure) equal the total combined number of d electrons and conduction electrons. In the absence of any field, this number is divided equally between electrons of spin up (plus) and those of spin down (minus). Thus, for each band at any given energy, the densities of states for the electrons of plus and minus spin ($N_+(E_F)$ and $N_-(E_F)$, respectively) are equal, each being half of the total density of states $N(E_F)$. Each band can therefore be subdivided into two equal spin sub-bands, as shown in Fig. 1b. Under these conditions the net spin moment is zero

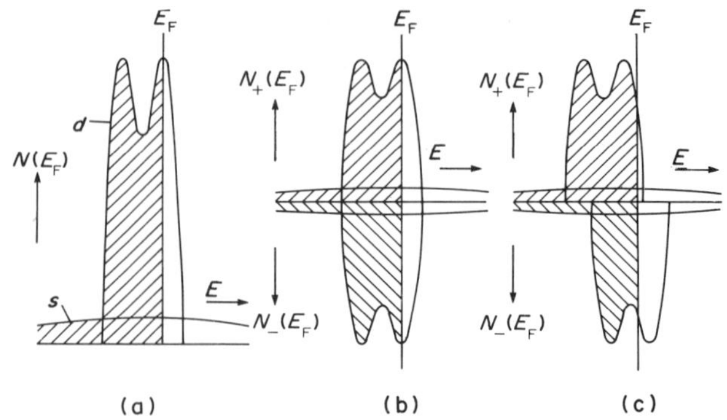

Figure 1
(a) Total density of states $N(E_F)$ as a function of energy E for d and s (conduction) electrons of a transition group atom in a metal; (b) same as (a) but with $N(E_F)$ subdivided into densities of states for electrons of spin plus and spin minus; (c) same as (b) but with spin sub-bands shifted in energy by the effective field. Shaded areas show electron states occupied up to the Fermi energy E_F

and, since the orbital moment of a transition group atom is normally quenched, the total moment of the atom is also zero.

However, there are interactions among the electrons that favor a shift in energy of one spin sub-band relative to the other. These are known from quantum mechanics as exchange interactions, which in an insulator give rise to an analogous shift of discrete energy levels in accordance with Hund's rule, as described earlier. In a metal, a spontaneous energy splitting of the spin sub-bands can occur only if the product of the exchange-interaction strength and the density of states at E_F exceeds a certain critical value. This threshold condition is satisfied most easily by the d electrons with their high densities of states and when it is satisfied the d electron spin sub-bands shift in opposite directions in energy, as shown in Fig. 1c. The split sub-bands continue to be fully occupied at energies up to E_F by the same total number of electrons. The net result, as shown by the shaded areas in the figure, is an imbalance between the numbers of electrons having opposite spin, which corresponds to a nonzero spin moment per atom.

The unsplit spin sub-bands in Fig. 1b represent the subcritical situation that prevails in most transition group metals, the important exceptions being chromium, manganese, iron, cobalt and nickel of the $3d$ series. With some of the same exceptions, this band picture also pertains locally to transition group atoms in a nonmagnetic metal host, such as a noble metal. In all these cases, an atomic magnetic moment does not form spontaneously. However, when a magnetic field H is applied to these materials, the spin sub-bands undergo a small relative shift of the type shown in Fig. 1c, resulting in a magnetic moment $\chi_p H$, where χ_p is the Pauli paramagnetic susceptibility, which is essentially constant with temperature. When the exchange interactions are weak, as in most of these materials, χ_p is simply proportional to $N(E_F)$, the density of states at the Fermi energy. In cases such as palladium and platinum (the end members of the 4d and 5d series), the product of $N(E_F)$ and the exchange interaction strength E_{exch}, although below the critical value for spontaneous band splitting, is large enough to cause an anomalous increase (a so-called exchange enhancement) of χ_p.

In the transition group metals at the end of the 3d series, that is, chromium through nickel, the $E_{exch}N(E_F)$ product actually exceeds the critical value, resulting in the split-band picture of Fig. 1c, even when H is zero. Hence, there is a spontaneous magnetic moment per atom, which is not necessarily integral in μ_B since the number of electrons in each spin sub-band is a time-averaged quantity that need not be an integer. The split-band picture of Fig. 1c is similarly relevant to the formation of stable magnetic moments by certain transition group atoms, even when they are extremely dilute impurities in, say, a noble metal. Classic examples are iron and manganese which as impurity atoms of a few parts per million in copper or gold are known to have well-defined moments. It is interesting to note that nickel as an isolated impurity atom in copper (or in any other nonmagnetic metal host) does not form a stable magnetic moment. However, in a nickel–copper alloy not so dilute in nickel, the local atomic environment begins to play an important role in moment formation; this and related effects will be discussed in Sect. 2.

2. Magnetic States of Materials

Thus far, it has been seen that only certain atoms are capable of producing magnetic moments and that such an atom, when located in a material, produces a moment only under suitable conditions which are different for insulators and metals. Given that these conditions are satisfied, the next question of interest is when and how are the atomic moments of a material aligned vectorially relative to each other. This is tantamount to asking when does such a material develop an ordered magnetic state and what are the general features of these states.

If there are no interactions between the atomic moments, the thermal energy causes them to fluctuate among all possible orientations without preference and unless a magnetic field is applied the magnetization of the material as a whole is zero. However, at any temperature T, application of an external magnetic field H produces a preferential orientation of the moments, resulting in a magnetization M which, with increasing H, rises at first linearly and then more slowly as it approaches a saturation value corresponding to complete alignment of all the moments parallel to the field. Moreover, the effects of H and T on the moment alignment oppose each other in such a way that M depends universally on the ratio H/T. For N atoms per unit volume, each with magnetic moment μ, this dependence can often be expressed as

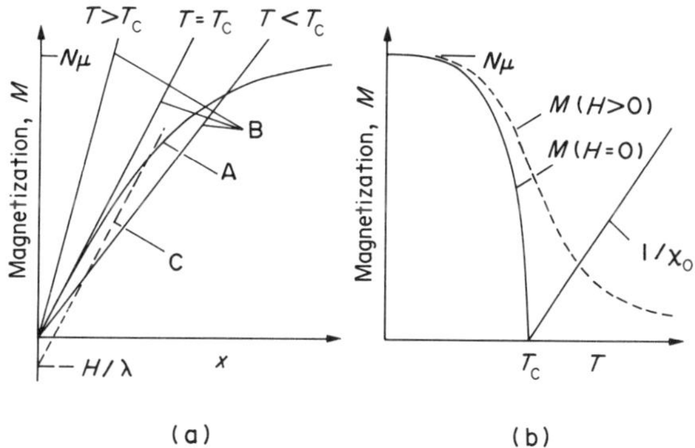

Figure 2
(a) Magnetization M as a function of parameter x ($=\mu H_{eff}/k_BT$, H_{eff} being the effective field) for a typical ferromagnet; different curves are discussed in text. (b) M in zero and nonzero field H and inverse susceptibility $1/\chi_0$ as functions of temperature T, as derived from the construction in (a)

$$M = N\mu B_S(x), \qquad x = \mu H/k_BT \tag{1}$$

where k_B is the Boltzmann constant and $B_S(x)$ is the Brillouin function (of argument x) for effective spin $S=\frac{1}{2}(n_+ - n_-)$, n_+ and n_- being the numbers of electrons per atom with positive and negative spin. For simplicity, the orbital moments are considered quenched, so that $\mu = 2S\mu_B$. Equation (1) is represented by curve A in Fig. 2a. It gives not only the isothermal variation of M with H described above but, since $B_S(x) \propto x$ for small x, it also states that at sufficiently low H,

$$\chi_0 \equiv (M/H)_{H\to 0} \propto 1/T \tag{2}$$

which is the well-known Curie law for the temperature dependence of the initial magnetic susceptibility χ_0. This entire behavior is said to be typically paramagnetic and is characteristic of many materials whose atomic moments are well defined but are essentially noninteracting (see *Paramagnetic Materials*).

The best examples of materials exhibiting typical paramagnetism are to be found among ionic salts in which transition group cations are very widely separated. More commonly in such salts, however, there is an appreciable exchange coupling between the moments of the magnetic cations via the intervening anions. In this process, known as superexchange, the outermost electron shell of an anion such as a halide, which normally is completely filled, is raised to an excited state that is magnetically polarized by and interacts simultaneously with the partly filled electron shells of a pair of neighboring magnetic cations. In most transition group ionic compounds the superexchange coupling is antiferromagnetic and the coupled magnetic moments tend to align antiparallel to each other, but there are several cases (e.g., $CrBr_3$, CrO_2) where the superexchange interactions are ferromagnetic and the tendency is towards a parallel moment alignment.

In the ferromagnetic cases, each ionic moment can be thought of as experiencing a so-called exchange field, which is parallel to any applied field H and is proportional to the magnetization through a coefficient λ whose magnitude reflects the combined strength of all the exchange interactions with the neighboring ionic moments. Thus, within this approximation, the total effective field on each magnetic ion is $H+\lambda M$ and its feedback effect on the parallel moment alignment that produces M can be determined by substituting it for H in Eqn. (1). The equation can then be split into the following two parametric expressions for M:

$$M = N\mu B_S(x) \tag{3a}$$

$$M = (k_B Tx/\mu - H)/\lambda \tag{3b}$$

where x acts as a dummy variable. The simultaneous solutions of these equations give the magnetization M at any temperature T and field H. They are shown graphically in Fig. 2a, where Eqn. (3a) is represented by curve A. For the case of $H=0$, Eqn. (3b) is represented by various straight lines labelled B, which emerge from the origin and have different slopes for different temperatures. There are clearly two regimes: the ferromagnetic regime at temperatures sufficiently low that line B intersects curve A at a nonzero value of M, corresponding to the spontaneous magnetization, and the paramagnetic regime at higher temperatures where the intersection occurs only at the origin. The two regimes are separated by the condition that line B is tangent to curve A at the origin, which takes place at and defines the Curie temperature T_C. The temperature dependence of M at zero H, obtained from this graphical construction, is shown in Fig. 2b and typifies the ferromagnetic state. Since the spontaneous magnetization decreases smoothly to zero as the temperature is raised to T_C, the change from the ferromagnetic state to the paramagnetic state is, thermodynamically speaking, a continuous (as opposed to abrupt) order–disorder transition, for which T_C is the critical temperature.

In a nonzero applied field, the variation of the magnetization with temperature is thermodynamically different from the zero-field behavior just described. This is seen in Fig. 2a, where Eqn. (3b) with H nonzero is represented by line C, which emerges from a negative M-axis intercept (equal to $-H/\lambda$); as the lines increase in slope with increasing temperature they continue to cross curve A at nonzero values of M. Thus, as shown in Fig. 2b, the magnetization in a constant nonzero field never goes to zero at finite temperatures. No transition occurs because, in a field, the ferromagnetic and paramagnetic states are qualitatively indistinguishable. The distinction between the states reappears when the field is reduced to an infinitesimal value. In that case, above T_C, x is vanishingly small and $B_S(x)$ becomes simply $(S+1)x/3S$, where S is the effective spin. Equations (3a) and (3b) can then be combined directly to eliminate x and obtain for the initial susceptibility at temperatures above T_C

$$\chi_0 = C/(T-\theta) \tag{4}$$

which is the Curie–Weiss law, where C (the Curie constant) $= N(S+1)\mu^2/3Sk_B$ and $\theta = TC = C\lambda$, the latter quantity thus reflecting the strength of the exchange coupling. Hence, as the temperature is lowered in the paramagnetic regime, χ_0 diverges at T_C, signalling the onset of the ordering transition to the ferromagnetic state. More specifically, $1/\chi_0$ decreases linearly to zero, as shown in Fig. 2b, and the inverse slope of this variation with temperature is C, which is often expressed as $N\mu_p^2/3k_B$, where μ_p (the effective paramagnetic moment) equals $[(S+1)/S]^{1/2}\mu$, μ being the fully aligned ferromagnetic moment per atom at zero temperature.

Although Eqns. (3a) and (3b) are broadly correct in the description they provide for the advent of ferromagnetism at T_C, they have been derived within the so-called mean-field approximation and thus have certain important limitations. In representing the exchange interactions on each magnetic atom by a mean effective field proportional to the magnetization M, no account is being taken of any correlations between the orientations of the magnetic moments of neighboring atoms. However, such correlations do exist and they play a vital role in various temperature regimes. First, it is found that the gradual decrease of M with increasing temperature at very low temperatures is more properly described in terms of spin waves (also known as magnons), which correspond to wavelike deviations of the atomic magnetic moments from a perfectly parallel alignment. Theoretically, spin waves are predicted to give rise to a decrease of M proportional to $T^{3/2}$, in contrast to the mean-field prediction of an exponential decay, and this famous $T^{3/2}$ law has been confirmed by many experiments. Second, magnetic correlations have a crucial effect on the detailed behavior of M and χ_0 just below and above T_C, respectively. For instance, $1/\chi_0$ is predicted to vary as $(T-T_C)^\gamma$, with the "critical" exponent γ having a value near 4/3, as opposed to the value of unity prescribed by mean-field theory (see Eqn. (4)). Indeed, the various exact calculations of the highly correlated behavior of a ferromagnet near its Curie point transition and their subsequent experimental confirmation have been an exciting modern development with important implications for the many analogous transitions in nature (condensation of fluids, atomic ordering in alloys, etc.).

When the superexchange coupling is antiferromagnetic, as it is in most transition group and rare-earth compounds, the ordered magnetic state that occurs below the critical temperature involves some nonparallel alignment of the ionic moments. In the simplest cases, the magnetic ions collectively subdivide into two interpenetrating crystallographically equivalent sublattices, where all the ionic moments in each sublattice are oriented parallel to each other and antiparallel to the ionic moments in the other sublattice. The resulting configuration of moments, as represented in Fig. 3a, satisfies all the antiferromagnetic interactions between nearest neighbor magnetic ions if these interactions are sufficiently dominant; any weaker interactions of longer range that would prefer another configuration are then said to be "frustrated." However, should the disruptive demands of the latter be strong enough, a compromise is reached and the ionic moments may form a modulated configuration of the type indicated in Fig. 3b, in which the moments are noncollinear (i.e., neither parallel nor antiparallel). In any case, a characteristic feature of any antiferromagnetic configuration (or state) is that the vector sum of all the moments is zero. Hence, the material as a whole has zero magnetization unless a field is applied and the moments respond by canting away from their zero-field directions. The susceptibility of a polycrystalline sample normally shows a sharp but finite peak at the antiferromagnetic ordering temperature, called the Néel temperature T_N. Above T_N, the paramagnetic susceptibility again obeys the Curie–Weiss law, but with θ in Eqn. (4) equalling $-T_N$ in the simple case of dominant antiferromagnetic interactions between sublattices. These characteristic properties of antiferromagnets are derived from mean-field theory but, as in the case of ferromagnets, magnetic correlations produce significant detailed effects at low temperatures (antiferromagnetic spin waves) and at temperatures near the magnetic ordering transition at T_N.

In certain ionic compounds such as the transition group ferrites and the rare-earth iron garnets, the antiferromagnetically coupled sublattices are not equivalent (see *Ferrites, Hard*; *Flux Pinning and the Summation of Pinning Forces*; *Rare-Earth Intermetallics*). Consequently, the ordered ionic moments below T_N do not add up vectorially to zero and there is a spontaneous magnetization. The configuration of moments for this so-called ferrimagnetic state in the case of magnetite (ferrous ferrite: $Fe^{2+}Fe_2^{3+}O_4^{2-}$) is represented schematically in Fig. 3c. The spinel structure of magnetite has two crystallographically distinct types of cation sites: the A sites occupied solely by Fe^{3+} ions and the B sites, which are twice as numerous as the A

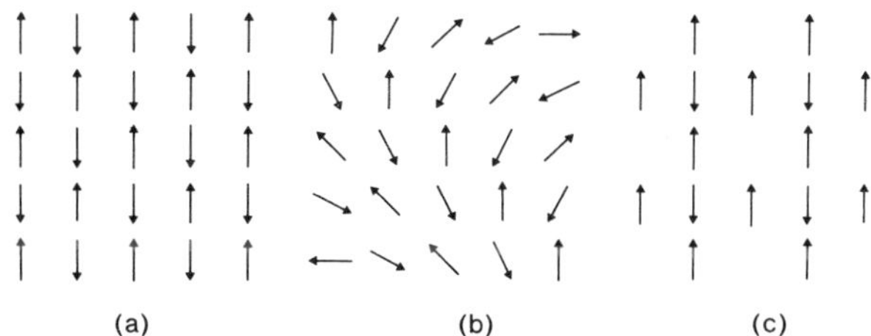

Figure 3
Two-dimensional representations of (a) a collinear antiferromagnetic structure, (b) a noncollinear (modulated) antiferromagnetic structure and (c) a collinear ferrimagnetic structure

sites and are occupied by Fe^{2+} and Fe^{3+} ions in equal numbers. Since the superexchange interactions between A and B sites are strongly antiferromagnetic, the A-site and B-site moments are aligned antiparallel to each other, the weakly antiferromagnetic A–A and B–B interactions being totally frustrated. Thus, the net magnetic moment per formula unit is simply equal to the moment of Fe^{2+}, namely $4\mu_B$. This moment value corresponds to a large spontaneous magnetization (at zero temperature) whose monotonic decrease to zero as the temperature is raised to T_N resembles the temperature dependence of the spontaneous magnetization of a ferromagnet below T_C, as typified in Fig. 2b. In the case of the rare-earth iron garnets, where ferric and rare-earth moments are aligned antiparallel to each other, the rare-earth moments are the larger of the two at low temperatures but, as they decrease more rapidly as the temperature rises, they are eventually exceeded in size by the ferric moments. Hence, the spontaneous magnetization actually goes through zero and reverses direction. The temperature at which this occurs (generally well below T_N) is called the compensation point or temperature (see *Magnetooptic Storage Media*). This exotic behavior is allowed within the context of ferrimagnetism but obviously has no counterpart in ferromagnetism.

In metallic materials, the exchange interactions between magnetic moments involve electrons that are itinerant rather than strictly localized at the atoms. As the conduction electrons move quite freely in a metal crystal, they can exchange interact with the moments of different magnetic atoms sequentially and thus produce an indirect exchange coupling between the atomic moments, known as the RKKY (Ruderman–Kittel–Kasuya–Yosida) coupling. This is the predominant coupling mechanism in nearly all the rare-earth metals and their intermetallic compounds and alloys, where the 4*f* electrons are too tightly bound to be able to participate directly in the interatomic exchange. Indirect exchange interactions via the conduction electrons, being effective over large interatomic distances, also provide the main coupling in dilute alloys of transition group atoms (with stable local moments) in nonmagnetic metal hosts. However, in concentrated transition group alloys and in the pure transition group metals, there is considerable overlap of the *d* electron orbital distributions of neighboring atoms; this gives rise to a direct exchange coupling, which in these cases predominates over the indirect coupling. It should be noted that this *d* electron orbital overlap makes it difficult to distinguish between the intra-atomic exchange that produces the atomic moments (as discussed in Sect. 1) and the direct interatomic exchange that couples these moments together. Nevertheless, it is convenient and reasonably justifiable to retain an operational distinction between these two modes of exchange.

In metals, as much as in insulators, the interatomic exchange coupling can be ferromagnetic (positive) or antiferromagnetic (negative). Indeed, the RKKY coupling via the conduction electrons oscillates in sign as a function of the distance between the interacting atomic moments. Consequently, the ordered magnetic states of the rare-earth metals and their intermetallic compounds run the gamut from simple ferromagnetism (as in pure gadolinium and various Laves-phase compounds) to simple antiferromagnetism, through a rich variety of complex magnetic structures, some of them modulated as illustrated in Fig. 3b. Regarding the direct exchange coupling that prevails in transition group metals, it can obviously be ferromagnetic as in iron, cobalt and nickel, or it can be antiferromagnetic, as in chromium, manganese and even iron when retained in the face-centered-cubic (fcc) structure (e.g., as a fine-particle precipitate in copper).

The solid-solution alloys of transition group metals are often magnetically complicated for two basic reasons, both of which derive from their atomic disorder giving rise to different atoms having different local environments. First, the *d* electrons of a transition group atom, having a fairly broad spatial distribution, explore the local environment and, therefore, the net magnetic moment that they form is affected by it. In the fcc nickel–iron alloys, for example, this effect is small but not negligible, particularly for the nickel atoms whose moments are thought to be enhanced when the number of their iron atom neighbors happens statistically to be large. It is consistent with this premise, incidentally, that atomic ordering at the Ni_3Fe composition results in a small increase of the spontaneous magnetization. In the nickel–copper system, this effect is very much stronger since a nickel atom forms no spontaneous moment at all unless it is surrounded by a large number of nickel atom neighbors, in which case the entire nickel-rich local region becomes what is known as a magnetic cluster, having a total giant moment of about $10\mu_B$. This phenomenon occurs especially in nickel–copper alloys of about 40 at.% nickel, which are almost but not quite ferromagnetic and in which the magnetic clusters behave as superparamagnetic entities.

The second complication posed by magnetic alloys stems from the local environment effect on the exchange interactions experienced by different magnetic atoms. This effect is most pronounced when the exchange coupling between two magnetic atoms may be ferromagnetic or antiferromagnetic, depending on the interatomic distance. Under this condition, since the magnetic atoms in a nonmagnetic host are situated randomly (or nearly randomly), the sum of all the exchange forces on each magnetic atom will be uniquely different in magnitude and direction. Hence, the atomic moments in the alloy will assume all possible orientations. In this so-called spin-glass state, all the atomic moments add up vectorially to zero in zero applied field, which resembles antiferromagnetism but without the translational symmetry of the magnetic atoms in the crystal. Since indirect exchange is spatially oscillatory, as described earlier, the spin-glass state has been found in various dilute magnetic alloys (e.g., manganese in copper or silver, iron in gold) and identified from certain characteristic properties, such as time-dependent remanent magnetizations below the so-called freezing temperature, at which the susceptibility shows a finite peak. In copper–manganese and silver–manganese, where spin-glass properties are observed at manganese concentrations as high as 25 at.%, it would appear that the direct exchange interactions between manganese atoms may also oscillate in sign with the interatomic distance. This would also explain the spin-glass behavior of other concentrated alloys such as palladium–manganese and nickel–manganese. In any case, the spin-glass state appears to be a fairly prevalent phenomenon in disordered magnetic alloys.

3. Magnetic Anisotropy in Materials

In most magnetic materials, the exchange interactions between the atomic magnetic moments are essentially isotropic, in that they depend primarily on the orientations of the moments relative to each other. Ordinary dipolar interactions, which are inherently anisotropic in that they also depend on the spatial locations of the atoms relative to their moment directions, are usually too weak to contribute significantly to any material properties. However, for reasons that involve the symmetry of the environment of the interacting magnetic atoms, the atomic moments generally have a preference of orientation and the resulting magnetic anisotropy effects a variety of properties. Since this is the subject of the article *Anisotropy in Magnetic Materials*, a brief discussion here of the underlying mechanisms is appropriate and sufficient.

Most magnetic materials are composed of small crystallites, within each of which there is a three-dimensional periodic arrangement of the magnetic atoms. In the simplest case, each magnetic atom has the same local environment in which the neighboring atoms (magnetic or otherwise) form a spatial configuration of a particular symmetry. This local crystalline environment imposes an electrostatic "crystal" field on the orbiting electrons of the magnetic atom and causes both its orbital magnetic moment and (via a spin–orbit coupling) its spin to have an orientational preference consistent with the symmetry of the local atomic configuration. This phenomenon, which thus produces energetically preferred (or "easy") directions for the magnetization of a ferromagnet or the sublattice magnetizations of an antiferromagnet, is known as magnetocrystalline anisotropy.

The simplest magnetocrystalline symmetry is one that gives a ferromagnet a single easy axis of magnetization. The anisotropy is then said to be uniaxial and, as such, it pertains to cobalt metal and many of its alloys, whose hexagonal crystal structure has a dominant symmetry axis normal to the basal plane. In general, the uniaxial anisotropy energy is expressible as

$$E_K(z) = -K_1\alpha_z^2 - K_2\alpha_z^4 - K_3\alpha_z^6 - \cdots \quad (5)$$

where α_z is the cosine of the angle between the magnetization vector $\boldsymbol{M}$ and the z axis of symmetry. Axial symmetry requires that only even powers of α_z appear in the expansion. Equation (5) is written with minus signs by convention, so that if all the anisotropy coefficients (K_1, K_2, etc.) are positive or if K_1 is dominant and positive (as in cobalt), $E_K(z)$ is minimized by $\alpha_z = \pm 1$ (i.e., by $\boldsymbol{M}$ directed in either sense along the z axis). In the opposite case of K_1 being dominant but negative, $E_K(z)$ is minimized by $\alpha_z = 0$ (i.e., by $\boldsymbol{M}$ lying in the plane normal to the z axis). There is an interesting (and not unknown) case where K_1 is positive and K_2 is sufficiently negative so that $|K_2| > K_1/2$, for which the minimum $E_K(z)$ state is described by $\alpha_z = \pm(K_1/2|K_2|)^{1/2}$, corresponding to a compromise situation in which $\boldsymbol{M}$ is canted at some angle ($\cos^{-1}\alpha_z$) relative to the z axis.

Though simplest, a uniaxial symmetry material has fairly low symmetry and in Eqn. (5) all even orders of α_z, the direction cosine of $\boldsymbol{M}$, are allowed to play a role. The situation is quite different for the highest (cubic) symmetry possessed by the crystal structures of many ferromagnetic materials (e.g., iron, nickel and their alloys). In this context, it is instructive to derive an expression for the cubic anisotropy energy from Eqn. (5), when truncated for simplicity to its leading three terms. Since cubic symmetry requires invariance with reference to three mutually orthogonal axes (x, y, z), a simple sum is taken of the expressions for $E_K(x)$, $E_K(y)$ and $E_K(z)$, each of them having the truncated form of Eqn. (5) with the same anisotropy coefficients. This sum is reduced by the use of the geometric identity $\alpha_x^2 + \alpha_y^2 + \alpha_z^2 = 1$, which connects the three direction cosines of the magnetization vector $\boldsymbol{M}$. The expression thus obtained for the cubic anisotropy energy is

$$E_K(x,y,z) = K'_0 + K'_1(\alpha_x^2\alpha_y^2 + \alpha_y^2\alpha_z^2 + \alpha_z^2\alpha_x^2) + K'_2\alpha_x^2\alpha_y^2\alpha_z^2 \quad (6a)$$

whose coefficients are related to those of Eqn. (5) by

$$\left.\begin{aligned} K'_0 &= -K_1 - K_2 - K_3 \\ K'_1 &= 2K_2 + 2K_3 \\ K'_2 &= -3K_3 \end{aligned}\right\} \quad (6b)$$

Note that each K' contains only the Ks of higher order, which are normally progressively weaker, and that the K'_0 term is constant and therefore innocuous. Hence, it can be seen why the magnetocrystalline anisotropy coefficients of cubic materials are usually much smaller in magnitude than those of related materials with uniaxial anisotropy. Nevertheless, the magnetic anisotropies of many cubic ferromagnetic materials are fairly substantial and it follows from Eqn. (6a) that, for K'_1 dominant, the easy directions of $\boldsymbol{M}$ are along the cube edges (e.g., $\alpha_x = 1$, $\alpha_y = \alpha_z = 0$) if $K'_1 > 0$ or along the body diagonals (e.g., $\alpha_x = \alpha_y = \alpha_z = 1/\sqrt{3}$) if $K'_1 < 0$. If $K'_1 < 0$, a third simple possibility exists if also $K'_2 > 2.25|K'_1|$, in which case the easy $\boldsymbol{M}$ directions are along the face diagonals (e.g., $\alpha_x = \alpha_y = 1/\sqrt{2}$, $\alpha_z = 0$).

Another more primitive source of magnetic anisotropy in a ferromagnet arises if the material is effectively a dispersion of small single-domain particles. If the particles are approximately ellipsoidal in shape with a single long axis, the demagnetization energy (derived from Maxwell's equations) is lowest when the magnetization vector $\boldsymbol{M}$ lies along the long axis. This so-called shape anisotropy is uniaxial and its energy has the form of Eqn. (5) with z being the long axis and K_1 (dominant and positive) representing a demagnetization coefficient. This equation also applies if the ellipsoidal particles are oblate and have a single short axis (z), in which case the dominant K_1 is negative and $\boldsymbol{M}$ will prefer to lie in the plane normal to z. There is no shape anisotropy if the particles are spherical. Moreover, if the particles are not dilutely dispersed, the magnetostatic interactions between them will effectively reduce the shape anisotropy. These considerations are important in the design of permanent magnets that make primary use of this phenomenon.

A very different type of magnetic anisotropy is one of the characteristic features of certain spin glasses (e.g., the copper-rich alloys of copper–manganese). When such a material is cooled from above its spin-glass freezing temperature in a fixed magnetic field H_{cool}, its hysteresis loop of magnetization M vs field H is found to be very narrow (nearly reversible) and peculiarly asymmetric with respect to the origin. For the two branches of the complete loop, the coercive fields (where $M = 0$) are both negative, such that both remanent magnetizations (at $H = 0$) are positive, the positive direction being that of $\boldsymbol{H}_{cool}$. Thus, $\boldsymbol{H}_{cool}$ defines a single easy direction (rather than axis) of magnetization and the spin glass in its field-cooled metastable state is said to have a unidirectional anisotropy. In this case, the lowest-order term in the anisotropy energy expression is linear in α_z, in contrast to the even powers of α_z in Eqn. (5). The unidirectional anisotropy is thought to derive from a weak spin–orbit coupling acting in consort with the isotropic exchange interactions whose frustration (together with the atomic disorder) produces the random arrangements of magnetic moments in spin glasses.

See also: Magnetic Materials: An Overview; Magnetic Units and Material Specification; Measurements in Magnetic Materials

Bibliography

Berkowitz A E, Kneller E (eds.) 1969 *Magnetism and Metallurgy*. Academic Press, New York

Binder K, Young A P 1986 Spin glasses: experimental facts, theoretical concepts, and open questions. *Rev. Mod. Phys.* 58: 801–976

Chikazumi S 1964 *Physics of Magnetism*. Wiley, New York

Morrish A H 1965 *The Physical Principles of Magnetism*. Wiley, New York

Rado G T, Suhl H (eds.) 1963–73 *Magnetism: A Treatise on Modern Theory and Materials*, Vols. 1–5. Academic Press, New York

White R M 1970 *Quantum Theory of Magnetism*. McGraw–Hill, New York

Zeiger H J, Pratt G W 1973 *Magnetic Interactions in Solids*. Clarendon, Oxford

J. S. Kouvel
[University of Illinois, Chicago, Illinois, USA]

Magnetic Properties of Superconductors

In addition to the restriction of the temperature regime to below the critical temperature T_c, superconductors only tolerate magnetic fields up to a distinct value. If this critical magnetic field is exceeded, superconductivity is destroyed and electric resistance appears as in a normal conductor. Like the critical temperature T_c, the critical magnetic field H_c varies between the different superconducting materials. At low magnetic fields all superconductors expel magnetic flux from their interior. This phenomenon is referred to as the Meissner effect and represents the most fundamental property characterizing superconductors. At higher magnetic fields magnetic flux penetrates into superconductors forming a characteristic magnetic domain structure. Whereas in a type I superconductor the flux carrying domains contain many flux quanta, in a type II superconductor a lattice of single flux quanta is generated (see *Flux Line Lattice*). In the presence of an electric transport current, motion of the magnetic flux structure can occur as a result of the Lorentz force of the current. This process of flux motion causes a resistive voltage and represents the most important mechanism limiting the dissipation-free transport of electric current in superconductors (see *Resistive Transition and Flux Flow in Superconducting Materials*).

1. Meissner Effect, Penetration Depth and Coherence Length

The most fundamental property of a superconductor is its ability to expel magnetic flux from its interior, as displayed in the Meissner effect. Although the conduction of electric current without power dissipation has provided the name for the phenomenon of "superconductivity," the Meissner effect is more fundamental. In the absence of the Meissner effect, a material with infinite electric conductivity would have the peculiar property that, for a given value of temperature and magnetic field, its state would depend on the path along which the values of these two thermodynamic variables had been reached. It is the Meissner effect which removes this dependence on the thermodynamic path and which establishes equilibrium thermodynamics for the superconducting phase transition. Of course, the magnetic flux expulsion is achieved by means of electric shielding currents flowing without power dissipation. From this it is seen that infinite electric conductivity necessarily follows from the observation of the Meissner effect, whereas the opposite conclusion is not possible.

Denoting the free-energy densities in zero magnetic field in the normal and the superconducting phases by $f_n(T)$ and $f_s(T)$, respectively, the difference between them can be found from equilibrium thermodynamics, yielding

$$f_n(T) - f_s(T) = \frac{\mu_0 H_c^2(T)}{2} \tag{1}$$

Here, μ_0 is the vacuum permeability and $H_c(T)$ is the thermodynamic critical magnetic field. For a value $H_c = 10\ \mathrm{kA\,m^{-1}}$, the difference between the energy densities of the normal and superconducting phases amounts to $4 \times 10^{-5}\ \mathrm{J\,cm^{-3}}$.

The perfect diamagnetism displayed by superconducting materials is expressed in the relation

$$M = -\mu_0 H \tag{2}$$

between the magnetization M and the magnetic field H. In a type I superconductor with a geometry where demagnetization effects can be ignored, Eqn. (2) holds up to the critical field $H_c(T)$. At $H = H_c(T)$ the magnetization abruptly drops to the value near zero of the material in the normal state. The temperature dependence of the critical field $H_c(T)$ in good approximation follows the empirical relation

$$H_c(T) = H_c(0)\left[1 - \left(\frac{T}{T_c}\right)^2\right] \tag{3}$$

Looking closely at the surface region of a superconductor in the Meissner state, it is found that the magnetic flux expulsion is not perfect and that the magnetic field penetrates a small but finite distance into the superconductor. This distance is called the penetration depth and typically amounts to a few tens of nanometers. It is within this surface layer that the shielding supercurrents are flowing. The first theoretical description of this magnetic field penetration was provided by the London theory, yielding an exponential decay of the local magnetic flux density $\boldsymbol{h}(x)$ within the superconductor with increasing distance x from the surface:

$$\boldsymbol{h}(x) = \boldsymbol{h}(0)\exp\left(-\frac{x}{\lambda_L}\right) \tag{4}$$

The London penetration depth λ_L is given by

$$\lambda_L = \left(\frac{m}{\mu_0 e^2 n_s}\right)^{1/2} \tag{5}$$

where m, e and n_s are the electron mass, the elementary charge and the single-electron number density of the superconducting electrons, respectively. From the appearance of the density n_s in the denominator of Eqn. (5), λ_L is expected to be temperature dependent and to diverge for $T \rightarrow T_c$. Similarly, a strong increase in the penetration depth is expected in the case of a strongly inhomogeneous superconductor, such as superconducting network or tunnel junction, where the superconducting parts are only weakly connected and the spatially averaged value of n_s is correspondingly reduced. The experimental data on the temperature dependence of the penetration depth λ can be described to a good approximation by the empirical relation

$$\lambda(T) = \lambda(0)\left[1 - \left(\frac{T}{T_c}\right)^4\right]^{-1/2} \tag{6}$$

The experimental values $\lambda(0)$ of the penetration depth at zero temperature are up to five times larger than the quantity $\lambda_L(0)$ obtained from Eqn. (5). This deviation has been explained by Pippard in terms of the superconducting coherence length ξ_0 and by extending the London theory accordingly (see *Superconducting Materials: Physical Properties*). The length ξ_0 describes the minimum distance over which the superconducting properties (e.g., the density n_s) can change and indicates the spatial rigidity of the superconducting wave function. It can be estimated from the uncertainty principle yielding

$$\xi_0 \simeq \frac{\hbar v_F}{\Delta} \simeq \frac{\hbar v_F}{k_B T_c} \tag{7}$$

where $\hbar$ is Planck's constant divided by 2π, v_F is the Fermi velocity, Δ is the superconducting energy gap and k_B is Boltzmann's constant. From the BCS theory (see *Superconducting Materials: BCS and Phenomenological Theories*), a more accurate result can be obtained:

$$\xi_0 = 0.18\frac{\hbar v_F}{k_B T_c} \tag{8}$$

For pure metals typical experimental values of ξ_0 range around 100 nm (see *Coherence Length, Proximity Effect and Fluctuations* for values of ξ_0 in specific materials).

The result shown in Eqn. (5) is only valid for $\xi_0 \ll \lambda_L$. However, this condition is usually not satisfied. For a pure metal, $\xi_0 \gg \lambda$ and the penetration depth is

$$\lambda = \lambda_L\left(\frac{\xi_0}{\lambda_L}\right)^{1/3} \tag{9}$$

2. *Type I and Type II Superconductors*

It is only in type I superconductors that a direct transition from the Meissner state to the normal state and an abrupt drop of the magnetization to zero take place at the critical field $H_c(T)$. However, there also exists a second class of superconductors, referred to as type II superconductors, for which, in increasing magnetic field, the Meissner state is followed by the "mixed state" before the normal state is reached. In this mixed state, single-quantum magnetic flux lines enter the superconductor forming a flux line lattice representing a mixture between the superconducting and the normal phases. The properties of the flux line lattice are discussed in more detail in the article *Flux Line Lattice*. The value of magnetic field at which the mixed state starts to form is called the lower critical field H_{c1}. For $H > H_{c1}$ with increasing magnetic field, the areal density n of the flux lines and the magnetic flux density

$$B = n\Phi_0 \tag{10}$$

increases within the superconductor. Here the quantity

$$\Phi_0 = \frac{h}{2e} = 2.07 \times 10^{-15}\,\mathrm{T\,m^2} \tag{11}$$

represents the magnetic flux quantum. The mixed state extends up to the magnetic field value H_{c2}, called the upper critical field. The thermodynamic critical field H_c defined from the difference between the free-energy densities of the normal and superconducting phases according to Eqn. (1) lies between H_{c1} and H_{c2}: $H_{c1} < H_c < H_{c2}$. Figure 1a shows magnetic flux density vs magnetic field and Fig. 1b magnetization vs magnetic field for a type II superconductor.

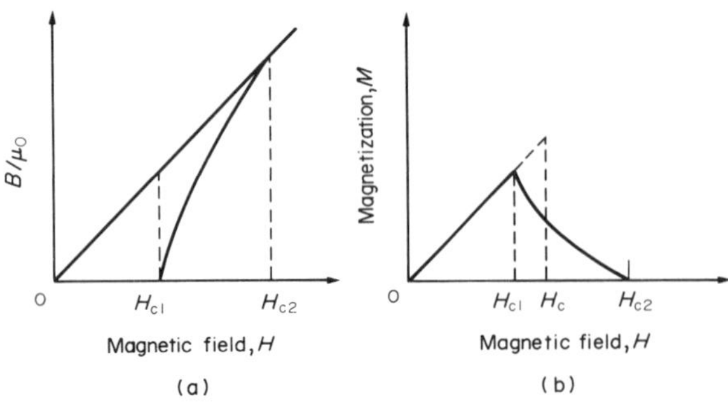

Figure 1
(a) Magnetic flux density/μ_0 and (b) magnetization vs magnetic field for a type II superconductor with a demagnetization factor $D = 0$

A phenomenological understanding of type I and type II superconductivity has been provided by the Ginzburg–Landau (GL) theory. Here, the superconducting property is described by an order parameter which is taken as a complex wave function:

$$\Psi = |\Psi| \exp(i\varphi) \tag{12}$$

The absolute value $|\Psi|$ as well as the phase φ can be space and time dependent. The squared amplitude $|\Psi|^2$ is identified as the Cooper pair density or the superconducting energy gap. The GL theory is restricted to the temperature regime close to T_c where the order parameter is small. By expanding the free-energy density in powers of the order parameter, including a term containing the spatial variations of Ψ and of the magnetic vector potential $\boldsymbol{A}$, and by applying a standard variational procedure for minimization, the two differential equations of the stationary GL theory are obtained (see *Superconducting Materials: Physical Properties*). One equation has the form of the Schrödinger equation for an eigenvalue of the energy, whereas the other equation represents the quantum mechanical description of a current. The GL theory has been further extended to also treat time-dependent phenomena.

One of the simplest problems that can be treated by the GL theory concerns the spatial deformation of the order parameter. As the characteristic length scale, the temperature-dependent coherence length $\xi(T)$ is found to be related to the Pippard coherence length ξ_0 of Eqns. (7) and (8). For a pure material with $t = T/T_c$

$$\xi(t) = 0.74\xi_0 \left(\frac{1}{1-t}\right)^{1/2} \tag{13}$$

The magnetic penetration depth $\lambda(T)$ can also be found from the GL theory, yielding the expression

$$\lambda(t) = \frac{\lambda_L(0)}{\sqrt{2}} \left(\frac{1}{1-t}\right)^{1/2} \tag{14}$$

for a pure material. Here $\lambda_L(0)$ is the London result in Eqn. (5) at zero temperature. The temperature dependence shown in Eqns. (13) and (14) is generally valid, independent of the purity of the material. Since the two characteristic lengths $\xi(T)$ and $\lambda(T)$ show the same proportionality to $[T_c/(T_c - T)]^{1/2}$ near T_c, it is useful to introduce the dimensionless GL parameter

$$\kappa = \frac{\lambda(T)}{\xi(T)} \tag{15}$$

Another important quantity that can be derived from the GL theory is the wall energy associated with the interface separating a superconducting from a normal domain. This wall energy per unit surface is written as $[\mu_0 H_c^2(T)/2]\delta$. The characteristic length δ is called the wall energy parameter. Since the transition of the order parameter from its value in the superconducting domain to zero in the normal domain can take place only over the distance $\xi(T)$, the existence of the interface causes a loss in superconducting condensation energy approximately given by $[\mu_0 H_c^2(T)/2]\xi(T)$ per unit area. However, due to the magnetic flux penetration into the superconducting domain over the distance $\lambda(T)$, in this layer, no gain and, hence, no loss in condensation energy is possible. From this, the expected result for the wall energy parameter is

$$\delta \simeq \xi(T) - \lambda(T) \tag{16}$$

As an important consequence of Eqn. (16) it can be seen that the wall energy can be positive or negative, depending on which of the two lengths $\xi(T)$ and $\lambda(T)$ is larger and, hence, on the magnitude of κ.

The exact numerical result obtained from the GL theory is as follows: $\delta = 0$ for $\kappa = 1/\sqrt{2}$; $\kappa < 1/\sqrt{2}$ gives $\delta > 0$ (positive wall energy) and type I superconductivity, whereas $\kappa > 1/\sqrt{2}$ gives $\delta < 0$ (negative wall energy) and type II superconductivity. It is the negative wall energy in type II superconductors that favors the formation of the mixed state with the finest possible spatial dispersion of the magnetic flux, each magnetic domain containing only a single flux quantum.

Equations (13) and (14) refer to the clean limit where the electron mean free path l is large compared with the coherence length ξ_0. The electron mean free path limits the electric conductivity in the normal state. The opposite situation with $l \ll \xi_0$ is referred to as the dirty limit. In the dirty limit, instead of Eqns. (13) and (14),

$$\xi(t) = 0.85 \left[\frac{\xi_0 l}{(1-t)}\right]^{1/2} \tag{17}$$

and

$$\lambda(t) = \frac{\lambda_L(0)}{[2(1-t)]^{1/2}} \left[\frac{\xi_0}{1.33 l}\right]^{1/2} \tag{18}$$

For the GL parameter of Eqn. (15),

$$\kappa = 0.96\lambda_L(0)/\xi_0 \quad \text{(clean limit)} \tag{19}$$

and

$$\kappa = 0.715\lambda_L(0)/l \quad \text{(dirty limit)} \tag{20}$$

The clean and dirty limits can also be expressed in terms of the gap frequency $\omega_g = 2\Delta/\hbar$ and the scattering time $\tau = l/v_F$ through the relations $\omega_g\tau \gg 1$ and $\omega_g\tau \ll 1$, respectively.

In general, pure metals lead, indium and tin are type I superconductors, whereas alloys and impure materials are type II superconductors. The recently discovered high-T_c oxide superconductors are of extreme type II with large values of κ (up to several hundred).

The following are useful relations for the lower and upper critical fields in a type II superconductor:

$$H_{c1} \simeq \frac{\Phi_0}{4\pi\mu_0\lambda^2}\ln\kappa \simeq \frac{H_c}{\sqrt{2}\kappa}\ln\kappa \tag{21}$$

$$H_{c2} = \frac{\Phi_0}{2\pi\mu_0\xi^2} = \sqrt{2}\kappa H_c \tag{22}$$

3. *Demagnetization and Magnetic Flux Structures*

Complete flux expulsion in the simple form of the Meissner effect only takes place when the dimensions of the superconductor perpendicular to the applied magnetic field are small such that demagnetization effects can be neglected. However, in general, flux expulsion leads to a magnetic field enhancement at the edges of the superconductor. As a consequence, the field H_e at the edge is given by

$$H_e = H/(1-D) \tag{23}$$

where D is the demagnetization factor (see *Demagnetizing Factors*). For a thin wire or plate oriented parallel to the applied field H, $D \simeq 0$. If the wire and plate are oriented perpendicular to H, $D = \frac{1}{2}$ and $D \simeq 1$, respectively. The field enhancement due to the factor $(1-D)^{-1}$ restricts the Meissner effect to the regimes $H < H_c(1-D)$ and $H < H_{c1}(1-D)$ in type I and type II superconductors, respectively.

For a type I superconductor, in the regime $H_c(1-D) < H < H_c$, magnetic domains are generated in which the local magnetic flux density is equal to $\mu_0 H_c$. These normal domains are separated by regions that are locally in the Meissner state. This situation is referred to as the intermediate state of a type I superconductor and is shown schematically in Fig. 2. The first theoretical treatment of this domain structure was given by Landau. It involves the minimization of the total free energy taking into account three contributions:

(a) the wall energy of the interface between the superconducting and normal domains;

(b) the excess energy contained in the nonuniform magnetic field outside the superconductor; and

(c) an excess energy due to the broadening of the normal domains near the surface of the superconductor.

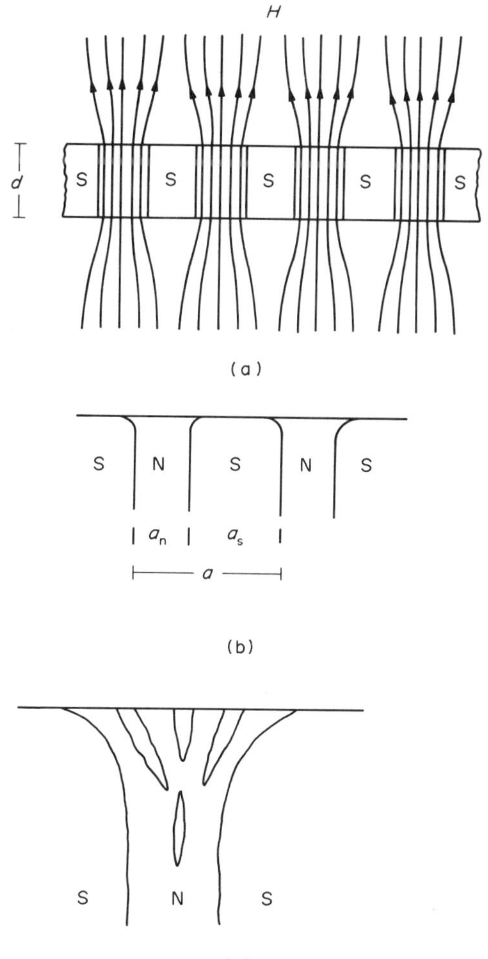

Figure 2
Intermediate state of a type I superconductor: (a) schematic of the normal domains penetrated by magnetic flux and separated from each other by the superconducting phase S, (b) nonbranching model with periodicity length a and (c) branching model with magnification increased to indicate detailed structure

From this Landau domain theory it is possible to calculate the periodicity length a indicated in Fig. 2b, representing the characteristic length scale of the domain structure. In its simplest form, the theory yields

$$a = a_n + a_s = \left[\frac{\delta d}{f(H/H_c)}\right]^{1/2} \qquad (24)$$

where δ is the wall energy parameter given in Eqn. (16), d is the thickness of the plate measured along the magnetic field direction as shown in Fig. 2a and $f(H/H_c)$ is a numerical function of the reduced field H/H_c. An expression analogous to Eqn. (24) is obtained for the domains caused by magnetic ordering as, for example, in a ferromagnet (see *Domains and Domain Walls*).

The Landau theory provides a good semiquantitative understanding of the experimental observations for type I superconductors. In its simplest form it is restricted to the two-dimensional case, neglecting any variation along the direction perpendicular to the plane of the diagram in Fig. 2. In the intermediate state there are long laminar normal domains or flux tubes with nearly circular cross section and both kinds of domains contain many flux quanta. A typical domain pattern observed in a type I superconducting lead film is shown in Fig. 3 for different perpendicular magnetic fields. As can be seen from Eqn. (24), the domain size decreases with decreasing thickness d of the sample. Eventually, below a critical thickness of the specimen film, the normal domains are reduced to single flux quanta and the material behaves as a type II superconductor.

If the magnetic field is oriented exactly perpendicular to the surface of a large plate or film, the long normal or superconducting laminae are statistically oriented in the sample plane, as shown in Fig. 3. If the magnetic field is inclined having a finite component within the surface plane of the sample, this isotropic symmetry is broken and the laminae become oriented parallel to this field component. This effect was first demonstrated by Sharvin and this Sharvin geometry has been very useful for accurately determining the periodicity length of the magnetic domain structure.

The Landau model discussed so far deals with simply connected, straight laminae of normal phase embedded in the superconducting phase and is referred to as the nonbranching model. A refinement of this model has been carried out by Landau assuming that branching of the normal laminae into smaller parts takes place near the sample surface. The branches are separated from each other by inclusions of the superconducting phase. This Landau branching model is shown schematically in Fig. 2c.

In the case of a finite demagnetization coefficient, the mixed state of a type II superconductor is established in the regime $H_{c1}(1-D) < H < H_{c2}$.

4. Fluxoid and Magnetic Flux Quantum

As previously mentioned for type II superconductors, the mixed state consists of a lattice formed by single-quantum magnetic flux lines. In the following, magnetic flux quantization in superconductors will be derived from general arguments.

The initial requirement is that the complex order parameter of Eqn. (12) must be single valued at any point in the superconductor. Therefore, the phase $\varphi(\boldsymbol{r})$ must change only by integer multiples of 2π following a complete loop along the path of integration, yielding

$$\oint \nabla\varphi . d\boldsymbol{s} = N2\pi \qquad (25)$$

where $d\boldsymbol{s}$ denotes the path element of integration and N is an integer. In a magnetic field the generalized momentum of the Cooper pairs is given by

$$2m\boldsymbol{v}_s = \hbar\nabla\varphi - 2e\boldsymbol{A} \qquad (26)$$

where $\boldsymbol{v}_s$ is the Cooper pair velocity. Inserting Eqn. (26) into Eqn. (25) and multiplying by $(2e)^{-1}$ gives

$$\frac{1}{2e}\oint (2m\boldsymbol{v}_s + 2e\boldsymbol{A})d\boldsymbol{s} = N\frac{h}{2e} \qquad (27)$$

The left-hand side of Eqn. (27) is referred to as the fluxoid introduced by F. London. If the dimensions of the superconducting object are sufficiently large, an integration path can be taken along which $\boldsymbol{v}_s = 0$. Then applying Stokes' theorem during the integration gives the magnetic flux quantum of Eqn. (11).

Fluxoid quantization is the macroscopic analog of the quantization of angular momentum in an atomic system. It clearly characterizes a superconductor as a macroscopic quantum system, an aspect first emphasized by F. London.

The first experimental demonstrations of magnetic flux quantization in a superconductor were reported in 1961 by Doll and Näbauer and by Deaver and Fairbank. To illustrate the magnitude of φ_0 it is noted that 1 cm^2 penetrated perpendicularly by the earth's magnetic field contains about 10^6 flux quanta.

The structure of an isolated flux line carrying a single flux quantum and representing the building block of the mixed state is shown schematically in Fig. 4. In the center of the flux line the order parameter (density of the superconducting electrons) reaches zero, the depression of the order parameter extending radially up to the distance ξ (Fig. 4a). The magnetic flux density reaches a maximum value in the center and vanishes approximately exponentially beyond a radius of the order of the penetration depth λ (Fig. 4b). The flux density in the center is about equal to H_{c1} given by Eqn. (21). Crudely speaking, a flux line can be taken as a tube of normal phase with radius ξ embedded in the superconducting phase. The density j_s of the circular supercurrents generating the field $\boldsymbol{h}(\boldsymbol{r})$ of the flux line reaches its maximum near the radius λ and vanishes in the normal core (Fig. 4c). Since flux lines are generated by circular supercurrents, they are often also called vortex lines. Isolated vortex lines with the structure shown in Fig. 4 and with negligible

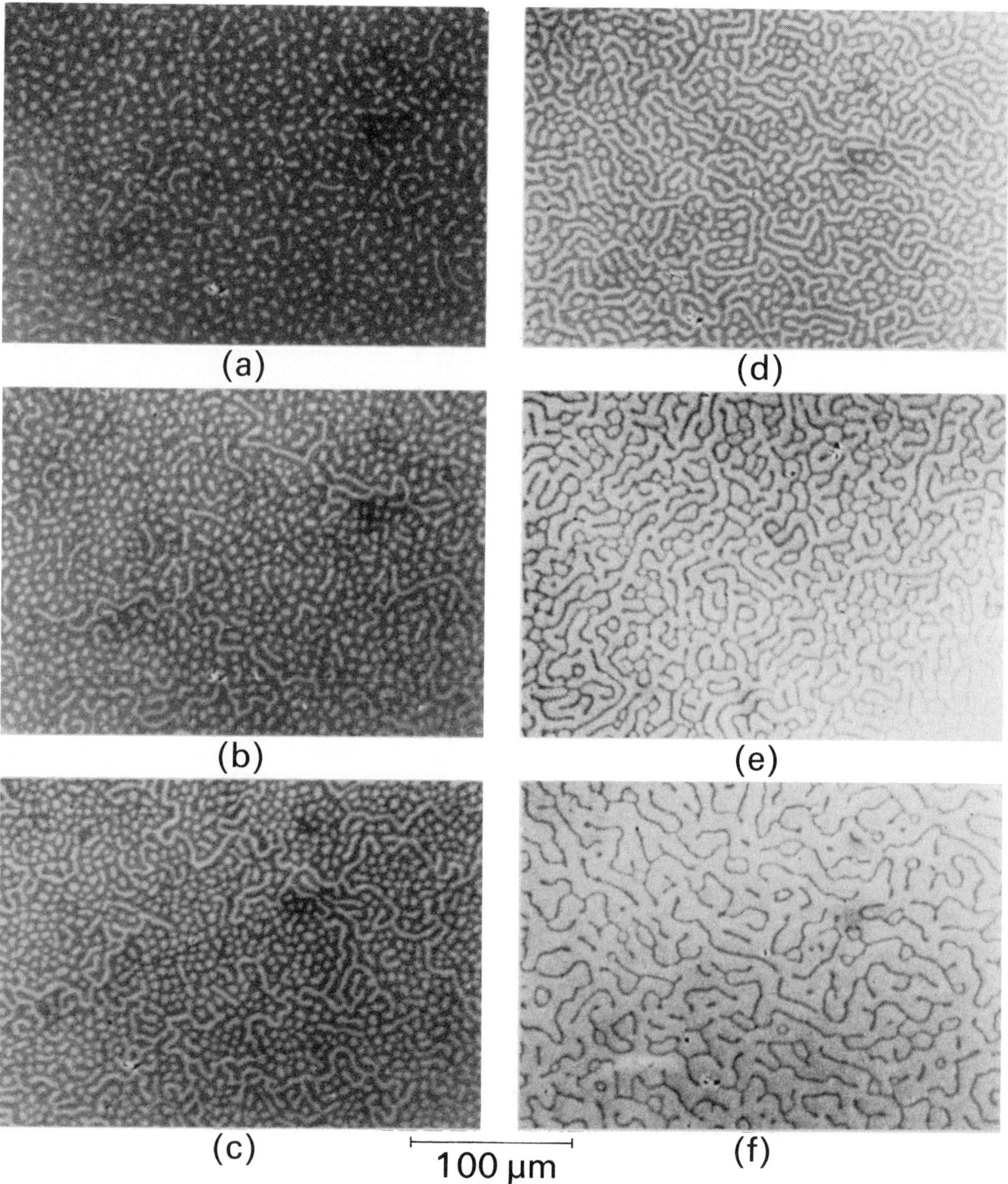

Figure 3
Intermediate-state structure of a lead film in a perpendicular magnetic field displayed magnetooptically for the following field values: (a) 7.6 kA m^{-1}, (b) 10.5 kA m^{-1}, (c) 14.2 kA m^{-1}, (d) 17.4 kA m^{-1}, (e) 27.8 kA m^{-1}, (f) 32.7 kA m^{-1}; normal phase is bright, $T = 4.2$ K, film thickness = 9.3 μm

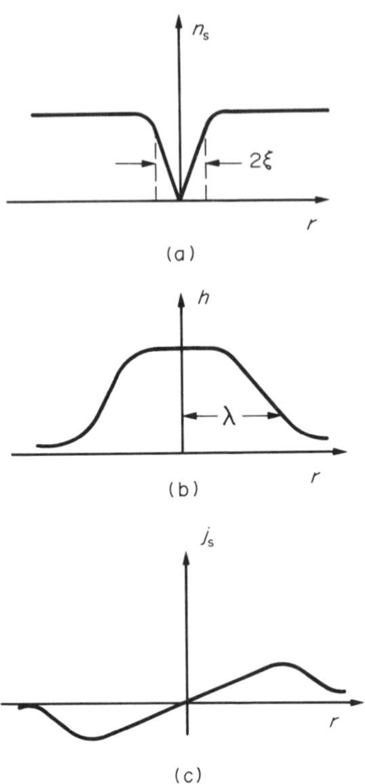

Figure 4
Structure of an isolated flux line: (a) density of the superconducting electrons, (b) magnetic flux density and (c) supercurrent density vs the distance from the vortex axis

interaction with other vortex lines are realized at magnetic fields only slightly larger than H_{c1}, where the flux line density is sufficiently small.

In a superconducting tunnel junction with an applied magnetic field parallel to the barrier, single-quantum vortex lines are also formed. However, due to the strongly weakened superconductivity in the barrier region, these vortex lines, referred to as Josephson vortices, are much larger than in a bulk superconductor (see Eqn. (5) and the appearance of the density n_s in the denominator). When it is necessary to distinguish Josephson vortices from the flux vortices of a type II superconductor the latter are termed Abrikosov vortices after A. A. Abrikosov who predicted the flux vortex lattice in 1957.

5. *Experimental Observation of Magnetic Flux Structures*

The magnetization of a superconducting specimen can be measured using different techniques (ballistic method, electronic integration method, vibrating-sample magnetometer) (see *Measurements in Superconducting Materials*). However, magnetization measurements only yield global information integrated over the whole sample volume and do not resolve the microscopic configuration of the magnetic flux structure within the superconductor. On the other hand, detailed knowledge about the microscopic magnetic flux configuration is important because of many aspects. As an example, it is noted that in the presence of an electric current the magnetic flux structure can be set in motion in a superconductor due to the action of the Lorentz force. This process of flux motion results in a resistive voltage and represents the most important mechanism limiting the dissipation-free transport of electric current in superconductors. A detailed discussion of this subject can be found in the article *Resistive Transition and Flux Flow in Superconducting Materials*. Clearly, our understanding of the flux motion process benefits from an accurate knowledge of the magnetic flux structure. In Sects. 5.1–5.5, the most important experimental techniques used for investigating magnetic flux structures in superconductors with high spatial resolution will be briefly discussed.

5.1 Neutron Diffraction

Neutron diffraction represents the only method yielding microscopic information on the magnetic flux structure in the interior of a bulk superconductor. The method is based on the interaction between the magnetic moment of the neutrons and the modulated magnetic flux density in the superconductor. Usually, the incoming neutrons are directed perpendicular to the direction of the vortex lines. According to the Bragg equation,

$$2a_{hk} \sin \tfrac{1}{2}\theta_{hk} = \lambda_n \tag{28}$$

maxima of the scattered neutron intensity appear at certain scattering angles θ_{hk}. Here, a_{hk} is the flux line lattice spacing between the hk lattice planes and λ_n is the wavelength of the neutrons. Since the flux line lattice spacing is typically 100 nm and since reasonable neutron beam intensities can be obtained up to about 1 nm wavelength, it can be seen from Eqn. (28) that small scattering angles must be resolved. Hence, these experiments have been performed using a special small-angle scattering apparatus.

Neutron diffraction experiments have yielded important information on the crystallographic perfection of the vortex lattice in type II superconductors and on the correlations between the vortex lattice and the atomic lattice of the superconducting crystal. In addition, the microscopic field distribution $\boldsymbol{h}(\boldsymbol{r})$ of an individual vortex line has been determined by means of the form factor, which represents the normalized Fourier transform of the two-dimensional field distribution $\boldsymbol{h}(\boldsymbol{r})$ of a flux line.

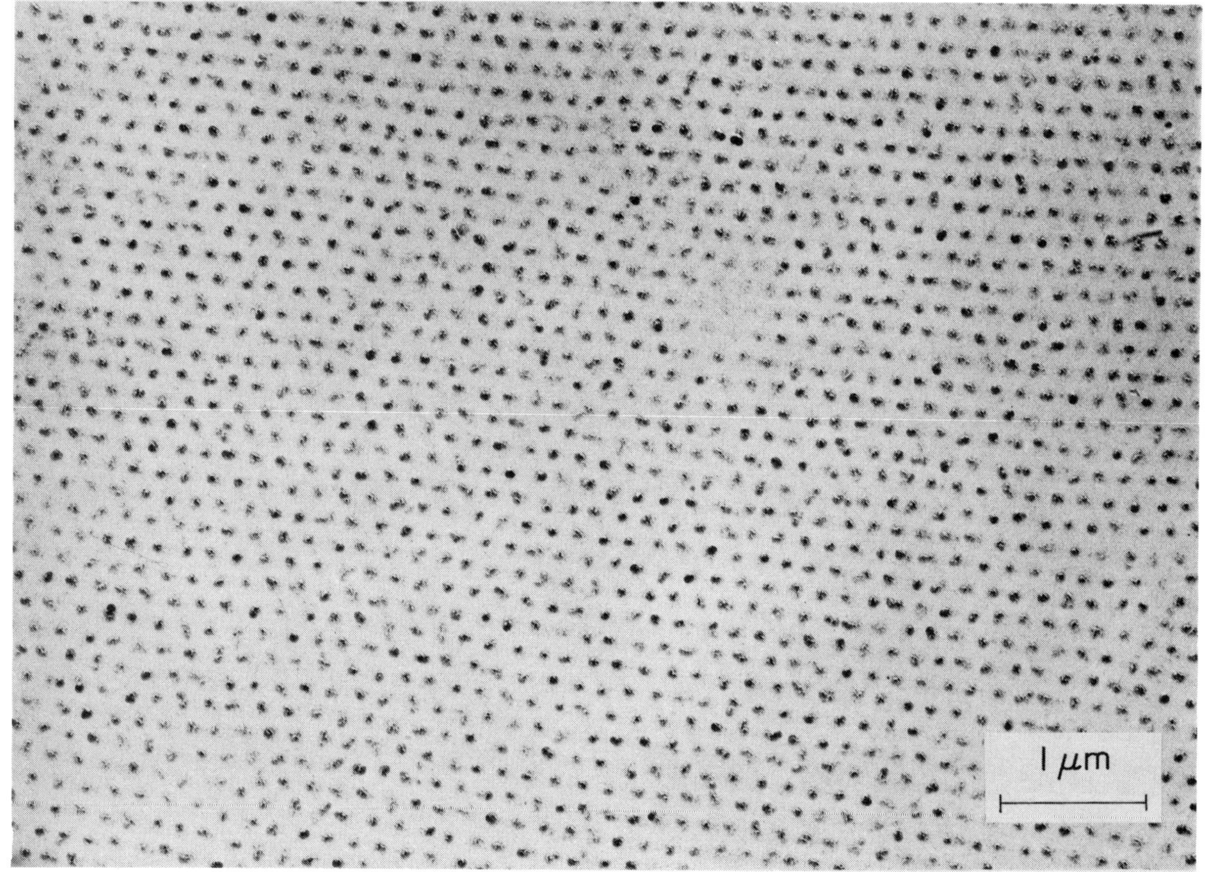

Figure 5
Bitter image of the triangular vortex lattice in a niobium disk of 0.5 mm thickness and 4 mm diameter for a perpendicular magnetic field of 78.8 kA m^{-1} at 1.2 K

5.2 Bitter Technique

In contrast to neutron diffraction, the Bitter technique and the magnetooptic method discussed in Sect. 5.3 only probe the spatial inhomogeneity of the magnetic field protruding from the surface of the superconductor.

In the early 1930s, F. Bitter successfully utilized small ferromagnetic particles for decorating the magnetic domain structure in ferromagnetic materials (see *Magnetization Imaging*). If the particles are sprinkled on the specimen surface, they are attracted to the regions with the largest value of the local magnetic field protruding from the sample. In 1966, Träuble and Essmann adopted this method for the two-dimensional imaging of the mixed state in a type II superconductor. To obtain sufficient spatial resolution they had to use ferromagnetic particles with diameters as small as 8 nm and electron transmission microscopy in combination with a carbon replica method. The preferential deposition of the extremely small particles at the specimen surface requires the additional condition

$$\boldsymbol{m}.\nabla \boldsymbol{H} > 3k_B T \tag{29}$$

where $\boldsymbol{m}$ is the magnetic moment of a particle and $\boldsymbol{H}$ is the spatially varying magnetic field at the sample surface. Hence, low temperatures are favorable for this decoration. A spatial resolution as high as about 10 nm has been achieved. As an example of the application of this technique, Fig. 5 shows the triangular vortex lattice in the mixed state of a niobium disk observed by the high-resolution Bitter method.

The decoration experiments of Träuble and Essmann provided the first direct experimental verification of the existence of the flux line lattice in a type II superconductor. Subsequently, this method has yielded important information on various types of lattice defects appearing in the vortex lattice. More recently, the method has also been applied successfully to the two-dimensional imaging of the flux line

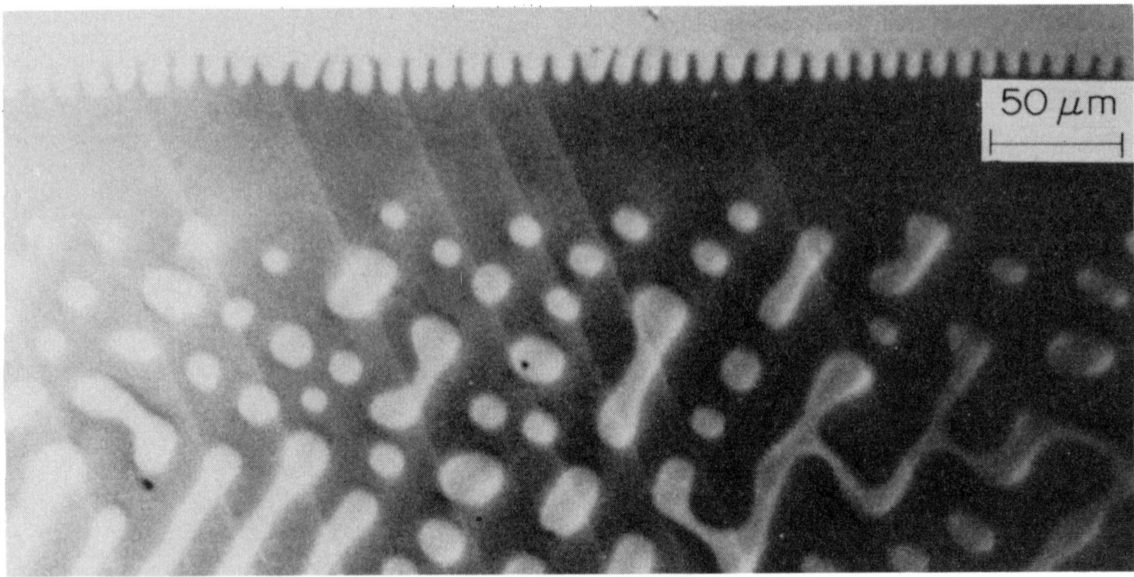

Figure 6
Magnetooptic image of the intermediate state of a monocrystalline mercury plate with 106 μm thickness in a perpendicular magnetic field $H = 0.29H_c$; $T = 1.7$ K, normal phase is bright

configurations in high-T_c oxide superconductors. Usually, scanning electron microscopy (SEM) is used for inspection of the decorated specimen surface.

5.3 *Magnetooptic Imaging*

Faraday rotation of the polarization vector of a beam of polarized light can serve for two-dimensional imaging of the magnetic flux structure at the surface of a superconductor. In this case, the surface of the superconducting sample carries a highly reflecting film which is subsequently covered with a suitable magnetooptic material. Usually, the applied magnetic field is oriented perpendicular to the sample surface. Since the inhomogeneous magnetic flux structure within the superconductor extends some distance beyond the sample surface, the magnetic field in the magnetooptic material is also spatially modulated. A beam of polarized light, passing perpendicularly through the magnetooptic layer and being reflected back, undergoes Faraday rotation (see *Magnetooptics*). The total rotation angle is given by

$$\theta_F = VB2t \tag{30}$$

where V is the Verdet constant, B is the local flux density in the magnetooptic layer and t is the thickness of the magnetooptic layer. It is noted that during both passages of the beam through the magnetooptic material (before and after reflection) Faraday rotation takes place in the same direction. Therefore, twice the thickness t appears in Eqn. (30). By passing the reflected light beam through an analyzer, a magnetooptic contrast is obtained which images the spatial modulation of the magnetic field in the magnetooptic material (see *Magnetooptic Materials*).

An important advance in this magnetooptic method was achieved by Kirchner in 1968 when he discovered that films of the rare-earth compounds EuO, EuS and EuSe of only 100–200 nm thickness yield sufficient Faraday rotation and magnetooptic contrast. Using such films, a spatial resolution limit of about 0.5 μm has been obtained, this limit being determined by the wavelength of the light. The magnetic resolution of this method is about 2 kA m^{-1}.

The pattern of the intermediate-state structure of a lead film shown in Fig. 3 was imaged magnetooptically. As another example, Fig. 6 displays the intermediate state of a monocrystalline mercury plate near the sample edge. The edge runs horizontally close to the top of the figure. A configuration of flux tubes and extended normal laminae can be seen exhibiting a pronounced anisotropy resulting from the anisotropy of the crystal lattice. The flux-free region running horizontally along the upper sample edge originates from the Gibbs free-energy barrier for magnetic flux penetration. The typical fingerlike magnetic flux pattern on the outside of this barrier region can also be seen clearly.

The overall pattern of trapped magnetic flux in a type II superconductor due to flux pinning effects can also be observed well by the magnetooptic technique. Similar to the situation of the domain structure in a ferromagnet, the total free energy of the system is lowered by forming several domains with opposite polarity instead of a single-domain configuration. The

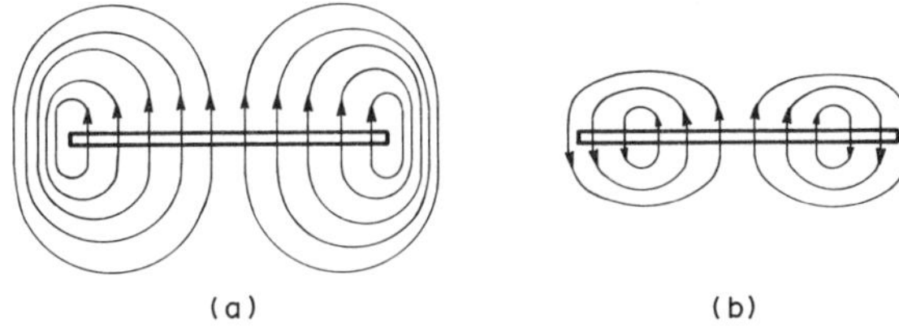

Figure 7
Origin of the domain structure of trapped magnetic flux in a superconductor: by replacing (a) the large single domain by (b) several smaller domains of both polarities, the magnetic field energy is reduced at the expense of the wall energy associated with the domain interface

geometrical shape of the resulting domain structure is closely related to the symmetry and overall shape of the sample. This principle is outlined in Fig. 7 and the domain structure of the trapped magnetic flux observed magnetooptically in niobium foil is shown in Fig. 8. Of course, the spatial resolution is insufficient for imaging the individual flux quanta in the mixed state of the domains.

The spatial resolution of the magnetooptic method is almost two orders of magnitude lower than that of the Bitter technique and the flux line lattice in a type II superconductor cannot be resolved magnetooptically. However, magnetooptic experiments have yielded important information on the magnetic flux structures in type I superconductors, particularly those in the form of thin films. In contrast to the Bitter technique, the magnetooptic method yields high time resolution for studying dynamic phenomena. Using a stroboscopic setup, a time resolution of better than 100 ns has been achieved magnetooptically.

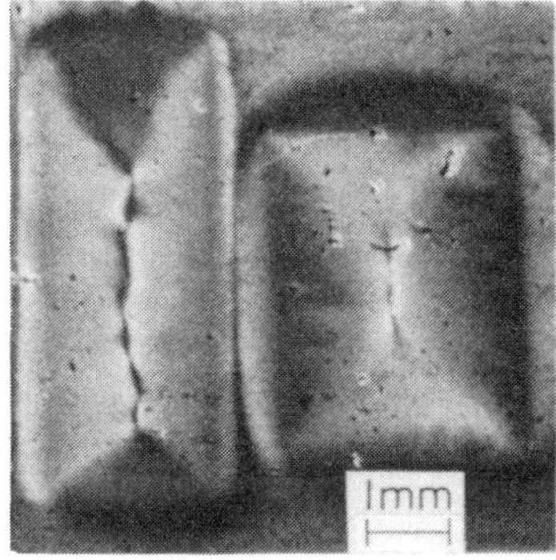

Figure 8
Trapped flux pattern observed magnetooptically in two rectangular niobium foils of 18 μm thickness after the magnetic field cycle $0 \rightarrow 0.13\,T \rightarrow 0$, $T = 4.2\,K$ (left, annealed; right, not annealed)

5.4 Low-Temperature Scanning Electron Microscopy

By scanning a current-biased superconducting thin film with the electron beam of a scanning electron microscope (SEM) and by recording the beam-induced resistive voltage change $\delta V(x, y)$ along the specimen as a function of the coordinate point (x, y) of the electron beam focus on the sample surface, the two-dimensional imaging of spatial structures in the specimen is possible. For such experiments an SEM equipped with a low-temperature stage is necessary and this technique is referred to as low-temperature scanning electron microscopy (LTSEM). The two-dimensional imaging process is based on the small thermal perturbation (local heating effect) of the specimen due to the beam irradiation. The maximum spatial resolution achieved by LTSEM for superconducting thin films is about 1 μm. This method has been applied successfully for imaging the incomplete Meissner effect due to the Josephson penetration depth and the different vortex states of superconducting tunnel junctions in a barrier-parallel magnetic field. In addition, trapped vortices in tunnel junctions have been studied using this method.

5.5 Scanning Tunnelling Microscopy

Scanning tunnelling microscopy (STM) represents an experimental technique developed recently for investigating the structural surface properties of solids yielding a spatial resolution on the atomic scale. Due to this extremely high resolution, STM appears highly promising for imaging the vortex lattice near the surface of a type II superconductor and the first results were reported during 1989. STM experiments for studying magnetic flux structures in superconductors with extremely high spatial resolution are only in their initial stages, but interesting results can be expected in the near future.

See also: Domains and Domain Walls; Flux Line Lattice; Magnetization Imaging; Resistive Transition and Flux Flow in Superconducting Materials; Transport Currents and Force-Free Current Flow

Bibliography

Bednorz J G, Müller K A 1988 Perovskite-type oxides—the new approach to high-T_c superconductivity. *Rev. Mod. Phys.* 60: 585–600

Binning G, Rohrer H 1987 Scanning tunneling microscopy—from birth to adolescence. *Rev. Mod. Phys.* 59: 615–25

Clem J R 1988 Granular and superconducting-glass properties of the high-temperature superconductors. In: Müller J, Olsen J L (eds.) 1988 *High-Temperature Superconductors and Materials and Mechanisms of Superconductivity*. North-Holland, Amsterdam, pp. 50–5

de Gennes P G 1966 *Superconductivity of Metals and Alloys*. Benjamin, New York

Hess H F, Robinson R B, Dynes R C, Valles J M Jr, Waszczak J V 1989 Scanning-tunneling-microscope observation of the Abrikosov flux lattice and the density of

states near and inside a fluxoid. *Phys. Rev. Lett.* 62: 214–16
Huebener R P 1979 *Magnetic Flux Structures in Superconductors.* Springer, Berlin
Huebener R P 1981 Motion of magnetic flux structures. In: Gray K E (ed.) 1981 *Nonequilibrium Superconductivity, Phonons and Kapitza Boundaries.* Plenum, New York, pp. 621–64
Huebener R P 1988 Scanning electron microscopy at very low temperatures. In: Hawkes P W (ed.) 1988 *Advances in Electronics and Electron Physics*, Vol. 70. Academic Press, New York, pp. 1–78
Tinkham M 1975 *Introduction to Superconductivity.* McGraw-Hill, New York

R. P. Huebener
[Universität Tübingen, Tübingen, Germany]

Magnetic Separation

Magnetic separation is achieved by a combination of a magnetic field and a field gradient that generates a force on magnetizable particles such that paramagnetic and ferromagnetic particles move towards the higher magnetic field regions and the diamagnetic field particles move towards the lower magnetic field regions. The force $\boldsymbol{F}_m$ on a particle is given by

$$\boldsymbol{F}_m = \chi(V_p/2)[\nabla(\boldsymbol{B}.\boldsymbol{B})]/\mu_0 \quad (1)$$

where χ is the magnetic susceptibility of the particle, V_p is the particle volume, $\boldsymbol{B}$ is the magnetic field, ∇ is the field gradient operator and μ_0 is the constant $4\pi \times 10^{-7}\,\mathrm{H\,m^{-1}}$.

Electromagnets in conjunction with an iron circuit have been used to generate a magnetic field in an air gap. Field gradients are produced by shaping the poles or by using secondary poles. Secondary poles consist of pieces of shaped ferromagnetic material introduced into the air gap. There are a large number of machines employing such principles. These machines have been described in detail in the classic works by Taggart (1944) and Svoboda (1987). In this article it is more appropriate to discuss the recent penetration of superconducting magnets and of rare-earth permanent magnet materials into magnetic separation technology. There are many proposals for the use of superconductors and rare-earth materials in magnetic separation but only those machines that have been developed commercially will be discussed here.

There are three ways in which particle separation can be achieved using magnetic methods:

(a) where the difference in magnetic properties between the particles to be separated is sufficiently large;

(b) where the material, although not sufficiently magnetic, can be attached to something that is sufficiently magnetic for separation to be achieved; and

(c) when magnetic ions are to be separated from solution—a chemical or a biochemical treatment is required to produce a magnetic precipitate which can either be extracted itself or attached to a magnetic particle.

1. High-Gradient Superconducting Separators

In the early 1970s the so-called high-gradient magnetic separators were developed which made it possible to extract weakly magnetic colloidal particles from a fluid, liquid or gas stream. High-gradient magnetic separation (see Kolm 1971) is a process in which magnetizable particles are extracted onto the surface of a fine ferromagnetic wire matrix which is magnetized by an externally applied magnetic field. The process, which is used to improve the brightness of kaolin clay, was developed for, and in conjunction with, the kaolin industry in the USA (Kolm 1971, Marston et al. 1971, Oder and Price 1973). This process allows weakly magnetic particles of colloidal size to be manipulated on a large scale at high processing rates so there is, in addition to the clay industry, a large number of potential applications in fields as diverse as the cleaning of human bone marrow, nuclear fuel reprocessing, sewage and waste water treatment, industrial effluent treatment, industrial and mineral processing, extractive metallurgy and biochemical processing.

These kaolin separators consist of an iron-bound solenoid, weighing typically 250 t, which provides a magnetic field within the solenoid of up to 2 T and using typically 0.4–0.5 MW of electric power. The solenoidal space contains a fine ferromagnetic wire matrix that acts as a secondary pole system and which occupies 5–10% of the solenoidal space. The radius of the wire is chosen to meet the needs of the separation process and, for kaolin, type 430 stainless steel is used with a strand radius of commonly 70–80 µm. Field gradients as high as $10^5\,\mathrm{T\,m^{-1}}$ can be achieved. This kind of separator has been described by Marston et al. (1971). The use of a finely divided matrix was previously suggested by Frantz (1937). In this method, a slurry containing paramagnetic particles is passed through the matrix which captures the particles and holds them while the magnetic field is on. Eventually the efficiency of the trapping process becomes reduced by the accumulation of the captured particles. These trapped particles can be released and the efficiency of the matrix restored by switching off the magnetic field or by withdrawing the matrix from the magnetic field and washing the particles from the matrix. The cost per tonne of the clay processing can contain a 60% component due to the magnet power consumption; there is, therefore, a great potential for cost reduction if superconducting coils are used instead of the copper

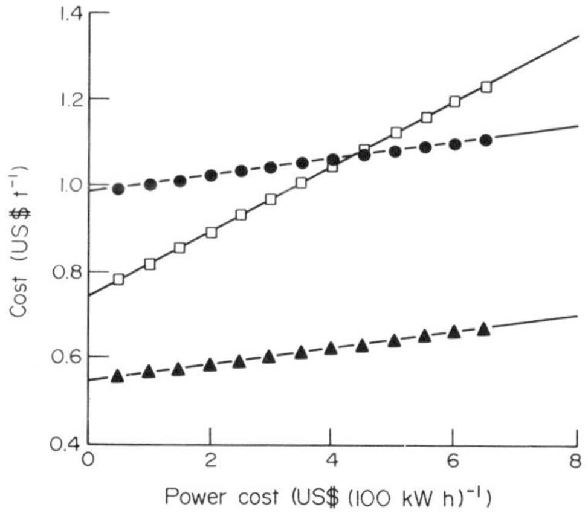

Figure 1
Comparison of cost vs power cost for a conventional separator (□), for the first superconducting separator installed in 1986 (●) and for the improved larger-capacity machine installed in 1989 (▲); costs include the capital cost of the machine depreciated over ten years

coils. A cost comparison for the treatment of kaolin is illustrated in Fig. 1 for a conventional machine, a superconducting machine and an improved superconducting machine.

In 1986 the first large-scale superconducting separator of this type was built and installed by Eriez Magnetics, Erie, Pennsylvania, USA at a kaolin clay plant of J.M. Huber Corporation in Georgia, USA and is shown in Fig. 2. The performance of this plant was so satisfactory that a number of other plants have been built with considerably improved economics.

English China Clays (ECC) pioneered a reciprocating canister system (RCS) in which a superconducting magnet, in the form of an open solenoid, is used alternately by two separation systems (Windle 1977), a method with the great advantage that it avoids switching the superconducting magnet on and off, which costs time and energy. When one system is working on separation the other is being cleaned. A separator of this type was built by Cryogenic Consultants Ltd, London, and is illustrated in Fig. 3. This machine was installed and has been operating in Cornwall, UK at the kaolin plant of Goonvean and Rostowrack since May 1989. The RCS has a number of features that lead to lower processing costs than the switch-on switch-off systems, namely higher magnetic fields obtainable with the open solenoid, higher process feed velocities, short dead time while switching between the two separation systems, lower refrigeration requirements and a radial feed for the kaolin slurry (Watson 1978). The RCS machine is also more compact than the switch-on switch-off system at the same processing capacity and product quality.

Figure 2
Superconducting separator produced by Eriez Magnetics and installed at a J. M. Huber kaolin plant in Georgia, USA; this plant began operation in May 1986

Figure 3
Superconducting reciprocating canister system on test at Cryogenic Consultants Ltd prior to installation in the kaolin plant of Goonvean and Rostowrack, Cornwall, UK

2. *Superconducting and Permanent Magnet Drum Separators*

In the open-gradient separation system the magnetic field and field gradient are produced in a separation region by using specially shaped electromagnets or permanent magnets so that when particles fall through or are guided through the separation region they are deflected by the magnetic force which varies with their magnetic susceptibility and volume according to Eqn. (1). An example of this type of separator is the drum separator. This separator has been used extensively in the minerals industry. Consider a drum within which there are stationary electromagnets producing a magnetic field and a field gradient on a part

Figure 4
Superconducting drum separator built by KHD Humboldt Wedag, Cologne, Germany; the separator is shown concentrating dry magnesite from a Greek source and operating at $80\,t\,h^{-1}$ over a length of 0.8 m

of the drum surface in such a way that part of the drum surface attracts paramagnetic particles by the force described by Eqn. (1). If the drum rotates so that the paramagnetic particles moving with the drum are removed from the magnetic field–field gradient region they can be released from the surface of the drum. Nonmagnetic particles are not held on the drum and, moving solely under the forces of gravity, fluid drag or drum centrifugal forces, can be directed away from the magnetics and therefore separated from them. Drum separators can be used to process wet or dry materials but the configurations are very different and are described by Taggart (1944) and Svoboda (1987).

Permanent magnets have almost completely replaced electromagnets as the field generating elements in drum separators. The magnetic fields and gradients are produced by using three to seven stationary magnets of alternating polarity. The design of drum separators has been closely associated with the development of permanent magnet materials (Svoboda 1987). For example, using barium ferrite, an increasing energy product from $20\,kJ\,m^{-3}$ to $25\,kJ\,m^{-3}$ has allowed an increase in the drum wall thickness which, in turn, allows the length of the drum to be increased from 1.5 m to 3 m. A combination of strontium and barium ferrite has been responsible for the development of drums 4 m in length. Similarly, the magnetic field available at the surface of the drum has increased from 0.16 T to 0.22 T. Samarium–cobalt and neodymium–boron–iron materials hold out great promise for this application as these rare-earth materials have residual magnetism approximately two times greater than that of materials such as alnico. The material cost is the controlling factor. If a higher field and a higher field gradient are required at the surface of the drum, it is usual to return to electromagnetic coils within the drum. In the UK drum separators are manufactured by Box-Mag Rapid Ltd, Birmingham and Eriez Magnetics, Caerphilly.

A superconducting drum separator has been developed by KHD Humboldt Wedag, Cologne, Germany (see Fig. 4); in this machine fields as high as 3.2 T and with field gradients of $100\,T\,m^{-1}$ have been achieved at the surface of a drum of diameter 1.2 m and length 1 m. This is achieved by the use of four racetrack coils of niobium–titanium which operate at 4 K and which form a 120° sector. This sector can be moved through 100° so that both wet and dry separation can be achieved. The drum is constructed from carbon-fiber-reinforced plastic and can be driven at continuously variable velocities between 2 rpm and 30 rpm.

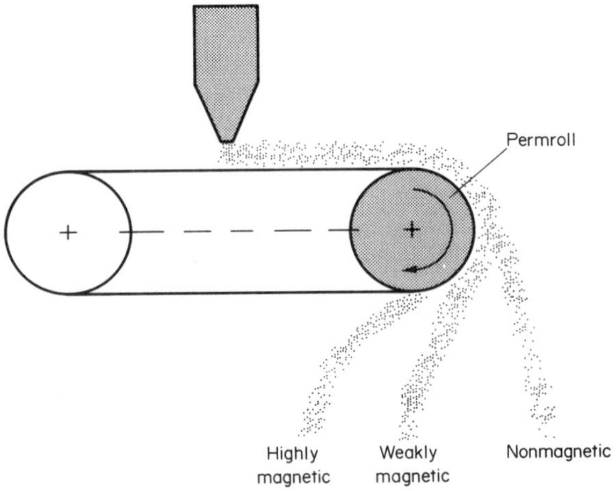

Figure 5
Operation of the Permroll dry magnetic separator; the material is fed from a hopper onto the moving belt enabling various magnetic components to be separated

3. Permanent Magnet Roll Separator

A recent development in permanent magnet separators has been that of the Permroll separator by E.L. Bateman Ltd in the UK. The active part of this separator is a cylindrical roller consisting of disks of samarium–cobalt or neodymium–boron–iron permanent magnets interleaved with mild steel disks. The magnet in this configuration generates a field of up to 1.6 T on the surface of the roll and field gradients of the order of $300\,\mathrm{T\,m^{-1}}$. The roll is covered by a thin endless belt which is held tightly by another roller, parallel to the Permroll. When the Permroll is rotated the endless belt moves continuously over the two rollers. If material is introduced on the top of the moving belt, magnetic particles will be held onto the belt when they pass around the Permroll while the nonmagnetic or weakly magnetic particles will be thrown off, as illustrated in Fig. 5. This machine has proved useful as a dry separator but is not nearly as effective for wet separation.

See also: Magnetic Materials: An Overview; Superconducting Magnets; Superconducting Materials: Large-Scale Applications

Bibliography

Frantz S G 1937 Magnetic separator. US Patent No. 2,074,085
Kolm H H 1971 Magnetic device. US Patent No. 3,567,026
Marston P G, Nolan J J, Lontai L M 1971 Magnetic separator and magnetic separation method. US Patent No. 3,627,678
Oder R R, Price C R 1973 Brightness beneficiation of kaolin clays by magnetic treatment. *Tappi* 56: 75–8
Svoboda J 1987 *Magnetic Treatment of Minerals*. Elsevier, Amsterdam
Taggart A F 1944 *Handbook of Mineral Dressing*. Wiley, New York
Watson J H P 1973 Magnetic filtration. *J.Appl. Phys.* 44: 4209–13
Watson J H P 1978 Magnetic separators, apparatus and method (radial flow). US Patent No. 4,124,503 (British Patent No. 1,530,296, Czech Patent No. 205,022)
Windle W 1977 Magnetic separation, method and apparatus (reciprocating canister). US Patent No. 4,124,503 (British Patent Nos. 1,469,765, 1,599,824 and 1,599,825, Czech Patent No. 180,638 DIV P)

J. H. P. Watson
[University of Southampton, Southampton, UK]

Magnetic Storage Media

Information can be recorded in the form of changes in the strength and direction of the remanent magnetization of the storage material. If the recording is to be permanent, the recorded state of the material must be stable and resistant to change; that is, recording materials have high remanent magnetizations, high coercivities and almost rectangular hysteresis loops. They resemble the materials from which permanent magnets are made except that recording materials have lower coercivities since their magnetic remanence must be changeable by the (limited) magnetic field available from the recording head.

In contrast, the magnetic state of the materials used in the "writing" and "reading" heads must change easily and quickly in response to changes in the head current during writing and changes of the magnetic flux (coming from the recorded medium) during reading. Head materials then, are characterized by very low remanent magnetization and coercivity, but very high saturation magnetization and permeability. The permeability must remain high with increasing frequency to at least the highest frequency to be recorded and read. Since both the coercivity and the remanent magnetization are low in head materials the hysteresis loop has a very small (ideally zero) area. This article is concerned only with those magnetic recording materials that may take the form of continuous thin films or dispersions of submicroscopic, single-domain particles of metallic oxides or metals. The materials used in heads and in magnetooptical recording are discussed in the articles *Recording Head Technology* and *Magnetooptic Storage Media*.

1. Magnetic Recording

Recording is effected by means of a recording head in which the strength and polarity of the external fields are controlled by electrical currents flowing in a conductor that passes around the magnetic circuit of the head. During reading, the recorded pattern of magnetization in the storage medium is inductively coupled to the reading head. Changes in the magnetization pattern passing by the gap in the reading head become changes in the magnetic flux circulating around the head and are detected as voltage changes in a coil of wire wound around the head. No processing of the recorded medium is needed (in the sense that a photographic film must be developed) and so information that has been recorded magnetically may be verified immediately. Since the recording process involves changing the direction of remanent magnetization in the storage medium, it is possible to erase or to update old information by simply recording the new message in its place. Remaining traces of the old message will be reduced to less than 5 dB of the original level. The cycle of writing–erasure–rewriting can be repeated indefinitely since it is an electronic process rather than one involving the movement of atoms or a chemical change. The recorded information is remarkably stable: it is almost unaffected by time, temperature, pressure, ionizing radiation or any physical changes other than the application of strong magnetic fields (greater than the coercivity of the medium) or the destruction of the medium.

Two recording methods are used: analog and digital. In analog recording of audio, video and instrumentation signals, the recorded pattern of magnetization continuously reproduces the varying amplitude and frequency of the incoming signal. Since the relationship between the magnetization and the field from the recording head is otherwise nonlinear, a high-frequency bias tone of constant amplitude is added to the signal to be recorded. Digital recording is used mainly to store data, but is increasingly being used to store audio and image signals as well. Using this method the information is encoded as a sequence of ones and zeros and so only two levels (or two directions) of magnetization are needed. Each binary digit or "bit" is stored in the transition region where one direction of magnetization changes to the opposite direction. Analog or digital recording place different requirements on the storage medium but in either case the readback signal must be large (in comparison with the noise level) over the range of frequencies of the recorded signal.

2. *Properties of Magnetic Recording Materials*

The essential property of magnetic recording materials is the remanent magnetization M_r that persists after the recording field has been removed. For reliable recording M_r must be high enough for the readback signal to exceed random noise by at least 20 dB. However, in a magnetized body having free poles, a demagnetizing field is always present whose strength depends on the value of M_r and on the shape of the body, and whose direction is opposite to that of M_r—hence the name "demagnetizing" field (see *Demagnetizing Factors*). Demagnetization is obviously undesirable. Its magnitude and effects can be reduced by (a) making the magnetized regions thin in directions perpendicular to M_r or (b) making M_r smaller with respect to the coercivity H_c of the medium (i.e., the magnetic field required to reduce the magnetization to zero). H_c should also be high enough to ensure that the recording is not erased either by self-demagnetization or by stray magnetic fields, but not so high that difficulty is experienced in rewriting or erasing the recording medium.

A practical storage material then must have balanced values of M_r, H_c and thickness. The optimal values are determined empirically. For archival storage M_r and H_c should be independent of time, temperature and pressure. The magnetic materials clearly must not react chemically with the atmosphere, the substrate, the binder materials in which magnetic particles are dispersed, or with any lubricant or overcoat material.

The nonmagnetic properties of the recording material are just as important as the magnetic ones and they are generally much more difficult to measure and to control. The dispersion of magnetic particles is a good example. A uniform dispersion can be defined precisely but verified only qualitatively, yet a poor dispersion can spoil the recording performance of the best particles; it can adversely affect both M_r and H_c as well as increase the noise level of the coating through its internal inhomogeneity and its surface roughness. The chemical and physical nature of the surface of the particle affects the strength of the cohesive bond between the particle and the binder polymers and thus affects the durability of the recording medium. The electrical resistivity of the medium must be low enough to prevent the accumulation of electrostatic charges and, in applications where the position of a tape or disk is determined by means of a beam of light, the transmissivity of the medium must be suitably low.

3. *Magnetic Particles*

Magnetic recording coatings can be continuous thin films (usually alloys) or dispersions of submicroscopic magnetic particles in a polymeric matrix. At present, particulate materials dominate flexible media; for example, tapes, of which around 90% use some form of iron oxide particle. Hard disks now use metallic thin films to a large (and increasing) extent.

Single-domain particles (i.e., particles without domain walls which must therefore reverse their direction of magnetization by a rotation of the electron spin) generally have their longest dimension in the range 0.1–1.0 μm. Particles smaller than 0.02–0.03 μm are superparamagnetic and thus are useless for recording. Particles larger than 1 μm are multidomain and usually have coercivities that are too low to support high recording densities. The coercivity of a single-domain particle can arise from shape anisotropy or from magnetocrystalline anisotropy or, as is often the case, from a combination of the two anisotropies (see *Anisotropy in Magnetic Materials*). Magnetoelastic anisotropy is a third possibility but one that is practically unimportant.

Shape anisotropy is uniaxial in most particles and so an aligned array of particles having dominant shape anisotropy would show $\sin 2\theta$ torque curves. Magnetocrystalline anisotropy in materials of cubic structure is, in principle, associated with six, eight or 12 equivalent easy directions of magnetization depending on the easy directions being parallel to $\{100\}$, $\{111\}$ or $\{110\}$ directions. Crystalline anisotropy can be very sensitive to changes in temperature (a distinct disadvantage) in contrast to shape anisotropy which is quite insensitive to temperature.

The five major types of particulate recording materials are described in Sects. 4–8; their properties are summarized in Table 1. Thin-film recording materials are covered in Sects. 9 and 10.

Table 1
Typical properties of magnetic particles

Property	Particle				
	γ-Fe_2O_3	Co–Fe_2O_3	CrO_2	BaO.6Fe_2O_3 + Co, Ti	Fe
Saturation magnetization, M_s (kA m^{-1})	336–340	350–355	340–390	380	870–1100
$\frac{M_r}{M_s}$	0.5	0.5–0.8	0.5	0.6–0.7	0.23–0.52
Curie temperature, T_C (K)	[863]	[863]	390–400	590	1040
Magnetocrystalline anisotropy constant, K_1 (J m^{-3})	-4.64×10^3	-5 to $+100 \times 10^3$	$+2.5 \times 10^4$	$+3.3 \times 10^5$	$+4.4 \times 10^4$
Saturation magnetostriction, λ_S	-5×10^{-6}	-5 to -15×10^{-6}	$+1 \times 10^{-6}$	$+10 \times 10^{-6}$	$+4 \times 10^{-6}$
Coercivity, H_c (kA m^{-1})	20–28	43.8–60	36–53	26–157	30–131
Switching field distribution, $\frac{\Delta H}{H_c}$	0.26–0.61	0.30–0.6	0.35–0.6	0.16–0.6	0.5–0.75
Temperature coefficient of coercivity, $\Delta H_c/H_c$/°C (20–70 °C)	-1×10^{-3}	impreg. -2.4×10^{-3} doped -10×10^{-3}	-5×10^{-3}	$+3.1 \times 10^{-3}$	-0.6×10^{-3}
Specific surface area (BET) (m^2 g^{-1})	20–40	20–40	25–37	15–31	26
Density, ρ	4.60	4.80	4.88–4.95	5.28	5.8
Crystal structure	cubic $a_0 = 0.25$	cubic	tetragonal $a = 0.4218$ nm $c = 0.9182$ nm	hexagonal $c = 2.32$ nm $a = 0.589$ nm	bcc $a = 0.2861$ nm
Ferromagnetic or ferrimagnetic	ferri	ferri	ferro	ferri	ferro
Particle size (μm)	needles $l = 0.3$ $d = 0.06$	needles: $l = 0.3$, $d = 0.06$ equiaxed: $l = 0.2$	needles $l = 0.5$ $d = 0.05$	hex. platelets dia = 0.08–0.1 thick = 0.001–0.025	needles $l = 0.3$ $d = 0.06$

4. γ-Fe_2O_3

The usual method of preparing particles of γ-Fe_2O_3 has evolved from the methods originally used to make particles for pigments. First, small yellow acicular particles of α-(FeO)OH (goethite) are precipitated from $FeSO_4$ using NaOH. These particles are then used as seeds to grow larger particles of goethite. At the end of this step both the size and the aspect ratio of the final particle are determined. The goethite precursor particles are then dehydrated to produce hematite (α-Fe_2O_3) which in turn is reduced to magnetite (Fe_3O_4). These are high-temperature steps that involve the evolution of water and can result in porous particles of inferior properties unless carefully controlled. Finally, the magnetite particles are carefully reoxidized to give γ-Fe_2O_3. All of the steps are critical to the properties of the final particle and have been the subject of numerous patents since the early 1970s. Many variations of the basic process have been described (Bate 1980).

The structure of γ-Fe_2O_3 is cubic. However, the lattice contains vacancies which have been reported to be arranged on a tetragonal superlattice for which $c/a = 3$ where c is the vacancy spacing and $a = 0.833$ nm. The particles as used in recording media are typically acicular of length 0.2–0.7 μm and have an aspect ratio ranging from 3:1 to 10:1. The particles are not usually single crystals but their crystallites share a common ⟨110⟩ axis. The intrinsic properties of γ-Fe_2O_3 are:

(a) saturation magnetization per unit mass $\sigma_s = 74\ \mathrm{A\,m^2\,kg^{-1}}$;

(b) Curie temperature $T_C = 861$ K;

(c) magnetocrystalline anisotropy constant $K_1 = 4.64\ \mathrm{kJ\,m^{-3}}$; and

(d) saturation magnetostriction $\lambda_s = 5 \times 10^{-6}$.

The principal extrinsic property is the coercivity H_c (the field value of the points on the major hysteresis loop where the magnetization $M = 0$). It depends on the distribution of the particle size and acicularity, the packing density and the degree of alignment within the coating. It should be noted that measurements of coercivity made on samples of the powder consisting of irregular clumps of particles are not likely to give the same result as measurements made on a coating of well-dispersed particles. Commercial powders can be obtained having coercivities of 19.9–31.8 $\mathrm{kA\,m^{-1}}$ while the coercivity of the tapes made from these powders is typically about 5% below that of the original powder.

Particles of magnetic tapes and of rigid disks are almost always oriented by the application of a magnetic field during the coating process. In tapes designed for longitudinal recording, that is, the direction of the tracks is parallel to the edge of the tape the particles are usually oriented parallel to the tape and track edge. In tapes used for video casette recorders, the tracks are at an angle of almost 6° to the tape edge and the average direction of the long axis of the particles is parallel to the track rather than to the tape edge. In rigid disks for use in computers the particles are oriented circumferentially. In flexible disks, the disks are punched out from a wide web of coated material. For practical reasons it is not possible to orient the particles. The coercivity of an oriented coating depends *inter alia* on the angle between the average direction of orientation and the direction of measurement.

The switching field of individual particles of γ-Fe_2O_3 has been measured by Knowles (1981) and found to be much higher than the average switching field for a dispersion of particles. Coercivity is an average value of the switching field of particles that have a variety of shapes and sizes and orientation and which may exist as single particles or as clumps containing many particles. The parameter that is used to estimate these variations is the switching field distribution (SFD). It can be measured by differentiating the major hysteresis loop in the vicinity of $H = H_c$ and determining the width at half-amplitude (ΔH) of the differentiated function and dividing by the coercivity H_c. Thus, SFD $= \Delta H/H_c$. A desirably narrow distribution would have SFD $\leqslant 0.2$.

The remanent magnetization M_r of a coating is a key property in determining the amount of flux from the magnetized regions on the disk or tape that can be coupled with the reading head. It depends on a number of factors:

(a) the saturation magnetization M_s (in $\mathrm{kA\,m^{-1}}$), the magnetic moment per unit volume, where $M_s = \rho\sigma_s$, σ_s being the saturation magnetization per unit mass and ρ being the density;

(b) the degree of orientation of the particles (typical orientation ratios are within the range 1.5 to 3.0);

(c) the number n of equivalent easy directions of magnetization within a single crystal of the recording material (e.g. if the easy axis is ⟨100⟩, $n = 6$, if the easy axis is ⟨111⟩, $n = 8$ and if the easy axis is ⟨110⟩, n = 12; the higher the value of n the greater is the remanent magnetization as a fraction of the saturation magnetization); and

(d) the size of the particles (and thus whether they are multidomain, single domain, or superparamagnetic).

Nonmagnetic properties are just as important as magnetic properties in determining satisfactory total performance of the recording material. The important non magnetic properties include the distribution of the shape and size of the particles (determined by transmission microscopy aided by automatic image analysis), specific surface area (determined by gas adsorption and ranging typically from 15 $\mathrm{m^2\,g^{-1}}$ to 30 $\mathrm{m^2\,g^{-1}}$ and porosity (determined by oil adsorption, a typical value being 50 ml per 100 g). The surface chemistry of the particle controls the strength of the bond between the particles and the polymeric binder; it thus plays an important part in determining the durability of the coating—a key property for the user. Unfortunately, there are few practical techniques by which the particle surface may be characterized (Bhushan 1990).

Particles of γ-Fe_2O_3 are relatively easy to make, inexpensive (US\$3.50 kg^{-1} at 1991 prices) and very stable with respect to changes in temperature, pressure and time. Unfortunately, the highest achievable coercivity of 31.8 kA m^{-1} is not high enough either for recording audio and video signals of short wavelength or for recording data at high bit density.

Magnetic particles of Fe_3O_4 are produced as the penultimate step in the preparation of γ-Fe_2O_3 particles. They have higher values of H_c and M_s than γ-Fe_2O_3 particles, but are not used for recording because of a combination of physical and chemical instabilities, real or believed. However, it is established that the recording performance of γ-Fe_2O_3 particles can be improved by reducing them back to Fe_3O_4 and then reoxidizing to Fe_2O_3. The cycle may be repeated several times and usually ends with Fe_2O_3 particles containing less than 0.2% FeO. If, however, the cycle of oxidation–reduction–oxidation is stopped at a composition such that the FeO content is about 20%, very high coercivities (>37.4 kA m^{-1}) can be obtained. The origin of the increased coercivity is thought to lie in the residual stresses that exist in particles having a core of Fe_3O_4 and an outer shell of Fe_2O_3. The surface of this type of particle has been found to act curiously with phosphate groups giving rise to a temporary increase of coercivity to $\lesssim 63$ kA m^{-1}. The remarkable increase in coercivity is, unfortunately, only a temporary one. Washing the particles brings the coercivity back to ordinary levels and demonstrates that the surface state of a magnetic particle can play a very important part in determining the switching field of the particle.

5. *Cobalt-Modified Iron Oxide Particles*

Bickford et al. (1957) found that the magnetocrystalline anisotropy constant of magnetite (Fe_3O_4) could be reversed in sign and tripled in magnitude by substituting 4% of the Fe^{2+} with Co^{2+}. Unfortunately their results showed that the temperature dependence of coercive force also increased greatly. It was later found that by substituting 2–3% of $CoSO_4$ and then using the standard method of making particles of γ-Fe_2O_3, the result was equiaxed (almost spherical) particles of Fe_2O_3 containing 2–3% of cobalt. The coercivity was attractively high but unfortunately the temperature dependence of H_c was high also. The coercivity was also found to be pressure sensitive, another very undesirable attribute particularly in a magnetic particle for use in tapes. The addition of manganese, zinc and other metals has been patented as a means of reducing the temperature sensitivity that is thought to involve the migration of the Co^{2+} ions among the lattice defects. The increased pressure sensitivity is believed to arise because the magnetostrictive constant λ_s increases with increasing cobalt in Fe_2O_3. The λ_s of Fe_3O_4, in contrast, decreases with increasing cobalt content (Flanders 1976).

Fortunately, it is now possible to combine the high coercivity introduced by cobalt atoms with the stability and the acicularity of standard γ-Fe_2O_3 by confining the cobalt atoms to a surface layer of the particles. One method is described by Imaoka et al. (1978). It involves dispersing acicular iron oxide particles in an aqueous solution of cobalt sulfate and then adding sodium hydroxide to precipitate cobaltous hydroxide onto the surface of the particles. Warming the treated particles to 100 °C for several hours decomposes the hydroxide and allows the cobalt ions to penetrate a short distance into the surface of the particles and so to form a layer of cobalt ferrite. Particles of Fe_2O_3, Fe_3O_4, Fe_2O_3–Fe_3O_4 and α-(FeO)OH have all been used successfully as starting materials.

The remarkable contrast in properties that results from adding a small amount of cobalt (2–3 wt%) to the surface of iron oxide particles, compared with the effect of distributing the same amount of cobalt uniformly throughout the particle, was demonstrated by Kishimoto et al. (1984). They studied the temperature dependence of coercivity and the time, temperature and field dependence of the remanence of particles containing 3.6–7 wt% of cobalt. The particles were first surface impregnated and characterized, then heated in air at a higher temperature and for a longer time. This had the effect of driving the cobalt atoms deeper into the particle. The conclusions were as follows.

(a) Iron oxide particles containing a certain amount of cobalt that is confined to a surface layer show a smaller dependence of H_c on temperature than when the same amount of cobalt is uniformly distributed throughout the particle.

(b) Similar results were obtained when the time, temperature and field dependence of the remanent magnetization were studied on the same samples. The cobalt-impregnated particles were more resistant to demagnetization by an applied field than were the cobalt uniformly doped particles.

(c) the coercivity was the same when a fixed percentage of cobalt ions (3.6%, 4.5%, 7%) was added regardless of whether they were confined to the surface of the particle or distributed throughout its volume.

There is no completely satisfactory explanation of these results. However, this inability to understand the behavior of cobalt-surface-impregnated particles has not affected the sales of the particles for use in video tapes, flexible disks and hard disks.

It is now possible to make cobalt-modified iron oxide particles with coercivities from $32\,kA\,m^{-1}$ to around $80\,kA\,m^{-1}$ (Sharrock 1989). Prices range from US\$7 kg^{-1} for materials used in video cassettes to US\$25 kg^{-1} for advanced applications; for example, S-VHS video format tapes.

6. CrO_2 Particles

Chromium dioxide particles, introduced in the mid-1960s, were the first magnetic recording particles to be made commercially having a coercivity significantly higher ($37.4\,kA\,m^{-1}$) than that of γ-Fe_2O_3 particles. The compound CrO_2 does not occur naturally and is usually made by the oxidation of Cr_2O_3 with excess CrO_3, under pressure and in the presence of water. To reduce the pressure and temperature requirements and to modify the magnetic properties (increased coercivity and Curie temperature), additives such as antimony, tellurium or iron are normally used. Chromium dioxide is unusual among the magnetic oxides in that it is ferromagnetic rather than ferrimagnetic. In the pure form, $\sigma_s = 100\,A\,m^2\,kg^{-1}$ and $T_c = 390\,K$. The latter value is rather too close to room temperature to ensure that the recorded data has archival properties, but does enable the tapes to be copied by thermal transfer. By means of the additive elements, Curie temperatures in the range 341–441 K are obtained. The saturation magnetization per unit mass, however, is reduced to about $70\,A\,m^2\,kg^{-1}$ by the addition of antimony and tellurium.

Particles of chromium dioxide are strikingly different in appearance from those of the iron oxides. They are very uniform, clean (free from side growths and other excrescences) and are highly acicular. Aspect ratios of 20:1 or 30:1 are not uncommon. The structure is tetragonal with $a_0 = 0.4423\,nm$ and $c_0 = 0.2917\,nm$. The long axis of the particle is parallel to the c axis but there is disagreement on whether the easy axis of magnetization is along that axis or canted as much as 40° from it.

Chromium dioxide particles are used in audiotapes and in the tape cartridges used in the IBM3480 tape system. The particles have always been expensive (partly because the high pressures used in the manufacturing process necessitate batch processing). The cost in 1989 was US\$14 kg^{-1}

7. Metal Particles

The attraction of using metal particles for recording materials is that they widen the range of magnetization levels and coercivities. Pure iron has a saturation magnetization per unit mass of $178\,A\,m^2\,kg^{-1}$ (cf. $\sim 74\,A\,m^2\,kg^{-1}$ for γ-Fe_2O_3, Co-γ and CrO_2). In particles whose dominant anisotropy is due to the particle shape, the coercivity increases linearly with the magnetization. Thus, single-domain particles of pure iron should have a coercivity roughly 2.5 times that of particles of iron oxide having the same aspect ratio; that is, $H_c \simeq 65\,kA\,m^{-1}$. In fact, even higher coercivities are found; particles of iron are commercially available having $H_c = 91.5\,kA\,m^{-1}$.

Such high coercivities mean that this factor is no longer a limiting one in the achievement of high recording densities. The limiting factor has moved from the medium to the recording–erasing head (see *Recording Head Technology*). The search now is for head materials having vanishingly small remanence and coercivity, high resistivity (to reduce losses due to eddy currents) and high saturation magnetization.

Pure iron particles are highly reactive and some method of passivating them must be found. The preferred method is to expose them to a mixture of nitrogen and oxygen. Coating the particles with inorganic compounds such as chromates or organic compounds (e.g., sodium oleate) has also been used (Lee et al. 1987).

The final product is an acicular particle having a core of iron and an outer shell of Fe_3O_4 and Fe_2O_3. The saturation magnetization of the passivated particles is thus lower than that of pure iron (e.g., $150\,A\,m^2\,kg^{-1}$ compared with $178\,A\,m^2\,kg^{-1}$).

Three principal methods have been used to make iron particles. They are, in order of increasing cost:

(a) reduction in hydrogen of particles of γ-(FeO)OH (synthetic goethite) at 300–400 °C,

(b) reduction of ferrous ions in a solution of sodium borohydride, and

(c) vacuum deposition of iron into an atmosphere of argon or helium.

The cost of metal particles is high ($\sim$US\$75–85 kg^{-1}), but is likely to be somewhat lower as the particles become widely used, for example, in digital audiotape (DAT) machines.

8. Barium Ferrite Particles

The existence of ferrites having a hexagonal structure, for example, barium ferrite $BaO.6Fe_2O_3$, has long been known (see *Ferrites, Hard*). Their high magnetocrystalline anisotropy led to their widespread use as inexpensive magnets in, for example, refrigerator door gaskets. The use of barium ferrite in magnetic recording was delayed because the coercivity ($\sim 200\,kA\,m^{-1}$) was too high for the available head materials. However, the introduction of a process for substituting cobalt and titanium atoms for iron led to a practical method of reducing the coercivity to recordable levels at a very small cost in M_s (Kubo et al. 1982).

In barium ferrite the magnetocrystalline easy axis is the hexagonal axis. The particles are obtained in the form of hexagonal platelets whose easy directions of shape anisotropy lie in the plane of the particle. Thus, for a random assembly of barium ferrite platelets the coercivity is:

$$H_c = 0.48\left(\frac{2K}{\mu_0 M_s} - NM_s\right)$$

where K is the magnetocrystalline anisotropy constant, N is the demagnetizing factor of the platelet along the hexagonal axis and μ_0 is the permeability of free space. The composition of the particle is $BaFe_{12-2x}Co_xTi_xO_{19}$. When $x = 0.8$, the $0.48(2K/\mu_0 M_s)$ term is $229\,kA\,m^{-1}$ and $0.48\,NM_s = 153\,kA\,m^{-1}$. Thus, $H_c = 229 - 153 = 76\,kA\,m^{-1}$ which matches the writing and erasing capability of current head materials.

The effect of increasing temperature on the coercivity of barium ferrite particles is unusual. The magnetization M_s decreases with increasing temperature and so $2K/\mu_0 M_s$ increases and NM_s decreases. Thus the coercivity *increases* with increasing temperature.

The platelets of barium ferrite tend to lie with their hexagonal axis perpendicular to the substrate; that is, the easy axis of magnetization is perpendicular. Barium ferrite particles are presently being used in the Toshiba high-density flexible disk.

The magnetic properties of particles used for recording are summarized in Table 1.

9. Particles Versus Thin Films

An alternative method of making magnetic coatings suitable for recording is to coat a suitable substrate directly with a thin film of metal by electroplating, chemical plating, vacuum evaporation or sputtering. The advantages of particulate coatings can be summarized as follows:

(a) The magnetic properties of the coating are determined principally, but not entirely, by the magnetic properties of the magnetic particles and these properties can be determined before coating.

(b) The nonmagnetic properties are determined mainly by the binder polymers and other nonmagnetic constituents. Thus, the nonmagnetic properties of particulate coatings can be optimized and controlled independently of their magnetic properties.

(c) The particulate coating process lends itself to the coating of very wide webs (e.g., 1–2 m) at low cost.

(d) The choice of available particles is extremely wide and is limited only by the magnetic properties available in the head materials.

The disadvantages of particulate coatings can be summarized as follows.

(a) The magnetic particles in a coating occupy only 40% of the volume of the coating in tapes and flexible disks and only 20% in hard disks. The magnetic properties are thus diluted and the recording performance is diminished.

(b) It is difficult to coat flexible, particulate media thinner than about 1 μm and hard disks thinner than about 0.25 μm, at high yields.

(c) The processes by which magnetic particles are manufactured often lead to agglomeration of the particles. The agglomerates must be subdivided without breaking the particles.

(d) Changing the magnetic properties of the coating means using a different particle which then must be characterized in terms of its surface chemistry and its interaction with the binder. This can be a lengthy process.

(e) Magnetic orientation of particles is inefficient or difficult or both.

10. Thin-Film Recording Surfaces

The promise that thin films have always offered to the magnetic media developer has been a method of achieving a thin coating consisting of 100% magnetic material, with uniform, consistent magnetic properties. The properties could, in principle, be fine tuned to particular recording applications by turning a knob on the coating processor. The concept, of course, also included nonmagnetic undercoats and overcoats. All three layers, undercoat, magnetic layer and overcoat (for corrosion and wear resistance) could, in principle, be deposited in one pass through the coating machine. Machines much more complicated than this have been in use for many years in the manufacture of photographic film. There were attempts to make thin-film computer tapes by IBM, NCR and others during the early 1960s, but it is only recently that thin-film recording products have appeared in the form of high-performance audiotapes and videotapes (vacuum deposited) and thin-film disks, $\leqslant 130$ mm diameter, which at first were plated and now are sputtered. With a single exception, the easy axes of commercial thin-film disks lie in the plane of the coating.

The magnetic layer (50–100 nm thick, $H_c = 50$–$100\,kA\,m^{-1}$) is deposited on an underlayer of plated nickel–phosphorus whose purpose is to provide a smooth, passive substrate and to encourage the desired crystallographic orientation of the magnetic layer.

A thin film of large area whose magnetization is normal to the plane has a demagnetizing field $H_d = -M_s$ and the energy associated with the shape anisotropy of the thin film will be $\frac{1}{2}\mu_0 M_s^2$. The magnetization is thus encouraged to rotate into the plane of

Table 2
Typical properties of magnetic film media

Material	Deposition process[a]	Orientation[b]	Dominant crystal structure[c]	M_s (kA m^{-1})	H_c (kA m^{-1})	$S_s S^*$	K_u (10^5 J m^{-3})
Co	OIE	IPA	hcp	1100–1400	61–120		4 (bulk)
	NIE, SP	IPI	hcp	1100–1400	30–60[d]		
Fe	OIE	IPA	bcc	1600	60–90		0.3–3.0
	SP	IPI	bcc	1600	10		
Ni	OIE	IPA	fcc	400	20–28		
	SP	IPI	fcc	400			
Co–Ni	OIE	IPA	hcp/fcc	800–1200	30–70		
Co–Fe	OIE	IPA	bcc	1400–1600	60–120	0.9, 0.9	
Co–Sm	NIE[e]	IPA	amorphous	500–1000	33–55	1.1	
Co–P	EL, EP	IPI	hcp	800–1100	36–96	0.9, 0.9	
Co–Re	SP	IPI	hcp/fcc	500–750	18–58	0.9, 0.9	
Co–Pt	SP	IPI	hcp/fcc	800–1400	→140		
Co–NiP	EI, EP	IPI	hcp/fcc	600–1000	40–120	0.8, 0.8	
Co–30 at.% Ni:N_2	SP	IPI	hcp/fcc	650	80	0.95	
Co–Ni:O_2	OIE	IPA	hcp/fcc	300–400	80	0.7–0.8	
Co–Ni–Pt	SP	IP*I	hcp/fcc	800–900	60–70	0.9, 0.97	
Co–Ni–W	SP	IPI	hcp/fcc	450	30–50	0.8, 0.8	
Fe_3O_4	NIE, SP	IPI	is	400	17–32		
γ-Fe_2O_3:Co	SP	IPI	is	200–250	40–100	0.8, 0.8	
γ-Fe_2O_3:O_s	SP	IPI	is	240	160	0.8, 0.8	
Co–18 at.%Cr	SP	⊥	hcp	300–550	80–100 (⊥)		− 1.0
Co–20 at.%Cr	SP	⊥	hcp	400	65–95 (⊥)		0.15
Co–22 at.%Cr	SP	⊥	hcp	300–340	80–105 (⊥)		0.4

Source: Köster and Arnoldussen (1987)
a OIE = oblique incidence evaporation; NIE = normal incidence evaporation; SP = sputtered; EL = electroless plated; EP = electroplated b IPA = in-plane anisotropic; IPI = in-plane isotropic; ⊥ = perpendicular c hcp = hexagonal close-packed; fcc = face-centered cubic; bcc = body-centered cubic; is = inverse spinel d Coercivities for Co films deposited on Cr or W underlayers e Co–Sm values for films deposited in the presence of an orienting field

the film and it will do so unless prevented by some additional anisotropy; for example, magnetocrystalline anisotropy K_u having a preferred direction normal to the film plane and for which $K_u > \frac{1}{2}\mu_0 M_s^2$. This inequality is not satisfied for pure cobalt films. However, if the cobalt is diluted with a nonmagnetic metal such as chromium, a composition can be found (~20 at.% chromium) for which $K_u > \frac{1}{2}\mu_0 M_s^2$ and the film then has an easy axis perpendicular to its plane. Thus, films of $Co_{0.8}Cr_{0.2}$ and so on, can, in principle, be used to make perpendicular recording materials. Much experimental work has been devoted to the preparation and study of rf sputtered films of cobalt–chromium having a chromium content in the range 15–22 at.% (Arnoldussen 1986).

A film of 20–80 nm of sputtered carbon is used as a protective overcoat against corrosion and unscheduled contacts with the head. An organic lubricant is also applied topically. The problem here is to apply a film of the right material and thick enough to protect the magnetic layer without being so thick that it reduces the recording performance of the system. Among the materials used have been sputtered rhodium with a chromium adhesion layer and a liquid lubricant, SiO_2 with solid polar lubricants, TiC, TiN, SiC, Cr_2C_3 and Al_2O_3 (Hintermann 1984).

The factors in a thin-film recording material that should lead to high signal levels in the reading head are understood. The magnetization should be high to deliver a large amount of flux to the head. The coercivity should also be high, especially in films where the magnetization lies in the plane of the film so that the effects of demagnetization are minimized. The coating should be thin, again to reduce the effects of demagnetization. However, the factors controlling noise in thin films were understood only recently (Belk et al. 1985). In contrast to particulate media, where the intensity of magnetization is low intrinsically (within a particle) and is further reduced by the low packing fraction of particles within the coating (~0.2–0.4), the magnetization intensity in thin films can be almost equal to that of the iron or cobalt. Adjacent crystallites are coupled to one another by strong magnetostatic forces but also by strong quantum mechanical exchange forces. The latter operate only over short distances, approximately one atomic spacing, but in thin metallic films, the distances between crystallites are very small. Exchange coupling in ferromagnetic materials leads to the spins being parallel over considerable distances. The coupling results in a coarse-grained magnetic structure and leads to high noise levels. The answer is to separate the spins in adjacent particles by a distance that is small, but is large enough to prevent or reduce exchange coupling between grains. These conditions exist in films of cobalt–chromium and cobalt–platinum–chromium (Yogi et al. 1990) and films of the latter material were used in a successful demonstration of reliable recording at 158 000 bits in^{-1}, 7470 tracks per inch or 1.180×10^9 bits in^{-2} (1.83×10^6 bits mm^{-2}) (Yogi et al. 1990).

The magnetic properties of film media used for recording are summarized in Table 2.

See also: Ferrites, Hard; Magnetic Materials: An Overview; Magnetooptic Storage Media; Recording Head Technology

Bibliography

Arnoldussen T C 1986 Thin-film recording media. *Proc. IEEE* 74(11): 1526–39

Bate G 1980 Recording materials. In: Wohlfarth E P (ed.) 1980 *Ferromagnetic Materials, a Handbook on the Properties of Magnetically Ordered Substances*, Vol. 2. North-Holland, Amsterdam, pp. 381–507

Belk N R, George P K, Mowry G S 1985 Noise in high-performance, thin-film longitudinal magnetic recording media. *IEEE Trans. Magn.* 21: 1350–5

Bhushan B 1990 *Tribology and Mechanics of Magnetic Storage Devices*. Springer, Berlin

Bickford L R, Brownlow J M, Penoyer R F 1957 Magnetocrystalline anisotropy in cobalt-substituted magnetite single crystals. *Proc. IEEE B* 104(5): 238–44

Flanders P J 1976 Impact-induced demagnetization, magneto-striction and coercive force in Co-doped iron oxide tapes. *IEEE Trans Magn.* 12: 770–2

Hintermann H E 1984 Thin solid films to combat friction, wear and corrosion. *J. Vac. Sci. Technol. B* 2(4): 816–22

Imaoka Y, Umeki S, Kubota Y, Tokuoka Y 1978 Characteristics of cobalt absorbed iron oxide tapes. *IEEE Trans. Magn.* 14: 649–54

Kishimoto M, Amemiya M, Hayama F 1984 Magnetic annealing and relaxation in cobalt-substituted acicular iron oxide particles. *J. Appl. Phys.* 55: 2272–4

Knowles J E 1981 Magnetic properties of individual acicular particles. *IEEE Trans. Magn.* 17: 3008–13

Köster E, Arnoldussen T C 1987 Recording media. In: Mee C D, Daniel E D (eds.) 1987 *Technology*, Magnetic Recording, Vol. 1. McGraw-Hill, New York, pp. 98–243

Kubo O, Ido T, Yokoyama H 1982 Properties of Ba–ferrite particles for perpendicular magnetic recording. *IEEE Trans. Magn.* 18: 1122–4

Lee T-H D, Hu S, Madudid N 1987 Stability studies of iron particles. *IEEE Trans. Magn.* 23: 2280–2

Sharrock M P 1989 Particulate magnetic recording media: a review. *IEEE Trans. Magn.* 25: 4374–89

Yogi T, Ching T, Nguyen T A, Ju K, Gorman G L, Castillo G 1990 Longitudinal media for 1-GB/in² areal density. *IEEE Trans. Magn.* 26: 2271–6

G. Bate
[Santa Clara University, Santa Clara, California, USA]

Magnetic Units and Material Specification

1. Magnetic Quantities and Symbols

It was the prerogative of the nineteenth-century electromagneticians to suggest names and symbols for the magnetic quantities. Maxwell was responsible for the codification of many of these and influenced the terminology adopted by his contemporaries. He used the symbol H for the "magnetizing force," now called the magnetic field strength, and the symbol B for the magnetic induction or magnetic flux density. In this tradition, which originated with Faraday, the primary magnetic field was H. The term "magnetic induction" alluded to the status of B as a dependent variable. Thus, the "coefficient of induced magnetization" χ (called magnetic susceptibility by William Thomson in 1872) was the ratio of the magnetization M to H, not M to B, and the "coefficient of induction" μ (named permeability by Thomson) was the ratio of B to H.

This philosophy has not been universally accepted. The field within magnetized matter is not H but B; H has its source only in conduction currents while B is produced by both conduction and Amperian (atomic) currents and is therefore more general. Those emphasizing the primacy of B have an excellent point but are at a historical disadvantage. In magnetostatic problems, the magnetic pole approach has traditionally been used in direct analogy with electrostatics. Fictitious surface poles are taken as sources of H, and B is the sum of H and M. In this traditional approach, H internal to a magnetized sample originates not only from external field sources but also from the self-field of the magnetization. Therefore, H, in the equations for B that follow, is the *internal* field, equal to the external or applied field H_a corrected by the demagnetizing field H_d:

$$H = H_a + H_d$$

H_d depends on M and on the geometric demagnetizing factor N (see *Demagnetizing Factors*):

$$H_d \equiv -NM$$

Ellipsoids, including spheres, have uniform M in a uniform H_a. If H_a is along one of the principal axes, M is in the direction of H_a. For nonellipsoidal bodies, M is uniform only in the limit $\chi = 0$, applicable to weakly magnetic and saturated ferromagnetic materials. The demagnetizing factors for three orthogonal axes of any uniformly magnetized body sum to unity. For certain shapes, such as long cylinders in axial fields or toroids in azimuthal fields, $N = 0$.

The measurement of susceptibility requires the application of H_a and the measurement of M. The susceptibility dM/dH_a is characteristic of the *sample* and may be termed the *external* susceptibility χ_{ext}. The *internal* susceptibility χ, characteristic of the *material*, is dM/dH. The two susceptibilities are related:

$$\chi = \chi_{ext}/(1 - N\chi_{ext})$$

$$\chi_{ext} = \chi/(1 + N\chi)$$

2. Magnetic Units

An expression of the relationship between B, H and M is sufficient to establish the system of units in use. In the rationalized meter–kilogram–second (mks) systems, there are at least two recognized relationships, depending on whether one wishes to refer to magnetization M or to magnetic polarization J (sometimes the symbol I, for intensity of magnetization, is used):

$$B = \mu_0(H + M) \quad (1)$$

$$B = \mu_0 H + J \quad (2)$$

where μ_0 is the permeability of vacuum equal to $4\pi \times 10^{-7}\,\mathrm{H\,m^{-1}}$. The structure of Eqn. (1) is such that B is analogous to the electric field E. In Eqn. (2), H is analogous to E. The word rationalized refers to the absence of the irrational number 4π in Eqns. (1) and (2) and in Maxwell's equations.

In the *Système International d'Unités* (SI), which evolved from the rationalized meter–kilogram–second–ampere (mksa) system, the defining relationship is Eqn. (1). However, magnetic polarization J is also a recognized quantity. The units of B and J are webers per square meter ($\mathrm{Wb\,m^{-2}}$) or teslas (T). The units of H and M are amperes per meter ($\mathrm{A\,m^{-1}}$), although H was sometimes expressed as ampere turns per meter in mksa.

In the Gaussian and the electromagnetic unit (emu) systems, which are unrationalized centimeter–gram–second (cgs) systems,

$$B = H + 4\pi M \quad \text{(cgs)}$$

where B is in gausses (G) and H is in oersteds (Oe) (a unit that is dimensionally and numerically equivalent to G). The magnetization, when written as $4\pi M$, is also in gausses and may be thought of as a field arising from the magnetic moment. When magnetization is expressed simply as M (the magnetic moment m per unit volume) its units are $\mathrm{erg\,G^{-1}\,cm^{-3}}$. In terms of base units, $\mathrm{erg = cm^2\,g\,s^{-2}}$ and $\mathrm{G = cm^{-1/2}\,g^{1/2}\,s^{-1}}$; therefore, $\mathrm{erg\,G^{-1}\,cm^{-3}}$, the units for M, are dimensionally but not numerically equivalent to G. It is often a source of confusion that $4\pi M$ and M have different units. M is conventionally expressed as $\mathrm{emu\,cm^{-3}}$. The designation "emu," however, is merely an indicator that electromagnetic units are in use; it is not a unit.

Magnetic susceptibility χ is dM/dH in both cgs and SI. In cgs, the ambiguity in the units for M extends to χ. In terms of base units, χ is a dimensionless quantity. However, χ is usually expressed as emu, $\mathrm{emu\,cm^{-3}}$ or $\mathrm{emu\,cm^{-3}\,Oe^{-1}}$. Occasionally, cgs susceptibility is written as $4\pi\chi$ (dimensionless), which is equal to SI susceptibility (dimensionless). The quantities M and χ are based on a unit volume of material. Consistent with SI, but not recognized as important SI quantities, are the mass (specific) magnetization σ and mass (specific) susceptibility χ_ρ.

Differential permeability μ is dB/dH in both cgs and SI. In cgs, it is dimensionless and

$$\mu = 1 + 4\pi\chi \quad \text{(cgs)}$$

In SI, permeability has the dimensions of μ_0. The relative permeability is defined as

$$\mu_r \equiv \mu/\mu_0 = 1 + \chi$$

and is numerically equal to cgs permeability.

The use of SI, rather than cgs, is recommended by all international standards organizations. One of its advantages is the painless unification of magnetic quantities with practical electrical units—the ampere and the volt. In comparison, the unit for current in emu is the abampere. In the Gaussian system, which uses electrostatic units (esu) for the electrical quantities, it is the statampere. Another advantage of SI is that each magnetic quantity has defined units, so that the dimensional balance of equations is always apparent. Because SI is a rationalized system, the ambiguity between M and $4\pi M$ does not occur. Disadvantages of SI for magnetics are the dual recognition of M and J, the awkward magnitudes of values for H and the fact that $\mu_0 \neq 1$. Table 1 is a list of the magnetic quantities, their accepted names and symbols, their recommended SI units and the conversion of cgs units to SI units.

Although other consistent metric systems of units are occasionally used in magnetism, the SI and cgs systems are those in common use. Some authors use "practical" units, combining length in centimeters, current in amperes and magnetic field strength in oersteds. While this might be of some utility to uncritical users of formulae, the equations do not balance dimensionally. Consequently, the haphazard use of such mixed units is discouraged. English units in electromagnetism, such as kilolines per square inch, are obsolete.

3. Magnetic Hysteresis Loop

The static magnetic properties of a material are represented by a plot of either magnetization or magnetic flux density as a function of field (M–H or B–H curves). Ferromagnetic materials exhibit hysteresis, a word first used by Ewing in 1881 to mean "lag behind." The volume energy dissipation per field cycle w is the area of the hysteresis loops, in units of joules per cubic meter. As shown in Fig. 1, M–H and B–H loops have different contours. For samples with a nonzero demagnetizing factor, the loop shapes will depend on whether H_a or H is plotted. Despite the different shapes of these hysteresis loops for a given sample, the energy loss is the same for each representation:

$$w = \mu_0 \oint H_a\, dM = \mu_0 \oint H\, dM = \oint H_a\, dB = \oint H\, dB$$

The following parameters are identified in Fig. 1a: saturation magnetization M_s, remanent (not "remnant") magnetization M_r and intrinsic coercivity (or intrinsic coercive force) H_{ci}, H_{cm} or ${}_mH_c$. The squareness of the loop may be represented by the ratio of M_r to M_s. The field H_s required to saturate the magnetization is not labelled. Also not labelled is the remanent coercivity H_r or ${}_rH_c$ ($H_r > H_{ci}$); when H is reduced from a value of H_r to zero, M_r is zero. As discovered by Lord Rayleigh in 1886, the M–H_a hysteresis loop of a ferromagnetic sample with a nonzero demagnetizing factor will be sheared at an angle $\tan^{-1}(1/N)$. When the H_a axis is rescaled to the internal field, the loop becomes unsheared. In either representation, the value of H_{ci} is the same.

In Fig. 1b, the parameters are remanent (or residual) induction B_r (also known as remanence or retentivity) and coercivity (or coercive force, the term used by Poisson) H_c, H_{cB}, or ${}_bH_c$. The value measured for coercivity depends on the rate at which the hysteresis loop is cycled. Static coercivity is measured with instruments such as a vibrating-sample magnetometer. Dynamic coercivity, which usually has a measured value somewhat larger than static coercivity, is measured at 50 Hz or 60 Hz with instruments such as a B–H looper. After M saturates, B is linearly proportional to H, so there is no "saturation induction" (although the term is used anyway). Note that $\chi = \chi(H)$ and $\mu = \mu(H)$. The differential permeability $\mu(H)$ is the slope of the line tangent to the curve at point (B, H); the average permeability $\bar{\mu}(H)$ is the slope of the line from the origin to point (B, H). The initial curves in Fig. 1a, b begin at the origin with the sample in a virgin state achieved by thermal or alternating-field demagnetization.

4. Magnetic Material Specification

Commercial magnetic materials are characterized by many of the parameters discussed previously as well as several others.

4.1 Hard Magnetic Materials

Hard magnetic materials include permanent magnets, recording media and hard ferrites. For these materials, a metallurgical attempt is made to have large H_c and B_r. A further important magnetic parameter for the specification of these materials is the maximum energy product $(BH)_{max}$ along the "demagnetization curves" in the second and fourth quadrants of the B–H loop. Ferromagnetic materials become paramagnetic above their Curie temperature T_C. In addition to T_C, the maximum operating temperature is specified. Common permanent magnets are Al–Ni–Co, rare-earth–cobalt compounds and Nd–Fe–B.

Materials used as recording media may be particulate (γ-ferric oxide, cobalt-modified γ-ferric oxide, chromium dioxide, barium ferrite, iron particles) or thin film (cobalt–nickel, cobalt–chromium, cobalt–phosphorus). These materials are read with inductive

Table 1
Magnetic quantities recognized under SI

Symbol	Quantity	SI units[a]	Conversion from Gaussian and cgs emu to SI[b]
Φ	magnetic flux	Wb, V s	1 Mx → 10^{-8} Wb
B	magnetic flux density, magnetic induction (Φ m^{-2})	T, Wb m^{-2}	1 G → 10^{-4} T
U_m	magnetic potential difference	A	1 Gb → $10/(4\pi)$ A
H	magnetic field strength (U_m m^{-1})	A m^{-1}	1 Oe → $10^3/(4\pi)$ A m^{-1}; 1 γ → $10^{-2}/(4\pi)$ A m^{-1}
m	magnetic moment	A m^2, J T^{-1}	1 erg G^{-1}[c] → 10^{-3} A m^2
M	(volume) magnetization (m m^{-3})	A m^{-1}	1 erg G^{-1} cm^{-3}[d] → 10^3 A m^{-1}; 1 G → $10^3/(4\pi)$ A m^{-1}
σ	(mass) magnetization	A m^2 kg^{-1}	1 erg G^{-1} g^{-1}[e] → 1 A m^2 kg^{-1}
j	magnetic dipole moment ($\mu_0 m$)	Wb m	1 erg G^{-1}[c] → $4\pi \times 10^{-10}$ Wb m
J	magnetic polarization ($\mu_0 M$)	T, Wb m^{-2}	1 erg G^{-1} cm^{-3}[d] → $4\pi \times 10^{-4}$ T
χ, κ	(volume) susceptibility (M/H)	dimensionless	1 emu cm^{-3} Oe^{-1}[f] → 4π
χ_ρ	(mass) susceptibility	m^3 kg^{-1}	1 emu g^{-1} Oe^{-1}[g] → $4\pi \times 10^{-3}$ m^3 kg^{-1}
χ_{mol}	(molar) susceptibility	m^3 mol^{-1}	1 emu mol^{-1} Oe^{-1}[h] → $4\pi \times 10^{-6}$ m^3 mol^{-1}
μ	permeability (B/H)	H m^{-1}, Wb A^{-1} m^{-1}	1 → $4\pi \times 10^{-7}$ H m^{-1}
μ_r	relative permeability ($\mu/\mu_0 = 1 + \chi$)	dimensionless	$\mu \rightarrow \mu_r$
w, W	(volume) energy density (BH, $\mu_0 MH$)	J m^{-3}	1 erg cm^{-3} → 10^{-1} J m^{-3}
N, D	demagnetizing factor	dimensionless	1 → $1/(4\pi)$

a Wb = weber, V = volt, s = second, T = tesla, m = meter, A = ampere, J = joule, kg = kilogram, H = henry b Gaussian units are the same as cgs emu for magnetostatics; Mx = maxwell, G = gauss, Gb = gilbert, Oe = oersted, γ = gamma c Often expressed as emu d Often expressed as emu cm^{-3} e Often expressed as emu g^{-1} f Often expressed as emu cm^{-3} or as a dimensionless quantity g Often expressed as emu g^{-1} or as cm^3 g^{-1} h Often expressed as emu mol^{-1} or as cm^3 mol^{-1}

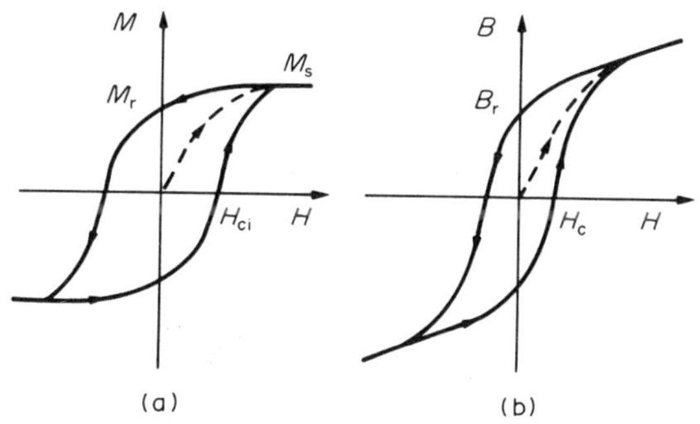

Figure 1
(a) Magnetization M and (b) magnetic flux density B as functions of magnetic field strength H (the initial curves are shown as broken lines)

or magnetoresistive heads. Important material specifications are M_r and H_c. The magnetocrystalline anisotropy constant K_u and the anisotropy field H_k,

$$H_k = 2K_u/\mu_0 M_s$$

are relevant to media with perpendicular anisotropy. Other magnetic specifications relate specifically to reading and writing performance. Another class of materials with perpendicular anisotropy is used for magnetooptic recording. These are written thermomagnetically and read using the Kerr effect (see *Magnetooptic Storage Media*). The Kerr rotation angle θ_K is specified. Typical materials are terbium–iron and Tb–Fe–Co.

4.2 Soft Magnetic Materials

Soft magnetic materials include transformer steels, ferrites, amorphous alloys and recording heads. These materials are designed to minimize hysteretic loss. Thus, H_c and B_r are small but μ is large. The important magnetic parameters for material specification are the initial permeability μ_i (μ at $B = 0$), the maximum permeability μ_{max}, B_r, H_c, M_s and T_C. Sometimes values for B are given for several values of H. Saturation magnetostriction λ_s is sometimes given. Because these materials are used in ac applications where eddy currents dissipate energy, the resistivity of the material is specified. In ac applications, permeability is a complex quantity, given by

$$\mu = \mu' + i\mu''$$

where μ' is the real (dispersive, inductive) part and μ'' is the imaginary (absorptive, resistive, loss) part. The loss tangent

$$\tan\delta \equiv \mu''/\mu'$$

is often specified at a given frequency, where δ is the phase angle between B and H. The quality factor Q is the reciprocal of $\tan\delta$. The relative loss factor is $\mu_0 \tan\delta/\mu'$. Total losses consist of eddy current, hysteresis and residual losses.

Write heads are soft magnetic materials designed for high-frequency use. Typical materials are ferrites, nickel–iron and amorphous alloys, often in the form of thin films. Important parameters are μ, H_s and H_k. Read heads may be inductive or magnetoresistive or use the Hall effect.

4.3 Paramagnetic and Diamagnetic Materials

For the material specification of paramagnetic and diamagnetic materials, magnetic susceptibility is given at the temperature of interest. Pure copper is diamagnetic but often contains iron impurities that contribute a large paramagnetic susceptibility. Aluminum is paramagnetic. Liquid oxygen is strongly paramagnetic. Austenitic steels (AISI types 304, 316 and 321) are paramagnetic but may partially transform to ferromagnetic martensite on thermal cycling to, or mechanical straining at, cryogenic temperatures. Most organic materials are diamagnetic. Lide (1991) gives tables of the molar susceptibilities of the elements and many inorganic and organic compounds.

See also: Demagnetizing Factors; Magnetic Materials: Basic Concepts; Measurements in Magnetic Materials

Bibliography

Bennett L H, Page C H, Swartzendruber L J 1976 Comments on units in magnetism. In: Becker J J, Lander G H, Rhyne J J (eds.) 1976 *Magnetism and Magnetic Materials—1975*, AIP Conference Proceedings, Vol. 29. American Institute of Physics, New York, pp. xix–xxi (reprinted 1978. *J. Res. Natl. Bur. Stand.* 83: 9–12)

Brown W F Jr 1962 *Magnetostatic Principles in Ferromagnetism*. North-Holland, Amsterdam, pp. 1–13

Brown W F Jr 1984 Tutorial paper on dimensions and units. *IEEE Trans. Magn.* 20: 112–17

Brown W F Jr, Frank N H, Kemble E C, Michener W H, Murdock C C, Webster D L 1950 The teaching of electricity and magnetism at the college level. *Am. J. Phys.* 18: 1–25; 69–88

Cohen E R, Giacomo P 1987 Symbols, units, nomenclature and fundamental constants in physics. *Physica A* 146: 1–68

Cullity B D 1972 *Introduction to Magnetic Materials*. Addison-Wesley, Reading, MA, pp. 49–61

International Electrotechnical Commission 1973 *Magnetism*. IEC, Geneva, Publication No. 50(901), Chap. 901

International Electrotechnical Commission 1986 *Magnetic Materials. Part 8: Specifications for Individual Materials*, IEC 404–8. IEC, Geneva, Sects. 1, 6

International Organization for Standardization 1982 *ISO Standards Handbook 2: Units of Measurement*, 2nd edn., ISO 31/V–1979. ISO, Geneva, pp. 85–108

Jackson J D 1975 *Classical Electrodynamics*, 2nd edn. Wiley, New York, pp. 811–21

Lide D R (ed.) 1991 *CRC Handbook of Chemistry and Physics*, 72nd edn. CRC Press, Boca Raton, FL, pp. 9.33–9.104

Maxwell J Clerk 1892 *A Treatise on Electricity and Magnetism*, 3rd edn., Vol. 2. Clarendon, Oxford, pp. 47–78 (reprinted 1954, Dover, New York)
Shadowitz A 1988 *The Electromagnetic Field*. Dover, New York, pp. 294–341
Silsbee F B 1962 Systems of electrical units. *J. Res. Natl. Bur. Stand.* 66C: 137–83 (reprinted 1963, National Bureau of Standards, Washington, DC, Monograph 56)

R. B. Goldfarb
[National Institute of Standards and Technology, Boulder, Colorado, USA]

Magnetization Imaging

The magnetic domain structure plays an important role in determining the properties of ferromagnetic and ferrimagnetic materials and it is therefore necessary to be able to study the domain structure experimentally. This article discusses the methods available for imaging domains. Indirect evidence for the existence of domains is provided by the Barkhausen effect discovered in 1919, which showed that changes of magnetization due to changes in applied field took place by discontinuous steps (see *Barkhausen Effect*). These steps were first attributed to sudden changes in the magnetization direction of the domains, but were later shown to be caused mainly by abrupt jumps of domain walls. The first images of magnetic domains were obtained in the 1930s. The date is significant, as it coincides approximately with the development of the theory of the subdivision of magnetic materials into domains. There are now a large number of techniques for imaging domains, many of which are also used for the imaging of magnetic structure in superconductors (see *Magnetic Properties of Superconductors*). The techniques known up to the mid-1960s are reviewed by Carey and Isaac (1966).

1. *The Bitter Technique*

The first method, the earliest to be developed, makes use of magnetic liquids, which are colloidal suspensions of very small magnetic particles, usually of a ferrite such as Fe_3O_4, in a carrier liquid, which may be water or an organic solvent (see *Fluids, Magnetic*). In a nonuniform magnetic field, the particles move towards the positions where the magnetic field is strongest. Near the surface of a ferromagnet, stray fields are present, which are usually strongest along the lines where domain walls meet the surface. When a drop of magnetic liquid is placed on the surface, the particles tend to delineate the domain walls, making the domain pattern observable under a microscope. When the method was first used, in the early 1930s (notably by Bitter, after whom the method is named), the importance of correct preparation of the specimen surface was not realized. Mechanically polished surfaces of metals are heavily strained and the strains produce an irregular "maze" pattern of domains which is quite unlike the domain structure in the interior of the specimen. In order to eliminate the strains, the surface must usually be electropolished. In earlier days (until about the early 1970s), the magnetic liquid had to be prepared by the experimenter, but nowadays a variety of suitable liquids with different ranges of particle sizes are available commercially. Figure 1 shows an example of a Bitter pattern.

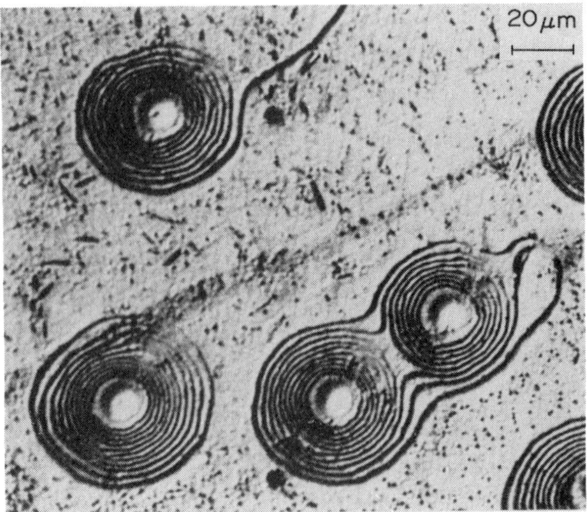

Figure 1
Bitter pattern of circular walls around bubble domains in an Sm–YIG garnet coated with a 10 nm thick permalloy layer (after Puchalska and Jouve 1977. © 1977 IEEE, New York. Reproduced with permission)

2. *Polarized Light Techniques*

The second optical microscopy technique makes use of polarized light. When linearly polarized light is either reflected from a magnetized surface, or transmitted through a magnetized medium, the plane of polarization is rotated. The angle of rotation depends on the component of magnetization in the direction of the light beam (see *Magnetooptics*). If the light is reflected from, or transmitted through, two domains magnetized in opposite directions, the plane of polarization rotates in the opposite sense in the two domains. Using an analyzer, it can be set to extinguish the light from one of the domains while allowing some light from the other to pass through. The images of the domains therefore have different intensities. As the angles of rotation are small, the correct setting of the analyzer is usually very near, but not exactly at, the crossed position. The effects of the magnetization on the transmitted and reflected light are called the Faraday and Kerr effects respectively.

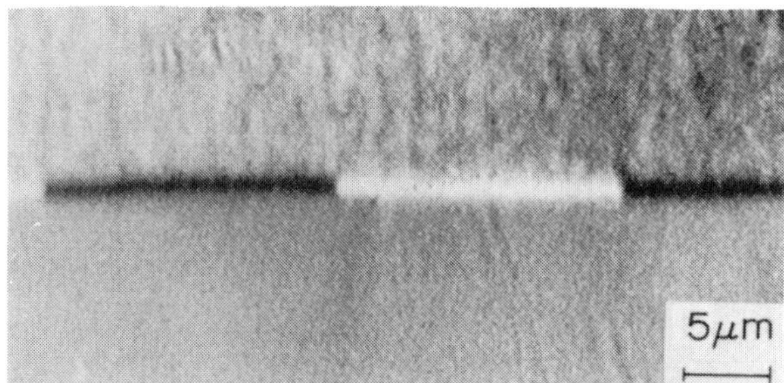

Figure 2
Digitally enhanced Kerr effect of a Néel wall with segments of opposite polarity (after Niedoba et al. 1991. © American Institute of Physics, New York. Reproduced with permission)

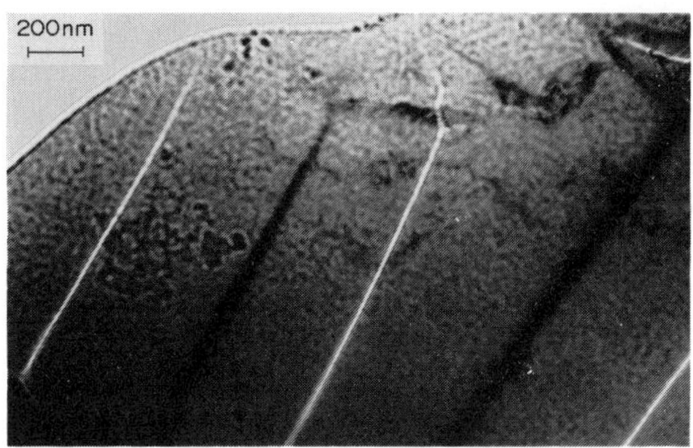

Figure 3
TEM Fresnel image of domains in a cobalt specimen

As the Faraday effect is obviously only suitable for transparent materials, it is restricted to very thin metal films or to fairly thin slices of oxide magnetic materials. The most suitable materials to study using the Faraday effect are garnets, which are good electrical insulators and are therefore fairly transparent. As the effect depends on the component of magnetization in the direction of the light beam, the method works best on materials in which the domains are magnetized normal to the surface. If the magnetization is parallel to the surface, oblique incidence must be used.

Similar considerations apply to the Kerr effect. Normal incidence can only be used for materials in which the magnetization has a component normal to the surface. As the demagnetizing energy favors the magnetization to be parallel to the surface, this configuration (called the polar Kerr effect) can only be used for materials with a strong enough anisotropy to overcome the demagnetizing effect. In other cases, the longitudinal Kerr effect can be used, in which the illumination is oblique and the plane of incidence is parallel to the magnetization. The different types of Kerr effect were summarized by Rave et al. (1987), who developed a method of processing the images digitally. Digital processing allows the contrast to be enhanced by subtracting two images of the same area in one of which the domain structure is eliminated by the application of a saturating field; Fig. 2 shows an example.

3. *Transmission Electron Microscopy*

As electrons carry electric charges, they experience a force when they move through a region in which a magnetic induction $\boldsymbol{B}$ is present. The force $\boldsymbol{F}$ is given by

$$\boldsymbol{F} = -|e|\boldsymbol{v} \times \boldsymbol{B}$$

where e and v are the charge and the velocity of the electron, respectively. As electrons pass through a magnetic specimen, this force causes them to be deflected by a small angle, given by

$$\beta = \frac{e\lambda Bt}{h}$$

for electrons travelling a distance t in a region where the magnetic induction has a component B perpendicular to the motion of the electrons; λ is the electron wavelength and h is Planck's constant. The small deflection can be used to image magnetic domains.

Specimens suitable for transmission electron microscopy (TEM) are usually up to a few hundred nanometers thick. For strongly magnetic materials such as iron or cobalt, the deflection of the electrons may be up to about 10^{-4} rad, in different directions in different domains. To image the domain structure, the specimen must be shielded from the strong field of the objective lens in order to prevent this field from destroying the domain structure. If the specimen is viewed slightly out of focus, the images of different domains are displaced sideways in different directions depending on the deflection of the electrons in the specimen, which in turn depends on the magnetization directions. Domain walls then appear as either bright or dark lines, depending on whether the images of the two domains on either side move towards or away from each other. Moreover, a wall that is bright in an overfocused image becomes dark in an underfocused image and vice versa. This effect can be used to distinguish domain walls from other features that may be seen in the image. If the illumination is sufficiently coherent, interference fringes may be seen along bright domain walls (see Fig. 3); the method is therefore called the Fresnel method.

The domain structure may also be revealed by displacing the objective aperture so that it intercepts the electrons that have passed through one set of domains, while still allowing other electrons to pass through (Foucault method). This method produces

Figure 4
Scanning electron microscope type 2 contrast image of domains in iron: both domain contrast and wall contrast are visible (after Fathers and Jakubovics 1977. © Elsevier Science Publishers, Amsterdam. Reproduced with permission)

images in which the domains themselves, rather than the domain walls, are revealed as bright or dark areas. These techniques have been reviewed by Jakubovics (1975) and Chapman (1984).

4. Scanning Electron Microscopy

There are a number of different ways in which domains can be imaged by scanning electron microscopy (SEM), two of which have been used for detailed domain studies. One of the methods (type 1 contrast) is based on the deflection of secondary electrons as they pass through stray fields near the specimen surface. The other method (type 2 contrast) makes use of deflections of the incident electrons after they enter the specimen. Depending on the magnetization direction, the electrons can be deflected towards or away from the surface, and the number of backscattered electrons produced can therefore vary from one domain to another. Contrast between the domains occurs only if the electrons are incident obliquely. However, a related mechanism produces contrast at domain walls, which is visible even at normal incidence. These techniques have been reviewed by Tsuno (1988). An example of type 2 contrast is shown in Fig. 4. Digital enhancement of the images is possible, as in the case of the Kerr effect, which enables domains to be imaged in materials with relatively low magnetization.

5. Scanning Transmission Electron Microscopy

The magnetization distribution may be imaged in a scanning transmission electron microscope fitted with a suitable detector, by the method of differential phase contrast (DPC). This method has the great

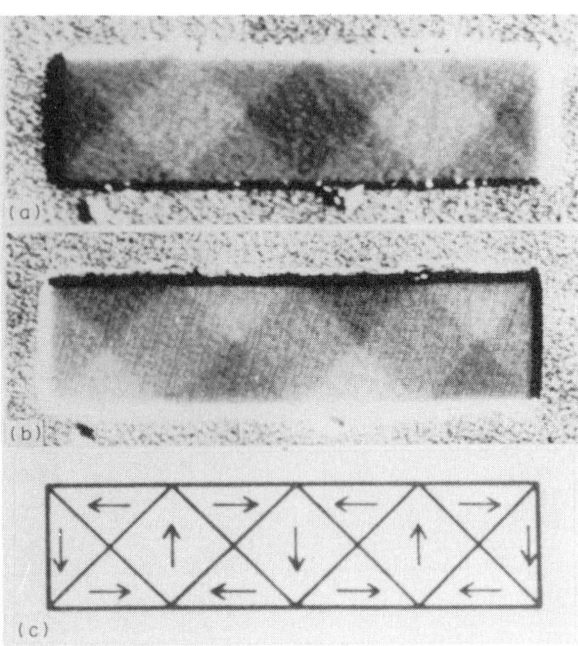

Figure 5
DPC images of a 3 μm × 0.75 μm permalloy particle of thickness 60 nm: the image intensity is related to the component of magnetization (a) perpendicular and (b) parallel to the long dimension of the particle; the two images enable the magnetization distribution (c) to be determined (courtesy of J. N. Chapman)

advantage that the image intensity is directly related to a selected component of the magnetic induction in the specimen. A quadrant detector is normally used, which enables a complete map of the magnetization distribution to be obtained at high resolution (see Fig. 5). Using a field emission electron source, the resolution is sufficient to enable the structure of domain walls to be studied. For further details see Chapman (1984) and Tsuno (1988).

6. Electron Holography

Electron holography (Tonomura 1987) can also be used for mapping the magnetic induction at a resolution similar to that of the scanning transmission electron microscopy (STEM) technique. However, this method requires very complex instrumentation and the upper limit of specimen thickness is lower than that for STEM.

7. Scanning Electron Microscopy with Polarization Analysis

Electrons emitted from the surface of a magnetic specimen have an unbalanced spin distribution, the

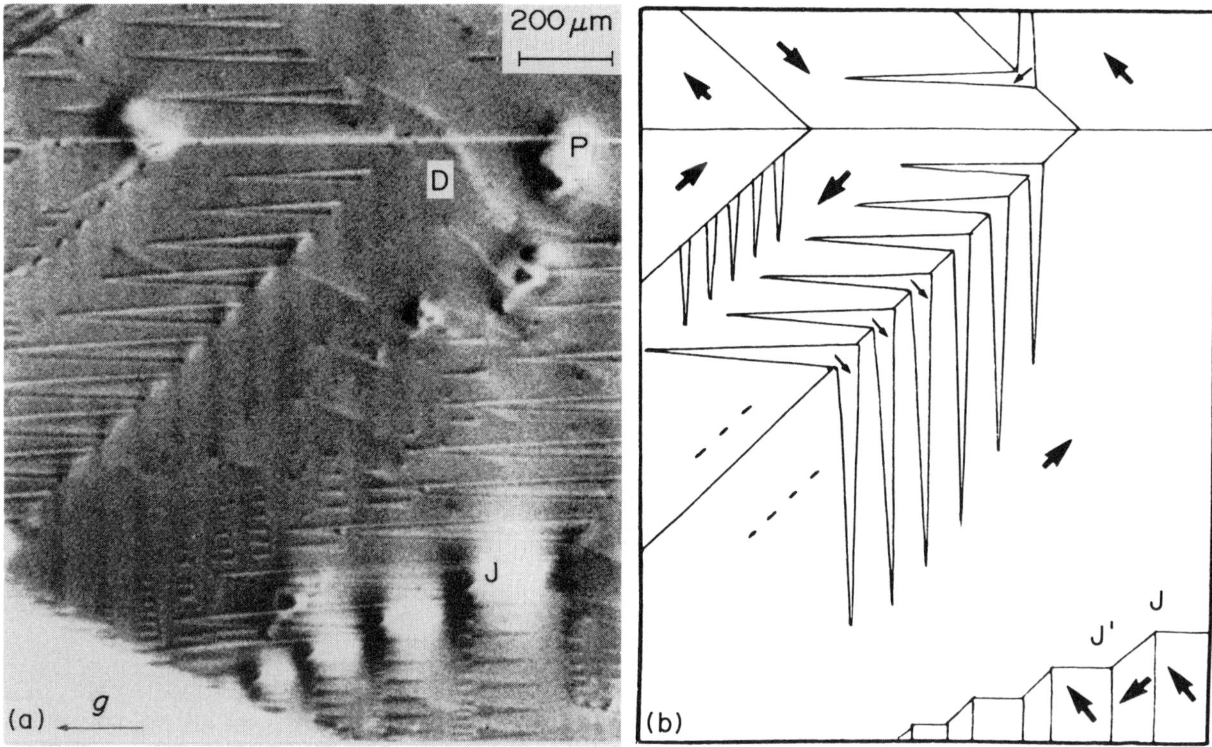

Figure 6
(a) X-ray topographic image of domains in an Fe–3.5%Si specimen of thickness 100 μm and (001) orientation, with (b) diagram of the domain structure: all 180° walls as well as those 90° walls that are in the vertical direction on the page are invisible, but triple wall junctions (J) show strong contrast; a precipitate (P) and a dislocation (D) are also visible (after Miltat 1976. © Taylor and Francis, Basingstoke, UK. Reproduced with permission)

majority of electrons having spins antiparallel to the magnetization. A scanning electron microscope fitted with a polarization analyser can therefore be used to map the magnetization distribution (Scheinfein et al. 1990). The scanning electron microscopy with polarization analysis (SEMPA) method is capable of the high resolution needed for the study of the structure of domain walls but, unlike STEM and electron holography, it is useful for mapping the magnetization close to the surface of bulk materials.

8. *Magnetic Force Microscopy*

In magnetic force microscopy (MFM), a very fine magnetic needle (or a very small magnetic sphere attached to a nonmagnetic needle) is moved parallel to, and very close to, the specimen surface and the force on the needle is measured. As the force depends on the magnetic field at the tip, the method enables the stray field to be mapped, from which the magnetization distribution near the surface can be deduced (Rugar et al. 1990).

9. *X-Ray Topography*

Unlike electrons, x rays are not deflected by the magnetization, but they are very sensitive to small changes in the interatomic spacing or in the orientation of atomic planes. In magnetic materials, such changes occur as a result of magnetostriction, which causes the domains to be spontaneously strained. If two neighboring domains are magnetized in opposite directions, they are strained in the same direction, and are therefore not distinguished from each other by x-ray topography. However, if the angle between the magnetization directions is not 180°, the domains are strained in different ways and therefore diffract x rays differently. X-ray topography enables domains to be imaged in quite thick specimens. However, x rays are very sensitive to crystal defects such as dislocations and grain boundaries and, therefore, x-ray topography is restricted to single crystals with low dislocation densities; an example is shown in Fig. 6. As x-ray topography does not produce magnified images, the photographs must subsequently be enlarged by optical methods and the resolution is therefore similar to that of optical microscopy techniques—much poorer than that of electron optical techniques. X-ray topography generally requires very long exposures, of the order of several hours, if a conventional x-ray source is used. However, exposure times can be reduced considerably by using intense sources based on synchrotron particle accelerators. A special feature of x-ray topography is that it can be used to image domains in antiferromagnetic materials. A detailed description of domain imaging by x-ray topography has been given by Polcarova (1969) and references to later work can be found in Tanner (1976).

10. Neutron Topography

The principles of neutron topography are similar to those of x-ray topography, except that as neutrons have a magnetic moment they interact not only with lattice distortions, but also with the magnetization, and can therefore detect 180° domain walls (Schlenker and Baruchel 1978).

11. Moving Domain Walls

The motion of domain walls can be studied by applying magnetic fields while the specimen is being imaged. In principle, fields can be applied in conjunction with any of the imaging techniques described. However, some techniques are more suitable for studies of domain wall motion than others. In the case of the Bitter technique, the magnetic particles cannot follow domain walls if they move too rapidly and the pattern becomes fixed once the carrier liquid has evaporated. Electron optical techniques suffer from the disadvantage that electrons are deflected by the applied magnetic fields, causing shifts and distortions in the illumination and the image. These shifts and distortions must be compensated by applying similar fields in the opposite direction away from the specimen. Even so, the magnitude of fields that can be applied is limited, partly by the fact that the compensation is never perfect and partly by the restricted space available in the specimen chamber for the coils necessary to produce the fields. The techniques that are most easily adapted to dynamic studies are the Faraday and Kerr effects and x-ray and neutron topography, because images obtained by these techniques are unaffected by the applied fields. However, topography techniques are unsuitable for studies of rapidly changing domain structures because of the long exposure times needed.

See also: Domains and Domain Walls; Magnetic Properties of Superconductors

Bibliography

Carey R, Isaac E D 1966 *Magnetic Domains and Techniques for Their Observation*. English Universities Press, London

Chapman J 1984 The investigation of magnetic domain structures in thin foils by electron microscopy. *J. Phys. D* 17: 623–47

Fathers D J, Jakubovics J P 1977 Methods of observing magnetic domains by scanning electron microscopy. *Physica B* 86–88: 1343–4

Jakubovics J P 1975 Lorentz microscopy and applications (TEM and SEM). In: Ruedl E, Valdrè U (eds.) 1975 *Electron Microscopy in Materials Science; Third Course of the International School of Electron Microscopy*, Pt. 4. Commission of the European Communities, Luxembourg, pp. 1303–403

Miltat J E A 1976 Fir-tree patterns. Elastic distortions and application to x-ray topography. *Philos. Mag.* 33: 225–54

Niedoba H, Heyderman L J, Gupta H O, Puchalska I B, Hubert A 1991 Micromagnetics of two-layer laminated films with superimposed domains. *J. Appl. Phys.* 69: 5865–7

Polcarova M 1969 Applications of x-ray topography to the study of magnetic domains. *IEEE Trans. Magn.* 5: 536–44

Puchalska I B, Jouve H 1977 Comparison between planar domain walls associated with bubbles in implanted and permalloy coated garnets. *IEEE Trans. Magn.* 13: 1178–80

Rave W, Schäfer R, Hubert A 1987 Quantitative observation of magnetic domains with the magneto-optical Kerr effect. *J. Magn. & Magn. Mater.* 65: 7–14

Rugar D, Mamin H J, Guethner P, Lambert S E, Stern J E, McFadyen I, Yogi T 1990 Magnetic force microscopy: general principles and application to longitudinal recording media. *J. Appl. Phys.* 68: 1169–83

Scheinfein M R, Unguris J, Kelley M H, Pierce D T, Celotta R J 1990 Scanning electron microscopy with polarisation analysis (SEMPA). *Rev. Sci. Instrum.* 61: 2501–26

Schlenker M, Baruchel J 1978 Neutron techniques for the observation of ferro- and antiferromagnetic domains. *J. Appl. Phys.* 49: 1996–2001

Tanner B K 1976 *X-Ray Diffraction Topography*. Pergamon, Oxford, pp. 119–22

Tonomura A 1987 Applications of electron holography. *Rev. Mod. Phys.* 59: 639–69

Tsuno K 1988 Magnetic domain observation by means of Lorentz electron microscopy with scanning technique. *Rev. Solid State Sci.* 2: 623–58

J. P. Jakubovics
[University of Oxford, Oxford, UK]

Magnetocaloric Effect

The temperature of thermally isolated paramagnetic material can be raised by the application of a strong magnetic field; this phenomenon typifies the magnetocaloric effect. The effect is easily understood using classical thermodynamic theory as follows. During isothermal magnetization at temperature T_1, the heat ΔQ released from the spin system is given by its magnetic entropy change ΔS_J, which is determined by the degree of spin alignment as

$$\Delta Q = \Delta S_J T_1 \quad (1)$$

Therefore, under adiabatic conditions, since the magnetic material is heated by ΔQ as in Eqn. (1), its temperature rises.

Recent applications of the magnetocaloric effect, for example in magnetic refrigeration and as magnetic regenerator material, have attracted the attention of many investigators in the cryogenic engineering field. These technologies are very effective in raising the efficiency of helium liquefaction and can be expected to be developed further in the near future.

This article explains the basic concepts of the magnetocaloric effect, that is, magnetic entropy and spin temperature, and then outlines the applications mentioned.

1. *Basic Concept*

To simplify the description of the magnetocaloric effect, an ideal paramagnetic spin system composed of N noninteracting spins J will be used.

1.1 *Magnetic Entropy*

In the paramagnetic state the $(2J+1)$ energy states are degenerate and each spin can arbitrarily occupy one of the $(2J+1)$ states, so that the total number of available states is given by $(2J+1)^N$. Therefore, the magnetic entropy S_J in the paramagnetic state can be obtained from Boltzmann's theory as

$$S_J = Nk_B \ln(2J+1) \qquad (2)$$

where k_B is Boltzmann's constant.

Then, as the magnetic field B_0 is applied to the spin system, the degenerate spin states are resolved into $(2J+1)$ nondegenerate states determined by the Zeemann exchange interaction as shown in Fig. 1. If the external field B_0 satisfies the condition that the Zeemann splitting energy $g\mu_B B_0$ is much larger than the thermal agitation energy $(\sim k_B T)$, all the spins align parallel to the external field (the perfect ferromagnetic alignment). In this case, since each spin occupies only the ground state $(m_J = -J)$, as shown in Fig. 1,

$$S_J = k_B \ln(1^N) = 0 \qquad (3)$$

If a magnetic phase change from the paramagnetic phase to the ferromagnetic phase occurs, the magnetic entropy change is given by Eqn. (2). If this change is performed under isothermal conditions at temperature T_1, the spin system expels heat as given by

$$\Delta Q = Nk_B T_1 \ln(2J+1) \qquad (4)$$

reducing its magnetic entropy. Since the variation mentioned is a thermodynamically reversible process, in the inverse process the magnetic material absorbs the same heat ΔQ.

1.2 *Spin Temperature*

As temperature increases, the thermal energy of the spins $(\sim k_B T)$ increases until comparable with the

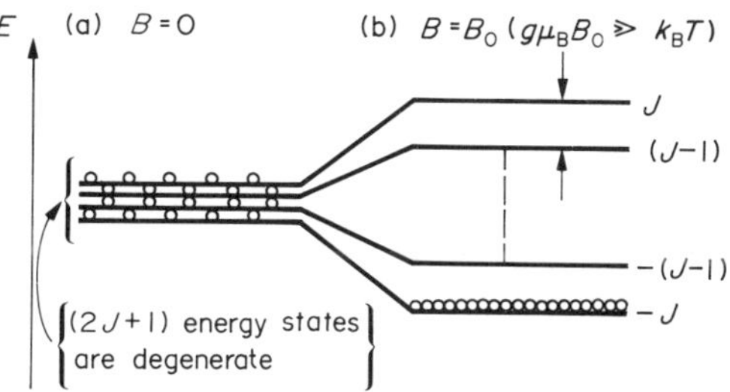

Figure 1
Energy scheme of spin in the ideal paramagnet: ○ indicates the population of spins. (a) External magnetic field is zero. The $(2J+1)$ energy states are degenerate and spins occupy each state equivalent probability.
(b) External magnetic field is B_0 where $g\mu_B B_0 \gg k_B T$. All spins occupy the ground state $(m_J = -J)$

Zeeman splitting energy $g\mu_B B$. Each spin can then occupy not only the ground state but also the excited states. The spin distribution in each energy state is given by the Boltzmann distribution function and, therefore, the spin population of the ith energy state is given by

$$n_i = n_0 \exp(-g\mu_B m_i B/k_B T) \qquad (5)$$

where m_i is a magnetic quantum number of the ith state. Therefore, the temperature of the spin system is defined by the spin distribution over the available energy levels under thermal equilibrium.

2. *An Example of Real Magnetic Material*

In most magnetic materials the spins are coupled by magnetic exchange interactions. This situation can be modelled in terms of a hypothetical internal magnetic field B_{int} applied to each spin. However, since the magnetic transition caused by the magnetic exchange interaction is a cooperative phenomenon, it occurs suddenly at the critical temperature T_c which is determined by B_{int}. Thus, the phase transition process near T_c is quite different from spin alignment caused by an external magnetic field.

EuS is a typical three-dimensional Heisenberg ferromagnet whose T_c is 16.1 K. Its saturation moment depends on temperature as shown in Fig. 2a. Below T_c the order parameter of spins suddenly increases and at 0 K there is perfect ferromagnetic spin alignment.

In Fig. 2b, the specific heat of EuS is shown for different applied magnetic fields. Since the Debye temperature of EuS is 230 K, the magnetic specific heat $C_J(T, B)$ is superimposed on the lattice specific heat $C_l(T)$ above 10 K.

Figure 2c shows the total entropy S_{tot} of EuS calculated using the thermodynamic relation

$$S_{tot} = \int_0^T (1/T)(C_J + C_l)dT \qquad (6)$$

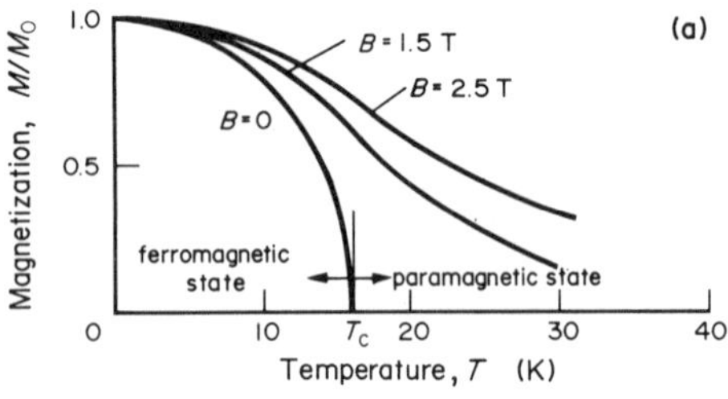

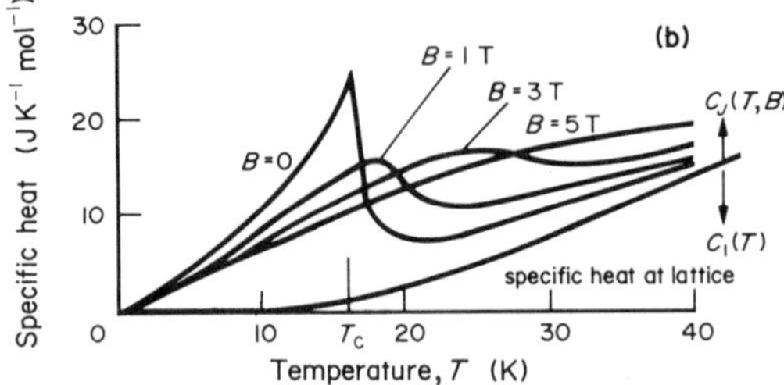

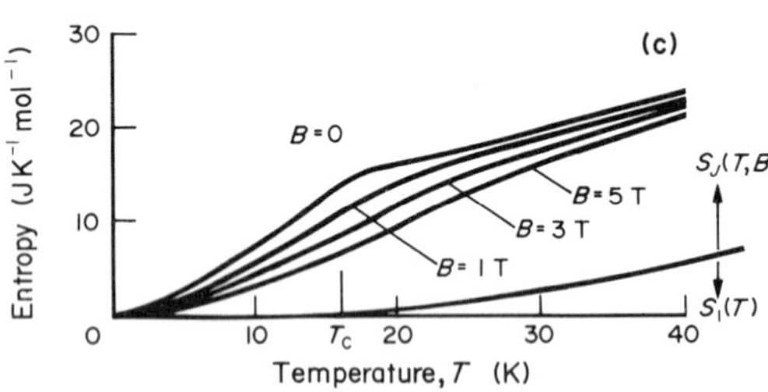

Figure 2
The magnetic and caloric properties of the three-dimensional Heisenberg ferromagnet EuS: (a) the temperature dependence of the magnetic moment in various magnetic fields, (b) the temperature dependence of the specific heat in various magnetic fields, and (c) the temperature dependence of the entropy estimated from the results of (b) and Eqn. (6)

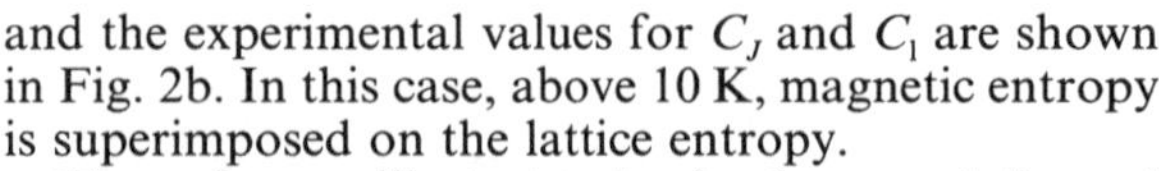
and the experimental values for C_J and C_l are shown in Fig. 2b. In this case, above 10 K, magnetic entropy is superimposed on the lattice entropy.

These figures illustrate clearly the correlation of spin alignment, magnetic entropy and specific heat.

3. *Applications*

A well-known example of the application of the magnetocaloric effect is in adiabatic demagnetization cooling (Giauque and MacDougall 1933). This depends on the two different processes as shown in Fig. 3. During the isothermal magnetization process (A→B) at temperature T_1 the spin system expels heat ($\Delta Q = \Delta S_J T_1$) to reduce its entropy. Then, the applied

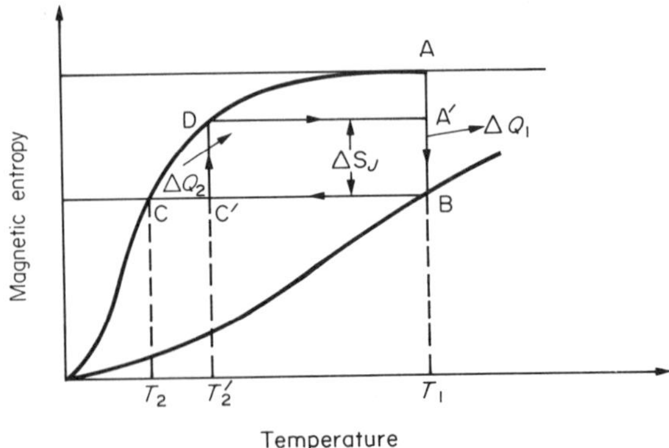

Figure 3
Schematic diagram of the adiabatic demagnetization process. The upper and lower lines correspond to the magnetic entropy curves at magnetic fields of zero and 6 T, respectively. Path A→B is the isothermal magnetization process. Magnetic material expels heat $\Delta Q = \Delta S_J T_1$. Path B→C is the adiabatic demagnetization process. The temperature of the material decreases along path B→C as magnetic field decreases. The rectangle A′BC′D shows the Carnot magnetic refrigeration cycle

magnetic field B is reduced under adiabatic conditions ($\Delta Q = 0$). Since this is a constant entropy process, the temperature of the spin system decreases along the path B→C in Fig. 3 with decreasing magnetic field. This cooling method is very useful in the range below 4.2 K and is used for the production of very low temperatures of the order of microkelvin (Pobell 1982).

A promising application of this effect is magnetic refrigeration in which magnetic materials are used as the working substance. The simplest example is the Carnot magnetic refrigerator whose conceptual lay-

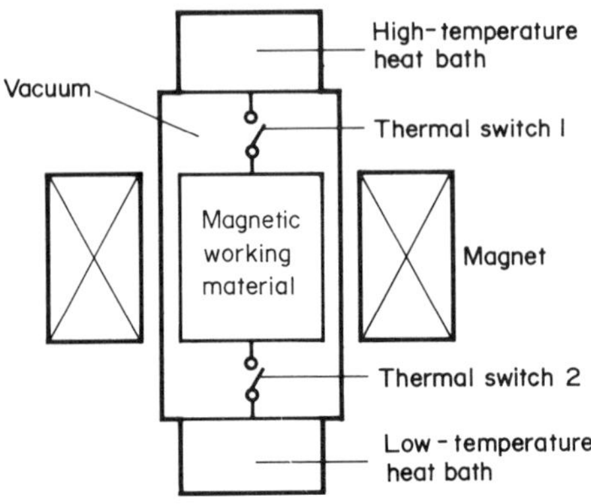

Figure 4
Conceptual Carnot magnetic refrigerator

out is given in Fig. 4. The refrigeration cycle is composed of four processes: the isothermal magnetization (A′ →B), the adiabatic demagnetization process (B→C′), the isothermal demagnetization cooling process (C′ → D) and the adiabatic magnetization process (D → A′) shown in Fig. 3.

This cycle is very useful in the range below 20 K, where the lattice entropy is almost zero. In this range, two types of magnetic refrigerators have been developed (Beranger et al. 1981, Hakuraku and Ogata 1986, Nakagome et al. 1986, Hashimoto et al. 1988a). One is used for production of superfluid ^4He (from 4.2 K to 1.8 K) and the other for ^4He liquefaction from 20 K to 4.2 K. It has been verified that these test machines have very high efficiency (~20–30% Carnot) in comparison with the low-power gas refrigerator (~1–5% Carnot), although the refrigeration power is low at only about 1–3 W.

In this range the most promising magnetic refrigerants are the heavy rare-earth garnets, $Gd_3Ga_5O_{12}$ (GGG) and $Dy_3Al_5O_{12}$ (DAG) (see *Rare-Earth Intermetallics*; Hashimoto et al. 1988a). As shown in Fig. 5, GGG has a high refrigeration power for a low-temperature range and DAG has a wide temperature span. The entropies of $Gd_3(Ga_{1-x}Al_x)_5O_{12}$ (Kimura et al. 1989) and $DyPO_4$ (Barclay 1980) have also been investigated.

Above 20 K, where the lattice entropy is comparable with the magnetic entropy, an Ericsson magnetic refrigerator has been investigated (Matsumoto et al. 1988). In the high-temperature range, where the thermal activation energy of spins (~k_BT) is large, heavy rare-earth ferromagnetic materials must be used. Some properties of typical refrigerants from 77 K to 10 K are shown in Table 1. Several kinds of refrigerants have been investigated, for instance, gadolinium (Hashimoto et al. 1981) at room temperature, R_3Al_2 (Wohlfarth 1980) from approximately 300 K to 10 K and RAl_2 (Hashimoto et al. 1981) from 200 K to 10 K, where R indicates gadolinium, terbium, dysprosium, holmium or erbium.

A magnetic material with large magnetic specific heat in the low-temperature range below 10 K has been applied as the regenerative material for a small refrigerator in the ^4He temperature range and it is possible to reach the approximate 2 K range by using the magnetic regenerator materials whose specific heats are shown in Fig. 6. Promising new magnetic

Table 1
Properties of some typical refrigerants

Refrigerant	Curie temperature, T_C (K)	Effective Bohr magneton number, μ_{eff}	Magnetic entropy change $\Delta S_{J(max)}$ at T_C ($J K^{-1} mol^{-1}$) at 4 T	at 5 T
$ErAl_2$	11.7	9.6	8.3[a] / 7.0[b]	7.6[b]
$HoAl_2$	26.8	10.8	5.7[a] / 5.7[b]	6.3[b]
$Dy_{0.5}Ho_{0.5}Al_2$	46.0	11.7	4.2[b]	4.8[b]
$DyAl_2$	55.9	11.7	3.6[b]	4.1[b]

a From specific heat b From magnetization

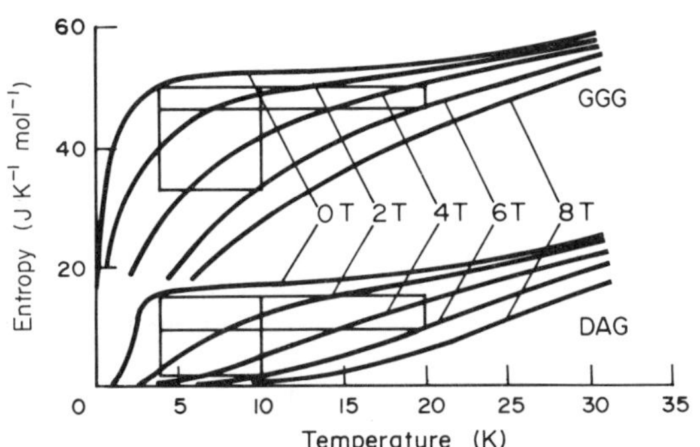

Figure 5
Variation of the entropy of $Gd_3Ga_5O_{12}$ (GGG) and that of $Dy_3Al_5O_{12}$ (DAG) as a function of temperature in various magnetic fields; two kinds of Carnot cycle (from 20 K to 4.2 K and from 10 K to 4.2 K) are shown

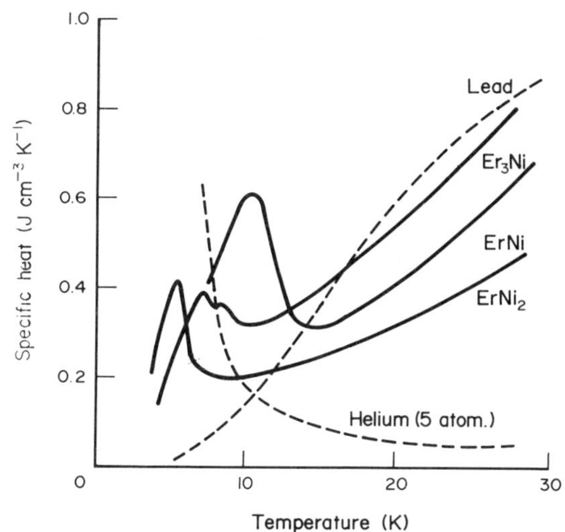

Figure 6
The specific heats of several magnetic regenerator materials: for comparison the specific heats of helium gas and lead are also shown

regenerator materials such as $(Er_{1-x}Dy_x)Ni_2$ (Hashimoto et al. 1988b), (Gd–Er)Rh (Buschow 1980), ErNi (Hashimoto et al. 1988b), Er_3Ni (Sahashi et al. 1990) and so on are being investigated.

See also: Magnetic Materials: Basic Concepts; Rare-Earth Intermetallics

Bibliography

Barclay J A 1980 A 4 K to 20 K rotational cooling magnetic refrigerator capable of 1 mW to >1 W operation. *Cryogenics* 20: 467–71

Beranger R, Mardion G B, Claudet G, Delpuech C, Lacaze A F, Lacaze A A 1981 A gadolinium gallium garnet double acting reciprocating magnetic regenerator. *Adv. Cryog. Eng.* 27: 703–9

Buschow K H J 1980 *Rare Earth Compounds*, Ferromagnetic Materials, Vol. 1. North-Holland, Amsterdam, pp. 279–414

Giauque W F, MacDougall D P 1933 Attainment of temperatures below 1° absolute by demagnetization of $Gd_2(SO_4).8H_2O$. *Phys. Rev.* 43: 768

Hakuraku Y, Ogata H 1986 A magnetic refrigerator for superfluid-helium equipped with a rotating superconducting magnet system. *Jpn. J. Appl. Phys.* 25: 140–6

Hashimoto T, Kuzuhara T, Sahashi M, Inomata K, Tomokiyo A, Yayama H 1987 New application of complex magnetic materials to the magnetic refrigerant in an Ericsson magnetic refrigerator. *J. Appl. Phys.* 62: 3873–8

Hashimoto T, Li R, Matsumoto K, Sahashi M, Yayama H, Tomokiyo A 1988b Recent progress in the magnetic materials for a regenerator in the range from 4.2 K to 20 K. *Proc. Int. Cryogenic Materials Conf.*, Schenyang, pp. 667–76

Hashimoto T, Numazawa T, Shiino M, Okada T 1981 Magnetic refrigeration in the temperature range from 10 K to room temperature: the ferromagnetic refrigerants. *Cryogenics* 21: 647–53

Hashimoto T, Yazawa T, Li R, Kuzuhara T, Nakagome H, Takahashi M, Sahashi M, Inomata K, Tomokiyo A, Yayama H 1988a Recent progress in magnetic refrigeration studies. *Adv. Cryog. Eng.* 33: 733–41

Kimura H, Numazawa T, Sato M, Maeda H 1989 Crystal structure and thermal conductivity of $Gd_3(Ga_{1-x}Al_x)_5O_{12}$ garnets. *Jpn. J. Appl. Phys.* 28: 1644–7

Li R, Ogawa M, Hashimoto T 1990 Magnetic intermetallic compounds for cryogenic regenerator. *Cryogenics* 30: 521–6

Matsumoto K, Ito T, Hashimoto T 1988 An Ericsson magnetic refrigerator for low temperature. *Adv. Cryog. Eng.* 33: 743–50

Nakagome N, Kuriyama T, Ogiwara H, Yazawa T, Hashimoto T 1986 A rotating magnetic refrigerator for helium liquefaction. *Proc. 11th Int. Cryogenics Engineering Conf.*, Berlin, pp. 246–55

Pobell F 1982 The quest for ultralow temperatures: what are the limitations? *Physica B* 109–10: 1485–98

Sahashi M, Tokai Y, Kuriyama T, Nakagome H, Li R, Ogawa M, Hashimoto T 1990 New magnetic material R3T system with extremely large heat capacities used as heat regenerator. *Adv. Cryog. Eng.* 35: 1175–90

Wohlfarth E P 1980 *Ferromagnetic Materials*, A Handbook on the Properties of Magnetically Ordered Substances, Vols. 1–3. North-Holland, Amsterdam

T. Hashimoto
[Tokyo Institute of Technology,
Tokyo, Japan]

Magnetoelastic Phenomena

In experiments with magnetic solids, changes in stiffness are found when a magnetic field is applied. These effects have their origin in the magnetostrictive response of a magnetic material (see *Magnetostrictive Materials*). A measure of the longitudinal stiffness of a long slender body is Young's modulus E—the larger the stiffness the larger the value of E. The decrease in the value of Young's modulus $E(H)$ caused by rotation of the magnetic moment in a solid is called the ΔE effect. The externally applied magnetic field H is used to rotate the magnetic moment. The ΔE effect has become important because it has recently become possible to reversibly change the value of $E(H)$ by a factor of ten in certain amorphous ferromagnetic materials with an applied field three times that of the earth (Wun-Fogle et al. 1986). This has led to quite definitive experiments in the field of nonlinear dynamics where important nonlinear phenomena, chaotic and nonchaotic, have been observed for the first time (Savage et al. 1990). There is active research to exploit the new materials in applications such as accelerometers, pressure sensors, strain gauges, torque sensors and security devices (see *Sensors and Devices, Magnetic*; Hernando et al. 1988). With the theory presented here, predictions can be made for the use of the amorphous materials in applications. In the past, a thermodynamic approach has been used for such predictions. The thermodynamic approach conceals the underlying physics in partial derivatives that are treated as constants to be determined experimentally. Here, large-amplitude stress–strain theory is used to describe the underlying physics (Savage and Spano 1982).

1. Theory of the ΔE Effect

The theory of magnetoelasticity is a description of the interaction of the elastic and magnetic subsystems of magnetic materials. The ΔE effect is an excellent example of such a magnetoelastic interaction and is shown in Fig. 1. Amorphous ferromagnetic iron alloys present an opportunity for the simplest and most complete description of the ΔE effect (see *Metallic Glasses*). This article therefore begins with a discussion of the ΔE effect in transversely field-annealed amorphous ribbon material which is shown with its domain structure in Fig. 2. In this simplified model

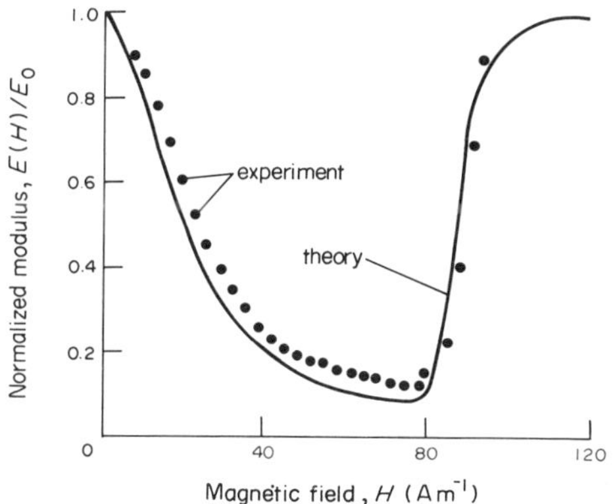

Figure 1
Normalized modulus $E(H)/E_0$ vs H of transversely field-annealed amorphous material of approximate composition $Fe_{80}B_{15}Si_5$: E_0 is Young's modulus without magnetoelastic interactions; the decrease is caused by the stress rotating the magnetic moment

the domain walls play no role in the generation of longitudinal magnetization M. The total magnetization is the sum of equivalent longitudinal components in each domain due to moment rotation. It is therefore only necessary to consider what goes in one domain when the applied magnetic field is in the usual (longitudinal) direction.

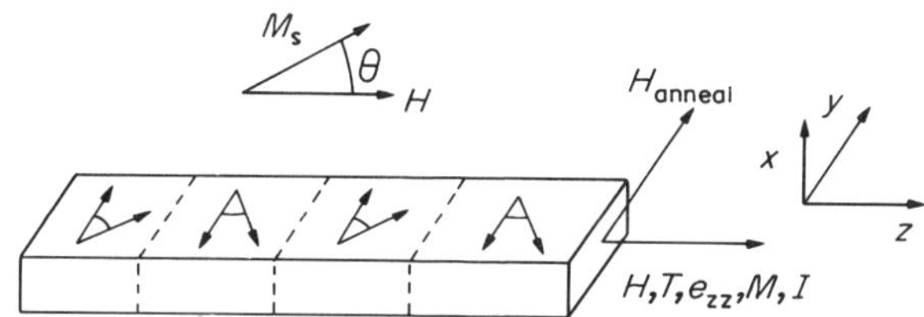

Figure 2
Magnetic structure of transversely field-annealed amorphous material of approximate composition $Fe_{80}B_{15}Si_5$. The annealing field (not the applied field) is in the plane of the ribbon and perpendicular to the longitudinal axis. The ribbon is held in the annealing field at around 475 °C for 10 min. As the temperature is lowered below the Curie point the presence of exchange forces and the transverse-annealing field of $10^3\,\mathrm{A\,m^{-1}}$ or more generates a magnetic anisotropy K of about $40\,\mathrm{J\,m^{-3}}$. The value of K represents the magnetic energy necessary to rotate the moment from the transverse to the longitudinal direction. The longitudinal variables are the moment M, applied field H (not the annealing field), transport current I, stress T and strain e_{zz}. The saturation moment is M_s. As H is increased from zero the moments in each domain rotate and the sample becomes longer (shorter) with positive (negative) magnetostriction. The change in length per unit length at magnetic saturation ($M = M_s$) is $3\lambda_s/2$. In this material the value of λ_s is around 35×10^{-6}. The current I is only used for the portion of the text concerning current excitation

In the derivation of an expression for Young's modulus in a linear elastic solid the elastic energy is minimized with respect to the strains e_{ij}. The differentials of the energy with respect to the strains are set equal to zero except for the differential with respect to the longitudinal strain e_{zz} which is set proportional to the applied longitudinal stress T (Fig. 2). The result is Hooke's law, $T = E_0 e_{zz}$. E_0 is an expression for Young's modulus in terms of the elastic constants c_{ij} of the solid. To derive an expression for the ΔE effect the same thing is done except that now there are more terms in the energy (see *Micromagnetic Calculations*). The Zeeman energy $-MH$ (due to the forces tending to align the magnetic moment with the field) must be included, as must the anisotropy energy K (due to atomic forces tending to align the magnetic moment with a certain material axis) and the magnetoelastic energy (due to atomic forces resulting from the interaction between the magnetic and elastic systems). It is assumed that there is no nonzero component of the magnetization out of the plane of the ribbon. Then $M = M_s \cos\theta$ where θ is the angle between the longitudinal axis and the magnetic moment vector M_s (Fig. 2). With this assumption the Zeeman energy is $-HM_s\cos\theta$, the anisotropy energy is $K\cos^2\theta$ and the magnetoelastic energy (Chikazumi 1978) becomes

$$E_{me} = -(3\lambda_s/2)(c_{11} - c_{12})[e_{xx}(\sin^2\theta - \tfrac{1}{3}) + e_{zz}(\cos^2\theta - \tfrac{1}{3}) - e_{yy}/3 + e_{xz}\sin\theta\cos\theta]$$

with the elastic energy

$$E_e = c_{11}(e_{xx}^2 + e_{yy}^2 + e_{zz}^2)/2 + c_{12}(e_{xx}e_{yy} + e_{yy}e_{zz}e_{zz}e_{xx}) + (c_{11} - c_{12})(e_{xy}^2 + e_{xz}^2 + e_{yz}^2)/2 - e_{zz}T$$

Minimizing the total energy with respect to the e_{ij} gives

$$0 = -(3\lambda_s/2)(c_{11} - c_{12})(\sin^2\theta - \tfrac{1}{3}) + c_{11}e_{xx} + c_{12}(e_{yy} + e_{zz}) \quad (1)$$

$$0 = -(3\lambda_s/2)(c_{11} - c_{12})(-\tfrac{1}{3}) + c_{11}e_{yy} + c_{12}(e_{zz} + e_{xx}) \quad (2)$$

$$T = -(3\lambda_s/2)(c_{11} - c_{12})(\cos^2\theta - \tfrac{1}{3}) + c_{11}e_{zz} + c_{12}(e_{xx} + e_{yy}) \quad (3)$$

Adding Eqns. (1–3) yields

$$T = (c_{11} + 2c_{12})(e_{xx} + e_{yy} + e_{zz}) \quad (4)$$

Since $e_{xx} + e_{yy} + e_{zz}$ is the volume strain it is apparent that linear magnetostriction has no volume effects. Solving Eqn. (3) for $e_{xx} + e_{yy}$ and inserting into Eqn. (4) yields

$$e_{zz} = T/E_0 + (3\lambda_s/2)(\cos^2\theta - \tfrac{1}{3}) \quad (5)$$

where $E_0 = (c_{11} - c_{12})(c_{11} + 2c_{12})/(c_{11} + c_{12})$. The $-\lambda_s/2$ term in the strain can be dropped because this only alters the zero in the strain measurement.

Equation (5) is called Hooke's law for an isotropic magnetoelastic solid. The result contains the usual expression for Young's modulus E_0 in an isotropic solid. The usual strain T/E_0 is incremented by the magnetostrictive strain $(3\lambda_s/2)\cos^2\theta$, the change in length of the sample caused by rotation of the magnetic moment. The cause of the ΔE effect can be seen in Eqn. (1). When a small stress δT is imposed on the sample with H being a constant greater than zero, the moment rotates to lower the energy. Thus an additional strain to $\delta T/E_0$, a magnetostrictive strain, is observed and since Young's modulus $E(H) \equiv (\delta T/\delta e)_H$, a lower or "softer" modulus is observed.

The energy must also be minimized with respect to the direction of the moment (i.e., with respect to θ). The result is: $\sin\theta[M_sH - \cos\theta(2K - 3\lambda_sT)] = 0$, which has solutions $\sin\theta = 0$ or

$$M_sH = (2K - 3\lambda_sT)\cos\theta \qquad (6)$$

$\sin\theta = 0$ is magnetic saturation, that is, the moment lies completely in the longitudinal direction (along H). The other solution implies that M is a linear function of the magnetic field with a slope of $M_s^2/(2K - 3\lambda_sT)$. The slope is called the susceptibility χ and has almost the same value as the permeability μ in the materials being discussed. The stress appears as a magnetic anisotropy. The stress is externally applied, giving an externally controllable susceptibility or permeability. How this can be exploited in devices will be shown in Sect. 2.

This simplified model has a serious problem. The susceptibility becomes infinite for a tensile value of $T = 2K/(3\lambda_s)$. Although Eqn. (6) is good for small stresses, no singularity is observed experimentally. The reason is that most small defects remove the singularity. In this example, defects will be simulated by imposing a small shear stress $T_{xz} = S$ on the sample. This adds a term $-Se_{xz}$ to the elastic energy. The resulting modification of Eqn. (6) is

$$M_sH = (2K - 3\lambda_sT + 6\lambda_sS\cot 2\theta)\cos\theta \qquad (7)$$

$\sin\theta = 0$ is no longer a solution. The sample never truly saturates no matter how high the field. The moment only rotates close to the longitudinal direction when T or H are large. The possibility of a singularity in the susceptibility no longer exists. This improvement over the simple rotational model is necessary for large-amplitude stress–strain considerations. Also, the susceptibility is closer to what is observed in magnetization experiments.

Equations (5) and (7) can now be normalized with $h = H/H_k$, $t = 3\lambda_sT/2K$, $e = e_{zz}/(3\lambda_s/2)$ and $s = 3\lambda_sS/2K$. H_k is the anisotropy field. It is the field at which, in the simple rotational model ($S = 0$) with $T = 0$, the sample saturates. These normalizations give

$$e = 2t/D + \cos^2\theta \qquad (8)$$

$$h = (1 + t + 2\cot 2\theta)\cos\theta \qquad (9)$$

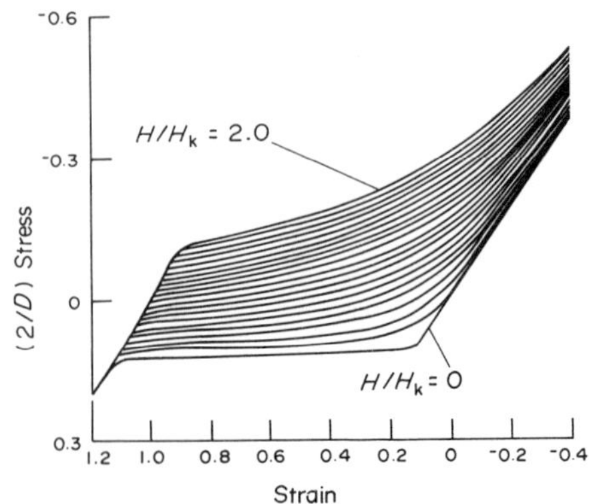

Figure 3
Magnetoelastic strain vs stress/$(D/2)$ in dimensionless units for a range of values of the applied field: values of $D = 17$ (from static measurements) and $s = 0.01$ were used; a nonzero value of s reduces the ΔE effect

For this model $h \geqslant 0$, and $0 \leqslant \theta < \pi/2$. The value of $D = (3\lambda_s)^2E_0/2K$ is a measure of the ΔE effect. The larger the value of D the larger the change in Young's modulus $E(h)$. In the amorphous ribbons considered here the value of E can be changed by a factor of ten, which is much larger than has ever been achieved in a crystalline solid. The reason is the extremely low value of the anisotropy constant $K \simeq 40\,\mathrm{J\,m^{-3}}$ In good material, values of s should be small. In the materials considered here $s \simeq 0.01$ and reduces the theoretical value of the change in $E(h)$ from about 17 to 10. The large-scale stress–strain relations can now be obtained with the magnetic field as a parameter. An infinite series can be derived:

$$t = [E(h)/E_0](e - e_0) + \sum_2^\infty a_n(h)(e - e_0)^n$$

where the magnetostrictive strain $e_0 = e_0(h)$ is the value of e at $t = 0$. The $a_n(h)$ are completely determined (Savage et al. 1990). The form is instructive, however, the stress–strain relations can be most easily obtained numerically. The relations are shown in Fig. 3. The ΔE effect or the determination of the field dependence of $E = \delta t/\delta e$ at constant h around $e = e_0$ can now be determined. The results are shown in Fig. 1. The excellent agreement with experiment is also shown.

The coupling factor k_{33} can now be defined by the relationship $E(h) = E_0(1 - k_{33}(h)^2)$. Large values of k_{33} imply a large ΔE effect. In fact, if k is the maximum value of k_{33} then, in the simple rotational model ($s = 0$), $D = k^2/(1 - k^2)$. Values of the ΔE effect and coupling factor are shown in Table 1 for different materials. If the sample is constrained so that it cannot elongate, then the observed susceptibility is $\chi_e = \chi(1 - k_{33}^2)$ where χ is the unconstrained value of

Table 1
Coupling factor and ΔE effect in various ferromagnetic materials: in the amorphous ribbon material the large values of k_{33} are highly dependent on composition, annealing conditions and surface quality; the addition of cobalt leads to a higher magnetic anisotropy energy, hence, lower values of k_{33}; the higher value in the last entry is due to an improvement in surface quality

Ferromagnetic material	Maximum coupling factor, k_{33}	$(E_0 - E(H)^a_{min})/E(H)^a_{min}$
Iron	~0	~0
Nickel	0.32	0.11
Amorphous ribbon $Fe_{67}Co_{17}B_{14}Si_2$	0.72	1.08
Amorphous ribbon $Fe_{78}B_{12}Si_{10}$	0.82	2.03
Amorphous ribbon $Fe_{81}B_{13.5}Si_{3.5}C_2$	0.95	9.25
Amorphous ribbon $Fe_{81}B_{13.5}Si_{3.5}C_2$	0.97	10.0

a $E(H)_{min}$ in the minimum measured value of Young's modulus

the susceptibility as defined previously (Wun-Fogle et al. 1986). Engineers desire large values of k_{33} in applications.

2. Applications

As shown, the magnetic permeability $\mu = \mu_0 + \chi$ is a function of the applied stress. The denominator of the χ in the simple rotational model is $(2K - 3\lambda_s T)$. Because the magnetic anisotropy energy K is very small the effects of small stresses are very large. This is equivalent to saying that effects of small strains are very large because T can be replaced by its equivalent in Eqn. (8). This characteristic has been exploited in the construction of a magnetoelastic strain gauge. With this device, strains are measured by measuring values of μ. Figures of merit around 4×10^5 were observed. This is three orders of magnitude higher than in present strain gauges (Wun-Fogle et al. 1987, Savage et al. 1988). Good temperature characteristics are expected.

Suppose a magnetic field $H = H_0 + h_{ac} \sin \omega t$ is imposed on a sample similar to that of Fig. 2. Because the field is varying, $E(H)$ is varying. This is equivalent to having a time-dependent spring constant. With a system like this a subharmonic bifurcation is expected, that is, the appearance of a response with period two (a frequency of $\omega/2$) if the ΔE effect is large enough. This experiment has been done by placing a pickup coil around the sample and observing the induced voltage. As h_{ac} is increased the period two voltage suddenly appears at a threshold value of h_{ac}. The threshold value of h_{ac} was in good agreement with the theory presented here (Savage et al. 1988). This experiment forms the basis for acoustic parametric amplification at sonic frequencies.

Another interesting topic is the electrical self-impedance of the amorphous ribbon. If an ac current, $I \sin \omega t$, is supplied from a current source with constant amplitude I, the voltage drop V along the length divided by I is the self-impedance Z. The value of Z at large ω is complex and is a function of the magnetic permeability of the ribbon. The magnetic permeability is a function of the longitudinal stress so that V is a function of the stress, or equivalently, the strain. To calculate Z, Maxwell's equations must be solved in a conductor where the current density $j = \sigma E$ where σ is the conductivity and E is the electric field. The calculation can be simplified by assuming the ribbon width W is very large so that the only nonzero component of the vector $\boldsymbol{H} = (H_x, H_y, H_z)$ is $H_y(x)$, that is, edge effects can be ignored. It is assumed that the domain walls run orthogonally to the direction shown in Fig. 2 and the simple rotational model so that μ is a function of stress only. Maxwell's equations become

$$\frac{\partial^2 H_y}{\partial x^2} = \sigma \frac{\partial B_y}{\partial t} = -\mathrm{i}\omega\sigma\mu H_y$$

after Fourier transforming. This has the solution $H_y = I/(2W)\sin(dx/a)/\sin d$, where $d = (\mathrm{i}\omega\sigma\mu a^2)^{1/2}$. Then $E = j/\sigma = Id\cos(dx/a)/(2Wa\sigma \sin d)$. This yields the solution $Z = R_0 d \cot d$, where R_0 is the dc resistance. Thus, measurements of Z yield the stress or strain state of the ribbon. In applications this can be an advantage because no bulky solenoids are necessary. Miniaturization is usually a problem with magnetoelastic devices. This kind of excitation has been used to explore the torsional magnetoelastic resonances in Laves-phase rare-earth iron crystalline alloys (Terfenol) (see *Magnetostrictive Materials*). Information on the field dependence of the torsional elastic modulus similar to the data presented in Fig. 2 on Young's modulus was obtained by Savage and Abbundi (1978).

3. Future Developments

The ideas presented in this article have just begun to be exploited in physics and engineering. In physics, an experiment that controls chaos is now being effected. A chaotic system, in some respects, can be thought of as a system randomly visiting a series of unstable periodic motions that are at frequencies equal to the driving frequency divided by integers. With the experiment of a ribbon in chaotic motion it may be possible, by appropriately varying H, to lock the system onto one of the unstable periodic motions. Thus an unstable, chaotic system can be turned into a stable one with (perhaps) the desired motion. A new form of amorphous ferromagnetic material is now available as amorphous wires. These have significant advantages over ribbon geometry and could be the basis for a torque transducer that would have a big impact on the automobile industry.

See also: Magnetic Materials: Basic Concepts; Magnetostrictive Materials; Sensors and Devices, Magnetic

Bibliography

Chikazumi S 1978 *The Physics of Magnetism.* Krieger, Huntington, NY, p. 167

Hernando A, Vasquez M, Barandian J M 1988 Metallic glasses and sensing applications. *J. Phys. E* 21: 1129–39

Savage H T, Abbundi R 1978 Perpendicular susceptibility, magnetomechanical coupling and shear modulus in $Tb_{.27}Dy_{.73}Fe_2$. *IEEE Trans. Magn.* 5: 545–7

Savage H T, Adler C L 1988 Effects of magnetostriction in amorphous ferromagnets. *Mater. Sci. Eng.* 99: 13–18

Savage H T, Ditto W L, Braza P A, Spano M L, Rauseo S N, Spring W C III 1990 Crisis-induced intermittency in a parametrically driven, gravitationally buckled, magnetoelastic amorphous ribbon experiment. *J. Appl. Phys.* 67: 5619–23

Savage H T, Spano M L 1982 Theory and application of highly magnetoelastic Metglas 2605SC. *J. Appl. Phys.* 53: 8092–7

Wun-Fogle M, Clark A E, Hathaway K B 1986 Permeability in frozen high magnetomechanical coupling amorphous ribbons. *J. Magn. & Magn. Mater.* 54–57: 893–4

Wun-Fogle M, Savage H T, Clark A E 1987 Sensitive, wide frequency range magnetostrictive strain gauge. *Sensors Actuators* 12: 323–31

H. T. Savage
[Instituto de Magnetismo Aplicado, Madrid, Spain]

Magnetooptic Materials

In addition to any nonmagnetic effects, light passing through or reflected from magnetically ordered materials interacts with the magnetization resulting either in changes to the polarization state and intensity of the light or sometimes to the magnetization. Certain paramagnetic and diamagnetic materials in a magnetic field often show similar, but much weaker, behavior towards light (see *Magnetooptics*). These changes are expressed in terms of a number of magnetooptic effects, of which the most familiar and important is magnetooptic rotation, the rotation of the axis of linear polarization of light on transmission through a magnetic material. In diamagnetic and paramagnetic materials it is usually called Faraday rotation. The behavior of this rotation in magnetooptic materials is discussed in Sect. 1. Several other prominent magnetooptic effects are also defined and related to materials. Throughout, the emphasis is on the magnetooptic properties of materials with a spontaneous magnetization, that is, ferromagnets and ferrimagnets.

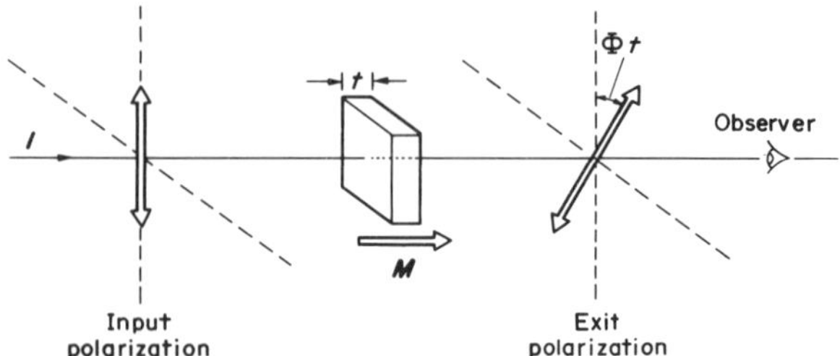

Figure 1
Magnetooptic rotation; the axis of linearly polarized light is rotated on transmission through a material with magnetization along the propagation direction. In a common sign convention the rotation is taken as positive in this case: $\boldsymbol{M}$ and $\boldsymbol{l}$ are parallel and anticlockwise rotation is seen by the observer

1. Magnetooptic Rotation and Ellipticity

Magnetooptic rotation (MOR) is most clearly seen in crystals that would be cubic in the absence of a magnetization (e.g., the ferrimagnetic garnets). Lower-symmetry crystals often have strong birefringences which complicate the observations. As indicated in Fig. 1, the axis of linear polarization is rotated after passage through a crystal with a magnetization $\boldsymbol{M}$. The specific rotation Φ of a material at some wavelength is given as the rotation per unit thickness for a sample that is magnetically saturated along the line of sight. The observed rotation φ is proportional to the thickness t of the sample and to the projection of $\boldsymbol{M}$ on $\boldsymbol{l}$, a unit vector along the direction of propagation; that is, $\varphi = \Phi t \boldsymbol{M}.\boldsymbol{l} |\boldsymbol{M}|^{-1}$. If $\boldsymbol{M}$ is perpendicular to $\boldsymbol{l}$, the rotation is zero; if $\boldsymbol{M}$ is parallel to $\boldsymbol{l}$, the rotation has one sign; but if $\boldsymbol{M}$ is reversed, it has the opposite sign. In this antireciprocity it differs fundamentally from the reciprocal rotations (termed optical activity or natural rotary power) which are attributable to the spatial arrangement of atoms in a molecule or a crystal. A common sign convention is illustrated in Fig. 1. Because of the great possibility of error and of other sign conventions, great care should be exercised in comparing the signs given in different papers.

On solving Maxwell's equations for the propagation of light in the simple case of a medium with a magnetization and no other anisotropy, it is found that light travels in two noninteracting normal modes and that these are the positive and negative circular polarizations. In general, the two complex refractive indices $N_\pm = n_\pm - ik_\pm$ are different. The difference between the dispersive terms $n_\pm$ is called magnetic circular birefringence (MCB) and is directly equivalent to MOR. The difference between the absorptive terms $k_\pm$ is called magnetic circular dichroism (MCD) and is equivalent to the ellipticity of the transmitted

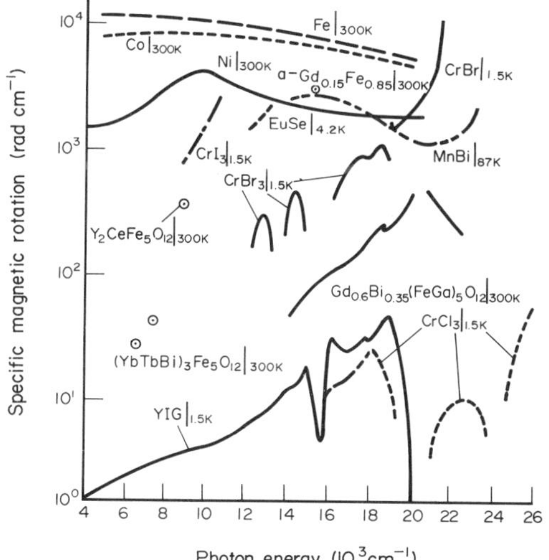

Figure 2
Summary of specific magnetic rotation as a function of photon energy for a variety of materials. Fe, Ni, Co data from Clemens and Jaumann 1963 *Z. Physik* 173: 135–48; $CrBr_3$, $CrCl_3$, CrI_3 data from Dillon et al. 1966 *J. Phys. Chem. Solids* 27: 1532–49; EuSe data from Suits et al. *J. Appl. Phys.* 37: 1391–7; MnBi data from Chen et al. 1962 *J. Appl. Phys.* 36: 1261–2; $Gd_{0.65}Bi_{0.35}(FeGa)_5O_{12}$ data from Hansen et al. 1981 *IEEE Trans. Magn.* 17: 3211; $Y_2CeFe_5O_{12}$ data from Gomi et al. 1990 *Jpn. J. Appl. Phys.* 29: L99–100; $(YbTbBi)_3Fe_5O_{12}$ data from Nakajima et al. 1988 *IEEE Trans. Magn.* 24: 2565

light (MOE). Light transmitted through magnetooptic materials shows both rotation and ellipticity, though the emphasis here is on the MOR.

In simple ferromagnetic crystals where there is only one magnetic lattice, the constant Φ is essentially proportional to $M(T)$, the spontaneous magnetization as it varies with temperature; that is, $\Phi = AM$. In the more complicated case of a material with multiple magnetic sublattices, such as a rare-earth–iron garnet or the rare-earth–transition metal amorphous materials used in magnetooptic recording, there are proportionality constants for each sublattice and the proportionality between Φ and the total magnetization is lost. Note particularly that in ferromagnets and ferrimagnets the MOR effects are principally associated with spontaneous magnetization rather than with the field.

Figure 2 shows how the absolute value of specific magnetic rotation of a broad range of materials varies with photon energy. Note that the discontinuities in the chromium trihalide data arise from changes in sign. Values for the ferromagnetic metals iron and cobalt reach up to over 10^4 rad cm^{-1} over a wide range. The intermetallic MnBi and rare-earth–transition metal amorphous metals such as a-$Gd_{0.15}Fe_{0.85}$ also show large values. However, because they are metals, they are exceedingly opaque and transmit measurable light only through very thin sections, perhaps 100 nm in thickness. It is their magnetooptic properties in reflection, that is, the magnetooptic Kerr effects, that are primarily of interest. Nonconducting magnetic compounds have sharper spectral structure and may be orders of magnitude less absorbing. Data are shown here for a few diverse classes of compounds: magnetic garnets, europium chalcogenides and chromium trihalides.

Some of these materials show magnetic order at low temperatures; for instance, EuSe becomes ferrimagnetic at 4.6 K and then ferromagnetic at 2.8 K, while EuO becomes ferromagnetic at 66.8 K. For applications in imaging the magnetic structure of superconductors, it is often necessary to use a paramagnetic material; for instance, EuS which becomes ferromagnetic at 16.3 K is mixed with 10–60 wt% EuF_2 to suppress its Curie temperature below 4.2 K (Huebener 1979; see also *Magnetic Properties of Superconductors*).

Magnetooptic effects are to be understood in terms of electric and magnetic dipole transitions within the energy level structure of the material. In the visible and near ir magnetic dipole effects arise from electron spin resonances in the microwave region. This magnetic contribution to overall rotation is featureless and, at photon energies above the near ir, much smaller than that due to electric dipole transitions. Because of the spontaneous magnetization of the Fe^{3+} ions in the iron sublattices the lowest lying members of the ground multiplet $^6S_{5/2}$ are disproportionately occupied. Transitions between these ground states and excited states split by spin–orbit interaction occur at different energies for positive and negative circularly polarized radiation and have different strengths because of the occupation differences. Both the splittings of the excited states and the occupation differences determine magnetooptic rotation. The general upward trend of rotation with photon energy in the ferrimagnetic iron garnets arises from transitions to excited states in the visible and near uv. Compositions that contain bismuth show large rotations because the large spin–orbit coupling of the heavy bismuth ion is mixed into the excited states and results in increased splittings. While the lowest-lying crystal field transitions of Fe^{3+} are nearly spin forbidden and have little absorption, higher energy transitions are allowed and produce strong absorption structure.

Superimposed on the magnetooptic spectroscopy of the dense Fe^{3+} sublattices just sketched is that of the trivalent rare-earth ions. The rare-earth multiplets are split by both crystal field and exchange field. The structures of both absorption and rotation associated with transitions within and between these rare-earth multiplets are different for each of the rare earths, and sharper than those of the iron sublattices.

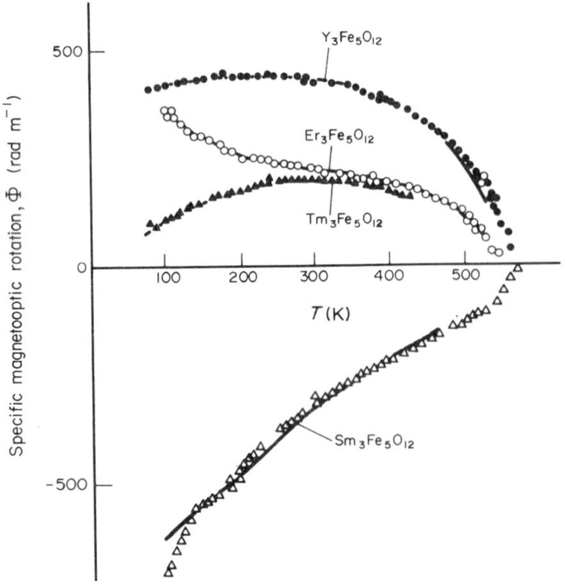

Figure 3
Variation of the spontaneous magnetooptic rotation with temperature for four magnetic garnets at $\lambda = 1152$ nm (data from Hansen and Krumme 1984)

Since the discovery and the subsequent availability of single crystals in 1956, the ferrimagnetic garnets have been the materials whose magnetooptic properties have been the most thoroughly studied, used and considered for transmission applications. In the absence of magnetization they would be cubic and typically have Néel temperatures well above room temperature. Techniques for growing high-quality single crystals have been continuously refined. Most importantly, they comprise a crystal chemical system with great substitutional flexibility so that extensive tailoring of physical properties is possible (see *Rare-Earth Intermetallics*).

Figure 3 shows the temperature variation of rotation at fixed photon energy for several ferrimagnetic garnets. For $Y_3Fe_5O_{12}$ there are two contributions to the rotation, that arising from the Fe^{3+} ions on octahedral sites and that from those on tetrahedral sites. Although there are three tetrahedral ions and two octahedral ions in each formula unit, the net octahedral contribution is larger. The temperature variations of the two sublattice magnetizations are the same. However, in the case of the rare-earth garnets, there is a third rare-earth sublattice on dodecahedral sites and its magnetization falls off much more quickly than that of the iron sublattices. For this case

$$\Phi = (A^e + A^m)M_a + (D^e + D^m)M_d + (C^e + C^m)M_c$$

where M_a, M_d, and M_c are the magnetizations of octahedral, tetrahedral and dodecahedral sites; A^e, A^m and so on are the corresponding coefficients for electric and magnetic contributions. The steeper slopes on the $\Phi(T)$ curves at lower temperatures reflect the contributions of the rare-earth sublattices.

Yttrium–iron garnet $Y_3Fe_5O_{12}$ (YIG) is the starting point for substitutions among the ferrimagnetic garnets. In single crystals of YIG, Φ varies from perhaps 3 rad cm^{-1} in the near ir to 30–40 rad cm^{-1} in the visible. However, the absorption coefficient α (as in the Beer–Lambert law, $I = I_0 \exp(-\alpha t)$) also increases from essentially zero for wavelengths longer than 1.2 μm to exceedingly large values in the green. Since rotation is only of interest insofar as detectable light can be transmitted, an important magnetooptic figure of merit is the rotation–absorption ratio (i.e., Φ/α). This dimensionless quantity is given in mrad dB^{-1} and it too has a spectral variation. For a truly transparent material it would be infinite. For iron, which has a large specific rotation but is nearly opaque, it is about 2 mrad dB^{-1} over a wide spectral range. For YIG at room temperature it falls from values above 2×10^4 mrad dB^{-1} in the ir to about 20 mrad dB^{-1} in the visible red and about 3 mrad dB^{-1} in the green. In thin sections of YIG and related compounds (0.015 cm and thinner) it is readily possible to use the MOR to see the distribution of magnetization (i.e., domain structure) using visible light. The high values of the figure of merit in the near ir make possible practical magnetooptic transmission-line devices, such as isolators and circulators, in the important wavelength range 1.1–1.5 μm.

The value of Φ at a specific wavelength varies widely for compositions within the garnet system. A few examples are given in Table 1. Note particularly that specific rotations much larger than those of YIG were obtained with substitution of the rare-earth ion Pr^{3+} and with the nonmagnetic ion Bi^{3+}.

Since the contribution of the rare-earth ions to MOR depends on details of their individual energy level schemes, there is no systematic variation across the rare-earth series. Thus, the large specific rotations

Table 1
Magnetooptic rotation Φ of various magnetic garnets at 1.06 μm and room temperature

Material	Φ (rad cm^{-1})
$Y_3Fe_5O_{12}$	+ 4.9[a]
$Gd_3Fe_5O_{12}$	+ 1.1[a]
$GdPr_2Fe_5O_{12}$	− 19.6[a]
$Y_2BiFe_5O_{12}$	− 43.6[b]
$Y_2CeFe_5O_{12}$	− 38.4[c]
$Pr_3Fe_5O_{12}$	− 55.8[d]
$Nd_3Fe_5O_{12}$	− 24.2[d]

a Wemple et al. 1973 *Appl. Phys. Lett.* 22: 331–3 b Takeuchi 1974 *Jpn. J. Appl. Phys.* 13: 2059–60 c Gomi et al. 1990 *Jpn. J. Appl. Phys.* 29: L99–100 d Dillon et al. 1987 *J. Mag. Soc. Jpn.* 11: S241–4.

of praseodymium- and neodymium-containing iron garnets must be regarded as empirical. The transitions responsible for the enhancement associated with cerium can be located accurately, but their assignments have not been discussed. On the other hand, the large contribution of Bi^{3+} is understood in terms of an admixture of its large spin–orbit splitting into an excited state. Large magnetooptic rotations correspond to large spin–orbit splittings of the excited state.

Ferromagnetic and ferrimagnetic compounds are quite rare and it is difficult to guess *a priori* which compounds might exhibit attractive magnetooptic properties, that is, high magnetooptic figure of merit, magnetic ordering at room temperature and high symmetry. Day (1979) has suggested that it is best to look for compounds in which the metal ion is in a lower oxidation state and in which the ligands have a high electronegativity. Compounds with spin-forbidden transitions in the visible are especially likely candidates. To date, the application of these criteria has led only to compounds with Curie temperatures below 77 K.

2. Other Magnetooptic Effects

2.1 Magnetic Linear Birefringence

If, instead of the orientation in Fig. 1, $\boldsymbol{M}$ is perpendicular to $\boldsymbol{l}$, there is an optical birefringence associated with the magnetization. This magnetic linear birefringence (MLB) is expressed by the difference between refractive indices for light polarized parallel and perpendicular to $\boldsymbol{M}$; that is, $\Delta n = n_{\parallel} - n_{\perp}$. Note that Δn is dimensionless. The effect is even in $\boldsymbol{M}$ and thus does not change sign when $\boldsymbol{M}$ is reversed. This MLB varies considerably with composition and temperature and is sometimes large enough that it must be considered in applications. Furthermore, the MLB has been found not to vary with wavelength over the range 1.4–2.1 μm. Figure 4 displays MLB values measured in this wavelength interval with $\boldsymbol{M}$ along [100] at 300 K for a number of rare-earth–iron garnets. In the figure they are plotted against the magnetostriction constant λ_{100} and it will be seen that there appears to be a correlation except for the case of $Sm_3Fe_5O_{12}$. This correlation is seen for 77 K and 300 K data with fields along both [111] and [100], except for the terbium and samarium garnets which seem to be highly anisotropic.

2.2 Magnetooptic Kerr Effects

Light reflected from a magnetic material also interacts with the magnetization. In the simplest arrangement, linearly polarized light at normal incidence is reflected from a surface magnetized normally. The reflected light is elliptically polarized with its major axis slightly rotated. This rotation and ellipticity are called polar Kerr effects, and represent but one of three traditional

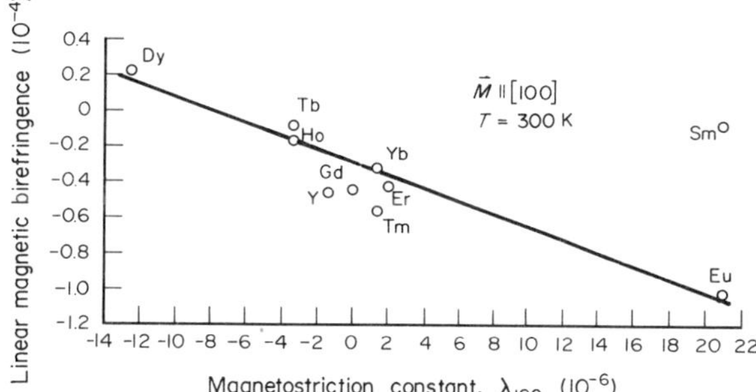

Figure 4
Magnetic linear birefringence with $\boldsymbol{M}$ along [100] at 300 K plotted against the corresponding magnetostriction for various rare-earth–iron garnets (data from Dillon et al. 1970 *J. Appl. Phys.* 41: 4613–19)

geometries. In Table 2 reported values of the polar Kerr rotation are shown for a number of materials. Again, it is necessary to be cautious about comparing the signs of rotation given in different papers. At room temperature, a rotation of 15 mrad is exceedingly large but at cryogenic temperatures values near 175 mrad have been observed.

2.3 Faraday Rotation in Materials Containing Paramagnetic and Diamagnetic Ions

Transparent paramagnetic or diamagnetic materials in parallel and perpendicular magnetic fields may also

Table 2
Magnetooptic materials with high polar Kerr rotation

Material	T (K)	λ (nm)	φ_k (mrad)	Remarks
Fe[a]	300	633	− 7.2	
Co[a]	300	633	− 6.1	
Ni[a]	300	633	− 2.3	
MnBi[b]	300	633	− 9.8	low temperature phase
PtMnSb[b]	300	633	16	strong λ dependence
$USb_{0.8}Te_{0.2}$[c]	10	1.5×10^3	157	strong λ dependence, polished (111) face, $H = 4$ T
Amorphous $U_{0.51}Sb_{0.49}$[d]	10	633	59	mild λ dependence $H = 4$ T
Amorphous $Gd_{0.17}Tb_{0.10}Fe_{0.73}$[e]	300	1.45×10^3	8.9	

a Buschow et al. 1983 *J. Magn. Magn. Mat.* 38: 1 b van Engen et al. 1983 *Appl. Phys. Lett.* 42: 202 c Schoenes and Reim 1986 *J. Magn. Magn. Mat.* 1371: 54–7 d McElfresh et al. 1990 *Appl. Phys. Lett.* 57: 730 e Hansen et al. 1987 *J. Magn. Soc. Jpn.* 11: S257

Table 3
Magnetooptic materials with high Faraday rotation in the visible

Material	Verdet constant ($rad\,m^{-1}T^{-1}$) At 405 nm	At 600 nm
Crystal		
$LiTbF_4$[a] (295 K)	-3.58×10^2	-1.40×10^2
$Tb_3Al_5O_{12}$[b] (295 K)	-6.57×10^2	-2.1×10^2
$Tb_3Al_5O_{12}$ (77 K)	-19×10^2	-6.7×10^2
$Cd_{0.55}Mn_{0.45}Te$[c] (295 K)		$+38\times10^2$
Glasses at 295 K		
$(Tb_2O_3)0.85(La_2)4.73(B_2O_3)$[d]	-3.5×10^2	-0.5×10^2
$(EuO)0.44(Al_2O_3)1.5(B_2O_3)$[e]		-0.44×10^2
Corning code 8363[f](PbO glass)		$+0.29\times10^2$
Hoya Fr-5		
Terbium[g](borosilicate glass)	-2.2×10^2	-0.79×10^2

a Weber et al. 1978 *J. Appl. Phys.* 49: 3464–9 b Rubinstein et al. 1964 *J. Appl. Phys.* 35: 3069–70 c Turner et al. 1983 *Appl. Optics* 22: 3152–54 d Rubinstein et al. 1964 *J. Appl. Phys.* 35: 2338–40. e Shafer and Suits 1964 *J. Am. Ceram. Soc.* 49: 261–4 f Borelli 1964 *J. Chem. Phys.* 41: 3289–93 g Hoya Faraday rotator glasses, Hoya Optics, Fremont, CA

show magnetooptic rotations, ellipticities and linear birefringences. For ordinary laboratory fields at room temperature these are much smaller than the corresponding effects associated with spontaneous magnetizations. The field-dependent rotation is termed the Faraday rotation and starts out linearly with field B; that is, $\Phi_F = VBt$, where V is the Verdet constant, which varies with photon energy, usually increasing steeply as $h\nu$ increases toward the energy of the operative transition (see *Magnetooptics*). The appropriate SI unit is $rad\,m^{-1}T^{-1}$ although the Verdet constant is still frequently expressed in $min\,Oe^{-1}\,cm^{-1}$ where $1\,min\,Oe^{-1}cm^{-1} \equiv 290.89\,rad\,m^{-1}T^{-1}$. Faraday rotation is used in optical isolators, polarization modulators and other devices. Table 3 gives illustrative values of V for several compositions that appear promising for some of these applications.

3. Applications

3.1 Observation of Magnetic Domains

The magnetooptic effects make it possible to see magnetic domains in a polarizing microscope (and also the magnetic structure of superconductors (Huebener 1979)) (see *Magnetic Properties of Superconductors*; *Magnetization Imaging*). The contrast arises from the differing magnetooptic rotations of domains or flux structures whose magnetization vectors project differently on the line of sight. Furthermore, the MLB often enables the differentiation of magnetizations that lie in different directions in the plane. Thus, researchers have been able to see these domains for various geometries and combinations of magnetic properties as well as their response to changing field, temperature and stress. One outstanding example is the development of sophisticated magnetic bubble technology which depended crucially on the magnetooptic observation of domain behavior (see *Bubble Domain Materials*).

In a similar way, the magnetic Kerr effects make it possible to see many details of domains on the surface of opaque materials such as metals and intermetallic compounds.

3.2 Magnetooptic Memories

Computer memories based on magnetooptic effects are now coming into wide use. A description of this recording technology and the necessary material characteristics is given in the article *Magnetooptic Storage Media*. It is merely noted here that the media currently used are amorphous rare-earth–transition metal films such as the amorphous $Gd_{0.17}Tb_{0.10}Fe_{0.73}$ in Table 2 with a rotation of only 9 mrad. A large and vigorous industry is being based on a Kerr rotation of only a half degree!

3.3 Isolators

The antireciprocity of MOR makes possible optical isolators and a variety of related devices (see *Magnetooptics*). Isolators appear to be necessary in the design of high-bit-rate fiber-optic transmission systems where they are used to prevent instabilities in laser diodes caused by unwanted reflections. Single-crystal magnetic garnets, such as YIG or compositions containing bismuth, are the magnetooptic materials used in current ir designs. Materials of Table 3 are used in the visible.

3.4 Faraday Effect Sensors

There is increasing interest in sensor operation at high voltages or in the presence of substantial electromagnetic interference (Day and Rose 1988). Recent dramatic increases in sensitivity in the dilute magnetic semiconductor materials such as CdMnTe are extending the scope for applications (Butler 1987).

3.5 Experimental Uses

MOR has been used in studies of magnetic phase diagrams, magnetic phase transitions and critical phenomena in magnetic materials. This depends on the close proportionality between rotation and magnetization. The usual magnetic properties of a material, such as magnetization, anisotropy and magnetic resonance behavior, only involve the ground state. Magnetooptic properties, specifically the spectral variation of rotation and ellipticity on transmission and reflection, also depend on energy levels well above the ground state. Thus, they provide powerful tools and tests for exploring the energy level structure of magnetic materials.

See also: Magnetic Materials: An Overview; Magnetooptic Storage Media; Magnetooptics; Rare-Earth Intermetallics

Bibliography

Butler M A 1987 Wavelength and temperature dependence of the Faraday effect in $Cd_{1-x}Mn_xTe$. *Solid State Commun.* 62: 45–7

Day G W, Rose A H 1988 Faraday effect sensors: the state of the art. *SPIE Fiber Optic Laser Sensors* 985: 138–50

Day P 1979 New transparent ferromagnets. *Acc. Chem. Res.* 12: 236–43

Dillon J F Jr 1978 Magnetooptical properties of magnetic garnets. In: Paoletti A (ed.) 1978 *Physics of Magnetic Garnets*. North-Holland, Amsterdam, pp. 379–416

Freiser M J 1968 A survey of magnetooptic effects. *IEEE Trans. Magn.* 4: 152–61

Hansen P, Clausen C, Much G, Rosenkranz M, Witter K 1989 Magnetic and magnetooptical properties of rare-earth transition-metal alloys containing Gd, Tb, Fe, Co. *J. Appl. Phys.* 66: 756–67

Hansen P, Krumme J -P 1984 Magnetic and magnetooptical properties of garnet films. *Thin Solid Films* 114: 69–107

Huebener R P 1979 *Magnetic Flux Structures in Superconductors*. Springer, Berlin

Robinson C J, Suzuki T, Falco C M (eds.) 1989 *Materials for Magneto-Optic Data Storage*. Materials Research Society, New York

Suits J C 1972a Faraday and Kerr effects in magnetic compounds. *IEEE Trans. Magn.* 8: 95–105

Suits J C 1972b Faraday effects. *American Institute of Physics Handbook*, 3rd edn. McGraw–Hill, New York, Sect. 5, pp. 219–21

Wettling W 1976 Magneto-optics of ferrites. *J. Magn. & Magn. Mater.* 3: 147–60

J. F. Dillon Jr
[AT&T Bell Laboratories, Murray Hill, New Jersey, USA]

Magnetooptic Storage Media

Magnetooptic recording is a new information storage technology for computers and other electronic apparatus which make use of digital information. Like other forms of optical recording technology, it uses a finely focused laser beam to write and read information on a disk, much like the familiar compact disk which is used for digital audio. However, unlike the compact disk which has all the information imprinted on it at the factory, a magnetooptic disk may be repeatedly written, read, erased and rewritten by the user.

Magnetooptic recording media may be removed from the disk drive used to record and read them. In this sense they are like floppy disks which are also used to store information for computers. On the other hand, magnetooptic storage media provide a much higher density of information storage than floppy disks. About 650 Mbytes or about 325 000 typewritten pages can be stored on a single 5.25 in. (130 mm) diameter magnetooptic disk and even higher densities are planned. Thus, in terms of storage density, magnetooptic media are similar to rigid disks. This new storage technology is thus expected to assume a portion of the present US$5 × 10^{10} worldwide data storage market and compete with floppy and rigid magnetic disk and tape drives.

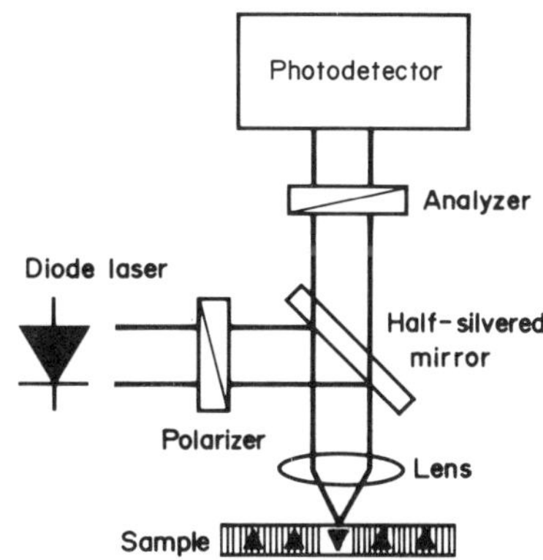

Figure 1
Apparatus for writing and reading magnetooptic storage media

Magnetooptic storage media consist of a thin film of magnetic material deposited on a transparent substrate—usually glass or plastic. The thin film is made in such a way that it is energetically most favorable for the magnetization to lie perpendicular to the plane.

Information is written onto magnetooptic storage media using a submicrometer focused beam of light from a laser and a magnetic field applied perpendicular to the film plane, as shown in Fig. 1. Heat generated by absorption of the light causes the temperature in the submicrometer region to increase. This rise in temperature in turn causes the coercive force, or magnetic field required to reverse the magnetization, to decrease. When the coercive force falls below the magnitude of the applied magnetic field, the magnetization will align with the magnetic field. Thus, by controlling the direction of the magnetic field and the pulses from the laser, the magnetization in the magnetooptic storage medium may be controllably written into up and down magnetized regions, referred to as magnetic domains. This pattern of up and down magnetization can be used to represent digital information.

To read information from magnetooptic recording media the laser beam intensity is reduced so that it will not alter the state of magnetization in the media and the Kerr magnetooptic effect is used (see *Magneto-*

optic Materials; *Magnetooptics*). This effect causes the plane of polarization of light reflected from a magnetic material to be rotated on reflection. The direction of the rotation (clockwise or counterclockwise) is dependent on the direction of magnetization (up or down) in the material. By using a polarizer in the incident beam of light from the laser and an analyzer (a second polarizer) in the reflected beam, as in Fig. 1, the difference in polarization of light reflected from upward pointing magnetization and downward pointing magnetization is converted to a change in reflected light intensity which can be sensed with a photodetector.

1. Amorphous Rare-Earth–Transition Metal Thin Films

The magnetooptic storage media most commonly made consist of amorphous alloys of rare-earth elements (typically gadolinium and terbium) and transition metal elements (typically iron and cobalt). Most commonly they are deposited onto substrates by some form of sputtering process.

A sputtering system consists of a vacuum chamber into which a target and the substrates to be coated are placed. The target consists of the magnetic material to be deposited onto the substrates and is biased with a strong negative voltage (typically about 1000 V) relative to the ground potential of the vacuum chamber itself. When an inert gas, such as argon, at low pressure is leaked into the vacuum chamber, the large voltage potential in the system causes electrons to be stripped from the argon atoms and a plasma is formed. The positive argon ions are then accelerated towards the negatively biased target and strike it with high energy. The transfer of momentum from the argon ions then causes atoms or clusters of atoms of the target material to be sputtered away from the target. These sputtered atoms diffuse away from the target and, if the substrate is in close proximity to the target, it is coated with a thin film of the target material.

Rare-earth–transition metal thin films deposited by such sputtering techniques are typically amorphous; that is, they have no long-range crystalline order. This fact has proven important to their performance as low-noise recording media. Other polycrystalline magnetooptic materials have larger Kerr magnetooptic rotations but, in spite of the higher signal, have a poorer signal-to-noise ratio. Noise in a large-grain polycrystalline material is much greater than in an amorphous material because the domain boundaries tend to follow the crystallite boundaries and because each crystallite produces a slightly different magnetooptic effect. In an amorphous material the nonuniformities are on an atomic scale and therefore noise from them is averaged by domains on a micrometer scale.

In spite of their amorphous structure, the rare-earth–transition metal thin films exhibit anisotropic properties such that the lowest energy orientation for the magnetization is perpendicular to the film plane. The origins of this perpendicular anisotropy are the subject of debate and investigation. Short-range pair ordering, single-ion anisotropy produced by non-s-state rare-earth ions, stress-induced anisotropies produced during the growth process and bond orientational anisotropy have all been suggested as playing roles (see *Anisotropy in Magnetic Materials*).

Perpendicular anisotropy is necessary for a good magnetooptic storage medium. To obtain the largest Kerr magnetooptic effect and therefore the largest signal during readback, the magnetization should be perpendicular to the film plane. In addition, to have stable submicrometer domains, strong perpendicular anisotropy is necessary.

The rare-earth and transition metal elements most frequently used in magnetooptic storage media exhibit negative exchange coupling so that the magnetic moments of the rare-earth atoms are opposite to those of the transition metal atoms, thus forming a ferrimagnet. Consequently, the net magnetization of the alloy is the difference between the magnetization produced by the rare-earth elements and the magnetization produced by the transition metal elements, as illustrated in Fig. 2. Since the rare earths have a larger magnetic moment than the transition metals at low temperatures but the transition metals have the larger moment at high temperatures, it is possible to adjust the composition of the magnetooptic storage media so that at low temperatures the net magnetization is parallel to the rare-earth magnetization, while at high temperatures it is parallel to the transition metal magnetization. Thus, at some intermediate temperature, the rare-earth and transition metal magnetizations are equal and opposite, providing zero net magnetization. This temperature is referred to as the compensation temperature.

As the compensation temperature is approached, the coercive force of the rare-earth–transition metal alloys tends toward infinity, as shown in Fig. 2. This is because the energy per unit volume of sample coupled to the net magnetization $\boldsymbol{M}$ by an externally applied magnetic field $\boldsymbol{H}$ is $\mu_0.\boldsymbol{M}.\boldsymbol{H}$. As $\boldsymbol{M}$ tends towards zero, $\boldsymbol{H}$ must increase towards infinity to couple in the energy required to reverse the magnetization. On the other hand, as the temperature of these materials is raised far above the compensation temperature, thermal energy causes the magnetization to be more easily changed in direction. Consequently, at high temperatures, the coercive force decreases and approaches zero at the Curie temperature of the material where the thermal energy overcomes the exchange coupling causing the magnetic moments to align. By locating the compensation temperature near room temperature and the Curie temperature near the temperature to which the film is heated by the laser beam used

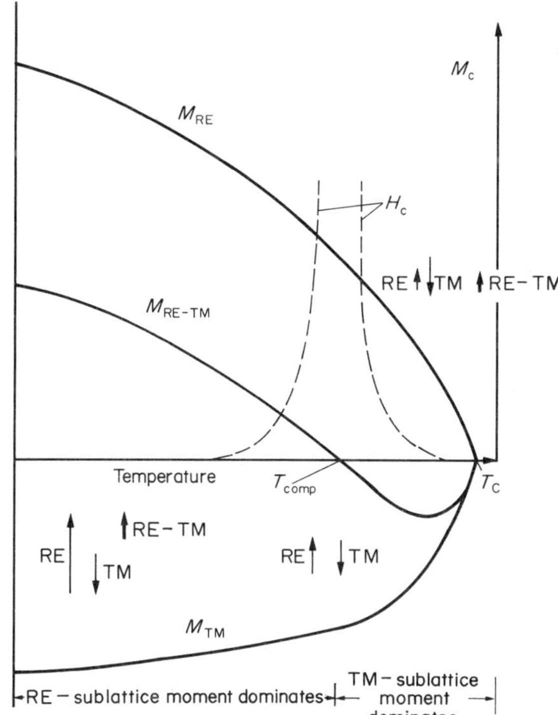

Figure 2
The net magnetization, coercive force and the magnetization contributions of rare-earth (RE) and transition metal (TM) constituents of a rare-earth–transition metal amorphous alloy as a function of temperature

for writing (150–300 °C), a suitably temperature-dependent coercivity can be achieved so that writing occurs with moderate laser powers (10 mW).

The Kerr magnetooptic effect in rare-earth–transition metal alloys is mainly determined by the transition metal components, at least for light in the infrared and visible wavelength spectrum. This is because the Fermi level of the 4*f* electrons responsible for the magnetism in the rare earths lies too deep for them to be excited by photons of light with infrared or visible wavelengths, whereas the Fermi level of the 3*d* electrons responsible for the magnetism in the transition metals is much shallower and easily reached by photons with such wavelengths.

The fact that the magnetooptic effect is due primarily to the transition metal magnetization and not the net magnetization of the alloy is the reason that data may be read out of magnetooptic media at or near the compensation temperature. Since the transition metal magnetization varies smoothly across the compensation temperature and approaches zero only at the Curie temperature, the signal produced by the Kerr magnetooptic effect does so also.

Table 1
ANSI specifications for digital information interchange on 130 mm rewritable optical disk cartridge

Carrier-to-noise ratio, CNR (dB)	Storage temperature T_{min} (°C)	Storage temperature T_{max} (°C)	Applied field, H_{max} ($A\,m^{-1}$)	Reflectance R_{min}	Reflectance R_{max}	Figure of merit F_{min}	Figure of merit F_{max}
>45	−20	+55	48000	0.1	0.34	0.0017	0.0052

Rare-earth–transition metal alloys are subject to corrosion and must be protected from the environment. This is accomplished in magnetooptic media by using a thin dielectric which is impervious to water vapor and oxygen, such as Si_3N_4 or AlN, between the substrate and magnetic film. To protect the top surface of the film, a similar dielectric and a reflective metal overcoat, such as aluminum, are frequently used. Extensive testing has shown that the rare-earth–transition metal media manufactured in this way have lifetimes in excess of 10 years.

Modern magnetooptic disks use rare-earth–transition metal thin films for the storage medium. Manufactured materials include GdTbFe and TbFeCo with the total rare-earth content near 20%. The standards agreed on by manufacturers of magnetooptic drives and media specify a number of parameters for the media. The American National Standards Institute (ANSI) standards for magnetooptic disk media specify, among other things, the minimum carrier-to-noise ratio (CNR), the maximum ambient field which the disk should be exposed to during storage (H_{max}), the minimum and maximum temperatures during storage (T_{min} and T_{max}), the minimum and maximum reflectances (R_{min} and R_{max}) and the minimum and maximum magnetooptic figures of merit (F_{min} and F_{max}). The ANSI standards for these parameters are given in Table 1. Indirectly, these standards place minimum requirements on the media. The maximum ambient field to which the disk should be exposed $H_{max} = 48\,000\ A\,m^{-1}$ may be interpreted as the minimum coercive force H_c for the media over the temperature range at which it can be stored (−20 °C to 55 °C). The magnetooptic figure of merit is calculated as

$$F = R \sin \varphi_K \cos 2\varepsilon$$

where φ_K is the Kerr rotation and ε is the ellipticity of the reflected beam. Note that this figure of merit relates to the signal level, rather than the shot-noise-limited signal-to-noise ratio, and is appropriate as a standard in drives having a noise floor higher than the shot noise limit. If the ellipticity is small, then, utilizing the maximum and minimum values of R, it is possible to calculate the minimum and maximum values of φ_K as 0.3° and 3°, respectively.

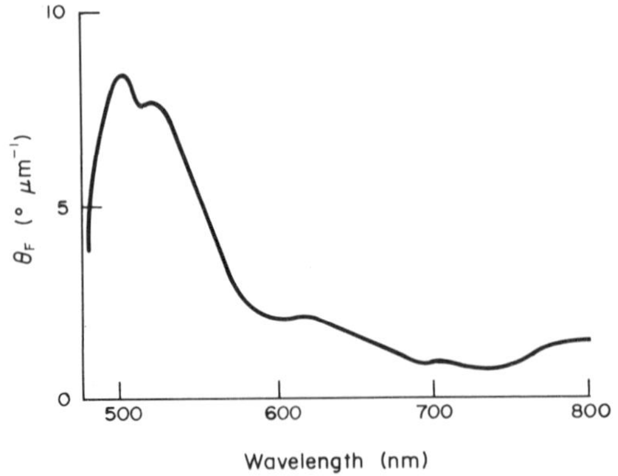

Figure 3
Typical Faraday rotation spectrum of bismuth, gallium: DyIG film on glass substrate (after Gomi and Abe 1989)

An alternative figure of merit $FOM_{S/N} = R\varphi_K^2$ is useful for characterizing magnetooptic materials rather than the drive and is the technically correct expression for the signal-to-noise ratio.

For typical bare rare-earth–transition metal media, $\varphi_K \simeq 0.3°$ and $R \simeq 0.6$. Using dielectric layers to increase φ_K and decrease R, typical manufactured media might have $\varphi_K = 1$–$2°$ and $R = 0.2$.

2. *Magnetic Oxides*

Although rare-earth–transition metal magnetooptic media are in production and available for purchase, other materials are under investigation for use in future media. One class of materials which has received considerable interest as potential magnetooptic media is magnetic oxides. Magnetic oxides such as garnets have relatively large magnetooptic effects, particularly in the short-wavelength spectrum as shown in Fig. 3. Furthermore, being oxides, they are naturally resistant to corrosion. The large magnetooptic effect in the short-wavelength spectrum is advantageous for future higher-density magnetooptic recording applications. Since the dimensions of a focused laser beam scale with the wavelength, it is expected that future magnetooptic recording systems will utilize shorter wavelengths than currently used.

Garnets have been the most extensively investigated magnetic oxide for magnetooptic recording media (see *Bubble Domain Materials*). Garnets, like the rare-earth–transition metal materials, are ferrimagnetic and can be made to exhibit a compensation temperature near room temperature so that a high room-temperature coercivity is achieved as indicated in Fig. 4. They also have Curie temperatures which can be adjusted in the 150–260 °C range as indicated by where the Faraday rotation goes to zero in Fig. 5. Thus, domains can be written in them with practical laser powers. Bismuth-doped garnets such as $Bi_2Dy_1Fe_{3.8}Al_{1.2}O_{12}$ produce large magnetooptic effects and single-crystal bismuth-doped garnets epitaxially grown on single-crystal gadolinium gallium garnet substrates have been shown to produce excellent carrier-to-noise ratios, comparable with the best obtained with rare-earth–transition metal films (Shono et al. 1989). However, large carrier-to-noise ratios have not been demonstrated in films deposited on glass or other inexpensive substrates.

The reason for the poor carrier-to-noise ratio in the garnet media deposited onto glass substrates is that they are polycrystalline with crystallite sizes near 100 nm. The randomly oriented crystallites give rise to noise, both because each crystallite produces a slightly different magnetooptic effect and because the domain boundaries used to define the recorded bits tend to follow the irregular grain boundaries. Current research on these materials is focused on reducing this grain noise.

In addition to the grain noise problem, garnets have an additional drawback. When they are deposited by sputtering or pyrolysis, they are in a nonmagnetic amorphous state and generally require high-temperature processing to be converted to the magnetic crystalline state. The temperatures involved preclude the use of plastic substrates, although glass remains a possibility.

Another magnetic oxide which has received attention as a potential magnetooptic recording medium is hexaferrite. Barium hexaferrite (see *Ferrites, Hard*) has been the most extensively investigated, because it offers strong perpendicular anisotropy. Doping with cobalt and titanium to form compounds such as $BaFe_{12-2x}Co_xTi_xO_{19}$ has been shown to increase the

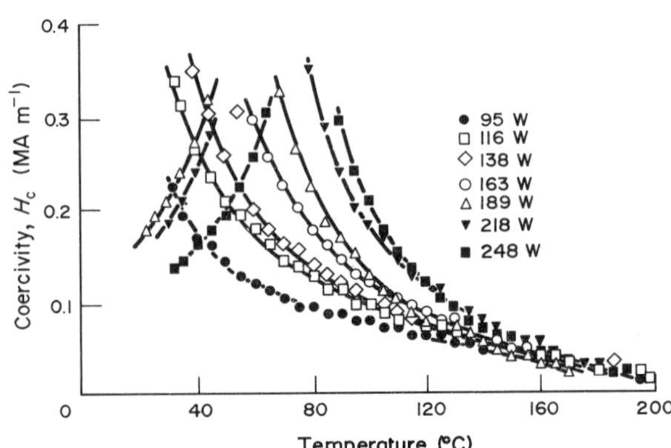

Figure 4
Temperature dependence of the coercivity for GdBiFeAl garnet films deposited at radio-frequency powers from 95 W to 248 W at an argon bleeding pressure of 10 mTorr (after Eppler and Kryder 1989)

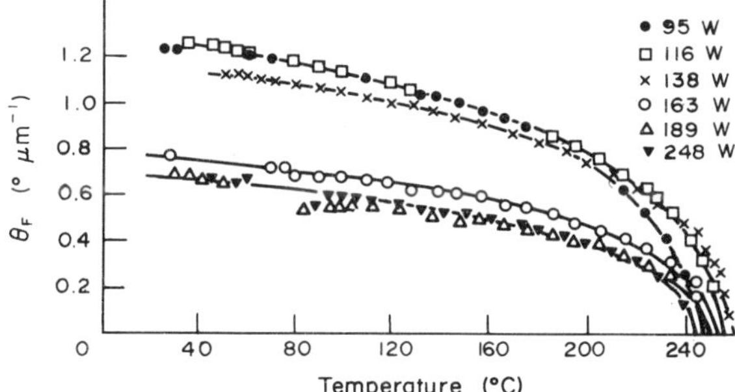

Figure 5
Temperature dependence of the Faraday rotation for GdBiFeAl garnet films deposited at radio-frequency powers between 95 W and 248 W and an argon bleeding pressure of 10 mTorr (after Eppler and Kryder 1989)

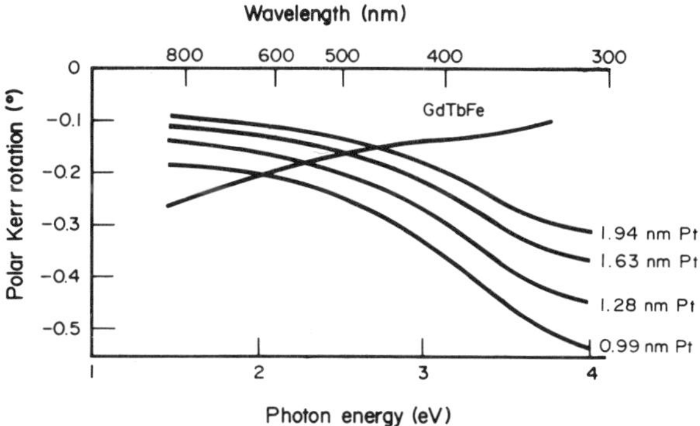

Figure 6
Kerr rotation spectra for several cobalt–platinum layered structures made of approximately 0.36 nm-thick cobalt layers laminated with various thicknesses of platinum and for GdTbFe (after Zeper et al. 1989)

Kerr rotation (Nakamura et al. 1987). Like the rare-earth–transition metal alloys and the garnets, barium hexaferrite is ferrimagnetic. Barium hexaferrite, like the garnets, is polycrystalline when deposited on high-temperature glass or quartz substrates. Consequently, it suffers from the same grain noise problem as the garnets. However, it has a smaller magnetooptic effect and requires even higher temperature processing than the garnets. Hence, garnets appear more attractive for future magnetooptic media applications than hexaferrites.

3. *Cobalt–Platinum and Cobalt–Palladium Multilayers*

A relatively new class of materials attracting increasing attention for future magnetooptic media applications comprises cobalt–platinum and cobalt–palladium multilayers (see *Thin Films and Multilayers, Magnetic*). These materials, like the garnets, exhibit increasing magnetooptic effects at shorter wavelengths of light, as shown in Fig. 6, and therefore are of interest for future higher density recording systems (Zeper et al. 1989). Although not an oxide, they are substantially more corrosion resistant than rare-earth–transition metal films.

Unlike the rare-earth–transition metal and garnet materials, which are ferrimagnetic, the cobalt–platinum and cobalt–palladium multilayers are ferromagnetic. In spite of this, they exhibit moderately high coercivity at room temperature and Curie temperatures ranging from about 250°C to 400°C (see, for example, in Fig. 7).

The perpendicular anisotropy in these multilayer materials is attributed to interfacial cobalt atoms which are more easily magnetized perpendicular to the plane and possibly also to stresses between layers. Since the anisotropy arises from interfacial effects between the cobalt and platinum layers, the cobalt layers must be extremely thin. A typical multilayer consists of 0.3 nm-thick layers of cobalt interlaced with 1 nm-thick layers of platinum or palladium. The number of layer pairs deposited to make a magnetooptic recording medium is in the range 10–25 (Zeper et al. 1989).

In spite of the multilayer structure of these media, they are relatively easily fabricated. The multilayers with the highest anisotropy and coercivity have been made using vapor deposition techniques, but ultrahigh vacuum is not a requirement. Sputtering has been used and may be favored for manufacturing because it offers lower deposition rates and therefore better control over the layer thicknesses, but thus far

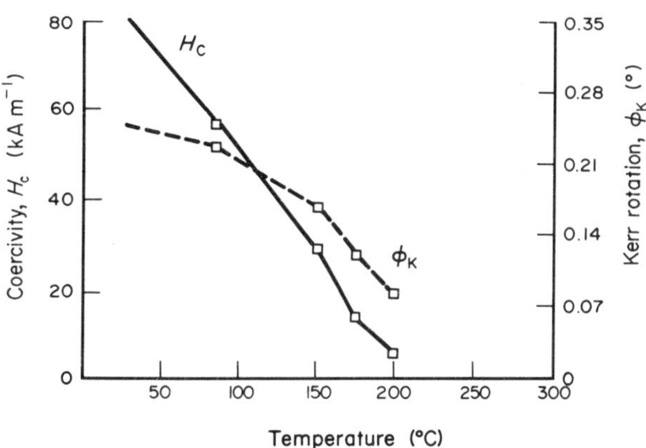

Figure 7
Coercivity and magnetooptic Kerr rotation as functions of temperature for cobalt–platinum multilayers

it has not produced anisotropy as strong or coercivity as high as vapor deposition.

It is believed that the reason for the reduced anisotropy in sputtered multilayers is that the interface between the cobalt and the other materials is not well defined. In sputtering, high-energy neutral atoms of the sputtering gas recoil from the target and bombard the substrate causing intermixing of the deposited atoms at the interface. By using high sputtering pressures and sputtering gases with lower recoil energies than argon, progress is being made on creating media with high anisotropy by sputtering (Carcia and Zeper 1990, Highmore et al. 1991).

The multilayers offering the highest Kerr magnetooptic effects are cobalt–platinum. The best carrier-to-noise ratios reported for cobalt–platinum multilayer media are over 60 dB and are comparable to the very best rare-earth–transition metal media (Lin 1990). The excellent carrier-to-noise ratio, combined with the enhanced intrinsic corrosion resistance and improved performance of these materials at shorter wavelengths, makes them very promising candidates for future magnetooptic recording media.

See also: Bubble Domain Materials; Magnetooptic Materials; Magnetooptics

Bibliography

Carcia P F, Zeper W B 1990 Sputtered Pt/Co multilayers for magneto-optic recording. *IEEE Trans. Magn.* 26: 1703–5

Eppler W, Kryder M H 1989 The effects of the sputtering conditions on bismuth doped gadolinium iron garnet films. *IEEE Trans. Magn.* 25: 3743–5

Gomi M, Abe M 1989 Sputtered iron garnet films for magneto-optical storage. *Mater. Res. Soc. Symp. Proc.* 150: 121–30

Highmore R J, Shih W C, Somekh R E, Evetts J E 1991 Sputter deposition of cobalt–palladium multilayers. *J. Vac. Sci. Technol. A* 9(4): 2123–7

Kryder M H 1985 Magneto-optic recording technology. *J. Appl. Phys.* 57: 3913–18

Kryder M H 1989 The design and fabrication of magneto-optic recording materials. *Mater. Res. Soc. Symp. Proc.* 150: 3–14

Lin C J 1990 Magneto-optical recording on evaporated Co/Pt multilayer films. *IEEE Trans. Magn.* 26: 1700–2

Nakamura H, Ohmi F, Kaneko Y, Sowada Y, Watoda A, Machida H 1987 Cobalt–titanium substituted barium ferrite films for magneto-optical memory. *J. Appl. Phys.* 61: 3346–8

Shono K, Kuroda S, Kano H, Koshino N, Ogano S 1989 Magneto-optical recording of sputtered garnet films crystallized during deposition. *Mater. Res. Soc. Symp. Proc.* 150: 131–5

Zeper W B, Greidanus F J A M, Carcia P F 1989 Evaporated Co/Pt layered structures for magneto-optical recording. *IEEE Trans. Magn.* 25: 3764–6

M. H. Kryder
[Carnegie Mellon University, Pittsburgh, Pennsylvania, USA]

Magnetooptics

When plane-polarized light travels through a medium in the direction parallel to an applied magnetic field, the plane of polarization of the light is observed to rotate. This phenomenon was discovered by Michael Faraday in 1845 and is now known as the Faraday effect. It is a universal effect occurring for all materials, anisotropic as well as isotropic, unlike optical activity. Magnetooptical effects are related to dipole transitions within the energy level structure of the material. The variation of optical rotation with wavelength is characteristic of a material and can diverge at wavelengths close to strongly absorbing transitions. When the Faraday rotation is not observed for a material by an experimenter, it usually means that, besides being weaker than the sensitivity of the measurement, the rotation has been masked by the material's birefringence, either natural or strain induced. For most materials the magnetic-induced circular index difference $(n_{-} - n_{+})$ is about 10^{-5} which is nearly three orders of magnitude below the birefringence usually associated with anisotropic crystals. Therefore, most of the Faraday measurements have been made on optically isotropic materials: gases, liquids, glasses and cubic crystals. In fact, accurate measurements of the Faraday rotation in cubic crystals have been hindered by the effects of strain birefringence.

1. Verdet Constant

The magnetic rotation Φ of the plane of polarization is proportional to the path length l of light through the material and the component of the magnetic induction B along the propagation direction:

$$\Phi = V\int \boldsymbol{B}.d\boldsymbol{l}$$

where $\boldsymbol{dl}$ is the differential vector along the direction of propagation. The proportionality constant V depends on the material and the wavelength of light, and is called the Verdet constant of the material. The sense of the magnetic rotation, left or right handed, is determined by the direction in which light passes through the material. This differs from natural optical activity where the sense of rotation bears a fixed relation to the direction of light propagation through the medium. In that case, when a beam of light is reflected back through an optically active medium, the rotation reverses its absolute sense in space and the net rotation is zero in such a double traversal of the medium. For the Faraday effect, the sense of the rotation is determined by the magnetic field direction so that the rotations are cumulated instead of cancelled by passing the light backward and forward through the medium. Many magnetooptic devices take advantage of this nonreciprocal property.

2. Diamagnetic Materials

Faraday rotation is produced by a magnetically induced double refraction for right- and left-circularly polarized light. The diamagnetic Faraday effect is related to the longitudinal Zeeman effect where the rotation angle is determined by the dispersion curves associated with the resonance frequencies of the Zeeman effect. Using the classical explanation for the Zeeman effect, Becquerel (1897) derived the basic formula for the diamagnetic Verdet constant, accurately giving its wavelength dependence. A complete quantum mechanical treatment of the Faraday effect for atoms shows that the diamagnetic Verdet constant is given exactly by the Becquerel formula when it is multiplied by a factor called the magnetooptic anomaly γ. This magnetooptic anomaly factor remains constant over the visible and near-ultraviolet spectrum and has a value between zero and unity depending on the material. The modified Becquerel formula is

$$V = -\gamma \frac{e}{2mc} \lambda \frac{dn}{d\lambda}$$

where e and m are the electronic charge and mass, c is the velocity of light and λ is the wavelength. The Verdet constant V is positive for diamagnetic materials because $dn/d\lambda$ is negative, where positive rotation is that of a right-handed screw advancing in the direction of the magnetic field. When $\lambda\, dn/d\lambda$ is known for a diamagnetic material, then only γ is needed to determine the Verdet constant. In a wavelength range far from absorption resonances, the Verdet constant is proportional to the inverse square of the wavelength, similar to Cauchy's law for indices of refraction. The Verdet constant of diamagnetic materials is nearly temperature independent, with small changes due to shifts in the effective bandgap. Table 1 lists Verdet constants at 589.3 nm (NaD line) for some diamagnetic crystals, along with their magnetooptic anomaly.

Table 1
Verdet constants at 589.3 nm (NaD) for various diamagnetic crystals

Crystal	Verdet constant (rad T^{-1} m^{-1})	γ
ZnS	65.7	0.91
KI	20.4	0.78
NH_4Br	14.5	0.69
KBr	11.6	0.74
NH_4Cl	10.8	0.72
NaCl	10.2	0.91
MgO	9.9	0.81
KCl	8.1	0.85
C (diamond)	6.7	0.28
Al_2O_3 (*c* axis)	6.1	0.64
$CaCO_3$ (*c* axis)	5.4	0.49
$NaClO_3$	4.9	0.67
SiO_2 (*c* axis)	4.9	0.78
$NH_4Al(SO_4)_2.12H_2O$	3.8	0.54
$KAl(SO_4)_2.12H_2O$	3.5	0.53
CaF_2	2.6	0.64
LiF	2.6	0.76

3. Paramagnetic Materials

When paramagnetic ions such as the transition metals and the rare-earth metals are present in a material, their magnetooptic effect generally dominates the ever-present diamagnetic effect. The paramagnetic Verdet constant for an ion with one dominant transition at wavelength λ_0 is approximately given by

$$V_{\text{para}} = K/(\lambda_0^2 - \lambda^2)$$

where K is inversely proportional to temperature. In the long-wavelength limit, the paramagnetic constant varies as $1/\lambda^2$ but with a sign opposite to that for diamagnetic rotation. Weber (1981) has tabulated Verdet constants for numerous paramagnetic glasses and crystals for optical applications (see *Magnetooptic Materials*). Trivalent terbium ions exhibit a strong paramagnetic rotation in the red and near-infrared spectra and are particularly useful for magnetooptic applications in either glasses (terbium borosilicate glass, Hoya FR-5) or crystals (terbium aluminum garnet).

The paramagnetic crystal terbium gallium garnet (TGG) is currently the material best suited to the widest range of Faraday rotator applications. TGG has a large Verdet constant of $-39.56\,\text{rad}\,T^{-1}m^{-1}$ with low absorption ($0.15\,m^{-1}$) at 1060 nm. The 500–1500 nm transmission range of TGG is not as broad, however, as for diamagnetic materials because of the visible and mid-infrared absorptions of the paramagnetic terbium ions in TGG.

4. Kerr Effect Materials

The Kerr magnetooptic effect is the change of polarization of light when reflected from a magnetic material. John Kerr (1877) first observed this polarization change when plane-polarized light reflected from the pole of a magnet displayed an ellipticity and a rotation of the plane of polarization dependent on the strength of magnetization of the mirror surface. This magnetic Kerr effect is to be distinguished from the more commonly known Kerr electrooptic effect used in Kerr cell light modulators and switches. Although the Kerr magnetooptic effect is small (usually $<0.5°$ rotation), it has become important due to its prominence in optical readout of information stored in magnetized thin films (Fujiwara 1989).

When a beam of linearly polarized light is incident upon a metal surface, the reflected beam is, in general, elliptically polarized. When a magnetic field is applied to a ferromagnetic surface, the reflected beam gains an additional ellipticity, apart from that due to ordinary reflection, and the plane of polarization is slightly rotated by an angle φ_k called the Kerr rotation. Three forms of the Kerr magnetooptic effect are produced by altering the orientation of the magnetization relative to the plane of incidence and to the plane of the reflecting surface. The polar form of the Kerr effect is when the magnetization is normal to the mirror surface, and longitudinal and transverse forms are when the magnetization is parallel to the mirror and, respectively, parallel and perpendicular to the plane of incidence (Lissberger 1961). The polar form gives the largest Kerr angle and is the readout method used in magnetooptical data storage. The ellipticity of magnetic origin is found to be negligible for the purpose of optical display. Materials used in early investigations were iron, nickel and cobalt, pure and alloyed, both as bulk mirrors and thin films (Ballantyne 1964). The common materials used nowadays for magnetooptical data storage disks are alloys of these previous metals with the rare-earth metals gadolinium, terbium, holmium and dysprosium in thin films. Many Kerr materials are plotted in Fig. 1 showing a correlation of Kerr rotation with Curie point. Fast writing and erasing requires a low Curie point for the Kerr material, while high Kerr rotations needed for good reading performance exist only for high Curie points. This means that, in general, high speed and good reading are contradictory to each other. Recently, however, two exchange-coupled layers of Kerr materials were found to solve this problem. For example, in Fig. 1, a magnetooptic layer A of high write/erase speed is exchange coupled to another layer B having good reading performance. The functions of write/erase and reading are separated in layers A and B, with the best operative combinations used, where the magnetization of each layer is conveyed to each other by the exchange coupling. The layered medium provides the high performance level of point C, an impossible achievement for a single layer (Fujiwara 1989). Recent developments in Kerr materials for magnetooptic data storage are given by Robinson et al. (1989).

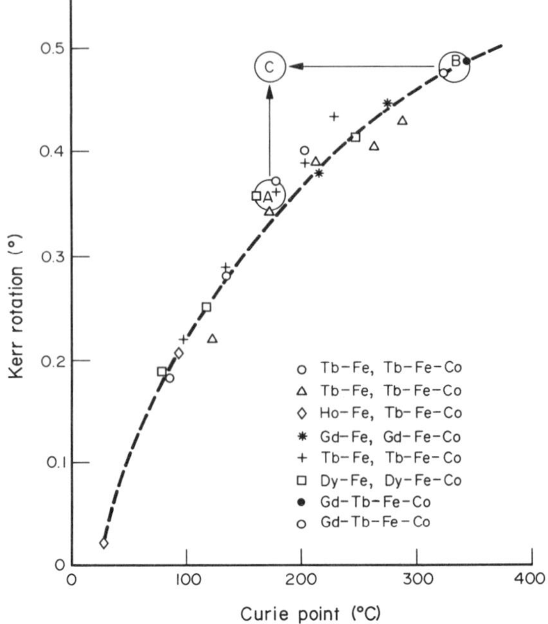

Figure 1
Correlation of Kerr rotation with Curie point for magnetooptic materials: when material A is exchange coupled to material B, the resulting layered medium C has high performance in both recording sensitivity (low Curie point) and reading performance (high Kerr rotation)

5. *Faraday Magnetooptic Applications*

The most extensive application of magnetooptic materials has been for optical isolators. An isolator is a nonreciprocal one-way transmission device. The basic isolator, shown in Fig. 2, consists of a material in a magnetic field causing 45° rotation placed between two polarizers. The acceptance planes of the polarizers are oriented at 45° to each other so that light passed by the first polarizer will be passed undiminished by the second. However, in the reverse direction, due to the nonreciprocity of the Faraday rotation, light admitted to the isolator by the second polarizer will be blocked by the first. Sometimes the available size of good optical quality material and the attainable magnetic field strengths dictate multiple passes through the Faraday material. The design of a triple-pass isolator shown in Fig. 2 further emphasizes the accumulative nature of the nonreciprocal Faraday

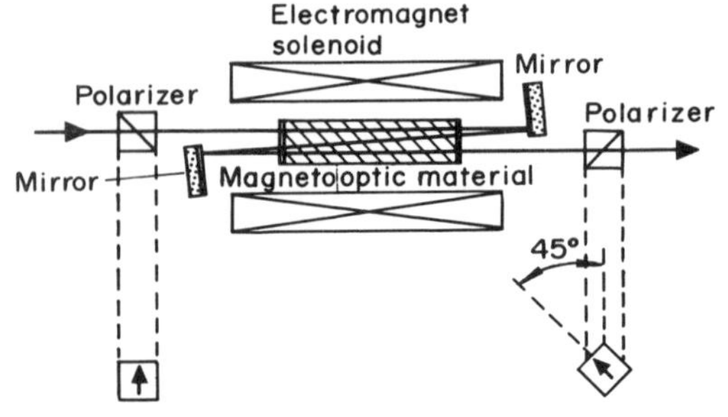

Figure 2
Optical isolator

rotation. Numerous isolator designs have been advanced using diamagnetic, paramagnetic and ferromagnetic materials for operation in the visible and infrared up to 10 μm. Permanent magnets and electromagnets have been used to provide a uniform axial magnetic field. Isolators are principally used in laser systems. They are inserted between laser amplifiers to suppress self-oscillation and to protect a laser from its laser radiation reflected from a target. Faraday isolators have also been built using silica fibers as the active medium (Turner and Stolen 1981). In addition, fast optical switches and modulators have been constructed using the Faraday effect. Faraday rotation of laser light in optical fibers has been used as an optical diagnostic method for high-current measurements and magnetic field measurements; currents in the milliampere range have been measured (Day and Rose 1988). Faraday effect sensors are faster than other types of optical sensor for magnetic fields, their speeds being limited only by the transit time of light through the device; also, they have particular advantages when working at high voltage or in the presence of substantial electromagnetic interference. The principal disadvantage of Faraday effect sensors has been their lack of sensitivity. Research, however, has led to dramatic gains in sensitivity, and further gains are in prospect in the development of dilute magnetic semiconductor materials (Butler 1987).

See also: Magnetooptic Materials; Magnetooptic Storage Media

Bibliography

Ballantyne J M 1964 Kerr magnetooptic effect in thin cobalt films. *J. Opt. Soc. Am.* 54: 1352–3

Becquerel H 1897 On an explanation applicable to the phenomena of Faraday and Zeeman. *Compt. Rend.* 125: 679

Butler M A 1987 Wavelength and temperature dependence of the Faraday effect in $Cd_{1-x}Mn_xTe$. *Solid State Commun.* 62: 45–7

Chen D 1971 Magnetooptic materials. In: Pressley R J (ed.) 1971 *Handbook of Lasers*. CRC Press, Boca Raton, FL, p. 460

Darwin C G, Watson W H 1927 The constants of the magnetic dispersion of light. *Proc. R. Soc. London, Ser. A* 114: 474–90

Day G W, Rose A H 1988 Faraday effect sensors: the state of the art. *SPIE Fibre Opt. Laser Sensors* 985: 138–50

Fujiwara T 1989 Layered media make good magnetooptical storage disks. *Laser Focus World* November: 109–16

Kerr J 1877 On rotation of the plane of polarization by reflection from the pole of a magnet. *Phil. Mag.* 3: 321–43

Lissberger P H 1961 Kerr magnetooptic effect in nickel–iron films. *J. Opt. Soc. Am.* 51: 948–66

Papp A, Harms H 1980 Magnetooptical current transformer. *Appl. Opt.* 19: 3729–45

Robinson C J, Suzuki T, Falco C M (eds.) 1989 *Materials for Magnetooptic Data Storage*, Materials Research Society Symposium Proceedings, Vol. 150. MRS, Pittsburgh, PA

Sansalone F J 1971 Compact optical isolator. *Appl. Opt.* 10: 2329–31

Turner E H, Stolen R H 1981 Fiber Faraday circulator or isolator. *Opt. Lett.* 6: 322–3

Van Vleck J H 1932 *Theory of Electric and Magnetic Susceptibilities*. Oxford University Press, Oxford

Weber M J 1981 *Faraday Rotator Materials*. Lawrence Livermore Laboratory Report No. M-103. National Technical Information Service, Springfield, VA

Zeiger H J, Pratt G W 1973 *Magnetic Interactions in Solids*. Clarendon, Oxford

L. G. De Shazer
[McDonnell Douglas, St. Louis, Missouri, USA]

Magnetostrictive Materials

Magnetostriction refers to any change in dimensions of a magnetic material caused by a change in its magnetic state. This effect has been recognized since the late nineteenth century. Frequently, magnetostriction is brought about by a change in magnitude or direction of an applied magnetic field, but it may also result from a change in temperature. Important types of magnetostriction are: (a) Joule magnetostriction, a change in linear dimension parallel to an applied magnetic field; (b) transverse Joule magnetostriction, a change in dimension perpendicular to an applied magnetic field; and (c) volume magnetostriction. Two less common magnetostrictive effects are the Guillemin effect, a bending of the magnetic material, and the Wiedemann effect, a twisting of the magnetic material.

Magnetostrictive effects are reciprocal. In addition to the change in length caused by a change in the magnetic field, a magnetostrictive material undergoes a change in its magnetic state (magnetization) under the influence of an applied mechanical stress. Measurements are generally performed in the presence of a small biasing field. A stress is applied and the resulting change in magnetization is measured.

Until recently, observed magnetostrictive changes of all types have been quite small, only a few parts per million. For iron, nickel and cobalt, maximum changes per unit length are of the order of 10×10^{-6}, 30×10^{-6} and 100×10^{-6}, respectively. Alloys of these metals yield magnetostrictions within this same range. Recently, with the advent of the rare-earth metals, much larger strains ($\sim 2000 \times 10^{-6}$ at room temperature, $\sim 10\,000 \times 10^{-6}$ at cryogenic temperatures) have been observed. Although the magnetostrictive dimensional changes are small, even for the rare earths, they are driven by strong interatomic forces, similar to those responsible for thermal expansion. Thus, magnetostrictive materials, very much like piezoelectric materials, find application in transducer devices, for example, in echo ranging and ultrasonic sound generation. Here, very strong forces but not necessarily large movements are required. For some applications, on the other hand, it is desirable to minimize magnetostrictive effects. For example, in electrical transformers magnetostriction is a source of mechanical loss and hum.

1. Joule Magnetostriction

In the most common type of magnetostriction, Joule magnetostriction, the dimensional change is associated with the distribution of distorted magnetic domains (small regions of fixed magnetic moments within a magnetic body). In the demagnetized state, the domains are distributed such that the net external magnetization of the body is zero. In a material exhibiting Joule magnetostriction, each domain is distorted by interatomic forces in such a way as to minimize the total energy. If the magnetostriction is positive, the dimension along the magnetization direction within the domain is elongated, while the dimension perpendicular to it is shortened, keeping the volume constant. In a body with negative magnetostriction, the dimension along the direction of magnetization is shortened, while that perpendicular to it is elongated. Thus, in a magnetic body consisting of many magnetostrictively distorted domains, overall dimensions change (a) when the magnetization within a domain is rotated by an external field and (b) when domains with magnetization close to the magnetic field direction grow at the expense of those with magnetization further away from the magnetic field. The shape of the magnetostriction curve during the magnetization process is illustrated in Fig. 1a. At saturation the body is a single large domain. Dimensional changes in the magnetically saturated material occur with field rotation (Fig. 1b); the fractional change in length of an isotropic polycrystal during rotation is given simply by

$$\Delta l/l = \tfrac{3}{2}\lambda_s(\cos^2\theta - \tfrac{1}{3})$$

where λ_s is the saturation magnetostriction and θ is the angle between the field and the measurement direction. Note that the maximum length change $3\lambda_s/2$ results not from the application of a magnetic field but from the rotation of a saturation field from a direction perpendicular to the measurement direction to a direction parallel to the measurement direction. Values of λ_s quoted in the literature are often obtained by this 90° rotation at a saturating field rather than by the application of a magnetic field.

The room-temperature magnetostrictions of some common magnetic materials are listed in Table 1. Iron has a very small magnetostriction. Alloys in the iron–nickel systems are both positive (iron rich) and negative (nickel rich), with zero magnetostriction occurring near 80% nickel. Alloys in the iron–cobalt system exhibit the largest known strains of the transition metals, 100–150 parts per million, depending on the metallurgical treatment. The largest magnetostrictions by far have been found in the rare-earth metals and compounds. The rare earths terbium and dysprosium possess magnetostrictions near 1% at cryogenic temperatures. RFe_2 compounds with R = Tb or Sm possess the largest known magnetostrictions at room temperature. These compounds offer a fertile field for future investigation, since magnetization, magnetostriction and magnetic anisotropy can be tailored by various rare-earth substitutions.

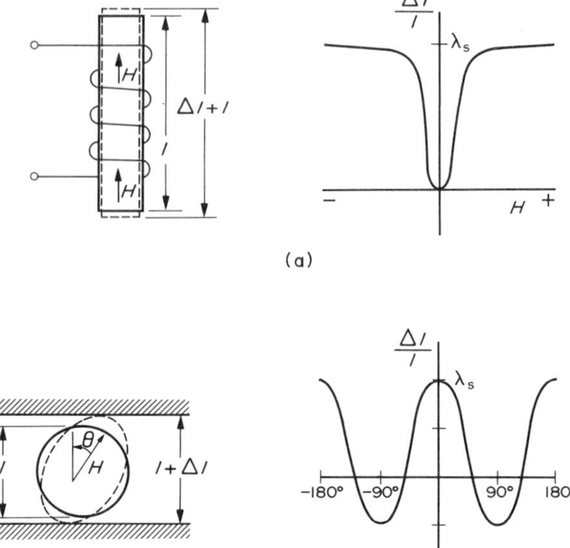

Figure 1
Joule magnetostriction caused by (a) application of a magnetic field and (b) rotation of a saturating magnetic field

Table 1
Room-temperature saturation magnetostriction (in parts per million) of some common polycrystalline magnetic materials

Material	λ_s	Material	λ_s
Iron	−9[a]	$TbCo_3$	65[d]
Nickel	−35[a]	$SmFe_2$	−1500[d]
Cobalt	−62[a]	$TbFe_2$	1753[d]
60%Co–40%Fe	68[b]	Fe_3O_4	40[e]
60%Ni%40%Fe	25[c]	$CoFe_2O_4$	−110[e]

a Bozorth (1951) b Williams 1932 *Rev. Sci. Instrum.* 3: 675 c Masiyama 1931 *Sci. Rep. Tohoku Univ.* 20: 574 d Clark 1974 *AIP Conf. Proc.* 18: 1015 e Smit and Wijn 1959 *Ferrites*. Wiley, New York

Table 2
Room-temperature magnetostriction constants (in parts per million) of some magnetic materials having cubic symmetry

Materials	λ_{100}	λ_{111}
Iron[a]	24	−22
Nickel[b]	−51	−23
Fe_3O_4[c]	−20	80
$Co_{.8}Fe_{2.2}O_4$[d]	−590	120
$NiFe_2O_4$[e]	−46	−22
$Dy_3Fe_5O_{12}$[f]	−14	−8
$SmFe_2$[g]		−2100
$TbFe_2$[h]		2460
$DyFe_2$[g]		1260

a Gersdorf et al. 1962 *J. Phys. Soc. Japan* 17 (B1): 342 b Bozorth and Hamming 1953 *Phys. Rev.* 89: 865 c Bickford et al. 1955 *Phys. Rev.* 99: 1210 d Bozorth et al. 1955 *Phys. Rev.* 99: 1788 e Smith and Jones 1966 *J. Appl. Phys.* 37: 1001 f Clark et al. 1966 *J. Appl. Phys.* 37: 1324 g Clark et al. 1977 *Physica B* 86–88: 73; Clark et al. 1977 *AIP Conf. Proc.* 29: 192 h Clark et al. 1975 *AIP Conf. Proc.* 24: 670

Magnetic materials in single-crystal form possess magnetostrictions which reflect the symmetry of the crystal. Changes in length at magnetic saturation depend on the measurement direction as well as the initial and final direction of the magnetization in the crystal. For crystals of cubic symmetry, such as iron and nickel, the fractional change in length is given by an expression containing two magnetostriction constants:

$$\Delta l/l = \tfrac{3}{2}\lambda_{100}((\alpha_x^2\beta_x^2 + \alpha_y^2\beta_y^2 + \alpha_z^2\beta_z^2) - \tfrac{1}{3}) + 3\lambda_{111}(\alpha_x\alpha_y\beta_x\beta_y + \alpha_y\alpha_z\beta_y\beta_z + \alpha_z\alpha_x\beta_z\beta_x)$$

Only two constants, λ_{100} and λ_{111}, are required to describe the magnetostriction along any direction in the crystal with magnetic field applied along any other direction (β_x, β_y and β_z are the direction cosines of the measurement with respect to the perpendicular x, y and z crystal axes; α_x, α_y and α_z are the direction cosines of the magnetization with respect to the same perpendicular axes). The magnetostriction constants are seldom equal and frequently have opposite signs. Thus, the single-crystal form of the magnetic material almost always has a particular direction in which the magnetostriction is larger than that found in its polycrystalline average. The magnetostriction constants for a few common single-crystal metals and oxides are listed in Table 2. Magnetostriction constants greater than 0.2% are found in the cubic rare-earth–iron compounds $TbFe_2$ and $SmFe_2$. Other important magnetic materials have hexagonal symmetry. For crystals of this lower symmetry, such as cobalt and the heavy rare-earth elements, the magnetostriction expression contains four constants, λ_1^α, λ_2^α, λ^γ and λ^ε:

$$\Delta l/l = \lambda_1^\alpha(\alpha_z^2 - \tfrac{1}{3})(\beta_x^2 + \beta_y^2) + \lambda_2^\alpha(\alpha_z^2 - \tfrac{1}{3})\beta_z^2 + \lambda^\gamma[\tfrac{1}{2}(\alpha_x^2 - \alpha_y^2)(\beta_x^2 - \beta_y^2) + 2\alpha_x\alpha_y\beta_x\beta_y] + 2\lambda^\varepsilon(\alpha_y\alpha_z\beta_y\beta_z + \alpha_z\alpha_x\beta_z\beta_x)$$

Table 3 lists the constants for the rare earths terbium, dysprosium and gadolinium at low temperature, as well as hexagonal cobalt and MnBi at room temperature (RT). Single crystals of the hexagonal rare earths terbium and dysprosium possess the largest known magnetostrictions, but these elements are magnetic only at low temperatures. Each of the single-crystal magnetostriction constants represents a crystalline distortion that occurs as the magnetization is rotated. Figure 2 depicts the particular distortion modes of the cubic and hexagonal constants cited above.

2. ΔE Effect

Closely associated with Joule magnetostriction is the change in Young's modulus with the degree of magnetic saturation (the ΔE effect). When a stress is applied to two adjacent magnetostrictively distorted domains, the "wall" separating these regions moves in such a manner as to minimize the restoring force. If the restoring force can be reduced further by domain rotation, this rotation also takes place. When a compressive stress is applied to a body with positive magnetostriction, the domains with magnetization perpendicular to the stress axis grow. In a body with negative magnetostriction, the domains with magnetization parallel to the stress axis grow. The opposite is true when a tensile stress is applied. In all cases the modulus is reduced from the intrinsic modulus, never increased. The maximum ΔE effect depends on (a) the square of the saturation magnetostriction and (b) the ease of domain wall motion and domain rotation. Typical changes are of the order of a few percent.

Table 3
Magnetostriction of some hexagonal crystals (in parts per million)

Materials	T (°C)	λ^γ	λ^ε	λ_1^α	λ_2^α
Cobalt[b]	RT	50	−233	70	−110
MnBi[c]	RT	−45		37	−50
Gadolinium[a,c]	−269	105	34	143	−150
Gadolinium[a,d]	−173	14	4	42	−110
Terbium[a,e]	−269	8900	15000	−2600	9000
Dysprosium[a,f]	−269	9000	5500		

a Nonmagnetic at room temperature b Bozorth 1954 *Phys. Rev.* 96: 311 c Williams et al. 1957 *J. Appl. Phys.* 28: 445 d Bozorth 1962 *J. Phys. Soc. Japan* 17: 1669 e Rhyne and Legvold 1965 *Phys. Rev.* 138: A507; DeSavage and Clark 1965 *Proc. 5th Rare Earth Res. Conf.*, Ames, IO f Clark et al. 1965 *Phys. Rev.* 138: A216

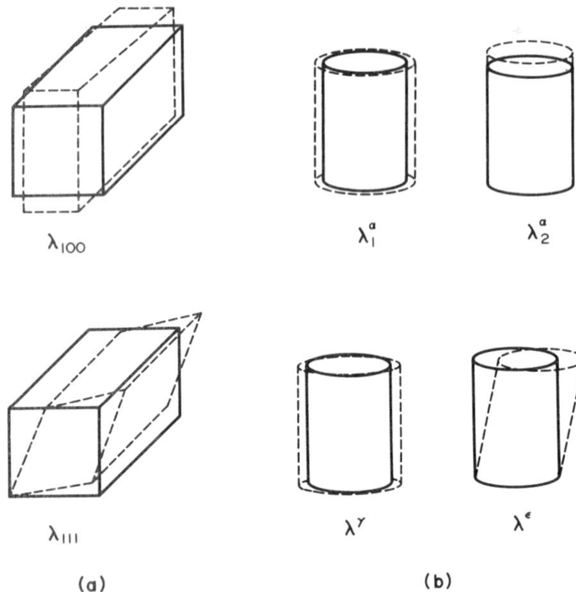

Figure 2
Single-crystal magnetostriction modes: (a) cubic and (b) hexagonal

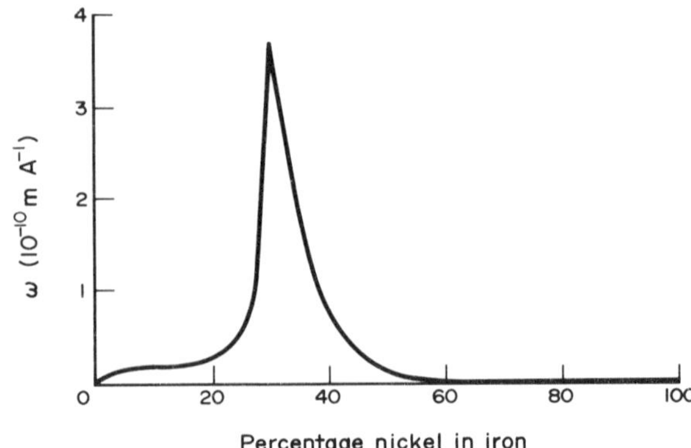

Figure 3
Volume magnetostriction of iron–nickel alloys at room temperature (after Bozorth 1951)

However, two new classes of magnetostrictive materials have recently emerged which possess very high ΔE effects. The first class, RFe_2 compounds with large magnetostrictions, displays $\Delta E/E$ effects greater than 100% (Clark 1980). The second class, magnetostrictive amorphous ribbons, in which domain walls move easily, possesses $\Delta E/E$ effects greater than 1000% (Wun-Fogle et al. 1986). The ΔE effect is described in more detail in *Magnetoelastic Phenomena.*

3. Volume Magnetostriction

Magnetically induced changes in volume (ΔV) of a body are usually observed in high fields and at temperatures near magnetic ordering (Curie) temperatures. The source of the volume magnetostriction in most materials is the dependence of the magnetic interaction (exchange field) on the separation between the atoms. It is not a domain effect. The magnetization of a body generally increases or decreases slightly with the separation of its magnetic atoms. When this is so, the volume of the body depends on its magnetization. This is known as volume magnetostriction. If the body expands with an increase in magnetization, the volume magnetostriction is defined as positive.

The magnetization can be changed in two ways: (a) by temperature and (b) by an applied magnetic field. The largest changes in magnetization with temperature occur near the Curie temperature and thus the largest volume magnetostriction occurs near this temperature. It is this volume magnetostriction that gives rise to the technically important Invar alloys, which possess zero or low thermal expansion. In these alloys a positive volume magnetostriction expansion compensates the intrinsic thermal expansion, yielding a small or zero change in volume with temperature (Bozorth 1951, Chikazumi 1964). An applied field has a strong effect on the magnetization and magnetostriction only near the Curie temperature. Far from the Curie temperature, an applied field causes a small change in magnetization and, therefore, only a very small change in volume. The change is proportional to magnetic field H and is often written as $\Delta V/V = \omega H$, where ω is a constant. For cobalt and iron $\omega \simeq 6 \times 10^{-12}\,\mathrm{m\,A^{-1}}$; for nickel, $\omega \simeq -0.6 \times 10^{-12}\,\mathrm{m\,A^{-1}}$ (see Carr 1959). Figure 3 illustrates the volume magnetostriction for the iron–nickel alloy system at room temperature. The sharp rise in magnetostriction near 30% nickel is a result of the low Curie temperature of this alloy.

See also: Anisotropy in Magnetic Materials; Electromagnetic–Acoustic Transducers; Magnetic Materials: Basic Concepts; Magnetoelastic Phenomena; Micromagnetic Calculations

Bibliography

Bates L F 1961 *Modern Magnetism*, 4th edn. Cambridge University Press, London

Bozorth R M 1951 *Ferromagnetism.* Van Nostrand, New York

Carr W J Jr 1959 *Magnetostriction. Magnetic Properties of Metals and Alloys.* American Society for Metals, Cleveland, OH

Chikazumi S 1964 *Physics of Magnetism.* Wiley, New York

Clark A E 1980 Magnetostrictive rare earth–Fe_2 compounds. In: Wohlfarth E P (ed.) 1980 *Ferromagnetic Materials: A Handbook on the Properties of Magnetically Ordered Substances*, Vol. 1. North-Holland, Amsterdam, Chap. 7, pp. 531–89

Clark A E, Savage H T 1975 Giant magnetically induced changes in the elastic modulus of $Tb_{0.3}Dy_{0.7}Fe_2$. *IEEE Trans. Sonics Ultrason.* 22: 50–2
Cullity B O 1972 *Introduction to Magnetic Materials.* Addison-Wesley, Reading, MA
Wun-Fogle M, Clark A E, Hathaway K B 1986 Permeability in "frozen" high magnetomechanical coupling amorphous ribbons. *J. Magn. & Magn. Mater.* 54–57: 893–4

A. E. Clark
[US Naval Surface Warfare Center, Silver Spring, Maryland, USA]

Magnetotransport Phenomena

An applied magnetic field produces a change in the electrical resistivity of a metal and also induces a voltage in the direction perpendicular to the field and the current. The former effect is the magnetoresistance and the latter is the Hall effect. The magnetoresistance and the Hall effect are the principal magnetotransport phenomena. In normal metals they are due to the Lorentz force that curves the electron trajectories. In magnetic metals and alloys, these so-called "normal effects" also exist but are generally small with respect to additional effects arising from the interaction of the conduction electrons with the *d* or *f* electrons which carry the magnetization. Considering the simple case of a system described by a localized moment model, the central problem is that of the *k–d* or *k–f* interaction between the conduction electrons (k) and the localized moments (d or f): the magnetotransport effects arise from the scattering of the conduction electrons by their interaction with localized moments oriented by the applied field.

A description of the scattering of the electrons by their interaction with localized moments and of the resulting magnetotransport effects is simplest in the case of the dilute magnetic alloys. These alloys are described in Sect. 1, followed by the more complicated case of the concentrated magnetic systems in Sect. 2 and the technologically important case of magnetic thin films and multilayers in Sect. 3.

1. Dilute Magnetic Alloys

The dilute magnetic alloys are normal metals containing a small concentration of magnetic impurities carrying a localized moment. In the low concentration limit, the interactions between impurities can be neglected and the system can be regarded as a collection of independent paramagnetic impurities. Typically, a piece of gold containing less than 1 at.% of gadolinium impurities can be regarded as a dilute magnetic alloy down to the liquid helium temperature range.

The simplest expression of the interaction between conduction electrons and magnetic impurities is

$$H_0 = (V - 2\Gamma \boldsymbol{s}.\boldsymbol{J})\delta(\boldsymbol{r}) \tag{1}$$

Equation (1) includes a potential term V and an exchange term Γ, proportional to the exchange constant $\mathscr{J}$ (see *Rare-Earth Intermetallics*; *Rare-Earth Metals and Alloys*), $\delta(\boldsymbol{r})$ is the Dirac delta function, $\boldsymbol{r}$ is the position vector of the conduction electron relative to the localized moment, $\boldsymbol{s}$ is the conduction electron spin and $\boldsymbol{J}$ is the impurity angular momentum. In zero field, when the impurity moments are randomly oriented, the electron scattering by this interaction gives the following contribution to the resistivity in the Born approximation:

$$\rho = \frac{2\pi m N(\varepsilon_F)c}{ne^2\hbar}[V^2 + \Gamma^2 J(J+1)] \tag{2}$$

where $N(\varepsilon_F)$ is the density of states at the host metal Fermi surface (per atom and per spin direction), c is the relative impurity concentration, m is the electron mass and n is the number of conduction electrons per unit volume. The second term of Eqn. (2),

$$\rho_m = \frac{2\pi m N(\varepsilon_F)c\Gamma^2 J(J+1)}{ne^2\hbar} \tag{3}$$

is the spin disorder resistivity or magnetic resistivity arising from exchange scattering.

An applied field orders the impurity spins and lowers the resistivity; this is the negative magnetoresistance of the dilute magnetic alloys. The origin of this effect is twofold: first, the spin–flip scattering is progressivly frozen out and second, the scattering potential becomes different for the spin ↑ and spin ↓ conduction electrons ($H_{0\uparrow(\downarrow)} \sim V \pm \Gamma J$ for $J_z = +J$). This implies that one spin orientation will be less scattered than the other, giving rise to a sort of short-circuit effect. The latter effect is more important. For $V \gg J$ the expression for the negative magnetoresistance is written (Beal-Monod and Weiner 1968) as

$$\frac{\Delta\rho}{\rho} = -4\left(\frac{\Gamma}{V}\right)^2\left[\langle J_z^2\rangle - \frac{1}{4}|\langle J_z\rangle| \times \left(\coth\left(\frac{\alpha}{2}\right) - \frac{1}{2}\frac{\alpha}{\sinh^2(\alpha/2)}\right)\right] \tag{4}$$

where $\alpha = g\mu_B H/k_B T$, with μ_B the Bohr magneton and g the gyromagnetic ratio, and J_z is the component of $\boldsymbol{J}$ along the field axis. An approximate expression for Eqn. (4) is simply

$$\frac{\Delta\rho}{\rho} = -4\left(\frac{\Gamma}{V}\right)^2\langle J_z^2\rangle \tag{5}$$

which accounts for the experimentally established proportionality between $\Delta\rho$ and M^2.

This negative magnetoresistance is isotropic, that is, it does not depend on the relative orientation of the field with respect to the current. Experimentally, a strictly isotropic negative magnetoresistance is found

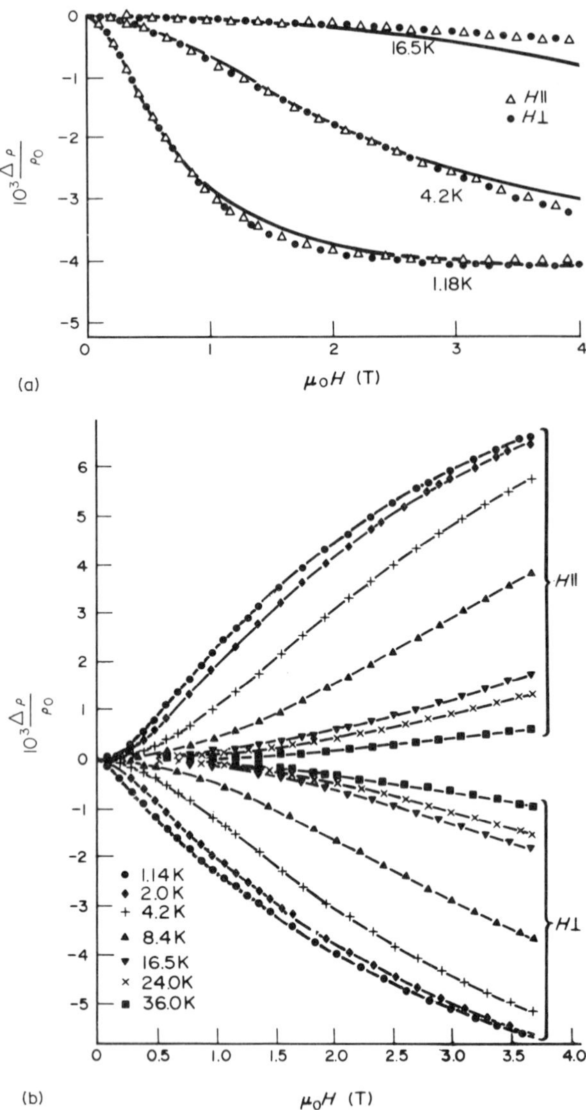

Figure 1
(a) Isotropic negative magnetoresistance of an Au–0.5%Gd alloy; as Gd^{3+} is an *s* ion with no electric quadrupole moment, the Au–Gd alloys show only the isotropic negative magnetoresistance arising from exchange scattering; (b) anisotropic magnetoresistance of an Au–0.98%Tb alloy at several temperatures; for non-*s* impurities, the magnetoresistance predominantly arises from quadrupole scattering, which gives positive and negative magnetoresistance in longitudinal and transverse fields respectively (Fert et al. 1977)

when the magnetic impurities carry only a spin angular momentum, as is the case for the *s*-ion gadolinium impurities. The example of a gold–gadolinium alloy is shown in Fig. 1a, with experimental data for longitudinal and transverse fields both giving a single fit with Eqn. (5).

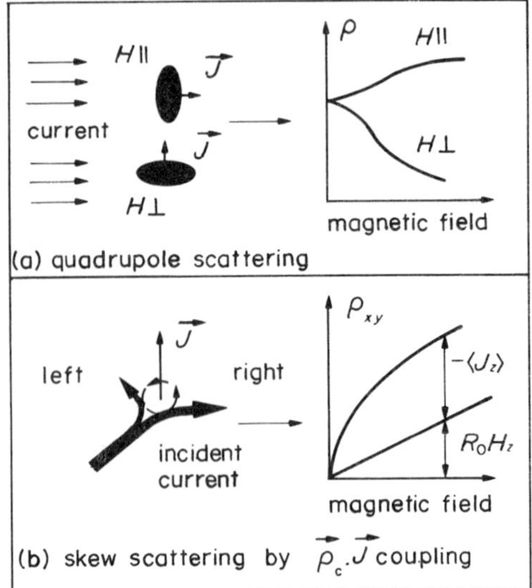

Figure 2
(a) Quadrupole scattering and (b) skew scattering, and the resulting contributions to the magnetoresistance and to the Hall resistivity

Equation (1) is a reasonable approximation for transition metal impurities but not for the rare earths, which generally carry a significant orbital angular momentum. In the presence of orbital magnetism, another important interaction term depending on the orientation of the impurity momentum $\boldsymbol{J}$ is the Coulomb interaction of the conduction electrons with the nonspherical distribution of charge on the magnetic impurity. This interaction exists only for non-*s* ions and can be approximated by its quadrupole term

$$H_Q \simeq D\tilde{O}_2(\boldsymbol{l}).\tilde{O}_2(\boldsymbol{J}) \tag{6}$$

where $\tilde{O}_2$ is a quadrupole tensor operator and $\boldsymbol{l} = (\boldsymbol{r} \times \boldsymbol{p})/\hbar$ is the orbital angular momentum of the conduction electron. This interaction produces magnetoresistances of opposite signs in longitudinal and transverse fields,

$$\Delta\rho_{\parallel} = -2\Delta\rho_{\perp} \sim D\left[\langle J_z^2 \rangle - \frac{J(J+1)}{3}\right] \tag{7}$$

The quadrupole scattering mechanism and the resulting magnetoresistance are illustrated in Fig. 2a, and an experimental example is shown in Fig. 1b. The anisotropic magnetoresistance observed in Fig. 1b for terbium impurities is not seen for gadolinium impurities in Fig. 1a because the quadrupole interaction, Eqn. (6), cancels out for *s*-ion rare earths.

The typical magnetoresistance curves of Fig. 1 are found for noble metals containing rare-earth impurities. For other host metals, the normal magnetoresistance may be nonnegligible, and the overall magnetoresistance is harder to analyze. For transition metal impurities, the orbital contribution to the magnetism is generally much smaller than for rare earths (the orbital momentum is said to be quenched) (see *Magnetic Materials: Basic Concepts*). Consequently, the spin–spin exchange term of Eqn. (1) is generally predominant, which gives an almost isotropic negative magnetoresistance. For many transition metal dilute alloys, an additional complication is the Kondo effect which will be presented at the end of this section.

Proceeding now to the Hall effect of dilute magnetic alloys, in addition to the normal or ordinary Hall effect due to the Lorentz force, written as $R_H H_z$ where R_H is the ordinary Hall coefficient, the Hall effect includes a contribution from the skew scattering of the conduction electrons by the magnetic impurities. Skew scattering is characterized by a left–right asymmetry as is represented in Fig. 2b. When a magnetic field orients the impurity moments, the electrons undergo preferential deflection to one side. The resulting contribution to the Hall resisitivity is, in first approximation, proportional to concentration c of the impurities and to their polarization $\langle J_z \rangle$. (The Hall resistivity ρ_H or ρ_{xy} relates the Hall transverse electric field E_y to the longitudinal current density i_x by $E_y = \rho_H i_x$.)

For rare-earth impurities, the skew scattering arises from terms of the k–f interaction that depend in an odd order on the electron orbital angular momentum ($\boldsymbol{l} = (\boldsymbol{r} \times \boldsymbol{p})/\hbar$). This includes orbital exchange terms proportional to $\boldsymbol{l}.\boldsymbol{J}$ and, with a more complicated mechanism, spin–orbit terms proportional to $\boldsymbol{l}.\boldsymbol{s}$ (Chien and Westgate 1980, Fert 1988). For transition metal magnetic impurities, the skew scattering can be explained by the spin–orbit splitting of the d levels (Fert et al. 1981). An experimental example of skew scattering contribution to the Hall resistivity is shown in Fig. 3. A fairly good fit is obtained by taking a Brillouin function for the variation of $\langle J_z \rangle$ (the discrepancy at the lowest temperatures is probably due to interaction effects) (see *Magnetic Materials: Basic Concepts*). Finally, at relatively large concentrations of magnetic impurities, another scattering asymmetry, the scattering with side jump, can also contribute to the Hall effect of the magnetic alloys (Berger 1970, Fert 1988).

For most transition metal magnetic impurities and for cerium or ytterbium impurities in a number of metal hosts, the exchange interaction with the conduction electron spins is due to k–d or k–f mixing and is antiferromagnetic, that is, Γ is negative in Eqn. (1). In this case, the impurity spin is progressively compensated by the conduction electron spins as the temperature decreases (see *Paramagnetic Materials*). This is the Kondo effect (Kondo 1969, Wilson 1975). A nonmagnetic singlet state is reached when T is much smaller than the Kondo temperature T_K. The order of magnitude of T_K is

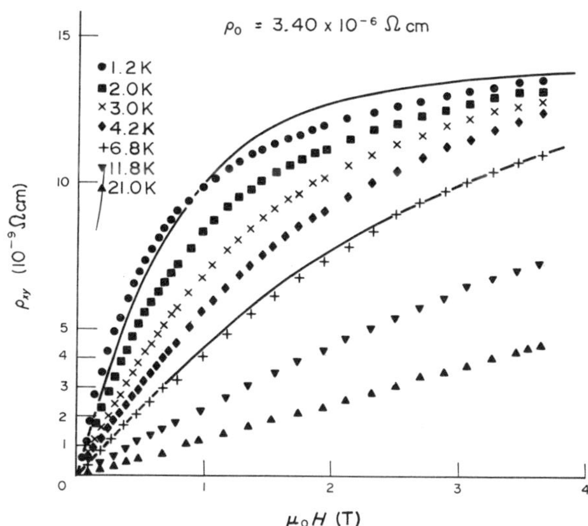

Figure 3
Skew scattering contribution to the Hall resistivity vs applied field for an Au–0.5%Gd alloy; the solid lines are Brillouin functions for $J = 7/2$, which represents the polarization $\langle J_z \rangle$ for perfectly paramagnetic impurities (after Fert and Friederich 1976)

$$T_K \simeq \frac{\Delta}{k_B} \exp\left(-\frac{1}{N(\varepsilon_F)|\Gamma|}\right) \tag{8}$$

where Δ is the bandwidth. The experimental values of T_K are scattered between a few millikelvin and a few hundred kelvin. As the temperature decreases, the magnetic resistivity increases and reaches a plateau for $T \ll T_K$. The variation of the magnetic resistivity for $T \gg T_K$ is given by the perturbation model of Kondo (1969)

$$\rho = \rho_m[1 + 4\Gamma N(\varepsilon_F)\ln(k_B T/D)] \tag{9}$$

The temperature-independent term ρ_m is the term of second order in Γ obtained in the first-order Born approximation and given by Eqn. (2). The second-order Born approximation gives the second term enclosed in the square brackets, the second-order correction is proportional to $\Gamma^3 \ln(k_B T/D)$. This term is generally a small correction to ρ_m but it is temperature dependent. As Γ is negative in Kondo systems, the resistivity increases as the temperature decreases.

The temperature dependence of the resistivity of copper–manganese and copper–iron Kondo alloys is shown in Fig. 4. The minimum is due to the competition between Kondo and phonon contributions to the resistivity. Below 10 K the Kondo term is predominant and a logarithmic variation is observed. A plateau at low temperatures is observed in copper–iron but not for copper–manganese because the Kondo temperature of copper–manganese is well below 1 K.

Figure 4 illustrates also the negative magnetoresistance of Kondo alloys. A frequent erroneous statement is that this negative magnetoresistance is due to the reduction of the Kondo logarithmic term by the applied field. In fact the Kondo term proportional to $\Gamma^3 N(\varepsilon_F)$ is generally much smaller than the term in Γ^2 of the first-order Born approximation, so that the magnetoresistance includes only relatively weak Kondo contributions and is still approximately described by Eqn. (4). Beal-Monod and Weiner (1968) have calculated the magnetoresistance terms of second and third order in Γ. They have applied their results to the interpretation of the data of Fig. 4 and found that a reasonably good fit can be obtained with only the non-Kondo terms. This is true for most transition metal Kondo alloys, except at very high fields where the Kondo terms account for the variation as ln H observed in some systems (Berman and Kopp 1973). In contrast, for cerium and ytterbium Kondo systems and especially for the category of the heavy fermion systems the contribution of the Kondo effect to the resistivity and the magnetoresistance can be much larger (see *Heavy Fermion Metals*).

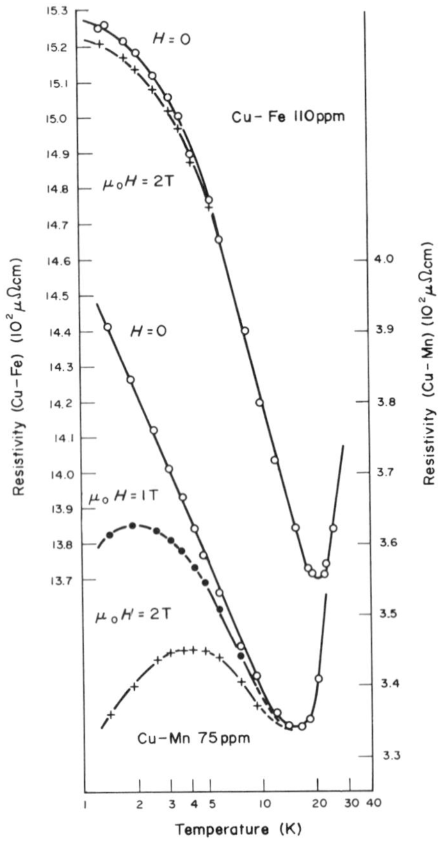

Figure 4
Resistivity of Cu–Mn and Cu–Fe alloys vs temperature in several applied fields; the positive normal magnetoresistance has been subtracted (after Monod 1967)

A contribution from the Kondo terms to the Hall effect has also been predicted by More (1969). However, it turns out that, besides the ordinary Hall effect, the main contribution arises from the non-Kondo skew scattering terms described previously. This skew scattering contribution is huge for the category of heavy fermion systems (Fert and Levy 1987).

2. *Ferromagnetic Metals and Alloys*

2.1 Magnetic Contributions to the Resistivity and the Resistivity Tensor in a Single-Domain Material

In the zero temperature limit, when all the magnetic moments of an ideal ferromagnet are aligned, the exchange interaction of the conduction electrons with the ordered spin forms a periodic potential. As a periodic potential cannot be a source of scattering, the contribution from the exchange to the resistivity is exactly zero (of course, for real ferromagnets, there is a residual resistivity due to impurities, defects, etc.). With an increase in temperature, some spin disorder is induced, the exchange potential is no longer periodic and there is a spin-disorder contribution to the resistivity, as illustrated in Fig. 5a (schematic) and Fig. 5b (experimental examples). This spin-disorder term increases as the temperature increases and its maximum value ρ_M is reached at T_c. In the first approximation, ρ_M can be regarded as the exchange resistivity induced by a collection of independent magnetic impurities and, from Eqn. (2) with $c = 1$, can be written as

$$\rho_M = \frac{2\pi m N(\varepsilon_F)}{ne^2\hbar} J(J+1)\Gamma^2 \qquad (10)$$

The variation of the spin-disorder resistivity from zero (at zero kelvin) to ρ_M (above T_c) can be written as (Kasuya 1956)

$$\rho_M(T) = \frac{\rho_M}{J(J+1)}(J - |\langle J \rangle_T|)(J + 1 + |\langle J \rangle_T|) \qquad (11)$$

where $\langle J \rangle$ is the mean value of J at temperature T. Eqn. (11) is obtained in a mean field approximation that does not take into account the short-range correlations. At low temperatures and at the approach of T_c, calculations taking into account spin waves or short-range order are more appropriate (de Gennes and Friedel 1958, Goodings 1963). In itinerant ferromagnets (transition metals), the temperature dependence of the band structure also contributes to the variation of the resistivity with temperature.

In the simple picture presented, it is assumed that the interaction of the conduction electrons with the magnetic moments includes only the spin–spin exchange term of Eqn. (1). Of course, as shown in Sect.

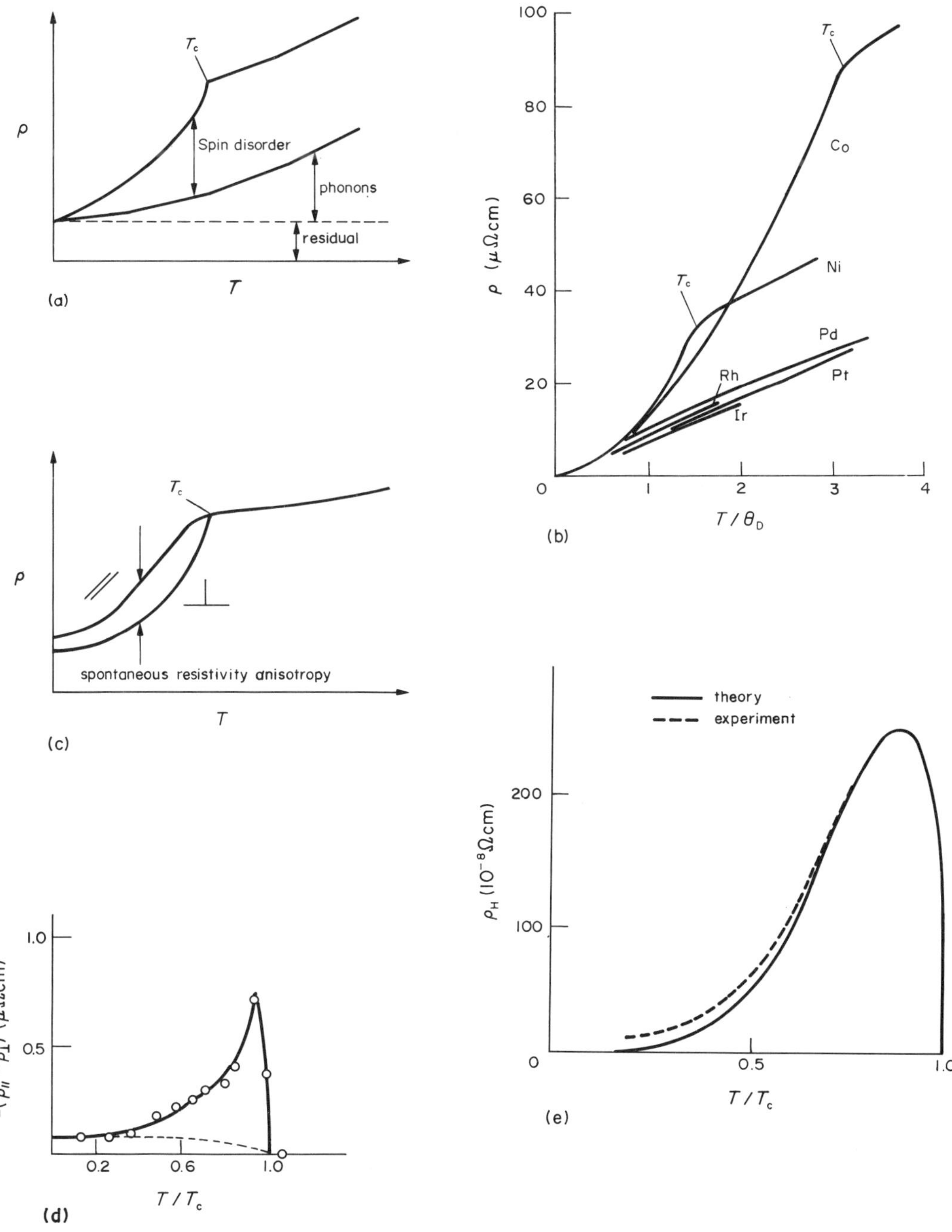

Figure 5
(a) The contributions to the resistivity of a ferromagnet; (b) resistivity of several transition metals vs T/θ_D; only cobalt and nickel are ferromagnetic; θ_D is the Debye temperature; (c) schematic variation of longitudinal and transverse resistivities for a single-domain ferromagnet; (d) the spontaneous resistivity anisotropy versus temperature for the ferromagnetic compound $PrAl_2$ (after Sablik et al. 1983); and (e) extraordinary (or spontaneous) Hall resistivity of iron versus temperature (after Kondo 1962)

1 for the dilute magnetic alloys, this interaction includes also anisotropic terms, for example those arising from electric-quadrupole-type or spin–orbit interactions. When such terms are taken into account, the resistivity of a single domain is no longer isotropic and the following two effects are obtained.

(a) The resistivity is different according to whether the electron current is parallel or perpendicular to the magnetization. This is the spontaneous resistivity anisotropy.

(b) For a current along the x axis with the magnetization along the z axis, there is a nonzero electric field along the y axis. This is the spontaneous (or extraordinary or anomalous) Hall effect.

The spontaneous resistivity anisotropy and the spontaneous or extraordinary Hall effect can be expressed using the following form of the spontaneous resistivity tensor for a monodomain polycrystal having its magnetization along the z axis:

$$[\rho_{ij}] = \begin{bmatrix} \rho_\perp & -\rho_{EH} & 0 \\ \rho_{EH} & \rho_\perp & 0 \\ 0 & 0 & \rho_\parallel \end{bmatrix} \qquad (12)$$

This form of the resistivity tensor corresponds to the following expression of the electric field $\boldsymbol{E}$:

$$\boldsymbol{E} = \rho_\perp \boldsymbol{j} + (\rho_\parallel - \rho_\perp)(\hat{\alpha}.\boldsymbol{j})\hat{\alpha} + \rho_{EH}\hat{\alpha} \times \boldsymbol{j} \qquad (13)$$

where $\boldsymbol{j}$ is the current density and $\hat{\alpha}$ is the unit vector in the direction of the magnetization. For the dependence of the resistivity on the angle θ between the field and the current, this gives

$$\rho = \frac{\rho_\parallel + 2\rho_\perp}{3} + (\rho_\parallel - \rho_\perp)(\cos^2\theta - \tfrac{1}{3}) \qquad (14)$$

The transverse and longitudinal resistivities $\rho_\perp$ and $\rho_\parallel$ are plotted schematically as a function of temperature in Fig. 5c. The difference between $\rho_\parallel$ and $\rho_\perp$ is the spontaneous resistivity anisotropy. It disappears above T_c. The temperature dependence of the spontaneous resistivity anisotropy has been calculated for rare-earth ferromagnetic compounds by Sablik et al. (1983) in a model taking into account the interaction of the electrons with the electric quadrupole of the rare-earth ions. The experimental temperature dependence of $(\rho_\parallel - \rho_\perp)$ in the ferromagnetic compound $PrAl_2$ is shown in Fig. 5d. The maximum value of $\rho_\parallel - \rho_\perp$ amounts to about 5% of the mean resistivity.

The term ρ_{EH} in Eqn. (12) is the extraordinary or spontaneous Hall resistivity. It increases above its residual value as the spin disorder increases and decreases back to zero at T_c, as illustrated by the experimental results for iron in Fig. 5e. This behavior is found for most ferromagnetic metals and compounds but a more complicated variation is observed for terbium and dysprosium. An example of calculation of the temperature dependence of the extraordinary Hall resistivity induced by spin disorder is that described by Kondo (1962). There is no simple relation between extraordinary Hall resistivity and magnetization.

The expression of the resistivity tensor, Eqn. (12), has implicitly been written for a single domain in zero field. In the presence of a magnetic field in the z direction, the variation of the resistivity tensor can be written as

$$[\Delta\rho_{ij}] = \begin{bmatrix} \Delta\rho_\perp & -\Delta\rho_H & 0 \\ \Delta\rho_H & \Delta\rho_\perp & 0 \\ 0 & 0 & \Delta\rho_\parallel \end{bmatrix} \qquad (15)$$

There are several contributions to the change of ρ_{ij}. First, as in a nonmagnetic metal, there is a normal magnetoresistance and an ordinary Hall effect due to the effect of the Lorentz force on the electron trajectories. In a first approximation the normal magnetoresistance scales with the ratio B/ρ, where B is the magnetic induction (in a magnetic material, $B = \mu_0(H_0 + H_D + M)$ where H_0 is the applied field, H_D is the demagnetizing field and M is the magnetization) and ρ is the resistivity for $B = 0$. This is known as Kohler's rule and implies the existence of Kohler's functions $G_\parallel$ and $G_\perp$ defined by $\Delta\rho_\parallel/\rho = G_\parallel(B/\rho)$ and $\Delta\rho_\perp/\rho = G_\perp(B/\rho)$ which vary from metal to metal but are nearly independent of temperature and of impurity type and concentration. The Lorentz contribution to $\Delta\rho_H$ is just the ordinary Hall resistivity $\rho_{OH} = R_H B$ where R_H is the ordinary Hall coefficient. In addition to these normal effects directly induced by the Lorentz force, an applied field can also increase the ferromagnetic order (forced magnetization), which leads to a reduction of the spin-disorder resistivity and an enhancement of the extraordinary Hall resistivity.

Two types of effects induced by the application of a magnetic field on a ferromagnetic material can now be predicted.

(a) The effects simply related to the orientation of the magnetization in the direction of the applied field. All these effects can be predicted by Eqn. (14) expressing the electric field as a function of the relative orientation of the magnetization and the current density $\boldsymbol{j}$.

(b) The effects corresponding to field-induced changes of the resistivity tensor arising from the Lorentz force (normal magnetoresistance and ordinary Hall effect) and from forced magnetization effects.

All that has so far been presented holds in a polycrystal (macroscopically isotropic medium for which, by symmetry, the resistivity tensor has the form of Eqn. (12)). In a single-domain single crystal the resistivity depends not only on the relative orientation of the current and the magnetization, but also on their orientation with respect to the crystal axes. For example, in a cubic crystal with $\boldsymbol{M}$ in the direction

$(\alpha_1, \alpha_2, \alpha_3)$ and $\boldsymbol{j}$ in the direction $(\beta_1, \beta_2, \beta_3)$, the resistivity can be written as (Campbell and Fert 1982)

$$\begin{aligned}\rho = \rho_0\Big[& 1 + k_1(\alpha_1^2\beta_1^2 + \alpha_2^2\beta_2^2 + \alpha_3^2\beta_3^2 - \tfrac{1}{3}) \\ & + k_2(2\alpha_1\alpha_2\beta_1\beta_2 + 2\alpha_2\alpha_3\beta_2\beta_3 + 2\alpha_3\alpha_1\beta_3\beta_1) \\ & + k_3(S-1) + k_4(\alpha_1^4\beta_1^2 + \alpha_2^4\beta_2^2 + \alpha_3^4\beta_3^2 + \tfrac{2}{3}S - \tfrac{1}{3}) \\ & + k_5(2\alpha_1\alpha_2\alpha_3^2\beta_1\beta_2 + 2\alpha_2\alpha_3\alpha_1^2\beta_2\beta_3 \\ & + 2\alpha_3\alpha_1\alpha_2^2\beta_3\beta_1)\Big] \qquad (16)\end{aligned}$$

where $S = \alpha_1^2 + \alpha_2^2 + \alpha_3^2$.

2.2 Magnetoresistance

Figure 6 represents typical magnetoresistance curves of a ferromagnetic metal in the case $\rho_\parallel > \rho_\perp$. Figure 6a is for the case (the low-temperature range in general) where the resistivity change is predominantly due to the spontaneous resistivity anisotropy (at low fields) and to the normal magnetoresistance. The zero field resistivity is an average of $\rho_\parallel$ and $\rho_\perp$. At low fields the resistivity change is mainly due to the orientation of the magnetization in the field direction. At higher fields, after saturation of the magnetization in the field direction, only the positive normal magnetoresistance remains. To separate the spontaneous resistivity anisotropy from the normal magnetoresistance, it is necessary to extrapolate the normal magnetoresistance back to $B = 0$, that is, $H_0 + H_D + M = 0$, which gives the spontaneous resistivities $\rho_\parallel(B=0)$ and $\rho_\perp(B=0)$. Their difference is the spontaneous resistivity anisotropy.

At the approach of T_c, as the spin-disorder resistivity is large, the negative magnetoresistance arising from field-induced spin ordering is considerable and generally more important than the positive normal magnetoresistance. This gives the behavior of Fig. 6b: the low field change is still mainly due to the spontaneous resistivity anisotropy, while, after saturation of the magnetization, the resistivity decrease is due to the negative contribution from spin ordering.

The metallic ferromagnetic materials are used to make magnetoresistive sensors because their spontaneous resistivity anisotropy gives rise to low-field magnetoresistance. In particular, for soft ferromagnetic materials, very small fields can orient the magnetization and induce significant resistivity changes. The spontaneous resistivity anisotropy—which determines the resitivity changes—is particularly important in nickel-based alloys. In fact, the relative anisotropy of the resistivity, $\Delta\rho/\rho = (\rho_\parallel - \rho_\perp)/\rho$, is not very large in pure nickel (about 2% at low temperatures) but is strongly enhanced by some impurities, cobalt and iron particularly. The most efficient impurities are those that scatter very differently the spin ↑ and spin ↓ electrons. If the ratio of spin ↑ and spin ↓ currents is α, it turns out that, at least at low temperatures and for α

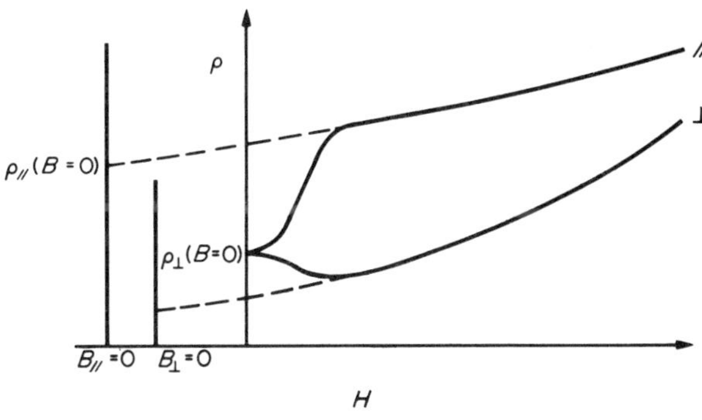

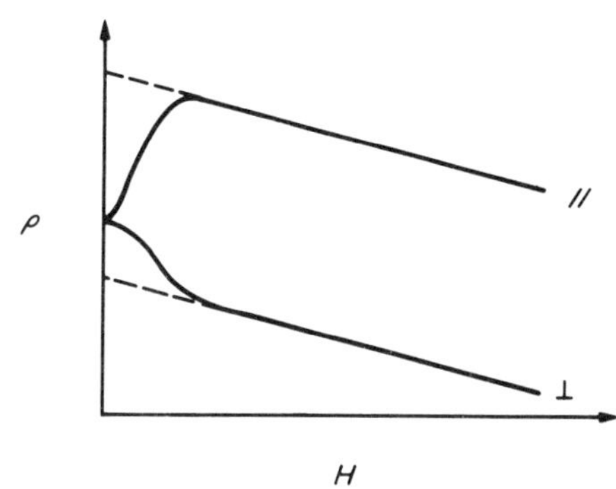

Figure 6
Schematic magnetoresistance curves for a ferromagnet: the variation during the orientation of the magnetization at low field is predominantly due to the spontaneous resistivity anisotropy; after saturation of the magnetization, the variation is due either to (a) the positive normal magnetoresistance or (b) the reduction of the spin-disorder resistivity

much larger than unity, the spontaneous resistivity anisotropy is roughly proportional to $\alpha - 1$, as illustrated in Fig. 7. This behavior can be explained in the extension (Campbell and Fert 1982, Malozemoff 1986) of the model of Smit (1951).

2.3 Hall Effect

Once the magnetization is aligned in the direction of the applied field, the Hall electric field is given by

$$\boldsymbol{E}_H = (\rho_{EH} + \Delta\rho_H)\hat{\alpha} \times \boldsymbol{j} \qquad (17)$$

The extraordinary (or spontaneous) Hall resistivity ρ_{EH} is the off-diagonal term of the spontaneous resistivity tensor in Eqn. (12) and, if any forced extraordinary contribution is neglected, $\Delta\rho_H$ reduces to the ordinary Hall resistivity ρ_{OH} defined in Sect. 2.1 and is generally proportional to B. This gives the typical behavior of Fig. 8: after saturation of the magnetization and the extraordinary contribution, the linear variation is due to the ordinary Hall effect. The extraordinary Hall resistivity is derived from the extrapolation shown in the figure. An example of the temperature dependence of the extraordinary Hall resistivity is shown in Fig. 5e.

The extraordinary Hall effect of the ferromagnetic metals is due to skew scattering (Kondo 1962) and scattering with side jump (Berger 1970). Both scattering asymmetries arise from the spin–orbit coupling and also, for the rare earths, from orbital exchange. In first approximation, the contribution from skew scattering is proportional to the resistivity ρ, while the contribution from the side jump is proportional to ρ^2. This explains why the skew scattering is observed in low-resistivity dilute alloys (see Sect. 1) and in ferromagnetic metals at relatively low temperatures, while the side jump seems to be largely predominant when the resistivity is high, that is, for pure metals at higher temperatures and in concentrated alloys at any temperature (Chien and Westgate 1980, Campbell and Fert 1982).

2.4 Planar Hall Effect

When the current density is in the x direction and the voltage is measured along the y direction, Eqn. (13) predicts the following transverse electric field when $\boldsymbol{M}$ is in the xy plane:

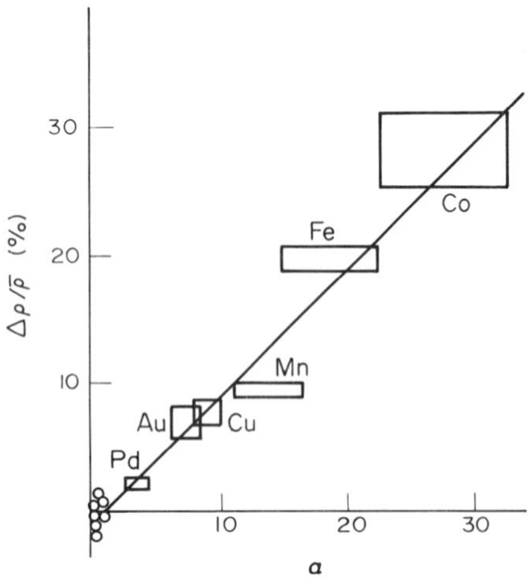

Figure 7
The spontaneous resistivity anisotropy ratio of nickel containing different types of impurity plotted as a function of the ratio α of the spin ↑ and spin ↓ cross sections for each of these impurities. The data are for 4.2 K and the rectangles correspond to the error bars on $\Delta\rho/\rho$ and α; $\Delta\rho = \rho_{\parallel} - \rho_{\perp}$; $\bar{\rho} = (1/3)\rho_{\parallel} + (2/3)\rho_{\perp}$ (after Campbell and Fert 1982)

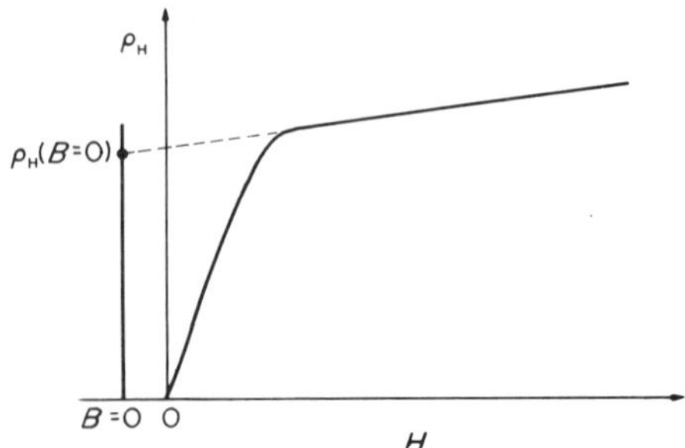

Figure 8
Schematic variation of the Hall resistivity as a function of the applied field for a ferromagnet: the linear variation at high field is due to the ordinary Hall resistivity, which is derived by extrapolating the linear variation back to $B = 0$, $\rho_{EH} = \rho_H(B \rightarrow 0)$

$$E_y = \tfrac{1}{2}(\rho_{\parallel} - \rho_{\perp})\sin 2\theta\, j_x \qquad (18)$$

where θ is the angle between $\boldsymbol{j}$ and $\boldsymbol{M}$. This phenomenon is known as the planar Hall effect, but it has nothing to do with the true or off-diagonal Hall effect and is a mere consequence of the resistivity anisotropy. The transverse voltage is maximum for $\boldsymbol{M}$ at 45° from the current.

3. Magnetic Thin Films and Multilayers

The magnetotransport properties are often used as an interesting tool for investigating the magnetic properties of thin films and multilayers (see *Thin Films and Multilayers, Magnetic*). In particular, the angular dependence of the spontaneous resistivity anisotropy, Eqn. (14), can be useful in determining the orientation

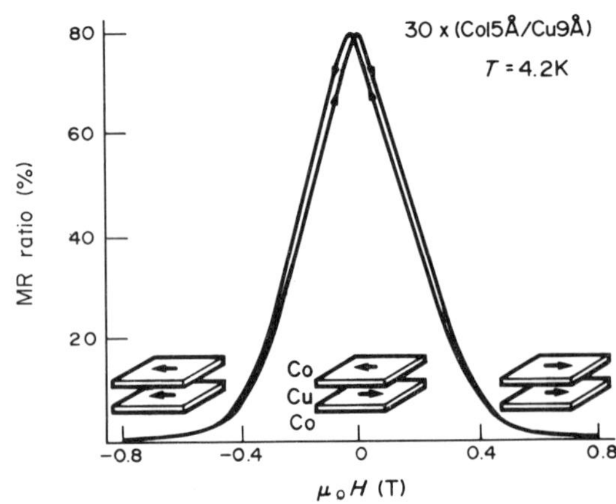

Figure 9
Magnetoresistance curves of a (Co 15 Å/Cu 9 Å) × 30 multilayer at 4.2 K: the magnetoresistance ratio, $r = (\rho_{max} - \rho_{min})/\rho_{min}$, amounts to 78% (after Mosca et al. 1991)

of the magnetization in a given field and in deriving the anisotropy constants (Dahlberg et al. 1988). In addition, the magnetic multilayers exhibit a specific effect known as the giant magnetoresistance or spin-valve effect. This phenomenon has been observed in multilayers with an antiferromagnetic exchange coupling between the magnetic layers across nonmagnetic layers: iron/chromium (Baibich et al. 1988, Binasch et al. 1989), copper/ruthenium (Parkin et al. 1990) and cobalt/copper (Mosca et al. 1991). An example of experimental curves for cobalt/copper is shown in Fig. 9: the resistivity decreases by almost a factor of two when an applied field overcomes the antiferromagnetic exchange and aligns the magnetizations of all the cobalt layers. The giant magnetoresistance can be accounted for when it is assumed that the majority and minority spin electrons are differently scattered by the impurities or defects in the magnetic layers or by the roughness of the interfaces. When the magnetizations of all the layers are aligned, the electrons of a given spin direction are systematically more weakly scattered in all the magnetic layers and, from a sort of short-circuit effect by these electrons, the resistivity is low. At zero field, the magnetizations are in alternating orientations and each spin direction is alternately weakly and strongly scattered, giving rise to a higher resistivity. Theoretical models have been worked out by Camley and Barnaś (1989) and Levy et al. (1990).

The giant magnetoresistances of the magnetic multilayers are higher than those of bulk materials by an order of magnitude (at room temperature, at least), which is very promising for applications in magnetoresistive sensors, magnetic recording heads and so on (Dibbern 1986). However, for such applications, it is necessary to obtain the resistivity change in very small fields. For multilayers with antiferromagnetic interlayer exchange, a relatively high field is needed to overcome the exchange and induce the resistivity change. As an alternative approach, multilayer structures have been prepared in which the ferromagnetic layers are prepared with alternating coercivities. This is accomplished by using different thicknesses for odd and even ferromagnetic layers, or by using alloys with different compositions. In such a system the magnetization of the softer layers can reverse while that of the hard layers is kept fixed, which results in a magnetoresitance effect at a field fixed by the coercivity of the soft layers. Interesting results have been obtained in several groups (Binasch et al. 1989, Dupas et al. 1990, Dieny et al. 1991, Shinjo 1991).

Bibliography

Baibich M N, Broto J M, Fert A, Nguyen Van Dau F, Petroff F, Etienne P, Creuzet G, Friederich A, Chazelas J 1988 Giant magnetoresistance of (001) Fe / (001) Cr magnetic superlattices. *Phys. Rev. Lett.* 61: 2472–5

Beal-Monod M T, Weiner R A 1968 Negative magnetoresistance in dilute alloys. *Phys. Rev.* 170: 552–9

Berger L 1970 Side-jump mechanism for the Hall effect of ferromagnets. *Phys. Rev. B* 2: 4559–66

Berman R, Kopp J 1973 The magnetoresistance of gold–iron alloys in high fields. *J. Phys. F* 3: 847–55

Binasch G, Grünberg P, Saurenbach F, Zinn W 1989 Enhanced magnetoresistance in layered magnetic structures with antiferromagnetic interlayer exchange. *Phys. Rev. B* 39: 4828–30

Camley R E, Barnaś J 1989 Theory of giant magnetoresistance effects in magnetic layered structures with antiferromagnetic coupling. *Phys. Rev. Lett.* 63: 664–7

Campbell I A, Fert A 1982 Transport properties of ferromagnets. In: Wohlfarth E P (ed.) 1982 *Ferromagnetic Materials*, Vol. 3. North-Holland, Amsterdam, pp. 747–804

Chien C L, Westgate C R (eds.) 1980 *The Hall Effect and its Applications*. Plenum, New York

Dahlberg E D, Riggs K, Prinz G A 1988 Magnetotransport: an ideal probe of anisotropy energies in epitaxial films. *J. Appl. Phys.* 63: 4270–5

de Gennes P G, Friedel J 1958 Anomalies de résistivité dans certains métaux magnétiques. *J. Phys. Chem. Solids* 4: 71–7

Dibbern 1986 Magnetic field sensors using the magnetoresistive effect. *Sensors and Actuators* 10: 127–40

Dieny B, Speriosu V S, Parkin S S P, Gurney B A, Wilhoit D R, Mauri D 1991 Giant magnetoresistance in soft ferromagnetic multilayers. *Phys. Rev. B* 43: 1297–300

Dupas C, Beauvillain P, Chappert C, Renard J P, Trigui F, Veillet P, Velu E 1990 Very large magnetoresistance effects induced by antiparallel magnetization in two ultrathin cobalt films. *J. Appl. Phys.* 67: 5680–2

Fert A 1988 The Hall effect in magnetic materials. In: Tabahashi M, Maekawa S (eds.) 1988 *Physics of Magnetic Materials*. World Scientific, London, pp. 503–14

Fert A, Asomoza A, Sanchez D H, Spanjard D, Friederich A 1977 Magnetotransport properties of noble metals containing rare-earth impurities. I. Quadrupole scattering by rare-earth impurities in gold. *Phys. Rev. B* 16: 5040–51

Fert A, Friederich A 1976 Skew-scattering by rare-earth impurities in silver, gold, and aluminum. *Phys. Rev. B* 13: 397–411

Fert A, Friederich A, Hamzic A 1981 Hall effect in dilute magnetic alloys. *J. Magn. & Magn. Mater.* 24: 231–57

Fert A, Levy P M 1987 Theory of the Hall effect in heavy fermion compounds. *Phys. Rev. B* 36: 1907–16

Goodings D A 1963 Electrical resistivity of ferromagnetic metals at low temperatures. *Phys. Rev.* 132: 542–58

Kasuya T 1956 Electrical resistance of ferromagnetic metals. *Prog. Theor. Phys.* 16: 58–63

Kondo J 1962. Anomalous Hall effect and magnetoresistance of ferromagnetic metals. *Prog. Theor. Phys.* 27: 772–92

Kondo J 1969 Theory of dilute magnetic alloys. In: Seitz F, Turnbull D, Ehrenreich H (eds.) 1969 *Solid State Physics*, Vol. 23. Academic Press, New York, pp. 183–281

Levy P M, Zhang S F, Fert A 1990 Electrical conductivity of magnetic multilayered structures. *Phys. Rev. Lett.* 65: 1643–6

Malozemoff A P 1986 Anisotropic magnetoresistance with cubic anisotropy and weak ferromagnetism: a new paradigm. *Phys. Rev. B* 34: 1853–63

Monod P 1967 Magnetic field dependence of the Kondo resistivity minimum in CuFe and CuMn alloys. *Phys. Rev. Lett.* 19: 1113–17

More R 1969 Hall anomaly in dilute magnetic alloys. *Solid State Commun.* 7: 237–8
Mosca D H, Petroff F, Fert A, Schroeder P A, Pratt W P Jr, Laloee R 1991 Oscillatory interlayer coupling and giant magnetoresistance in Co/Cu multilayers. *J. Magn. & Magn. Mater.* 94: 11–15
Parkin S S P, More N, Roche K P 1990 Oscillations in exchange coupling and magnetoresistance in metallic superlattice structures: Co/Ru, Co/Cr, Fe/Cr. *Phys. Rev. Lett.* 64: 2304–7
Rohrer H 1968 Magnetoresistance of dilute alloys. *Phys. Rev.* 174: 583–94
Sablik M J, Pureur P, Levy P M 1983 Quadrupole scattering in $PrAl_2$. *Phys. Rev. B* 28: 3890–6
Shinjo T 1991 Interface magnetism. *Surf. Sci. Rep.* 12: 49–98
Smit J 1951 Magnetoresistance of ferromagnetic metals and alloys at low temperatures. *Physica* 16: 612–27
Wilson K G 1975 The renormalization group: critical phenomena and the Kondo problem. *Rev. Mod. Phys.* 47: 773–840

A. Fert and D. K. Lottis
[Université Paris-Sud, Orsay, France]

Manganese–Aluminum–Carbon Alloys

Mn–Al–C permanent magnets are of considerable interest because they are free of cobalt. Although cast isotropic alloys have been used, the magnetic properties are generally poor, with $(BH)_{max} \simeq 6.4\,kJ\,m^{-3}$. However, an anisotropic, high-energy-product material can be made by warm extrusion, with values of $(BH)_{max}$ as high as $56\,kJ\,m^{-3}$.

A typical Mn–Al–C alloy contains 70% manganese, 29.5% aluminum and 0.5% carbon. The cast cylindrical ingot is first solution treated at 1100°C, quenched to 500°C, tempered at 600°C and extruded at 700°C. A further increase in coercivity is obtained by aging the extruded material at 700°C. Table 1 shows some typical magnetic properties.

Metallurgically, alloys in this composition range exist as the hexagonal-close-packed ε phase at high temperature. This phase is transformed, during controlled cooling or quenching followed by isothermal annealing, to the metastable magnetic τ phase, which has an ordered AuCuI tetragonal structure. In the transformation, the ε phase is first changed to an ordered orthorhombic ε' phase via diffusion, which is then changed to the τ phase via a martensitic reaction. The extrusion orients the c axis of the τ phase in the axial (extrusion) direction. Since the τ phase exhibits a very large magnetocrystalline anisotropy of around $1\,MJ\,m^{-3}$ with the c axis as the easy magnetization direction, the optimum magnetic properties of the extruded alloy are in the extrusion direction.

Table 1
Mn–Al–C magnets

Alloy (wt%, balance Mn)	B_r (T)	H_c ($kA\,m^{-1}$)	$(BH)_{max}$ ($kJ\,m^{-3}$)
29.5 Al, 0.5 C (isotropic)	0.28	72	6.4
29.5 Al, 0.5 C (isotropic, by melt extraction)	0.34	104 (132)[a]	12.0
29.5 Al, 0.5 C (anisotropic, by extrusion)	0.61	216 (240)[a]	56
29.3 Al, 0.5 C, 0.7 Ni (anisotropic, after upsetting)	0.46	230 (326)[a]	33

a Intrinsic coercivity H_{cJ}

The carbon addition slows down the rate of formation of the τ phase and at the same time stabilizes it against decomposition into the equilibrium MnAl rhombohedral and β-Mn phases. The best magnetic properties have been obtained when the carbon concentration exceeds its solubility limit in the τ phase. This limit is about 0.4% at 70% manganese and 0.6% at 72% manganese. In the supersaturated condition, the coercivity increase developed by the extrusion appears to arise from the precipitation of Mn_3AlC carbides which restrict grain growth. The coercivity of MnAl is degraded by the presence of stacking faults. Such stacking faults are apparently suppressed by a fine grain size.

Other recent developments in the Mn–Al–C permanent magnet alloys include the following.

The addition of 2–4% copper to the Mn–Al–C ternary decreases the pressure required for warm extrusion, with the consequent advantage of prolonging die life. Further improvements have been claimed to result from adding a small amount of titanium (0.01–0.05%) to the quaternary.

The addition of a small amount of nickel to Mn–Al–C is said to raise the value of $(BH)_{max}$ to $64\,kJ\,m^{-3}$ with an increase in coercivity.

The upsetting (forging) of previously extruded cylinders of Mn–Al–C alloys at 660°C led to the development of a planar magnetic anisotropy with an easy magnetic plane in the plane of the upset disk, that is, normal to the previous extrusion direction. The properties are isotropic in the plane. They are shown in Table 1. One advantage of this type of processing is the preparation of magnets of larger area as compared with extruded material. On the other hand, the properties appear to be lower.

By quenching manganese–aluminum and Mn–Al–C alloys rapidly from the melt using a melt extraction technique, it is possible to prepare isotropic permanent magnets superior to the previously best isotropic materials. Optimum magnetic properties for two alloys are shown in Table 1. The binary alloy was aged at 450°C for 25 min while the ternary alloy was aged at 550°C for 5 min, after rapid quenching from the melt. There is a potential for preparing powdered manganese–aluminum or Mn–Al–C magnets using this process.

The Mn–Al–C alloys can be machined readily. The raw materials are relatively low in cost. The rather large intrinsic coercivity is an advantage. On the other hand, the Curie temperature is low (~ 300 °C) and the warm extrusion technique is expensive and presents limitations in size and shape of the magnets. However, the latter problem might be overcome with techniques such as upsetting, and new alloy development may ease the problem of warm working.

At present, Mn–Al–C magnets are in limited production for internal use and are not available commercially.

See also: Magnetic Materials: An Overview; Magnetic Materials: Basic Concepts.

Bibliography

Kojima S, Kojima K, Mitani S 1979 Permanent magnet Mn–Al–C alloy. US Patent No. 4,133,703 (9 January 1979)

Ohtani T, Kato N, Kojima S, Kojima K, Sakamoto Y, Konno I, Tsukahara M, Kubo T 1977 Magnetic properties of Mn–Al–C permanent magnet alloys. *IEEE Trans. Magn.* 13: 1328–30

Sakamoto Y, Ibata A, Kojima S, Ohtani T 1980 New MnAlC permanent magnets exhibiting macroscopically-plane magnetic anisotropy. *IEEE Trans. Magn.* 16: 1056–8

Willens R H 1980 Melt extracted MnAl and MnAlC. *IEEE Trans. Magn.* 16: 1059–61

Zijlstra H, Haanstra H B 1966 Evidence by Lorentz microscopy for magnetically active stacking faults in MnAl alloy. *J. Appl. Phys.* 37: 2853–6

G. Y. Chin[†]
[AT&T Bell Laboratories, Murray Hill, New Jersey, USA]

J. H. Wernick
[Bellcore, Red Bank, New Jersey, USA]

Measurements in Magnetic Materials

The investigation, understanding and use of magnetic materials requires the measurement of magnetic properties. Superconductors also possess important magnetic properties, as a consequence of the exclusion of magnetic flux from fully superconducting materials (the Meissner effect). This article describes the measurement of basic magnetic properties and of important engineering magnetic properties, emphasizing mainly techniques used at very low frequency (dc), power frequencies and audio frequencies (up to about 20 kHz).

1. Magnetization

The fundamental experimental quantity in magnetic materials is the magnetization, defined as the magnetic moment per unit volume. In SI units, magnetization is measured in $\mathrm{A\,m^2\,m^{-3} = A\,m^{-1}}$. Alternatively, magnetization can be measured in tesla ($\mathrm{Wb\,m^{-2}}$), in which case it is called magnetic polarization. In cgs (electromagnetic) units, using the convention that magnetic moment is measured in emu (electromagnetic units, $\mathrm{erg\,Oe^{-1}}$), magnetization is in $\mathrm{emu\,cm^{-3}}$ (see *Magnetic Units and Material Specification*). When the volume of the sample may change, as with temperature, it is convenient to measure the moment per unit mass rather than per unit volume; this quantity is usually given the symbol σ and is equal to the magnetization per unit volume multiplied by the density, in appropriate units. Values of σ are numerically equal in SI and cgs units.

Magnetization is in general a strongly nonlinear function of the applied field H and, in sufficiently high field, reaches a limiting value called the saturation magnetization, M_s. The saturation magnetization is temperature dependent, in general decreasing from a maximum value at 0 K to a zero value at the Curie temperature. At fixed temperature, magnetization vs field is both nonlinear and hysteretic; a plot of M vs H for a cycle of field from $+H \rightarrow 0 \rightarrow -H \rightarrow 0 \rightarrow +H$ is called a magnetic hysteresis loop and contains much of the information needed to characterize a magnetic material.

1.1 Force Methods

A magnetic moment m in a gradient field ∇H experiences a force $\boldsymbol{F} = m\nabla H$. A gradient field can be produced by tapering the pole pieces of an electromagnet; a sample is suspended in the gap of the magnet, where it is subject to an average field H and a field gradient ∇H. The sample is suspended from a force-measuring device, commonly a chemical balance or an electrobalance, and the measured force gives directly the magnetic moment if the field gradient is known (see Fig. 1). This method is variously known as the Faraday or the Curie method; the apparatus is often called a magnetic balance. It is capable of measuring small moments and, since the only connection to the sample is a support wire, the method is well suited to measurements at low and especially at high temperatures. It is often used for the determination of Curie temperatures (see Sect. 5) and for measurements of susceptibility (the slope of the M vs H curve) as a function of temperature. Using a balance with a sensitivity of 0.1 mg and a field gradient of $400\ \mathrm{kA\,m^{-2}}$ (typical for tapered pole pieces), the limiting sensitivity is about $10^{-6}\ \mathrm{A\,m^2}$ (10^{-3} emu). The system is usually calibrated by measuring a sample of known magnetic properties, frequently pure nickel. Accuracy is limited by the fact that it is difficult to make the field gradient uniform over a large volume and constant over a range of field and it is also difficult to measure the field gradient accurately. These problems are avoided if the sample is made large enough so that a uniform cylinder of sample extends from the point of maximum field to a region of essentially zero field. In this case (called the Gouy

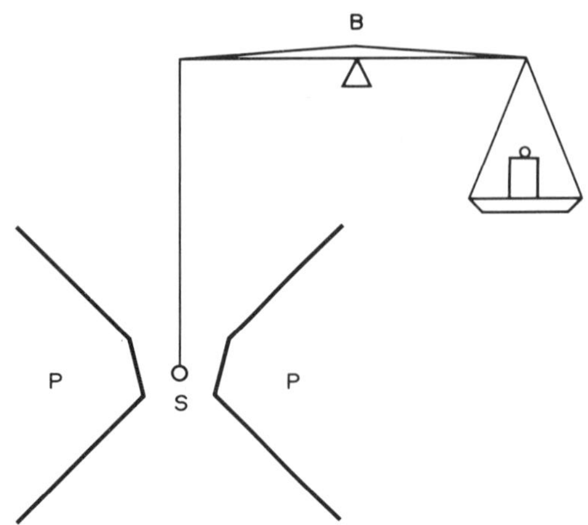

Figure 1
Magnetic balance: P, electromagnet pole piece with tapered gap; S, sample; and B, balance or other force-measuring device

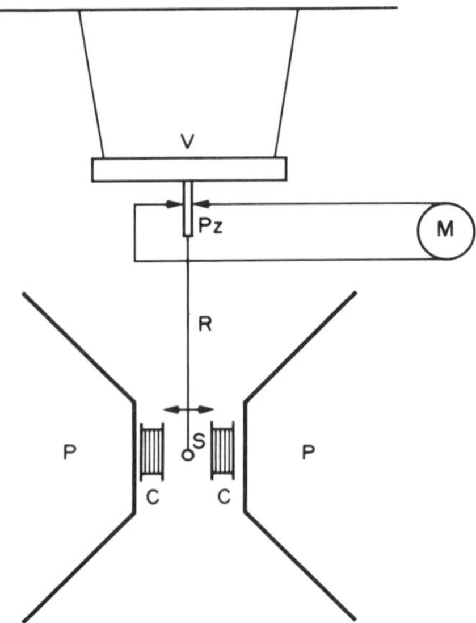

Figure 2
Alternating-gradient magnetometer: P, electromagnet pole piece; C, drive coil to create alternating-field gradient in horizontal direction; S, sample; R, support rod; Pz, piezoelectric crystal motion detector; V, vibration-damping support table; and M, meter (usually lock-in amplifier)

method), the force on the sample does not depend on the shape of the field gradient.

The field gradient in the arrangement of Fig. 1 vanishes as the field approaches zero, so this method is not useful for low field measurements. If water-cooled field gradient coils are used instead of tapered pole pieces, the field gradient becomes independent of the average field, can be varied or reversed and the volume of uniform gradient can be made larger. Furthermore, the gradient and therefore the sensitivity can be increased by about a factor of ten. However, the gradient coils require an increase in the magnet gap and, consequently, reduce the maximum average field.

If the gradient field is produced by coils in the magnet gap and is made ac instead of dc, the sample is subjected to an alternating force, which can be made either parallel or perpendicular to the dc field of the magnet. The displacement caused by this force can be detected by mounting the sample on a thin support arm which is attached to a piezoelectric crystal; the ac voltage output of the piezoelectric is detected and amplified by a lock-in amplifier (see Fig. 2). If the frequency of the gradient field is tuned to the mechanical resonant frequency of the support arm and the sample, the amplitude of the vibration is increased by the Q of the resonant system, which is of order 100. This device has been called the alternating-gradient magnetometer, and can have a sensitivity of 10^{-10} A m^2 (Flanders 1988, Frey et al. 1988); a commercial version is available for room-temperature use.

1.2 Induction Methods

If the magnetic flux density $B=\mu_0(H+M)$ ($B=H+4\pi M$ in cgs units) threading a coil of N turns and area A is changed by an amount dB, a voltage is generated in the coil of magnitude $E=-kNA(dB/dt)$, where k depends on the system of units in use. A device capable of integrating this voltage gives an output signal proportional to the change in B. Before electronic instrumentation was available, the integration was done by a ballistic galvanometer or by a related electromechanical instrument called a fluxmeter; with the availability of electronic instrumentation the integration is done by an operational amplifier circuit and the complete instrument is called an electronic fluxmeter or simply a fluxmeter. Note that the fluxmeter measures B, not M. For many engineering purposes, B is actually the more useful quantity and M can always be found from B if the field H is known. Alternatively, two identical coils can be connected in series opposition and placed in identical fields. One coil contains the sample and measures $(H+M)$; the other contains no sample and measures only H. The difference is then the magnetization M.

Note also that the fluxmeter measures only *changes* in B, so that in order to make a measurement, the flux density B threading the coil must be changed, either by changing H or by moving the sample relative to the coil. The sensitivity of the fluxmeter is limited by drift in the output, and is about 10^{-6} Wb turns (100 maxwell turns), corresponding to 0.01 mT (0.1 G) with a coil of 1000 turns and 1 cm^2 area. Fluxmeters are most often used when the sample is in a closed magnetic circuit, either in the form of a complete ring or as a part of a ring with the rest of the magnetic path comprising a yoke of high-permeability magnetic material. In the latter case, the measurement of the field presents certain difficulties (see Sect. 3).

Instead of electronic analog integration of the induction voltage, the voltage signal can be digitized, stored and digitally integrated. When the samples are large and strongly magnetic, either method will serve; however, analog integration is better for small or weakly magnetic samples.

If the sample is magnetized with an ac field and the induction voltage is simply amplified rather than integrated, the ac incremental susceptibility ($\Delta M/\Delta H$) is measured directly. This method is limited to relatively low fields (since high fields cannot be cycled rapidly), but is useful for determining Curie temperatures of ferromagnetic and ferrimagnetic materials and the temperature dependence of susceptibility of weakly magnetic materials and superconductors. A lock-in amplifier is usually used to detect the ac induction signal at the drive frequency; it can be used to give both the real and imaginary parts of the susceptibility. A dc field superimposed on the ac drive field can give additional useful information in measurements on superconducting samples (see *Measurements in Superconducting Materials*).

1.3 Measurement of Field Produced by Sample

A magnetized sample not in a closed magnetic circuit produces a field in the surrounding space. The magnitude of this field at any point is proportional to the magnetization of the sample. Various methods have been used to measure this field; the principal difficulty is to separate the field produced by the sample from the applied field causing the magnetization of the sample. One method is to choose a measuring point where the field from the sample has a large component perpendicular to the applied field and measure the perpendicular field, usually with a Hall-effect sensor. Since the applied field is usually large compared with the sample field, careful orientation of the field sensor is required.

The most common solution is to oscillate or vibrate the sample at a fixed frequency. The sample field is then ac, not dc, and can be easily separated from the dc applied field. The instrument based on this principle is the vibrating-sample magnetometer, or VSM (also sometimes called a Foner magnetometer after its inventor). It is widely used for measurements on small samples at high fields. The sample is usually oscillated perpendicular to the applied field, either by a loudspeaker cone or by a mechanical cam (see Fig. 3). In the former case the frequency is about 100 Hz and the amplitude is about 0.1 mm; in the latter case the frequency is about 20 Hz and the amplitude about 2 mm. The loudspeaker system needs some kind of feedback control to maintain the vibration amplitude constant and, generally, is limited to samples of less than about 100 mg. In both cases, the ac field produced by the sample motion is sensed by symmetrical sets of coils placed in the gap of the magnet and the signal is amplified by a lock-in amplifier locked to the frequency of sample vibration. Both types of VSM are available commercially, with accessories to permit measurements from 4 K to about 1000 K.

The limiting sensitivity of the VSM is usually quoted as 10^{-8} A m^2 (10^{-5} emu); the practical limit in normal operation is about ten times larger.

1.4 SQUID Measurements

The highest available sensitivity for the measurement of magnetization is provided by the SQUID (superconducting quantum interference device) magnetometer (see *SQUID Design, Fabrication and Application*). The SQUID itself, incorporating a superconducting Josephson junction, usually operates in liquid helium, so a SQUID magnetometer is often used in conjunction with a superconducting magnet. This means that applied fields below a few hundred amperes per meter are not easily available. The sample is passed slowly through a set of coils wound from superconducting material, which are connected to the SQUID. The total flux passing through the coils is proportional to the SQUID output voltage. The field must be accurately constant during the measurement and, in practice, several minutes elapse between measurements if either the field or the temperature is changed. A complete hysteresis loop or a temperature sweep from 4 K to 300 K then requires several hours and commercial systems are set up for automatic unattended operation. The claimed limiting sensitivity is 10^{-12} A m^2 (10^{-9} emu); as in the case of the VSM, a practical limit is about a factor of ten higher. SQUID systems are beginning to become available with superconducting junctions based on high-temperature oxide superconductors which can operate at liquid nitrogen temperatures (77 K). The limiting sensitivity at 77 K is less than in liquid helium (4 K). A detailed discussion is given in the article *SQUID Design, Fabrication and Application.*

2. Measurement of Magnetic Field

The standard general-purpose field-measuring instrument uses the Hall effect in a small plate of a semiconducting material. Usually known as a gaussmeter, this instrument can measure fields from levels comparable with the earth's field ($\sim$ 10 A m^{-1}) up to at least 5 MA m^{-1} and often 15 MA m^{-1} with accuracies of about 1%. The sensing element can be as small as 1 mm^2 and both dc and ac fields can be measured. Special Hall elements are made for use at cryogenic temperatures and can be repeatedly cycled to low temperatures without damage (Rubin and Sample 1980).

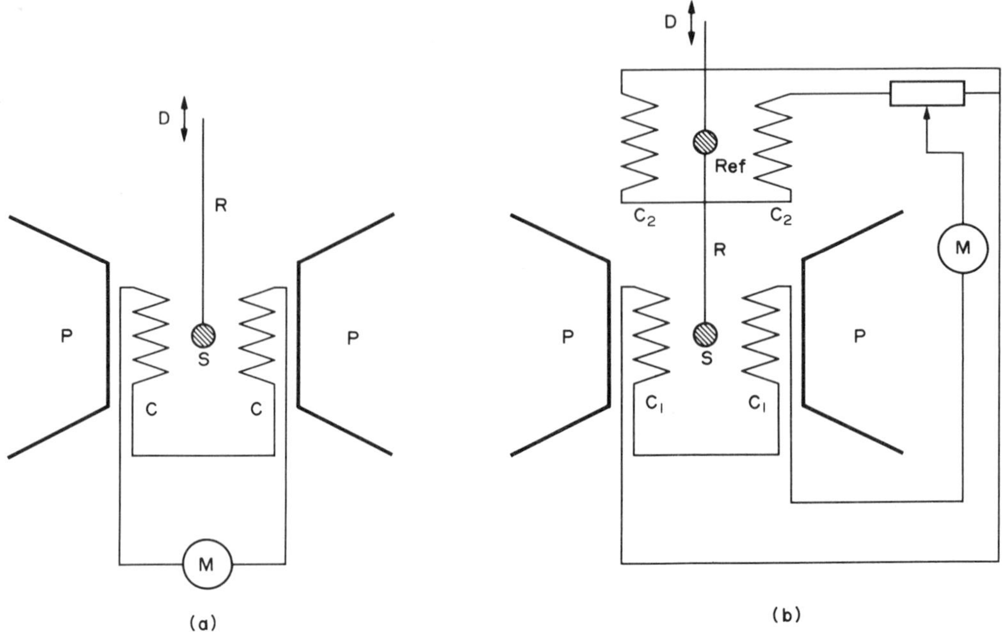

Figure 3
(a) VSM and (b) alternate version in which the signal from sample S is balanced against a signal from a reference sample (Ref) of fixed magnetic moment: P, electromagnet pole piece; R, sample support rod; C, sensing or pickup coils (orientation and location may vary); D, sample motion (mechanism not shown); and M, meter (usually lock-in amplifier)

A simple copper coil connected to a fluxmeter can be used to measure changes in the flux through the coil, which is equivalent to the average field in the coil. Because of drift in the integrator output, this system is useful primarily when the field is cycled fairly rapidly, as in the measurement of hysteresis loops. However, if the signal is large (a big coil with many turns in a strong field), drift may not be a problem. Coils of special geometries can be used to minimize the errors due to fields that are nonuniform over dimensions comparable with the size of the coil (Zijlstra 1967 Pt. 2, pp. 3–7). If the coil is oscillated or rotated in the field, an ac voltage is generated whose magnitude is proportional to the field. Commercial instruments called rotating-coil gaussmeters have been built using this principle, but are less convenient to use than Hall-effect instruments.

The resonance frequency of various atomic nuclei depends linearly on the applied magnetic field, with a proportionality constant known to high accuracy. The measured resonance frequency then establishes the value of the applied field (see *Nuclear Magnetic Resonance Imaging*). For successful nuclear magnetic resonance (NMR) measurements, the field must be highly uniform both in space and time (although in modern instruments the size of the probe has been much reduced); consequently this method is used mainly to calibrate other sensors.

For fields lower than the earth's field, the most common field-measuring instrument is the fluxgate magnetometer, which measures the applied dc field required to make the ac hysteresis loop of a high-permeability magnetic strip exactly symmetric (see *Sensors and Devices, Magnetic*). At this point, the net dc field on the material is zero, so the applied dc field is equal and opposite to the (preexisting) field to be measured. Fields of this level are rarely used in the measurement of magnetic materials.

3. Demagnetizing Fields

As noted, any magnetized sample not in a closed magnetic circuit creates a field in the surrounding space. This field exists within the sample itself, as well as in the surrounding space, and inside the sample is always opposite in direction to the sample magnetization. It is therefore called the demagnetizing field. This is a perfectly real field, not a mathematical abstraction, and is often far from negligible. Proper and accurate correction for the demagnetizing field is a persistent problem in many magnetic measurements.

If the sample is an ellipsoid and is uniformly magnetized, the demagnetizing field is uniform throughout the sample and is linearly related to the magnetization of the sample by a proportionality constant known as the demagnetizing factor (D in SI, N in cgs), which depends on the sample shape. The demagnetizing factor can be calculated from formulae given in textbooks, and tabulated values and graphs are also widely available (see *Demagnetizing Factors*). The maximum value of the demagnetizing field, for an infinite plate magnetized perpendicular to its surface, is unity in SI or 4π in cgs. The maximum possible demagnetizing field is therefore numerically equal to M_s (SI) or $4\pi M_s$ (cgs); this corresponds to over 1.6 MA m^{-1} (20 kOe) for iron samples.

Ellipsoidal samples are difficult to make and to mount in experimental equipment and are not used as often as they should be. The demagnetizing factor parallel to the axis of a rod decreases with increasing length/diameter and, for a disk sample along a diameter, decreases with increasing diameter/thickness, so long thin rods or thin disks are often used as samples with the demagnetizing factor assumed equal to that of the equivalent ellipsoid. The "equivalent" ellipsoid is usually taken to be the ellipsoid of the same aspect ratio as the sample, although this is not strictly correct.

If the demagnetizing correction needs to be made with high accuracy, two general methods are available. Since the tangential fields just outside and just inside the surface of a sample are equal, a direct measurement of the parallel field at the surface of the sample gives the true field acting inside the sample: that is, the applied field minus the demagnetizing field. This field can be measured with a Hall probe, or with a coil and fluxmeter. Since the field cannot be measured exactly at the sample surface, and is usually changing rapidly with distance from the surface, it may be necessary to measure at several points, each a known distance from the surface, and extrapolate to the surface. Of course, the demagnetizing field may not be uniform within the sample.

Alternatively, use can be made of the fact that for a high-permeability, low-coercive-field material, the slope of the magnetization curve at low fields is governed almost entirely by the demagnetizing field. In such a material, the magnetization adjusts itself to a value such that the demagnetizing field is equal and opposite to the applied field, making the true internal field zero. Then the slope of the curve of magnetization vs applied field at low fields is numerically equal to the reciprocal of the demagnetizing factor. The demagnetizing factor thus determined for a high-permeability sample should apply to other samples of the same shape measured in the same apparatus.

To eliminate the demagnetizing field, a homogeneous ring sample can be used. The field is applied by a coil wound onto the ring and can be calculated exactly from the circumference of the ring, the number of turns and the current. Note that the field is higher at the inside diameter than at the outside diameter, because of the difference in circumferences; for minimum uncertainty in field, the radial thickness of the ring should be small compared with the ring diameter. The maximum field is limited by heat generated in the windings to about 8 kA m^{-1} (100 Oe). In a composite ring sample, comprising a sample to be measured clamped in a high-permeability yoke, the yoke material ideally has much higher permeability and cross-sectional area than the sample, so that all the ampere turns of magnetomotive force act on the sample and the field can be calculated from the known ampere turns and the length of the sample. In practice, this is often an inadequate approximation and one of the experimental methods described previously must be used to determine the true applied field.

4. *Image Effect*

When measurements are made in the gap of an electromagnet, the presence of large volumes of high-permeability magnetic material in the pole pieces acts to distort the field surrounding the sample. In the limiting case of semi-infinite slabs of uniformly permeable material, the effect can be described mathematically by magnetic "images" of the sample existing in the pole pieces at the mirror image positions. At high fields, the material of the pole pieces approaches magnetic saturation, the permeability decreases towards unity and the images tend to disappear. The calibration constant relating the magnetization of the sample to the output signal of the magnetometer then changes with field. This usually results in an apparent decrease in magnetization with increasing field at high fields; this phenomenon is what is usually meant by the image effect. The image effect may be minimized by increasing the magnet gap, at the cost of lower maximum field, or simply by limiting the maximum field to values where the image effect remains sensibly constant. Placing the magnetization sensor as far as possible from the pole pieces is also helpful, but is often incompatible with placing a furnace or low-temperature vessel in the magnet gap.

A field-dependent calibration can be made using a sample whose magnetization is independent of field (or varies with field in a known way) over a range of fields at high field. A better solution in principle is to use a copper coil carrying a known current, which acts like a cylindrical sample of fixed magnetic moment. However, the magnetic moment produced by such a coil is small compared with the moment of a strongly magnetized sample of the same size and shape.

5. *Curie Temperature*

The Curie temperature is the temperature at which a ferromagnetic or ferrimagnetic material loses its magnetization and becomes paramagnetic. The equivalent temperature in an antiferromagnet is called the Néel temperature. An approximate value for the Curie temperature can be obtained by measuring the magnetization (by any method) as a function of temperature. However, in high fields the transition from ferromagnetic to paramagnetic behaviour is often not sharply defined, so that either a low-field measurement or an extrapolation to low field is necessary. The low-field susceptibility (0.1–10 A m^{-1}) often shows a maximum at or near the Curie temperature; this is known as the Hopkinson effect (Hopkinson 1889) and is strongly dependent on the sample microstructure.

6. *Magnetocrystalline Anisotropy*

Magnetocrystalline anisotropy describes the fact that the energy of a magnetic crystal depends on the direction of the saturation magnetization relative to the crystallographic axes. Other magnetic anisotropies, due to shape, stress and so on, can also exist. Accurate measurement of the crystal anisotropy requires a single-crystal sample, although some information can be extracted from measurements on an aligned powder sample or a textured polycrystal. Anisotropy is expressed as a series of trigonometric terms, each multiplied by a constant called an anisotropy constant (see *Anisotropy in Magnetic Materials*).

The most satisfactory experimental method for the determination of anisotropy is the measurement of the torque acting on an ellipsoidal (or disk-shaped) sample as the sample is rotated about its axis of symmetry in a saturating applied field perpendicular to this axis. An instrument made for this purpose is called a torque magnetometer or simply a torquemeter. The shape of the torque curve reflects the crystallographic symmetry of the sample and the magnitude of the torque is related to the magnitude of the anisotropy. A difficulty is that the direction of the magnetization relative to the field and the crystal axes is not known directly, but must be deduced from the value of the saturation magnetization (assumed constant at high fields) and the measured torque. Generally, the apparent anisotropy increases slightly with field and some kind of extrapolation is required. Details of the method and the mathematics that apply to various crystal structures and to different crystallographic planes in the plane of the disk sample are given by Chikazumi (1964), Cullity (1972) and Pearson (1979).

A variety of torque-measuring methods can be used (Pearson 1979). If the samples and torques are large, rigid systems with deflections measured by strain gauges can be used. Small torques associated with small samples or small values of the anisotropy require sensitive torque-measuring systems using fine filaments. The wide range of torque values that may be encountered makes the development of commercial instruments difficult and most instruments are purpose built.

If a magnetic field produces a torque on a disk-shaped sample, there must be a component of magnetization perpendicular to the applied field and a measurement of this perpendicular magnetization is equivalent to the measurement of the torque. Various schemes for the measurement of perpendicular magnetization have been used (Pearson 1979); they can have the advantage of being able to be continuously read, which is convenient for following the changes in anisotropy with temperature or field. They have the disadvantage that the measured signal decreases with increasing field, so that extrapolation to high fields is somewhat uncertain.

Values of the anisotropy constants can also be obtained by measuring the magnetization curves (M vs H) in various crystallographic directions, and comparing the energy required to reach magnetic saturation. Physically, this is not strictly equivalent to rotating the saturation magnetization from one crystallographic direction to another but, in practice, the difference is not usually significant.

The dependence of the ferromagnetic resonance frequency or field on crystallographic direction can also be interpreted to give anisotropy constants.

Some rare-earth elements and compounds have very strong anisotropy, so that fields higher than those available from laboratory electromagnets ($\sim$1.6 MA m^{-1} or 20 kOe) are necessary for their measurement. Superconducting magnets or Bitter magnets (see Sect. 8) may be used, except their usual configuration makes torque measurements difficult. Very high pulsed fields with pulse widths of many milliseconds can be produced by discharging a large capacitor bank through a reinforced coil; equipment of this kind is made commercially for magnetizing permanent magnets and can also be used with appropriate instrumentation for some kinds of high-field measurements.

7. *Magnetostriction*

Magnetostriction describes the change in dimensions of a magnetic sample with changes in the state or direction of magnetization. From an engineering point of view, the quantity of interest is the change in length of a sample when its magnetization is changed, typically from plus to minus saturation or near saturation. Scientifically, the measured values are the changes in dimension of a single crystal as the saturation magnetization is rotated to various crystallographic directions. The experimental considerations are similar in both cases (see *Magnetoelastic Phenomena*; *Magnetostrictive Materials*; Lee 1979).

Magnetostrictive strains are generally small, of the

order of 10^{-5}. If the available samples are large enough, electric resistance strain gauges provide the simplest measurement technique. They can be driven with either ac or dc currents, and the manufacturer's calibration constants are sufficient for accuracy of about 1%. Most strain gauge alloys show significant magnetoresistance, which must be balanced with a dummy gauge not attached to the sample but in the same field, or otherwise corrected for.

Linear variable differential transformers (LVDTs) are position sensors with sufficient sensitivity for magnetostriction measurements and are useful for ribbon or wire samples that are too thin or too narrow for strain gages. LVDTs have the advantage that the measuring instrument can be held at room temperature when the sample is heated or cooled. Very small samples may be measured using capacitance methods, where changes in the length of the sample vary the spacing of a pair of capacitor plates.

If long samples (1 m) are to be measured, as on steel transformer sheets, commercial phonograph pickups have sufficient sensitivity and are easy and convenient to use.

8. *Use of Superconducting Solenoids or Bitter Magnets*

The traditional sources of magnetic fields are simple copper coils for low fields (0.01 T, 8 kA m^{-1} or 100 Oe) and iron-core electromagnets for high fields (2 T, 1.6 MA m^{-1} or 20 kOe). Much higher fields are attainable in water-cooled reinforced copper coils (Bitter magnets, available at only a few locations in the world), high-field superconducting solenoids or pulsed coils. Superconducting magnets are commercially available in the range 7–20 T (see *Superconducting Magnets*). Bitter magnets can reach 25 T and Bitter magnets combined with superconducting magnets ("hybrid magnets") can reach 35 T. (Practical superconducting magnets currently require operation in liquid helium although the application of high-temperature superconductors in small coils has already been demonstrated.) There are no fundamental differences between magnetic measurements in electromagnets and in solenoids or coils, beyond the obvious fact that access to the sample and any sample motion are generally parallel to the applied field in a solenoid and perpendicular to the field in an electromagnet. This geometry makes torque measurements difficult, but not impossible. (Both superconducting and water-cooled solenoids can be built as split coils to permit radial or transverse access to the highest-field region, at the cost of lower maximum field.) Pulsed fields may require special measurement techniques, especially if the pulse time is very short.

The fact that superconductors exclude magnetic flux at low fields but not completely at high fields gives rise to an image effect in superconducting magnets analogous to that in electromagnets.

9. *Magnetic Measurements on Superconductors*

Superconductors in low fields exclude all magnetic flux and, therefore, have a relative susceptibility of -1 in SI units or a susceptibility of $-1/4\pi$ in cgs units. Magnetic measurements on superconducting samples can provide useful information on critical fields, flux trapping and pinning, and so on (see *Measurements in Superconducting Materials*; Goldfarb and Spomer 1990). All the basic measurement techniques for magnetic materials can be used for superconducting materials, remembering that the magnetization is frequently negative (opposite to the applied field) and therefore the demagnetizing field is positive (in the same sense as the applied field).

As noted previously, ac susceptibility measurements are useful in deducing the properties of inhomogeneous superconducting samples. For measurements on superconducting cables, it may be desirable to have substantial transport current flowing through the sample during the measurement of magnetic properties; this makes the VSM and force methods generally unsatisfactory, and some system in which the sample remains stationary must be used.

10. *Soft Magnetic Materials*

Sheet or strip materials for use in power-frequency devices such as transformers and motors are usually specified primarily on the basis of power loss, or the input electrical energy lost as heat when the material is magnetically cycled through a symmetrical hysteresis loop of specified maximum flux density and frequency. The standard test procedure for loss measurements has been the same for many years: a group of identical rectangular strips is assembled into a square frame which is then operated as a transformer under no-load conditions. The power loss in the transformer is measured, usually with a sensitive and accurate wattmeter, with a correction applied (if necessary) for the Joule heating in the windings. This is known as the Epstein test; the results are quoted in watts per kilogram (or watts per pound) at a specified maximum flux density, usually 1 T, 1.5 T or 1.8 T (10 kG, 15 kG or 18 kG), and a specified frequency, normally 50 Hz or 60 Hz. The output voltage waveform of the Epstein frame should be sinusoidal, which means that the flux change with time in the material is sinusoidal; this condition requires a very nonsinusoidal input current, a condition that is increasingly difficult to meet as the maximum flux density is increased.

Because the Epstein test uses a relatively large amount of material, and assembling the test frame is a slow manual operation, there is a strong tendency to supplement or replace the Epstein test with some form of single-strip test, using a single strip of test material and a high-permeability yoke. However, no generally accepted single-strip test has been developed.

Sometimes, the iron loss (or core loss) is "separated" into hysteresis loss, measured at very low frequency or "dc," and assumed to be independent of frequency; and eddy current loss, assumed to vary as the square of the frequency and inversely as the electrical resistivity. If the measured hysteresis loss plus the calculated eddy current loss is less than the measured total ac loss (as is always the case), the difference is called the anomalous loss. This simple linear separation of losses is almost entirely without physical meaning (see *Power Loss in Magnetic Materials*).

Another common specification for high-quality transformer material is the peak flux density at an applied field of 800 A m^{-1} (10 Oe). This is normally measured at 50 Hz or 60 Hz and is a measure of the crystallographic orientation of the steel; higher values are better.

For many applications, either the initial permeability μ_i or the maximum permeability μ_{max} is important; permeability is defined as B/H. Other tests are used on high-permeability materials that are more or less directly related to their intended application as relays, high-frequency transformers, shielding materials, and so on.

11. Permanent Magnet Materials

Permanent magnets are always used to create a magnetic field in space and so always operate in their own demagnetizing field. Their behavior in the second quadrant of the hysteresis loop ($H < 0$, $B > 0$) is therefore of primary interest. Since high fields are required for measurements, ring samples cannot be used; samples are either clamped in the gap of an electromagnet, which operates as a yoke, or are measured as isolated samples in a VSM or similar device, with a correction for the demagnetizing field.

The common measure of quality of a permanent magnet is the maximum energy product $(BH)_{max}$, which is the maximum value of the product of the flux density B multiplied by the (true) field H, measured in the second quadrant of the hysteresis loop and taken as a positive value. The units are joules per cubic meter in SI (gauss oersteds, usually MG Oe, in cgs). The volume of permanent magnet material required to produce a specified level of field in a specified volume of space is inversely proportional to the maximum energy product (if the geometry of the magnet is correctly chosen).

In permanent magnet materials, the distinction between the ordinary coercive field (or coercive force) and the intrinsic coercive field becomes important. The ordinary coercive field H_c is the (negative) field required to reduce the flux density B from positive saturation to zero; the intrinsic coercive field H_{ci} is the field required to reduce the magnetization M to zero. In a quality modern permanent magnet, M remains almost constant at about 10^6 A m^{-1} (1000 emu cm^3) in negative fields up to 2 MA m^{-1} (24 kOe) or more. Under these circumstances, H_{ci} is twice as large as H_c.

The remanent induction B_r at $H = 0$ is another important quantity for permanent magnet materials. For some applications, the reversible and irreversible changes in magnetic properties with temperature need to be measured, usually in the temperature range from room temperature to 150 °C.

See also: Barkhausen Effect; Magnetic Materials: An Overview; Magnetic Units and Material Specification; Measurements in Superconducting Materials; Power Loss in Magnetic Materials; Sensors and Devices, Magnetic

Bibliography

American Society for Testing and Materials 1964 *Magnetic Testing—Theory and Nomenclature*. ASTM Technical Publication 371. ASTM, Philadelphia, PA

American Society for Testing and Materials 1970 Direct Current Measurements of Soft Magnetic Materials. ASTM Technical Publication 371 S1. ASTM, Philadelphia, PA

American Society for Testing and Materials 1991 *Annual Book of ASTM Standards*, Pt. 44. ASTM, Philadelphia, PA

Chikazumi S 1964 *Physics of Magnetism*. Wiley, New York, pp. 15–37; 128–47

Cullity B D 1972 *Introduction to Magnetic Materials*. Addison-Wesley, Reading, MA, Chaps. 1,2; pp. 215–33

Flanders P J 1988 An alternating gradient magnetometer. *J. Appl. Phys.* 63: 3940–6

Foner S 1967 Special magnetic measurement techniques. *J. Appl. Phys.* 38: 1510–19

Frey Th, Jantz W, Stibal R 1988 Compensated vibrating reed magnetometer. *J. Appl. Phys.* 64: 6002–8

Goldfarb R, Spomer R L 1990 Magnetic characteristics and measurements of filamentary Nb–Ti wire for the superconducting supercollider. *Adv. Cryog. Eng. (Mater.)* 36: 215–22

Hopkinson J 1889 Magnetisation of iron at high temperature. *Proc. R. Soc. (London)* 45: 318–21

Lee E W 1979 Magnetostriction. In: Kalvius G M, Tebble Robert S (eds.) 1979 *Experimental Magnetism*. Wiley, New York, pp. 137–222

McGuire T R, Flanders P J 1969 Direct current magnetic measurements. In: Berkowitz A I, Kneller E (eds.) 1969 *Magnetism and Metallurgy*. Academic Press, New York

Oguey H J 1969 Alternating current magnetic measurements. In: Berkowitz A I, Kneller E (eds.) 1969 *Magnetism and Metallurgy*. Academic Press, New York

Pearson R F 1979 Magnetic anisotropy. In: Kalvius G M, Tebble Robert S (eds.) 1979 *Experimental Magnetism*. Wiley, New York, pp. 225–65

Rubin L G, Sample H H 1980 Hall effect magnetometers for high magnetic fields and temperatures between 1.5 and 300 K. In: Chien C L, Westgate C R (eds.) 1980 *The Hall Effect and its Applications*. Plenum, New York, p. 463

Zijlstra H 1967 *Experimental Methods in Magnetism*. North-Holland–Wiley, New York

C. D. Graham Jr
[University of Pennsylvania, Philadelphia, Pennsylvania, USA]

Measurements in Superconducting Materials

Materials may be in the form of single crystals, polycrystalline pellets or cast samples, thin films or commercial wires. The characterization of the superconducting properties of a material involves a number of different measurements; the types of measurement appropriate for a particular material depend on the size, shape and homogeneity of the sample available. Most materials exhibit some inhomogeneity and some, for instance granular materials (see *Granular Superconductors*) or materials with high critical current densities (see *Critical Current Density*), have intrinsic inhomogeneity. The characterization of inhomogeneous materials is complex; even the simplest of measurements require careful interpretation.

Emphasis is placed on techniques for the characterization of extreme type II superconducting materials (see *Magnetic Properties of Superconductors*). Two types of measurement can be used to derive a number of different parameters. The first is a resistive measurement in which a current is passed through a sample and the voltage measured. The second is an inductive measurement which gives information on the magnetic moment of the sample. This article begins by describing these two techniques and the precautions to be taken if reliable results are to be obtained. The measurement of individual material parameters is then described.

1. Resistive Measurements

In a resistive measurement a current is passed through the sample and the voltage between two points in the current path measured. There are two separate regimes. In the first regime, very low currents are used to measure the critical temperature T_c. For this purpose, the geometry is not very important and the main errors arise from thermoelectric effects and poor contacts. Thermoelectric effects can be eliminated by chopping or reversing the input current to check the voltage zero. More sensitive measurements can be made by using an ac drive current and a lock-in amplifier. Inductive pickup can be recognized because it has a different phase and frequency dependence from the resistive component and it can then be balanced out if necessary. In the early days of high-T_c oxide materials, many anomalous and irreproducible results were announced. These have been attributed, not entirely convincingly, to cracking or microscopic voltaic cells. An unambiguous value for the resistivity in these circumstances can best be obtained by measuring all permutations of voltage and current leads. The details are described by van der Pauw (1961) and Montgomery (1971). The technique is laborious and cannot be used for transient effects. When a material is inhomogeneous, the resistive transition is a measure of the first percolative superconducting path, the fraction of superconducting material at the transition depends on the dimensionality of the sample and a comparison of the resistive and inductive transitions can give quantitative information on the nature of the sample inhomogeneity (Evetts 1983).

The second regime is the high-current regime used for measuring critical current densities. Since low voltages are involved, thermoelectric effects are also important, but the main difficulty is the heating of the sample by the transport current and the tendency of flux flow to start at the current contacts and spread to the rest of the sample (see *Resistive Transition and Flux Flow in Superconducting Materials*). Because of the self-field of the current, the most meaningful measurements are carried out in an external field, applied perpendicular to the current, which is much larger thant the self-field of the current. Measurements in parallel fields are of considerable interest but most applications involve perpendicular fields (see *Transport Currents and Force-Free Current Flow*). To ensure that flux flow starts well away from the contacts the center section should be narrowed; alternatively, the center section is sometimes heated, but this makes the temperature of the measurement uncertain. For measurements in a solenoid, wire samples are usually bent into a U shape so that the contacts can be out of the high-field region and flux flow starts at the central section of the wire where the field is transverse to the sample.

2. Inductive Measurements

2.1 DC Magnetization

The magnetization curve gives a great deal of information about a superconducting material. Most standard methods can be used but some are more suitable than others (see *Measurements in Magnetic Materials*). In extreme type II materials the large value of the Ginzburg–Landau parameter κ poses particular problems. At low applied fields, the magnetization curve contains detail up to a few millitesla with a relatively high susceptibility, while the high-field end of the curve can involve very small magnetizations at external fields of up to 100 T. No measurement system is ideal for both.

The principle is nearly always to change the flux in a search coil due to the sample magnetization; for three geometries the signal can be directly related to the magnetization whatever the current distribution in the sample. If the sample is in a long solenoidal coil or if it is small compared with the distance to the search coil windings, the signal is directly proportional to the moment of the sample, whatever its shape. In addition, if the sample is a long cylinder of arbitrary cross section placed inside two opposing coils, the signal when it is moved is proportional to the magnetization per unit length. These geometries involve loose inductive coupling between sample and search coil, which tends to increase spurious signals and noise, and they are not those of optimum sensitivity, for which the coils should closely follow the specimen outline.

For a close-coupled geometry, mutual inductance tables can be used or the system can be calibrated with a bulk type I supeconductor. Although the coupling factor changes somewhat as currents penetrate a type II superconductor, this is not a very large correction and a diamagnetic calibration sample is generally adequate. For bulk samples of simple geometry, the system is self-calibrating since below the lower critical field H_{c1} the sample can be assumed to be perfectly diamagnetic, although the actual initial slope of the magnetization curve will depend on the demagnetizing factor for the sample (see *Demagnetizing Factors*).

A number of methods are used to produce an inductive signal. The commonest is probably to move the sample out of the search coils or between two coils. The displacement should be kept to a minimum since, if the magnetization curve is hysteretic, small variations in the external field can take a sample from one branch of the hysteresis curve to the other (see *Critical State Model*). Movement into a region of different field will trap the new field and bring flux back into the coils. A change of a few millitesla is sufficient to give spurious results if hysteresis is low, so that a small sample displacement at a higher frequency is preferred, as in a vibrating sample magnetometer (VSM). SQUID systems, although sensitive, use large displacements and extreme field homogeneity is needed if the sample is hysteretic.

It is possible to keep the sample stationary and integrate the signal as the applied magnet field is swept. This is not very sensitive over a large field range since it is necessary to balance out of the large signal due to the applied field. The field distribution from a superconducting solenoid depends on both the field and history so that the balance is not constant.

Demagnetizing the sample by heating it above T_c is effective, but time consuming, particularly for high-T_c materials, and it is important that the heating does not move the search coils in the magnet field. The measurement of changes in magnetization caused by flux creep (see *Flux Creep Phenomena*) present fewer problems since the change can be picked up directly as a voltage in the search coils. The signal is measured by an integrator which tends to drift, but use of a SQUID detector will avoid this problem. However, most flux creep experiments must be made in a fairly large external field so that flux creep in the magnet is the dominant error. Commercial instruments have poor coupling between sample and search coils and magnet creep is a major limitation.

Techniques which measure the force, or torque, on a sample have not been used much but are sensitive and direct, although delicate. They have been of considerable value in measurements on high-T_c materials (Hagen and Griessen 1989, Hamzic et al. 1990). The Faraday balance as normally supplied is not suitable as it puts a large and variable field gradient across the sample. However, by putting the sample in a uniform main field and adding a small gradient field with external coils which can be chopped and detected with a lock-in amplifier, a sensitive magnetometer can be made which does not move the sample significantly. The gradient coils should not move the flux vortices in the sample by more than the pinning interaction distance for linear displacement, as will be discussed later.

Torque measurements have the same advantages as a Faraday balance in that it is not necessary to move the sample position in the magnet field. Torque magnetometers cannot measure the reversible component of the magnetization, but they can measure anisotropy and flux vortex pinning forces, although the geometry can make quantitative interpretation difficult, particularly if both are present.

2.2 AC Techniques

The response of a superconductor to an alternating field allows a number of parameters to be measured. Many of the same considerations apply as to dc measurements but ac techniques, particularly with the addition of a lock-in amplifier, can be extremely convenient and sensitive. The simplest configuration is to measure the change in impedance of a coil containing the sample. This may be small and greater sensitivity can be obtained by looking for the third harmonic, since a superconductor can be distinguished from other magnetic materials by the generation of a strong third-harmonic signal.

However, most methods use a mutual inductance so that the pickup coil voltage is independent of the much larger drive coil voltage. For quantitative measurements the drive coil should provide a uniform field over the sample and the pickup coils should be such that the voltage is easily related to the magnetization, as for dc magnetic measurements. It is generally necessary to use a balancing network to cancel out unwanted signals due to mutual inductance in the leads, capacitive effects and ground loops. The system can be balanced either when the sample is normal or when it is superconducting, according to the measurement. If it is balanced when normal, the balance need only be set once since it is independent of sample size. However, if it is desired to observe small penetrations of flux in a mainly superconducting sample, it is necessary to balance in the fully superconducting state.

The apparatus is generally a compromise between conflicting requirements. Making the system small reduces temperature differences, but also reduces sensitivity. Any metals used to reduce temperature gradients carry eddy currents which produce spurious signals. These out-of-balance signals tend to be the main limitation rather than noise or sensitivity. They arise from dimensional changes in the probe, from conventional soldered joints, from thermometer cans too close to the coils and from eddy currents in cans round the system and the superconducting magnet. This last factor is very nonlinear at high drive amplitudes and there is also a danger of quenching the magnet if it is close to its maximum field. To minimize external eddy currents the drive coil can be wound as two coils of the same number of area turns excited in opposition, although this reduces the available amplitude. A technique to limit the effects of thermal expansion is to wind drive coils on quartz tubes and put the search coils outside them. The signal is then independent of changes to the search coil dimensions and the drive coils can be wound to be more nearly identical since they are of thicker wire (Cooper et al. 1990).

The signal from the search coils can be treated in a number of ways depending on the parameter being investigated. These are discussed in Sect.3.

3. *Individual Parameters*

The parameters described in this section may not have a single value in a material because of material anisotropy (see *Anisotropy in Superconducting Materials*) or because of inhomogeneity. Thus, high-T_c sintered ceramic materials can be treated as composites of two quite different superconductors with two separate sets of parameters characteristic of the grains and grain boundaries (see *Granular Superconductors*). In other materials, different parameters can vary continuously as, for instance, in a proximity-coupled composite structure.

3.1 Critical Temperature

This section describes measurements of T_c in zero magnetic field.

(*a*) *Resistive measurements.* In a homogeneous superconductor the resistance goes to zero over a few millikelvin and provides an accurate measurement of T_c. The conventional definition is the temperature to reach half the normal state resistivity. What is actually measured is the temperature at which the critical current is equal to the measuring current, so a check should be made that the curve is independent of the measuring current. This is normally the case for currents of a few milliamperes. A broad transition or a finite resistance at low temperatures is indicative of inhomogeneity, although curvature at the high-temperature end of the transition may be due to fluctuations. If the proportion of superconductor is less than the percolation threshold, the relative reduction in resistance gives an indication of the proportion of superconductor present. Although, in principle, a network of fine superconducting threads could give zero resistance, this seems to occur rather rarely; grain boundary phases are usually less superconducting than the bulk, although a notable exception is seen in dilute solutions of iron in titanium (see *Superconducting Materials: An Overview*).

(*b*) *Inductive measurements.* As the sample goes through T_c it becomes diamagnetic and the size of the signal at any temperature depends on the ratio of the magnetic penetration depth λ at that temperature to the diameter of the sample. The half-way point of the transition is when λ is approximately equal to the radius of the sample. This point will be close to the peak in the imaginary part of the susceptibility or in the third harmonic if the amplitude is large enough. Small samples will therefore have wide transitions even if they are homogeneous. The amplitude dependence must be checked since small applied fields can induce current densities greater than the critical current density.

The induced signal is proportional to the product of the amplitude and the frequency and greater sensitivity can be obtained at high frequencies; up to several megahertz is used for oxide superconductors. However, it is important that the normal state resistivity is such that the skin depth is much greater than the sample size; this limits the frequency to less than 1 kHz for bulk metallic samples. The dc magnetization measured by first cooling in zero field and then raising the temperature in a small applied field gives the same measurement as the inductive techniques extrapolated to zero frequency.

(*c*) *Specific heat anomaly.* This is a very reliable method of measuring T_c since it is independent of the shape of any inhomogeneous regions and measures a bulk property. However, it is time consuming and needs large samples and so is not very suitable for standard testing.

(*d*) *The effect of magnetic fields.* This is considered in more detail in Sect. 3.7. In general, the transition is broadened. This may be due to a greater inhomogeneity in the upper critical field H_{c2} than in T_c. However, there are other reasons for broadening even in homogeneous samples. For a type II superconductor in the mixed state, zero resistance is only achieved by the pinning of vortices (see *Flux Pinning and the Summation of Pinning Forces*). If the pinning energy is less than k_BT, thermally activated flux flow gives a linear resistivity and the ac susceptibility will be slightly paramagnetic with a value $1/2\kappa^2$. The inductive transition only appears at the irreversibility line where thermal activation stops (see Sect. 3.14). In conventional superconductors this may be very close to T_c but in oxide superconductors the two temperatures may be separated by tens of degrees. Even if the transition is close to T_c the inductive signal in the mixed state is determined by the pinning penetration depth λ_{pin}, an effective penetration depth determined by the pinning centers (see Sect. 3.10). In reversible materials this can be more than 100 μm; even quite massive specimens may be of a size comparable to λ_{pin} with a correspondingly broad transition to less than perfect diamagnetism. The detailed interpretation of the inductive transition in an applied field is still the source of some controversy.

3.2 Superconducting Fraction

The term superconducting fraction is generally taken to mean the proportion of a fully diamagnetic signal measured either in an inductive transition or in a sample warmed in a small field (i.e., less than H_{c1} at low temperatures). It will give the proportion of superconductor present only in the following rather unusual circumstances. The superconducting phase must consist of particles that are large compared with the magnetic penetration depth which do not form an interconnected structure. Large samples with a connected network of superconductor will give perfect diamagnetism even with very large volume fractions of normal material. Conversely, homogeneous superconductors can give a very small diamagnetism if the particle size is smaller than the magnetic penetration depth and the grains are decoupled. Since newly prepared materials are often well away from the optimum composition with a correspondingly large penetration depth, it is easy to assume inhomogeneity where none exists. Interpretation of the inductive signal should be made in conjunction with resistive measurements to make sure they are consistent. A large diamagnetic signal is almost certainly indicative of the presence of a superconductor, although care must be taken in measurements using bridge systems to make sure that it is not a decrease in paramagnetism which is being observed. A discussion is presented by Campbell et al. (1991).

3.3 Meissner Fraction

The Meissner effect is the expulsion of flux which occurs when a superconductor is cooled in a field (see *Magnetic Properties of Superconductors*). Although sometimes quoted as the best test of a superconductor, it is in fact a complex process and many good superconductors have almost no Meissner effect. In a type II superconductor, significant flux begins to move towards the nearest surface when the temperature is low enough for the external field to be below H_{c1}. The extent to which it can get out is determined by flux pinning, or the critical current density J_c, at this temperature. The Meissner fraction is the magnetic moment expressed as a fraction of the full diamagnetic moment; it is a complex combination of the applied field, the sample size and the strength of the pinning and its variation with temperature. Little useful information can be obtained from it (Campbell et al. 1991).

3.4 Magnetic Penetration Depth

Different techniques have been used to measure the magnetic penetration depth λ according to the nature of the samples available. If the material is available as small particles which can be sized, the susceptibility as a function of particle size can be used to give λ. Alternatively, the size distribution of a random sized sample can be measured and the susceptibility as a function of temperature can be fitted to give the best value of λ (Cooper et al. 1989). A similar technique is the measurement of the imaginary part of the microwave impedance. The impedance of a thin-film strip line of the material has also been used (Fiory et al. 1988, Drabeck et al. 1989). The properties of the vortex state near H_{c1} and the value of H_{c1} are directly related to λ. If hysteresis is small, the shape of the reversible magnetization curve at fairly low fields can be expressed in terms of λ, although paramagnetic contributions make this difficult in some oxide materials (Mitra et al. 1989). The rms field fluctuations of the vortices can be directly related to λ and values obtained from positron spin relaxation have been widely quoted (Harshman et al. 1989). There are, however, other components to the rms field fluctuations caused by pinning effects which affect the value of λ derived. Finally, a value for λ can be calculated from a knowledge of H_{c1} and κ using the Ginzburg–Landau theory.

Problems arise in all methods that use the theory of flux vortices if the particle size is comparable with the penetration depth since, in these circumstances, the apparent values of H_{c1} and the magnetization curve are affected by the proximity of the sample surface. Conventional superconductors have small penetration depths and can usually be treated as large samples. However, the oxide superconductors can have large penetration depths, particularly when off stoichiometry, and are often only available as small independent grains, so that the situation when λ is comparable with the particle size frequently occurs.

3.5 Thermodynamic Critical Field

The thermodynamic critical field H_c is the field at which the magnetic energy equals the condensation energy; it is therefore isotropic. In a type I superconductor H_c can be found directly from the magnetization curve but in a type II superconductor the only direct measurement is from the specific heat. A review of specific heat measurements can be found in Junod (1989). The thermodynamic critical field can also be found indirectly from measurements of H_{c1} and H_{c2} using the Ginzburg–Landau theory.

3.6 Lower Critical Field

The lower critical field H_{c1} is the field at which flux vortices first penetrate a long superconducting cylinder. It is difficult to measure because pinning effects prevent flux entry until considerably larger fields. Pinning can occur both due to bulk effects and surface effects. The latter pose a particular problem since there is no significant penetration until the surface barrier is overcome. Bulk penetration leads to a detectable curvature of the magnetization curve above H_{c1} and, if J_c is proportional to $1/B$, extrapolation of $(H-M)^{1/2}$ to zero should give H_{c1}. A more sensitive technique is to use the ac susceptibility either as a function of signal amplitude or external field. Both real and imaginary parts increase rapidly when flux first penetrates. Although bulk penetration may be at fields considerably higher than H_{c1} due to surface pinning effects, losses and penetration can be detected below H_{c1} because of field concentrations at asperities. In general, the point of first penetration of flux is much less well defined than is commonly suggested.

Surface barriers can be removed by diffusing an element into the surface which progressively depresses T_c as the concentration increases, so that the surface is spread out over several λ. This has only been effective in pure niobium using oxygen and in PbTl using thallium (Evetts 1970). Without this treatment flux first penetrates the surface at about 1.5 times H_{c1} in long cylinders. The penetration of other shapes depends on demagnetizing effects (see *Demagnetizing Factors*) and has not been investigated very thoroughly.

An estimate of the effect of pinning can be obtained from the low-field magnetic hysteresis curve. If the barrier is symmetric for the entry and exit of flux, H_{c1} will be half way between the peak of the increasing curve and the peak of the decreasing curve. A better measurement can be made by fitting the shape of the curve as calculated by Abrikosov in fields above H_{c1} where it is reasonably reversible and extrapolating back to the irreversible region. Bulk pinning effects are reduced if the sample size is reduced, so that fitting curves is easier for small samples. However, they must not be too small, since for particles smaller than the penetration depth the field at which the first vortex enters is determined by the particle size rather than H_{c1}.

3.7 Upper Critical Field and Ginzburg–Landau Coherence Length

The upper critical field H_{c2} is the field at which the vortex cores overlap and bulk superconductivity disappears. It is therefore directly related to the Ginzburg–Landau coherence length ξ. Two problems arise: first, the fields involved can be very large and, second, the transition is second order so that if κ is large the change in properties is very small. In high-T_c oxides fluctuations smooth out the discontinuity making it difficult to define H_{c2} precisely. The reversible magnetization is linear near H_{c2} so extrapolation of the magnetization curve to zero magnetization is a good way of determining H_{c2}. Since the slope is $1/(2\kappa^2-1)$, the change in slope may be difficult to detect, especially if there is a paramagnetic moment. A clearer transition for large κ is generally seen if the field is kept constant and the temperature varied. Extrapolation of the linear section to meet the normal state susceptibility gives the best value for H_{c2} (Welp et al. 1989).

Resistive measurements may be the only possible technique at very high fields which are invariably pulsed so that the measurement time is limited but, as pointed out previously, this measures the irreversibility line rather than H_{c2}. Another complication with resistive measurements is that resistanceless surface currents can be carried up to H_{c3}, the critical field for the nucleation of surface superconductivity. The currents are not large on rough surfaces and can be eliminated by plating with a normal metal. However, this does mean that a resistive transition may be rather uncertain and current dependent.

3.8 Surface Nucleation Field

The boundary condition on the order parameter at a vacuum or insulating surface allows surface superconductivity to exist in fields parallel to the surface up to $1.7H_{c2}$. Except in the case of very-low-κ material, the magnetization is not detectable but the resistive transition in a field parallel to a smooth surface can be used to measure H_{c3}. The possibility of thermal activation of the point vortices involved leading to an irreversibility line between H_{c2} and H_{c3} has not been investigated.

3.9 Resistive Transition and Flux Flow Resistivity

The movement of a flux vortex is lossy, even without pinning, and it experiences a viscous drag, just like flux in a normal metal. The resulting linear resistivity is called the flux flow resistivity ρ_f and, according to the theory of Bardeen and Stephen, decreases approximately linearly from the normal state value at H_{c2} to zero at zero field (see *Resistive Transition and Flux Flow in Superconducting Materials*). There is no satisfactory theory for the voltage–current (VI) characteristics of materials with pinning but the commonest assumption is that $E=\rho_f(J-J_c)$ (i.e., the flux flow resistivity is given by the slope of the VI characteristic well above J_c).

A second technique is to use low-amplitude oscillations which do not unpin the flux lines and measure the frequency response. In principle, this should be a better technique but experimental measurements are few and it is not clear to what extent these ac measurements agree with dc measurements. There are two possible regimes, separated by the pinning frequency above which viscous forces dominate pinning forces (Gilchrist and Monceau 1968). This requires the skin depth to be smaller than the pinning penetration depth λ_{pin} and losses due to normal electrons must be considered. At lower frequencies a model using a linear restoring force from the pinning wells plus a viscous force due to the flux flow resistivity can be used (Gittleman and Rosenblum 1966). Measurements of the real and imaginary parts of the susceptibility give values for the two parameters.

3.10 Pinning Penetration Depth

In the mixed state the flux line is pinned in a pinning well which gives a linear restoring force for small displacements. The result is that an ac signal decays exponentially from the surface and penetrates a distance λ_{pin} (Campbell and Evetts 1972). This depth is considerably larger than the usual magnetic penetration depth or London penetration depth and can be measured by measuring the ac susceptibility as a function of external dc field. (In a bridge balanced below T_c in zero field the signal is proportional to $\lambda_{pin}-\lambda$.) For independent pinning centers this would give the curvature of the foot of the pinning well but collective effects make the interpretation in terms of the microstructure complex. The signal remains linear until the vortex displacement is sufficient to unpin flux lines at the pinning interaction distance (see Sect. 3.11). The critical state model cannot be applied in samples that have a dimension comparable to λ_{pin} in the direction of penetration of the critical state front (see *Critical State Model*).

3.11 Pinning Interaction Distance

The pinning interaction distance d_{pin} is the distance flux lines can move before they become unpinned. It is typically about a tenth of the vortex spacing (Campbell 1971, Campbell and Evetts 1972). The vortex movement at the surface can be derived directly from the susceptibility since the voltage in the search coils is proportional to the flux change in the sample. If the vortices at the sample surface move a distance y when the applied flux density is B, the flux change is By per unit length of surface (see *AC Losses in Superconducting Materials*). Note that there is an upper limit to the movement of the magnetic lines of force which applies even in the normal state. For a cylindrical sample of radius a and an oscillating field of amplitude b, the amplitude of the flux vortex movement at the sample surface is $ab/2B$. If this is less than about a tenth of the vortex spacing, the response of the system will always be linear and the critical state model does not apply (see *Critical State Model*).

An alternative technique for investigating the displacement of the flux lattice for an applied force less than the critical depinning force depends on measuring the voltage in a resistive measurement for an ac current with peak amplitude below the critical current. The force–displacement response for pinned vortices has been measured using this technique by McLean and results are given by Evetts (1983). The initial slope of the force–displacement response α is related to d_{pin} and λ_{pin} by the relations

$$\alpha = \frac{BJ_c}{d_{pin}} = \frac{1}{\mu_0}\left(\frac{B}{\lambda_{pin}}\right)^2$$

3.12 Critical Current Density

Values of critical density J_c can be measured either by direct transport currents or inductively. Zero-field critical currents can only be a rough guide to the quality of a material since they are affected by the field generated by the transport current; J_c is said to be self-field limited.

(*a*) *Direct measurements.* These use the same four-terminal technique as resistive transitions but with much higher currents. Heating at the current contacts is a major problem and spurious results are common unless care is taken over the design of current contacts and the location of the voltage contacts. If a normal conducting sheath is present, current transfer to the superconductor may take place over a length long enough to overlap the voltage contacts, giving a finite voltage below J_c. In sintered materials it is important to know whether the voltage contacts span grain boundaries or are so close that they only measure the single-grain critical current. For practical materials it is usually necessary to carry out the experiments in a liquid cryogen, since gaseous surroundings do not remove heat fast enough. Heating problems should be suspected if the critical current densities become independent of magnetic field or show hysteresis in the current applied although variations in internal trapped flux can cause hysteresis in J_c in granular superconductors. One way of overcoming heating problems is to pulse the applied current. However, there is a limit to how short the pulse can be which is set by the need for flux to penetrate to the center of the wire before a true resistive voltage can be seen. This only occurs when the skin depth calculated for the pulse duration is greater than the sample size. The conductivity used to calculate the skin depth should be the average for the moving flux and can be amplitude and frequency dependent.

In homogeneous bulk samples well below the irreversibility line the critical current density is not very dependent on the criterion used (see *Critical Current Density*). For instrumental convenience 1 μV is the most usual criterion. In typical samples this corresponds to a field of about $1\,\mu V\,cm^{-1}$. However, the limitations of patterning thin films mean that the same voltage criterion in measurements on thin-film samples often implies electric fields two or three orders of magnitude higher. For magnet applications where wires may be kilometers in length a resistivity criterion may be more appropriate. In inhomogeneous material such as sintered oxides and multifilamentary superconductors, the choice of a lower electric field criterion can lower the critical current density by a very large factor. In these materials the VI characteristic can be approximated by a power law over several decades of voltage; such materials are often specified in terms of an n value, where n is an empirical power derived by the best fit of the VI characteristic to the expression $V \propto I^n$. For magnet applications which require extremely low electric fields in composite conductors the value of n is very important for design purposes; in such conductors the n value may be related directly to the spatial inhomogeneity in J_c (Plummer and Evetts 1987).

(*b*) *Inductive techniques.* A number of techniques derive J_c by analysis of the signal from a search coil of the same type as is used to measure inductive transitions. To obtain valid results the vortex movement must be sufficient to build up a critical state (i.e., $b \gg (\Phi_0 B)^{1/2}/a$ where a is the radius of the sample and Φ_0 is the flux quantum). The nonlinearity caused by flux pinning can be used in a number of ways to determine the critical current density.

If the critical current density is assumed constant, the Bean critical state model can be used to predict the waveform, and hence the third harmonic, or the real and imaginary parts of the susceptibility in terms of J_c. The value can then be chosen to give the best fit to the results (Bean 1964). This technique can be extended by assuming a variation of J_c with B such as $J_c \propto 1/B$ and fitting the parameters to the signal at various amplitudes. If viscous forces are to be ignored, the frequency must be such that the skin depth obtained from the flux flow resistivity is much greater than the sample size. Frequencies between 100 Hz and 1000 Hz are commonly used. Parameters to describe the difference between intergrain and intragrain currents in oxide superconductors can also be fitted (Müller and Pauza 1989, Takacs and Gömöry 1989).

Another group of measurements measures the flux in the sample directly (Campbell 1969, Rollins et al. 1974). This can be done either by integrating the signal in a computer or from the signal from a lock-in amplifier which gives the flux directly. This requires a broader band detector than the harmonic analysis, so increasing the noise, but it has the major advantage that the total flux in the sample is much easier to interpret and it is not necessary to fit a number of parameters in complex expressions which may be based on invalid assumptions. A comparison of the techniques has been made by Kroeger et al. (1975).

For an external field considerably greater than the amplitude of the ac drive field it can generally be assumed that J_c is a constant. Differentiation of the signal from the pickup coils with respect to the amplitude then gives the profile of J_c across the sample. This is particularly useful in distinguishing surface pinning from bulk pinning. In zero applied field a similar analysis yields the critical current density as a function of B for much lower values of B than can be achieved in transport measurements because of the self-field of the current. The technique can be extended to give the complete force–displacement curve for the pinned lattice (Campbell 1971, Campbell and Evetts 1972).

A problem with high-T_c materials is that it is difficult to find out the effective size of the region in which currents are flowing. The most convincing way is to break the sample into smaller sizes to see if the magnetization changes, but this is destructive and may change the properties. It has recently been pointed out that if the demagnetizing factor is large, the slope of the differential susceptibility can be used to give the size of the current loops and this appears to be a useful technique (Angadi et al. 1991). High-T_c materials are also anisotropic and it must be appreciated that inductive techniques will always involve currents flowing in more than one direction; if J_c varies with the direction of current flow a quantitative analysis is difficult.

(*c*) *DC hysteresis.* The critical current density can be calculated from the hysteresis in the dc magnetization using the critical state model (see *Critical State Model*). The method is less sensitive than techniques based on measurement of the ac susceptibility but since the dc magnetization should always be measured it is convenient to derive J_c from such measurements. The same problems arise in deciding on the effective size of the sample in the case of granular high-T_c materials; an unambiguous conclusion can only be obtained from a series of measurements where the sample is progressively reduced in size.

3.13 Flux Creep

Flux creep is generally accepted to be a result of the thermal activation of flux lines past pinning centers (see *Flux Creep Phenomena*). It leads to a logarithmic decay of the magnetization if the activation energy U is much greater than $k_B T$ and to complete reversibility and zero critical current if U is comparable with or less than $k_B T$. The first measurements were made by measuring the decay of a field trapped in a hollow cylinder by measuring the voltage of a coil inside it (Kim et al. 1962, Beasley et al. 1969). Alternatively, the coil can be wound round the outside of a solid specimen. To obtain meaningful values it is important that the critical state has penetrated to the center of the sample. Any technique which measures magnetization can be used to measure flux creep by measuring the magnetization as a function of time, although the most sensitive techniques measure just the change in magnetization.

Measurements are easiest in zero applied field but it is important to know how activation energies vary with field. Since the magnet may creep at a similar rate to the specimen, it is not easy to distinguish between the two. In addition, small changes in magnet field, caused either by magnet overshoot during ramping to the measurement field or inhomogeneity over the region where the sample is moved, affect the flux creep rate at the sample surface. Torque magnetometry has proved a successful technique, largely eliminating problems of magnet overshoot and inhomogeneity. Simple theory predicts that $\Delta M/M = (k_B T/U)\ln(t)$, where M is the magnetization and t is time, so that what is derived is an effective activation energy (see *Flux Creep Phenomena*). Statistical effects make this difficult to interpret in any simple way but it gives a good guide as to the creep rate in magnet applications.

3.14 The Irreversibility Line

The term irreversibility line $B_{irr}(T)$ is a new parameter introduced for the oxide superconductors. When the activation energy for flux creep is comparable to $k_B T$, pinning is ineffective and the magnetization of the sample becomes completely reversible (see *Critical Current Density*). The resistivity is linear for small currents and although it is less than the flux flow resistivity the material is useless for practical purposes. No diamagnetism is observed, apart from a skin effect at high frequencies, and the low-frequency ac susceptibility is slightly paramagnetic. The irreversibility line is not precisely defined and can be taken as the point where the magnetization curve becomes reversible, the point of zero resistance, the start of the diamagnetic transition or the peak in the lossy component of the susceptibility. All these occur at similar points but research is continuing into their relationship. Of these techniques, the least reliable is the closing of the dc hysteresis curve since this is ill defined and can be obscured by inhomogeneity in the magnet. However, looking for distortion in the response function of a SQUID magnetometer can give good results. It can be seen that most methods of detecting the superconducting transition in a magnetic field detect the irreversibility line and not H_{c2}.

The existence of such a line should not have been unexpected since the energy wells which pin flux lines must reach a depth comparable with $k_B T$ before they go to zero at H_{c2}. However, the deep wells and low T_c values encountered in conventional superconductors meant that this occurred very close to H_{c2} and the phenomenon was not observed or looked for. In high-T_c materials the energy wells are smaller relative to $k_B T$; the irreversibility line is easily distinguished from H_{c2} and can be a major problem in practical applications. In sintered high-T_c materials there are two such lines, one for the intergrain currents and one for the intragrain currents.

3.15 AC Losses

Measurements of ac losses are of two types: calorimetric and electromagnetic.

(*a*) *Calorimetric methods*. In one technique the heat generated by losses in the superconductor is measured by the increase in boil-off of the cryogenic coolant. This gives an unambiguous loss measurement but for accurate results it is necessary for the loss to be considerably greater than the background boil-off. Even with a calorimeter round the sample to collect the gas this may be difficult to achieve because of the boil-off due to conduction, particularly if electrical leads are attached to the sample. The background boil-off is usually dependent on the level of cryogen present and the cooling due to increased boil-off can change the conditions markedly. The most reliable technique is to use a null method in which the power to a resistor adjacent to the sample is reduced in such a way as to keep the boil-off constant. The reduction in power to the resistor is then equal to the dissipation in the superconductor.

This technique is used for large volumes of material and particularly for magnet coils, where the large inductive voltage makes electrical measurements inaccurate. It is generally limited to temperatures for which a liquid cryogen can be used, although in principle the rise in temperature of a gaseous coolant might be used. The ultimate sensitivity is about 1 mW.

Another technique is to thermally isolate the sample and measure the rate of temperature rise. This has not been used much as temperature gradients and the continously changing temperature make the results difficult to analyze. However, a modification of this method in which the sample is connected to the helium bath by a thermal resistance and the temperature rise is measured in the steady state can give good results. Schmidt and Specht (1990) have used the technique to measure losses with a resolution of 10 nW. This method is not suitable if losses in the presence of a transport current are required because of the heat leak down the current leads.

(*b*) *Electrical methods*. These are convenient and sensitive provided there is only one source of power, either a current in a wire or a magnetization loss. If the power is coming from two sources such as a solenoid and a transport current supply, it is necessary to measure both magnetization and longitudinal voltage. This is true even if one of them is not oscillating since a wire carrying an ac current in an oscillating magnetic field will have a voltage induced along its length with a consequent power loss from the supply, leading to a dynamic resistance (see *AC Losses in Superconducting Materials*). There can be considerable problems in balancing out inductive signals from the two sources and a criterion for balance based on the expected shape of the waveform may have to be used. Calorimetric methods are probably more reliable in this situation.

If the losses are purely magnetic, the sample is placed in an oscillating field and the magnetic response is measured by one of the techniques described previously. It is then possible to measure the area of a hysteresis loop on the M–H curve. However, a more sensitive method is to measure the imaginary part of the susceptibility. For this technique, it is only the out-of-phase component of the fundamental which gives the loss, so a tuned filter is needed to eliminate harmonics. The main problem is in balancing out signals which are not due to the sample. It is important to do the balancing in such a way that there is not a large in-phase component since, if this is present, a very small error in setting the phase leads to a large error in the loss. Therefore, if the penetration of the signal is small, the bridge should be balanced in the superconducting state and the phase set in the normal state. If the penetration is large, the bridge should be balanced in the normal state and the phase set from the signal in the superconducting state. This works for losses in a field when the losses in zero field can be assumed negligible and this state is used for calibration and phase setting. For zero-field losses it is helpful to be able to move the sample out of the coils to set the phase. A sign that the balance is correct if the loss is hysteretic is the absence of a linear signal at low amplitudes. The lossy component of the susceptibility due to hysteresis is always proportional to at least the square of the drive signal since the sample must be driven round a Rayleigh loop.

For samples carrying a current in zero applied field the voltage in phase with the current again gives an unambiguous measure of the loss. Although voltages are smaller than for susceptibility measurements, careful design can lead to very small inductive signals for straight samples and the method is very sensitive. However, for solenoids, the large inductive signal will require a very accurate setting of the phase for reliable results.

See also: Critical Current Density; Critical State Model; Magnetic Properties of Superconductors; Measurements in Magnetic Materials

Bibliography

Angadi M A, Caplin A D, Laverty J R, Shen Z X 1991 Nondestructive determination of the current-carrying length scale in superconducting crystals and thin films. *Physica C* 179: 479–87

Bean C P 1964 Magnetisation of high field superconductors. *Rev. Mod. Phys.* 36: 31–9

Beasley M R, Labusch R, Webb W W 1969 Flux creep in type II superconductors. *Phys. Rev.* 181: 682–700

Campbell A M 1969 The response of pinned flux vortices to low frequency fields. *J. Phys. C* 2: 1492–1501

Campbell A M 1971 The interaction distance between flux lines and pinning centres. *J. Phys C* 4: 3186–98

Campbell A M, Blunt F J, Johnson J D, Freeman P A 1991 Quantitative determination of percentage superconductor in a new compound. *Cryogenics* 31: 732–7

Campbell A M Evetts J E 1972 Flux vortices and transport currents in type II superconductors. *Adv. Phys.* 21: 199–428

Cooper J R, Chu C T, Zhou L W, Dunn B, Grüner G 1989 Determination of the magnetic field penetration depth in superconducting YBaCuO: deviations from the Bardeen Cooper Schrieffer laws. *Phys. Rev. B* 37: 638–41

Cooper J R, Forro L, Collin G, Henry J Y 1990 T^2 behaviour in the low temperature penetration depth of oxygen depleted YBaCuO crystals. *Solid State Commun.* 75: 737–41

Drabeck L, Carini J P, Grüner G, Hylton J, Char K, Beasley M R 1989 Power law temperature dependence of the electrodynamic properties in oriented $YBa_2Cu_3O_{7-\delta}$ and $Y_2Ba_4Cu_8O_{16-\delta}$ films. *Phys. Rev. B* 39: 785–8

Evetts J E 1970 Flux pinning by phase boundaries in type II superconductors. *Phys. Rev. B* 2: 95–6

Evetts J E 1983 The characterization of superconducting materials: conflicts and correlations. *IEEE Trans. Magn.* 19: 1109–19

Fiory A T, Hebard A R, Mankiewich P M, Howard R E 1988 Renormalisation of the mean-field superconducting penetration depth in epitaxial YBaCuO films. *Phys. Rev. Lett.* 61: 1419–22

Gilchrist J le G, Monceau P 1968 The pinning frequency of type II superconductors. *Philos. Mag.* 18: 237–50

Gittleman J I, Rosenblum B 1966 Radio frequency measurements for subcritical currents. *Phys. Rev. Lett.* 16: 734–7

Hagen C W, Griessen R 1989 Distribution of activation energies for thermally activated flux motion in high T_c superconductors: an inversion scheme. *Phys. Rev. Lett.* 62: 2857–60

Hamzic A, Fruchter L, Campbell I A 1990 Non-activated magnetic relaxation in a high T_c superconductor. *Nature* 345: 515–16

Harshman D R, Schneemeyer L F, Waszczak J V, Aeppli G, Cava R J, Batlogg B, Rupp L W, Ansaldo E J, Williams D L 1989 Magnetic penetration depth in single-crystal $YBa_2Cu_3O_{7-\delta}$. *Phys. Rev. B* 39: 851–4

Junod A 1989 Specific heat of high-temperature superconductors: a review. In: Ginsberg D M (ed.) 1989 *Physical Properties of High Temperature Superconductors*, Vol. 2. World Scientific, Singapore, pp. 13–120

Kim Y B, Hempstead C E, Strnad A R 1962 Critical persistent currents in hard superconductors. *Phys. Rev. Lett.* 9: 306–9

Kroeger D M, Koch C C, Charlesworth J P 1975 A comparison of methods for measuring flux gradients in type II superconductors. *J. Low Temp. Phys.* 19: 493–512

Mitra S, Chu J H, Lee W C, Johnston D C, Kogan V G 1989 Magnetic field penetration depth of polycrystalline $(Y,Cd)Ba_2Cu_3O_7$, grain aligned YBaCuO and single crystal $Bi_2Sr_2CaCu_2O_8$. *Phys. Rev. B* 40: 2674–7

Montgomery H C 1971 Methods for measuring electrical resistivity of anisotropic materials. *J. Appl. Phys.* 42: 2971–5

Müller K H, Pauza A J 1989 Susceptibility of high temperature superconductors in a critical state model. *Physica C* 159: 717–26

Plummer C J G, Evetts J E 1987 Dependence of the shape of the resistive transition on composite inhomogeneity in multifilamentary wires. *IEEE Trans. Magn.* 23: 1179–82

Rollins R W, Küpfer H, Gey W 1974 Magnetic field profiles in type II superconductors with pinning using a new ac technique. *J. Appl. Phys.* 45: 5392–8

Schmidt C, Specht E 1990 AC loss measurements on superconductors in the microwatt range. *Rev. Sci. Instrum.* 61: 988–92

Takacs S, Gömöry F 1989 Study of flux flow in high T_c superconductors. *Physica C* 162: 681–2 van der Pauw L J 1961 Determination of resistivity tensor and Hall tensor of anisotropic conductors. *Philips Res. Rep.* 16: 187–95

Welp U, Kwok W K, Crabtree G W, Vandervoort K G, Liu J Z 1989 Magnetic measurements of the upper critical field of YBaCuO single crystals. *Phys. Rev. Lett.* 62: 1908–11

A. M. Campbell
[University of Cambridge, Cambridge, UK]

Metallic Glasses

Metallic glasses are metal alloys that have no long-range atomic order. There are two ways of achieving such materials. A suitable alloy can be quenched from the melt so rapidly ($\sim 10^6\,K\,s^{-1}$) that the random liquidlike structure is retained in the solid. A second method involves deposition by sputtering or evaporation. Because of the lack of long-range ordering, the bulk magnetic crystal anisotropy averages to zero, resulting in low magnetic losses and high magnetic permeability. These are features that characterize "soft" magnetic alloys.

The age of metallic magnetic glasses started with the prediction by Gubanov in 1960 that amorphous solids would be ferromagnetic in contrast to the previously held view that they would be paramagnetic. However, it was not until 1965 that the existence of weak ferromagnetic behavior in amorphous alloys of cobalt–gold was demonstrated. In 1967 the strongly ferromagnetic properties of $Fe_{75}P_{15}C_{10}$ were discovered. This is a typical soft magnetic alloy with high saturation magnetization and low coercivity. It was made by a splat-cooling technique in which individual drops of molten alloy were cooled between two piston faces. Finally, in 1969 the continuous casting of amorphous ribbons was reported leading to the realization of the full technological potential of these alloys.

There have been a number of reviews covering the general field of rapidly quenched metals, but also some covering specifically the amorphous magnetic alloys. Useful reviews in the literature include those by Alben et al. (1977), Graham and Egami (1978), Luborsky (1980a) and O'Handley (1987). Several books on amorphous metallic alloys are also available (Anantharaman 1983, Hasegawa 1983, Kaneyoshi 1983, Luborsky 1983, Moorjani and Coey 1984).

In this article, only those alloys produced by continuous casting from the melt will be considered. It is these materials that currently show the largest level of application and on which the greatest degree of research and development has been expended.

Alloys of interest include two classes: the transition metal–metalloid type, such as (Fe,Co,Ni)–(B,Si,C,P,...), and the transition metal–Zr, Hf type. It is the former type that shows the greater potential for application.

1. Magnetic Moments and Curie Temperatures

Amorphous magnetic alloys are best described by the itinerant electron model of ferromagnetism (see *Magnetic Materials: Basic Concepts*). These alloys are characterized by their density-of-states curves and the effective interaction between the itinerant electrons. A central feature of this theory is that the rigid band model is not generally applicable. The material is assumed to have a unique and uniform density of states with the alloying influencing only the relative occupation of the available energy levels. More sophisticated calculations consider coherent potentials on different atomic sites and local environment effects. In this way local density of states can be obtained and it is these which determine the magnetic properties of alloys. For the amorphous state the calculations are more difficult because they depend on models involving the disorder. A significant feature of the ferromagnetism in amorphous alloys is the possibility of noncollinear moments. This arises from the competition between the exchange force, which ranges from strong ferromagnetic to antiferromagnetic with the spread in local topological order, and the local crystal anisotropy. The correlation length can be of the order of the interatomic spacing, and this has consequences for technical parameters such as magnetostriction. This topic has been reviewed by Gibbs (1991). The evidence that the magnetic properties of these amorphous alloys obey the itinerant electron model may be summarized as follows:

(a) The saturation magnetization M_s follows typical Slater–Pauling plots as a function of outer electron concentration with a systematic shift along the abscissa which depends on the ratio of the metalloid to the transition metal. These shifts are most simply discussed in terms of charge transfer from the metalloid to the metal atoms. Measurements of M_s for iron-based alloys with boron, carbon, silicon, germanium, aluminum and phosphorus show an approximate trend in terms of the effective valency of the added metalloid atoms, thought to be also the result of charge transfer. Typical results are shown in Fig. 1.

(b) The Curie temperature T_C has also been shown to have a simple variation with outer electron concentration for both amorphous and crystalline alloys. The results on the same alloys as in the preceding paragraph can be discussed semiquantitatively in terms of the charge transfer, local environment and atom size, which are all concepts in the itinerant electron model. A phenomenological model has been used to correlate the compositional dependence of T_C. Typical results are shown in Fig. 2.

(c) The spin-wave stiffness D, defined by the ratio of the spin-wave energy at small wave vectors by $E(k) = Dk^2 + ...$, may be calculated from the

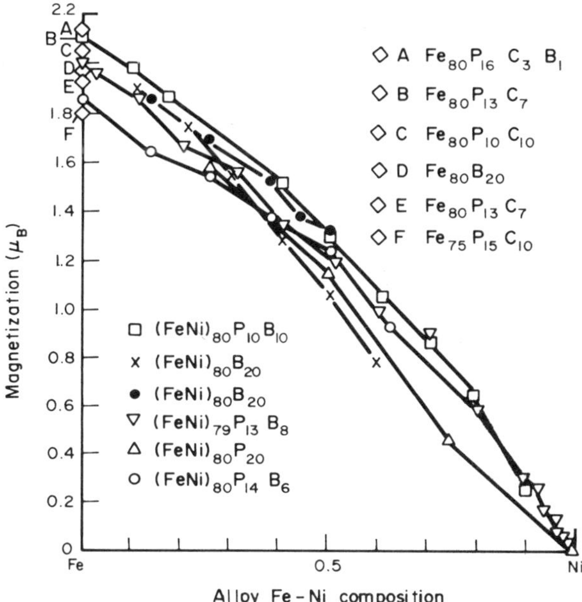

Figure 1
Magnetization as Bohr magnetons per metallic atom at 0 K for various iron–nickel amorphous alloys (after Graham and Egami 1978)

temperature dependence of magnetization from the coefficient of the $T^{3/2}$ law (see *Magnetic Materials: Basic Concepts*). It has been found (Luborsky et al. 1980b) that by plotting D against metalloid content linear plots are obtained with slopes that depend on the metalloid. A universal curve is obtained for all metalloids when D is plotted against T_C. The qualitative theoretical analysis of this trend is based on the itinerant electron model.

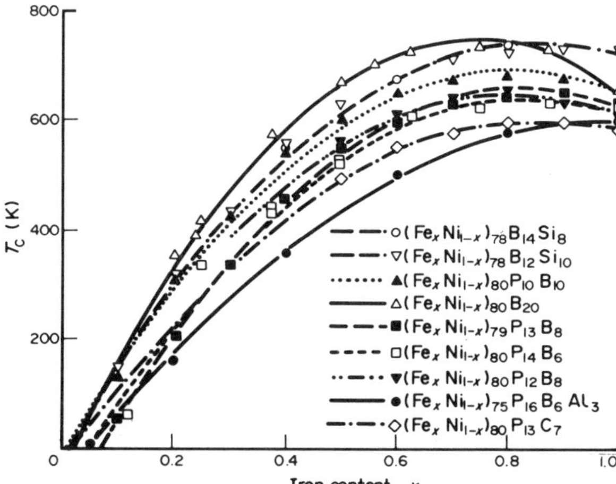

Figure 2
Curie temperatures of iron–nickel amorphous alloys as a function of iron content; curves calculated (after Luborsky 1980b)

(d) The Curie–Weiss constant, obtained from the susceptibility measured above T_C, can be used to calculate the number of magnetic carriers q_c. This is, in general, larger than the carriers q_s calculated from M_s. As T_C decreases, the ratio q_c/q_s for amorphous as well as crystalline alloys increases.

(e) The Arrott plots of high-field magnetization, M^2 versus H/M, show a curvature for amorphous alloys which is absent in magnetically homogeneous alloys. This curvature can be explained in terms of the spatial fluctuations in M by the itinerant model.

(f) The invar effect, which includes a large negative pressure derivative of T_C, large high-field susceptibility and volume magnetostrictions, and resulting thermal expansion and elastic constant anomalies, is also best explained by the itinerant electron model.

2. *Magnetostriction*

A split-band model reproduces the essential results of CPA-calculated densities of states and adds to it a perturbation calculation of the angular momentum at the Fermi energy in order to predict the compositional dependence of magnetostriction particularly for the $\lambda = 0$ compositions. A comparison of the calculated and observed $\lambda = 0$ line in the Fe–Co–Ni–B amorphous alloys is shown in Fig. 3. The calculated line describes the slope and intercept of the observed $\lambda = 0$ line on the iron–nickel side of the diagram. The observed shift of the nonmagnetostrictive composition from 18% iron in crystalline iron–nickel to approximately 7% iron in amorphous Fe–Ni–B is well predicted by the split-band model with a charge transfer to the transition metal d bands of 1.6 e per boron atom.

The temperature dependence of λ (O'Handley 1978) reflects the nature and range of magnetic interactions. The temperature dependence of λ for $Fe_{100-x}B_x$ alloys is dominated by one-ion interactions. Nevertheless, λ in iron-rich amorphous alloys shows strong dependences on the type of metalloid. This presumably occurs because the spins and orbits are strongly affected by the local environment. It is suggested that the striking differences in λ between iron-rich amorphous alloys and polycrystalline α-iron are due primarily to the presence of metalloid nearest neighbors and the higher TM–TM coordination in the amorphous alloys. The presence or absence of long-range order is apparently insignificant.

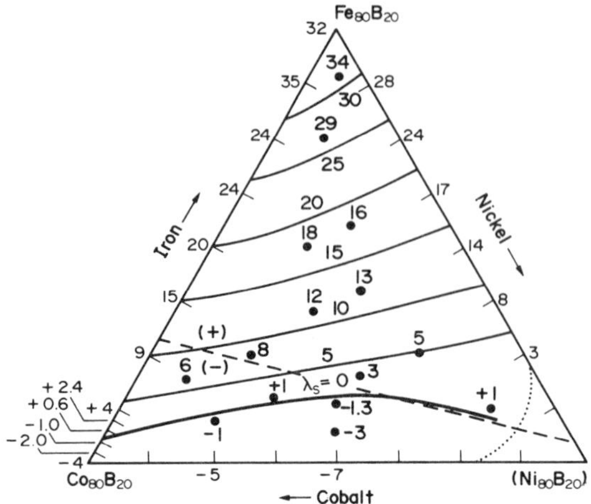

Figure 3
Magnetostriction of $(Fe_xCo_yNi_{1-x-y})_{80}B_{20}$ amorphous alloys: solid lines for average values of λ; dashed line for $\lambda = 0$ from split-band theory; dotted line indicates the limit of fabricability of the amorphous phase (after O'Handley and Berger 1978)

3. *Induced Anisotropy*

In an ideal amorphous alloy there should be no anisotropy. However, anisotropic microstructures can arise due to structural or compositional fluctuations, strain–magnetostriction interaction anisotropy, directional order anisotropy or exchange anisotropy. These may arise from field or stress annealing, from the casting process or from subsequent cold working. All of these sources of anisotropy have been reported for amorphous alloys. We will consider here only the field-induced and strain-induced directional order anisotropy K_u.

As in crystalline alloys, the amorphous alloys also exhibit atomic ordering under the influence of a magnetic field or stress, applied at a temperature below the Curie temperature. The magnitude of K_u is expected to be a function of the anneal temperature θ and the magnetization M_θ at θ, M_T at the measurement temperature and M_0 at 0 K, as given by $K_u = Af(c)(M_T/M_0)^2(M_\theta/M_0)^2/\theta$ where A is a constant which depends on the atomic arrangement and the range of interactions considered and $f(c)$ is a function of the concentration of the ordering species. For nonideal solutions, the effective nonmagnetic interaction energy may be included as described by Luborsky and Walter (1977). Reported values for K_u in various amorphous alloy systems are shown in Fig. 4. The data of Luborsky and Walter (1977) and of Miyazaki and Takahashi (1978) do follow the predicted behavior assuming a nonmagnetic interaction energy. Note that, at $x = 0$ and 1, where pair ordering is not possible, there is still measured an induced anisotropy. This has been attributed to interstitial ordering.

The kinetics of the reorientation of K_u have received considerable attention because of the relevance to the stability of the materials in devices and because of the insight it provided into atomic rearrangements. As shown in Fig. 5 the temperature at which the kinetics become noticeable is much lower in amorphous alloys than in annealed crystalline alloys. The mechanism of this enhancement in the time constant is attributed to the larger free volume in the amorphous materials and the existence of a broad spectrum of activation energies for structural relaxation (Gibbs et al. 1983, Evetts 1985). Thus, magnetic aftereffect or relaxation phenomena are also much more noticeable in amorphous alloys (Hodson and Evetts 1986).

The measured magnitude of macroscopic anisotropy is now believed to depend on two factors (Gibbs 1991). There are local structural units that have anisotropies related to equivalent crystalline structures. These units can rotate under the action of temperature and external field to give a degree of alignment of local easy axes. The greater the alignment the greater the macroscopic anisotropy. Superimposed on this comes the atomic ordering discussed already. This will determine the final (equilibrium) magnitude of the macroscopic anisotropy.

4. *Coercive Force and Remanence*

The reversal of magnetization in amorphous alloys occurs by wall motion. Thus, the value of the coercive field H_c is controlled by wall pinning (see *Coercivity and Domain Wall Pinning*). Experiments on ribbons of

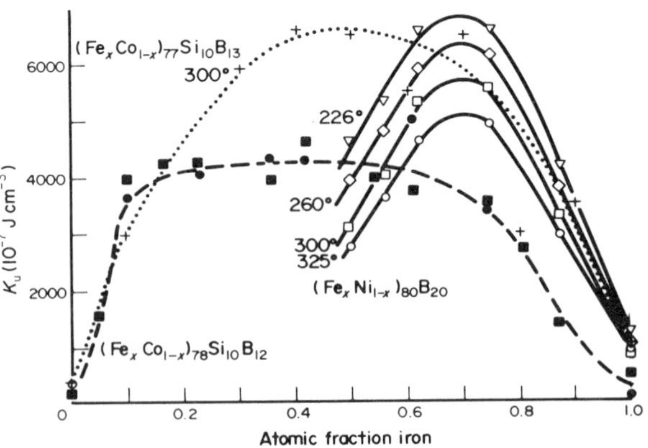

Figure 4
Magnetically induced anisotropy in various amorphous alloys at different anneal temperatures: solid symbols (after Fujimori et al. 1977) obtained from maximum values on heating for 15 min at successively higher temperatures (●) and from cooling from 380 °C at 1.6 °C min^{-1} (■); open symbols after Luborsky and Walter (1977); crosses after Miyazaki and Takahashi (1978)

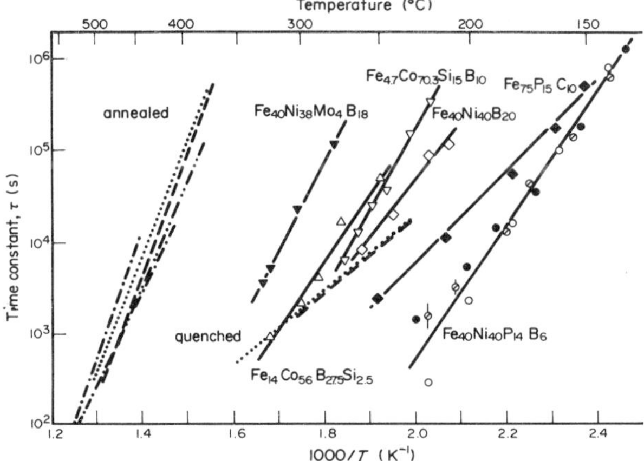

Figure 5
Time constants for the reorientation of the induced anisotropy in amorphous alloys (solid lines) and in crystalline alloys (broken lines): --- Fe–60Ni; ···· Fe–85Ni; –·– Fe–50 to 80Ni; —··— Fe–84Ni; —·— Fe–20Ni–60Co. After Luborsky 1980a, except for ○ (after Chambron and Chamberod 1980); ▽ (after Guo et al. 1981); △ (after Fujimori et al. 1978) and ▼ (after Hasegawa 1978)

varying thicknesses (Luborsky et al. 1980a) have indicated that wall pinning appears to occur principally at the surfaces. Under these conditions $\mu_0 H_c = 4S(AK)^{1/2}/dM_s$ where S is a constant determined by the nature of the surface roughness, A is the exchange constant, K is the anisotropy and d is the ribbon thickness. Contributions to H_c may also arise from fluctuations in K giving $\mu_0 H_c = (2K/M_s)f$ where $f = \Sigma \delta K/K$.

In as-cast ribbons H_c typically falls between 0.8 A m^{-1} and 8 A m^{-1} (10 mOe and 100 mOe), the value at the low end corresponding to alloys with low or zero magnetostriction. Values of H_c as low as 0.08 A m^{-1} (1 mOe) have been reported in $\lambda = 0$ alloys. However, even alloys with high λ will yield low H_c after a stress-relief anneal which reduces K. Values of 0.4–0.8 A m^{-1} (5–10 mOe) have been measured in Fe–B–Si alloys. If this annealing is carried out in a field or under the influence of a stress then the magnetization ratio B_r/B_s increases, typically from an as-cast value of 0.3–0.6 up to a value of 0.8–1.0 after annealing.

Table 1
Losses and exciting power reported for amorphous iron-based alloys at high induction

Alloy	60 Hz, sin B loss (W kg^{-1}) at 1.4 T	60 Hz, sin B loss (W kg^{-1}) at 1.5 T	Exciting power (V A kg^{-1}) at 1.4 T	Exciting power (V A kg^{-1}) at 1.5 T
$Fe_{82}B_{12}Si_6$	0.58	0.76	1.2	4.1
$Fe_{18}B_{13.5}Si_{3.5}C_2$	0.29	0.39	0.30	1.1
$Fe_{81}B_{13.5}Si_{3.5}C_2$	0.31	0.38	2.5	7.8
$Fe_{81}B_{13.5}Si_{3.5}C_2$	0.24	0.29	0.26	0.33
$Fe_{80.8}B_{15.9}Si_{2.4}C_{0.9}$	0.15	0.18	0.19	0.24
Fe–3.2%Si[M-3H(Hi-B)]	0.97	1.1	1.05	1.25
Fe–3.2Si[M-5]	1.0	1.2	1.1	1.45

5. Losses, Exciting Power and Permeability

The major impetus provided by amorphous alloys towards application in magnetic devices comes from their demonstrated extremely low magnetic losses and low exciting power in transformers (Table 1). In work on the effect of toroid diameter, losses of a factor of 20 lower than Fe–3.2% Si have been demonstrated while in 0.5 kV A model transformers losses and exciting power which were a factor of 3 lower than Fe–3.2% Si have been achieved using clamping, winding and insulating procedures amenable to production. The origin of these extremely low losses arises from the same factors contributing to the low H_c, namely the absence of major wall pinning centers and low anisotropy after stress-relief annealing.

As with the losses, annealing which reduces K increases μ. However, the initial permeabilities tend to be somewhat lower than the best crystalline alloys, namely the 4–79% molybdenum permalloy or supermalloy, but the maximum permeabilities are somewhat higher. Various special anneals have been developed for alloys in which T_C is greater than the crystallization temperature in order to maximize the permeability.

6. Applications

Amorphous alloys offer a unique combination of physical properties that make them particularly interesting for applications (Luborsky and Johnson 1980). For example, amorphous alloys can combine high mechanical hardness and wear resistance with high permeability, high saturation magnetization and zero magnetostriction. This combination of mechanical and magnetic properties is not found in crystalline materials, where mechanical hardness and magnetic hardness go together. An obvious area of application here is in recording heads, particularly with thin-film or multilayer material.

The topological disorder gives rise to an increase in electrical resistivity (typically 150 μΩ cm) and this factor is a significant contribution to the low losses under ac excitation. The distribution transformer market could see major reductions in loss with a switch to amorphous alloy cores in place of the iron–silicon now used. The US and Japanese distribution networks are currently taking delivery of amorphous-metal-cored transformers at a rapidly increasing rate. However, the historical level of investment in iron–

silicon technology, and recent advances in the processing of these crystalline alloys, especially the increase in silicon content and laser scribing, has continued to improve the performance of iron–silicon. However, the iron–silicon alloys still have higher losses than the amorphous alloys.

The high permeability coupled with the excellent as-cast mechanical performance make amorphous metal suitable for magnetic shielding applications. There is no material degradation with mechanical work. The sheet can be formed or woven to suit many applications. However, amorphous alloys do embrittle after heat treatment and this can limit their durability in service.

The large magnetostrictions ($\lambda_s \simeq 40$ ppm) available can be exploited in stress, torque and knock sensors. There have also been proposals for magnetometers with amorphous alloys as the active element. These have relied on the magnetostrictive or ΔE effects. Other applications include antitheft devices and proximity detectors.

At present, the major commercial applications are in power, high-frequency and pulse-compression transformers. Iron-based alloy is being cast on a continuous basis in sheet up to 20 cm wide and 25 μm thick. It is less clear whether there will be commercial-scale production of specialized alloys for sensors and transducers. The deposition routes may prove to be more significant here.

Metallic glasses are metastable with respect to crystallization and other relaxations of the structure. For optimum performance in most applications a heat treatment is required. Extensive studies have indicated maximum operating temperatures of 120 °C over 20-year time periods after heat treatment.

The degrees of freedom in alloy design in metallic glasses give them a unique place in the family of soft magnetic materials. Their commercial impact will continue to grow, and some of the traditional roles for permalloy, iron–silicon and certain ferrites will be overtaken by these new materials.

See also: Magnetic Materials: An Overview; Magnetic Materials: Basic Concepts; Magnetoelastic Phenomena; Sensors and Devices: Magnetic

Bibliography

Alben R, Budnick J I, Cargill G S 1977 Magnetism in metallic glasses. In: Leamy H J, Gilman J J (eds.) 1977 *Metallic Glasses*. American Society for Metals, Metals Park, OH, Chap. 12

Anantharaman T R 1983 *Metallic Glasses: Production, Properties and Applications*. Trans Tech Publications, Switzerland

Chambron W, Chamberod A 1980 Kinetics of induced magnetic directional order in a $Fe_{40}Ni_{40}P_{14}B_6$ amorphous alloy. *Solid State Commun.* 33: 157–9

Evetts J E 1985 Level systems and kinetic processes in amorphous solids. In: Steeb S, Warlimont H (eds.) 1985 *Rapidly Quenched Metals*. North-Holland, Amsterdam, pp. 607–10

Fujimori H H, Morita H, Obi Y, Ohta S 1977 On the magnetically induced anisotropy in amorphous ferromagnetic alloys. In: Levy R A, Hasegawa R (eds.) 1977 *Amorphous Magnetism II*. Plenum, New York, pp. 393–402

Fujimori H, Ohta S, Masumoto T, Nakamoto K 1978 Magnetic after-effect on soft magnetic properties of amorphous ferromagnetics. In: Cantor B (ed.) 1978 *Rapidly Quenched Metals III*, Vol. 2. Metals Society, London, pp. 232–9

Gibbs M R J 1991 The effect of large spread in moment direction on anisotropy and magnetostriction in amorphous alloys. *Proc. 5th. Int. Conf. Physics of Magnetic Materials*. World Scientific, London

Gibbs M R J, Evetts J E, Leake J A 1983 Activation energy spectra and relaxation in amorphous materials. *J. Mater. Sci* 18: 278–88

Graham C D Jr, Egami T 1978 Magnetic properties of amorphous alloys. *Annu. Rev. Mater. Sci.* 8: 423–57

Gubanov A I 1960 Quasi-classical theory of amorphous ferromagnetics. *Sov. Phys. Solid State* 2: 468

Guo H Q, Shen B G, Lin D M, Pan S T 1981 Magnetic annealing and crystallization kinetics of the metallic glass $Fe_5Co_{70}Si_{15}B_{10}$. *J. Magn. & Magn. Mater.* 23: 156

Hasegawa R 1978 Reorientation kinetics of the magnetization in a glassy ferromagnet. *J. Appl. Phys.* 49: 5610–11

Hasegawa R 1983 *Glassy Metals: Magnetic, Chemical and Structural Properties*. CRC Press, Boca Raton, FL

Hodson P D, Evetts, J E 1986 Polarization "crossover" of field induced magnetic anisotropy in ferromagnetic amorphous alloys. *J. Magn. & Magn. Mater.* 59: 81–5

Kaneyoshi T 1983 *Amorphous Magnetism*. CRC Press, Boca Raton, FL

Luborsky F E 1980a Amorphous ferromagnets. In: Wohlfarth E P (ed.) 1980 *Ferromagnetic Materials*, Vol. 1. North-Holland, Amsterdam, pp. 451–530

Luborsky F E 1980b Composition dependence of the Curie temperatures of amorphous alloys. *J. Appl. Phys.* 51: 2808–10

Luborsky F E 1983 *Amorphous Metallic Alloys*. Butterworths, London

Luborsky F E, Johnson L A 1980 Application of magnetic amorphous alloys. *J. Phys. (Paris)* 41: C8.820–6

Luborsky F E, Liebermann H H, Walter J L 1980a Effect of ribbon thickness, composition and process changes on the properties of rapidly quenched metal–metalloid alloys. *Proc. Conf. Metallic Glasses*, Vol. 1. Hungarian Central Research Institute for Physics, Budapest, Paper P-10, p. 203

Luborsky F E, Walter J L 1977 Magnetic anneal anisotropy in amorphous alloys. *IEEE Trans. Magn.* 13: 953–6

Luborsky F E, Walter J L, Liebermann H H, Wohlfarth E P 1980b The effect of temperature on magnetic saturation of some amorphous alloys. *J. Magn. & Magn. Mater.* 15–18: 1351–4

Luborsky F E, Walter J L, Wohlfarth E P 1980 The saturation magnetization, Curie temperature and size effect of amorphous iron alloys. *J. Phys. F* 10: 959–66

Miyazaki T, Takahashi M 1978 Magnetic annealing effect of amorphous $(Fe_{1-x}Co_x)_{77}Si_{10}B_{13}$ alloys. *Jpn. J. Appl. Phys.* 17: 1755–63

Moorjani K, Coey J M D 1984 *Magnetic Glasses*. Elsevier, Amsterdam

O'Handley R C 1978 Magnetostriction of transition metal/metalloid glasses: temperature dependence. *Phys. Rev. B* 18: 930

O'Handley R C 1987 Physics of ferromagnetic amorphous alloys. *J. Appl. Phys.* 62: R15–49

O'Handley R C, Berger L 1978 Evidence supporting the split-band model from magnetostrictions of transition metal/metalloid glasses. In: Lee M J G, Perz J M, Fawcett E (eds.) 1978 *Proc. Int. Conf. Physics Trans. Metals.* Institute of Physics, Bristol, UK

F. E. Luborsky
[General Electric Company, Schenectady, New York, USA]

M. R. J. Gibbs
[University of Bath, Bath, UK]

Micromagnetic Calculations

Micromagnetic calculations are used to compute a continuous variation in magnetization on a scale as small as an order of magnitude larger than the size of a magnetic unit cell. The static parameters that describe the medium include exchange, magnetocrystalline anisotropy and the local magnetic field. The dynamic properties include gyromagnetic ratio and damping. The aim of these calculations is to be able to compute all these parameters from solutions of Schrödinger's equation. The objective of these calculations then is to explain a variety of macroscopic phenomena in the presence of an applied field, including coercivity (see *Coercivity and Domain Wall Pinning*) and Barkhausen noise (see *Barkhausen Effect*).

1. Statics

The magnetostatic states are computed (Morrish 1965) by minimizing the energy of the system, which is the sum of the exchange energy, the anisotropy energy and the Zeeman energy:

$$W = \int_V \{A[(\nabla M_x)^2 + (\nabla M_y)^2 + (\nabla M_z)^2] + Kf(\theta,\varphi) + \mu_0 \boldsymbol{H}.\boldsymbol{M}\}\, dV \quad (1)$$

where A is the exchange constant, K is the anisotropy constant, $f(\theta, \varphi)$ is the angular variation of the anisotropy energy, $\boldsymbol{M}$ is the magnetization and $\boldsymbol{H}$ is the total internal field, which is the sum of the externally applied field $\boldsymbol{H}_A$ and the demagnetization field $\boldsymbol{H}_D$, which is due to the variation in the magnetization in the medium and its shape. Thus, the Zeeman energy includes the shape anisotropy.

Both the anisotropy energy and the exchange energy can be described in terms of effective fields. For uniaxial anisotropy $Kf(\theta, \varphi) = K_u \sin^2 \theta$ and it is possible to define an anisotropy field $\boldsymbol{H}_K$ by

$$\mu_0 \boldsymbol{H}_K = \frac{K_u \cos\theta}{M} \mathbf{1}_K \quad (2)$$

where $\mathbf{1}_K$ is a unit vector along the easy axis of the crystal. Note that, in order for an anisotropy field to describe a unidirectional energy, this field has to be a function of the direction of the magnetization. The concept of the anisotropy field is useful only for small deviations of the magnetization from the easy axis, for which this field is almost constant. The anisotropy energy can then be written as

$$W_K = \int_V \mu_0 \boldsymbol{M}.\boldsymbol{H}_K\, dV \quad (3)$$

Since the magnitude of the magnetization at a point in the medium is constant,

$$\nabla^2(\boldsymbol{M}.\boldsymbol{M}) = 2[(\nabla M_x)^2 + (\nabla M_y)^2 + (\nabla M_z)^2] + 2(\boldsymbol{M}.\nabla^2\boldsymbol{M}) = 0 \quad (4)$$

Therefore, an exchange field $\boldsymbol{H}_{ex}$ can be defined by

$$\mu_0 \boldsymbol{H}_{ex} = -A\nabla^2 \boldsymbol{M} \quad (5)$$

so that the exchange energy is given by

$$W_{ex} = -\int_V \mu_0 \boldsymbol{H}_{ex}.\boldsymbol{M}\, dV \quad (6)$$

If these fields are defined, then the energy can be computed using

$$W = -\int_V \mu_0 \boldsymbol{M}.\boldsymbol{H}_T\, dV \quad (7)$$

where $\boldsymbol{H}_T$ is the sum of the external field, the demagnetizing field, the anisotropy field and the exchange field:

$$\boldsymbol{H}_T = \boldsymbol{H}_A + \boldsymbol{H}_D + \boldsymbol{H}_K + \boldsymbol{H}_{ex} \quad (8)$$

2. Methods of Solution of Static Problems

With the general availability of fast computers, numerical techniques have become increasingly important in solving these partial differential equations. The numerical techniques used to solve these equations involve obtaining the solution at a collection of specified points. If these points fall on a regular grid, then the finite-difference method can be used to obtain the equations governing the relationship between solutions at the different points. On the other hand, if they do not, then the finite-element method must be used. The latter method is used when trying to solve device problems in the case where the geometry is complex. The solution at other points is then obtained by interpolation.

If the magnetization of the material can be characterized by a nonlinear single-valued function of the applied field, then very powerful general-purpose finite-element programs can be used that are capable of analyzing any media shape. For hysteretic media, it is necessary to write *ad hoc* programs, in which case it is advisable for simplicity to use regular meshes. In certain cases, especially when the region of solution is unbounded, the boundary-element method may be preferred.

Since these equations are nonlinear, iterative techniques have to be used to solve them. The methods are driven by the residual error, that is, the amount by which the trial solution does not satisfy the equations. The simplest of these methods is the Gauss–Seidel method, where all the points are sampled in a specific order for a given iteration and the solution is adjusted at each point so that its corresponding equation is satisfied. If a system of linear equations has diagonal dominance, then this method will converge. This method is very slow and so other methods, such as the successive overrelaxation (SOR) method, have been used to accelerate convergence. The SOR method involves magnifying the change that is necessary to adjust the solution so that the equation is satisfied by an overrelaxation factor before making the change. A problem with this method is that it can become unstable since the system is nonlinear. In this case, the overrelaxation factor has to be reduced to the point that no speed increase is obtained.

Two problems that occur with iterative methods are choosing a satisfactory initial trial solution and knowing when to stop. The choice of initial solution markedly affects not only the time required to obtain the solution but also which solution is obtained when, as is usually the case, more than one solution is possible. It has been found that a good choice of the initial solution can be obtained from classical solutions obtained from problems with simplified geometry. These problems fall into two categories to be discussed later: continuous media and particulate media.

The choice of a terminating criterion for the iterative method is affected by the conditioning of the problem. It can be shown that, for residual driven methods for linear problems, the residual error, the absolute error and the amount that a solution is changed on a given iteration are proportional to each other. Thus, if the solution does not change on a given iteration, both the residual and the absolute errors must be zero. The problem is that the range of proportionality constants (i.e., the conditioning numbers) is many orders of magnitude. It is therefore not possible to conclude that, if on a given iteration the solution changes by a certain percentage, the error is even the same order of magnitude as that percentage.

A related problem is the mode in which the trial solution approaches the real solution in successive iterations. In these problems, certain modes, which are referred to as stiff-convergent modes, converge very quickly, while others converge extremely slowly. For the latter modes, a very small change in the solution occurs at each iteration so that it can be mistakenly concluded that a solution has been found although the real solution is still very far away. Special techniques have to be used to accelerate the convergence of these modes when they occur (Torfeh-Isfahani and Della Torre 1983, Della Torre 1984, Della Torre and Perlov 1991).

3. Dynamics

When an input, such as the applied field, is changed, the static states also change. An important question concerns the choice of state of the system after a change in applied field. An attempt may be made to resolve this by taking small steps in the input and by following the state. This approach becomes thwarted when that particular state becomes unstable. If there is only one stable state of the system for that particular input, then there is no problem; however, choosing the correct state is a matter of concern when there are several possibilities because, although a state may be energetically favorable, there might not be any physical path to it. The only way to resolve this problem is to solve the dynamics of the system, which is a much longer calculation.

A variety of effects determines the dynamic behavior of magnetization changes: eddy currents, gyromagnetic resonance, aftereffect and accommodation. The first of these occurs in conductors where electric currents can be induced in the magnetic material, due to the change in magnetization, which shield the interior of the material from the applied field. Gyromagnetic resonance is due to the spin angular momentum of the electrons, which causes the magnetic moment to precess about a magnetic field. The moment will eventually align with the field due to damping. This behavior is described by the Gilbert form of the Landau–Lifschitz equation:

$$\frac{d\boldsymbol{M}}{dt} = \gamma(\boldsymbol{M} \times \boldsymbol{H}) - \frac{\alpha}{M}\left(\boldsymbol{M} \times \frac{d\boldsymbol{M}}{dt}\right) \tag{9}$$

where $\boldsymbol{H}$ is the total effective field, γ is the gyromagnetic ratio, and α is the phenomenological damping constant. Aftereffect is the slow change in magnetization as a function of time after the applied magnetic field has changed. Accommodation refers to the gradual shift in position of a minor hysteresis loop during repeated transversals.

The phenomenological damping constant is not well understood and its value appears to differ depending on how it is measured. The factors that contribute to it are radiation from the precessing dipole and losses during domain wall motion. Thus, if it is measured using resonance linewidth techniques with no domain wall motion, a different value is obtained to that when wall motion is involved.

4. Methods of Solution of Dynamic Problems

The solution of dynamic problems involves an additional degree of freedom, namely time. These problems have to be solved either in the time domain or the frequency domain. Thus, in addition to all the equations for static problems, it is necessary, in the time domain, to solve a temporal differential equation or, in the frequency domain, to solve for the harmonics that are generated by the nonlinearities.

In the time domain, the temporal dependence in the differential equation can be obtained by the Runge–Kutta method. The more efficient predictor–corrector methods, such as the Adams–Moulton method, cannot be used, since it is usually necessary to vary the step size continuously due to the nonlinearities.

5. Domains and Domain Walls

In a perfect crystal the magnetization tends to form magnetic domains in order to minimize the exchange energy (see *Domains and Domain Walls*; *Magnetization Imaging*). Furthermore, these domains tend to be aligned with the easy axis of magnetization in order to minimize the magnetocrystalline anisotropy energy. The domains are separated by domain walls whose width is determined by balancing the increase in exchange energy due to the change in magnetization with the increase in magnetocrystalline anisotropy energy due to the rotation of the magnetization from the easy axis. For an infinite Bloch wall between two regions in which θ takes on the values of 0 and π, respectively, and in which φ is equal to 0 if there is no applied or demagnetizing field, then the variation in θ is given by

$$\frac{d\theta}{dz} = \left(\frac{Kf(\theta,0)}{A}\right)^{1/2} \tag{10}$$

For thin films this has to be modified due to the demagnetizing field induced by the Bloch wall at the film surfaces. In this case, if the easy axis of the medium lies in the plane, then a Néel wall forms, in which φ is equal to π. This occurs despite the fact that this wall would normally have a higher energy than a Bloch wall due to its internal demagnetizing field.

These domain walls can translate in a perfect crystal with a minimal applied field. Any imperfections can impede this motion. A general discussion of analytic calculations of coercivity for different crystal imperfections is given in the article *Coercivity and Domain Wall Pinning*, where the problem of solving the micromagnetic equations numerically for the coercivity is discussed for the case of inclusions in the microstructure. Two models have been proposed to account for the coercivity caused by inclusions: the Kersten–Néel model, in which the equilibrium position of the wall is centered at the inclusion because the energy is lower there, and the wall-repulsion model, in which the wall is repelled by the magnetic moment induced by the omission of magnetized material in the inclusion. The latter model has been calculated numerically for two cases. In the first case the size of the inclusion is much larger than the thickness of the domain wall so that the wall can be assumed to be of zero thickness. In the second case the inclusion is small compared with the thickness of the domain wall so that the structure of the domain wall has to be taken into account. In garnets and materials with higher ratios of magnetization to wall energy, it has been shown that the wall repulsion model gives answers that are approximately correct, but the Kersten–Néel model predicts values that are wrong by several orders of magnitude. In either case, the effect of the inclusion is to make the wall motion erratic, since the wall has to overcome the potential well due to the inclusion. Once this is overcome, the wall moves quickly to the next inclusion. This effect is known as a Barkhausen jump and can be detected as noise as the magnetization changes (see *Barkhausen Effect*).

In order to solve these problems numerically, it is necessary to limit the scope of the calculation in order to obtain answers in a reasonable time. Thus, the techniques used are a blend of classical micromagnetic and pure numeric. For example, when an inclusion is approximated by a sphere, it is then assumed that the surrounding region is uniformly magnetized so that its effects are given by a dipolar field. Also, the variation in magnetization in the domain wall is approximated by the variation in the magnetization of a plane domain wall which can be computed classically. Even with these types of approximations, the solution cannot be obtained in a reasonable time without using the techniques described by Torfeh-Isfahani and Della Torre (1983), Della Torre (1984) and Della Torre and Perlov (1991).

Another source of coercivity is the change with position in material parameters, such as exchange. This can be caused by a gradual variation in composition which is especially important in amorphous materials. The domain wall prefers to be situated where the exchange constant is lower and, hence, the exchange energy is lower. The wall will then have a different shape than it would have in the rest of the medium. In order to propagate, the wall has to change its shape and therefore increase its energy, thus creating a potential well which must be overcome by the application of a sufficiently high field. A one-dimensional analysis (Della Torre and Perlov 1991) which illustrates this point is plagued by the problem of stiff convergence.

When a field is applied to a medium containing a domain wall, magnetization change takes place by wall motion, which does not occur until the applied field exceeds the coercivity. In conducting materials, eddy currents shield the wall from the applied field. Thus, the velocity of the wall is such that the eddy currents induced by this rate of change in magnetization reduce the field at the wall to just above the coercive field. At higher velocities in conducting materials and at all velocities in insulating materials the gyromagnetic effects are important.

An analysis of this behavior shows that, in Bloch walls for velocities below a certain limiting velocity, the Walker velocity, the wall appears to be viscous and, thus, its velocity is linearly proportional to the field at the wall. Above the Walker velocity, steady-state solutions break down and spin-wave analysis must be used. This behavior can be explained by the distortion in the domain wall that takes place when a magnetic field is applied parallel to the magnetization of one of the domains. This causes a precession of the moment in the center of the domain wall tending to change it from a Bloch wall to a Néel wall. The demagnetizing field present in the Néel wall causes the spins to precess in a way that translates the wall at a constant velocity. This velocity is directly proportional to the demagnetizing field and saturates at the Walker velocity when the wall is entirely a Néel wall and cannot increase further. If the applied field is increased further there is no further change in wall structure until it is large enough to nucleate other magnetization patterns. The Walker velocity for garnets is of the order of 50 m s^{-1} but varies considerably with composition.

6. Particulate Media

Particulate media that are useful in recording consist of an assembly of fine particles (see *Magnetic Storage Media*). These may be individual particles or a continuous film in which islands have been defined by the segregation of an element that greatly reduces the exchange coupling between islands. The critical field for magnetization in these particles is determined by a combination of shape and magnetocrystalline anisotropy. In order for these particles to be useful in a recording medium they must have a discrete number of states. A two-state particle is usually referred to as a single-domain particle, although its magnetization is not strictly uniform, especially at the ends of the particle.

The Stoner–Wohlfarth model assumed that the shape of the particle was a perfect ellipsoid. In this case, if the particle is placed in a uniform field, it can be magnetized uniformly since the demagnetizing field is also uniform. Thus, magnetization reversal can be accomplished by rotating the uniform magnetization against the demagnetizing field for the so-called coherent reversal mode. The calculated field at which magnetization reverses was much larger than that which had been observed, so a nucleation theory was developed that used perturbation methods. Large-scale numerical calculations were impossible at the time, making it impossible to carry out complete reversals. It was then shown that incoherent reversal modes could nucleate at smaller applied fields, which incidentally were much closer to the observed values. The fanning, buckling and curling modes are derived from various models (Mee 1986).

These calculations were all based on ellipsoidal shapes and not the jagged shapes of realistic particles. Numerical micromagnetic models have been applied to particles whose shapes were simple rectangular parallelepipeds and similar rectangular parallelepipeds with pyramidal ends. It was found that all acicular particles tended to be uniformly magnetized in the central region but that magnetization was nonuniform at the ends in order to reduce the demagnetizing energy. For small particles, this nonuniform magnetization at the ends tended to fan out in the so-called flower state. For larger particles, the magnetization at the ends tended to approximate the classical curling modes; however, the central region of the particle was still uniformly magnetized. Particles that were larger still were multidomain, but still useful for recording, since there were still only a few states separated by a potential barrier. These domains were columnar in shape and parallel to the long axis of the particle.

Magnetization reversal in all of these particles is accomplished by the flipping mode; that is, a magnetization reversal nucleates at one or both ends and then propagates towards the center of the particle. This reversal might reverse the entire particle or just a column inside the particle.

7. Nucleation

A particular micromagnetic problem of interest is the nucleation problem, that is, finding the inception of a magnetization reversal. In addition to the fact that the problem of finding a classical solution for the intermediate states is extremely difficult, so that the choice of an initial solution for the numerical problem is problematic, there are the numerical problems associated with characterization of the nucleation event. An associated problem is that a reversal mode might require cooperative phenomena in order to nucleate. An example of this was discussed in the nucleation of reversals in magnetooptic films (Perlov et al. 1990) where the film was divided into cubes and it was found that under certain conditions reversing a single cube increased energy while reversing several cubes together decreased energy and was thus a possible reversal mode. This type of cooperative phenomenon is similar to stiff-convergent modes previously discussed.

Another problem in the numerical analysis of nucleation is the breaking up of symmetry. An example of this occurs in the reversal modes of fine particles when they are characterized by a mathematically perfect shape, which of course does not happen for real particles. When relaxing a high-energy configuration, there is no preferred direction for the reversal and thus nucleation does not occur until much higher fields than would ordinarily be required. This problem can be overcome by the order in which the points are sampled. This intellectually unsatisfying technique does resolve the issue but is somewhat sensitive to the choice of relaxation order.

The effect of temperature in the calculation of magnetization patterns has not been adequately addressed by current analyses. When the nucleation threshold approaches the product of Boltzmann's constant and the absolute temperature, it is possible to obtain a discontinuous change in the magnetization pattern. It is necessary to develop new techniques to include these effects.

Finally, any analyses must include sufficient degrees of freedom to permit all possible nucleation modes if the correct nucleation threshold is to be obtained. This was classically observed by the unrealistically high field predicted for the reversal critical field by the Stoner–Wohlfarth model which permitted only coherent reversals. The curling mode under certain conditions requires lower nucleation fields and thus dominates the reversal process. Similarly, for particles with complex mathematical shapes other reversal modes, such as the flipping mode, that have even lower nucleation fields are possible.

See also: Coercivity and Domain Wall Pinning; Domains and Domain Walls; Magnetic Materials: Basic Concepts; Magnetization Imaging

Bibliography

Della Torre E 1984 Adaptive mode pushing. *IEEE Trans. Magn.* 20: 1906–8

Della Torre E, Perlov C M 1991 A one-dimensional model for wall motion coercivity in magnetooptic media. *J. Appl. Phys.* 69: 4596–8

Mee C D 1986 *The Physics of Magnetic Recording*. North-Holland, Amsterdam

Morrish A H 1965 *The Physical Principles of Magnetism*. Wiley, New York

Perlov C M, Della Torre E, Birecki H 1990 Magnetic domain behavior in magneto-optic recording. *J. Appl. Phys.* 67: 4444–6

Torfeh-Isfahani M, Della Torre E 1983 Mode pushing and stiff convergent problems. *IEEE Trans. Magn.* 19: 2527–30

E. Della Torre
[George Washington University,
Washington, DC, USA]

Microwave Applications of Superconducting Materials

The microwave applications of superconductors can be divided into two main categories; active devices (based on the Josephson effect) and passive devices, the latter forming the basis of this article. In general, incorporating superconductors into passive devices utilizes the lower surface resistance R_s over conventional conductors leading to improved device characteristics. A superconductor has a finite R_s at finite temperature and frequency. For a normal metal, R_s is proportional to $\omega^{1/2}$ while, for a superconductor in the local limit at frequencies well below the gap frequency (typically a few THz), R_s is approximately proportional to ω^2 (see *High-Frequency Impedance of Superconducting Materials*). Therefore, at a fixed temperature, these different frequency dependencies result in a superconductor being more lossy than a normal metal above a particular frequency. This frequency is known as the crossover frequency and the metal used as the comparator is usually copper.

The crossover frequency of $YBa_2Cu_3O_{7-\delta}$ (YBCO) with copper at 77 K at low rf fields for thick-film material is of the order of a few tens of gigahertz and, for thin-film material, is of the order of a few hundreds of gigahertz (Piel and Müller 1991), corresponding to values of R_s at 77 K and 10 GHz of around 5 mΩ for thick films, and around 0.01 mΩ for good quality thin films; the corresponding R_s for copper is around 10 mΩ. YBCO will prove useful for practical applications below these crossover frequencies. However, these figures have to be compared with crossover frequencies well in excess of 500 GHz for low-T_c superconductors like Nb_3Sn at the same reduced temperature T/T_c; the surface resistance of niobium can be reduced to a few tens of nΩ at 4.2 K and 10 GHz by careful surface preparation (see *Superconducting Cavities*). In addition, the rf field handling capabilities of high-T_c materials are presently poor; this point will be discussed further in Sect. 4.

Microwave passive device applications that benefit from replacing normal metals with superconductors can be divided into three main categories.

1. Resonators and Filters

There are three major forms of loss that limit the quality factors of resonators—ohmic loss, dielectric loss and radiation loss. The losses are additive, such that

$$\frac{1}{Q} = \frac{1}{Q_c} + \frac{1}{Q_d} + \frac{1}{Q_r} \qquad (1)$$

where Q is the overall unloaded quality factor and Q_c, Q_d and Q_r are the quality factors due to the ohmic (conductor), dielectric and radiation losses, respectively. Additional forms of loss mechanism can be easily incorporated into Eqn. (1). In the majority of resonator microwave applications using conventional conductors, it is the ohmic loss that limits the device performance. The advantage of incorporating superconductors is apparent if the resonator operates well below the crossover frequency. The dielectric and radiation losses can usually be reduced by careful design and choice of materials. Resonators such as cylindrical cavities, coaxial cavities and helical cavities can be used to characterize superconductors in the frequency range 100 MHz to 100 GHz, where the superconductor forms all or part of the cavity.

One of the most important types of resonator for applications is the planar resonator, which is made in the form of a planar circuit fabricated from thin or thick-film material. There are many possible types of this resonator based on stripline, microstrip, coplanar or other microstrip geometries (see Fig.1). In the case of low-T_c superconductors, the dielectric used is usually sapphire (Al_2O_3), which exhibits very low microwave loss, particularly at low temperatures. For epitaxial thin films of high-T_c materials, MgO and $LaAlO_3$ are commonly used in preference to sapphire since, although the microwave properties of sapphire are by far superior, there are problems due to lattice mismatch and Al^{3+} poisoning the superconductor. The development of thin dielectric buffer layers for sapphire substrates is a very active area of research (see *Oxide Superconductors: Thin-Film Deposition*).

As illustrated in Fig.1, a typical resonator consists of a length of planar transmission line which is open circuit at both ends, resonance occurring when the device length is an integral number of half wavelengths. Coupling can be achieved using capacitive gaps or loops. Ground plane contact is made either by direct soldering or by spring loading to the resonator container. Figure 1 schematically illustrates a typical package for a coplanar resonator with capacitive coupling. For cryogenic purposes it is desirable to have the whole package hermetically sealed. The performance of a number of different types of planar resonator is shown in Table 1.

These resonators are extensively used in microwave circuits and they form the basis for the frequency selective network in oscillators and filters. When used in oscillators, a good resonator will reduce the sideband phase noise and when used in filters will either reduce the insertion loss, increase the quality factor Q, or both. High unloaded resonator Qs greatly improve the performance of oscillators. Around 10 GHz, low-T_c materials can offer more than 20 dB reduction in sideband noise over competing technologies such as crystal oscillators. Considerable improvement is expected with the use of improved designs and higher-Q resonators.

A number of microstrip filters have been produced using both high-T_c and low-T_c superconductors, all showing significant improvement over filters made from normal metals. For some applications, the reduction in insertion loss of even fractions of a decibel are important, for example, when using a filter to receive signals from deep space probes. Such filters are generally placed at the front end of any receiving system in order to reduce out-of-band noise so that it does not mix into the signal band in the initial amplification stages.

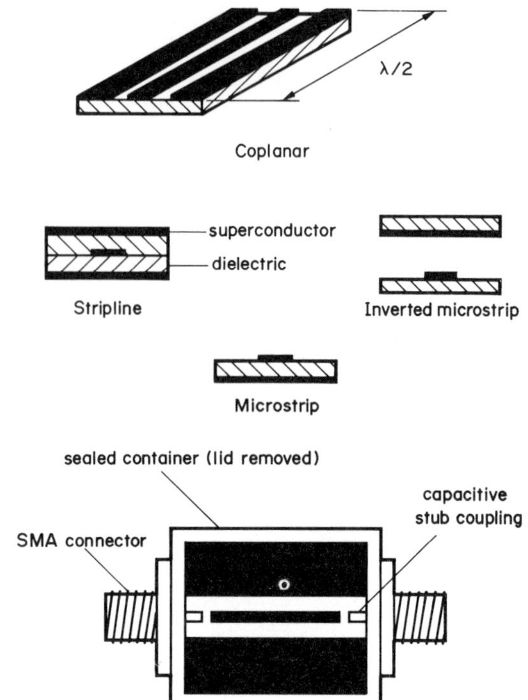

Figure 1
Planar resonator geometries and packaging

Table 1
Resonator Q for various planar resonator geometries and materials

Geometry	Material	Frequency (GHz)	Q	*T* (K)
Microstrip	lead	14.3	2×10^5	4.2
Stripline	lead	7.8	6×10^5	4.2
Stripline	YBCO	2.8	2×10^4	4.2
Coplanar	YBCO	9.1	3.3×10^3	4.2
Coplanar	YBCO	8.8	1.3×10^3	77
Stripline	Niobium	1.1	2.5×10^5	4.2
Microstrip	TlBaCaCuO	2.6	3×10^4	77

Source: Lancaster (1990)

2. Delay Lines

Owing to the low surface resistance, very low loss and low dispersion superconducting transmission lines can be manufactured. A delay line is a very long length of transmission line, usually in the form of a spiral, patterned onto a substrate with a corresponding ground plane. Superconducting delay lines have been studied by MIT Lincoln Laboratories since the early 1980s (Withers and Ralston 1989). Their importance lies not in the simple application of delaying a microwave signal but, once a delay has been established, in it being possible to construct a transversal filter. This is accomplished by tapping off the signal at points along the line using backward wave couplers. The types of transversal filter that can be constructed in this way include linear phase filters and dispersive filters.

These techniques have been used for many years in the case of surface acoustic wave (SAW) filters to produce filters for pulse compression radar systems, among other applications. However, SAW devices are not able to produce bandwidths greater than a few gigahertz, whereas superconductor delay lines are envisaged to produce bandwidths of several tens of gigahertz, although with much shorter delays for a given area.

The degree of sophistication achieved in superconducting delay line technology is illustrated by the hamming-weighted dispersive chirp filter of Dilorio et al. (1989), the results of which are shown in Fig.2. The filter consists of a niobium film spiral (of diameter 50 mm) on a silicon wafer with a niobium film ground plane. Agreement between theory and experiment is excellent. At present, these filters are mainly limited by the short delay due to the limited length of line that can be fabricated on a wafer while maintaining isolation requirements, although this situation is likely to improve.

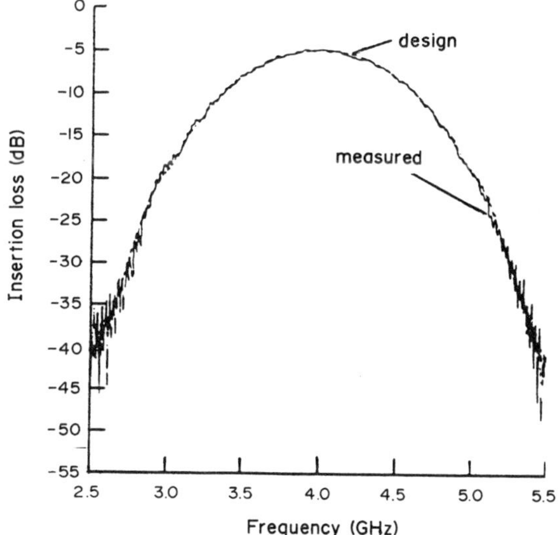

Figure 2
Design parameters and measured characteristic for amplitude response of the hamming-weighted, superconducting chirp filter of DiIorio et al. (1989). The bandwidth was 2.6 GHz; the dispersion was 37.5 ns; the insertion loss was 5 dB; and the center frequency was 4 GHz (© 1989 IEEE, New York. Reproduced with permission)

3. Antennas

Hansen (1990) has identified six areas of application for superconducting antennas: those of electrically small antennas, antenna arrays, millimeter-wave-length arrays, antenna matching networks, phasers for electronic scanning and travelling wave array feeds. The performance of any antenna system is governed by the antenna ohmic loss resistance R_l, the radiation resistance R_r and the antenna reactance X. In many applications, R_r is much larger than R_l and so the choice of antenna material has little effect on the overall efficiency. However, in applications such as small antennas or superdirective antenna arrays constructed out of a normal metal like copper, R_l can be large compared with R_r, leading to a reduction in efficiency. Since R_l is proportional to R_S (the constant of proportionality being a simple geometrical factor), replacing conductors with superconductors leads to an increase in efficiency. It should be noted, however, that reducing R_l can considerably increase Q of the antenna; there is a trade-off between Q and efficiency which must be considered carefully depending on the application (Hansen 1990, Lancaster et al. 1991).

There have been a number of low-T_c and high-T_c superconducting antennas demonstrated; small loops and dipoles have been shown to improve the antenna efficiency considerably. The first high-T_c superconducting antenna was demonstrated at Birmingham University, UK, in 1987. This small dipole antenna showed an improvement in gain over a similar copper antenna at 77 K of 6 dB (Lancaster 1990).

4. Future Developments

The three main areas of application of superconducting materials to passive microwave devices have been discussed. Superconductors have been proposed for use in many other applications, for example in matching networks, signal combiners, duplexers, screening and microwave switches. Little work has been performed on these applications. In addition, devices based on kinetic inductance effects (e.g., Pond 1987) may hold prospects for tunable resonators and filters.

With the newly discovered high-T_c superconductors, the field is wide open. Crossover frequencies at

77 K for thin films are continually improving. Until recently, R_S for both bulk and thin film YBCO degraded very rapidly as a function of applied rf magnetic field owing to the granular nature of the samples, even for rf fields less than 10^{-4} T. This is very important for very-high-Q devices where, even if the input power is low and the coupling is weak, high surface rf fields can be produced. Some of the very best YBCO thin films, with a small number of grain boundaries, show very little degradation up to rf fields of around 10^{-2} T at 4.2 K (Piel and Müller 1991), as would be expected since the superheating field (at which flux enters the material) is an order of magnitude higher than this. The prospect of high-Q and high-field thin-film devices is continually improving, although at 77 K the bulk material looks destined only for low-field applications up to around 10 GHz.

See also: High-Frequency Impedance of Superconducting Materials; Microwave Materials and Devices, Magnetic; Superconducting Cavities

Bibliography

DiIorio M S, Withers R S, Anderson A C 1989 Wideband superconductive chirp filters. *IEEE Trans. Microwave Theory Tech.* 37: 706–10

Hansen R C 1990 Superconducting antennas. *IEEE Trans. Aerospace Electron. Syst.* 26: 345–55

Hartwig W H, Passow C 1975 RF superconducting devices. In: Newhouse V L (ed.) 1975 *Applied Superconductivity*, Vol. 2. Academic Press, New York, pp. 541–639

Lancaster M J 1990 Passive microwave device applications of high T_c superconductors. *Proc. 2nd Int. Symp. New Materials and Applications*, IOP Conference Series No. 111. Institute of Physics, Bristol, UK, pp. 273–84

Lancaster M J, Wu Z, Huang Y, Maclean T S M, Zhou X, Gough C, Alford N McN 1991 Superconducting antennas. *Supercond. Sci. Technol.* (in press)

Piel H, Müller G 1991 The microwave surface impedance of high-T_c superconductors. *IEEE Trans. Magn.* 27: 854–7

Pond J M, Claassen J H, Carter W L 1987 Measurements and modelling of kinetic inductance microstrip delay lines. *IEEE Trans. Microwave Theory Tech.* 35: 1256–62

Withers R S, Ralston R W 1989 Superconductor analog signal processing devices. *Proc. IEEE* 77: 1247–63

M. J. Lancaster and A. Porch
[University of Birmingham,
Birmingham, UK]

Microwave Materials and Devices, Magnetic

Magnetic materials have been employed at microwave frequencies—initially in the range 1–20 GHz but later up to 60 GHz and beyond—since the late 1940s. Their uses have been mainly in communication and measurement systems at these frequencies. These applications are made possible by the fact that the Lamor precession of a magnetic moment in a static magnetic field can interact with microwaves to produce attenuation, phase shift, Faraday rotation and, most importantly, nonreciprocal effects in which waves travelling in opposite directions are affected differently (see *High-Frequency Properties of Magnetic Materials*).

The step which led to the widespread microwave use of magnetic materials was the development at Philips Laboratories, Eindhoven, between 1933 and 1946 (under adverse conditions), of high-resistivity (around $10^6\,\Omega\,\mathrm{m}$), magnetite-based ferrites which, unlike ferromagnetic metals, had a substantial skin depth (see *Ferrites, Soft*).

The frequency of operation of a ferrite sample in a microwave device is determined in principle by the *effective* applied magnetic field H_0, since the natural Lamor angular frequency is given by

$$\omega_0 = \gamma H_0 \tag{1}$$

where γ, the gyromagnetic ratio, is $2\pi \times 2.8 \times 10^{10}$ rad s^{-1} T^{-1} (2.8×10^{10} Hz T^{-1}). Thus, for a field of 0.3 T, the resonance would occur at 8.4 GHz.

1. Microwave Materials

1.1 Spinel Ferrites

The original ferrites were of the basic formula $MeFe_2O_4$ where Me is a divalent ion, usually magnetic. The Me^{2+} ion will, in effect, replace the Fe^{2+} ion in magnetite (Fe_3O_4), thereby creating an insulator. Typically, Me^{2+} would be Ni^{2+}, Mn^{2+} or Co^{2+} or the nonmagnetic ions Zn^{2+} or Mg^{2+}—or some combination of these. An unusual case is when $Me^{2+} \equiv (Li^{+}_{0.5}\,Fe^{3+}_{0.5})$ which gives the most useful lithium ferrite with the highest ferrite Néel temperature of 635 °C.

These magnetite-based ferrites, such as $NiFe_2O_4$ and $MnFe_2O_4$, have the structure of the cubic mineral spinel $MgAl_2O_4$.

1.2 Garnet Ferrites

Between 1952 and 1956 a new class of ferrites was developed with the formula $Me_3Fe_5O_{12}$ where Me is generally a trivalent rare-earth ion (see *Rare-Earth Intermetallics*). These ferrites were orthomorphic with the mineral garnet ($Ca_3Fe_2Si_3O_{12}$). The classic example of the garnet ferrites is $Y_3Fe_5O_{12}$ (yttrium–iron–garnet (YIG)). In the pure state, YIG has only trivalent, *s*-state ions so that its losses—magnetic and dielectric—are particularly low; however, because small amounts of magnetic rare-earth ions such as Ho^{3+} and Er^{3+} give high losses, YIG is usually manufactured from 99.9999% pure Y_2O_3. To avoid the high cost of YIG while keeping losses low, an alternative, CalVan BIG ($Ca_{2x}V_xBi_{3-2x}Fe_{5-x}O_{12}$), was developed.

1.3 Hexagonal Ferrites

Also in the 1950s, another group of ferrites was developed, with hexagonal structures. Barium hexaferrite ($BaO(Fe_2O_3)_6$, $BaFe_{12}O_{19}$ or BaM) has a uniaxial anisotropy which, in the single crystal, gives an effective field of about 1.9 T so that there is a self-resonance (or natural ferromagnetic resonance (NFMR)) at about 50 GHz. BaM, as a permanent-magnet material, is also known by the Philips trade name of Ferroxdure.

Another hexagonal ferrite—the W type—has a (smaller) uniaxial anisotropy. Two others—Y type and Z type—are known as ferroxplana materials because of their planar anisotropy; the NFMR of the planar ferrites occurs at 2 GHz and below.

2. Resonance Phenomena

Because of the precessional motion of the magnetization of a ferrite which occurs in a static field $\boldsymbol{H}_0 = \boldsymbol{i}_z H_0$ when a radio-frequency field $\boldsymbol{h} = \boldsymbol{h}_0 \exp(\mathrm{j}\omega t)$ is imposed, the permeability has the form of the Polder tensor so that the radio-frequency flux density $\boldsymbol{b}$ is given by

$$\begin{bmatrix} b_x \\ b_y \\ b_z \end{bmatrix} = \begin{bmatrix} \mu & -\mathrm{j}\kappa & 0 \\ \mathrm{j}\kappa & \mu & 0 \\ 0 & 0 & \mu_0 \end{bmatrix} \begin{bmatrix} h_x \\ h_y \\ h_z \end{bmatrix} \qquad (2)$$

where μ_0, the permeability of free space, is $4\pi \times 10^{-7}\,\mathrm{H\,m^{-1}}$ for H_0 in amperes per meter and is unity for H_0 in tesla or for cgs units. The components of the tensor μ and κ can be written for small signals, when only transverse variations of the magnetization are important, as

$$\left.\begin{aligned} \mu &= \mu' - \mathrm{j}\mu'' = \mu_0 \left(1 + \frac{\omega_\mathrm{m}\omega_0}{\omega_0^{*2} - \omega^2}\right) \\ \kappa &= \kappa' - \mathrm{j}\kappa'' = \mu_0 \left(\frac{\omega_\mathrm{m}\omega}{\omega_0^{*2} - \omega^2}\right) \end{aligned}\right\} \qquad (3)$$

where $\omega^*_0 = \omega + \mathrm{j}\Delta\omega/2$, ω_0 being from Eqn. (1) and $\Delta\omega$ being the width of the resonance line at the points where μ'' (or κ'') falls to one-half of its peak value; $\omega_\mathrm{m} = \gamma M_\mathrm{s}$ (with the ferrite magnetization M_s taken in the same units as H_0).

Away from resonance, Eqn. (3) can be written with

$$\omega_0{}^* \simeq \omega_0, \quad \mu'' = \kappa'' \simeq 0$$

Near resonance, provided $\Delta\omega \ll \omega_0$,

$$\kappa' \simeq \tfrac{1}{2} \frac{\omega_0(\omega_0 - \omega)}{(\omega_0 - \omega)^2 + (\Delta\omega/2)^2}, \quad \mu' \simeq \mu_0 + \kappa'$$

$$\kappa'' \simeq \mu'' \simeq \tfrac{1}{4} \frac{\omega_\mathrm{m}\Delta\omega}{(\omega_0 - \omega)^2 + (\Delta\omega/2)^2} \qquad (4)$$

The width $\Delta\omega$ of the resonance line observed from the variation of μ'' or κ'' as the frequency varies at constant field is related to the width ΔH of the line observed as H_0 varies at constant frequency:

$$\Delta\omega = \gamma\Delta H \qquad (5)$$

2.1 Resonant Field

If the field $\boldsymbol{H}_0 = \boldsymbol{i}_z H_0$ acts on a finite sample along a symmetry axis, there will be a demagnetizing field $\boldsymbol{H}_\mathrm{d}$ such that

$$\begin{bmatrix} H_{\mathrm{d}x} \\ H_{\mathrm{d}y} \\ H_{\mathrm{d}z} \end{bmatrix} = -\begin{bmatrix} N_x & 0 & 0 \\ 0 & N_y & 0 \\ 0 & 0 & N_z \end{bmatrix} \begin{bmatrix} M_x \\ M_y \\ M_z \end{bmatrix} \qquad (6)$$

where M_x, M_y and M_z are the components of $\boldsymbol{M}_\mathrm{s}$ and the demagnetizing factors N_x, N_y and N_z depend on the configuration of the sample relative to $\boldsymbol{H}_0$. The actual resonance frequency (which replaces Eqn. (1)) is given by the Kittel equation (See *High-Frequency Properties of Magnetic Materials*):

$$\omega_0^2 = [\gamma H_0 + (N_x - N_z)\omega_\mathrm{m}][\gamma H_0 + (N_y - N_z)\omega_\mathrm{m}] \qquad (7a)$$

or, since $\omega_\mathrm{m} = \gamma M_\mathrm{s}$

$$\omega_0 = \gamma\{[H_0 - (N_z - N_x)M_\mathrm{s}] \times [H_0 - (N_z - N_y)M_\mathrm{s}]\}^{1/2} \qquad (7b)$$

For a sphere, $N_x = N_y = N_z$ ($= \frac{1}{3}$ for H_0 and M_s in the same units) so that $\omega_0 = \gamma H_0$; thus, spheres are generally used for the measurement of ferrite parameters.

For a disk in the xy plane, $N_x = N_y = 0$, $N_z = 1$ and

$$\omega_0 = \gamma(H_0 - M_\mathrm{s}) < \gamma H_0$$

For a disk in the xz plane, $N_x = N_z = 0$, $N_y = 1$ and

$$\omega_0 = \gamma[H_0(H_0 - M_\mathrm{s})]^{1/2} > \gamma H_0$$

2.2 Resonance Linewidth

The linewidth ΔH_s for a *single crystal* of a ferrite depends on

(a) the perfection of the crystal,
(b) the quality of the surface polish,
(c) the number of Fe^{2+} ions, and
(d) the presence of ions such as Co^{2+} or Ho^{3+} which have significant spin–orbit coupling.

In the case where (a) and (b) are the only contributions, the linewidth will be due to the scattering of spin waves (or magnons) and will relate to the spin wave linewidth ΔH_s.

Stoichiometric, highly polished YIG crystals grown from a fluxed melt have the lowest linewidth, about $1\,\mathrm{A\,m^{-1}}$ (10^{-6} T). Single crystals of spinel ferrites will have $\Delta H_\mathrm{s} \simeq 400\,\mathrm{A\,m^{-1}}$ while BaM crystals have $\Delta H_\mathrm{s} \simeq 4000\,\mathrm{A\,m^{-1}}$, largely due to residual Fe^{2+} ions.

Polycrystalline (or ceramic) ferrites have higher linewidths which can be written

$$\Delta H = \Delta H_s + \Delta H_p + \Delta H_{an} \tag{8}$$

where $\Delta H_p \simeq M_s p$ is the contribution due to the fraction p of the sample made up of pores (p is the "porosity") and ΔH_{an} is the contribution due to misalignment of the anisotropy fields of individual crystallites. (For a cubic crystal, $\Delta H_{an} \simeq 1.7 \,|\, K_1 \,|/M_s$ where K_1 is the first-order anisotropy constant and has an approximate value of $16\,kA\,m^{-1}$ for $NiFe_2O_4$.) For BaM, ΔH_{an} for a ceramic specimen depends on how well the crystallites have been aligned during preparation and can vary between $8\,kA\,m^{-1}$ and $600\,kA\,m^{-1}$. For 5% porosity, ΔH_p will typically be $16\,kA\,m^{-1}$.

3. Microwave Devices

3.1 YIG-Tuned Filter

For resonance at 10 GHz, a YIG sphere can have a Q factor $\omega_0/\Delta\omega$ (or $H_0/\Delta H$) as high as 3×10^5. If two coils are wound orthogonally around the sphere and it is subjected to a static but variable field at 45° to the two coils, a tunable narrow band pass filter (as required in swept-frequency generators) will have been created, since a signal into one coil will only be detected by the other when the sphere is at resonance.

3.2 Resonance Isolator

When an electromagnetic wave travels along a rectangular waveguide in its fundamental TE_{01} mode, the H field is in the plane of the broad dimension of the guide. At a point one-quarter of the way across the guide the H field will be circularly polarized: the sense of this polarization will reverse if the direction of propagation is reversed.

If a piece of ferrite is placed approximately one-quarter of the way across the guide and a field is applied to cause precession of the magnetization in the plane of the H field, then there will only be significant interaction between the H field and the precession if the sense of rotation is the same.

The electromagnetic wave is said to be travelling in the "forward" direction if there is little interaction. In the opposite, "reverse" direction there is interaction and there will be attenuation of the wave which will be greatest at resonance. The reverse attenuation can be made, depending on the size of the ferrite and its linewidth, as high as 40 dB at resonance for a forward or insertion loss of 0.3 dB.

The device is said to be an isolator because any signal source would be isolated by it from reflections of power back down the guide.

3.3 The Y or Junction Circulator

If a ferrite specimen (usually cylindrical) is placed at the junction between three waveguides (of whatever type) meeting at 120° to each other and if there is a field applied normal to the plane of the guides, then, by matching the three ports into the junction to give zero reflection, the ferrite can be made to *circulate*: that is, the power in at port one will be almost all transmitted through port two while port three is isolated and so on cyclically for power in at ports two and three. (Clearly the device can be used as an isolator when it will have low insertion loss and high isolation.)

3.4 Other Devices

Other ferrite microwave devices include power limiters (based on the fact that spin waves are generated exponentially above a certain critical power level), magnetostatic surface wave (MSSW) variable delay lines (based on the generation on thin films of MSSWs which have frequency—and/or field—dependent velocities) and Faraday rotation devices.

These latter devices depend on the fact that microwaves propagating through a ferrite medium parallel to a nonresonant applied field $\boldsymbol{H}_0$ will suffer a rotation of the plane of polarization by

$$180(\varepsilon_r)^{1/2}\left[\left(1+\frac{\omega_m}{\omega_0-\omega}\right)^{1/2} - \left(1+\frac{\omega_m}{\omega_0+\omega}\right)^{1/2}\right]$$

degrees in travelling a distance equal to a free-space wavelength, where ε_r is the relative permittivity. A 90° Faraday rotation can be used to switch an antenna between two senses of polarization and a 45° rotation can be used to produce a circulator.

4. Molecular Engineering

The properties of microwave ferrites can be changed by varying the ionic composition.

Fe^{2+} ions may be suppressed in spinel ferrites and BaM by replacing about 2% of the Fe^{3+} ions by the buffer ion Mn^{3+} which takes up spare electrons.

The magnetization of ferrites may be reduced by replacing Fe^{3+} ions by Al^{3+}: $Y_3AlFe_4O_{12}$ has a room-temperature magnetization of 0.05 T compared with 0.175 T for YIG. Such reduction increases the power handling capacity and lowers the minimum frequency of operation (and, unhelpfully, reduces the Néel temperature—in the example, from 287 °C to 172 °C). On the other hand, Zn^{2+} in place of Ni^{2+} in $NiFe_2O_4$ increases the magnetization by 0.004 T for each percent of Zn^{2+} added.

In BaM, $(Co^{2+}_{0.5}\,Ti^{4+}_{0.5})$ in place of an Fe^{3+} ion reduces the anisotropy field, while Al^{3+} increases it and raises the NFMR frequency.

See also: Ferrites, Soft; High-Frequency Properties of Magnetic Materials

Bibliography

Baden-Fuller A J 1987 *Ferrites at Microwave Frequencies*. Peregrinus, London

Lax B, Button K 1962 *Microwave Ferrites and Ferrimagnetics*. McGraw-Hill, New York
Von Aulock W H 1965 *Handbook of Microwave Ferrite Materials*. Academic Press, New York

A.J. Pointon
[Portsmouth Polytechnic, Portsmouth, UK]

Millimeter-Wave Detection

It is natural to look to superconductivity for sensitive detection of millimeter- and submillimeter-wavelength electromagnetic radiation. Superconductivity is intrinsically a low-noise phenomenon, with characteristic energy Δ (for elemental superconductors) falling in this frequency range. Indeed, the most sensitive such detectors are heterodyne mixers utilizing nonlinear quasiparticle tunnelling (rather than Josephson pair tunnelling) in superconductor–insulator–superconductor (SIS) junctions. In 1990, 10–15 astronomical observatories worldwide were routinely using SIS mixers from 43 GHz to 360 GHz.

The SIS mixer is the only superconducting device that has had significant practical applications at these or higher frequencies. Nevertheless, great efforts have gone to the development of other superconductor-based detector components; the most promising will be described. The theory and experimental development of SIS mixers and receivers was reviewed by Tucker and Feldman (1985), and Richards and Hu (1989) review the entire subject matter of this article but at greater depth.

1. SIS Mixers

An extraordinary nonlinearity is created by the gap in available energies for quasiparticles on both sides of an SIS tunnel barrier. In an ideal SIS junction at absolute zero, no dc current can flow below the threshold voltage $V_g = 2\Delta/e$, while above V_g the dc current is near the normal state value. This perfectly sharp nonlinearity is smeared over some range δV_g in real SIS junctions (see Fig. 1a). An SIS mixer consists of such a junction embedded in a coupling structure. The mixer behaves classically if radiation with quantum energy $\hbar\omega/e$ smaller than δV_g is applied. But when $\hbar\omega/e$ exceeds δV_g the SIS junction instead responds to individual quanta by photon-assisted tunnelling. This leads to phenomena that are forbidden or inaccessible in classical resistive mixers, such as conversion gain, radiation-induced negative resistance and sensitivity limited only by irreducible quantum noise.

1.1 Theory

The quantum theory of mixing appeared fully formed in the work of Tucker (1979), shortly after the first SIS mixer experiments. Subsequent papers have merely explored the implications of Tucker's theory. In simplest terms, the tunnel barrier is characterized by a time-dependent response function containing the physics of the quasiparticle tunnelling. Since this function is causal its Fourier transform $j(\omega)$, the frequency-dependent response function, must be complex with real and imaginary parts related through a Kramers–Kronig transform. It turns out that $\text{Im}\, j(\omega)$ is equal to the dc I–V characteristic $I_{dc}(\hbar\omega/e)$ of the tunnel junction, which is dissipative. $\text{Re}\, j(\omega)$ is thus a nonlinear reactance, determined by $I_{dc}(V)$ through the Kramers–Kronig transform, which is an intrinsic property of the tunnelling process. The noise properties of the tunnel barrier can also be expressed in terms of I_{dc} through the fluctuation–dissipation theorem. Thus, the entire physical response of a *particular* quasiparticle tunnel junction is contained in its measured dc I–V characteristic such as Fig. 1a, and this gives the theory an extraordinary predictive power.

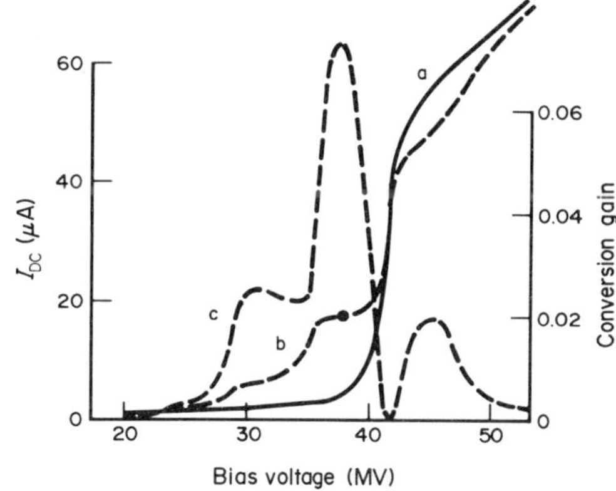

Figure 1
(a) The dc I–V curve of an unpumped series array of $N = 14$ SIS junctions, (b) the I–V curve pumped at 114 GHz, and (c) the conversion gain (badly mismatched). Note that $14\hbar\omega/e = 6.6$ mV; the best dc bias point for gain is given by the dot on curve (b)

Tucker wedded the mathematics describing quasiparticle tunnelling, much of which was known in other contexts, with the formalism of classical mixer theory, which describes the relationship of the mixing element to its environment. The final equations are exceedingly complex and so it is common to make the simplifying assumptions that the local oscillator (LO) voltage is sinuosidal and that the intermediate frequency (IF) is an extremely low frequency. These assumptions are quite well satisfied in most experiments, and the simplified theory has been successful in predicting the behavior of SIS mixers both in general and in great detail for individual mixers.

1.2 Phenomena

Photon-assisted tunnelling results in a (pumped) dc I–V curve consisting of a series of steps of width $\hbar\omega/e$ ($N\hbar\omega/e$ for the series array of Fig. 1b), which can be understood as follows. At dc voltages between V_g and V_g–$\hbar\omega/e$ quasiparticles gain sufficient energy to tunnel across the barrier by absorbing one photon, giving a large current. At lower dc voltages a quasiparticle must simultaneously absorb two or more photons to tunnel and this less likely event gives a smaller but still finite current. The I–V curve

$$I_{dc}(V_{dc},V_{rf}) = \sum_{n=-\infty}^{\infty} J_n^2(\alpha)I_{dc}(V_{dc}+n\hbar\omega/e) \quad (1)$$

is the sum of contributions from the unmodulated I–V curve each displaced in voltage by integer multiples of the quantum energy $\hbar\omega/e$ and weighted by Bessel functions of the reduced LO voltage $\alpha = eV_{rf}/\hbar\omega$. This expression has been known since 1963. In the classical theory the pumped I–V curve is smooth, a continuous integral over the unpumped curve. The LO power determines the junction rf current, given by

$$I_{rf}(V_{dc},V_{rf}) = \sum_{n=-\infty}^{\infty} \frac{2n}{\alpha} J_n^2(\alpha)I_{dc}(V_{dc}+n\hbar\omega/e) \quad (2)$$

Since an experimental I–V curve is taken by sweeping V_{dc} while holding LO power constant, it is possible to have α decrease as V_{dc} increases in Eqn. (1) to the extent that the differential resistance on a step is negative.

A heterodyne mixer is a linear frequency converter. A weak incoming signal at frequency ω_s is added to the large-amplitude LO at ω to produce a total input that is quasistatically modulated at the beat frequency $\omega_{IF} = |\omega_s - \omega|$. Therefore, the IF output is proportional to the signal and the signal-to-IF-power gain (see Fig. 1c) depends on the extent to which the dc bias current changes with small changes in LO power. In the simplified theory, Eqns.(1) and (2), together with the formalism of classical mixer theory, are sufficient to calculate the conversion gain of an SIS mixer. The gain can be and often is greater than unity. This was surprising at first because in classical theory it is prohibited. That prohibition, however, results from an intrinsic symmetry: the response of a classical mixer is independent of frequency. If this symmetry is broken by only an infinitesimal amount the conversion gain can, in principle, be arbitrarily large.

The LO power required for a single-junction SIS mixer, roughly $(\hbar\omega/e)^2/2R_N$, is quite small, several nanowatts at 100 GHz if the normal resistance R_N is chosen for convenient impedance matching. Because of this, SIS mixers saturate at relatively low signal levels; in one extreme case a mixer was severely saturated by a 50 K thermal input. The use of series arrays of junctions, as in Fig. 1, increases the saturation power as N^2 and ameliorates this problem. A series array should in principle act like an equivalent single junction, but in fact some series arrays have considerable excess noise.

The geometrical capacitance of an SIS junction tends to short circuit the desired nonlinearity and this constrains the junction size to be quite small, measured, at most, in micrometers. Before such small junctions were available an effective SIS mixer was not possible. Junctions with the product $\omega R_N C \gtrsim 1$, however, generally result in a fairly large mixer conversion loss. The benefit of larger capacitance, which should be tuned out at the signal frequency, is presumably in shorting undesirable higher harmonic-frequency effects.

The predominant internal noise source of an SIS mixer is shot noise, which is always small because the current scale of SIS junctions is much less than that of most semiconductor devices. External thermal noise coupled into the IF output can also be significant. If these are minimized, by using a near-ideal SIS junction either at very low temperature or with proper external impedances, the noise of an SIS mixer can approach the quantum limit given by Heisenberg's uncertainty principle, corresponding to the noise energy of half a photon added to the input signal.

1.3 Current Status

SIS mixers for the 100 GHz region have become quite sophisticated, achieving broad bandwidths, conversion gain and mixer noise temperatures of a few times the quantum limit. The most common are waveguide mounted, with several mechanical tuners, and are operated at 4.2 K with about unity gain. Lead-alloy SIS junctions, once standard, are giving way to the excellent characteristics of Nb–Al_2O_3–Nb junctions (see *Superconducting Device Fabrication*). Newer designs utilize planar integrated tuning elements to provide optimum embedding terminations.

There are greater difficulties at higher frequency. Even smaller junction areas and correspondingly higher current densities are required. Submillimeter waveguide is lossy and unwieldy and many quasioptical designs are thus being explored. Mechanical tuning of impedance is impracticable. Interference from Josephson tunnelling effects, even if partially suppressed by a magnetic field, makes a submillimeter mixer more susceptible to saturation and also to mechanical vibration. Nevertheless, the future appears bright. Büttgenbach et al. (1988) demonstrated an extremely broadband 100–760 GHz quasioptical SIS receiver with good performance in the laboratory, which had fairly large conversion loss to reduce saturation. NbN SIS junctions with V_g almost twice that of Nb will certainly be used in the near future, resulting in much less Josephson interference. The design of integrated tuning elements, ideal for quasi-optical mixers, is advancing.

As yet, SIS mixers have no significant applications beyond astronomy, where the benefits of low noise far outweigh the disadvantage of cooling to 4 K. The new high-T_c superconductors might seem to offer a solution, if SIS junctions can be developed. The noise of a carefully designed mixer, shot noise, would be independent of operating temperature. Also, the high-frequency cutoff for SIS mixing, $\nu_{max} = 2eV_g/\hbar$, scales as T_c. The problems, however, are formidable. For millimeter wavelengths a high-T_c SIS junction must be of quite high quality to have $\delta V_g \lesssim \hbar\omega/e$. For submillimeter wavelengths the current difficulties are largely material independent and would remain (ν_{max} for NbN is ~3 THz), except that a high-T_c mixer would have no Josephson interference.

2. Other Superconducting Devices

The SIS direct detector, essentially an SIS mixer operated without an LO, uses the nonlinear quasiparticle I–V curve to rectify an rf signal. The current responsivity, the induced current per unit signal power $S = \Delta I_{dc}/P_{sig}$, is easily determined from Eqns. (1) and (2) taking $\alpha \to 0$. If $I_{dc}(V_{dc} + \hbar\omega/e) \gg I_{dc}(V_{dc})$ the responsivity reaches its quantum-limited maximum $S = e/\hbar\omega$, one electron out for each signal photon in, which is readily approached in practice. The noise of the SIS direct detector is shot noise from the bias point leakage current and $1/f$ noise; for realistic junctions the noise is worse than that of competing devices and this has precluded applications. Since SIS junctions can be easily formed in planar arrays, the potential remains for multielement SIS direct detector receivers.

The Josephson-effect mixer uses a junction with such low capacitance, for example a point contact, that there is no hysteresis on the I–V curve. When sufficient LO power is applied to depress the critical current I_c to about half of its full value, a small signal at a nearby frequency sees a real impedance and produces oscillations at the difference frequency. The gain peaks at dc voltages where the dynamic resistance is large, most strongly at $V \simeq \hbar\omega/4e$, and can exceed unity. These mixers tend to be noisy. The best noise temperature is roughly 40 times either the physical temperature or $\hbar\omega/2k_B$, mostly attributable to many high-frequency noise components down-converted by the strong Josephson nonlinearity. Although Josephson mixing using grain-boundary junctions in high-T_c superconductors is readily observed, even worse sensitivity will result from operation above 4.2 K.

A Josephson junction biased with dc current I less than I_c is a nonlinear inductor, with inductance $L^{-1} = 2e(I_c^2 - I^2)^{1/2}/\hbar$. As such, it can be used as a reflection parametric amplifier. Early experiments encountered a peculiar broadband noise output rising in direct proportion to the square of the gain. It is now understood that a shunt inductor can eliminate that noise. Yurke et al. (1988) demonstrated a Josephson parametric amplifier at 19.4 GHz cooled to 0.1 K which had the remarkably low noise temperature of 0.28 K for one signal phase, and which produced a squeezed state from 4.2 K thermal noise. There is no intrinsic reason why good performance should not be expected at any millimeter wavelength. Note, however, that reflection amplifiers require circulators to separate output from input, and these are difficult to build at high frequencies. Also, as with all Josephson devices, the saturation threshold of this amplifier is inconveniently low.

Most of the detecting devices mentioned here require external microwave radiation, of modest power, to operate. The Josephson junction, itself a voltage-controlled oscillator, is naturally suggested for the source. The maximum power available from a single junction is quite small, however, and the linewidth is large (thus, internally pumped Josephson mixers and parametric amplifiers have been less successful than externally pumped counterparts). One approach to these problems is to use arrays of phase-locked junctions: the available power increases as N^2 and the linewidth decreases as N. Wan et al. (1989) designed a Josephson oscillator that forces 40 junctions to oscillate in phase by spacing them each one wavelength apart along a superconducting microstrip and, for the first time, coupled usable power, about 1 μW at 380 GHz, into an external load. The existence of such an oscillator opens the prospect of an entire superconducting receiver on a single chip.

It is still too early to predict the success of using high-T_c superconductors for the active element of a millimeter-wave detector. For most of the devices considered here, a high-T_c counterpart must await the development of high-T_c SIS junctions. An alternative is to employ the particular properties of high-T_c films for detection. Higher operating temperature will in general result in higher noise levels, but for many potential applications this is less important than the simplification of the cryogenics.

See also: Josephson Effects in Weak Links; Superconducting Device Fabrication; Tunnelling Spectroscopy

Bibliography

Büttgenbach T H, Miller R E, Wengler M J, Watson D M, Phillips T G 1988 A broadband low noise SIS receiver for submillimeter astronomy. *IEEE Trans. Microwave Theory Tech.* 36: 1720–6

Richards P L, Hu Q 1989 Superconducting components for infrared and millimeter-wave receivers. *Proc. IEEE* 77: 1233–46

Tucker J R 1979 Quantum limited detection in tunnel junction mixers. *IEEE J. Quantum Electron.* 15: 1234–58

Tucker J R, Feldman M J 1985 Quantum detection at millimeter wavelengths. *Rev. Mod. Phys.* 57: 1055–113

Wan K, Jain A K, Lukens J E 1989 Submillimeter wave generation using Josephson junction arrays. *Appl. Phys. Lett.* 54: 1805–7

Yurke B, Kaminsky P G, Miller R E, Whittaker E A, Smith A D, Silver A H, Simon R W 1988 Observation of 4.2 K equilibrium-noise squeezing via a Josephson-parametric amplifier. *Phys. Rev. Lett.* 60: 764–7

M. J. Feldman
[University of Rochester, Rochester, New York, USA]

Multifilamentary Superconducting Composites

Type II superconductors in the mixed state (i.e. the state in which most practical superconductors normally function) contain a lattice of flux lines, or fluxoids, which must be pinned on imperfections such as precipitates, dislocations, twin boundaries or grain boundaries if the material is to support a significant resistanceless current at relatively high magnetic fields (see *Critical Current Density*; *Flux Pinning and the Summation of Pinning Forces*). In NbTi the flux is primarily pinned on α-titanium precipitates which are produced by repeated applications of cold work and heat treatments (see *Niobium–Titanium Alloys*; Larbalestier et al. 1985). More recently, attempts have been made to produce a network of pinning sites artificially (Motowildo et al. 1990). In the case of Nb_3Sn, a brittle compound which is difficult to cold work, the flux is principally pinned on grain boundaries (see *A15 Compound Superconductors*). Again, attempts are being made to produce Nb_3Sn multifilamentary wires with artificial pinning centers (Flükiger 1991).

Since, in practical operation, the superconductor is subjected to local fluctuations in the mechanical and electromagnetic environment, it is necessary to design practical conductors in such a way that these local pertubations do not lead to total, and perhaps catastrophic, loss of superconductivity in the device in which they are used. To help to reduce this possibility, the superconductor is surrounded by a highly thermally conductive medium such as high purity copper. In addition, in order to reduce the extent of flux motion and to ensure the proximity of the copper, it is desirable to produce the superconductor in the form of fine filaments. Therefore, the structure desired is a composite one consisting of fine filaments in a copper matrix.

The design and fabrication techniques for the production of superconducting multifilamentary composites have developed along slightly different lines from those of other composites, such as those to be used in high-strength applications. The reason for this is that, in the case of the latter, at least one of the components is hard and frequently brittle. In the case of the superconductors, however, the extreme ductility and low work-hardening coefficient of niobium, niobium–titanium alloys and copper enable enormous amounts of reduction to be carried out without frequent intermediate anneals.

The same basic fabrication techniques (Hillmann 1981, Collings 1986) are used for the production of both the ductile multifilamentary niobium titanium (NbTi) and the hard intermetallic compound of niobium tin (Nb_3Sn). These two materials constitute the bulk of the commercial superconductors currently in use: NbTi for low-field and Nb_3Sn for higher field and somewhat higher temperature applications. In the case of Nb_3Sn, the two components (niobium and tin) are prevented from reacting to form the hard compound until the fabrication procedure is completed. In many cases the higher field product is fabricated into the coil required in the device before the relatively brittle Nb_3Sn is formed. These basic fabrication procedures are billet assembly, evacuation, welding, consolidation, extrusion, wire drawing, twisting, insulating and cabling. The details of the fabrication techniques employed and the design criteria used in the manufacture of multifilamentary superconductors are dependent on the device in which the conductor will be used.

1. NbTi Strands

1.1 Strands with less than 200 Filaments

These strands are made by gun drilling solid copper billets and inserting relatively large diameter NbTi rods into the holes, after carefully cleaning and etching both of the components (see Fig. 1a). The billets are evacuated, sealed by welding on a "nose" and a "tail," extruded and drawn to finished size. Unless the wire is drawn to very small dimensions, the filaments are usually of a large diameter; that is, greater than 20 μm.

Few manufacturing problems are encountered in the fabrication of this type of material when the correct production and quality control procedures are in place. Long piece lengths, which are required for magnetic resonance imaging (MRI) applications, are relatively easily achieved.

Typical applications are small and intermediate size magnets such as low-field (MRI) magnets and the ratio of copper to superconductor ranges from 1.3:1 to 7:1.

1.2 Strands with 200–2500 Filaments

When the number of strands exceeds 200, it is no longer practical to use drilled billets. One method frequently used with this number of filaments is to insert round rods into hexagonal tubes. These are stacked into 250 mm or 300 mm diameter cans and then processed in a manner similar to that illustrated in Fig. 1b. This approach is sometimes termed the "Fermi kit" approach as it was used to manufacture the strand for the Fermi tevatron dipoles and quadrupoles. Filamentary diameters down to 8 μm can be achieved with this technique in wires of diameter between 0.5 mm and 1 mm (see Fig. 2).

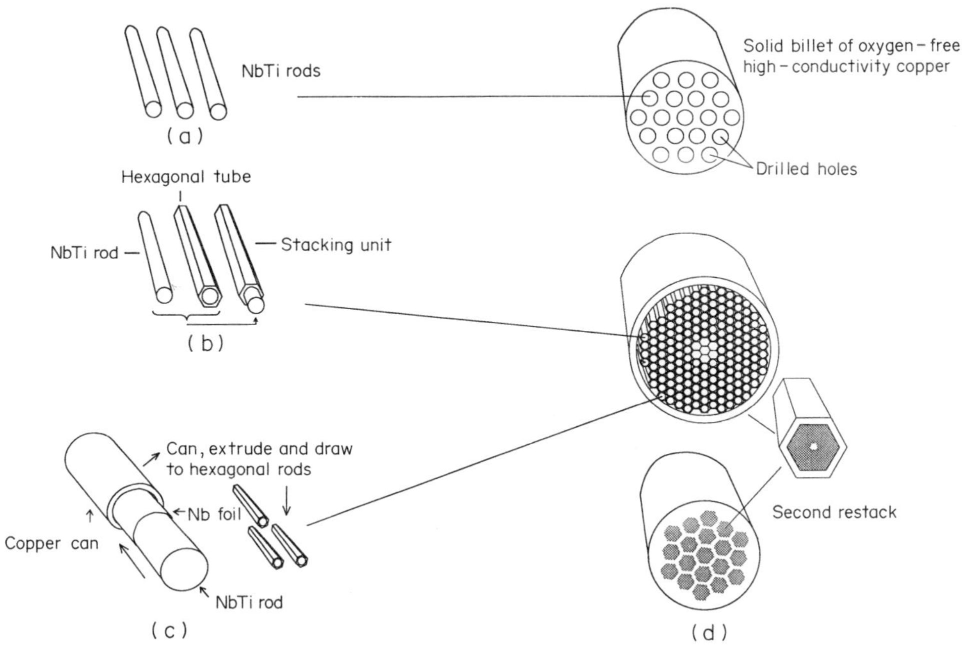

Figure 1
Fabrication methods for NbTi with the number of filaments being (a) less than 200, (b) 200–2500, (c) 2500–25 000 and (d) greater than 25 000

While the method of assembly is relatively economical, it has significant drawbacks when the capability to have smaller filament diameters combined with the highest current density (J_c) is required. The relatively thick walls of the available hexagonal tubes lead to filament "sausaging." The many surfaces involved at the assembly stage are difficult to clean and this may lead to piece length problems.

Conductors for a wide range of applications, including dipoles, quadrupoles and MRI magnets, have been made using this assembly technique. The values of J_c achieved can be high (~ 3000 A mm^{-2}), if the conductor dimensions are such that the filament diameters are greater than 15 μm. If, however, the filament diameter is below 10 μm then J_c values of 2000–2500 A mm^{-2} are more common. In the early work using this approach piece length problems were severe, but these have now been reduced to tolerable levels.

1.3 Strands with 2500–25 000 Filaments

Largely in response to the increasingly demanding specifications for dipole and quadrupole magnet conductors for such proposed accelerators as the superconducting super collider (SSC) and the large hadron collider (LHC) attempts have been made to improve the values of J_c in fine filamentary NbTi wires. Considerable advances were made in the late 1980s (Gregory 1989, Kanithi et al. 1989). The best way of meeting these high J_c fine-filament requirements has been to first make a monofilament in which the NbTi core is surrounded by a wrap of foil of material such as niobium and then by sufficient copper (or a copper–manganese alloy) to give the correct spacing between the filaments (see Fig. 1c). After extrusion and drawing these monofilaments are either restacked as round rods or passed through a hexagonal finishing die and then stacked. The latter approach gives the most perfect array but, in some cases, the round rods are easier and quicker to straighten and stack. The largest number of individual hexagonal monofilaments reported to have been stacked into a 300 mm diameter SSC type billet is 22 900 (Gregory et al. 1989). This makes a 0.65 mm diameter strand with 2.5 μm diameter filaments and was a prototype for an advanced outer cable conductor for the SSC–HEB (high energy booster) magnets. It had Cu–0.5 wt%Mn between the filaments in order to ensure that no proximity-effect coupling existed at the finished wire diameter (see Fig. 3). Further discussion of proximity coupling and pair-breaking may be found in the articles *Coherence Length, Proximity Effect and Fluctuations* and *Granular Superconductors*.

This technique is now frequently used for both the inner (~ 7000 filaments) and the outer (~ 4600 filaments) 6 μm diameter filamentary material specified for the SSC cables. Piece length problems have been encountered during the development work and seem

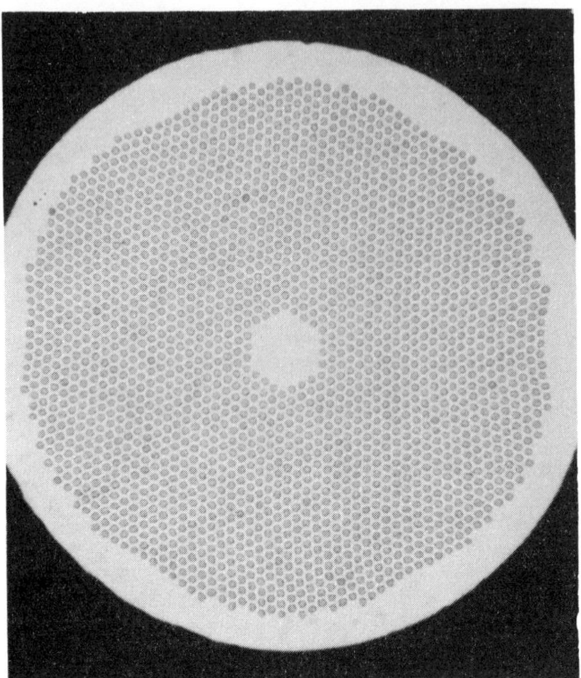

Figure 2
Cross section of a tevatron strand with approximately 2070 filaments, made by the Fermi kit technique

to be more severe as the number of filaments is increased. It appears that, if the necessary precautions are taken, the specified piece lengths can be achieved reliably and reproducibly. While it is possible to make at least the outer conductor with 2.5 μm diameter filaments by this method, as mentioned above, the piece length and values of J_c obtained have been lower than those obtained with the larger filaments. Due to this, and because the inner SSC conductor with 2.5 μm diameter filaments requires approximately 40 000 filaments, more refined packing steps are being explored. If these fail, one method that is not ideal but may have to be used for practical reasons, is to employ a further restacking step (see Fig. 1d) for both these very fine filament materials, as has been done for some time when making conductors for ac applications (see *AC Applications of Superconducting Materials*).

While the principal application for strands made by this technique is advanced accelerator magnets, the knowledge gained in this work can be applied to the development of high-J_c conductors for a wide range of applications.

1.4 Strands with more than 25 000 Filaments

Since the conductors under consideration for the LHC are of an overall larger diameter than those specified for the SSC, and because small filaments are generally desirable to reduce magnetization effects, wires with greater than 25 000 strands are likely to be specified in the future, if they can be developed with the high values of J_c and the long piece lengths required. The need for submicrometer-diameter filaments has long been recognized when ac applications have been considered but the high J_c value and long piece lengths requirements have not been stressed for these applications in the past. The approach used has been to make a subelement containing several hundred filaments by an approach such as that described in Sect. 1.3, draw to a relatively large size, finish with a hexagonal die and again restack (see Fig. 1d), then extrude and draw to final size. One of the disadvantages of this approach is that the filaments on the outer layers of the second restack are not uniformly supported on all sides and can, therefore, sausage more readily than those in the centers of each group.

2. Nb_3Sn Strands

Many of the design and fabrication factors which apply to NbTi conductors must also be taken into account when Nb_3Sn strands are to be made (Suenaga and Clark 1980). There are three main commercial

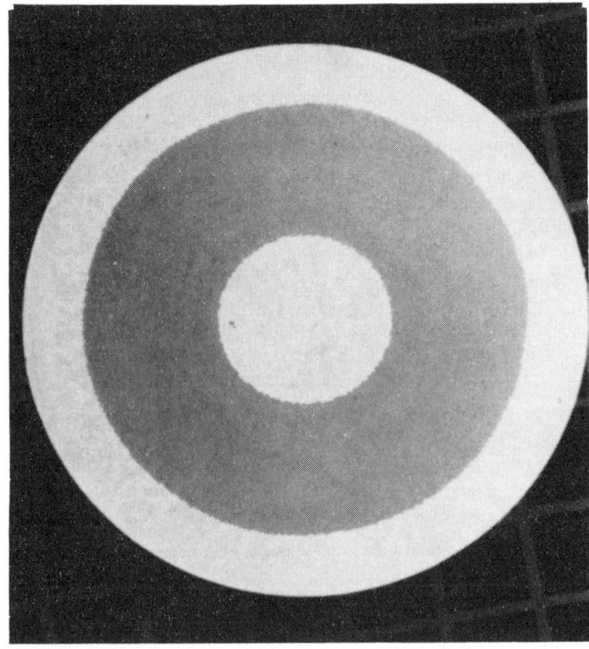

Figure 3
Cross section of an approximately 23 000 filament conductor made by the single-stack technique (courtesy of Supercon Inc.)

fabrication methods which will be discussed in this section. There is also a Dutch process based on niobium-clad $NbSn_2$ powder filaments (Elen et al. 1981) which has not yet been developed to a full commercial stage. The filament size achievable by this method appears to be significantly larger than that made by the other methods.

2.1 The Bronze Method

The process most frequently used in the past is one of "double stacking." A drilled 13.5 wt% tin bronze billet, similar to that shown in Fig. 1a, is packed with rods of pure niobium or niobium containing small percentages of titanium or tantalum. After extrusion, this is restacked into a copper can with a barrier layer of either niobium or tantalum to separate the copper stabilizer from the tin in the bronze. The principal drawback to this process is that the bronze work hardens rapidly and requires frequent annealing steps, which are expensive and tend to result in prereaction and the formation of brittle Nb_3Sn during fabrication. Although recent work using the process has overcome the technical problems, the relatively high cost of the process has led to attempts to develop alternative methods for commercial production of Nb_3Sn multifilamentary wire.

2.2 The Internal Tin Process

The internal tin process, developed by the IGC Advanced Superconductors Inc. (Schwall et al. 1983) and illustrated schematically in Fig. 4, does not require the repeated intermediate anneals. The product is free from prereaction and the method is more economical than some of the others. The process as originally developed required the production of two components: a composite tube and a stabilizer tube. The composite was made in a tubular form by a process similar to that shown in Fig. 1b and contained many filaments of niobium, or alloyed niobium, in a copper matrix. The stabilizer was made in a similar manner and contained a diffusion barrier of tantalum or niobium. After hot extrusion, the composite tube was filled with tin, or a high-tin alloy, and drawn down to a suitable size for restacking. Recently, this process has undergone considerable modification to increase piece length and improve the J_c which can be achieved for given ac losses (Gregory et al. 1991). Figure 5a shows a strand with a tantalum barrier and Nb–7.5 wt.%Ta strands made by drawing down a restacked assembly. A conductor of this type has been supplied as one of the prototype designs for the toroidal coils for NET (next European torus). Figure 5b shows the cross section of a conductor made by the tubular tin source modification of the internal tin method. Such a conductor has high current density but is, at the present time, only available in small diameter ($\leqslant$ 0.5 mm) wires. Both these conductors are shown before reaction to form the intermetallic compound Nb_3Sn.

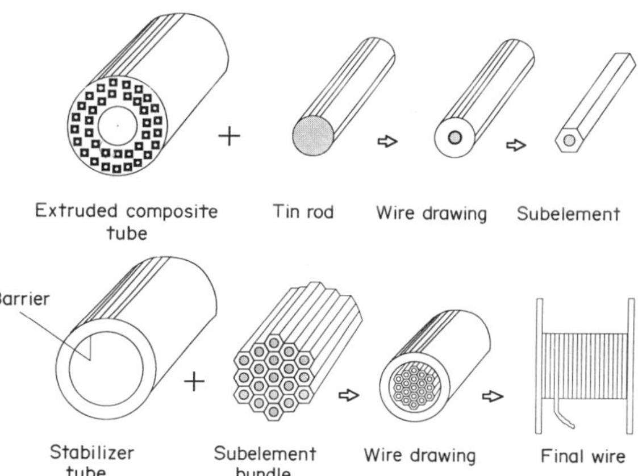

Figure 4
Fabrication steps in the internal tin process for making Nb_3Sn

2.3 The Modified Jelly Roll Process

The modified jelly roll (MJR) process, developed by Teledyne Wah Chang Albany (TWCA) (McDonald et al. 1983) and illustrated schematically in Fig. 6, consists of rolling expanded niobium, or niobium with small percentages of tantalum or titanium foil, interleaved with copper on a tin rod. A vanadium–niobium barrier layer separates the jelly roll configuration from the stabilizing copper. Many tin-core MJR units are restacked to reduce the tin diffusion distance from the core of each jelly roll and Fig. 7 shows such a multisubelement conductor.

3. Design Factors

3.1 Copper for Stability and Quench Protection

The desirability of having the filaments embedded in a material of high-conductivity copper has been alluded to in the introduction. For a more detailed explanation of the concepts of cryostability, adiabatic stability and dynamic stability see *Superconducting Magnets*.

From the point of view of the composite conductor design, it is important to provide copper in sufficient amounts, and in the correct distribution, so that a low-resistance current path is provided around any length of superconductor that has entered the normal state. The copper also conducts heat away from a "hot spot" and thus lowers the temperature of the superconductor by allowing it to return to the resistanceless state.

The type of stabilization and the degree of quench protection that the magnet designer requires will determine the ratio of copper to superconductor requested. This, together with the filament and conductor size, are frequently fixed in the specification

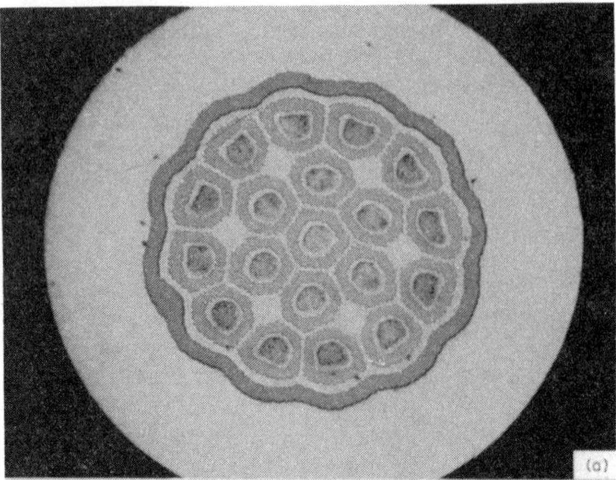

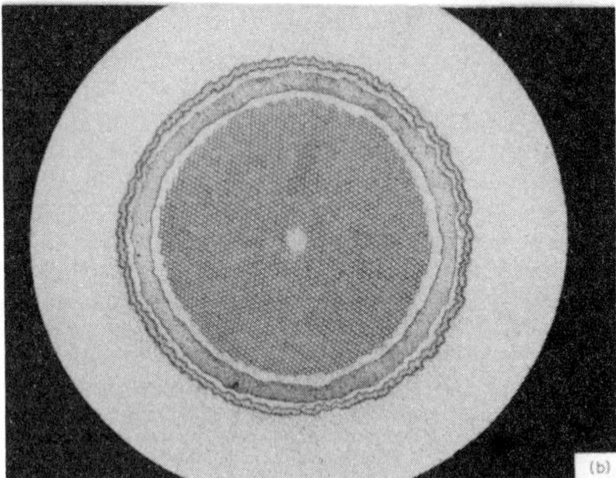

Figure 5
(a) Cross section of internal tin (tin core) Nb_3Sn conductor containing 19 subelements and (b) cross section of internal tin (tubular tin source) Nb_3Sn conductor

along with the current-carrying capability. The distribution of the copper in the cross section of the wire is not specified as frequently. While uniform distribution of the copper would intuitively provide the best protection, little work has been done on the effects of copper distribution on stability.

The importance on J_c of closely spaced filaments is outlined in Sect. 3.2 and, obviously, if the total amount of copper is to remain constant that which is removed from between the filaments must be located elsewhere in the conductor cross section. Changes in this copper distribution can have profound effects on the mechanical properties. This is particularly the case in the fabrication of multifilamentary NbTi. Here, the filament array becomes harder with each successive heat treatment while the copper is repeatedly annealed and remains relatively soft. These hardness differences markedly affect composite fabricability.

In the case of Nb_3Sn, the filaments are surrounded by relatively high-resistance bronze after the reaction to form the compound. It is therefore necessary to provide high-purity copper either in the center and/or on the outside of the filament array, or distributed within it. Such copper areas have to be surrounded by barrier layers to prevent their contamination by the tin in the bronze. The larger the number of discrete copper areas the more effective they will be from the point of view of providing stability but the greater the barrier volume introduced into the cross section becomes. This increase will lower the overall current-carrying capacity of the conductor.

From this section, it can be seen that the provision of copper for stability and quench protection in both NbTi and Nb_3Sn multifilamentary conductors can markedly affect mechanical behavior and thus fabricability and performance characteristics.

3.2 Improvements in Current-Carrying Capacity

The design challenge, particularly in NbTi conductors, has been to obtain the highest J_c at relatively high fields in fine filamentary material (see *Critical Current Density*). J_c values were raised by 70% (from $1800\ \mathrm{A\ mm^{-2}}$ to $3000\ \mathrm{A\ mm^{-2}}$ at a field of 5 T) and filament sizes reduced by a factor of two or three (from 8–15 μm to 5 μm) during the late 1980s. These improvements were achieved by optimizing the heat treatment (Larbalestier et al. 1985, Li and Larbalestier 1987), reducing the filament sausaging and preventing the formation of the copper–titanium intermetallics (see *Niobium–Titanium Alloys*; Gregory 1988, 1989).

NbTi with a high degree of homogeneity is now used in the designs requiring high J_c in fine filaments. In addition, thin niobium barriers surround the NbTi filaments (see Fig. 1c) to ensure that no intermetallics

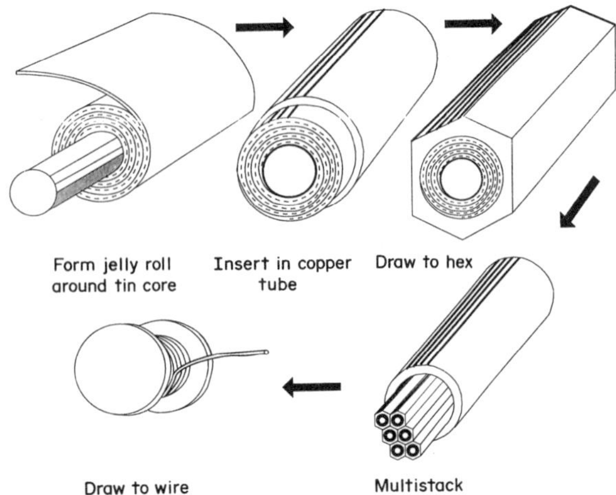

Figure 6
Fabrication steps in the MJR process for making Nb_3Sn

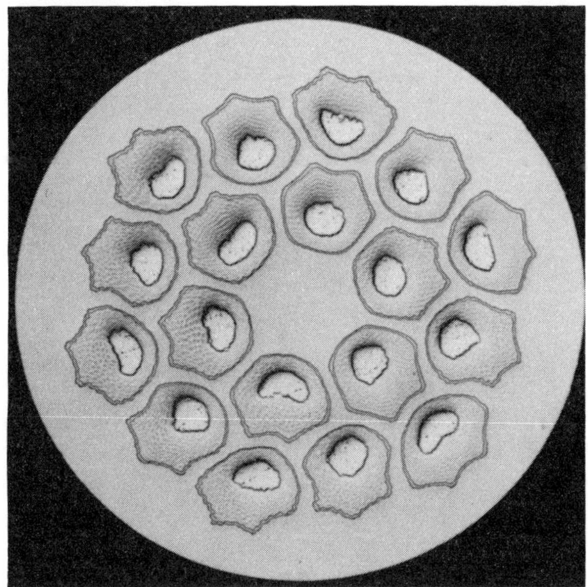

Figure 7
Cross section of MJR conductor containing 18 subelements before reaction to form the intermetallic compound (courtesy of TWCA)

form between the titanium in the NbTi and the copper in the matrix. These barriers allow the manufacturer to employ a wider range of fabrication and heat-treatment procedures than would otherwise be possible.

The creation of a uniform and small spacing between the filaments, particularly in the case of filaments of small diameter, has been shown to significantly decrease sausaging (Gregory 1988). In order to achieve the small spacing the procedure of inserting round NbTi rods into copper tubes with a hexagonal exterior configuration, normally employed for the manufacture of larger filamentary material (see Fig. 1b), was abandoned in the techniques for making the finer filamentary material. A "double extrusion" (see Fig. 1c) was adopted instead. This enabled thinner copper layers to be developed on the monofilaments as well as allowing the manufacturer further cold-work and heat-treatment flexibility. The use of double stacking (as distinct from double extrusion) has been discontinued in many cases for filament numbers below 10 000, because of the desirability of maintaining a uniform filament array and a constant spacing between the individual filaments. In most billet designs, as the copper between the filaments is minimized to reduce sausaging, large areas of copper must be carefully accommodated elsewhere in the cross section, as a fixed copper-to-superconductor ratio must be maintained.

The small spacing between the filaments can lead to electrical coupling, particularly at low fields and, hence, much of the advantage of subdividing the NbTi (reduced magnetization and losses under cyclic fields) is eliminated. The replacement of a small amount of low-resistivity copper between the filaments by copper–manganese alloy has solved the coupling problem (Collings 1988, Gregory 1988).

When billets are assembled, particularly by the "single-stacking" techniques, sophisticated cleaning and assembly techniques have to be employed in order to ensure that good bonding is achieved in subsequent fabrication because of the many surfaces involved. This is believed to be of particular importance in the achievement of the long piece lengths which are of great importance if economical fabrication is to be performed.

4. Conductors for Large Devices

Large magnets require conductors to carry currents far in excess of those that can be transported by the single strands of a small diameter wire described previously in this article and, therefore, for this type of application a multistrand cable or a composite monolith is required. The cables for the SSC dipoles are of the keystoned Rutherford-type flat cable, developed at the Rutherford laboratory in the UK. The inner and outer grades have 30 and 36 strands, respectively. An earlier design with 23 strands is shown schematically in Fig. 8a.

As has been previously mentioned for low-field MRI magnets, a single strand containing relatively few filaments (see Fig. 1a) and with a copper-to-superconductor ratio of around 7:1 is frequently used. Some of the higher field (1–2 T) MRI magnets, however, use a "wire-in-channel" type of conductor made by continuously soldering a strand that contains relatively few filaments and a low copper-to-superconductor ratio (1.35:1) into a high-purity copper channel. The overall copper-to-superconductor ratio of the resulting composite is approximately 15:1. This wire-in-channel approach (see Fig. 8b) enables a relatively high current to be obtained in a large cross-sectional conductor. In addition, long lengths of conductor can be relatively easily produced at low cost. Both these advantages are particularly valuable in MRI and many other applications.

Another technique of adding additional stabilizer to a high-current-density core is to add aluminum by extrusion cladding (Kanithi et al. 1988). Practical aluminum-stabilized conductors based on continuous cladding of multifilamentary NbTi–Cu composites have been successfully developed. Many geometries can readily be fabricated to take advantage of the high-purity aluminum with a resistivity at room temperature 2000 times that at 4.2 K and these are commercially available at competitive prices.

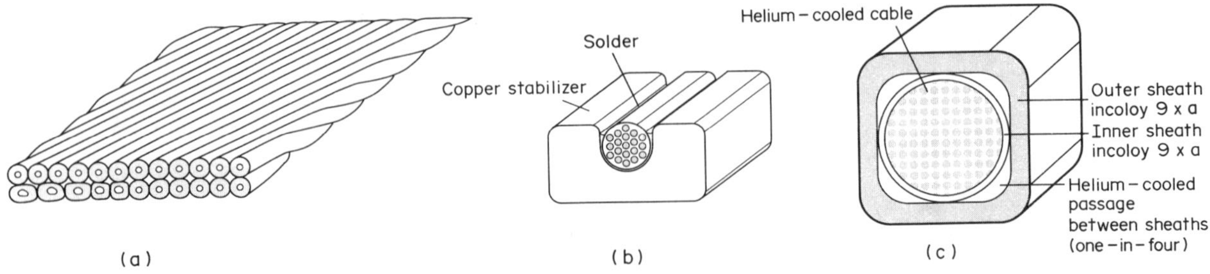

Figure 8
Methods of fabrication for large conductors for high-current applications: (a) Rutherford cable, (b) wire-in-channel conductor and (c) CICC

Soldering has also been used to attach an aluminum stabilizer to a normal copper-matrix superconductor. Conductors of this type are used primarily for detector magnets for accelerators.

While not many large Nb_3Sn coils have been made in the US, the notable exception is the Westinghouse LCP coil (Dresner et al. 1988). This employed one of the first cable-in-conduit conductors (CICCs), a design where the liquid helium flows through the conduit instead of the whole coil being immersed in a bath of the cryogen. A similar conductor design was used in the 12 T program at the Lawrence Livermore laboratory. Both conductors were composed of early bronze strands exhibiting low current densities. Figure 8c shows a schematic cross section of a more recent design of a CICC (Hoenig et al. 1988).

In Europe, a considerable amount of work has been done, particularly in Germany, Switzerland, the Netherlands and Italy, on the design of large A15 coils for the insert to a large test coil named "Sultan" in Switzerland and for the toroidal and poloidal coils for NET. Many of these conductors are of the forced flow type, some with special helium conduits. In Japan, the development of large-scale Nb_3Sn conductors has made considerable progress. Toroidal coils have been developed successfully for use in the test module coil in JAERI and in the demonstration poloidal coil including a Nb_3Sn pulse coil (DPC-EX), using a forced flow conductor that is currently under development.

5. *Practical High-Temperature Superconductors*

The discovery of high-temperature superconductors in 1986 was widely acclaimed as one of the most important scientific breakthroughs of the twentieth century. This was confirmed by an immediate award of the Nobel prize for physics to the inventors (Bednorz and Müller 1986). A flurry of activity unprecedented in material science, and perhaps in any scientific endeavor, followed. The aim of this activity was to raise T_c even further. It led to the discovery of LaSrCuO, $T_c = 40$ K (Kishio et al. 1987), YBaCuO, $T_c = 95$ K (Wu et al. 1987), BiSrCaCuO, $T_c = 85$–115 K (Maeda et al. 1988) and TlBaCaCuO, $T_c = 125$ K (Sheng and Herman 1988). This quest still continues although it is now several years since a confirmed increase in T_c above that of the thallium-based superconductor has been reported.

A new generation of large-scale practical superconductors does not automatically appear as soon as a material with a markedly improved T_c is found. Many other accompanying properties are required, including first an ability to carry high current in high fields and, second, a fabrication procedure to manufacture fine multifilamentary composites.

From the very beginning it was realized that the oxide superconductors are relatively brittle and could present fabrication problems for the production of long lengths of multifilamentary composites. This fact was rapidly overshadowed by the more serious one that high J_cs were difficult to achieve, particularly in bulk materials. For this reason, work concentrated on thin films where good properties appeared to be easier to attain. In this electronic device area, applications seemed to be more realistically achievable. For bulk practical applications concentration has been on the bismuth-based high-temperature superconductor composites and, to a lesser extent, on the thallium-based composites. For such applications, J_cs of the order of 1000 A mm^{-2} at the operating temperature and magnetic field are required. At present, there are problems in achieving such J_c values in wires or tapes at 77 K because of weak links or flux creep but, recently, very significant progress has been achieved in this area by many investigators. Various techniques have been used to process oxide superconductors into wires and tapes, but the one in most frequent use is the powder-in-tube method (see *Oxide Superconductors: Ceramic Processing*). The appropriate powders are mixed, calcined, sintered and ground. They are then put into silver tubes and drawn to round wires. These are then rolled or pressed into tapes as thin as 0.1 mm. The properties appear to be highly dependent on the cycles of working and heat treating employed. The most outstanding results at 77 K have been on the bismuth-based 2223 phase (Sato et al. 1991). They reported J_cs at 77.3 K of 470 A mm^{-2} at 0 T, and 110 A mm^{-2} at 1 T.

From a practical point of view, the most interesting data was first presented in 1989 by Vacuumschmelze (Tenbrink et al. 1991a) at 4.2 K and 26 T. They obtained 150 A mm^{-2} in Bi-2212/Ag wires. Recently, Tenbrink et al. (1991b) reported that the Bi-2223 compound is better suited for application at 77 K due to its higher T_c and especially due to the weaker magnetic field dependence of J_c at this temperature (Sato et al. 1991, Tenbrink et al. 1991a). The Bi-2212 compound may, however, be the choice for application at 4.2 K for magnetic fields greater than 20 T due to its much simpler processing conditions. Both bismuth-based compounds seem to show little dependency of J_c on field from 5 T to 23 T and values as high as 1030 A mm^{-2} at 23 T have been reported (Sato et al. 1991) in Bi-2223/Ag tapes, although others appear to have had difficulty in achieving such high values.

Such tapes have been restacked using the "double-stacking" technique into a wire containing 1296 filaments (Sato et al. 1991). Also prototypes of coils and current leads have been produced. One coil generated 0.014 T at 77 K and 0.088 T at 4.2 K. Critical current properties were measured with back-up fields of up to 0.1 T at 77 K and 3.7 T at 4.2 K.

6. Commercial Applications

The basic superconducting properties of J_c, H_{c2} and T_c do not in themselves determine solely which materials are used commercially. The ability to economically produce the composites in the form of multifilamentary wire is mandatory for many applications, although tapes have experienced limited usage in the past. Long lengths of these wire composites with reliable properties, invariant along the entire length, are required. The result of this is that NbTi, a ductile material with a very low work hardening capability, is used at 4.2 K for most applications. Where NbTi cannot meet all the field or temperature requirements, Nb_3Sn, a brittle intermetallic compound, is generally employed although the ternary NbTiTa at 1.8 K is being considered for some applications.

The discovery of high-temperature superconducting materials has not yet resulted in significant bulk applications, but recent property improvements have suggested that these materials may have significant potential in the future.

See also: A15 Compound Superconductors; AC Losses in Superconducting Materials; Critical Current Density; Flux Pinning and the Summation of Pinning Forces; Granular Superconductors; Niobium–Titanium Alloys; Oxide Superconductors: Ceramic Processing

Bibliography

Bednorz J G, Müller K A 1986 Possible high-T_c superconductivity in the Ba–La–Cu–O system. *Z. Phys. B* 64: 189–93

Collings E W 1986 *Applied Superconductivity Applications*, Vol. 2. Plenum, New York

Collings E W 1988 Stabilizer design considerations in fine filament Cu–NbTi composites. In: Clark A F, Reed R P (eds.) 1988 *Advances in Cryogenic Engineering*, Vol. 34. Plenum, New York, pp. 867–78

Dresner L, Fehling D T, Lubell M S, Lue J W, Luton J N, McManamy T J, Shen S S, Wilson G T 1988 Stability tests of the Westinghouse coil in the international fusion superconducting magnet test facility. *IEEE Trans. Magn.* 24(2): 1779–92

Elen J D, Schinkel J W, van Wees A C A, van Beynen C A M, Hornsveld E M, Stahlie T, Veringa H J, Verkaik A 1981 Development of stabilized Nb_3Sn wire containing a reduced number of filaments. *IEEE Trans. Magn.* 17(1): 1002–5

Flükiger R 1991 Bronze processed Nb_3Sn multifilamentary wires with artificial Ta pinning centers in filaments consisting of jelly rolled Nb/Ta sheets. In: Reed R P, Fickett F R (eds.) 1991 *Advances in Cryogenic Engineering*, Vol. 38. Plenum, New York

Gregory E 1988 Recent advances in commercial multifilamentary NbTi wires in the United States. In: Reed R P. Xing Z S, Collings E W (eds.) 1988 *Cryogenic Materials '88*, Vol. 1. International Cryogenics Materials Conference Publishing, Boulder, CO, pp. 361–71

Gregory E 1989 Conventional wire and cable technology. *Proc. IEEE* 77(8): 1110–23

Gregory E, Kreilick T S, Wong J, Collings E W, Marken K R Jr, Scanlan R M, Taylor C E 1989 A conductor with uncoupled 2.5 µm diameter filaments, designed for the outer cable of SSC dipole magnets. *IEEE Trans. Magn.* 25(2): 1926–9

Gregory E, Ozeryansky G M, Zeitlin B A 1991 Improvement of the structure and properties of internal tin Nb_3Sn conductors. In: Reed R P, Fickett F R (eds.) 1991 *Advances in Cryogenic Engineering*, Vol. 38. Plenum, New York

Hillmann H 1981 Fabrication technology of superconducting material. In: Foner S, Schwartz B B (eds.) 1981 *Superconductor Material Science, Metallurgy, Fabrications and Applications*, NATO Advanced Study Institute Series. Plenum, New York, pp. 275–388

Hoenig M O, Steeves M M, Gibson C R 1988 The selection of a 30 kA ohmic heating coil conductor. *IEEE Trans. Magn.* 24(2): 1452–4

Kanithi H C, King C G, Zeitlin B A 1989 Fine filament NbTi conductors for the SSC. *IEEE Trans. Magn.* 25(2): 1922–5

Kanithi H C, Phillips D, King C, Zeitlin B A 1988 Development and characterization of aluminum clad superconductors. *IEEE Trans. Magn.* 24(2): 1029–32

Kishio K, Kitazawa K, Kanbe S, Yasuda I, Sugü N, Takagi H, Uchida S I, Fueki K, Tanaka S 1987 New high-temperature superconducting oxides—$(La_{1-x}Sr_x)_2\ CuO_{4-\delta}$ and $(La_{1-x}Ca_x)_2\ CuO_{4-\delta}$. *Chem. Lett.* 429–32

Larbalestier D C, West A W, Starch W, Warnes W, Lee P, McDonald W R, O'Larey P, Hemachalam K, Zeitlin B A, Scanlan R M, Taylor C 1985 High critical current densities in industrial-scale composites made from high homogeneity Nb 46.5 wt%Ti. *IEEE Trans. Magn.* 21(2): 269–72

Li C G, Larbalestier D C 1987 Developments of high critical current densities in niobium 46.5 wt% titanium. *Cryogenics* 27(4): 171–7

McDonald W K, Curtis C W, Scanlan R M, Larbalestier D C, Marken K Jr, Smathers D B 1983 Manufacture and evaluation of Nb_3Sn conductors fabricated by the MJR method. *IEEE Trans. Magn.* 19(3): 1124–7

Maeda H, Tanaka Y, Fukutomi M, Asano T 1988 A new high-T_c oxide superconductor without a rare-earth element. *Jpn. J. Appl. Phys.* 27: L209–10

Motowidlo L R, Kanithi H C, Zeitlin B A 1990 NbTi superconductors with artificial pinning structures. In: Reed R P, Fickett F R (eds.) 1990 *Advances in Cryogenic Engineering*, Vol. 36A, Plenum, New York, pp. 311–16

Sato K, Kato T, Masuda T, Hagata M, Iwata K, Mitsui T 1991 High-J_c silver-sheathed Bi-based superconducting wires. *IEEE Trans. Magn.* 27(2): 1231–8

Schwall R E, Ozeransky G M, Hazelton D W, Cogan S F, Rose R M 1983 Properties and performance of high current density Sn-core process MF Nb_3Sn. *IEEE Trans. Magn.* 19(3): 1135–8

Sheng Z Z, Herman A M 1988 Superconductivity in the rare-earth-free Tl–Ba–Cu–O system above liquid-nitrogen temperature. *Nature* 332: 55–8

Suenaga M, Clark A F 1980 *Filamentary A-15 Superconductors.* Plenum, New York

Tenbrink J, Wilhelm M, Heine K, Krauth H 1991a Development of high-T_c superconductor wires for magnet applications. *IEEE Trans. Magn.* 27(2): 1239–46

Tenbrink J, Wilhelm M, Heine K, Krauth H 1991b Processing of high-T_c superconductor wires for magnetic application. In: Reed R P, Fickett F R (eds.) 1991 *Advances in Cryogenic Engineering*, Vol. 38. Plenum, New York

Wu M K, Ashburn J R, Toring C J, Hor P H, Meng R L, Gao L, Huang Z J, Wang Y Q, Chu C W 1987 Superconductivity at 93 K in a new mixed-phase Ya–Ba–Cu–O compound system at ambient pressure. *Phys. Rev. Lett.* 58: 908–10

E. Gregory
[IGC Advanced Superconductors,
Waterbury, Connecticut, USA]

N

Nanocrystalline Soft Magnetic Materials

Nanocrystalline materials are single-phase or multiphase polycrystals, the crystal size of which is in the nanometer range, typically 5–50 nm in at least one dimension. They can be produced by processes such as compaction of nanometer-size powders, deposition techniques or by crystallization from the amorphous state.

The magnetic properties of fine particle systems have so far been found to be typical of high-coercive materials (see *Elongated-Single-Domain Magnets*). The reduction of particle size is a well-known approach to subdividing a material into single-domain particles, which increases coercivity H_c towards a maximum controlled by the anisotropies present (Luborsky 1961). Although it is known that H_c decreases again for very small particles, the regime in which the grain size is much smaller than the single-domain particle and domain wall sizes has remained relatively unexplored.

Only recently has it become evident that smallest grain sizes also offer the opportunity of tailoring low coercive materials with good soft magnetic properties. The most prominent example is devitrified glassy FeCuNbSiB (Yoshizawa et al. 1988), which has a homogeneous ultrafine grain structure of bcc FeSi with grain sizes of typically 10 nm. Further examples are devitrified glassy alloys based on CoNbB (O'Handley et al. 1985) or FeZrB (Suzuki et al. 1990) and multilayered structures like Fe/CoNbZr (Dirne et al. 1988); further examples are to be expected since the field is rapidly evolving.

The basic mechanism underlying the soft magnetic properties of nanocrystalline materials is the suppression of magnetocrystalline anisotropy due to ferromagnetic exchange interaction. The situation resembles the case of amorphous alloys where it is well established that atomic-scale local anisotropy is randomly averaged out so that there is no overall anisotropy contribution to the magnetization process (see *Metallic Glasses*).

This article focuses on nanocrystalline FeCuNbSiB alloys produced via crystallization from the amorphous state. They are of particular interest because, apart from their suppressed magnetocrystalline anisotropy, their low or vanishing saturation magnetostriction leads to soft magnetic properties comparable with those of permalloys and cobalt-based amorphous alloys. The general features exhibited by FeCuNbSiB alloys, however, also apply to other nanocrystalline magnetic materials.

1. Microstructure

The alloy composition originally proposed and subsequently studied most widely is 73.5 at.%Fe–1 at.%Cu–3 at.%Nb–13.5 at.%Si–9 at.%B (Yoshizawa et al. 1988). It can be considered as a typical FeSiB metallic glass composition with additions of copper and niobium. If formed by a single-roll high-rate strip casting process into ribbons of 15–25 μm thickness it solidifies in the amorphous state. A subsequent anneal at $T_a = 500$–$600\,°C$ produces a homogeneous, ultrafine grain structure of bcc FeSi with grain sizes D that are typically around 10 nm and random texture, embedded in a residual amorphous (minority) matrix (see Fig. 1). These features are the basis for the excellent soft magnetic properties indicated by high values of the initial permeability μ_i (up to 100 000) and correspondingly low values of coercivity H_c. Annealing above 600 °C leads to the formation of iron borides and coarsening which deteriorates the soft magnetic properties significantly.

The formation of the nanocrystalline structure is ascribed to the combined effects of copper and niobium or other group V or VI elements such as molybdenum, tantalum or tungsten (Yoshizawa et al. 1990). Copper is thought to enhance the nucleation of the bcc grains. Local clustering associated with bcc-like topological short-range order seems to be a likely explanation. The addition of niobium increases the chemical short-range order and, thus, stabilizes the residual amorphous regions. This inhibits the

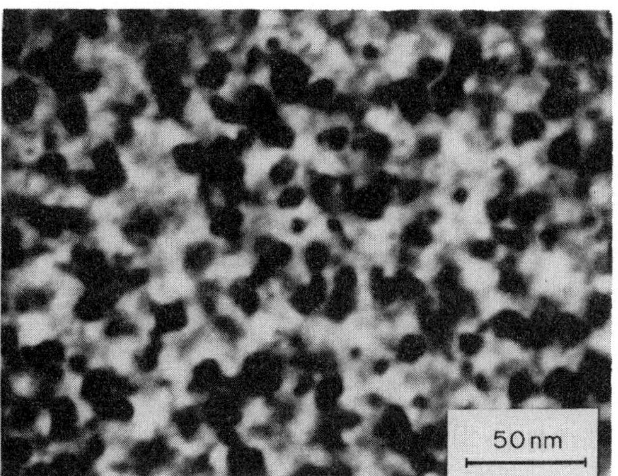

Figure 1
Transmission electron micrograph of 73.5 at.%Fe–1 at.% Cu–3 at.%Nb–13.5 at.%Si–9 at.%B annealed for 1 h at 540 °C

formation of borides and, at the same time, impedes coarsening of the bcc grains.

The basic conditions of formation were found to be determined by the onset of crystallization on temperature and composition. The nanocrystalline state can be attained only by (a) alloying additions which lead to two clearly separated stages of crystallization with onset temperatures T_{x1} and T_{x2} and (b) annealing at $T_{x1} < T_a < T_{x2}$ such that only the phase forming above T_{x1} is crystallized.

The nanocrystalline state of devitrified glassy alloys is characterized by crystallites of, typically, 5–20 nm in size embedded in a residual amorphous matrix. The crystalline volume fraction v_{cryst} ranges from about 50% to 80% depending on alloy composition and heat treatment. Generally, v_{cryst} decreases with decreasing grain size, which can be understood from the increased number of interfaces. A simple geometric estimate relates the grain size D and the crystalline volume fraction v_{cryst} to the thickness δ of the amorphous interfacial layer by $\delta \simeq (1 - v_{cryst})D/3$. Typically, this yields a thickness $\delta \simeq 1$–2 nm. These features are somewhat similar to nanocrystalline materials produced by compaction of nanometer-size powders which, however, are characterized by an intergranular linkage that has a wide spectrum of interatomic spaces with practically no short-range order. This is in marked contrast to nanocrystallized glassy alloys where the interfacial layer is in a well-relaxed amorphous state with pronounced short-range order.

The two-phase nature of nanocrystalline glassy alloys is also seen in the temperature dependence of saturation magnetization which reveals two magnetic phase transitions, typically at $T_{C1} \simeq 200$–400 °C and $T_{C2} \simeq 600$–650 °C reflecting the local Curie temperatures T_C of the amorphous matrix and the bcc grains, respectively (Herzer 1989).

Since direct chemical analysis (e.g., by electron beam microprobe) is inaccurate on the nanometer scale, lattice parameter and intrinsic magnetic properties have been evaluated to determine the composition. Boron is practically insoluble in bcc iron ($\ll$0.01 at.%); the solubilities of copper and niobium are low (< 0.2 at.% copper, < 0.1 at.% niobium at 550 °C). If these or even lower values apply to the FeSi solid solution, too, the bcc phase consists essentially of FeSi. Curie temperature and lattice parameter measurements indicate an enhanced silicon concentration up to 20–22 at.% silicon for the bcc grains, suggesting that, effectively, all silicon has partitioned into the bcc grains. Accordingly, the amorphous interfacial phase mainly consists of FeNbB enriched with boron (25–30 at.%) and niobium (5–10 at.%).

2. Magnetic Properties

The relation between microstructural and magnetic properties is shown in Fig. 2 where the coercivity H_c of nanocrystalline iron-based alloys has been plotted versus the average grain size D together with the data for amorphous and conventional polycrystalline soft magnetic materials. The $1/D$ dependence for large grain sizes reflects the classical rule that best soft magnetic properties require very large grains (see *Nickel–Iron Alloys*). Low coercivities, however, are also found for amorphous alloys ("grain size" of the order of atomic distances; see *Metallic Glasses*) and in the nanocrystalline FeCuNbSiB alloys for grain sizes D less than 15 nm. The steep variation of coercivity for grain sizes between about 10 nm and 40 nm demonstrates the close proximity of soft and hard magnetic behavior in the same alloy system.

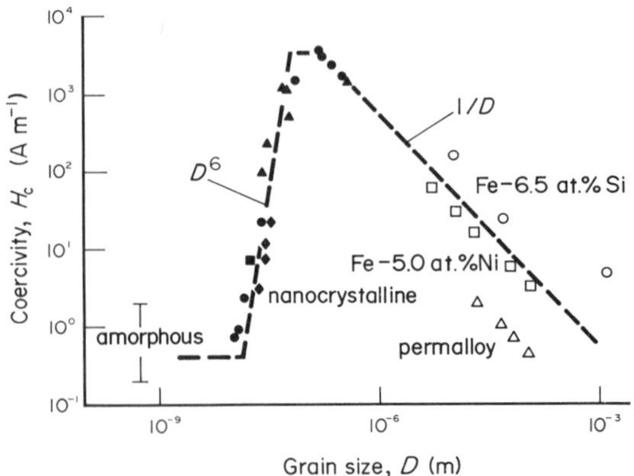

Figure 2
Coercivity versus grain size for various soft magnetic metallic alloys: ▲, FeNbSiB (Herzer 1990); ●, FeCuNbSiB (Herzer 1990); ◆, FeCuVSiB (Sawa et al. 1990); and ■, FeZrB (Suzuki et al. 1990)

The grain size dependence of magnetic properties originates from the counterplay of local magnetocrystalline anisotropy energy and ferromagnetic exchange interaction. The ferromagnetic exchange interaction tends to align the magnetic moments, preventing the magnetization from following the easy directions of each individual grain if the grain size becomes too small. As a consequence the effective anisotropy for the magnetic behavior is an average over several grains and, thus, reduced in magnitude. The critical scale where the exchange interaction starts to dominate is given by the ferromagnetic exchange length

$$L_{ex}^0 = (A/K_1)^{1/2} \tag{1}$$

where A denotes the exchange stiffness and K_1 is the local magnetocrystalline energy constant. The exchange length L_{ex}^0 is about 35 nm for bcc Fe–20 at.%Si ($K_1 \simeq 8$ kJ m^{-3}, $A \simeq 10^{-11}$ J m^{-1}). The degree to which the magnetocrystalline anisotropy is finally averaged out has been successfully addressed in terms of the so-called random anisotropy model originally proposed for amorphous ferromagnets (Alben et al. 1978). As a result the average anisotropy $\langle K \rangle$ relevant to the magnetization process is given by (Herzer 1989, 1990)

$$\langle K \rangle \simeq K_1 (D/L_{ex}^0)^6 = K_1^4 D^6 / A^3 \quad (2)$$

which holds as long as the grain size is smaller than the ferromagnetic exchange length. An estimate for grain sizes $D = 10$ nm, using the material parameters of bcc Fe–20 at.%Si yields $\langle K \rangle = 4\,\mathrm{J\,m^{-3}}$ which is almost negligible compared with the local magnetocrystalline anisotropy of $K_1 = 8000\,\mathrm{J\,m^{-3}}$. High-resolution Kerr effect studies of nanocrystalline 73.5at.%Fe–1at.%Cu–3at.%Nb–13.5at.%Si–9at.%B indeed reveal very wide domain walls of about 2 μm thickness indicative of the low effective anisotropy of the material (Schäfer et al. 1991). The generalization of Eqn. (2) to other dimensions yields a $D^{2/3}$ and D^2 dependence of $\langle K \rangle$ for one- and two-dimensional structures, respectively, and may be applied, for example, to multilayers (cf. Dirne et al. 1988, Herzer 1990).

The average anisotropy $\langle K \rangle$ essentially determines the soft magnetic properties of the material. In the regime where D is less than L_{ex}^0, H_c and inverse initial permeability $1/\mu_i$ are directly proportional to $\langle K \rangle$ irrespective of the detailed mechanisms of the magnetization reversal. Coercivity, for example, is explicitly given by (Herzer 1990)

$$H_c = p_c \frac{\langle K \rangle}{J_s} \simeq p_c \frac{K_1^4 D^6}{J_s A^3} \quad (3)$$

where J_s is the saturation magnetization and p_c is a dimensionless prefactor with, typically, $p_c \simeq 0.1$–0.2. The behavior of the initial permeability μ_i, being essentially inversely proportional to H_c, can be described analogously (Herzer 1990). As schematically indicated in Fig. 2, the reduction of H_c with decreasing grain size D is, in practice, limited to a lower bound due to more long-range inhomogeneities such as surface defects, induced anisotropies and internal mechanical stresses.

The suppression of magnetocrystalline anisotropy requires that the grains are coupled by exchange interaction via the interfacial phase. Since the amorphous interfacial component has a significantly lower Curie temperature than the bcc grains, this coupling is reduced with increasing temperature in terms of the reduction of the magnetic moment in the interfacial phase. As a consequence, coercivity increases and permeability decreases with increasing temperature, the reverse of the temperature dependence seen in ferromagnetic metallic glasses. The effect becomes most evident when the amorphous matrix becomes paramagnetic. Coercivity then increases by orders of magnitude (Herzer 1989) and the domain structure changes from wide domains to a pattern of small irregular domains (Schäfer et al. 1991), indicating that the exchange interaction between the crystallites is largely interrupted.

An additional feature of the FeCuNbSiB alloys is that the originally high positive saturation magnetostriction of $\lambda_s \simeq +2.3 \times 10^{-5}$ decreases significantly on the formation of the nanocrystalline state by about one order of magnitude. The magnetostriction in the nanocrystalline state decreases with increasing silicon content and passes through zero at about 16 at.% silicon. The behavior of λ_s can be understood from the high silicon content of the bcc grains and the balance in magnetostriction among the bcc grains and the residual amorphous matrix, that is,

$$\lambda_s \simeq v_{cryst} \lambda_s^{FeSi} + (1 - v_{cryst}) \lambda_s^{am} \quad (4)$$

The analysis of the experimental data for λ_s and v_{cryst} typically yields $\lambda_s^{am} \simeq +2.0$–$2.5 \times 10^{-5}$ for the residual amorphous phase and $\lambda_s^{FeSi} \simeq -5 \times 10^{-6}$ to -10^{-5} for the bcc FeSi grains, which is in reasonable agreement with the data of amorphous FeNbB alloys and bcc Fe–20 at.%Si, respectively.

A low saturation magnetostriction is essential to avoid magnetoelastic anisotropies arising from internal mechanical stresses. Thus, the increase of initial permeability on the formation of the nanocrystalline state is closely related to a simultaneous decrease of λ_s. However, the best soft magnetic properties are not necessarily to be found on the $\lambda_s = 0$ line but for compositions where the total amount of magnetoelastic and effective magnetocrystalline anisotropy is minimized.

3. *Soft Magnetic Properties Relevant to Application*

Low effective magnetocrystalline anisotropy and low or vanishing magnetostriction are the key to superior soft magnetic properties. There are only a few alloy systems that exhibit this combination of properties: the permalloys, Sendust, manganese–zinc ferrites, the amorphous cobalt-based alloys and, the nanocrystalline iron-based alloys. Among these, the nanocrystalline alloys offer the highest saturation magnetization, $J_s = 1.2$–1.3 T, which makes them an interesting candidate for soft magnetic application.

Coercivities of less than $1\,\mathrm{A\,m^{-1}}$ and initial permeabilities of more than 100 000 can be achieved which compares favorably with permalloys (see *Nickel–Iron Alloys*). The hysteresis loop is usually rounded with remanence-to-saturation ratios of about 0.5. However, as with other soft magnetic materials, magnetic field annealing provides a technique for tailoring the shape of the hysteresis loop. Remanence-to-saturation ratios ranging from about 0.05 to 0.90 may be achieved this way (Yoshizawa and Yamauchi 1989).

Nanocrystalline material exhibits a relatively high electrical resistivity of about 1.35 μΩ m comparable to that of amorphous alloys. Combined with the low ribbon thickness of, typically, 20 μm this yields a favorable frequency dependence of permeability and low eddy current losses up to frequencies of 100 kHz and more (Yoshizawa and Yamauchi 1989).

The soft magnetic properties of the new nanocrystalline iron-based alloys are very similar to those of amorphous, near-zero magnetostrictive cobalt-based alloys (see *Metallic Glasses*). The advantages of the nanocrystalline material, however, are a higher saturation magnetization and a significantly better thermal stability of the soft magnetic properties.

There is no doubt that nanocrystalline iron-based alloys will supplement the spectrum of existing soft magnetic materials in the near future. Possible applications already envisaged are tape wound cores for common mode chokes, saturable reactors, high-frequency transformers and magnetic heads.

See also: Measurements in Magnetic Materials

Bibliography

Alben R, Becker J J, Chi M C 1978 Random anisotropy in amorphous ferromagnets. *J. Appl. Phys.* 49: 1653–8

Dirne F W A, den Broeder F J A, Toolboom J A M, de Wit H J, Witmer C H M 1988 Soft-magnetic properties and structure of Fe/CoNbZr multilayers. *Appl. Phys. Lett.* 53: 2386–8

Herzer G 1989 Grain structure and magnetism of nanocrystalline ferromagnets. *IEEE Trans. Magn.* 25: 3327–9

Herzer G 1990 Grain size dependence of coercivity and permeability in nanocrystalline ferromagnets. *IEEE Trans. Magn.* 26: 1397–402

Luborsky F E 1961 High coercive materials. *J. Appl. Phys.* 32: S171–83

O'Handley R C, Megusar J, Sun S-W, Hara Y, Grant N J 1985 Magnetization process in devitrified glassy alloy. *J. Appl. Phys.* 57: 3563–5

Sawa T, Takahashi Y 1990 Magnetic properties of FeCu (3*d* transition metals) SiB alloys with fine grain structure. *J. Appl. Phys.* 67: 5565–7

Schäfer R, Hubert A, Herzer G 1991 Domain observation on nanocrystalline material. *J. Appl. Phys.* 69: 5325–7

Suzuki K, Kataoka N, Inoue A, Makino A, Masumoto T 1990 High saturation magnetization and soft magnetic properties of bbc Fe–Zr–B alloys with ultrafine grain structure. *Mater. Trans. JIM* 31: 743–6

Yoshizawa Y, Oguma S, Yamauchi K 1988 New Fe-based soft magnetic alloys composed of ultrafine grain structure. *J. Appl. Phys.* 64: 6044–6

Yoshizawa Y, Yamauchi K 1989 Effects of magnetic field annealing on magnetic properties in ultrafine crystalline Fe–Cu–Nb–Si–B Alloys. *IEEE Trans. Magn.* 25: 3324–6

Yoshizawa Y, Yamauchi K 1990 Fe-based soft magnetic alloys composed of ultrafine grain structure. *Mater. Trans. JIM* 307–14

G. Herzer
[Vacuumschmelze, Hanau, Germany]

H. Warlimont
[Am Weinberg, Freigericht, Germany]

Neodymium–Iron–Boron Alloys

Permanent magnets based on the $Fe_{14}Nd_2B$ phase were introduced in 1983 and are currently the strongest magnets available. They are the latest stage of the remarkable twentieth-century progress in permanent magnets, shown in Fig. 1 in terms of maximum energy product (the most commonly used measure of magnet quality) and intrinsic coercivity (the reverse magnetic field required to reduce net magnetization to zero).

For many centuries, carbon steels were the only improvement in permanent magnet materials over lodestones. Starting in about 1890, improved understanding of the metallurgy of alloy steels and, later, control of solid-state precipitation in iron-based alloys via heat treatment, led to substantial improvements. This path of materials development culminated in the alnicos, which retain today a large fraction of the magnet market. However, increasing availability of rare-earth metals in recent decades has led to a new development path based on compounds between 3*d* transition metals and rare earths. The iron–rare earth magnets represent the third "generation" of rare-earth magnets, the first and second generations being based on Co_5R and $(Co, Fe)_{17}R_2$ phases, respectively (see *Cobalt–Rare-Earth Compounds*). As seen in Fig. 1, the Fe–Nd–B magnets have provided a fourfold increase in energy product over the alnicos—and an increase of 50-fold over carbon steels, the most popular magnet a century ago.

1. Basic Properties of $Fe_{14}R_2B$

The crystal structure of $Fe_{14}R_2B$ (see Fig. 2) is complex; the tetragonal unit cell contains four formula units (68 atoms), six nonequivalent iron sites and two nonequivalent rare-earth sites. There is a six-layer stacking sequence along the *c* axis. The first and fourth layers are mirror planes containing iron, neodymium and boron atoms, and the other layers are puckered nets containing only iron atoms. Each boron atom is at the center of a trigonal prism formed by six iron atoms, three above and three below the Fe–R–B planes. (Such trigonal prisms are common to many transition metal–metalloid compounds, including FeB and Fe_3C.) For $Fe_{14}Nd_2B$, $c = 1.22$ nm and $a = 0.88$ nm. Both lattice parameters decrease with increasing atomic number of the rare earth, the usual "lanthanide contraction."

In Fig. 3, the saturation magnetization of various $Fe_{14}R_2B$ phases is shown in comparison with those of Co_5R, $Co_{17}R_2$ and $Fe_{17}R_2$ phases. Net magnetizations for compounds formed with heavy rare earths from gadolinium through ytterbium are below those for nonmagnetic rare earths such as yttrium because the magnetic moments of the rare-earth atoms are coupled antiparallel to the moments of iron or cobalt. In $Fe_{14}Nd_2B$, the neodymium and iron are coupled in

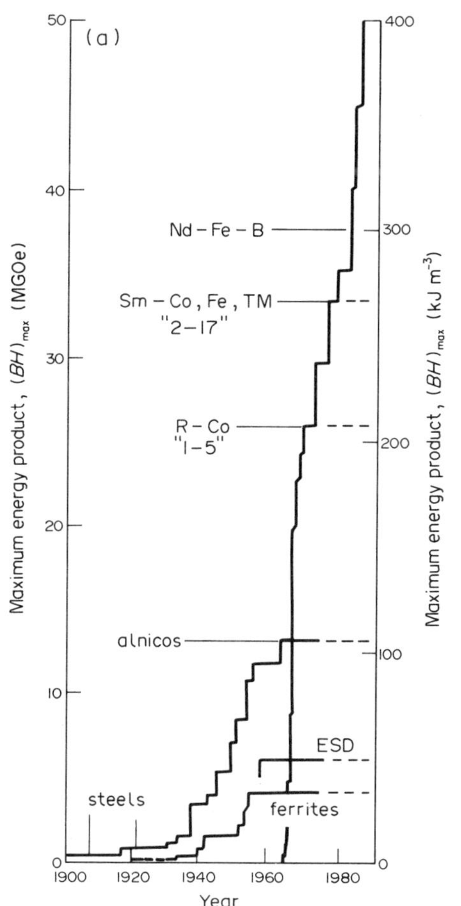

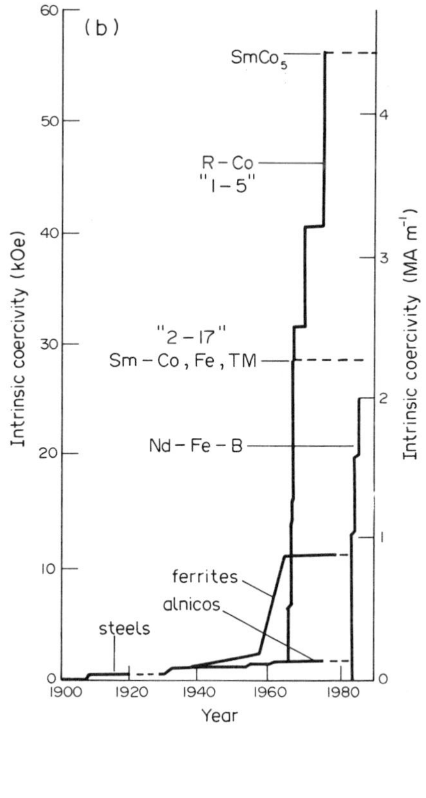

Figure 1
Chronology of magnet development since 1900: best reported laboratory values for (a) the maximum energy product and (b) intrinsic coercivity (after Strnat 1988)

parallel, resulting in a net magnetization of approximately 1.6 T, higher than that of any $Co_{17}R_2$ or Co_5R phase. Since the upper limit to the energy product is proportional to M_s^2, the higher magnetization of $Fe_{14}Nd_2B$ allowed the progress shown in Fig. 1.

To achieve a high energy product, a magnet must have both a high M_s and a high resistance to demagnetization in reverse fields, that is, a high coercivity. In rare-earth magnets, the basic origin of coercivity is a high uniaxial magnetocrystalline anisotropy that opposes rotation of the magnetization out of the low-energy direction or easy axis. In most $Fe_{14}R_2B$ compounds, the tetragonal *c* axis is the "easy" axis. The anisotropy field H_a, the transverse field required to rotate the magnetization from the easy axis to a transverse ("hard") direction, is shown in Fig. 4 for various $Fe_{14}R_2B$ compounds as a function of temperature. Although intrinsic coercivity is a highly structure-sensitive property, it seldom exceeds 10–20% of the anisotropy field, so that anisotropy field is a good measure of the potential of a material for coercivity. For nonmagnetic rare-earth ions (Y^{3+}, La^{3+}, Ce^{4+} and Lu^{3+}) and spherical ions (Gd^{3+}), the anisotropy field of $Fe_{14}R_2B$ at room temperature is about 2 MA m^{-1}. This reflects the contribution of the iron sublattice to anisotropy. For $Fe_{14}Nd_2B$, the Nd^{3+} ions make a substantial additional contribution, yielding a total value of about 6 MA m^{-1}. Although substantially below the value of about 25 MA m^{-1} for Co_5Sm, the anisotropy field of $Fe_{14}Nd_2B$ is sufficiently high to permit the development of high coercivity. Of the other $Fe_{14}R_2B$ compounds, only $Fe_{14}Pr_2B$ has the combination of high magnetization and high anisotropy field required to make it a promising candidate for permanent magnets.

A third important parameter for permanent magnet materials is the Curie temperature T_C. As shown in Fig. 5, the $Fe_{14}R_2B$ phases have substantially higher

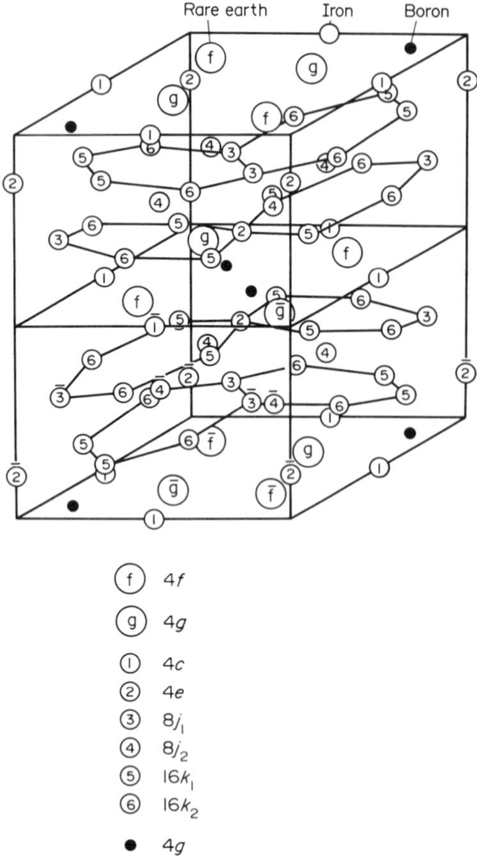

Figure 2
Schematic drawing of the crystal structure of $Fe_{14}R_2B$ (after Sagawa et al. 1987)

Curie temperatures than the $Fe_{17}R_2$ phases, the only binary Fe–R phases with high magnetization. However, they remain far inferior to Co_5R and $Co_{17}R_2$. This is a major weakness that limits the temperature range of application of Fe–Nd–B magnets.

2. *Sintered Magnets*

Most Fe–Nd–B magnets are made by a process analogous to that used for Co_5R and ferrite magnets. Castings are crushed and milled to powder below 10 µm in size, sufficiently small that most particles are single crystals. Powders are then aligned in a magnetic field, pressed, sintered at about 1100 °C and rapidly cooled. A postsintering heat treatment is usually used to enhance coercivity. To enhance densification via liquid phase sintering, magnet compositions are more neodymium rich and boron rich than the stoichiometric compound and the final microstructure

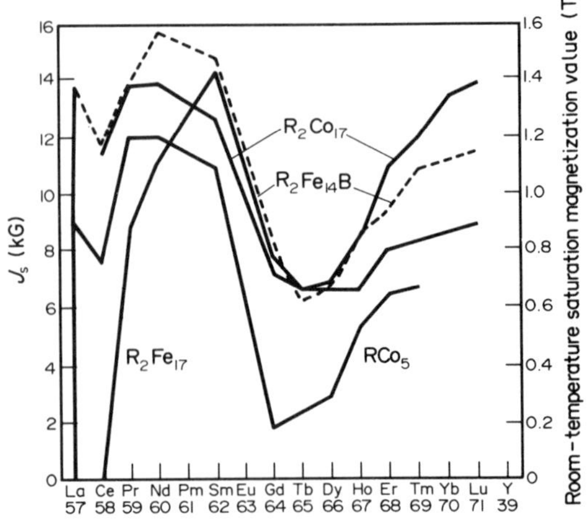

Figure 3
Room-temperature saturation magnetization values of various transition metal–rare-earth (–boron) compounds (after Strnat 1988)

consists of $Fe_{14}Nd_2B$ grains separated by a thin neodymium-rich grain-boundary phase, plus inclusions of $Fe_4Nd_{1.1}B_4$. These secondary phases are nonmagnetic at room temperature. The boron-rich phase is believed to have no beneficial effect on magnetic properties, but the neodymium-rich phase appears to enhance coercivity. On the other hand, it decreases resistance to corrosion and oxidation.

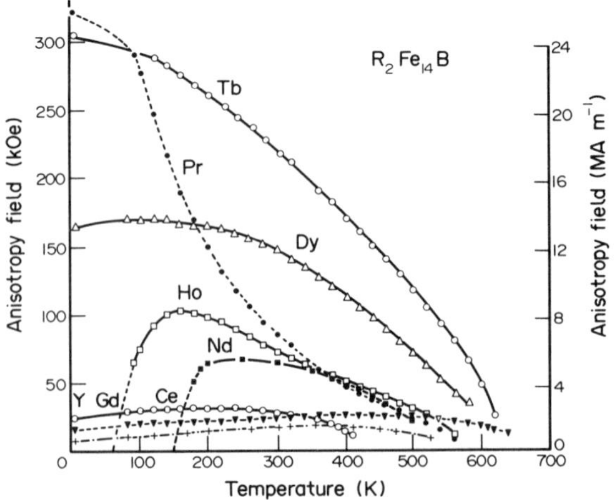

Figure 4
Anisotropy field of various $Fe_{14}R_2B$ compounds as a function of temperature (after Sagawa et al. 1987)

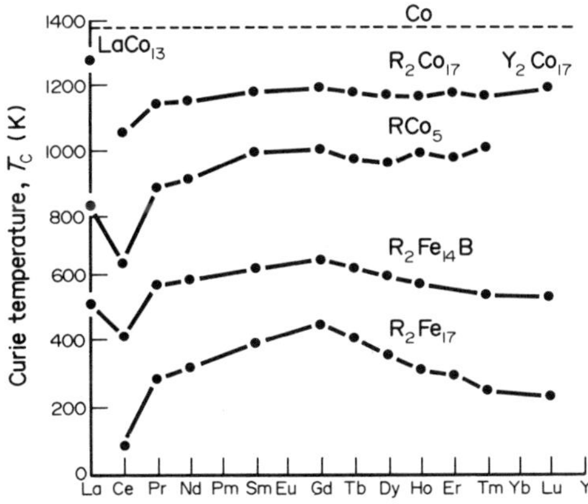

Figure 5
Curie temperatures of various transition metal–rare-earth (–boron) compounds (after Strnat 1988)

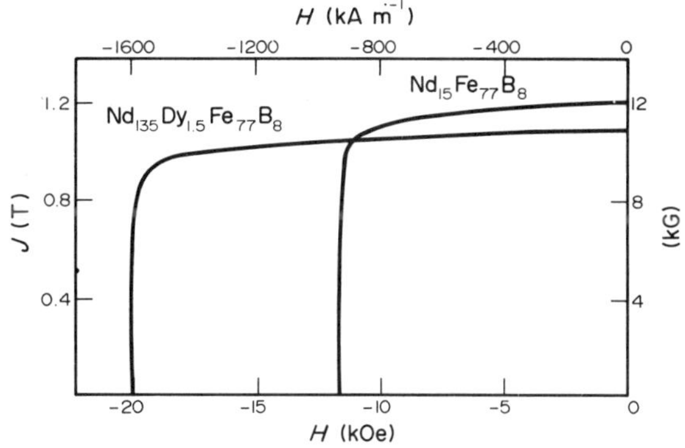

Figure 6
Magnetization vs reverse magnetic field for sintered Fe–Nd–B and Fe–Nd–Dy–B magnets, showing the enhancement in coercivity produced by dysprosium substitution (after Sagawa et al. 1984)

By minimizing the presence of secondary phases, maximum energy products of up to 400 kJ m^{-3} have been reported for aligned sintered magnets, but commercial Fe–Nd–B magnets typically cover the range 200–320 kJ m^{-3}. Intrinsic coercivity approaches 1 MA m^{-1} and remanence, the net flux density remaining after removal of the magnetizing field, is about 1.2 T.

With increasing temperature, remanence of Fe–Nd–B sintered magnets decreases by about 0.12% °C^{-1} and intrinsic coercivity by about 0.6% °C^{-1}. This decrease is much larger than for Co_5R and $(Co, Fe)_{17}R_2$ magnets and results primarily from the lower Curie temperature. The decrease of coercivity with increasing temperature is a particular problem in many applications, since demagnetizing fields then cause irreversible losses of magnetization. Various alloying additions have been studied as a means of improving the thermal stability of Fe–Nd–B magnets. Alloying substitutions of cobalt for iron raise the Curie temperature, but also tend to lower the coercivity. As a result, cobalt additions alone do little to enhance thermal stability. Substitutions of dysprosium for neodymium raise the coercivity (see Fig. 6) by increasing the anisotropy field (see Fig. 4). This improves thermal stability, but with a loss of magnetization and energy product. Numerous other alloying additions, including aluminum, gallium and magnesium, have been reported to enhance coercivity, either through enhancement of anisotropy or through effects on microstructure. Magnetization reversal in aligned sintered magnets occurs by nucleation and growth of reversed magnetic domains, with the neodymium-rich grain-boundary phase enhancing coercivity by pinning domain walls and thereby resisting the propagation of magnetic reversal from grain to grain. Subtle microstructural changes induced by alloying additions may reduce reverse-domain nucleation or enhance grain-boundary pinning, allowing coercivity to reach a higher fraction of the anisotropy field.

3. Melt-Spun Magnets

Commercial Fe–Nd–B magnets have also been produced by melt spinning, a process in which an alloy is rapidly solidified by directing a molten stream against a cold rotating substrate. An inert gas is used to avoid oxidation. The resulting ribbon is 15–30 μm thick and may be either amorphous or microcrystalline depending on the substrate surface velocity, which controls the cooling rate. Optimum coercivity for Fe–Nd–B occurs at an intermediate quench rate that yields grains less than 100 nm in diameter, about 100 times finer than the grains in sintered magnets. Melt-spun ribbons are more nearly single-phase $Fe_{14}Nd_2B$ than sintered magnets, but may contain a thin neodymium-rich phase along the grain boundaries. Optimally quenched ribbon is crushed to coarse powder, blended with polymer and formed into bonded magnets by various molding techniques. The resulting magnets have a volume fraction of $Fe_{14}Nd_2B$ of about 80% and are isotropic, limiting the remanence to about 0.65 T and the maximum energy product to about 64 kJ m^{-3}.

Alternatively, magnets of nearly 100% $Fe_{14}Nd_2B$ can be produced by the hot pressing of melt-spun powder. Since some grain growth occurs at the pressing temperature, overquenched ribbon is used so that optimum grain size and coercivity are obtained after pressing. Such hot-pressed isotropic magnets typically have remanences approaching 0.8 T and maximum energy products of about 100 kJ m^{-3}. For both these and isotropic bonded magnets, intrinsic coercivities are about 1.3 MA m^{-1}. The relative decrease of coercivity with increasing temperature is somewhat less than for sintered magnets. Demagnetization curves for isotropic bonded and hot-pressed magnets at various temperatures are shown in Figs. 7a and 7b, respectively.

4. *Alignment by Hot Working*

Deformation of hot-pressed Fe–Nd–B magnets in compression at temperatures above 700 °C, called "die upsetting," produces substantial alignment of the *c* axis along the compression direction. After this treatment, the magnet contains platelike grains lying normal to the compression direction, typically 60 nm in thickness and 300 nm in diameter. The alignment produced by die upsetting can increase the remanence from 0.8 T to 1.2 T (Fig. 7c). As a result, the maximum energy product is increased to over 250 kJ m^{-3}. Intrinsic coercivity, however, is decreased and becomes more temperature dependent. As with sintered magnets, alloying additions are being studied as a means of improving temperature capability.

Die upsetting can also be used as a route to produce anisotropic bonded magnets. Die-upset magnets are reground to powder, blended with polymer, magnetically aligned and then molded into bonded magnets. As with isotropic melt-spun powder, the coercivity is related to the grain size, which is much finer than the powder size. Thus coercivity is not severely degraded by powdering or by chemical reactions with binder materials. Melt-spun magnets also contain less neodymium-rich phase than most sintered magnets and are therefore less subject to corrosion and oxidation.

For some applications, a magnet with easy axes distributed randomly in a plane is more desirable than either a fully isotropic or a uniaxially aligned magnet. Such a texture has been produced in $Fe_{14}Nd_2B$ by hot extrusion of compacted powder through circular dies. This process imposes radial compressive forces and, as with die upsetting, the *c* axes preferentially align parallel to the compression forces. Other variations on standard production methods of sintered and melt-spun magnets are described in a comprehensive review by Buschow (1988).

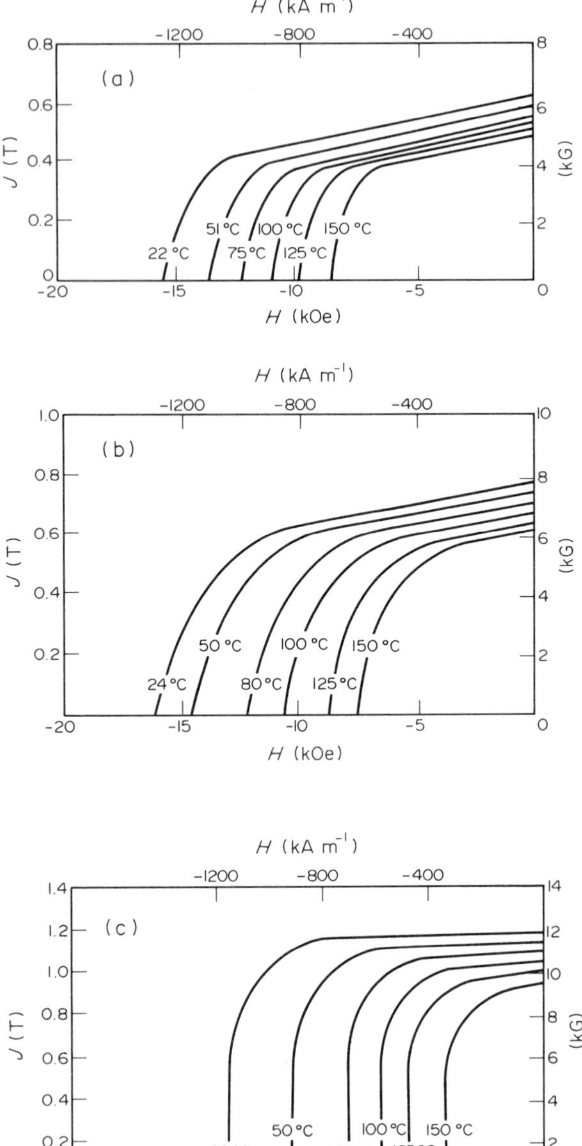

Figure 7
Magnetization vs reverse magnetic field of melt-spun Fe–Nd–B magnets at various temperatures: (a) isotropic, epoxy-resin-bonded, (b) isotropic, hot-pressed, (c) anisotropic, hot-formed (after Croat 1988)

5. *Current Status*

Despite their high cost, rare-earth magnets have found many applications where their high coercivities and energy products produce design advantages, including electric motors, magnetic bearings, earphones and microwave power devices. Because of higher energy products and because iron and neodymium are more abundant than cobalt and samarium, Fe–Nd–B magnets are likely to replace Co_5Sm in applications where high thermal stability and high coercivity are not required. However, broader applicability of Fe–Nd–B will require improvements in thermal stability, coercivity and corrosion resistance. The relative importance of sintered and melt-spun magnets remains unclear and will be determined by relative cost as well as by property differences.

Stimulated by the unexpected discovery of $Fe_{14}Nd_2B$, researchers have also begun an intensive search for other ternary and higher-order compounds with sufficiently high values of magnetization, anisotropy and Curie temperature to be useful as permanent magnets. Most attention is focused on alloy systems containing iron, cobalt and rare earths, in addition to other elements. Although a higher energy product via higher magnetization is one goal of such studies, higher coercivity, better thermal stability and lower cost are equally or more important.

More recently alloys based on the compound $(Fe, Ti)_{17}Sm_5$ have shown exceptional coercivities as high as $5\,MA\,m^{-1}$ (Katter et al. 1990). These materials have been prepared by melt spinning as well as by mechanical alloying and in thin-film form by sputtering. One disadvantage of this material is its low saturation moment at about 0.3 T. Developments on these and other materials have been summarized by Stadelmaier and Henig (1991). The most promising newly discovered material is the interstitial nitrided intermetallic $Fe_{17}Sm_2N_{3-\delta}$ (Coey and Sun 1990). It has a higher Curie temperature (470 °C) than $Fe_{14}Nd_2B$ with comparable saturation moment and anisotropy field. If a practical processing technique to produce dense aligned magnets with good coercivity can be developed, this may become the next addition to the growing family of commercial rare-earth magnets.

See also: Cobalt–Rare-Earth Compounds; Rare-Earth Intermetallics; Rare-Earth Metals and Alloys

Bibliography

Buschow K H J 1986 New permanent magnet materials. *Mater. Sci. Rep.* 1: 1–63

Buschow K H J 1988 Permanent magnet materials based on 3*d*-rich ternary compounds. In: Wohlfarth E P, Buschow K H J (eds.) 1988 *Ferromagnetic Materials*, Vol. 4. Elsevier, Amsterdam

Coey J M D, Sun H 1990 Improved magnetic properties by treatment of iron-based rare earth intermetallic compounds in ammonia. *J. Magn. & Magn. Mater.* 87: L251–4.

Croat J J 1988 Neodymium–iron–boron permanent magnets prepared by rapid solidification. *J. Mater. Eng.* 10: 7–13

Katter M, Wecker J, Schultz L, Grossinger R 1990 Preparation of highly coercive Sm–Fe–Ti by rapid quenching. *Appl. Phys. Lett.* 56: 1377–9

Sagawa M. Fujimura S, Yamamoto H, Matsuura Y, Hiraga K 1984 Permanent magnet materials based on the rare earth–iron–boron tetragonal compounds. *IEEE Trans. Magn.* 20: 1584–9

Sagawa M, Hirasawa S, Yamamoto H, Fujimura S, Matsuura Y 1987 Nd–Fe–B permanent magnetic materials. *Jpn. J. Appl. Phys.* 26: 785–800

Stadelmaier H H, Henig E T 1991 Permanent magnet materials developments during the last 18 months. *JOM* 43 (2):32–5

Strnat K J 1988 Rare earth–cobalt permanent magnets. In: Wohlfarth E P, Buschow K H J (eds.) 1988 *Ferromagnetic Materials*, Vol. 4. Elsevier, Amsterdam

J. D. Livingston
[Massachusetts Institute of Technology, Cambridge, Massachusetts, USA]

Nickel–Iron Alloys

Soft magnetic nickel–iron alloys with diverse properties are used in many fields of electrical and electronic engineering. Their structure-dependent magnetic properties, such as coercivity, permeability and hysteresis-loop shape, depend on the basic magnetic constants and on the microstructure. The basic magnetic constants—crystalline anisotropy K_1, magnetostriction λ and uniaxial anistropy K_u—are determined by the alloy composition and by an annealing treatment which leads to a definite degree of Ni_3Fe order. The microstructure is determined by the details of the production procedure—melting, cold working and intermediate and final annealing. The most important microstructural parameters are the grain size and texture as well as the density, size and chemical nature of nonmagnetic inclusions.

1. Basic Magnetic Constants

Figure 1 shows the basic magnetic constants for the binary (fcc) nickel–iron alloys, as well as the saturation polarization J_s and the Curie temperature T_C. The five technically useful composition ranges are also indicated. The crystalline anisotropy K_1 depends not only on alloy composition but also on the degree of long- and short-range order of Ni_3Fe, which is stable below 600 °C. The magnetostriction constants λ_{111} and λ_{100} depend mainly on the composition and are affected by the ordering process only in the vicinity of Ni_3Fe. The uniaxial anisotropy K_u is caused by directional short-range order of iron and nickel atoms. During annealing below the Curie temperature in a magnetic field, a uniaxial magnetically preferred direction parallel to the applied field is induced. The hysteresis loop squares up in the field anneal direction and flattens in the perpendicular direction. An annealing treatment without a magnetic field leads to a constricted loop (perminvar effect).

2. Microstructure

For alloys with a nickel content of 70–81%, $K_1 \leqslant 0$, that is, the preferred magnetic direction is $\langle 111 \rangle$. For this reason the textures commonly found in nickel–iron alloys, $\{100\}\langle 001 \rangle$ and $\{210\}\langle 001 \rangle$, must be

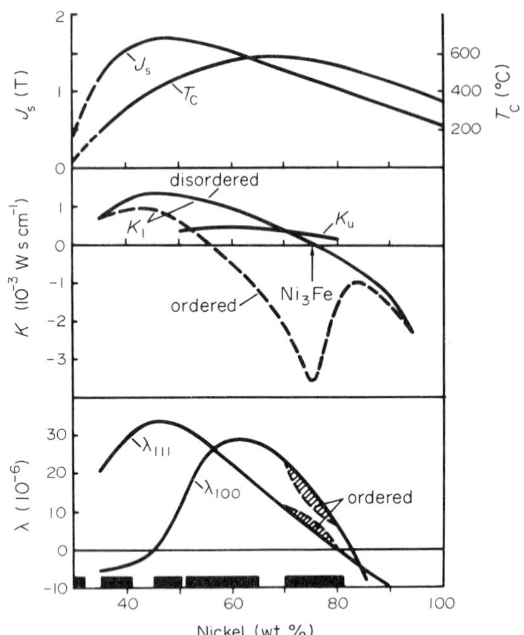

Figure 1
Saturation polarization J_s, Curie temperature T_C, uniaxial anistropy K_u (after annealing at 450 °C), crystalline anistropy K_1 and magnetostriction λ_{111} and λ_{100} in the binary nickel–iron system; ▬, concentration ranges of alloys in technical use

avoided and an essentially random texture must be achieved. On the other hand, for materials in the 45–65% nickel range, where $K_1 > 0$ and thus the preferred direction of magnetization is ⟨100⟩, the aforementioned textures are used to optimize the magnetic properties.

The processing methods which lead to the various microstructures will now be described.

Cube texture, {100}⟨001⟩, is obtained by a high degree of cold rolling (90–99%) followed by annealing at temperatures of 900–1050 °C. This texture is more pronounced the higher the degree of final cold work and the lower the intermediate annealing temperature prior to final cold working. Figure 2a shows a schematic diagram of this behavior. However, if the material is annealed at 1100–1200 °C, secondary recrystallization occurs which leads to a large grain size. The conditions for secondary recrystallization as well as for attaining the cube texture are a low intermediate annealing temperature and as high a degree of cold working as possible. Normal secondary-recrystallized material contains grains with different orientations. However, under certain circumstances, especially with laminations of 0.05–0.1 mm, the formation of grains with {210}⟨001⟩ orientation is preferred (Fig. 2b).

The boundaries between cube texture and isotropic microstructure or the annealing temperature at which secondary recrystallization begins depend on the deoxidizers in the melt, such as aluminum, magnesium and calcium. The conditions shown in Fig. 2 refer to molten material which has been weakly deoxidized with manganese and silicon.

To attain an isotropic microstructure, which is required for high-nickel alloys and for tapes of greater thickness, a low degree of final cold working of 50–80% and a high intermediate annealing temperature of 800–1000 °C are required (Fig. 2). Final annealing is generally carried out in oxygen-free dry hydrogen with a dew point below −40 °C.

Suitable combinations of the basic constants and the microstructure result in hysteresis loops of different shapes. These are characterized by different remanence values: a round loop with remanence

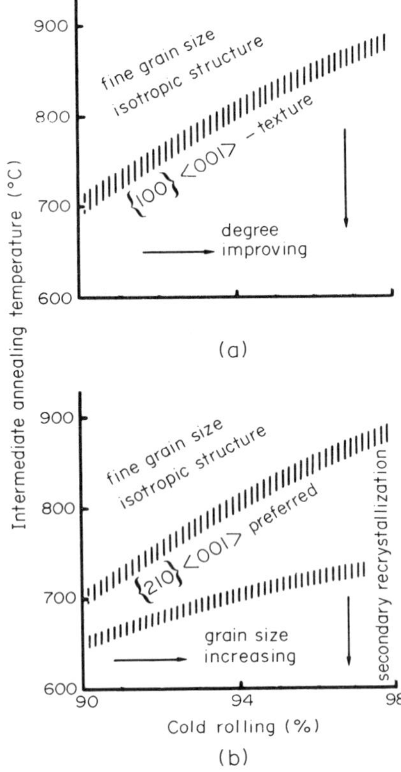

Figure 2
Influence of intermediate annealing temperature and degree of final cold working on the microstructure of nickel–iron alloys after final annealing at (a) 900–1050 °C and (b) 1100–1200 °C

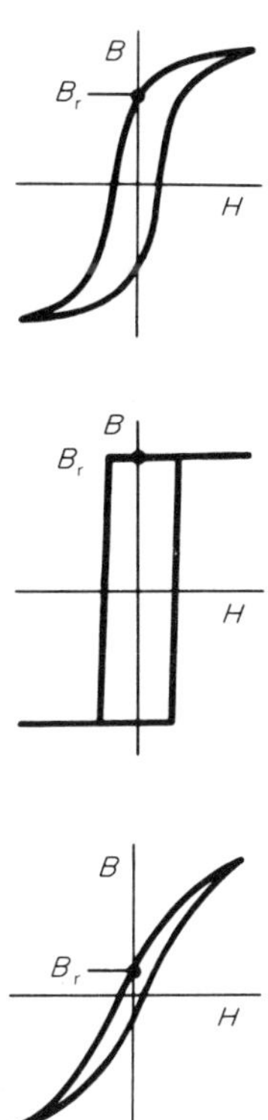

Figure 3
Typical hysteresis loops

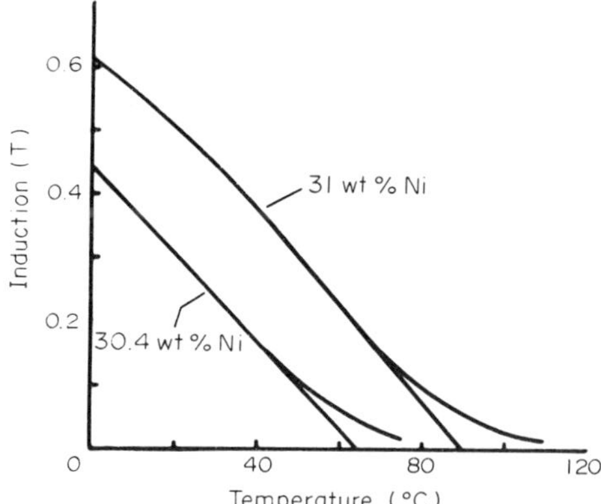

Figure 4
Induction as a function of temperature for alloys with about 30% nickel ($H = 10\,\mathrm{kA\,m^{-1}}$)

values 50–70% of the saturation induction B_s, a square loop with 90–98% and a flat loop with 5–30% (see Fig. 3).

3. *Technically Useful Alloys*

3.1 30% Nickel Alloys

In alloys with approximately 30% nickel the Curie temperature is close to room temperature. Materials with a Curie temperature of 50–120 °C and therefore a strong temperature dependence of the saturation induction are used for temperature compensation applications (see *Temperature Compensation Materials*), such as in permanent magnet designs. To attain a definite Curie point, a high degree of accuracy in the alloy composition is required. Changes as small as 0.1% in the nickel content or 0.01% in the carbon content result in a shift of the Curie point by about 5 °C. Figure 4 shows the induction–temperature curves of two alloys with different nickel content.

3.2 35–40% Nickel Alloys

In the 35–40% nickel alloy range the crystalline anisotropy K_1 decreases with nickel content and the squareness ratio B_r/B_s is reduced, leading to a round hysteresis loop. This round loop, in combination with a large electrical resistivity (40% nickel, $\rho = 60\,\mu\Omega\,\mathrm{cm}$; 48% nickel, $\rho = 45\,\mu\Omega\,\mathrm{cm}$) and a fine-grained isotropic microstructure, results in a lower core loss than that of conventional alloys with 48% nickel. For example, at 40% nickel, p_{Fe} (0.1 T, 20 kHz) $= 9\,\mathrm{W\,kg^{-1}}$, compared with $14\,\mathrm{W\,kg^{-1}}$ at 48% nickel, for a tape thickness of 0.05 mm. Alloys with about 37% nickel are in technical use. However, alloys with about 40% nickel whose saturation induction is only 5% lower than that of 48% nickel are also of interest (see Fig. 5 and Table 1).

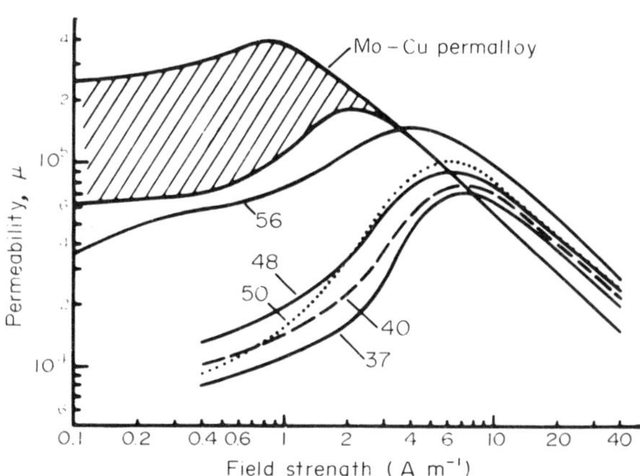

Figure 5
Permeability as a function of field strength of nickel–iron alloys with a round hysteresis loop (toroidal tape-wound cores, tape thickness 0.05–0.1 mm, 50 Hz) (see Table 1 for heat treatment and microstructure)

Table 1
Magnetic properties of technical alloys

Alloy (wt%)	Heat treatment[a]	Microstructure	Saturation induction, B_c (T)	Remanence ratio, B_r/B_s	Curie temperature (°C)	Coercivity (static) (A m^{-1})	Permeability (50 Hz) μ_4 (10^3)	Permeability (50 Hz) μ_{max} (10^3)	Tape thickness (mm)
30NiFe	650 °C	fine grained	temperature-dependent saturation induction		50–120	50			1
37NiFe	1100–1200 °C	isotropic	1.30	0.60	250	7	8	70	0.05–0.1
40NiFe			1.48	0.62	310	6	10	80	
48NiFe	1000 °C	{100}⟨001⟩	1.55	0.97	440	12		50	0.05
48NiFe	1100–1200 °C	medium fine	1.55	0.75	440	4	12	80	0.1
50NiFe		second. recryst. isotropic	1.52	0.72	470	5	10	90	0.1
56NiFe	1200 °C + ThT (longit. field) for high μ	fine grained second. recrystal.	1.5	0.55	540	1.2	60	120	0.1
65Ni 2.5MoFe	1200 °C + ThT (longit. field) for square loop	isotropic	1.25	0.93	520	2		250	0.05
50NiFe	1200 °C + ThT (transv. field)	second. recryst. {210}⟨001⟩	1.52	0.05–0.1	470	5	pulse permeability (Fig. 7)		0.05
65Ni 2.5MoFe	1100 °C + ThT (transv. field)	isotropic	1.25	0.05–0.15	520	7			0.05
80NiMo Fe or 77NiMo CuFe	1000–1200 °C + AT for high μ	isotropic	0.78	0.7		0.3–1.2	70–300	180–400	0.05–0.1
	1100 °C + AT for square loop	isotropic	0.8	$B_r/B_m = 0.92$ $H_m = 0.1$ mA cm^{-1}	400	0.7		250	0.05
	900–100 °C + ThT (transv. field)	isotropic	0.78–0.8	0.05–0.3		1	pulse permeability (Fig. 7)		0.03–0.1
+ Nb,Ti	1000 °C + AT for high μ	isotropic	0.5	0.6	280	1.5	50	120	0.15

a ThT = thermomagnetic treatment below T_C; AT = annealing treatment above T_C

3.3 45–50% Nickel Alloys

The composition of 45–50% nickel alloys is in the range of maximum saturation polarization J_s with $K_1 > 0$ and the 'easy' magnetic axis ⟨100⟩. The main differences in the magnetic properties of technical materials are due to different microstructures.

A square loop is obtained by the formation of a cube texture. Applications are in toroidal cores and core laminations for magnetic amplifiers, chokes and transformers (Table 1).

Technical materials with round hysteresis loops, smaller coercivity and higher permeability may have either a secondary-recrystallized or a fine-grained isotropic microstructure developed by primary recrystallization. Here, secondary-recrystallized material with a {210}⟨001⟩ texture shows a higher permeability than the primary-recrystallized material. Compared with the secondary-recrystallized material, the fine-grained material has a lower initial permeability but a slightly higher maximum permeability (Table 1). Toroidal cores and core laminations with secondary-recrystallized microstructure are used in current transformers and ground-fault interrupters, for example, whereas fine-grained isotropic material is used for rotor and stator laminations and relay elements. The behavior of the materials after the final heat treatment is different. The secondary-recrystallized materials show a prompt increase in permeability only after annealing above the secondary-recrystallization temperature of 1100–1200 °C. On the other hand, the permeability of the primary-recrystallized material increases steadily with annealing temperature and grain size so that this material exhibits higher permeability values after annealing at temperatures below about 1100 °C.

To improve formability, materials are often processed with chip-breaking strong deoxidizers such as aluminum, magnesium and selenium. The microstructure of these materials is fine-grained isotropic. As a result of inclusions associated with these additives, the coercivity is higher than that of the aforementioned vacuum-melted materials weakly deoxidized with silicon and manganese.

The coercivity of nickel–iron alloys with medium nickel content and relatively high crystalline anisotropy is very sensitive to defects such as grain boundaries, nonmagnetic inclusions and internal stresses.

The measured coercivity H_c may be interpreted as the sum of three parts:

$$H_c = H_{c0} + H_{cg} + H_{ci} \tag{1}$$

where H_{cg} represents the influence of grain boundaries, H_{ci} the influence of nonmagnetic inclusions and H_{c0} the coercivity caused by internal stresses.

H_{cg} is a linear function of the reciprocal grain diameter $1/d_g$:

$$H_{cg} \simeq 3\gamma_w / J_s d_g \tag{2}$$

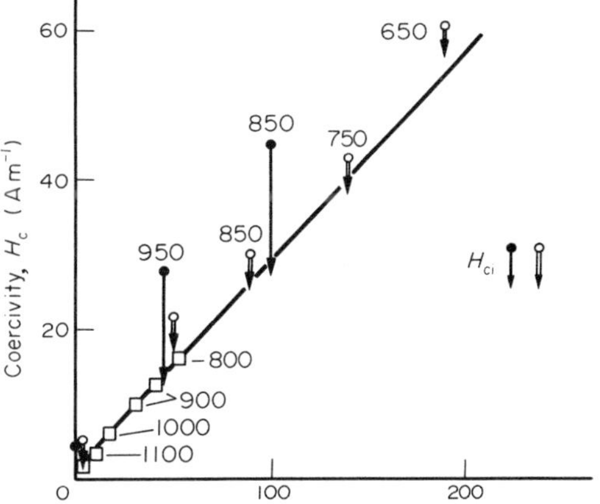

Figure 6
Coercivity of 50% nickel–iron as a function of reciprocal grain diameter: ●, vacuum-melted, deoxidized with silicon and manganese; ○, air-melted, deoxidized with calcium and magnesium; □, sintered (high-purity); annealing temperatures (°C) are indicated

where the wall energy γ_w is proportional to $K_1^{1/2}$. The coercivity caused by the nonmagnetic inclusions is essentially determined by particles whose diameter corresponds to the Bloch wall thickness. The Bloch wall thickness is proportional to $K_1^{-1/2}$, from which a value of 0.2 μm can be derived for 50% nickel. This means that submicroscopic particles are particularly effective in raising the value of H_c.

Experimental investigations on alloys with 47.5% nickel and a tape thickness of 0.5–1 mm yield the following empirical relation:

$$H_c = 0.8 + 0.29/d_g + 2.8 \times 10^{-4} N_{sub} \tag{3}$$

where H_c is in amperes per meter, d_g is in millimeters and N_{sub} is the number of particles per square millimeter.

Figure 6 shows the measured H_c values versus the reciprocal grain diameter, as obtained by annealing at various temperatures, for melted material with weaker (silicon, manganese) and stronger (calcium, magnesium) deoxidation as well as for a highly pure sintered alloy. If the contribution H_{ci} (length of arrows in Fig. 6) of the submicroscopic particles (0.02–0.5 μm) is subtracted from the measured coercivity, then all the results can be fitted to the line for the sintered material, which has a very low concentration of particles.

The nonmagnetic inclusions are mainly oxides and sulfides. The concentration of the submicroscopic particles in strongly deoxidized material is almost independent of annealing temperature ($N_{sub} = 1.4 \times 10^4$ mm^{-2}). In weakly deoxidized material it decreases with increasing annealing temperature (N_{sub}(950 °C) $= 5.5 \times 10^4$ mm^{-2} and N_{sub}(1150 °C) =

$1.2 \times 10^4\ \mathrm{mm}^{-2}$). The sintered material has a very low concentration of $10^3\ \mathrm{mm}^{-2}$.

Consideration of various contributions to H_c is technically important. For example, the contribution of particles to the coercivity of weakly deoxidized material after annealing at 1150 °C is 75%. Thus, the reduction in the concentration of particles during melting and final annealing is an important consideration in alloys with a medium nickel content.

3.4 50–65% Nickel Alloys

A particularly strong effect of thermomagnetic treatment below T_C is attained for 50–65% nickel alloys because K_1 decreases to zero and K_u rises to high values. Two characteristic states may be obtained, depending on the field annealing temperature. A square loop with high static maximum permeability is obtained by annealing far below the Curie temperature ($T_C - T_A = 130$ °C) where $K_u \gg K_1$. Since such a material has only a few 180 ° Bloch walls, the magnetization reversal is accompanied by strong eddy current effects which lower the initial permeability, particularly under dynamic conditions. If a higher annealing temperature is selected, about 60 °C below T_C, K_1 is larger and K_u smaller. This results in a less-square-shaped loop and a lower static maximum permeability but a greater number of Bloch walls and, hence, smaller eddy current effects. The permeability at 50 Hz is higher in the range up to medium excitation than that of the square-loop material.

From a technical point of view there are two materials of importance. Alloys with about 55% nickel attain high initial permeability by magnetic field annealing at low temperature. The microstructure is very important. The higher permeability values occur either with the {210}⟨001⟩ texture or with fine-grained secondary recrystallization (Fig. 5, Table 1). Fine-grained isotropic microstructure or coarse-grained secondary recrystallization is unfavorable.

A square-loop alloy with 65% nickel and 2.5% molybdenum (balance iron) with a fine-grained isotropic microstructure has also been developed commercially. Both the fine-grained isotropic microstructure and an increase in electrical resistivity due to the molybdenum addition contribute to the favorable dynamic magnetization reversal behavior of this alloy. Furthermore, a small dynamic coercivity with a moderately high value of $B_r/B_s(\sim 0.92)$ can be achieved by controlled cooling in the temperature range where Ni_3Fe ordering occurs (Table 1).

If the same alloy with 65% nickel and a fine-grained microstructure is subjected to magnetic field annealing perpendicular to the measurement direction, a hysteresis loop with low remanence and relatively high pulse permeability ($\mu_p = \Delta B/\mu_0 \Delta H \sim 4000$) is obtained (see Fig. 7), where ΔB and ΔH are the unipolar excursion range of induction and of field strength, respectively.

Higher values of pulse permeability can be obtained by transverse field annealing of alloys in the range 47–65% nickel with a {100}⟨001⟩ or {210}⟨001⟩ texture. An example is shown in Fig. 7 for the 50% nickel alloy.

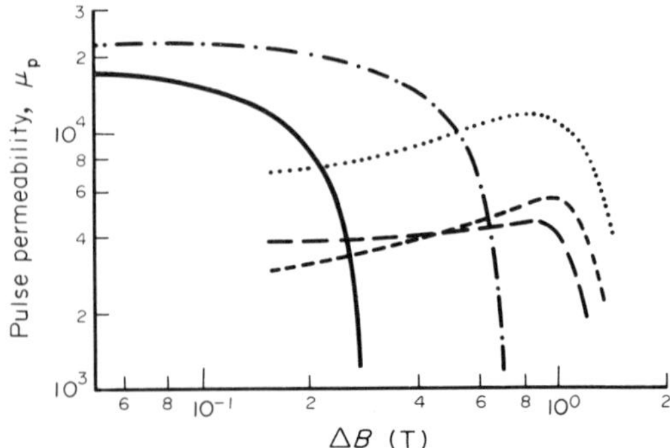

Figure 7
Pulse permeability μ_p as a function of unipolar excursion range of induction ΔB. (a) Alloys annealed in a transverse magnetic field (tape thickness 0.05 mm, pulse duration 50 μs); — —, 65%Ni–2.5%Mo–Fe, isotropic microstructure; ---, 50%Ni–Fe, isotropic microstructure; ····, 50%Ni–Fe, {210}⟨001⟩ texture. (b) Molybdenum–copper permalloys with isotropic microstructure (tape thickness 0.03 mm, pulse duration 10 μs): ————, annealed for high μ above T_c; —·—, annealed in a transverse field

In addition to pulse permeability, the magnetization reversal losses become important in applications with high pulse-repetition rate. The lowest losses occur in the 65% Ni–2.5%Mo–Fe alloy with fine-grained isotropic microstructure, with p_{Fe}(0.3 T, 10 kHz) = 14 W kg^{-1}. The 50% Ni–Fe alloy with a {210} texture and high pulse permeability is less favorable ($p_{Fe} = 22$ W kg^{-1}).

3.5 70–81% Nickel Alloys

Alloys with a high nickel content show the highest permeability values, with initial permeability μ_i of around 150 000 and μ_4(4 mA cm^{-1}) of around 300 000 for 50 Hz excitation in 0.05–0.1 mm-thick tapes. Highest permeability occurs when the crystalline anisotropy K_1 and the magnetostriction λ are both zero. This cannot be obtained in binary nickel–iron alloys ($K_1 = 0$ for 75% nickel, $\lambda_s = 0$ for 81% nickel). To meet the requirement of $K_1 = 0$, $\lambda = 0$ with the same composition, elements such as molybdenum, copper and chromium must be added. Moreover, the crystalline anisotropy is influenced by short- and long-range ordering. Figure 8 shows the zero of K_1, λ_s and λ_{111} as affected by annealing treatment. High initial permeability is reached at the intersections of the lines $K_1 = 0$ and $\lambda_s = 0$. The condition $\lambda_s = 0$ is achieved with decreasing nickel content (81% to 71%) and by increasing the addition of copper (0% to 14%). To remain on the $K_1 = 0$ line during cooling from 550 °C,

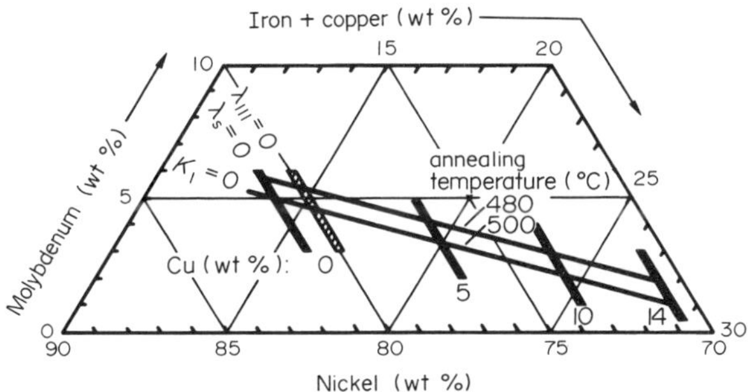

Figure 8
Lines for $K_1 = 0$ and λ_{111}, $\lambda_s = 0$ in a Ni–Fe–Cu–Mo system (permalloys with high nickel content)

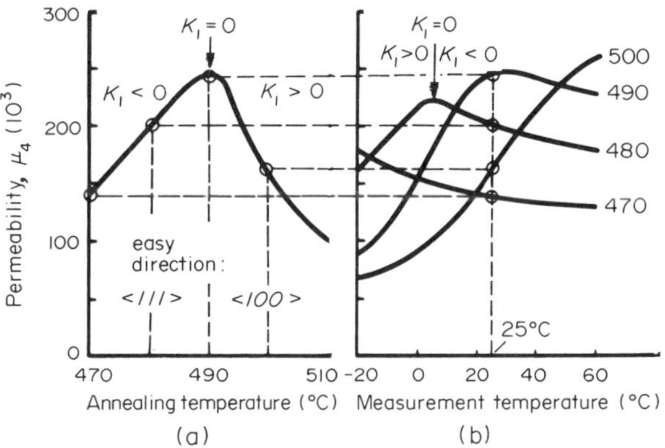

Figure 9
Permeability μ_4 at 50 Hz as a function of vs (a) annealing and (b) measurement temperatures for 77%Ni–5%Cu–4%Mo–Fe: (a) measurement temperature 25 °C; (b) annealing temperature (°C) indicated

the molybdenum content must be increased from 1.2% to 5% with an increasing nickel content (71% to 81%). The $K_1 = 0$ line shifts to a higher molybdenum content as the annealing temperature decreases. A similar diagram can be constructed to show the effect of cooling rates.

The intersection of $K_1 = 0$ and $\lambda = 0$ can be predicted theoretically for ternary systems (nickel + iron + an additional element).

Apart from the particularly high initial permeability associated with a round hysteresis loop, a square loop can be produced in high-nickel alloys. The so-called "spontaneous square loop" with $B_r/B_s \simeq 0.8$ and $B_r/B_m \simeq 0.9$ with $H_m = 2–5H_c$ occurs when the magnetostriction is zero in the preferred magnetic direction; for example, $\lambda_{111} = 0$ if $K_1 < 0$.

The correlation presented in Fig. 8 refers to measurements at room temperature and if other annealing or measurement temperatures are applied the zero K_1 line will shift. As an example, for one of the technical alloys with $\lambda_{111} = 0$, the influence of annealing temperature T_A and measurement temperature T_M on the permeability μ_4 is presented in Fig. 9. The μ_4–T_M curves show a maximum for a small negative value of K_1 near zero, because $\lambda_{111} = 0$. The permeability μ_4 drops faster below the maximum because $K_1 > 0$ and $\langle 100 \rangle$ becomes the preferred direction, while the magnetostriction λ_{100} is not zero. For $K_1 < 0$ the drop in μ_4 as a function of T_M is less, because here the preferred direction is $\langle 111 \rangle$ and the magnetostriction λ_{111} in the preferred direction is zero. Since $|K_1|$ increases with rising measurement temperature, the loop becomes increasingly square. The permeability at medium excitation is then decreased, although at a modest level. Thus, the temperature dependence of the permeability at medium excitation can be reduced near room temperature by annealing at a temperature T_A below the maximum of the μ–T_A curve, thus shifting the maximum of the μ–T_M curves to lower temperatures. However, a certain fall in permeability at 25 °C must be accepted.

A shift in the maximum of the μ–T_M curves into the range of cryotechnical importance can be achieved by a lower molybdenum content and suitable annealing temperatures.

Pulse applications and the development of ground-fault interrupters which react to both ac and dc pulses have led to important developments in high-nickel alloys. By annealing below the Curie temperature in a magnetic field perpendicular to the measurement direction, a hysteresis loop with low remanence B_r is attained. To obtain a high pulse permeability at as low a remanence as possible, a high-permeability alloy with $K_1 = 0$ and $\lambda = 0$ must be used. The Curie temperature of such an alloy is about 400–420 °C. Thus, the thermomagnetic treatment must be carried out at relatively low temperatures (250–380 °C), where the rate of short-range order is slow. To increase the rate it is advantageous to quench rapidly from higher temperatures (e.g., from the annealing temperature to set $K_1 \simeq 0$) to increase the quenched-in vacancies. Figure 7 shows the pulse permeability μ_p of one such alloy (molybdenum–copper permalloy annealed in a transverse field).

The first reports on permalloys with additions of niobium and titanium were published in 1972. These alloys attain a Vickers hardness (HV) of more than 200 after optimum heat treatment without noticeable loss of permeability. The HV of permalloys without additives is about 120. The increase in hardness is caused by coherent precipitation of γ'-Ni_3(Nb, Fe) or Ni_3(Nb, Ti, Fe) whose diameter is small in comparison with the Bloch wall thickness. These alloys have found use for magnetic heads in tape recorders due to their high permeability and superior wear resistance. The basic composition of the alloys is the same as that of the common alloys with 70–80% nickel and addi-

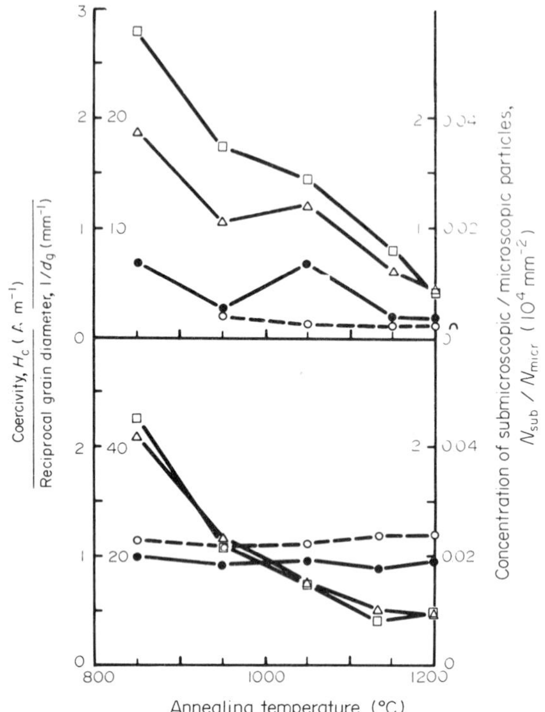

Figure 10
Influence of annealing temperature on properties of vacuum-melted, weakly deoxidized (upper graph) and air-melted, strongly deoxidized (lower graph) permalloys: □, reciprocal grain diameter; △, coercivity; ● and ○, concentrations of submicroscopic and microscopic particles, respectively

tions of molybdenum and copper. The partial replacement of iron by hardening additives, particularly niobium, tantalum, titanium and aluminum leads to a reduction in saturation induction (Table 1).

In molybdenum permalloy, quite differently from 50%Ni–Fe, a strong increase in coercivity due to 1–20 μm particles is expected. If it is assumed that $|K_1| = 3–10\ \mathrm{W\,s\,m^{-3}}$ for molybdenum permalloy, the resulting Bloch wall thickness, and thus the critical particle diameter, is 1–3 μm. In melted alloys with either weak deoxidation (silicon, manganese) or strong deoxidation (calcium, magnesium), different effects of the nonmagnetic inclusions on H_c have been observed.

Figure 10 (upper) shows H_c, $1/d_g$ and particle concentration as functions of annealing temperature for 1 mm-thick tapes of weakly deoxidized material. The rise in H_c between 950 °C and 1050 °C, in spite of the increase in grain size, is due to the increase in the number of submicroscopic particles. The following empirical equation has been derived from the data:

$$H_c = 0.1 + 0.05/d_g + 0.5 \times 10^{-4} N_{sub} \quad (4)$$

where the units are the same as for Eqn. (3).

The results for strongly deoxidized material (Fig. 10, lower) show almost identical trends of coercivity and reciprocal grain size, whereas N_{sub} and N_{micr} are practically independent of the annealing temperature.

The weakly deoxidized material shows abnormal behavior in that the coercivity is not influenced by the number of microscopic particles. Here, in contrast to conventional ideas, nonmagnetic inclusions smaller than the Bloch wall thickness do affect the coercivity. However, the only difference in the two materials is in the chemical nature of their particles. The manganese–silicate (mixed oxides) inclusions observed in the weakly deoxidized material apparently create stress fields as a result of unequal thermal expansion with the matrix and these, in turn, interact more strongly with the Bloch walls. On the other hand, the MgO and MgS inclusions in strongly deoxidized material lead to considerably fewer additional stresses because their thermal expansion coefficients are closer to that of the nickel–iron matrix.

In spite of the unexpected behavior of weakly deoxidized materials, they are nevertheless selected for highest-permeability applications. The reason is that these materials show faster grain growth than the strongly deoxidized materials, since the former can be purified at high annealing temperatures. Moreover, this type of annealing reduces the particle concentration considerably, in particular for thin tape of about 0.1 mm thickness.

See also: Magnetic Materials: An Overview; Metallic Glasses

Bibliography

Boll R 1979 *Soft Magnetic Materials*. Heyden, London

Chen C W 1977 *Magnetism and Metallurgy of Soft Magnetic Materials*. North-Holland, Amsterdam

Chin G Y 1971 Review of magnetic properties of Fe–Ni alloys. *IEEE Trans. Magn.* 7: 102–13

Chin G Y, Wernick J H 1980 Nickel–iron alloys. In: Wohlfarth E D (ed.) 1980 *Ferromagnetic Materials, A Handbook on the Properties of Magnetically Ordered Substances*, Vol. 2. North-Holland, Amsterdam, pp. 123–68

Couderchon G, Tiers J F 1982 Some aspects of magnetic properties of Ni–Fe and Co–Fe alloys. *J. Magn. & Magn. Mater.* 26: 196–214

English A T, Chin A T 1967 Metallurgy and magnetic properties control in permalloy. *J. Appl. Phys.* 38: 1183–7

Pfeifer F, Boll R 1969 New soft magnetic alloys for applications in modern electrotechnics and electronics. *IEEE Trans. Magn.* 5: 365–70

Pfeifer F, Kunz W 1982 Crystalline and amorphous alloys for sophisticated electronic devices. *IEEE Trans. Magn.* 18: 1406–11

Pfeifer F, Radeloff C 1980 Soft magnetic Ni–Fe and Co–Fe alloys—some physical and metallurgical aspects. *J. Magn. & Magn. Mater.* 19: 190–207

F. Pfeifer
[Vacuumschmelze, Hanau, Germany]

Niobium–Titanium Alloys

The niobium–titanium system contains the important alloys of Nb–46–50 wt%Ti, from which the vast majority of all superconducting magnets are constructed (see *Superconducting Materials: Large-Scale Applications*). Although the critical temperature T_c and upper critical field B_{c2} of these niobium–titanium alloys are only moderate ($T_c \sim 9.3$ K, B_{c2} (4.2 K) $\sim$ 11 T, B_{c2} (2 K) $\sim$ 14 T), their very high critical current densities ($\sim 3 \times 10^9$ A m^{-2} at 5 T, 4.2 K) and great strength and ductility render them outstandingly useful as engineering materials. The useful range of niobium–titanium permits operation up to 9 T at 4.2 K. This can be extended to 12 T (or even 13 T by partially replacing niobium by about 15–25 wt% tantalum) by cooling with superfluid liquid helium to approximately 1.8 K.

Niobium–titanium alloys are type II superconductors (see *Magnetic Properties of Superconductors*) and in order for them to carry large enough current densities for magnet applications, fluxoid motion must be inhibited by developing strong flux pinning centers (see *Flux Pinning and the Summation of Pinning Forces*). In commercial niobium–titanium superconductors, strong pinning is achieved by periodic heat treatment during wire processing which produces a very dense array of normal (i.e., nonsuperconducting) α-titanium precipitates. Further wire drawing distorts the α-titanium precipitates into extremely thin, folded ribbons ($\sim$1 nm thick) which are separated from each other by only about 5 nm. Niobium–titanium is almost always fabricated in a multifilamentary form such that the filaments (normally 5–50 μm in diameter) are dispersed in a normal metal matrix (generally pure copper). Typically, the number of filaments is of the order of 50–5000, the wire diameter is 0.5–2 mm and the Cu:Nb–Ti ratio of the order of 1:1 to 2:1. Such composites are made on a scale of roughly 200 kg and wire lengths can exceed 10 km at final size (see *Multifilamentary Superconducting Composites*).

1. Phases and Microstructures

1.1 Equilibrium Phases

Above 882 °C there is complete liquid and solid miscibility between niobium and titanium, allowing a single-phase substitutional body-centered cubic (bcc) alloy to be formed. Commercial alloying is most commonly performed by consumable electrode vacuum arc melting using niobium and titanium composite electrodes or by electron beam melting of high-purity ingots. The large differences in the freezing points of titanium and niobium (1670 °C for titanium and 2467 °C for niobium) can result in localized solute enrichment if the process is not carefully controlled. An inclusion-free, high-homogeneity alloy is required to produce fine filament high critical current density (J_c) wire. Below 882 °C precipitation of the hexagonal close-packed (hcp) α-titanium phase can occur. The position of the $\alpha + \beta/\alpha$ phase boundary is uncertain in the alloy composition range used for superconducting applications. It lies in the vicinity of 600 °C, but the actual equilibrium values have not been determined due to the slow diffusion rates at these low temperatures (the absolute reduced temperature, T/T_m, $\simeq 0.4$, where T is the temperature and T_m is the melting point, both measured in kelvin). For this reason also the β-niobium–titanium single phase can be retained at room temperature, even by relatively slow cooling through the two-phase region.

1.2 Cold-Worked Microstructures

Precipitation of the α-titanium phase is required in order to produce high critical current density in conventionally processed niobium–titanium composites (Pfeiffer and Hillman 1968, Larbalestier and West 1984) and the heat treatments used to precipitate the α-titanium are carried out between 375 °C and 420 °C, well below temperatures at which equilibrium conditions can be obtained. In addition to the α-titanium phase, several metastable phases are also possible, most importantly in this composition range, the hexagonal ω phase. Although the ω phase can produce strong flux pinning, the phase substantially hardens the alloy, thus reducing wire drawability. This phase is thus to be avoided, for example by applying extremely high cold work strains to the alloy prior to precipitation heat treatment (Buckett and Larbalestier 1987). The amount of cold work can be represented by the true strain, ε_T, which is given by

$$\varepsilon_T = \ln\frac{A_0}{A} = 2\ln\frac{d_0}{d} \tag{1}$$

where d_0 and A_0 are the starting diameter and cross-sectional area of the alloy rod at the final recrystallization anneal, respectively, and d and A are the diameter and area of the niobium–titanium after cold work. In commercial practice (see *Multifilamentary Superconducting Composites*), warm extrusions are often used in the fabrication process and prior cold work performed before an extrusion process may be lost. Both extrusion and wire drawing produce a strong [011] wire texture in the β-niobium–titanium and result in plane strain deformation with flow tending to be confined to the (011) plane. The plane strain condition, which becomes more pronounced as the drawing strain increases, results in a highly distorted transverse cross-sectional microstructure in which the β-niobium–titanium grains bend around each other. By a true strain of 5–7, however, a high-angle grain structure of regular, small-diameter (50–100 nm) grains has formed. Higher prestrains (the prestrain ε_p is the total cold work strain prior to precipitation heat treatment) homogenize and refine the microstructure further. For a Nb–47 wt%Ti alloy (from which the vast majority of composites are made), a typical heat treatment (10 h to 80 h at between 375 °C and 420 °C) will produce the equilibrium α-titanium phase at grain boundary intersections (most frequently triple points). Increasing the titanium content of the alloy increases the prestrain required to suppress ω and the Widmanstätten form of α-titanium, both of which precipitate on a fine scale within the β-niobium–titanium grains. Both of these phases significantly increase filament hardness and reduce drawability. The increased strain increment required to suppress these phases is approximately linear with respect to composition within the range Nb–44–62 wt%Ti. An increase in ε_p of roughly 0.77 is required for each wt%titanium increase in composition (Lee et al. 1989). The restrictively high strains required to produce triple-point α-titanium precipitation in high-titanium alloys (up to 12 for alloys containing over 55 wt% titanium) limit their use. The amount of precipitation is also dependent on the alloy composition; the quantity of α-titanium produced falls steeply when the titanium content is reduced below 45 wt% titanium and such alloys cannot produce enough α-titanium to develop the highest J_c values. This factor is particularly strong in the range Nb–47 wt%Ti to Nb–53 wt%Ti where the quantity of precipitate can increase by more than 50 vol.%. Finally, it is important to note that a chemically homogeneous alloy is very desirable in order to produce the uniform precipitate morphology characteristic of the highest J_c.

1.3 The Multiple Heat Treatment Process

The maximum amount of precipitate that can be produced by a single heat treatment in Nb–47 wt%Ti alloy is approximately 11 vol.%. This is insufficient to produce the highest J_c. By adding additional strain ($\Delta\varepsilon_T \simeq 1$)–draw–heat treatment cycles, additional precipitate can be formed. This process can be performed 3–6 times advantageously. In Nb–47 wt%Ti alloys at least three heat treatments are necessary to produce the approximately 20 vol.% of α-titanium precipitate required to produce high J_c ($J_c > 3 \times 10^9$ A m^{-2} at 5 T, 4.2 K). Increasing the duration and temperature (within the 375 °C to 420 °C range) of heat treatment increases the quantity of precipitate but with diminishing returns as the titanium content of the β-niobium–titanium matrix is depleted (Lee and Larbalestier 1988). The limit to α-titanium precipitate formation by this process occurs when the β-niobium–titanium composition drops to between 36 wt% and 37 wt% titanium. Increasing the temperature and times of heat treatments also increases the dimensions of the precipitates and, as niobium–titanium wire is not processed bare but is sheathed in a stabilizer, any reaction with the stabilizer is also increased. In order to reduce this reaction, a diffusion barrier is often applied to the niobium–titanium surface (see *Multifilamentary Superconducting Composites*).

After final heat treatment, the α-titanium precipitates are rather uniformly distributed and are generally equiaxed in transverse cross section with average diameters ranging from approximately 80 nm to 200 nm. In longitudinal cross section the precipitates extend along the drawing axis. The aspect ratio depends on the heat treatment conditions and typically ranges from 4 to 15. In order to produce a high-J_c microstructure the wire must be further strained to reduce the microstructural dimensions to those of the fluxoid lattice (see *Flux Line Lattice*).

1.4 Optimized, Final Wire Microstructures

When the niobium–titanium is drawn after the final heat treatment, the α-titanium precipitates gradually distort into densely folded sheets. The folding of the α-titanium dramatically increases the interfacial density of the microstructure and increases the rate at which the precipitate thickness decreases with increasing strain (Lee and Larbalestier 1987). As the strain from final heat treatment ε_f is increased, the J_c of the wire increases until a peak is reached, typically for a final strain of 4–5. The strain required to reach maximum J_c increases with the size of precipitate after final heat treatment and the field at which J_c is measured, but is decreased by nonuniformity in the niobium–titanium filament cross section (which also reduces the magnitude of J_c). At peak J_c, the average precipitate thickness and spacing are of the order of 1 nm and 5 nm, respectively, with the folded sheets of α-titanium extending for micrometers along the length of the wire. These precipitate densities are very high and generally higher than the fluxoid density, whose spacing is of the order of $(\varphi_0/B)^{1/2}$, where $\varphi_0 = 2 \times 10^{-15}$ Wb is the flux quantum and B is the field. The microstructural pin density and the fluxoid density are contrasted in Fig. 1.

In each of the processing steps outlined in this section, considerable strain is required; since there are finite limits to the dimensions of both the starting material (e.g., maximum extrusion diameters of 300–350 mm) and final wire size (e.g., 0.5–1 mm diameter) there is a limit to the strain space available (~ 12). This frequently limits the number of heat treatments that can be applied and also limits the range of alloy compositions that can be optimized.

2. Mechanical Properties

Niobium–titanium alloy is not used bare but is sheathed (during wire fabrication) in a stabilizer, most commonly copper. Copper provides the appropriate thermal and electrical stability, as well as good drawing properties. When pulsed or alternating fields are needed, a copper–nickel or mixed copper and copper–nickel stabilizer is used to reduce coupling losses.

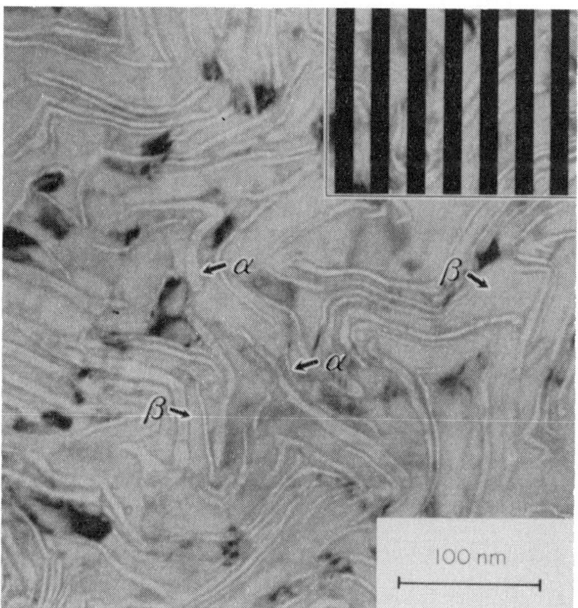

Figure 1
Transmission electron micrograph of part of a niobium–titanium (Nb–46.5 wt%Ti) filament in transverse cross section; α-titanium precipitates appear lighter due to their relatively low atomic number compared to the β-niobium–titanium matrix. Inset is a representation of the equilibrium fluxoid (black) spacing at 5 T, 4.2 K

More rarely, high-purity aluminum is used for applications where a high transparency to radiation or low conductor density is required. For the vast majority of applications, the sheathed niobium–titanium is assembled into a multifilamentary composite, consisting of the stabilizer and the superconductor (see *Multifilamentary Superconducting Composites*). The elastic and plastic behavior of niobium–titanium alloys are therefore examined here in terms of their behavior within a composite.

The high strength and excellent ductility of niobium–titanium composites have been crucial elements in their success. The mechanical properties of both the individual components and the composites have been extensively reviewed (Koch and Easton 1977). Providing that the precipitation of significant amounts of ω and Widmanstätten α-titanium phases is avoided, niobium-titanium composites remain ductile throughout the extensive wire-making process even when around 20 vol.% of nanometer spaced precipitates are present. The composite mechanical properties can be approximated by applying the rule of mixtures to the properties of the components and thus the wire properties will be affected by the copper–superconductor ratio. The properties of the components will depend on the processing of the wire, for instance whether a final copper anneal is applied to increase the residual resistivity ratio (RRR) of the stabilizer or whether a twist is applied to the wire. The ultimate tensile stress (UTS) is much higher for niobium–titanium (~2200 MPa at 4.2 K) than for copper (750 MPa at 4.2 K), and thus the UTS for the conductor as a whole increases with increased volume of superconductor, whereas the elastic modulus, which is approximately 10% less for niobium–titanium (83 GPa at 4.2 K) than for copper, decreases with increasing superconductor fraction. If the copper is annealed, the heavily cold-worked state of the niobium–titanium is retained but the work hardening of the copper is lost.

In magnets the principal mechanical concern is the response of the conductor to stresses arising from fabrication, magnet winding, differential thermal contraction and magnet energization. The relative contributions of these stresses have been reviewed by Ekin (1981). Typically these stresses are less than 200 MPa and they do not produce any significant degradation in J_c (Bray and Ekin 1989).

3. Normal State Physical Properties

The normal state physical properties and the superconducting properties are strongly connected: in the GLAG (Ginzburg–Landau–Abrikosov–Gorkov) theory (see *Superconducting Materials: Physical Properties*), the nonparamagnetically limited upper critical field at 0 K in the dirty limit is given by

$$\mu_0 H^*_{c2}(0) = 3.1 \times 10^3 \gamma \rho_n T_c \qquad (2)$$

where ρ_n is the normal state resistivity (in Ωm) and γ is the electronic specific heat coefficient (in $\mathrm{J\,m^{-3}\,K^{-2}}$). The T_c in turn is related to the density of states $N(E_F)$ (as is γ) and is also related to the Debye temperature θ_D and the electron–phonon interaction parameter λ_{ep}.

The electrical resistivity of the niobium–titanium alloy is a strong function of composition with ρ_n increasing with titanium content (see Fig. 2). The relationship between resistivity and composition for annealed alloys deviates from the expected parabolic one due to the instability of the β-phase lattice (this also manifests itself in a negative $d\rho/dT$ at high titanium contents). This is the same lattice instability that permits athermal ω-phase formation.

The electronic specific heat coefficient peaks between approximately 34 wt% titanium and 26 wt% titanium, as does $N(E_F)$. The peak value for γ is approximately $1.1 \times 10^3\ \mathrm{J\,m^{-3}\,K^{-2}}$, but the value declines slightly towards the useful composition range, such that for Nb–47–50 wt%Ti the value is approximately $10^3\ \mathrm{J\,m^{-3}\,K^{-2}}$. The Debye temperatures of niobium–titanium alloys are in the 240 K to 290 K range and the electron–phonon interaction parameter ranges from 0.75 to 0.90.

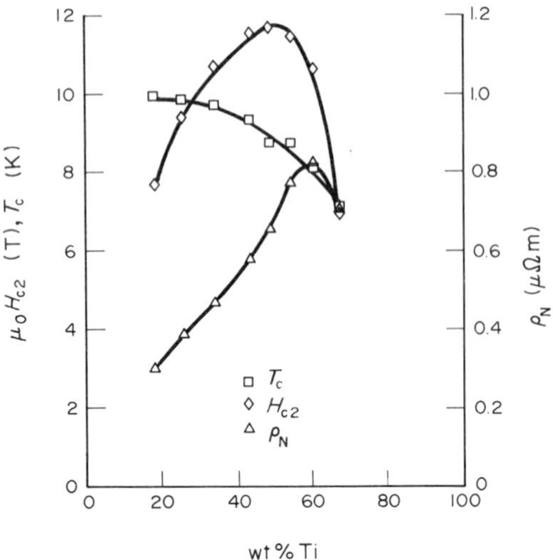

Figure 2
Variation of ρ_n (Moffat 1985), T_c and H_{c2} (Muller 1988) with composition β-niobium–titanium alloys: the T_c was obtained inductively

In the composition range of practical concern (40–50 wt% titanium) the upper critical field is paramagnetically limited (see *Superconducting Materials: Physical Properties*). The Pauli paramagnetic moment χ_P peaks at approximately 55 wt% titanium.

The thermal conductivity of a niobium–titanium alloy depends on the composition and metallurgical state and at 4.2 K ranges from approximately 0.1 W m^{-1} K^{-1} to 1 W m^{-1} K^{-1}. The integrated thermal contraction from 273 K to 4 K for niobium–titanium alloys is approximately 0.17%.

These and other properties of niobium–titanium alloys are extensively reviewed in the two volumes by Collings (1986).

4. Superconducting Properties

4.1 Transition Temperature and Upper Critical Field

In Fig. 2, the T_c, H_{c2} and ρ_n values are compared for annealed single-phase niobium–titanium alloys. There is a broad peak in T_c, which reaches a maximum value of approximately 10 K in the 18–25 wt% titanium range. The T_c begins to drop steadily when the titanium content is increased beyond about 40 wt% titanium. The peak in $\mu_0 H_{c2}$ occurs between 44 wt% titanium and 47 wt% titanium at a value of approximately 11.5 T at 4.2 K, rising to roughly 14.5 T at 1.8 K. (Note, however, that H_{c2} is quite sensitive to metallurgical state and the criterion chosen and may easily vary by the order of $\pm$ 0.5 T in a given alloy.) When the titanium content is increased beyond Nb–47 wt%Ti, the increased ρ_n contribution to H_{c2} is more than offset by the decline in T_c and γ.

In optimized niobium–titanium wires, the alloy condition is not the equilibrium, annealed two-phase state. Rather the alloy is very heavily cold worked, with a typical total strain of approximately 12 (an area reduction of 99.9994%) and has between 18 vol.% and 20 vol.% of α-titanium phase. This heavy precipitation shifts the β-matrix composition to between 42 wt% titanium and 39 wt% titanium. The shift to lower titanium would be expected to lower ρ_n (see Fig. 2), increase T_c and slightly reduce H_{c2}; furthermore, as the Ginzburg–Landau parameter κ is proportional to ρ and γ (see *Superconducting Materials: Physical Properties*), and γ is not a strong function of composition, κ is also expected to decline. In fact, the resistivity is lowered approximately 20% by the precipitation, somewhat more than might be expected simply due to the shift in β composition (Meingast et al. 1989); at 20 vol.% titanium the low resistivity of the α-titanium precipitates ($\sim$20 μΩ cm) must be taken into account. The T_c is increased by roughly 5%, whereas the H_{c2} changes little after precipitation. All these changes are observed immediately after precipitation when the dimensions of both the superconducting β-phase grains and normal α-titanium precipitates are significantly larger than the superconducting coherence length ξ. T_c, H_{c2} and ρ_n undergo further significant changes as the precipitate thickness is reduced to less than ξ during the final wire drawing process. As the wire is drawn, ρ_n increases slightly and T_c decreases until, at peak J_c, their values are approximately those of the nonprecipitated material. Furthermore, the value of κ increases from approximately 40 immediately after the final heat treatment to 44–45 for typical final wire sizes. The increase in ρ_n is matched by a decrease in H_c, resulting in H_{c2} being roughly independent of strain. Thus the values of T_c, H_{c2} and ρ after straining to peak J_c are approximately those of the initial single-phase alloy, despite the very different nature of the microstructures (Meingast et al. 1989).

4.2 Critical Current Density

Filamentary composites undergo a progressive transition to the dissipative state as the current is increased (see *Resistive Transition and Flux Flow in Superconducting Materials*) and consequently an arbitrary choice must be made as to what point in the transition defines the critical current density J_c. I_c and J_c are not in any case fundamental physical properties of superconductors since they are defined by the balance between the Lorentz force ($\boldsymbol{F}_L = \boldsymbol{J} \times \boldsymbol{B}$) pushing on the vortices and the summation of the local microstructural pinning forces $\boldsymbol{f}_p$ which oppose $\boldsymbol{F}_L$. A widely accepted criterion for niobium–titanium composites defines I_c or J_c at that point at which the composite resistivity reaches 10^{-14} Ω m. This represents the condition under which a typical high-current-density magnet might quench; this criterion will be used throughout this section. However, the generally rather less sensitive criterion of $1\,\mu V\,cm^{-1}$ is also frequently used (but beware that this criterion can overestimate J_c defined at $10^{-14}\,\Omega\,m$ by 10–50%).

The most important factors controlling J_c are the alloy composition and its thermomechanical processing treatment. A typical process for Nb–47 wt%Ti would involve extruding a multifilamentary Cu:Nb–Ti composite from about 300 mm to 75 mm diameter, then wire drawing to approximately 25 mm diameter. At this point ($\varepsilon_p \simeq 5$) heat treatment can begin. The wire is further drawn, heat treated, drawn and heat treated again. Typically three or four such heat treatment draw cycles are employed. Final optimization involves a large final drawing strain ($\varepsilon_f \simeq 4$–5). Heat treatment times and temperatures lie in the range 10–80 h at 375–420 °C (Li and Larbalestier 1987). The longer in duration and higher in temperature the heat treatment, the higher is the intrinsic J_c of which the composite is capable. The effect of this processing on the microstructure is described fully in Sects. 1.2 and 1.3; the same heat treatment conditions that increase the intrinsic J_c primarily increase the volume of α-titanium precipitated. In practice, however, there is always a compromise between raising the intrinsic J_c to its highest limit and degrading the filament quality, thus producing an extrinsic limit on J_c.

As the wire is drawn from its final heat treatment size, J_c initially shows a marked increase until it reaches a maximum. The magnitude of the peak and its position in terms of final cold work strain for a given field and temperature are dependent on this balance between intrinsic flux pinning at the α-titanium precipitates and extrinsic limits produced by factors such as chemical and dimensional inhomogeneities. Filament nonuniformity ("sausaging") frequently has the greatest impact since chemical inhomogeneities are rare in modern commercial niobium–titanium alloys. The techniques used to limit filament nonuniformity are described in the article *Multifilamentary Superconducting Composites*.

Since, in practice, it is impossible to produce high-J_c wire without some degree of filament sausaging, I_c in each filament becomes limited by the cross-sectional area of the necked regions. In order to quantify the level of extrinsic limitation, the sharpness of the resistive transition is an important criterion. This can be determined by the index n, which models the resistive transition $V \propto I^n$ where V is the voltage (see *Resistive Transition and Flux Flow in Superconducting Materials*). The higher the value of n, the sharper is the transition. Low n values are primarily due to voltage generated by the transfer of current from locally necked filaments to their neighbors. In the intrinsic limit (where J_c is determined by flux pinning interactions) n is very high, for example, 60–90 at 5 T (4.2 K) falling to less than 20 for conductors in which filament sausaging controls J_c (Warnes and Larbalestier 1986). As filament sausaging increases, n tends to become small and independent of field. The highest J_c composites yet produced have J_c values at 5 T (4.2 K) of approximately $3.7 \times 10^9\,A\,m^{-2}$ and an n value of approximately 40 (20 at 8 T). The corresponding standard deviation in filament cross-sectional area is typically 4–8%. If filament uniformity could be maintained, J_c could be further improved. Unfortunately, the processing required to produce intrinsically high J_c values often contributes to sausaging. Increasing the length and temperature of heat treatment increases the reaction between copper and niobium–titanium that initiates much of the sausaging. In addition, multiple heat treatments increase the likelihood of drawing instabilities due to softening of the stabilizer relative to the superconductor. Sausaging also increases with final drawing strain, particularly when ε_f exceeds 4.

Careful studies in which sausaging is minimized indicate that, for fields greater than 1.5 T, the peak in the bulk pinning force F_p occurs for final drawing strains exceeding 5. At this high level of cold work, the average α-titanium precipitate thickness is typically between 1 nm and 2 nm and the center to center separation is roughly 5 nm. The precipitate thickness is thus much smaller than the coherence length (ξ (4.2 K) = 5.5 nm) and the precipitate separation is much smaller than the equilibrium fluxoid spacing which is 22 nm at 5 T. Whether this condition represents the maximum J_c is doubtful, since this value is only of order 5% of the maximum possible (the depairing J_c) (Meingast and Larbalestier 1989).

Alloys in the composition range of roughly Nb–46–49 wt%Ti are at a far more advanced state of J_c optimization than alloys outside this range; alloys with compositions below Nb–46 wt%Ti produce insufficient precipitation for high J_c values and alloys with higher titanium contents than 49 wt% titanium have only recently been brought under microstructural control. Recent studies, however, indicate that the optimum alloy composition for obtaining maximum J_c depends on the field and temperature of operation. At 4.2 K, the higher the desired field, the lower should be the titanium content; for a field of 7 T the optimum composition lies in the range 46.5–49 wt% titanium whereas at 2 T the optimum composition is roughly 10 wt% titanium higher (McKinnell et al. 1989).

5. Future Developments

The major driving force for advanced conductor development has been and still is for high-energy physics applications. The largest such project is the superconducting super collider (SSC) which will demand large production scale quantities of high J_c ($> 2.75 \times 10^9\,A\,mm^{-2}$ at 5 T, 4.2 K) conductor with fine filaments (< 6–9 μm diameter). The emphasis for this application is the reduction in extrinsic J_c limitation by improving the production techniques for conventional niobium–titanium alloy conductor.

For very-high-field applications, such as focusing quadrupoles and the planned large hadron collider (LHC) at CERN, there has been some development of Nb–Ti–Ta ternary alloys. The addition of tantalum to niobium–titanium suppresses the paramagnetic limitation of the upper critical field. The enhancement in critical field is small at 4.2 K (0.3 T) but large at 1.8 K where H_{c2} values of 15.5 T can be obtained (1.3 T better than niobium–titanium). The ternary conductors are still very much under development but J_c values in excess of 2×10^9 A m^{-2} at 10 T (2 K) have been obtained (McInturff et al. 1990). With increased microstructural control and the impact on the flux pinning performance of niobium–titanium base superconductors, it is likely that further development of both ternary and quaternary alloys will take place in the near future.

An interesting recent development has been the development of artificial pinning center (APC) composites (Dorofejev et al. 1985, Motowidlo et al. 1990). Such composites artificially introduce a normal pinning center into the superconducting filament. Once the APC filament has been assembled and extruded, it can be restacked like a conventional filament. An APC composite, in principle, requires no heat treatment and the ratio of pinning material to superconductor can be precisely controlled. At present the best APC composites can outperform similar conventionally processed material up to a field of about 5 T with J_c values approaching 7.5×10^9 A m^{-2} at 3 T for material with 25% of niobium pinning center in a Nb–47 wt%Ti matrix (Kanithi et al. 1992). The APC composites have unfortunately shown poor workability and low J_c values at fields above 5 T. However, a study on conventionally processed monofilament (Cooley et al. 1991) has indicated that by making the pinning centers more efficient, a substantial increase in J_c for niobium–titanium composites can still be achieved. They rolled a conventionally processed Nb–47 wt%Ti monofilament into a tape with an aspect ratio of 9:1 producing a strong alignment of the α-titanium ribbons parallel to the broad face of the tape. The alignment of the α-titanium ribbons increased J_c at 5 T(4.2 K) from 2.61×10^9 A m^{-2} to 5.2×10^9 A m^{-2} when the field was applied parallel to the ribbons (a 40% improvement over the highest previous value for niobium–titanium). The increased pinning efficiency of the aligned α-titanium ribbons when parallel with the field was at the expense of the pinning efficiency away from this orientation (the J_c with the field perpendicular to the ribbons was similar to material without precipitates). The important implication of this experiment is, however, that there is a wide range over which the pinning efficiency and thus the J_c can be improved by further innovations in niobium–titanium processing.

See also: A15 Compound Superconductors; Critical Current Density; Multifilamentary Superconducting Composites

Bibliography

Bray S L, Ekin J W 1989 Effect of room temperature stress on the critical current of NbTi. *J. Appl. Phys.* 65: 684–7

Buckett M I, Larbalestier D C 1987 Precipitation at low strains in Nb46.5 wt%Ti. *IEEE Trans. Magn.* 23: 1638–41

Collings E W 1986 *Applied Superconductivity, Metallurgy and Physics of Titanium Alloys*, Vol. 1: Fundamentals, Vol. 2: Applications. Plenum, New York

Cooley L D, Jablonski P D, Lee P J, Larbalestier D C 1991 Strongly enhanced critical current density in Nb47 wt%Ti having a highly aligned microstructure. *Appl. Phys. Lett.* 58: 2984–6.

Dorofejev G L, Klimenko E Yu, Frolov S V 1985 Current carrying capacity of superconductors with artificial pinning centers. In: Marinucci C, Weymuth P (eds.) 1985 *Proc. 9th Int. Conf. Magnet Technology*. Swiss Institute for Nuclear Research, Villigen, Switzerland, pp. 564–6

Ekin J W 1981 Mechanical properties and strain effects in superconductors. In Foner S, Schwartz B B (eds.) 1981 *Superconductor Material Science: Metallurgy, Fabrication and Applications*. Plenum, New York, pp. 455–510

Kanithi H C, Valaris P, Motowidlo L R, Zeitlin B A, Scanlan R M 1992 Further developments in NbTi superconductors with artificial pinning centers. (Paper presented at ICMC–CEC 1991, Huntsville, AL). *Adv. Cryog. Eng.* 38

Koch C C, Easton D S 1977 A review of mechanical behavior and stress effects in hard superconductors. *Cryogenics* 17: 391–413

Larbalestier D C, West A W 1984 New perspectives on flux pinning in niobium–titanium composite superconductors. *Acta Metall.* 32: 1871–81

Lee P J, Larbalestier D C 1987 Development of nanometer scale structures in composites of Nb–Ti and their effects on the superconducting critical current density. *Acta Metall.* 35: 2526–36

Lee P J, Larbalestier D C 1988 Determination of the flux pinning force of α-Ti ribbons in Nb46.5 wt%Ti produced by heat treatments of varying temperature, duration and frequency. *J. Mater. Sci.* 23: 3951–7

Lee P J, McKinnell J C, Larbalestier D C 1989 Microstructure control in high Nb–Ti alloys, *IEEE Trans. Magn.* 25: 1918–21

Li Chengren, Larbalestier D C 1987 Development of high critical current densities in niobium 46.5 wt% titanium. *Cryogenics* 27: 171–7

McInturff A D, Carson J, Larbalestier D, Lee P, McKinnell J, Kanithi H, McDonald W, O'Larey P 1990 Ternary superconductor NbTiTa for high field superfluid magnets. *1990 IEEE Int. Magnet Conf.*, Digest of Technical Papers. Brighton Polytechnic, Brighton, UK, p. AP15

McKinnell J C, Lee P J, Larbalestier D C 1989 The effect of titanium content on the pinning force in Nb44 wt%Ti to Nb62 wt%Ti. *IEEE Trans. Magn.* 25: 1930–3

Meingast C, Larbalestier D C 1989 Quantitative description of a very high critical current density Nb–Ti superconductor during its final optimization strain: II. Flux pinning mechanisms. *J. Appl. Phys.* 66: 5971–83

Meingast C, Lee P J, Larbalestier D C 1989 Quantitative description of a high J_c Nb–Ti superconductor during its final optimization strain, I: Microstructure, T_c, H_{c2} and resistivity. *J. Appl. Phys.* 66; 5962–70

Moffat D L 1985 Phase transitions in the titanium–niobium binary alloy system. Ph.D. thesis, University of Wisconsin–Madison
Motowidlo L R, Kanithi H C, Zeitlin B A 1990 NbTi superconductors with artificial pinning structures. *Adv. Cryog. Eng.* 36: 311–6
Muller H 1988 The upper critical field of niobium–titanium. Ph.D. thesis, University of Wisconsin–Madison
Pfeiffer I, Hillmann H 1968 Der Einfluss der Struktur auf die Supraleitungseigenschaften von NbTi50 und NbTi65. *Acta Metall.* 16: 1429–39
Warnes W H, Larbalestier D C 1986 Critical current distributions in superconducting composites. *Cryogenics.* 26: 643–53

P. J. Lee and D. C. Larbalestier
[University of Wisconsin–Madison, Madison, Wisconsin, USA]

Nitride and Carbonitride Superconductors

In the early 1930s niobium carbide was found to be a superconductor and then in 1941 Aschermann and co-workers reported superconductivity in niobium nitride (δ-NbN) with a transition temperature, T_c of 16 K. In 1953 niobium carbonitride was found by Matthias to have a T_c of 17.8 K. At this time a series of related compounds was discovered based around the transition metal (vanadium, molybdenum, tantalum, tungsten, zirconium) carbonitrides, all with the cubic B1 (NaCl) structure. The variation of T_c was measured as a function of composition across many binary combinations of these transition metals with carbon and nitrogen. Because the materials are isostructural and allow continuous substitutions with respect to each of the constituents, these studies gave an insight into the mechanisms of superconducting behavior as a function of the other normal state properties. This situation was unique, other classes of high-T_c material such as the A15 compounds (see *A15 Compound Superconductors*) were far more prone to structural disorder which did not allow such smooth interpolation from one metal compound system to another. In common with transition metal solid solutions and A15 compounds the variation of T_c as a function of the valence electrons per atom (e/a) is in agreement with the empirical observation that T_c shows two maxima for e/a close to 4.7 and 6.7 (see *Superconducting Materials: An Overview*).

1. Structure and Superconductivity

These materials exhibit characteristic instability at low temperatures. Stoichiometric NbC is metastable at ambient temperatures and in the stable phase only 80% of the carbon sites are occupied. Figure 1 shows a proposed equilibrium phase diagram for the niobium–nitrogen system which is complex and has proved to be very difficult to establish accurately. δ-NbN is structurally stable only at elevated temperatures and off stoichiometry. Acoustic phonon dispersion data indicate a phonon softening characteristic for this class of superconductor and the structural instability has been observed in tunnelling data (Phillips 1989).

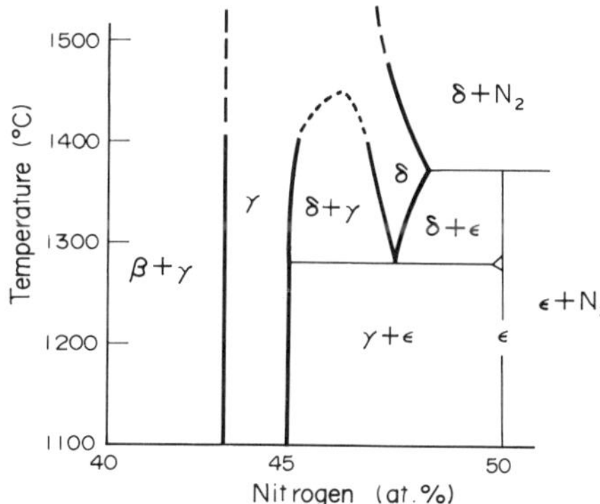

Figure 1
The proposed phase diagram for niobium–nitrogen (after Toth 1971)

The nitrides and carbonitrides are strong coupling superconductors characterized by a low density of states and a large electron–phonon interaction, similar to that in lead. The high T_c is linked to the occurrence of structural instability, both being a consequence of the strong electron–phonon interaction. Variations in T_c are due to variations of the density of states and models for T_c are based on the McMillan theory of strong coupling superconductors (see *Superconducting Materials: BCS and Phenomenological Theories*). Among the nitrides and carbides, niobium nitride is the binary system with the highest T_c, and the highest overall T_c is that of niobium carbonitride. MoN has been predicted to have a T_c of 29.3 K; however, the highest value achieved is 15.8 K in a mixed phase cubic B1 and hexagonal MoN thin film annealed under high pressure. Figure 2 shows the variation in T_c as the composition of mixed nitrides and carbides is varied.

2. Thin Films

The metastability of these materials makes fabrication difficult; bulk samples must be made at high temperatures and then quenched. However, thin films can be prepared relatively easily with vacancies on the nonmetal sites. Nitride and carbonitride films may be prepared by several different deposition techniques (e.g., chemical vapor deposition, reactive evaporation, postdeposition nitriding) but reactive sputtering is one of the most suitable methods, offering control over composition and film structure. Both rf and dc

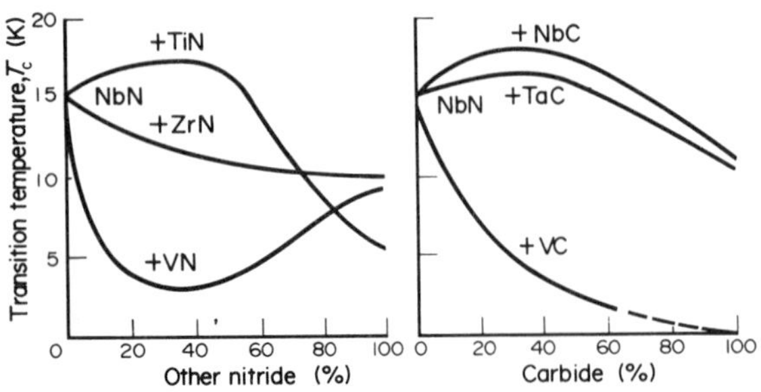

Figure 2
Critical temperature vs composition for nitride and carbide additions to NbN (after Hulm and Blaugher 1972)

sputtering from metal targets in an argon–nitrogen gas mixture have been widely reported. The deposition geometries used by different workers show many variations, but it is often found that optimum film formation occurs with the substrates placed close to the target (e.g., 20–30 mm) and at low gas pressures. This implies that some interaction with the active plasma region and/or high energies in the arriving atoms aids nitride formation.

NbN films can be deposited successfully at lower substrate temperatures than other high-T_c materials. Many workers have found that the carbonitride films resulting from C_2H_2, CH_4 or cyanogen additions during deposition are more readily formed than the pure nitride. Figure 3 shows the variation of the transition temperature as a function of methane addition during deposition. When grown without any external substrate heating polycrystalline films form with a columnar microstructure, but T_c values as high as 16.4 K have been observed by Villegier et al. (1985). By increasing the deposition temperature and using

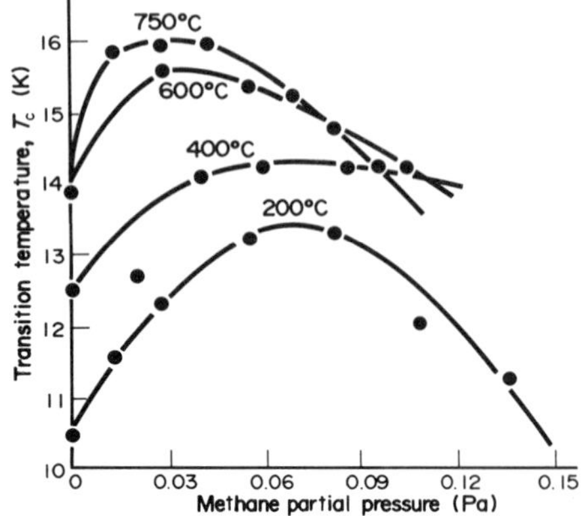

Figure 3
The transition temperature vs methane partial pressure for rf magnetron sputtered NbN films at various substrate temperatures (after Cukauskas et al. 1985)

single-crystal substrates (e.g., MgO) epitaxial growth may be promoted.

Very large critical currents have been reported in NbN films by Gavaler et al (1981): $> 2 \times 10^6$ A cm^{-2} for a 100 nm-thick film at 4.2 K in zero field and well over 10^5 A cm^{-2} at 20 T. However, as film thickness increases, the critical current density may be reduced by more than an order of magnitude. The coherence length is very short because of the low Fermi velocity and high T_c, typically 4–5 nm, but the penetration depth is relatively large (200–300 nm) due to the low electron density. Film resistivity may vary between about 60 μΩ cm and more than 300 μΩ cm and drops with temperature significantly only if the films are single crystal (e.g., by 20–40% for good epitaxial films between 300 K and 20 K). For polycrystalline films the resistance ratio is generally close to, or less than, unity. This corresponds to a very short electron mean free path which leads in turn to a very small Ginzburg–Landau coherence length and a high zero-field upper critical field, $B_{c2}(0)$, which can be in the range 25–50 T (see *Magnetic Properties of Superconductors*). Another feature of these films is that their characteristic columnar morphology can lead to pronounced anisotropy in transport properties.

3. Junction Development and Device Technology

Niobium nitride and carbonitride, with their high transition temperatures, high superconducting energy gap, chemical stability and refractory nature are very promising materials for use in superconductor–insulator–superconductor (SIS) tunnel junctions (see *Thin Films, Multilayers and Devices, Superconducting*). They offer very high stability on thermal cycling and storage and are less susceptible to damage (e.g., radiation damage) than other high-T_c materials. Devices based on NbN or NbCN may be used in closed cycle cryocoolers operating at around 10 K and may be thought of as a bridging technology between conventional niobium devices, which need liquid helium as a cryogen, and high-T_c ceramic superconducting devices using, for example, $(Ba,K)BiO_3$ or $YBa_2Cu_3O_{7-\delta}$. The high superconducting energy gap promises an extended frequency range of application, well in excess of 1 THz.

The ability to deposit high-T_c films at ambient temperatures has meant that full SIS trilayer structures (see *Superconducting Device Fabrication*) could be produced at an early stage so that junction processing techniques could be developed and tunnelling characteristics measured. The short coherence length has considerable consequences for tunnelling phenomena, making it important that there should be very-high-quality material immediately adjacent to the tunnel barrier. Although SIS junctions have been made using native niobium oxide as the barrier layer with a NbN counterelectrode (see, for example,

Villegier et al. 1985) the energy gap measured in such structures is low (4.0–4.2 meV) implying that the counterelectrode energy gap is reduced, presumably due to oxygen interdiffusion at the barrier–counterelectrode interface.

Artificial tunnel barriers based on an oxidized aluminum or magnesium layer have advantages as substrates for counterelectrode deposition due to their greater thermodynamic stability. An amorphous MgO barrier deposited on a polycrystalline NbN base electrode with a NbN counterelectrode has yielded devices with a gap of 5.0 meV (Shoji 1991). In principle the deposition of a single-crystal epitaxial NbN base electrode leads to a smoother substrate for barrier deposition, a more uniform energy gap and a shorter magnetic penetration depth, although problems with barrier layer coverage have been reported by Talvacchio and Braginski (1987). Shoji (1991) has fabricated SIS junctions using an epitaxial base which showed gaps as high as 5.8 meV and a steep subgap characteristic, indicating high uniformity of the gap, although the subgap structure suggested that excess leakage was due to multiparticle tunnelling through locally thinner areas. Epitaxial barrier growth at high temperatures (700 °C for Al_2O_3 and > 900 °C for MgO) results in layer growth rather than island growth giving smoother barriers of more uniform thickness; however, higher processing temperatures lead to greater possibility of interaction at the interfaces.

A mature NbN-based junction fabrication technology has been developed using similar processing techniques to those developed for niobium-based devices, for example the selective niobium etch process (SNEP) (see *Superconducting Device Fabrication*). LeDuc et al. (1991) have reported small-area junctions (0.1 μm^2) made using electron beam lithography, and the fabrication of large scale integrated (LSI) circuits using NbN–MgO–NbN tunnel junctions has been demonstrated by Aoyagi et al. (1991).

4. Conductor Development

The notable strain and radiation tolerance of NbN together with excellent high-field properties make it a promising candidate for application in wires or tapes for very-high-field magnets (Ekin et al. 1982). Although a number of different fabrication routes have been investigated and significant lengths of both single-filament and multifilamentary conductor prepared, these materials have not yet found commercial application.

The earliest conductors were made by chemical conversion of precursor carbon filaments at temperatures between 1100 °C and 1600 °C (Pike et al. 1975, Smith et al. 1975). A multifilamentary conductor was fabricated by passing a carbon yarn consisting of about 720 6 μm filaments through a furnace in an atmosphere of $NbCl_5$, N_2, and H_2 or CH_4. A reaction time of 2–5 min resulted in a layer of mixed NbC and NbCN phases 0.2–1.0 μm thick. Critical current densities were in excess of 10^6 A cm^{-2} in zero field but dropped to less than 10^5 A cm^{-2} in fields greater than 3 T. More recently, continuous tapes and fibers with higher critical current densities (about 10^5 A cm^{-2} at 15 T and 10^4 A cm^{-2} at 20 T) have been fabricated by reactive sputtering onto a variety of substrates (Dietrich et al. 1987, Kampwirth et al. 1987). Although the results are promising the remarkably high critical current densities seen in very thin films have not been reproduced in the thicker coatings required for practical conductors. For this reason conductors based on NbN and NbCN have so far been unable to compete with commercial NbTi and Nb_3Sn conductors (see *Critical Current Density*).

See also: Superconducting Materials: An Overview; Superconducting Materials: Physical Properties

Bibliography

Aoyagi M, Nakagawa H, Kurosawa I, Takada S 1991 Josephson LSI fabrication technology using NbN/MgO/NbN tunnel junctions. *IEEE Trans. Magn.* 27:3180–3

Aschermann G, Friederich E, Justi E, Kramer J 1941 Supraleitfähige Verbindenungen mit extrem hohen Sprungtemperaturen (NbH und NbN). *Physik. Zeit.* 42:349–60

Cukauskas E J, Carter W L, Qadri S B, Skelton E F 1985 Niobium nitride thin films for use in Josephson junctions. *IEEE Trans. Magn.* 21:505–8

Dietrich M, May D, Windte V, Schauer W 1987 High field NbN superconductor on a fiber basis. *IEEE Trans. Magn.* 23:991–4

Ekin J W, Gavaler J R, Gregg J 1982 Effect of strain on the critical current and critical field of B1 structure NbN superconductors. *Appl. Phys. Lett.* 41:996–8

Gavaler J R, Santhanam A T, Braginski A I, Ashkin M, Janoko M A 1981 Dimensional effects on current and field properties in NbN films. *IEEE Trans. Magn.* 17: 573–6

Hulm J K, Blaugher R D 1972 In: Douglass D H (ed.) 1972 *Superconductivity in d- and f-band metals*. AIP, New York

Kampwirth R T, Capone D W, Gray K E, Ho H, Chumley S 1987 High field properties of NbN conductors on practical substrates. *IEEE Trans. Magn.* 23:995–8

LeDuc H G, Judas A, Cypher S R, Bumble B, Hunt B D, Stern J A 1991 Submicron area NbN/MgO/NbN tunnel junctions for SIS mixer applications. *IEEE Trans. Magn.* 27:3192–5

Matthias B T 1953 Transition temperatures of superconductors. *Phys. Rev.* 92:874–6

Phillips J C 1989 *Physics of High-T_c Superconductors*. Academic Press, San Diego, CA

Pike G E, Mullendore A W, Schirber J E, Napier J 1975 Superconducting properties of thin film niobium carbonitrides on carbon fibres. *IEEE Trans. Magn.* 11:185–8

Shoji A 1991 Fabrication of all-NbN Josephson tunnel junctions using single crystal NbN films for the base electrodes. *IEEE Trans. Magn.* 27:3184–7

Smith W D, Lin R Y, Coppola J A, Economy J 1975 Preparation and properties of multifilament niobium carbonitride superconductor. *IEEE Trans. Magn.* 11:182–4
Talvacchio J, Braginski A I 1987 Tunnel junctions fabricated from coherent NbN/MgO/NbN and NbN/Al_2O_3/NbN structures. *IEEE Trans. Magn.* 23:859–62
Toth L E 1971 *Transition Metal Carbides and Nitrides.* Academic Press, New York
Villegier J C, Vieux-Rochaz L, Goniche M, Renard P, Vabre M 1985 NbN tunnel junctions. *IEEE Trans. Magn.* 21:498–504

Z. H. Barber, R. E. Somekh and J. E. Evetts
[University of Cambridge, Cambridge, UK]

Nondestructive Evaluation Using Magnetism

Magnetic methods of nondestructive evaluation utilize the response of magnetic materials to static or slowly varying magnetic fields. They find wide application both on the production line and in the field and can often provide a sensitive, rapid and relatively inexpensive test method. Uses include measurement of material properties, such as carbide content or hardness, detection of harmful impurities or defects, and measurement of coating thickness.

Magnetic methods are generally confined to ferromagnetic materials, such as ferritic steels and nickel or cobalt alloys, or to materials containing a ferromagnetic component, such as ferrite in nonmagnetic stainless steel welds or magnetic corrosion products.

Magnetic inspection can be conveniently divided into two types: the detection of magnetic leakage fields and the measurement of magnetic properties. A leakage field is a disruption in the normal magnetic flux pattern around a magnetized part due to discontinuities in that part. The leakage fields from defects, such as cracks, will have strong gradients which can attract magnetic particles sprayed onto the part, giving a visible indication of that defect. Leakage fields can also be detected using magnetic recording tape and measuring probes such as moving coils, Hall effect devices, flux gate magnetometers (Förster probes) (see *Sensors and Devices, Magnetic*), SQUID magnetometers (see *SQUID Design, Fabrication and Application*) and nuclear magnetic resonance (NMR) magnetometers (see *Nuclear Magnetic Resonance Imaging*).

In addition to the detection of discontinuities, magnetic measurements can give information about the metallurgical condition of the material. The magnetic properties which can be measured include coercive force, remanence, saturation magnetization, permeability, overall hysteresis loop shape and Barkhausen noise (see *Barkhausen Effect*). Uses of magnetic property measurement include sorting of steel, detection of internal stresses, determination of ferrite content in stainless steel welds and castings, determination of grain orientation and detection of ferromagnetic impurities.

1. Leakage Fields

A defect, such as a blowhole or crack, in a magnetized material causes the surrounding magnetic field to differ from that in a defect-free material. This difference is referred to as a leakage field. For example, consider a cylindrical defect below the surface of a magnetic material with permeability μ_1 in an applied field of magnitude H_0 (see Fig. 1). It is assumed that the defect is filled with a material, such as air or vacuum, with a permeability near μ_0 which is much less than μ_1. For this situation, the magnetic leakage field is a dipole field and may be obtained analytically. The tangential, H_t, and perpendicular, $H_\perp$, leakage field components are illustrated in Fig. 1. The parameter m is the dipole strength and for a linear magnetic material is given approximately by $2H_0a^2$, where a is the radius of the defect. As seen in Fig. 1, the leakage field strength at the surface falls off inversely with the square of the defect depth h. For a nonlinear material such as steel, the measured values from a cylindrical defect are about twice this value and the apparent depth is less than the true depth.

Although the leakage fields shown in Fig. 1 are for an idealized case (i.e., linear magnetic material and cylindrical defect), real leakage fields look like more or less distorted representations of the fields plotted in

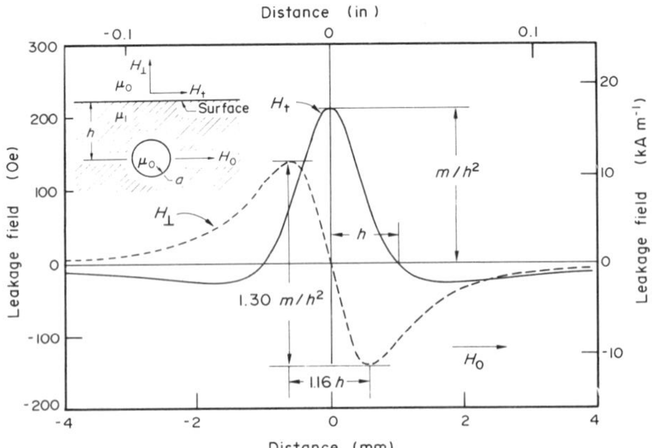

Figure 1
Magnetic leakage field calculated for cylindrical defect in a linear isotropic magnetic material. The geometry of the defect is shown in the inset; the axis of the cylinder is parallel to the sample surface. The distance scale refers to distance along the surface, with zero being at the surface directly over the hole axis. The numerical values are those appropriate for a 10^{-3} m radius cylinder with axis at a depth of 10^{-3} m and an applied field H_0 of 6.4 kA m^{-1}

the figure. Added to the leakage fields are large but slowly varying bias fields. For example, the tangential field will have the added value H_0, because boundary conditions require that this field be the same directly above and below the surface of the sample. If H_0 is not strictly parallel to the sample surface, the magnetization of the sample will have a component in the perpendicular direction, resulting in a bias field in that direction because boundary conditions require the perpendicular component of the B field to be continuous.

2. Leakage Field Measurement

Methods used for detecting and measuring leakage fields include Hall effect probes, moving coils, magnetic tapes, flux gate magnetometers and NMR magnetometers. The magnetic particle method is discussed separately in Sect. 3. NMR magnetometers offer both high sensitivity and extreme accuracy. Their use at present is generally confined to calibration.

Hall effect probes with high sensitivity and small size are currently readily available. Sizes range down to about 10^{-3} m diameter for the active area. Hall elements as small as 100×25 μm have been specially fabricated and used for leakage field detection. Even smaller probes are possible and have been used for magnetic domain studies. The sensitivity of Hall probes depends on a large number of factors and can be of the order of 1 A m^{-1}. The probe diameter must not be much larger than the leakage field width h, as seen in Fig. 1, or the defect becomes difficult to detect. To cancel the effect of bias fields, two closely spaced probes are often used, with the output being the difference in field between the two probes.

A moving-coil detector utilizes a small, many-turned coil. The voltage output of such a coil is proportional to the rate of change of magnetic flux lines through the coil area. The coil can be moved with a uniform velocity over the sample surface, or it may be rotated or vibrated. The coil diameter must not be much greater than the leakage field width or, as with the Hall probe, detection sensivity will be reduced. Moving-coil detectors can be made with sensitivities comparable to Hall probes, but generally must be larger.

Flux gate magnetometers (also called second-harmonic detectors or Förster probes) are very sensitive and can accurately measure fields down to about 0.01 A m^{-1} (the earth's field is approximately 17 A m^{-1}). Because of their great sensitivity they are often used to measure residual leakage fields (fields existing after the applied field H_0 is turned off).

The magnetic tape method consists of applying a magnetic tape to the surface of a magnetized material. A pattern of the magnetic flux at the surface remains on the tape, which is subsequently scanned by a tape reader. The information obtained can then be processed and displayed on an oscilloscope, chart recorder or other display device for interpretation.

3. Magnetic Particle Methods

The magnetic particle method of detecting leakage fields consists of applying finely divided ferromagnetic particles to a suitably magnetized sample. The particles will tend to accumulate in the regions of large magnetic field gradients found at leakage fields from defects and other discontinuities. These accumulations, often called indications, are then observed and interpreted. Both surface and near-subsurface defects can be detected. Nonmagnetic coatings up to 100 μm generally do not affect the sensitivity. Interpretation includes the ability to sort out nonrelevant indications such as intentional discontinuities, cross-sectional changes and so-called magnetic writing.

With suitable magnetization and particle-application techniques, almost any size and shape part made of ferromagnetic material can be inspected. The technique can be adapted to both rapid 100% production line inspection using automated equipment and to field testing using portable general-purpose equipment. Procedures for magnetization, particle application and interpretation must be carefully worked out for each type of part. In establishing a test procedure for a part, consideration must be given to (a) the magnetization method; (b) the size, shape, location and orientation of the defects to be detected; (c) the direction and magnitude of magnetization; (d) the type of magnetic particles to be used; (e) the method of particle application; (f) the illumination present during observation; and (g) acceptance and rejection criteria.

Magnetization can be accomplished by passing a current directly through the part (direct method), or by passing a current through a cable threaded through the part or wrapped around the part in the form of a coil (indirect method). Permanent magnets, induced currents and electomagnetic yokes can also be used. When using the direct or indirect method, one of three types of current may be chosen. Direct current from full-wave-rectified three-phase alternating current, direct current from half-wave-rectified single-phase alternating current or unrectified alternating current (ac). Because of the skin depth effect, ac is generally used when only surface defects are being sought. In this case, lower current can often be used, because the current is confined to a layer near the surface. When subsurface defects are important, or when larger currents are required, full-wave-rectified three-phase alternating current or its equivalent is used.

Greatest sensitivity is obtained when the magnetic particles are applied to the part while the magnetization current is flowing. This is referred to as the continuous method. Precise timing is necessary because the large currents can cause overheating when allowed to flow for more than several seconds. For material with a high coercive force and when greatest sensitivity is not desired, the particles may be applied after the magnetization current has been turned off. This is referred to as the residual method.

Proper choice of magnetizing level is a critical factor. With too small a magnetizing field, important detectable flaws will not be revealed. With too high a field, the number of nonrelevant indications will increase, possibly masking indications from real defects. For each part type and inspection situation the optimum current level is generally established through trial and error, using parts containing known or artificial defects. A number of rules of thumb have been developed to guide the establishment of proper levels. For example, using full-wave-rectified three-phase alternating current and direct magnetization on rod-shaped samples, the current level should be 25–35 kA m^{-1} of part diameter for diameters up to 0.125 m, 20–25 kA m^{-1} for diameters of 0.125–0.380 m and no less than 12 kA m^{-1} for diameters over 0.380 m. This implies a tangential component of magnetic field at the surface of the part of 2–5.5 kA m^{-1}. The upper limit would be preferred in the ideal case, while the lower limit recognizes the difficulty of working with large currents, along with the fact that smaller defects are less important on larger diameter parts. The use of a gaussmeter to determine whether critical areas of the part under test are actually receiving this amount of applied field has proven advantageous. (The gaussmeter should be placed directly on the surface and oriented to determine the magnetic field parallel to the surface, i.e., with the plane of the Hall effect element perpendicular to the surface.)

Magnetic particles may be applied by blowing dry particles (at low velocity) across the part under test or by flowing a wet suspension of particles over the part. Particles are available with or without a fluorescent coating. Nonfluorescent particles are available in a variety of colors to provide contrast against various backgrounds.

Other types of particles for specialized applications have been developed. Wet particles in aerosol cans are available. Magnetic paints detect leakage fields by using the reflective properties of platelike magnetic pigments when they are deflected by a leakage field. Some of these paints are useable under water. Polymerizable rubberlike formulations which contain magnetic particles have been developed. During polymerization, the particles will concentrate in the region of a leakage field, leaving a permanent record of that field. They are especially useful for inspecting for cracks in inaccessible locations such as bolt holes.

Figure 2 illustrates the inspection of a rod using direct circular magnetization to detect defects parallel to the cylinder axis and coil magnetization to detect defects perpendicular to the cylinder axis. Only two such perpendicular directions of magnetization are required, since sufficient sensitivity is generally obtained for defects whose dimensions lie at angles between 45° and 135° to the field direction. Current prods can be used for detecting defects in sheet material. This method has found widespread use in the inspection of welds for cracks.

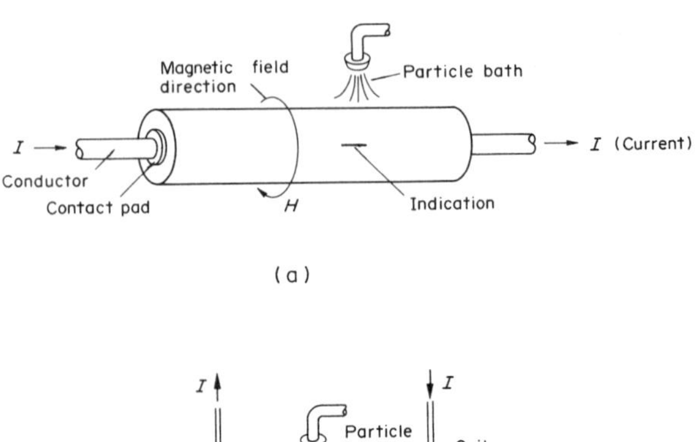

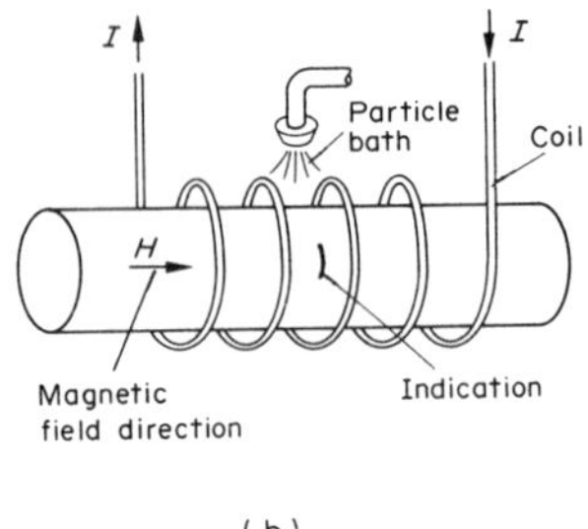

Figure 2
Inspection of a rod using (a) direct circular magnetization (head shot) to detect longitudinal defects, and (b) longitudinal magnetization with a coil (coil shot) to detect defects lying parallel to the circumference. Defects inclined up to 45° from the horizontal will be detected by the head shot, the remainder by the coil shot

4. Magnetic Properties

The magnetic properties of a ferromagnetic material can be represented on a hysteresis loop, as shown in Fig 3. The shape of this loop is influenced by the values of initial permeability, saturation induction, remanence and coercive force, which are dependent on the metallurgical history of the material. Such details as grain structure and orientation, internal stress, alloy content, hardness and thermomechanical history all affect these details. As with many other NDE methods, correlations, such as between coercive force and hardness, that might exist are not general, but must be established for each particular material and its particular thermomechanical history.

The properties with perhaps the greatest range of variation in ferromagnetic materials are coercive force

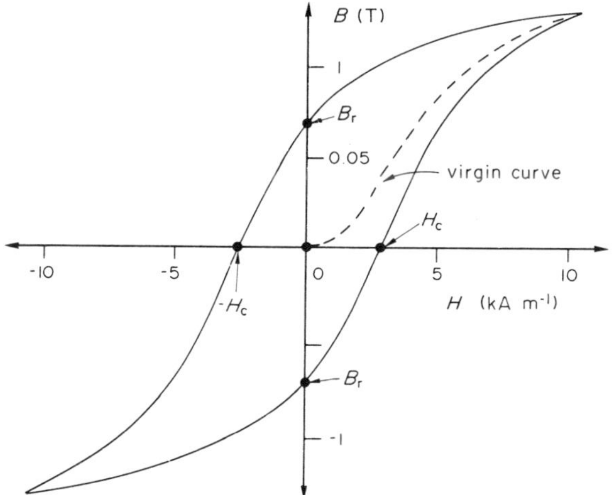

Figure 3
Hysteresis loop (solid) and initial magnetization curve (dotted) for a ferromagnetic material; H_c is the coercive force and B_r the remanence

and permeability. In permeability measurements, the shape of the test specimen has a large influence on the measurements because of demagnetizing effects (see *Demagnetizing Factors*). The coercive force has the advantage that it is independent of sample shape and can be readily measured as a function of temperature with suitable instrumentation.

Electromagnetic testing (often classed as an eddy current test) involves tracing out a B–H hysteresis curve at low frequencies (10–400 Hz). The curve obtained is compared in various ways to standard patterns and, based on experience, parts which are out of specification can be sorted. The remote field electromagnetic inspection technique, based on the diffusion of electromagnetic energy through a ferromagnetic material, has proven useful for the detection of external casing corrosion even when access is available only inside the casing.

5. *Future Developments*

Progress continues to be made in correlating magnetic properties to material properties such as internal stress, grain size and hardness. Recent decreases in the cost of computing power should lead to increased automation in both gathering and interpreting such information and should improve the certainty of the correlations through multivariate analysis. Improvements in specifications, magnetic inks and automation are increasing the utility of magnetic particle inspection as a reliable and cost-effective method of locating deleterious defects in steel parts and structures.

See also: Electromagnetic Forming; Inks, Magnetic; Measurements in Magnetic Materials; Nuclear Magnetic Resonance Imaging; Sensors and Devices, Magnetic; SQUID Design, Fabrication and Application

Bibliography

American Society for Testing and Materials 1981 Standard recommended practice for magnetic particle examination, ASTM E709–80. *Metallography, Non-Destructive Testing*, 1981 *Annual Book of ASTM Standards*, Part 2. ASTM, Philadelphia, PA, pp. 626–67

American Welding Society, Committee on Filler Metal 1974 *Standard Procedures for Calibrating Magnetic Instruments to Measure the Delta Ferrite Content of Austenitic Stainless Steel Weld Metal*, Standard Procedure AWS A4.2–74. American Welding Society, Miami, FL

Bainton K F 1977 Characterizing defects by determining magnetic leakage fields. *Non-Destr. Test. (Guildford, Engl.)* 10: 253–7

Betz C E 1967 *Principles of Magnetic Particle Testing*. Magnaflux Corporation, Chicago, IL

Bozorth R M 1953 *Ferromagnetism*, 2nd edn. Van Nostrand, New York

Bray D E, Stanley R K 1989 *Nondestructive Evaluation — A Tool for Design, Manufacturing and Service*. McGraw-Hill, New York

Carr M J, Gavert R B, Moore R L, Wawrousek H W, Westbrook J H 1976 *Critical Surveys of Data Sources: Electrical and Magnetic Properties of Metals*, NBS Special Publication 396–4. US Government Printing Office, Washington, DC

Förster F 1981 Theoretical and experimental results on the magnetic stray flux technique. *Materialprüf.* 21: 372–8

Jiles D C 1990 Review of magnetic methods for nondestructive evaluation. *NDT Int.* 23: 83–92

Kwun H, Burkhart G L 1987 Nondestructive measurement of stress in ferromagnetic steels using harmonic analysis of induced voltage. *NDT Int.* 20: 167–71

Lord W 1985 *Electromagnetic Methods of Nondestructive Testing*, Vol. 3. Gordon and Breach, London

Lord W, Hwang J H 1977 Defect characterization from magnetic leakage fields. *Br. J. Non-Destr. Test.* 19: 14–18

Lovejoy D J 1980 The nature of magnetic inks and powders. *Br. J. Non-Destr. Test.* 22: 242–5

McMaster R C (ed.) 1959 *Nondestructive Testing Handbook*, Vol. 2 Ronald Press, New York, pp. 30.1–34.24

Metals Handbook, 8th edn., Vol. 11 1976 Magnetic particle inspection. American Society for Metals, Metals Park, Ohio, pp. 44–74

Metals Handbook, 8th edn., Vol. 11 1976 Electromagnetic sorting and testing. American Society for Metals, Metals Park, Ohio, pp. 93–105

Schmidt T R 1984 The remote field eddy current inspection technique. *Mater. Eval.* 42: 225–30

Schmidt J T, Skeie K, McIntire P (eds.) 1989 *Magnetic Particle Testing*, Nondestructive Testing Handbook, Vol. 6. *American Society for Nondestructive Testing*

Stratton J A 1941 *Electromagnetic Theory*. McGraw-Hill, New York

L. J. Swartzendruber
[National Institute of Standards and Technology, Gaithersburg, Maryland, USA]

Nonequilibrium Superconductivity

In a superconductor under equilibrium conditions the density of unpaired electrons (i.e., quasiparticles) is constant. The quasiparticle recombination rate is balanced exactly by the rate at which thermal energy breaks Cooper pairs. Equilibrium may be disturbed in a variety of ways, for instance by current injection, microwave or infrared radiation or by irradiation with x rays or nuclear particles. A nonequilibrium superconductor displays a number of complex effects. Under steady-state conditions the superconducting gap may be depressed due to the creation or injection of excess quasiparticles. Correspondingly, quasiparticle extraction can lead to extreme gap enhancement even at temperatures above the equilibrium critical temperature (Blamire et al. 1991a). The nonequilibrium state may also involve charge imbalance when the number of quasielectrons does not equal the number of quasiholes. When the disturbing influence is removed the superconductor recovers from the nonequilibrium state at an overall rate that depends on the interaction of a number of different scattering and relaxation rates for charge carriers and phonons.

Most studies of nonequilibrium superconductivity enjoy a strong symbiotic relationship with tunnelling and phonons (see *Thin Films, Multilayers and Devices, Superconducting*; *Tunnelling Spectroscopy*). Tunnel junctions provide the most fruitful microscopic measurements of properties of the nonequilibrium state and often create it. The excess energy in nonequilibrium states is primarily removed by phonons and thus the coupling to phonons and their escape was recognized by Rothwarf and Taylor (1967) as being crucial. Reviews of these topics are given by Gray (1981) and Langenberg and Larkin (1986). The escape of phonons through Kapitza boundaries at the surface of the superconductor is discussed by Kaplan (1979).

For superconductors at $T>0$, thermal energy breaks Cooper pairs into quasiparticles with energies E greater than the pair binding energy 2Δ, where Δ is the energy gap. Nonequilibrium states usually increase or otherwise alter the energy/momentum distribution of the quasiparticle density N from its equilibrium composition and energy distribution $N(T)$. Rothwarf and Taylor (1967) summarize the behavior for an injection rate I_0 and a phonon escape rate Γ_{es} as

$$\frac{\partial N}{\partial t} = I_0 + 2\Gamma_B N_\omega - RN^2 \qquad (1)$$

$$\frac{\partial N_\omega}{\partial t} = \frac{RN^2}{2} - \Gamma_B N_\omega - \Gamma_{es}[N_\omega - N_\omega(T)] \qquad (2)$$

where N_ω is the density of phonons with energy $\hbar\omega$ greater than 2Δ, $N_\omega(T)$ is its equilibrium density at temperature T, Γ_B is the pair-breaking rate by such phonons and R is the coefficient of the pairwise recombination rate of quasiparticles into Cooper pairs. Thermal equilibrium implies that $2\Gamma_B N_\omega(T) = RN(T)^2$, but note also that $N(T)$ is divided equally between quasielectrons ($k>k_F$) and quasiholes ($k<k_F$), where k is the momentum and k_F is the Fermi momentum: any inequality is a charge imbalance.

When the perturbation from equilibrium is small, equilibrium relaxation rates can be measured. These include quasiparticle recombination, charge imbalance relaxation and inelastic scattering. The rates generally agree well with theoretical predictions based on the BCS theory of superconductivity and other measurements (see *Superconducting Materials: BCS and Phenomenological Theories*). For larger perturbations, the nonlinearities of Eqns. (1) and (2) are manifest together with other unexpected phenomena.

1. Close to Equilibrium

For $\Delta N = N - N(T) \ll N(T)$, Eqns. (1) and (2) can be linearized. In steady state, ΔN is related to the quasiparticle recombination rate, $\Gamma_R(T) = RN(T)$, by

$$\Delta N = I_0(1 + \Gamma_B \Gamma_{es})/2\Gamma_R \qquad (3)$$

Note that $\Gamma_{es} \leqslant s/d$, where s is the phonon (sound) velocity and d is the distance to a thermal reservoir. For $\Gamma_B \ll \Gamma_{es}$, ΔN directly measures Γ_R, but for $\Gamma_B \gg \Gamma_{es}$, ΔN measures $\Gamma_R\Gamma_{es}/\Gamma_B$, which has the same temperature dependence but the dependence on the electron–phonon scattering rate, characterized by $1/\tau_E$, the normal-state scattering rate at T_c, cancels in the ratio Γ_R/Γ_B. In the latter case, which is hard to distinguish experimentally, the injected energy resonates between ΔN and ΔN_ω with their magnitudes limited only by the small Γ_{es}. This problem can be overcome by measuring: Γ_R very near T_c; charge relaxation, since phonon readsorption will not lead to a charge imbalance; the quasiparticle inelastic scattering rate; or the lifetime broadening of Δ in thermal equilibrium, which works if $\Gamma_R \simeq \Delta/\hbar$.

Very near to T_c, Γ_Q, the charge relaxation rate, and Γ_R are also limited by Δ which decreases to zero at T_c. The reason is that charge relaxation and Cooper pairing only occur for final states within around Δ of the Fermi level, but inelastic scattering processes occur over an energy range of approximately k_BT: thus, $\Gamma_Q = 0.79\Delta/k_BT\tau_E$ and $\Gamma_R = 0.27\Delta/k_BT\tau_E$. Because $\Delta \ll k_BT$ near T_c, there are no special 2Δ phonons and Γ_R is measured directly. Results of $\Gamma_R(T)$ and $\Gamma_Q(T)$ for aluminum are shown in Fig. 1, together with theory: however, following Eqn. (3), $\Gamma_R(T) = RN(T) \propto N(T)/\tau_E$ has been scaled at low temperatures by $\Gamma_B/\Gamma_{es} = 30$ to account for poor phonon escape.

Determination of the quasiparticle inelastic scattering rate Γ_s exploits the energy dependence of I_0 and

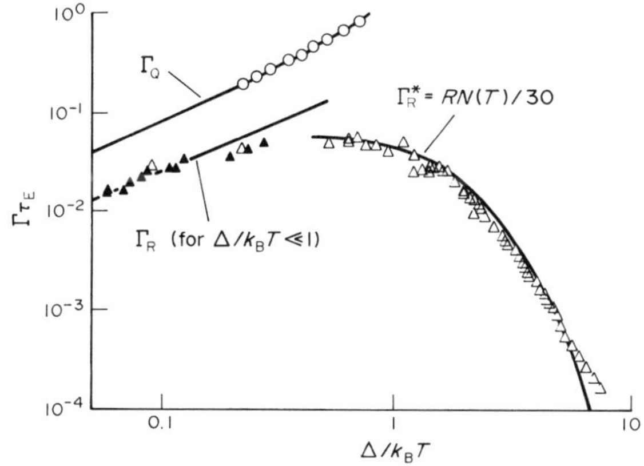

Figure 1
Experimental determinations of the rates of quasiparticle recombination (triangles) and charge relaxation (circles) for aluminum, normalized to $1/\tau_E$. Open symbols are steady-state measurements and solid symbols are real-time decays. The lines are theoretical dependences with the low-temperature Γ_R scaled by $\Gamma_B/\Gamma_{es} = 30$ to account for poor phonon escape (see Eqn. (3))

ΔN attainable with tunnel junctions and one finds $\Gamma_s(E) \simeq E^3$, as expected from the T^3 dependence of the electron–phonon scattering. In aluminum the results of all these measurements and theory agree within a factor of two, although they depend systematically and strongly on purity, through variation in T_c and localization effects. They extrapolate to a pure-limit value consistent with theoretical estimates of $\tau_E \simeq 40$ ns. The temperature-dependent rates Γ_R and Γ_Q can be found using Fig. 1. For example, for pure aluminum Γ_R has its maximum value of $10^6\,s^{-1}$ for $\Delta/k_BT \simeq 0.5$, or $T \simeq 0.97\,T_c$, and falls to $5 \times 10^3\,s^{-1}$ at lower temperatures $\Delta/k_BT \simeq 7.5$, or $T \simeq 0.23\,T_c$. In general, τ_E decreases with T_c (i.e., $\Delta(0)$); for example, it is 0.3 ns for tin, approximately 0.02 ns for lead and niobium and about 0.2 ps for the high-T_c superconductor $Tl_2Ba_2Ca_2Cu_3O_{10}$ (Eesley et al. 1990). The impurity effect also increases $1/\tau_E$ in $Pb_{90}Bi_{10}$ by a factor of about five over pure lead.

Because superconductors are a charged superfluid, the frequency of collective, soundlike modes is shifted to the plasma frequency and the motions are damped by lattice defects. However, charge imbalances result in propagating waves in a nonequilibrium superconductor. An imbalance is produced by an inflow of normal current, which for electroneutrality induces an outflow of supercurrent and a shift in the pair potential. This in turn acts as a restoring force to produce currents in the opposite direction. These heavily damped, propagating oscillations have been observed.

2. *Far from Equilibrium*

The first effect of a large injection rate is the nonlinear saturation of ΔN according to Eqn. (1). The determination of Γ_R from this nonlinearity agrees with the previous methods. In addition, spatial instabilities can occur in thin-film tunnel junctions because quasiparticle injection begins abruptly at $V_i = 2\Delta$. As such, there are values of V_i just less than $2\Delta_0$ for which the junction area splits into two regions, one with intense injection which reduces the undisturbed Δ_0 to $\Delta' < V_i/2$ and another which has no injection because $\Delta_0 > V_i/2$. Experiments (Gray and Willemsen 1978, Iguchi and Langenberg 1980, Blamire et al. 1991b) are in good agreement with theory (Schön and Tremblay 1979).

The energy spectroscopy of tunnel junctions permits determination of the nonequilibrium distribution function $f(E)$, which corresponds to the Fermi function in equilibrium. The gross features of $f(E)$ in aluminum are well fitted by

$$f(E) = \frac{1}{1 + \exp(E - \mu^*)/k_BT^*} \tag{4}$$

where μ^* and T^* are the effective chemical potential and temperature of the nonequilibrium quasiparticles, respectively. With increasing tunnel junction injection (and thus V_i), T^* steadily increases above T and μ^* drops more sharply from about $\Delta_0/2$ to slightly negative values. The smooth decrease in measured Δ' is reproduced by the BCS gap equation using the nonequilibrium $f(E)$ (see *Superconducting Materials: BCS and Phenomenological Theories*). It is interesting that $T^* > T_c$ is possible because $\mu^* < 0$ and it is observed for $\Delta' < \Delta_0/2$. For $V_i \gg 2\Delta_0$ and $\Delta' < k_BT^*$, charge imbalance is expected and manifests itself as a different $f(E)$, μ^* and T^* for quasielectrons and holes.

For $V_i < 2\Delta$, tunnel junctions redistribute the energies of quasiparticles without injecting new ones. This redistribution is operationally equivalent to photon (microwave) or phonon adsorption with $\hbar\omega = eV_i$. A gap enhancement results from substituting the nonequilibrium $f(E)$ in the BCS gap equation and measurements using microwave, phonon and tunnel junction injection have verified the theory quantitatively. Enhancement of the critical current is also found with microwaves.

Microbridges between superconducting banks can be realized by small constrictions (Dayem type) or by interposing a thin normal-metal layer which becomes superconducting by the proximity effect (SNS type). At finite voltages, Josephson oscillations occur and the nonequilibrium analysis requires numerical solutions of the time-dependent Ginzburg–Landau equation. This procedure is a vast improvement over the resistively shunted junction model and correctly includes the effects of microwave irradiation. In addition, superconducting and normal probes, closely spaced on each side of a microbridge, have measured directly the difference in quasiparticle and Cooper pair potentials resulting from the charge imbalance of the nonequilibrium state for currents above the critical value.

The thermal propagation of normal regions (hotspots) that bridge a current-carrying conductor is solvable analytically for sufficiently long lengths L_n by linearizing the thermal properties. The power dissipated per unit area for the steady-state solution P_0, which separates propagating from collapsing hotspots, is independent of L_n. Numerical simulations using realistic, temperature-dependent properties agree in that limit, but also allow extension to smaller L_n, for which $P_0 \simeq 1/L_n^2$. Generating small-L_n hotspots in narrow films with x rays has verified calculations of the propagation velocity (Gabutti et al. 1990) and collapsing hotspots.

Direct measurements of phonons are possible using tunnel junctions both to create excess phonons and to measure them (see *Tunnelling Spectroscopy*). Examples include: longitudinal and transverse modes and their absorption by defects in single crystals; coupled phonon–quasiparticle modes in superconducting lead; Kapitza conductance; and the dynamics of phonons in liquid helium.

3. Theory

Nonequilibrium superconductors may be described from two different theoretical points of view. One characterizes the nonequilibrium state in terms of a Boltzmann equation for quasiparticles: it is more intuitive but is limited to situations where variations are small on time scales of the order $\hbar/\Delta$ and length scales of the order of $\hbar v_F/\Delta$, where v_F is the Fermi velocity. A second, more general approach is based on the semiclassical Green function. It is also valid when the width of states is so significant that the concept of quasiparticles loses its meaning. This approach allows one to include pair breaking to all orders in the scattering rates and is particularly better for cases like the thermoelectric effect associated with a supercurrent and a temperature gradient. However, in most cases, the Boltzmann equation is a sufficiently accurate approach which is conceptually simpler.

4. Devices

There are a number of important superconducting devices in which operation is not intimately dependent on a nonequilibrium state. These include Josephson switching devices for digital applications, phase-locked arrays of Josephson junctions for microwave sources up to several GHz and SQUID magnetometers (for further details on these see *SQUID Design, Fabrication and Application*; *Superconducting Devices and Electronics*).

Millimeter-wave detectors and mixers can be made using the Josephson effect, but tunnel junctions using quasiparticle characteristics have been more successful (see *Millimeter-Wave Detection*). The key feature is the effect of photon-assisted tunnelling on the $I(V)$ near the sharp increase in current at the onset of pair breaking at $V = 2\Delta$ (or $\Delta_1 + \Delta_2$ for dissimilar electrodes). The induced steps at $\pm n\hbar\omega$ provide detection and the addition of a pump signal provides a mixer.

Three-terminal devices offer important advantages for digital electronics circuit design. Mentioned here are device concepts based on nonequilibrium superconductivity, which consist of stacked back-to-back tunnel junctions with a common electrode. They include devices based on: tunnelling through the superconducting gap; injected quasiparticles recirculating in an output junction; and gap suppression. Injection-controlled weak links have also been studied, together with the potentially important hybrid superconductor–semiconductor devices.

Particle detector concepts using superconductors include energy resolution from the enhanced quasiparticle current ($\propto \Delta N$) of a tunnel junction and position-sensitive devices using either the normal-to-superconductor transition or switching in a Josephson junction. Tunnel junctions operating at around 100 mK have achieved better resolution ($\sim$50 eV) than semiconducting devices and are being pursued for a number of applications such as x ray detectors. Thin-film strips have also successfully demonstrated extremely high spatial resolution for x rays and are being investigated (Gabutti et al. 1990) as detectors of minimum ionizing particles. The interaction of a high-energy ($\gg k_B T$) particle results in a complex, spatial- and time-dependent nonequilibrium state.

See also: Josephson Effects in Weak Links; Superconducting Materials: BCS and Phenomenological Theories; Superconducting Materials: Physical Properties; Thin Films, Multilayers and Devices, Superconducting; Tunnelling Spectroscopy

Bibliography

Blamire M G, Kirk E C G, Evetts J E, Klapwijk T M 1991 a Extreme critical-temperature enhancement of Al by tunneling in Nb/A1O$_x$/A1/A1O$_x$/Nb tunnel junctions. *Phys. Rev. Lett.* 66: 220–3

Blamire M G, Kirk E C G, Somekh R E, Evetts J E 1991 b Novel hysteresis effects in Nb/A1O$_x$/A1/A1O$_x$/Nb tunnel junctions. *J. Appl. Phys.* 69: 2376–81

Eesley G L, Heremans J, Meyer M S, Doll G L, Liou S H 1990 Relaxation time of the order parameter in a high-temperature superconductor. *Phys. Rev. Lett.* 65: 3445–8

Gabutti A, Gray K E, Wagner R G, Ono R H 1990 Granular aluminum superconducting detector for 6 keV x-rays and 2.2 MeV beta sources. *Nucl. Instrum. Methods A* 289: 274–82

Gray K E 1981 *Nonequilibrium Superconductivity, Phonons, and Kapitza Boundaries*. Plenum, New York

Gray K E, Willemsen H W 1978 Inhomogeneous state of superconductors by intense tunnel injection of quasiparticles. *J. Low Temp. Phys.* 31: 911–25

Iguchi I, Langenberg D N 1980 Diffusive quasiparticle instability toward multiple-gap states in a tunnel-injected nonequilibrium superconductor. *Phys. Rev. Lett.* 44: 486–9

Kaplan S B 1979 Acoustic matching of superconducting films to substrates. *J. Low Temp. Phys.* 37: 343–65

Langenberg D N, Larkin A I 1986 *Nonequilibrium Superconductivity*. North-Holland, Amsterdam

Rothwarf A, Taylor B N 1967 Measurement of recombination lifetimes in superconductors. *Phys. Rev. Lett.* 19: 27–31

Ruggiero S T, Rudman D A 1990 *Superconducting Devices*. Academic Press, Boston, M A

Schön G, Tremblay A-M 1979 Nonequilibrium superconducting states with two coexisting energy gaps. *Phys. Rev. Lett.* 42: 1086–9

K. E. Gray
[Argonne National Laboratory, Argonne, Illinois, USA]

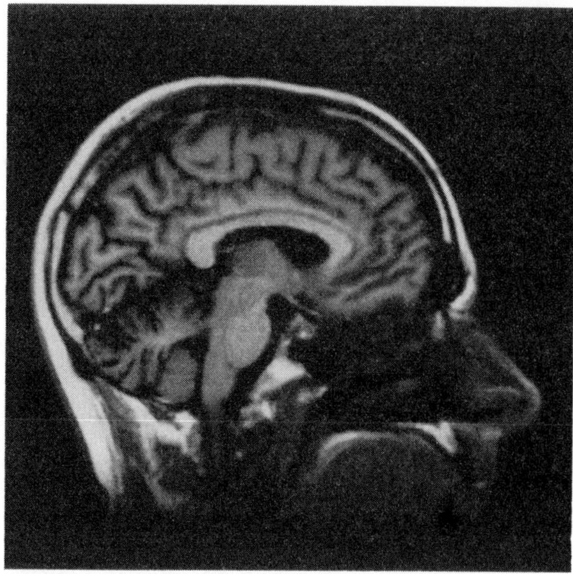

Figure 1
Magnetic resonance image of a sagittal slice, 3 mm thick, through the head of an adult, measured at 1.5 T

Nuclear Magnetic Resonance Imaging

The phenomenon of nuclear magnetic resonance (NMR) was discovered in 1946 and its importance recognized by the award of a Nobel Prize in 1952. Following a short development, largely by physicists, its potential was recognized in the 1950s by chemists for many of whom NMR spectroscopy (MRS) is now their most versatile analytical tool; since the early 1980s MRS has increasingly been used to study oligopeptides and small proteins taking advantage of the fact that it can be used for noncrystalline substances. More recently, it has become feasible to use the method to map the spatial distribution of many chemicals, particularly water, and NMR imaging (MRI) has already become a powerful medical imaging tool, capable of producing images such as that of the human head shown in Fig. 1.

Most magnets used for NMR are based on superconducting solenoids. For MRS the room-temperature bore size is generally in the range 50–150 mm and the magnetic field strength B_0 from 4–14.5 T. Although the same magnets can be used for MRI, the fact that larger objects are generally of interest means that magnets with bore size from 300–1000 mm and strength from 0.5–4.7 T are more commonly used. The field homogeneity at the magnet isocenter for a sphere 20% of the bore diameter has to exceed one part in 10^7 for MRS; for MRI one part in 10^5 is required for 50% bore diameter. Niobium–titanium wire is generally used for magnets up to about 7 T and niobium–tin above that; higher-field MRS magnets often have an inner winding of the latter enclosed within an outer jacket of the former (see *Superconducting Magnets*). The thermal performance of the cryogenic enclosures of such magnets is now so outstanding that their running costs are invariably lower than those of a comparable electromagnet; liquid helium refill intervals invariably exceed two months and cryopump cooling systems can ensure indefinite hold times.

1. General Introduction

NMR depends on the fact that the nuclei of many atoms are weakly magnetic; ^{1}H, ^{13}C, ^{31}P, ^{19}F and ^{23}Na are the most commonly studied. As a result, when a sample, generally a solution, which contains them is placed within a powerful, homogeneous magnetic field, such nuclei tend to become aligned with the field, much like the needle of a magnetic compass in the earth's field. If the sample is then bathed in radio waves of the appropriate frequency (which depends on the value B_0 of the static magnetic field and the magnetogyric ratio of the nucleus being studied) the orientation of some of the nuclear spins is changed. This results in a macroscopic magnetization vector which precesses about the B_0 field direction; as it does so, it induces a voltage in a suitably placed radiofrequency (rf) receiver coil, which is then digitized and, after suitable data processing, is displayed as a plot of magnetization intensity vs the resonance frequency.

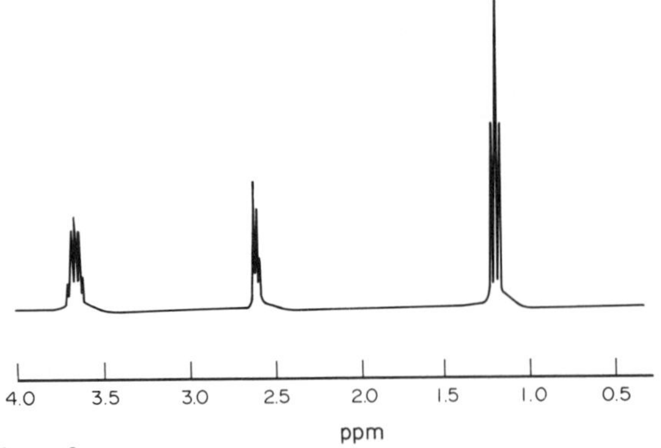

Figure 2
High-resolution proton NMR spectrum of ethyl alcohol measured at 7 T (300 MHz)

In current practice, this process is generally achieved using the "pulse Fourier transform" technique. A pulse of rf energy a few microseconds in duration serves to excite some of the nuclei into resonance by promoting them from a lower to a higher spin state. The signal from the resultant precessing magnetization vector is then digitized as the "free-induction" decay signal (FID); its intensity corresponds to the amount of NMR signal elicited from the sample and decays exponentially with time due to nuclear relaxation. When the FID is subjected to Fourier transformation the plot of magnetization intensity vs "time" is changed to one of magnetization intensity vs "frequency," which is the conventional display mode of NMR spectroscopy. The frequencies in the spectrum correspond to the differences between the frequency of the rf transmitter used and the natural precessional frequencies (Larmor frequency) of all of the nuclei excited by the rf pulse.

The resultant NMR spectrum shows a separate resonance for each set of nuclear spins that have a distinctive chemical environment: the relative peak intensities reflect the relative proportions of the different nuclei in the sample, and their relative frequencies depend on their molecular electronic environment. Thus, the proton NMR spectrum of ethyl alcohol comprises three separate bands with intensity ratios of 1:2:3 (see Fig. 2), corresponding to the HO, CH_2 and CH_3 resonances. The latter two resonances show fine structure due to "scalar spin coupling" if the magnetic field used is sufficiently homogeneous (1 part in 10^8).

Because the total amount of induced magnetization is so small, the signal-to-noise (S/N) ratio of an NMR spectrum is generally low. Fortunately it can be enhanced by averaging over the NMR responses elicited by a series of suitably spaced rf pulses; since the enhancement of signal is proportional to the number of pulses n and that of the noise is proportional to $n^{1/2}$, the overall S/N ratio increases as $n^{1/2}$.

2. Magnetic Resonance Imaging

The first reported use of magnetic resonance imaging (MRI) (Lauterbur 1973) determined the spatial location of two test tubes containing water. Consider the NMR response from the water protons in a sample comprising two water-filled test tubes (see Fig. 3). In a homogeneous B_0 field a single response is detected since all the protons have a chemically and magnetically identical environment. However, if a linear magnetic field gradient is applied, the NMR "spectrum" will correspond to a projection of the spin density of the sample onto the axis defined by the direction of that gradient, since the resonance frequency now reflects the combined magnetic field due to B_0 and that of the gradient G. In this case the projection induced by G_x, a gradient in the x direction, shows the separate profiles of the two tubes, whereas that due to G_y, a gradient in the y direction, shows them in silhouette; a sine/cosine combination of $G_x + G_y$ enables the sample to be viewed from any direction. The image of the object is then computed from a series of these individual views using the same algorithm for back-projection image reconstruction as x-ray computerized axial tomography (CAT scan).

Lauterbur's method, dubbed zeugmatography, has now been largely displaced by variants of the "two-dimensional Fourier transform" method (Kumar et al. 1975). In common with Lauterbur's method, the spatial-frequency relationship along one axis is imposed by a static field gradient; however, the second spatial dimension is now effectively "line scanned" by measuring a series of NMR responses, each for a different value of G_y. Following a two-dimensional Fourier transformation the resultant data set shows the distribution of magnetization projected onto the plane defined by the orientation of G_x and G_y; the image shown in Fig. 1 was obtained by this method (Edelstein et al. 1980).

So far the third spatial dimension has been neglected in this discussion. Generally, images are obtained either from one slice or a series of parallel

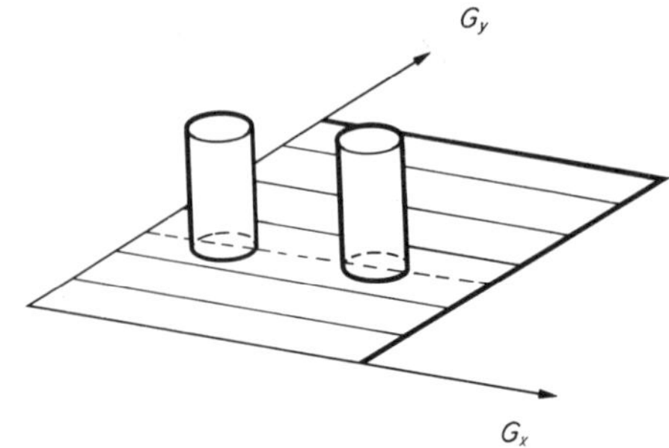

Figure 3
Projected views of two test tubes of water obtained by use of linear magnetic field gradients

slices, generated by imposing a third linear field gradient G_z orthogonal to the other two; the location of each slice is then defined by the frequency of a narrow-bandwidth, frequency-selective rf pulse used to excite the NMR responses, and its slice thickness by the frequency spread of that rf pulse. However, it is also possible to generate a full three-dimensional data set, which can be viewed from any direction, thereby providing visualization of oblique slices.

Two nuclear parameters are of prime importance both to the measurement of an NMR image and to the subsequent interpretation of the data contained therein. These are the spin–lattice relaxation time (T_1 value) and the spin–spin relaxation time (T_2 value). Both of these reflect the motion of the molecule being studied and it is these which enable MRI to provide such a rich source of information concerning the spatial distribution of structural, chemical and physicochemical properties of the sample; these include internal compartmentalization, concentration, local viscosity, translational self-diffusion coefficients and, hence, pore structure, mass transport either by perfusion or flow, and temperature distribution, or that of acidity (pH), to name but a few. For samples that contain several different chemical species, it is possible not only to map their separate spatial distribution, as in oil–water imaging (see Sect. 4), but also their environmental properties. The insight which can be obtained by interpretation of these parameters for multiphase structures is unique and merits further discussion of how these NMR parameters are reflected in an image.

The perceived intensity for each voxel in an image depends on the total magnetization signal intensity, which is dependent on two sets of variables. First, there are those related to the sample itself, which include the concentration of proton-containing chemicals and the T_1 and T_2 values. Second, there are the data acquisition parameters of which two are of paramount importance: the spin echo time (TE value) which is the delay time between stimulation of the NMR response and its subsequent detection; and the relaxation delay (TR) which is the delay allowed between successive line scans to enable the nuclear spins to recover from the effects of the previous rf pulse. For nuclei which have short T_2 values it is difficult not to lose signal intensity during the TE value; for short TE values the image intensity will be enhanced for those regions where the nuclei have long T_2 values. For short TR values, spins with longer T_1 values will become progressively "saturated" with a consequential loss of signal intensity from that region of the image and an enhancement from the region having shorter T_1 values.

The effects of flow also have a pronounced effect on signal intensity. Consider a slice image taken transverse to the direction of fluid flow; if the image is recorded immediately after slice excitation, the full NMR signal intensity of the fluid will be visualized; however, as the delay between excitation and visualization is increased, there will be a progressive loss of signal intensity as the excited fluid leaves the image plane and is replaced by fluid that has not been subjected to NMR excitation; in the limit, all the signal intensity will be lost. This and other methods can produce maps which show quantitatively the distribution of flow velocities.

3. *Medical Applications of MRI*

Most clinical MRI systems use superconducting solenoid magnets with an access bore diameter of 1 m operating at field strengths between 0.1 T and 2.0 T. Above 1.5 T it is possible to obtain separate images from water (muscle) and fat, but generally a composite proton image is obtained. The cost of an instrument increases with field strength; so too does the cost of its installation since, by law, human exposure to a B_0 field of greater than 5×10^{-4} T (5 G) must be controlled and the fringing field of a 2 T, 1 meter-bore magnet extends to about 20 m. This has prompted the development of shielded magnets in which the return flux is constrained and, hence, contained within a smaller volume room; this can be achieved "passively" by use of suitable iron in the form of either a walled enclosure or a large surrounding sleeve; alternatively "active" shielding can be produced by a second superconducting solenoid surrounding that of the magnet itself.

The image in Fig. 1, although obtained in the mid-1980s, still represents the state of the art in terms of spatial resolution; slice thickness is 5 mm and in-plane resolution is submillimeter. The exquisite contrast between the different soft tissues reflects the choice of TR and TE values used for the data-acquisition sequence. As a result of this control of image contrast, it is possible to distinguish many diseased tissues, such as tumors. Since the gradients can be applied at any angle with respect to the patient, slice images can be obtained from any arbitrary direction, optimized for the particular diagnosis. Because the overall NMR signal intensity is so low, many examinations are made using a surface coil which is placed near the tissue of interest; by restricting the field of view this restricts the noise and, hence, optimizes the perceived signal.

4. *Nonmedical MRI*

Although medical studies continue to dominate all aspects of MRI, apparatus for studies of physically smaller objects is available and is progressively attracting the attention of those in need of new insight into the structure of objects that contain water, hydrocarbon (fat) or mixtures thereof. The magnets used for such systems are generally smaller in size (room-temperature bore from 50 mm to 600 mm) than those of clinical units and operate at higher field strengths (typically from 2.4 T to 12 T). Consequently, the S/N level will support far higher spatial resolution for proton imaging, with a slice thickness down to 0.1 mm and in-plane resolution to 0.01 mm; furthermore, chemical shift resolved imaging is not only possible, but is mandatory if water–fat mixtures are studied. Imaging studies of less receptive nuclei such as deuterium, ^{13}C, ^{29}Si, ^{31}P and ^{59}Co are also feasible.

Areas to which MRI has already been applied include studies of the following:

(a) determination of the distribution of oil and/or water in porous rock and, by measurement of the T_1 and T_2 relaxation times, characterization of the pore structure of the rock;

(b) mapping the spatial distribution of fat and/or water in food produce and foodstuffs—importantly, it is possible to study quantitatively the movement of water or fat during food processing and storage;

(c) imaging of polymers and of polymer composites which is largely restricted to studies of the more mobile components, such as monomers, plasticizers or ingressed liquids—however, in all those areas it provides unique insight; and

(d) imaging of small laboratory animals which although equivalent in principle to MRI of man, the smaller physical scale necessitates substantially higher spatial resolution—once this is achieved, MRI will become an ideal tool for evaluating animal models for disease in man, which is of great importance for the pharmaceutical industry.

5. Concluding Remarks

Looking ahead it is exciting to speculate what might be achieved using high-T_c superconductors. Besides the obvious saving in cryogens that could come from higher operating temperatures, any relaxation in engineering specifications could create the possibility for new magnet geometries; of those, the most exciting would be a flat-bed magnet, on which a patient or a large sheet of material could be positioned. Currently most MRI magnets are superconducting solenoids that use multifilamentary wires based on low-T_c superconductors.

A second possible application for high-T_c superconductors could be in superconducting preamplifiers and radio-frequency probes. Operating at liquid nitrogen temperatures such devices would offer the prospect of increasing the effective detector sensitivity of NMR, thereby reducing one of the remaining problems which limits the applicability of NMR to bio-medical studies.

See also: Superconducting Magnets; Superconducting Materials: Large-Scale Applications

Bibliography

Edelstein W A, Hutchinson J M S, Johnson G, Redpath T 1980 Spin warp NMR imaging and application to human whole body imaging. *Phys. Med. Biol.* 25: 751–6

Kumar A, Welti D, Ernst R R 1975 NMR Fourier zeugmatography. *J. Magn. Reson.* 18: 69

Lauterbur P C 1973 Image formation by induced local interactions: examples employing nuclear magnetic resonance. *Nature* 242: 190–1

L. D. Hall
[University of Cambridge, Cambridge, UK]

O

Optical Properties of Superconductors

The most elementary property of a superconductor, for which the phenomenon is named, is the ability to carry a dc current without resistance. This property is not preserved at nonzero frequency, and a significant resistance is observed, both at low frequencies (see *AC Losses in Superconducting Materials*) and in the microwave region (see *High-Frequency Impedance of Superconducting Materials*). This article is concerned with the sharp increase in absorption that is observed at the higher frequencies of the far infrared.

1. Density of States and Absorption Threshold

A qualitative understanding of both the microwave and the infrared response to an electromagnetic field can be based on the simplest form of the microscopic (BCS) theory, which is described in the standard texts (e.g., Tinkham 1975, Tilley and Tilley 1990). In the normal state of a metal, the density of electron states at the Fermi surface varies only slowly with energy, and on the energy scale considered here can be taken as a constant $N(0)$. In the superconducting state, however, the density of states is reduced near the Fermi surface. The expression derived from the simplest form of BCS theory (see *Superconducting Materials: BCS and Phenomenological Theories*), in which the electron–phonon interaction is replaced by an effective electron–electron interaction, and the Fermi surface is assumed spherical, is

$$N(E)\begin{cases} =0 & |E|<\Delta \\ =\dfrac{N(0)E}{(E^2-\Delta^2)^{1/2}} & |E|>\Delta \end{cases} \qquad (1)$$

where E is the energy measured from the Fermi surface. In effect, Eqn. (1) describes how the states that in the normal state occupy the region $|E|<\Delta$ are removed and piled up outside that region, with an inverse square root singularity at $E=\Delta$.

At zero temperature, absorption of an infrared photon of angular frequency ω requires the creation of two quasiparticles of almost opposite wave number $\pm k$. Since the minimum quasiparticle energy is Δ, this means that there is an absorption threshold described by

$$hf \geqslant 2\Delta(0) \qquad (2)$$

where $\Delta(0)$ indicates the zero-temperature value. In the simple theory,

$$2\Delta(0) = 3.52\, k_B T_c \qquad (3)$$

For strong-coupling superconductors, in which the electron–phonon interaction cannot be treated so simply, the value of the BCS gap $2\Delta(0)$ exceeds that given by Eqn. (3); values of $2\Delta(0)/k_B T_c$ for some elemental superconductors are given in Table 1.

Equations (2) and (3) give for the threshold frequency or gap frequency $f_G = 2\Delta(0)/h$, the numerical value $f_G = 73.35\, T_c$(GHz) where T_c is measured in kelvins. Thus, for T_c in the range 1–20 K, which applies to conventional superconductors, the absorption threshold is in the far infrared. It is common practice in infrared spectroscopy to express frequency in units of cm^{-1}, that is, the value of the inverse wavelength λ, where $f\lambda = c = 3\times10^{10}\,\mathrm{cm\,s^{-1}}$. Thus the threshold frequency given by Eqns. (2) and (3) may be expressed alternatively as $f_G/c = 2.445\, T_c\,\mathrm{cm}^{-1}$.

For $T>0$, there is some thermal excitation of quasiparticles, so that absorption can occur without a threshold. In addition, as mentioned, the energy gap $2\Delta(T)$ decreases, eventually to zero at $T=T_c$. Thus, as T increases the absorption threshold loses its sharpness and moves to lower frequencies. Since these effects are conspicuous in the microwave properties, they are not discussed further in this article but are considered in the article *High-Frequency Impedance of Superconducting Materials.*

It should be mentioned that magnetic impurities have a pair-breaking effect which results in a nonzero density of states for $|E|<\Delta$ as well as a decrease in T_c. Addition of magnetic impurities therefore removes the sharpness of the absorption edge, even at low temperatures.

Table 1
Some elemental superconductors, with values of $2\Delta(0)$ (determined from tunnelling measurements) and corresponding gap frequencies f_G in gigahertz and per centimeter

Superconductor	T_c (K)	$2\Delta(0)/k_BT_c$	$f_G=2\Delta(0)/h$ (GHz)	$f_G/c=2\Delta(0)/hc$ (cm^{-1})
Aluminum	1.19	3.37	83.6	2.79
Mercury	4.16	4.60	399	13.3
Indium	3.4	3.63	257	8.57
Niobium	9.2	3.84	736	24.5
Lead	7.2	4.29	643	21.4
Tin	3.72	3.46	268	8.93
Tantalum	4.4	3.60	330	11.0

2. Theoretical Background

The theoretical framework for the quantitative treatment of infrared absorption may be sketched by starting with the observation that the dc response of a simple alloy superconductor with cubic symmetry is a diamagnetic response of the supercurrent $\boldsymbol{J}$ to the vector potential $\boldsymbol{A}$, described by the London equation

$$\boldsymbol{J} = -(\mu_0 \lambda^2)^{-1} \boldsymbol{A} \tag{4}$$

This is a nondissipative response, since it corresponds to a pure imaginary conductivity. Elimination of $\boldsymbol{J}$ between Eqn. (4) and the Maxwell equation $\nabla \times \boldsymbol{B} = \mu_0 \boldsymbol{J}$ leads to

$$\nabla^2 \boldsymbol{B} = \boldsymbol{B}/\lambda^2 \tag{5}$$

This describes the Meissner effect, namely the exclusion of magnetic flux from a bulk specimen.

Equation (4) is not general enough for most pure superconductors; it must be replaced by the nonlocal Pippard equation

$$\boldsymbol{J}_\alpha(\boldsymbol{r},\omega) = -\sum_\beta \int K_{\alpha\beta}(\boldsymbol{r}-\boldsymbol{r}',\omega) A_\beta(\boldsymbol{r}',\omega)\, d^3r' \tag{6}$$

Here, two simple generalizations have been made: a dependence on frequency ω has been included and, mainly with a view to later discussion of the high-T_c oxides (which are orthorhombic), so have tensor suffices α and β. More important, the relation has been written as a nonlocal one, with a kernel $K_{\alpha\beta}(\boldsymbol{r}-\boldsymbol{r}', \omega)$. A key part is played by the range ξ of the kernel (i.e., the distance $|\boldsymbol{r}-\boldsymbol{r}'|$ beyond which it is effectively zero). In fact, it is found that

$$\xi^{-1} = \xi_0^{-1} + l^{-1} \tag{7}$$

where $\xi_0 = \hbar v_F/\pi\Delta(0)$ (the intrinsic coherence length) is a property of the pure material, v_F is the Fermi velocity and l is the low-temperature mean free path for electron scattering.

It follows from Eqn. (5) that the characteristic distance for variations of $\boldsymbol{A}$ is λ. Hence, if $\lambda \gg \xi$, $A_\beta(\boldsymbol{r}', \omega)$ can be taken out of the integral in Eqn. (6), since it cannot vary appreciably over the range ξ, and Eqn. (6) reduces to the local form

$$J_\alpha(\boldsymbol{r}, \omega) = -\sum_\beta Q_{\alpha\beta}(\omega) A_\beta(\boldsymbol{r}, \omega) \tag{8}$$

with

$$Q_{\alpha\beta}(\omega) = \int K_{\alpha\beta}(\boldsymbol{r}-\boldsymbol{r}', \omega)\, d^3r' \tag{9}$$

Materials with $\xi > \lambda$, satisfying Eqn. (6), are called Pippard superconductors, while those with $\xi < \lambda$, satisfying Eqn. (8), are called London superconductors. In view of Eqn. (7), alloys (with sufficiently small l) are of London type, and in addition the high-T_c oxides are London superconductors. It is seen that London superconductors can exhibit only tensorial anisotropy in their electromagnetic properties, while Pippard superconductors can exhibit more subtle anisotropies since the direction of $\boldsymbol{r}-\boldsymbol{r}'$ relative to the crystal axes is of importance in Eqn. (6).

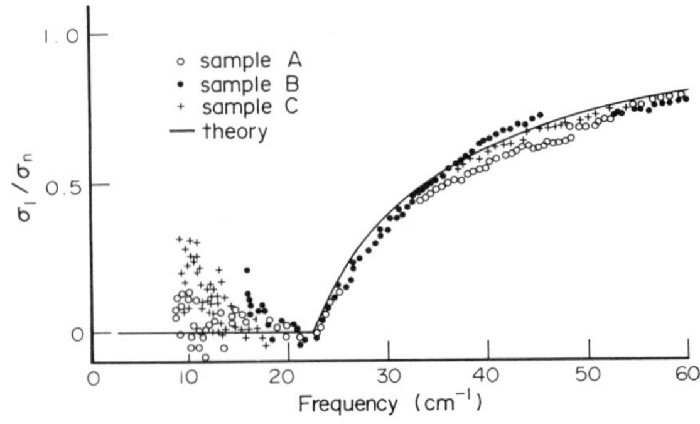

Figure 1
Real part of conductivity σ_1, normalized to normal-state value σ_n, derived from measurements on superconducting lead films (after Palmer 1966)

3. Complex Conductivity

The kernels $K_{\alpha\beta}(\boldsymbol{r}-\boldsymbol{r}', \omega)$ or $Q_{\alpha\beta}(\omega)$ are the obvious interface between theory and experiment; it is more convenient to work in terms of the Fourier transform $K_{\alpha\beta}(\boldsymbol{q}, \omega)$, and use is often made of the complex conductivity $\sigma_{\alpha\beta}(\boldsymbol{q}, \omega) = -K_{\alpha\beta}(\boldsymbol{q}, \omega)/i\omega$, a relationship which follows from substitution of $\boldsymbol{E} = -\partial \boldsymbol{A}/\partial t = i\omega \boldsymbol{A}$ in Eqn. (6). The details of the calculation of $K_{\alpha\beta}(\boldsymbol{q}, \omega)$ from the microscopic theory are given by Abrikosov et al. (1963) and Rickayzen (1964), for example. Figure 1 shows the real (i.e., dissipative) part of the conductivity derived from experiments on lead films. The absorption threshold at $2\Delta(0)$ is very clear, and it is seen that the agreement with theory for $\hbar\omega > 2\Delta(0)$ is very good.

4. High-Temperature Superconductors

Since late 1986, enormous activity has been generated in the field of high-T_c superconductivity. A general introduction is given by Tilley and Tilley (1990), and various articles in this Concise Encyclopedia give more detailed accounts of specialized aspects. Specimen preparation and characterization is a difficult matter in these materials. For this reason, there is not yet detailed agreement between the optical data published by different groups, and discussion has to be somewhat tentative.

The main classes of high-T_c superconductors are oxides containing puckered copper–oxygen planes (see *Oxide Superconductors: Solid-State Chemistry*).

The crystal structures are orthorhombic, but with only a slight distortion from tetragonal; for example, $YBa_2Cu_3O_{7-\delta}$ (denoted YBCO) has unit cell dimensions $a = 0.3859$ nm, $b = 0.3920$ nm and $c = 1.1843$ nm (Bell 1988). Most results are discussed with the simplifying assumption that the crystal is tetragonal, with a and b axes equal. The stoichiometric composition is O_7 and the physical properties are very sensitive to the oxygen deficiency. Thus $YBa_2Cu_3O_6$ is a nonsuperconductor, with antiferromagnetic ordering of the copper moments and a Néel temperature $T_N = 450$ K, while in $YBa_2Cu_3O_{6.4}$ antiferromagnetism and superconductivity coexist at low temperature, with $T_N < T_c = 33$ K (Brewer et al. 1988).

In all the compounds the copper–oxygen planes lie in the *ab* plane, and the function of the oxygen deficiency is to introduce holes, which are the carriers involved in the superconductivity. Given this structure, it is to be expected that physical properties are highly anisotropic (see *Oxide Superconductors: Physical Properties*). In particular, the normal-state resistivity is much higher along the *c* axis (ρ_c) than in the *ab* plane (ρ_{ab}) and, in addition, ρ_c increases with decreasing T (semiconductorlike) while ρ_{ab} decreases with decreasing T (metallike). Nevertheless, both components of resistivity drop to zero simultaneously at the same critical temperature T_c, indicating true three-dimensional ordering.

The results on flux quantization (Gough et al. 1987) show unequivocally that in YBCO the carriers are paired, as in BCS theory, and the same is presumed to be true of the other high-T_c compounds. However, the mechanism, or mechanisms, underlying the pairing is still not clear. The magnetism, or incipient magnetism, of the compounds suggests that spin fluctuations may be important, and these would favor p-wave pairing, in which the gap Δ becomes a function of position on the Fermi surface, with regions of zero gap. One of the main motivations for infrared studies in the gap region is to find evidence for or against low-gap areas and, in addition, in some orientations the phonon spectrum can be studied.

The first specimens of the high-T_c materials were polycrystalline ceramics. Infrared measurements on these were interpreted as evidence of a gap, but obviously could not address the question of angular variation. Techniques for preparing single-crystal specimens were developed soon after the initial discovery of the compounds. Single crystals of YBCO have the form of flat platelets, with the faces parallel to the *ab* plane. They are therefore of a convenient shape for normal-incidence reflectivity measurements with the electric field in the *ab* plane. Reflectivity measurements show two features that may be associated with energy gaps: an absorption onset at about 140 cm^{-1} and a reflectivity shoulder at just under 500 cm^{-1}; an example of the latter is shown in Fig. 2. For a T_c of 90 K these two frequencies correspond to about 2.1 k_BT_c and 8 k_BT_c.

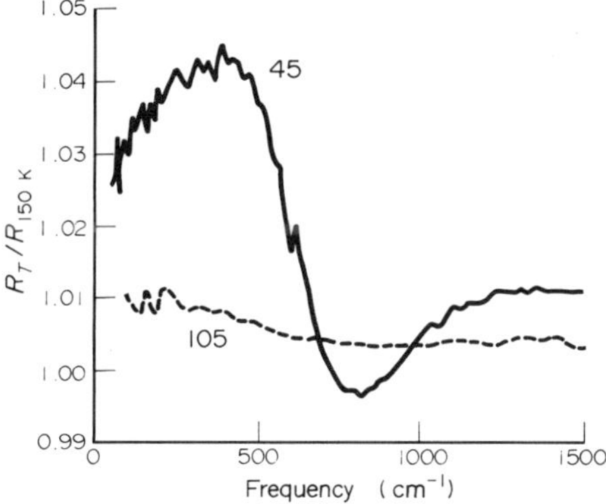

Figure 2
Frequency dependence of normal-incidence far-infrared reflectivity of a YBCO crystal with propagation along *c* axis and electric field in *ab* plane. Curves shown for $T = 45$ K ($T < T_c$) and $T = 105$ K ($T > T_c$). For the lower-temperature curve, $T/T_c \simeq 0.5$, and $\Delta(T)$ is very nearly equal to its zero-temperature value $\Delta(0)$ (after Collins et al. 1989)

Collins et al. (1989) interpret the results shown in Fig. 2 as evidence for an energy gap with a very high value of $2\Delta/k_BT_c$. They maintain this interpretation in later work (Schlesinger et al. 1990) using polarized light in which an interesting difference appears between the reflectivities with the E field along the *a* and *b* axes. However, this interpretation is disputed by other groups. For example, Kamarás et al. (1990) measure reflectivity from YBCO films with thicknesses in the range 200–500 nm deposited on $SrTiO_3$ substrates. They transform their results via a Kramers–Kronig analysis into a frequency-dependent conductivity. Figure 3 shows the conductivity for various temperatures above and below T_c; the former are corrected by subtraction of a Drude term. It is seen that although the 140 cm^{-1} and 500 cm^{-1} features are present in the superconducting phase, they can also be seen in the normal phase.

The work of Forro et al. (1990) is notable for two technical advances. First they prepared free-standing $Bi_2Sr_2CaCu_2O_x$ crystals, so that they were able to carry out transmission measurements and interpret their data without the complication of allowing for a substrate. Second, they used a bright synchrotron-radiation infrared source. Figure 4 shows their results for the ratio of the transmissions in the superconducting and normal phases. The expectation is that this ratio should show a strong peak at a frequency just above 2Δ, as indicated by the two theoretical curves in Fig. 4. The striking aspect of the results is the absence of such a peak; this is tentatively ascribed by Forro et al. (1990) to breaking of Cooper pairs by magnetic impurities, which is well known (e.g., Parks 1969) to produce the phenomenon of gapless superconductivity.

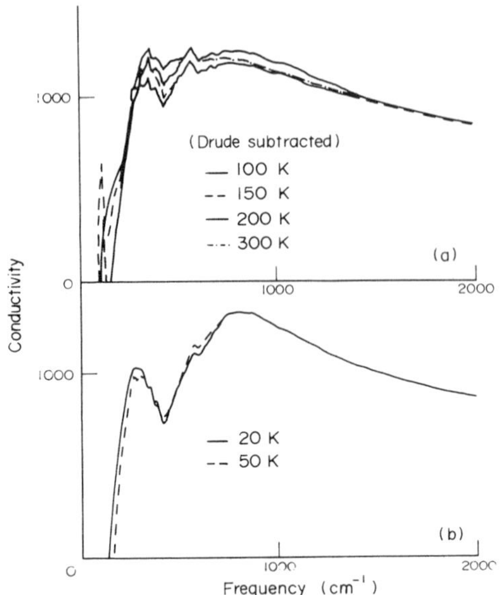

Figure 3
(a) Frequency-dependent conductivity of a $T_c = 89$ K YBCO film above T_c with Drude contribution subtracted; and (b) conductivity below T_c (after Kamarás et al. 1990)

If the reflectivity shoulder shown in Fig. 2 is substantiated as evidence for a gap $2\Delta/k_BT_c \simeq 8$, this will be consistent with photoemission data (Imer et al. 1989), but larger than the values found in some tunnelling measurements (e.g., Gurvitch et al. 1989) and in Raman scattering (Cooper et al. 1988, Hackl et al. 1988, Friedl et al. 1990). The value of 8 is of course much larger than the value 3.52 of the BCS theory, and much larger than is found in any low-temperature superconductor (Table 1). However, it is inside the range that is allowed by the theory of strong-coupling phonon-mediated superconductivity, and a theoretical discussion along these lines is given by Dolgov et al. (1990).

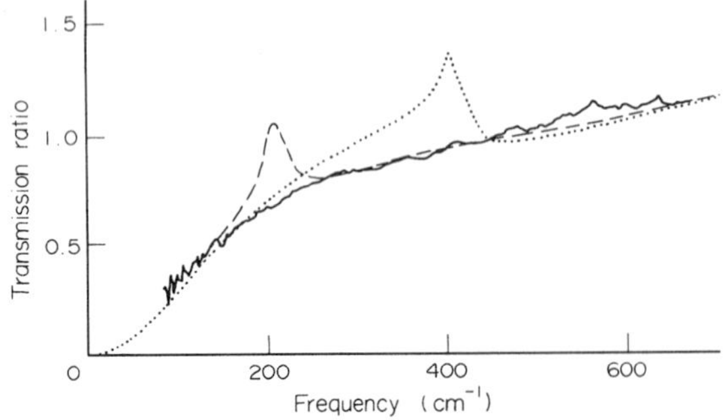

Figure 4
Ratio of transmissions at 25 K and 90 K of a $Bi_2Sr_2CaCu_2O_x$ film with thickness $d = 135$ nm and $T_c = 76$ K. The dashed and dotted lines are theoretical curves assuming $2\Delta = 3.5\,k_BT_c$ and $2\Delta = 7.0\,k_BT_c$ (after Forro et al. 1990)

Bibliography

Abrikosov A A, Gor'kov L P, Dzyaloshinski I E 1963 *Methods of Quantum Field Theory in Statistical Physics.* Prentice-Hall, Englewood Cliffs, NJ

Bell J M 1988 Model for the order-disorder transition in $YBa_2Cu_3O_y$. *Phys. Rev. B* 37: 541–4

Brewer J H, Ansaldo E J, Carolan J F, Chaklader A C D, Hardy W N, Harshman D R, Hayden M E, Ishikawa M, Kaplan N, Keitel R, Kempton J, Kiefl R F, Kossler W J, Kreitzman S R, Kulpa A, Kuno Y, Luke G M, Miyatake H, Nagamine K, Nakazawa Y, Nishida N, Nishiyama K, Ohkuma S, Riseman T M, Roehmer G, Schleger P, Shimada D, Stronach C E, Takabatake T, Uemura Y J, Watanabe Y, Williams D Ll, Yamazaki T, Yang B 1988 Antiferromagnetism and superconductivity in oxygen-deficient $YBa_2Cu_3O_x$. *Phys. Rev. Lett.* 60: 1073–6

Collins R T, Schlesinger Z, Holtzberg F, Feild C 1989 Infrared evidence for gap anisotropy in $YBa_2Cu_3O_7$. *Phys. Rev. Lett.* 63: 422–5

Cooper S L, Klein M V, Pazol B G, Rice J P, Ginsberg D M 1988 Raman scattering from superconducting gap excitations in single-crystal $YBa_2Cu_3O_{7-\delta}$. *Phys. Rev. B* 37: 5920–3

Dolgov O V, Golubov A A, Shulga S V 1990 Far-infrared properties of high T_c superconductors. *Phys. Lett. A* 147: 317–22

Forro L, Carr G L, Williams G P, Mandrus D, Mihaly L 1990 Far-infrared transmission study of single-crystal $Bi_2Sr_2Ca_1Cu_2O_x$ superconductors. *Phys. Rev. Lett.* 65: 1941–4

Friedl B, Thomsen C, Cardona M 1990 Determination of the superconducting gap in $RBa_2Cu_3O_{7-\delta}$. *Phys. Rev. Lett.* 65: 915–8

Gough C E, Colclough M S, Forgan E M, Jordan R G, Keene M, Muirhead C M, Rae A I M, Thomas N, Abell J S, Sutton S 1987 Flux quantization in a high-T_c superconductor. *Nature* 326: 855

Gurvitch M, Valles J M Jr, Cucolo A M, Dynes R C, Garno J P, Scheenmeyer L F, Waszczak J V 1989 Reproducible tunneling data on chemically etched single crystals of $YBa_2Cu_3O_7$. *Phys. Rev. Lett.* 63: 1008–11

Hackl R, Glaser W, Muller P, Einzel D, Andres K 1988 Light-scattering study of the superconducting energy gap in $YBa_2Cu_3O_7$ single crystals. *Phys. Rev. B* 38: 7133–6

Imer J-M, Patthey F, Schneider W-D, Baer Y, Petroff Y, Zettl A 1989 High-resolution photoemission study of the low-energy excitations reflecting the superconducting state of Bi–Sr–Ca–Cu–O single crystals. *Phys. Rev. Lett.* 62: 336–9

Kamarás K, Herr S L, Porter C D, Tache N, Tanner D B, Etemad S, Venkatesan T, Chase E, Inam A, Wu X D, Hegde M S, Dutta B 1990 In a clean high-T_c superconductor you do not see the gap. *Phys. Rev. Lett.* 64: 84–7

Palmer L H 1966 Ph.D. thesis, University of California, Berkeley

Parks R D (ed.) 1969 *Superconductivity.* Dekker, New York

Rickayzen G 1964 *Theory of Superconductivity*. Wiley, New York
Schlesinger Z, Collins R T, Holtzberg F, Feild C, Blanton S H, Welp U, Crabtree G W, Fang Y, Liu J Z 1990 Superconducting energy gap and normal-state conductivity of a single-domain $YBa_2Cu_3O_7$ crystal. *Phys. Rev. Lett.* 65: 801–4
Tilley D R, Tilley J 1990 *Superfluidity and Superconductivity*. Hilger, Bristol, UK
Tinkham M 1975 *Introduction to Superconductivity*. McGraw-Hill, New York

D. R. Tilley
[University of Essex, Colchester, UK]

Organic and Molecular Ferromagnets

Unlike common ferromagnets such as iron, nickel, cobalt or even cobalt–samarium and Nb–Fe–B, it is extremely rare for a discrete molecule, especially an organic or polymeric material, to have cooperative magnetic interactions. Since the mid-1980s a few such materials have been prepared which demonstrate that ferromagnetic interactions can be present in a molecular-based system. The reports have stimulated studies as to the origins as well as to the properties of molecular/organic ferromagnets. Three routes have led to highly magnetic molecular/polymeric materials: (a) chains of isolated alternating radical cations and radical anions, (b) covalently bonded chains containing alternating different metal ions, and (c) covalently bonded chains involving metal ions and nitroxide radicals.

Bulk ferromagnetic behavior has been reported for the linear chain electron transfer salt $[Fe(C_5Me_5)_2]^{\cdot+}[TCNE]^{\cdot-}$ where $[Fe(C_5Me_5)_2]$ is decamethylferrocene (see Fig. 1) and [TCNE] is tetracyanoethylene (see Fig. 2). Crystals of $[Fe(C_5Me_5)_2]^{\cdot+}[TCNE]^{\cdot-}$ are composed of parallel linear chains of alternating radical $[Fe(C_5Me_5)_2]^{\cdot+}$ cations $D^{\cdot+}$ and $[TCNE]^{\cdot-}$ anions $A^{\cdot-}$ without any close interion interactions (see Fig. 3). This material spontaneously becomes a ferromagnet below a Curie (transition) temperature T_C of 4.8 K. Its saturation magnetization M_s of 16.3 A m^2 mol^{-1} (1.63 × 10^4 emu G mol^{-1}) (36% greater than that of iron metal (on either a per mole or iron basis)) and its 80 kA m^{-1} coercive field are much greater than that of iron. Consistent with the magnetic order and a transition to ferromagnetic behavior at T_C the measured critical exponents are $\alpha = 0.09$, $\beta \simeq 0.5$, $\gamma = 1.21$ and $\delta = 4.4$. Ferromagnetic ordering was also detected by neutron diffraction. Zero-applied-field Zeeman split ^{57}Fe Mössbauer spectra are observed at low temperatures. The $[TCNE]^{\cdot-}$ acts as an internal dipolar field inducing the Zeeman splitting with extraordinarily large internal fields (> 30 MA m^{-1}). The Mössbauer data confirms that all iron is low-spin Fe^{III} and that iron metal is not present. Ferromagnetic interactions are lost on replacing either ion radical with similarly structured diamagnetic ions demonstrating the necessity of having radicals at both the donor and acceptor sites.

Replacement of manganese for iron leads to the formation of $[Mn(C_5Me_5)_2]^{\cdot+}[TCNE]^{\cdot-}$ with a T_C of 8.8 K and an M_s of 20 A m^2 mol^{-1} (2 × 10^4 emu G mol^{-1}). The 2.0 K coercive field is 100 kA m^{-1} for $[Mn(C_5Me_5)_2]^{:+}[TCNE]^{\cdot-}$. A mean-field Heisenberg model containing two different spin sites has been developed for $[M(C_5Me_5)_2][TCNE]$ (M = Fe, Mn). This enables the scaling of T_C for these iron and manganese magnets. For a specific set of exchange interactions, T_C for the manganese magnet should exceed that of the iron magnet by 87% in excellent agreement with the observed value of 83%.

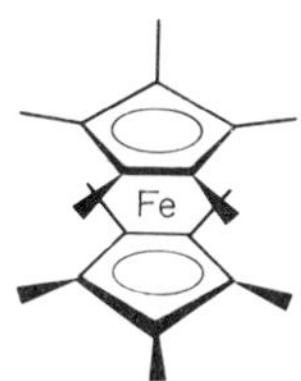

Figure 1
Tetracyanoethylene structure

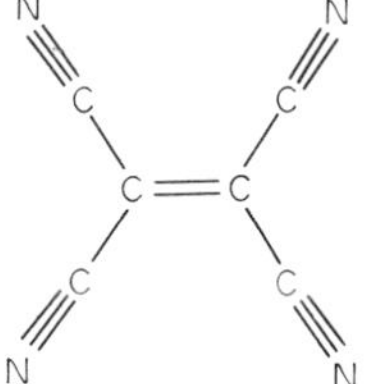

Figure 2
Decamethylferrocene structure

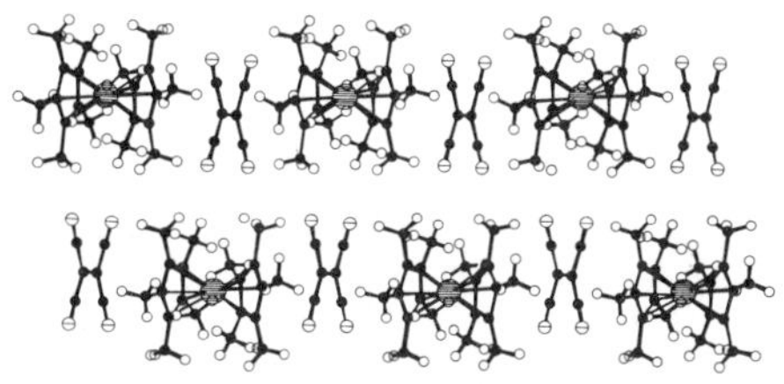

Figure 3
Alternating ... $D^{\cdot+}A^{\cdot-}D^{\cdot+}A^{\cdot-}$ ... structure of $[Fe(C_5Me_5)_2]^{\cdot+}[TCNE]^{\cdot-}$ D = $Fe(C_5Me_5)_2$ and A = TCNE); hydrogens have been omitted for clarity

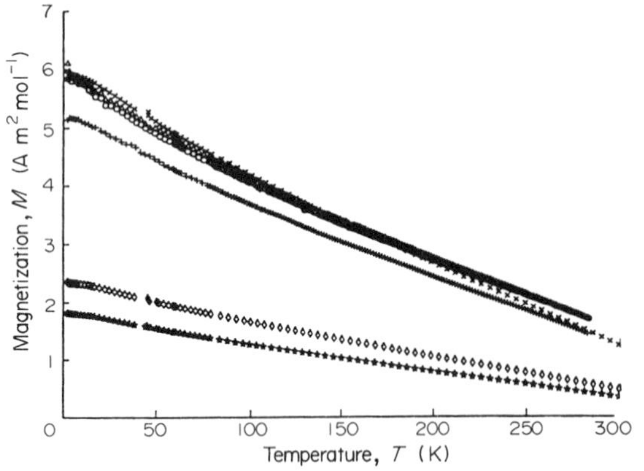

Figure 4
Magnetization M as a function of temperature T at 1.56 MA m^{-1} (○), 1.28 MA m^{-1} (△), 1.42 MA m^{-1} (+), 0.16 MA m^{-1} (×), 0.04 MA m^{-1} (◇) and 0.01 MA m^{-1} (☆) for $V(TCNE)_x \cdot y(CH_2Cl_2)$ (after Manriquez et al. 1991. © AAAS, Washington, DC. Reproduced with permission)

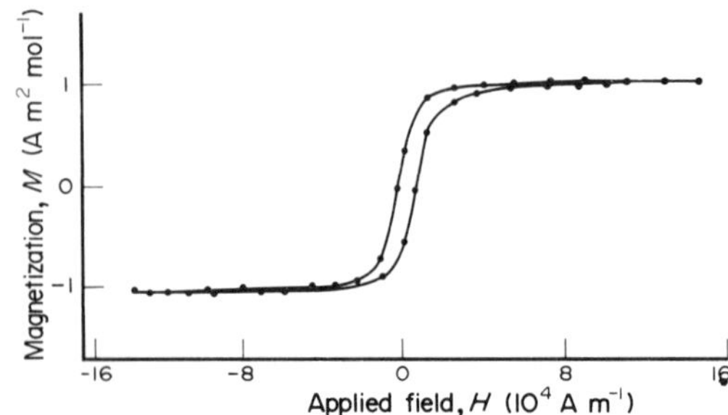

Figure 5
Hysteresis $M(H)$ of $V(TCNE)_x \cdot y(CH_2Cl_2)$ at room temperature. The data was taken on a vibrating sample magnetometer (after Manriquez et al. 1991. © AAAS, Washington, DC. Reproduced with permission)

Research aimed towards high-T_C molecular-based magnets was extended to the reaction of bis(benzene) vanadium and TCNE. This reaction results in the formation of a disordered, air-sensitive, thermally unstable black solid material of nominal $V(TCNE)_x \cdot yCH_2Cl_2$ ($x \sim 2$; $y \sim \frac{1}{2}$) composition which exhibits a field-dependent magnetic susceptibility and magnetization M (see Fig. 4). The unusual nearly linear increase of M with decreasing T most likely reflects the contribution of two or more spin sublattices of this ferrimagnet. Hysteresis with a coercive field of 5 kA m^{-1} is observed at room temperature (see Fig. 5). This is the first example of molecular/organic-based material with a critical temperature above room temperature. The critical temperature exceeds 350 K, the thermal decomposition temperature of the sample. The linear extrapolation of the temperature at which the magnetization should vanish leads to an estimation of a T_C of 400 K. From spin wave theory and data taken below 20 K a spin wave dispersion coefficient has been estimated for the vanadium-based magnet. This also leads to an estimation of T_C to be 400 K. Due to the disordered nature and variable composition of the magnetic material the disordered structure has yet to be elucidated. The TCNEs, however, probably bind up to four different vanadiums via σ–N bonds enabling the formation of a three-dimensional network structure. The disorder leads to random anisotropy which results in a correlated spin-glass or wandering axis ferrimagnet behavior depending on the amount and type of solvent present, temperature and applied magnetic field.

Bulk ferrimagnetic behavior arising from the antiferromagnetic coupling of $S = \frac{5}{2}$ Mn^{II} alternating with either $S = \frac{1}{2}$ Cu^{II} or $S = \frac{1}{2}$ nitronyl nitroxides has been reported for several systems containing covalently bonded chains. The solid $Cu^{II}Mn^{II}$[oxamidobis(benzoato)]$\cdot H_2O$ (see Fig. 6) exhibits an ordering temperature T_C of 14 K, a saturation magnetization of 23.4 A m^2 mol^{-1} (2.34×10^4 emuG mol^{-1}) and a coercive field of 5 kA m^{-1} whereas Mn^{II} (hexafluoroacetylacetonate)$_2$NITEt (NITEt = 2-ethyl-4,4,5,5-tetramethyl-4,5-dihydro-1*H*-imidazole-1-oxy1-3-oxide or ethyl nitronyl nitroxide) (see Fig. 7) exhibits a T_C of 8.1 K, an M_s of 19.4 A m^2 mol^{-1} (1.94×10^4 emu G mol^{-1}) and a coercive field of 37.5 kA m^{-1}.

The origins of the magnetic behavior of $[Fe(C_5Me_5)_2]^{\cdot +}[TCNE]^{\cdot -}$ are important for the design of new ferromagnetic materials with higher saturation magnetization and critical temperatures. Stabilization of ferromagnetic coupling via the configurational admixture of a charge transfer excited state with the ground state is the basis of the present understanding of the spin exchange in this system. Essentially, if an alternating chain of radicals, $\cdots D^{\cdot +}A^{\cdot -}D^{\cdot +}A^{\cdot -}\cdots$ (where D = donor = $[Fe(C_5Me_5)_2]^{\cdot +}$; A = acceptor = $[TCNE]^{\cdot -}$), exists, such that on virtual electron transfer from either D or A, but not both, either D or

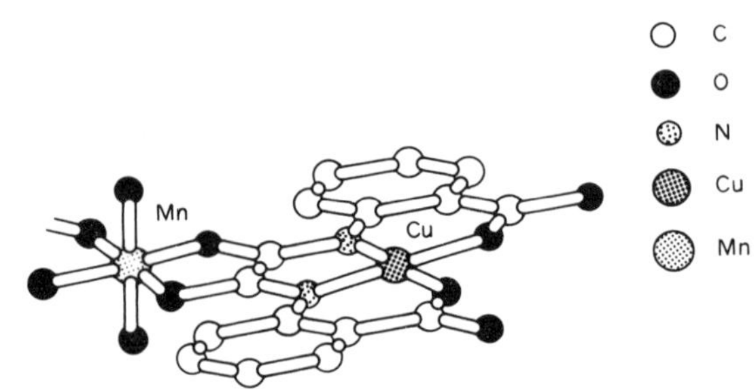

Figure 6
$Cu^{II}Mn^{II}$[oxamidobis(benzoato)]$\cdot H_2O$ structure

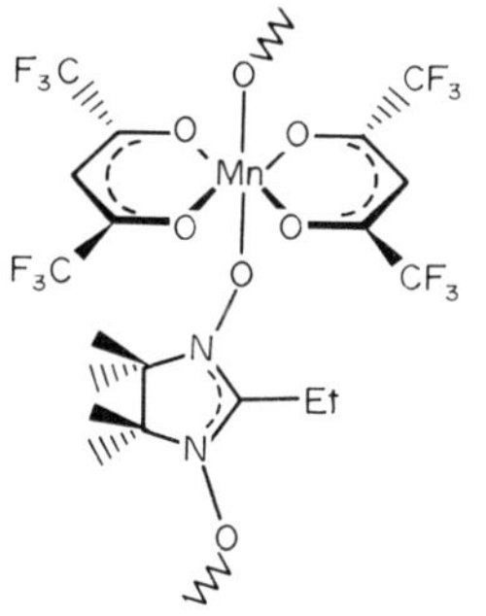

Figure 7
Mn^{II}(hexafluoroacetylacetonate)$_2$NITEt structure

A becomes a triplet, then ferromagnetic coupling could be stabilized. That is, on removing one electron from either D or A and the remaining electrons having a moment equivalent to two electron moments aligning parallel, the ferromagnetic coupling could be stabilized. Since $[Fe(C_5Me_5)_2]^{\cdot +}[TCNE]^{\cdot -}$ has the prescribed structure and triplet $[Fe(C_5Me_5)_2]^{2+}$ could form on virtual electron transfer from $D^{\cdot +}$ to $A^{\cdot -}$, this model can account for the observed ferromagnetism. Note that if either the donor or acceptor is spinless (diamagnetic) the system has only weak exchange confirming the importance of both radicals. Electron transfer between $[Fe(C_5Me_5)_2]^{\cdot +}$ on adjacent chains also leads to ferromagnetic alignment while electron transfer between $[TCNE]^{\cdot -}$ on adjacent chains leads to antiferromagnetic alignment. The competition between ferromagnetic and antiferromagnetic alignment may account for the ferromagnetic ground state in $[Fe(C_5Me_5)_2]^{\cdot +}[TCNE]^{\cdot -}$ and $[Mn(C_5Me_5)_2]^{\cdot +}$-$[TCNE]^{\cdot -}$ (discussed earlier), the metamagnetic ground state in $[Fe(C_5Me_5)_2]^{\cdot +}[TCNQ]^{\cdot -}$ (TCNQ = 7,7,8,8-tetracyano-*p*-quinodimethane) and the antiferromagnetic ground state in $[Ni(C_5Me_5)_2]^{\cdot +}[TCNE]^{\cdot -}$. Likewise, as discussed earlier.

The presence of $\cdots D^{\cdot +}A^{\cdot -}D^{\cdot +}A^{\cdot -}\cdots$ chains within the crystal leads to important anisotropy effects. For example, specific heat measurements of $[Fe(C_5Me_5)_2]^{\cdot +}[TCNE]^{\cdot -}$ are consistent with only around 4% of the magnetic entropy results from the three-dimensional ordering of the spins at 4.8 K. The majority of the entropy is involved in magnetic correlation that persists along the stacks to temperatures of the order of the intrachain exchange energy ($J_{intra} \simeq$ 25 K) for $[Fe(C_5Me_5)_2]^{\cdot +}[TCNE]^{\cdot -}$. One consequence of this quasi-one-dimensionality is that the three-dimensional ordering temperature of 4.8 K is dramatically reduced by the introduction of spinless sites. For example, replacement of 2.5% of the $S=\frac{1}{2}$ $[Fe(C_5Me_5)_2]^{\cdot +}$ sites with spinless $S=0$ $[Co(C_5Me_5)_2]^{+}$ leads to a 43% reduction in T_C. For potential polymer ferromagnets, where the anisotropy of the exchange (J_{inter}/J_{intra}) may be larger than that for the metallocenes the suppression of T_C for oligomers may be even more dramatic.

See also: Organic Superconductors: Anisotropic Molecular Metals

Bibliography

Caneschi A, Gattesche D, Sassoli R, Rey P 1989 Toward molecular magnets: the metal-radical approach. *Acc. Chem. Res.* 22: 392–8

Chakraborty A, Epstein A J, Lawless W N, Miller J S 1989 specific heat of the molecular ferromagnet decamethylferrocenium tetracyanoethylenide: $[DMeFc]^{\cdot +}[TCNE]^{\cdot -}$. *Phys. Rev. B* 40: 1422–4

Dixon D A, Suna A, Miller J S, Epstein A J 1991 Models for magnetic coupling in metallocene based donor/acceptor complexes. In: Gatteschi D, Kahn O, Miller J S, Palacio F (eds.) 1991 *NATO ARW Molecular Magnetic Materials E* 198. Kluwer, Dordrecht, The Netherlands, pp. 171–90

Gatteschi D, Kahn O, Miller J S, Palacio F 1991 Proceedings of the conference on molecular magnetic materials. In: Gatteschi D, Kahn O, Miller J S, Palacio F (eds.) 1991 *NATO ARW Molecular Magnetic Materials E* 198. Kluwer, Dordrecht, The Netherlands, pp. 1–411

Manriquez J M, Yee G T, McLean R S, Epstein A J, Miller J S 1991 A room temperature molecular/organic-based magnet. *Science* 252: 1415–17

Miller J S, Dougherty D A 1989 Proceedings of the symposium on ferromagnetic and high spin molecular based materials. *Mol. Cryst. Liq. Cryst.* 176: 1–562

Miller J S, Epstein A J 1988 Ferromagnetic exchange in molecular solids. *NATO Adv. Study Inst., Ser. B* 168: 159–74

Miller J S, Epstein A J 1990 Organic based magnets. *Chemtech* 20: 168–73

Miller J S, Epstein A J, Reiff W M 1988a Ferromagnetic molecular charge-transfer complexes. *Chem. Rev.* 88: 201–20

Miller J S, Epstein A J, Reiff W M 1988b Molecular ferromagnets. *Acc. Chem. Res.* 21: 114–20

Miller J S, Epstein A J, Reiff W M 1988c Molecular/organic ferromagnets. *Science* 240: 40–7

Miller J S, Yee G T, Manriquez J M, Epstein A J 1991 Room temperature cooperative behavior in a molecular-based material. *Proc. Nobel Symp. NS-81. Conjugated Polymers and Related Materials: The Interconnection of Chemical and Electronic Structure*. Oxford University Press, Oxford

Yee G T, Manriquez J M, Dixon D A, Narayan K S, Epstein A J, Miller J S 1991 Decamethylmanganocenium tetracyanoethenide, $^3[Mn(C_5Me_5)_2]^{:+}[TCNE]^{\cdot -}$, a molecular ferromagnet with an 8.8 K T_c. *Adv. Mater.* 3:309–11

Joel S. Miller
[E.I. Du Pont de Nemours, Wilmington, Delaware, USA]

A. J. Epstein
[The Ohio State University, Columbus, Ohio, USA]

Organic Superconductors: Anisotropic Molecular Metals

Although the vast majority of organic compounds are electrically insulating, there is a smaller class of compounds whose electrical conductivities are in the range $100–1000\,\Omega^{-1}\,cm^{-1}$ at room temperature. These organic conductors or "molecular metals" achieve metallic delocalization through intermolecular contact of extended π-electron systems. About 30 of the latter class of compounds are superconductors at low temperatures. At present the maximum transition temperature T_c is around 13 K; this represents a rapid improvement since the initial discovery of superconductivity in the organics at $T_c \simeq 1$ K (Jérome et al. 1980) and activity in this field has been reviewed by Ishiguro and Yamaji (1990), Saito and Kagoshima (1990), Jérome (1991), Schegolev (1991), Underhill (1992) and Williams et al. (1991).

The molecular units forming the known organic superconductors are shown in Fig. 1. All of these molecules contain sulfur or selenium atoms and can be regarded as derivatives of the TTF parent molecule (Fig. 1a). Conducting molecular crystals are formed when these large planar molecules are stacked face-to-face to form columns. Between these columns or chains are other molecular units which act as electron donors or acceptors; thus, the overlapping π orbitals of the molecular chains from a partially filled conduction band.

The directional nature of the π orbitals causes considerable anisotropy in both normal state and superconducting properties. Estimates of the bandwidths ($4t$, where t is the tight-binding transfer integral) in the various symmetry directions for these materials are given in Table 1. These are estimated principally from the polarized free-carrier reflectivities (plasma frequency) and from band structure calculations. Typical bandwidths in the direction of best overlap are low, being no more than 1 eV, and values in the directions perpendicular to this can be up to 3000 times smaller. Thus, by comparison with ordinary metals, organic crystals have lower electron density and much narrower bandwidths and this has at least two consequences. First, the cohesive energy of a particular crystal structure is not affected by the "metallic" bonds. Instead it arises from a combination of coulomb forces (as in ionic solids) and steric effects (which are nearly always important in organic materials). Thus, for any given highly conducting organic solid, there are several closely related but often nonconducting derivatives. Second, electron–electron interactions are only weakly screened and the interaction energy tends to be comparable with the largest electron bandwidths, so there is a strong tendency to form localized electronic states and Mott–Hubbard insulators.

Crystals of organic superconductors are often prepared either by coprecipitation from solutions which are allowed to mix slowly by diffusion or by an electrochemical method. In a typical small-scale electrochemical process a current of 1–10 μA is passed

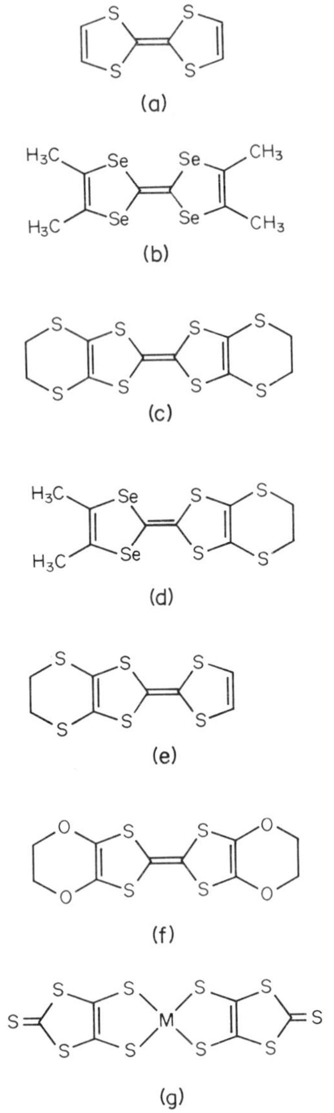

(a)

(b)

(c)

(d)

(e)

(f)

(g)

Figure 1
Chemical structure of molecular units of the known organic superconductors: (a) the "parent" donor molecule tetrathiafulvalene (TTF), (b) bis-tetramethyltetra-selenafulvalene (TMTSF), (c) bis-ethylendithia–tetrathiafulvalene (BEDT–TTF), (d) dimethyl-(ethylendithio) diselena–dithiafulvalene (DMET), (e) methylenedithia–tetrathiafulvalene (MDT–TTF), (f) bis(ethylenedioxy)–tetrathiafulvalene (BEDO–TTF) and (g) metal (M = Ni, Pd) complexes of dimercaptoisotrition (dmit)

Table 1
Normal state properties of some superconducting molecular metals

	Conductivity at 300 K[a] (Ω^{-1} cm^{-1})	Tight-binding transfer integral, t (meV)		Electronic density of states $n(E_F)$ determined from the following (states eV^{-1} molecule^{-1})		
		Experimental[a]	Theoretical[a]	Band structure calculations	Low-temperature specific heat	Magnetic susceptibility
$(TMTSF)_2ClO_4$	800 10 0.2[c]	258 25 0.9[c]	240–366 29–24	1.88–1.23	2.2	4.4 (300 K) 2.4 (60 K)
$(TMTSF)_2PF_6$	800 10 0.025[c]	264 23 0.1	231–365 33–26	1.94–1.23		4.4 (300 K) 2.4 (60 K)
β_L-(BEDT–TTF)$_2I_3$	25–50[b] 0.1[c]	100,130[b] 0.5[c]	~125	1.8–2.2	5.1	7.1 (300 K) 5.6 (20 K)
κ-(BEDT–TTF)$_2$–Cu(SCN)$_2$	20[b]	100, 180[b]	~125	3.7	5.2	7.5 (300 K)

Source: Jérome and Schulz (1982), Ishiguro and Yamaji (1990), Delhaes (1991), Williams et al. (1991)
a Data for three perpendicular directions b In-plane data(two perpendicular directions) c Out-of-plane, c-axis data (e.g., Cooper et al. 1986)

between platinum electrodes through a solution containing the organic molecules and appropriate counter-ions. The crystals grow slowly over a period of 10 d or so on one of the electrodes. This process favors growth of the conducting polytypes of the crystal and also gives single crystals of high purity.

Many organic crystals which are good conductors at room temperature undergo a sharp metal–semiconductor transition on cooling below room temperature. Some examples are shown in the resistivity–temperature plots of Fig. 2. In many cases the transition arises from a small periodic distortion of the conducting chains—a Peierls distortion or charge density wave (CDW). The resultant semiconductor has a small bandgap typically of the order of 10 k_BT_p where k_B is Boltzmann's constant and T_p the transition temperature. However, in some cases (e.g., the $(TMTSF)_2X$ or Bechgaard salts, see Fig. 1b for TMTSF structure) the small gap arises from the setting up of weak antiferromagnetic order (the spin density wave) and is typically 3–4 k_BT_p. The Peierls and spin density wave states are both associated with the one-dimensional chain structure of the material. They tend to be suppressed by any modification (e.g., the application of hydrostatic pressure) which increases the overlap of electronic wavefunctions on neighboring chains and makes the electronic band structure more two- or three-dimensional.

Measurements under hydrostatic pressure have proved to be very useful in suppressing these semiconductor phases and in identifying new classes of organic superconductors. If one member of a family is found to be superconducting under applied pressure, then usually a related ambient pressure superconductor can be synthesized, using, for example, smaller counter-ions.

1. TMTSF Salts

The crystal structure of one of the Bechgaard salts, the ambient pressure superconductor $(TMTSF)_2ClO_4$, is shown in Fig. 3. The large planar TMTSF donor molecules are stacked face-to-face to form the highly conducting axis. Intermolecular contacts in one of the transverse directions (normal to the ClO_4^- sheets) are relatively strong and this material is only weakly susceptible to the one-dimensional electronic instabilities discussed previously. This structure is maintained for a large number of monovalent anions. The control of unit cell volume through selection of the anion and by the application of hydrostatic pressure establishes phase diagrams of the type shown for the PF_6^- salt in Fig. 4. For large cell volumes the low-temperature ground state is a small-gap spin density wave semiconductor; this is considered to be driven by the one-dimensional instability of the electron gas. At lower cell volumes, a superconducting state is found, with a maximum value of T_c of around 1.2 K. For the PF_6^- salt, this is found above 0.8 GPa (8 kbar) (Jérome et al. 1980), while the ClO_4^- salt is superconducting at ambient pressure (Bechgaard et al. 1981).

It is noted that the anions are situated at the corners of the monoclinic cell at points which have inversion symmetry, so that there are important differences

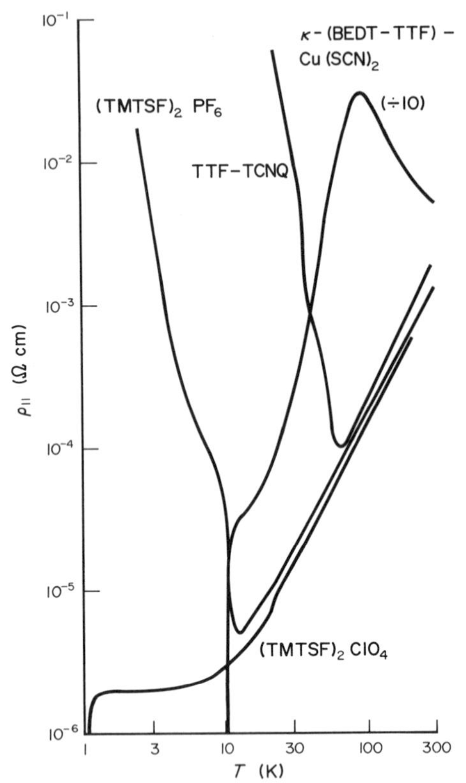

Figure 2
Resistivity vs temperature for several molecular metals

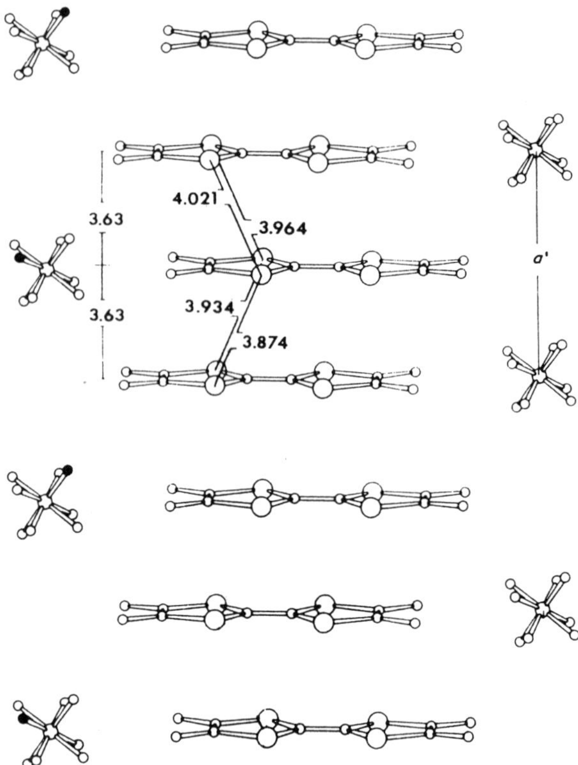

Figure 3
Structure of $(TMTSF)_2ClO_4$

between salts formed with noncentrosymmetric, tetrahedral anions such as ReO_4^- and ClO_4^- and octahedral ones such as PF_6^- and AsF_6^-. The former can undergo an anion ordering phase transition which lowers symmetry and this can introduce gapping at the Fermi energy which is not driven by the one-dimensional instability of the electron gas. Control of this in the ClO_4^- salt is important in determining the low-temperature properties.

Despite the low value of T_c the superconducting properties of the Bechgaard salts have been extensively investigated. Being an ambient pressure superconductor, the ClO_4^- salt is of particular interest and some important parameters are shown in Table 2. Because of the low dimensionality of this class of materials there have been suggestions that T_c is strongly depressed from its mean field value by one-dimensional fluctuations (Jérome and Schulz 1982). However, the prevailing view is that fluctuations are not very important and it is possible to model the heat capacity jump (Garoche et al. 1982) at T_c in terms of standard BCS theory (see *Superconducting Materials: Physical Properties*). The possibility of novel pairing mechanisms (Bourbonnais 1987) remains an open question.

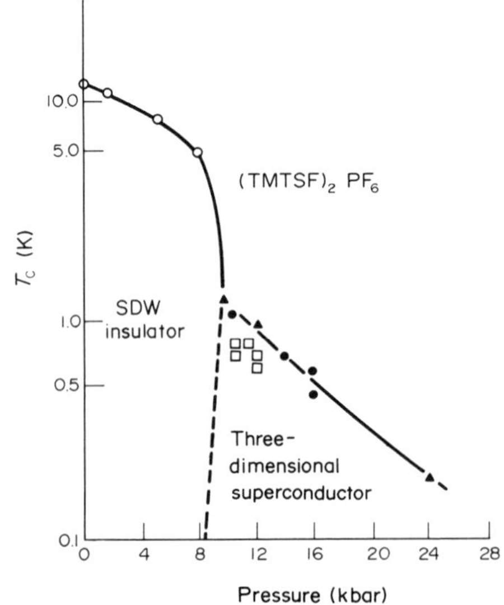

Figure 4
Phase diagram for $(TMTSF)_2PF_6$ (after Jérome and Schulz 1982)

Table 2
Some superconducting properties of anisotropic molecular metals

	T_c (K)	$H_{c1}(0)$ (mT)	$H_{c2}(0)$ (T)	$\xi(0)$ (nm)	$\lambda(0)$ (nm)
$(TMTSF)_2ClO_4$	1.2	0.02[a]	6.35[a]	140[d]	500[d]
		0.1	0.98	20	2800
		1	0.06	1.2	80×10^3
β_L-(BEDT–TTF)$_2$I$_3$	7.5–8	0.5[b]	40[b]	10[d]	250[d]
		5[c]	4[c]	1	2500[e]
κ-(BEDT–TTF)$_2$–Cu(SCN)$_2$	10.4	1.5[b]	19[b]	18[d]	400[d]
		4.5[c]	1.3[c]	0.96	1500[e]

Source: Jérome and Schulz (1982), Bulaevskii (1988)
a Magnetic field along each of the three principal axes, the chain axis corresponds to minimum H_{c1} and maximum H_{c2}, etc. b Field applied in the conducting plane c Field applied perpendicular to the conducting plane d Values of ξ and λ for the principal axes, the largest value of ξ and smallest value of λ, correspond to the chain axis (or conducting plane for the two-dimensional salts), etc. e Values estimated from H_{c1}; for this direction the values determined from the Meissner fraction may be several orders of magnitude larger (e.g., Takahashi et al. 1991)

The magnetic field–temperature phase diagram for the superconducting Bechgaard salts is particularly rich. In brief, application of a magnetic field makes the carrier motion more one-dimensional and enhances the formation of the spin density wave. However, competing periodicities between the Fermi surface vectors and the magnetic field length give rise to a cascade of spin density wave phases (Jérome 1991) and these fulfill the conditions for observation of the quantum Hall effect (Cooper et al. 1989, Hannahs et al. 1989). A further exotic phenomenon is the ability of the spin density wave to slide with respect to the lattice and to thus provide a new channel for electrical conduction (Jérome 1991, Tomić et al. 1991); analogous effects are well established for charge density wave materials.

2. *BEDT–TTF Salts*

The donor molecule BEDT–TTF, shown in Fig. 1c, provides the largest class of organic superconductors. Though the central fulvalene section of the molecule is planar, the ethylene groups at the ends do not lie in the plane; this has the effect of restricting the face-to-face contact within the molecular stacks and dimerized structures are often formed. In contrast, interstack interactions via sulfur–sulfur contacts are often strong and many of these salts therefore show essentially two-dimensional electronic structure, with poor intermolecular contact in the third direction through the counter-ion sheets. This has the effect of limiting (and largely preventing) the one-dimensional electronic instabilities that characterize, for example, the TMTSF salts, and many of the BEDT–TTF salts are metallic to low temperatures and are sometimes superconducting. BEDT–TTF salts show very extensive polytypism; for example, there are seven known phases of the I_3^- salt, of which four show superconductivity.

Superconductivity was first discovered in (BEDT–TTF)$_2$ReO$_4$ under hydrostatic pressure at 2 K (Parkin et al. 1983), but many of the interesting superconductors are formed with linear anions such as I_3^- and AuI_2^-. The *β* phase of (BEDT–TTF)$_2$I$_3$ has a well-defined stack structure but with strong interstack interactions within the molecular sheets. It shows a T_c at 1.5 K and the closely related phase stabilized at hydrostatic pressures of above 0.05 GPa has a T_c as high as 8 K (Yagubski et al. 1984). Extensive work on salts formed with related anions resulted in the discovery of the *κ*-phase (BEDT–TTF)$_2$Cu(NCS)$_2$ which has a T_c of 10.4 K (Uramaya et al. 1988). The structure is shown in Fig. 5 and it can be seen that the stack structure is lost, with a two-dimensional arrangement of BEDT–TTF pairs within sheets of the donors which are separated by anion sheets which form a three-dimensional network of threefold coordinated Cu(I). The electronic properties of this salt are effectively isotropic in the plane. Further compounds with closely related structures and with higher values of T_c have since been found; (BEDT–TTF)$_2$Cu(N(CN)$_2$Br shows a T_c of 11.6 K (Kini et al. 1990) and the corresponding salt with chlorine in place of bromine shows an onset for T_c at 12.8 K at a pressure of 0.03 GPa (it is semiconducting at atmospheric pressure) (Wang et al. 1991). The number of superconductors formed as charge transfer salts of BEDT–TTF continues to grow and was given as 20 by Williams et al. (1991).

The normal state properties of the BEDT–TTF salts reflect the two-dimensional character of the electronic structure; for example, the electrical conductivity and the free-carrier reflectivity are effectively isotropic within the plane of the organic sheets and show poor metallic behavior in the direction transverse to these sheets (Table 2). De Haas–Shubnikov magnetoresistance oscillations have been measured in several of these materials (including the β-phase I_3^- salt (Kang et al. 1989) and the κ-phase $Cu(SCN)_2$ salt (Oshima et al. 1988)) and these direct measurements of the Fermi surface are in good agreement with two-dimensional band calculations. One-dimensional electronic instabilities are not therefore in evidence. However, the feature that apparently distinguishes these metals from conventional inorganic metals is the narrowness of the conduction bands formed from intermolecular π-electron overlap, with typical bandwidths as determined from the optical free-carrier reflectivity of around 0.5 eV. At this level of band delocalization, electron–electron interactions are expected to be strong and there is much interest in determining the role of these interactions. The experimental evidence (see Table 1) is that the paramagnetic susceptibilities of the normal state are high and are enhanced over the values expected for noninteracting electrons. Similarly, effective masses determined in de Haas–Shubnikov experiments are higher than the calculated band masses by a factor of about four. Possibly related to this are the remarkably high normal state in-plane resistivities; for the case of (BEDT–TTF)$_2$Cu(NCS)$_2$ the room temperature conductivity is no higher than 20 $\Omega^{-1}\,cm^{-1}$ and this even falls on cooling to 100 K, as is evident in Fig. 2.

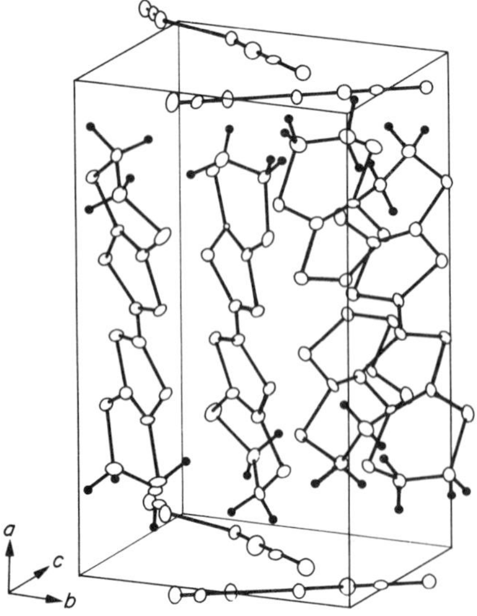

Figure 5
Structure of (BEDT–TTF)Cu(SCN)$_2$

The superconducting properties, collated in Table 2, are well investigated and the anisotropy of the coherence lengths is, as expected, very large. There has been considerable speculation about the factors that control the strength of the superconducting pairing mechanism. It is clear that within a given structural class there is a correlation between the unit cell volume and the value of T_c (Williams et al. 1991). T_c is also found to fall very quickly with applied hydrostatic pressure (Schirber et al. 1986), consistent with this chemical trend. The reason for this correlation is less clear. The unit cell volume controls the bandwidth through the size of the intermolecular overlaps and narrow bands are expected with high values of the density of states at the Fermi energy $N(E_F)$ and of the Pauli susceptibility. However, calculated variations of $N(E_F)$ with volume do not show large variations between salts and Williams et al. (1991) use a BCS pairing model and argue that the unit cell volume controls T_c through the lattice softness. It is also necessary to consider the role of electron–electron interactions in these very-narrow-band metals. Though the BEDT–TTF salts differ from the TMTSF salts in not commonly showing a common phase boundary with a magnetic phase (the spin density wave), it does seem clear that the optimum conditions for superconductivity are those just to the metallic side of a Mott–Hubbard metal–insulator transition. In this regard, (BEDT–TTF)$_2$Cu(N(CN)$_2$Cl is particularly interesting in that modest pressure is required to drive this salt from semiconductor to superconductor and there is some evidence that the semiconducting phase is also a spin density wave state (Wang et al. 1991). Although there has been much speculation about the role of exotic pairing mechanisms in the organic superconductors, firm evidence for such mechanisms has been hard to find. However, there has been some evidence, in the form of a power law low-temperature variation of the penetration depth (Takahashi et al. 1991), that the superconducting gap for (BEDT–TTF)$_2$Cu(NCS)$_2$ may vanish at certain points in the Brillouin zone, though this work has been challenged by Klein et al. (1991).

3. Other Charge Transfer Salts

There are a great many other organic donor and acceptor molecules that can be used to form metallic charge transfer salts, but only a small number of these have been found to be superconductors. Three derivatives of TMTSF and BEDT–TTF are in this class (Figs. 1d–f; there are seven superconducting salts of DMET, with a maximum value of T_c of 1.9 K, and one each for MDT–TTF ($T_c = 4.5$ K) and BEDO–TTF ($T_c = 1$ K) (see Williams et al. 1991 for references). Organometallic complexes have been investigated in which the planar structure necessary to allow stacking and intermolecular π-electron delocalization is achieved though square–planar ligand coordination around nickel, palladium, platinum and gold (Underhill 1991). The M(dmit)$_2$ complex shown in Fig. 1g is an example of these. Charge transfer salts similar in structure to the all-organic structures discussed previously are formed and the strong intermolecular interactions are those between the π-electron systems of the ligands. These give comparable bandwidths and comparable anisotropies to those of the TMTSF and BEDT–TTF salts (Underhill et al. 1991). Several can be stabilized as metals at low temperatures (Underhill et al. 1991), indicating that the electronic structure can be sufficiently two-dimensional to avoid one-dimensional electronic instabilities, and some are superconductors under applied hydrostatic pressure: (TTF)[Ni(dmit)$_2$]$_2$ at 1.6 K (0.7 GPa) (Brossard et al. 1986), two polytypes of (TTF)[Pd(dmit)$_2$]$_2$ at 6.4 K (2.07 GPa) and 1.7 K (2.18 GPa) (Brossard et al. 1989), and Me$_4$N[Ni(dmit)$_2$]$_2$ at 5 K (0.7 GPa) (Kobayashi et al. 1987).

See also: Organic and Molecular Ferromagnets; Superconducting Materials: An Overview

Bibliography

Bechgaard K, Carneiro K, Olsen M, Rasmussen F B, Jacobsen C S 1981 Zero-pressure organic superconductor di-(tetramethyltetraselenafulvalinium)-perchlorate [(TMTSF)$_2$ClO$_4$]. *Phys. Rev. Lett.* 46: 852–5

Bourbonnais C 1987 Electronic correlations in organic superconductors. In: Jérome D, Caron L (eds.) 1987 *Low Dimensional Conductors and Superconductors*, NATO Advanced Study Institutes Series. Plenum, New York, pp. 155–83

Brossard L, Ribault M, Busseau M, Valade L, Cassoux P 1986 A new type of molecular superconductor—TTF[Ni-(Dmit)$_2$]$_2$. *C. R. Acad. Sci., Ser. II* 302: 205–10

Brossard L, Ribault M, Valade L, Cassoux P 1989 Pressure-induced superconductivity in molecular TTF(Pd(dmit)$_2$)$_2$. *J. Phys. (Paris)* 50: 1521–34

Bulaevskii L N 1988 Organic layered superconductors. *Adv. Phys.* 37: 443–70

Cooper J R, Forró L, Korin-Hamzić B, Bechgaard K, Moradpour A 1986 Magnetoresistance of the organic conducting tetramethyltetraselenafulvalene salts (TMTSF)$_2$-ClO$_4$ and (TMTSF)$_2$PF$_6$—search for the coherent diffusive transition or localization effects with increasing temperature. *Phys. Rev. B* 33: 6810–14

Cooper J R, Kang W, Auban P, Montambaux G, Jérome D, Bechgaard K 1989 Quantized Hall-effect and a new field-induced phase-transition in the organic superconductor (TMTSF)$_2$PF$_6$. *Phys. Rev. Lett.* 63: 1984–7

Delhaes P 1991 Some recent trends in π cation-radical salts based on tetrathiafulvalenes. *Synth. Met.* 41–43: 1977–83

Garoche P, Brusetti R, Bechgaard K 1982 Influence of the cooling rate on the superconducting properties of the organic solid di-(tetramethyltetraselenafulvalinium)-perchlorate (TMTSF)$_2$ClO$_4$. *Phys. Rev. Lett.* 49: 1346–9

Hannahs S T, Brooks J S, Kang W, Chiang L Y, Chaikin P M 1989 Quantum Hall effect in a bulk crystal. *Phys. Rev. Lett.* 63: 1988–91

Ishiguro T, Yamaji K 1990 *Organic Superconductors*, Series in Solid State Sciences, Vol. 88. Springer, Berlin

Jérome D 1991 The physics of organic superconductors. *Science* 252: 1509–14

Jérome D, Mazaud A, Ribault M, Bechgaard K 1980 Superconductivity in the synthetic organic superconductor (TMTSF)$_2$PF$_6$. *J. Phys. Lett. (Paris)* 41: L95–L98

Jérome D, Schulz H J 1982 Organic superconductors and superconductors. *Adv. Phys.* 31: 299–490

Kang W, Montambaux G, Cooper J R, Jerome D, Batail P, Lenoir C 1989 Observation of giant magnetoresistance oscillations in the high-T_c phase of the two-dimensional organic conductor β-(BEDT–TTF)$_2$I$_3$. *Phys. Rev. Lett.* 62: 2559–62

Kini A M, Geiser V, Wang H H, Carlson K D, Williams J M, Kwok W K, Vandervoort K G, Thompson J E, Stupka D L, Jung D, Whangbo M-H 1990 A new ambient-pressure organic superconductor, κ-(ET)$_2$-Cu[N(CN)$_2$]Br, with the highest transition temperature yet observed. *Inorg. Chem.* 29: 2555–7

Klein O, Holczer K, Gruner G, Chang J J, Scalapino D, Wudl F 1991 The electrodynamics of the superconducting state of κ-[BEDT–TTF]$_2$Cu[NCS]$_2$. *Synth. Met.* 41–43: 2063–7

Kobayashi A, Kim H, Sasaki Y, Kato R, Kobayashi H, Mariyama S, Nishio Y, Kajita K, Sasaki W 1987 The first molecular superconductor based on PI-acceptor molecules and closed-shell cations, [(CH$_3$)$_4$N][Ni(dmit)$_2$]$_2$ low-temperature x-ray studies and superconducting transition. *Chem. Lett.* 1819–22

Oshima K, Mori T, Inokuchi H, Uramaya H, Yamochi H, Saito G 1988 Shubnikov–De Haas effect and the Fermi-surface in an ambient-pressure organic superconductor [bis(ethylenedithiolo)tetrathiofulvalene]$_2$Cu(NCs)$_2$. *Phys. Rev. B* 38: 938–41

Parkin S S P, Engler E M, Schumaker R R, Lagier R, Lee V Y, Scott J C, Greene R L 1983 Superconductivity in a new family of organic conductors. *Phys. Rev. Lett.* 50: 270–3

Saito G, Kagoshima S (eds.) 1990 *The Physics and Chemistry of Organic Superconductors*, Proceedings in Physics, Vol. 51. Springer, Berlin

Schegolev I F 1991 Organic superconductors. *Europhys. News* 22: 12–15

Schirber J E, Azevedo L J, Kwak J F, Venturini E L, Leung P C W, Beno M A, Wang H H, Williams J M 1986 Shear-induced superconductivity in β-di[bis(ethylenedithio)-tetrathiafulvalene] triiodide [β-(BEDT–TTF)$_2$I$_3$]. *Phys. Rev. B* 33: 1987–9

Takahashi T, Kanoda K, Akiba K, Sakoa K, Watabe M, Suzuki K, Saito G 1991 Unconventional superconductivity in the organic superconductor κ-[BEDT–TTF]$_2$Cu[NCS]$_2$. *Synth. Met.* 41–43: 2005–10

Tomić S, Cooper J R, Kang W, Jérome D, Maki K 1991 The influence of chemical impurities and x-ray induced defects on the single-particle and spin-density wave conductivity in the Bechgaard salts. *J. Phys. (Paris)* 1: 1603–25

Underhill A E 1992 Molecular metals and superconductors. *J. Mater. Chem.* 2: 1–11

Underhill A E, Clarke R A, Marsden I R, Allan M, Friend R H, Tajima H, Naito T, Tamura M, Kuroda H, Kobayashi A, Kobayashi H, Canadell E, Ravy S, Pouget J-P 1991 Structural and electronic properties of Cs[Pd(dmit)$_2$]$_2$. *J. Phys. Condensed Matter* 3: 933–54

Uramaya H, Yamosho H, Saito G, Nozawa K, Sugano T, Kinoshita M, Sato S, Oshima K, Kawamoto A 1988 A new ambient pressure organic superconductor based on

BEDT–TTF with T_c higher than 10 K (10.4 K). *Chem. Lett.* 55–6

Wang H H, Carlson K D, Geiser U, Kini A M, Schultz A J, Williams J M, Montgomery L K, Kwok W K, Welp U, Vandervoort K G, Boryshuk S J, Strieby Crouch A V, Kommers J M, Watkins D M, Schirber J E, Overmyer D L, Jung D, Novoa J J, Whangbo M-H 1991 New κ-phase materials, κ-$(ET)_2Cu[N(CN)_2]X$; X=Cl, Br and I. The synthesis, structure and superconductivity above 11 K in the Cl (T_c=12.8 K, 0.3 kbar) and Br (T_c=11.6 K) salts. *Synth. Met.* 41–43: 1983–90

Williams J M, Schultz A J, Geiser U, Carlson K D, Kini A M, Wang H H, Kwok W-K, Whangbo M-H, Schirber J E 1991 Organic superconductors—new benchmarks. *Science* 252: 1501–8

Yagubski E B, Schegolev I F, Laukhin V N, Kononovitch P A, Kartsovnik M V, Zvarykina A V, Buravov L I 1984 Normal-pressure superconductivity in an organic metal $(BEDT–TTF)_2L_3$[bis(ethylene dithiolo) tetrathiofulvalene triiodide]. *JEPT Lett.* 39: 12–16

R. H. Friend and J. R. Cooper
[University of Cambridge, Cambridge, UK]

Organic Superconductors: Doped C_{60} (Fullerene)

Graphite has long been known to form intercalation complexes with a wide range of electron donor and acceptor groups. There is substantial interaction between the graphite sheets and the intercalant groups between these layers, and the charge transfer to or from the graphite sheets gives metallic properties due to the partial occupation of the carbon π bands. Some of the intercalation complexes formed with alkali and related metals have been found to be superconductors, although mostly with low values of the superconducting transition temperature (e.g., < 0.2 K for C_8K). Superconductivity in these materials has been reviewed by Dresselhaus and Chaiken (1986). Very recently, much higher values of T_c, possibly above 50 K, have been found in similar complexes of shell-structure carbon allotropes with similar π electron systems.

Shell-structure carbon molecules were proposed by Kroto et al. (1985) to account for the preponderance of certain molecular masses in experiments in which graphitic carbon was laser-beam evaporated. Two masses were particularly stable, corresponding to C_{60} and C_{70}, and the structure proposed by Kroto et al. for C_{60} is the truncated icosahedron shown in Fig. 1 and formed from 20 regular hexagons and 12 regular pentagons. This structure is now well established, and the molecule which was termed Buckminster Fullerene by Kroto et al. has been variously known as fullerene, soccerene and as a Bucky-ball. C_{70} is simply related to C_{60} by the separation of the two halves of the C_{60} molecule by a ring of ten extra carbon atoms to form an elongated spheroid (US football or UK rugby football). The unexpected stability of these and other shell molecules has considerable implications for the interpretation of spectra due to interstellar material, and also for the understanding of the formation of soot particles, as has been discussed by Kroto (1988).

Figure 1
Structure of C_{60}

These fullerenes were available in very small quantities until Krätschmer et al. (1990) reported that it was possible to form relatively large quantities, of up to grams, by the simple device of striking an arc between two graphite rods held in a helium gas environment. A sizable fraction of the "soot" collected from the sides of the evaporation chamber is composed of fullerenes, and these can be solvent extracted (in for example benzene) and chromatographically separated into molecular weight fractions. Solid films are conveniently prepared by vacuum sublimation. C_{60} is the dominant fraction produced in this way, and has been very extensively studied by a large number of groups. This article will be restricted to C_{60}.

The electronic structure of the C_{60} molecule is controlled by the delocalized molecular orbitals formed predominantly from the p_z orbitals at each carbon, although since the coordination about any given carbon is not quite planar, the separation of carbon valence electrons into sp^2 and p_z is not exact. As expected, the energy gap between highest filled and lowest unfilled orbitals is relatively low, at about 1.8 eV, and many groups have now shown that the C_{60} molecule can be readily oxidized or reduced. It is the reduced forms of C_{60} that have shown superconductivity at temperatures now in excess of 30 K.

Alkali metal doping of C_{60} was reported by Haddon et al. (1991) to produce metallic compounds, through charge transfer from the alkali metal to the C_{60} cluster, and very shortly afterwards superconductivity in the potassium-doped material was reported with an onset for the superconducting transition temperature T_c at

18 K (Hebard et al. 1991). Values of T_c have been rising steadily, with T_c for the rubidium-doped material at 28 K (Rosseinsky et al. 1991) and for a mixed rubidium- and cesium-doped material at 33 K (Tanigaki et al. 1991). Although the materials first reported contained only a small fraction of superconducting material, as determined by flux exclusion measurements, careful attention to uniform doping has produced bulk superconductors, and detailed information about the structure and superconducting properties of these materials is rapidly becoming available.

The stoichiometry of the superconducting phases is M_3C_{60}, and the structure can be described in terms of the occupation of interstitial sites of the fcc lattice formed by the undoped C_{60} which has a spacing between the centers of adjacent clusters of 1.00 nm (Heiney et al. 1991). The unit cell parameter of K_3C_{60} is only slightly expanded from that of C_{60}, from 1.411 nm to 1.424 nm, and both octahedral and tetrahedral interstitial sites are occupied by the alkali metal (Stephens et al. 1991). Measurements of superconducting properties for this material show it to be a type-II superconductor, with values for the penetration depth of 240 nm and for the coherence length of 2.6 nm (Holczer et al. 1991).

The reasons for these remarkably high values of T_c are still a matter for speculation, but there are some chemical trends which seem well established. It seems that there is a unique superconducting phase of stoichiometry in M_3C_{60} and that this is found with either the C_{60} or the M_6C_{60} phase in materials which contain less or more than this ratio of metal to cluster, so that the degree of band filling is the same for all superconducting phases. It appears that the factor controlling T_c is the unit cell parameter, with the expanded lattices formed with the larger alkali metal ions producing the higher values of T_c. This correlates with the large negative pressure coefficient for T_c reported for K_3C_{60}, of $-5\,K\,GPa^{-1}$ (Sparn et al. 1991). This chemical/pressure-dependent trend is very much as observed in the superconducting organic charge transfer salts based on bis-ethylendithia–tetrathiafulvalene (BEDT–TTF) and it is very likely the case that the metallic electrons in the C_{60} compounds are also in narrow conduction bands. In distinction to the BEDT–TTF salts, and also the cuprate superconductors, these C_{60} compounds have three-dimensional electronic properties.

This unexpected class of materials, which on passing a T_c of 25 K gained the epithet of high-temperature superconductors, are certain to be developed rapidly. They have already shown that superconductivity at comparatively high temperatures is a more widespread phenomenon than previously suspected.

See also: Organic and Molecular Ferromagnets; Organic Superconductors: Anisotropic Molecular Metals

Bibliography

Dresselhaus G, Chaiken A 1986 Superconductivity in graphite intercalation compounds. In: Dresselhaus M S (ed.) 1986 *Graphite Intercalation Compounds.* Plenum, New York, pp. 387–406

Haddon R C, Hebard A F, Rosseinsky M J, Murphy D W, Duclos S J, Lyons K B, Miller B, Rosamilia J M, Fleming R M, Kortan A R, Glarum S H, Makhija A V, Muller A J, Eick R H, Zahurak S M, Tycko R, Dabbagh G, Thiel F A 1991 Conducting films of C_{60} and C_{70} by alkali metal doping. *Nature* 350: 320–2

Hebard A F, Rosseinsky M J, Haddon R C, Murphy D W, Glarum S H, Palstra T T M, Ramirez A P, Kortan A R 1991 Superconductivity at 18 K in potassium-doped C_{60}. *Nature* 350: 600–1

Heiney P A, Fischer J E, McGhie A R, Romanow W J, Denenstein A M Jr, McCalauley J P III, Smith A B, Cox D E 1991 Orientational ordering transition in C_{60}. *Phys. Rev. Lett.* 66: 2911–14

Holczer K, Klein O, Grüner G, Thompson J D, Diederich F, Whetten R L 1991 Critical magnetic fields in the superconducting state of K_3C_{60}. *Phys. Rev. Lett.* 67: 271–4

Krätschmer W, Fostiropoulos K, Huffman D R 1990 The infrared and ultraviolet absorption sprectra of laboratory-produced carbon dust—evidence for the presence of the C_{60} molecule. *Chem. Phys. Lett.* 170: 167–70

Kroto H W 1988 Space, stars, C_{60} and soot. *Science* 242: 1139–46

Kroto H W, Heath J R, O'Brien S C, Curl R F, Smalley R E 1985 C_{60} Buckminsterfullerene. *Nature* 318: 162–3

Rosseinsky M J, Ramirez A P, Glarum S H, Murphy D W, Haddon R C, Hebard A F, Palstra T T M, Kortan A R, Zahurak S M, Makhija A V 1991 Superconductivity at 28 K in Rb_xC_{60}. *Phys. Rev. Lett.* 66(21): 2830–3

Sparn G, Thompson J D, Huang S-M, Kaner R B, Diederich F, Whetten R L, Grüner G, Holczer K 1991 Pressure-dependence of superconductivity in single-phase K_3C_{60}. *Science* 252: 1829–31

Stephens P W, Mihaly L, Lee P L, Whetten R L, Huang S-M, Kaner R, Diederich F, Holczer K 1991 Structure of single-phase conducting K_3C_{60}. *Nature* 351: 632–4

Tanigaki K, Ebbesen T W, Saito S, Mizuki J, Tsai J S, Kubo J, Kuroshima S 1991 Superconductivity at 33 K in $Cs_xRb_yC_{60}$. *Nature* 352: 222–3

R. H. Friend and J. R. Cooper
[University of Cambridge, Cambridge, UK]

Oxide Glasses as Magnetic Materials

All oxide glasses can be rendered magnetic by including magnetic ions (the transition metal and rare-earth ions) in the amorphous matrix. The magnetic behavior of an oxide glass indirectly reflects the glass structure, and it is for this reason that magnetic studies of oxide glasses are important. In some cases the studies are complementary to other structural probes using structure-sensitive physical properties of these magnetic ions such as the color of glasses (optical absorption) and paramagnetic resonance. Magnetic oxide glasses are not presently used to replace crystalline magnetic oxides (i.e., ferrites).

Metallic ferromagnetic glasses, however, are potentially of technical importance in low hysteresis loss applications (see *Metallic Glasses*).

The magnetization M of oxide glasses is the resultant of the diamagnetism of the inner shell electrons of the matrix and the paramagnetism of a wide variety of magnetic entities that may exist in the glasses. The complete range of magnetic behavior can be observed depending on the particular circumstances.

(a) *Diamagnetism*. The (temperature-independent) magnetization M is in opposition to the applied field H and is linearly dependent. It is customary to write

$$M = \chi H \tag{1}$$

with the value of the magnetic susceptibility $\chi \simeq -10^5$ (SI).

(b) *Paramagnetism*. The magnetic moment M is in the direction of the applied field. It is usually temperature dependent. At small fields ($H < kT/\mu_0\mu$ where μ is the magnetic moment of the contained species in Bohr magnetons), Eqn. (1) is a good approximation with

$$\chi = C/(T - \theta) \tag{2}$$

where C and θ are constants.

(c) *Superparamagnetism*. This is due to crystalline ferromagnetic or ferrimagnetic particles (see *Magnetic Materials: Basic Concepts*).

(d) *Spin glass behavior*. The lack of long-range order causes random variations in the exchange interaction between a given magnetic moment and its magnetic neighbors.

1. Diamagnetic Behavior

Diamagnetism in an oxide glass results from the combination of contributions of different atoms:

$$\chi_{\text{dia}} = -\frac{\mu_0 e^2}{6m} \Sigma\, n_i \langle r_i^2 \rangle \tag{3}$$

where n_i is the number of atoms of the ith component per unit volume and $\langle r_i^2 \rangle$ is the effective mean square radius. The larger atoms contribute most to the diamagnetism.

2. Paramagnetic Behavior

Paramagnetism in oxide glasses arises from unpaired spins of intrinsic "defect centers" or from paramagnetic ions intentionally introduced for coloring or other purposes. Generally Eqn. (2) holds; the so-called Curie behavior occurs when the temperature-independent coefficient $\theta = 0$, Curie–Weiss behavior for positive θ and Curie–Néel for negative θ. The constant C is known as the Curie constant in all these situations and, in the first approximation, is proportional to the square of the magnetic moment of the contained magnetic species and to their concentration.

The susceptibility in Eqn. (2) is the low-field, high-temperature approximation to the number of magnetic species per unit volume multiplied by the average component of the magnetic moment of each in the direction of the applied field H. This average is numerically less than μ owing to the disordering effects of temperature. The value and sign of θ reflects the tendency of neighboring moments to align mutually parallel or antiparallel.

A typical intrinsic defect in SiO_2 glass is the E' center, consisting of an oxygen vacancy which has trapped one electron. This defect should contribute to the static magnetic susceptibility with a Curie constant that depends only on the electron spin in the first approximation. The paramagnetism from such intrinsic defects is small, since their concentrations are small, and dc magnetic measurements on them and related centers have not been performed. E' centers can be readily detected by electron paramagnetic resonance (EPR) and by optical absorption. Such defects in glass have received considerable attention in view of their importance in optical waveguide performance.

As an example of a glass intentionally containing paramagnetic ions, the xMnO. $(1-x)$ [$19TeO_2$, PbO_2] glass system displays some of the magnetic behavior typical of these systems particularly well. It is usual to invert Eqn. (2) and to plot χ^{-1} versus T to obtain a straight line from which C and θ may be found, as shown in Fig. 1.

The salient features displayed here are an almost ideal Curie behavior (Eqn. (2) with $\theta = 0$) for small concentrations of magnetic ions, followed by an abrupt onset of Curie–Néel behavior at concentrations above around 5–10% magnetic oxide, in which case the experimental points do not extrapolate to zero. The Curie constant is proportional to the magnetic ion concentration, while the Néel temperature θ increases linearly with concentration as shown in Fig. 2. The value of C for a given magnetic ion depends on the strength and symmetry of the surrounding ligand field which reduces the magnetic moment to some value between the free-ion value and the spin-only value. For oxide glasses, θ is always negative.

The different valence states are variously determined using optical absorption, by Mössbauer spectroscopy, by EPR or by wet chemical methods. The valence concentration ratio c (fraction of total ions in the high valence state) is determined mainly by the melting conditions such as temperature and oxygen pressure (activity) and less by the host glass composition. Since an electron, or polaron, can move from a low-valence-state ion to a neighboring high-valence-state ion by thermally activated tunnelling, or so-called "hopping," there is a subtle relation between

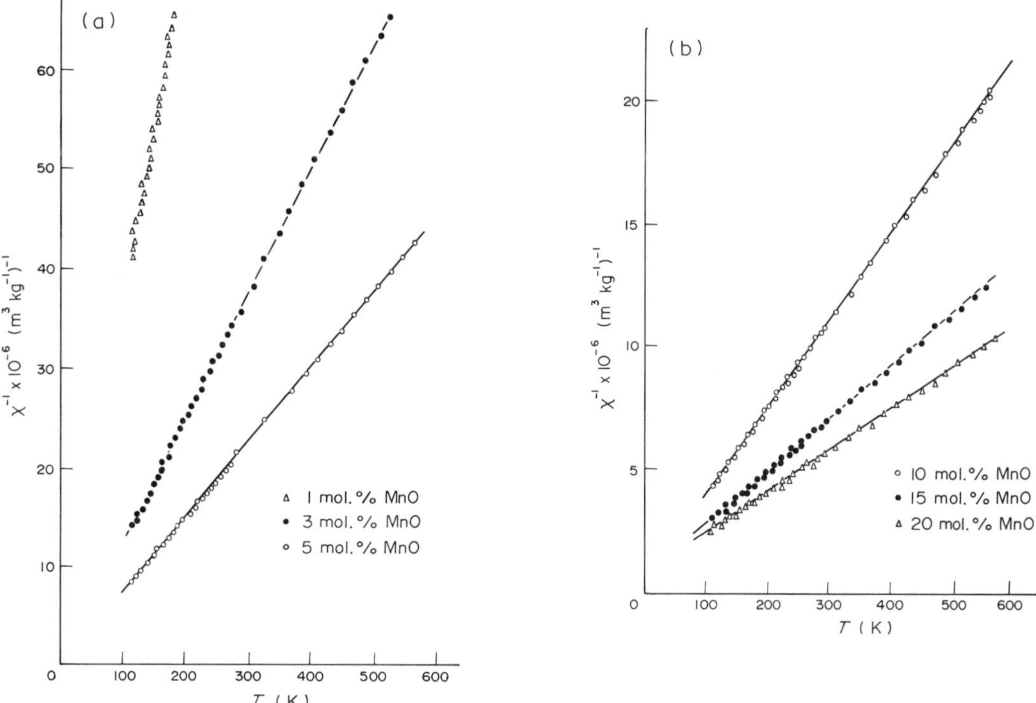

Figure 1
Temperature dependence of reciprocal susceptibility for glasses with MnO content (a) <5 mol.% and (b) >5 mol.% (after Ardelean et al. 1980 *Solid State Commun.* 33: 653–5)

the electronic conductivity and the magnetic susceptibility induced at the same time by defects or transition metal ions in the glass.

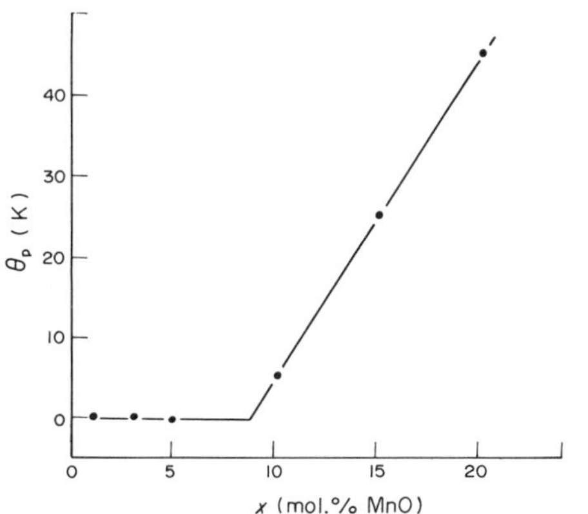

Figure 2
Composition dependence of the paramagnetic Curie temperature (after Ardelean et al. 1980 *Solid State Commun.* 33: 653–51)

The magnetic ions may also exhibit different coordination in different glasses. The dependence of the coordination number on host glass composition is illustrated by the Ni^{2+} ion in alkali borate glasses, where at 20 mol.% alkali, tetrahedral coordination in addition to octahedral coordination sets in. This results in a change in the value of the Curie constant C accompanied by changes in the color of the glass as the ligand field changes together with the orbital contribution to the magnetic moment. As may be expected, different coordinations can exist simultaneously.

A negative θ (Néel temperature) indicates a tendency for antiparallel alignment, which is also the case in crystalline magnetic oxides (ferrites). In concentrated glass (~50 mol.% magnetic ion, a composition easily realized without crystallization in the phosphate system, for example) where intermagnetic ion distance and number of nearest neighbors are comparable to those of their counterpart crystals, a comparable Néel temperature (~100 K), resulting from the short-range interaction, is often attained. Effective-field theories and cluster calculations have been developed to predict the Néel temperatures in this concentrated regime. At low temperatures, antiferromagnetic oxide glasses do not order abruptly; rather, progressively larger clusters form with small resultant moment so that a maximum in the χ versus T curves is observed (see Sect. 4).

In some oxide glasses, most notably those containing iron, the Curie constant is much reduced from the spin-only value. This low value of C is evidence for closely spaced pairs, triplets and quartets, and so on, of antiferromagnetically coupled atoms, which are formed in the glass during melting and/or subsequent heat treatment. Conversely, very large magnetic moments per ion have been observed in very dilute glasses.

3. Superparamagnetism

Superparamagnetism occurs when an increased concentration of ions and/or the appropriate heat treatment enlarges and orders the atom clusters in some systems so that small crystalline ferrimagnetic particles are formed in the glass. In this case the magnetic moment per particle is given by $\mu = M_s V$ where M_s is the saturation magnetization of the particles and V is the volume of the particle. Thus $\mu_0 \mu H$ can become equal to kT at moderate field values, and the magnetic system is then said to be superparamagnetic and has a Langevin function response rather than that of Eqn. (2). These are ferritic glass ceramics (Bonnenfant et al. 1982) which are analogous to ferroelectric glass ceramics in many ways. Some glasses may be prepared containing ferrometallic particles which are also superparamagnetic. Glasses containing larger ferrimagnetic particles may have practical potential as magnetic material for data storage. Small antiferromagnetic crystalline particles may also exhibit large positive magnetic moments. This arises from incomplete magnetic compensation in crystallites containing an odd number of magnetic planes.

4. Spin Glass Behavior

Spin glass behavior results when, in a predominantly antiferromagnetic situation, the lack of long-range structural order leads to a situation called "frustration" (see *Paramagnetic Materials*). Consider three magnetic ions arranged linearly (representing crystalline order) compared with the ions arranged on a triangle (representing disorder). In the first case the lowest energy state, the so-called ground state, is easily identified as in Fig. 3. In the second case, it is not possible to satisfy nearest-neighbor antiparallel spin arrangement between all the spins simultaneously (i.e., the system is frustrated). In this case the energy of the ground state is higher than that of the crystalline case, and so is the degeneracy. These are the two characteristic features of a spin glass, namely

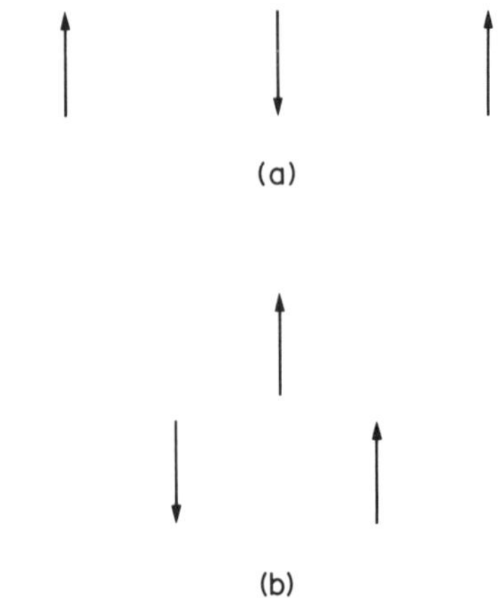

Figure 3
Schematic representation of frustration in a spin glass (b). In (a), representing a crystal, antiferromagnetic alignment between all nearest neighbours can be simultaneously realized. In contrast, this is not possible in a spin glass (b). There are two ways the lowest energy state in (a) can be realized; the degeneracy is thus two. In (b), the degeneracy is six. This larger degeneracy of the ground state is one characteristic of spin glasses

random structure and/or random exchange interaction, and frustration in the highly degenerate ground state.

At high temperatures, the magnetic behavior of a spin glass follows a Néel behavior as before. As the temperature is lowered, the magnetic terms dominate and the spin system undergoes a phase transition into the spin glass state, where ultimately, the spins become frozen into one of the many frustrated ground states, with very small if not zero net magnetic moment. In the spin glass state, any given spin has a nonzero Edwards and Anderson order parameter $Q(t) = [\langle S(0)\rangle \langle S(t)\rangle]$ where $\langle\ \rangle$ is a thermal average and [ ] is a configuration average, reflecting a tendency to a frozen state. This static but random configuration of spin glass below T_g (the spin glass transition temperature) is in contrast to the high-temperature paramagnetic state in which the orientation of a given spin fluctuates very rapidly, and $Q(t) = 0$.

First, a cusp appears in the magnetic susceptibility versus temperature curve, which becomes blunter at higher measuring fields. The specific heat exhibits a rounded maximum which shifts to higher temperatures in increasing applied magnetic field. Below T_g the zero field magnetization depends on the magnetic field during cooling. The archetypical spin glass is the dilute copper–manganese alloy—a spin glass because of the random distance between the magnetic manganese ions. Spin glass behavior has been observed in metal alloys, aluminosilicate, and fluorophosphate glasses, spinels and other insulating and semiconducting crystals. Magnetic spin glasses are of interest not only for the theoretical complexity of the theory, but also their relation to other disordered systems, such as dipolar glasses, neural networks and intergranular Josephson junctions (see *Granular Superconductors*).

See also: Magnetic Materials: Basic Concepts; Paramagnetic Materials

Bibliography

Bonnenfant A, Freidt J M, Maurer M, Sanchez J P 1982 Magnetic and structural properties of iron-based oxide glasses, Fe_2O_3-BaO-Ba_2O_3, from ^{57}Fe Mössbauer spectroscopy. *J. Phys.* (*Paris*) 43: 1475–87

Chowdhury D 1986 *Spin Glasses and Other Frustrated Systems.* Princeton University Press, Princeton, NJ

Hooper H, de Graaf A (eds.) 1973 *Amorphous Magnetism.* Plenum, New York

Levy R, Hasegawa R (eds.) 1977 *Amorphous Magnetism,* Vol. 2. Plenum, New York

Morrish A H 1965 *The Physical Principles of Magnetism.* Wiley, New York

R. K. MacCrone
[Rensselaer Polytechnic Institute,
Troy, New York, USA]

Oxide Superconductors: Ceramic Processing

The recent discovery of high-T_c ceramic superconductors has led to unprecedented excitement and intense research effort due to their potential for significant technological applications. For the new superconductors (e.g., Y–Ba–Cu–O, Bi–Sr–Ca–Cu–O and T1–Ba–Ca–Cu–O) to be useful in bulk forms, they must be fabricated into desirable shapes (e.g., wires) and should carry sufficiently high electrical currents. Both the fabrication and the current-carrying capacity are intimately related to how these mechanically brittle, ceramic materials are processed. In this article, various fabrication and processing techniques for the ceramic superconductors are discussed.

1. Fabrication of Bulk Ceramic Superconductors

Extensive knowledge and techniques are available for the processing of various ceramics. Essentially similar processing methods may be employed to synthesize and shape the high-T_c superconductors into bulk geometry (Murphy et al. 1988). Fine powders of the materials are prepared, which are then compacted into preforms and finally densified by high-temperature sintering heat treatment.

1.1 Powder Preparation

The most common method for the preparation of superconductor powders begins with mechanical mixing of the constituent oxides or oxide precursors such as nitrates or carbonates in appropriate ratios (for the Y–Ba–Cu–O system, the stoichiometric ratio to aim at is $Y_1Ba_2Cu_3O_7$). Laboratory grinding and mixing devices such as a mortar and pestle can be used for small batches. For large batches, the constituents can be mixed in a water slurry using ball mills or shear mixers. The mixture is filtered and dried. The powder is then calcined (baked) at 900–950 °C for chemical mixing. Repeated grinding and calcining is often needed to acquire desirable chemical homogeneity. Typical particle size is 1–10 μm in diameter. Various alternative preparation techniques have also been developed, for example coprecipitation of the ceramic powders from aqueous solutions, sol–gel processing, freeze drying and others. These methods aim at achieving high chemical homogeneity and fine particle size for easy sintering.

1.2 Shaping

The superconductor powder is compacted into desired shapes before sintering. Dry processing, which consists of filling a metal die with powder and pressing at 100–500 MPa, is the most widely used technique. A small amount of organic binder such as polyvinyl alcohol is often added to the powder for lubrication and improved strength of the compact after pressing or extrusion.

In tape casting, a larger amount of binder and solvent is added to make a slurry with the viscosity of heavy paint. The slurry is spread out as an even layer of 0.05–0.5 mm thickness over a carrier film and is allowed to dry. This tape is mechanically flexible before sintering and thus can be peeled off, cut or shaped. During sintering, the binder is burned off and the superconductor particles are consolidated into a brittle ceramic sheet.

Screen printing also employs the slurry technique. The slurry is forced through a patterned mesh screen and printed on selected areas on a substrate as, for example, a conductive circuit line which interconnects devices. Similar technology has been used in the production of multilayer ceramic capacitors and thick-film hybrid integrated circuits. A proper substrate material has to be chosen, as it should withstand the subsequent high-temperature sintering treatment and should not poison the ceramic superconductor through extensive chemical interdiffusion.

1.3 Sintering

During sintering heat treatment, the particle–particle contact points are bonded and progressive consolidation (densification) takes place through diffusion of atoms. Typical sintering treatment for the $YBa_2Cu_3O_7$ superconductor consists of heating the shaped body slowly to 900–960 °C, holding it at the temperature for

5–20 h and cooling slowly to room temperature. A continuous flow of oxygen gas is used to facilitate the removal of organic binders and other impurities during sintering, as well as to allow a sufficient uptake of oxygen during the cooling stage.

In $YBa_2Cu_3O_7$, the copper atoms have an average valence of 2.33 +. The seven oxygen atoms are particularly crucial for superconductivity (see *Oxide Superconductors: Solid-State Chemistry*). When the oxygen content is reduced from seven to six (for example by heat treating in a reducing atmosphere), the resultant $YBa_2Cu_3O_6$ compound (with copper atoms at 2+ valence state) exhibits insulator behavior. This is because the electrons are now localized and are not free to move around. As the oxygen content is increased towards seven, the added oxygen atoms now pull enough electrons away from the copper atoms in the adjacent CuO_2 plane to yield superconductivity with $T_c \simeq 90$ K. The compound is often denoted as $YBa_2Cu_3O_{7-\delta}$ to reflect this variability in oxygen stoichiometry. The thermodynamic equilibrium favors the higher level of oxygen content at lower temperatures ($YBa_2Cu_3O_{6.4}$ at ~900 °C as compared with $YBa_2Cu_3O_7$ at ~400 °C). Well-sintered and dense Y–Ba–Cu–O requires a longer time for oxygen diffusion and, hence, a slower cooling rate should be employed.

The Bi–Sr–Ca–Cu–O and Tl–Ba–Ca–Cu–O superconductors can be sintered at lower temperatures and they exhibit a less severe dependence of superconductivity on oxygen content than Y–Ba–Cu–O. There are other techniques for densifying the superconductor powders such as hot isostatic pressing (HIP) and explosive compaction; however, these are not as commonly used as the sintering method. Sintered Y–Ba–Cu–O typically exhibits a polycrystalline microstructure with an average grain size of 3–20 μm in diameter. The critical current density is unacceptably low, usually below about 1000 A cm^{-2} at 77 K because of the weak-link problem which will be discussed in Sect. 3.

2. Wire Fabrication

Commercially useful bulk superconductors for magnet or power transmission applications require shaping into composite multifilamentary wires containing both the superconductor and a normal metal. Such geometry is desirable for several important reasons: (a) mechanical protection of the brittle superconductor core against the Lorentz force and other unavoidable stresses during operation and handling, (b) parallel electrical conduction path to serve as a bypass in case of failure in the superconductor, (c) heat sinking, (d) minimization of energy loss by magnetic flux motion, (e) ac loss reduction, and (f) protection from atmospheric environments.

The high-T_c ceramic superconductors are brittle and cannot be drawn into desirable, fine-wire geometry. However, several different approaches to single-filament Y–Ba–Cu–O, Bi–Sr–Ca–Cu–O or Tl–Ba–Ca–Cu–O wires have been reported: metal-clad composite wire by the powder-in-tube method, and metal-core composite wire and bare ceramic wires by the slurry extrusion, oxide melt-spin and metal melt-spin methods.

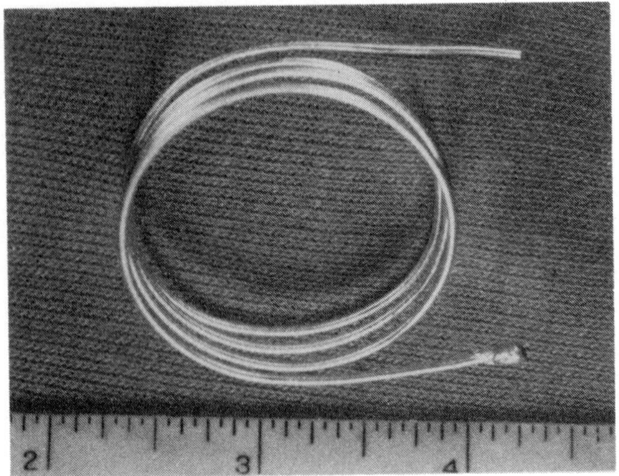

Figure 1
Metal-clad composite wire

2.1 Metal-Clad Composite Wire

The choice of proper cladding metal is critical as most of the common metals interact with the ceramic superconductors at sintering temperature and degrade the superconducting properties. Silver and gold were found to be benign in contact with the superconductors. Silver is relatively inexpensive and could be economically acceptable as a thin-walled clad metal or a thin-layered diffusion barrier.

The fabrication of silver-clad wire consists of filling a silver tube with powders of $YBa_2Cu_3O_7$ (or other types of ceramic superconductor), drawing to a much smaller diameter and sintering at around 900 °C. Once sintered, the superconductor inside the silver cladding is brittle and therefore any additional shaping such as roll flattening or solenoid winding has to be carried out before sintering. An exemplary wire is shown in Fig. 1. Photomicrographs revealing the cross section of the composite wire (~1.5 mm diameter) are shown in Fig. 2.

2.2 Metal-Core Composite Wire

Metal-core composite wire is basically the opposite of the metal-clad configuration and has been devised to cope with the oxygen supply problem in metal-clad Y–Ba–Cu–O. Through the exposed superconductor on the outside of the metal core, oxygen can easily diffuse in or out. A typical processing consists of coating of a silver wire or ribbon with a superconduc-

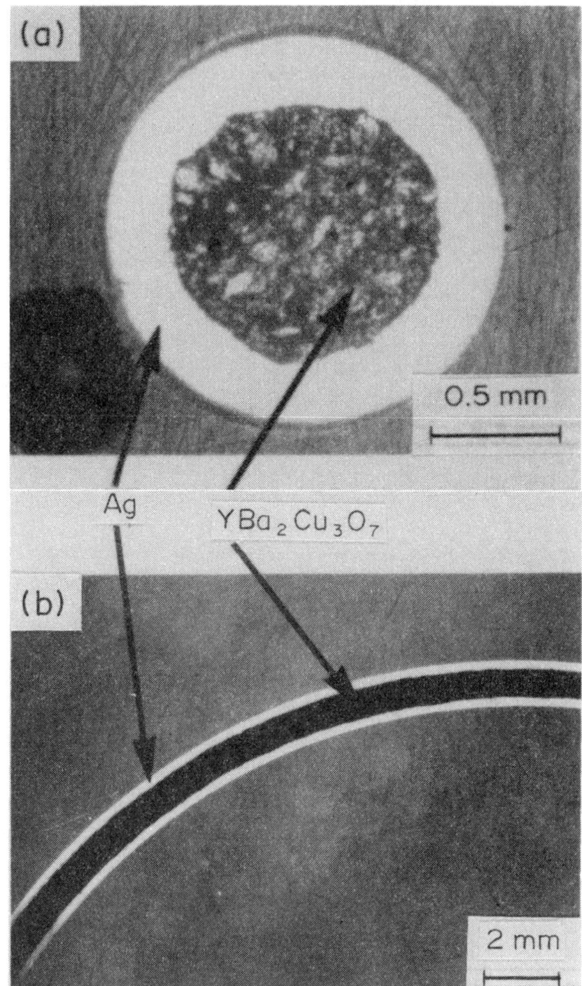

Figure 2
Cross-sectional photomicrographs of silver-clad composite wire superconductor: (a) transverse section, (b) longitudinal section

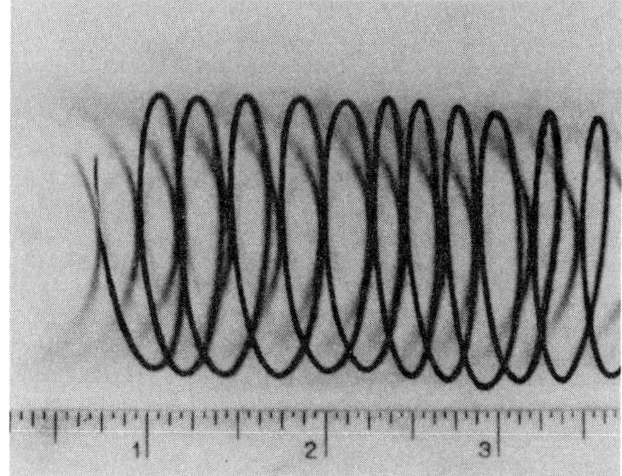

Figure 3
Metal-core composite wire

tor slurry (a mixture of powder, organic binder and solvent), drying, winding and sintering. An exemplary metal-core composite wire prepared this way is shown in Fig. 3 (~1 mm diameter with ~0.5 mm core of silver).

2.3 Bare Wire

Bare ceramic wires can be made by a number of different methods. A superconductor slurry with a properly controlled viscosity can be extruded through a die into a wire or cast into tapes. As described earlier, these materials are then dried, cut or shaped, and sintered. Ceramic wires can also be prepared using a melt-spinning method. A stream of molten superconductor oxide or a precursor alloy melt (e.g., Eu–Ba–Cu) is poured onto the surface of a rapidly spinning metal drum, producing a thin continuous ribbon of solidified material. The melt-spun ribbons are then heat treated or oxidized to render proper chemistry and microstructure. For stabilization, a suitable metal coating may have to be added to the surface of these bare ceramic wires. Other high-temperature processing such as plasma spraying on a wire- or ribbon-shaped substrate can also be used to prepare the superconducting wires or ribbons.

While the wire fabrication of the ceramic superconductors has been successfully demonstrated using a number of different processing methods, one common problem encountered in most of these wires is the low current-carrying capacity caused by grain boundary weak links. A suitable ceramic processing has to be employed to increase the critical current density J_c as discussed in the following.

3. Ceramic Processing for Improved J_c

3.1 Grain Boundary Weak Links

Progress towards major applications of the new high-T_c superconductors has been hindered by low critical currents and their severe degradation in magnetic fields (see *Critical Current Density*). The cause of the low J_c in polycrystalline ceramic superconductors is now understood in terms of the presence of Josephson-type weak links at grain boundaries, across which the flow of electrical current is severely limited especially in a magnetic field. While the exact nature of the grain boundary weak links is still unknown, they are probably related to the deviation in chemistry, altered crystal structure or presence of impurity atoms, microcracks or anisotropy-related weak connections at grain boundaries.

3.2 Melt-Textured Growth

The weak-link problem can be greatly reduced by elimination or minimization of high-angle grain boundaries in the paths of transport current. This can be achieved by a preferential alignment of grains parallel to the *ab* conduction plane, for example by a melt-textured growth of the bulk Y–Ba–Cu–O superconductor. In the melt-texture processing, a sintered $YBa_2Cu_3O_7$ precursor sample is partially melted by decomposition at about 1010–1200 °C, is directionally solidified (2–10 °C h^{-1} cooling rate) in the presence of a temperature gradient to crystallize and grow parallel and aligned plates of the superconductor phase and is then given an oxygenation treatment. This procedure is schematically illustrated in Fig. 4. The optical microstructure of the melt-textured material (see Fig. 5) reveals near-perfect alignment of the superconducting phase in the *ab* direction along the length of the sample. The blocky, isolated particles are the second-phase Y_2BaCuO_5, which do not interfere with the supercurrent flow except for the volume fraction effect. The transport current is passed along the direction of the horizontal axis in Fig. 5 and does not encounter the weak links (high-angle grain boundaries).

As a result of grain alignment in the melt-textured Y–Ba–Cu–O, significantly improved J_c values and greatly reduced field dependency of J_c are achieved as shown in Fig. 6. The figure includes recent J_c data obtained by a number of researchers using melt-textured growth or related melt processing. $J_c(H)$ values for melt-textured Y–Ba–Cu–O, in the range of 10^4–10^5 A cm^{-2} at 77 K, are improved by orders of magnitude over the values for the randomly grained Y–Ba–Cu–O, especially in magnetic fields. Melt-texture processing of long, continuous wires will be needed to make it commercially useful. Some progress has been made in the area of continuous melt texturing. Successful fabrication of wire about 1 m long has been reported.

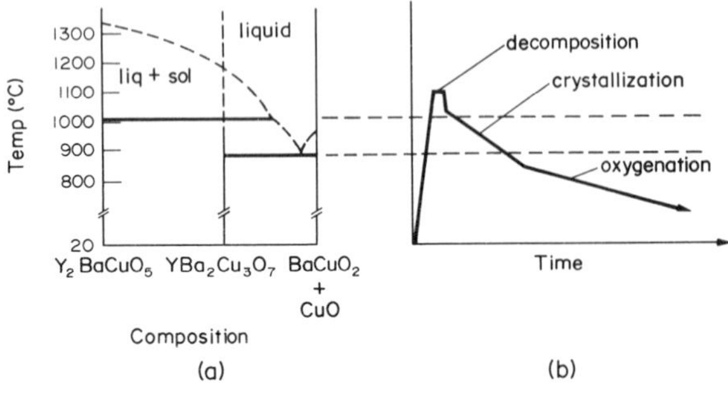

Figure 4
Schematic illustration of melt-texture processing: (a) qualitative sectional phase diagram, (b) three basic steps shown as a temperature–time curve

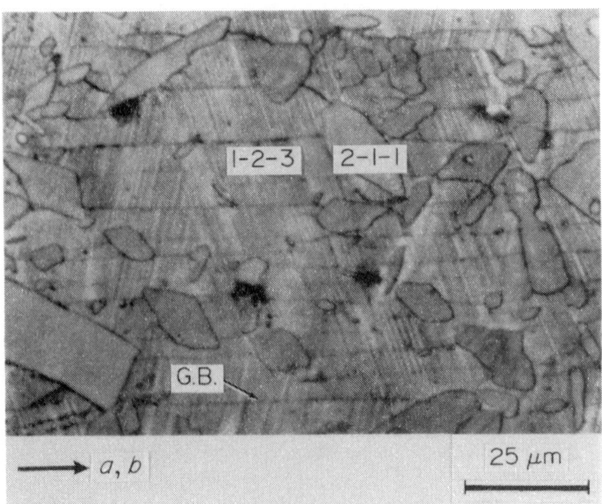

Figure 5
Microstructure of Y–Ba–Cu–O processed by melt-textured growth showing the parallel aligned grains separated by low-angle grain boundaries (1-2-3, $YBa_2Cu_3O_7$; 2-1-1, Y_2BaCuO_5; G. B., grain boundary)

The melt-textured samples show large magnetic hysteresis when a magnetic field is applied at 77 K. Because of the large positive magnetization left over

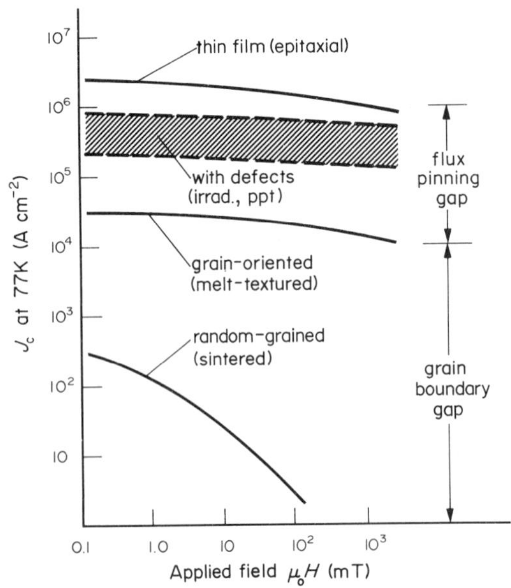

Figure 6
Comparative transport critical current density ($J_c(H)$ curves) for grain-aligned and random-grained Y–Ba–Cu–O at 77 K

Figure 7
Photograph showing the melt-textured Y–Ba–Cu–O pellet attracting nails

after the field is removed (trapped flux due to persistent current), the material exhibits a magnetlike behavior as shown in Fig. 7. Recent work by researchers at the Nippon Steel Corporation, Japan, demonstrates that the trapped flux in melt-processed Y–Ba–Cu–O superconductors could generate a magnetic field in excess of 1 T at 77 K. These bulk oxide magnets, while the stability of the magnetic moment in a variable field is still an issue, may be useful for levitation, bearings or some special permanent magnet applications.

3.3 Other Methods for Grain Alignment

While melt-textured-growth processing has been the most successful method in accomplishing a near-perfect alignment of grains, there are other possible processing schemes which have shown some success in inducing various degrees of texture in the high-T_c superconductors. These include (a) magnetic alignment of superconductor powders before sintering (using the anisotropic paramagnetic susceptibility in $YBa_2Cu_3O_7$), (b) mechanical alignment of plate-shaped particles especially in the Bi–Sr–Ca–Cu–O system (using the tendency of anisotropic fracture of the superconductors along the *ab* plane and the tendency of their alignment during pressing, wire drawing or cold-rolling processing), (c) press forging (high-temperature alignment of grains under applied stress), and (d) substrate-induced texture (based on the superconductor–substrate reaction at the interface). The superconductor material near the interface (especially on a silver substrate) experiences more extensive melting and layerlike growth (*c*-axis texture), which is aided by the two dimensionality of the interface.

The techniques based on magnetic or mechanical alignment have demonstrated somewhat limited effectiveness in texture formation; however, the technique based on the substrate-induced texture has been successful in introducing a strong *c*-axis texture and producing high transport J_c values especially in Bi–Sr–Ca–Cu–O-type superconductor wires/ribbons. The Bi–Sr–Ca–Cu–O (sometimes modified by lead doping) wires/ribbons can be prepared by at least three different processing methods: (a) silver-clad wire processing (as described earlier) followed by roll flattening, (b) doctor blade coating of silver ribbon with superconductor–binder slurry, and (c) spray coating of silver ribbon with diluted slurry followed by cold rolling. The composite ribbons are then subjected to a partial melt processing at around 850–900 °C. An exemplary processing sequence (spray coating/rolling technique) is schematically illustrated in Fig. 8. Some exemplary microstructures from the textured Bi–Pb–Sr–Ca–Cu–O ribbons are shown in Fig. 9. The layered growth along the *ab* basal plane of the superconductor and the *c*-axis texture is evident.

High transport critical current densities of around 5.4×10^4 A cm^{-2} ($H=0$) and 1.2×10^4 A cm^{-2} ($\mu_0H=1$ T) at 77 K were reported for the silver-clad Bi–Pb–Sr–Ca–Cu–O ribbons by researchers in Japan. Their prototype superconducting solenoids made from the ribbons generated about 0.1 T field at 77 K and about 0.9 T field at 4.2 K. Bi–Sr–Ca–Cu–O-based ribbons prepared by all three methods exhibit very high J_c at 4.2 K of around 10^5 A cm^{-2}, even in high magnetic fields approaching 30 T. The conventional metallic superconductors such as niobium–titanium or Nb_3Sn are not capable of carrying high currents in such a high field (see *Critical Current Density*).

The high J_c values of around 10^5 A cm^{-2} in Bi–Sr–Ca–Cu–O are basically maintained up to a temperature of near 20 K; however, a severe deterioration in J_c is observed at above 30 K if the applied field is greater than about 1 T especially if the direction of the field is perpendicular to the *ab* plane of the superconductor, as shown in Fig. 10. The curve $B_{in}(B,T)$ in the B,T field defines an irreversibility line; for values of B and T above the irreversibility line the transport critical current density is zero. This behavior is attributed to the thermally activated flux creep or flux cutting, or thermally induced modification of the flux line lattice, which is known to be much more severe in the Bi–Sr–Ca–Cu–O-type superconductors than in Y–Ba–Cu–O

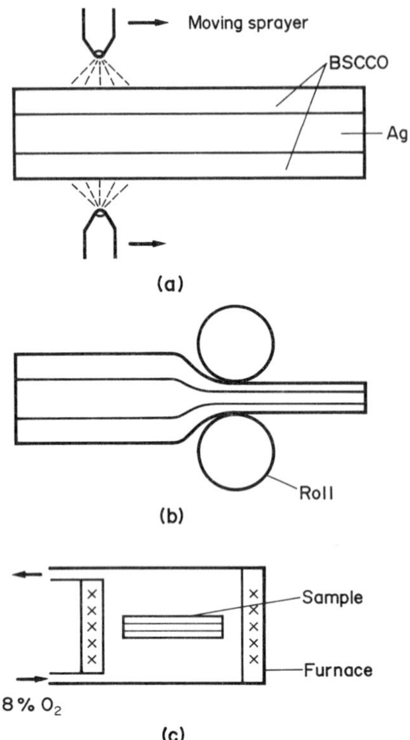

Figure 8
Schematic illustration of processing steps for Bi–Pb–Sr–Ca–Cu–O ribbons: (a) spray coating, (b) cold rolling and (c) partial melt processing

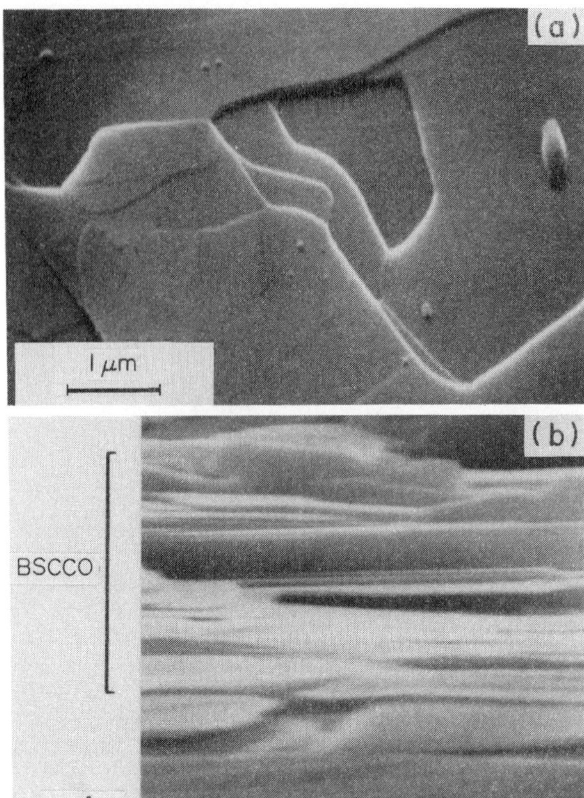

Figure 9
Microstructures of the textured Bi–Pb–Sr–Ca–Cu–O ribbon: (a) top view, (b) cross-sectional view

(see *Flux Creep Phenomena*; *Flux Line Lattice*). Further improvements in materials behavior, especially in flux pinning characteristics, are needed in order to reduce the flux creep problem and make the material more useful at higher operating temperatures.

3.4 Flux Pinning

Many of the superconductor applications depend on the ability of the material to carry sufficiently large currents, that is, J_c more than about 10^5 A cm^{-2} at 77 K in high magnetic fields of several tesla (Fitzgerald 1988, Wolsky et al. 1989). While the weak-link problem in bulk, high-T_c superconductors can be resolved through grain alignment, the high-field J_c still seems to be limited to 10^4–10^5 A cm^{-2} at 77 K, as shown in Fig. 6. Epitaxial thin films of $YBa_2Cu_3O_7$ often exhibit $J_c(H)$ of more than 10^6 A cm^{-2}. Apparently it is the low density of flux pinning sites in the bulk Y–Ba–Cu–O that limits $J_c(H)$ to 10^4–10^5 A cm^{-2}. The lack of sufficient pinning is even more evident in Bi–Sr–Ca–Cu–O and Tl–Ba–Ca–Cu–O. Effective flux pinning requires the presence of defects with at least one dimension on a size scale comparable to the superconducting coherence length (about 0.4–3.0 nm in Y–Ba–Cu–O) (see *Flux Pinning and the Summation of Pinning Forces*). Several processing techniques can induce fine defects and improve the flux pinning as described in the following.

It has been demonstrated that irradiation with fast neutrons improves the flux pinning in high-T_c superconductors. J_c (at 77 K, $H \simeq 0.9$ T) as high as 0.6×10^6 A cm^{-2} has been obtained, which approaches those in epitaxial thin films (see *Radiation Effects on Superconducting Materials*). Proton irradiation has also proven effective in flux pinning enhancement. The irradiation techniques are not as practical as is desired for large-scale industrial applications because of the inconvenience and radioactivity issues.

Phase-transformation-induced defects may be utilized for flux pinning enhancement. An example is the decomposition of the $YBa_2Cu_4O_8$ precursor compound into $YBa_2Cu_3O_7$ containing a high density of stacking defects and dislocations. About tenfold

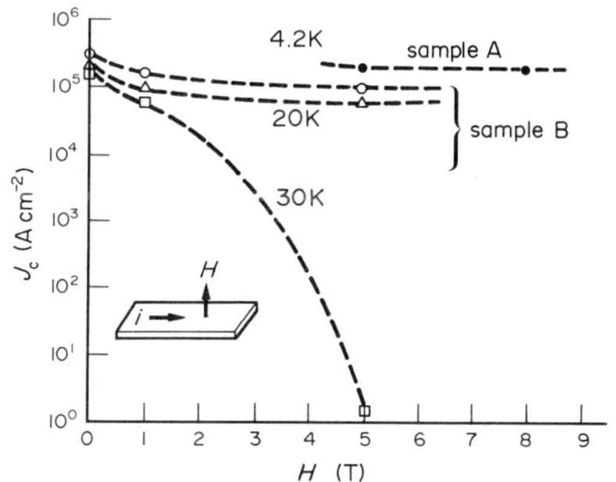

Figure 10
Transport J_c vs H in textured Bi–Pb–Sr–Ca–Cu–O ribbons

improvement in flux pinning has been reported. The dispersion of nonsuperconducting second-phase inclusions is also useful for pinning enhancement. For effective pinning, the particle size of the inclusions will probably have to be made small (of the order of 100 nm or smaller).

Chemical doping may also be employed to locally disturb the crystal structure of the superconductor at the unit cell or subunit cell level and create coherence-length-scale defects. The chemical substitution may be performed on the yttrium, barium, copper or oxygen site. Because the chemical doping technique is less likely to interfere with various required processing steps such as sintering, texturing and wire fabrication compared with other pinning-enhancement techniques, it is inherently more desirable and deserves further attention. The development of commercially viable processing methods that will generate a large density of microscopic defects for effective flux pinning remains as a challenge in the materials science of high-T_c superconductors.

See also: Flux Pinning and the Summation of Pinning Forces; Measurements in Superconducting Materials; Oxide Superconductors: Electron Microscopy; Oxide Superconductors: Physical Properties; Oxide Superconductors: Solid-State Chemistry; Superconducting Materials: An Overview; Superconducting Materials: Large-Scale Applications

Bibliography

Fitzgerald K 1988 Superconductivity: fact vs fancy. *IEEE Spectrum* 25: 30–1

Jin S, Graebner J E 1991 Processing and fabrication techniques for bulk high T_c superconductors: a critical review. *Mater. Sci. Eng. B* 7: 243–60

Murphy D W, Johnson D W, Jin S, Howard R E 1988 Processing techniques for the 93 K superconductor $Ba_2YCu_3O_7$. *Science* 241: 922–30

Wolsky A M, Giese R F, Daniels E J 1989 The new superconductors: prospects for applications. *Sci. Am.* 260: 60–9

S. Jin
[AT&T Bell Laboratories, Murray Hill, New Jersey, USA]

Oxide Superconductors: Electron Microscopy

Defects are known to play a critical role in the current-carrying capacity of conventional superconductors and are equally important in high-T_c oxide superconductors (see *Flux Pinning and the Summation of Pinning Forces*). The only way that the nature and spatial distribution of such defects can be determined is by transmission electron microscopy (TEM). The advantage of TEM is that it is possible to combine various imaging modes with electron diffraction and other analytical techniques such as x-ray energy-dispersive spectrometry (EDS) in order to characterize defects structurally and chemically. Important defects in oxide superconductors include point defects, impurities, dislocations, stacking faults, twins, grain boundaries and second phases. In this article the various techniques and their application to the study of defects in the oxide superconductors will be described. As each new oxide superconductor has been discovered over the past few years, it has been subjected to intense structural and microstructural analysis by TEM. A wealth of information about crystal chemistry, phase transformations and defect structures has emerged from such studies, both as fundamental studies in their own right and as attempts to correlate with the superconducting transition temperature T_c and the critical current density J_c.

1. Experimental Techniques

Specimen preparation is critical. For electron transparency, thicknesses of about 100 nm are required and surfaces need to be free of contamination and damage. Even thinner specimens (~10 nm) are desirable for high-resolution electron microscopy (HREM) and electron energy loss spectroscopy (EELS) while thicknesses approaching 1 μm can be tolerated for regular diffraction contrast microscopy, especially at higher voltages (see later). Such specimens are made commonly by two techniques—crushing and ion milling. Gentle crushing with a mortar and pestle is a rapid and convenient method of producing electron-transparent particles. However, the spatial relationship between the features observed in individual particles is lost. Ion milling is performed on a thin section around 3 mm in diameter with an argon ion beam at

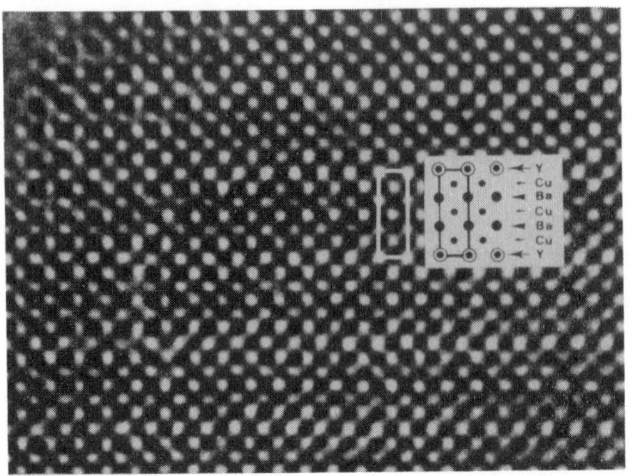

Figure 1
HREM image of Y123 along [100]: a one-to-one correspondence exists between the projected structure (inset) and the white (or black) dot configuration

about 5 kV and an incidence angle of approximately 15°. Specimen cooling is used to minimize ion damage. The method is more tedious but the relationship between the features (grain boundaries, different phases, etc.) is retained.

A wide range of electron microscopy techniques is available. The traditional approach is bright-field/dark-field amplitude contrast microscopy in which contrast is produced through variation in the local diffracted intensity. For example, a dislocation is seen as a line due to the bending of the Bragg planes by the elastic strains around the dislocation. Obtaining diffraction patterns from the same regions is a simple push-button operation, either by selected-area diffraction (SAD) or by convergent beam electron diffraction (CBED). Consequently, crystal structure information can be obtained from areas smaller than 10 nm. The traditional accelerating voltage has been 100 kV but 200–400 kV microscopes have now become routine. Such microscopes can be optimized either for high resolution (HREM) or for analytical electron microscopy (AEM). HREM utilizes phase contrast microscopy whereby an image is produced by allowing the transmitted beam to interfere with many diffracted beams; resolutions of better than 0.2 nm are commonly available so that the image can be thought of as a projection of the atomic structure along the direction of the electron beam (e.g., see Fig. 1). In AEM, the electron beam can be focused to a small probe ($\sim$1 nm) and the various signals produced can be used for analysis. For example, the characteristic x rays of each element can be analyzed by EDS for chemical analysis (e.g., Y123 in Sect. 2.2); additional chemical and crystallographic analysis can be obtained from EELS and CBED techniques. The electron probe can also be scanned using scanning transmission electron microscopy (STEM). A recent development in STEM techniques is high-resolution Z-contrast imaging. In this technique, a high-angle annular detector is used; since heavy (high-Z) elements scatter more strongly into high angles, then atomic images can be obtained with strong compositional sensitivity, provided the scanned electron probe is of atomic dimensions. An example of Z-contrast use is given in Sect. 4 for a superlattice. SEM is used for imaging the surface of bulk material and an example of its usefulness is given in Sect. 3 for investigating bismuth compounds.

2. *Defects in $YBa_2Cu_3O_{7-x}$*

The structure of $YBa_2Cu_3O_{7-x}$ (hereafter referred to as Y123) is based on the perovskite unit cell with the smaller copper ions occupying the corners of the unit cell and the larger yttrium or barium ions occupying the body-centered position; the oxygen ions occupy the edge centers. The Y123 unit cell consists of three stacked perovskite unit cells (see *Oxide Superconductors: Solid-State Chemistry*). In the fully oxygenated condition ($x \simeq 0$), the unit cell is orthorhombic with $a = 0.382$ nm, $b = 0.388$ nm and $c = 1.168$ nm ($\cong 3b$). The small orthorhombicity is due to the fact that there are copper–oxygen chains aligned along the b axis with vacant oxygen sites in between at the $\frac{1}{2}00$ positions. An HREM picture of the structure is shown in Fig. 1, obtained by imaging along [100]. The projected structure is shown in the inset and there is clearly a one-to-one relationship between the cation positions and the black/white dots of the HREM image. (The atoms can be imaged as black or white simply by changing the defocus under phase contrast conditions.) The oxygen ions are less obvious because they scatter the electrons much less than the heavier cations; however, they can be imaged under ideal conditions.

The orthorhombic structure can be transformed to a tetragonal structure either by heating or by electron irradiation at voltages above 100 kV. In both cases the oxygen sublattice becomes disordered so that the copper–oxygen chains no longer exist. Intermediate states of order exist, especially for oxygen-poor materials ($0.63 < x < 0.69$), which reveal superlattice reflections in SAD patterns and also have lower values of T_c.

2.1 Twins

The most prominent defects in Y123 are twin boundaries (see Fig. 2). These are formed during cooling from high temperatures or during oxygenation at intermediate temperatures, when the tetragonal structure transforms to orthorhombic. The shape change resulting from the transformation is accommodated by the formation of twins. The twins form in two sets of colonies (Fig. 2b) parallel to either (110) or $(1\bar{1}0)$.

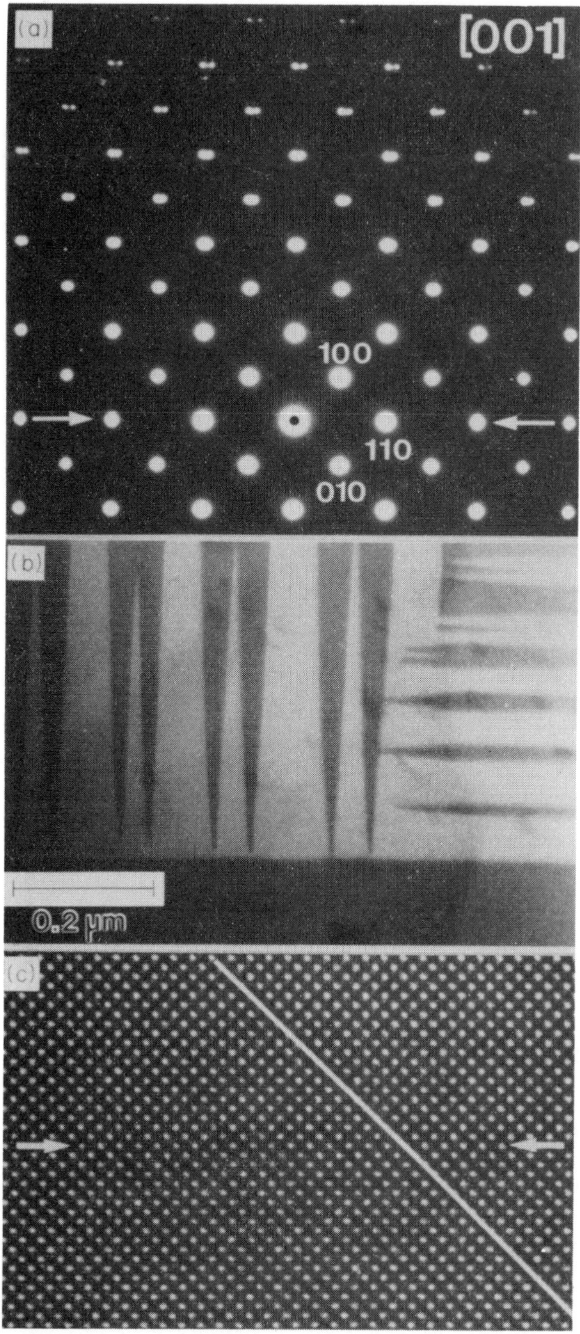

Figure 2
Twins in Y123: (a) [001] SAD pattern taken over (110) twin interface, (b) the two sets of colonies and (c) an HREM image

The *b* and *a* axes are interchanged or reflected across each twin boundary. SAD patterns across (110) twin boundaries show spot splitting along the [110] reciprocal lattice direction which increases with increasing distance from the origin (Fig. 2a). This spot splitting is due to the small difference between *a* and *b*. The twin spacing is about 100 nm and the twin ends are lenticular in shape; the bright-field contrast in Fig. 2b is due to the small orientation difference between the matrix and twins. The small orientation change can also be revealed in HREM images; if a straight edge is aligned along the white dots in Fig. 2c, an approximately 2° rotation is evident across the twin boundary (marked with a white line).

2.2 Grain Boundaries

Grain boundaries in Y123 are widely regarded as the "weak links" which limit J_c in bulk polycrystalline material. A typical grain boundary triple point is shown in the HREM image in Fig. 3. Grain boundaries are frequently parallel to the *ab* plane of one of the grains due to the plate morphology developed during grain growth. The crystal planes in each of the three grains in Fig. 3 appear to remain unchanged up to the atomic core of the boundaries. In other cases, second phases are frequently observed at grain boundaries and triple points (e.g., CuO, Y_2BaCuO_5, $BaCuO_2$). These are easily detected not only by SAD and CBED, but also by EDS. An x-ray spectrum from Y123 is shown in Fig. 4; the yttrium, barium and copper peaks are labelled and are seen to have intensities roughly in the ratio 1:2:3, respectively. Such data can easily be quantified and used to identify second-phase material at grain boundaries. HREM analysis of carefully prepared stoichiometric Y123 shows that the grain boundaries are generally clean and that therefore the weak link behavior is intrinsic.

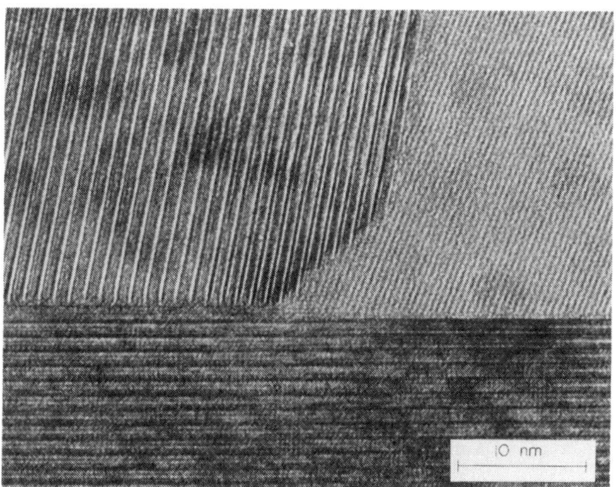

Figure 3
HREM image of grain boundaries in Y123

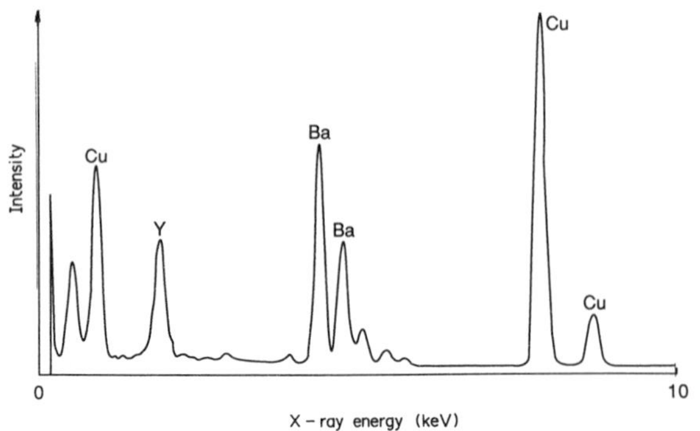

Figure 4
EDS spectrum (intensity vs x-ray energy) for Y123 showing that the intensities of yttrium, barium and copper peaks are roughly in the ratio 1:2:3

Figure 5
Stacking fault in Y123 imaged (a) along [010] and (b) along [100]; the fault is due to the insertion of an extra CuO layer between two BaO layers with an additional shear of $\frac{1}{2}b$ in the fault plane

2.3 Dislocations and Stacking Faults

The most common defects in the interior of grains are stacking faults and their associated partial dislocations. Perfect dislocations are relatively rare unless deliberately introduced by deformation. Planar defects are observed on (001) planes and are associated with the insertion of an extra copper–oxygen plane. These extra planes are arrowed in the HREM images in Fig. 5. Figure 5a shows that the lattice is displaced only along [001] when the fault is viewed along [010] while Fig. 5b shows that there is an additional shear of $\frac{1}{2}$[010] when the fault is viewed along [100]. It is easily deduced from the HREM images that the extra copper–oxygen layer is inserted between two barium–oxygen layers. The faults are terminated along their periphery by partial dislocations corresponding to such displacements. Regular arrangements of planar faults along the c axis can produce the Y247 and Y248 structures.

An example of a mixed phase region is shown in Fig. 6. The planes that show bright white contrast in Fig. 6 are CuO planes. The single CuO planes, corresponding to the Y123 structure, and the double CuO planes, corresponding to the Y248 structure, are seen clearly. The 1.36 nm spacing corresponds to half the Y248 c parameter. The other features indicated are (1) a stacking fault where a CuO single layer changes to a double layer, (2) a step in a CuO layer and (3) a partial dislocation (arrowed) where a single layer changes to a double layer with another partial of opposite sign separated by a single Ba–Y–Ba stacking sequence. The movement of these defects facilitates phase transformation between the 123, 247 and 248 structures.

The phenomenon of the intergrowth of structures with different stacking sequences is called microsyntaxy. In fact, many of the new high-T_c compounds discovered since 1986 were often present as microsyntactic intergrowths in parent compounds prior to their isolation as separate compounds, especially for the bismuth- and thallium-based superconductors described in Sect. 3.

Perfect dislocations are observed in Y123 specimens deformed at high temperatures (not shown). Standard diffraction contrast of these dislocations shows that they have Burgers vectors of [100] and [010] (the shortest lattice vectors). Occasionally [110] Burgers vectors are also observed. There have been no [001] dislocations reported except in small-angle tilt boundaries parallel to (001) planes which necessitates the c-axis Burgers vector.

3. Modulated Structures and Intergrowths in Bismuth- and Thallium-Based Superconductors

Many of the superconducting cuprates containing bismuth and thallium can be described in terms of the structural series

$$(AO)_m B_2 Ca_{n-1} Cu_n O_{2+2n+\delta}$$

where A = bismuth or thallium and B = strontium or barium, respectively. The members of the series are

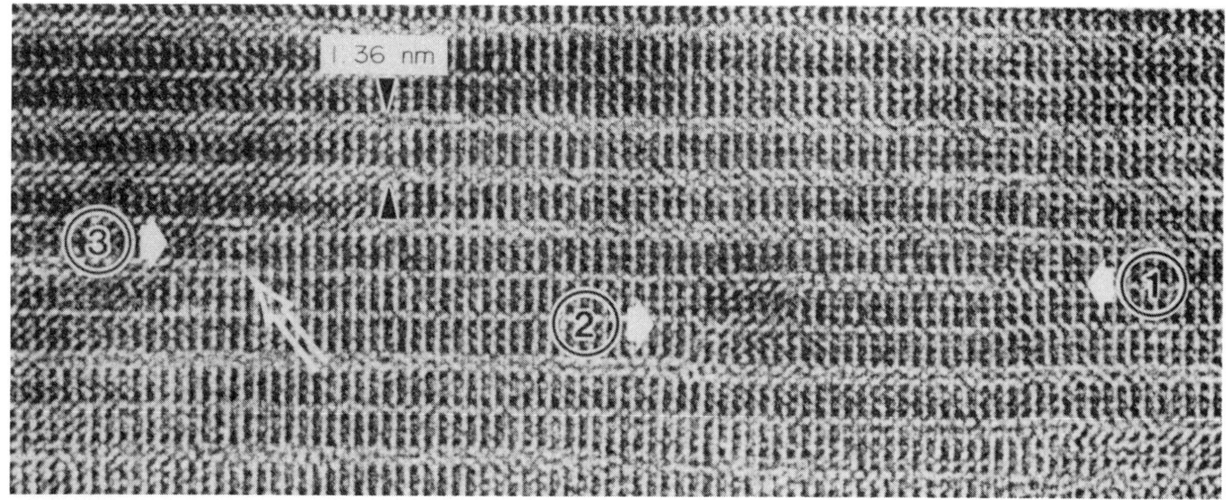

Figure 6
HREM image of mixed Y123 and Y248. The single bright CuO planes correspond to Y123 while the double bright CuO planes correspond to Y248. The numbers and arrows point to stacking faults and partial dislocations separating the two structures

often referred to as 2201, 2212, 2223, 2234 and so on. For example, Bi-2201 is $Bi_2Sr_2CuO_6$, Bi-2212 is $Bi_2Sr_2CaCu_2O_8$, Tl-2201 is $Tl_2Ba_2CuO_6$ and Tl-2223 is $Tl_2Ba_2Ca_2Cu_3O_8$. These are all layered structures consisting of perovskitelike units interlaced with rock-salt-like units. Because of the relationship between the members of the structural series, phase purity is hard to establish and microsyntactic intergrowths between the different members are common. A consequence of the layered nature of the structures is that they tend to be "micaceous" (i.e., they break up easily into flakes because of the weak bonding between the layers). This is especially true of the bismuth compounds and is illustrated vividly in the SEM image of Fig. 7. This has consequences not only for the structural integrity of practical superconductors but also for the difficulty of performing TEM on platelet cross sections.

Incommensurate modulations seem to be the norm in both the bismuth and thallium structures. An HREM image and corresponding SAD pattern from Bi-2201 ($T_c = 20$ K) are shown in Fig. 8 viewed along [100]. The Bi-2201 has an underlying tetragonal structure ($a = 0.54$ nm, $c = 2.4$ nm). The [100] SAD pattern in Fig. 8b shows satellite reflections due to modulations in the base structure in the (100) plane. The rows

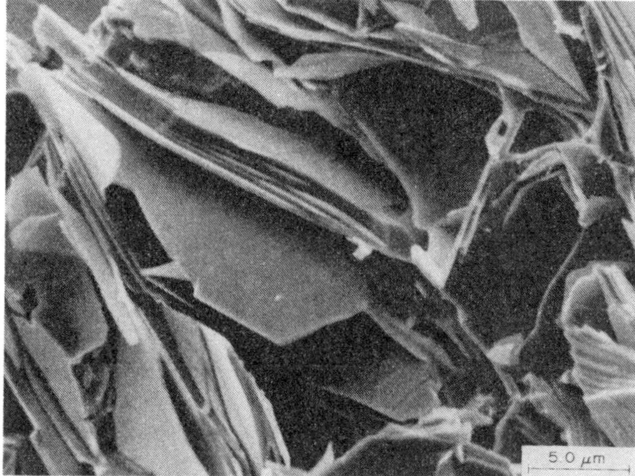

Figure 7
SEM image of Bi-2212 illustrating its micaceous (platelike) morphology

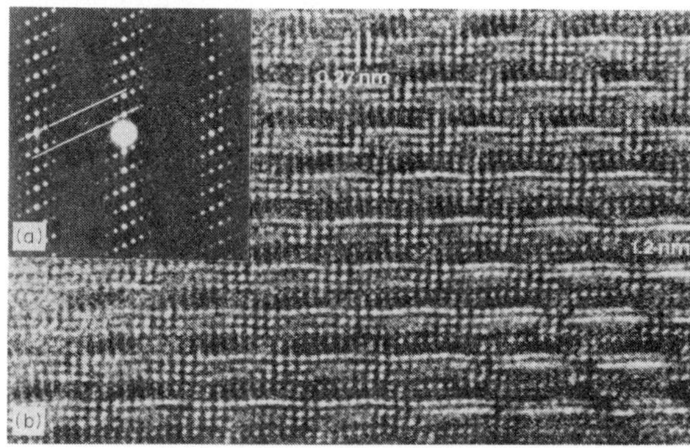

Figure 8
(a) SAD pattern and (b) corresponding HREM image of the modulated structure in $Bi_2Sr_2CuO_6$

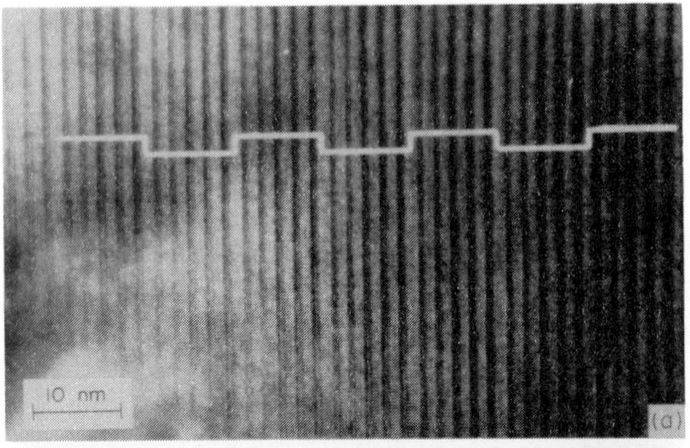

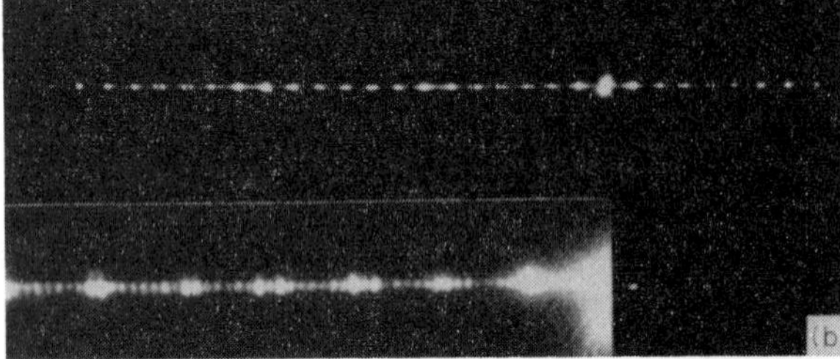

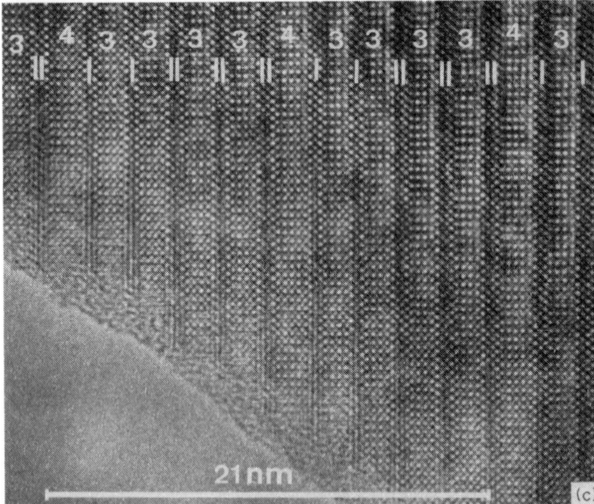

Figure 9
(a) "Low-magnification" HREM image of long-period modulation in $Tl_{2-x}Ba_2Ca_2Cu_3O_{10}$ showing 21 nm periodicity (white line). (b) Corresponding SAD pattern of the [100] reciprocal row of spots with enlargement (inset) showing a large number of closely spaced spots.
(c) HREM image of the modulated structure where single and double white bars indicate single and double thallium–oxygen layers and 3 and 4 denote n in the formula $Tl_mBa_2Ca_nCu_{n+1}O_{2+m+2n}$

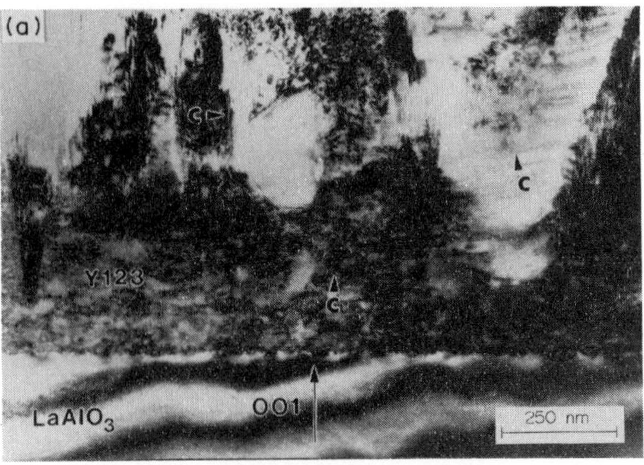

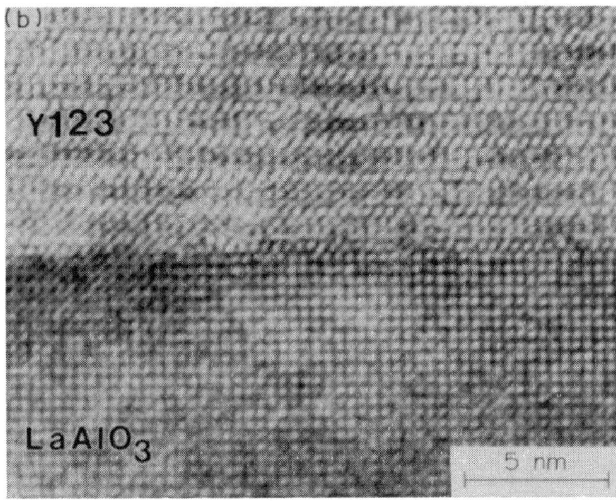

Figure 10
(a) Cross-sectional view of Y123 film grown by laser ablation on (001) $LaAlO_3$, showing the epitaxial grain near to the interface and misoriented grains further away (note the high density of defects in the epitaxial grain);
(b) HREM image of interface region

of satellite reflections do not point in a rational direction in the base structure, nor does their spacing correspond to a commensurate periodicity in the structure. The modulation is therefore called incommensurate. This modulation can also be seen in the HREM image (Fig. 8a) where the crests of the waves can be seen to align parallel to the rows of satellite reflections. The modulation reduces the symmetry from tetragonal to monoclinic. Similar types of modulation are observed in Bi-2212 and Bi-2223. Here the modulations are along the [010] direction with a wavelength of 4.8 times the b parameter. The origin of the modulation is thought to reside in the bismuth–oxygen double layer which locally has the rock salt structure and does not match well with the underlying perovskite structure. The wavelength of the modulation can be changed by substitutions

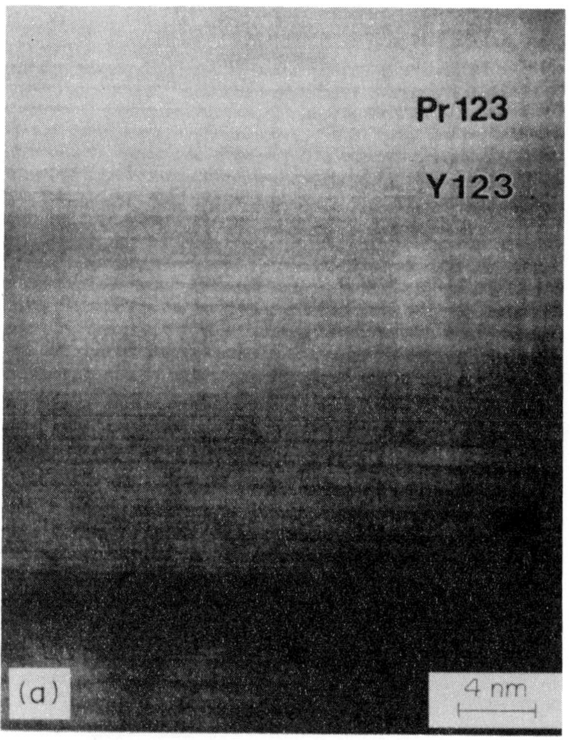

Figure 11
Z-contrast image of a Y123–Pr123 superlattice showing (a) wavy compositional modulation and (b) compositionally abrupt, stepped interfaces at higher magnification (courtesy of S. J. Pennycook)

(e.g., lead substituting for bismuth increases the wavelength while substituting yttrium for calcium decreases it). Modulations are also found in the isostructural thallium compounds but the satellite reflections have a weaker intensity and longer wavelength.

Microsyntactic intergrowths are common, especially in the thallium compounds. Variations in stoichiometry are easily accommodated by variations in stacking along the *c* axis. For example, to produce the Tl-2223 structure from Tl-2212 it is only necessary to insert extra layers of CaO and CuO and the only difference between Tl-1223 and Tl-2223 is that the thallium–oxygen single layer is replaced by a thallium–oxygen double layer. An example is shown in Fig. 9. Figure 9a is a lower magnification HREM image of $Tl_{2-x}Ba_2Ca_2Cu_3O_{10}$, showing a long period modulation with a 21 nm periodicity. Figure 9b is a [001] diffraction pattern showing spots corresponding to the various periodicities in the structure; the inset enlargement shows the large number of finely spaced spots from the 21 nm periodicity. Figure 9c shows a higher magnification HREM image of the 21 nm periodicity; the single and double bars indicate single and double thallium–oxygen layers, respectively, while 3 and 4 indicate *n* in the general formula $Tl_mBa_2Ca_{n-1}Cu_nO_{2+m+2n}$. The structure therefore consists of ordered syntactic intergrowths of the 1223, 1234, 2223 and 2234 structures.

4. *Thin Films*

The highest critical current densities have been found in epitaxially grown thin films of Y123. These films are highly textured with their *c* axis perpendicular to the substrate. Their high J_c values are due to the absence of "bad defects" (grain boundaries) and the presence of "good defects" which pin the vortices (see later). Films have been grown on a variety of substrates, particularly single crystals based on the perovskite structure, such as $SrTiO_3$, $LaGaO_3$ and $LaAlO_3$. A wide range of physical and chemical deposition techniques have been successful, including both *in situ* techniques (requiring no subsequent annealing) and *post situ* techniques (which do require heat treatment after deposition) (see *Oxide Superconductors: Thin-Film Deposition*).

TEM examination of cross-sectional specimens is valuable for understanding epitaxy and growth morphology. An example is shown in Fig. 10 for Y123 deposited by laser ablation onto (001) $LaAlO_3$. The cubic perovskite parameter of $LaAlO_3$ is 0.379 nm which is close to the *a* and *b* parameters of Y123. ($LaAlO_3$ actually has a small rhombohedral distortion but it is cubic at the 750 °C deposition temperature where Y123 is tetragonal and where the epitaxy is established.) Figure 10a shows that, for the first 500 nm or so, the Y123 grows with its *c* axis parallel to the [001] substrate normal. The epitaxy at the interface is shown in the HREM image in Fig. 10b. Thereafter, grains tend to nucleate with the *c* axis flipped through 90° or with random orientations. This may be due to the misfit strain between $LaAlO_3$ and Y123. However, the absence of high-angle grain boundaries in the epitaxial layer close to the substrate is the reason for the high J_c values. This region also shows a high density of defects (Fig. 10a) particularly planar faults and associated partial dislocations. It is likely that these defects (and perhaps the interface itself) may increase J_c by pinning the vortices.

Thin-film superlattices involving the superconductors have also been grown successfully, for example Y123–Pr123 where Pr123 is insulating. Atomically sharp interfaces are essential in superlattices and this can usefully be investigated using the *Z*-contrast technique described in Sect. 1. A *Z*-contrast image of such a superlattice grown by laser ablation is shown in Fig. 11. The praseodymium layers are heavier and therefore brighter than the yttrium layers. The compositional modulation is wavy (Fig. 11a) but at higher magnifications is seen to be atomically abrupt with the waviness being accommodated by abrupt steps in the interface (Fig. 11b).

See also: Oxide Superconductors: Physical Properties; Oxide Superconductors: Solid-State Chemistry; Superconducting Materials: An Overview

Bibliography

Busek P R, Cowley J M, Eyring L 1988 *High-Resolution Electron Microscopy and Associated Techniques.* Oxford University Press, Oxford

Eibl O 1990 Crystal structure of $(Bi,Pb)_2Sr_2Ca_{n-1}Cu_nO_{4+2n+\delta}$ high-T_c superconductors. *Physica C* 168: 215–38

Hiraga K, Shindo D, Hirabayashi M, Kikuchi M, Kobayashi N, Syono Y 1988 Crystal structures of Tl–Ba–Ca–Cu–O superconducting phases studied by high resolution electron microscopy. *Jpn. J. Appl. Phys.* 27: 1848–51

Iijima S, Ichihashi T, Shimakawa Y, Manako T, Kubo Y 1988 Intergrowth structures in superconductor Tl–Ba–Ca–Cu–O oxides. *Jpn. J. Appl. Phys.* 27: 1054–7

Joy D C, Romig A D Jr, Goldstein J I 1986 *Principles of Analytical Electron Microscopy.* Plenum, New York

Matsui Y, Maeda H, Tanaka Y, Horiuchi S 1988 Possible model of the modulated structure in high-T_c superconductors in a Bi–Sr–Ca–Cu–O system revealed by high-resolution electron microscopy. *Jpn. J. Appl. Phys.* 27: L372–L375

Mitchell T E, Davis P K 1988 Electron microscopy of ceramic superconductors. *J. Electron Microsc. Technol.* 8: 245

Pennycook S J, Jesson D E 1991 High resolution *Z*-contrast imaging of crystals. *Ultramicroscopy* 37: 14–38

Ramesh R, Hwang D M, Barner J B, Nazar L, Ravi T S, Inam A, Dutta B, Wu X D, Venkatesan T 1990 Defect structure of laser deposited Y–Ba–Cu–O thin films on single crystal MgO substrate. *J. Mater. Res.* 5: 704–16

Raveau B, Michel C, Hervieu M, Groult D, Provost J 1990 What about structure and non-stoichiometry in superconductive layered cuprates. *J. Solid State Chem.* 85: 181–201

Van Tandeloo G, Broddin D, Zandbergen H W, Amelinckx S 1990 Detwinning mechanism, twinning dislocations and planar defects in $YBa_2Cu_3O_{7-\delta}$. *Physica C* 167: 627–39

Yvon K, François M 1989 Crystal structures of high-T_c oxides: the years 1987 and 1988. *Z. Phys. B* 76: 413–44

Zandbergen H W, Gronsky R, Thomas G 1988 On the structure of $(CuO)_2$ double layers in superconducting $YBa_2Cu_3O_7$. *Nature* 331: 596–9

Zandbergen H W, Gronsky R, Van Tendeloo G 1989 Atomic structure of grain boundaries and surfaces in $YBa_2Cu_3O_{7-\delta}$ and $LaBaCaCu_3O_{7-\delta}$. *J. Supercond.* 3: 103–11

T. E. Mitchell
[Los Alamos National Laboratory, Los Alamos, New Mexico, USA]

R. Gronsky
[University of California, Berkeley, California, USA]

G. Van Tendeloo
[Universiteit Antwerpen, Antwerp, Belgium]

Oxide Superconductors: Physical Properties

Since the late 1980s, few areas of condensed matter physics have received as much research attention as oxide superconductivity. These efforts have resulted in unprecedented improvements in the physical properties of these materials. Superconducting transition temperatures as high as 125 K and critical current densities in the 10^{11} A m^{-2} range have been achieved. These remarkable developments provide the first real opportunity for widespread application of superconductivity (Clark 1988, Geballe and Hulm 1988).

The superconducting oxides belong to a large group of compounds which contain a chalcogenide or an element of row VI of the periodic table. A transition series element such as copper (Vonsovsky et al. 1982) is also frequently present. They may be classified by composition into two broad categories: the copper oxides (Poole et al. 1988) and the noncopper oxides. The noncopper spinels, $LiTiO_x$ oxides, and the perovskites $BaPbBiO_x$ were the first oxide superconductors to be discovered. The early materials were remarkable because not only were they superconducting, but they also had among their number the highest extant transition temperatures. For example, the superconducting transition temperature T_c is as high as 13.7 K in $LiTi_2O_4$ and 13 K in $BaPb_{1-x}Bi_xO_6$. Among the noncopper oxides the highest T_c, in the 30 K range, is found in potassium-substituted BaBiO, that is $(Ba_{1-x}K_x)BiO_3$.

Most of the highest-T_c materials are based on copper oxides. These form relatively stable compounds particularly when compared with the metastable phases of the technologically well known A15 superconductors (see *A15 Compound Superconductors*). The copper oxides are derived from highly correlated single-valence Mott insulators with composition between the insulating and metallic range. The first copper oxide superconductors were derived from the lanthanate La_2CuO_4, often known as "214" (Bednorz and Müller 1986). Pure 214 is an electrical insulator, but it may be doped to create free charge carriers and give rise to electrical conduction when about 10% of the trivalent lanthanum atoms are substituted by divalent barium, calcium or strontium. These doped compounds can superconduct with T_c up to about 40 K. Another copper oxide, $Tl_2Ba_2Ca_2Cu_3O_{10}$, is the highest-temperature (125 K) superconductor (Sheng and Hermann 1988). Many others have values of T_c between 80 K and 125 K (Cava 1990). For this reason, the copper oxide materials are known as high-T_c, high-temperature superconductors or HTSs. All the classic properties of superconductivity (Parks 1969, Rose-Innes and Rhoderick 1978) are evident in the oxides. As in the conventional low-temperature superconductors (LTSs), the clarity of these features is influenced by sample quality and preparation conditions (Roberts 1978, University of South Carolina 1991).

One parameter of particular importance for the high-T_c oxides is the superconducting coherence length ξ (Batlogg 1989, Tinkham and Lobb 1989, Kresin and Wolf 1990), which is a measure of the distance over which the superconducting order remains the same. It is also the spatial extent of a charge carrier pair in the condensed or superconducting state. Within a region of size ξ in space, there is phase coherence of the superconducting state similar to that of a de Broglie wave packet of a quantum particle. In terms of the average carrier separation $n^{-\frac{1}{3}}$, where n is the number of free charges in a unit volume, ξ may be obtained as follows:

$$\xi \sim (E_F/k_B T_c)n^{-1/3} \\ \sim n^{1/3}(h^2/m^* k_B T_c) \quad (1)$$

where E_F is the Fermi energy, k_B is Boltzman's constant and h is Planck's constant. In the LTSs, ξ is about 1 μm and at any instant the center-of-mass coordinates of about a million carriers lie within the coherence sphere, which is a sphere of radius ξ. In the high-T_c materials, small n, large $m^* \sim 10 m_e$ and high transition temperature conspire to severely reduce ξ to the order of 1 nm; so in these oxide materials the coherence sphere contains about one pair of carriers. Due to the Heisenberg uncertainty principle, such localization in the space coordinates renders the momentum of the pair uncertain and causes the charges to be more like pairs in physical or real space; which raises the question, how can so few particles produce the strong pairs that are necessary for high T_c and what is the correct theoretical description?

The short coherence length causes high-T_c superconductivity to be more susceptible to defects and impurities. It also encourages fluctuation effects which enormously broaden the critical region of the superconducting transition.

Anisotropy, or the dependence of physical properties on crystal direction, is another feature common to oxides. Fluctuations are further assisted by two-dimensional anisotropy. Strong fluctuations, both above and below T_c, can be observed in many physical properties of these materials (Salamon 1989, Salamon et al. 1990).

1. Crystal and Electronic Structure

The highest-T_c oxides are often related structurally to the perovskites. In simple perovskites, the chemical formula of which is ABO_3, the metals A and B, and the oxygen atoms are arranged in a cubic layer sequence. The crystal structures of the high-T_c oxides are complex variations of the layering scheme (see *Oxide Superconductors: Solid-State Chemistry*; Poole et al. 1988, Yvon and Francois 1989, Hazen 1990). Three typical examples of these structures will now be described.

In the highest-T_c materials, as shown in Fig. 1a, $2n+1$ atomic planes comprising $n+1$ sheets of CuO_2 plus n intercalating sheets of CaO, are sandwiched by several boundary layers of TlO (or BiO) and BaO (or SrO). These boundary layers which effectively separate the CuO regions are common to most materials and cause many of the physical properties of these solids to be anistropic, as already mentioned.

The K_2NiF_4 type, or the T structure, of $La_{2-x}M_x$-CuO_{4-y} (M is Ba or Sr) can be seen in Fig. 1b. This structure has just a single CuO_2 plane. The boundary layers contain lanthanum, M and oxygen; the dopant alkaline-earth element M goes substitutionally into these boundary layers without forming distinct oxide planes of its own.

The so-called "123" compounds $RBa_2Cu_3O_{7-\delta}$, where R is Y, La, Nd, Sm, Eu, Gd, Ho, Er, Tm, Yb or Lu have BaO planes, not calcium planes between the three copper–oxygen sheets. The central sheet comprises partially filled chains of CuO orientated along the crystalline b axis. These chains contribute to electrical conduction and cause the lattice constant a

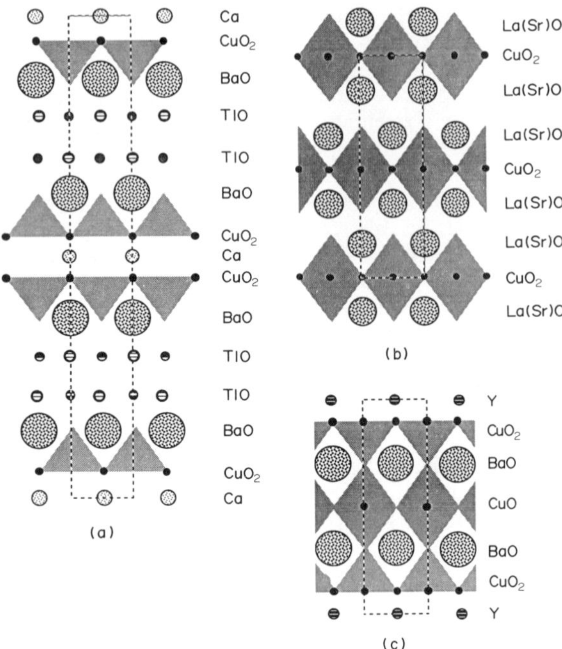

Figure 1
(a) Diagram of the atomic arrangement in the highest-T_c superconductors, such as the TlBaCaCuO- or BiSrCaCuO-based copper oxides. The structure shown is that of $Tl_2Ba_2CaCu_2O_8$. The unit cell marked by the dashed box contains two copper–oxygen layers (CuO_2) separated by the calcium ions. (b) Atomic layer scheme of the K_2NiF_4 or 214 structure of the lanthanum-based (LaBaSrCuO) oxides. (c) Schematic of the 123-oxide structure. Diagram is for Y123, but is true for any R123 oxide. The small Y^{3+} ions effectively isolate the unit cell in the *c*-axis direction

to be slightly shorter than *b* and produce an orthorhombic distortion. Finally, only R–O planes form the boundary layers. A diagram of the atomic planes observed in this structure is shown in Fig. 1c.

Hundreds of other superconducting compounds have been synthesized by substitution and hybridization of the structures described. However, the basic atomic arrangements remain similar. Table 1 provides a compilation of the ideal unit cell data for several oxide superconductors. The real materials are generally oxygen deficient which makes the structure less perfect. For all of these materials, crystal structure is of low symmetry. The highest-T_c conventional superconductors, on the other hand, tend to have high symmetry.

A distinctive feature of the high-T_c oxides is the planar arrangement of the atomic layers of CuO_2. Often these are derived from the CuO_6 octahedra of perovskites. Cu^{2+} is a well-known Jahn–Teller ion; the Jahn–Teller effect distorts the octahedra by stretching the axial copper–oxygen bonds to 0.24 nm and reducing the other four bonds to 0.19 nm, creating the square-planar geometry of the copper with the four nearest oxygens. The atoms form a nearly two-dimensional square grid. These layers play a central role in the high-T_c oxides. The charge carriers responsible for the electronic properties reside in these planes.

The copper oxides form layered structures by Jahn–Teller-type distortion because of the orientation of the electronic orbitals in O^{2-} and Cu^{2+}. Oxygen has four external electrons in the $2p$ shell. It accepts two more to complete the six states of $2p$ required by O^{2-}. Copper has 11 external electrons, one in the $4s$ shell and a full $3d$ shell with ten electrons; as Cu^{2+}, copper loses the single $4s$ electron and another one from the $d_{x^2-y^2}$ orbital, leaving nine electrons or a single hole in the full $3d$ shell. The two-lobed, figure-of-eight-shaped $2d$ orbital of oxygen, overlaps with the four-lobed, clover-leaf pattern $3d$ (hole) copper orbital, and forms a two-dimensional copper–oxygen network in the horizontal direction. A half-filled band arises from this antibonding hybridization of the oxygen $2p_x$ and $2p_y$ (σ) with the copper x^2-y^2 orbitals (Kitazawa et al. 1988). Electronically, the planar copper–oxygen arrangement produces a two-dimensional band structure. For this half-filled case, the Fermi surface will be roughly the diagonal boundary of the square-shaped region with rounded corners, inscribed by the lines joining the reciprocal space k points (π/a, 0) and (0, π/a).

The electronic band structure of a real oxide superconductor is much more complicated (Pickett 1989). In such materials states at the Fermi surface may arise from the oxide bonds of more than one kind of element. An example of the energy bands closest to the Fermi level E_F in $Tl_2Ba_2CaCu_2O_8$ along several high-symmetry directions is shown in Fig. 2. The inset indicates these directions on the reduced Brillouin zone for the body-centered tetragonal structure of space group I4/mmm appropriate for this material (Table 1). The copper–oxygen bands not only cross the E_F but stay close to it and thereby define the Fermi surface over large patches. The region XPG_3 due to partially empty copper–oxygen bands creates a hole pocket (Fig. 2). In the L–Z direction the Fermi surface is due to a partially filled thallium–oxygen band which creates an electron pocket. A feature frequently observed in the layered oxides is also visible in Fig. 2; this is the small dispersion or curvature of the bands along L–Z which causes the Fermi surface to be flat in the k_z direction. This is due to the nearly two-dimensional nature of the metal oxide bonds in these materials. Although the conduction properties are determined by the carriers in the broad bands, plus those in the pockets, all of which are sensitive to deviations from ideal atomic positions and stoichiometry, on the whole the electronic band structure adequately describes the behavior of many

Table 1
Oxide superconductor crystal structures

Composition	Space group	*a* (nm)	*b* (nm)	*c* (nm)	$\alpha = \beta = \gamma$
$Ba_{0.71}K_{0.29}BiO_3$	Pm$\bar{3}$m	0.42932	0.42932	0.42932	90°
$Bi_{1.88}Sr_{2.08}Ca_{0.97}Cu_{2.18}O_{7.98}$	A2aa	0.54095	0.54202	3.09297	90°
$Bi_2Sr_2CuO_6$	Amaa	0.5362	0.5374	2.4622	90°
$La_{1.85}Sr_{0.15}CuO_4$	I4/mmm	0.37793	0.37793	1.322660	90°
$Nd_{1.32}Ce_{0.27}Sr_{0.41}CuO_{3.93}$	P4/nmn	0.38563	0.38563	1.24842	90°
$Nd_{1.85}Ce_{0.15}CuO_4$	I4/mmm	0.395	0.395	1.207	90°
$Pb_2Sr_{2.24}Nd_{0.76}Cu_3O_8$	Cmmm	0.5435	0.5463	1.5817	90°
$Tl_{1.1}Ba_{1.88}Ca_{1.90}Cu_3O_9$	P4/mmm	0.3853	0.3853	1.5913	90°
$Tl_{1.17}Ba_2NCa_{0.83}Cu_2O_{6.75}$	P4/mmm	0.38566	0.38566	1.2754	90°
$Tl_{1.84}Ba_{2.1}CuO_{5.70}$	A2aa	0.54684	0.54727	2.32382	90°
$Tl_{1.94}Ba_{1.96}Ca_{1.07}Cu_{1.96}O_{8.21}$	14/mmm	0.38558	0.38558	2.92596	90°
$Tl_2Ba_2Ca_2Cu_3O_{10}$	I4/mmm	0.38503	0.38503	3.588	90°
$Y_2Ba_4Cu_7O_{14.6}$	Ammm	0.3851	0.3869	5.029	90°
$YBa_2Cu_{2.86}O_{6.62}$	P4/mmm	0.3869	0.3869	1.1723	90°
$YBa_2Cu_3O_{6.91}$	Pmmm	0.38177	0.38836	1.16827	90°
$YBa_2Cu_4O_8$	Ammm	0.38411	0.38718	2.7240	90°

real oxide superconductors. However, such a one-electron band description may fail because it ignores electron–electron correlation; for example, it predicts even the undoped compound La_2CuO_4 to be a metal.

The failure of the one-electron approximation to explain why La_2CuO_4 is an insulator underscores the importance of electronic correlation. Correlation restricts double occupancy of electrons; the electrons stay away from positions which are occupied. In single-valence solids, there is one electron at each atom; all sites are occupied, therefore, the electrons have nowhere to move and the solid is forced to be an insulator. By this reasoning, the CuO compounds are insulators because the electrons in the top half-filled band are strongly correlated. It has been argued that before the more complex superconducting oxides can be understood, the role of correlation on the properties of the relatively simple oxide insulators needs to be explained. In many new theoretical models of high-temperature superconductivity single occupancy and charge immobility are essential features.

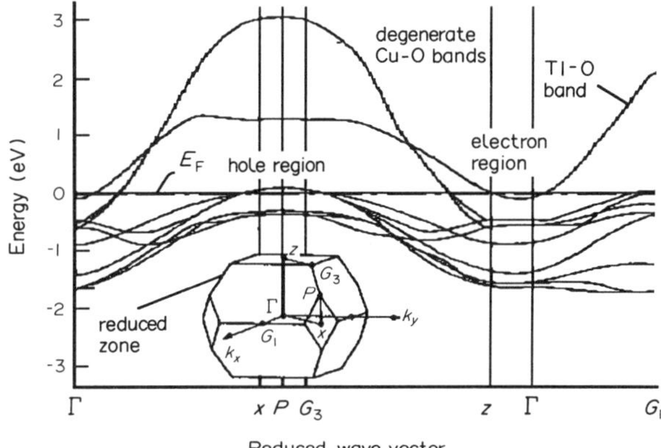

Figure 2
Electronic band structure diagram for a double copper–oxygen layer oxide $Tl_2Ba_2CaCu_2O_8$ (only the states closest to the Fermi level E_F are shown; most of the copper–oxygen states stay close to E_F giving rise to broad energy bands; the antibonding thallium–oxygen states also fall on the Fermi surface and produce an electron region)

2. *Antiferromagnetic Interactions*

Many of the oxides contain localized magnetic moments. For instance, in copper oxides the Cu^{2+} ions possess moments. There may also be moments due to elements such as the magnetic rare earths R. All these moments interact and give rise to interesting magnetic properties. This section will outline some of these features starting with the rare earths.

Low-temperature Mössbauer, neutron diffraction and specific heat (C_p) data for the magnetic rare-earth (R) materials of formula $RBa_2Cu_3O_{7-\delta}$ indicate that the R^{3+} ion possesses nearly theoretical moments. These interact antiferromagnetically and order in a long-range three-dimensional Néel state. The Néel temperature T_N is generally much lower than T_c. The small value (~ 1 K) of T_N implies that the strength of interaction between the R^{3+} moments is weak. Although ordering may take place in spite of the superconducting transition, both crystal field and charge carrier effects on the magnetic ordering can be detected. Generally the specific heat peak is sharper and is at a lower temperature for the superconducting compositions.

The Cu^{2+} ions also show antiferromagnetic instability, but the Cu^{2+} Néel state does not coexist with superconductivity. Quantitatively, when T_N for Cu^{2+} moments are nonzero T_c tends to be zero and vice versa. Furthermore, the interplay between superconductivity and the Cu^{2+} moments is very complex. The correlation effects responsible for poor electronic conduction, discussed earlier, also determine the magnetic properties. In some theoretical models of these correlated oxides the charge carriers are believed to be responsible for both magnetism and superconductivity. The possibility of a magnetic route to high-temperature superconductivity has contributed to the interest in Cu^{2+} magnetism.

An immense amount of experimental data exists relating to the Cu^{2+} ions (Ginsberg 1989). In La_2CuO_4 the copper ion exhibits a moment of roughly 0.5 μ_B, a value much less than the 1.1 μ_B expected for Cu^{2+}. Zero point fluctuation effects are believed to be responsible for this reduction. Below the Néel point (300 K), these moments order antiferromagnetically in three dimensions; the in-plane nearest-neighbor coupling parameter is around 116 meV and the net interlayer coupling is around 2 μeV (Birgeneau and Shirane 1989).

The in-plane interaction is almost five orders of magnitude stronger than the interplane interaction; therefore, spin fluctuations continue in the copper–oxygen planes well above the three-dimensional ordering temperature T_N. Two-dimensional fluctuation is strong with a correlation length of around 40 nm at the Néel point. The spin dynamics follow those of conventional magnetic excitations with spin wave velocity of around 0.085 eV nm. Between room temperature and 5 K, this velocity increases by about 10%. The Cu^{2+} ions in other nonsuperconducting copper oxides, such as oxygen-deficient $YBa_2Cu_3O_{6+x}$, also order antiferromagnetically. In this case, the moment is a little higher at around 0.66 μ_B and for $x = 0$ T_N is around 450 K.

In the conducting or doped samples, T_N is reduced. Neutron scattering, μSR, NMR and other measurements show the copper ions disorder and fluctuate vigorously as free carriers are introduced. The long-range ordered moment as indicated by neutron diffraction also decreases rapidly; however, the local moment as measured by μSR does not change at such a fast rate. As the density of the charge carriers increases, random disorder also increases. The long-range Néel order is destroyed as the magnetic correlation length becomes comparable to the unit cell size. At still higher carrier densities, superconductivity is observed; the NMR and μSR time-averaged moments vanish as the moments fluctuate many times in the timescale of the measurements. Even in the superconducting range ($T_c < 50$ K), two-dimensional correlations exist and the total magnetic scattering cross section of the neutrons remains comparable to that of the insulating composition.

3. Some Normal State Properties

One unusual normal state or nonsuperconducting property of the high-T_c oxides is anisotropic conduction. These compounds are metallic along the crystalline *ab* directions but can be poorly conducting in the perpendicular or *c* direction. Even in the easy directions conductivity is low, or the values of ρ, the resistivity, are moderately high; typically, at room temperature, ρ is in the range 1–10 μΩ m. This rather large ρ is primarily due to the low carrier concentration and large effective mass. In YBaCuO single crystals at room temperature the carrier relaxation rate is similar to that in pure copper.

The resistivity $\rho(T)$ of the high-T_c copper oxides is often a linear function of temperature (see Fig. 3), that is,

$$\rho(T) = \alpha + \beta T \tag{2}$$

The temperature coefficient or slope $\beta = d\rho/dT$, is characteristic of the material; in many cases, β is about 10^{-2} μΩ m K^{-1}. For a set of similar specimens, α is roughly constant. The residual or extrapolated zero-temperature resistance may vary both in sign and in magnitude. In the best samples (i.e., those with the sharpest and highest transition), residual resitivity can be negligibly small. Table 2 provides the values of resistivity for several oxides.

Linearity of resistivity is particularly obvious in materials with a large temperature region over which normal state properties can be examined. For example, in $Bi_2Sr_2CuO_6$, this region is $40\text{ K} < T < 700\text{ K}$. Exceptions to linear resistivity do occur, such as in the low-T_c ($T_c \sim 22$ K) $Nd_{2-x}Ce_xCuO_{4-y}$ and in oxygen-deficient $YBa_2Cu_3O_{6+x}$; in the oxygen-poor material, ρ is not linear, even though T_c may be

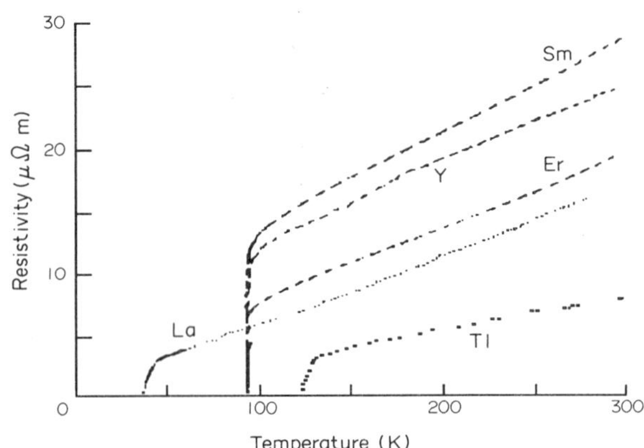

Figure 3
Temperature dependence of resistivity $\rho(T)$ for several ($La_{1.85}Sr_{0.15}CuO_4$, $SmBa_2Cu_3O_{7-x}$, $YBa_2Cu_3O_{7-x}$, $E_2Ba_2Cu_3O_{7-x}$ and $Tl_2Ba_2Ca_2Cu_3O_{10-x}$) bulk oxide materials: even these polycrystalline samples clearly display the linear $\rho(T)$ characteristics of the high-T_c oxides

Table 2
Electrical resistivity

Composition	ρ(300 K) (μΩ m)	$\rho(T)$ (μΩ m)	T (K)
$Ba_{0.6}K_{0.4}BiO_4$	900	112	272
$Ba_2EuCu_3O_x$	7.4	17.5	96
$Bi_2Sr_2CaCu_2O_x$ (*a* axis)	15	8	125
$Bi_2Sr_2CaCu_2O_x$ (*b* axis)	6.5	3	125
$Bi_2Sr_2CaCu_2O_x$ (*c* axis)	1100	650	125
$EuBa_2Cu_3O_{7-x}$	7	4	100
$La_{1.8}Sr_{0.2}CuO_4$	12	16	44
$La_{1.85}Sr_{0.15}CuO_4$	27	8.5	50
$Nd_{1.85}Ce_{0.15}CuO_{4-x}$ (*a,b* axis)	45		
$Nd_{1.85}Ce_{0.15}CuO_{4-x}$ (*c* axis)	750	2000	80
$Tl_2Ba_2Ca_2Cu_3O_x$	7	4.4	200
$YBa_2Cu_3O_{7-x}$ (*a,b* axis)	4.5	2	100
$YBa_2Cu_3O_{7-x}$ (*c* axis)	130	180	100

as high as around 60 K. Also, some alloys and metal oxides such as $ZrTe_3$ and rhenium and niobium oxides show linear $\rho(T)$, but are not high-temperature superconductors.

In metals, resistivity can be separated into an impurity part and an intrinsic part given by the Matthiessen rule:

$$\rho = \rho_{\text{imp}} + \rho'(T) \quad (3)$$

where the impurity contribution ρ_{imp} is assumed to be independent of temperature and is equal to α in Eqn. (2); ρ_{imp} is directly proportional to the impurity concentration and $\rho'(T)$ is the intrinsic resistivity of the pure metal; electron–phonon interaction is the common source of the intrinsic resistance.

For phonon scattering, $\rho'(T)$ follows a nonlinear relationship with temperature, usually expressed by the semiempirical Bloch–Gruneisen formula. This behavior may be split into several different ranges of temperature. In the Matthiessen's phase or the low-temperature region, $\rho'(T)$ tends to zero and the total resistivity becomes close to ρ_{imp} or α, hence ρ is independent of T. The intermediate region, that is, for temperatures up to about a tenth of the Debye temperature Θ_D, $\rho'(T)$ is of the order of T^{α} with α around 5. In the high-temperature region ($T > \Theta_D$), the value of α tends to unity.

The high-T_c materials tend not to show the temperature-independent behavior of the Matthiessen phase. At low temperatures, when impurities are introduced, the HTSs either remain superconducting or become insulating. Linear resistivity over a large temperature range may indicate weak electron–phonon coupling, but as will be pointed out, most other properties tend to indicate considerable interaction between the electrons and phonons. Linear resistivity is also evidence against the presence of electronic excitations with energy $h\Omega_{\text{excit}}$. Because such excitations would give rise to additional scattering causing nonlinearity of $\rho(T)$, at least for temperatures in the $T \simeq h\Omega_{\text{excit}}/k_B$ range (Cohen et al. 1990), a vast amount of theoretical work has been devoted to the linear $\rho(T)$ and its connection to high T_c.

For T above T_c the resistivity in the c direction $\rho_c(T)$ is generally larger than ρ_{ab}, that in the ab plane and ρ_c may even be larger than the localization or Mott limit h/e^2k_F where k_F is the electron wave number at the Fermi surface. In addition, ρ_c shows a power-law-type temperature dependence. In many of the HTSs, ρ_c decreases with increasing temperature, and has an extrapolated divergence at absolute zero; in others, metallic or $d\rho_c/dt > 0$ behavior is observed. The ratio ρ_c/ρ_{ab} or the resistivity anisotropy factor is both material and temperature dependent. In the highly anisotropic materials, like $Bi_2Sr_2CaCu_2O_x$ at room temperature, this ratio can be as high as 10^4 and m_c^*/m_{ab}^* is around 50.

In all the oxides, the highest T_c tends to correlate with an optimal density of charge carrier concentration n. In the low-temperature noncopper oxides, such maximization of T_c with n can be attributed to screening effects. In the copper oxides, there is also an optimal number of copper–oxygen planes which results in the highest T_c. For example, in both the bismuthates and thallates, three copper–oxide planes separated by two calcium layers or the compositions $Bi_2Sr_2Ca_2Cu_3O_{10-x}$ and $Tl_2Ba_2Ca_2Cu_3O_{10-x}$ produce the highest T_c.

4. *Resistive and Calorimetric Transitions*

In a physically perfect and pure LTS material, the width δT_c of the superconducting transition may be a few microkelvins. Due to thermodynamic fluctuations, even for the purest superconductors, δT_c increases as the square root of T_c. By this reasoning, δT_c for a "100 kelvin" material will be around 50 μK. In the purest defect-free well-oxygenated high-T_c single crystal the experimentally observed transition width is about 100 mK. For impure specimens, δT_c can be a substantial fraction of T_c.

In an external magnetic field H, the transition region broadens immensely even for pure high-T_c samples (Palstra et al. 1988, Iye 1989), two examples of which are shown in Fig. 4; as may be seen, the shape of the resistivity temperature curve ρ (H, $T < T_c$) in a magnetic field is material dependent. The onset temperature remains unaffected, but the resistive midpoint is drastically lowered by H. The transition width δT_c, defined as $\delta T_c(H) = T_c - T_c(H)$ is a function of T_c, and generally, $\delta T_c(H)$ increases with the zero field T_c. For low-T_c oxides such as the SmCeCuO materials with T_c in the 20 K range, $\delta T_c(H)$ could be insignificant.

Most LTSs also display broadening and material-sensitive magnetoresponse. There are, however, some

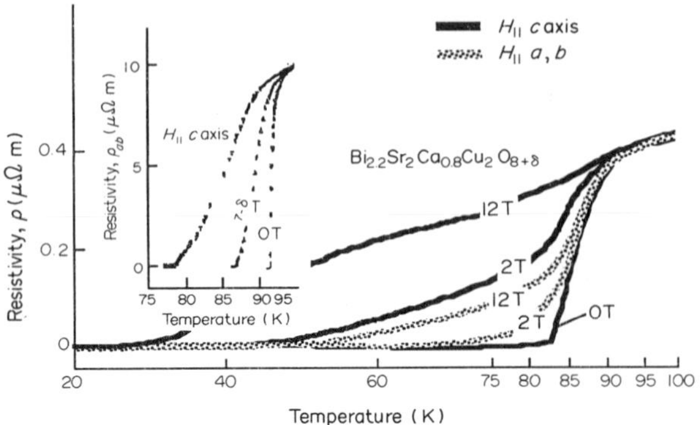

Figure 4
The broadening of the high-T_c resistive transition by external magnetic field for two types of single-crystal oxides: above T_c the magnetoresistance is small; below T_c it is material and orientation dependent as can be seen from the bismuth data; the value of the applied field is indicated

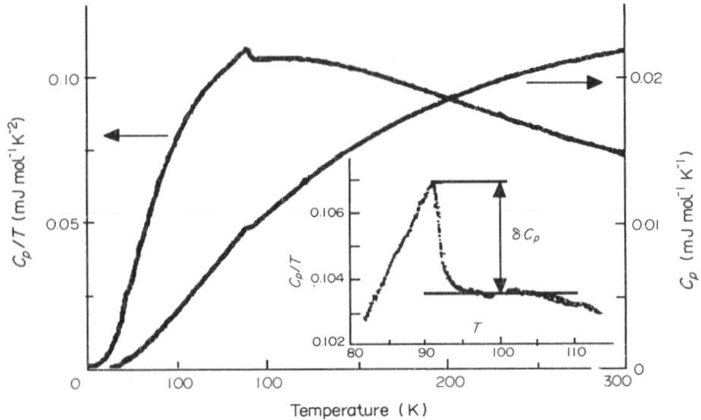

Figure 5
Specific heat C_p, C_p/T and the jump in C_p at the high-T_c transition for $YBa_2Cu_3O_{7-x}$: the inset is a magnified diagram of the jump δC_p

differences: for instance, thermal activation is much less important in conventional materials (Campbell and Evetts 1972); in high-T_c oxides, motion of the magnetic flux lines is readily affected by temperature. Because T_c and, hence, the associated thermal energy, is up to about ten times higher and since ξ the coherence length is extremely short, the pinning energy, or vortex condensation energy, $U \sim \xi^x$ $(2 < x < 3)$ can be small. Even for H less than the upper critical field H_{c2}, high temperatures and a small U encourage energy dissipative flux motion, giving rise to electric resistance and broadening of the transition. The curvature of $\rho(H, T)$ causes experimental values of many important materials parameters such as $T_c(H)$, $J_c(H)$ and H_{c2} to be dependent on operational definitions. Because magnetoresistance measurement is one of the popular methods of determing H_{c2}, $\rho(H, T)$ is a topic of intense research activity (see *Resistive Transition and Flux Flow in Superconducting Materials*).

A jump or anomaly at T_c in the specific heat C_p as a function of temperature is a signature of the second-order phase transition associated with superconductivity (see *Superconducting Materials: Physical Properties*). Measurement of $C_p(T)$ is the most stringent test for bulk superconductivity and sample homogeneity. It is also an indicator of the fraction of specimen volume that is actually superconducting. In most of the oxides, specific heat jumps at T_c have been observed (Junod 1990); for $YBa_2Cu_3O_x$ this behavior is shown in Fig. 5. The magnitude of the C_p jump at T_c $\delta C_p(T_c)$, where $\delta C_p(T_c) = [C_p \text{ (super)} - C_p \text{ (normal)}]$ represents the entropy change due to the transition. Ignoring the phonon contribution in YBaCuO, δC_p is found to be close to the entropy of all the electrons, implying all carriers in the oxides contribute to superconductivity. Phonon renormalization corrections reduce this estimate to about 10% of the total carrier concentration (Salje 1990).

Theoretically, the ratio of the $\delta C_p(T_c)$ jump of the normal-state specific heat at T_c is a measure of the coupling or interaction between the carriers, irrespective of the type of excitation responsible for superconductivity. In the weak coupling limit, the BCS model predicts

$$\delta C_p/C_p \text{ (normal)} = 1.43 \tag{4}$$

Larger values of this ratio are indicative of strong coupling. For high-T_c materials the experimental values of $\delta C_p/C_p$ (normal) are around 2–3.5, that is, in the mildly strong coupling region.

Away from the transition, external magnetic fields do not influence $C_p(T)$ strongly, but just below T_c the behavior is similar to that of the resistive transition (Salamon et al. 1990). This calorimetric broadening has been related to anisotropy, dimensional crossover, criticality, finite size scaling and, ultimately, to the short coherent length. The exact interpretation of both the $\rho(H, T)$ and $C_p(T, H)$ results may be debatable but the similarity between the dissipative response $\rho(T)$ and equilibrium response C_p is striking and may indicate a common origin of both the resistive and calorimetric broadenings. Typical values of some Debye temperatures Θ_D, the electronic specific heat γ and $\Delta C_p/T_c$ for several materials are provided in Table 3.

5. *Diamagnetism of the Superconducting State*

The London penetration depth λ is large in the HTSs, so the Ginzburg–Landau parameter $\kappa = \lambda/\xi$ is large; for example, κ is around 100 in $Sm_{1.85}Ce_{0.15}Cu_{4-y}$ and around 70 in $YBa_2Cu_3O_{7-y}$. Furthermore, the carrier

Table 3
Thermodynamic parameters of some oxide superconductors

Composition	T_c (K)	Θ_D (K)	γ (mJ mol^{-1} K^{-2})	$\Delta C/T_c$
$Ba_{0.6}K_{0.4}BiO_{3+y}$	30	210		1
$BaPb_{0.8}Bi_{0.2}O_3$	7			5
$Bi_2Sr_2Ca_2Cu_3O_x$	105	246	1	37
$Bi_2Sr_2CaCu_2O_x$	90	237	±2	40
$GdBa_2Cu_3O_x$	90	313	6	40
$La_{1.8}Sr_{0.2}CuO_4$	35	396	4	17
$La_{1.85}Ba_{0.15}CuO_{6.7}$	30	405	6	18
$La_{1.85}Ca_{0.15}CuO_4$	22	450	3	2
$La_{1.85}Sr_{0.15}CuO_4$	35	372	1	17
$Nd_{1.85}Ce_{0.15}Cu_3O_{3.98}$	20		50	
$Tl_2Ba_2Ca_2Cu_3O_x$	115	300	60	20
$Tl_2Ba_2CaCu_2O_x$	100	254	6	35
$YBa_2Cu_3O_7$	90	375	4	60

mean free path is also much longer than ξ. These two conditions make the HTSs type-II superconductors in the clean limit, a rather unusual behavior for a poor conductor.

Levitation of a magnet above a flat high-T_c disk is a popular demonstration of superconductivity. Another spectacular effect is the inverted levitation or suspension of a magnet below a superconducting oxide. Both stability and suspension in these materials are evidence for flux trapping and pinning.

Within the interior of the oxides the trapped flux form conventional magnetic vortices. Each of the vortices contains one fluxon or a single unit of a magnetic flux, $\Phi_0 = hc/2e = 2.07 \times 10^{-15}$ Wb. This value of flux quantization indicates charge pairing in the oxides to be the same as in conventional superconductors. Generally, defects and impurities provide pinning sites for the flux vortices. In the oxides because of the short coherence length even smaller-scale irregularities such as oxygen vacancies may act as traps. Twin boundaries in the orthorhombic materials such as $YBa_2Cu_3O_{7-x}$ also trap vortices. When special care is taken to avoid defects, a well-formed flux line lattice can be observed (see *Flux Line Lattice*). Sometimes, peculiarities such as anisotropy in the lattice parameter and approximately oval-shaped vortices are noticeable; they arise from the anisotropy of λ (see *Anisotropy in Superconducting Materials*). Table 4 shows the critical field and related data for a few superconducting oxide single crystals.

Flux pinning produces hysteresis and a wide variety of diamagnetic behavior in all type-II materials including the oxides. At low temperatures, part of the complexity is manifested as a difference in the zero-field-cooled magnetization M_{ZFC} and the field-cooled magnetization M_{FC}. The difference δM between FC and ZFC responses is the net amount of trapped flux. When temperature is raised, δM decreases as the M_{ZFC} and M_{FC} curves converge asymptotically (Malozemoff 1990). The merging point T^* is a common experimental parameter. It represents the demarcation between the high-temperature region of easy flux motion, and the less mobile low-temperature region. At temperatures above T^* ($T^* < T < T_c$), near equilibrium flux arrangements are approached in the timescale of the measurement and M_{FC} and M_{ZFC} are equal within the instrumental resolution of the measurements (Campbell and Evetts 1972). Figure 6 shows M_{FC}, M_{ZFC} δM (marked by shading) and the merging point T^* in a thallium–copper oxide material for two values of applied field. In some cases, even in the reversible region, both time dependence (Datta et al. 1988a) and trapped or remanant moments persist (Malozemoff 1990). The interpretation of electrical resistance is complicated by flux motion; similar concerns also apply to magnetic measurements.

The trajectory of T^* in the H–T plane has been given various names, for example the H_{c2} line, the depinning line and the vortex glass-transition line. The $T^*(H)$ curve may be more accurately described as the irreversibility line (Malozemoff 1989); calculations based on this curve can greatly underestimate H_{c2}. For

Table 4
Coherence length and magnetic properties

Material	Penetration depth λ(0) (nm)	λ_c/λ_{ab}	$\mu_0 H_{c2}$(0 K) (T)	$\mu_0 H_{c2}/\delta T$ (T K^{-1})	ξ(0 K) (nm)
$Bi_2Sr_2CaCu_2O_{8-y}$	300 (*ab* plane)		280 (∥) 32 (⊥)	5(∥) 0.6 (⊥)	~3 (∥) 0.4 (⊥)
$La_{1.9}Sr_{0.1}CuO_{4-y}$	100		50	2	2
$Nd_{1.85}Ce_{0.15}CuO_{4+y}$			5	0.5	
$Sm_{1.85}Ce_{0.15}CuO_{4+y}$	460 (*ab* plane)	4	24 (∥) 7 (⊥)	1.8 (∥) 0.5 (⊥)	6 (∥) 2 (⊥)
$Tl_2Ba_2Ca_2Cu_3O_{10-y}$	200	3	120	1.5	
$YBa_2Cu_3O_{7-y}$	150 (*ab* plane)	6	670 (∥) 120 (⊥)	10 (∥) 3 (⊥)	2 (∥) 0.4 (⊥)

this reason, some of the H_{c2} values provided in Table 4 may be underestimates. In many oxides $(1 - T^*)/T_c$ is a power-law function of H; the inset of Fig. 6 shows such behavior. Here, the irreversibility line is shown as the solid line and is a fit to $(1 - T^*)/T_c \sim H^x$, with $x \sim \frac{2}{3}$. The irreversibility line is one of the important experimental features of the magnetic phase diagram of the oxide superconductors. The interpretation of flux flow resistivity is complicated by flux motion; similar concerns also apply to magnetic measurements.

The dynamics of magnetic flux vortices in the oxides have been extensively studied. Changes towards and away from the Bean-type critical state is nonlinearly diffusive (Beasley et al. 1969). Nonlinearity of the laws of vortex motion give rise to a period doubling route to chaos and magnetic disorder. Such deterministic chaos has been observed in $YBa_2Cu_3O_{7-x}$ (Datta et al. 1988b).

Close to the critical state another characteristic of flux diffusion is the logarithmic time dependence of the magnetic moment, a behavior predicted by the Anderson equation (Anderson and Kim 1964):

$$M(t, T) = A(T) - S(T)\ln(t - \tau), \qquad t > \tau \qquad (5)$$

where the parameters A and S are often field dependent and τ is a fiduciary time (see *Flux Creep Phenomena*). Over a large range of field, temperature and time, the $M(t)$ data for most oxides roughly follow the Anderson behavior. Because of low pinning force and high temperature, flux motion is faster in the high-T_c

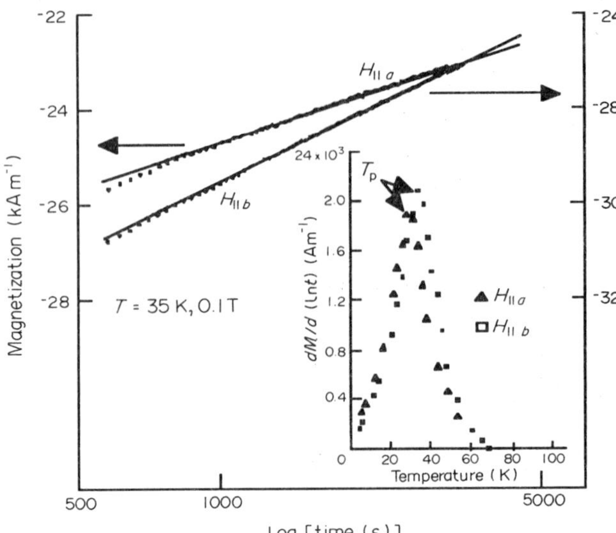

Figure 7
$YBa_2Cu_3O_{7-x}$ logarithmic creep of trapped field into a twin-free $YBa_2Cu_3O_{7-x}$ single crystal: the vortices move faster when they are parallel to the copper–oxygen chains (b axis) as indicated by the larger slope of the magnetization vs ln t (the inset shows the peak of the initial slope)

materials than those measured in the low-temperature superconductors. In single crystals, flux motion is anisotropic. For twin-free $YBa_2Cu_3O_7$, flux vortices are trapped more strongly when the field is oriented along the crystalline b axis (Rice, et al. 1990), this data is shown in Fig. 7. The slope $S(T)$ of the relaxation curve is a metric parameter and is a measure of the pinning strength. $S(T)$ does not increase monotonically with temperature as in the simple flux creep model. Instead, $S(T)$ shows a maximum at T_p. The value of this peak temperature T_p depends on the choice of the time when $S(T)$ is determined. Often, T_p is a small fraction of T_c. Such a peak can be due to two competing factors: the increase in the escape attempt frequency of the flux and the decrease in the superconducting circulation currents as temperature is raised; as a result the net relaxation rate first increases to a maximum and then decreases at higher temperatures.

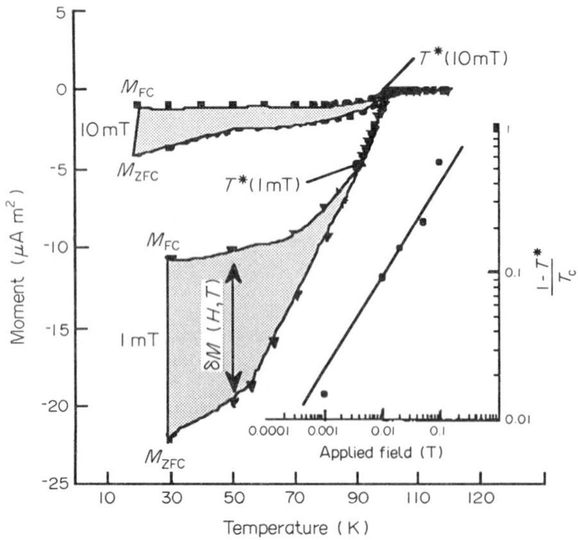

Figure 6
Type-II superconductivity in a high-T_c oxide ceramic specimen of $Tl_2Ba_2CaCu_2O_x$: the amount of trapped flux δM is marked by shading; for $T > T^*$, $M_{ZFC} \simeq M_{FC}$ and T^* rapidly approaches T_c as the field is raised (the inset shows the algebraic behavior of the irreversibility line $T^*(H)$)

6. Electron–Phonon Interaction

The quanta of lattice vibrations referred to as phonons determine many physical properties of solids. Interaction between electrons and phonons is the mechanism for pairing in the BCS model of superconductivity. The classic test for a phonon mechanism is the isotope effect or the decrease of T_c as the isotopic mass M of the superconductor is increased. Mathematically, $T_c \sim \Theta_D \sim (M)^{-\alpha}$; ideally $\alpha = \frac{1}{2}$. In the low-

Table 5
Oxygen isotope effect in oxide superconductors

Material	T_c (K)	^{18}O (conc. % of total O)	α
$Ba_{0.625}K_{0.375}BiO_3$	30	96	0.41
$BaBi_{0.25}Pb_{0.75}O_3$	11	60	0.46
$Bi_2Sr_2Ca_2Cu_3O_{10-y}$	110	98	0.02
$La_{1.85}Sr_{0.15}CuO_4$	21	75	0.54
$Tl_2Ba_2Ca_2Cu_3O_{10-y}$	120		0.6
$Y_{0.5}Pr_{0.5}Cu_3O_7$	31	86	0.49
$Y_{0.6}P_{0.4}Cu_3O_7$	46	86	0.29
$Y_{0.8}Pr_{0.2}Cu_3O_7$	76	85	0.08
$YBa_2Cu_3O_7$	92	90	0.05

temperature oxide superconductors such as BaPbBiO this behavior is confirmed. However, in other materials, the isotope effect is progressively less prominent with increasing T_c.

For example, in YBaCuO with T_c in the 90 K range, isotopic influence of all the four elements has been found to be negligible. Several values of α for oxygen isotope substitution are provided in Table 5. There is an opposite ^{18}O correlation in several oxides between T_c and Θ_D (i.e., lower-T_c oxides generally have higher Θ_D). Also, in comparison with LTSs, T_c values of the oxide superconductors are large with respect to the values of the electronic specific heat or Sommerfield constant γ.

Temperature dependences of elastic properties provide information on the electron–phonon interaction. When a traditional metal becomes superconducting, many of its elastic parameters such as sound, velocity, bulk and shear moduli, soften or decrease in value.

Elastic properties of complex materials such as the A15s (Testardi, 1977) and the oxide superconductors, behave in a complicated manner. In such situations temperature dependence of a macroscopic parameter might not be a clear indicator of superconductivity. But when background trends are included, then even some high-temperature superconductors indicate elastic softening (Datta et al. 1988b). Softening of microscopic vibration of the copper ions can be seen in $Bi_2Sr_2CaCu_2O_8$ (Mook et al. 1990). Figure 8 shows the temperature dependence of both the bulk and microscopic softening for two oxides. It should be noted that premonitory reductions start well above the transition. Softening is indicated by a decrease in the physical properties from those expected (solid curves) in the normal state.

Thermal conductivity is another indicator of the electron–lattice interaction. In most solids $\kappa(T)$ is a measure of the heat energy transported by charge carriers plus that by the phonons. In a normal metal, phonon contribution is negligible, the charge carriers are electrons and the electrons carry most of the heat

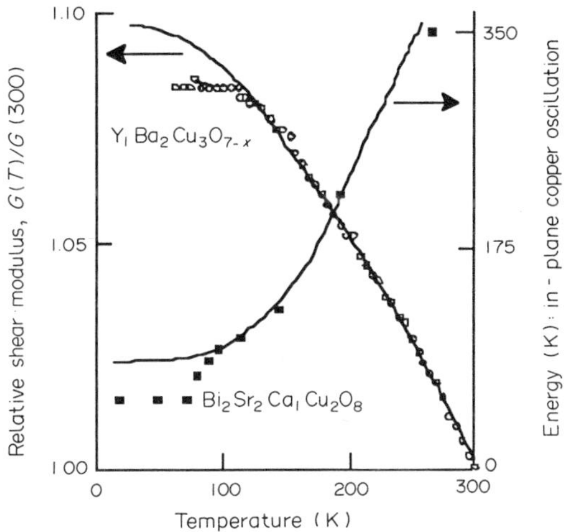

Figure 8
Elastic softening associated with oxide superconductivity: solid curves are extrapolated normal-state behavior; both bulk shear modulus and microscopic elastic energy show a decrease as superconductivity sets in

current; below the superconducting transition, $\kappa(T)$ falls drastically as the charge carrier contribution decreases exponentially. At low temperatures, only the small phonon contribution is left. In contrast, in many HTSs, an increase in $\kappa(T)$ is observed just below T_c (Uher 1990). Thermal conductivity of several HTS materials as a function of temperature is shown in Fig. 9. The increase in $\kappa(T)$ below the superconducting transition for these materials are marked by arrows. In these oxides, as in many alloys, the phonon contribution is comparable with that due to the charge carriers. At the transition, when the charge carriers start to condense out, the phonon mean free path becomes longer and so the net $\kappa(T)$ increases. The characteristic room-temperature values of some bulk elastic and thermal properties for a typical high-T_c oxide $YBa_2Cu_3O_7$ are shown in Table 6.

Pressure dependence of T_c indicates the influence of the lattice, change in the carrier density and role of the phonons on superconductivity (see *Superconducting Materials Under Pressure*). In the oxides pressure tends to increase the transition temperature: $dT_c/dP \simeq 0.3\ K\ kbar^{-1}$ in $La_{1.85}Sr_{0.15}CuO_4$; $dT_c/dP \simeq 0.1\ K\ kbar^{-1}$ in $YBa_2Cu_3O_7$; and $dT_c/dP \simeq 0.6\ K\ kbar^{-1}$ in $YBa_2Cu_4O_8$. The effect is larger in the materials with smaller T_c (Poole et al. 1988, Phillips 1989).

Although the role of phonons in the superconductivity of these materials is not yet clear, the experimental facts described are strong evidence that electrons and phonons do interact and significantly influence many physical properties of the high-T_c oxides.

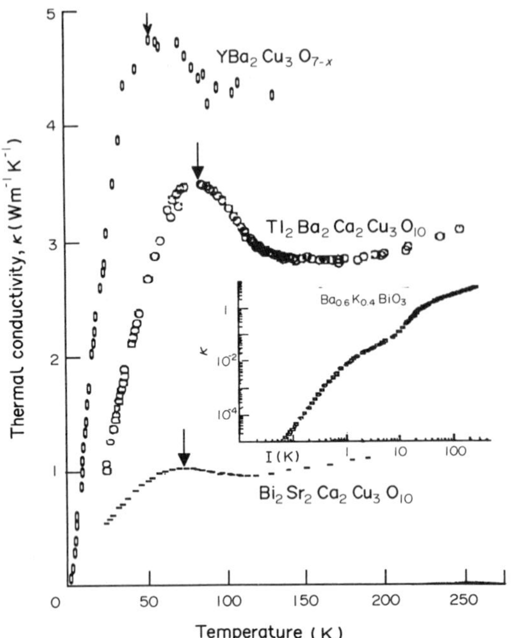

Figure 9
Temperature dependence of thermal conductivity $\kappa(T)$ for several polycrystalline oxides superconductors: in all materials for $T \lesssim 300$ K, κ is small (~ 1 W m^{-1} K^{-1}); the increase in κ following the superconducting transition is marked by arrows

7. *Electron–Hole Symmetry*

Electron counting is a convenient method of estimating band filling and the sign of the charge carriers. By this reckoning, carriers in most cuprate superconductors are expected to be positive, that is hole or *p* type. Many chemical and physical parameters agree with this prediction and indicate a low density of carriers ($\sim 10^{25}$–10^{27} m^{-3}).

Table 6
Bulk, elastic and thermal properties of a common high-T_c oxide ($YBa_2Cu_3O_{7-y}$) at 300 K

Ideal (x-ray) density (kg m^{-3})	6.38×10^3
Thermal conductivity (W m^{-1} K^{-1})	4
Thermal expansivity (K^{-1})	2×10^{-5}
Gruneisen parameter	3
Longitudinal wave speed (km s^{-1})	3.5–5.3
Shear wave speed (km s^{-1})	2–3
Shear modulus (GPa)	20–60
Poisson's ratio	0.26–0.3
Young's modulus (GPa)	50–150
Bulk modulus (GPa)	30–100

In some of the oxides, low carrier density permits electron doping for negative or *n*-type conduction. Such electron-doped materials may be evidence for electron–hole symmetry of high-T_c superconductivity. The T'-phase materials with composition $R_{2-x}Ce_xCuO_{4-y}$ where R is Pr, Nd, Sm or Eu and $R_{1-x}Th_xCuO_{4-y}$ where R is Pr, Nd or Sm, and a few others with copper or oxygen doping are likely examples of electron-type superconductors.

These T' materials show T_c as high as 25 K, which is slightly less than that observed in the T phase $La_{2-x}Ba_xCuO_4$, to which the T' materials are related (Tokura et al. 1989, Maple 1990). Some of the important properties of the T' materials are that they are the only class of copper–oxygen material in which substitution of copper and/or oxygen improves the superconducting properties and where praseodymium and cerium are significant components and, most importantly, they are likely to be electron type. Electron doping follows from a straightforward electron counting argument. However, carrier assignment in these complex solids is not just a matter of valence consideration. The difficulty is that different experimental techniques are sensitive to different aspects of the carriers and can give conflicting results. For example, x-ray absorption spectroscopy shows the expected increase in the negative charge (Cu^+) with doping; but electron energy loss spectroscopy suggests an increase in oxygen 2*p* holes. The Hall coefficient R_H is not a clear guide because in many oxides R_H decreases with increasing temperature (Mott 1990). In $Nd_{2-x}Ce_xCuO_{4-y}$, R_H indicates electron-type behavior between room temperature and T_c; in other T' materials, the sign of R_H changes from negative at room temperature to positive for temperatures below 100 K.

In simple conductors the sign of the Seebeck coefficient $S(T)$ or thermoelectric power is equal to the carrier sign, but in the oxides two specimens of the same material may exhibit opposite signs. Figure 10 shows such conflicting $S(T)$ behavior for two oxides. Two other features visible in Fig. 10 are (a) as in conventional metals, S goes to zero for temperatures below T_c—this indicates that the charges do not carry entropy; and (b) the value of S is small, which is appropriate for high-carrier-density conductors like metals. The sign reversal behavior, however, is similar to other low-carrier-density materials such as semiconductors where even small concentrations of impurities may reverse the sign.

The electron-type materials exhibit flux motion appropriate for weak pinning, but do not show significant broadening of the transition in a magnetic field (Almasan et al. 1991). Another point of departure between *n* and *p* doping concerns pressure effects. For the *n*-type material, pressure reduces T_c which is opposite behavior to that for the *p*-type material. The effect is more pronounced in oxides with a lower T_c such as the *n*-type compounds.

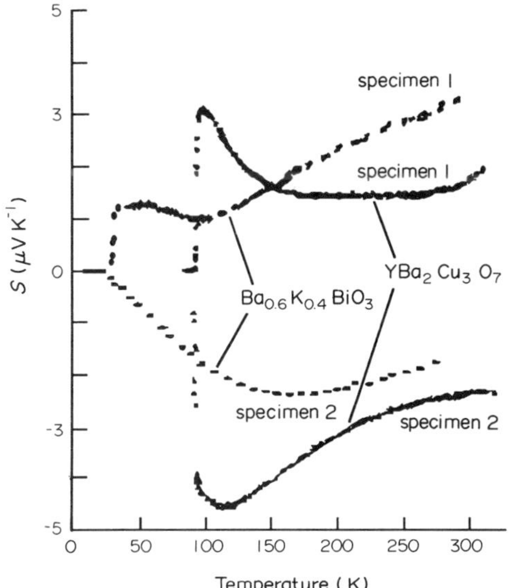

Figure 10
Thermoelectric coefficient S for two types of oxide superconductors: in all cases $S\ (T < 300\ K)$ is small, characteristic of good metals with a lot of carriers, but the sign of S may be opposite from one specimen to another as in low-carrier-density materials

8. *Physical and Chemical Perturbations*

An enormous amount of experimental data on the effects of physical and chemical alterations in the oxides may be found in the literature (Poole et al. 1988, Ginsberg 1989, Phillips 1989).

It is observed that the perovskite-type crystal structure is necessary to reach high values of T_c. Nevertheless, such a structure does not guarantee superconductivity. For example, $PrBa_2Cu_3O_7$ has the 123 structure but is not a superconductor. Strict stoichiometry is not of paramount importance—chemical substitution, within limits, does not destroy superconductivity. Perturbation of oxygen and copper sites produces the greatest changes. In some of the electron-doped materials, partial substitution of both copper and oxygen are required to obtain the highest T_c. For the hole compounds, replacement of copper or oxygen markedly reduces T_c. Both oxygen stoichiometry and the local atomic arrangement or ordering of the oxygen atoms contribute to determine T_c. Atomic ordering is also a key factor for other superconductors such as the A15 materials. In conventional LTSs, magnetic impurities break the superconducting (Cooper) pairs and depress T_c. For low concentrations, the rate of decrease in T_c with paramagnetic impurity of spin S and concentration x is given by the Abrikosov–Gorkov (AG) result:

$$\frac{dT_c}{dx} \simeq -\frac{\pi^2 N(E_F) f^2 (g-1)^2 J(J+1)}{2k_B} \tag{6}$$

where g is the g factor, f is the exchange coupling constant and J is the total angular momentum. Other superconducting properties are also strongly affected by paramagnetism. In the oxides such antagonism between magnetism and superconductivity is not apparent, particularly if the magnetic ion occupies noncopper sites.

In copper oxides, normal state properties may be influenced by paramagnetic impurities, but the principal superconductive parameters such as T_c, J_c and $d(H_{c2})/dT$ are not sensitive to magnetic substitution (Poole et al. 1988, Ginsberg 1989, Phillips 1989). Nickel and zinc are periodic table neighbors of copper; substituting copper by either reduces T_c. Zinc depresses T_c more rapidly than nickel: in $YBa_2Cu_3O_{7-y}$, $dT_c/dx \simeq -12$ K at.%$^{-1}$ for zinc and $dT_c/dx \simeq -3$ K at.%$^{-1}$ for nickel. This is surprising because nickel ions are magnetic and should be more effective in pair breaking than zinc. Furthermore, contrary to the AG formula, dT_c/dx is not independent of concentration, even in the dilute samples.

The Debye temperature Θ_D is also decreased sharply on substitution of zinc for copper, or oxygen removal from $YBa_2Cu_3O_{7-y}$. In the rare-earth (123) materials, the copper–oxygen chain sites complicate substitution studies because substitution can take place on both the chains and the planes. Due to valence requirements, most copper replacement processes also involve unintentional changes of oxygen stoichiometry. Hence, the role of oxygen needs to be included to properly interpret the copper substitution results.

Both dT_c/dx and the overall shape of the T_c versus x curves, depend on sample treatment, such as thermal annealing. It is believed that the sample treatment effects may be due to local ordering of the atoms and the defects. In short-coherence-length materials like the high-T_c oxides, superconductivity is likely to be influenced by intermediate and short-range order.

9. *Spectroscopy*

Optical and Raman spectroscopy are invaluable in studying low-energy properties of solids (Timask and Tanner 1989, Thomsen and Cardona 1990). Qualitatively, optical conductivity $\sigma(\Omega)$ of many oxides is observed to be temperature dependent and is consistent with a Drude-like contribution. The Drude part agrees with the measured dc conductivity $\sigma(0)$; the optical data also indicates a strongly frequency (ω) dependent effective mass m^* and lifetime of the quasiparticles. For $YBa_2Cu_3O_{7-x}$, $m^*/m \sim 1$ at high frequencies ($\omega > 0.35/h$ eV) and $m^*/m \sim 7$ in the low-frequency ($\omega < 2k_BT/h$) range. Figure 11 shows optical reflectivity data for YBaCuO and BiSrCaCuO

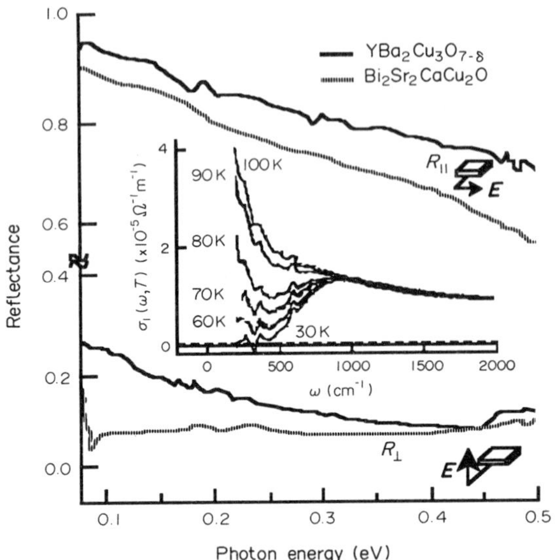

Figure 11
Variation of optical reflectivity with the polarization of electric vector as a function of the photon energy in two types of high-T_c oxides: the opening of the gap in $YBa_2Cu_3O_{7-x}$ may be observed ($\omega \sim 500\ cm^{-1}$) in $\sigma_1(\omega, T)$; the real part of infrared conductivity as the specimen is cooled through T_c (~ 90 K) is shown in the inset

crystals with two electric field polarizations. The inset shows $\sigma_1(\omega, T)$ of $YBa_2Cu_3O_7$ in the infrared region for several temperatures. As a function of temperature, the real part of $\sigma_1(\omega, T)$ shows a gaplike behavior ($T < T_c$). Due to the opening of the gap, σ_1 falls to zero at low ($\sim 500\ cm^{-1}$) frequencies (see also *Optical Properties of Superconductors*).

In the superconducting phase of conventional metals, electronic scattering is suppressed for energies below E_g. This gives rise to a sharp onset and peak in the electronic Raman spectra. In the high-T_c oxides, the Raman signal rises gently with energy; a peak but no clear onset is observed (Cooper and Klein 1990). The peak energy E_p is not sensitive to the sample T_c and may indicate the formation of real-space pairs with energy E at temperatures well above the superconductive condensation.

Below T_c, a BCS superconductor exhibits a "Hebel–Slichter coherence peak" in the spin-lattice relaxation rate of the nuclear magnetic resonance (NMR) spectrum. At low temperatures, this rate is exponential with an activation energy $\Delta = E_g/2$. Neither the coherence peak nor activation has been observed in the NMR spectra of the high-T_c materials (Pennington and Slichter 1990).

Inside a superconductor, an electron with kinetic energy less than the gap energy E_g is unstable. At the interface between a normal metal and a BCS superconductor, a slow electron condenses in momentum space by pairing with a time- and spin-reversed electron from the metal. The resulting Cooper pair in the superconductor moves forward; the empty state or hole goes exactly backwards into the metal. This back flow or the hole current is called Andreev reflection. At voltages below the gap, the net conductance ideally doubles by this process. Andreev reflection corroborates the zero-momentum nature of the pairs and can be utilized to determine E_g. Partial Andreev signals have been observed in several oxides.

Table 7
Superconducting gap and coupling parameters

Material	T_c (K)	Δ (meV)	$2\Delta/K_BT_c$
$Ba_{0.6}K_{0.4}BiO_{3-x}$	25	4	4
$Bi_{1.7}Pb_{0.3}Sr_2CaCu_2O_x$	96	26	6
$La_{1.85}Sr_{0.15}CuO_4$	32	7	5
$Nd_{1.85}Ce_{0.15}CuO_{4-x}$	22	4	4
$Tl_2Ba_2CaCu_2O_x$	100	25	6
$YBa_2Cu_3O_7$	90	25	7

Tunnelling is another technique that has been very successful in the study of conventional LTS superconductors (see *Tunnelling Spectroscopy*). Extremely short coherence length ξ, anisotropy and surface contamination problems make tunnelling measurements difficult in the oxides. However, the point contact method is often effective; this procedure is not orientation selective and the sharp counter electrode can probe below the dead top surface. Point contact measurements of the low-temperature oxides such as $Ba_{1-x}K_xBiO_3$, electron-doped samples like $Nd_{2-x}Ce_x$-CuO_{4-y} and the highest-T_c materials BiPbSrCuO and TlBaCuO provide a wealth of information. Normal state density of states $N_n(E)$, superconducting state density of states $N_s(E)$, gap E_g and Eliashberg function have been obtained from these data. The values of $2\Delta/k_BT_c$, obtained by these measurements, are listed in Table 7; observe that the values tend to be smaller for the lower-T_c oxides and in all cases $2\Delta/k_BT_c$ lies in the mild-to-strong coupling range. In the $YBa_2Cu_3O_x$ materials, the copper–oxygen chains tend to complicate tunnelling results. In many junctions of $YBa_2Cu_3O_x$, a prominent feature at high bias is a V-shaped conductance centered at zero voltage (Gurvitch et al. 1989).

Anisotropic tunnelling requires special samples. Oriented junctions of YBaCuO, with very small contact area, clearly exhibit anisotropy. Figure 12 shows tunnelling behavior for oriented YBaCuO–YBaCuO and Pb–Ag–YBaCuO superconductor–normal–superconductor (SNS) junctions (Lee et al. 1990). In the (001) and (103) oriented junctions, supercurrents are not well defined despite low contact resistance ($\sim 1\ m\Omega$ at 4.2 K). For the (100) orientation, large

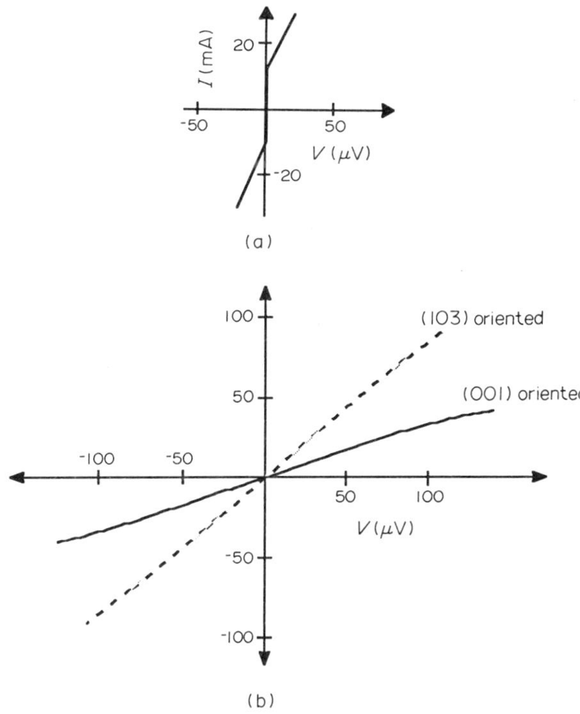

Figure 12
Anisotropic junction behavior of small-area Pb–Ag–YBaCuO contacts: (a) tunnelling clearly takes place along (100) direction but seems to be absent in the other direction (b), even for this low-resistance, small-area SNS junction

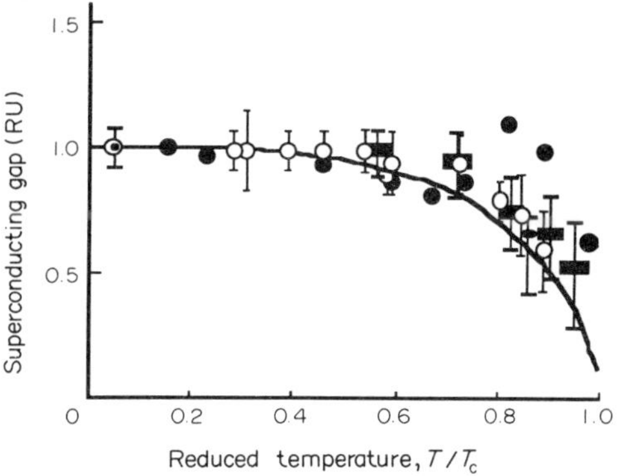

Figure 13
Temperature dependence of high-T_c order parameter: a selection of YBaCuO data obtained by a variety of experimental techniques show overall self-consistency; open circles represent tunnelling, solid dots are Raman and squares are obtained by optical measurements; the solid curve is a guide to the eye; notice the differences at low temperatures—the open circles indicate saturation but the solid dots show some temperature dependence (after University of South Carolina 1991)

supercurrents with very low resistivity of $3 \times 10^{-17}\,\Omega\,m^2$ are observed.

The anisotropy observed in many physical properties is attributable to clean-limit superconductivity and asymmetry of the Fermi surface. In nonconventional superconductivity, the paired state shows lower symmetry than the crystal structure and the Fermi surface. Symmetry of the pairing state can be determined from the low-temperature behavior of the London penetration depth λ, the gap Δ and other superconducting parameters. In the conventional LTS materials, the order parameters saturate exponentially at low temperatures ($T/T_c < 0.3$). For heavy fermion and other nonconventional superconductors with nodes on the Fermi surface, the order parameter $\Delta(T)$ approaches $\Delta(0)$ as a power law in T.

In HTS materials, λ shows a strong coupling behavior at intermediate temperatures. The $\lambda(T)$ data at low temperatures for these materials do not presently preclude a nonconventional behavior. A broad selection of high-T_c gap vs temperature data from different measurement techniques is shown in Fig. 13. On the whole the data corroborate each other but, due to scatter, low-temperature behavior cannot be determined with certainty.

Photoemission spectroscopy (PES) is very effective in determining the Fermi liquid properties of the oxides. For example, this technique reveals that the quantum states at the Fermi level of $Bi_2Sr_2Ca_1Cu_2O_x$ can be decomposed into a 3:7 mixture of $3d$ and $2p$ orbitals. Fermi surfaces of the normal state are clearly seen by PES. A sharp edge similar to the behavior in platinum metal is observed. On the whole, the PES data are consistent with theoretical band structure predictions. When the sample is cooled below T_c, the edge of the density of states $N(E)$ moves to lower energies, as the energy gap E_g of around 30 meV opens. $N(E)$ of the superconducting state shows a peak and a trough at the low-energy side of the edge. This oscillatory peak in the density of states is similar to that in lead. It implies the oxide material, like lead, may also be a strongly coupled superconductor.

The oxide superconductors are a class of low carrier density solids with maximum T_c values in the triple digit range; this makes these materials the highest-temperature superconductors. Along with superconductivity, they exhibit a rich variety of electrical, thermal, structural and magnetic behaviors and may soon find extensive applications.

Bibliography

Almasan C C, Seaman C L, Dalichaouch Y, Maple M B 1991 Irreversibility line and magnetic relaxation in a $Sm_{1.85}Ce_{0.15}CuO_{4-y}$ single crystal. *Physica C* 174: 93–100

Anderson P W, Kim Y B 1964 Hard superconductivity: theory of the motion of Abrikosov flux lines. *Rev. Mod. Phys.* 36: 39–43

Batlogg B 1990 Selected experiments on high T_c cuprates. In: Bedel K, Coffey D, Meltzer D, Pines D, Schrieffer J R (eds.) 1990 *High Temperature Superconductivity: The Los Alamos Symposium*. Addison-Wesley, Reading, MA, pp. 37–82

Beasley M R, Labusch R, Webb W W 1969 Flux creep in type-II superconductors. *Phys. Rev.* 181: 682–700

Bednorz J G, Müller K A 1986 Possible high-T_c superconductivity in the Ba–La–Cu system. *Z. Phys.* 64: 189–93

Birgeneau R J, Shirane G 1989 Neutron scattering studies of structural and magnetic excitation in lamellar copper oxides: a review. In: Ginsberg 1989, pp. 151–212

Campbell A M, Evetts J E 1972 Flux vortices and transport currents in type II superconductors. *Adv. Phys.* 12: 199–421

Cava R J 1990 Superconductors beyond 1–2–3. *Sci. Am.* 261: 42–9

Clark J 1988 Small-scale analog applications of high transition-temperature superconductors. *Nature* 333: 29–35

Cohen M L, Penn D R, Barbee T W 1990 Limitations on electronic mechanisms for high temperature superconducting oxide. *Solid State Commun.* 75: 971–4

Cooper S L, Klein M V 1990 Light scattering studies of the low frequency excitation spectra of high temperature superconductors. *Commun. Solid State Phys.* 15: 99–104

Datta T, Almasan C, Estrada J, Violet C E, Gubser D U, Wolf S A 1988a Magnetic behavior of YBaCuO superconductors prepared under different oxygen anneals. *J. Appl. Phys.* 63: 4204–6

Datta T, Almasan C, Gubser D U, Wolf S A, Toth L E 1988c Period doubling and magnetic chaos in superconducting $YBa_2Cu_3O_7$. *Physica C* 153: 332–3

Datta T, Ledbetter H M, Violet C E, Almasan C, Estrada J 1988b Re-entrant softening in perovskite-like superconductors. *Phys. Rev. B* 37: 7502–5

Geballe T H, Hulm J K 1988 Superconductivity—the state that came in from the cold. *Science* 239: 367–75

Ginsberg D M (ed.) 1989 *Physical Properties of High Temperature Superconductors*, Vol. 1. World Scientific, Singapore

Gurvitch M, Valles J M, Cuculo A M, Dynes R, Garno J P, Scheemeyer L F, Waszczak J V 1989 Reproducible tunneling data on chemically etched single crystals of YBaCuO. *Phys. Rev. Lett.* 63: 1008–11

Hazen R M 1989 Crystal structures of high temperature superconductors. In: Ginsberg 1989, pp. 121–98

Iye Y 1989 Transport studies on high T_c oxides. In: Kamimura H, Oshiyama A (eds.) 1989 *Mechanism of High Temperature Superconductivity*. Springer, Berlin, pp. 263–74

Junod A 1989 Specific heat of high temperature superconductors: a review. In: Ginsberg 1989, pp. 13–120

Kitazawa K, Takagi K, Kishio K, Hasegawa T, Uchisda S, Tajima S, Tanaka S, Fueki K 1988 Electronic properties of cuprate superconductors. *Physica C* 153: 9–14

Kresin V Z, Wolf S A 1990 *Fundamentals of Superconductivity*. Plenum, New York

Lee M, Lew D, Eom C B, Geballe T H, Beasley M R 1990 Anisotropic proximity coupling in small $YBa_2Cu_3O_7$–normal–Pb junctions. *Appl. Phys. Lett.* 57: 1152–4

Malozemoff A P 1989 Macroscopic magnetic properties of high temperature superconductors. In: Ginsberg 1989, pp. 71–150

Malozemoff A P 1990 The magnetic phase diagram of high T_c superconductors. *Mater. Res. Soc. Bull.* 15(6): 50–4

Maple B M 1990 Electron-doped high T_c superconductors. *Mater. Res. Soc. Bull.* 15(6): 60–7

Mook H A, Mostroller M, Harvey J A, Hill N W, Chakourmakos B C, Sales B C 1990 Observations of phonon softening at the superconducting transition in $Bi_2Sr_2CaCuO_8$. *Phys. Rev. Lett.* 65: 2712

Mott N F 1990 The Hall effect above T_c in the superconductor $YBa_2Cu_3O_{7-\delta}$ (YBCO). *Phil. Mag. Lett.* 62: 273–5

Palstra T T M, Batlogg B, Schneemeyer L, Waszczak J V 1988 Thermally activated dissipation in BiSrCaCuO. *Phys. Rev. Lett.* 61: 1662–5

Parks R D (ed.) 1969 *Superconductivity*, Vols. 1, 2. Dekker, New York

Pennington C H, Slichter C P 1989 Nuclear resonance studies of YBaCuO. In Ginsberg: 1989, pp. 269–367

Phillips J C 1989 *Physics of High-T_c Superconductors*. Academic Press, New York

Pickett W E 1989 Electronic structure of the high-temperature oxide superconductors. Rev. *Mod. Phys.* 61: 433–512

Poole C P, Datta T, Farach H A 1988 *Copper Oxide Superconductors*. Wiley, New York

Rice J P, Ginsberg D M, Rabin M W, Vandervoot K G, Crabtree G W, Claus H 1990 Anistropic magnetic relaxation, hysteresis and Meissner fraction in untwinned single-crystal YBaCuO produced without applying stress. *Phys. Rev. B* 41: 6532–40

Roberts B W 1978 Properties of selected superconductive materials, supplement. National Bureau of Standards Technical Note 983. US Government Printing Office, Washington, DC

Rose–Innes A C, Rhoderick E H 1978 *Introduction to Superconductivity*, 2nd edn. Pergamon, Oxford

Salamon M B 1989 Thermodynamic properties: fluctuations and anisotropy of high temperature superconductors. In: Ginsberg 1989, pp. 39–70

Salamon M B, Inderhees S E, Rice J P, Ginsberg D M 1990 Heat capacity of untwinned $YBa_2Cu_3O_{7-\delta}$ in magnetic fields: dimensional crossover near T_c. *Physica A* 168: 283–90

Salje 1990 Phonon renormalization, specific-heat anomaly and why some aspects of mean-field theory can be relevant for high-T_c superconductivity. *Phil. Mag. Lett.* 62: 277–81

Sheng Z Z, Hermann A M 1988 Superconductivity in the rare-earth-free system above liquid-nitrogen temperature. *Nature* 332: 55–6

Testardi L R 1977 Structural instability of A-15 superconductors. *Phys. Acoust.* 13: 29–47

Thomsen C, Cardona M 1989 Raman Scattering in high-T_c superconductors. In: Ginsberg 1989, pp. 409–507

Timusk T, Tanner D B 1989 Infared properties of high-T_c superconductors. In: Ginsberg 1989, pp. 339–407

Tinkham M, Lobb C J 1989 Physical properties of the new superconductors. *Solid State Phys.* 42: 91–132

Tokura Y, Tagaki H, Uchida S 1989 A superconducting oxide compound with electrons as the charge carriers. *Nature* 337: 345–7

Uher C 1990 Thermal conductivity of high-T_c superconductivity. *J. Supercond.* 3: 337–89

University of South Carolina 1991 *Superconductor Data Archive*. Institute of Superconductivity, University of South Carolina, Columbia, SC

Vonsovsky S V, Izyumov Y A, Kurmaev E Z 1982 *Superconductivity of Transition Metals*. Springer, Berlin

Yvon K, Francois M 1989 Crystal structure of high-T_c oxides. *Z. Phys. B* 76: 413–44

T. Datta
[University of South Carolina, Columbia, South Carolina, USA]

Oxide Superconductors: Solid-State Chemistry

The oxide superconductors with highest superconducting transition temperature T_c are all based on CuO_2 sheets, and there are now a great many phases in this family. Another class of high-temperature superconductors can be represented as (Ba, A)(Bi, Pb)O_3. Although the upper T_c in this cubic class is presently only about 35 K, this class is potentially preferred for certain applications. There are also many other oxide superconductors with lower T_c.

Pervasive interrelated themes of the oxide superconductors are defects, disorder, inhomogeneity and thermodynamic instability (Sleight 1991). These interrelationships are not well understood because there is no consensus on a theory for high-temperature superconductivity. Nonetheless, the thermodynamic instabilities create big challenges for synthesis. The large concentration of defects greatly complicate attempts at complete structural definition, and the inherent inhomogeneities cause confusion in the interpretation of physical property measurements.

1. Superconductivity Below 15 K

1.1 NbO and TiO

Superconductivity was discovered in oxides in the early 1960s. These were oxides based on titanium and niobium (see Table 1). The structures of NbO and TiO are closely related to the sodium chloride structure. However, NbO and TiO have both cation and anion vacancies with respect to the NaCl structure. In the case of NbO, these "defects" are well ordered yielding a compound with a rather small range of composition. The coordination of both niobium and oxygen is fourfold, square planar. Despite the ordered structure, the oxygen–niobium ratio for NbO varies from about 0.96 to 1.02. The highest T_c is obtained close to the 1.0 ratio. In the case of TiO, the concentrations of cation and anion vacancies are much more variable. Thus, this material exhibits a considerable range of composition with the oxygen–titanium ratio varying from about 0.8 to 1.3. There is, however, only a small change in T_c over this range.

Table 1

Oxide superconductors with T_c below 15 K (those with an oxygen-to-metal ratio less than one are not listed; superconductivity discovered between 1964 and 1990)

Compound	T_c (K)	Compound	T_c (K)
NbO	1.4–1.6	$Cs_{0.30}WO_3$(h)	4.8
TiO	0.7–1.1	$Li_{0.30}WO_3$(h)	2.2
$SrTiO_{3-x}$	0–0.5	Na_xWO_3(h)	5.4
$Na_{0.3}WO_3$(t)	0.5	$(NH_4)_{0.33}WO_3$	3.2
$Na_{0.2}WO_3$(t)	3.0	K_xMoO_3(t)	4.2
$Ca_{0.1}WO_3$(h)	3.4	K_xReO_3(h)	3.6
$Sr_{0.08}WO_3$(h)	4.0	$Li_{0.9}Mo_6O_{17}$	~2
$Ba_{0.14}WO_3$(h)	2.2	$Li_{0.45}NbO_2$	5.5
$In_{0.11}WO_3$(h)	2.8	Ag_7O_8X	1.4
$Tl_{0.30}WO_3$(h)	2.1	$LiTi_2O_4$	13.7
$K_{0.30}WO_3$(h)	5.7	$BaBi_{1-x}Pb_xO_3$	13
$Rb_{0.30}WO_3$(h)	6.6		

t Tetragonal h Hexagonal

1.2 $SrTiO_3$-Based Superconductors

The first report of superconductivity in the perovskite structure was for phases based on $SrTiO_3$ (Table 1). Stoichiometric $SrTiO_3$ is, of course, an insulator. Electron carriers can be produced by oxygen vacancies or by suitable cation substitutions. Superconductivity is observed for carrier concentrations of about 10^{24}–$10^{26}\,m^{-3}$.

1.3 Oxide Bronzes

The T_c for oxides began its creep above 4.2 K with the discovery of superconductivity in the oxide bronzes. These are a family of compounds with the formula A_xMO_3 where M is W, Mo or Re; A is a variety of cations including the alkalies, alkaline earths and NH_4^-. Those that have been found to be superconducting are given in Table 1.

The structures of the oxide bronzes are closely related to that of perovskite. The M cation is in octahedral coordination to oxygen. The octahedra link by corners to form three-dimensional structures. The A cations occupy interstitial sites of these networks. One variation on this family is actually the perovskite structure (i.e., AMO_3). However, in the bronzes, the A cation content is variable and the octahedra can link up in ways very different than in the perovskite structure. Superconductivity is, in fact, never observed in the oxide bronzes based on the simple, cubic perovskite structure. Only when the octahedra are linked in such a way to produce tetragonal or hexagonal structures is it observed.

An oxide system formally analogous to the oxide bronzes is the WO_3–ReO_3 solid solution (Sleight and Gillson 1966). The cubic perovskite structure exists over a wide range of the solid solution, but there is no A cation. Very good metallic properties are observed for ReO_3 ($5d^1$), and WO_3 ($5d^0$) is an insulator. At no point in this solid solution does superconductivity occur above 4.2 K. This is consistent with the lack of superconductivity in the cubic bronzes of the type A_xMO_3 where M is W, Mo or Re. The tentative conclusion is, therefore, that cubic site symmetry for transition metal cations is not favorable for superconductivity. However, there is no such requirement for *s*-band materials such as (Ba, K)BiO_3.

1.4 $Li_{0.9}Mo_6O_{17}$

A series of compounds of the type $A_{0.9}Mo_6O_{17}$ exist where A may be K, Na or Li. Superconductivity is observed only when A is Li (Table 1). Both $K_{0.9}Mo_6O_{17}$ and $Na_{0.9}Mo_6O_{17}$, which are not isostructural with $Li_{0.9}Mo_6O_{17}$, exhibit charge density wave behavior instead.

1.5 Li_xNbO_2

$LiNbO_2$ is an insulator, but the lithium content can be decreased by reaction with bromine. A series of Li_xNbO_2 phases so prepared are conducting and some are superconducting. A T_c of 5.5 K is reported for $Li_{0.45}NbO_2$, and a T_c of 5.0 K for $Li_{0.50}NbO_2$. The structure of these phases is made up of layers of edge-sharing NbO_6 trigonal prisms. The Li^{1+} cations are found between these layers. There are many examples of reduced oxides of niobium which possess metallic properties, and the usual coordination of niobium to oxygen in these phases is octahedral. In the two examples of superconductivity in niobium oxides, the coordination of niobium to oxygen deviates strongly from octahedral (i.e., square planar or trigonal prismatic).

1.6 Ag_7O_8X Phases

Superconductivity is observed at rather low temperatures (Table 1) in phases of the type Ag_7O_8X where X may be NO_3^-, HF_2^-, ClO_4^- or BF_4^-. Despite the low T_cs, this system is of considerable interest due to the position of silver immediately under copper in the periodic table. These silver oxide phases contain mixed valent silver just as all the copper oxide superconductors contain mixed valent copper. Also, the coordination of silver is basically fourfold, square planar just as it tends to be in the copper oxide superconductors. However, the silver–oxygen square-planar units link to form clusters, and these clusters in turn link to form a cubic structure. The resulting Ag–O–Ag bond angles deviate much further from 180° than is found for Cu–O–Cu bond angles in the superconducting copper oxides. All attempts to prepare complex silver oxides with the same structures as the superconducting copper oxides have failed.

1.7 $LiTi_2O_4$

The first oxide discovered with a T_c above 10 K was in the system $Li_{1+x}Ti_{2-x}O_4$. This material has the cubic spinel structure, but the site symmetry for titanium is not cubic. The properties vary with x, and the maximum T_c of about 13.7 K is observed close to $x = 0.0$.

1.8 $BaBi_{1-x}Pb_xO_3$ System

Superconductivity at about 13 K was discovered in the $BaBi_{1-x}Pb_xO_3$ system in 1975. The end members of this solid solution both have the perovskite structure, but with somewhat curious properties. Insulating properties might have been expected for $BaPbO_3$ ($Pb^{IV} = 6s^0$) just as in fact observed for $BaSnO_3$ ($Sn^{IV} = 5s^0$). For $BaBiO_3$ ($Bi^{IV} = 6s^1$), metallic properties might have been expected due to a half-filled 6*s* band. In fact, the properties for $BaPbO_3$ and $BaBiO_3$ are the reverse of these expectations. Metallic properties are observed for $BaPbO_3$ which is presumed to be a semimetal with overlap of the conduction and valence bands. Semiconductivity is observed for $BaBiO_3$ due to the disproportionation of Bi^{IV} into Bi^{III} and Bi^V. (An equivalent description of this situation is a charge density wave in $BaBiO_3$.) Semiconductivity is observed over most of the $BaPbO_3$–$BaBiO_3$ solid solution, and EXAFS studies (Boyce et al. 1991) show discrete Bi^{III} and Bi^V exist but are disordered over the octahedral sites. Metallic properties are observed when x is greater than about 0.7 and the T_c maximum is observed for x about 0.725. It was observed in 1975 that the superconducting properties were somewhat improved by addition of potassium (i.e., $Ba_{1-y}K_y$-$Bi_{1-x}Pb_xO_3$). More than ten years later, superconductivity close to 30 K was found in $Ba_{1-y}K_yBiO_3$ (see Sect. 3).

2. Cuprate Superconductors

Before the 1980s, low-temperature superconductivity was known in copper sulfides, but superconductivity was completely unknown in oxides of copper. There is now a very large family of superconductors which may be represented as $A'_xX_y.A_{n-1}Cu_nO_{2n}$. The value of n equals the number of CuO_2 sheets stacked consecutively. Between any two CuO_2 sheets which are immediately adjacent to each other there are A cations, most commonly Ca and/or rare earths. There is usually no oxygen between these adjacent sheets. Since the oxidation state of copper is always close to two, the CuO_2 sheets always carry a negative charge. Thus, the net charge on the interlayer assembly, A'_xX_y, must always be positive as it is for the A cations. This requirement may be met in many different ways: La_2O_2, Bi_2O_2, Tl_2O_2, Sr_2Cl_2 and so on.

2.1 Infinite Layer Phase

The simplest structure with consecutively stacked CuO_2 sheets is found in the (Ca, Sr)CuO_2 system. This phase contains only divalent copper; thus, this phase is an antiferromagnetic insulator. Since the A'_xX_y assembly is completely lacking for this phase, the value of n is now infinity. When prepared at ambient pressures, the infinite layer phase exists in the $Ca_{1-x}Sr_xCuO_2$ system only for values of x close to 0.15. However, at high pressure, values of x up to 1.0 may be obtained for this structure. In fact for samples prepared at 60 kbar, even the $Sr_{1-x}Ba_xCuO_2$ system possesses the infinite layer structure for values of x up to about 0.85 (Takano et al. 1989). These high-pressure studies have also revealed the superconductivity in (Ba, Sr)CuO_2 compositions at temperatures as high as 90 K. However, the actual composition of the superconducting phase is unknown.

High-pressure conditions have also been used to prepare $Sr_{0.85}Nd_{0.15}CuO_2$ with the infinite layer structure. This is an n-type material with a T_c of 40 K.

2.2 A_2CuO_4-Based Superconductors

(*a*) *The T structure*. The La_2CuO_4 structure (Fig. 1a) is a one-layer structure (i.e., n = 1). There are four short copper–oxygen bonds (~0.19 nm) within the CuO_2 sheets. The copper–oxygen bonds to the interlayer oxygen (~0.25 nm) bring the total coordination number of copper up to six. Above 250 °C, the La_2CuO_4 structure is tetragonal and the CuO_2 sheets are flat. Below 250 °C, the La_2CuO_4 structure is orthorhombic and the Cu–O–Cu bond angles in the CuO_2 sheets are no longer 180°.

In stoichiometric La_2CuO_4, all copper is divalent and antiferromagnetic insulating behavior is observed. When the average oxidation state of copper is sufficiently increased, the insulating state is destroyed and superconductivity results. Various substitutions may be used to increase the average oxidation state of copper. Most commonly, a lower-valent cation is substituted for lanthanum. This may be barium, strontium, calcium, sodium or potassium (see Table 2). An alternate method of oxidizing the CuO_2 sheets is to simply place extra oxygen between the CuO_2 sheets (i.e., La_2CuO_{4+x}).

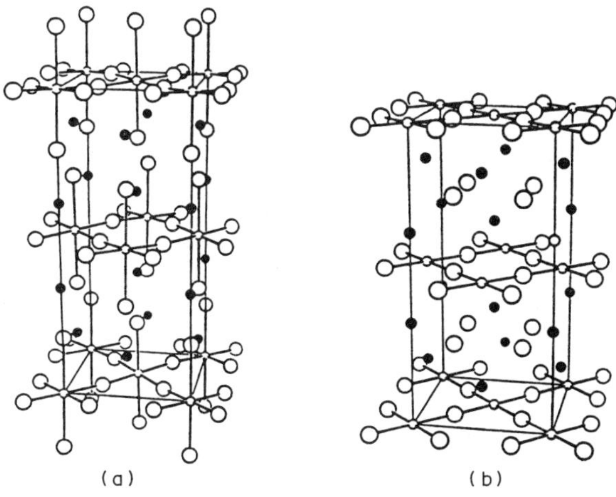

Figure 1
(a) Structure of La_2CuO_4: the small open spheres are copper; the large open sphěres are oxygen; and the filled spheres are lanthanum. (b) Structure of R_2CuO_4 phases where R is Pr, Nd, Sm or Gd: the small open spheres are copper; the large open spheres are oxygen; and the filled spheres are the R cations

(*b*) *The T′ structure*. For certain rare earths other than lanthanum, a different structure (Fig. 1b) exists for the A_2CuO_4 composition. The same CuO_2 sheets are present. The symmetry is basically tetragonal although small distortions from this symmetry may sometimes be present. The A cations occupy the same positions as lanthanum cations in the La_2CuO_4 structure. However, the interlayer oxygen occupies very different positions in the T and T′ structures. The oxygen coordination to copper is now regarded as strictly fourfold, square planar. For samples prepared at ambient pressure, this structure is found for rare-earth cations from praseodymium through gadolinium. For samples prepared at high pressures, this structure is also found for the rare cations yttrium, dysprosium, holmium, erbium and thulium (Okada et al. 1990).

Antiferromagnetic insulating properties are observed for the T′ phases of the type $A_2^{III}CuO_4$. Substitutions aimed at increasing the average oxidation state of copper have failed to produce superconductivity. However, substitutions which decrease the average oxidation state of copper do destroy the antiferromagnetic insulating state, and superconductivity is observed (Table 2). A higher valent cation can be substituted for the rare earth (e.g., $Nd_{2-x}A_x^{IV}CuO_4$ where A^{IV} is Ce or Th) or some fluorine can be substituted for oxygen (i.e., $Nd_2CuO_{4-x}F_x$).

(*c*) *T* phases*. The third variation on the A_2CuO_4 theme may be regarded as a one-to-one intergrowth of the T and T′ phases. These phases may all be represented by the formulae $(R, R')_{2-x}Sr_xCuO_4$ where R and R′ are two of the lighter rare earths (Table 2). The partial substitution of strontium produces p-type conductivity and superconductivity.

2.3 $A_3Cu_2O_6$ Phases

Phases of the type $A_2^{III}A^{II}Cu_2O_6$ with $n = 2$ were known before the discovery of superconductivity in cuprates. These phases have a structure similar to La_2CuO_2 but with double CuO_2 layers. The A^{II} cation is Sr or Ca, and A^{III} can be a variety of rare earths. Oxidation of the CuO_2 sheets is accomplished primarily by partial replacement of A^{III} by A^{II} (i.e., $La_{2-x}Sr_{1+x}CaCu_2O_6$). These phases apparently can also contain either oxygen vacancies or oxygen interstitials.

Many early attempts to produce superconductivity in this family failed. Finally, superconductivity at 60 K was induced by treating samples with high-pressure oxygen. The reason for the difficulty in producing superconductivity in these phases is

Table 2
Cuprate superconductors (superconductivity discovered between 1986 and 1991)

Compound	T_c (K)	Compound	T_c (K)
$(Sr, Nd)_2CuO_2$	40	$Tl_2Ba_2CaCu_2O_8$	110
$(La, Ca)_2CuO_4$ [T]	18	$Tl_2Ba_2Ca_2Cu_3O_{10}$	125
$(La, Sr)_2CuO_4$ [T]	38	$Tl(Ba, La)_2CuO_5$	40
$(La, Ba)_2CuO_4$ [T]	35	$TlBa_2CaCu_2O_7$	90
$(La, Na)_2CuO_4$ [T]	40	$TlBa_2Ca_2Cu_3O_9$	110
$(La, K)_2CuO_4$ [T]	38	$TlBa_2Ca_3Cu_4O_{11}$	122
La_2CuO_{4+x} [T]	38	$TlSr_2(Ca, R)\ Cu_2O_7$	90
$(Nd, Ce)_2CuO_4$ [T′]	24	$Bi_2Sr_2(R, Ce)_2Cu_2O_{10+x}$	30
$(Pr, Ce)_2CuO_4$ [T′]	17	$Pb_2Sr_2RCu_3O_8$	70
$(Sm, Ce)_2CuO_4$ [T′]	15	$Pb_2(Sr, La)_2Cu_2O_6$	32
$Nd_2Cu(O, F)_4$ [T′]	27	$PbBaSr(Y, Ca)Cu_3O_{8-x}$	65
$(Nd, Ce, Sr)_2CuO_4$ [T*]	30	$Tl(Sr, La)_2CuO_5$	48
$(La, Sm, Sr)_2CuO_4$ [T*]	20	$(Tl_{0.5}Pb_{0.5})Sr_2CaCu_2O_7$	85
$(La, Sr)_2CaCu_2O_6$	60	$(Tl_{0.5}Pb_{0.5})Sr_2Ca_2Cu_3O_9$	120
$RBa_2Cu_3O_7$	94	$(Tl, Bi)Sr_2CaCu_2O_7$	95
$Nd_2Ce_{0.67}Sr_{0.33}BaCu_3O_{9-x}$	38	$(Tl, Bi)Sr_2Ca_2Cu_3O_9$	115
$Bi_2Sr_2CuO_{6+x}$	9	$(Pb_{0.5}Sr_{0.5})Sr_2(Y_{0.5}Ca_{0.5})Cu_2O_{7-y}$	100
$Bi_2Sr_2CaCu_2O_{8+x}$	94	$(Pb_{0.5}Ca_{0.5})Sr_2(Y_{0.5}Ca_{0.5})Cu_2O_{7-y}$	80
$Bi_2Sr_2Ca_2Cu_3O_{10+x}$	110	$(Pb, Cu)(Eu, Ce)_2(Sr, Eu)_2Cu_2O_x$	25
$Tl_2Ba_2CuO_6$	90	$(Pb_{0.65}Cu_{0.35})Sr_2(Y_{0.7}Ca_{0.3})Cu_2O_7$	25

unclear. Many phases have been prepared in this family which appear to have a carrier concentration favorable for superconductivity, but superconductivity is not observed. One possible explanation is that this phase seems to show some tendency to accept oxygen between the immediately adjacent CuO_2 layers. This locally raises the coordination number of copper to six, now with five short copper–oxygen distances. Possibly, this unusual coordination for copper adversely affects superconductivity.

2.4 *$RBa_2Cu_3O_7$ and Related Phases*

(*a*) *$RBa_2Cu_3O_7$ phases.* T_c was raised above the boiling point of liquid nitrogen with the discovery of superconductivity in the $n = 2$ phase $YBa_2Cu_3O_7$. Most rare earths may be substituted for yttrium giving a series of the type $RBa_2Cu_3O_7$. This phase does not, however, exist for R = Ce. In the case of R = Pr, the structure exists but superconductivity is not observed.

Actually, the basic formula for this class of materials is $RBa_2Cu_3O_6$. This structure (see Fig. 2a) contains double CuO_2 sheets with R cations between the two sheets. The double sheets are held together by barium cations and O–Cu–O "sticks." The oxidation state of copper is one in the sticks and two in the sheets. The usual antiferromagnetic insulating behavior is observed for this situation. The $RBa_2Cu_3O_6$ phases can be oxidized by annealing or slow cooling in an oxygen-containing environment. The $RBa_2Cu_3O_7$ phases prepared in this way are kinetically stable, but they are not actually thermodynamically stable at any condition of pressure and temperature. The maximum T_c is observed close to the O_7 composition, and phases of this type with higher oxygen contents are not readily obtained. The ideal structure of the $RBa_2Cu_3O_7$ phases is shown in Fig. 2b. Between the double CuO_2 sheets, there are now copper oxygen chains as well as the barium cation. The chains are actually not straight as shown. Instead, they are presumed to be zigzagged but this distortion is not well ordered in three dimensions.

As a single crystal of $RBa_2Cu_3O_6$ is oxidized to $RBa_2Cu_3O_7$, the chains may not form so as to all run parallel to each other as shown in Fig. 2b. Such crystals are twinned, and there are two possibilities. In a given plane perpendicular to the *c* axis, the chains may all be parallel to each other but perpendicular to those in an adjacent plane of chains. Such incoherent twinning introduces strain into the lattice which is proportional to the difference in the dimension of the *a* and *b* axes of the orthorhombic unit cell for $ABa_2Cu_3O_7$. Thus, this type of twinning is more common when R is large where *a* and *b* are more similar, and the strain from twinning is not so great.

In the more common type of twinning, there are different domains within one sheet of chains. The chains are all parallel to each other within one domain, but they are perpendicular to those in an adjacent domain. No strain is created at the boundary between these domains. In fact, this domain boundary created by coherent twinning relieves strain created during the oxidation process.

A description of the oxidation state of copper is complex in the $RBa_2Cu_3O_7$ phases. The coordination

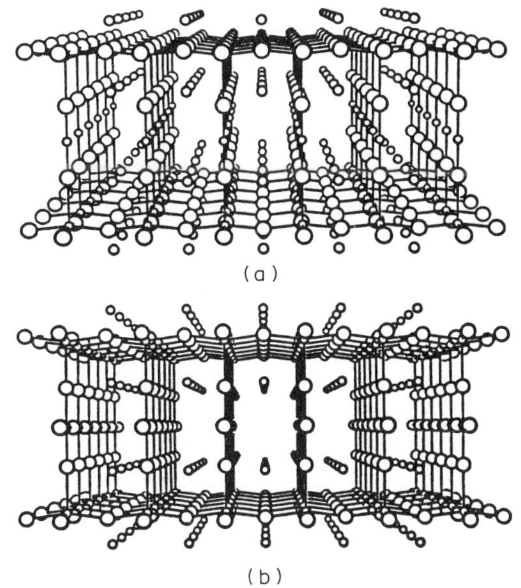

Figure 2
Structures of (a) $YBa_2Cu_3O_6$ and (b) $YBa_2Cu_3O_7$: the small connected spheres are copper; the large connected spheres are oxygen; and the isolated spheres are barium (inside sandwich) or yttrium (outside sandwich)

of copper belonging to the O–Cu–O sticks of the $RBa_2Cu_3O_6$ phase has now been brought up to fourfold, square planar. It is, thus, primarily this copper which has been oxidized. Therefore, a description has nearly been arrived at where the copper oxidation state in the chains would have become three, and the copper oxidation state in the CuO_2 sheets would remain at two. However, such a material would remain an antiferromagnetic insulator. Instead, the material contains mixed valency $Cu^{II,III}$ on both the chains and sheets. The disorder caused by mixed valency on the chains possibly prevents the ordering of the zigzag distortion of the chains.

As the oxygen content decreases below O_7, T_c falls. Most plots of T_c vs oxygen content show a plateau region with a T_c close to 60 K. However, if T_c is plotted against carrier concentration in the CuO_2 sheets (i.e., percent Cu^{III} in the sheets), no such plateau is observed. Instead, T_c decreases monotonically as the carrier concentration is decreased. Sharp, complete superconducting transitions have been observed for single crystals of $YBa_2Cu_3O_{6+x}$ for any T_c from 95 K down to about 20 K (Veal et al. 1990).

The lack of superconductivity in $PrBa_2Cu_3O_7$ is almost certainly due to an insufficient number of carriers in the CuO_2 sheets. The praseodymium cations are directly bonded to the CuO_2 sheets but not to the copper–oxygen chains. Presumably, a small amount of oxidation of Pr^{III} to Pr^{IV} occurs. This decreases the oxidation state of copper in the CuO_2 sheets but not in the more distant chains. The number of carriers in the CuO_2 sheets is thus reduced to the point where superconductivity is not supported.

For phases between $RBa_2Cu_3O_6$ and $RBa_2Cu_3O_7$, the properties are highly dependent on the sample preparation conditions. This is because T_c and carrier concentration in the CuO_2 sheets depend not only on oxygen content but also on the nature of the ordering of the interstitial oxygen. Various superstructures for phases intermediate between $RBa_2Cu_3O_6$ and $RBa_2Cu_3O_7$ have been observed by electron microscopy. However, only one of these superstructures is reasonably well characterized. This is the one for the $RBa_2Cu_3O_{6.5}$ composition where every other chain is presumed missing relative to the $RBa_2Cu_3O_7$ composition. Some superstructures which had been assumed to be related to oxygen ordering may in fact have other origins (Werder et al. 1991).

(*b*) *$RBa_2Cu_4O_8$ phases*. This phase was first discovered as an intergrowth in some thin films of $YBa_2Cu_3O_7$. It has subsequently been prepared single phase with various rare earths. This phase is not readily prepared under the conditions normally used to synthesize the $RBa_2Cu_3O_7$ phases. Instead, the formation of $RBa_2Cu_4O_8$ phases is favored by high oxygen pressure and low temperature. Normally, the T_c of phases is about 80 K. However, substitutions of calcium for yttrium have led to a stabilized $Y_{0.9}Ca_{0.1}Ba_2Cu_4O_8$ phase with an enhanced T_c of 90 K (Miyatake et al. 1990).

The structure of the $RBa_2Cu_4O_8$ phases is the same as $RBa_2Cu_3O_7$ phases except that the chain is now a double copper–oxygen chain formed from square-planar units with share edges. The average oxidation state of copper in $RBa_2Cu_4O_8$ phases is 2.25 whereas it is 2.33 in $RBa_2Cu_3O_7$ phases. The $RBa_2Cu_4O_8$ phases do not have a broad range of oxygen stoichiometry as do the $RBa_2Cu_3O_{6+x}$ phases. Thus, the oxygen anneal is not required for $RBa_2Cu_4O_8$ phases, and these phases tend to be less susceptible to oxygen loss under reducing conditions. Higher oxygen pressure is required to form $RBa_2Cu_4O_8$ phases, but this is because the $RBa_2Cu_3O_x$ phases form closer to $RBa_2Cu_3O_6$ than to $RBa_3Cu_3O_7$.

(*c*) *$R_2Ba_4Cu_7O_{15}$ phases*. This structure may be regarded as a one-to-one intergrowth of $RBa_2Cu_3O_7$ and $RBa_2Cu_4O_8$. The CuO_2 sheets remain as double sheets containing the rare-earth cations. Alternating with the double CuO_2 sheets are sheets with single copper–oxygen chains and double copper–oxygen chains. Other intergrowths between $RBa_2Cu_3O_7$ and $RBa_2Cu_4O_8$ are also possible.

2.5 $(A'O)_mA_{1+n}Cu_nO_{2+2n}$ Phases

(*a*) *$(BiO)_2Sr_2Ca_{n-1}Cu_nO_{2+2n}$ phases*. The structures for phases of this type are shown in Fig. 3. The number of consecutively stacked CuO_2 sheets is again indicated by *n*. Between the consecutively stacked

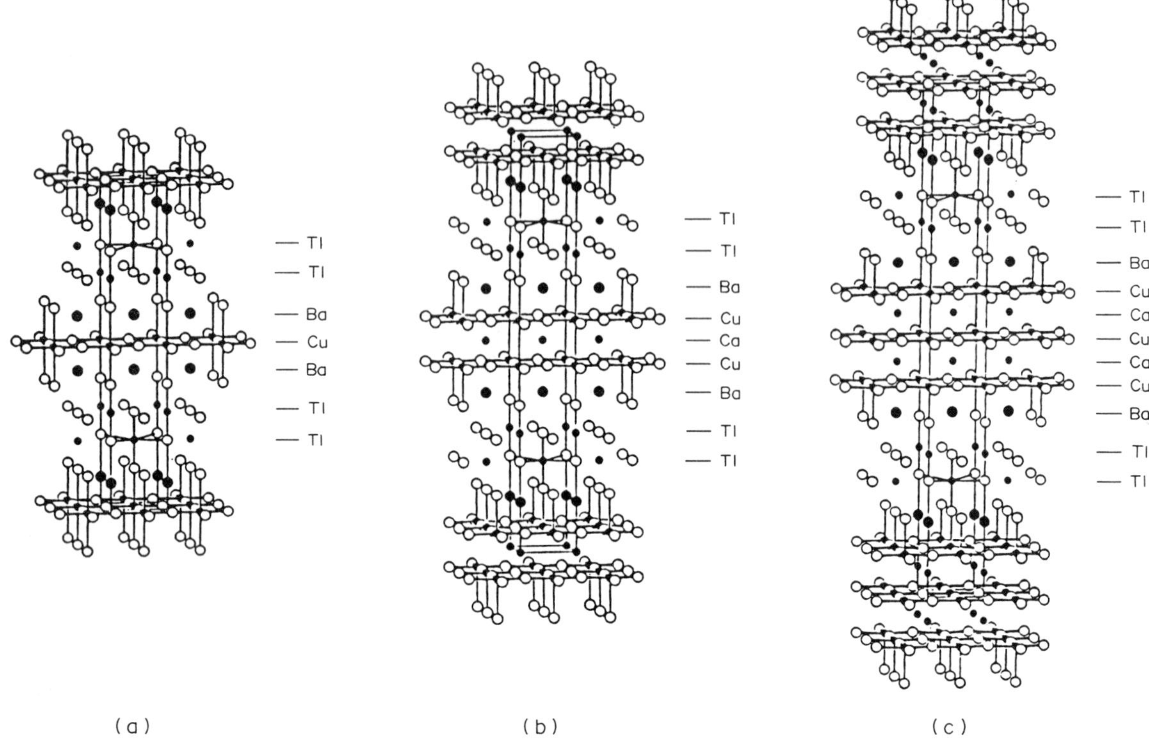

Figure 3
Structures of $(TlO)_2Ba_2Ca_{n-1}Cu_nO_{2n+2}$ for (a) $n = 1$, (b) $n = 2$ and (c) $n = 3$

CuO_2 sheets are mainly Ca^{II} cations, but other cations seem to also always be present. Phases of this type are readily prepared for n equal to one and two. The n equal to three phase is very difficult to prepare. Formation of this phase is facilitated by addition of lead, only some of which actually substitutes into the superconducting phase. Phases with n equal to four through eight have been prepared only by artificially structuring (Kanai et al. 1989). As n increases, T_c first increases and then decreases.

The $(BiO)_mSr_2Ca_{n-1}Cu_nO_{2+n}$ formulation is only an ideal formulation, and the structures shown in Fig. 3 are also highly idealized. It is doubtful that the properties of these materials can be understood within the framework of these ideal formulas and structures. We know that actually a significant range of composition exists for each n value. This range of composition in general does not even include the ideal composition. If the ideal compositions could be made with the ideal structures, these phases would presumably all be antiferromagnetic insulators.

For an $n = 2$ phase, a more accurate structural representation of a $Bi_{2.01}Sr_{1.84}Ca_{0.90}Cu_2O_{8+x}$ composition would be $[Bi_{1.90}][Sr_{1.7}Ca_{0.10}Bi_{0.50}][Ca_{0.80}Sr_{0.14}Bi_{0.06}]$-$Cu_2O_{8+x}$. The average copper oxidation state becomes greater than two mainly due to deficiencies of the A^{II} cations Ca^{II} and Sr^{II}. The bismuth site seems to contain only bismuth cations, but these sites may not be fully occupied. The oxidation state of bismuth is primarily trivalent, but small amounts of pentavalent bismuth may be present. The strontium site appears to contain some vacancies as well as some Ca^{II} and Bi^{III} cations. The calcium site appears to be fully occupied, but some Sr^{II} and Bi^{III} have substituted for Ca^{II}. The CuO_2 sheets are free of defects within our capability to find such defects by x-ray and neutron diffraction techniques.

The oxygen content of these phases is variable for any fixed cation composition, and T_c increases as oxygen is removed. It is generally assumed that the labile oxygen lies within the Bi_2O_2 double layer. It has also been frequently assumed that this extra oxygen serves to oxidize copper. There is no direct proof for either of these assumptions. It should be noted that extra oxygen in $Pb_2Sr_2(Ca,R)Cu_3O_{8+x}$ phases actually causes the copper oxidation state in the CuO_2 sheets to decrease (see Sect. 2.6).

The $(BiO)_mSr_2Ca_{n-1}Cu_nO_{2+2n}$ phases contain large concentrations of point defects in the form of vacancies, interstitials and antisite disorder. However, the deviation from ideality does not end here. All of these phases have complex superstructures which are only partially understood. These superstructures are normally not commensurate with the subcell lattice, and they seem to be caused by a size mismatch between the Bi_2O_2 double sheet and the CuO_2 sheets (see *Oxide Superconductors: Electron Microscopy*). We may assume that the highly covalent bonding of the CuO_2 sheets leads to a rather rigid behavior for these sheets. That is, the copper–oxygen distance in the sheets cannot be readily compressed to accommodate bismuth–oxygen distances which should be much shorter than in the idealized structure. Thus, the distortion caused by the size mismatch is mainly accommodated within the Bi_2O_2 double sheet. Both bismuth and oxygen of the Bi_2O_2 double sheet are strongly displaced from their ideal positions on fourfold axes. In conjunction with these displacements, a wave develops perpendicular to the *c* axis which further destroys the fourfold symmetry reducing even the subcell to orthorhombic symmetry. The periodicity of the modulation is frequently found to be about 4.7 for phases with the nominal $Bi_2Sr_2CaCu_2O_8$ composition. Actually, the modulation can be commensurate at 5.0 *a* if the sample is prepared under mildly reducing conditions (Gai et al. 1990). It is likely that the extra oxygen is sited at particular places along the modulation, but this has not been proven for these phases or the structurally related $Bi_2Sr_3Fe_2O_9$ phase.

Given the large structural mismatch between the CuO_2 layers and the Bi_2O_2 double layer, it is strange that these phases should form at all. In fact, they appear to be phases which form only at high temperatures, close to melting temperatures in these systems. They are not actually thermodynamically stable phases at room temperature and below.

The bonding within the Bi_2O_2 double layer which holds the two BiO layers together is very weak. This is presumably because the lone pairs of Bi^{III} electrons (from 6*s*–6*p* hybridization) are pointed in the direction of the adjacent BiO layer. Thus, the $Bi_2A_{1+n}Cu_nO_{4+2n}$ crystals are very readily cleaved perpendicular to the *c* axis, much like mica. Of all the cuprate superconductors, these phases are the most two dimensional in character. Iodine can be intercalated between the BiO sheets of the Bi_2O_2 double layer giving an even more anisotropic material (Xiang et al. 1990). This iodine is presumably uncharged; thus, the carrier concentration and T_c are only slightly changed.

(*b*) *$(TlO)_mA_{1+n}Cu_nO_{2+2n}$ phases*. Phases of this type raised T_c to the maximum observed values. The oxidation state of thallium is always close to three but may be reduced somewhat in certain cases. This series is very closely related to the $Bi_2Sr_2Ca_{n-1}Cu_nO_{4+2n}$ series described earlier. The maximum value of *n* obtained without resorting to artificially structuring is four. Again T_c first rises and then falls for increasing *n*. A significant difference between the bismuth and thallium series is that the latter can be obtained with *m* equal to either one or two (i.e., either a TlO or a Tl_2O_2 layer stacking with the CuO_2 sheets). For the $m=1$ series, we have the ideal formula $TlA_{1+n}Cu_nO_{3+2n}$. The average oxidation for copper is now greater than two if the A cations are all divalent. Such phases are metallic, but not superconducting, and are considered to be overdoped (i.e., too much Cu^{III}). These can be made superconducting by bringing down the oxidation state of copper (e.g., $Tl(A^{II}_{1-x}A^{III}_x)_{1+n}Cu_nO_{3+2n}$ where A^{III} is a rare-earth cation).

The $Tl_2A_{1+n}Cu_nO_{4+2n}$ series is more closely analogous to the $Bi_2Sr_2Ca_{n-1}Cu_nO_{4+2n}$ series. There are, however, important structural variations. The bonding within the Tl_2O_2 double layer is distinctly different than that within the Bi_2O_2 double layer. One difference is that the bonding between adjacent TlO layers is much stronger than between adjacent BiO layers. Thus, the very easy cleavage perpendicular to the *c* axis which exists for the $Bi_2Sr_2Ca_{n-1}Cu_nO_{4+2n}$ phases is not found for any of the $(TlO)_mA_{1+n}Cu_nO_{2+2n}$ phases. Another distinction is with regard to superstructures. The $(TlO)_mA_{1+n}Cu_mO_{2+2n}$ phases do not have superstructures analogous to those found for all the $Bi_2A_{1+n}Cu_nO_{4+2n}$ phases. Furthermore, the basic subcell remains tetragonal for nearly all the $(TlO)_nA_{1+n}Cu_nO_{2+2n}$ phases.

The situation with regard to defects, nonstoichiometry, distortions, disorder and stability is very similar for the Tl–A–Cu–O and Bi–A–Cu–O superconductors. Neither series appears to be thermodynamically stable at room temperature and below. Both series contain rather massive quantities of point defects. Atoms in the Bi_2O_2, Tl_2O_2 and TlO layers are displaced off their ideal sites on the fourfold axes. Although these displacements are highly correlated in all cases, the thallium phases do not normally show superstructures indicative of long-range order.

(*c*) *Mixed phases*. In addition to the phases just described, there is also the $(A'O)_mA_{1+n}Cu_nO_{2+2n}$ phase where A′ is Bi and Pb, Tl and Pb, Bi and Tl, Pb and Sr, or Pb and Ca (Table 2). Sometimes these appear to be rather obvious solid solutions. However, in other cases the ratio of the two A′ cations is essentially fixed.

2.6 $Pb_2Sr_2(Ca, R)Cu_3O_{8+x}$ Phases

The structure for $Pb_2Sr_2(Ca, R)Cu_3O_{8+x}$ phases resembles that of $YBa_2Cu_3O_{6+x}$ in that there are O–Cu–O sticks between double CuO_2 sheets for $x=0.0$. However, there is a very important difference. As *x* increases for the $Pb_2Sr_2(Ca, R)Cu_3O_{8+x}$ phases, both lead and copper of the O–Cu–O sticks are oxidized, but the average oxidation state of copper in the sheets decreases (Marezio et al. 1990). Thus, the hole carrier concentration in the CuO_2 sheets decreases and T_c decreases. The T_c for these phases approaches 80 K.

Sharp, complete transitions have been observed in single crystals.

2.7 Miscellaneous Cuprate Superconductors

Other cuprate superconductors are also listed in Table 2. In some cases, the structures of these phases are still not well determined.

3. $Ba_{1-x}A_x^{1+}BiO_3$ Superconductors

Although superconductors of the type $Ba_{1-x}K_x$-$Bi_{1-y}PbO_3$ were prepared in the 1970s, the synthesis condition used at that time did not yield superconductors of the type $Ba_{1-x}A_x^{1+}BiO_3$. Very special synthesis conditions are required to prepare $Ba_{1-x}A_x^{1+}BiO_3$ superconductors, just as for the cuprate superconductors.

A T_c of 34 K can be routinely obtained in the $Ba_{1-x}K_x^{1+}BiO_3$ system, and a T_c of 37 K has been reported in the (Ba, K, Rb)BiO_3 system. Although this T_c is considerably lower than for many cuprate superconductors, these bismuthate superconductors are of considerable interest because they can be prepared at temperatures as low as 250 °C, they possess a cubic crystal structure and therefore completely isotropic properties, and they readily produce hysteretic Josephson junctions (Huang et al. 1990), something not observed for the high-T_c cuprates.

As discussed in Sect. 1.8, $BaBiO_3$ is a diamagnetic insulator due to the disproportionation of Bi^{IV} into Bi^{III} and Bi^{V}. As potassium is substituted for barium (i.e., $Ba_{1-x}K_x^{1+}BiO_3$), the average oxidation state of bismuth increases. Initially, the Bi^{IV} continues to be disproportioned into Bi^{III} and Bi^{V}. However, at an x value of about 0.25 superconductivity appears and there is now only one kind of bismuth. The *6s* electron of bismuth can now be assumed to be delocalized.

4. Future Developments

Undoubtedly, new superconducting oxides will continue to be discovered. Despite the fact that a theory for high-temperature superconductivity is lacking, the rules for where to search for such materials seems rather clear (Sleight 1987). Varying composition close to a metal–insulator boundary has always been fruitful territory for the discovery of superconductors. However, there are restrictions on this approach. It would appear that oxide superconductors never have a transition metal site of cubic symmetry (Sleight 1988), but this rule does not apply to post-transition metals. This rule appears to be related to a requirement of no more than one unpaired electron per cation. Mixed valency is required for superconductivity, but again the requirement of no more than one unpaired electron for each of the valent states involved must be met.

High-T_c superconductors always seem to be thermodynamically unstable at room temperature and below (Sleight 1991). Some have only very narrow stability ranges at a particular pressure/temperature condition. For others, there appears to be virtually no condition of temperature and pressure where the phase is truly thermally stable. Thus, we must be very clever and very lucky to find the right synthesis condition to prepare such compounds.

Of course, the potential of materials other than oxides can be considered as high-T_c superconductors. However, such materials may also require different metals than are employed for high T_c in oxides. The most fruitful area for finding high-T_c superconductors has involved materials containing Cu^{II}/Cu^{III} mixed valency. The oxidation state of three for copper is only barely possible in oxides. It is readily achieved in fluorides, but it is not expected to exist in nitrides, sulfides, chlorides, and so on. All fluorides seem to be too ionic to support delocalized electron behavior. Thus, there would appear to be nowhere to search for superconductivity based on Cu^{II}/Cu^{III} mixed valency except oxides. Of course, there is also superconductivity based on Cu^{II}/Cu^{I} mixed valency, and here there is less difficulty in defining nonoxide systems to explore. In fact, CuS is a superconductor containing Cu^{II}/Cu^{I} mixed valency because some of the sulfur in this compound has dimerized.

Bibliography

Boyce J B, Bridges F G, Claeson T, Geballe T H, Li G G, Sleight A W 1991 Local structure of $BaBi_xPb_{1-x}O_3$ determined by x-ray absorption spectroscopy. *Phys. Rev. B* 44: 6961–72

Gai P L, Subramanian M A, Sleight A W 1990 In: Daly D A 1990 (ed.) *Proc. 12th Int. Conf. Electron Microscopy*, San Francisco, CA, p. 36–9

Huang Q, Zasadzinski J F, Gray KE, Richards D R, Hinks D G 1990 Ideal tunneling characteristics in $Ba_{1-x}K_xBiO_3$ point-contact junctions with Au and Nb tips. *Appl. Phys. Lett.* 57: 2356–8

Kanai M, Kawai T, Kawai S, Tabata H 1989 Low-temperature formation of multilayered Bi(Pb)–Sr–Ca–Cu–O thin films by successive deposition using laser ablation. *Appl. Phys. Lett.* 54: 1802–4

Marezio M, Santora A, Capponi J J, Hewat E A, Cava R J, Beech F 1990 Crystal structure, atomic ordering and charge location in $Pb_2Sr_2Y_{1-x}Ca_xCu_3O_{8-\delta}$. *Physica C* 169: 401–12

Miyatake T, Koshizuka N, Tanaka S 1990 *Oyo Butsari* 59: 601 Okada H, Takano M, Takeda Y 1990 Synthesis of Nd_2CuO_4-type Y_2CuO_4, Dy_2CuO_4, Ho_2CuO_4, Er_2CuO_4, Tm_2CuO_4 under high pressure. *Physica C* 166: 111–14

Sleight A W 1987 Rules for high-temperature superconductivity. *Chemtronics* 2: 116–19

Sleight A W 1988 Chemistry of high-temperature superconductors. *Science* 242: 1519–27

Sleight A W 1991 Synthesis of oxide superconductors. *Phys. Today* 44(6): 24–30

Sleight A W, Gilson J L 1966 Preparation and properties of alkali rhenium bronzes and a WO_3–ReO_3 solid solution. *Solid State Commun.* 4: 601–2

Veal B W, Paulikas P, You H, Shi H, Fang Y, Downey J W 1990 Observation of temperature-dependent site disorder in $YBa_2Cu_3O_{7-\sigma}$ below 150 °C. *Phys. Rev. B* 42: 6305–16

Werder D J, Chen C H, Espinosa G P 1991 Origin of the $2\sqrt{2}AX_2\sqrt{2}A$ superstructure in Y–Ba–Cu–O. *Physica C* 173: 285–92

Xiang X D, McKernan S, Vareka W A, Zettl A, Corkill J L, Barbee T W, Cohen M L 1990 Iodine intercalation of a high-temperature superconducting oxide. *Nature* 348: 145–7

A. W. Sleight
[Oregon State University, Corvallis, Oregon, USA]

Oxide Superconductors: Thin-Film Deposition

During the rapid growth of research associated with the new high-critical-temperature oxide superconductors (HTSs) thin films have played a varying yet always vital role. The importance of thin films was generally recognized when it became clear that grain boundaries have such a deleterious effect on the critical transport currents. Thus, highly oriented films prepared in the early days were a key to giving encouragement that genuinely useful ceramic superconducting material could be made. In the future the main role for thin films will be as part of passive or active devices and they will continue to be used for scientific studies related to these potential applications. Microwave filter applications are being commercially evaluated and the first use of a thin-film dc SQUID (superconducting quantum interference device) operating in liquid nitrogen has been demonstrated for magnetocardiograph analysis (Miklich et al. 1991).

In an overview of the preparation of thin films of these materials it is important to define first the key requirements of a mature technology for future applications. The main need is the ability to deposit smooth epitaxial films with the desired crystal structure and the correct level of oxygenation. However, deposition of a single superconducting layer is not sufficient for most requirements, it is usually necessary to prepare advanced epiheterostructures that involve successive deposition of several epitaxial layers which may be superconducting, insulating or metallic. Other objectives are the deposition of ultrathin films, the formation of controlled grain boundaries by the use of specially prepared substrates and the reduction of the processing temperatures without degradation of important superconducting properties. Longer term, one can envisage the need to prepare perfect single-crystal films in which all three axes of the film are uniquely oriented.

The scale and intensity of research on thin-film deposition of HTSs has been unprecedented and the sheer weight of published material makes assessment and the crosslinking of different authors' ideas very difficult. An indexed bibliography has been collated by Talvacchio (1989–1991). HTS thin films are now being consistently produced with good epitaxy and critical current densities in excess of 10^7 A cm^{-2} at 4 K and in the range 10^6–10^7 A cm^{-2} at 77 K. It is evident that the technical challenges which have been overcome since 1987 have lifted the science and technology of the thin-film deposition of oxides and complex structures to a new level of sophistication and achievement.

In the following description $YBa_2Cu_3O_{7-\delta}$ (YBCO) is discussed in the major part of the article and other ceramic superconductors are addressed in Sect. 2.5.

1. Fundamental Issues

Although there have been studies of epitaxy relationships for the growth of some simple oxides (e.g., MgO, ZrO_2) the general problem of the vapor deposition of epitaxial thin films of oxides with large unit cells had not been addressed prior to 1987. The epitaxial growth of metals and semiconductors from the vapor phase requires deposition temperatures that are a small fraction of their melting points T_m, usually 0.2–$0.4T_m$ and 0.3–$0.45T_m$, respectively (the higher range for semiconductors reflecting their higher covalency). The typical range used for superconducting oxide films is 0.7–$0.85T_m$ which seems high by comparison, probably reflecting the chemical dissimilarity of the constituent oxides. Complex oxides with large unit cells, for instance multicomponent garnets, have been prepared by liquid phase epitaxy (LPE); because they are grown from the melt at low levels of supersaturation so as to be close to equilibrium, they can be grown with relative ease. Already some promising work on the LPE of $Bi_2Sr_2Ca_1Cu_2O_{8+x}$ compounds has resulted in impressive crystal perfection when compared with corresponding vapor-phase-deposited films (Balestrino et al. 1990).

1.1 Phase Stability and Oxygen Supply

In view of the relatively high homologous deposition temperatures (T/T_m) used in preparing this material, the bulk phase diagram gives a close indication of the stability limits of the elemental combinations as a function of temperature and oxygen partial pressure (see Fig. 1). However, thin-film growth is generally a nonequilibrium process and the kinetics of the nucleation process can play an important role in determining which phases appear in a growing film. In addition, one can expect a different set of kinetics to be operative depending on the activity of oxygen present; active oxygen can considerably aid the formation of the correct phases. This raises the general problem of how to interpret Fig.1 in terms of the degree of activity of the oxygen that is present for the particular deposition configuration used.

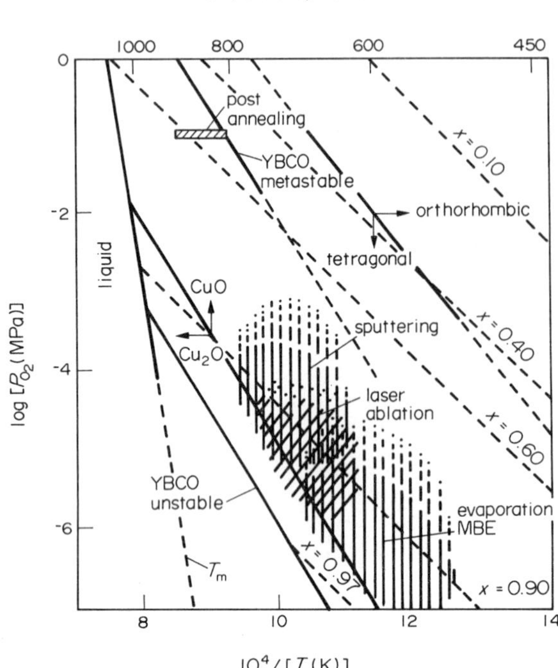

Figure 1
Oxygen and partial pressure diagram of $YBa_2Cu_3O_{7-x}$ showing phase boundaries and Cu_2O–CuO phase line. Also shown are the regimes of thin-film deposition (after Feenstra et al. 1991)

These materials are not only complex oxides but also have an unusual oxygen configuration which does not correspond to many of the more standard oxides. An interesting aspect of this is the general observation that as the oxygen chemistry is changed from $O_{6.1}$ to $O_{7.0}$ the melting point increases by 100 °C. The behavior of the oxygen is critical throughout the consideration of thermodynamic and kinetic aspects of the growth of this material as a thin film. At present, with regard to the growth of YBCO, one can see two distinct regimes which merge into one another; there is a growth regime in which the oxygen is in a low concentration and the metal species are quite mobile at low temperatures (630 °C to 700 °C), while at higher oxygen concentrations the metal species are less mobile (the majority being in some oxidized state) and a higher temperature is needed to form the required phase. Recent work by Feenstra et al. (1991) on the effect of oxygen pressure on the stability of YBCO gives considerable insight into the whole thin-film deposition process.

1.2 Diffusion and Kinetics

Any thin-film growth phenomenon depends on diffusional processes. Routbort et al. (1991) have measured the diffusivity of copper in YBCO, finding that its diffusion coefficient is 10^{-12} $cm^2\,s^{-1}$ at 805 °C, but comment that the diffusion of yttrium and barium are much slower. The activation energy for surface diffusion tends to be about half that for the bulk; making this assumption leads to the estimates of diffusion coefficients and diffusion lengths (for 10 s on the surface and about 1 h in the bulk) shown in Table 1. Although mainly constructed from estimates, these numbers illustrate the key mobilities and show how the formation of outgrowths is likely to be dependent on growth temperature and indicate why films will appear much smoother at lower temperatures simply because the mobilities of the yttrium and barium are low and will be feature-size limiting.

Table 1
Diffusion coefficient in $cm^2\,s^{-1}$ (and diffusion length) on the surface and in bulk for yttrium, copper and barium

Temperature	Surface (10 s on film surface)			Bulk (4000 s in bulk)		
	Yttrium	Barium	Copper	Yttrium	Barium	Copper
800 °C	10^{-9} (1 µm)	10^{-7} (10 µm)	10^{-5} (100 µm)	10^{-19} (0.2 nm)	6×10^{-16} (20 nm)	10^{-12} (0.6 µm)
700 °C	10^{-10} (0.3 µm)	10^{-8} (3 µm)	2×10^{-6} (30 µm)	5×10^{-22} (0.02 nm)	10^{-17} (2 nm)	10^{-14} (60 nm)
600 °C	4×10^{-12} (60 nm)	10^{-9} (1 µm)	3×10^{-7} (20 µm)	10^{-24}	10^{-19} (0.2 nm)	10^{-15} (20 nm)

Another important kinetic consideration in the growth of these materials as a function of temperature is the likelihood of phase separation if the stoichiometry is not perfect. Again this depends on the nucleation processes for the unwanted phase. At lower temperatures there is more likelihood of cation disorder (due to the effect on ordering kinetics of low bulk and surface diffusion coefficients). While layer-by-layer growth has been observed, it is not clear whether fully ordered films can be grown at these relatively low temperatures and there is some evidence for cation disorder with yttrium substitution on barium sites (Matijasevic et al. 1991).

1.3 Nucleation and Growth

The standard theories of single-crystal thin-film growth (reviewed by Venables et al. 1984) suggest three types of growth; layer-by-layer growth (Frank–van der Merwe), capillary growth in which islands nucleate, grow and coalesce (Volmer–Weber) and an intermediate mechanism in which the film starts layer by layer and then becomes islandlike (Stranski–Krastanov). Layer-by-layer growth tends to occur when there is a good lattice match between the growing film and the substrate and also a favorable chemistry which prevents strong chemical reaction which could disrupt such an interface. The nucleation rate depends on the arrival rate of atoms, the activation energies of desorption and surface diffusion and the activation energy ΔG^* associated with formation of a critical nucleus of radius r^*:

$$r^* \propto \sigma / \Delta G_v$$

where σ is the surface energy and ΔG_v is the free energy difference between the solid and vapor phases. In a simple capillary model of nucleation

$$\Delta G^* \propto \sigma^3 / \Delta G_v^{\ 2}$$

with ΔG_v given by $L\Delta T/T$, where L is the latent heat and ΔT is $T_m - T$, the degree of undercooling. The main consequences of this are that at lower temperatures nuclei are smaller, there is a larger driving force to form them and the nucleation rate is higher. However, one does not expect a complex unit cell to nucleate at all at a low temperature because the critical radius is smaller than the complete unit cell.

If growth proceeds the strain due to lattice mismatch between film and substrate increases and at some point misfit dislocations are likely to occur. For $SrTiO_3$ with its 1.2% lattice misfit one expects misfit dislocations to appear with film thicknesses of order 10 nm, while on MgO the lattice mismatch is so great (8%) that one expects misfits at thicknesses of order 1 nm or one unit cell. Recent atomic resolution scanning tunnelling microscopy (STM) studies by, for example, Schlom et al. (1991) and Lang et al. (1992) are leading to a general understanding of the nucleation and growth processes of these materials on a range of substrates. At low temperatures, particularly with low levels of oxygen, one sees layer-by-layer growth as observed by Terashima et al. (1990) using reflection high-energy electron diffraction (RHEED), though there is some evidence that after a few unit cells the structure ceases to be perfectly flat. Some RHEED interpretations are that the initial nuclei are in fact full unit cells as the YBCO streaks form after only 0.4 nm of film has been grown. At higher temperatures, but still less than 740 °C, and with a more oxygen-rich deposition technique such as laser ablation or sputtering one sees more island growth on substrates that are not perfectly lattice matched. Thus, islands have been observed by Miyazawa et al. (1992) for an early stage of growth of YBCO on $SrTiO_3$; however, they also observed that YBCO growing on $NdGaO_3$, and particularly on a vicinally cut substrate, showed a near layer-by-layer growth which is consistent with the better lattice matching of YBCO with $NdGaO_3$ (0.4% compared with 1.2% for $SrTiO_3$).

As the temperature is raised to the 740–760 °C growth regime STM observations show screw dislocation growth (see Fig. 2). These may have their origin in the underlying misfit dislocations threading through to the film surface as screw dislocations. The fact that

Figure 2
A high-resolution image of the surface of a YBCO film obtained using scanning tunnelling microscopy (STM). Part of a single-screw dislocation growth spiral can be seen, the individual ledges are of order 20 nm wide and one unit cell high (after Schlom et al. 1992)

they do not appear at lower temperatures could be because they are more closely packed and annihilate one another. The explanation given by Schlom et al. (1992) is that they originate from the intergrowth of islands in the coalescence stage, as seen in other systems. Screw dislocation growth is one of the characteristic growth mechanisms of a solid from the melt and its occurrence is not surprising in view of the closeness to the melting temperature.

1.4 Orientation Effects

A consensus exists that for the growth of YBCO on $SrTiO_3$ substrates by laser ablation and sputtering films tend to grow with *c*-axis orientation (that is, with the *c* axis perpendicular to the substrate surface) in the temperature region of around 750 °C to 830 °C while at about 100 °C lower they grow with *a*-axis orientation (with the *a* axis perpendicular to the substrate surface) (see, for example, Inam et al. 1991). Evaporated films tend to be *c*-axis oriented even at the lower temperatures, although a transition from *c*-axis, through mixed orientation, to *a*-axis growth has been reported on $SrTiO_3$. The prevalence of *a*-axis nucleation at lower temperatures may be explained by the fact that an *a*-axis nucleus would be smaller than one of *c*-axis material and thus could occur at a lower temperature. Overall the resulting orientation of a given film will depend on a combination of the nucleation phenomena and the subsequent growth processes in which there is an apparent higher surface mobility on the *c* planes leading to fast growth in the *a*, *b* directions.

For *a*-axis growth, because of the lower temperature, films are formed which are rather imperfect from a structural point of view. To overcome this problem some workers (for example, Linker et al. 1992) have used a technique of starting the growth at this lower temperature for the nucleation process (for *a* axis) and then raising the temperature for the main part of the film growth. Inam et al. (1991) have used $PrBa_2Cu_3O_{7-\delta}$ for the initial "template" layer. YBCO films with a (110) orientation may be grown, particularly on (110) $SrTiO_3$ and $LaSrGaO_4$, and it is possible to produce YBCO with its *c* axis lying in a single and unique direction in the plane of the film. Habermeier et al. (1992) have made high-quality films of this type by starting film growth at a low substrate temperature and subsequently increasing the deposition temperature. Such "*c*-axis directed" films are of potential importance for application in devices.

The orientation of films grown on vicinal substrates depends on the substrate type and lattice misfit; films grown on vicinal $SrTiO_3$ have an epitaxial orientation locked to the substrate for vicinal angles as large as 6°, while in the case of vicinal MgO the *c* axis tends to be perpendicular to the plane of the substrate.

1.5 Substrates and Buffer Layers

An ideal substrate must be well lattice matched to give good epitaxy and should have a low loss and low dielectric constant at high frequencies, while also being chemically stable, reasonably priced and potentially available in large-diameter wafers. One should also consider buffers on a variety of substrates either for high-frequency applications or as part of some integrated electronic device (semiconducting or optoelectronic). The area has been reviewed by Scheel (1991).

Since YBCO has a perovskitelike structure, it was natural that the earliest widely used substrate was $SrTiO_3$. YBCO films on yttria-stabilized zirconia (YSZ) have never achieved quite as high critical currents as those films on MgO and $SrTiO_3$ due to poor epitaxy, but nonetheless YSZ still represents one of the best buffers on silicon and, more recently, has been used by Fork et al. (1991) on silicon on sapphire (SOS). MgO, despite its lattice mismatch (8%), has shown itself to be a very compatible substrate and is widely used particularly for lower temperature growth techniques such as evaporation; careful pretreatment of the surface can lead to very-high-quality films. It has a very low dielectric constant (11 at 10 GHz), although not as low as that of Al_2O_3 (7 at 10 GHz); however, Al_2O_3 unfortunately reacts too strongly with YBCO. Attempts at combining these two materials with MgO as a buffer layer on Al_2O_3 have met with some success and other materials are now being evaluated as potential buffer layers on Al_2O_3.

A large area of substrate research is associated with the replacement of $SrTiO_3$ by other perovskitelike materials which have a low dielectric constant and low loss tangent together with a good lattice match. At a deposition temperature of 700 °C the lattice mismatch between $SrTiO_3$ and YBCO is +1.2%, that is, the $SrTiO_3$ unit cell is larger; the percentages quoted in the following are defined similarly. There is a complete family of perovskites potentially of use. At present those being closely considered are slightly distorted simple perovskites most like $SrTiO_3$ such as the aluminate $LaAlO_3$ (−2.6%) and gallates such as $LaGaO_3$ (+0.4%), $NdGaO_3$ (−0.27%) and $PrGaO_3$ (−0.02%). Next comes the series with the K_2NiF_4 structure (with the structure of the original LaSrCuO superconductor) such as $LaSrAlO_4$ (−3.1%) and $LaSrGaO_4$ (−0.5%). All these $SrTiO_3$ substitutes have dielectric constants in the range 24–28 at high frequencies. $NdGaO_3$ appears as quite close to an ideal in that the lattice mismatch at the deposition temperature is small and yet the film is still put under a small compressive stress at room temperature.

Certain tunnel junction devices depend for their operation on the growth of biepitaxial films with a single grain boundary of specific orientation (see *Thin Films, Multilayers and Devices, Superconducting*). This has been acheived by lithographically patterning a "seed" layer which alters the orientation of subsequently grown layers, the edge of the seed layer then defines the position of a grain boundary in the active layer. An example of an epiheterostructure similar to those already being incorporated into devices is shown in Fig.3.

1.6 Film Structure and Morphology

HTS thin films and heterostructures are now being consistently produced with good epitaxy; typically with a mosaic spread of 0.10° for films with the *a* axis in the plane of the substrate and a full-width half maximum (005) of 0.17° for films where the *c* axis is normal to the substrate. This is still large in comparison with the substrates implying that there is room for further improvement. If the material is grown exactly on stoichiometry and perfectly epitaxially then very smooth films will result. Most of the extensive morphological studies have been made on films deposited by coevaporation and these have shown a very consistent pattern of morphologies as the composition of the films deviate from stoichiometry. For instance, if films are copper poor they tend to have a pitted surface, while if copper rich, precipitate particles are seen. An example of the surface morphology variations with film composition as observed by Baudenbacher et al. (1992) is illustrated in Fig.4. They combine their studies of the film surface morphology with RHEED of the film growth mechanism and report that copper-poor films tend towards island

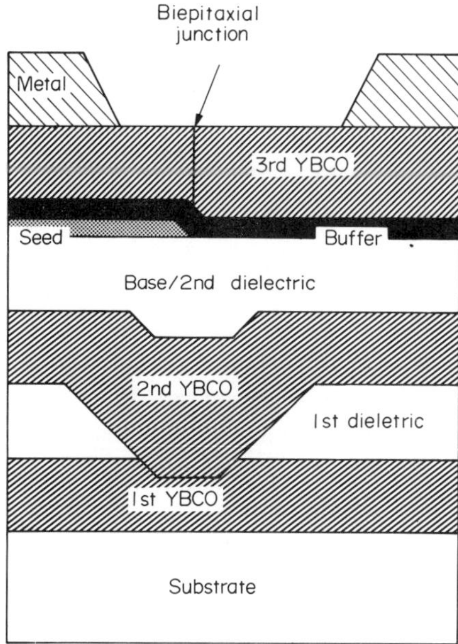

Figure 3
A schematic cross section of a multilayer epiheterostructure film for application in an integrated SQUID sensor, incorporating buffer layers, a "seed" layer, an insulating crossover and a controlled biepitaxial layer with a single grain boundary junction (courtesy of Conductus Inc.)

growth and that pinholes are observed prior to coalescence of the nucleated islands. In contrast, copper-rich films grow layer by layer in the early stages, followed by three-dimensional growth resulting in a pronounced grain structure and outgrowths. Outgrowths with *a* axis normal to the substrate may nucleate at second-phase particles as observed by Ramesh et al. (1991), at regions of film that are off stoichiometry, at lattice imperfections or at steps or defects on the substrate surface. After nucleation these outgrowths often develop a characteristic lenticular morphology due to relatively fast growth in the *b* direction as compared with the *c* direction.

2. *Deposition Methods*

The preparation of single-crystal films grown epitaxially on a single-crystal substrate requires high temperatures. At first this was achieved by a technique of postdeposition annealing (*ex situ* preparation), a method already generally in use for making other textured oxide films. Different substrates worked with varying degrees of success but all showed more chemical interaction at the substrate–film interface than was desirable and, in addition, the films exhibited a high level of surface roughness. It was rapidly recognized that *in situ* preparation would be a superior method. Here the film is deposited directly with the required crystal structure onto a hot substrate (Somekh et al. 1987). Such an *in situ* technique can be performed at a lower deposition temperature since it relies on surface, rather than bulk, diffusion. In addition to reducing the adverse effects of film–substrate interaction this enables the juxtaposition of fine-scale epitaxial insulating barriers and insulation layers which are an essential requirement for device applications. However, groups that have persisted with *ex situ* methods have shown that, despite the solid-state diffusion processes involved, films can be produced at comparable temperatures to many of the *in situ* methods by using lower oxygen pressures during annealing (see, for example, the work of Feenstra et al. 1991).

Essentially the first step of the *ex situ* route is to use the same deposition methods as described in the following for *in situ* films except that they are deposited onto substrates at much lower temperatures (300 °C to 400 °C) in a microcrystalline or an amorphous form. The film is then annealed at high temperatures to form an epitaxial film with the required crystal structure and orientation; the final anneal stage depends on which HTS is being prepared. In the case of YBCO a final oxygenation anneal is necessary after the film has been formed in the tetragonal state. This is typically performed at between 350 °C and 470 °C which implies from the thermodynamics that few films are made with a full O_7 oxygenation level. There is considerable evidence that this last stage may be improved by using some sort of active gas (Edwards et al. 1990).

All techniques used to grow these materials must bring the correctly proportioned material to a growth surface with the appropriate oxygen concentration and activity. The different deposition methods differ by virtue of the differing oxygen environments. Nearly all methods use some form of activated oxygen in their own environment. All methods, though giving reasonably well-oriented material at low temperatures (650 °C to 730 °C), tend to produce better material at 750 °C to 820 °C, which one can assume relates to the need to order the atoms in their structure. Most workers agree on the basic requirements to produce good films; however, the exact nature of the nucleation and growth phenomena are still being investigated and the recipe followed by each group reflects the peculiarities of each fabrication facility. Although the techniques used for making these films are the standard thin-film fabrication processes, small process control deviations can easily lead to loss of epitaxy and the nucleation of unwanted phases. Furthermore in a given laboratory the film deposition is often optimized for a particular property such as the critical current, microwave surface resistance or normal state resistivity.

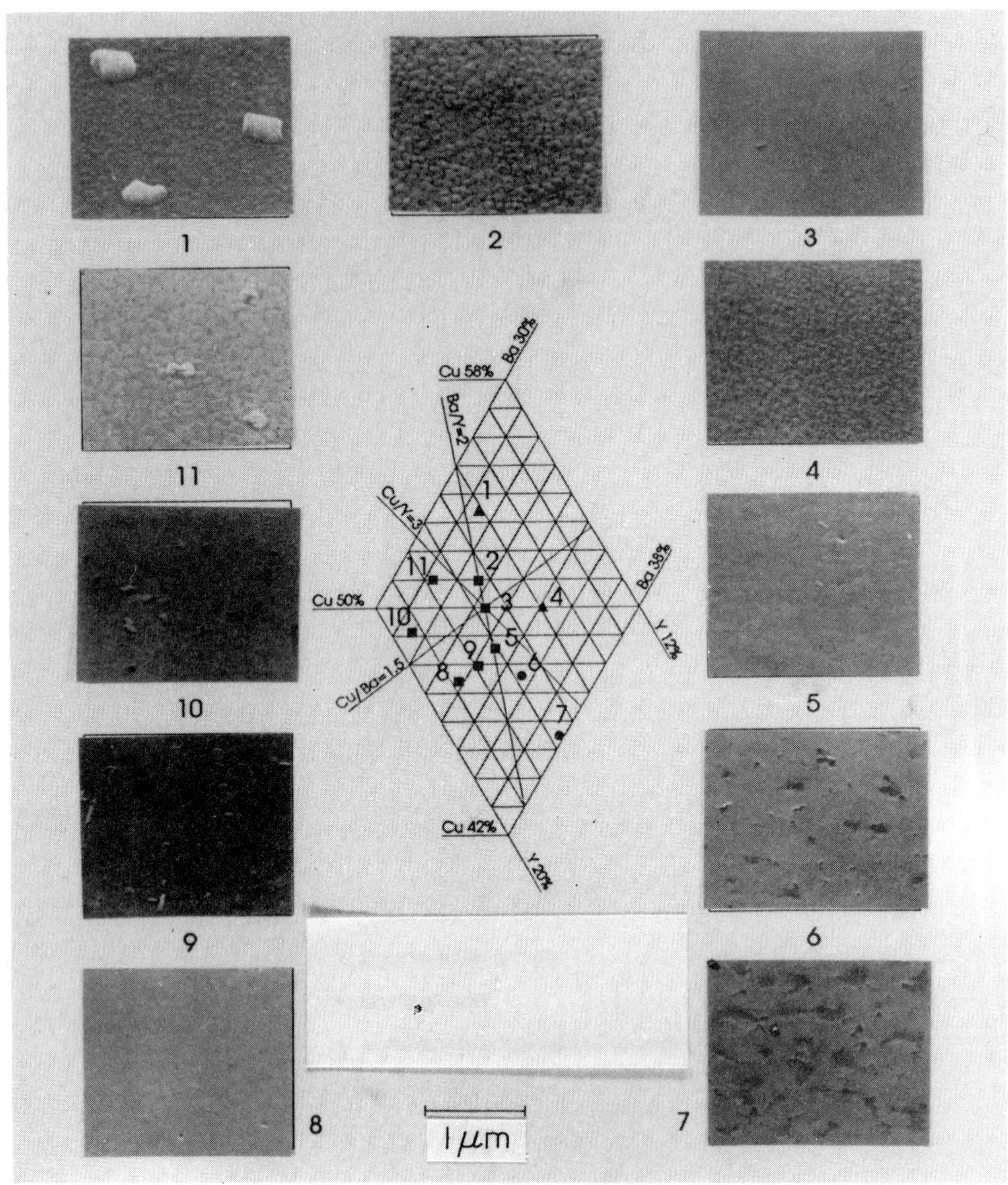

Figure 4
Surface morphologies of a series of YBCO films coevaporated under the same conditions but with different elemental fluxes (■, layer-by-layer growth; ●, island growth; ▲, transition to three-dimensional growth) (after Baudenbacher et al. 1992)

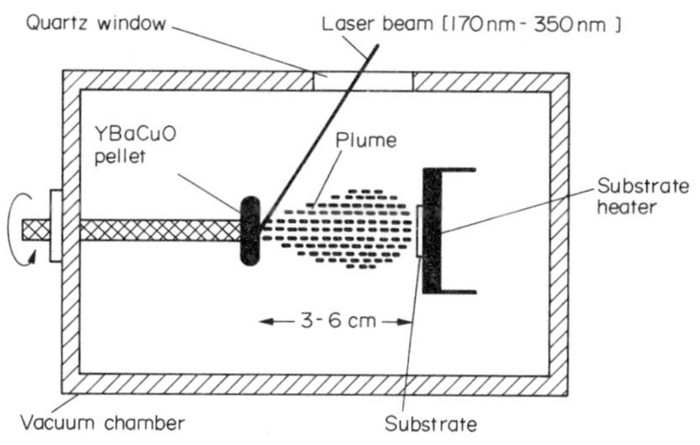

Figure 5
A schematic diagram showing the configuration of a laser deposition system (after Wu et al. 1988)

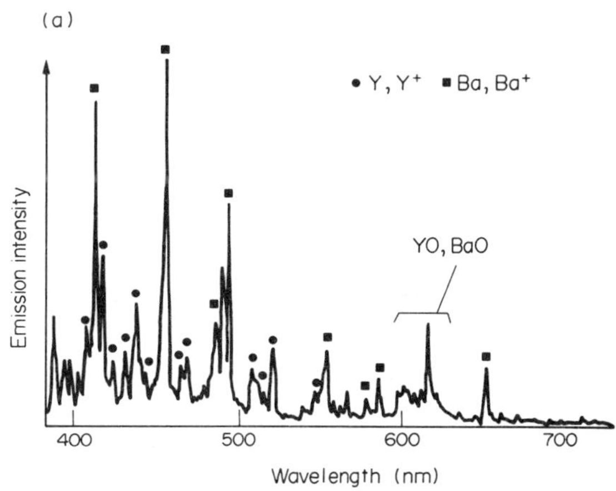

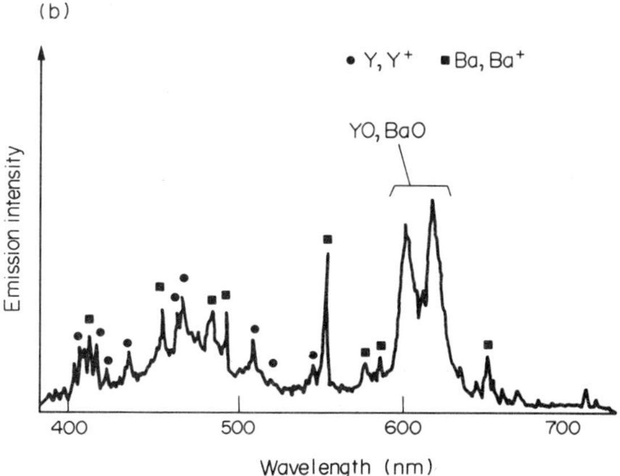

Figure 6
Spectroscopic analysis of the plume emanating from a target, (a) close to (20 mm) and (b) far from the target (50 mm) showing significant increase in the oxidation of the atomic species (after Yoshida et al. 1991)

2.1 Laser Ablation

This technique consists of focusing an accurately controlled excimer laser onto the surface of a rotating (sometimes water-cooled) target. The substrate to be coated is placed as shown in Fig. 5, at a distance on the edge but just within the visible part of the laser plume. An argon–oxygen gas mixture is used with a typical pressure of 20–50 Pa. With an excimer laser the short 170–260 nm wavelength (photon energy 5–7 eV) is able, for a modest energy of 0.1–2 J per pulse (typically of 10–40 ns duration) and an energy density of 1–4 J cm^{-2}, to create a powerful plume which glows brilliantly as it spreads from the target at initially supersonic velocities of 40 km s^{-1}. It travels some 40–60 mm from the target and a copious supply of nascent oxygen has been observed in the plume, as well as oxidized metal atoms (see Fig. 6). Although most of the oxygen in the film originates from the target the background oxygen pressure is also important; thus, as the deposition rate is increased from 0.3 μm min^{-1} to 3.7 μm min^{-1} the oxygen pressure has to be increased to maintain a T_c of 90 K (Yoshida et al. 1991).

The technique requires careful optimization since the composition of the film produced varies somewhat as a function of laser power density, target to substrate distance and the oxygen ambient. A particular problem is the appearance of micrometer-size and smaller droplets on the surface of the grown film; these originate from debris ejected from the region of impact of the laser pulse. A variety of procedures can help to reduce this problem so that the droplet density can be reduced by several orders of magnitude from about 10^5 cm^{-2} to less than 10^2 cm^{-2}. Such procedures include polishing the target between deposition runs, defocusing the laser spot in an optimal way or mechanically chopping part of the plume from the target, besides ensuring that the target does not rise too much in temperature.

In general terms laser ablation has proved the simplest technique to deposit YBCO films with a superconducting transition above 85 K, it can also deposit other HTSs successfully as well as the insulating layers required in the deposition of epiheterostructures for devices. Laser ablation has the potential to produce films very rapidly with consistently high current-carrying capacity and for the majority of present research applications it seems a very potent technique.

2.2 Sputter Deposition

Sputter deposition has the deserved reputation as being the technique for preparing thin films of alloys and complex materials and it was the natural method to try when these materials were first discovered. The problem of composition control in the presence of resputtering due to oxygen ions generated in the plasma had been recognized some 10 years before the discovery of HTSs and has been fully investigated by Rossnagel and Cuomo (1988). Resputtering can be minimized by thermalizing the energetic species either by working at very high pressure (which also provides a particularly rich oxygen environment) or by what has come to be called "off-axis" sputtering at a modestly high sputtering pressure, which forces all the atoms emanating from the target to have one or two energy-reducing collisions before arriving at the substrate.

Different combinations and techniques have been tried: mixed metal targets, separate oxide targets, separate metal targets, stoichiometric targets and multiple nonstoichiometric targets. Sputtering from fully oxygenated stoichoimetric targets has come to be seen as the most straightforward method either with off-axis techniques using a cylindrical magnetron, facing magnetrons or a single magnetron, or on axis with a very high sputtering pressure. All these configurations seem capable of producing films with T_c over 90 K and critical current densities of 2×10^6 to 8×10^6 A cm^{-2} at 77 K, although problems with reproducibility are common. Gavaler et al. (1991) have made a systematic study and find that in general off-axis sputtering can be more reproducible than on-axis sputtering and that dc sputtering is better than rf sputtering. They suggest that the high level of thermalization is not only needed to prevent resputtering of cations but is also concerned with the loss of a weakly held species such as atomic oxygen from the surface of the growing film.

Sputtering can produce films of slightly higher structural quality and does not suffer from the droplet problem of laser ablation, but it seems inherently slow (typically 0.3–1.5 nm min^{-1}) because of the high levels of thermalization required and the limits on power density set by the low thermal conductivity of the targets; this makes the technique expensive. However, it is being widely used as the workhorse technique to produce films for many programs because it is reliable and relatively easy to run automatically or semiautomatically.

2.3 Evaporation and MBE

The pioneering work of Hammond (1975, 1978) and others in the 1970s and early 1980s laid the foundations on which ultrahigh-vacuum evaporation systems have been based. Various forms of spectroscopic techniques have been developed for the independent accurate control of several sources simultaneously in an oxygen environment. Evaporation must operate with a background pressure of less than 10^{-2} Pa and yet high oxygen levels are required for the formation of YBCO, as seen from Fig. 1. This conflict has been resolved by using various forms of active oxygen often injected close to the substrate. Sources of active oxygen based on ozone generators, electron cyclotron resonance (ECR) or an rf plasma have been developed. A review of this subject has been given by Humphreys et al. (1990). Growth rates are currently of the order 5–10 nm min^{-1}; although evaporation and MBE have the potential for high-rate deposition (600 nm min^{-1}) the supply of oxygen and the maintenance of an adequate base pressure would become major problems.

Evaporation and MBE techniques are quite difficult undertakings but are scientifically rewarding because one may easily incorporate *in situ* analysis, particularly RHEED, which can give a continuous signature of the perfection of the structure being grown. Also, direct control of the composition may be relatively easily incorporated. Though individual atom layered structures (sometimes called ALE) have been attempted with both laser ablation and sputtering, evaporation presents the obvious route for film growth with this sort of atomic precision.

2.4 Chemical Vapor Deposition Techniques

Metallorganic chemical vapor deposition (MOCVD) and plasma assisted MOCVD (PAMOCVD) represent techniques that have considerable potential for reasonable deposition rates and large area coverage. The main means of composition control in this technique is by adjusting the rate of evaporation of the precursor into a constant stream of an oxygen-rich carrier gas which then passes over the (heated) substrates in the susceptor in a separate chamber. For HTSs the main challenge is to develop suitable volatile precursors. Recently simplified methods have been developed in which the precursors are evaporated directly in the same chamber (Schulte et al. 1992) or a premixed compacted powder is used (Hiskes et al. 1991). There tends not to be active control, so the process depends on a high level of reproducibility of the flow and evaporation from the sources. Once again, by reducing the oxygen partial pressure the deposition temperature is reduced; high-pressure, high-temperature techniques are at the moment being compared with (and replaced by) lower-temperature techniques as used by Zhao et al. (1991) which make use of lower pressures of oxygen, activated in some way.

The best films being produced have critical current densities of order 3×10^6 A cm^{-2} and T_c close to 90 K which compares well with other deposition processes. MOCVD and its variants are seen as the techniques which could coat large areas at very high deposition rates (up to 50–100 nm min^{-1}) for commercial application.

2.5 Deposition of Other HTSs

Although most effort has been concentrated on the deposition of YBCO there is increasing scientific and technological interest in the preparation of high quality films of other oxide superconductors. The bismuth- and thallium-based compounds (see *Oxide Superconductors: Solid-State Chemistry*) are almost all routinely prepared by *ex situ* methods involving cold deposition followed by a heat treatment. The reason for this approach is that at present no one has found it easy to overcome the problem of the thallium reevaporation. In the more complex bismuth compounds, besides reevaporation, there is in addition the problem that the thermodynamic driving forces for the various phases are very similar so that it is difficult to grow the material as a single phase. Wasa et al. (1991) and Kawai et al. (1992) have tried very careful layer-by-layer growth but although the correct structure seems to have resulted few workers have succeeded in producing high-T_c phases.

$BaKBiO_3$ is a cubic material which has been prepared by a few groups for tunnelling experiments. The technique which has been widely published is coevaporation although the relative volatility of the metallic elements makes its deposition difficult. Hellman et al. (1991) report T_c in the range 20–25 K.

3. Concluding Remarks

Remarkable advances have been made in the science and technology of oxide thin-film deposition since the discovery of HTSs in 1986. Each of the four main deposition methods has a role to play although laser ablation is undoubtedly the simplest technique to establish for a laboratory not previously involved in precision thin-film deposition.

From a technological viewpoint objectives in the near future will be to deposit fully oriented smooth films and epiheterostructures on large-area substrates with suitable buffer layers. In the case of YBCO the main problems are to control oxygenation and manage the tetragonal to orthorhombic transition without introducing an undesirable twin stucture. For other HTSs and in particular the bismuth and thallium phases the main objective must be to achieve full *in situ* deposition of smooth films and heterostructures.

There remain several real fundamental problems to be addressed. As is often the case, the technology can easily outstrip the scientific understanding and the growth issues of these complex materials are still only beginning to be considered on a fundamental level. How does nucleation really start, with the full unit cell or a precursor to it? What species are mobile if different levels of oxygen are present (e.g., Y or YO_x)? What initiates the nucleation of an unwanted phase if the composition is slightly off stoichiometry? How are the phase limits on the surface modified by the oxygen activity? Even in well-researched areas such as the silicon–germanium system there are still problems associated with a full understanding of interfaces so it is not surprising that these new oxide systems present many difficult challenges.

See also: Oxide Superconductors: Electron Microscopy; Oxide Superconductors: Physical Properties; Oxide Superconductors: Solid-State Chemistry; Thin Films, Multilayers and Devices, Superconducting

Bibliography

Balestrino G, Foglietti V, Marinelli M, Milani E, Paoletti A, Paroli P 1990 Transport critical current density in epitaxial $Bi_2Sr_2CaCu_2O_{8+x}$ films: effects of the substrate twinning. *Appl. Phys. Lett.* 57: 2359–61

Baudenbacher F, Hirata K, Berberich P, Kinder H, Assman W 1991 Smooth YBCO films on MgO with good superconducting properties. In: Correra 1992, pp. 365–70

Correra L 1992 *High Temperature Superconductor Thin Films*. North-Holland, Amsterdam

Edwards J A, Chew N G, Goodyear S W, Satchell J S, Blenkinsop S E, Humphreys R G 1990 Optimisation of $YBa_2Cu_3O_7$ thin films for multilayers. *J. Less-Common Met.* 164, 165: 414–21

Feenstra R, Lindemer T B, Budai J D, Galloway M D 1991 Effect of oxygen pressure on the synthesis of $YBa_2Cu_3O_{7-x}$ thin films by post-deposition annealing. *J. Appl. Phys.* 69: 6569–85

Fork D K, Ponce F A, Tramontana J C, Newman N, Phillips J M, Geballe T H 1991 High critical current densities in epitaxial $YBa_2Cu_3O_{7-\delta}$ thin films on silicon-on-sapphire. *Appl. Phys. Lett.* 58: 2432–4

Gavaler J R, Talvacchio J, Braggins T T, Forrester M G, Greggi J 1991 Critical parameters in the single target sputtering of YBCO. *J. Appl. Phys.* 70: 4383–91

Habermeier H-U, Lourenco A A C S, Leibold B, Kircher J, Friedl B, Lu G 1991 Preparation and properties of YBCO thin films with the *c*-axis aligned in the film plane. In: Correra 1992, pp. 343–52

Hammond R H 1975 Electron beam evaporation synthesis of A15 superconducting compounds: Accomplishments and prospects. *IEEE Trans. Magn.* 11: 201–9

Hammond R H 1978 Synthesis and physical properties of superconducting compound films formed by electron beam co-deposition of the elements. *J. Vac. Sci. Technol.* 15: 382–5

Hellman E S, Hartford E H, Gyorgy E M 1991 Epitaxial $Ba_{1-x}K_xBiO_3$ films on MgO: nucleation, cracking, and critical currents. *Appl. Phys. Lett.* 58: 1335–7

Hiskes R, DiCarolis S A, Young J L, Laderman S S, Jacowitz R D, Taber R C 1991 Single source metalorganic chemical vapor deposition of low microwave surface resistance $YBa_2Cu_3O_7$. *Appl. Phys. Lett.* 59: 606–7

Humphreys R G, Satchell J S, Chew N G, Edwards J A, Goodyear S W, Blenkinsop S E, Dosser O D, Cullis A G 1990 Physical vapour deposition techniques for the growth of $YBa_2Cu_3O_7$ thin films. *Supercond. Sci. Tech.* 3: 38–52

Inam A, Ramesh R, Rogers C T, Wilkens B, Remschnig K, Hart D, Barner J 1991 Properties of epitaxially grown *a*-axis oriented $YBa_2Cu_3O_{7-x}$-$PrBa_2Cu_3O_{7-y}$ heterostructures. *IEEE Trans. Magn.* 27: 1603–6

Kawai T, Kanai M, Matsumoto T, Tabata H, Horiuchi K, Kawai S 1991 Superconducting artificial lattices grown by laser MBE. In: Correra 1992, pp. 287–94

Lang H P, Frey T, Sum R, Guntherodt H-J 1991 Atomic resolution and nanostructure of $Y_1Ba_2Cu_3O_{7-\delta}$ laser-ablated epitaxial thin films on various substrates studied by scanning tunneling microscopy. In: Correra 1992, pp. 789–94

Linker G, Brecht E, Geerk J, Meyer O, Rauschenbach B, Reiner J, Remmel J, Ritschel Ch, Wang R L 1991 Growth of *a*-axis oriented YBaCuO and EuBaCuO thin films. In: Correra 1992, pp. 359–64

Matijasevic V, Rosenthal P, Shinohara K, Marshall A F, Hammond R H, Beasley M R 1991 Reactive coevaporation of YBaCuO superconducting films. *J. Mater. Res.* 6: 682–98

Miklich A H, Wellstood F C, Kingston J J, Clarke J 1991 Thin-film YBCO magnetometer. *Nature* 352: 482–3

Miyazawa S, Sasaura M, Mukaida M 1991 Evidence of two-dimensional epitaxial growth of $YBa_2Cu_3O_x$ thin films on lattice-matched $NdGaO_3$ substrates. In: Correra 1992, pp. 741–6

Ramesh R, Inam A, Hwang D M, Sands T D, Chang C C, Hart D L 1991 Surface outgrowth problem in *c*-axis oriented Y–Ba–Cu–O superconducting thin films. *Appl. Phys. Lett.* 58: 1557–9

Rossnagel S M, Cuomo J J 1988 Negative ion effects during magnetron and ion beam sputtering of $YBa_2Cu_3O_x$. In: Harper J M E, Colton R J, Feldman L C (eds.) 1988 *Thin Film Processing and Characterization of High-Temperature Superconductors*, American Institute of Physics Conf. Proc. 165. AIP, New York, pp. 106–13

Routbort J L, Rothman S J, Chen N, Mundy J N, Baker J E 1991 Site selectivity and cation diffusion in $YBa_2Cu_3O_{7-\delta}$. *Phys. Rev. B* 43: 5489–97

Scheel H J, Berkowski M, Chabot B 1991 Substrates for high-temperature superconductors. *Physica C* 185–189: 2095–6

Schlom D G, Anselmetti D, Bednorz J G, Broom R, Catana A, Frey T, Gerber Ch, Guntherodt H-J, Lang H P, Mannhart J, Muller K A 1992 Screw dislocation mediated growth of sputtered and laser-ablated $YBa_2Cu_3O_{7-\delta}$ films. *Z. Phys. B* 86: 163–75

Schulte B, Maul M, Becker W, Elschner S, Schlosser E G, Haussler P, Adrian H 1991 A new MO-CVD technique for the preparation of $Y_1Ba_2Cu_3O_{7-\delta}$ thin films. In: Correra 1992, pp. 325–30

Somekh R E, Blamire M G, Barber Z H, Butler K, James J H, Morris G W, Tomlinson E J, Schwarzenberger A P, Evetts J E 1987 High superconducting transition temperatures in sputter-deposited YBaCuO thin films. *Nature* 326: 857–9

Talvacchio J 1989–1991 Bibliography of high-T_c superconductivity. *J. Supercond.* 2(1): 1–210; 3(1): 1–153; 4(2): 75–169

Terashima T, Bando Y, Ijima K, Yamamoto K, Hirata K, Hayashi K, Kamigaki K, Terauchi H 1990 Reflection high-energy electron diffraction oscillations during epitaxial growth of high-temperature superconducting oxides. *Phys. Rev. Lett.* 65: 2684–7

Venables J A, Spiller G D T, Hanbücken M 1984 Nucleation and growth of thin films. *Rep. Prog. Phys.* 47: 399–459

Wasa K, Adachi H, Hirochi K, Ichikawa Y, Matsushima T, Setsune K 1991 Thin film processing for high-T_c superconductors of the Bi-system. *J. Mater. Res.* 6: 1595–604

Wu X D, Venkatesan T, Inam A, Chase E W, Chang C C, Jeon Y, Croft M, Magee C, Odom R W, Radicati F 1988 Versatility of pulsed laser deposition technique for preparation of high T_c superconducting thin films. *SPIE High-T_c Superconductivity: Thin Films and Devices* 948: 50–65

Yoshida N, Okuda S, Takano S, Nagata M, Yamamoto T, Hara T, Okaniwa K 1991 Y-Ba-Cu-O superconducting film deposition by the laser ablation method. *Sumitomo Electr. Tech. Rev.* 31: 49–54

Zhao J, Chern C S, Li Y Q, Norris P, Gallois B, Kear B, Wu X D, Muenchausen R E 1991 Compositional effects on plasma-enhanced metalorganic chemical vapor deposition of $YBa_2Cu_3O_{7-x}$ thin films. *Appl. Phys. Lett.* 58: 2839–41

R. E. Somekh, Z. H. Barber and J. E. Evetts
[University of Cambridge, Cambridge, UK]

P

Paramagnetic Materials

Materials can exhibit paramagnetic behavior for a variety of reasons. The most important factor is whether or not a material contains magnetic moments on constituent atoms in the absence of a magnetic field. All materials that undergo, at some temperature, any form of magnetic ordering are, above that temperature, paramagnetic in their response to a field, as are also those magnetically ordered or frozen materials that possess no spontaneous magnetization below that temperature.

There is no complete agreement on SI units for susceptibility. In this article the dimensionless volume susceptibility $\kappa=\mu_0 M/B_0$ is used, giving a mass susceptibility χ in units of $m^3\,kg^{-1}$.

1. Paramagnetism of Electrons in Metallic Energy Bands

The paramagnetism of electrons in metallic energy bands is the simplest form of response to an applied field and exists in all metallic materials. However, in some, the net response may be dominated by a larger diamagnetic response of closed shells (e.g., copper, silver, gold) or an additional term (Landau–Peierls diamagnetism) in the band susceptibility that becomes large when energy bands are full or nearly so (e.g., bismuth, antimony, γ-copper–zinc).

Band paramagnetism (sometimes called Pauli paramagnetism) arises simply from the energetically favorable slight repopulation of the spin-up (parallel to the field) and spin-down bands when the field is applied. Although normally small (volume susceptibility $\sim 10^{-6}$ in dimensionless SI units), it is enhanced when the relevant energy band is narrow, as in the transition metals. In such materials the temperature dependence is also greater, since significant variations in the electronic density of states may be taking place on an energy scale of the order of $k_B T$. However, Since $\mu_B B$, where μ_B is the Bohr magneton, is much smaller, a field dependent susceptibility is not normally found. Data for various metals are shown in Fig. 1. It should be emphasized that a simple calculation of the susceptibility from accurate band structure calculations or experimentally determined densities of electron states will always give too small a value for the susceptibility. This is due to the enhancement produced by exchange interactions even when these are not strong enough to produce spontaneous magnetic ordering. In palladium, where this enhancement is large, an increase in exchange can be mimicked by alloying with isoelectronic nickel to follow the transition to ferromagnetism at about 2.5% nickel.

2. Paramagnetism of Materials with Local Atomic Moments

Except for atoms or ions with only closed shells of electrons, all atoms and ions possess magnetic moments that can respond to an applied field by virtue of their possession of a net electronic angular momentum, specified by the quantum number J. The ratio (the magnetomechanical ratio) of the magnetic moment to the electronic angular momentum is dependent on the relative contributions of spin (S) and orbit (L) to the value of J. (In the few instances where S and L cancel to give $J=0$, a weak temperature-independent paramagnetism (Van Vleck paramagnetism) is produced by the role of the applied field in mixing small amounts of excited states with nonvanishing J into the nonmagnetic ground state.) It

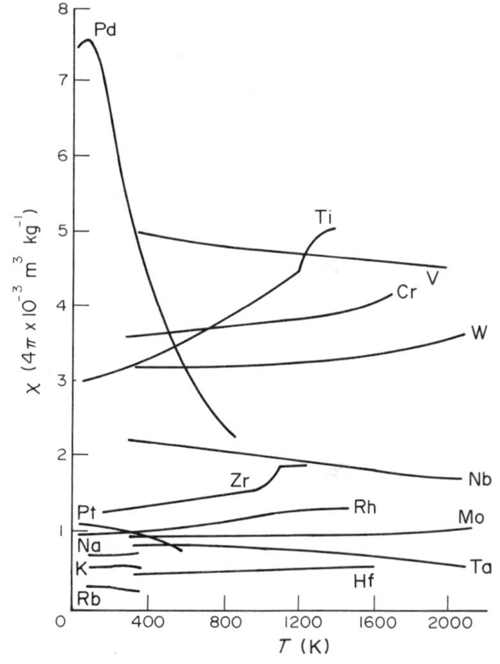

Figure 1
Mass susceptibility vs temperature for various metals (after Kittell 1971). The equivalent cgs units of susceptibility are 10^{-6} emu (i.e., erg $Oe^{-2}\,g^{-1}$)

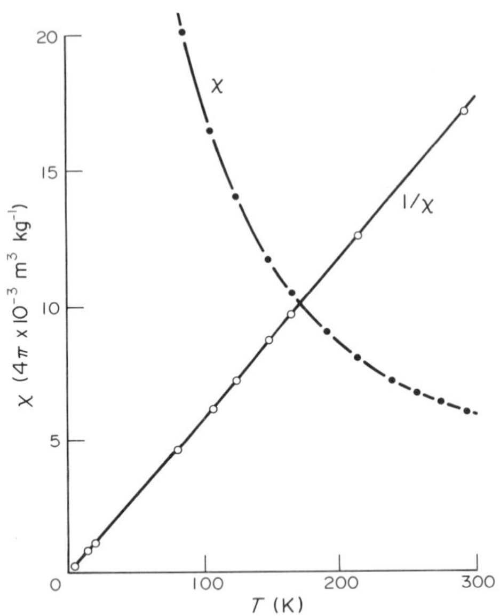

Figure 2
Mass susceptibility and its reciprocal as a function of temperature for $CuSO_4.5H_2O$ (Curie law) (after Crangle 1977)

must be recognized, however, that for most atoms the very nature of the chemical or metallic bonding destroys these moments in the formation of solid elements and compounds. Thus, such moments persist in solid materials only on atoms or ions possessing incomplete electron shells only slightly perturbed by the formation of the materials. In practice, this restricts the materials to be considered under this heading to compounds and alloys of 3*d* transition elements, 4*f* rare-earth elements and 5*f* actinide elements. Where the perturbations to the atomic moments μ can be ignored and in the absence of indirect coupling of such moments to one another by conduction electrons or other atoms, the resultant susceptibility (the Curie law susceptibility) is $N\mu_0\mu^2/3k_BT$ where N is the number of atomic moments per unit volume (see Fig. 2).

Fuller accounts of rare-earth-containing materials can be found in the articles *Rare-Earth Metals and Alloys* and *Rare-Earth Intermetallics*. Consideration here will be restricted to the general features that govern the sizes of the moments and to a brief account of some of the special features of the environment of these moments in some materials that modify their response. Since the magnetomechanical ratio is twice as large for spin as for orbital angular momentum, the magnetic moment has to take into account the relative contributions of the two so that μ is equal to g times the total angular momentum vector where g is the Landau splitting factor $[(3J(J+1)+S(S+1) - L(L+1))/2J(J+1)]$. As the expectation value of the angular momentum is, in quantum mechanics, $(J(J+1))^{1/2}$, the effective moment μ_{eff} that replaces μ in the Curie law is $g(J(J+1))^{1/2}\mu_B$. For a given atom or ion, J can then be found by applying Hund's rules to give the values of S and L that specify the ground state for the particular configuration (e.g., $3d^9$ for Cu^{2+} giving $S=\frac{1}{2}$ and $L=2$ or $4f^3$ for Nd^{3+} giving $S=\frac{3}{2}$ and $L=6$) and applying the rule that makes $J=L-S$ for a less than half full shell and $L+S$ for a more than half full shell. Curie's law can be shown to be valid only for small values of $\mu_0\mu_{eff}H/k_BT$ and at very low temperatures or high fields the magnetization must obviously approach saturation at $gJ\mu_B$. In a field of 1 T this behavior only becomes significant below about 2 K.

Atomic moments derived on this basis often fail to explain the observed behavior of crystals containing 3*d* and 4*f* elements, even in the absence of conduction electrons that can be expected to introduce complications. Such deviations have to be ascribed to the nonspherical crystalline electric field experienced by the 3*d* and 4*f* shells. The effect is naturally much stronger for the former because of the greater radial extent of the 3*d* wave functions in transition metals. Often it is strong enough for it to be a good approximation to regard the orbital contribution as quenched, leaving μ_{eff} equal to $2(S(S+1))^{\frac{1}{2}}\mu_B$. For 4*f* shells the effect is weaker and above 100–200 K the full J contribution is found, although at the lowest temperatures a reduced moment is found corresponding to an eigenstate of the full Hamiltonian. For ions with even numbers of 4*f* electrons this may often be a nonmagnetic singlet and a temperature-independent Van Vleck susceptibility is then observed, analogous to that found for a $J=0$ singlet of the spin–orbit coupling.

3. Local Moments in Dilute Magnetic Alloys

For more concentrated magnetic systems conduction electrons play an important role in coupling local moments to give a variety of magnetically ordered states, but for dilute alloys they produce subtler effects which have only recently been given a satisfactory description. Such effects are important in the transport properties (see *Magnetotransport Phenomena*) and in superconductivity as well as in the paramagnetic response (Dugdale 1978). As first pointed out by Friedel for dilute alloys of 3*d* metals, local 3*d* states will hybridize to some degree with conduction electron states. For sufficiently strong hybridization (as for nickel, cobalt and iron in solid solution in aluminum) no local moment is found and a weakly temperature-dependent response is observed, rather as in metallic palladium where the hybridization of conduction and 4*d* electrons also plays a role. For weak hybridization more subtle many-body effects can be approximated by an effective exchange J_{eff} between the local moment and the conduction electrons that is of negative sign, so that below some Kondo temperature the moment is compensated and the susceptibility becomes temperature independent. (This temperature is around 10 K for iron in copper, but below 1 K for iron in gold, so that at higher iron concentrations in the latter system iron–iron interactions lead to a magnetically ordered or frozen state, the spin glass.) For most 4*f* ions the hybridization effect is too weak to counter the normal exchange coupling, but for some with less stable 4*f* configurations, cerium in particular, the Kondo effect can operate.

4. Kondo Lattices or Heavy Fermion Compounds

Unlike compounds of 3*d* metals, more concentrated atomically ordered cerium intermetallic compounds can show the Kondo effect because 4*f*–4*f* overlap effects remain unimportant (see *Heavy Fermion Metals*; *Rare-Earth Intermetallics*). In $CeCu_6$, for example, the high-temperature behavior of the resistivity and susceptibility is that of a noninteracting set of 4*f* moments with Kondo scattering of the conduction electrons leading to a negative temperature coefficient of resistance. At low temperatures, coherence of the hybridized 4*f*–conduction electron states leads to a maximum in the resistivity, which falls rapidly below 10 K, but has little effect on the susceptibility which shows the low-temperature approach to a constant value typical of individual Kondo moments, while remaining paramagnetic. Some other Kondo lattice systems (e.g., $CeAl_2$ and U_2Zn_{17}) show antiferromagnetic ordering at low temperatures, often with reduced moments, some become superconducting (e.g., $CeCu_2Si_2$ and UPt_3) and very remarkably, URu_2Si_2 first becomes antiferromagnetic (17.5 K) and then superconducting (1 K). It has been suggested that such behavior has links with high-temperature cuprate superconductors, since variations in doping or oxygen deficit can change the cuprates from antiferromagnetic to superconducting (see *Oxide Superconductors: Physical Properties*), with the possibility of overlap. In these materials there is clearly again a balance between local 3*d* states on copper atoms (with possible local moments) and hybridized states, involving here oxygen 2*p* states.

5. Paramagnetic Response of Ordered or Frozen Moments

Two categories of materials which have no spontaneous magnetization show a paramagnetic response in an applied field which is not that of conduction electrons or noninteracting atomic moments. These are antiferromagnets below their ordering (Néel) temperature T_N and spin glass materials (metallic or nonmetallic) below the spin glass temperature T_g. Theoretical arguments still continue about whether this temperature, marked by a peak in the ac susceptibility, should be regarded as a true magnetic phase transition (Binder and Young 1986). The two types differ, however, in that a zero-field magnetization can be induced in spin glasses, either by cooling in a moderate to large dc field which is removed well below T_g or by cycling a dc field up to large values below that temperature. The susceptibility of single crystals of antiferromagnetic material below T_N (see Fig. 3) is strongly anisotropic, since applied fields can produce little effect at low temperatures if applied parallel to the direction of moment alignment when $\mu B \ll k_B T$, while they can induce a perpendicular

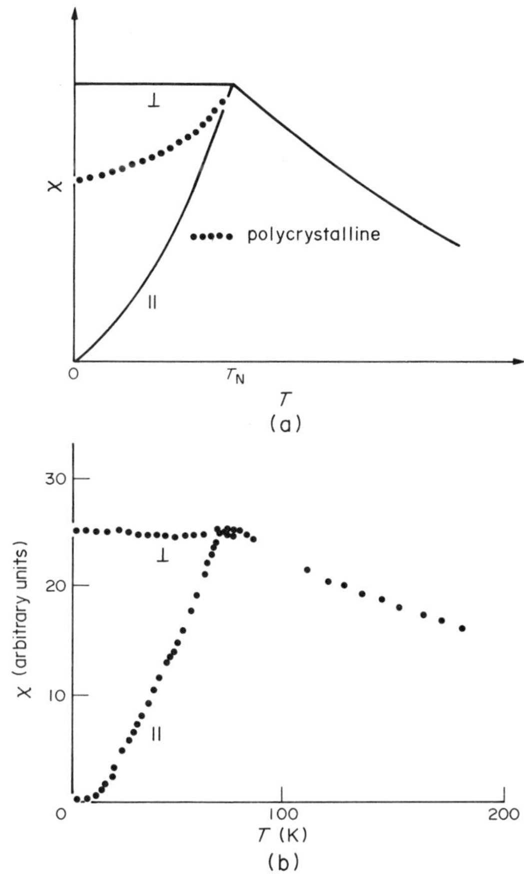

Figure 3
Schematic presentation of (a) the anisotropy in the temperature variation of the antiferromagnetic susceptibility and (b) experimental data for a crystal of MnF_2 (after Myers 1990)

magnetization by tilting the moments in both sublattices. Above T_N the susceptibility of an antiferromagnet is $N\mu_0\mu^2/3k_B(T-\theta)$ like that of a ferromagnet above T_c; for the former θ is negative, while for the latter it is positive and $\simeq T_c$.

Bibliography

Binder K, Young A P 1986 Spin-glasses—experimental facts, theoretical concepts, and open questions. *Rev. Mod. Phys.* 58: 801–976

Crangle J 1977 *The Magnetic Properties of Solids.* Arnold, London

Dugdale J S 1978 *The Electrical Properties of Metals and Alloys.* Arnold, London

Kittell C 1971 *Introduction to Solid State Physics.* Wiley, New York

Myers H P 1990 *Introductory Solid State Physics.* Taylor and Francis, London

Van Vleck J H 1932 *Theory of Electric and Magnetic Susceptibilities.* Oxford University Press, Oxford

B. R. Coles
[Imperial College of Science, Technology and Medicine, London, UK]

Powder Cores, Magnetic

A metal powder core, one formed by the compression of insulated metal particles, is a natural extension of the subdivision of a magnetic material from bulk to laminations to wire to powder. Since eddy current losses increase with the smallest dimension perpendicular to the flux path, metal powder cores are very advantageous at high frequencies where these losses are most important. The effective relative permeability of these cores ranges from 5 to 550. The lower effective permeability leads to increased stability of this permeability against temperature variations and dc bias conditions. In addition to the linearity in the initial permeability region, the shearing of the hysteresis loop by the distributed air gap prevents saturation at high power levels. This, in turn, prevents high currents in the windings and subsequent overheating.

Powdered iron cores were used for telephone loading coils from 1918 and for radio cores from 1933. Improved materials such as powdered permalloy (1925) and finally powdered molybdenum-substituted permalloy (1940) were developed for these applications. In 1937, Sendust, an Al–Si–Fe alloy, was invented in Japan. Recently, powder cores of 50:50 nickel–iron were introduced for power applications.

1. Iron Powder Cores

Iron for powder cores is prepared by chemical reduction, mechanical disintegration, electrolytic deposition or the carbonyl process. The first two are the least costly and the last is the most expensive. The powder is insulated by superficial oxidation, phosphating with phosphoric acid or coating with water glass or another inorganic insulant. In some cases, an additional binder is added. The powder is pressed at varying pressures up to 700 MPa (50 ton in^{-2}). The core is then heat treated to cure the resin or for stabilization when inorganic insulants are used. A coat of paint is then applied to prevent oxidation or moisture penetration.

The properties of iron powder cores depend greatly on the chemical purity, particle size and shape of the iron powder used. High-purity larger-particle high-density powders give higher permeabilities at low and medium frequencies. Smaller or layered particles (such as those prepared by the carbonyl process) give lower permeabilities at low frequencies but permit operation with low losses at high frequencies. High-density cores, achieved by lower insulation percentages and higher forming pressures, have higher permeabilities. The highest relative permeability is typically about 100 and is effective to about 100 kHz. On the other hand, carbonyl iron cores with relative permeabilities of 5–10 are usable up to several hundred megahertz. For low permeability cores, the temperature coefficient of initial permeability is of the order of 10–50 ppm °C^{-1}, while for higher permeability cores it is several hundred ppm per degree. A similar situation exists with dc bias. An iron powder core with a relative permeability of 20 retains that value at a superimposed dc field of 1600 A m^{-1} (20 Oe), while one with a relative permeability of 60 decreases 15% under the same conditions.

As stated previously, the earliest application of iron powder cores was in telephone loading coils. These cores are used to compensate for the increased capacitance in long transmission lines. Other applications include if and rf coils, noise filter coils, FM coils, oscillator coils, permeability tuning, waveguide antenna coils and television coils. Recent uses are in power applications for switched mode power supplies, switching regulators, pulse transformers and light-dimming SCR circuits.

As a variation, iron powder can be rolled into flake and cores pressed with the flake oriented in the pressing plane. These cores have comparatively high relative permeability (about 90) and can be made from relatively inexpensive material.

2. Permalloy and Molybdenum-Permalloy Powder Cores

Between 1929 and 1940, powder cores were made of permalloy, an 80% Ni–Fe alloy (see *Nickel–Iron Alloys*) which was an improvement over existing powder iron cores and led to a reduction in core size.

Legg and Given (1940) reported the development of a further improved material, 2–81 permalloy powder cores (2% molybdenum, 81% nickel, balance iron). The molybdenum increases the resistivity of the alloy, reducing eddy currents. In addition, the permeability is increased. This material has been in continuous use since then with improvements being made mainly in insulation and powder processing. Early cores were of toroidal shape and this shape is still used for loading coils. Other shapes with air gaps, such as pot cores, offer simple inductance adjustments.

Powder for molybdenum-permalloy cores is manufactured somewhat differently from iron powder. One method includes embrittlement of the material by the use of additives such as sulfur. The alloy is melted and cast into an ingot, hot rolled to a thickness of about 6 mm ($\frac{1}{4}$ in), granulated and milled. The powder must be heat treated before pressing. The material is sieved into size fractions, with the +120 mesh (+0.125 mm) being used for higher permeability cores and the −400 mesh (−0.037 mm) powder for low permeability cores. Insulants such as magnesium oxide, sodium silicate and clay have been used but the insulation procedure is usually a manufacturing secret. The amount of insulation applied will partially determine the permeability. The cores are pressed at pressures up to 1400 MPa (100 ton in^{-2}). The cores must be further heat treated in hydrogen at about 650 °C to relieve the pressing strains. To eliminate the shorting out effect of surface smearing of the metal during pressing, acid etching and oxidation are used to improve the properties. The cores are dried and coated with a protective paint.

3. *Properties of Molybdenum-Permalloy Powder Cores*

The excellent properties of molybdenum-permalloy material are derived from the simultaneous minimization of the magnetocrystalline anisotropy and magnetostriction giving high permeability. Hysteresis losses are very low, eliminating unwanted harmonics. These harmonics are deleterious as they give rise to cross talk in telephone circuits. For use as a powder core, another useful property is the malleability, allowing the powder to conform during the pressing operation so that the air gap can be minimized, again increasing permeability. Because of the variety of processing conditions, relative permeabilities from 14 to 300 can be obtained. The high-permeability cores are used for loading coils and low-frequency coils. Cores with relative permeabilities of 14 and 26 are used for high-frequency filters. A core with a relative permeability of 125 will retain that permeability to a flux density of 0.3 T (3000 G). After applying a dc magnetization to saturation, the core with relative permeability 14 changes only 0.02% in inductance. Temperature stability is another critical feature in many applications. Aside from the basic increase in temperature stability due to low permeability, molybdenum-permalloy cores can be stabilized in another manner. This is done by adding to the basic 2–81 powder a fraction of low-Curie-point powder (often a 12% molybdenum alloy). The superposition of the two temperature courses of permeability flattens out the resultant curve. Another type of core has a linear variation of permeability with temperature for use in *LC* filters with polystyrene capacitors.

An important factor affecting stability is the effect of moisture. Since the intergranular insulation is oxidic, moisture penetrating the core will swell the insulation and increase the air gap, lowering the permeability. Therefore, it is important that the core be protected after drying. This protection is in the form of a varnish or paint. In addition to the air gap effect, the adsorbed moisture will decrease the dielectric constant and increase the losses. The varnish or paint will often be required to pass a voltage breakdown test of 1250 V or higher.

Other applications include coils for audio frequency, high- and low-pass filters (high-Q applications), multiplexing filters, noise filters, decade inductors, noise suppression filters and rf interference filters.

Flake cores made from molybdenum-permalloy powder can achieve relative permeabilities up to 550. This flake is rolled, insulated in the same way as the molybdenum-permalloy powder and pressed after orientation of the flakes in the pressing plane.

4. *Sendust Powder Cores*

Masumoto invented Sendust in 1937, naming it after Sendai where it was developed (see *Iron–Aluminum and Iron–Aluminum–Silicon Alloys*). It was designed to be a substitute for molybdenum permalloy. The composition is about 10% silicon, 5% aluminum and the remainder iron. The bulk material has an initial relative permeability of about 35 000.

Because Sendust is so brittle it can be cast and broken up to powder quite easily. The resistivity can be increased by additives such as molybdenum. Because of its hardness, it is a difficult material to press.

To overcome problems of pressing. Sendust powder has been heated and rolled to flake. Relative permeabilities of 230 have been achieved. The basic resistivity of this material is higher than molybdenum permalloy, so that use at higher frequencies is possible.

Sendust powder cores can be made with relative permeabilities of 60 and 100. The eddy current losses are lower than molybdenum permalloy, but the other losses are higher. Despite the low raw materials cost, Sendust cores have not achieved widespread use over the years.

5. 50:50 Nickel–Iron Powder Cores

The 50:50 nickel–iron alloy has a saturation magnetization nearly double that of the 2–81 alloy. In addition, the lower nickel content makes it a lower-cost material. This combination makes the powder from this alloy a candidate for lower-frequency power applications, primarily in switching power supplies. The losses are not as low as 2–81 molybdenum permalloy but lower than powdered iron at lower frequencies.

6. Comparison of Powder Cores with Other Magnetic Materials

In comparison to laminated metal cores, tape-wound cores or cut cores, the permeability of the metal powder core is significantly lower. At low frequencies, cores of silicon–iron are the choice because of high saturation and low cost. As the frequency goes into the high kilohertz region, the metal strip of silicon–iron or nickel–iron must be rolled quite thin to maintain low eddy current losses. This raises the material cost, especially if nickel–iron material is used for high permeability at low losses. If high permeability is not needed and losses and stability are important, there is a frequency range in which a specific metal powder core (iron permalloy or molybdenum permalloy) would definitely be the choice.

Ferrites have higher toroidal permeabilities. Because of high resistivity, eddy currents are low, permitting high-frequency operation. The stability can also be obtained with gapping of ferrite pot cores. To get the same permeability, large gaps are needed. This leads to increased leakage and stray-field loss, but the ferrite pot cores can screen this field.

Disadvantages of ferrites are their low saturation values and lower stability of permeability to temperature, dc bias and dc shock such as lightning. In Europe and Japan where telephone transmission lines are either buried or otherwise not subject to temperature extremes, or are short and hence require no loading, these are not problems. In the USA, where rural transmission lines are above ground and subject to temperature extremes and lightning, the situation is different. It is not surprising to see ferrites replacing molybdenum-permalloy powder cores for loading almost everywhere but in the USA.

See also: Magnetic Materials: An Overview; Magnetic Materials: Basic Concepts

Bibliography

Chin G Y, Wernick J H 1980 Soft magnetic metallic materials. In: Wohlfarth E P (ed.) 1980 *Ferromagnetic Materials*. North-Holland, Amsterdam

Heck C 1974 *Magnetic Materials and Their Applications*. Crane, Russack & Co., New York

Legg V E, Given F J 1940 Compressed powdered molybdenum Permalloy for high quality inductance coils. *Bell Syst. Tech. J.* 19: 385–406

Owens C D 1956 *Stability Characteristics of Molybdenum Permalloy Powder Cores*, Monograph Number 2576. Bell Telephone System, New York

A. Goldman
[Spang and Co., Butler, Pennsylvania, USA]

Power Loss in Magnetic Materials

One of the most important criteria used in the assessment of magnetic materials is that of power loss under magnetization conditions. This energy loss manifests itself as heating of the material and the investigation of methods of reducing this loss, thereby producing financial savings, has been a major concern in research and development in the field of magnetic materials since the end of the nineteenth century.

1. Hysteresis Loss

The general behavior of a ferromagnetic material in a magnetic field may be understood by referring to Fig. 1. When an increasing magnetic field is applied to a material in a demagnetized condition, the resulting increase in flux density B is shown by the initial magnetization curve OA. This process is not reversible: if the magnetic field is reduced, the induction decreases along the path AB, so that at the point B where the applied field has been reduced to zero, an induction OB, known as the remanence, remains in the material. To reduce the remanence to zero, an applied field strength OC must be applied in the reverse direction. This reverse field is called the coercive force. The relation between the flux density B and applied field H for a complete cycle is given by the closed loop. This phenomenon of irreversible magnetization is called hysteresis (from the Greek for lag, which describes the phase shift between B and H).

The total energy input to the magnetic material per unit volume to make the magnetization traverse the whole loop is measured by the area enclosed by the loop:

$$E = \oint H dB \quad (\mathrm{J\,m^{-3}})$$

This loss, when measured in slowly varying magnetic fields, is termed the static hysteresis loss.

2. Eddy Current Losses

In rapidly varying fields, energy losses that are much greater than the static hysteresis loss are observed. These are the result of eddy currents set up in the material. Eddy current losses are inevitable in conducting ferromagnetic materials and are one of the main causes of the phase shift (lag) between the magnetization and the external fluctuating magnetic field. According to Faraday's law of induction, any change in the magnetic flux in a conducting medium produces an emf, which in turns sets up electrical eddy currents. These eddy currents flow parallel to the surface of the sample and perpendicular to the direction of induction. Eddy currents thus induced in the material give rise to energy loss through I^2R heating. Lenz's law states that the magnetic field of these currents opposes the external field causing them. Thus the eddy currents have a shielding effect, so that the amplitude of the external field (and corresponding magnetization) becomes damped, decreasing from the surface inwards. This effect is known as the skin effect, which causes the flux in the material not to penetrate fully towards the centre of the magnetic material.

It is important that the amount of flux carried by the material is maximized by achieving full flux penetration. Flux penetration decreases as permeability μ and frequency f increase and increases as the resistivity ρ of the material increases, for a constant material thickness.

The classical analysis of eddy current loss makes certain simplifying assumptions, namely that the permeability μ is constant and independent of magnetic field H and that it has the same value at all points in the sheet. Thus for a sinusoidal applied field when all other quantities are sinusoidal and applying Maxwell's equations, the eddy current losses for a thick sheet, in which full flux penetration occurs, are given by

$$P_e = (\pi d f B_{max})^2/6\rho$$

where d is the thickness of the material, B_{max} is the peak induction, f is the frequency of the applied field and ρ is the resistivity of the material. It can therefore be seen that to minimize the power loss it is necessary to minimize the thickness d of the magnetic material by using it in a laminated form. A decrease in the eddy current loss can also be achieved by increasing ρ. For example, the addition of silicon to iron produces a material with higher resistivity (Fe–3 wt%Si), which is the principal material used in transformers and low-loss power frequency devices (see *Silicon Steels*).

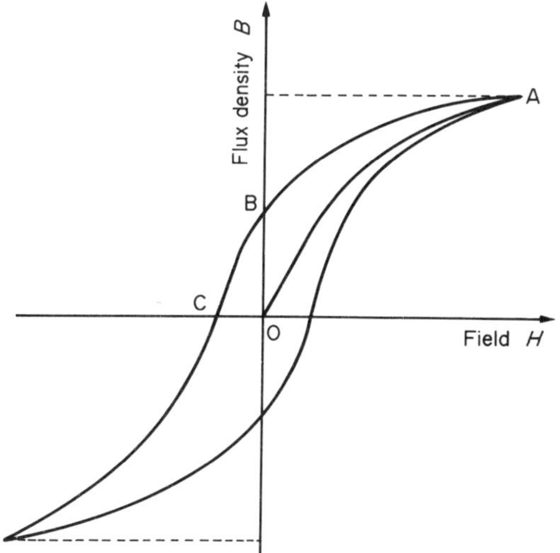

Figure 1
Hysteresis loop of a magnetic material, showing the relationship between the flux density B and the applied magnetizing force H

3. *Separation of Losses*

In order to obtain a better understanding of the total power loss it is useful to divide the total power loss into its constituent losses. There are two approaches:

(a) the classical method of loss separation, which ignores the domain structure of the material, and
(b) a micromagnetic approach, which relates the losses to physical phenomena which occur in magnetic materials under magnetization.

The static hysteresis loss k_1 is determined from the area of the hysteresis loop when it is measured at very low frequency (0.001 Hz) or preferably at dc over a long period. Classically the contribution from this loss source at frequency f is taken to be

$$W_h = k_1 f$$

that is, proportional to frequency. It has been shown that the classical eddy current loss is proportional to the square of the frequency:

$$W_e = k_2 f^2$$

Therefore the total power loss

$$P_t = W_h + W_e = k_1 f + k_2 f^2$$

and the total energy loss per cycle is

$$\frac{P_t}{f} = k_1 + k_2 f$$

Therefore, on the classical picture the loss per cycle against frequency characteristic is linear with its intercept at a point with a value equal to the static hysteresis loss (see Fig. 2).

Actual loss measurements at frequencies below 50 Hz show that linear variation of the total power loss with frequency does not occur: the relationship follows a curve with a concave downward trend at low frequencies. It can be seen from Fig. 3 that the measured total loss is greater than the addition of the hysteresis loss and the classical eddy current loss. The difference between the measured and calculated losses is called the anomalous loss. The anomalous loss is often described in terms of the "anomaly factor:"

anomaly factor =

$$\frac{\text{classical eddy current loss} + \text{anomalous loss}}{\text{classical eddy current loss}}$$

$$= 1 + \frac{\text{anomalous loss}}{\text{classical eddy current loss}}$$

It should be noted that the use of the term "anomalous loss" is anachronistic since this loss is really only anomalous when the loss is viewed classically as a summation of static and eddy current losses. It is now appreciated that an accurate estimation of the total loss depends on a knowledge of the domain structure and the dynamic micromagnetic response of the system.

4. *Anomalous Loss (Excess Eddy Current Loss)*

Anomalous loss has received much attention in recent years since it can be responsible for as much as 50% of the total power loss in silicon–iron at power frequencies. The anomalous loss is more accurately described as the excess eddy current loss since it results from a lack of knowledge of the true conditions under which a magnetic material operates. The classical eddy current loss is calculated from Maxwell's equations, which assume that the material is homogeneous and has a constant permeability. Neither of these assumptions is correct due to the existence of domain walls and the metallurgical nature of the material. Several different sources of anomalous loss have been suggested:

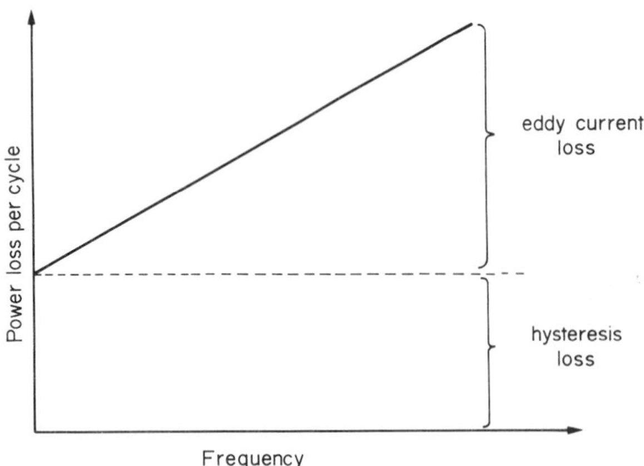

Figure 2
Classical method of separating power loss into constituent losses

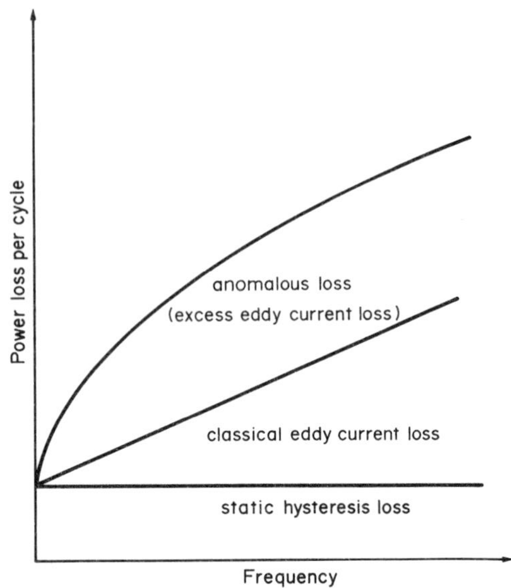

Figure 3
Schematic diagram of actual power loss (upper curve) and separation into constituent losses

(a) occurrence of domain walls with the resulting nonsinusoidal, nonuniform and nonrepetitive domain wall motion;

(b) the nucleation and annihilation of domain walls;

(c) lack of flux penetration which results in domain wall ruckling (a form of bowing) through the thickness of the material;

(d) nonsinusoidal and localized variation of flux density; and

(e) interaction between grains, grain size, grain orientation and other metallurgical inhomogeneities.

The anomalous loss is only an appreciable percentage of the total loss when the material is magnetized at low frequency, because the classical eddy current loss dominates at higher frequencies. The most obvious method of reducing the power loss for a material to be operated at high frequency is to use a material with a high resistivity. Thus, it is preferable to use a ceramic material, such as a soft ferrite, rather than a metallic material.

5. *Variation of Power Loss with Magnetization Conditions*

The classical eddy current loss is dependent upon the square of the product of frequency and flux density

$$P_e \propto f^2 B^2$$

Therefore, the total power loss of a material is also dependent upon frequency and flux density and it is necessary to quote these parameters when stating the total power loss of a magnetic material. It is normal to quote power loss under alternating flux conditions but it is not unusual for more complex flux conditions to occur in applications. For example, in the root of the stator teeth of rotating machines or close to the corners of transformers two alternating fluxes meet at right angles, thereby producing rotational flux conditions. The power loss under these rotational loss conditions exceeds values under alternating flux conditions by as much as a factor of three.

It is normal to quote power loss under sinusoidal flux conditions but the nonlinear relationship between B and H and other parameter nonlinearities produce nonsinusoidal conditions. The harmonics of the fundamental flux present in nonsinusoidal fluxes produce higher values of power loss than are measured under pure sinusoidal conditions. In most applications, magnetic materials are subject to mechanical stresses which change the power loss of the material. In general, tensile stress does not appreciably change the level of power loss, but the existence of compressive stresses increases the power loss by as much as an order of magnitude. Typical values of the power loss under various magnetization conditions are stated in manufacturers' information sheets and in some of the references in the bibliography (see *Magnetic Materials: An Overview*; *Metallic Glasses*; *Silicon Steels*).

6. Measurement of Power Loss

There are many experimental techniques utilized for measuring the power loss of magnetic materials, such as thermometric methods, wattmeter methods, the Hall effect and Poynting vector techniques, bridge methods and measuring the B/H relationship, which are fully described in the literature (see *Measurements in Magnetic Materials*). However, care must be taken when interpreting power loss measurements because in many cases special specimens have to be produced in order that a specific experimental technique can be used.

See also: Domains and Domain walls; Magnetic Materials: An Overview

Bibliography

Bates L F 1961 *Modern Magnetism*, 4th edn. Cambridge University Press, Cambridge

Bozorth R M 1951 *Ferromagnetism*. Van Nostrand Reinhold, New York

Brailsford F 1966 *Physical Principles of Magnetism*. Van Nostrand Reinhold, New York

Chikazumi S 1964 *Physics of Magnetism*. Wiley, New York

Crangle J 1977 *The Magnetic Properties of Solids*. Arnold, London

Cullity B O 1972 *Introduction to Magnetic Materials*. Addison-Wesley, Reading, MA

Lameraner J, Staf M 1966 *Eddy Currents*. Iliffe, London

Luborsky F E 1983 *Amorphous Metallic Alloys*. Butterworth, London

Martin D H 1967 *Magnetism in Solids*. Iliffe, London

K. J. Overshott

[Gwent College of Higher Education, Newport, UK]

R

Radiation Effects on Magnetic Materials

Exposure of magnetic materials to radiation environments eventually causes deterioration of their structure-sensitive magnetic properties such as coercivity, permeability, induction and remanence. In rare cases, the intrinsic properties (saturation magnetization and the Curie temperature) and even the form of magnetism could be changed.

Radiation effects may not be detected until the fluence of the irradiating agent reaches a minimum level, which varies widely with the type (soft or hard) and composition of the magnet and with the species and energy of the bombarding particles as well. In general, magnetic properties are much less susceptible to detectable radiation effects than are the electronic properties of semiconductors. Figure 1 compares the lowest fluence levels of detectable radiation effects for various classes of engineering materials. Note that the lowest level of neutrons is of the order of 10^{12} n cm^{-2} for transistors, 10^{15} n cm^{-2} for permanent magnets of the Nd–Fe–B type, 10^{16} n cm^{-2} for silicon diodes and 10^{17} n cm^{-2} for the softest magnetic alloys. The lowest levels for magnetic materials are, therefore, two to five orders of magnitude greater than those for semiconductors.

The effects of radiation on magnetic materials were first actively investigated with neutrons in research reactors in the period 1958–1964. Later studies exploited other energetic particles (protons, alpha particles and heavier ions) and laser beams as irradiation media.

The bombarding agents that produce the most pronounced effects are neutrons, particularly fast

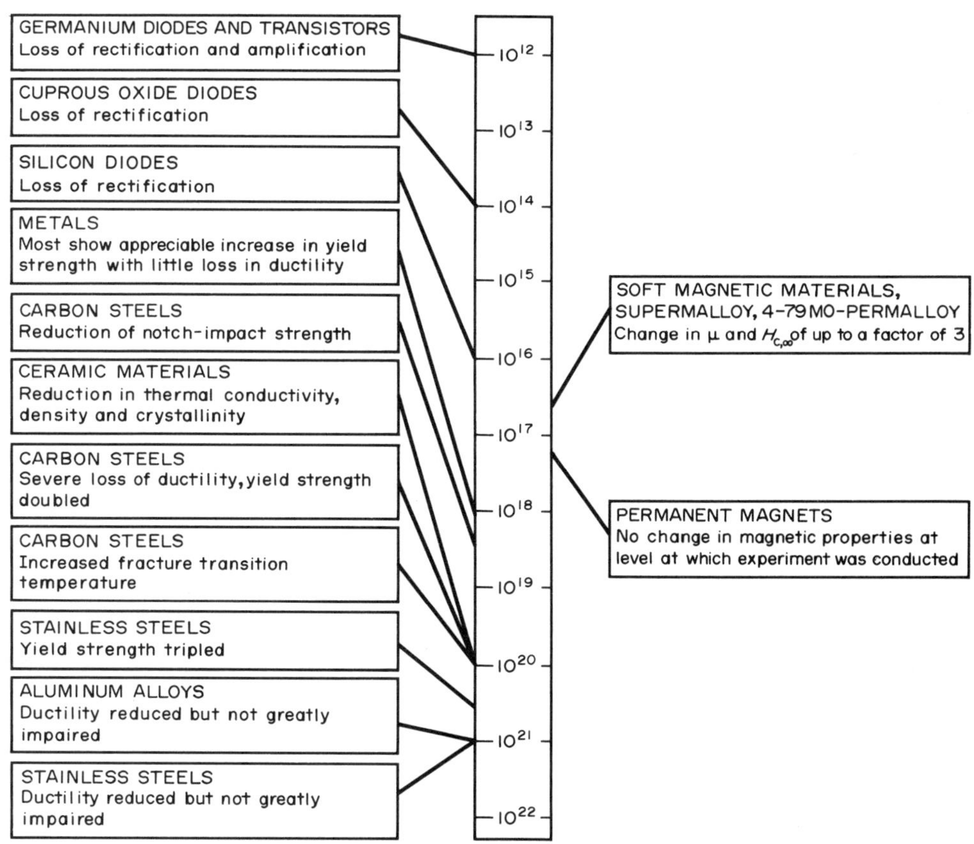

Figure 1
Comparison of the lowest level of neutron fluence (n cm^{-2}) at which various classes of engineering materials are affected. The permanent magnets refer mainly to the Alnico type. The newer Nd–Fe–B magnets are susceptible to neutron damage at only 10^{15} n cm^{-2} (courtesy of D. I. Gordon and R. S. Sery)

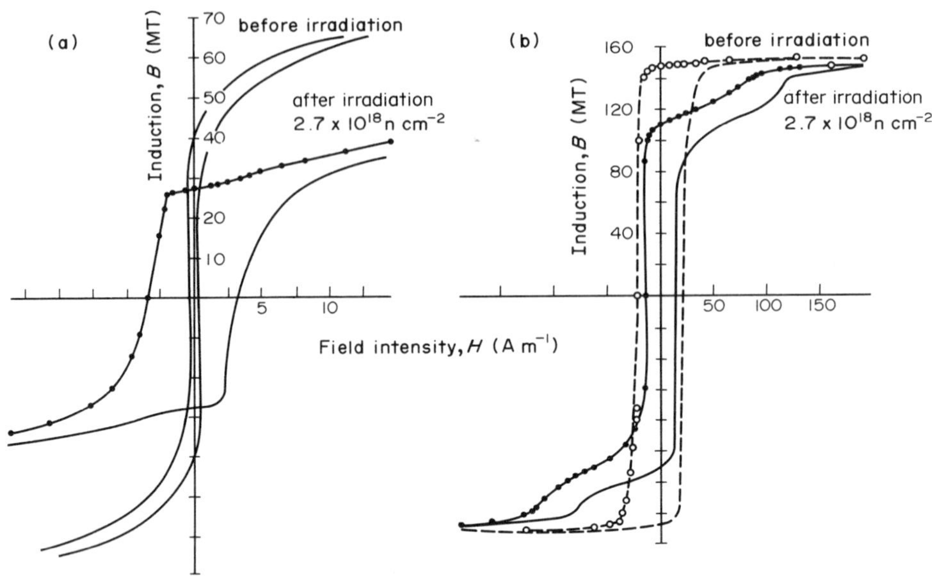

Figure 2
Effect of neutron irradiation on the hysteresis loop: (a) the increase in coercivity and the decrease in retentivity cause a considerable shape change towards squareness; and (b) the emergence of constricted loops (courtesy of D. I. Gordon)

neutrons with energy greater than 0.1 MeV. Experiments with neutron irradiations of moderate fluences (up to 10^{19} n cm^{-2} at 70–115 °C) clearly show that fast neutrons cause severe damage to the structure-sensitive magnetic properties of the softest magnetic materials only when coercivity is less than 4 A m^{-1} (0.05 Oe). The damage is insignificant to less-soft magnets and practically nil to hard magnets. (Magnetic materials are generally classified into soft and hard types on the basis of their coercivity (H_c) values. Although the classification is not sharp, materials with H_c greater than 8000 A m^{-1} (100 Oe) are labelled hard.)

Two examples of the effect of neutron irradiation on the hysteresis loop are depicted in Fig. 2. Both the shape and size of the loop can be severely changed, leading to an increase in coercivity, a reduction in remanence and an increase in the associated energy loss. At 2×10^{18} n cm^{-2} ($E > 0.5$ eV), the relative permeabilities (μ'_{20}) at a magnetizing field of 1600 A m^{-1} (20 Oe) and the maximum relative permeabilities (μ'_{max}) of Supermalloy (15%Fe–5%Mo–79%Ni) and 4–79 Mo-permalloy (15%Fe–4%Mo–79%Ni) were reduced by 90%; their H_c increased more than eightfold. Also, the hysteresis loops of Supermalloy, grain-oriented 50%Ni–Fe and other nickel-base alloys were seriously distorted. As far as high-frequency energy losses were concerned, cores composed of nickel ferrites, 2–81 Mo-permalloy (2%Mo–81%Ni–Fe) dust, and Sendust (10%Si–5%Al–Fe) exhibited significant increases, but carbonyl iron-dust cores showed no increase. For exposures to neutron fluences greater than 1.3×10^{18} n cm^{-2} the deterioration of the structure-sensitive magnetic properties proceeded at a decelerating rate, implying that there was probably a saturation level of fluence beyond which no further damage to the properties would occur.

Several important iron-base alloys, including untextured 3.5%Si–Fe, 3%Si–1%Al–Fe, and 2V Permendur (2%V–49%Co–Fe), were in fact found to be radiation resistant. Even for grain-oriented 3%Si–Fe, only μ'_{20} suffered an 18% drop at 2×10^{18} n cm^{-2}; all other properties, such as μ'_{max}, H_c and remanence B_r were not affected. Contrary to the usual adverse effect of radiation, laser irradiation renders a more than 10% reduction of the core loss of the grain-oriented silicon–iron. This beneficial effect is attributed to a refinement of the 180° domain wall spacing and other changes in the domain structure.

The most damaging effect of irradiation with neutrons or energetic ions is the production of atomic displacements; neutron irradiations have been observed to alter the intrinsic magnetic properties of compounds and of ordered, concentrated alloys. For example, irradiation to levels of 5.4×10^{18} thermal and 1.2×10^{17} fast neutrons per square centimeter at 50 °C resulted in 3% and 15% decreases in the saturation magnetization of γ-Fe_2O_3 and Fe_3O_4, respectively. Similar irradiations also led to changes in the first anisotropy constant and magnetostriction of a variety of soft magnetic alloys and to the conversion from antiferromagnetism in the ordered state to superparamagnetism in the partially disordered state in Pt_3Fe.

Neutrons, especially thermal neutrons, induce radioactivity in both soft and hard magnets. The induced radioactivity may present some problems in handling and storage of magnets, especially those containing cobalt because of the large cross section (37 b) for the reaction

$$^{59}_{27}\mathrm{Co} + ^{1}_{0}\mathrm{n} \rightarrow [^{60}_{27}\mathrm{Co}^{*}]$$

from which the compound nucleus undergoes decay into two isomeric forms, one with a half-life of 5.27 years. This irradiation effect is particularly unfortunate because cobalt is an important constituent in many prominent hard magnets, notably the rare-earth–cobalt compounds (such as $SmCo_5$ and Sm_2Co_{17}) and the alnico alloys.

The remarkable improvement in the maximum energy product, $(BH)_{max}$, of rare-earth–cobalt compounds ($SmCo_5$, Sm_2Co_{17}, etc.) and rare-earth–iron–boron ($R_2Fe_{14}B$, where R is a rare-earth element, notably neodymium) has resulted in new applications for permanent magnets, such as generating permanent magnetic lenses for accelerator and beam-transport systems. In many of these areas, the magnets are subjected to high levels of irradiation by energetic ions or neutrons.

The work of Blackmore shows that the Nd–Fe–B magnets are very sensitive to ionizing radiation contained in a proton beam and lose essentially all of their magnetization at a dose of 7×10^7 rad (with a conversion factor of 1 rad = 6.24×10^7 MeV g^{-1}, a proton irradiation of 1 μA h^{-1} at 500 MeV corresponds to an average dose of 4.5×10^8 rad to the sample) and more than 50% at 4×10^6 rad. However, samarium–cobalt permanent magnets exhibit significant demagnetization only when irradiated at dose levels of 10^9–10^{10} rad. This large difference in radiation resistance is partly ascribed to the much higher Curie temperatures of the latter compounds (e.g., 724 °C for $SmCo_5$ vs 312 °C for $Nd_2Fe_{14}B$). Neutron irradiation in a reactor at Julich, Germany, to neutron fluence of 8×10^{19} n cm^{-2} ($0 < E < 7$ MeV) reveals that the sensitivity of Sm_2Co_{17} to demagnetization is a factor of ten lower than that of $SmCo_5$. For Nd–Fe–B permanent magnets, similar reactor irradiation detected 5% and 10% losses of remanence for fast neutron fluence of 10^{15} n cm^{-2} at 77 °C and 153 °C, respectively. This loss of remanence is fully recoverable by remagnetization, however, which also gives rise to a 20% increase in the coercive force.

See also: Magnetic Materials: An Overview; Magnetic Materials: Basic Concepts; Radiation Effects on Superconducting Materials

Bibliography

Chen C W 1986 *Magnetism and Metallurgy of Soft Magnetic Materials*. Dover, New York, pp. 436–77

Dautreppe D 1969 Radiation effects in magnetic materials. In: Dienes G J (ed.) 1969 *Studies in Radiation Effects in Solids*, Vol. 3. Gordon and Breach, New York, pp. 1–151

Thompson M W 1969 *Defects and Radiation Damage in Metals*. Cambridge University Press, Cambridge

C. W. Chen
[Lawrence Livermore National Laboratory,
Livermore, California, USA]

Radiation Effects on Superconducting Materials

Superconductors not only are essential in many technological applications such as magnets, but also represent interesting systems in which to investigate basic physical and metallurgical phenomena such as electron–phonon interactions, the effects of impurity additions and the role of metallurgical treatments such as cold working or precipitate formation (Luhman and Dew-Hughes 1979, Scanlan 1980, McKinnel et al. 1989). The effect of irradiation on superconducting materials plays an important role in both of these areas (Brown et al. 1978, Luhman and Dew-Hughes 1979, Snead and Luhman 1985). Furthermore, the superconducting magnets in fusion reactor devices will be exposed to fast-neutron and γ irradiation fields and it is necessary to determine the effect of the irradiation on the operating characteristics of the magnet and, therefore, on the individual components (Brown 1981). The microscopic behavior of materials is frequently investigated by introducing defects and observing changes. Irradiation experiments are useful for studying superconductors since the metallurgy (e.g., impurity level, dislocation structure) can be held constant and the irradiation conditions (particle type, energy, fluence and irradiation temperature) can be varied to result in well-characterized types of damage.

The basic properties characterizing a superconductor include its critical temperature T_c, the critical magnet fields, especially the upper critical field H_{c2} since type-II superconductors are being considered in all practical cases, and the critical current density J_c (see *Critical Current Density*; *Flux Pinning and the Summation of Pinning Forces*; *Magnetic Properties of Superconductors*). All of them are affected by radiation, although through different mechanisms. Whereas changes of T_c depend entirely on the significance of atomic ordering for the actual value of T_c prior to irradiation, the upper critical fields H_{c2} are directly correlated with the normal state resisitivity ρ_n and, hence, generally increase with the introduction of point defects or point defect clusters as long as the transition temperature T_c is not degraded too much. On the contrary, the changes of critical current density depend largely on the preirradiation defect structure and the extent to which it has been optimized for flux pinning. As a general rule, radiation-induced defects must have sizes of the order of the superconducting coherence length (~30 nm in niobium, 4 nm in Nb_3Sn, and < ~1 nm in high-T_c superconductors at 4.2 K) in order to act as effective pinning centers. Hence, isolated or point defects such as those introduced by bombardment with 1–10 MeV electrons have no effect on J_c in metallic superconductors.

With the advent of high-temperature superconductors, radiation effects have been studied extensively, much along the lines described at the beginning of this chapter. With regard to critical current densities, the granular nature of most of these materials (see *Granular Superconductors*) imposes severe limitations on the assessment of intrinsic (intragranular) property changes, which can be obtained at present only indirectly from magnetization measurements on single crystals and to some extent from transport measurements on epitaxial films. In addition, the material characterization prior to irradiation is indeed very limited at the present stage.

In the following, the study of radiation effects in superconducting materials will be discussed for each of the three basic properties (T_c, H_{c2}, J_c) separately, although there are of course correlations among them. For each property the discussion will be subdivided into the four classes of superconductors: elements, alloys, compounds and high-T_c materials.

1. Critical-Temperature Changes

Critical-temperature changes after irradiation are related to changes in the normal state resistivity ρ_n and are generally much larger in compounds than in elemental or alloy superconductors. The large values of T_c in compound superconductors (up to ~23 K) are due to the ordered crystal structure which, for example, results in long chains of A atoms in the A_3B compounds in the A15 crystal structure. These structures can be disordered during irradiation in which, for example, A and B atoms exchange positions, resulting in a drop in T_c. During the defect production process many more atoms can be replaced (in an exchange of sites) than are displaced (moved off a site). This is especially true for high-energy events (cascades) such as those produced by fast neutrons, which can result in large regions (~10 nm diameter) of disorder in which the ratio of replaced to displaced atoms can be of the order of 10–20.

The T_c of elemental superconductors (~9 K in niobium) is, of course, not sensitive to site-exchange disorder since all atoms are the same. Therefore, only the displaced atoms (or impurities due to implantation or transmutation) have an effect on T_c changes, thus making this effect much smaller than in compounds. Although the changes are small, the effect has been used to understand better some of the basic physical properties of these materials such as the energy gap or the phonon spectrum.

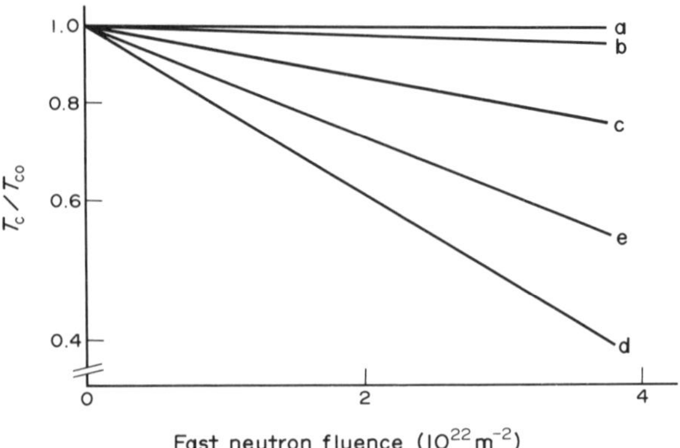

Figure 1
Schematic illustration of critical temperature T_c changes during irradiation normalized to unirradiated values T_{c0} for: (a) elements, alloys and various compounds; (b) Laves phase, B1 phase; (c) A15; (d) Chevrel phase; and (e) high-T_c superconductors

The T_c of alloy superconductors (~9 K in NbTi) is also very insensitive to irradiation, although some decreases have been measured. The physical reason for this decrease is not completely clear, but may be due to the removal of the energy-gap anisotropy by the introduction of additional scattering centers.

The T_c of high-temperature superconductors is known to depend critically on the degree of order, especially with regard to the positions of the oxygen atoms. As a consequence, radiation-induced disordering, which is particularly pronounced for the lightest atom of the compound (i.e., the oxygen atom), leads to large depressions of T_c and eventually to a metal–insulator transition. Results obtained from a variety of charged particle irradiations show perfect scaling of the T_c depression with calculated displacements per atom.

Figure 1 is a schematic representation of the changes in T_c in various types of superconductors. The T_c of elemental and alloy superconductors changes very little, whereas the T_c of compound superconductors decreases more rapidly and depends strongly on the crystal structure.

2. Critical-Field Changes

Radiation-induced changes of the upper critical fields H_{c2} are interesting in their own right only at low particle fluences, where the changes of the basic material parameters, such as T_c or the density of states at the Fermi level, are still small. In this fluence range, increases of H_{c2} by 2–10% have been reported for compounds (e.g., Nb_3 Sn, NbN) and elemental superconductors. They can be explained quantitatively by the radiation-induced increase of the normal state resistivity. Alloy superconductors with their characteristic high preirradiation resistivities do not show this effect and H_{c2} tends to follow the T_c depression also at low fluences (Weber et al. 1988). No information about radiation-induced H_{c2} changes is available for the high-temperature superconductors at present.

3. Critical-Current Changes

The critical current has been the subject of extensive investigation since it is this property that determines the applicability of a material for use in a superconducting magnet. A higher J_c results in more compact, cheaper and higher-field magnets. Whereas T_c is a fundamental property of superconductors, J_c depends strongly on the microstructure of the material. For example, cold working, addition of second-phase precipitates and variation of grain size have been used to vary J_c significantly. It is the ability of these inhomogeneities to pin the flux lines that result in J_c changes. In order to determine the effect of radiation-induced defects on J_c, it is necessary to determine the effect of the initial flux-pinning structures on J_c and to understand the interaction of the radiation-induced defects with the initial pinning structures. Furthermore, since there are many flux lines and many defect structures, it is necessary to know how to sum the various interactions; for example, the motion of a single flux line can be influenced by many defect structures. J_c changes at a fixed low field (< 10 T) are shown schematically in Fig. 2. There is also a strong magnetic field dependence of J_c changes, which is shown for some alloy and compound superconductors in Fig. 3. Whereas T_c changes depend on the defect concentration or degree of disorder and not critically on the defect configuration (e.g., cascades or point defects), J_c changes depend critically on the configuration of the defects as well as on their concentration.

Most of the information on fundamental defect–flux line interactions has been determined in elemental superconductors (Brown et al. 1978). Very pure, defect-free samples have been irradiated under conditions chosen to introduce a well-known defect structure (e.g., point defects or voids). Models based on superconducting properties agree fairly well with experiments in predicting the single defect–flux line interaction and the scheme for summing these interactions (see *Niobium–Titanium Alloys*). Since the samples are generally prepared to minimize initial flux pinning, J_c is usually greatly increased by irradiation and depends critically on the irradiation conditions and the resulting defect structures.

The details of the effect of irradiation on J_c changes in alloy superconductors are fairly complicated because of the various metallurgical treatments (resulting in dislocation-cell-wall or precipitate pinning) that have been used to increase the unirradiated value of J_c (i.e., J_{c0}) (Weber 1982). In particular,

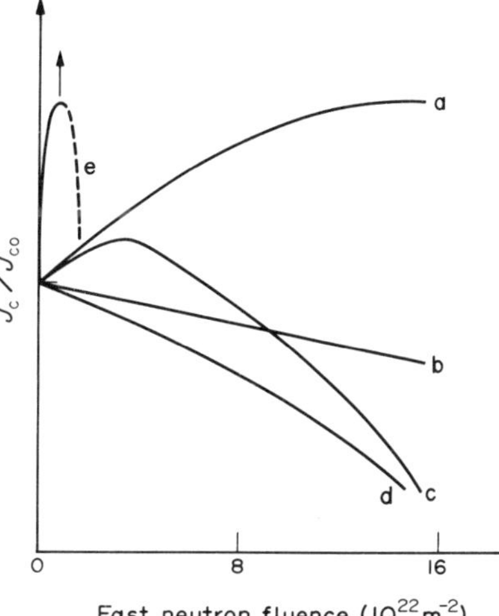

Figure 2
Schematic illustration of critical current J_c changes normalized to unirradiated value J_{c0} at fixed low field for: (a) low-J_{c0} material; (b) alloys with high J_{c0}; (c) A15 compounds with high J_{c0} for fast-neutron irradiations at 4 K; (d) A15 compounds with high J_{c0} for irradiation at $\gtrsim$300 K; and (e) high-T_c single crystals

$Nb_{1-x}Ti_x$ (with ~0.4–0.5%) has been extensively investigated since it is the most widely used superconducting magnet material for magnetic fields $\lesssim$ 8–10 T (see *Flux Pinning and the Summation of Pinning Forces*). Low-J_{c0} materials show an increase in J_c during irradiation due to the introduction of flux pinning structures. Most experiments have been done on high-J_{c0} (i.e., commercially interesting) material for which J_c decreases during irradiation. The largest observed decrease has been about 20% and the decrease with irradiation fluence has been found to depend critically on the metallurgical history, including degree of cold working, presence of titanium-rich precipitates and annealing treatments. For the specific applications of NbTi as the superconductor for fusion reactor magnets, it is generally accepted that this will not be the limiting component with regard to radiation-induced degradation since damage in the insulators and stabilizers poses more serious problems (Brown 1981).

The magnitude and sign of radiation-induced changes in compound superconductors also depend critically on the initial metallurgical state and irradiation temperature. The A15 compounds (e.g., Nb_3Sn, Nb_3Ge, V_3Ga) have been studied most extensively because of the sensitivity of T_c to irradiation and the fact that they are used as conductors for magnets with fields > 10 T (see *A15 Compound Superconductors*). It has been shown for the A15 compounds that radiation can dramatically increase J_c in low-J_{c0} material,

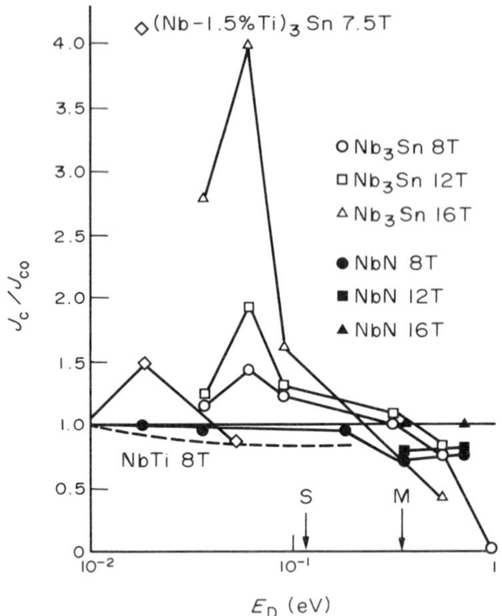

Figure 3
Critical current J_c changes normalized to unirradiated value J_{c0} for NbTi, Nb_3Sn, $(Nb–1.5\,wt\%Ti)_3Sn$ and NbN, at various high magnetic fields. The data are plotted versus damage energy (i.e., the average energy transferred to each atom by fast neutron irradiation). S and M indicate the lifetime damage at the magnet location of two fusion reactor designs (after Weber 1989)

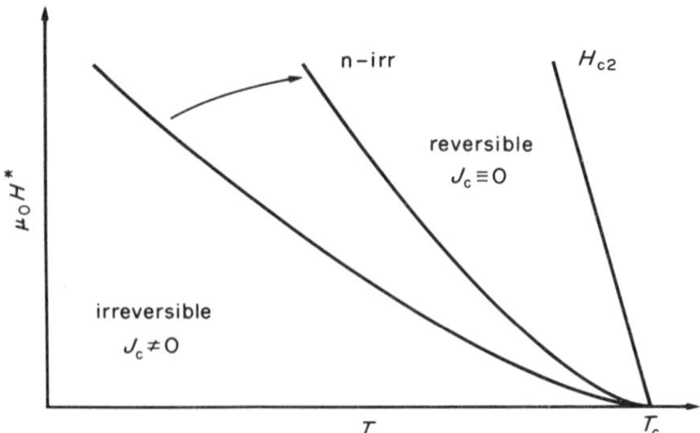

Figure 4
Schematic illustration of the shift of the irreversibility line in $YBa_2Cu_3O_{7-\delta}$ superconductors following neutron irradiation

but also in high-J_{c0} material at comparatively low fluences, especially at high magnetic fields. The latter has been successfully ascribed to the radiation-induced H_{c2} enhancement discussed in Sect. 2. Since J_c goes to zero at H_{c2}, increases of H_{c2} by one or two tesla will lead to pronounced J_c enhancements at fixed high magnetic field as shown in Fig. 3. Finally, after high-fluence irradiation, under all conditions, J_c decreases due to the decrease in T_c. Among the other compound superconductors, NbN films have recently been shown to be practically unaffected by neutron irradiation to fluences of greater than about $10^{23}\,m^{-2}$ and involving low and ambient temperature irradiation conditions (Herzog et al. 1991).

A similarly detailed understanding of radiation effects on J_c in high-temperature superconductors is presently not at hand. Largely different conditions prevail depending on which form of material is investigated. In the case of sintered ceramics, the transport J_c values always decrease because of the destruction of the weak link structure and consquently the loss of intergrain conductivity. Magnetization measurements on the same materials show, however, significant broadening of the hysteresis loops. This indicates an enhancement of J_c within the grains (intragrain J_c). Epitaxially grown thin films show in general very high critical current densities prior to irradiation ($\sim 7 \times 10^{11}\,A\,m^{-2}$ at 4.2 K, $\sim 7 \times 10^{10}\,A\,m^{-2}$ at 77 K). Hence, their defect structure is very well developed and presumably nearly optimized. As a consequence, transport measurements on neutron-irradiated thin films show only small (Schindler et al. 1990) or no (Schalk et al. 1992) enhancements of J_c at low fluences and significant degradations later on. The most detailed information is available for $YBa_2Cu_3O_{7-\delta}$ and $Bi_2Sr_2CaCu_2O_8$ single crystals. Neutron and some ion irradiations have shown significant increases of J_c (Sauerzopf et al. 1991, Weber and Crabtree 1991) at fluences where the T_c depression is not too large ($\sim 1–2 \times 10^{22}\,m^{-2}$, $E > 0.1$ MeV, for YBCO). Transmission electron microscopy (TEM) observations of radiation-induced defects as well as damage calculations of the energy distribution of the recoil atoms (Kirk et al. 1990, Kirk and Weber 1992) begin to shed some light on the flux-pinning mechanisms in these materials. Fast neutrons have been shown to produce a significant number of defect cascades with a diameter of around 10 nm (10^{22} cascades per m^3 at a fluence of $2 \times 10^{21}\,m^{-2}$), which are certainly able to act as strong flux-pinning centers. Detailed experiments on the fluence dependence of J_c at various fields and temperatures have shown, first, that the relative J_c enhancements are larger at higher temperatures, second, that the intrinsic anisotropy of J_c is strongly reduced on irradiation and, third, that the magnetization relaxation rates (due to thermally assisted flux flow and flux creep) are strongly reduced on irradiation (Lensink et al. 1991) (see *Flux Creep Phenomena*). Particularly important are recent results pertaining to considerable enhancements of the "irreversibility" line (see *Critical Current Density*), that is, the field and temperature values defining the boundary between flux pinning and flux motion, following neutron irradiation (see Fig. 4) (Kritscha et al. 1990, Küpfer et al. 1991, Sauerzopf et al. 1992). This demonstrates in a clear way that pinning centers can be introduced into the oxide superconductors, which are suitable for flux pinning even at high temperatures. Details of the experimental results seem to be understandable on the basis of significant interactions between the newly created defects and the preirradiation defect structure, but further work will be needed to clarify many aspects of flux pinning and current transport in these materials. In conclusion, it should be mentioned that interesting "model" systems for the study of flux line–defect interactions are provided by amorphous "tracks" produced by GeV-heavy ion irradiation of thin single crystals and films (Civale et al. 1991).

Acknowledgement

This work was performed under the auspices of the US Department of Energy, Division of Materials Sciences, Office of Basic Energy Sciences, under Contract W-31-109-ENG-38.

See also: Critical Current Density; Flux Pinning and the Summation of Pinning Forces; Magnetic Properties of Superconductors; Superconducting Materials: Large-Scale Applications

Bibliography

Brown B S 1981 Radiation effects in superconducting fusion-magnet materials. *J. Nucl. Mater.* 97: 1–14

Brown B S, Freyhardt H C, Blewitt T H (eds.) 1978 Proceedings of the International Discussion Meeting on Radiation Effects on Superconductivity. *J. Nucl. Mater.* 72: 1–300

Civale L, Marwick A D, Worthington T K, Kirk M A, Thompson J R, Krusin-Elbaum L, Sun Y, Clem J R, Holtzberg F 1991 Vortex confinement by columnar defects in $YBa_2Cu_3O_7$ crystals: enhanced pinning at high fields and temperatures. *Phys. Rev. Lett.* 67: 648–51

Herzog R, Weber H W, Kampwirth R T, Gray K E, Gerstenberg H 1991 Low-temperature neutron irradiation of magnetron-sputtered NbN films. *J. Appl. Phys.* 69: 3172–5

Kirk M A, Frischherz M C, Liu J Z, Greenwood L R, Weber H W 1990 The structure of neutron-irradiation damage in $YBa_2Cu_3O_{7-x}$. *Philos. Mag. Lett.* 62: 41–9

Kirk M A, Weber H W 1992 Electron microscopy of radiation-induced defects in high-temperature superconductors. In: Narlikar A V (ed.) 1992 *Studies of High Temperature Superconductors*, Vol. 10. Nova, New York

Kritscha W, Sauerzopf F M, Weber H W, Crabtree G W, Chang Y C, Jiang P Z 1990 Flux pinning enhancement and shift of the irreversibility line by fast neutron irradiation of single crystal $Bi_2Sr_2CaCu_2O_8$. *Europhys. Lett.* 12: 179–84

Küpfer H, Keller C, Meier-Hirmer R, Salama K, Selvamanickam V, Tartaglia G P 1991 Critical current and relaxation of oriented grained $YBa_2Cu_3O_{7-x}$ after fast neutron irradiation. *IEEE Trans. Magn.* 27: 1369–74

Lensink J G, Griessen R, Wiesinger H P, Sauerzopf F M, Weber H W, Crabtree G W 1991 Fast neutron irradiation effects on magnetization relaxation in YBCO single crystals. *Physica C* 185–189: 2287–8

Luhman T, Dew-Hughes D (eds.) 1979 *Metallurgy of Superconducting Materials*, Treatise on Materials Science and Technology, Vol. 14. Academic Press, New York

McKinnel J C, Lee P J, Larbalestier D C 1989 The effect of Ti-content on the pinning force in Nb–44 wt%Ti to Nb–62 wt%Ti. *IEEE Trans. Magn.* 25: 1930–3

Sauerzopf F M, Wiesinger H P, Kritscha W, Weber H W, Crabtree G W, Liu J Z 1991 Neutron-irradiation effects on critical current densities in single-crystalline $YBa_2Cu_3O_{7-\delta}$. *Phys. Rev. B* 43: 3091–100

Sauerzopf F M, Wiesinger H P, Weber H W, Crabtree G W, Frischherz M C, Kirk M A 1992 Anisotropic shift of the irreversibility line by neutron irradiation. *Supercond. Sci. Technol.* 5: S105–8

Scanlan R M 1980 Superconducting materials. *Annu. Rev. Mater. Sci.* 10: 113–32

Schalk R M, Weber H W, Barber Z H, Davies C E, Evetts J E, Somekh R E, Kim D H 1992 Critical currents in sputtered YBCO films. *Supercond. Sci. Technol.* 5: S224–7

Schindler W, Roas B, Saemann-Islenko G, Schultz L, Gerstenberg H 1990 Anisotropic enhancement of the critical current density of epitaxial $YBa_2Cu_3O_{7-x}$ films by fast neutron irradiation. *Physica C* 169: 117–22

Snead C L, Luhman T 1985 Radiation damage and stress effects in superconductors: materials for high-field applications. In: Johnson R A, Orlov A N (eds.) 1985 *Physics of Irradiation Effects in Crystals*. North-Holland, Amsterdam

Weber H W 1982 Neutron irradiation effects on alloy superconductors. *J. Nucl. Mater.* 108/109: 572–84

Weber H W 1989 Neutron damage of superconductors for fusion magnets. *Kerntechnik* 53: 189–96

Weber H W, Crabtree G W 1991 Neutron irradiation effects in high-T_c single crystals. In: Narlikar A V (ed.) 1991 *Studies of High Temperature Superconductors*, Vol. 9. Nova, New York

Weber H W, Khier W, Wacenovsky M, Hoch H 1988 Radiation-induced changes of critical fields in NbTi superconductors. *Adv. Cryog. Eng.* 34: 1033–9

H. W. Weber
[Technische Universität Wien, Vienna, Austria]

B. S. Brown
[Argonne National Laboratory, Argonne, Illinois, USA]

Rare-Earth Intermetallics

Intermetallic compounds formed by combining rare-earth metals with other metals are a fascinating class of materials. There are several hundred such compounds, displaying a large variety of different crystal structures. Their magnetic ordering temperatures range from below liquid helium temperature to above 1000 K. The corresponding magnetic structures comprise simple ferromagnets, antiferromagnets and ferrimagnets but also complicated noncollinear spin structures including magnetic structures incommensurate with the crystal lattice. Coexistence of magnetism with superconductivity is observed in cases where the coupling between the localized 4*f* moments is low. Nuclear-enhanced magnetism is observed in cases where the crystal field interaction has led to a nonmagnetic singlet or triplet state. Apart from the fundamental physics side many of the rare-earth intermetallics find technological applications as permanent magnet materials, high-magnetostriction materials, Invar alloys and hydrogen storage materials.

The rare-earth (RE) elements are mutually very similar with regard to their outer electron shells. For this reason there is little difference in their chemical properties. If one of the rare-earth elements forms a given type of compound with another metal then generally also the remainder of the rare-earth elements behave similarly. Intermetallic rare-earth-based compounds are therefore characterized by isotypic series, the second component being generally one of the *d* transition metals or one of the *s*, *p* elements of rows IA, IIA, III and IV of the periodic table. Interest in these materials has been very strong from the side of crystal chemistry and fundamental physics as well as from the side of technological applications. Composition and structural details of all these compounds can be found in the compilation of Villars and Calvert (1985). An analytical approach to describing the occurrence of binary compounds in terms of formation enthalpies has been given by de Boer et al. (1988).

In spite of the similarity in chemical behavior the intermetallic compounds display a large variety in physical phenomena. This is due to the incomplete 4*f* shell which is gradually filled up when passing through the lanthanide series. The electronic configuration of the 4*f* electrons is characterized by a total spin quantum number S and a total orbital quantum number L, the ground term multiplet being described by a total angular momentum $J = L + S$ for the light rare-earth elements and $J = |L - S|$ for the heavy rare-earth elements. It is this variety in 4*f* configuration that is mainly responsible for the wealth of magnetic properties found in the rare-earth intermetallics.

The 4*f* shell is situated well to the interior of the rare-earth atoms so that there is virtually no overlap between the 4*f* wave functions. This implies that any magnetic coupling between the 4*f* moments cannot be a direct one. Consequently the magnetic ordering temperatures of compounds in which rare earths are combined with nonmagnetic elements are comparatively weak. A different situation arises in the majority of compounds consisting of rare-earth and 3*d* elements in which the magnetic ordering temperatures are determined by the interplay of interactions between 4*f* and 3*d* moments and the mutual interactions between the 3*d* moments.

1. Compounds of Rare Earths with Nonmagnetic Elements

The coupling between the localized 4*f* moments in these compounds is commonly considered to proceed indirectly via polarization of the conduction electrons. In the Rudermann–Kittel–Kosuya–Yoshida (RKKY) model (for a detailed description see Vonsovskii 1974) it is assumed that the conduction spin $\boldsymbol{s}$ at the position $\boldsymbol{r}$ interacts with the 4*f* spin $\boldsymbol{S}$ at $\boldsymbol{R}$ through a perturbative nonlocal exchange interaction (see *Rare-Earth Metals and Alloys*) of the form

$$H_{sf} = \mathscr{J}(\boldsymbol{R}-\boldsymbol{r})\boldsymbol{S}.\boldsymbol{s}$$

which may be approximated by the local expression

$$H = -\Gamma\Omega\boldsymbol{s}.\boldsymbol{S}\delta(\boldsymbol{r}-\boldsymbol{R}) \quad (1)$$

where Ω is the atomic volume and where the *s–f* exchange integral $\Gamma \equiv \Gamma(\boldsymbol{q})$ depends in general on the difference $\boldsymbol{q} = \boldsymbol{k} - \boldsymbol{k}'$ of orbital states of the conduction electron before and after scattering. A perturbation approach up to the second order shows that the conduction electron polarization of the rare-earth spin is long range and oscillatory in character. Each rare-earth spin experiences an exchange field produced by the conduction electron polarization due to the surrounding spins and orients its direction accordingly. When expressed in terms of mean field theory, the following expression for the asymptotic Curie temperature T_c is found:

$$T_c = -\frac{3\pi Z^2\Gamma^2}{4k_B E_F}(g_J - 1)^2 J(J+1)\sum_{n\neq m} F(2k_F R_{nm}) \quad (2)$$

where E_F and k_F are, respectively, the Fermi energy and Fermi wave vector, Z is the number of conduction electrons, g_J is the Landé factor and k_B is Boltzmann's constant. The dependence of T_c on the type of crystal structure is contained in the summation of $F(2k_F R_{nm})$ over all rare-earth atom distances R_{nm} to the central atom considered, with $F(x) = (x \cos x - \sin x)/x^4$.

The results of investigations of the magnetic properties of numerous intermetallic compounds have been reviewed by Buschow (1980). These compounds give rise to ferromagnetic order ($T_c > 0$) as well as to antiferromagnetic order ($T_c < 0$). In general it proved difficult to analyze the magnetic properties exclusively in terms of Eqn. (2) owing to the unavailability of experimental values for k_F and E_F, while the use of the free-electron values for k_F and E_F does not generally lead to the right answer. For this reason results of magnetic measurements were frequently combined with other types of measurements able to provide experimental information on the spatial extent of the conduction electron polarization, keeping k_F (E_F) as adjustable parameters. These measurements comprised, for instance, the determination of the Knight shift and/or hyperfine field at suitable atoms in the lattice and the determination of the magnetic structure by neutron diffraction. A frequently employed technique is also magnetic dilution and the study of the changes in T_c in isotypic compounds in which the conduction electron concentration is changed continuously by replacing the nonmagnetic component by an element of different valence (see the review of Buschow (1980).

An important general rule which follows from Eqn. (2) is that the values of T_c in the series of isotypic compounds scale across the lanthanides with the de Gennes factor $G \equiv (g_J - 1)^2 J(J+1)$, the highest values being found for gadolinium located in the middle of the lanthanides, tailing off towards the end and the beginning of the series.

A further mechanism which determines the magnetic properties of rare-earth compounds is the interaction of the electrostatic field generated by the charges of the surrounding ions with the aspherical charge distribution associated with the $4f$ charge cloud of rare earths for which $L \neq 0$. This so-called crystalline electric field (CEF) interaction lifts the $2J+1$ fold degeneracy of the ground state. The Hamiltonian H_c describing the CEF interaction is most conveniently expressed in terms of operator equivalents (for a review, see Hutchings (1964)):

$$H_c = \sum_{n,m} B_{nm} O_{nm} \tag{3}$$

where $B_{nm} = \theta_n \langle r^n \rangle A_{nm}$ represent the structure-dependent crystal field intensities and O_{nm} are operators of J_x, J_y and J_z listed for all orders n, m by Hutchings (1964), together with the multiplicative factor θ_n. Values of the expectation values for the $4f$ radius are represented by $\langle r^n \rangle$. Values for A_{nm} were often obtained by means of point charge calculations where the electrostatic field at the rare-earth site is calculated by summing the effective point charges of the surrounding ions. These crystal field interactions determine the directions of the moments in the magnetic structures imposed by the (isotropic) RKKY interaction. The ground state moment in the magnetically ordered state is given by $M = -m_J g_J \mu_B = J g_J \mu_B$ when the crystal field interaction is much smaller than the magnetic interaction. But when the crystal field interaction is much larger than the magnetic interaction a situation may arise where the magnetic quantum number m_J of the lowest crystal field and exchange split $2J+1$ manifold is different from $m_J = -J$. This is the reason why the the magnetic moments in magnetic rare-earth compounds fall sometimes appreciably below the free-ion value $J g_J \mu_B$. In some cases, when the crystal field interaction is particularly strong, magnetic ordering is suppressed when the lowest crystal field split state is a nonmagnetic singlet or triplet for which $\langle m_J \rangle = 0$.

2. *Special Features of Systems with Localized 4f Electrons*

It follows from Eqn. (3) that the strength of the magnetic interaction in systems of localized $4f$ moments depends on the valence electron concentration (through the interconnected parameters Z, k_F and E_F) and on the crystal structure (through the summation over R_{nm}). Therefore, the extraordinary situation exists that suitable variation of these parameters can make the magnetic coupling extremely small even though the magnetic $4f$ moments keep their normal magnitudes. This offers the following four possibilities.

First, the specific heat released at the magnetic ordering transition depends on the magnitude of the moment but not on the coupling strength. By choosing a system with large moments (e.g., gadolinium) and by varying the magnetic coupling strength via k_F and the crystal structure one may obtain materials with a large specific heat contribution in a desired temperature interval.

An example of how tailoring the Curie temperature can lead to materials with the desired properties is given in Fig. 1 which shows some results of specific heat measurements (see *Magnetocaloric Effect*). In cryogenic engineering the temperature range of liquid helium is usually easily accessible. However, there is a lack of materials that have a large specific heat just above the boiling point of helium (i.e., in the range 4–10 K). Such materials are required for thermal shields, buffer heat capacities or heat regenerators. It can be seen from the results shown in Fig. 1 that a number of pseudobinary compounds of the type $Gd_{1-x}Er_xRh$ have a large specific heat contribution in the desired range, which compares favorably with that of lead (broken line).

Second, by tailoring the exchange interaction strength between the localized moments a situation can be obtained where the electronic interaction between superconducting Cooper pairs compares favorably with the magnetic exchange interaction and leads to the occurrence of (reentrant) superconductivity and magnetic ordering in one and the same compound. A survey of rare-earth-based superconducting materials has been given by Fischer (1990).

Third, if the magnetic exchange interaction can be made sufficiently small and a particularly favorable crystal field interaction is present, it is possible to observe nuclear-enhanced phase transitions or mixed electronic nuclear phase transitions. Some of the compounds exhibiting such features find applications in nuclear-enhanced cooling devices.

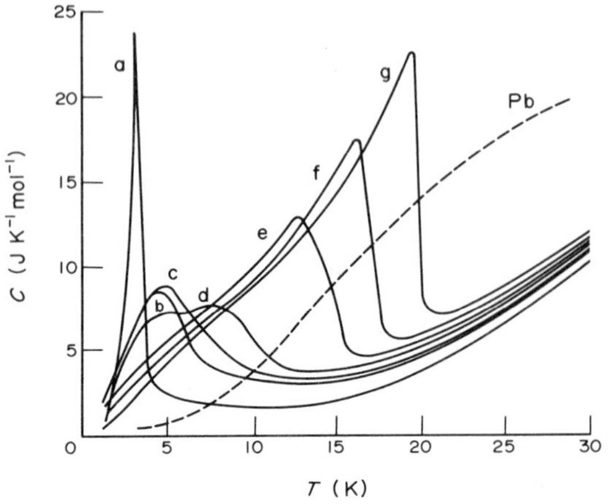

Figure 1
Temperature dependence of the specific heat C in various CsCl-type compounds of rare earth with rhodium: (a) ErRh, (b) $Gd_{0.4}Er_{0.6}Rh$, (c) $Gd_{0.5}Er_{0.5}Rh$, (d) $Gd_{0.6}Er_{0.4}Rh$, (e) $Gd_{0.8}Er_{0.2}Rh$, (f) $Gd_{0.9}Er_{0.1}Rh$, and (g) GdRh; the dashed line denotes the temperature dependence of the specific heat in lead

Finally, low magnetic coupling between the localized moments and a negative value of the *s–f* exchange integral of Eqn. (1) are prerequisites for the occurrence of Kondo anomalies.

The Kondo effect orginates from the elastic scattering of itinerant electrons by an ion having a net magnetic moment. But the virtual bound state associated with the localized electrons responsible for the magnetic moment has to be situated not far below the Fermi level, that is, the magnetic moment is close to its instability. The scattering process can proceed with or without spin flip of the itinerant electrons, the spin flip process becoming predominant when the temperature is low enough. The negative exchange interaction between the spin s of the itinerant electron and the spin S of the ion (Eqn. (1)), in a static picture, gives rise to a screening of the local spin by a surrounding cloud of conduction electrons with opposite spin. This situation can lead to the disappearance of magnetism at very low temperature. In other words if $S=\frac{1}{2}$ (in the case of Ce^{3+}) this interaction removes the degeneracy of the system giving rise to a nonmagnetic singlet ground state and to an excited triplet state.

The characteristic temperature below which this process becomes predominant is the Kondo temperature T_K given by

$$T_K = \frac{\Delta}{k_B}\exp - \left[\frac{1}{N(E_F)|\Gamma|}\right]$$

where Δ is the bandwidth and $N(E_F)$ is the density of states at the Fermi level. Typically $\Gamma N(E_F) \simeq 0.1$ and T_K is of the order of a few kelvin. At high temperatures the influence of this process decreases and the moment recovers the characteristics of the free ion as observed from the high-temperature slope of the reciprocal susceptibility. The clearest experimental signature of the Kondo effect is found in the electrical resistivity which shows a logarithmic increase when the temperature decreases in the range near and below T_K. The other scattering processes give rise to the normally observed increase of the resistivity with temperature. Therefore, the total resistivity of the material under consideration shows a minimum (see *Magnetotransport Phenomena*). More details can be found in the review by Brandt and Moshchalkov (1984).

3. Compounds of Rare Earths with 3d Elements

Of the binary systems between rare-earth and 3*d* elements, those with nickel are the most prolific in the formation of intermetallics. For instance, the erbium–nickel system is composed of 11 intermetallic compounds ranging from Er_3Ni to Er_2Ni_{17}. Owing to hybridization effects between the rare-earth valence electrons and the 3*d* electrons the nickel atoms do not carry a magnetic moment in compounds with nickel concentrations lower than $RENi_3$.

The number of RE–cobalt compounds is lower than that of RE–nickel compounds. Also in this case the 3*d* moment is lost although in some cases it persists up to much lower 3*d* atom concentrations (RE_4Co_3) than found in the RE–nickel compounds. In some compounds, such as $RECo_2$, the presence of a moment on the cobalt atom depends on the spin-polarizing influence of the rare-earth component: YCo_2 is a Pauli paramagnet whereas $GdCo_2$ is a ferromagnet with $T_C = 404$ K.

The number of compounds is even less in binary systems of rare earths and iron or manganese, the lowest 3*d* concentration being reached in the series $REFe_2$ and $REMn_2$. The 3*d* atoms in all RE–iron and RE–manganese compounds carry a magnetic moment.

The manganese atoms have the tendency to mutually antiparallel alignment. As a consequence there is antiferromagnetic or ferrimagnetic order in the 3*d* sublattice of RE–manganese compounds. For this reason the manganese sublattice generally contributes little to the overall magnetization. The situation is different in the rare-earth compounds with nickel, cobalt or iron. The 3*d* moments in these compounds mutually couple parallel, leading to a ferromagnetic 3*d* sublattice. The total magnetization is determined by the type of coupling between the rare-earth and 3*d* sublattices. In general, this coupling is ferromagnetic for the light rare-earth elements, which leads to relatively high overall magnetizations. For the heavy rare-earth elements the RE–3*d* intersublattice coupling is antiferromagnetic. This leads to ferrimagnetic compounds with lower overall magnetizations than found in the corresponding compounds with light rare-earth elements. The situation is basically the same in ternary compounds RE–3*d*–X in which rare-earth and 3*d* elements are combined with a nonmagnetic element X.

In order to describe the magnetic properties of RE–3*d* or RE–3*d*–X systems in a more quantitative sense, mean field theory is commonly used in which the sublattice magnetization M_{RE} is coupled to the 3*d* sublattice magnetization M_{3d} via the molecular field constant $n_{RE\text{-}3d}$. The 3*d* sublattice anisotropy is included in this approach via standard phenomenological expressions which take the form $K_1 \sin^2\theta$ for uniaxial compounds, where K_1 is the lowest-order anisotropy constant and θ is the angle between the unique crystallographic direction and the magnetization. The magnetic properties at any given temperature and applied field H_{appl} are then obtained by solving the coupled equations

$$H_{RE} = \sum_{nm} B_{nm}O_{nm} - (n_{RE\text{-}3d}\langle \boldsymbol{M}_{3d}\rangle + \boldsymbol{H}_{appl}).\boldsymbol{M}_{RE} \quad (4)$$

$$E_{3d} = K_1 \sin^2\theta - (n_{RE\text{-}3d}\langle \boldsymbol{M}_{RE}\rangle + \boldsymbol{H}_{appl}).\langle \boldsymbol{M}_{3d}\rangle \quad (5)$$

while minimizing the total energy

$$E_{tot} = E_{3d} + E_{RE} - n_{RE\text{-}3d}\boldsymbol{M}_{RE}.\boldsymbol{M}_{3d} \quad (6)$$

The free energy of the rare-earth sublattice E_{RE} is obtained by summing the individual rare-earth atom contributions $F_{RE} = -k_B T \ln Z_{RE}$ (Z_{RE} is the partition function) resulting from diagonalization of the Hamiltonian in Eqn. (4). More details of this procedure are described by Coey and Li (1991).

The upshot of this procedure is that, owing to the presence of a strong rare-earth sublattice anisotropy contained in the crystal field term $\Sigma B_{nm}O_{nm}$, the ultimate magnetic structure need not be a collinear one. Furthermore there may be changes in magnetization direction as a function of temperature. Examples of structural changes that have been satisfactorily explained with the two-sublattice model expressed by Eqns. (4–6) are shown in Fig. 2. Comprehensive reviews of the magnetic properties and their theoretical explanation are given for binary RE–3*d* compounds by Franse and Radwański (1992) and for ternary RE–3*d*–X compounds by Coey and Li (1991) and Szytula (1991)

4. *Permanent Magnet Materials*

Permanent magnet materials based on rare earths and 3*d* transition elements owe their favorable magnetic properties both to the rare-earth sublattice and the 3*d* sublattice. The former is the main contributor to the magnetocrystalline anisotropy whereas the latter is mainly responsible for the sufficiently high magnetic ordering temperatures. Both properties are linked together by the comparatively strong magnetic coupling between the rare-earth and 3*d* sublattices, expressed by the coefficient $n_{RE\text{-}3d}$ in Eqns. (4, 5).

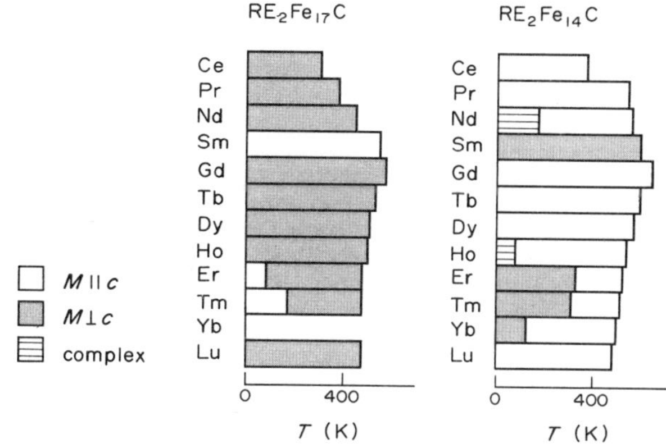

Figure 2
Temperature dependence of magnetic structures observed for the series $RE_2Fe_{14}C$ and $RE_2Fe_{17}C$

Powerful magnet materials are based on compounds such as $SmCo_5$ and $Nd_2Fe_{14}B$ and their favorable hard magnetic properties are discussed in more detail in the articles *Cobalt–Rare-Earth Compounds* and *Neodymium–Iron–Boron Alloys*, respectively. Other permanent magnet materials are based on the compounds $SmFe_{11}Ti$, $Sm_2Fe_{17}C$ and $Sm_2Fe_{17}N_{2.7}$.

An accurate description of the magnetocrystalline anisotropy and the magnetic coupling between 4*f* and 3*d* moments is available for the series $RE_2Fe_{14}B$, $RE_2Co_{14}B$, $RE_2Fe_{14}C$, $REFe_{11}Ti$, $RE_2Fe_{17}C$ and $RE_2Fe_{17}N_{2.7}$ owing to the results of numerous investigations, using modern techniques such as NMR, different types of Mössbauer spectroscopy, neutron diffraction and high-field magnetization measurements. The magnetocrystalline anisotropy was found to be rather well describable in terms of the two-sublattice model (Eqns. (4, 5)), which offers an explanation for the various types of magnetic structures found in these series. Band structure calculations have proved to be most helpful in the understanding of the different magnitudes of the iron moments associated with the various crystallographic 3*d* sites in the corresponding unit cells. Of particular interest are compounds of the series RE_2Fe_{17} which do not qualify for permanent magnets owing to their small crystalline anisotropy and low Curie temperatures. However, anisotropy as well as Curie temperature can be enhanced considerably by the introduction of carbon or nitrogen atoms into the interstitial sites available in the crystal structure. Experimental results of T_C for $RE_2Fe_{17}C$ and $RE_2Fe_{17}N_{2.7}$ can be compared with data for RE_2Fe_{17} in Fig. 3. A review of the magnetic

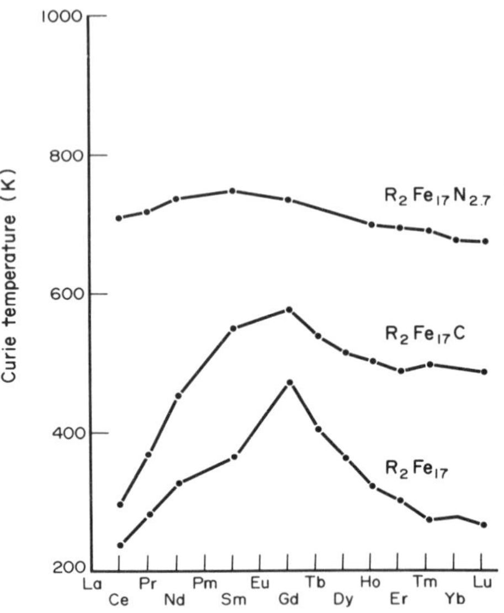

Figure 3
Changes of the Curie temperature in compounds of the series RE_2Fe_{17} after interstitial solution of carbon atoms (middle) and nitrogen atoms (top)

properties of compounds related to permanent magnet materials and their theoretical description is given by Buschow (1991).

5. High-Magnetostriction Materials

The large room-temperature magnetostriction observed in a number of rare-earth 3*d* compounds makes them suitable for applications as high-power magnetomechanical transducers. The rare-earth metals have intrinsically huge magnetostrictions. A drawback is their relatively low magnetic ordering temperatures. High values of the magnetostriction at room temperature have been obtained by combining rare-earth metals with 3*d* metals, leading to ordering temperatures that are sufficiently above room temperature. The most successful binary compounds are the Laves phases $REFe_2$. They derive their magnetostrictive properties basically from the large anisotropic 4*f* charge density of the rare-earth component. Owing to the strong magnetic coupling between the rare-earth and the iron sublattices (Eqns. (4–6)) the large intrinsic low-temperature magnetostriction of the rare-earth component is preserved at room-temperature. The favorable magnetostrictive properties are not restricted to crystalline materials but have also been observed in amorphous RE–Fe alloys. More details can be found in the article *Magnetostrictive Materials*.

6. Invar Alloys

Several types of alloys described in the literature have a small thermal expansion coefficient in the temperature range 273–573 K. These alloys are in general composed of three or more different 3*d* metals and are designated as Invar, Stainless Invar or Super Invar. A compilation of data and theoretical description regarding all these alloys has been given by Wassermann (1990). Invar-type behavior is also shown by rare-earth compounds. For instance, ternary RE–3*d*–X compounds of the types $RE(Fe,Co)_{11}Al_2$, $REFe_{10}V_2$, $RE_2Fe_{14}B$, $RE_2Fe_{14}C$ and $RE_2Fe_{17}C$ show pronounced Invar-like characteristics in their thermal expansion behavior. Examples of the thermal behavior of several of such compounds are shown in Fig. 4. An interesting aspect of this class of Invar alloys is that they are brittle compounds which lend themselves easily to pulverization in a mortar. The other Invar alloys mentioned are ductile and cannot be pulverized in a mortar. The advantage of rare-earth-based Invar alloys lies primarily in the possibility of using them in the manufacture of metallic components of complicated shape and low thermal expansion by means of powder compacting techniques, as for instance liquid phase sintering.

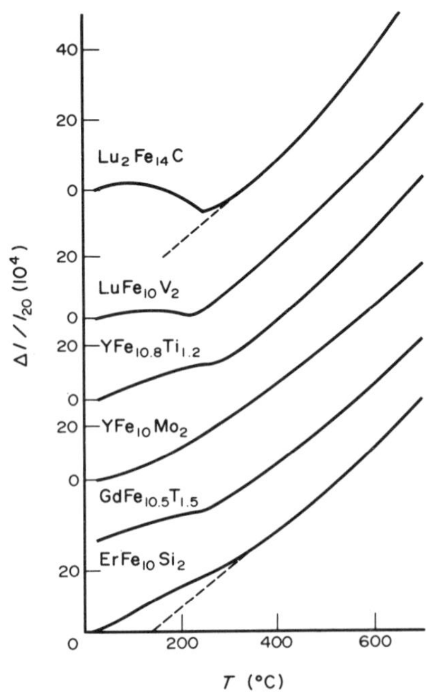

Figure 4
Thermal expansion behavior observed in several ternary rare-earth compounds

7. *Hydrogen-Induced Changes in Magnetic Properties*

A most fascinating property of rare-earth–3*d* metal compounds is their ability to absorb large amounts of hydrogen gas. This hydrogen absorption is reversible and generally occurs at moderate pressures. In many intermetallic compounds it is possible to reach hydrogen densities much higher than that of liquid hydrogen. The hydrogen is present in these materials in the form of screened protons and occupies interstitial hole sites in the various crystal structures. In most cases the total available number of hole sites is larger than that corresponding to the amount of hydrogen absorbed. The distribution of hydrogen atoms over the available hole sites is not static. Experiments involving nuclear magnetic resonance (NMR), neutron scattering and Mössbauer effect spectroscopy have shown that the hydrogen atoms interchange sites with jump rates which, at room temperature, lie in the range 10^{12}–$10^{13}\,s^{-1}$. (For a review see Buschow et al. (1982).)

The uptake of hydrogen gas in these intermetallic compounds often leads to spectacular changes in magnetic properties. Examples are known of hydrogen-induced paramagnetic-to-ferromagnetic, ferromagnetic-to-paramagnetic and antiferromagnetic-to-ferromagnetic transitions. All these changes and the various models proposed to explain them have been discussed in the review by Hilscher and Wiesinger (1991).

It is important to realize that the ternary hydrides are metastable with respect to the corresponding binary rare-earth hydrides and 3*d* metal. The ternary hydrides owe their existence to the limited diffusion of the metal atoms at temperatures below about 600 °C, which prevents phase separation. This feature has been employed advantageously to generate high coercive forces in permanent magnet materials. (For details see the review by Buschow (1991).) In fact, it has been shown that heating materials consisting primarily of $Nd_2Fe_{14}B$ in a hydrogen atmosphere, with subsequent vacuum annealing at temperatures higher than 750 °C, produces coercive powders which are suitable for the production of polymer-bonded magnets. Investigations performed at different stages of the process have confirmed that the Nd–Fe–B alloy disproportionates into iron, Fe_2B and neodymium hydride before recombining during the vacuum annealing stage to produce the fine-grained microstructure necessary for highly coercive properties. The grain size of the final microstructure depends strongly on the temperature at which the hydrogenation–disproportionation–desorption cycle is carried out.

See also: Cobalt–Rare-Earth Compounds; Magnetostrictive Materials; Neodymium–Iron–Boron Alloys; Rare-Earth Metals and Alloys

Bibliography

Brandt N B, Moshchalkov V V 1984 Concentrated Kondo systems. *Adv. Phys.* 33: 373–467

Buschow K H J 1980 Rare earth compounds. In: Wohlfarth E P (ed.) 1980 *Ferromagnetic Materials*, Vol. 1. North-Holland, Amsterdam, p. 297

Buschow K H J 1991 New developments in hard magnetic materials *Rep. Prog. Phys.* 54: 1123–1213

Buschow K H J, Bouten P C P, Miedema A R 1982 Hydrides formed from intermetallic compounds of two transition metals. *Rep. Prog. Phys.* 45: 937–1039

Coey J M D, Li Hong-Shuo 1991 Magnetic properties of ternary rare-earth transition-metal compounds. In: Buschow K H J (ed.) 1991 *Magnetic Materials*, Vol. 6. North-Holland, Amsterdam

de Boer F R, Boom R, Mattens W C M, Miedema A R, Niessen A K 1988 *Cohesion in Metals*. North-Holland, Amsterdam

Fischer Ø 1990 Magnetic superconductors. In: Buschow K H J, Wohlfarth E P (eds.) 1990 *Ferromagnetic Materials*, Vol 5. North-Holland, Amsterdam, p. 465

Franse J J M, Radwánski R J 1992 Binary rare-earth transition-metal intermetallic compounds. In: Buschow K H J (ed.) 1992 *Magnetic Materials*, Vol. 7. North-Holland, Amsterdam

Hilscher G, Wiesinger G 1991 Magnetism of hydrides. In: Buschow K H J (ed.) 1991 *Magnetic Materials*, Vol. 6. North-Holland, Amsterdam

Hutchings M T 1964 Point charge calculation of energy levels of magnetic ions in crystalline electric field. *Solid State Phys.* 16: 227–74

Szytula A 1991 Magnetic properties of ternary intermetallic rare earth compounds. In: Buschow K H J (ed.) 1991 *Magnetic Materials*, Vol. 6. North-Holland, Amsterdam

Villars P, Calvert L D 1985 *Pearson's Handbook of Crystallographic Data for Intermetallic Phases*. American Society for Metals, Metals Park, OH

Vonsovskii S V 1974 *Magnetism*. Wiley, New York

Wassermann E F 1990 Invar: moment-volume instabilities in transition metals and alloys. In: Buschow K H J, Wohlfarth E P (eds.) 1990 *Ferromagnetic Materials*, Vol. 5. North-Holland, Amsterdam, p. 237

K. H. J. Buschow
[Philips Research Laboratories,
Eindhoven, The Netherlands]

Rare-Earth Metals and Alloys

The term rare earths is used to describe 17 elements—lanthanum and the 14 following elements, called the lanthanides, together with scandium and yttrium which have very similar chemical properties. Within this group, the elements lanthanum through to europium and gadolinium through to lutetium are known as the light and heavy rare earths, respectively. They are all mutually soluble in one another and hence a complete range of intra-rare-earth alloys exists. These elements began to be discovered in the late eighteenth century, by which time the term earth had come to mean an oxide that could not be reduced on heating either by hydrogen or carbon. Alumina and silica were examples of the common earths. When Karl Axel Arrhenius, a Swedish lieutenant, discovered an unusual black mineral in 1787 he named it ytterbite, after the nearby village of Ytterby, on an island not far from Stockholm. From this the Finnish chemist Johannes Gadolin extracted a substance which he called at first ytterbia, then yttria. Its chemical properties were different from any of the known earths and it had been found in a rare mineral. Thus it was called a "rare" earth. Subsequently ytterbium, terbium and erbium as well as yttrium were named after this Swedish village.

Scandium is significant because it had been predicted in 1871 by the Russian chemist Dmitri Mendeleev. When he first proposed the periodic table, he found it necessary to leave a blank at the place now occupied by scandium. However, as well as predicting the existence of an element in this position, he also successfully predicted some of its properties. The discovery of scandium eight years later established the credibility of Mendeleev's periodic table. However, one of its weaknesses was the lack of a logical place for the lanthanides and hence the difficulty of predicting how many there should be.

Around 1912 the English physicist Henry Moseley discovered a direct relationship between the frequencies of the x-ray emission spectra and the atomic number Z of an element. This predated quantum mechanics and so the reason for Moseley's relationship was not properly understood. Nevertheless, this was an extremely significant discovery from which it was deduced that there were only 14 lanthanide elements. It could even be predicted that while 13 of them were known, there was one missing at $Z=61$. Promethium does not occur naturally since it has no stable isotope. It was isolated from fission products only in 1945 and not prepared in metallic form until 1963.

The rare earths are not actually rare at all. There is more cerium (46 ppm) than tin and more neodymium (24 ppm) than lead in the earth's crust. Even the relatively scarce lutetium (0.8 ppm) is more abundant than mercury. However, the similarity of their chemical properties makes them difficult to separate.

1. Basic Magnetic Interactions

The origin of the magnetism of the lanthanide elements lies in the gradual filling of the $4f$ electronic shell inside the $5d$ and $6s$ electrons (see Fig. 1). The magnetic ground state of an atom with an incomplete $4f$ shell is determined by Hund's rules (1925). Subject to the exclusion principle, the total spin S and then the total orbital angular momentum L are maximized. The total angular momentum J is given by $L+S$ or $L-S$, depending on whether the shell is, respectively, more than or less than half filled. The magnetic moment of the free-ion ground state, given by $g\mu_B J$, is indicated in Table 1. Because of the relatively large spin–orbit coupling, only the lowest J multiplet is populated at room temperature and below (except for samarium).

When lanthanide atoms condense to form a solid, the outer $5d$ and $6s$ electrons form a conduction electron gas, while the well-shielded $4f$ electrons retain their magnetic properties. Apart from the case of cerium, there is no direct overlap of the $4f$ wave functions on different ionic sites. However, the $4f$ electrons are coupled indirectly via the conduction electrons. In this process, a conduction electron with spin $\boldsymbol{s}$ at position $\boldsymbol{r}$ interacts with the localized moment $\boldsymbol{S}$ at $\boldsymbol{R}_i$ with a strength determined by their separation. This s–f interaction has the Heisenberg form:

$$H_{sf} = -\mathcal{J}(\boldsymbol{R}-\boldsymbol{r})\boldsymbol{S}.\boldsymbol{s}$$

where $\mathcal{J}$ includes both local and nonlocal exchange interaction. Thus, when a conduction electron passes close to a $4f$ ion, there is a coupling between the spins and hence the memory of this is carried to the next localized moment. The net result is an indirect coupling between f electrons at $\boldsymbol{R}_i$ and $\boldsymbol{R}_j$ which, in its simplest form, may be expressed as

$$H_{ff} = -\sum_{R_{ij}} \mathcal{J}(\boldsymbol{R}_{ij})\boldsymbol{S}_i.\boldsymbol{S}_j$$

where $\boldsymbol{R}_{ij} = \boldsymbol{R}_i - \boldsymbol{R}_j$. This is known as the Ruderman–Kittel–Kasuya–Yoshida (RKKY) coupling. The Fourier transform of $\mathcal{J}(\boldsymbol{R}_{ij})$ is

$$\mathcal{J}(\boldsymbol{q}) = \sum_{R_{ij}} \mathcal{J}(\boldsymbol{R}_{ij})\exp(\mathrm{i}\boldsymbol{q}.\boldsymbol{R}_{ij})$$

and can be readily determined from neutron inelastic

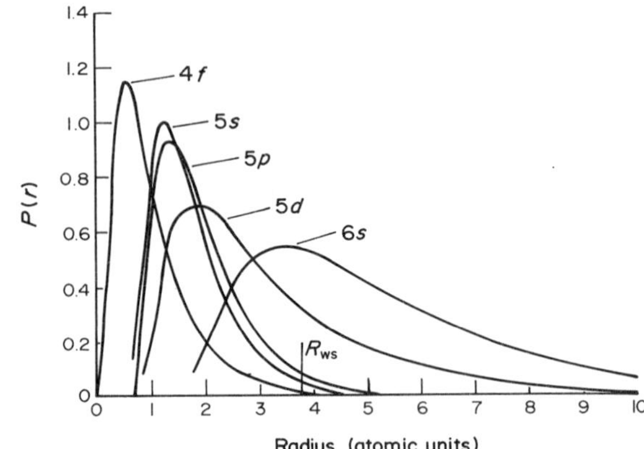

Figure 1
The radial part of the $4f$, $5s$, $5p$, $5d$ and $6s$ wave functions for gadolinium (R_{WS} is the Wigner–Seitz radius)

Table 1
Magnetic properties of the rare earths

Element (ion)	$4f^n$	L	S	J	g	$(g-1)^2J(J+1)$	$T_{N,C}$ (K)
Lanthanum (La^{3+})	0	0	0	0	0	0	a
Cerium (Ce^{3+})	1	3	$\frac{1}{2}$	$\frac{5}{2}$	$\frac{6}{7}$	0.18	13
Praseodymium (Pr^{3+})	2	5	1	4	$\frac{4}{5}$	0.80	0.05
Neodymium (Nd^{3+})	3	6	$\frac{3}{2}$	$\frac{9}{2}$	$\frac{8}{11}$	1.84	19.9
Promethium (Pm^{3+})	4	6	2	4	$\frac{3}{5}$	3.20	?
Samarium (Sm^{3+})	5	5	$\frac{5}{2}$	$\frac{5}{2}$	$\frac{2}{7}$	4.46	106
Europium (Eu^{2+})	7	0	$\frac{7}{2}$	$\frac{7}{2}$	2	15.75	94
Gadolinium (Gd^{3+})	7	0	$\frac{7}{2}$	$\frac{7}{2}$	2	15.75	293
Terbium (Tb^{3+})	8	3	3	6	$\frac{3}{2}$	10.50	230
Dysprosium (Dy^{3+})	9	5	$\frac{5}{2}$	$\frac{15}{2}$	$\frac{4}{3}$	7.08	178
Holmium (Ho^{3+})	10	6	2	8	$\frac{5}{4}$	4.50	133
Erbium (Er^{3+})	11	6	$\frac{3}{2}$	$\frac{15}{2}$	$\frac{6}{5}$	2.55	85
Thulium (Tm^{3+})	12	5	1	6	$\frac{7}{6}$	1.17	57
Ytterbium (Yb^{2+})	14	0	0	0	0	0	a
Lutetium (Lu^{3+})	14	0	0	0	0	0	a

a No magnetic moment

scattering experiments, as discussed later. This quantity $\mathcal{J}(\boldsymbol{q})$ is related to the s–f exchange and the susceptibility of the conduction electron gas, which can in principle be calculated:

$$\mathcal{J}(\boldsymbol{q}) = (2/N)|j(q)^2|\chi(q)$$

Strong spin–orbit coupling means that J rather than S is the constant of motion. Thus the exchange is due to the projection of S on J which is $(g-1)J$ and the indirect exchange interaction may be written as

$$H_{ff} = -\sum_{R_{ij}}(g-1)^2\mathcal{J}(\boldsymbol{R}_{ij})\boldsymbol{J}_i.\boldsymbol{J}_j$$

The strength of the exchange energies determines the temperature of magnetic ordering to a cooperative phase. As can be seen from Table 1, the ordering temperatures (T_N for an antiferromagnetic and T_C for a ferromagnetic transition) are below room temperature, in contrast to the iron group magnets.

Consideration will now be given to the second important interaction in the rare-earth metals. In the solid state, the $4f$ electrons experience the electric field of their neighbors—the crystalline electric field—and the eigenstates are linear combinations of the basis $|J, m_j\rangle$ states. The definitive description of this crystal field interaction in the rare earths was given by Stevens in 1952. It is still not possible to calculate these states from first principles, but they can be deduced by combining experimental evidence from a variety of measurements, such as heat capacity, magnetization and neutron inelastic scattering. The states in almost all of the pure metals are now well known. Figure 2 illustrates the crystal field level scheme for the $2J+1$ $(=9)$ states of praseodymium and the associated charge densities of the low-lying levels. Because the electric field is very inhomogeneous and the charge clouds are anisotropic, there is a strong tendency for the magnetic moments to align along particular crystallographic directions. This effect gives rise to a single-ion magnetic anisotropy which, in some cases, requires tens of tesla to overcome.

The majority of the magnetic properties of the rare-earth metals, alloys and compounds can be explained in terms of the two interactions mentioned, plus the Zeeman term (in a finite magnetic field) and the magnetoelastic interaction. The latter arises from the strain sensitivity of the exchange and crystal field interactions. Together they form the basis of the standard model of rare-earth magnetism. However, it is now clear that the $4f$ electrons in cerium and many cerium-based alloys and compounds have a strong itinerant character and thus the standard model cannot explain the magnetic properties of these materials. Extensions to the simple model are frequently required to explain the quantitative behavior of particular metals or alloys. For example, the exchange may be anisotropic, involving other interactions such

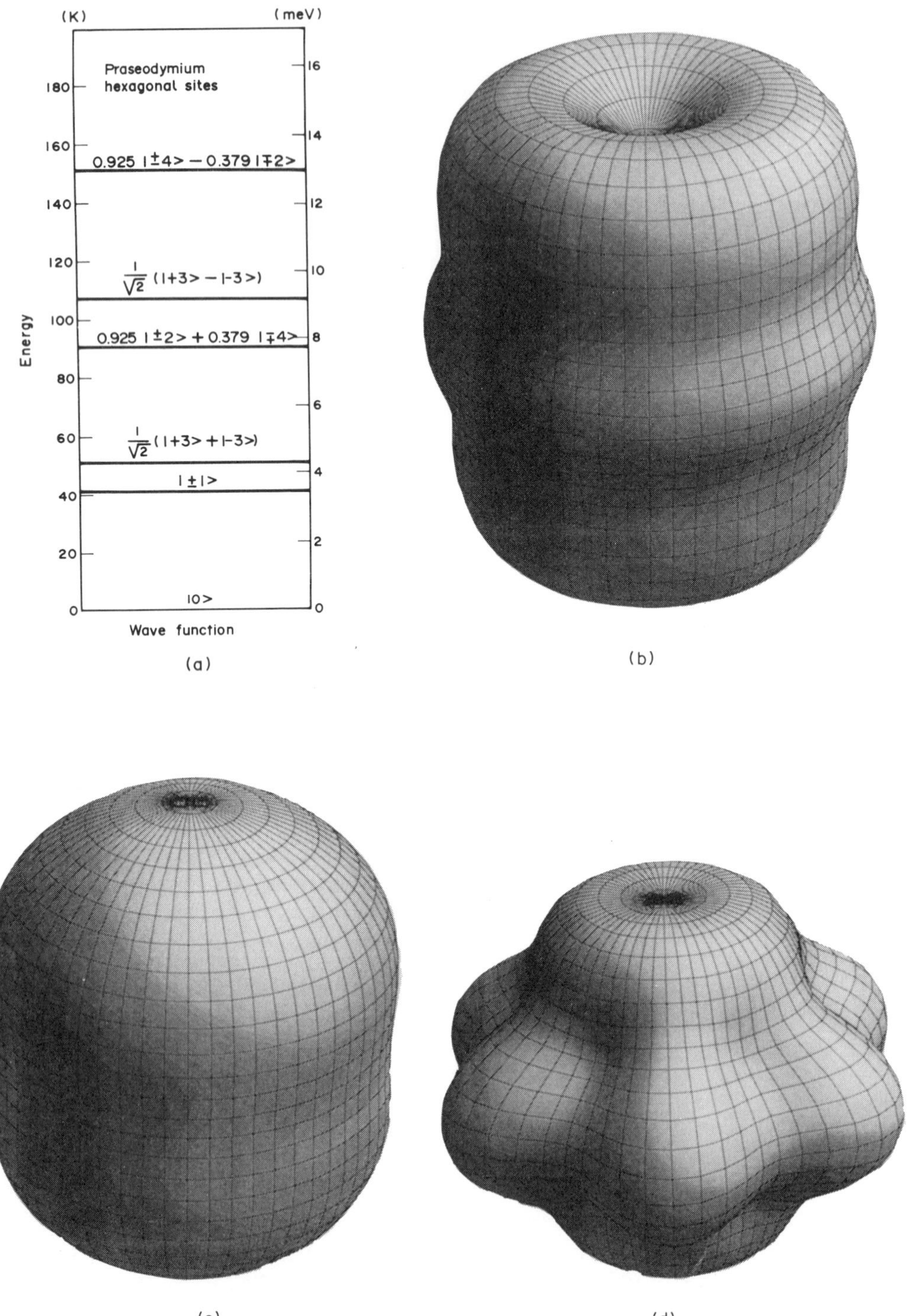

Figure 2
(a) The crystal field level scheme for praseodymium; the m_J part of the wave functions $|J=4, m_J\rangle$ are given.
(b), (c) and (d) The associated charge densities of, respectively, the ground state $|0\rangle$ and the low-lying excited states $|\pm 1\rangle$ and $|3s\rangle$

as dipolar or orbital coupling. The hyperfine coupling between the electronic and nuclear spins occasionally plays a vital role.

2. Magnetic Structures

When they are cooled below a critical temperature, most of the rare earths exhibit a transition to a magnetically ordered state, as discussed in Sect. 1. The first detailed studies of the magnetic structures were made in the late 1960s by Wallace Koehler and his collaborators at Oak Ridge National Laboratory using single crystals of the heavy rare-earth metals prepared at the Ames Laboratory, Iowa, by the group of Frank Spedding. The results obtained from their neutron diffraction work are shown schematically in Fig. 3. The simplest structure is that of gadolinium which is ferromagnetic below $T_C = 293$ K. In general these metals, which have the hexagonal close-packed crystal structure, exhibit a variety of antiferromagnetic phases in which the moments in a particular basal plane align parallel to each other, but this direction varies systematically along the hexagonal axis. For example, the helical structure found in terbium between $T_N = 230$ K and $T_C = 220$ K and in dysprosium between $T_N = 178$ K and $T_C = 86$ K may be described by

$$J_x = J \cos \boldsymbol{Q}.\boldsymbol{R}_i, \qquad J_y = J \sin \boldsymbol{Q}.\boldsymbol{R}_i$$

In the case of erbium and thulium, the uniaxial magnetic anisotropy favors alignment of the moments along the c axis rather than the basal plane and so the longitudinally modulated structure is found:

$$J_z = J \cos \boldsymbol{Q}.\boldsymbol{R}_i$$

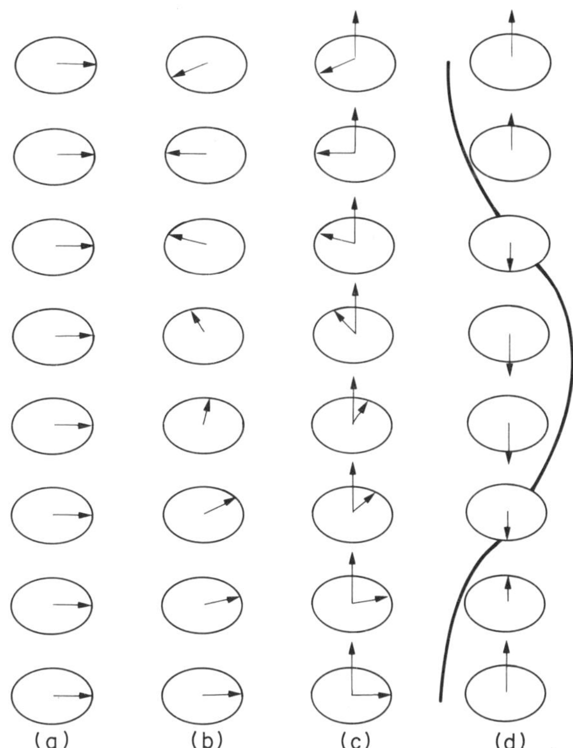

Figure 3
Magnetic structures of the heavy rare-earth elements (Gd, Tb, Dy, Ho, Er, Tm): (a) ferromagnet, (b) helix, (c) cone, (d) longitudinal modulation (after Koehler 1972)

The type of ordering is determined by the wave vector $\boldsymbol{Q}$ for which $\mathcal{J}(\boldsymbol{q})$ has a maximum. If $\boldsymbol{Q} = 0$ as in gadolinium, the ferromagnetic structure is found, whereas in terbium, dysprosium, holmium, erbium and thulium $\boldsymbol{Q}$ is nonzero and directed along the c axis. Below $T_N = 133$ K, the structure of holmium first has a simple helical behavior, but as the planar anisotropy increases with decreasing temperature, it becomes too costly in energy for the moments to maintain a constant interlayer turn angle. Using synchrotron x radiation, Gibbs et al. (1985) discovered that holmium exhibits a spin–slip structure. This comprises regions of commensurate spin layers with moments in two planes locked along successive easy directions, followed by a spin–slip in which the moments in only a single plane are directed along the next easy direction.

The structures of the light rare earths are considerably more complex. Neodymium displays a hierarachy of magnetic structures below $T_N = 19.9$ K. Since the moments and the ordering wave vector lie in the basal plane, the structures are longitudinally modulated. The degeneracy of symmetry-related wave vectors leads to the existence of multiple-$\boldsymbol{Q}$ structures. Most prominent are the double-$\boldsymbol{Q}$ structure and quadruple-$\boldsymbol{Q}$ structures discovered in neutron diffraction experiments (Forgan et al. 1989).

Praseodymium is interesting because the crystalline electric field results in the ground state of the $4f^2$ ions being a singlet. In such a case, the only contribution to the magnetic susceptibility is the Van Vleck term. The exchange interaction needs to exceed a certain threshold value in order to induce magnetic ordering. As discussed in Sect. 3, the RKKY coupling in praseodymium is only 92% of the critical value. However, the susceptibility of the nuclear spins is proportional to T^{-1} (i.e., Curielike) and the hyperfine interaction couples the ordered nuclear spins to the electronic moments in such a way that the $4f$–$4f$ exchange is enhanced over the threshold value. The coupled spin system thus becomes magnetically ordered only at $T_N = 50$ mK, the lowest ordering temperature found in the rare earths (McEwen and Stirling 1981, Møller et al. 1982).

3. Magnetic Excitations

The greatest insight into the microscopic mechanisms governing rare-earth magnetism has come from studies of the excitations of their spin systems. At low temperatures, the normal modes of the magnetically ordered structures are spin waves (see Fig. 4). The energy required to excite a spin wave varies with its wavelength and is most conveniently expressed in terms of the wave vector $\boldsymbol{q}$. For a simple isotropic ferromagnet, the dispersion is of the form

$$E(\boldsymbol{q}) = \hbar\omega(\boldsymbol{q}) = \boldsymbol{J}[\mathcal{J}(\boldsymbol{q}) - \mathcal{J}(\boldsymbol{0})]$$

$E(\boldsymbol{q})$ is normally in the range 1–20 meV and hence these magnetic excitations can be measured using the technique of inelastic neutron scattering. The exchange interaction may thus be measured quite directly. In anisotropic magnetic systems, the dispersion also depends on the crystal field parameters, permitting these to be determined as well.

The first observation of spin waves in a rare-earth metal was made at Risø, Denmark, in the mid-1960s. From their measurements on terbium, Hans Bjerrum Møller, Jens Houmann and Allan Mackintosh were able to deduce the exchange coupling $\mathcal{J}(\boldsymbol{q})$. Figure 5 shows the spin wave dispersion for terbium. Since the hexagonal close-packed lattice contains two atoms per unit cell, there are in general two modes, one with acoustic and one with optic character. Hybridization with the lattice vibrations (phonons) is also apparent. An energy gap at $\boldsymbol{q} = 0$ is clearly apparent: this arises from the energy required to turn the spins against the anisotropy forces and its magnitude provides a microscopic determination of these parameters, complementing bulk measurements.

Subsequently, spin waves in all the other heavy rare-earth metals have been studied. Natural gadolinium absorbs neutrons very strongly and it was therefore necessary to prepare a single crystal from isotopically purified ^{160}Gd before the spin waves in gadolinium could be measured. Figure 6 compares the exchange interaction along the c direction for all of the heavy rare earths. As can be seen, those elements with helical and modulated structures have their maximum in $\mathcal{J}(\boldsymbol{q})$ at the ordering wave vector $\boldsymbol{Q}$. The measured exchange interaction in Fig. 6 has been normalized by dividing by $(g-1)^2$. If the interaction was entirely of the RKKY type and the 4f wave functions and conduction electron susceptibilities $\chi(q)$ were the same for these elements, then these curves would be identical. Figure 6 demonstrates clearly that this is not the case. Significant differences are apparent, particularly involving those ions with large orbital angular momenta and an explanation of the variation across the series remains a theoretical challenge.

In the light rare earths, the situation is rather more complex. The exchange interaction is weaker and the crystal field interaction is stronger, as a result of the greater radial extension of the 4f wave functions. Indeed, it is more appropriate to describe the magnetic excitations in terms of a different model. In the paramagnetic phase, single-ion excitations between crystal field states separated by an energy gap Δ are coupled by the exchange interaction, producing propagating excitations whose dispersion is given by

$$E^2(\boldsymbol{q}) = [\Delta^2 - 4\alpha^2\Delta\mathcal{J}(\boldsymbol{q})]$$

where α is a dipolar matrix element between ground and excited states. It can be seen that the minimum energy of the excitations occurs where $\mathcal{J}(\boldsymbol{q})$ is greatest and that if $\mathcal{J}(\boldsymbol{q})$ becomes large enough for $\eta = 4\alpha^2\mathcal{J}(\boldsymbol{q})/\Delta$ to equal unity, a phase transition to an ordered state will take place. In praseodymium, the excitations reveal a deep minimum in the dispersion $E(\boldsymbol{q})$ from which it is concluded that $\eta \simeq 0.92$ for this element (Houmann et al. 1979). By applying pressure, it is possible to reduce Δ sufficiently to enable magnetic ordering to take place.

Recently another class of excitations has become accessible to experimental study by neutron scattering. The advent of neutron spallation sources (such as the UK ISIS facility), which produce neutrons at much higher energies than reactor sources, has enabled inelastic scattering studies to be extended into

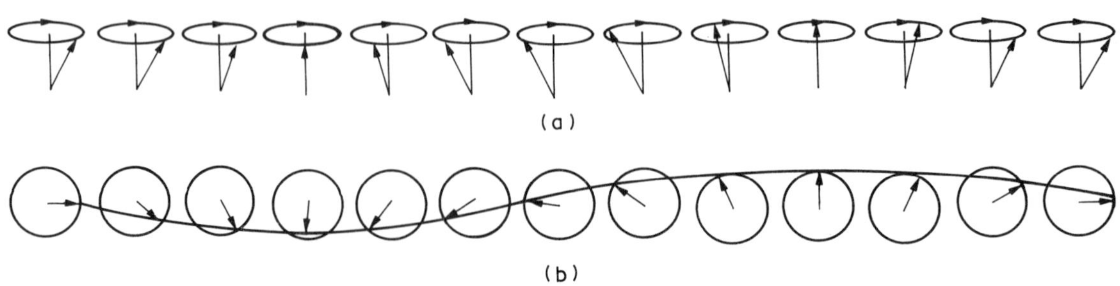

Figure 4
The precession of spins about the net magnetization direction in a spin wave, as seen from (a) side and (b) top; the wavelength is clearly apparent from (b)

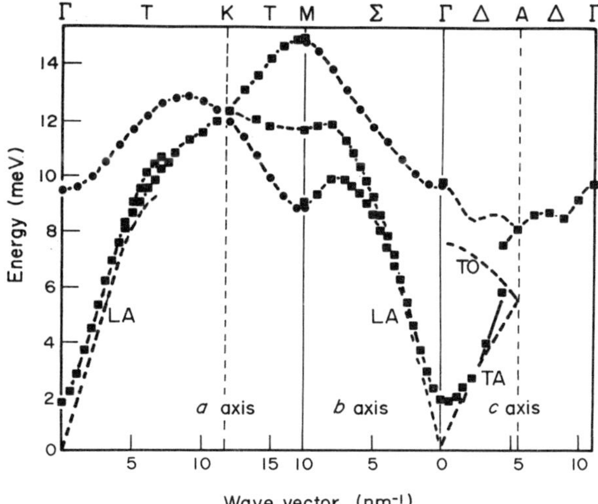

Figure 5
The spin wave dispersion in the ferromagnetic phase of terbium measured along the *a*, *b* and *c* directions: LA, TA and TO are longitudinal acoustic, transverse acoustic and transverse optic phonon modes, respectively (after Mackintosh and Møller 1972)

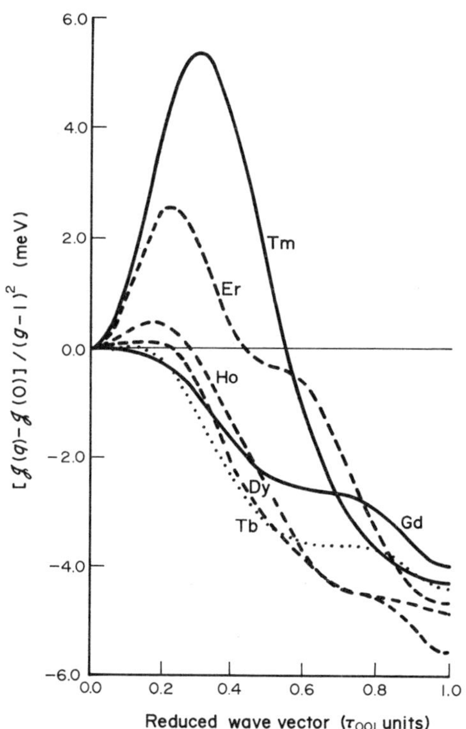

Figure 6
The exchange interaction, determined from spin wave measurements, along the *c* direction in the heavy rare-earth elements (after McEwen et al. 1991)

the eV rather than the meV range. Transitions to higher angular momentum states above the Hund's rule ground state may be studied in this way. A review of these experiments is given by McEwen (1988).

This article has given a brief, elementary description of the salient features of the magnetic structures and excitations of the rare earths. For further reading, see the comprehensive articles by Gschneidner and Eyring (1978) and the excellent monograph of Jensen and Mackintosh (1991).

Bibliography

Forgan E M, Gibbons E P, McEwen K A, Fort D 1989 Observation of a quadruple-q magnetic structure in neodymium. *Phys. Rev. Lett.* 62: 470–3

Gibbs D, Moncton D E, D'Amico K L, Bohr J, Grier B H 1985 Magnetic x-ray scattering studies of holmium using synchrotron radiation. *Phys. Rev. Lett.* 55: 234–7

Gschneidner K A Jr, Eyring L 1978 *Handbook on the Physics and Chemistry of Rare Earths*, Vol. 1. North-Holland, Amsterdam

Houmann J G, Rainford B D, Jensen J, Mackintosh A R 1979 Magnetic excitations in praseodymium. *Phys. Rev. B* 20: 1105–18

Jensen J, Mackintosh A R 1991 *Rare Earth Magnetism*. Oxford University Press, Oxford

Koehler W C 1972 Magnetic structures of rare earth metals and alloys. In: Elliott R J (ed.) 1972 *Magnetic Properties of Rare Earth Metals*. Plenum, London, pp. 81–128

McEwen K A 1988 High energy neutron spectroscopy of rare earths. *J. Magn. & Magn. Mater.* 76/77: 391–5

McEwen K A, Steigenberger U, Jensen J 1991 Spin waves, phonons, and crystal-field excitations in thulium. *Phys. Rev. B* 43: 3298–310

McEwen K A, Stirling W G 1981 Magnetic ordering in praseodymium at millikelvin temperatures. *J. Phys. C* 14: 157–65

Mackintosh A R, Møller H B 1972 Spin waves. In: Elliott R J (ed.) 1972 *Magnetic Properties of Rare Earth Metals*. Plenum, London, pp. 187–244

Møller H B, Jensen J Z, Wulff M, Mackintosh A R, McMasters O D, Gschneidner K A Jr 1982 Hyperfine interactions, magnetic impurities and ordering in praseodymium. *Phys. Rev. Lett.* 49: 482–5

K. A. McEwen
[Birkbeck College, London, UK]

Recording Head Technology

Since the 1940s, magnetic recording has been the universal technology for electronic information mass storage. Its presence is ubiquitous as audio tape, VCRs, floppy disks, computer tapes, hard disks, credit cards and so on. Sales of magnetic recording products exceeded US$50 × 10^9 in 1990 with projected strong growth. Since the 1940s, the areal density has risen to $> 10^8$ bits cm^{-2} on disks and to

2×10^7 bits cm^{-2} on tape. Thus, each bit uses about 1.0 μm^2 area.

The central problem in achieving a high information density with magnetic recording resides primarily in the materials area. The increase in the recording density requires media with higher coercivities and this, in turn, puts tougher requirements on head materials. With the development of video recording systems, response to high frequencies is required in addition to achieving very high areal densities.

Nearly all recording systems use an induction coil and magnetic core (inductive magnetic head) for writing, reading and erasing. Some computer systems also use magnetoresistive heads to read information.

This article describes the different types of recording and reproducing heads currently used in magnetic recording and the magnetic materials used in their fabrication. The trends in development of new materials for future heads will be discussed in the final section of this article.

1. Recording Heads

Magnetic recording heads can be divided into two major categories: inductive and magnetoresistive (MR). The inductive head can perform three functions: write, read and erase, while the MR head is a read-only head. In the case of the inductive head, the fringing field produced in the gap of the core creates a magnetic state in the medium that corresponds to signal currents in the head coil. During the read process, the head senses the states by intercepting magnetic flux that emanates from the magnetic medium. The MR heads utilize the magnetoresistive effect—change of the resistance of the thin layer of magnetic material as a function of the applied field. The inductive head senses dB/dt and its output is media-speed dependent, while the magnetoresistive head is sensing the tape flux and its output does not depend on media speed.

Different types of recording heads are schematically shown in Fig. 1. The conventional heads (Fig. 1a) have a bulky core made mostly from ferrite or laminated thick metal sheets and a wired coil. A modified version of a conventional head is the metal-in-gap (MIG) head where the layer of high-magnetization metal is deposited on one or both poles in the gap region. Ferrite and MIG heads are the most widely used heads in current recording devices. The most advanced head is a thin-film head (Fig. 1b), which uses magnetic metal (iron, cobalt, nickel or their alloys) as a magnetic core material and is produced entirely (including the coil) by a thin-film deposition process.

A high-frequency metal-laminated rotating head is shown in Fig. 1c, where the core, a multilayer of the magnetic metal–insulator, is sandwiched between two pieces of nonmagnetic ceramic. Regular wire is used to produce the coil.

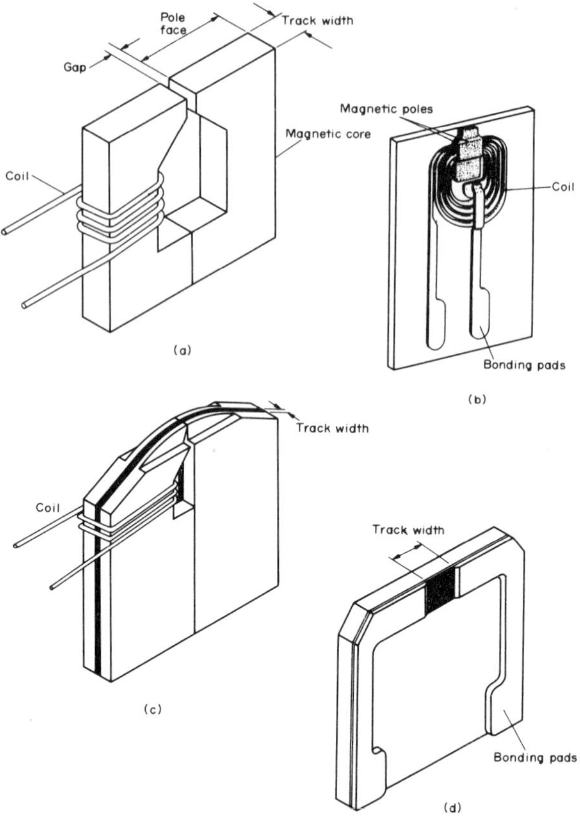

Figure 1
Magnetic recording heads: (a) ferrite type and MIG type, (b) thin film inductive, (c) metal laminated and (d) magnetoresistive

The last type of head, the MR head (Fig. 1d), is fully produced by thin-film technology and its use is growing. MR heads are currently used in tape drives such as the IBM 3480, Philips Data Logger and the recently introduced Philips Digital Compact Cassette system, where its low-speed characteristics are advantageous. In 1991 IBM introduced the MR head in its hard disk drive.

2. Material Requirements for Magnetic Heads

2.1 Inductive Head

The properties of recording head core or pole materials are listed in Table 1. The ability of the head to record on high-coercivity media is directly proportional to the saturation magnetization of the pole material. Since high recording density requires media with high coercivity, this, in turn, requires a larger

Table 1
Material requirements for magnetic recording heads

Property	Reason	Head type
Large saturation magnetization	large gap field	write
High permeability at all frequencies	high efficiency over wide frequency range	read/write
Small coercivity with low hysteresis losses	low thermal noise	write
Small, but nonzero uniaxial anisotropy	control of domain structure and permeability at high frequency	read
Low magnetostriction (negative, $\lambda_s < 0$)	low media contact noise, anisotropy control	read
High resistivity	minimize eddy current losses and improve high-frequency permeability	read/write
Wear resistant	long life	
Corrosion resistant	long life	
Good thermal and time stability	reliability	
Low forming effect	easy and reliable manufacturing process	

saturation flux density of the head material. Experience shows that B_s (T) (head) $\simeq 5 \times 10^{-10} H_c$ (A m^{-1}) (media) in the case of metals and B_s (T) $\simeq 10 \times 10^{-10}$ H_c (A m^{-1}) for ferrites.

Head efficiency is proportional to the pole permeability. High permeability over a wide frequency range is an obvious requirement for a high-frequency head. Uniaxial induced anisotropy is critical in maintaining the optimum domain structure and keeping high permeability over a wide frequency range. High resistivity of the material reduces the permeability rolloff due to eddy current losses. However, high resistivity and uniaxial anisotropy may not be sufficient to obtain high permeability at high frequency when poles are thicker than the skin depth. A multilayer structure must be used and the total required thickness is realized by sandwiching the magnetic material layers with an insulating spacer.

Another aspect of high-frequency operation is that it introduces hysteresis loss. Since the loss has a resistive character, it is a source of thermal noise.

During the in-contact recording, the heads experience a lot of stress that introduces strain. This, in turn, through magnetostriction λ_s, produces a variation in the magnetization which appears as noise. Therefore, it is important to keep λ_s small.

The resistance to wear, corrosion and material stability are also important requirements from the point of view of reliability of heads.

2.2 Magnetoresistive Head

The requirements for MR heads are listed in Table 2. The sensitivity of the MR head $(\delta\rho/\rho)/\delta H = \frac{4}{3}(\Delta\rho/\rho_0 H_{eff})$ depends on the size of the magnetoresistive effect $\Delta\rho/\rho$ (2.5% for currently used NiFe) and is inversely proportional to the effective anisotropy field $H_{eff} = H_k + H_d$, where H_k and H_d are magnetic uniaxial anisotropy and demagnetization fields, respectively. The dominant part of the effective anisotropy field, for small sensor geometries (element height $h \simeq$ 3–20 μm), is a demagnetization field $H_d = B_s(t/h)$, which is about 6.4 kA m^{-1} for the typical MR sensor (t is the film thickness $\simeq$ 40 nm). Since H_d is proportional to the magnetization, an increase in the $\Delta\rho/\rho_0$ effect, due to composition modification, may not bring an improvement in sensor performance if it is accompanied by an offsetting increase in M_s.

A uniaxial magnetic anisotropy is required to control the domain structure. Control of magnetostriction λ_s helps keep the stress-induced anisotropy, $H_\sigma = \frac{3}{2}\sigma\lambda_s$, low (for stress σ).

Table 2
Material requirements for magnetoresistive heads

Property	Requirement
Large magnetoresistance	large signal
Small magnetization	low demagnetization field and large sensitivity
Uniaxial anisotropy	control of domain structure
Small hard-axis coercivity	small MR hysteresis
Low magnetostriction	low stress-induced anisotropy
Low resistivity	large magnetoresistance $\Delta\rho/\rho$
Wear resistance	long life
Corrosion resistance	long life

3. Inductive Head Materials

3.1 Ferrites

The class of ferrites used in recording heads has the spinel structure with the general formula MFe_2O_4, where M is a divalent metal ion. Two types of ferrite are in general use: MnZn and NiZn compositions. The properties of both ferrites are strongly influenced by the Ni/Zn and Mn/Zn ratios and are listed in Table 3.

Most of the ferrites used are polycrystalline; however, as the required trackwidth gets narrower and is comparable with the grain size, single crystals must be used to avoid the Barkhausen noise problem. Polycrystalline ferrites are produced by cold or hot pressing of powders, with hot isostatic pressing being used to produce high-density material with good machinability and wear resistance. Single crystals are grown typically by the Bridgeman technique.

MnZn ferrite has a saturation flux density of 0.5 T, so the maximum media coercivity one can fully saturate with such a write head is $H_c = 50\,\mathrm{kA\,m^{-1}}$. The NiZn ferrite can be used with media where $H_c < 40\,\mathrm{kA\,m^{-1}}$. Therefore, ferrite heads can perform well on oxide media ($H_c < 48\,\mathrm{kA\,m^{-1}}$); however, they will perform very poorly on high coercivity media (e.g., metal particle media: $H_c \simeq 0.10$–$0.12\,\mathrm{MA\,m^{-1}}$) required for high performance and high-density recording. Additionally, ferrite permeability drops very quickly with frequency and the head cannot be used effectively at frequencies above 50 MHz. Also, the increase of head temperature, due to the hysteresis losses at high frequency, eliminates the ferrite head as a record head at these frequencies.

Thus, a ferrite head cannot be considered a practical solution for high-density, high-frequency recording. The development of hexaferrites could improve frequency characteristics, but, at present, this material is not to be considered practical as a record head material due to very low B_s (< 0.4 T).

Ferrite is very vulnerable to surface damage caused by the rubbing action of recording media. The contact layers of ferrite become magnetically "dead," resulting in output losses at higher recording frequencies. It has been shown experimentally that the (111) face of single-crystal ferrite is most susceptible to magnetic property decay. The permeability of ferrite is also influenced by a phenomenon known as disaccommodation, a time decay of permeability under constant operating conditions. This effect could be minimized by the use of additives and process control.

Ferrites are easy to form and their high hardness gives them good wear resistance (see *Ferrites, Soft*).

3.2 NiFe Alloys (Permalloy)

The term permalloy generally refers to a nickel–iron alloy with approximately 80 wt% nickel. Permalloy ($Ni_{81}Fe_{19}$) is the most common material used for inductive heads. Its reproducible, stable magnetic properties are well understood. Its saturation flux density is 1.0 T and the heads made of NiFe can easily be used to record on MR media. The induced anisotropy and, therefore, the domain structure also, can be controlled. Magnetostriction is a strong function of composition and the preparation conditions must be controlled to achieve $\lambda < 5 \times 10^{-7}$. Also, NiFe has a relatively low resistivity ($0.2\,\mu\Omega\,\mathrm{m}$) so, in bulk form it can only be used in a low-frequency range. In the thin-film form, the lamination should be very thin to maintain high permeability at high frequency.

In spite of all its good magnetic properties, permalloy is susceptible to corrosion and its wear rate is about 100 times higher than MnZn ferrite. However, it is still one of the most widely used head materials (see *Nickel–Iron Alloys*).

3.3 Sendust

Another widely used alloy is Sendust, $Fe_{85}Al_{5.4}Si_{9.6}$. As shown in Table 3, its saturation flux density is 1.1–1.3 T (see *Iron–Aluminum and Iron–Aluminum–Silicon Alloys*). Thin films of Sendust are widely used in MIG heads. Sendust sheets are also used as pole tips of audio and video heads.

The key advantage of Sendust is its high recrystallization temperature, which permits its use in heads assembled with high-temperature glass bonding. Sendust also has a higher resistivity than NiFe and, therefore, can perform better at high frequency. The as-prepared (as-cast or as-deposited, if a thin film) Sendust has magnetic properties unacceptable for head applications and annealing is required to achieve required magnetic softness. During the annealing process, there is a structure transformation and it is believed that the formation of a cubic DO_3 structure is required to achieve good magnetic properties.

Sendust is very difficult to fabricate both in thin-film form and sheet form, and its optimum properties are difficult to obtain. Also, its corrosion resistance is questionable; however, its wear resistance is superior to that of NiFe.

An undesirable side effect of using Sendust in MIG heads is the appearance of "ripple" in the read back. This effect is associated with the "pseudogap" produced by the thin, nonmagnetic layer formed at the Sendust–MnZn interface during the annealing–glass bonding process. This effect can be significantly reduced by depositing a thin layer of, for example, NiFe between the Sendust and ferrite or, as has been proposed by Sony, by using FeRuGaSi as an alternate. FeRuGaSi belongs to the same family of materials as Sendust, but is much easier to sputter and its corrosion and wear resistance are better than that of Sendust. Recently, Eastman Kodak Company developed a metal-laminated head for a 150 MHz ($250\,\mathrm{Mbits\,s^{-1}}$) recorder which utilizes a FeRuGaSi–SiO_2 multilayer.

Table 3
Properties of different head materials

Property	Ferrites		Crystalline alloys			Amorphous alloys		New materials	
	MnZn	NiZn	Permalloy	Sendust	Sofmax	CoZrNb	CoZrNb–N	FeN	Fe–Ni multilayer
Composition	$(MnZn)Fe_2O_4$	$(NiZn)Fe_2O_4$	$Ni_{81}Fe_{19}$	$Fe_{85}SiO_{9.6}Al_{5.4}$	$Fe_{68}Ru_8Ga_7Si_{17}$	$Co_{10-x}(ZrNb)_x$	$(CO_{89}Nb_{7.5}Zr_{3.5})_{88}N_{12}$		
Saturation flux density B_s(T)	0.5	0.33	1.0	1.1–1.3	1.2–1.4	1.4–1.8	1.3	2.0–3.0	1.7–2.4
Coercive force H_c (A m^{-1})	8	10	30	4	20	0.8	<10	80	<50
Permeability μ at 10 MHz (T m A^{-1})	5	3	13	13	16	19	12	2	38
Unaxial anisotropy field H_k (A m^{-1})			390	20	20	20–2000		400–800	300–800
Magnetostriction λ_s	10^{-6}–10^{-5}	10^{-6}–10^{-5}	10^{-7}–10^{-6}	$<10^{-6}$	$<10^{-6}$	10^{-7}–10^{-6}	10^{-6}	$<10^{-6}$	$<10^{-7}$
Electrical resistivity ρ($\mu\Omega$ m)	$>10^9$	$>10^9$	0.18	>0.8	>1.3	>1.2	0.6		
Annealing	no	no	no	yes	yes	yes (<675 K)	yes (<975 K)		
Curie temperature T_c (K)	403	423	723	773	850	>700			
Wear resistance	excellent	excellent	poor	better	good	good		unknown	
Corrosion resistance	excellent	excellent	poor	unknown	good	uncertain		good (?)	

3.4 Amorphous Alloys—Metallic Glass

The amorphous $(Fe,Co,Ni)_{1-x}$ $(M,Me)_x$ (where M = metal, Me = metalloid) alloys seem to be excellent soft magnetic materials for magnetic heads. These alloys have high saturation magnetization, low magnetostriction and high permeability with flat high-frequency dependence (see *Metallic Glasses*). Iron-based amorphous alloys also have high saturation flux density (> 1.7 T) but, unfortunately, also have high magnetostriction (10^{-5}). Amorphous alloys are produced in a form of ribbons 20–50 μm thick, using a rapid solidification method, or prepared in a form of thin films using vacuum deposition techniques. Most of the work on films for heads has been concentrated on cobalt–M alloys (M is one or several of zirconium, niobium, titanium, hafnium, etc.). CoZrNb alloys with high B_s and zero λ_s were used in MIG, metal-laminated and thin-film heads. Due to the amorphous nature of these materials, there is no magnetocrystalline anisotropy and the dominant anisotropy is a uniaxial magnetization-induced one. This creates a unique opportunity to adjust, by annealing, the anisotropy field $H_k = 2K_u/B_s$ and, therefore, permeability to the head materials. To ensure magnetization rotation rather than wall motion and, therefore, better high-frequency response, it is important that the anisotropy easy axis be perpendicular to the flux path. As the track becomes narrower, the demagnetization field becomes larger and a higher anisotropy field is required to minimize formation of closure domains. Amorphous materials are the only ones where H_k could be varied over a wide range by a simple annealing process. Amorphous ferromagnets are mechanically hard compared to NiFe and exhibit electrical resistance that is three to four times higher.

The thermal stability of amorphous alloys is one of the concerns preventing wider use of these materials. The crystallization temperature of the high-magnetization alloys is usually less than 780 K. However, by sputtering in the presence of nitrogen, a high saturation flux density (> 1.2 T), high crystallization temperatures (> 880 K) have been obtained in CoZrNb–CoZrNbN multilayers.

4. Magnetoresistive Heads Materials

As noted previously, a high-sensitivity MR head can be obtained when the overall anisotropy field, including the demagnetization field, is small but nonzero and the head materials have large $\delta\rho/\rho_0$. So far, the best material for the MR sensor is $Ni_{81}Fe_{19}$, where $\delta\rho/\rho_0$ is 2.5%, magnetostriction can be kept below 5×10^{-7} and uniaxial anisotropy can easily be induced during the deposition process.

To achieve high signal sensitivity, the MR element should be subjected to a bias field (hard-axis bias). For narrow-track MR heads, an additional bias field easy-axis bias must be applied along the length of the element (at the same time the easy axis of anisotropy) to prevent Barkhausen noise (see *Barkhausen Effect*). There are more than ten different hard-axis biasing techniques; however, the only one used in commercially available heads is called shunt bias, where the bias field is produced by current in an adjacent conductor layer. Two biasing methods that have drawn a lot of attention recently are thin-film permanent magnet biasing and exchange bias. In the first method, a uniquely shaped CoPt thin-film magnet is located a few micrometers behind the MR element and provides both biasing fields. The exchange coupling between NiFe and either antiferromagnetic $Fe_{0.5}Mn_{0.5}$ or ferrimagnetic TbCo is used in the other biasing scheme. Both techniques are very promising, but are not yet used in commercial devices.

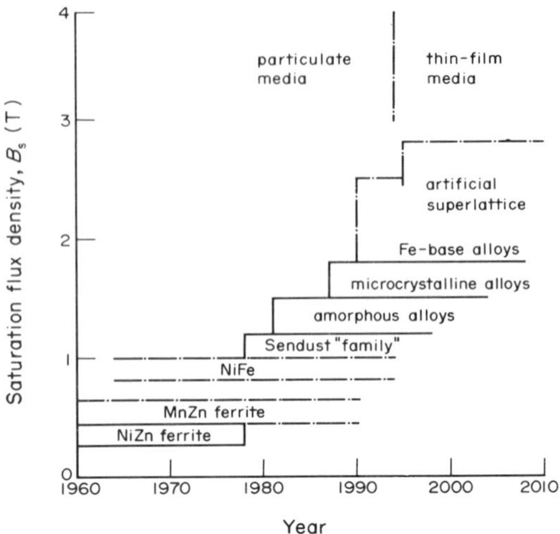

Figure 2
Trends in magnetic head materials development

5. Trends in Material Development

Figure 2 summarizes trends in magnetic head materials. As the media coercivity increases, the interest in head materials is shifted from MnZn ferrite to NiFe, Sendust and amorphous alloys. Future materials will be microcrystalline alloys, iron-based alloys and artificial superlattices, mostly iron-based. Most future materials will be prepared by the thin-film deposition techniques.

The highest saturation flux density obtained so far was found in iron–nitrogen, where B_s > 2.8 T has been observed. The addition of nitrogen to bcc (body-centered-cubic) iron results in a tetragonal expansion of the lattice with an increase in the average magnetic moment from 2.2 μB to 3 μB per iron atom. The magnetostriction of FeN is, however, large and its thermal stability and corrosion resistance are questionable.

Very interesting properties have been found in artificial superlattice multilayers (see *Thin Films and Multilayers, Magnetic*), such as iron–cobalt, iron–nickel and iron–CoFe multilayers, where each layer is between 2 nm and 20 nm thick. The saturation flux density of a multilayer can be as high as 2.2 T and its other properties, like H_k and λ_s, can be controlled independently by modifying the structure. Since the properties of the lattice depend on the thickness and fabrication of each of the layers, there is a unique opportunity for materials engineering. Superlattices have been built from crystalline and amorphous materials and a combination of both of them. For example, magnetostriction λ_s in cobalt–iron structures can be changed from $+15\times10^{-6}$ to -1×10^{-6} by varying the t_{Fe}/t_{Co} ratio, without change in the $B_s = 2.1$ T.

The artificial structures were also used in investigating giant-magnetoresistive-effect materials (see *Magnetotransport Phenomena*). When 3 nm-thick iron layers were separated by a thinner chromium layer, a giant magnetoresistance of more than 50% was observed. A similar effect has been observed in FeNi–Cu–FeNi (8%) and Co–Cu–FeNi (8–10%).

As the recording density increases, higher coercivity media will be used, the use of ferrites will be reduced and metals and alloys will dominate as head materials. The very long term future lies with multilayers and artificial structures in which properties can be engineered independently during the fabrication process.

See also: Magnetic Materials: An Overview; Magnetic Storage Media; Sensors and Devices, Magnetic

Bibliography

Ash K P, Wachenschwanz D, Brucker C, Olson J, Trcka M, Jagielinski T 1990 A magnetic head for 150 MHz, high-density recording. *IEEE Trans. Magn.* 26: 2960–5

Bajorek C, Thompson D 1975 Permanet magnet films for biasing of magnetoresistive transducers. *IEEE Trans. Magn.* 11: 1209

Bozorth R M 1973 *Ferromagnetism*. Van Nostrand Reinhold, New York, pp. 102–60

Cain W C, Markham D, Kryder M N 1989 Dual exchange biased NiFe–TbCo thin films for use in self-based magnetoresistive heads. *IEEE Trans. Magn.* 25: 3695–7

Chin G Y, Wernick J H 1980 Soft magnetic metallic materials: In: Wohlfarth E P (ed.) 1980 *Ferromagnetic Materials: A Handbook on the Properties of Magnetically Ordered Substances*, Vol. 2. North-Holland, Amsterdam, pp. 55–188

Doyle W, Jagielinski, T 1987 Magnetic materials in thin-film inductive heads. In: Gorzkowski W, Lachowicz H K, Szymezak H (eds.) 1987 *Physics of Magnetic Materials*. World Scientific, London, p. 491

Jagielinski T 1989 Trends in materials development for magnetic recording heads. *Mater. Sci. Eng. B* 3: 467–71

Jagielinski T, 1990 Materials for future high performance magnetic recording heads. *MRS Bulletin* 15: 36

Jeffers F 1986 High-density magnetic recording heads. *Proc. IEEE* 74: 1540–56

Jones R E, Mee C D, Daniel E D (ed.) 1982 *Recording Heads in Magnetic Recording*. McGraw–Hill, New York, pp. 244–337

Jorgensen F 1988 *The Complete Handbook of Magnetic Recording*. TAB, Blue Ridge Summit, PA

Luborsky F E 1983 *Amorphous Metallic Alloys*. Butterworth, London

Markham D, Jeffers F 1989 *Proc. American Electrochemical Society Meeting*. Electrochemical Society, Pennington, NJ

Shelledy F, Brock G 1975 A linear self-biased magnetoresistive head. *IEEE Trans. Magn.* 11: 1206–9

Shibaya M, Fukuda I 1977 Preparation by sputtering of thick Sendust film suited for recording head core. *IEEE Trans. Magn.* 13: 1029

Slick P I 1980 Ferrite for non-microwave applications. In: Wohlfarth E P (ed.) 1980 *Ferromagnetic Materials: A Handbook on the Properties of Magnetically Ordered Substances*, Vol. 2. North-Holland, Amsterdam, pp. 189–241

T. M. Jagielinski
[Eastman Kodak Company, San Diego, California, USA]

Reed Contact Materials

Dry-reed contacts are miniature hermetically sealed devices used for switching electrical voltages and currents. Production and operating characteristics place stringent requirements on the magnetic materials which comprise the switching elements of this class of contact.

1. Contact Description and Design Requirements

The dry-reed contact comprises two solid metal reeds hermetically sealed in a glass vial containing a controlled inert or reducing atmosphere. The metallic reeds are aligned and maintained in the glass vial so that their tips slightly overlap and therefore make contact when they are magnetically drawn together. The evolution of the dry-reed contact proceeded from non-self-latching designs, termed ferreeds, to the most recent self-latching designs, termed remanent reeds, or remreeds (see Fig. 1).

The ferreed has two soft magnetic reeds, which respond to the magnetic field of external hard magnetic plates. A current passing through electrical windings magnetizes the external plates which, in turn, draw the reed members together to complete an electrical path. The remanent magnetic field of the external plates maintains the reeds in a closed configuration until a reverse current is applied to the external windings to reverse the magnetic field. Iron–nickel alloys, principally 51 wt%Ni–49 wt%Fe, have gained universal acceptance as having the most satisfactory properties such as high permeability and good formability (see *Nickel–Iron Alloys*).

Remanent reed contacts were used extensively in the telecommunications industry in the 1970s and 1980s, but are now being replaced by solid-state switches. The design requirements call for a monolithic reed structure of a permanent magnetic material to eliminate the need for external plates. The typical geometry of a reed is shown in Fig. 2. The reed material must be sufficiently ductile to be drawn into wire 0.50–0.60 mm in diameter and maintain enough ductility to be flattened to form a reed blade (paddle). The operational requirements can be related to the magnetic performance of each sealed contact through its *B–H* loop (Archer et al. 1976) and, coupled with design requirements and restrictions, these define the magnetic requirements of the reed alloy. In terms of actual values these requirements are a coercive force H_c of 1.4–2.2 kA m^{-1} (18–28 Oe) on the round wire and 2–2.5 kA m^{-1} (25–31 Oe) on the flattened section, a minimum remanent flux φ_r of 2.9×10^{-7} Wb on the round wire and 3.6×10^{-7} Wb on the flattened section, and a minimum squareness B_r/B_8 of 0.85, where B_8 is the flux density measured at 8 kA m^{-1} (100 Oe).

Other requirements must be met. The alloy must be capable of being drawn into wire with a surface free from defects to create and maintain a hermetic seal. The magnetic properties must be stable during brief periods at high temperatures to withstand sealing in glass. The alloy must be readily solderable for external connections and must be compatible with the electrical contact metal which is applied to its flattened surface to provide low and stable contact resistance. Its thermal expansion coefficient must closely match that of the glass. Low magnetostriction is desirable to minimize contact susceptibility to scrub sticking when pulsed between remanence and saturation.

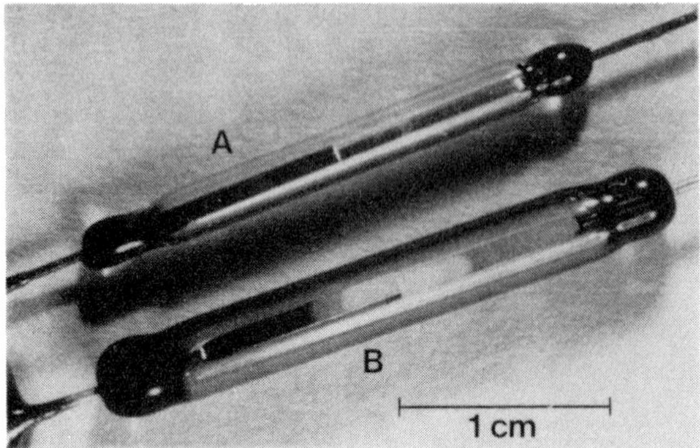

Figure 1
Typical remanent reed (A) and ferreed (B) electrical contacts

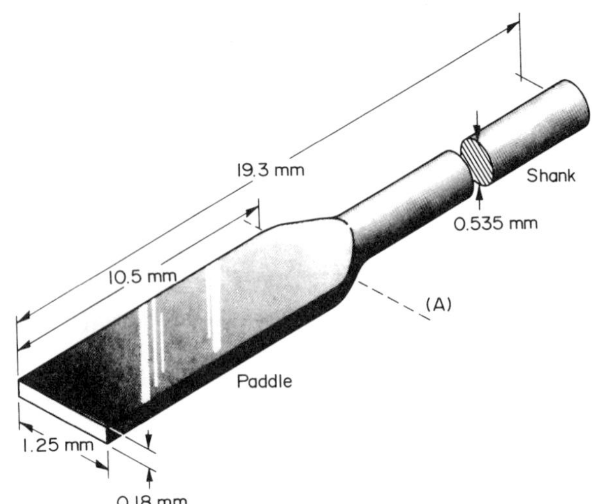

Figure 2
Physical dimensions of a typical reed blade

2. Potential Magnetic Alloys

The alloys which meet these requirements fall in the class commonly termed semihard magnetic materials which includes alloys having a coercive force in the range 1–8 kA m^{-1} ($\sim$10–100 Oe). Most are precipitation-hardenable alloys since they can be cold worked and formed while in the ductile solution-treated condition (see *Magnetic Alloys: Intermediate Coercivity Materials*). Then, in their final shape, they can be magnetically hardened to the desired degree by a precipitation anneal. These alloys may be divided into three basic groups: equiatomic iron–cobalt alloys, cobalt-rich alloys and iron-rich alloys. A partial list is given in Table 1 with values of pertinent magnetic parameters and other physical characteristics.

Of the equiatomic iron–cobalt group, the Fe–Co–V alloy termed Remendur has seen the widest application. Its coercive force is very easily controlled to meet the optimum value for a given reed contact design by varying the vanadium content. A rule of thumb is that the maximum H_c obtainable is equal to ten times the vanadium content in weight percent. The processing of Remendur is somewhat difficult, however. The equiatomic iron–cobalt alloys undergo an embrittling ordering reaction and critical quenches are necessary after heat treatments to maintain ductility. Remendur also has a two-phase structure even prior to the precipitation anneal and precise heat-treating temperatures are required to maintain the proper distribution of phases for optimum magnetic properties. Typical of all equiatomic iron–cobalt alloys, the magnetostrictive coefficient is high and this problem cannot be overcome with processing changes or small

Table 1
Alloys proposed as having possible application in reed contacts and their nominal properties

Alloy composition (%)	Coercive force, H_c (kA m^{-1})	Remanent flux density, B_r	Squareness ratio, B_r/B_s	Magnetostrictive coefficient ($\times 10^{-6}$)	Thermal expansion coefficient ($\times 10^{-6}$) 20–400 °C	Formability	Plateability	Solderability	Sealability	Aging temperature (°C)
49Co–48Fe–3V (Remendur)	2.4	1.8	0.90	50	10.2	good	good	good	good	610
38Co–39Fe–20Ni–3Nb	2.4	1.6	0.95	~45	9.8	good				600
Co–Fe–Ni–Al–Ti (Vacozet 655)	3.2	1.4	0.90	48	11	good	good	poor	good	550
85Co–12Fe–3Nb (Nibcolloy)	1.6	1.5	0.95	2	12.1	good	good	good	good	700
82Co–13Fe–5Mo	2.4	1.3	0.92	slightly positive	~12	good				650
84Co–12Fe–4Ti	1.8	1.4	>0.90	slightly positive	~12	good				650
82Co–12Fe–6Au (Coballoy)	1.1	1.6	0.85	small	12.5	good	good	good	good	550
89Co–10Fe–1Be	2.4		0.90	small	~12	good				800
54Fe–28Ni–18Co (Kovar)	<0.8	1.8	<0.70		5	good	good	good	good	
80Fe–16Ni–3A–1Ti	2.4	1.25–1.4	0.8–0.95	0.4	12.5	good	good	poor		600
78Fe–16Ni–5Cu–1W	2.8	1.4	0.95		13	good	good	good		550
80Fe–18Cu–2Mn	2.4	1.5	0.90		11.5	good	good			500
C–steel	1.6–3.2	1.0	<0.70	slightly negative	13	good	good	good	fair	
Cr–Co–C steel	1.6–4	1.0–1.25		small	13				fair	300–500
92Fe–8Mn	2.2	1.8	0.96	1.3		good			poor	400–450
Fe–7Mo–5Ni	2.4	1.9	0.97	small		good			good	600

alloying additions. It must be compensated for by making the contact metal immune to cold welding or sticking.

The other two equiatomic iron–cobalt alloys listed in the table are basically similar to Remendur except for the presence of nickel. This improves ductility to some degree but at the expense of a reduced remanent flux density. An alloy containing aluminum and titanium as the elements added to create a precipitate has also been reported to be difficult to solder. This presumably is due to tenacious oxides of aluminum and titanium formed during glass sealing.

The cobalt–rich alloys were developed because in the vicinity of 85–90% cobalt the magnetostriction is very low, the alloys have a single-phase fcc structure and are therefore ductile without special heat treatment, and they have a large value of crystal anisotropy with [111] being the "easy" direction of magnetization. Since wire drawing results in aligning the [111] direction along the wire axis, a very square hysteresis loop can be obtained.

Of these alloys, Nibcolloy has seen the widest use in reed contacts. It seems to provide adequate levels of H_c without too severe a loss in remanent flux density or an unacceptable decline in ductility as the percentage of the third element is increased. In some of the other alloys, too great a loss is suffered in these other properties in order to get the coercive force to adequate levels. Additions such as titanium and beryllium also are again suspect with respect to the influence of their oxides on solderability and plateability. A final point which has placed these alloys in some disfavor at various times is the economic and political situation involving cobalt. Its excessively high cost and questionable uninterrupted availability make users strongly prefer the less cobalt-rich choices.

These latter factors weigh heavily in favor of the iron-rich alloys. However, with few exceptions, suitable iron-rich alloys which provide an acceptable compromise of the various critical parameters have not been found. Typically, either the squareness ratio, remanent flux density or coercive force are too low or the thermal expansion coefficient is too high to be sealable to commercially available glass. Although none has been utilized in reed contact applications, several did hold some promise. The Fe–Cu–Mn alloy and the iron–manganese and Fe–Mo–Ni alloys have an excellent combination of magnetic parameter values. Their properties are developed through special processing involving deformation and heat treatment cycles. This affords structural refinement and particle alignment of a two-phase structure which, coupled with the martensitic transformation, produces the required properties. The iron–manganese alloy has shown potential problems in glass sealing due to the martensitic transformation but the Fe–Mo–Ni alloy shows promise in this respect. The other physical requirements have still to be evaluated.

See also: Iron–Cobalt Alloys; Magnetic Alloys: Intermediate Coercivity Materials; Magnetic Materials: An Overview

Bibliography

Archer W E, Olsen K M, Renaut P W 1976 Development of a remanent reed sealed contact. *Bell Syst. Tech. J.* 55: 511–35

Feiner A 1964 The ferreed. *Bell Syst. Tech. J.* 43: 1–14

Heck C 1974 *Magnetic Materials and Their Applications.* Crane-Russak, New York

Pinnel M R, Bennett J E 1973 The metallurgy of remendur: effects of processing variations. *Bell Syst. Tech. J.* 52: 1325–40

Pinnel M R 1976 Magnetic materials for dry reed contacts. *IEEE Trans. Magn.* 12: 789–94

Okada M, Kassai M, Suzuki Y, Sasaki T, Henmi Z 1972 Properties of semihard magnetic alloys of the Co–Fe–Nb system. *Trans. Jpn. Inst. Met.* 13: 391–5

M. R. Pinnel
[AT&T Bell Laboratories, Columbus, Ohio, USA]

Resistive Transition and Flux Flow in Superconducting Materials

Superconductors only carry an ideal zero-resistance current for sufficiently low values of the applied magnetic field and temperature and for current densities below some critical current density which itself depends on field and temperature. Dissipation in the superconducting state is most directly characterized in terms of the variation of electric field $\boldsymbol{E}$ with transport current density $\boldsymbol{J}$. In practice this can be measured by passing a current I through a gauge length of constant cross section and observing the appearance and variation of voltage V between two potential contacts. A measurement of the resistive transition may be performed at constant current and applied field as the temperature is varied through the superconducting transition temperature or, alternatively, for constant temperature and applied field with a varying current. Such a voltage–current (VI) characteristic is generally strongly nonlinear, the voltage varying as a high power of the current or exponentially with current.

The technological specification of a superconductor for a particular application depends on both the critical current density J_c defined at a particular voltage criterion, for instance $10^{-5}\,\mathrm{V\,m^{-1}}$, and on the shape of the VI characteristic defined loosely in terms of an n value derived by fitting a power law $V \propto I^n$ to the transition in the vicinity of the critical current. For commercial conductors n is generally in the range of 20–100 and varies with the applied magnetic field.

The physical mechanism for dissipation in a superconductor depends on whether the material is type I, type II or granular. Multifilamentary composites and multilayer structures bring additional complications. In each case the resistive state can be interpreted in terms of the motion of magnetic flux. The concept of current-induced motion of flux structures was introduced first by Shoenberg in 1952 for type I superconductors where the moving entities are the normal and superconducting domains of the intermediate state. In type II superconductors the flux lines or flux vortices move under the influence of the Lorentz force due to the transport current (see *Transport Currents and Force-Free Current Flow*); when $J < J_c$ thermal activation leads to flux creep or thermally activated flux flow (TAFF), when $J > J_c$ the flux flow velocity is determined by an effective flux flow viscosity related to the normal state resistivity. In the case of granular materials dissipation is determined by the collective phase-slip response of the weak link network which can be modelled in terms of the passage of flux quanta through the material.

Although the microscopic source of dissipation can either be eddy current damping in normal regions, quasiparticle currents in superconducting regions or nonequilibrium effects in weak links or tunnel junctions the overlying phenomenology of dissipation in different classes of material has a common basis. This is because dissipation can be treated quite generally according to the laws of irreversible thermodynamics as rate of entropy production expressed as the sum of the vector products of generalized forces and fluxes. When motion of the magnetic flux structure occurs in the presence of a temperature gradient and heat flux various thermoelectric and thermomagnetic cross effects can occur (e.g., Seebeck, Nernst, Ettingshausen effects). Measurement of such cross effects provides information of value in elucidating the mechanisms for flux flow and dissipation in different materials.

Although most low-frequency dissipation phenomena in bulk type II superconductors can be described adequately in terms of flux flow, flux pinning and occasionally flux cutting, the behavior of high-critical-temperature oxide superconductors seems to be more complex. They appear to exhibit a flux vortex structure; however, their resistive behavior is dominated by their extreme anisotropy; their properties have some of the features of granular materials and others that are characteristic of weakly coupled superconducting multilayer structures.

1. Flux Flow

Although the characteristics of the stationary flux vortex lattice are well known and can be quantitatively derived from microscopic theory (see *Flux Line Lattice*), the theoretical framework for treating moving flux vortices is less well established. The treatment of equilibrium structures is always simpler than those where there is dissipation, particularly if the latter involves turbulence. Fortunately the concept of the magnetic flux vortex appears extremely durable and it has been possible to develop a remarkably successful phenomenological theory of flux motion based on the idea of a Lorentz force which is balanced by pinning forces and a viscous drag force proportional to the flux vortex velocity. The basic assumption is that the vortex velocity at any instant is precisely that needed to balance through viscous forces any imbalance between the local driving force and the pinning potential field arising from local inhomogenity. The flux vortex is assumed to have no inertia.

1.1 Flux Vortex Velocity and Viscosity

Uniform vortex motion generates an electric field $\boldsymbol{E} = \boldsymbol{B} \times \boldsymbol{v}$ proportional to the vortex velocity $\boldsymbol{v}$ and the flux density (Josephson 1962). This expression can be derived quite generally by equating the dissipation per unit volume $\boldsymbol{J.E}$ to the product of the Lorentz force density and the velocity:

$$\boldsymbol{J}.\boldsymbol{E} = (\boldsymbol{J} \times \boldsymbol{B}).\boldsymbol{v} = \boldsymbol{J}.(\boldsymbol{B} \times \boldsymbol{v}) \qquad (1)$$

If η is defined as the viscous drag coefficient per unit length of vortex, the viscous drag per unit volume is $-n\eta\boldsymbol{v}$ where $n = B/\Phi_0$ is the vortex density and Φ_0 is the flux quantum; thus, the instantaneous dissipation may also be expressed by

$$\boldsymbol{J}.\boldsymbol{E} = n\eta v^2 \qquad (2)$$

In the absence of pinning inhomogeneities the vortex velocity is uniform and the average dissipation is also given by $n\eta(\langle \boldsymbol{v} \rangle)^2$ where $\langle \boldsymbol{v} \rangle$ is the mean vortex velocity. The resultant *EJ* characteristic is shown in Fig. 1a; if η is independent of vortex velocity the flux flow resistivity ρ_f defined by $E = \rho_f J$ is a constant:

$$\rho_f = B\Phi_0/\eta \qquad (3)$$

The flux flow resistivity is found to vary approximately in the low-temperature limit as $\rho_f = \rho_n(B/B_{c2}(0))$ where ρ_n is the normal state resistivity and $B_{c2}(0)$ is the upper critical field at zero temperature. On a simple picture the fraction of the total volume occupied by the normal cores of the vortices is B/B_{c2}, suggesting that dissipation is associated with normal cores and becomes equal to the normal state dissipation when the normal cores overlap and the material becomes normal at the upper critical field.

When the material contains pinning inhomogeneities the vortex velocity is no longer uniform $(\langle \boldsymbol{v}^2 \rangle) > (\langle \boldsymbol{v} \rangle)^2$ and the *EJ* curve becomes nonlinear. In the ideal case (Fig. 1b), there is an abrupt onset of resistance at J_c with a linear rise in electric field $E = \rho_f(J - J_c)$. In this ideal limit the additional dissipation due to the pinning forces is given by $(BJ_c)\boldsymbol{v} = EJ_c$. The dissipation due to pinning inhomogeneities can be viewed as arising from the nonuniformity imparted to the vortex motion by the pinning interactions

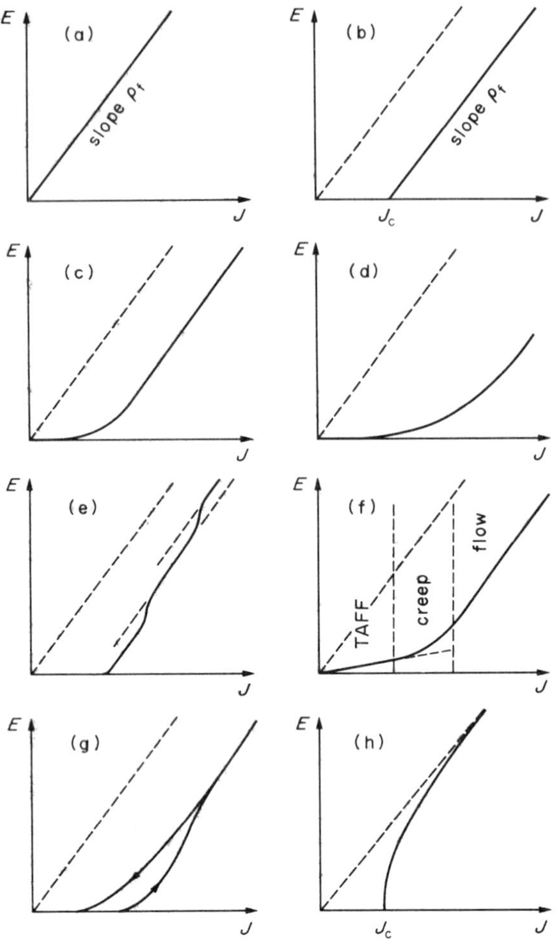

Figure 1
Schematic electric field vs current density (*EJ*) characteristics illustrating different features of flux flow: (a) ideal flow with constant flux flow resistivity and no flux pinning; (b) ideal flow with flux pinning; (c) typical homogeneous type II superconductor; (d) inhomogeneous superconductor or multifilamentary composite; (e) characteristic with fine structure; (f) effect of thermally activated depinning processes; at low temperatures the flux flow is linear with applied field, at higher temperature flux creep becomes exponential; (g) history effects in a defective flux lattice; (h) characteristic response for a regular pinning potential

$$EJ_c = \eta(\langle \mathbf{v}^2 \rangle - (\langle \mathbf{v} \rangle)^2) = \eta \langle \mathbf{s} \rangle^2 \tag{4}$$

where $\mathbf{s} = \mathbf{v} - \langle \mathbf{v} \rangle$ is the local deviation of the flux vortex velocity from its mean value.

Values of the vortex velocity of practical interest range from $10^{-15}\,\mathrm{m\,s^{-1}}$ in low-drift high-field solenoids to local vortex velocities as high as $10^4\,\mathrm{m\,s^{-1}}$ as a vortex becomes unpinned and moves at the velocity required to dissipate viscously the elastic stored energy released as it relaxes. At present, theory does not give predictions as to whether the phenomenological η is likely to remain constant over such a wide velocity range.

There are many factors including possible variations in η that can cause departure from the ideal behavior of Figs. 1a, b. The *EJ* characteristic is usually generally curved as in Figs. 1c, d or may contain fine structure as in Fig. 1e (Lumley et al. 1978). At finite temperatures there may be significant additional curvature due to thermal activation of vortices (Fig. 1f); this will be discussed in more detail in Sec. 2. Sources of particular deviations from ideality are difficult to identify. It can be seen, for instance, that inhomogeneities too small to cause pinning and contribute to the critical current (subthreshold pinning centres) will contribute to s and increase the dissipation during flux flow. Furthermore, since dissipation depends on the square of velocity, flux changes involving a few vortices moving a large distance are more dissipative than a large number of vortices moving a correspondingly shorter distance in the same time. Dissipation is therefore likely to depend on the defect structure of the vortex lattice. The defect structure of the lattice is itself dependent on the vortex lattice velocity, and at high velocities in isotropic superconductors the lattice becomes more perfect and rotates until a close-packed direction in the lattice is in the direction of vortex flow. Changes in the lattice structure also change the total pinning force and J_c leading to curvature and fine structure in the *EJ* characteristic and also to history effects that depend on changes in the vortex lattice defect structure (Fig. 1g) (Evetts and Mullock 1985). The dynamic response of a regular pinning array is also quite different. The *EJ* characteristic rises abruptly at the critical current $E = \rho_f J(1 - (J_c/J)^2)^{\frac{1}{2}}$ as indicated in Fig. 1h. The response of the vortex lattice is synchronous or coherent and the dissipation has parallels with that seen in an overdamped junction or weak link (see *Josephson Effects in Weak Links*) where the voltage arises from the regular passage of single flux quanta by a phase-slip process at the Josephson frequency $\omega = V/\Phi_0$. A similar response is seen for domain wall pinning by a regular pinning field (Rodbell and Bean 1956); in each case the primary field is "conservative" so that on applying a force the critical current or coercive force may be reached without dissipation.

A useful insight into the dissipation process when a single vortex moves is given by the Bardeen–Stephen model (Bardeen and Stephen 1965). A simplified model of the vortex was taken as a current vortex described by the London equations (see *Superconducting Materials: BCS and Phenomenological Theories*) with a normal core of radius ξ the coherence length. The vortex is taken to move freely in the absence of pinning forces and sufficiently slowly for it to move a distance small compared to ξ during the quasiparticle scattering time (see *Nonequilibrium*

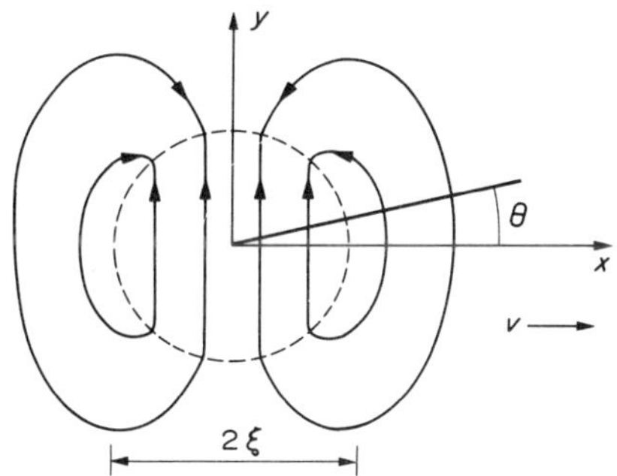

Figure 2
Normal electron flow in the vicinity of a vortex core radius ξ moving with velocity v in the x direction

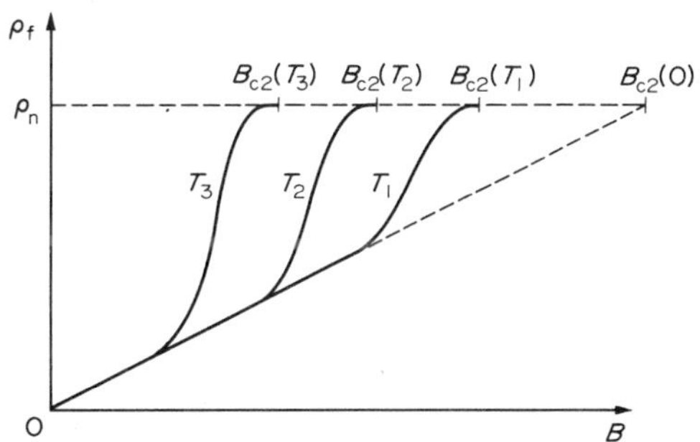

Figure 3
Variation of the flux flow resistivity ρ_f with flux density B for temperatures $T_3 > T_2 > T_1$

Superconductivity), so that the treatment of superconducting properties outside the normal core involves only equilibrium properties of the superconductor. At any position $\boldsymbol{r}$ near a moving vortex there will be a time-varying supercurrent density $\boldsymbol{J}_s$; if, for example, the vortex core crosses the point $\boldsymbol{r}$ the current $\boldsymbol{J}_s$ will fall to zero as the vortex approaches and then reverse direction as it moves away. The local electric field can then be derived from the second London equation, $\boldsymbol{E} = \mu_0 \Lambda d\boldsymbol{J}_s/dt$, and results in a uniform field inside the core perpendicular to the direction of motion of the vortex line with a corresponding dipole field outside the core (see Fig 2). From the London model the supercurrent density around an isolated core is $\boldsymbol{J}_s = \hbar ne\hat{\boldsymbol{\theta}}/2mr$ where $\hat{\boldsymbol{\theta}}$ is the azimuthal unit vector, so that for a vortex velocity v in the x direction

$$\boldsymbol{E} = \frac{v\Phi_0}{2\pi}\frac{\partial}{\partial x}\left(\frac{\hat{\boldsymbol{\theta}}}{r}\right) = \frac{v\Phi_0}{2\pi r^2}(\hat{\boldsymbol{\theta}} \cos\theta - \boldsymbol{r}\sin\theta)$$

$$E_{\text{core}} = \frac{v\Phi_0}{2\pi\xi^2} = vB_{c2}$$

The energy dissipation due to flow of normal currents in the core per unit length of core is

$$\pi\xi^2 E_{\text{core}}^2/\rho_n \tag{5}$$

If the dissipation due to normal current flow outside the core is of the same magnitude one has from Eqns. (5,2), for an isolated vortex

$$\boldsymbol{J}.\boldsymbol{E} = \eta v^2 = v^2\Phi_0 B_{c2}/\rho_n$$

whence $\eta = \Phi_0 B_{c2}/\rho_n$. Comparison with Eqn. (3) indicates that the model predicts $\rho_f = \rho_n(B/B_{c2})$ in agreement with experiments at low temperatures. Experimentally the variation of ρ_f diverges from the Bardeen–Stephen model at higher temperatures (see Fig 3). If the model is further developed to include real order parameter variations in the core there is little change in the predicted viscosity coefficient; a more complete description requires a solution of the time-dependent generalization of the Ginzburg–Landau theory (TDGL). The TDGL equations can be justified rigorously only for gapless superconductors (Gor'kov and Eliashberg 1968, Cyrot 1973); however, it appears that the TDGL theory can account reasonably for changes in the slope of $\rho_f(B)$ with B at different temperatures. Huebener (1979) gives a summary of solutions of the TDGL equations for a uniformly moving vortex lattice. The microscopic theory for ρ_f has been reviewed by Larkin and Ovchinnikiv (1986).

1.2 Thermoelectric and Thermomagnetic Cross Effects

When flux flow occurs in the presence of a temperature gradient and heat flux a number of thermoelectric and thermomagnetic cross effects occur. Some of these effects are large, up to three orders of magnitude larger than the corresponding effects in the normal state. They have their origin in the entropy difference between the superconducting condensate and normal electrons or quasiparticles. When a flux vortex moves it carries with it an "entropy of transport" s^* per unit length of vortex. In the presence of a thermal gradient the flux vortex driving force density becomes

$$(\boldsymbol{J} \times \boldsymbol{B} - ns^*\nabla T)$$

Typically, in the absence of pinning forces, the Lorentz force due to a current of $10\ \text{A cm}^{-2}$ could be compensated by a temperature gradient as small as $1\ \text{K cm}^{-1}$. If the transport current is zero, vortices will move under the influence of the thermal force leading to an electric field "transverse" to the temperature gradient; this is the Nernst effect. Correspondingly, the Ettingshausen effect is the heat flux (or entropy flux) in the absence of a temperature gradient resulting from flux flow transverse to an applied transport current. The Nernst and Ettingshausen effects are thus termed transverse effects. In addition there are "longitudinal" effects; the Seebeck effect is an electric field in the direction of a temperature gradient when the transport current is zero, while the Peltier effect is the heat current carried by the transport current in the direction of the applied electric field when the temperature gradient is zero.

A straightforward description of the interrelation of heat flux and flow of flux vortices may be obtained using the methods of irreversible thermodynamics (Evetts and Campbell 1968, Campbell and Evetts 1972, Huebener 1979). If the total irreversible entropy production σ is expressed as a sum over products of conjugate fluxes and forces

$$T\sigma = \sum \boldsymbol{J}.\boldsymbol{X}$$

the fluxes and forces are then connected by the phenomenological equation

$$J_i = \sum_j L_{ij} X_j$$

where the coefficients L_{ij} satisfy the Onsager reciprocity theorem. The transport current $\boldsymbol{J}$ and electric field $\boldsymbol{E}$ are an alternative conjugate flux–force pair to the flux of vortices $\boldsymbol{J}_{\mathrm{f}} = n\boldsymbol{v}$ and the Lorentz force $\boldsymbol{F}_{\mathrm{L}} = \boldsymbol{J} \times \boldsymbol{\Phi}_0$ on a single unit length of vortex line where $\boldsymbol{\Phi}_0$ is directed along the vortex line since

$$\begin{aligned} \boldsymbol{J}_{\mathrm{f}}.\boldsymbol{F}_{\mathrm{L}} &= (n\boldsymbol{v}).(-\boldsymbol{\Phi}_0 \times \boldsymbol{J}) \\ &= \boldsymbol{J}.(-n\boldsymbol{v} \times \boldsymbol{\Phi}_0) \\ &= \boldsymbol{J}.(\boldsymbol{B} \times \boldsymbol{v}) \\ &= \boldsymbol{J}.\boldsymbol{E} \end{aligned}$$

Thus, all cross effects involving E and J can be alternatively expressed as analogue cross effects in terms of the Lorentz force and flux flow.

It is important to appreciate that the phenomenological formalism of irreversible thermodynamics applies whenever irreversible entropy production can be expressed as the product of generalized forces and fluxes. For this reason the observation of cross effects in the transport properties of a superconductor does not necessarily imply that the transport has to be described in terms of the flow of flux vortices. Strong cross effects are observed in high-temperature oxide superconductors particularly for applied fields and temperatures above the irreversibility line where pinning effects are negligible (Hagen et al. 1990, Lengfellner et al. 1990, Zeh et al. 1990) and also in granular materials. In these materials the concepts of the flux vortex and flux flow are more complex and the interpretation of cross effects is still uncertain. The microscopic mechanisms leading to cross effects have their origin in the difference in entropy transported by quasiparticle currents and supercurrents and in the Josephson expression relating the rate of change of phase in the superelectron condensate to the electric field. These ingredients may be combined in various ways to produce cross effects in the different classes of superconductor.

In the case of pure flux flow the entropy of transport s^* has been calculated by Maki (1969) near T_{c} using TDGL theory:

$$s^*(T) \simeq \frac{\Phi_0}{T} \frac{(H_{\mathrm{c2}}(T) - H)}{(1.16(2\kappa^2 - 1) + 1)} L_{\mathrm{D}}(T)$$

where $L_{\mathrm{D}}(T)$ is a numerical factor close to unity near T_{c}.

1.3 Dissipation in Twisted Flux Structures

If the transport current $\boldsymbol{J}$ has a component that is parallel to $\boldsymbol{B}$ neighboring sheets of vortices become twisted with respect to each other; in this situation when the critical current is exceeded the flux motion is complex and may involve flux line cutting or more complex inhomogeneous current-induced dissipative states that may be periodic or chaotic (see *Transport Currents and Force-Free Current Flow*). The flux cutting process leads to large local flux line velocities which raise the dissipation above that due to the mean vortex velocity. When there is flux cutting the flux flow concept has to be applied with caution. Flux cutting occurs in very strong pinning materials where it is an alternative to pin breaking and it is also closely related to the phase-slip processes that occur in weakly coupled layer structure superconductors and in the more anisotropic high-temperature superconductors.

2. The Resistive Transition

An alternative measure of the onset of dissipation is the resistive transition obtained by varying the temperature at constant current. If the current is limitingly small and the applied field is zero the transition occurs sharply at the critical temperature for the material. For conventional low-temperature superconductors the intrinsic breadth of the transition, determined by fluctuations (see *Coherence Length, Proximity Effect and Fluctuations*), is typically of order a few millikelvins. If the applied field is raised the transition is displaced to lower temperatures defined by $B_{\mathrm{c2}}(T)$. Close to T_{c} the variation of $B_{\mathrm{c2}}(T)$ is linear with T; however, the breadth of the transition is not a strong function of B (see Fig. 4a).

The resistive transition for high-temperature oxide superconductors is more complex. First of all the transition breadth is much greater due to stronger fluctuations because the temperature is higher and because the number of electrons in a coherence volume is small. More characteristic still is the strong broadening of the transition observed in an applied field (Fig. 4b). This arises because the critical current becomes negligibly small for combinations of applied field and temperature that place the material above the irreversibility line (see *Critical Current Density*).

As the transition is traversed from high temperatures the initial curvature above T_{c} is due to fluctuations. Just below the transition the characteristic is

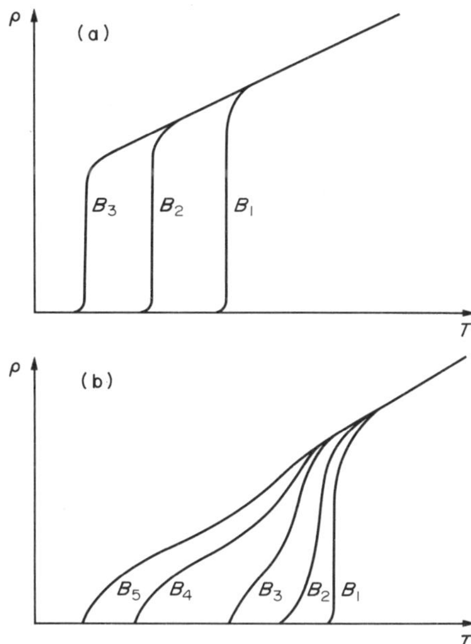

Figure 4
Resistive transition (ρ–T) for different values of the applied field: (a) low-temperature superconductor, $B_3 > B_2 > B_1$; (b) high-temperature superconductor, $B_{n+1} > B_n$

dominated by flux flow since the sample will in general be above the irreversibility line. The width of this region depends on the size of the measuring current, the position of the irreversibility line and the rate of increase of critical current density with reduction in temperature when the irreversibility line is crossed. When the critical current exceeds the measuring current the resistivity will be due to exponential flux creep at higher temperatures and thermally activated flux flow (TAFF) at lower temperatures (see *Flux Creep Phenomena*).

A resistive transition therefore displays at different points all the different dissipation or flux flow mechanisms that operate in a superconductor. If a series of resistive transitions are taken at closely spaced values of the applied field for a constant current density the regions where different dissipation or flow mechanisms dominate may be mapped on to the ρ–T field. A similar mapping process may be performed in the E–J field by plotting ln E vs ln J for a range of applied fields or temperatures (see Fig. 5). The value of mapping complex processes in materials engineering has been demonstrated by Ashby (see, for instance, Frost and Ashby 1982).

By mapping flux flow phenomena it should also be possible to identify phase changes in the vortex lattice if they have any relevance for the process that is being mapped. At present there is controversy as to whether

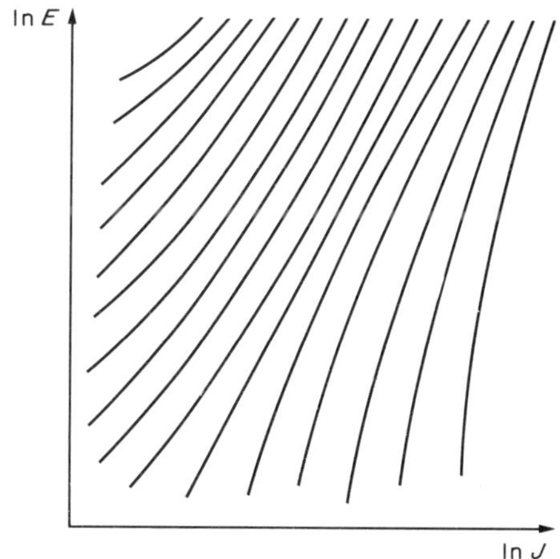

Figure 5
Schematic variation of ln E vs ln J for different values of the temperature

phenomena such as flux lattice melting, the glass transition or other phase transitions have an effect on the resistive transition. In general it is likely that phase transitions such as melting will have most effect on the resistive properties of two-dimensional materials where flux flow through weakly pinned channels can arise if melting occurs below the irreversibility line leading to nonuniform heterogeneous behavior with highly nonlinear EJ curves (see Sect. 3).

3. Heterogeneous Structures

Most real superconductors are to some extent inhomogeneous and indeed high-critical-current materials need to be inhomogeneous to ensure strong pinning. Thus, a niobium–titanium alloy wire with a critical current of $2 \times 10^5\,\mathrm{A\,cm^{-2}}$ must necessarily display inhomogeneity of at least 0.2 T in its upper critical field (Evetts 1983). Furthermore, multifilamentary composites and granular materials are clearly heterogeneous. Such heterogeneity is reflected in the VI characteristic of the material and for random heterogeneity leads to a strongly curved characteristic that can often be fitted approximately by a power law with a characteristic n value. Even if the power law approximation is poor it is useful to express the transition in terms of an n value that varies with I or J. When material has a low n value the value of the critical current density depends strongly on the voltage criterion specified. In commercial wires the critical current density, the voltage criterion and the n value must all be specified.

3.1 Multifilamentary Composite Conductors

Within a real composite conductor there are numerous sources of critical current inhomogeneity arising from variations in filament microstructure and cross section, fluctuations in the strain field and randomness in the geometrical packing of the filaments. It may therefore be deduced from the central limit theorem that the critical current distribution may to a good approximation be taken as normal or Gaussian. The resultant *EJ* characteristic depends on the precise geometry of the composite, in particular whether the filaments are resistively coupled or effectively isolated. In either case the *EJ* curve approximates very well to a power law; the departure from a simple power law is hard to detect when the curve is viewed over less than six decades of electric field (Plummer and Evetts 1987). For reasonable values of the composite parameters for an A15 composite it can be shown that the n value is directly related to the mean critical current $\langle J_c \rangle$ and its standard deviation σ as follows

$$n = (\langle J_c \rangle / \sigma)^{5/3} \text{ coupled filament limit}$$

$$n = 0.6(\langle J_c \rangle / \sigma)^{5/3} \text{ single filament limit}$$

An expression for the mean critical current that is insensitive to the composite parameters is

$$\langle J_c \rangle = J(10^{-5}\,\mathrm{V\,m^{-1}})^{(n/n-1)}$$

where $J(10^{-5}\,\mathrm{V\,m^{-1}})$ is the overall current density in $\mathrm{A\,m^{-2}}$ measured at $E = 10^{-5}\,\mathrm{V\,m^{-1}}$. Thus, for an n value of 27 a criterion of $10^{-5}\,\mathrm{V\,m^{-1}}$ would result in a typical high-current composite having a critical current $J_c = \frac{1}{2}\langle J_c \rangle$.

3.2 Granular Materials and Weak Link Networks

A granular material consists of an assembly of superconducting regions interconnected by a network of weak links (see *Granular Superconductors*). In a formal sense a granular material is equivalent to a weak link network (WLN) although the artificial WLNs that have been investigated are generally two dimensional (Falo et al. 1990) while granular materials are interconnected in three dimensions. It is evident that the resistive *EJ* transition of these materials is strongly curved. The characteristic is not, however, fitted well by a power law and if an "effective" n value is used it is also necessary to give $n(I)$ defined as the variation of n with I (Evetts et al. 1989). Various analytical expressions have been proposed for the form of the *EJ* curve; there is some consensus that a relation of the form $E \propto (J - J_c)^n$ offers a better fit than a simple power law to data for polycrystalline ceramic superconductors. However, it may also be significant that both the ceramic superconductors and WLNs formed from large arrays of tunnel junctions exhibit an effective n value proportional to I^{-2} when a power law fit is attempted.

3.3 Layer Structures and Anisotropic Superconductors

The systematic study of the resistive transition in layer-structure superconductors is in its early stages. Layer-structure superconductors include artificial proximity-coupled multilayers, intercalated compounds such as $NbSe_2$ and many of the high-temperature oxide superconductors. There is evidence that the transitions are broad and depend strongly on applied field and on the coupling between the different layers. The effect of the angle between the applied field and current does not follow the usual predictions of the Lorentz force picture (Kim et al. 1990). The mechanisms for flux transport and dissipation have not been elucidated. There is some controversy over whether the strong nonlinearity of the *EJ* characteristics arises from sample heterogeneity or from thermally activated processes (Palstra et al. 1990, Sun et al. 1991). It may be possible to unify these aspects by including thermally activated vortex cutting or phase slip which could lead to an additional melting transition that effectively decouples the superconducting layers of the structure, allowing strongly heterogeneous flux flow to occur that has some of the characteristics of flow in two-dimensional materials (Clem et al. 1990, White et al. 1991).

4. Concluding Remarks

The nature of the resistive transition in superconductors is of practical importance for applications and also of fundamental importance in the elucidation of the dynamic processes and dissipation mechanisms operative in different classes of superconducting material. Only in the simplest model systems is there at present a satisfactory quantitative description of flow phenomena. The high-temperature oxide superconductors have brought renewed interest in flow phenomena because flux creep and the irreversibility line transition which are obscure corrections to flux flow behavior in mainstream low-critical-temperature superconductors assume dominating importance in the new materials and are of critical relevance for their application. Consideration of the full range of experimental flux flow data including cross effects as a function of applied field and temperature will be necessary to discriminate between different models. In addition, measurements of Lorentz force effects, longitudinal effects and flux cutting, and flux flow noise provide important additional evidence for modelling.

See also: Critical Current Density; Flux Creep Phenomena; Flux Line Lattice; Magnetic Properties of Superconductors; Transport Currents and Force-Free Current Flow

Bibliography

Bardeen J, Stephen M J 1965 Theory of the motion of vortices in superconductors. *Phys. Rev. A* 140: 1197–1207

Campbell A M, Evetts J E 1972 Flux vortices and transport currents in type II superconductors. *Adv. Phys.* 21: 199–428

Clem J R, Coffey M W 1990 Viscous flux motion in a Josephson-coupled layer model of high T_c superconductors. *Phys. Rev. B* 42: 6209–16

Cyrot M 1973 Ginzburg–Landau theory for superconductors. *Rep. Prog. Phys.* 36: 103

Evetts J E 1983 The characterisation of superconducting materials—conflicts and correlations. *IEEE Trans. Magn.* 19: 1109–19

Evetts J E, Campbell A M, Dew–Hughes D 1968 Forces on flux vortices in an arbitrary configuration. *J.Phys. C* 1: 715–31

Evetts J E, Glowacki B A, Sampson P L, Blamire M G, Alford N McN, Harmer M A 1989 Relation of the *n*-value of the resistive transition to microstructure and inhomogeneity for $YBa_2Cu_3O_7$ wires. *IEEE Trans. Magn.* 25: 2041–4

Evetts, J E, Mullock S J 1985 Collective flux pinning and history effects in a defective flux vortex lattice. In: Matsushita T, Yamafuji K, Irie F (eds.) 1985 *Proc. Int. Symp. Flux Pinning and Electromagnetic Properties in Superconductors.* Matsukuma Press, Fukuoka, Japan, pp. 94–7

Falo F, Bishop A R, Lomdahl P S 1990 I–V characteristics in two-dimensional frustrated Josephson-junction arrays. *Phys. Rev. B* 41: 983–93

Frost H J, Ashby M F 1982 *Deformation-Mechanism Maps.* Pergamon, Oxford

Gor'kov L P, Eliashberg G M1968 Generalisation of the Ginzburg–Landau equations for non-stationary problems in the case of alloys with paramagnetic impurities. *Sov. Phys. JETP (Engl. Transl.)* 27: 328–34

Hagen S J, Lobb C J, Greene R L, Forrester M G, Talvacchio J 1990 Flux-flow Nernst effect in epitaxial $YBa_2Cu_3O_7$. *Phys. Rev. B* 42: 6777–80

Huebener R P 1979 Magnetic flux structures in superconductors. In: Cardona M, Fulde P, Queisser H J (eds.) 1979 *Solid-State Sciences*, Vol. 6. Springer, Berlin

Huebener R P 1981 Motion of magnetic flux structures. In: Gray K E (ed.) 1981 *Nonequilibrium Superconductivity, Phonons and Kapitza Boundaries.* Plenum, New York, Chap. 20, pp. 621–64

Josephson B D 1962 Possible new effects in superconductive tunnelling. *Phys. Lett.* 1: 251–3

Kim D H, Gray K E, Kampwirth R T, McKay D M 1990 Possible origins of resistive tails and critical currents in high-temperature superconductors in a magnetic field. *Phys. Rev. B* 42: 6249–58

Larkin A I, Ovchinnikov Y N 1986 Vortex motion in superconductors. In: Langenberg D N, Larkin A I (eds.) 1986 *Nonequilibrium Superconductivity*, Modern Problems in Condensed Matter Physics, Vol. 12. North-Holland, Amsterdam, pp. 493–552

Lengfellner H, Schnellbögl A, Betz J, Prettl W, Renk K F 1990 Nernst effect by laser-pulse heating in TlBaCaCuO superconducting thin films. *Phys. Rev. B* 42: 6264–7

Lumley J M, Evetts J E, Chao C H 1978 Flux flow measurements in irreversible type II superconductors. *J. Phys. (Paris) Lett.* 39: 393

Maki K 1969 Motion of the vortex lattice in a dirty type II superconductor. *J. Low Temp. Phys.* 1: 45–58

Palastra T T M, Batlogg B, van Dover R B, Schneemeyer L F, Waszczak J V 1990 Dissipative flux motion in high-temperature superconductors. *Phys. Rev. B* 41: 6621–32

Plummer C J G, Evetts J E 1987 Dependence of the resistive transition on composite inhomogeneity in multifilamentary wires. *IEEE Trans. Magn.* 23: 1179–82

Rodbell D S, Bean C P 1956 Some properties of the coercive force in soft magnetic materials. *Phys. Rev.* 103: 886–95

Shoenberg D 1952 *Superconductivity.* Cambridge University Press, Cambridge

Sun J Z, Eom C B, Lairson B, Bravman J C, Geballe T H 1991 Magnetic relaxation, current-voltage characteristics and possible dissipation mechanisms for high-T_c superconducting thin films of YBaCuO. *Phys. Rev. B* 43: 3002–8

Ullmaier H 1975 Irreversible properties of type II superconductors. *Tracts in Modern Physics*, Vol. 76. Springer, Berlin

White W R, Kapitulnik A, Beasley M R 1991 A model system for vortex motion in coupled two dimensional type-II superconductors. *Phys. Rev. Lett.* 66: 2826–9

Zeh M, Ri H C, Kober F, Huebener R P, Fischer J, Gross R, Müller H, Sermet T, Ustinov A V, Wener H G, Mannhart J 1990 Flux-flow induced Nernst effect in superconducting YBaCuO films. *Physica C* 167: 6–10

J. E. Evetts
[University of Cambridge, Cambridge, UK]

S

Sensors and Devices, Magnetic

Magnetic sensors and devices rely on exploiting a range of basic magnetic material properties for the transduction of a measurand such as magnetic field, force or torque into a useful signal. Table 1 is a summary of the magnetic phenomena that can be exploited for sensor applications. Detailed information on these phenomena can be found elsewhere in this Concise Encyclopedia.

To obtain the maximum material response in any given transducer arrangement, the relative orientation and coupling between input measurand (force, magnetic field, etc.) and magnetic properties must be optimized. Such optimization is primarily achieved by control of the magnitude and direction of the magnetic anisotropy, which is a complex function of material geometry, composition and thermomechanical history (see *Anisotropy in Magnetic Materials*). Changes in magnetic anisotropy are often detected through changes in such parameters as magnetic permeability. In crystalline materials the magnetocrystalline anisotropy dominates. Some alloys (e.g., the nickel–iron series) are capable of supporting a field-annealing-induced uniaxial anisotropy. Anisotropy can also be introduced by strain–magnetostriction coupling after mechanical deformation. In metallic glasses there is no macroscopic magnetocrystalline anisotropy and the easy direction can be controlled more readily by the material user.

The primary factor in sensor design is the intrinsic magnetic response of the sensor material. However, certain sensors may be required to work in environments that are mechanically, thermally or chemically hostile. These extra constraints dictate careful engineering design and perhaps a compromise on material choice. For example, with crystalline materials the general rule is that magnetic and mechanical hardness go together, or mechanical softness with magnetic softness. Mechanically hard materials are usually also brittle and wear resistant. Soft materials are ductile and have poor wear resistance. The majority of magnetic sensors employ soft magnetic materials, whose magnetic properties are easily degraded by mechanical work. The recent advent of the noncrystalline soft magnetic alloys, known generically as metallic glasses (see *Metallic Glasses*), has greatly aided sensor design and construction. Uniquely among soft magnetic materials, they possess excellent magnetic properties with high hardness, low wear resistance and, under certain circumstances, high ductility. New developments in materials preparation include thin-film technology and multilayer magnetic materials. These hold out the possibility of integrated and smart sensors in the future.

The magnetic elements in sensors are used in a wide range of forms: toroids, wires, rods, films on substrates and coatings. Some elements are essentially free standing, others are an integral part of a more complex device. Crystalline magnetic materials are

Table 1
Magnetic phenomena that can be exploited for sensors

Effect	Physical principle	Application areas
Joule	dimensional change in sample on magnetization—magnetostriction	force, torque, displacement, acoustic, magnetic field detection
ΔE	change in elastic modulus with the state of magnetization—magnetoelasticity	magnetic field detection
Matteucci	torsion-induced changes in magnetization—magnetoelasticity	torque
Wiedermann	with a magnetic field and an electric current in one direction, there develops a magnetization at right angles	torque, force
Villari	changes in magnetization on applying tension or compression—magnetostriction	force, acoustic
Sixtus–Tonks	fast change in magnetization due to Barkhausen jump	displacement, threshold magnetic field detection

available in all bulk forms and compositions have been optimized for transducer applications. Metallic glasses come in thin sheet (of thickness 20–40 μm) up to 0.2 m wide or as wires of 125 μm diameter, from commercial sources. The compositions available commercially have not been optimized for transducer applications, but still offer competition to crystalline materials. Both crystalline and noncrystalline materials can be deposited by sputtering or evaporation to make thin films.

1. Types of Sensor and System

All magnetic sensors involve the principle of transduction:

$$\Delta(\text{external parameter}) \rightarrow \Delta(\text{magnetic parameter}) \rightarrow \Delta(\text{electrical signal})$$

Magnetic sensors can be divided into two categories: primary and secondary. In primary sensors the external parameter is a magnetic field and the sensor and system are designed to give magnitude and/or directional information. For secondary sensor systems the external parameter may be something like a force or displacement which is translated into an electrical signal via some form of magnetic coupling. The sensor element can be regarded as dealing with the first step of the transduction process, while the system produces the second step. Whether or not a magnetic sensor system is appropriate for a given application depends on the availability of suitable materials, a satisfactory system including such factors as linearity, repeatability and resolution, and a competitive cost.

1.1 Primary Sensors

The general name for magnetic field sensing is magnetometry. The range of magnetic fields for which sensors are required spans seventeen orders of magnitude. Biomagnetic fields, such as those found around the cranium or torso, are at the femtotesla level (see *Biomagnetic Measurements*). The upper limit of interest is 100 T. The frequency range is also very wide. Geophysical and extraterrestrial measurements may last for some years with a lowest frequency component of 10^{-8} Hz, while in data transfer operations in the magnetic recording industry the frequency range may extend up to 1 GHz.

Two useful types of primary magnetic sensor rely on magnetotransport effects (see *Magnetotransport Phenomena*). Hall effect sensors and magnetoresistors both exhibit a change in electrical property (voltage across the device and electrical resistance, respectively) when there is a change in magnetic field in the region of the sensing element. Both forms of device are produced by thin-film technology and hence are ideally suited for inclusion in integrated systems where the signal processing is included in the same package.

Many standard laboratory field measurement systems employ Hall effect devices. The field probe can be configured to give measurement of axial or transverse magnetic fields, the nature of the Hall effect providing the vector response of the device. For typical systems the equivalent input magnetic field noise spectral density for a silicon-based Hall element at $I = 0.5$ mA is $3 \times 10^{-13}\,\mathrm{T^2\,Hz^{-1}}$ at 100 Hz and $10^{-15}\,\mathrm{T^2\,Hz^{-1}}$ at 100 kHz. The lowest nonlinearity ($\pm 3 \times 10^{-4}$) has been achieved in a GaAs-based device for fields below 1 T. Hall elements have a minimum detectable field of a few millitesla at dc decreasing to a few nanotesla at megahertz frequencies. The largest detectable fields are of the order of 50 T.

Magnetoresistors are either made from III–V semiconductors or from thin-film transition metal alloys (e.g., $Ni_{81}Fe_{19}$). At low frequencies their minimum detectable field is of the order of 100 pT. They saturate at fields in excess of 0.1 T. The high-frequency performance is reduced by eddy current effects and at 1 MHz the minimum detectable field is of the order of 1 μT. Magnetoresistors have good linearity and low temperature coefficients. Temperature stability is achieved using bridge circuits. Vector readings are possible by using shape anisotropy of the magnetoresistive strip. Two major application areas for these sensors are in security and validation. The magnetic stripe on tickets, identification passes or credits cards is usually read using magnetoresistive sensors. The output from the system (a few millivolts at a drive current of a few milliamperes) is independent of read speed and systems are guaranteed up to 3×10^5 card passes. Automatic vending machines can have coin validators installed based on magnetoresistive sensors. Again the output is independent of coin speed through the sensing system. The distinctive magnetic signatures of the different coins acceptable to the vending machine can be stored in the machine, for comparison with coins offered to the machine. Such systems can, for example, discriminate between British 5p and German 1DM coins. Other applications of magnetoresistive sensors can be found in noninvasive current and power measurement. The stray field from cables can be detected for currents between 10 A and a few kiloamperes. Magnetoresistors need a bias field to linearize the response, this being achieved by mounting the device in close proximity to a permanent magnet or by designing the current paths of the device to produce a biasing field.

The other major primary magnetic sensor is the flux gate. The magnetic element is usually made from a very soft magnetic material from the nickel–iron series of alloys. Flux gates using metallic glasses tend to have a poorer noise response owing to Barkhausen noise in the material. They are operated in two modes: absolute, where a single sensing element is used to give the local field magnitude and direction, or gradient, where two sensors are placed in close proximity to detect the magnitude and direction of field gradients. For a vector device a magnetic element is taken having a favorable shape anisotropy (usually a rod). Flux gates can be made on toroidal magnetic elements, but these tend to have a poorer vector response. In any flux gate sensor one winding produces a premagnetization of this element with ac current (sinusoidal, square wave or triangular wave) and a second winding acts as pickup. The premagnetizing field drives the element into saturation. The presence of a low-frequency (with respect to the premagnetizing field) or dc field introduces even harmonics into the signal from the pickup winding. Filtering allows detection of the second harmonic of the pickup signal and the amplitude of this second harmonic is a direct measure of the applied field. The schematic layout of a single-core flux gate probe is shown in Fig. 1a. The probe core is of magnetizable material with a high permeability and a low coercive force. The block diagram for a second-harmonic flux gate magnetometer with a phase-sensitive rectifier on the output is given in Fig. 1b. Flux gates cover a field range of 10^{-7}–10^{-3} T with a typical zero-level stability of 0.2 nT °C^{-1} and noise of 0.01 nT^2 Hz^{-1} in a 0–1 Hz bandwidth. A recent development in flux gate technology is the use of digital processing rather than analog. The field to be measured is compared with a reference field generated within the magnetometer. The reference field has a triangular form so as to obtain a linear relationship between the output and the field being measured. An electronic comparator experiences a change in pulse duty cycle as the external field and reference field coincide in magnitude. The output from the comparator is easily interfaced with microprocessors for advanced data handling. Such digital systems offer an improvement in the frequency range of the flux gate.

Other forms of magnetometry rely on the use of the ΔE effect (change in Young's modulus with applied field) or on magnetostrictive response (see *Magnetoelastic Phenomena*; *Magnetostrictive Materials*). Suitably field-annealed metallic glasses can show a large change in Young's modulus with applied field at fields close to zero. Hence the velocity of sound in the ribbon is a strong function of applied field (Squire and Gibbs 1987). By launching an ultrasonic wave into the ribbon and comparing the phase between the launched wave and the wave after travelling a measured distance down the ribbon, it is possible to detect applied fields of the order of 1 nT. Such a system is still at an early stage of development.

Magnetostrictive sensors have relied on coupling the magnetostrictive strain of annealed metallic glass, which is a strong function of the applied field, into a piezoelectric strip for conversion to an electrical signal (Mermelstein and Dandridge 1987) or into a fiber optic cable for conversion into a phase response in an interferometer (Yariv and Winsor 1980). The minimum detectable field has been of the order of a few picotesla. Severe problems of degradation of response have been reported, attributable to the bonding of the metallic glass to the substrates.

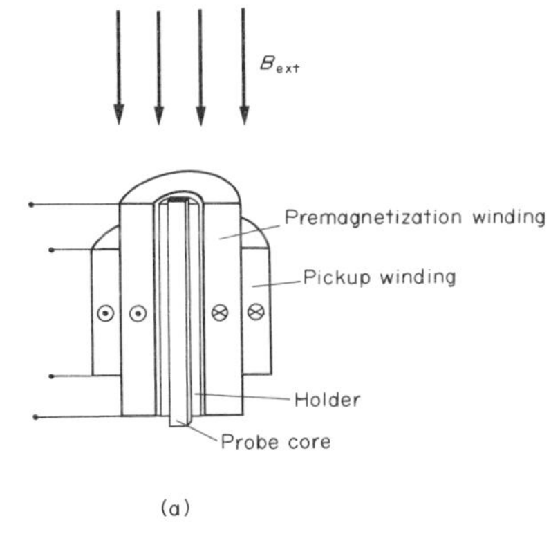

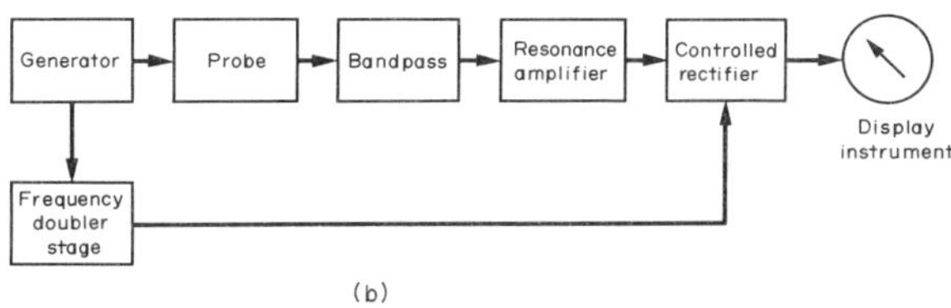

Figure 1
(a) Principal design of a flux gate probe and (b) single-core second-harmonic flux gate magnetometer (from Boll and Overshott 1989. ©VCH, New York. Reproduced with permission)

1.2 Secondary Sensors

The first category of secondary magnetic sensors to consider is that related to measurement of displacement or velocity. Applications cover proximity detection around a few nanometers to position indication over a few meters. The useful ranges for various linear displacement transducer technologies are indicated in broad terms in Fig. 2. The simplest form of position sensors are based on the principle of induction. For proximity detection the moving member has a small permanent magnet mounted on it. On the stationary member is mounted a small coil carrying an ac current. As the permanent magnet approaches the coil, the inductance changes and the sensor is activated either on a threshold principle or to give continuous readout. Advantages of this technique include the fact that the sensor can be physically small and also there is no physical contact between the moving part and the sensing coil. Noncontact switches for keypads, based on this proximity detection principle, can be produced for use in environmentally hostile areas.

If motion is to be detected (e.g., the rotary motion of gears or a shaft), inductive sensors based on

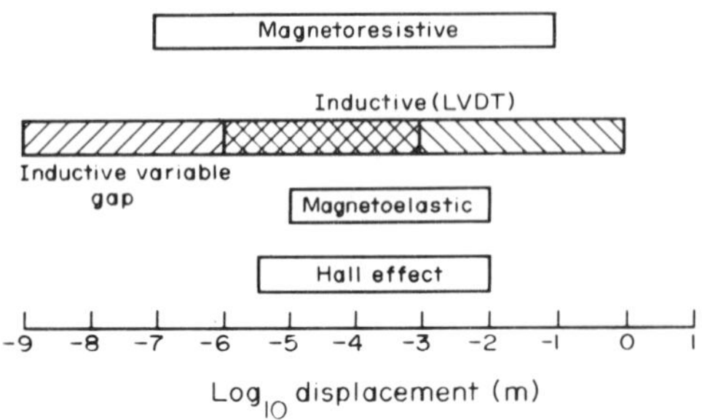

Figure 2
An indication of the useful ranges for various linear displacement transducer technologies (from Boll and Overshott 1989. ©VCH, New York. Reproduced with permission)

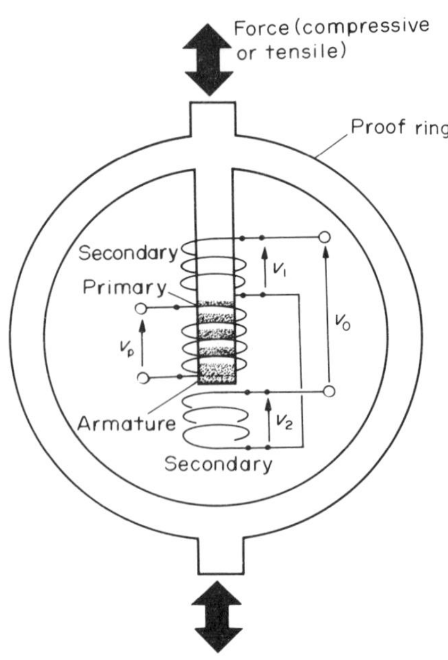

Figure 3
Proof-ring-based force transducer including an LVDT (from Boll and Overshott 1989. ©VCH, New York. Reproduced with permission)

Faraday's law can be used. If one or more permanent magnets are mounted on the rotating member, then an emf is induced in a coil as the magnet moves past. Revolution counting and rotary encoding are possible, again with no direct contact between moving parts and the sensing element. Angular velocities in the range 100–6000 rpm can be measured with a linearity of 0.01%. Angular velocity can also be determined using magnetoresistors or Hall elements in place of the detector coils around the rotating member. The output from the magnetic sensor can easily be digitized for interfacing with control and feedback electronics in a machine. Brushless dc motors use this principle in speed control. The unit cost of Hall effect or magnetoresistive devices is low owing to their ease of mass production. Wiegand or pulse wires can be used to detect rotary motion. Vicalloy wire or amorphous wire can be used to generate a voltage pulse every time the switching field on the wire is exceeded by the passage of a magnet at one end of the wire. Wire-based systems are easily interfaced to digital processors as the output is in voltage pulse form.

For numerically controlled machines, where computer control of a manufacturing process is required, a number of linear displacement and velocity systems have been proposed. One type has a magnetic track, similar to that on a credit card, with a recorded signal at a fixed wavelength which is traversed by read heads (Uemura and Kanegawa-ken 1969). Such devices can have resolutions down to 0.05 mm over a track length of 2 m. Traversing speed is high, typically several hundred meters per second. The ultimate resolution comes from the use of multiple heads and sophisticated signal processing. Linear displacement can also be detected using a linear variable differential transformer (LVDT) which is again an inductive device. Figure 3 shows how a proof ring coupled with an LVDT can be used as a force transducer for tensile or compressive loads. A soft magnetic core is attached to the moving part. Two secondary windings offset along the length of the magnetic core are the signal detectors. The core is magnetized at ac with a primary coil. As the core moves, the magnetic coupling between the secondaries varies. Null repeatability is excellent with these devices (better than 0.05%), with linearity at ±0.025% to 0.5% of the nominal displacement. Environmental sensitivity is low as the core and coils can be separately sealed. For the measurement of compressive forces or pressure an inductance bridge transducer can be used. Figure 4 is a schematic diagram of one such device. As the central diaphragm is displaced, the reluctance in each arm of the bridge will change and the signal output is a measure of the direction and magnitude of the force. A typical design can be used in the range of 0–3.5 MPa over a range of 0–100 °C. Displacement sensors based on the magnetoelastic properties of metallic glasses have been proposed. A toroidal core of metallic glass is deformed along a diameter by the moving member. The core is excited with a fixed amplitude ac current through a primary winding. The deformation causes a change in the material permeability and hence a change in the voltage induced in a secondary coil. The mechanical ruggedness of metallic glasses offers some advantages here, although the useful range of displacements is limited to 10^{-5}–10^{-2} m. The unit cost of these devices is high given the high cost of commercially available metallic glasses.

An important secondary magnetic sensor applica-

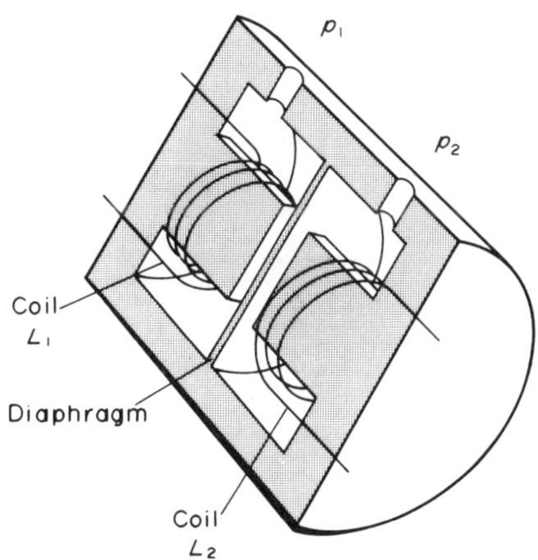

Figure 4
Schematic diagram of an inductance bridge transducer (from Boll and Overshott 1989. ©VCH, New York. Reproduced with permission)

tion is for mechanical transduction. Force, strain, torque and shock are all important parameters, particularly in such areas as the automotive industry where the modern vehicle may have upwards of 50 sensors and transducers built in. Magnetoelastic effects are important. The mechanical properties of the magnetic element are important as there must be direct linkage with the measurand. Load cells use the stress dependence of magnetic permeability as the transduction process. Cylinders of magnetic material such as iron–aluminum alloys are used, with the magnetic element being placed in compression by the applied force. Load cells based on magnetic sensing are tolerant of high overload conditions (say up to 300%) and are relatively insensitive to vibration. A typical working range is 0–10 MN. Linearity is 0.1% with 0.2% hysteresis. Semiconductor strain gauges have typical figures of merit (FOM) of 20 (with FOM defined as $(1/R)(\partial R/\partial \varepsilon)$ where R is the electrical resistance and ε the strain). More recently a magnetoelastic strain gauge was proposed with a figure of merit of 2×10^5 (Wun-Fogle et al. 1987) (with FOM now defined in terms of the strain dependence of the material permeability μ_r as $(1/\mu_r)(\partial \mu_r/\partial \varepsilon)$. A length of field-annealed metallic glass was bonded (rigidly or viscously) to the strained member. The metallic glass formed the core of a transformer. Under constant amplitude ac excitation, the output from the secondary was a direct measure of the applied strain, due to the strain–magnetostriction coupling. An advantage of this system over a semiconductor strain gauge is that there is no physical contact between the straining member and the sensor. Shock sensors operate on a similar principle. A toroidal core of metallic glass is excited at ac. The signal from a secondary winding can be passed to a threshold detector so that under shock excitation an output can be recorded. The strong mechanical properties of the metallic glasses combined with the soft magnetic properties are again important here.

Mechanical torque is conveniently detected using magnetic sensors. The simplest system uses the change in magnetic properties of the shaft being twisted. The torque couples with the magnetostriction of the shaft to produce a change in material permeability. A bridge arrangement of coils can be used for detection. A major drawback is that shaft materials must be mechanically strong and hence their magnetic properties are poor for transduction processes. A second approach is to coat or sleeve the shaft with a soft magnetic material to improve the sensitivity. The soft magnetic material is placed so that the easy axis is coincident with the principal stress axes of the twisted shaft (±45° to the shaft axis). A primary winding excites the soft magnetic material with an ac field. Secondary coils, often in a bridge circuit, detect the changes in permeability caused by the torque. Typically torques of 10–5000 N m can be sensed on shafts of 12–100 mm diameter. Problems are experienced with bonding the soft film in that the easy direction may be perturbed by the bonding stresses and hence not be in the preferred direction with a resulting loss in sensitivity. Such systems are also prone to dc fields which can cause offsets in the output signal.

2. *Future Trends*

Magnetic sensors have an important part to play as industry and commerce become increasingly automated. System design will move down routes that lead to low-power or self-powered sensors which are easily interfaced to signal processing systems. There are indications in Sect. 1 of the technologies that may lead to this in the future. New materials produced by deposition on substrates may well produce greater sensitivity and ease of integration. Magnetic sensors already have a market lead in areas such as linear and rotary position sensing and in security systems. New markets will open up as the materials and systems engineers work together to reduce unit costs.

See also: Magnetic Materials: An Overview; Magnetostrictive Materials; Measurements in Magnetic Materials; Metallic Glasses; SQUID Design, Fabrication and Application

Bibliography

Boll R, Overshott K J 1989 *Magnetic Sensors*, Sensors—A Comprehensive Survey, Vol. 5. VCH, New York

Hernando A, Vazquez M, Barandiaran J M 1988 Metallic glasses and sensing applications. *J. Phys. E* 21: 1129–39

Mermelstein M D, Dandridge A 1987 Low-frequency magnetic field detection with a magnetostrictive amorphous metal ribbon. *Appl. Phys. Lett.* 51: 545–7

Squire P T, Gibbs M R J 1987 Shear-wave magnetometry using metallic glass ribbon. *Electron. Lett.* 23: 147–8
Uemura S, Kanegawa-ken 1969 Magnetic heads utilized as a displacement measuring instrument. US Patent No. 3,597,749
Wun-Fogle M, Savage H T, Clark A E 1987 Sensitive, wide frequency range magnetostrictive strain gauge. *Sensors Actuators* 12: 323–31
Yariv A, Winsor H V 1980 Proposal for detection of magnetic fields through magnetostrictive perturbation of optical fibers. *Opt. Lett.* 5: 87–9

M. R. J. Gibbs
[University of Bath, Bath, UK]

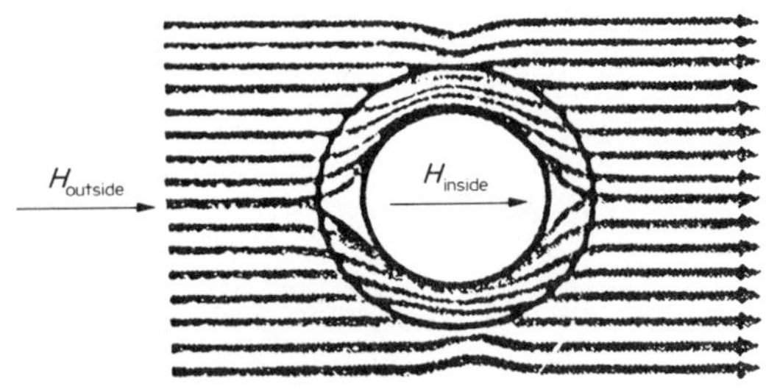

Figure 1
Schematic diagram of a typical magnetic shield

Shields, Magnetic

The correct functioning of an electrical circuit can be upset by stray magnetic fields from sources such as permanent magnets, power cables, transformers or motors. Shielding against an alternating electrical or electromagnetic field can be achieved by the use of a high-conductivity metal enclosure which reflects or restricts the field by setting up an eddy current barrier. These shields, typically constructed of copper or aluminum, become less effective once the frequency of the interference drops much below 10 kHz. For low-frequency or dc fields it is necessary to use a high-permeability soft magnetic material which provides a low-resistance path for the magnetic flux to pass around the device being shielded. The term magnetic shield is normally applied to this latter type of enclosure.

1. Material Selection

The properties of the materials normally employed for the construction of magnetic shields are given in Table 1. The high-permeability alloy, 80%Ni–20%Fe with additions of molybdenum and/or copper, is the normal choice of material to provide the most economic and effective shield (see *Nickel–Iron Alloys*). When the shield is required to operate in high fields and the operating flux level in the shield wall approaches 0.6 T, it may become necessary to employ materials such as 50%Ni–50%Fe or pure iron which have lower permeabilities but higher saturation flux densities. These materials tend not to be used on their own but in combination with 80%Ni–20%Fe in a multiple shield construction.

2. Shield Design

A number of factors need to be taken into account when designing a shield to achieve a particular screening factor. These include the size and shape of the screen, the thickness and permeability of screen material and the intensity of the field.

The shielding factor s is defined as the ratio

$$s = \frac{H_{\text{outside}}}{H_{\text{inside}}}$$

(see Fig. 1) and can be expressed in decibels as

$$\text{db} = 20 \log \frac{H_{\text{outside}}}{H_{\text{inside}}} \tag{1}$$

2.1 Shield Configuration

The shielding factor for various shapes of screen can be estimated using the following simplified equations:

$$\text{sphere } s = \frac{4}{3}\frac{\mu t}{D} \tag{2}$$

Table 1
DC magnetic properties of shield materials

Material	Initial permeability	Maximum permeability	Saturation flux density (T)	Coercive force ($A\,m^{-1}$)
$Ni_{77}Fe_{14}Cu_5Mo_4$	50000	200000	0.8	1.0
$Ni_{50}Fe_{50}$	6000	40000	1.6	8
Pure iron	300	5000	2.15	70

$$\text{cube } s = 0.7\frac{\mu t}{a} \tag{3}$$

$$\text{cylinder } s = \frac{\mu t}{D} \tag{4}$$

where t = sheet thickness, μ = sheet permeability, D = screen diameter and a = side length.

For open-ended shields, the length should exceed that of the component by an amount at least equal to that of the radius of the cylinder and the ratio of the length to the diameter of the shield should be greater than four. End caps are necessary if a short shield has to be employed. It is not always possible to orient the magnetic shield such that the interference field is perpendicular to the axis. Where the field impinges directly into the open end of a cylinder the shielding factor can be as low as 20% that of the perpendicular case.

These design considerations relate to dc fields; under ac conditions the shielding factor is generally greater due to the effect of the induced eddy currents which act to reduce field penetration. The improvement is not as great as might be expected, for the material permeability also drops with increasing frequency.

2.2 High-Intensity Fields

Generally the shield operates at flux levels well below the saturation level of the shield material. In high-field environments the designer must ensure that the flux in the shield wall does not approach saturation, otherwise the material permeability and thus the shielding factor will be poor. The flux in the shield wall can be calculated using the approximate equation

$$B = \mu_0 H_{\text{outside}} \frac{D}{t} \tag{5}$$

If the flux level does start to approach the saturation level of the shield material then there is the option of using a thicker material or a multiple shield, where the single shield is replaced by a number of concentric shields. In general, the use of multiple shields is more efficient than increasing the wall thickness. The screening factor of such multiple shields can be calculated using the following equations:

$$\text{sphere } s = s_1 + s_2 + s_1 s_2 \left(1 - \frac{D_1^3}{D_2^3}\right) \tag{6}$$

$$\text{cylinder } s = s_1 + s_2 + s_1 s_2 \left(1 - \frac{D_1^2}{D_2^2}\right) \tag{7}$$

$$\text{cube } s = s_1 + s_2 + 0.8\, s_1 s_2 \left(1 - \frac{a_1^3}{a_2^3}\right) \tag{8}$$

To attenuate very high fields, it may be necessary to employ a higher saturation/lower permeability material for the outer shield and to use more than two concentric layers.

The use of a multiple shield construction is also the usual means of achieving very low residual fields for very sensitive scientific or medical measurements. For such applications an additional technique known as "shaking" is frequently employed. This involves applying a continuous high-frequency field to the shield wall and results in an improved shielding factor.

3. Shield Fabrication

The usual methods of fabricating magnetic shields are deep drawing, spinning, or forming and welding using the parent alloy as a filler rod. After all the mechanical forming operations have been completed, the shield must be heat treated to develop the magnetic properties and, hence, the design performance. This treatment involves heating the shield, suitably supported, to a temperature of 1050 °C for a period of 2–4 h. A furnace atmosphere of dry hydrogen or cracked ammonia must be maintained throughout the heating and cooling cycles.

For research and development, design evaluation or small production runs, which do not justify the tooling expense of a fully fabricated shield, pre-annealed foil can be employed. This foil can be wrapped around the device and the number of wraps can be adjusted to provide the thickness necessary to achieve the required attenuation.

In order to achieve a very low remanence field it is important that the shield should be demagnetized. This is normally achieved by toroidally winding a number of turns around the shield through which is passed a mains-frequency current sufficient to saturate it. The current is then slowly reduced to zero.

See also: Magnetic Materials: An Overview; Measurements in Magnetic Materials

Bibliography

Cohen D 1967 Enhancement of ferromagnetic shielding against low-frequency magnetic fields. *Appl. Phys. Lett.* 10: 67–9

Kelha V, Peltonen K, Rantala B 1980 The effect of shaking on magnetic shields. *IEEE Trans. Magn.* 16: 575–8

Mager A 1981 The Berlin magnetically shielded room. *Biomagnetism*. Walter de Gruyter, Berlin, pp. 51–78

Wadey W G 1956 Magnetic shielding with multiple cylindrical shells. *Rev. Sci. Instrum.* 27: 910–16

R. V. Major
[Telcon, Crawley, UK]

Silicon Steels

Silicon steels are soft magnetic alloys widely used in transformers, rotating power equipment and relays. Their magnetic properties are derived from properties present in pure iron. Elemental silicon is added to increase resistivity and permeability and to decrease magnetostriction and magnetocrystalline anisotropy. Additionally, silicon stabilizes the alpha (ferrite) phase to higher temperatures, allowing the special heat treatments necessary to develop a preferred orientation. The commercially important properties of silicon steels include high permeability and low core loss.

Silicon steels fall into two general categories.

(a) There are nonoriented or isotropic products in which the grains are randomly oriented relative to the rolling and transverse directions of the sheet.

(b) There are grain-oriented or anisotropic products which develop a preferred crystallographic texture which leads to reduced core loss and increased permeability compared with nonoriented products. In such products the individual grains are aligned so that their easy axis of magnetization, the [001] direction, lies nearly parallel to the rolling direction of the sheet. Two types of preferred orientations can be developed in silicon steels: the cube-on-face or (100)[001] orientation, in which the (100) plane lies in the sheet plane, and the cube-on-edge (110)[001] Goss orientation in which the (110) plane lies in the sheet plane. The Goss texture is the only texture commercially produced on a significant scale.

Before a detailed discussion of oriented silicon steels, it is appropriate to describe briefly the nonoriented silicon steels. Nonoriented silicon steels are to be distinguished from the so-called motor lamination steels that are widely used in fractional horsepower motors and are the largest single tonnage electrical steel. Motor lamination steels generally contain less than 0.6 wt% total of silicon plus aluminum and are considered low-carbon steels (see *Iron and Low-Carbon Steels for Magnets*). Nonoriented silicon steels, on the other hand, contain over 1 wt% total of silicon and aluminum, and these steels are distinguished from oriented silicon steels by their properties which are nearly isotropic in the plane of the sheet and, thus, are used in applications where magnetic flux must be carried in almost all directions with nearly equal facility.

Nonoriented steels may be produced using much of the same equipment on which oriented silicon steels are produced. For example, they could be melted using a basic oxygen furnace, continuously cast and then hot rolled after soaking to achieve temperature uniformity. This can be followed by a heat treatment, removal of scale and cold rolling to final gauge, or the heat treatment may not be used in some cases. At this stage the steel may be annealed and coated and is ready for magnetic testing to verify the magnetic quality prior to sale to customer as a "fully processed" steel. In some cases the steel is cold rolled to within 6–10% of the desired final gauge, annealed and then rolled to final gauge at which point it can be coated with an inorganic coating, tested for magnetic quality and then sold as a "semiprocessed" steel. The small amount of cold work left in the steel as the result of the 6–10% reduction to final gauge enhances the punchability of the steel.

The total content of silicon and aluminum will range from about 1 wt% up to about 3.75 wt% in the nonoriented silicon steels. Those with lower alloy content have higher saturation levels and offer better permeability at high test inductions; whereas those with higher alloy content offer relatively low core losses, a feature required in high-efficiency rotating machinery. These steels are generally produced in gauges of 0.35 mm, 0.47 mm and 0.64 mm, the lighter gauges offering lower core losses and the heavier gauges offering efficiencies of assembly of machinery. The core loss and permeability traits of fully processed nonoriented silicon steel grades may be found in ASTM A 677–77 and in Table 1 of this article and for semiprocessed steel in ASTM A 683–77.

Nonoriented silicon steels may be sold as semiprocessed or fully processed products; the user of semiprocessed steel will provide a stress relief anneal of the stamped or sheared parts prior to assembly into machinery. Fully processed steel is in the fully annealed condition and the user does not anneal the steel after stamping or shearing. The semiprocessed product will usually provide the lowest core losses in the application because the stress relief anneal removes all magnetic damage done by the stamping or shearing operations. These steels may be sold with no added insulative coating on the surface or with inorganic (C-5 type) or organic coatings (C-3 type). The organic coating can only be supplied on fully processed steel since such coatings will not survive a stress relief anneal. There are also coatings of organic nature with inorganic fillers (C-6 type) that may be applied to steels that go into applications where rebuilds are a common occurrence. Such coatings are designed to provide good punchability and to maintain their insulative integrity through the burnout operations that are used in rebuilding motors, for example.

Integral horsepower motors and generators intended for continuous or near-continuous operation are generally provided with steels of higher alloy content and thinner gauges to minimize heat problems that can result from higher-loss steels and to provide highly efficient motors. There is some use of

Table 1
Properties of silicon steels

AISI designation	Type[a]	Thickness (mm)	Maximum core loss at 60 Hz, 1.5 T[b] ($W\,kg^{-1}$)	Saturation induction (T)	Nominal Si + Al content (wt%)
M47	NO	0.64	10.8	2.11	1.05
		0.47	10.14	2.11	
M45	NO	0.64	7.94	2.07	1.85
		0.47	6.72	2.07	
M43	NO	0.64	5.95	2.04	2.35
		0.47	5.07	2.04	
M36	NO	0.46	5.29	2.02	2.65
		0.47	4.52	2.02	
		0.36	4.19	2.02	
M27	NO	0.64	4.96	2.02	2.80
		0.47	4.19	2.02	
		0.36	3.97	2.02	
M22	NO	0.64	4.81	2.00	3.20
		0.47	4.08	2.00	
		0.36	3.70	2.00	
M19	NO	0.64	4.59	1.99	3.30
		0.47	3.84	1.99	
		0.36	3.48	1.99	
M15	NO	0.47	3.70	1.98	3.50
		0.36	3.20	1.98	
M6	CGO	0.35	1.46 (2.07)	2.00	3.15
M5	CGO	0.30	1.28 (1.83)	2.00	3.15
M4[c]	CGO	0.35	(1.79)	2.00	3.15
		0.28	(1.68)	2.00	
M3	CGO	0.28	1.09 (1.57)	2.00	3.15
		0.23	0.96 (1.44)	2.00	
M2	CGO	0.18	0.89 (1.40)	2.00	3.15
M2H	HP	0.30	(1.55)	2.00	2.90–3.15
		0.28	(1.46)	2.00	
M1H	HP	0.30	1.05 (1.41)	2.00	2.90–3.15
		0.28	1.02 (1.38)	2.00	
M0H	HP	0.30	1.03 (1.36)	2.00	2.90–3.15
		0.28	0.92 (1.30)	2.00	
		0.23	0.88 (1.21)	2.00	3.15

a NO = nonoriented products, $B_{800} \simeq 1.4$ T; CGO = conventional grain-oriented products, $B_{800} \simeq 1.8$ T; HP = high-permeability products, $B_{800} > 1.88$ T b Values in parentheses are at 1.7 T c Below M4, these are industry-accepted values and have not yet been established by AISI or ASTM

the higher-quality nonoriented silicon steels in small specialty transformers such as found in amplifiers. Higher core loss and thicker steels are used where core loss and heat are a secondary consideration and where higher flux-carrying capability may be desirable. For example, fractional horsepower or intermittently operating motors and pole pieces are likely applications for such lower-grade steels.

Nonoriented products are used in applications in which the magnetic flux is not forced along any one direction, as in rotating power equipment such as motors, alternators and generators. For applications such as power and distribution transformers, oriented silicon steel is cut to maximize magnetic-flux passage in the easy [001] direction.

1. Magnetic Properties

1.1 Permeability

Permeability is a measure of the ease with which the steel can be magnetized to a particular induction—the higher the permeability, the easier it is to magnetize the steel. In silicon steels, the perfection or sharpness of the final crystallographic texture determines its permeability in large applied fields and is measured by the induction obtained when the sample is magnetized to a point just beyond the "knee" of the magnetization curve. An applied field of 800 $A\,m^{-1}$ is used and the induction obtained is termed the material's B_{800} value. A perfectly oriented material—one with a perfect (110)[001] orientation—would be magnetically saturated at this point and would have a B_{800} value of 2.030 T. As grain misorientations increase (i.e., deviate from the perfect (110)[001] orientation), B_{800} values of 1.9 T, 1.8 T and 1.4 T are obtained for high-permeability grain-oriented, conventional grain-oriented and nonoriented silicon steels, respectively. Typical B_{800} values of 1.9 T, 1.8 T and 1.4 T result from grain misorientations of the order of 3°, 6° and 22.5°, respectively.

The key factor required to produce the desired grain orientation is the selective growth or secondary recrystallization of (110)[001] grains during a high-temperature anneal. This selectivity is obtained through a critical and complex processing sequence typically consisting of a large cold reduction, followed by a low-temperature recrystallization–decarburization anneal and finally a high-temperature texture-purification anneal during which secondary recrystallization occurs.

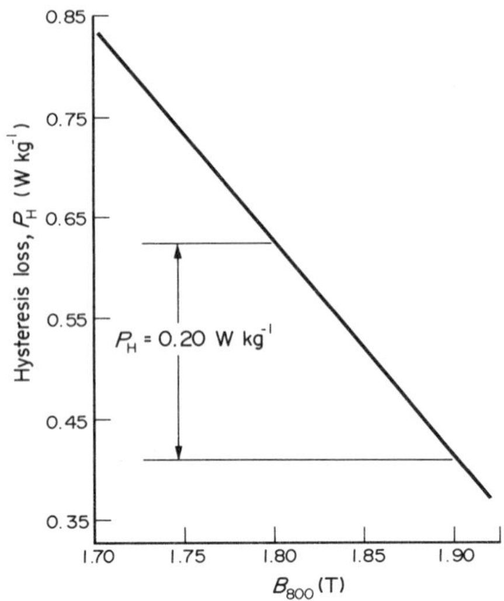

Figure 1
Variation in hysteresis losses with B_{800} in 3%Si–Fe (0.28 mm sheet thickness)

1.2 Core Loss

Core loss is the amount of energy dissipated during cyclic magnetization of soft magnetic materials and is ultimately related to changes in domain structure. Core loss therefore represents an inefficiency in the operation of electrical equipment and, although it cannot be eliminated entirely, it can be minimized through control of the domain structure of the magnetic material.

The core losses in silicon steels arise from two principal sources (see *Power Loss in Magnetic Materials*).

(a) Synchronous eddy current losses are, for example, eddy current losses resulting from domain-wall motion at the same frequency and in synchronization with the applied field. These losses are directly proportional to the square of induction and frequency, to the first power of sheet thickness and domain width, and inversely proportional to electrical resistivity.

(b) Losses due to nonuniform domain wall motion include conventional hysteresis losses as well as losses associated with the motion of supplementary domain structure and domain nucleation and annihilation events. Hysteresis losses are a function of crystallographic texture and domain-wall pinning, which depends on internal stresses, inclusions and precipitates.

All losses are dissipated by microeddy currents. Total core loss is merely the superposition of eddy currents arising from synchronous 180° domain-wall motion and nonuniform domain-wall motion.

Careful control of chemistry and stress relief annealing of the final product has led to reductions in hysteresis losses. However, additional reductions in hysteresis losses have resulted from improvements to crystallographic texture. Figure 1 shows the dependence of hysteresis losses on texture as measured by B_{800}. As can be seen in Fig. 1, an improvement in B_{800} from 1.8 T to 1.9 T decreases the hysteresis losses by 0.2 W kg^{-1}.

The dependence of total losses on hysteresis losses becomes smaller as texture improves and the losses become dominated by synchronous eddy current losses. If an eddy current model based on the existence of simple 180° domains is assumed, the synchronous eddy-current-loss component of silicon steels can only be reduced by increasing electrical resistivity, decreasing sheet thickness or decreasing the domain spacing. Elemental silicon, when added to pure iron, increases electrical resistivity linearly from 0.1 μΩ m at 20 °C to approximately 0.6 μΩ m at 20 °C and 4 wt%Si. Unfortunately, brittleness problems, particularly during cold rolling, generally preclude silicon alloying much above 3 wt% if conventional manufacturing processes are used. Spin casting, as is used in making metallic glasses or amorphous metals, has been used to produce iron alloys containing up to 6 wt%Si; however, such materials appear to be limited to a maximum thickness of about 0.04 mm.

Reducing sheet thickness to achieve lower core losses has been extensively studied, but losses increase below a certain sheet thickness. Minimum core losses in polycrystalline grain-oriented silicon steels are achieved at thicknesses of about 0.12 mm. The reasons for the increase in losses with further reduction in thickness are (a) an increase in the 180° domain-wall spacing as thickness is decreased and (b) an increased influence of surface roughness on the move-

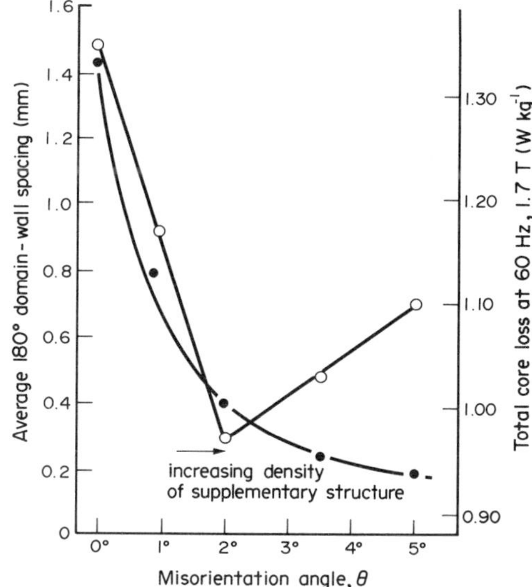

Figure 2
Dependence of 180° domain-wall spacing (●) and core loss (○) on misorientation angle θ (tilt of the [001] direction out of the crystal surface)

ment of domain walls as the surface-to-volume ratio increases with decreasing thickness. At thicknesses slightly larger than that for minimum losses, the practical lower limit of sheet thickness is determined by the expense and difficulty involved in production of this gauge strip and by the increased costs of fabricating transformers from the greater number of required laminations.

Domain observations in silicon–iron single crystals have indicated that domain-wall spacing can be reduced by several factors to yield lower eddy current losses. Minimum achievable wall spacings of the order of one-half the sheet thickness can be achieved by control of orientation, application of tensile stress or scribing. Figure 2 shows the dependence of domain spacing and total core loss on the tilt (misorientation angle) θ of the [001] direction out of the sheet plane. Values of θ less than 2° lead to an increase in 180° domain-wall spacing and a corresponding increase in losses. Values of θ greater than 2° result in little additional domain-spacing refinement and at the same time cause an increase in the density of flux closure domains (supplementary structure), thereby increasing core loss. Domain spacing decreases with increasing θ due to a buildup of magnetostatic (free-pole) energy on the sheet surface, which is offset by a reduction in wall spacing and an increase in supplementary structure.

Application of a uniaxial tensile stress along the projection of the [001] direction reduces the density of supplementary structure and thereby produces an additional reduction in 180° domain-wall spacing. However, the elimination of supplementary structure is only metastable for values of θ greater than 2°, and the supplementary structure reappears during subsequent magnetization. Hence, tensile stresses are only effective in reducing eddy current losses for misorientations of the order of 2° or less. Such behavior is observed in polycrystalline grain-oriented silicon steel strip as is shown in Fig. 3.

In recent years, the effectiveness of tensile stress in reducing core losses has been exploited by silicon steel manufacturers through the development of a stress-producing insulative coating or stress coating. Stress coatings have a lower coefficient of thermal expansion than the silicon steel. The stress coating is applied and then cured at a high temperature. On cooling, a residual tensile stress is induced in the steel due to the differences in contraction between the steel and the coating. Because of the elastic anisotropy of the steel, the greatest component of strain occurs in the rolling direction of the steel. This results in magnetoelastic effects equivalent to those produced by a uniaxial tensile stress applied along the rolling direction. More recently, similar magnetoelastic effects have been observed by lightly "scribing" the surface of silicon steel along the transverse direction.

There are now production scribing methods of effecting domain refinement. All of the techniques produce fine lines of disturbed material at angles nearly perpendicular to the easy direction of magnetization at intervals ranging from 3 mm to 15 mm. The disturbed material may be the result of thermal shock, cratering or plastic deformation, but it is always confined to very limited volumes of material. The

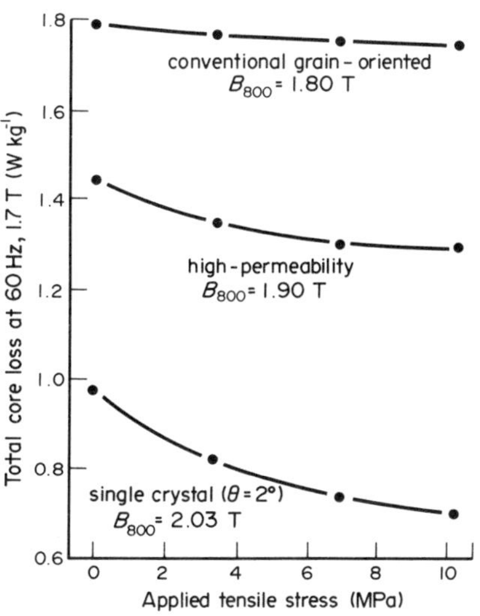

Figure 3
Effect of applied tensile stress on core loss in 3%Si–Fe

effect of such disturbed material is to create supplementary domains and to minimize the magnetostatic energy; additional antiparallel domain walls form thus reducing the domain-wall spacing and eddy current losses associated with their movement. Nippon Steel and Armco have developed techniques using lasers to accomplish domain refinement. Allegheny Ludlum has developed an electron beam method, British Steel has developed a spark ablation technique and Kawasaki Steel has developed a localized plasma flame heating method. These techniques are all satisfactory in stacked core transformer construction wherein no stress relief anneal is employed, but when used for wound core construction, the required stress relief anneal removes the benefits induced by these methods. Methods using localized plastic deformation to create fine lines of primary recrystallized grains have been developed to cause decreases in domain wall spacings; these treatments are heat resistant; that is, their domain refining effects survive the usual stress relief anneal temperatures. Such heat-resistant domain refinement results in steels with some of the very lowest core losses to be found.

1.3 Magnetostriction

Magnetostriction is the systematic expansion and contraction of ferromagnetic materials along the direction of the applied field (see *Magnetostrictive Materials*). This creates undesirable noise—the familiar transformer hum. Magnetostriction is measured by the net strain parallel to the applied field when the material is magnetized from the demagnetized condition to a given induction. The magnetostriction in grain-oriented silicon steels is a function of silicon concentration and the density of supplementary domain structure. In silicon steels the basic magnetostrictive constants approach zero at silicon levels of slightly greater than 6 wt%. However, as described already, grain-oriented silicon steels with such silicon levels are not produced using conventional manufacturing processes. Magnetostriction, therefore, can only be controlled by reducing the density of supplementary domain structures through control of grain orientation and residual stress levels. In general, the amount of supplementary structure and, hence, magnetostriction decreases with increasing degree of texture perfection.

2. *Metallurgical Processes*

Commercial processes developed make possible mass production of silicon steel with superior magnetic quality for the electrical industry. These processes, involving careful control of composition and a series of hot-rolling, cold-rolling and annealing operations, have evolved from significant progress in understanding the fundamental metallurgy of the iron–silicon system.

In a typical commercial process, during melting and refining, normally the major elements—silicon, carbon, manganese and sulfur for conventional silicon steel—are carefully controlled and residual impurities are kept as low as possible. After casting and solidification of ingots, or continuous casting as a slab, a high-temperature soaking, of at least 1320 °C, is required to dissolve MnS particles which precipitate during solidification. During the ensuing hot-rolling processes, finely dispersed MnS precipitates form—these are vital for (110)[001] Goss texture development. A hot-rolled band heat treatment may be used to ensure more uniform quality. Following cold reduction, such as a two-stage cold reduction with an intermediate anneal, the strip is then at its final gauge of 0.17–0.36 mm, with 50–70% final reduction. The cold-rolled strip must be given a final continuous anneal in a controlled wet atmosphere to reduce carbon levels to less than 0.003 wt%. This decarburization step eliminates magnetic aging in the final product and at the same time recrystallizes the cold-rolled structure to produce a uniform, equiaxed primary-grain structure with a weak (110)[001] texture. The coil is then coated with refractory oxide coating, such as MgO, and subjected to final texture anneal at 1200 °C in a controlled hydrogen atmosphere to ensure that the (110)[001] texture is fully developed and that the steel is thoroughly desulfurized. During this anneal, the MgO coating, which is used as an annealing separator to prevent sticking of adjacent coil wraps, reacts with a SiO_2 surface film formed during final continuous annealing and produces a Mg_2SiO_4 base coating. The base coating helps to reduce the total core loss of stacked lamination assemblies by reducing macroscopic eddy currents between adjacent laminations.

Recently, high-permeability grain-oriented steels have been developed that differ somewhat in composition and metallurgical processes. The first such product, commercialized by Nippon Steel, made use of AlN as the primary inhibitor. In addition, an 87% final cold reduction was used. More recently, Kawasaki Steel has introduced a high-permeability product. The composition and processing of such products are compared with conventional grain-oriented products in Table 2.

2.1 Secondary Recrystallization

The (110)[001] Goss texture in grain-oriented silicon steel is developed through secondary recrystallization, which is the selective growth of (110)[001] grains in a primary recrystallized matrix during the final texture anneal. The driving force for secondary recrystallization is the imbalance of grain-boundary surface tension between large grains and a surrounding finer-grain matrix. Secondary grains achieve their size advantage during the early stage of secondary recrystallization and subsequently grow into a stabilized primary-grain matrix, in which normal grain growth

Table 2
Typical composition and manufacturing of conventional and high-permeability silicon steels

Designation	Conventional grain-oriented	High-permeability grain oriented: Nippon Hi-B[a]	High-permeability grain oriented: Kawasaki RG-H[b]
Composition (%)			
C	0.03	0.05	0.020
Si	3.15	2.95	3.25
Mn	0.07	0.08	0.05
S	0.025	0.028	
Others		0.03 Al, 0.007 N	0.02 Se, 0.03 Sb
Ingot soak or slab reheat (°C)	1370	1370	1370
Hot-rolled band treatment (°C)	900 (optional)	1100, WQ[c]	900
Rolling stages	two	one	two
Final reduction (%)	55	88	65
Final anneal (°C)	1200	1200	900–1200

a HI-B is a trademark of Nippon Steel Corporation b RG-H is a trademark of Kawasaki Steel Corporation c WQ = water quenched

has been suppressed by an inhibitor, such as dispersed MnS and/or AlN or by solutes such as carbon, nitrogen and sulfur. When annealing in certain temperature ranges, usually 900–1200 °C, secondary recrystallization starts after an incubation period which decreases as temperature increases. The secondary-grain size varies from a few millimeters to a few centimeters, as shown in Fig. 4, indicating that a secondary grain consumes as many as 10^6 primary grains during secondary recrystallization and significantly reduces the total grain-boundary energy. It is easy to understand secondary recrystallization from this grain-boundary-energy concept, but why the final texture develops to such a high degree of orientation and how the selectivity develops at the early stage of the secondary recrystallization are still not clearly understood. However, improved techniques for measuring and understanding texture development, such as through the orientation distribution function (ODF), are helping to understand the evolution of the texture through the several cold reductions and anneals that are part of the processing of oriented silicon steels.

2.2 Inhibition System and Primary Texture

It is generally agreed that the inhibition system and primary texture are the two most important metallurgical factors necessary for good Goss-texture development in grain-oriented silicon steel. High-purity silicon–iron lacking inhibition cannot develop a Goss texture. May and Turnbull (1958) first demonstrated that small amounts of impurity additions, mainly manganese and sulfur, form dispersed MnS particles, which retard normal grain growth and maintain the driving energy for secondary recrystallization. Only small MnS particles, with radii less than about 26 nm, are effective in grain-growth inhibition. Figure 5 shows the dispersed MnS particles in a primary-grain matrix and interactions between a grain boundary and MnS particles in conventional grain-oriented silicon steel. Two-stage cold rolling with 50–70% final reduction produces a weak (110)[001] primary texture with grain sizes ranging from 15 μm to 20 μm after primary recrystallization. This weak (110)[001] component in the primary texture is necessary for the selective growth of (110)[001] grains during texture development. It has been proposed that (110)[001] grains store a larger strain energy during cold rolling than grains of other orientations, and thus (110)[001] grains recrystallize first and subsequently consume their neighboring high-energy regions to achieve their size advantage. This mechanism leads to the (110)[001] selectivity during secondary recrystallization and has been confirmed in conventional grain-oriented silicon steel. Final cold reduction greater than 70% cannot develop good Goss texture even with the same MnS particle-size distributions. In this case, it is thought that MnS particles are no longer sufficient to inhibit normal grain growth. Thus, the misorientation of the final texture appears to reach a limit of 6° in conventional silicon steels.

Taguchi and Sakakura (1966) first introduced high-permeability silicon steels which improve the misorientation of the Goss texture to 3°. This process is

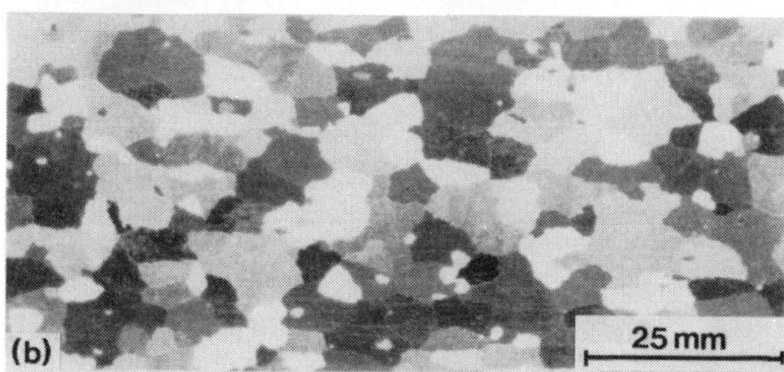

Figure 4
Grain structure in (a) high-permeability and (b) conventional grain-oriented 3%Si–Fe

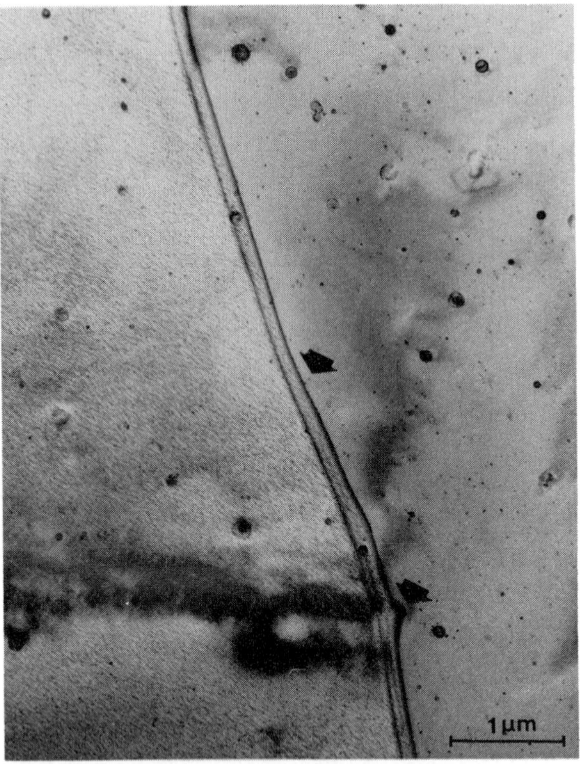

Figure 5
A grain boundary in conventional grain-oriented silicon steel pinned by MnS inclusions: the lower arrow points to a large particle; the upper arrow points to one of many sites pinned by the more important fine particles

characterized by single-stage heavy cold reduction with AlN as the primary inhibitor system. Rapid cooling after hot rolling enables AlN to precipitate during the partial alpha-to-gamma phase transformation in this alloy. Heavy cold reduction produces a strong (111)[112] component in the primary texture. Severe cold reduction leads to the formation of transition bands with (110) orientations embedded in (111) grains in the as-cold rolled state. The (110)[001] grains nucleate first from these (110) subgrains and achieve their size advantage during primary recrystallization. A special boundary relationship of $+35°$ about a common ⟨110⟩ axis exists between (110)[001] secondary nuclei and their surrounding (111)[112] matrix, providing the (110)[001] nuclei with the highest mobilities relative to the other possible nuclei in the primary recrystallized matrix.

Thus, the sharp (111)[112] primary texture enhances (110)[001] selectivity during secondary recrystallization and develops a sharper Goss texture. High-permeability silicon steels have a lower density of (110)[001] nuclei than conventional grain-oriented materials and, therefore, the high-permeability materials have a larger grain size after Goss-texture development (Fig. 4).

Solute segregation to grain boundaries can suppress grain-boundary migration in high-purity metals. Grenoble and Fiedler (1969) demonstrated that secondary recrystallization can occur by replacing the small dispersed inclusions in 3%Si–Fe with solute sulfur and nitrogen segregated on the grain boundaries.

Another process (Kawasaki) to produce high-permeability silicon steels makes use of a two-stage cold reduction with 65% final reduction in which the (110)[001] component in the primary texture is maximized. Sulfur, selenium and antimony all play a role in strengthening the inhibition system. A highly oriented Goss texture develops during a long soaking cycle in the final anneal at temperatures between 820 °C and 900 °C. The steel is subsequently purified at 1200 °C. This unique anneal indicates that the degree of (110)[001] texture development can be enhanced by decreasing the secondary recrystallization temperature.

3. Properties

Table 1 is a listing of the properties of standard grades of silicon steels as specified by the American Society for Testing and Materials (ASTM) and the American Iron and Steel Institute (AISI). Nonoriented products are tested on samples, half of which are sheared in the rolling direction and the remaining half are sheared across the rolling direction. Conventional grain-oriented and high-permeability materials are all tested in the rolling direction. Methods of effecting domain refinement using various scribing techniques can reduce the losses of the high-permeability steels shown in Table 1 by 6–10%.

See also: Iron and Low-Carbon Steels for Magnets; Magnetic Materials: An Overview

Bibliography

Beckley P 1987 Domain control by spark ablation. Proceedings of Conference on Hard and Soft Magnetic Materials. *Am. Soc. Met. Mater. Week.* October: 127–30

Benford J 1987 Low core losses via heat proof indigenous domain refinement. Proceedings of Conference on Hard and Soft Magnetic Materials. *Am. Soc. Met. Mater. Week.* October: 115–17

Fukuda B, Sato K, Sugiyama T, Honda A, Ito Y 1987 Loss reduction in grain oriented Si-steel sheets by new domain refining technique. Proceedings of Conference on Hard and Soft Magnetic Materials. *Am. Soc. Met. Mater. Week.* October: 107–9

Grenoble H E, Fiedler H C 1969 Secondary recrystallization of silicon-iron by solute-induced boundary restraint. *J. Appl. Phys.* 40: 1575–6

May J E, Turnbull D 1958 Secondary recrystallization in silicon iron. *Trans. Am. Inst. Min. Metall. Pet. Eng.* 212: 769–81
Nozawa T, Matsuo Y, Tanaka O, Kobayashi H, Zuroki K 1987 Domain refining techniques in grain-oriented silicon steel. Proceedings of Conference on Hard and Soft Magnetic Materials. *Am. Soc. Met. Mater. Week*. October: 91–106
Shilling J W, Houze G L 1974 Magnetic properties and domain structure in grain-oriented 3% Si–Fe. *IEEE Trans. Magn.* 10: 195–223
Shilling J W, Morris W G, Osborn M L, Rao P 1978 Orientation dependence of domain wall spacing and losses in 3% Si–Fe single crystals. *IEEE Trans. Magn.* 14: 104–11
Taguchi S, Sakakura A 1966 The effects of AlN on secondary recrystallization textures in cold rolled and annealed (001)[100] single crystals of 3% silicon iron. *Acta Metall.* 14: 405–23
Washko S D, Choby E G 1979 Evidence for the effectiveness of stress coatings in improving the magnetic properties of high permeability 3% Si–Fe. *IEEE Trans. Magn.* 15: 1586–91
Washko S D, Miller R F 1980 Sheet thickness effects on energy losses in 3% silicon–iron. *J. Magn. & Magn. Mater.* 19: 361–4
Washko S D, Morris W G 1980 Effects of coatings on domain wall spacing and core loss in 3% Si–Fe single crystals. *J. Magn. & Magn. Mater.* 19: 349–52

J. G. Benford
[Allegheny Ludlum, Brackenridge, Pennsylvania, USA]

S. D. Washko
[Allegheny Ludlum, Leechburg, Pennsylvania, USA]

SQUID Design, Fabrication and Application

Magnetic sensors are widely used in such diverse applications as medical research, geophysics, nuclear magnetic resonance and nondestructive evaluation of industrial products. For the detection of weak magnetic fields and field gradients, the superconducting quantum interference device (SQUID), coupled to an appropriate input flux transformer, offers unsurpassed sensitivity. A SQUID is a sensor of magnetic flux, providing an output voltage that is periodic in units of the flux quantum ($\Phi_0 \simeq 2 \times 10^{-15}$ Wb). Connected to appropriate feedback circuitry, it becomes a linear sensor of magnetic flux.

With suitable input transducers, SQUIDs can be extended to measure other quantities such as current, voltage, displacement and temperature. SQUIDs made from conventional superconductors such as niobium operating at liquid helium temperatures have been commercially available for many years and are by far the most widely used superconducting devices. Their range of potential applications is likely to become still greater with the advent of high-T_c materials such as $YBa_2Cu_3O_{7-\delta}$ (YBCO). These materials are already being used in prototype SQUID devices and good progress has been made towards the realization of practical instruments working at liquid nitrogen temperatures.

1 Basic Science

1.1 Flux Quantization in Rings Containing Weak Links

The operating principles of SQUIDs combine the properties of Josephson weak links or junctions (see *Josephson Effects in Weak Links*) with the phenomenon of quantization (in units of the flux quantum, $\Phi_0 = h/2e$, where h is Planck's constant and e the electronic charge (see *Magnetic Properties of Superconductors*)). This article discusses two types of device: the radio-frequency (rf) SQUID, in which a superconducting ring contains one weak link or tunnel junction and is inductively coupled to the coil of a resonant LC circuit, and the dc SQUID, in which the ring contains two Josephson elements and is operated with a constant bias current.

In both types of SQUID the application of magnetic field causes the circulation of supercurrents in the ring that self-adjust to maintain fluxoid quantization. However, the circulating supercurrent in either direction cannot exceed the critical current I_c of the Josephson element and, in a changing applied field, this results in a flux quantum entering or leaving the ring each time this limit is exceeded. Thus the ring is effectively a magnetic flux meter, whose properties and output are periodic in Φ_0. In a practical device one is able to detect changes in applied flux Φ (applied magnetic field multiplied by SQUID loop area) far smaller than Φ_0, limited by intrinsic or amplifier noise (see Sect. 1.5).

Designing SQUIDS to achieve such sensitivities requires careful attention to obtaining correct values for certain ring and junction parameters. For the SQUID ring, the most important quantity is its self-inductance L_s. For the Josephson elements, which are commonly represented by the resistively and capacitively shunted junction (RCSJ) model (see *Josephson Effects in Weak Links*), the significant parameters are I_c, the capacitance C and the shunt resistance R. Useful design parameters for SQUIDs are

$$\beta_c = 2\pi I_c R^2 C/\Phi_0 \quad (1)$$

and

$$\left.\begin{aligned} \beta_L &= 2L_s I_c/\Phi_0 \text{ (dc SQUID)} \\ \beta_L &= 2\pi L_s I_c/\Phi_0 \text{ (rf SQUID)} \end{aligned}\right\} \quad (2)$$

Although SQUIDS can be operated with a large range of values, optimum sensitivity for instrument design purposes is usually obtained for $\beta_c \leqslant 1$ (typically about 0.3) and $\beta_L \simeq 1$. In practice, for 4.2 K operation, values of L_s in the range 0.1–1 nH are used; corresponding values of I_c are 1–10 μA. One uses the largest possible value of R by making the junction area and hence C as small as possible. For oxide tunnel junctions, areas of 1–10 μm² are typical, with corresponding values of C between 0.05 pF and 0.5 pF.

1.2 RF and DC SQUIDs

If a superconducting ring containing one weak link as described in Sect. 1.1 is subjected to an applied magnetic flux Φ, an appropriate number of flux quanta enter the ring, leaving a small circulating current $I_s < I_c$. I_s varies periodically with Φ, the period being Φ_0.

The ring is operated as an rf SQUID by coupling it to the coil (self-inductance L_T) of an LC rf tuned circuit excited by an oscillator (see Fig. 1a). This induces a circulating rf supercurrent and it is found that the dissipation of rf energy depends on the average value of circulating current I_s. If Φ is steadily increased or decreased, the dissipation will pass through a series of maxima and minima and the tank circuit voltage amplitude will have a period Φ_0, corresponding to one flux quantum entering or leaving the SQUID loop. The rf signal across the tank circuit is amplified and rectified to produce a voltage proportional to the rf amplitude. This voltage is periodic in the flux applied to the SQUID as indicated in Fig. 1b.

In a dc SQUID (see Fig. 2a), the superconducting ring contains two weak links that are directly connected to an external static current source I. The division of I between the two junctions depends on the applied flux Φ and, in turn, the IV characteristic varies with a period Φ_0 between the two extrema shown in Fig. 2b. In operation the SQUID is biased with a current I_B, large enough to maintain a nonzero voltage for any applied flux; once again the voltage across the SQUID is periodic in Φ_0 (Fig. 2c).

The SQUID characteristics of Fig. 1b and 2c can be used directly as small-signal amplifiers to sense very small fractions of Φ_0 or to count flux quanta digitally. However, the periodic flux–voltage response is more

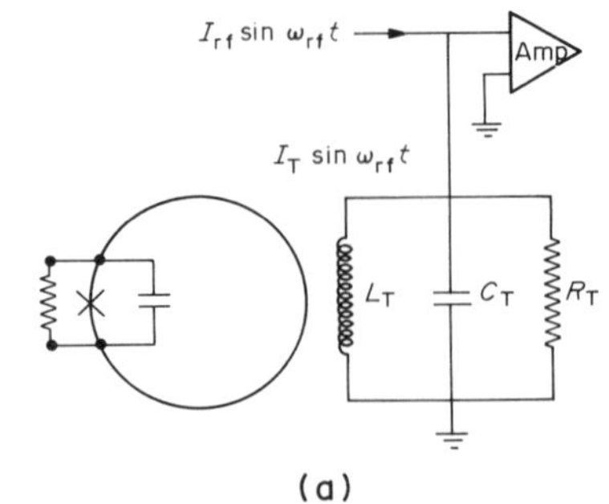

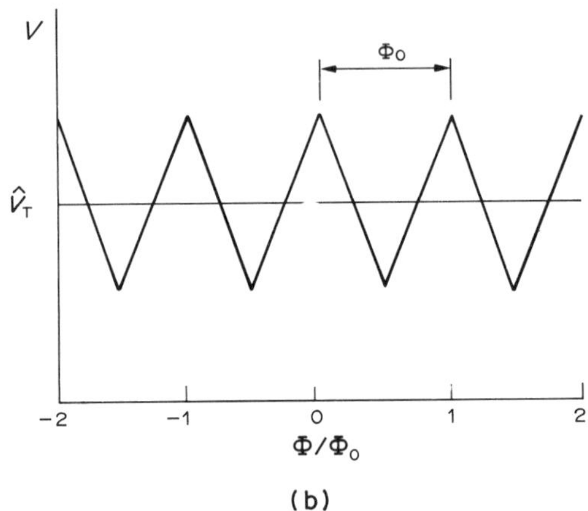

Figure 1
rf SQUID: (a) inductively coupled to a resonant tank circuit and (b) flux–voltage characteristic after rectification of rf signal ($\beta_L > 1$)

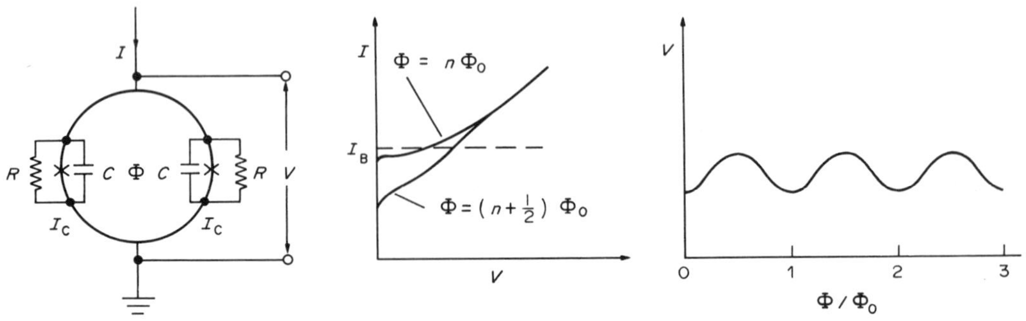

Figure 2
dc SQUID: (a) configuration, (b) IV characteristics and (c) flux–voltage characteristic at constant bias current I_B

usually linearized by incorporating the SQUID in a flux-locked loop. Here, flux modulation of amplitude approximately $\frac{1}{2}\Phi_0$ at a frequency of typically 100 kHz is applied to the SQUID and the output voltage is synchronously detected by a lock-in amplifier, whose output current is fed back through the modulation coil to null changes $\delta\Phi$ in the net low-frequency applied flux. Any change in Φ is thus cancelled by flux generated by the feedback current, which also generates the output voltage of the SQUID instrument: the SQUID therefore remains locked at a flux bias corresponding to one of the maxima or minima in Figs. 1b and 2c.

1.3 Fabrication Techniques

Early rf SQUIDs were made by machining niobium to produce rings with an internal area of about 1 mm^2. The weak link was established by careful adjustment of a finely pointed screw. The most commonly used commercially available rf SQUID consists of a toroidal cavity machined in niobium and uses a thin-film junction. The tank coil and flux input coil (see Sect. 2.1) are located within the cavity, which gives excellent flux coupling and magnetic screening from the external environment.

Most dc SQUIDs, on the other hand, are made from thin films. The most popular design incorporates two Nb–Al_2O_3–Nb tunnel junctions (see *Superconducting Device Fabrication*), typically 5 µm^2 in area, on either side of a narrow slit in a square niobium "washer" about 1 mm across (see Fig. 3). The washer and input leads are niobium, as are the connecting leads. Resistor films, designed to give the optimum value of R (Eqn. (1)), shunt the junctions. External fields are not applied directly to the SQUID: instead they are coupled into the SQUID loop by a current flowing in a superconducting spiral of typically ten to 50 turns deposited on top of the washer and separated from it by an insulating film, usually consisting of about 200 nm of SiO. The inductance of the washer is given approximately by $1.25\,\mu_0 d$ where d is the length of its inner edge; typical values are 0.05–0.5 nH. The inductance of the spiral input coil (n turns) is approximately n^2 times greater and is designed to match that of a pickup magnetometer or gradiometer (Sects. 2.1–2.2). The coupling coefficient between the washer and the spiral coil is typically 0.75–0.9.

1.4 Signal Amplitude, Dynamic Range and Frequency Limits

The performance of a SQUID is usually evaluated by first measuring its transfer function $V(\Phi) = \partial V/\partial\Phi$, in volts per flux quantum (i.e., twice the peak-to-peak voltage of the triangles in Fig. 1b). For the rf SQUID, the value of the transfer function is maximized when

$$k^2 Q_T \simeq 1 \quad (3)$$

where Q_T is the quality factor of the tank circuit and k is the coupling coefficient between the SQUID and the

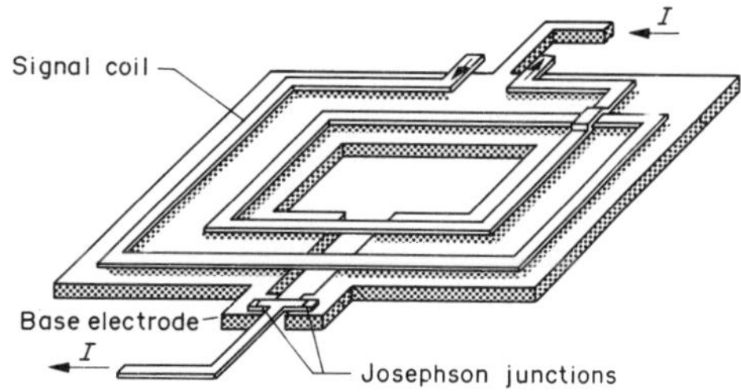

Figure 3
Schematic of thin-film "washer" dc SQUID

tank circuit coil. In such a case

$$V(\Phi) = 2\pi f_p (Q_T L_T / L_s)^{1/2} \quad (4)$$

where f_p is the rf pump frequency.

It can be seen that sensitivity can be increased by reducing L_s and increasing the operating frequency. In practice, one might operate with $f_p \simeq 20$ MHz, $Q_T \simeq 100$ and $k \simeq 0.1$.

For the dc SQUID, the transfer function is given approximately by

$$V(\Phi) \simeq R/L_s \quad (5)$$

where R is the shunt resistance defined in Eqn. (1). For a typical SQUID $V(\Phi) \simeq 10$–$100\ \mu V/\Phi_0$.

The maximum change in flux that a SQUID can measure depends on its mode of operation. In a flux-locked mode it is set by the maximum feedback flux that the electronics can generate and is typically 10^6–$10^7 \Phi_0\,Hz^{-1/2}$. However, some circuits are designed to reset whenever the applied flux changes by Φ_0, thus greatly extending the dynamic range provided there is circuitry to count the number of resets.

The frequency response of SQUIDs in a flux-locked loop extends, in principle, to one-half the modulation frequency (e.g., 50 kHz for a 100 kHz modulation). As a small-signal open-loop amplifier, the dc SQUID can be operated at frequencies from 0 Hz up to 200 MHz or higher (see Sect. 2.4).

1.5 SQUID Noise

The limiting flux sensitivity of a SQUID depends on various noise processes and, in general, two frequency ranges can be identified: an upper white noise region, in which the noise power per unit bandwith is independent of frequency, and a lower one (typically below a "corner" frequency in the range 0.2–10 Hz), in which noise increases with an approximately $1/f$ power spectrum.

At frequencies above the $1/f$ corner, dc SQUIDs are limited by white noise originating as Johnson noise in the junction shunt resistors. The noise in rf SQUIDs, which arises from the statistical fluctuations in the value of flux at which the devices change flux states, theoretically results in similar performance under optimum operating conditions, but in practice room-temperature rf amplifiers and components set a poorer limit than for dc SQUIDs.

The spectral noise density $S_V(f)$ is recorded in a well-shielded environment to determine the intrinsic noise of the SQUID and its associated electronics. From this, the spectral density of the flux noise can be calculated:

$$S_\Phi(f) = S_V(f)/V^2(\Phi) \quad (6)$$

which in the white noise region can be reduced by careful design to $10^{-12}\Phi_0^2\,\mathrm{Hz}^{-1}$ for dc SQUIDs and to about $10^{-9}\Phi_0^2\,\mathrm{Hz}^{-1}$ for rf SQUIDs. An alternative figure of merit is the noise energy per unit bandwidth

$$\varepsilon(f) = S_\Phi(f)/2L_s \quad (7)$$

which, by taking into account the SQUID inductance, gives a better criterion for comparing SQUIDs of different geometries. White noise energies for 4.2 K operation are typically $10^{-29}\,\mathrm{J\,Hz}^{-1}$ (rf SQUIDs) and $10^{-32}\,\mathrm{J\,Hz}^{-1}$ (dc SQUIDs), although the latter have achieved energy resolutions as small as $2\times 10^{-34}\,\mathrm{J\,Hz}^{-1}$ (i.e., $2\hbar$ or close to the limit expected from quantum fluctuations).

The low-frequency noise region is very important because some of the most demanding SQUID applications (e.g., geophysics and biomagnetism) require the detection of signals at frequencies of 0.5 Hz or lower, where $1/f$ noise dominates. It has been shown that this low-frequency noise arises from fluctuations in the critical currents of the weak links and from random movements of trapped flux in the superconducting material of which the SQUID is made. Noise arising from the former mechanism can be reduced by suitable modulation of the bias current and flux, but the latter can only be reduced (and then only to a limited extent) by careful choice of materials.

1.6 High-T_c SQUIDs

The advent of high-transition-temperature (high-T_c) superconductivity provided new impetus to research on SQUIDs and indeed SQUIDs were the first high-T_c devices. Some early devices were made from bulk material, but most of the subsequent effort has been focused on thin films. The major challenge has been to find a manufacturable process for the Josephson elements, especially because it has not proved possible to fabricate Josephson junctions with the near-ideal hysteretic characteristics shown by unshunted niobium–aluminum oxide structures (see *Superconducting Device Fabrication*). In the first SQUIDs, junctions were formed by patterning microbridges in polycrystalline high-T_c films, so that each bridge was crossed by one or more boundaries between adjacent grains; each grain boundary acts as a Josephson weak link (see *Thin Films, Multilayers and Devices, Superconducting*).

More recently, SQUIDs have been made with a wide variety of alternative junction structures, most of which show acceptable self-shunted RCSJ behavior. Some involve grain boundaries deliberately engineered by depositing a narrow film stripe over a 200 nm step etched into the substrate. Others use an epitaxial film of YBCO on a bicrystal in which the two parts of the substrate have been fused together with a known misorientation or on a substrate partially covered with a "seed" layer which causes a 45° in-plane rotation of the crystalline axes of the superconductor. These techniques produce a grain boundary at the edge of the seed layer. Other junctions have been made with normal barriers, such as silver or PrBaCuO.

Figure 4a shows an example of a dc SQUID which can operate satisfactorily at the boiling point of liquid nitrogen (77 K) or at even higher temperatures. The white noise energy $\varepsilon(f)$ is often in good agreement with the prediction of Eqns. (6) and (7) and values below $10^{-30}\,\mathrm{J\,Hz}^{-1}$ have been reported. On the other hand, at frequencies below a few hundred hertz, high-T_c SQUIDs exhibit large levels of $1/f$ noise. Much effort is being devoted to understanding the origin of this noise and to reducing it; significant reductions will be necessary if high-T_c SQUIDs are to have an impact on the more demanding applications such as magnetoencephalography.

In the case of high-T_c SQUIDs, fabrication of the superconducting multilayer structures needed to form the SQUID and the multiturn input coil poses new challenges. The multilayers involve the deposition of two epitaxial superconducting films with an intervening insulating film, with an option of making a connection between the superconductors via a window in the insulator. Multilayers of this kind have been fabricated successfully using a pulsed laser to deposit YBCO films and, for example, $SrTiO_3$ as the insulator (see *Thin Films, Multilayers and Devices, Superconducting*). The films are grown epitaxially on substrates maintained at 750 °C to 800 °C in a low pressure of oxygen. Figure 4b shows a five-turn coil that is pressed against the SQUID of Fig. 4a in a flip-chip arrangement. Although satisfactory performance has been achieved with such devices, for practical applications it is highly desirable to integrate the SQUID and flux transformer on a single chip as for low-T_c devices. Such integration is just becoming possible: an example is shown in Fig. 4c.

2. SQUID-Based Instruments

Both the dc and rf SQUIDs sense tiny changes in magnetic flux that in practical measurements are almost invariably generated by currents induced in an input coil tightly coupled to the SQUID. This coil can in turn be connected to a variety of different input circuits that determine the physical parameter

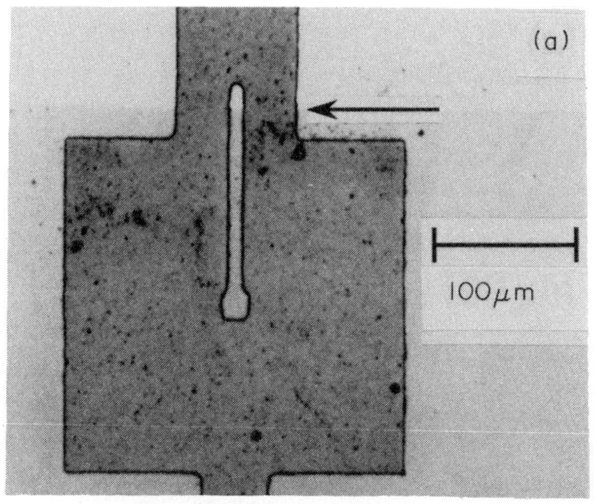

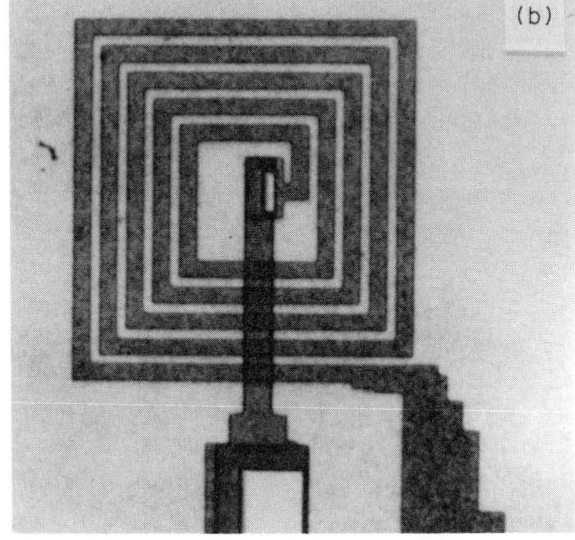

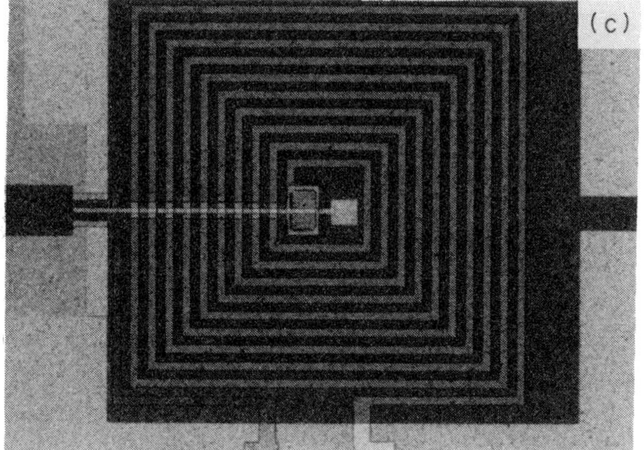

Figure 4
(a) Photograph of a high-T_c dc SQUID made with two biepitaxial junctions along the line indicated by the arrow: the region above the arrow was covered with a "seed" layer, the edge of which produced a 45° grain boundary in the YBCO film deposited subsequently (courtesy of Conductus Inc./Lawrence Berkeley Laboratory). (b) Five-turn YBCO flux transformer: the lower film of YBCO (running from the middle of the picture to the bottom) is insulated with a $SrTiO_3$ film from the upper film that is patterned into a spiral. A window in the $SrTiO_3$ is visible at each end of the lower film (courtesy of Lawrence Berkeley Laboratory). (c) A monolithic SQUID magnetometer involving biepitaxial Josephson junctions fabricated from multiple layers of the high-T_c superconductor YBCO and insulating oxides: in one design, 15 epitaxial layers are used (courtesy of L. P. Lee, K. Char, M. S. Colclough and G. Zaharchuk, Conductus Inc.)

measured by the SQUID such as magnetic field, magnetic susceptibility, displacement or nuclear spin precession.

2.1 The Flux Transformer: Magnetometers

One of the most important input circuits is the flux transformer, two types of which are illustrated in Figs. 5a, 5b. The magnetometer shown in Fig. 5a consists of a single-turn pickup loop connected in series with the multiturn input coil; the pickup coil may have more than one turn. For thin-film SQUIDs, the transformer is usually wound from niobium wire. The entire circuit is superconducting, so that the application of a magnetic field δB to the pickup loop requires, by flux quantization, the flow of a supercurrent to maintain the total magnetic flux in the transformer at its original value. As a result, a magnetic flux proportional to δB is applied to the flux-locked SQUID, and the feedback circuit, in turn, produces an output voltage proportional to δB.

The flux transformer can produce large enhancements of the magnetic field sensitivity of the bare

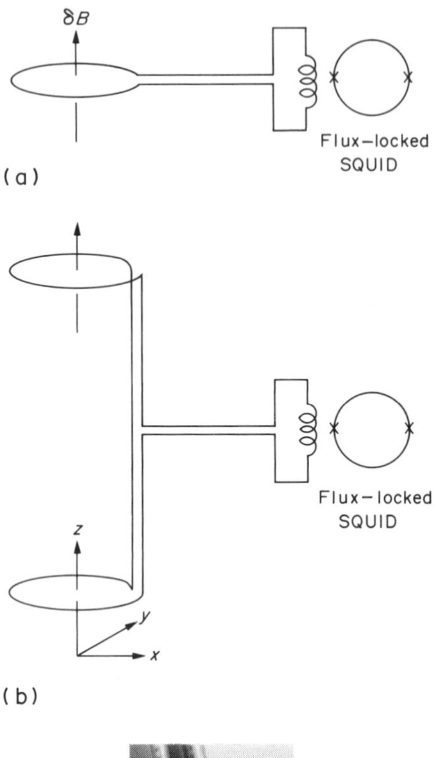

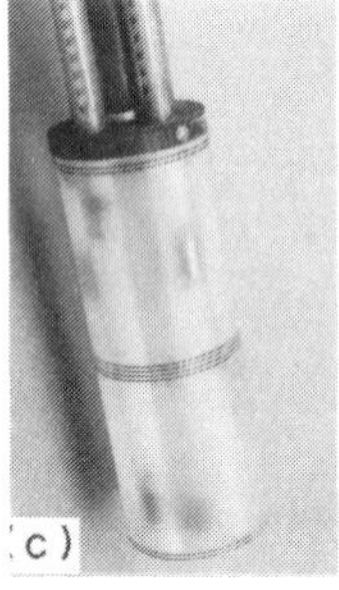

Figure 5
(a) Flux transformer configured as a magnetometer,
(b) flux transformer configured as a gradiometer and
(c) second-derivative gradiometer showing two turns of niobium wire at each end of the quartz cylinder and four turns in the middle (courtesy of Biomagnetic Technologies Inc.)

SQUID. Optimization involves making the inductances of the single-turn pickup loop and multiturn input coil approximately equal; thus, the area of the former A_p is much greater than that of the latter, which is comparable with the effective pickup area of the SQUID A_s. The gain in magnetic field sensitivity scales approximately as A_p/A_s; a resolution of 1–10 fT Hz$^{-1/2}$ is typical in the frequency range over which the SQUID noise is white.

Prototype high-T_c flux transformers have been made by patterning the second YBCO film in Fig. 4b into a single loop, typically 10 mm in size, that connects to the two ends of the spiral coil. When the spiral coil is coupled to the SQUID, the flux transformer can enhance the magnetic field sensitivity by as much as two orders of magnitude. The best performance of these early magnetometers is about 1 pT Hz$^{-1/2}$ at 1 Hz, improving to 50 fT Hz$^{-1/2}$ in the white noise region above 1 kHz. This performance, with the SQUID immersed in liquid nitrogen, has proved good enough to measure a magnetocardiogram. It is anticipated that significant improvements in high-T_c SQUID performance at low frequencies will be achieved in the near future as $1/f$ noise is reduced.

2.2 The Gradiometer

A widely used variant of the flux transformer is the first-derivative gradiometer shown in Fig. 5b. The two pickup loops are closely matched in area and orientation, often with the aid of superconducting trimming pads, and wound in opposite senses. Thus, application of a uniform magnetic field in any direction ideally produces no net change in the flux in the transformer and, hence, no output from the SQUID (it is assumed that the SQUID and input coil are surrounded by a superconducting shield that screens out any changes in external field). On the other hand, applying an axial magnetic field gradient of the form $\partial B_z/\partial z$ produces a proportional response from the flux-locked SQUID. With a baseline of the order of 0.1 m, a resolution of 10–100 fT m^{-1} Hz$^{-1/2}$ in the white noise frequency range is typical. The principle is readily extendable to other gradients. By winding the transformer to form two first-derivative gradiometers back to back, it is possible to make a second-order gradiometer which is sensitive to $\partial^2 B_z/\partial^2 z$. Off-diagonal gradiometers are particularly popular with thin-film SQUIDs because the SQUID and pickup loops can be integrated on a single chip to measure, for example, $\partial B_z/\partial x$ or $\partial^2 B_z/\partial x \partial y$. Gradiometers are of particular importance in biomagnetic measurements and systems containing as many as 37 channels are available commercially. Figure 5c shows a second-derivative gradiometer.

Integrated structures have grown more important as SQUID neuromagnetism has developed. Many parallel sensor channels (up to 100 have been suggested) are needed for rapid mapping over a surface such as a skull and it has been possible to produce many dc SQUIDs, and their gradiometer coils, on a single substrate.

Multichannel instruments will involve an excessive number of leads unless multiplexing of channels and incorporation of feedback electronics on the chip can be achieved. This has been addressed, for example, with a thin-film Josephson logic circuit, deposited on the same chip as the SQUID, that operates as an A–D SQUID signal converter, a digital store and a D–A feedback generator (see *Superconducting Devices and*

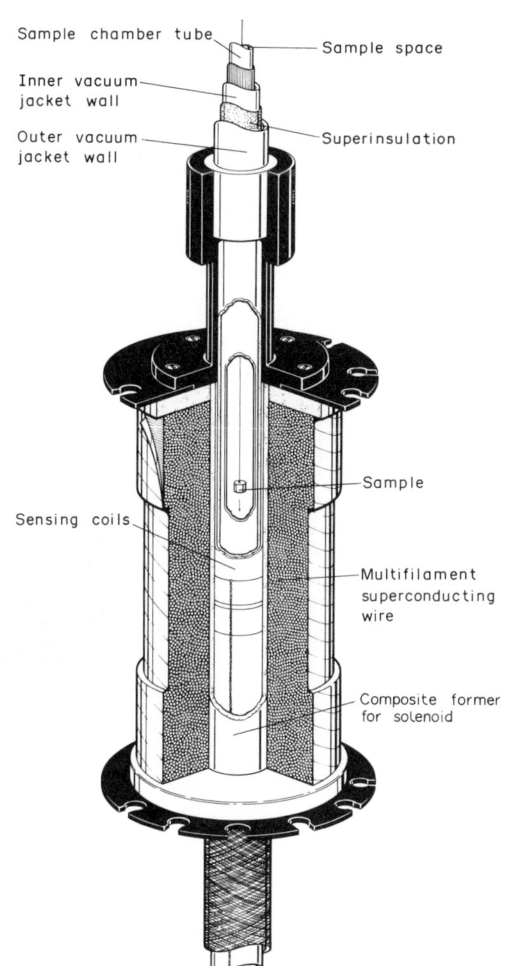

Figure 6
Cutaway drawing of a susceptometor showing sample above the uppermost coil of a second-derivative gradiometer (courtesy of Quantum Design Inc.)

Electronics). Performance is limited, but is adequate for magnetocardiography.

2.3 The Susceptometer

Another straighforward extension of the first-derivative gradiometer is to the measurement of magnetic susceptibility. A constant magnetic field is maintained along the z axis of the flux transformer in Fig. 5b and the sample under investigation is lowered through the upper pickup loop. The resultant imbalance of flux in the two loops is proportional to the susceptibility χ. Figure 6 shows a cutaway drawing of the cryogenic part of a commercially available system which provides a good example of the degree of sophistication available in SQUID-based instruments. The susceptometer is fully automated and enables the user to detect magnetic moments as small as $10^{-11}\,\mathrm{A\,m^2}$ (10^8 emu) in very stable fields as high as 5 T over the temperature range from 2 K to 400 K.

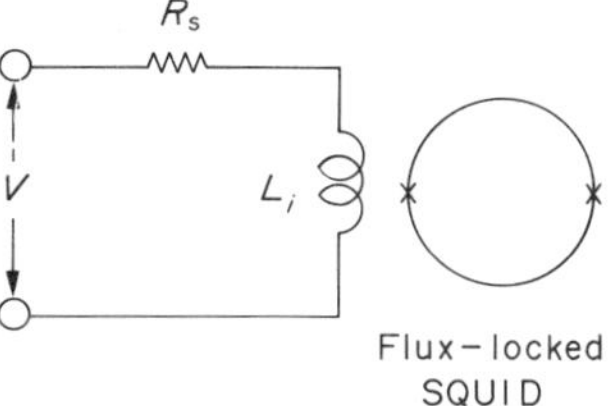

Figure 7
Configuration of voltmeter: the voltage V generates a current V/R_s that couples magnetic flux into the dc SQUID

2.4 Voltmeters and Amplifiers

Figure 7 shows a SQUID as a voltmeter. The quasistatic voltage V to be measured is applied to the input terminals and the resultant current in the input circuit produces an output from the flux-locked SQUID. In fact, it is often preferable to apply the feedback current directly to the series resistance R_s rather than as a flux to the SQUID, so that the voltmeter is null balancing; that is, it draws no current. Because the SQUID is a current-sensing device, its use is advantageous only for relatively low impedance circuits, say, below a few tens of ohms. The circuit is generally cooled to the same temperature as the SQUID and, in properly designed circuits, the resolution is generally limited by Nyquist noise generated in the resistance of the input circuit (e.g., $\sim 10^{-14}\,\mathrm{V\,Hz^{-1/2}}$ for $1\,\mu\Omega$ at 4.2 K). Examples of the use of such voltmeters include the measurement of flux creep in superconductors and of thermoelectric voltages.

The voltmeter circuit of Fig. 7 may readily be engineered to work up to radio frequencies of 100 MHz or higher. It is sometimes convenient to add a capacitance in series with the input coil to produce a resonance at the signal frequency. Operated at liquid helium temperatures, such amplifiers have achieved gains as high as 18 dB and noise temperatures as low as 1 K at a signal frequency of 100 MHz. A notable application of these amplifiers is to magnetic resonance.

2.5 Displacement Measurement

In Fig. 8, a superconducting plate is positioned close to a superconducting coil which is coupled to the SQUID input coil and supports a persistent current i_s. The Meissner effect, by excluding flux from the plate, distorts the self-field around the coil. During any motion of the plate, say towards the coil, the distortion changes and causes lines of force to cut the turns, creating an instantaneous voltage which changes i_s. A movement δx causes a current change δi_s, proportional to $i_s\delta x$, which can be detected by the SQUID.

A practical uniaxial sensor cell, for use in measurements of weak gravitational effects, consists of a superconducting niobium chamber which supports the superconducting plate (or proof mass) on a phosphor bronze spring of high compliance in the sensitive direction and high rigidity perpendicular to this. Two sensing coils, connected to each other and coupled to the SQUID, are placed one on either side of the plate. The persistent current i_s now circulates in the connected sensing coils. A proof mass displacement δx diverts a small current δi_s, proportional to $i_s\delta x$, into the SQUID input coil. With $i_s \simeq 1$ A, one can realize a transfer coefficient to the SQUID $\delta i_s/\delta x \simeq$ 1 A cm^{-1}, corresponding to a noise-limited sensitivity $\langle(\delta x_N)^2\rangle^{1/2} \simeq 10^{-16}\,\mathrm{mHz}^{-1/2}$ for a $2\times10^{-5}\Phi_0\,\mathrm{Hz}^{-1/2}$ SQUID. Two such cells, connected for gradiometry, are shown in Fig. 9.

Still higher sensitivity is possible, in a narrow bandwidth, if the natural frequency of the proof mass and its spring is the same as that of the body on which the chamber is mounted. The ratio of the resonant amplitude of the proof mass to that of the chamber (mass M) is then $(M/m)^{1/2}$. In the gravity wave detector (Sect. 3.3) the chamber is rigidly attached to a 5000 kg bar which oscillates freely at 1 kHz, while the proof mass (a niobium diaphragm) weighs only 1 g: the amplification factor of 2200 makes possible detection of bar amplitudes $\simeq 10^{-19}\,\mathrm{mHz}^{-1/2}$. Frequency matching is achieved by adjusting i_T: this alters the effective compliance of the spring (in this case the diaphragm itself) by adding a component associated with the compression of the stored magnetic field as the diaphragm moves.

For a quantum-limited SQUID the limit would become $10^{-19}\,\mathrm{mHz}^{-1/2}$, but as will be seen still higher

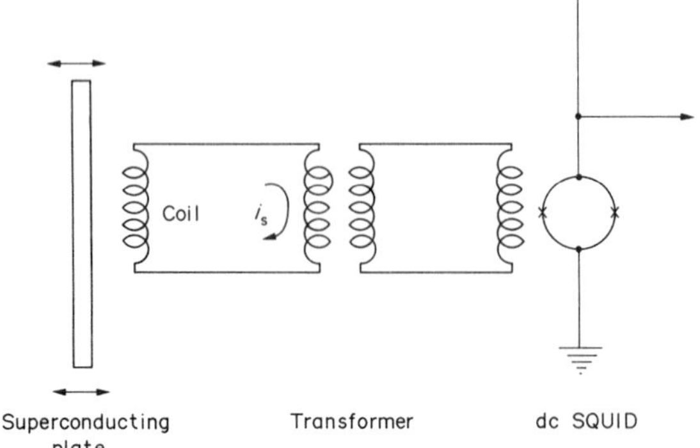

Figure 8
SQUID displacement sensor: movement of the plate induces a change in the persistent circulating current i_s; the change is coupled to the SQUID

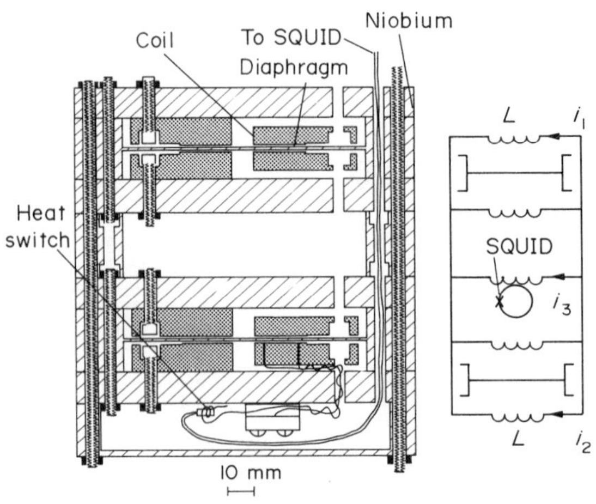

Figure 9
Double diaphragm gravity gradiometer: movement of a niobium diaphragm steers some of the current ciculating in the coils around it into the SQUID input transformer; the net input is proportional to the relative displacement of the two diaphragms

sensitivities may prove necessary. "Quantum nondemolition" measurement procedures have been proposed to achieve this further improvement in sensitivity. In such procedures position x is, in principle, measured with arbitrary precision, while the conjugate variable (in this case momentum) is, in effect, not measured at all.

3. Applications

One of the ongoing fascinations of SQUIDs is the amazingly broad range of their applications, of which a brief overview will be given. Biomagnetic measurements are described in detail in the article *Biomagnetic Measurements*.

3.1 Geophysics

Probably the single most important geophysical application is rock magnetometry—the identification of magnetic phases in rock samples. A commercially available magnetometer enables samples to be inserted into a horizontal access tube and the axial and transverse magnetization to be measured in the presence of an axial magnetic field. Alternatively, the ambient magnetic field can be reduced to a low level to enable the measurement of the remanent magnetism of samples. For example, it is possible to determine the Curie temperature of ferromagnets or the Néel temperature of antiferromagnets from the temperature dependence of the susceptibility. The concentration of a known ferromagnetic component can be estimated from the saturation susceptibility. Paramagnetic materials can be investigated in the presence of ferromagnetic phases by separating the saturating and nonsaturating magnetization in a high magnetic field and from the $1/T$ dependence of the susceptibility (T is the absolute temperature). Other interesting properties include anisotropy, which can be determined by rotating the sample, and the relaxation of magnetic remanence when the sample is transferred from high field to low field. Remanent magnetization is also of interest in paleomagnetism, which involves the study of the variation of the earth's magnetic field with time. Thermoremanent magnetization occurs, for example, in the baked clay of hearths, ovens and kilns and usually arises from the presence of magnetite (Fe_3O_4) or hematite (Fe_2O_3). Above the blocking temperature T_B, the magnetic domains orient in the earth's magnetic field; this orientation is frozen in when the material is last cooled below T_B. Thus, if one can date the structure, for example by reference to archeological chronology or by radiocarbon dating, one can obtain invaluable geophysical information. A similar method can be applied to core samples of lake sediments, which are deposited relatively rapidly and have provided information of the declination and inclination of the earth's magnetic fields over the last 11 or 12 millennia. Vastly longer timescales can be studied in deep-ocean sediments, which have provided important evidence for geomagnetic reversals.

A quite different geophysical application of SQUIDs is magnetotellurics, an electromagnetic survey technique which determines the resistivity ρ of the earth's surface as a function of depth. In this technique, one makes use of low-frequency (10^{-4}–10^2 Hz) electromagnetic energy generated in the ionosphere and magnetosphere or in thunderstorms. The incident fields are reflected by the earth, but components of the electric and magnetic fields $\boldsymbol{E}(\omega)$ and $\boldsymbol{B}(\omega)$ decay into the ground with a characteristic length δ that scales as $(\rho/\omega)^{1/2}$ (ω is the angular frequency). At 1 Hz, δ is typically between 1 km and 5 km. One measures simultaneously the horizontal components of $\boldsymbol{E}(\omega)$ and $\boldsymbol{B}(\omega)$ and uses them to estimate the impedance tensor $\mathbf{Z}(\omega)$ from the relation $\boldsymbol{E}(\omega) = \mathbf{Z}(\omega)\boldsymbol{B}(\omega)$. After rotating the tensor to align one of the axes along the direction of highest translational symmetry, it is possible to deduce estimates of the apparent resistivity of the ground as a function of frequency in the two orthogonal horizontal directions. Finally, one constructs a model describing the two resistivities as a function of depth. Although an excellent technique in principle, for many years it was plagued by problems of spurious noise sources that rendered the results unreliable. These problems were largely overcome by the introduction of three-axis SQUID magnetometers and, in particular, of the remote reference technique. Signals from the remote reference magnetometer, which is placed some distance away from the base site (typically 5–10 km), are cross-correlated with the local measurements of $\boldsymbol{E}(\omega)$ and $\boldsymbol{B}(\omega)$ to eliminate local noise sources, thereby retaining only signals that are coherent over long distances. In a complementary technique, known as controlled-source electromagnetic sounding, one applies large pulses of magnetic field to the ground and detects the response some distance away using SQUID magnetometers. Again, a remote reference magnetometer is used, in this case to cancel the naturally occurring magnetic field fluctuations. The results of these measurements are used to estimate the resistivity of the ground. Both the passive and active techniques are used in studies of the earth's crust and in surveying for natural resources.

SQUIDs have also been used in oceanographic measurements. One application has been the detection of the magnetic fields generated by internal ocean waves by means of an axial gradiometer suspended above the surface of the ocean. A three-axis SQUID magnetometer has also been operated successfully on the ocean floor at a depth of about 100 m.

The use of SQUIDs in many of the geophysical applications has been restricted by the difficulty of obtaining and transferring liquid helium in remote areas. The advent of high-T_c SQUID magnetometers, which are expected to become available in the near future, may have a considerable impact on this field because of the very long hold times (> 1 year) available with relatively small dewars of liquid nitrogen.

3.2 Magnetic Resonance

There have been numerous experiments using SQUIDs to detect nuclear magnetic resonance. In early work, the SQUID was used to detect the change in the static magnetic susceptibility when a signal was applied to the sample at the nuclear magnetic resonance (NMR) frequency. The development of SQUID-based tuned rf amplifiers, however, enabled the detection of NMR or nuclear quadrupole resonance (NQR) signals directly with a better signal-to-noise ratio than that available with conventional techniques. For example, a SQUID amplifier has been used to detect spin noise—the spontaneous emission of photons from an ensemble of ^{35}Cl nuclei, in which the populations of the two levels had been initially equalized—even though the lifetime for this process is about 10^6 centuries. A SQUID amplifier with an untuned input circuit has been used for the direct detection of low-frequency (< 100 kHz) NQR and NMR signals from molecules, such as the NQR spectrum of ^{14}N in NH_4ClO_3 and of ^2D in a tunnelling deuterated methyl group. Such techniques promise new windows on weak spin–spin and electric quadrupole interactions in solids.

3.3 Gravity Wave Detectors

Several projects are in hand to detect the gravitational radiation emitted during cataclysmic stellar events, such as supernovae, and a number involve SQUIDs. Gravity waves propagate at the speed of light, obey the $1/r^2$ law and cause instantaneous strains ($\delta L/L$) in bodies they encounter. For an event at the center of our galaxy, the distortion produced in a body on earth would be of order $\delta L/L \simeq 10^{-18}$–$10^{-19}$ and last for about 1 ms. There are only a few such galactic events per decade and to record 10 events per year one needs a range of 3 Mpc (9×10^{22} m) for which $\delta L/L \leqslant 10^{-22}$.

The project at Stanford University uses the 5000 kg aluminum bar displacement detector discussed in Sect. 2.5. A pulse of gravity waves excites a high-Q (5×10^6) longitudinal mechanical resonance at 1 kHz, which is amplified by the diaphragm of the SQUID displacement sensor. Although the thermal fluctuations of the bar at its operating temperature (4.2 K) are about 10^{-17} m Hz$^{-1/2}$, the bar retains a memory of resonant energy deposited in it for approximately Q cycles, making possible measurement of much smaller excited amplitudes. Currently, changes in amplitude corresponding to $\delta L/L \simeq 10^{-19}$ can be detected.

At this sensitivity there have been no confirmed events. (Unfortunately the nearby event of April 1988, for which $\delta L/L$ would probably have been 10^{-15}, occurred during an observational downtime of all the world's instruments.) It is planned to cool the Stanford bar to 30 mK, giving an antenna noise temperature of 30 mK/Q, at which only zero-point motional strain ($\delta L/L \simeq 3 \times 10^{-21}$) will be significant; the detector will be a dc SQUID with $\varepsilon(f) \simeq 20$ h. As mentioned in Sect. 2.5, to improve sensitivity still further will probably require quantum nondemolition measurement procedures.

3.4 Gravimeters and Gravity Gradiometers

The untuned SQUID displacement sensor of Sect. 2.5 can be used for the measurement of local geophysical gravitational effects, usually in mineral surveying or in inertial navigation instruments, though fundamental problems such as tests of the universal law of gravitation have also been addressed.

The following is a simple example from mineral surveying. It is easy to show, by calculating the local gravitational acceleration $g(h)$ at a small height h above the earth's surface, that a small nearby mass anomaly produces a much larger relative change in the gradient of $g(h)$ than in $g(h)$ itself. The average value of $dg(h)/dh$ at the earth's surface is 3000 Eötvös, where 1 Eötvös $= 10^{-9}$ s^{-2} or about $10^{-10}g(0)$ m^{-1}, whereas the local increase in that value due to a 21 m diameter ore body (or void) of density difference 3000 kg m^{-3} relative to the earth, located 100 m below the earth's surface, would be about 1 Eötvös. Gravity gradiometers are thus popular choices for mineral surveys and Fig. 9 shows how such a SQUID-based gradiometer can be realized using a combination of two diaphragm sensors rigidly mounted about 10 cm apart.

The two sensors share a SQUID, to which they are coupled in such a way that it measures the difference in the current produced by the displacements in the two cells. The electromagnetic contribution to the effective spring constant of one diaphragm is adjusted in an operation which trims the persistent circulating current by using a heat switch to introduce a $10^{-11}\,\Omega$ resistance into the current loop. In this way the two cells are balanced to one part in 10^9. In a vertical gravitational field $g(h)$ the relative displacements (sags) of the two diaphragms will be a measure of the difference in field between them. The apparatus shown achieved a displacement difference sensitivity of about 10^{-16} m Hz$^{-1/2}$, equivalent to the detection of a change of 70 Eötvös Hz$^{-1/2}$ in the local gravity gradient.

Improved sensitivities (0.1 Eötvös Hz$^{-1/2}$) have been achieved by replacing the diaphragms with proof masses mounted on high-compliance cantilever springs.

Basic applications include gravity surveying for mineral prospecting and also geodesy, for which a 10^{-3} Eötvös satellite-borne instrument has been proposed. There is also a possible use in inertial navigation where it is necessary to distinguish true accelerations from gravitational effects: this can be done by subtracting the spatially integrated output of a gravity gradiometer from the reading of an accelerometer, which measures the vector sum $(\boldsymbol{a}+\boldsymbol{g})$ of the spatial and gravitational acceleration.

A gradiometer of 0.2 Eötvös resolution has been used to test for possible corrections to Newton's inverse square law of gravity, $F = Gm_1m_2/r^2$, in which G is not constant but varies with distance. A possible form is

$$G(r) = G[1 + \alpha \exp(-\beta r)] \tag{8}$$

Laboratory measurements ($r \simeq 1$ m) have established an upper limit for the tensor trace

$$\frac{\partial g_x}{\partial x} + \frac{\partial g_y}{\partial y} + \frac{\partial g_z}{\partial z} \tag{9}$$

which should be zero for a strict inverse square law. From this a value of α less than 0.024 ± 0.036 has been obtained, showing that Newton's law holds to better than 2.5% over a 1 m scale. Further measurements, using tidal lakes and eccentric satellite orbits, have been proposed to test the law over larger length scales.

3.5 Nondestructive Evaluation

A growing use of SQUIDs is in the detection of defects in a variety of materials and structures, including electronic circuits (see *Nondestructive Evaluation Using Magnetism*).

(a) *Remote magnetometry.* This technique for the nondestructive evaluation (NDE) of mild steel plates involves a SQUID gradiometer rigidly fixed with respect to a superconducting coil producing a persistent mode field (20 mT) (see Fig. 10). Because the SQUID senses only changes in the flux linking it, there is no signal due to this field itself. However, as a mild steel plate is passed through the field beneath the magnet, the SQUID senses the very slight changes in the field produced by flaws such as cracks or even regions where the permeability is abnormal because of dislocations introduced by fatiguing (see Fig. 11). Cracks as small as 1 mm × 1 mm in cross section have been detected in this contact-free way at standoff distances as large as 6 cm and through protective coatings such as paint or coverings of rust or marine growth. Immersion of the plate in water does not affect the resolution of the technique.

This technique has been applied to stainless steels which have been thermally stressed, for example by heating for long periods above 400 °C. Under these circumstances, small amounts of ferrite precipitate out and can be detected easily in a field of 10 mT.

Magnetometric methods have also been used to monitor the magnetic realignment of domains in unmagnetized steel bars as they are stressed.

(*b*) *Remote galvanometry*. The sensing and locating of currents, perhaps flowing in faulty paths as a result of defects, have been the object of several SQUID applications. Simple magnetometry can, for example, detect a buried pipeline provided access is available at either end to pass the current along the pipe. The method can also detect current diverting flaws, such as holes or welds in the pipeline.

Another application of SQUIDs has been the study of electrolytic corrosion currents in nonvoltaic cells, in which both electrodes are made of the same metal and there is a single electrolyte. Such cells mimic corrosion situations in a number of industrial situations. The low-frequency current fluctuations in the cell caused by different parts of each electrode operating at various times as a local anode or cathode appear to characterize diffusion, hydrogen bubbling, crack propagation, etch pit deepening and other processes.

Remote SQUID galvanometry could possibly map current distributions in integrated circuits for fault finding. This is because the magnetic inverse problem,

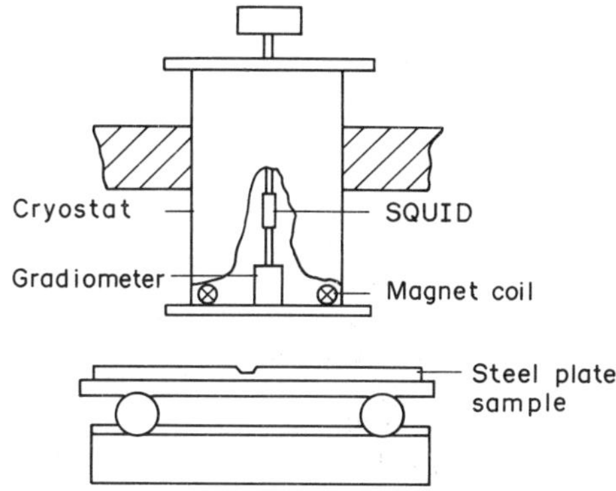

Figure 10
Nondestructive crack detection using SQUID gradiometer

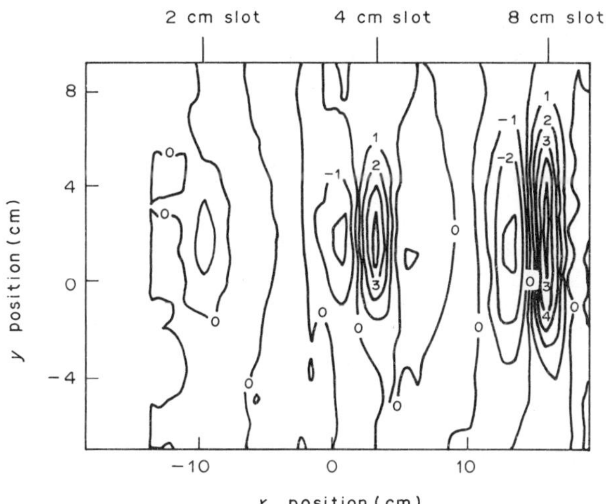

Figure 11
xy Contour plot of signal above three slots of varying length in a steel plate

which is very difficult to solve in three dimensions, has a unique solution in two dimensions: a map of a field $\boldsymbol{B}(x', y')$ made above a two-dimensional specimen, such as a plate, can be used to calculate details of the currents $\boldsymbol{J}(x, y)$ or magnetic moments $\boldsymbol{m}(x, y)$ in the specimen.

To be useful for integrated circuit mapping, spatial resolutions better than 10 μm will be necessary. Since the spatial resolution cannot be much smaller than the coil-to-specimen distance, this requirement has resulted in novel cryostat designs. Resolutions of a fraction of a millimeter have been achieved.

(*c*) *High-T_c SQUIDS*. NDE applications often do not require high SQUID sensitivity and are founded instead on the availability of gradiometers. Moreover they benefit from reduced standoff to improve spatial resolution. Pickup coils which need only be cooled to liquid nitrogen temperatures can exploit such reduced standoff. It follows that NDE might be a target application for the first high-T_c SQUIDs; indeed a simple step edge 77 K device has already been used to demonstrate remote galvanometry.

4. Future Developments

Low-T_c dc SQUIDs, in particular, are now highly developed and remarkable for their extraordinary flexibility; for example, the relatively simple design shown in Fig. 3 is the basis of both the most sensitive magnetometer available for frequencies as low as 10^{-4} Hz and the lowest noise rf amplifier operating at frequencies of 10^8 Hz or higher. Although these devices are sufficiently sensitive for most practical applications, modest improvements in sensitivity seem likely. Equally important will be the development of circuits for the flux-locked loop that are cheaper and more compact and that will facilitate the development of 100-channel systems for medical applications. Considerable progress has been made on high-T_c SQUIDs and input circuits since the late 1980s and the sensitivity of high-T_c magnetometers has advanced to the point where they are useful for practical applications. Significant further improvements in the near future may be expected. The advent of such devices operating in liquid nitrogen is likely to have substantial impact on systems operating in remote areas, for example for geophysical explorations or NDE. A concomitant growth in the commercial market is to be expected.

See also: Biomagnetic Measurements; Josephson Effects in Weak Links; Superconducting Devices and Electronics

Bibliography

Clarke J 1983 Geophysical application of SQUID. *IEEE Trans. Magn.* 19: 288–94

Clarke J 1989 Principles and applications of SQUIDS. *Proc. IEEE* 77: 1208–33

Clarke J 1989 In: Weinstock H, Nisenoff M (eds.) 1989 *Superconducting Electronics*, NATO-ASI Series. Springer, Berlin, pp. 87–148

Clarke J 1989 In: Ruggiero S T, Rudman D A (eds.) 1989 *Superconducting Devices*. Academic Press, London, pp. 51–99

Donaldson G B 1989 SQUIDS for everything else. In: Weinstock H, Nisenoff M (eds.) 1989 *Superconducting Electronics*, NATO-ASI Series. Springer, Berlin, pp. 173–207

Ryhanen T, Seppa H, Ilmoniemi R, Knuutila J 1989 SQUID magnetometers for low-frequency applications. *J. Low Temp. Phys.* 76: 287–386

Recent proceedings of the Applied Superconductivity Conference give excellent coverage and are reported in the following volumes of *IEEE Trans. Magn.*: 19 (1983), 21 (1985), 23 (1987), 25 (1989) and 27 (1991).

G. B. Donaldson
[University of Strathclyde, Glasgow, UK]

J. Clarke
[University of California, Berkeley, California, USA]

J. C. MacFarlane
[CSIRO, Lindfield, Australia]

Steels for Magnets

Magnet steels are carbon steels which contain about 1% carbon and various combinations of chromium, cobalt and tungsten. Since permanent magnet properties are developed in the martensitic structure obtained by quenching from the high-temperature austenitic phase, the magnet steels are also called martensitic steels. Straight carbon steel was the first man-made permanent magnet material. (Lodestone, a permanent magnet material, occurs in nature.)

Table 1
Magnet steels

Alloy (wt%, balance Fe)	B_r (T)	H_c ($kA\,m^{-1}$)	$(BH)_{max}$ ($kJ\,m^{-3}$)
2 Co, 4 Cr, 0.5 W, 1 C[a]	0.98	6.4	2.6
3 Co, 9 Cr, 1.5 Mo, 1 C[a]	0.72	10.3	2.8
6 Co, 9 Cr, 1.5 Mo, 1 C[a]	0.75	11.5	3.5
9 Co, 9 Cr, 1.5 Mo, 1 C[a]	0.78	12.7	4.0
15 Co, 9 Cr, 1.5 Mo, 1 C[a]	0.82	14.3	4.9
35 Co, 5 Cr, 4 W, 0.9 C[a]	0.90	20.0	7.6
3.5 Cr, 1 C[b]	1.03	4.8	2.4
3.25 Co, 4 Cr, 1 C[b]	0.97	6.4	3.0
18.5 Co, 3.75 Cr, 5 W, 0.75 C[b]	1.07	12.7	5.5
38 Co, 3.8 Cr, 5 W, 0.75 C[b]	1.04	18.3	7.8

a UK steels b US steels

The addition of carbon to iron lowers the saturation induction and hence the remanence B_r. However, the increased carbon also results in greater internal strain in the martensite structure, thereby raising the value of coercivity H_c. The product B_rH_c reaches a maximum value at about 0.8% carbon. The 1%-carbon steel develops values of $B_r \simeq 0.9\,T$, $H_c \simeq 4\,kA\,m^{-1}$ and $(BH)_{max} \simeq 1.6\,kJ\,m^{-3}$ after quenching from about 800 °C.

The addition of chromium, tungsten and molybdenum, singly or in combination, generally improves the processing of carbon steels by lowering the rate of cooling necessary for optimum magnetic hardening. In 1917 Takagi and Honda invented the well-known *Honda* or *KS steel* by adding cobalt. The cobalt addition increases all the permanent magnet properties: remanence, coercivity and energy product. Table 1 lists these values for several commercial magnet steels. The upper six alloys are typical of those made in the UK; the lower four are from the USA. The 35% cobalt steel has values of $B_r = 0.9\,T$, $H_c = 20\,kA\,m^{-1}$ and $(BH)_{max} = 7.6\,kJ\,m^{-3}$. The latter two values are about five times better than those of the plain carbon steels.

Magnet steels are produced by casting or forging and rolling ingot material. Since good permanent magnet properties depend on a fine dispersion of carbide particles, prolonged exposure around 1000 °C prior to quenching can coarsen the carbides and also result in a decarburized surface layer. The result is degradation of magnetic properties associated with what is known as "spoiling."

Magnet steel permanent magnets are now almost obsolete. They are used primarily in hysteresis motors. With the availability of other magnets that are superior in performance and cost, the magnet steels will continue to decrease in usage. The Magnetic Materials Producers Association (USA) lists four such alloys still in commercial production. It is believed that only the 3% and 35% cobalt alloys are still commercially available in the UK.

See also: Magnetic Materials: An Overview; Magnetic Materials: Basic Concepts

Bibliography

Gould J E 1971 *Cobalt Alloy Permanent Magnets.* Centre d'Information du Cobalt, Brussels

Hadfield D (ed.) 1962 *Permanent Magnets and Magnetism.* Iliffe, London

Heck C 1974 *Magnetic Materials and their Applications.* Crane, Russak & Co., New York

McCaig M 1977 *Permanent Magnets in Theory and Practice.* Wiley, New York

G. Y. Chin[†]
[AT&T Bell Laboratories, Murray Hill, New Jersey, USA]

J. H. Wernick
[Bellcore, Red Bank, New Jersey, USA]

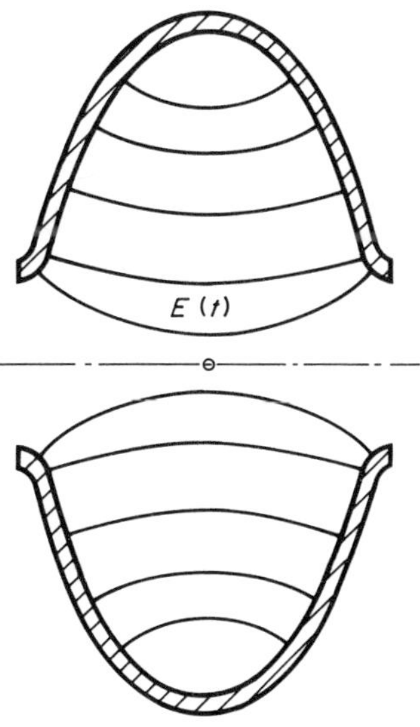

Figure 1
Modern elliptical electron accelerator TM_{010} mode cavity with suppressed multipacting with electric field lines shown; the diameter is around 50 cm for a 0.5 GHz cavity

Superconducting Cavities

Research on superconductors at radio frequencies was begun by London and Pippard in the 1940s using newly available radar equipment. In the 1960s Fairbanks proposed the application of superconductors with low rf (radio-frequency) losses in superconducting rf particle accelerators. This initiated much research which yielded actual working accelerators for electrons with elliptical cavities as shown in Fig. 1 and for heavy ions as shown in Fig. 2. In a parallel development modern microwave integrated circuit technology is also starting to exploit the low loss of superconductors at microwave frequencies (see *High-Frequency Impedance of Superconducting Materials*; *Microwave Applications of Superconducting Materials*).

1. Parameters For Superconducting Cavities

Superconductors are attractive to the rf engineer because the surface resistance R is reduced by more than five orders of magnitude easily compensating for the increased cooling costs ($\sim \times 100$). In addition, the small value of R allows new rf applications, for example, a continuous high-gain accelerator, new designs with smaller sizes or thin-film, integrated techniques.

The parameters used to describe the rf properties of cavities are the quality factor Q, the decay time τ, the resonance halfwidth $\Delta\omega/\omega_0$ and the geometry factor G and are related by

$$\mathrm{Q} = \omega_0 \tau = \frac{\omega_0}{\Delta\omega} = \frac{G}{R} = \frac{\omega_0 \int dV \mu_0 H^2}{R \oint ds H_{\parallel}^2} = \frac{\omega_0 \times \text{store energy}}{\text{wall losses}} \tag{1}$$

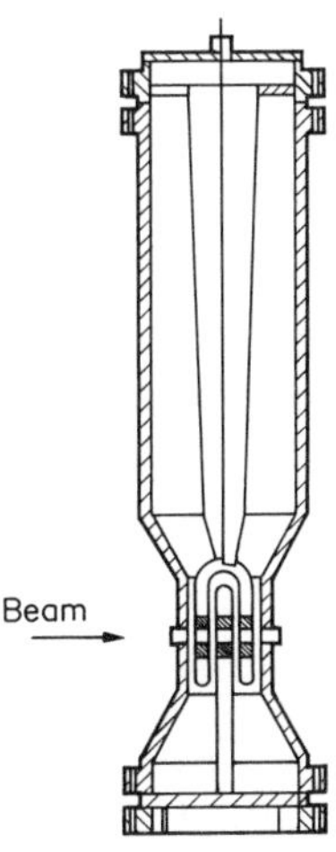

Figure 2
Split ring cavity of the Argonne National Laboratory for heavy ion acceleration at 50 MHz

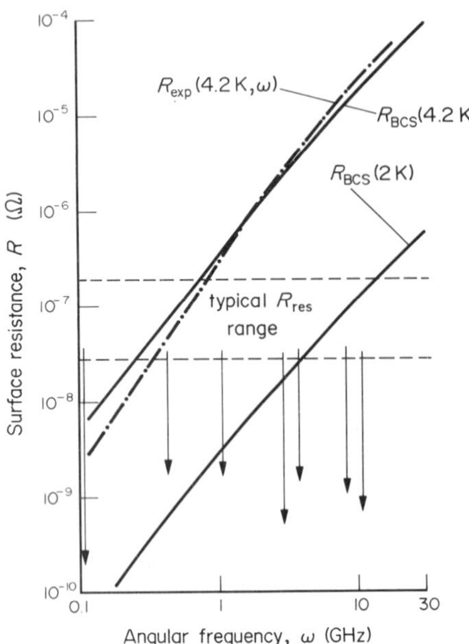

Figure 3
Summary of experimental (–·–·) and computed (——) surface resistance of niobium between 0.1 GHz and 30 GHz assuming $\Delta/k_BT = 1.85$. The $R(4.2\text{ K}, \omega)$ values show the crossover of experimental and computed values due to the smearing of the BCS singularity. With improved niobium quality (residual resistance ratio at 300 K (RRR) > 100) Δ/k_BT_c increases and the smearing is reduced. The residual losses show a large scatter depending on surface quality. With RRR > 100 niobium $R_{res} \simeq 10^{-8}\,\Omega$ are now obtained regularly and the arrows indicate best values

where ω_0 is the resonant angular frequency of the cavity, the geometry factor $G \simeq \omega_0\mu_0 d$ with d the smallest distance between current carrying surfaces and $H_{\parallel}(0) = \int j(x)\,dx$ describes the current flowing behind a surface. For TM_{010} mode cavities operating in the gigahertz range, values of $G \simeq 200\,\Omega$ and $Q \simeq 10^{10}$ are possible with superconducting niobium (see Fig. 3). Such small bandwidths $\Delta\omega/\omega_0 \simeq 10^{-10}$ need precisely regulated rf generators and protection against noise pickup and coupling to mechanical vibrations. Since the superconducting surface resistance R is very small compared to the surface reactance X (Halbritter 1974), the reactive shielding current dominates the surface impedance $Z = R + iX$ and the rf current, unless any normal conducting precipitates, fluxoid cores or lossy dielectrics cause rf residual losses R_{res} (Halbritter 1990).

For good power handling characteristics in accelerators with effective accelerating field E_{acc}, values of the ratio Z_p of the maximal surface fields H_p and E_p should be in the range

Table 1
Superconducting parameters: λ, ξ and l are typical values for the magnetic penetration depth, superconducting coherence length and electron mean free path, respectively

Material	T_c (K)	Δ (THz)	$\mu_0 H_c(0)$ (T)	λ_L (nm)	ξ_{FO} (nm)	l (nm)
Lead	7.2	0.5	0.07	28	175	≥10
Niobium	9.2	0.6	0.2	32	62	≤10
NbN	~17	~1	0.4	~80	~10	~1
YBCO	92	~6	~2	~140	~2	≥10

$$Z_p = \mu_0 \frac{E_p}{H_p} \simeq 0.1\text{–}0.3\ \text{MV m}^{-1}(\text{mT})^{-1}, \quad \frac{E_p}{E_{acc}} \simeq 2\text{–}5 \quad (2)$$

because, on one hand, the Meissner state of superconductors is stable only up to an "rf critical field" $H_{crit}(\omega, T)$ and, on the other hand, limits for E_p exist in multipactor loading or field emission loading.

The field $H_{crit}(\omega, T)$ where the cavity surface suddenly becomes partly normal conducting and Q degrades is related for smooth, homogeneous surfaces to the critical superheating field of the superconductor which is roughly equal to the thermodynamic critical field H_c (T). If the material contains local weak links with an enhanced penetration depth λ_y and leakage current and a reduced effective critical field for flux entry $H^*_{c1J}(\omega, T) \propto 1/\lambda_y^2$ then R_{res} is increased and H_{crit} is reduced. In practice, heating of surfaces, that is $T^* > T_{bath}$, is of crucial importance, because this accelerates the transition to the normal conducting state. An asterisk denotes the actual local conditions at the surface of the superconducting cavity material, the local properties of a real surface being well below the ideal values summarized in Table 1. To avoid small H_p fields any ridges parallel to the shielding current have to be avoided, that is, the currents have to be distributed as homogeneously as possible. Multipactor, or field emission, loading is more stringent for superconducting cavities, because electron impact heating initiates a transition to the normal state. To avoid enhanced field emission any protusion, especially dust, has to be eliminated (Halbritter 1985).

2. Superconducting Materials For RF Cavities

To obtain a low surface resistance (Halbritter 1974) a large energy gap 2Δ and critical temperature T_c are needed, with a consequent high thermodynamic critical field H_c (Table 1). More important for applications are the real accelerator conditions and the existence of surface defects enhancing rf residual losses R_{res} and degrading H_{crit}, for example:

(a) frozen-in flux has to be avoided by careful attention to shielding or proper cool down procedures;

(b) no defects should occur with low $(T_c^*, H_c^*) < (T_c, H_c)$ or other reductions of $H_{crit}(T, \omega)$, for example by local heating; and

(c) no weak links are acceptable that show large leakage currents and that trap field or allow the penetration of flux at fields $H^*_{C1J}(T, \omega) \ll H_c(T)$ and frequencies of interest (Halbritter 1990, 1992a,b).

For rf cavities for accelerators the following materials and preparation methods are used.

(a) Electroplating of lead or PbSn is performed on copper. These cavities are limited by the mechanical and chemical softness of lead and the low $T_c = 7.2$ K and $\mu_0 H_c(0) \simeq 80$ mT.

(b) Sputtering of niobium and NbN on copper has weak links and adherence as problems. However, sputtering of niobium and NbN for small-area strip line resonators is successfully applied in microelectronics and passive microwave devices.

(c) The workhorse is still a niobium sheath material of high purity. Fabrication is by careful machining or deep drawing, welding, UHV firing and cleaning with dust-free H_2O, CH_3OH. Problems are fluxoids, weak links caused by segregation of oxygen and hydrogen pickup with low $H^*_{C1J} \leq H_c$ and high R_{res}. The experimental $R(T, \omega)$ in Fig. 3 is in accord with the BCS theory with a reduced and smeared out energy gap related to oxygen pickup and oxygen segregation (Halbritter 1987).

3. Cavity Applications

Aside from miscellaneous small rf field applications, like stable oscillators or cavities for physics experiments, most superconducting cavities utilize the small rf losses to reach high fields in a continuous fashion. For higher power densities or higher E_{acc} values optimized cavity designs with minimal E_p and H_p and suitable surface treatments have been developed recently.

For fast particles with $v \simeq c$ (velocity of light) Fig. 1-type cavities are used assembled into a series (see Fig. 4) at frequencies around 1 GHz with $E_{acc} \simeq E_p/2$, $\mu_0 H_p/E_{acc} \simeq 4$ mT (MV m^{-1})$^{-1}$ and Q > 10^9. This design has good mechanical stability and allows for easy cleaning, reaching in a real unit $E_{acc} \simeq 20$ MV m^{-1} (Fig. 4).

For low-velocity accelerators ($v \leq c/2$), for example for heavy ions, various designs are in use. Such independently phased cavities operate around 0.1 GHz with $E_p/E_{acc} \simeq 4$ (see Fig. 2).

See also: High-Frequency Impedance of Superconducting Materials; Microwave Applications of Superconducting Materials

Figure 4
TM_{010} cavity of continuous electron beam accelerator facility (CEBAF) (Kojima 1990) which attained $E_{acc} \simeq 18$ MV m^{-1} at 2 K with Q $\simeq 10^{10}$ in an accelerator environment

Bibliography

Halbritter J 1974 On the surface resistance of superconductors. *Z. Phys.* 266: 209

Halbritter J 1985 On contamination of electrode surfaces and electric field limitations. *IEEE Trans. Electr. Insul.* 20(4): 671–81

Halbritter J 1987 On the oxidation and on the superconductivity of niobium. *Appl. Phys. A* 43(1): 1–28

Halbritter J 1988 Low-temperature oxidation of Nb and Nb compounds in relation to superconducting application. *J. Less-Common Met.* 139(1): 133–48

Halbritter J 1990 RF residual losses, surface impedance and granularity in superconducting cuprates. *J. Appl. Phys.* 68(12): 6315–26

Halbritter J 1992a Submillimeter surface impedance and relaxation, a case study of quasi two dimensional $YBa_2Cu_3O_x$. *J. Supercond.* 5 (in press)

Halbritter J 1992b On extrinsic effects in the surface impedance of cuprate superconductors by weak links. *J. Appl. Phys.* 71: 339–43

IEEE Trans. Magn. Proceedings of the Applied Superconductivity Conference. (1979) 15: 7–871; (1981) 17: 11–1108; (1983) 19: 851–1489; (1985) 21: 77–1212; (1987) 23: 343–1841; (1989) 25: 733–2591; (1991) 27: 814–3398

Kojima Y (ed.) 1990 *Proc. Workshop on RF Superconductivity*, KEK Report 89–21. KEK National Laboratory for High Energy Physics, Ibaraki, Japan

Kuntze M (ed.) 1980 *Proc. Workshop on RF Superconductivity*, KfK Report 3019. Kernforschungszentrum, Karlsruhe, Germany

Lengeler H (ed.) 1984 *Proc. Workshop on RF Superconductivity*. CERN, Geneva

Shepard K W (ed.) 1988 *Proc. Workshop on RF Superconductivity*, ANL–PHY–88–1. Argonne National Laboratory, Argonne, IL

J. Halbritter
[Kernforschungszentrum Karlsruhe,
Karlsruhe, Germany]

Superconducting Device Fabrication

Although rapid progress in the development of a thin-film technology for high-critical-temperature superconducting materials is bringing reliable high-T_c superconducting devices closer, the fabrication techniques available are still relatively primitive (see *Thin Films, Multilayers and Devices, Superconducting*).

Device technology for low-critical-temperature materials has, however, progressed steadily. In the early 1980s, it became clear that the lead alloy junctions that had been used in Josephson integrated circuits during the 1970s were not practicable, because the characteristics of the lead alloy junctions were unstable. Much work has since gone into replacing lead alloy electrodes with refractory materials such as niobium and NbN.

In 1983, an excellent Josephson junction with niobium electrodes was made by Gurvitch et al. (1983). It had a tunnel barrier of aluminum oxide ($Nb/AlO_x/Nb$). Since then, niobium junctions have been applied to many logic and memory circuits (Kotani et al. 1988, Hasuo and Imamura 1989, Suzuki et al. 1989). It is now possible to fabricate a circuit with tens of thousands of junctions (Kotani et al. 1989). The most promising techniques for superconducting devices use niobium Josephson junctions.

1. Josephson Circuit Fabrication

The Josephson junction is made of a thin (a few nanometers) tunnel barrier sandwiched between two superconducting films, the base and counter electrodes. Standard Josephson integrated circuits use $Nb/AlO_x/Nb$ Josephson junctions, molybdenum resistors, a niobium ground plane and niobium wiring usually in 9–11 layers on silicon substrates. Each layer is patterned through a photoresist mask by reactive ion etching (RIE).

A cross section of an integrated circuit with Josephson junctions is shown in Fig. 1. Josephson circuits are usually fabricated above a superconducting ground plane. Resistors that are used as loads or for damping in junction switching contact the base electrode. Via holes are made in insulators above junction counter electrodes. In some logic circuits wiring is the top layer, but usually two additional layers (insulators and control lines) are formed on the wiring.

Layer materials and thickness are listed in Table 1. Niobium, aluminum, molybdenum and SiO_2 are deposited by sputtering under the conditions in Table 2. SiO_x, used as a protective underlayer for molybdenum resistors during RIE of the niobium base electrode, is deposited by evaporating SiO in oxygen. All but the SiO_x layer are patterned by RIE. The reactive gases are listed in Table 3. Because aluminum and AlO_x are not etched by reactive CF_4 and CHF_3, the thin Al–AlO_x barrier is removed by argon sputter etching. The SiO_x layer used is patterned by liftoff.

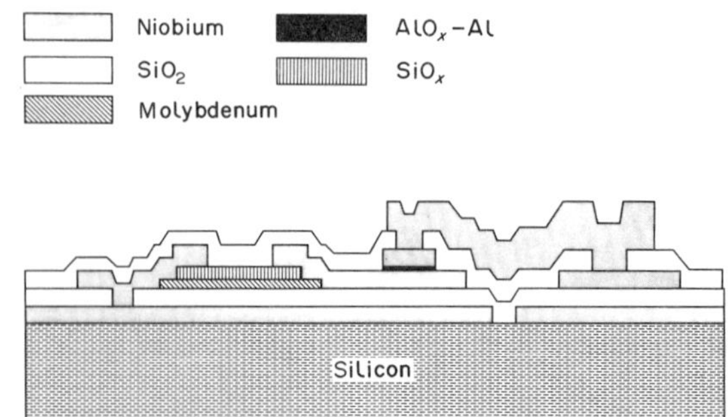

Figure 1
Cross section of integrated circuit with $Nb/AlO_x/Nb$ junction

The key process in Josephson integrated circuits is making high-quality junctions. However, several other elements are involved, including the insulation layer, the superconducting wiring, the resistors and the contacts between the two niobium electrodes.

2. $Nb/AlO_x/Nb$ Junctions

$Nb/AlO_x/Nb$ junctions have excellent *IV* characteristics and are reliable. The excellent controllability and reproducibility of these junctions makes them practicable for application in circuits with tens of thousands of junctions with a close tolerance to their critical parameters. Josephson junction characteristics depend on the formation of a uniform high-quality tunnel barrier on the niobium electrodes. An aluminum layer a few nanometers thick wets the niobium surface sufficiently and a uniform AlO_x layer can be formed on the aluminum layer. The discovery of the AlO_x barrier was the single most important factor in enabling the niobium junction to be put into application.

Table 1
Circuit layers

Layer	Material	Thickness (nm)
Ground plane	niobium	300
Insulation	SiO_2	300
Resistor	molybdenum	100
Resistor protection	SiO or SiO_x	100
Base electrode	niobium	200
Tunnel barrier	AlO_xAl	7
Counter electrode	niobium	100–200
Insulation	SiO_2	400
Wiring	niobium	600
Insulation	SiO_2	800
Control line	niobium	1000

Table 2
Conditions of metal and insulation layer sputtering

Material	Argon pressure (Pa)	Deposition rate ($nm\,min^{-1}$)
Niobium	1.3–2.3[a]	200
Aluminum	1.3	8
Molybdenum	0.67	130
SiO_2	1.3	8

a Target–substrate distance 55 mm

The fabrication of Josephson integrated circuits is described by Imamura et al. (1987) and Hasuo and Imamura (1989). The fabrication process is, of course, applicable to other circuits, such as superconducting quantum interference device (SQUID) chips (see *SQUID Design, Fabrication and Application*) and voltage-standard chips (see *Voltage-Standard Devices*).

The $Nb/AlO_x/Nb$ trilayer barrier is first deposited across the whole silicon substrate in the same vacuum run. This process is described as whole-wafer technology. The junctions are defined by trimming in the next patterning process. The quality of the *IV* characteristics depends greatly on the formation of a sharp interface between the niobium counter electrode and the AlO_x, and between the aluminum and the niobium base electrode. Optimized $Nb/AlO_x/Nb$ trilayer deposition has been reported in detail (Morohashi and Hasuo 1987).

Recently, the microstructure of the thin AlO_x barrier has been revealed by using cross-sectional TEM (Imamura and Hasuo 1991). The TEM photograph shown in Fig. 2 makes clear three features important for the junction barrier. First, both sputtered niobium and aluminum are polycrystalline films with columnar structures. Second, the 200 nm lower niobium surface is rough and wavy, but a thin aluminum film deposited on niobium planarizes the rough surface of the lower niobium. The 2 nm AlO_x barrier is formed on the planarized aluminum. Third, the upper niobium surface has a good crystalline structure just above the AlO_x barrier. These three features are thought to be essential for the high-quality characteristics of $Nb/AlO_x/Nb$ junctions.

Table 3
Conditions of metal and insulation layer patterning

Material	Reactive gas	Pressure (Pa)	Power density ($W\,cm^{-2}$)
Niobium	CF_4(5–20% O_2)	2.7–6.7	0.10
Aluminum	argon	0.7	0.15
Molybdenum	CF_4(5% O_2)	6.7	0.10
SiO_2	CHF_3(0–15% O_2)	2.0	0.20

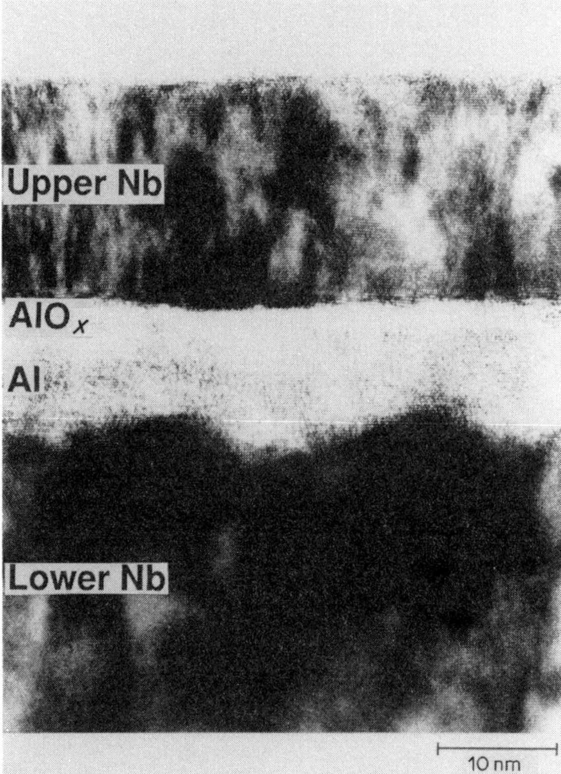

Figure 2
Lattice image of $Nb/AlO_x/Nb$ junction structure

An anodization profile method has been developed for analyzing the barrier structure (Imamura and Hasuo 1989). The anodization profile as shown in Fig. 3 is obtained by plotting dV/dt vs V during the anodization of the $Nb/AlO_x/Nb$ junction structure. An anodization voltage of 1 V corresponds to a thickness of 0.9 nm. The sharpness of the interface between the AlO_x barrier and niobium can be eva-

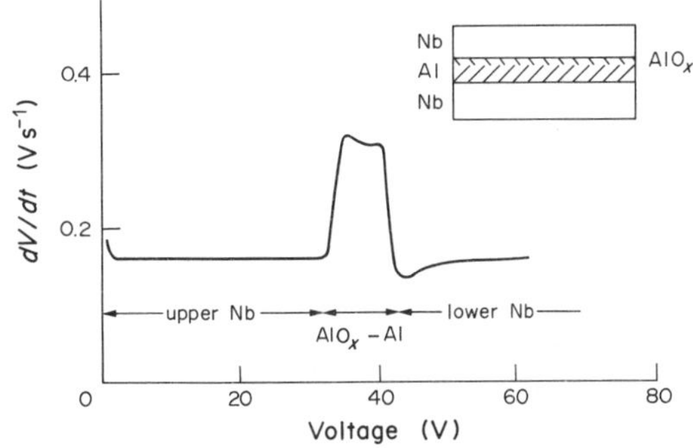

Figure 3
Anodization profile of $Nb/AlO_x/Nb$ junction structure

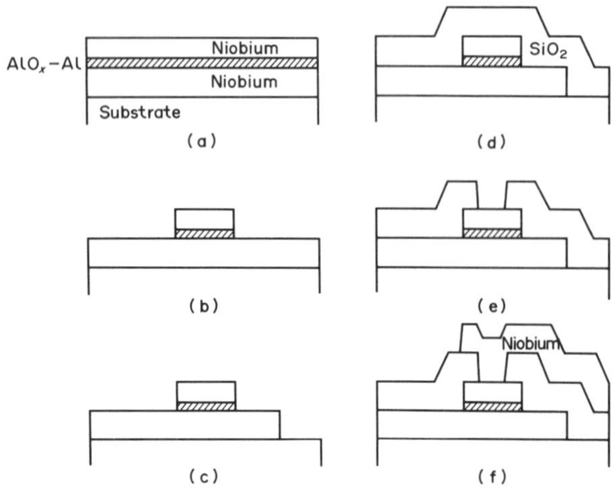

Figure 4
$Nb/AlO_x/Nb$ junction fabrication: (a) whole-wafer trilayer structure, (b) patterning of junction meter, (c) patterning of base electrode, (d) deposition of insulation layer, (e) "opening up" of contact via and (f) deposition of counter electrode

luated to an accuracy of less than 1 nm. Thus, the anodization profile is very effective for monitoring interdiffusion around the junction barrier. Another merit of the anodization profile is that the barrier quality can be estimated in process, that is, even at room temperature.

The $Nb/AlO_x/Nb$ junction is usually fabricated as shown in Fig. 4. Before trilayer deposition, substrates are etched by argon sputtering to contact the niobium ground plane through contact holes. The trilayer is then deposited by sputtering on the substrate attached to a water-cooled holder. To attain good thermal contact, a copper backing plate and indium foils are used as spacers between the substrates and the holder. The base pressure is less than 10^{-5} Pa. After the base electrode and aluminum barrier are deposited, argon with 10% oxygen is introduced into the chamber to form a thin 2 nm oxide on the aluminum. The critical current density of the junction J_c depends on the pressure (usually in the range 50–130 Pa) and oxidation time (typically 30–60 min); it is in the range 500–10 000 A cm^{-2}. For a design J_c of 2000 A cm^{-2}, junctions with J_c values in the range 1500–3000 A cm^{-2} can be reproducibly obtained. After oxidation, the niobium counter electrode is deposited on the AlO_x barrier. The trilayer is then trimmed to the device structure. The junction area is defined by patterning the niobium counter electrode. In RIE, AlO_x and aluminum act as good etch stops against CF_4 gas. The remaining AlO_x–Al barrier is removed by argon sputter etching. The lower niobium layer is then etched into a base electrode pattern. Junctions are covered with an SiO_2 insulation layer. Contact holes are formed in the SiO_2 by RIE, then niobium is deposited and patterned as a wiring layer.

The *IV* characteristics of fabricated junctions are shown in Fig. 5. The junction size is 10 μm × 10 μm. The gap voltage is 2.9 mV. The quality parameter V_m, which is defined as the product of the critical current and the subgap resistance at 2 mV, exceeds 60 mV. $Nb/AlO_x/Nb$ junctions are very stable in cycling between 4.2 K and 300 K and in storage at room temperature.

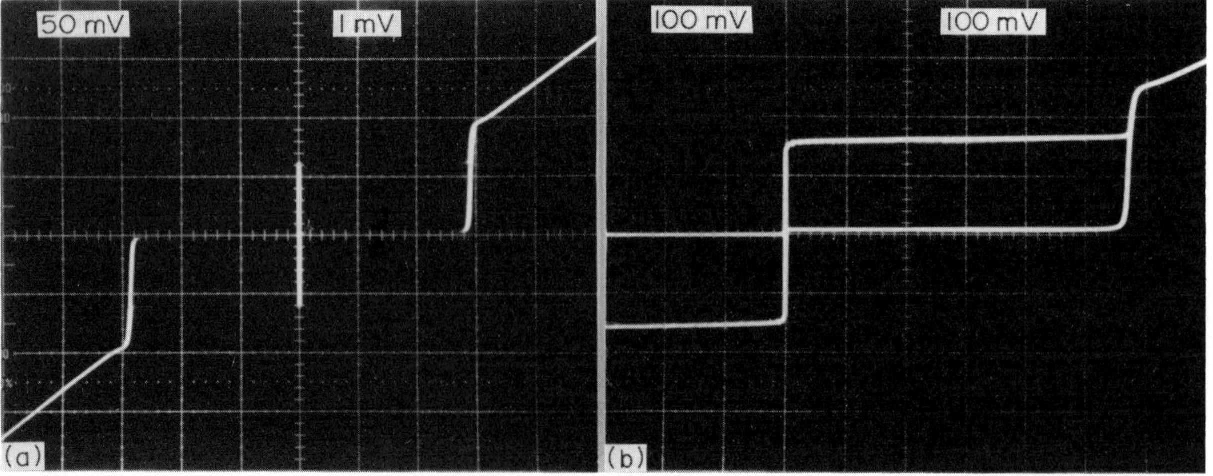

Figure 5
IV characteristics for (a) single junction (vertical scale: 1 mA div^{-1}, horizontal scale: 1 mV div^{-1}) and (b) 200 series-connected junctions (vertical scale: 1 mA div^{-1}, horizontal scale: 100 mV div^{-1})

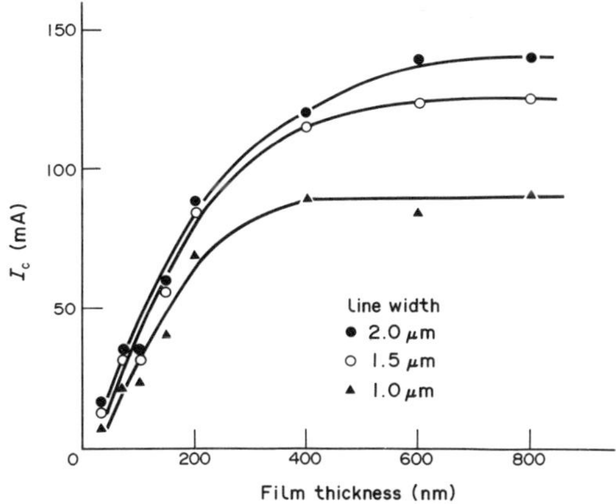

Figure 6
Superconducting critical current I_c vs film thickness for niobium wiring lines

3. SiO_2 Insulators

In lead alloy Josephson circuits, SiO and SiO_x films have been used for insulation layers, but these films are not good enough for niobium circuits. The breakdown voltage between two niobium electrodes with these insulators decreases to almost zero after the counter niobium electrode is deposited. This is because of the microcracks induced by heat or stress during refractory niobium deposition. The breakdown voltage in sputtered SiO_2 exceeds 200 V even after niobium is deposited as a counter electrode. Cooling substrates during SiO_2 sputtering proved that using sputtered SiO_2 does not deteriorate the junction characteristics, which is why it is used in Josephson circuits.

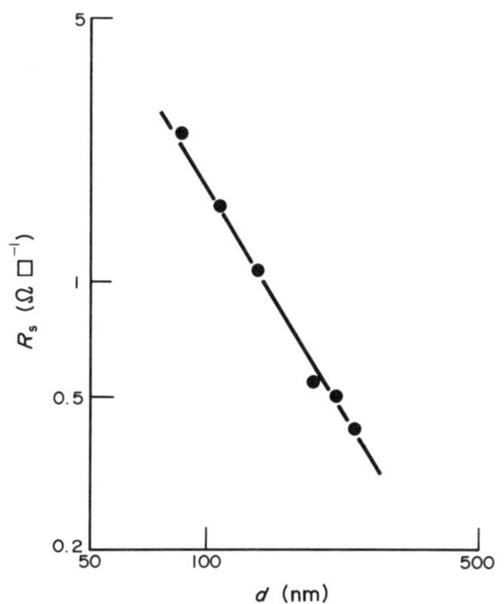

Figure 7
Sheet resistance R_s vs thickness d in molybdenum resistors

4. Superconducting Niobium Wiring

As circuits become more and more densely integrated, long narrow wiring lines are needed. The superconducting critical current I_c measured for niobium wiring is shown in Fig. 6. The lines are 1.0 μm to 2.0 μm wide and 10.88 mm long. For films more than 400 nm thick, I_c depends only on the line width and not the film thickness. For films thinner than 200 nm, however, I_c changes linearly with thickness, apparently due to the supercurrent flowing in the film surface up to the London penetration depth. The current flow in Josephson logic circuits is usually less than 1 mA. The measured I_c is more than ten times greater than the current level, so circuit malfunctions caused by normal transitions in niobium wiring cannot occur, making wiring integrity good enough for Josephson large-scale-integration circuit applications.

5. Molybdenum Resistors

Molybdenum film, deposited by sputtering, forms the resistors. Measured sheet resistance R_s is plotted as a function of thickness d in Fig. 7 (Imamura et al. 1987). R_s is proportional to $d^{-1.6}$. The spread of resistance on a wafer is only a few percent and the contact resistance between molybdenum and niobium is negligible.

6. Conclusion

Niobium Josephson junction fabrication technology progressed remarkably in the 1980s. It has become possible to apply the technique to integrated circuits with over 10 000 junctions per chip, because of the stable, reliable, reproducible and controllable characteristics of niobium junctions.

Niobium junction fabrication will play a key role in the creation of systems with low-temperature superconducting electronics, such as multichannel SQUID systems or Josephson computers.

See also: Superconducting Devices and Electronics; Thin Films, Multilayers and Devices, Superconducting; Voltage-Standard Devices

Bibliography

Gurvitch M, Washington M A, Huggins H A 1983 High quality refractory Josephson tunnel junctions utilizing thin aluminum layers. *Appl. Phys. Lett.* 42: 472–4

Hasuo S 1989 High-speed Josephson integrated circuit technology. *IEEE Trans. Magn.* 25: 740–9

Hasuo S. Imamura T 1989 Digital logic circuits. *IEEE Proc.* 8: 1177–93

Imamura T, Hasuo S 1989 Characterization of Nb/AlO$_x$–Al/Nb Josephson junctions by anodization profiles. *J. Appl. Phys.* 66: 2173–80

Imamura T, Hasuo S 1991 Cross-sectional transmission electron microscopy observation of Nb/AlO$_x$–Al/Nb Josephson junctions. *Appl. Phys. Lett.* 58: 645–7

Imamura T, Hoko H, Ohara S, Kotani S, Hasuo S 1987 Fabrication technology for Josephson integrated circuits with Nb/AlO$_x$/Nb junctions. In: Hara K (ed.) 1987 *Superconductivity Electronics*. Prentice-Hall, Englewood Cliffs, NJ, pp. 22–33

Kotani S, Fujimaki N, Imamura T, Hasuo S 1988 A Josephson 4b microprocessor. *Digest Technical Papers Int. Solid-State Circuits Conf.* IEEE, New York, pp. 150–1

Kotani S, Imamura T, Hasuo S 1989 A sub-ns clock Josephson 4b processor. *Digest Technical Paper Symp. VLSI Circuits*. IEEE, New York, pp. 23–4

Morohashi S, Hasuo S 1987 Experimental investigations and analysis for high-quality Nb/Al–AlO$_x$/Nb Josephson junctions. *J. Appl. Phys.* 61: 4835–49

Suzuki H, Fujimaki N, Tamura H, Imamura T, Hasuo S 1989 A 4K Josephson memory. *IEEE Trans. Magn.* 25: 783–8

T. Imamura and S. Hasuo
[Fujitsu Laboratories, Atsugi, Japan]

Superconducting Devices and Electronics

After some 25 years of development, the field of superconducting electronic devices remains subject to rapid change, while opinions differ considerably over the future of the area. The main reason for this controversy is the history of the field itself, which presents a curious mixture of bright achievements, false promises, and resounding failures.

On the one hand, a few superconducting electronic devices, notably SQUIDs (see *SQUID Design, Fabrication and Application*), SIS mixers (see *Millimeter-Wave Detection*) and dc voltage standard devices (see *Voltage-Standard Devices*) exhibit unparalleled performance and have become an indispensable part of modern practical electronics and instrumentation. Higher hopes still are pinned on some devices which have not yet left research laboratories. For example, key components of the recently invented rapid single-flux-quantum (RSFQ) digital family have been demonstrated to operate at clock frequencies beyond 100 GHz, some 100 times faster than semiconductor transistor devices (see Sect. 5).

On the other hand, the world market for electronics based on the low-T_c superconductor materials remains relatively small (of the order of US$100 million per year) because the need for cooling to helium temperatures ($T < 20$ K) leaves no place for devices with mediocre performance. Hopes for a rapid transfer to liquid nitrogen temperatures ($T \sim 70$–80 K) using the new high-T_c superconductors have proved to be unreasonable for most applications for a number of reasons which are discussed in Sect. 2. Other serious technical problems include the absence of a good superconductor analog of the semiconductor transistor and magnetic flux trapping in the superconducting thin-film electrodes.

The aim of this article is to present a picture of the current status and prospects of superconducting devices and electronics and to show how the next few years are likely to be decisive for the future development of the field. The whole spectrum of electronic devices using superconductors will be discussed, from the simplest to the most complex, with special emphasis on prospects for high-T_c superconducting materials and on new complex analog and digital devices (see Sects. 3–5).

1. Passive Components and Simple Devices

1.1 Superconducting Transmission Lines, Resonators and Antennas

Transmission lines, resonators and antennas make a use of a sharp drop in the ac resistance of superconductors below their transition point T_c, which is especially strong at frequencies much less than $2\Delta(T)/h$ (see *High-Frequency Impedance of Superconducting Materials*). For traditional materials with low critical temperatures such as lead and niobium, this upper frequency is of the order of 10^{12} Hz, so that the decrease in losses can be quite spectacular even in the microwave band if proper care is taken over the surface of the material. For example, the microwave surface resistance $R_s = \mathrm{Re}\,Z(\omega)$ of niobium at $T \simeq 4$ K can be made as low as a few nanoohms at 1 GHz, compared with a figure of few milliohms available for the best normal metals such as silver and copper.

This property can be used for the design of low-loss microwave transmission lines, high-quality resonators and antennas (for details, see *Microwave Applications of Superconducting Materials*). The high cost of helium refrigeration has restricted the practical use of low-T_c devices of this kind to a few specialist uses such as electron acceleration (Piel 1988; see also *Superconducting Cavities*), unique physical experiments (Braginskii et al. 1985) and preliminary attempts in analog signal processing (Withers and Ralston 1989). It is important to note, however, that superconducting microstrip lines are capable of passing picosecond waveforms over distances exceeding 1 cm, and in this capacity they are indispensable components of Josephson junction digital devices which are the subject of intensive development (see Sect. 5).

The high-T_c materials which can use much less-expensive nitrogen cooling allow refrigeration costs to be reduced. However, the values for the surface resistance R_s possible even for the best (quasimonocrystalline) bulk and thin-film samples of the high-T_c

superconductors are still relatively high (see *High-Frequency Impedance of Superconducting Materials*), and at microwave frequencies are only a few times lower than those of copper for similar temperatures ($T = 77$K). This is the reason why applications are highly dependent on whether the values of R_s can be brought down closer to the values following from Bardeen–Cooper–Schrieffer-type (BCS) theories. Note also that widely publicized prospects to use high-T_c superconductor interchip interconnections to speed up semiconductor transistor circuits cannot be considered as realistic (Malozemoff et al. 1987, Likharev 1990).

1.2 Electromagnetic Field Shields

Another possible application of high-T_c superconductors is for shielding sensitive electronic devices from external electromagnetic fields (noise, interference, etc.). Due to the Meissner effect it is sufficient to surround the volume to be protected by a thin film or layer of a thickness d only a few times larger than λ, the magnetic field penetration depth (see *Magnetic Properties of Superconductors*; *Superconducting Materials: Physical Properties*).

For quasimonocrystalline films of, say, YBaCuO, $d = 1$ μm would be sufficient for virtually perfect shielding; granular (ceramic) layers would require a larger d, but in all cases the thickness and weight of the superconducting shields would be much less than those of magnetic materials providing a comparable effect (see *Shields, Magnetic*). In which cases exactly this advantage will outweigh the problems of cooling is not yet clear.

1.3 Switches (Modulators)

If a signal transmission line has a superconducting component, it can be switched rapidly from the Meissner (low-loss) state to the purely normal or mixed state, providing large signal losses. It this way, signal attenuation can be controlled (modulated) effectively. Possible means of control include dc current, laser illumination and magnetic field. Moreover, switching can be induced by the signal itself when its amplitude exceeds a certain threshold (in which case the device is called either a power limiter or a noise discriminator).

Advantages of the best superconductor modulators include high speed (switching times below 1 ns for low-T_c thin films and probably faster than 10 ps for high-T_c thin films), low noise and a wide signal frequency band due to travelling-wave operation mode (see Likharev et al. 1990).

1.4 Bolometers

If the temperature T of a superconducting thin film is close to the critical temperature, even slight heating by incoming radiation can lead to a noticeable change of its dc resistance R, due to the high value of dR/dT. This effect is used in superconducting bolometers, which can be very sensitive detectors of electromagnetic radiation within a broad frequency range, from microwaves to visible light and beyond. This sensitivity (with the noise equivalent power (NEP) approaching, for example, 10^{-15} W Hz$^{-1/2}$ for a time constant ~ 1 s and $T \sim 1$ K) is especially valuable in the far infrared range where superconducting bolometers are second only to the best semiconductor bolometers.

Use of high-T_c superconductors implies higher operation temperatures and hence higher noise. Nevertheless, projected NEPs of a few 10^{-12} W Hz$^{-1/2}$ would make such bolometers the most sensitive nitrogen-temperature detectors in the far infrared. Infrared-imaging systems based on such bolometers could become useful in all the infrared range due to their potentially low cost.

1.5 Particle Detectors (Spectrometers)

Photons of higher frequencies (with energies E well beyond the superconductor gap Δ) as well as charged particles, when absorbed in a superconductor, create a "shower" of many ($\sim E/\Delta \gg 1$) nonequilibrium (overheated) electron quasiparticles. This phenomenon can be used for detection and spectrometry of x rays and ionizing particles (for details, see *Nonequilibrium Superconductivity* and Barone 1988).

2. Nonlinear and Active Devices

2.1 Superconductor Transistors

Much time and effort has been spent in attempts to find a superconductor analog of semiconductor transistors. One proposal was to inject nonequilibrium electrons (which would give an analog of the bipolar transistor), while another was electric-field control of the Cooper-pair transfer in two-dimensional electron gas layers (the so-called Josephson field-effect transistor, or JFET).

Unfortunately, neither of these has been successful; in the first case the voltage gain G is too close to unity to make the device useful, while in the second case, $G \ll 1$. Thus, practical superconducting electronics is left with only two useful nonlinear devices, SIS junctions and Josephson junctions, both of these being two-terminal devices rather that multiterminal ones.

2.2 SIS Quasiparticle Tunnel Junctions

The SIS junction is a tunnel junction between two superconductor electrodes in which the Josephson (Cooper-pair) tunnelling is either suppressed or not substantial. The remaining current I_q is due to tunnelling of single electrons and holes, and is a strongly nonlinear function of the applied voltage V, increasing sharply within some small range δV at $|V| \simeq 2\Delta(T)/e$. For frequencies below roughly $e\delta V/h$ (typically, $\sim 10^{10}$ Hz), the junction behaves exactly as a nonlinear resistor. At higher frequencies, the behavior is somewhat more complex due to quantum effects, but is nevertheless close to that of a nonlinear resistor. This strong nonlinearity is of value in a number of applications, notably SIS mixers (see Sect. 3.3).

2.3 Josephson Junctions

Tunnel junctions, as well as other types of weak link between superconductor electrodes (Likharev 1979) may all carry a considerable Cooper-pair "supercurrent" I_s and in this case are referred to as Josephson junctions (see *Josephson Effects in Weak Links*; Van Duzer and Turner 1981). In many cases, I_s can be described by simple equations

$$\left.\begin{aligned} I_s &= I_c \sin\varphi \\ \varphi &= \frac{2e}{\hbar}\int V\,dt \end{aligned}\right\} \qquad (1)$$

Despite their simplicity, these equations can describe quite complex dynamical behavior of the Josephson junctions (Likharev 1986). Most notable of the dynamic features are the ability of the Josephson junction, first, to pass a current ($-I_c < I < +I_c$) without any voltage drop and, second, to generate short (picosecond) "single-flux-quantum" (SFQ) pulses with the fixed area

$$\int V\,dt = \frac{h}{2e} = \Phi_0 \simeq 2.07\,\text{mV ps} \qquad (2)$$

If the dc voltage V across the junction is fixed, the frequency of these pulses is fixed as well:

$$f_J = \frac{2e}{h}V \qquad (3)$$

At large V, the pulses merge, and the junction generates nearly sinusoidal "Josephson oscillations" of the frequency given by Eqn. (3). One more important feature of the junction is its nonlinear reactance, which is reduced to just a parametric inductance $L_J = (h/2eI_c)[\cos\varphi(t)]^{-1}$ in the small-signal limit.

The complex dynamics allows the use of Josephson junctions in various devices with very different modes of operation, including active generation of quasi-sinusoidal and pulselike waveforms. The most important limitations of the junction performance are

(a) the "characteristic" frequency $f_c = (2e/h)I_cR_n$ (where R_n is the normal state resistance), which is fundamentally limited by the superconductor energy gap ($f_c < \pi\Delta(0)/h$);

(b) the "plasma" frequency $f_p = (1/2\pi)\,(L_JC)^{1/2}$ (typically, the performance of the junction degrades at frequencies above min$[f_c, f_p]$; junctions with $f_p > f_c$ are called overdamped, while those with $f_p < f_c$ are called underdamped); and

(c) the intrinsic thermal noise which can be characterized by the effective "fluctuation current" amplitude $I_T = (2e/h)\,k_BT$ (for various applications, the critical current I_c of the junction should be 10–500 times larger than I_T).

The final limitation is especially painful for the high-T_c Josephson junctions with their relatively low I_cR_n products (typically, below 1 mV at 77 K) and high operation temperatures T. The resulting large I_c and small R_n make the impedance matching of the junctions with both resistive loads (typically requiring $R \simeq R_n$) and inductive loops ($L \simeq L_J$) very difficult. A larger I_c also leads to increased power dissipation (scaling as $I_c^2R_n$), making it excessively large for some potential high-T_c VLSI circuits (see Sect. 5). Note also that the high-frequency performance of the high-T_c Josephson junctions can hardly be higher than that of their low-T_c counterparts, because the plasma frequency f_p can hardly be increased by transfer to these new materials (Likharev et al. 1990).

This is why the prospect for high-T_c superconductor electronic applications using conventional Josephson junction devices (with a possible exception of the SQUIDs) seems rather bleak. The reader will see, however, that this fact does not detract from the expected future of the field as a whole.

3. *Analog Devices*

3.1 SQUIDs

The celebrated SQUIDs (**s**uperconducting **qu**antum **i**nterference **d**evices) probably present the most important single achievement of superconducting electronics. The reader can find more details in the article *SQUID Design, Fabrication and Application*. Note also the recently proposed high-performance digital SQUIDs (see Sect. 4.3).

3.2 Samplers

Josephson junction samplers are able to measure periodically repeating electrical pulse waveforms. Because of lack of a good analog of the usual sample-and-hold circuits, superconductor samplers use Josephson junction comparators, where a sum of the measured waveform current, a special short sampling pulse, and the dc negative-feedback current (which also serves as output signal of the device) is periodically compared with the critical current I_c.

Practically implemented versions of the samplers (see for example Whiteley et al. 1989) have demonstrated time resolution of a few picoseconds and a sensitivity of a few microvolts. While the first figure is only a few times better than that for similar semiconductor devices, the sampler sensitivity is unique (due to the low intrinsic noise of the Josephson junctions). This is why these samples could become indispensable instruments for testing new ultrafast digital devices, especially the RSFQ-type logic circuits (see Sect. 5).

3.3 Microwave and Millimeter-Wave Receivers

Superconductor junction receivers of electromagnetic radiation can be used within a broad range of frequencies. Moreover, both the resistive nonlinearity of the SIS junctions and reactive nonlinearity of the Josephson junctions (Sect. 2) can be employed, each in several modes of operation (Kuzmin et al. 1981, Likharev 1986).

SIS mixers are the most popular, especially in the millimeter-wave band where they are the most sensitive receivers (see *Millimeter-Wave Detection*). Note, however, that Josephson junction self-oscillating mixers and quadratic detectors can be made even more sensitive in the same band (Kuzmin et al. 1981, Devyatov et al. 1986) by using optimal coherent multijunction arrays (Jain et al. 1984); this remarkable opportunity is still to be developed.

3.4 Oscillators

Josephson junction self-oscillation at the frequency given by Eqn. (3) suggests that Josephson junctions could be used as simple dc-voltage-tuned radiation sources. The major disadvantages are the low power (typically, below 10^{-9} W) available from a single junction and a relatively broad radiation line. Both of these drawbacks can be cured by a use of the coherent arrays (Jain et al. 1984, Likharev 1986), in which case the Josephson frequency generators can be used at least as on-chip local oscillators for SIS mixers and other superconductor junction receivers.

3.5 DC Voltage-Standard Devices and Multipliers

One more important analog application of Josephson multijunction systems is for dc voltage-standard devices. Simple arrays of up to about 20 000 junctions connected in series and phase-locked by an external radio-frequency reference can be successfully used to produce up to 10 V dc voltage (see *Voltage-Standard Devices*). The major drawbacks of this design are a need for a very-high-frequency reference source (typically, 70–100 GHz) with considerable output power and a low output dc current (a few tens of microamperes), which makes the device quite vulnerable to dc reference noise.

Recently, a very promising alternative device has been suggested (Semenov and Voronova 1989). Its core is the dc voltage multiplier based on the mutual locking (synchronization) of SFQ pulses (see Sect. 2.3) generated by ac-coupled Josephson junctions. Such locking provides exact equality of the dc voltages across the junctions, so that they can be dc-connected in series to form the basic cell of the device, the dc voltage doubler. A dc voltage multiplier can then be formed as a uniform array of the doubler cells. The n-fold multiplication of an input voltage requires a relatively large number ($N \simeq 6n$) of junctions; they can, however, be decomposed into several dc-connected circuits each of a moderate integration scale.

In addition to the potential importance of the multiplier as such, it is straightforward to turn it into a dc voltage-standard device by phase-locking its first junction using a low-power and relatively low-frequency reference source. An important potential advantage of the new standard is a possibly large output current (a few milliamperes). This promising device is yet to be implemented.

3.6 Digital-to-Analog Converters

The first junction of the dc voltage multiplier discussed in the previous section can be driven not only by a sinusoidal reference source but also by an arbitrary sequence of the voltage pulses (of proper amplitude). In this mode, the output dc voltage of the device is exactly proportional to the average rate of the pulses. Feeding this device from a usual (semiconductor) digital integrator gives a very precise digital-to-analog converter with the converter rate equal to a multiple of $2e/h$ (Semenov and Voronova 1989). One possible application of this digital-to-analog converter is discussed in Sect. 4.3.

4. Analog-to-Digital Converters

The natural quantization of the magnetic flux in superconductors makes Josephson junctions very convenient for analog-to-digital conversion (for a recent review, see Lee and Peterson 1989). The following three approaches are possible.

4.1 Parallel Converters

Critical current I_c of the two-junction interferometer (see *SQUID Design, Fabrication and Application*) is a Φ_0-periodic function of the applied flux Φ_x. Thus one can fix its bias current I at a value where a half of the Φ_0 period of Φ_x corresponds to $I > I_c$, while its remaining half corresponds to $I < I_c$. For underdamped junctions, the latter condition leads to switching of the interferometer into its resistive state with dc voltage $V \neq 0$ as soon as the bias current is turned on. Hence the voltage V can be interpreted as the single-bit binary code of Φ_x/Φ_0. Feeding a string of n such interferometers with the same pulsing bias current I, and with binary outputs of an analog divider ($\Phi_{xi} = \Phi_x/2^i$, where $i = 1,\ldots,n$), gives an n-bit parallel ("flash-type") analog-to-digital converter.

The maximum sampling frequency f_{max} of a parallel converter is limited by the time necessary to reset its underdamped junctions into their initial superconducting state, and can scarcely exceed a few gigahertz. The maximum bit number n_{max} is determined by the accuracy of the analog divider and by a parasitic "jitter" of the sampling pulses of the bias current. For these reasons, implemented versions of these converters are restricted to 6–8 bits. Serial counterparts of these devices (see the next section) seem to have better prospects.

4.2 Serial Converters

Serial converters work by counting the SFQ pulses generated by a Josephson junction. In advanced versions of the converter (see for example Filippenko et al. 1991, Miller et al. 1991) the analog signal flux Φ_x is applied to the interferometer with two overdamped Josephson junctions, so that its increase by Φ_0 causes generation of the SFQ pulse by one junction, while its decrease by Φ_0 causes generation by the other junction. These pulses can be picked up and counted by simple SFQ binary counters (also using overdamped Josephson junctions). Subtraction of the counts of these two channels yields the binary code of Φ_x/Φ_0. In order to achieve high performance, subtraction and digital filtering of the output signals should be fulfilled as quickly as possible.

These functions can be performed by the recently developed ultrafast RSFQ logic circuits (see Sect. 5.2). Estimates (Likharev and Semenov 1991) show that in this case the serial converter performance can be superior to those of both flash-type superconductor converters and the best semiconductor devices; for example, conversion of a 100 MHz signal with 16-bit accuracy seems quite feasible. By changing the number of pulse counting/filtering sections of the series converter, maximum signal frequency can be traded for better accuracy (or vice versa):

$$n_{\max} = \frac{3}{2}\log_2(\tau_a f_{\max})^{-1} \tag{4}$$

where $\tau_a = \pi/\min[f_c, f_p]$ is the so-called aperture time.

4.3 Low-Frequency Converters and Digital SQUIDs

At low frequencies (below ~10 MHz), Eqn.(4) yields such a large number of bits that the conversion accuracy is limited by the analog input components of the device. In this frequency range compensation-type converters become necessary. These devices use low-bit (but high-sensitivity) zero-point detectors to balance the signal with a negative-feedback-loop output.

The simplest zero-point detector of this kind can be the usual analog SQUID with a room-temperature analog-to-digital converter. Feeding the Josephson junction digital-to-analog converter described in Sect. 3.6 with the resulting digital signal, a high-precision, low-frequency analog-to-digital converter (or, in other words, a digital dc voltmeter) can be obtained which apparently has the capability to provide up to 30 correct bits of input dc voltage in less than about 1 ms (Semenov and Voronova 1989). Such voltmeters can be valuable for metrology and scientific instrumentation.

Another possibility is to use a dc SQUID together with a low-bit superconductor analog-to-digital converter. In this case the feedback loop can be implemented using the RSFQ technology and placed on the same chip with the SQUID and a necessary digital filter (Likharev and Semenov 1991). This device is most convenient for measurement of magnetic flux and can be considered as a digital SQUID. Analysis shows that this device (still to be implemented) can combine the high sensitivity of the usual SQUIDs with extremely high slew rates approaching $10^{10}\,\Phi_0\,\mathrm{s}^{-1}$. This feature would make such devices very convenient for applications in noisy environments.

5. Digital Devices

As early as in the 1960s it was clearly understood that Josephson junction integrated circuits could be very attractive for implementation of superfast digital signal processing, due to a unique combination of factors, including (see for example Likharev and Semenov 1991):

(a) the availability of compact superconducting microstrip lines capable of carrying picosecond waveforms with low dissipation and distortion, low crosstalk, and with very high speed (~100 μm ps^{-1}) by any on-chip distance;

(b) the ability to generate ultrashort pulses using Josephson junctions which can be impedance-matched with the transmission lines, thus enabling the ballistic propagation of the signals;

(c) low power dissipation (typically below 1 μW) of the junctions, making high device density and hence small propagation delays feasible.

This unique opportunity induced a lot of interest in the implementation of Josephson junction digital devices in the 1970s, with the famous IBM cryogenic computer project as a highlight. When this large-scale attempt was eventually dropped in 1983, after more than a decade of hardship, it was considered by many as a sign that the hopes for fast superconductor-aided computation had gone forever. However, since 1983 most of the problems responsible for the IBM project failure have been overcome.

The first of these problems, low reliability of the soft superconducting materials (e.g., lead alloys) used by IBM, was circumvented by development of the so-called all-niobium whole-wafer technology (see *Superconducting Device Fabrication*). This remarkable achievement has enabled several Japanese groups (including notably those of Fujitsu, Hitachi, NEC and ETL) to design and fabricate several single-chip digital LSI circuits with up to 20 000 Josephson junctions. Figure 1 shows an 8-bit digital signal processor (DSP) fabricated in early 1990 by Fujitsu. The DSP operates at 1 GHz calculating at 1 gigaoperation per second without parallel processing. The total power dissipation is 12 mW. A comparable silicon CMOS DSP operates with a clock frequency of 10 MHz and has a power consumption of 300 mW. A number of companies have demonstrated single-chip superconducting 4-bit microprocessors with clock rates over 750 MHz. Table 1 compares clock speed and power

Figure 1
The Fujitsu Josephson 8-bit digital signal processor (DSP) (size 5 mm × 5 mm). The DSP chip contains an 8-bit by 8-bit multiplier, a 13-bit ALU (Arithmetic logic unit) a 64-word by 24-bit instruction ROM (read only memory), a 16-word by 8-bit coefficient ROM, a sequencer and a shift register. It contains 6300 MVTL gates and 23 000 Josephson junctions 1.5 μm in diameter

dissipation for similar 4-bit microprocessors made in silicon and gallium arsenide. The speed of these circuits is, however, still quite low compared with the ultimate speed of the Josephson junctions and is less than an order of magnitude higher than that of the fastest room-temperature semiconductor circuits performing similar functions. The reason for the limitation on speed lies in the use of so-called latching logic which in one form or another is the basis for the original IBM project and current Japanese designs.

Table 1
Microprocessor characteristics (4-bit)

Device	Maximum clock (MHz)	Power (mW)
Si[a]	30	1.4
GaAs[b]	72	2.2
Josephson	770	0.005

Source: Hasuo and Imamura (1989)

5.1 Latching Logic

In latching logic, digital information is coded by the presence (binary 1) or absence (binary 0) of the dc voltage across an underdamped Josephson junction. This coding has allowed the design of logic and memory circuits with wide parameter margins (typically wider than ± 20%) and quite fast 0→1 switching times (a few picoseconds). An MVTL (modified variable threshold logic) OR gate using latching logic is shown in Figure 2, the gate has 1.2 μm minimum junction diameter, with a recorded 0→1 switching time of 1.5 ps per gate and a power consumption of 12 μW per gate. The reset (1→0 switching) of the circuits is, however, very slow and takes about 1 ns, due to certain peculiarities of Josephson junction dynamics (Likharev 1986). This is mostly the reset process which limits the operation speed of the latching logic circuits.

Another drawback of latching logic gates is their relatively high power dissipation which, in particular, ruins any hope for large-scale-integration circuits of

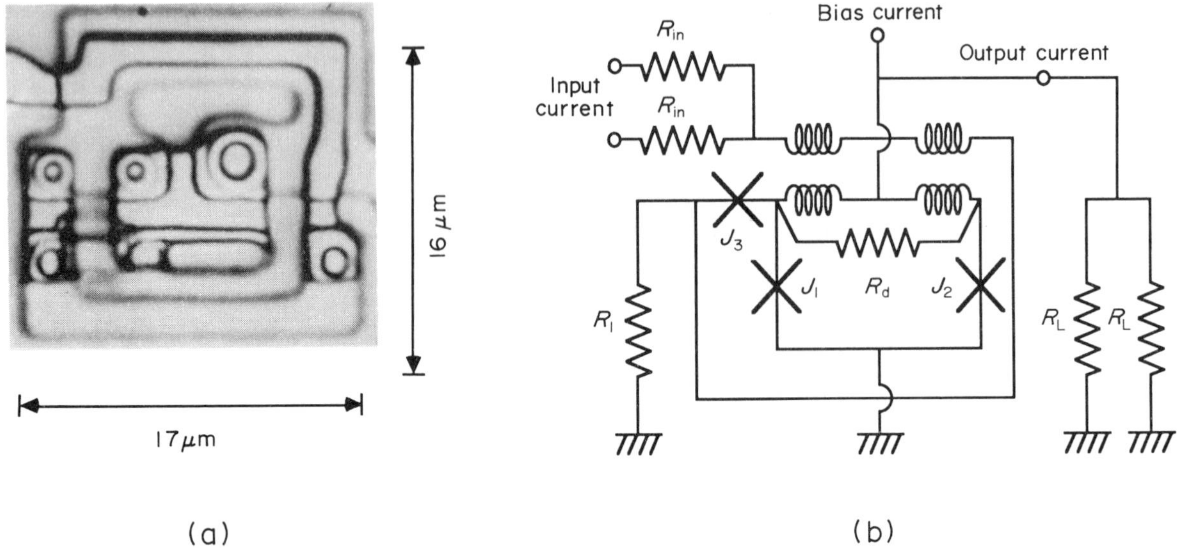

Figure 2
(a) Photomicrograph of an MVTL OR gate and (b) the equivalent circuit for the gate

this kind based on the high-T_c superconducting materials operating at liquid nitrogen temperatures.

5.2 Single-Flux-Quantum Circuits

Since the early 1970s, several attempts have been made to avoid latching logic problems by coding of digital bits by single flux quanta (SFQ). For a long time these attemps resulted in not much more than an exchange of the old problems for new ones. In particular, most SFQ circuits suffer from narrow parameter margins and an inability to communicate quickly with more than a few near neighbors. Most importantly, they need external global timing (typical for latching circuits as well) which limits their clock frequencies to well below 10 GHz.

Nevertheless, at least one approach proposed in 1985 and termed rapid single-flux-quantum (RSFQ) logic is apparently capable of overcoming these limitations (Likharev and Semenov 1991). In circuits of this family the information is stored in the form of the usual static SFQ, but is transferred in the form of short (picosecond) SFQ pulses (Sect. 2.3) which can propagate ballistically by any intrachip distance. Most importantly, the RSFQ circuits (using overdamped Josephson junctions) allow superfast local self-timing.

As a result, several low T_c (all-niobium) RSFQ circuits of small and moderate integration scales have been designed, fabricated, and tested successfully to operate at clock frequencies beyond 100 GHz, despite the relatively immature (5 μm) fabrication technology employed. Numerical simulations show that with a modern 1.5 μm technology the maximum frequency can be more than doubled.

The RSFQ circuits can be used practically, first of all, for single-chip digital-signal processing with very high rates (approaching, for example, one trillion 32-bit numbers per second, some two orders of magnitude larger than the speed available with the fastest semiconductor microprocessors). In particular, such processors would enable sophisticated methods of the digital filtering to be applied to compact radar and communication systems.

Another major possible application of the RSFQ circuitry is for analog-to-digital conversion. This technology alone seems capable of providing the high-speed digital-signal filtering necessary for implementation of the devices described in Sect. 4, in particular of the high-performance digital SQUIDs.

Note that a high-speed general-purpose computer is not among the primary applications of the RSFQ technology. Such a computer would require a large main memory ($\sim 10^9$ bits) which cannot be placed on the processor chip using current fabrication technology. Frequent processor-to-memory communication (typical for such a computer) would involve relatively slow interchip exchange which would slow down the pace of the RSFQ (or any other superfast) logic circuitry. This application will become possible only when the progress of packaging technology allows the passing of picosecond SFQ pulses from chip to chip.

6. Summary of Future Prospects

For superconductor electronics the last few years have been marked not only (and maybe not so much) by the advent of the high-T_c superconductors, but also by a notable series of inventions of new Josephson junction circuits using the SFQ pulses. Most of these new ideas are still to be implemented practically, first of all using well established all-niobium low-T_c fabrication technology. When it is done (presumably in the next few years), several new devices of an outstanding performance will be within reach, which will more than justify helium cooling.

On the other hand, high-T_c superconductors have most promise for use in the simplest electronic devices and components for which transfer to nitrogen refrigeration would bring significant cost savings. Use of these new materials for complex active devices may happen only after reliable technology for high-T_c Josephson junction fabrication is established, or if some completely new principles for their application in electronics are found.

See also: Josephson Effects in Weak Links; Superconducting Device Fabrication; Thin Films, Multilayers and Devices, Superconducting

Bibliography

Barone A (ed.) 1988 *Superconductive Particle Detectors*. World Scientific, Singapore

Braginskii V B, Mitrofanov V P, Panov V I 1985 *Systems with Small Dissipation*. Chicago University Press, Chicago, IL

Devyatov I A, Kuzmin L S, Likharev K K, Migulin V V, Zorin A B 1986 Quantum- statistical theory of microwave detection using superconducting tunnel junctions. *J. Appl. Phys.* 60: 1808–28

Filippenko L V, Kaplunenko V K, Khabipov M I, Koshelets V P, Likharev K K, Mukhanov O A, Rylov S V, Semenov V K, Vystavkin A N 1991 Experimental implementation of analog-to-digital converter based on the reversible ripple counter. *IEEE Trans. Magn.* 27: 2464–7

Hasuo S, Imamura T 1989 Digital circuits. *Proc. IEEE* 77: 1177–93

Jain A K, Likharev K K, Lukens J E, Sauvageau J 1984 Mutual phase locking in Josephson junction arrays. *Phys. Rep.* 109: 309–426

Kupriyanov M Yu, Likharev K K 1990 Josephson effect in high-temperature superconductors and structures. *Sov. Phys. Usp.* 33: 340–64

Kuzmin L S, Likharev K K, Migulin V V 1981 Microwave receivers using SQUIDs and Josephson junction arrays. *IEEE Trans. Magn.* 17: 822–5

Lee G S, Peterson D A 1989 Superconductive A/D converters. *Proc. IEEE* 77: 1164–76

Likharev K K 1979 Superconducting weak links. *Rev. Mod. Phys.* 51: 101–60

Likharev K K 1986 *Dynamics of Josephson Junctions and Circuits*. Gordon and Breach, New York

Likharev K K 1990 Progress and prospects of superconductor electronics. *Supercond. Sci. Technol.* 3: 325–37

Likharev K K, Semenov V K 1991 RSFQ logic memory family: a new Josephson-junction technology for subterahertz-clock-frequency digital systems. *IEEE Trans. Appl. Supercond.* 1: 1–26

Likharev K K, Semenov V K, Zorin A B 1990 New possibilities for superconductor electronics. In: Ruggiero S, Rudner D (eds.) 1990 *Superconducting Devices*. Academic Press, Boston, MA, pp. 1–49

Malozemoff A P, Galagher W J, Schwall R E 1987 Applications of high-temperature superconductors. In: Nelson D L, Whittingham T S, George T F (eds.) 1987 *Chemistry of High-Temperature Superconductors*. ASC, Washington, DC, pp. 280–306

Miller D L, Przybysz J X, Kang J, Hamilton C A, Burnell D M 1991 Josephson counting analog-to-digital converter. *IEEE Trans. Magn.* 27: 2761–4

Piel H 1988 Fundamental features of superconducting cavities for high-energy accelerators. In: Turner S (ed.) 1988 *Proc. CERN Accelerator School (Oxford, 1985)*. CERN, Geneva, pp. 736–65

Semenov V K, Voronova M A 1989 DC voltage multiplier: a novel application of synchronization in Josephson junction arrays. *IEEE Trans. Magn.* 25: 1432–5

Solymar L 1972 *Superconductive Tunneling and Applications*. Chapman and Hall, London

Van Duzer T, Turner C W 1981 *Principles of Superconducting Devices and Circuits*. Elsevier, New York

Whiteley R S, Kuo F, Radparvar M, Faris S M 1989 An all-NbN time domain reflectometer chip functional above 8 K. *IEEE Trans. Magn.* 25: 770–6

Withers R S, Ralston R W 1989 Superconductor analog signal processing devices. *Proc. IEEE* 77: 1247–63

K. K. Likharev
[State University of New York, Stony Brook, New York, USA]

Superconducting Magnets

Superconducting magnets can be generally recognized as comprising current-carrying windings that have no electrical resistance in operation. In principle, it is possible to build a magnet from superconducting material that is a copy of a conventional winding that creates an electromagnetic field, and a wide range of magnet types should be possible. The practice is rather different, although superconducting magnets with complex shapes and weighing many tonnes have been built (see *AC Applications of Superconducting Materials*; *Superconducting Materials: Large-Scale Applications*).

The properties of superconducting materials place constraints on the designs of magnets through the cost associated with engineering a suitable winding for the required magnetic-field profile while maintaining this in the superconducting condition. The point of considering using superconducting materials to create magnetic fields is that the field strength within any current-carrying circuit can be directly linked to the density of current in the circuit. The absence of resistance means very high current densities can be obtained, relative to conventional conductors. However, arbitrarily large current densities are not possible (see *Critical Current Density*). The industrial exploitation of superconducting magnets has been governed by what can be reliably achieved for given capital and operating costs set against the operational circumstances of the systems using the magnetic field. Because superconductivity only occurs at relatively low temperatures, special cryogenic arrangements are necessary to cool the winding and to prevent heat generation within the windings. Many of the practical restrictions on the exploitation of superconducting magnets result from this requirement.

High-strength magnetic fields are used mainly to influence matter for experimental or practical purposes. The source of the field interaction with matter is the electric charges associated with atoms, both in their outer electron configurations and their nuclei. Physical and chemical interactions between collections of atoms making up systems, either solid, liquid or gaseous, are driven by the energy difference between the systems. The energy content of each system has a component due to the impressed magnetic field that is a characteristic of the electronic and nuclear arrangements of the system. However, the magnetic energy component will be proportional to the intensity of the impressed field although the detailed relationship can be quite complex. Therefore, many situations arise in which the higher the impressed field, the greater is the magnitude of the response of the interacting systems and, in consequence, there is a requirement for sources of high magnetic field.

1. Importance of Current Density to Field Creation

In the SI system of units, magnetic field strength $\boldsymbol{H}$ is measured as ampere-turns per meter ($\mathrm{A\,m^{-1}}$) and in the cgs system as oersteds (Oe) where $1\,\mathrm{A\,m^{-1}} = 4\pi \times 10^{-3}\,\mathrm{Oe}$. The flux density $\boldsymbol{B}$ in SI units is measured in tesla (T) which are webers per square meter ($\mathrm{W\,m^{-2}}$), the weber is a unit of flux, and in the cgs system in gauss (G) where $1\,\mathrm{T} = 10^4\,\mathrm{G}$. The flux density $\boldsymbol{B}$ is related to the impressed field $\boldsymbol{H}$ by a characteristic constant μ, termed permeability, which is unique to the medium permeated by the field:

$$\boldsymbol{B} = \mu \boldsymbol{H} \tag{1}$$

In a vacuum $\mu_0 = 4\pi \times 10^{-7}\,\mathrm{Wb\,A^{-1}\,m^{-1}}$ (or henries per meter, $\mathrm{H\,m^{-1}}$) in SI units. For other materials, tables of relative permeability μ_r are available from measurement, where

$$\mu = \mu_0 \mu_r \tag{2}$$

The relative permeability in air is almost one.

The flux density $d\boldsymbol{B}$ due to current $\boldsymbol{I}$ flowing in an element of conductor at a point P is given by

$$d\boldsymbol{B} = \mu \boldsymbol{I} \times dl \frac{\sin\theta}{4\pi r^2} \tag{3}$$

(see Fig. 1a). For any continuous path, the contribution of all the elements dl can be summed for all the points P at which it is required to know the flux density or field strength. This may be done analytically for simple geometries, or numerically with computers for complex current paths and field zones (see *Design with Magnetic Materials: Computer Simulation*). Thus, in principle, electromagnets can be designed to produce any required field profile. For a simple loop, the field component in the z direction on the axis at P is given by

$$B_z = \mu I \frac{a^2}{2(a^2 + z^2)^{3/2}} \tag{4}$$

(see Fig. 1b). For a simple solenoid comprising a helical winding (like a spring) the field can be modelled by a stack of loops along the z axis; the contribution of the loops can be summed. The current density J in the conductor is the current I divided by the conductor cross-sectional area A; so J is measured in amperes per square meter. Therefore, the greater the current density in a conductor, the larger will be the flux density at the center line of the coil.

If the conductor is not superconducting, it will have a resistance to electric current and power will be dissipated. For a single loop the power W is:

$$W = 2\pi\rho a J^2 A \tag{5}$$

where a is the loop radius and ρ is the resistivity in ohm meters. Using resistive conductors, as the required field is increased greater efforts to cool the conductor are needed to extract the dissipated power and eventually a practical limit is reached beyond

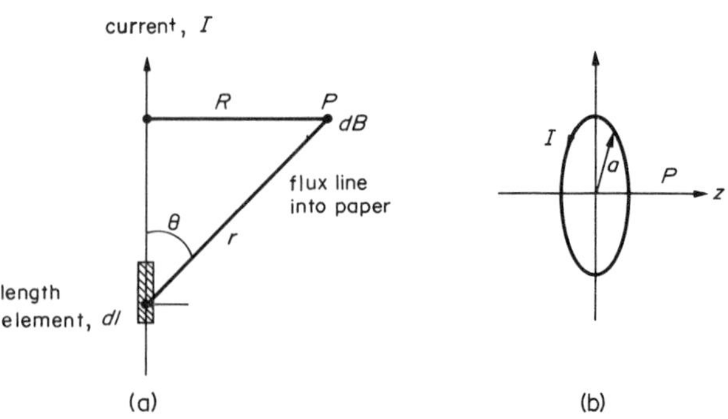

Figure 1
Relationship between magnetic field and current flow in a conductor on which a magnet design is based:
(a) magnetic field at point P due to a current element $l\,dl$; and (b) dimensions of a simple loop as an elemental building unit of a solenoid

which it will be necessary to employ a superconducting zero-resistance winding. Numbers of different ways of cooling exist and the magnet geometry plays an important part in defining "the economic" crossover value of J favoring a superconducting magnet. For comparison of flux densities, the flux density over the earth's surface is 0.000 05 T, a permanent "horseshoe" magnet between the poles is 0.1 T, near the surface of a rare-earth intermetallic permanent magnet the flux density is 0.4 T, large electromagnets in industrial also use 0.4 T and superconducting magnets may reach 18–20 T.

2. *Boundary Conditions on the Performance of Superconducting Magnets*

The value of field strength obtainable from a winding using superconducting materials is limited by the current density that the superconductor can support in the field of the magnet. As the field strength applied to the superconductor increases, so the maximum value of J decreases. The attainable current density in the actual winding may be less than the material is capable of under ideal conditions. This depends on the degree of success in designing the magnet structure and conductor layout to minimize the incidence of small internally generated heat pulses. Progress in the evolution of superconducting magnets can be charted by the refinement of designs to improve conductor stability in its working environment. The current density as a function of field and temperature is a fundamental characteristic of each superconducting material (see *Critical Current Density*).

It needs to be appreciated that the specific heat of all materials tends to decrease as the temperature of the material is reduced. Liquid helium is the refrigerant currently used for commercial purposes, and at its boiling point (bp) (4.2 K at atmospheric pressure) specific heats are about one-thousandth of their room-temperature value. At this temperature, it requires 10^3–10^4 J m^{-3} to raise the temperature by 1 K for materials commonly used to construct superconducting magnets. Conventional superconductors, or low-temperature superconductors (LTSs), only exhibit superconductivity in the temperature range 0–23 K, with each material having its own critical temperature within that range above which they will not conduct. The two most commonly used conductors are an alloy of niobium and titanium (see *Niobium–Titanium Alloys*) and an intermetallic compound of niobium and tin (see *A15 Compound Superconductors*). These materials came into widespread use because reliable processes for their fabrication in composite forms were evolved: their critical temperatures are approximately 9 K and 18 K, respectively. As with increasing the applied field, increasing temperature will reduce the current density until it is zero at the critical temperature. Commercial superconducting magnets are generally operated at temperatures close to the bp of liquid helium at atmospheric pressure. This gives a working temperature range between operating conditions and the critical temperature while using a practical, if exotic, refrigeration fluid. To be economic in the use of superconductor, magnets are usually operated at current densities 25–90% of their maximum at the magnet operating temperature. This safety margin can accommodate typically between 1 K and 0.1 K temperature rise during magnet operation. Because of the small specific heat, this means that minute quantities of heat are sufficient to cause a local loss of superconductivity which may cause the whole magnet to fail.

Sources of heat that may cause the magnet winding to lose its superconductivity can be grouped into three categories: environmental, mechanical and electromagnetic. Mechanisms that give rise to heating and design solutions to minimize heat effects will now be reviewed.

2.1 Environmental Sources of Heat

All superconducting magnets are housed in structures that shield the magnet from heat transferred from the environment by radiation, conduction and convection. The assembly of vessels and components is termed the cryostat and acts like the classic Dewar vessel. The size and shape of the cryostat are dictated by the dimensions, geometry and use of the magnet it contains, with various arrangements to gain access to the magnet chamber for refrigerant and to make electrical connections to the magnet. Figure 2 shows the layout of components for a medium-size magnet for nuclear magnetic resonance imaging (NMRI) (see *Nuclear Magnetic Resonance Imaging*).

A common element of cryostat design themes is to absorb most of the radiation from the room-temperature environment by boiling liquid nitrogen (bp 77 K) contained in a separate tank that is also linked to the magnet supports to thermally ground these. Reflective shields provide the magnet with its radiation environment and are either cooled by boiled-off helium gas passed through heat exchangers (negative thermal feedback) or local refrigeration engines are used to provide a low radiation temperature of 20–50 K. The whole system of vessels is evacuated and all magnet support members are made as long as possible, frequently employing low-conductivity composite materials, to minimize convection and conduction. Small magnets are nearly always immersed in a bath of helium liquid while very large magnets are sometimes built with integral cooling networks through which helium refrigerant can be pumped. Liquid helium is usually refrigerated by large liquifiers out of natural gas streams and supplied for transfer to magnets in transport Dewars, typically 50–5000 l capacity. Very large magnet installations often justify their own liquifier and work with a closed cycle of helium flow. Radiation gives rise to typically

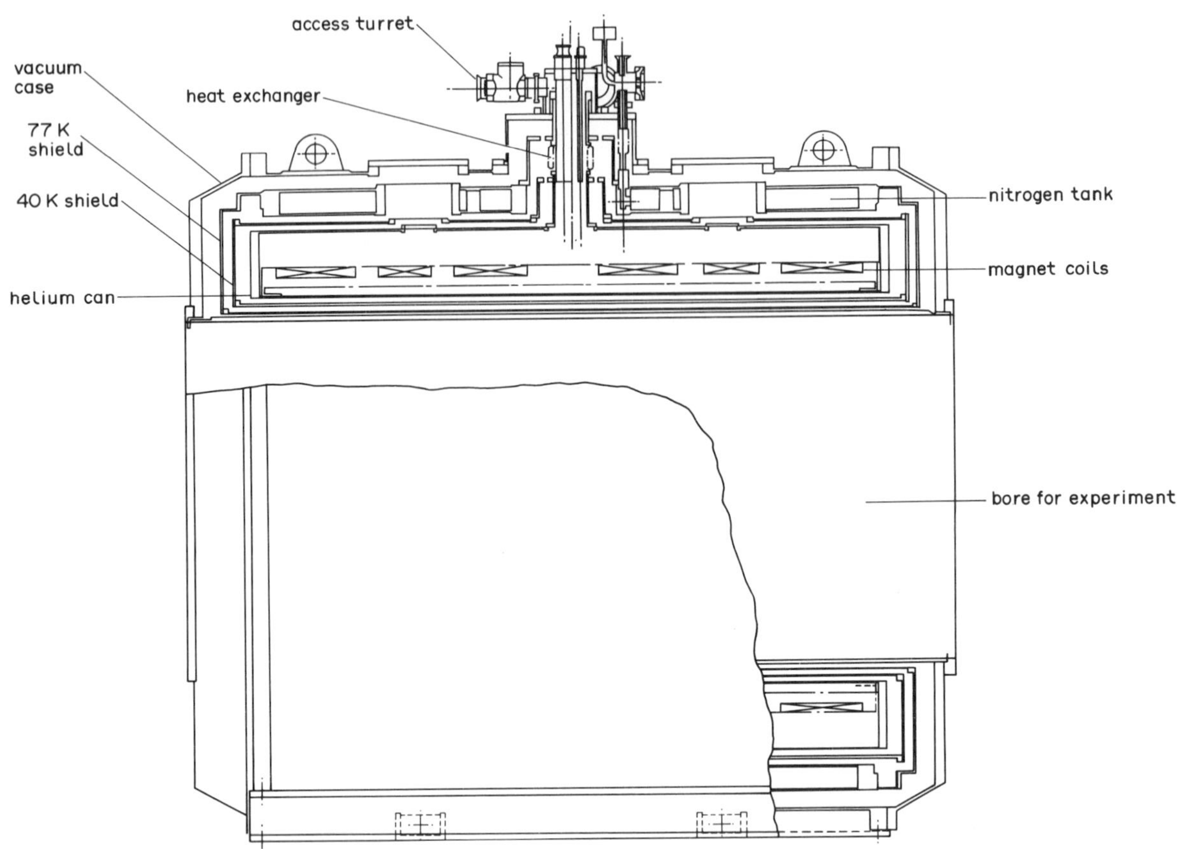

Figure 2
Principal components of a conventional superconducting MRI magnet (overall length 2.0 m, bore diameter 1.0 m)

20 W m^{-2} (going from 300 K to 4.2 K) and 0.1 W m^{-2} (going from 77 K to 4.2 K). It should be noted there are many detailed design considerations placed on cryostat structures by the need to consider the working objectives of the magnet.

2.2 Mechanical Sources of Heat

When a magnet suffers a local loss of the superconducting state in the winding, the normal or "hot" zone may spread throughout the whole winding. The magnet is said to have "quenched." In practice, superconducting magnets respond to the internal stress created in the winding by the electromagnetic forces with discrete movements of wire turns. The design aims to clamp turns as rigidly as possible to the position assigned to the turns during construction but, usually, very small movements are possible and the winding "settles in" to its exact operating set of positions. During settling in, bonding media between conductor turns will be distorted, possibly cracked and, because the movement of the wire turns does work, local heat pulses are generated.

Forces developed on a conductor turn per unit length are proportional to the product $\boldsymbol{H} \times \boldsymbol{I}$ and can amount to some hundreds of kilograms per meter of conductor length. The field varies in magnitude and direction within the windings, so producing a force pattern which for engineering purposes needs to be converted to a stress profile. This requires a complete knowledge of the force distribution and the effective material moduli of the components of the winding. Approximate analytical solutions can be obtained for simple geometries and finite-element methods using computers to provide solutions are employed for more complex winding shapes. In this latter approach, the solid body is divided into a series of small volumes and the stress–strain equation applied in each with boundary conditions for solution being matched at each boundary. Nevertheless, because of the small value of specific heat, it has been estimated by Wilson (1983) that the design of the winding must prevent sudden conductor movements of the order of 2 μm if the winding is to be electromagnetically efficient and reliable. The "settling in" process can give rise to the magnet quenching on a succession of runs up to field at progressively higher values of field on each run as the mechanically least-stable conductor elements "bed down." This process is referred to as "training."

2.3 Conductor Stability

If the superconductor in the magnet winding is operated at a current density less than the maximum possible in the applied field, a margin of operating safety exists to accommodate disturbances. However, the resistance of superconductors in the normal state is high which results in a small normal length of conductor heating up rapidly if current is maintained in the winding. All practical superconductors are used embedded in a good conventional conductor, like copper, to provide a low-resistance current path around the normal length of superconductor (see *Multifilamentary Superconducting Composites*). The copper also conducts heat away from the "hot zone" to the coolant and assists in lowering the temperature of the superconductor so that the winding recovers its resistanceless state after a localized discrete disturbance. Composite copper-to-superconductor volume ratios can be defined against operating current density and temperature to achieve recovery against a given size of heat pulse. An extreme is cryogenic stabilization, where the ohmic heating caused by carrying the magnet current in the copper is much less than the rate of cooling in the steady state. This latter condition requires refrigerant to be close to the composite everywhere in the winding and is only efficiently applied to large magnets.

2.4 Electromagnetic Sources of Heat

Superconducting materials are divided into two categories on the basis of the manner in which they exclude magnetic flux from their interior (the Meissner effect) (see *Magnetic Properties of Superconductors*). Conductors used for magnets are classed as type II which can exist with "threads" of magnetic flux penetrating the interior, with each thread surrounded by a supercurrent. These threads are termed fluxoids or flux vortices and have various energetic interactions with the crystal lattice of the superconductor and its natural defects (see *Flux Line Lattice*). These interactions are very important as there is a force $\boldsymbol{B} \times \boldsymbol{I}$ between a current flowing through the material and the fluxoids that, if it causes the fluxoids to move, will give rise to an apparent resistance in the material. The fluxoids need to be pinned to the lattice for the bulk of the material to behave as a resistanceless carrier of current. If a type-II conductor is placed in a magnetic field, it will tend to exclude magnetic flux by developing screening currents (analogous to eddy currents in conventional conductors under ac conditions) and a gradient in flux density will be established from a high value at the surface dropping towards zero at the interior (see Fig. 3).

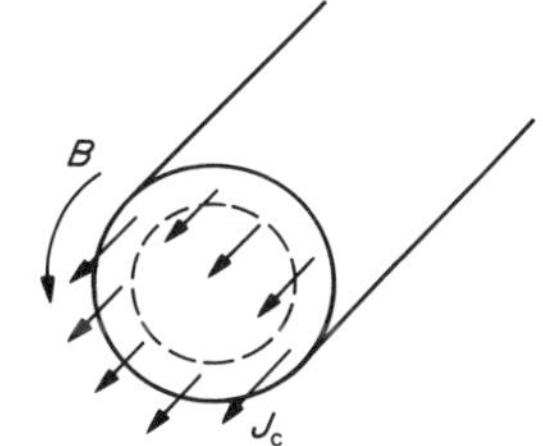

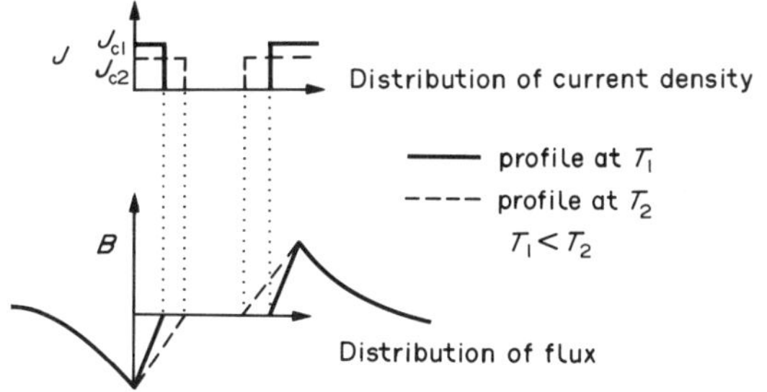

Figure 3
Superconductor carrying current and generating a "concentric self" field distribution of current and field across the section: J_c, the maximum current at the prevailing external surface field B and operating temperature T, is reduced by a rise in T

In fact, a gradient in magnetic field is always associated with a flow of current in any medium and a flux gradient must be maintained for transport current as well as screening current in the superconductor (the transport current being the current flowing from start to end of the magnet winding). There is no current in unpenetrated regions; however, if the current through the winding is increased, because the critical current density is constant, flux penetrates further into the superconductor (see *Critical State Model*). This flux motion generates heat within the conductor, which can raise the temperature of the conductor and, in turn, reduce the critical current density leading to further flux penetration (Fig.3). This is a positive feedback loop leading to loss of superconductivity.

Unstable flux penetration of a superconductor can be avoided by making the cross section of the superconductor (at right angles to current flow) very small (adiabatic stability). If each filament of superconductor is also surrounded by a good electrical and thermal conductor, so that movement of magnetic flux is slowed down and heat flow speeded up, the temperature excursion after a flux jump is minimized (dynamic stability). The form of a practical superconducting composite has now been defined as an array of fine filaments (10–50 μm diameter) running along the length of the composite conductor, where the matrix is copper which separates the filaments. Different designs of composite array are possible to match detailed performance requirements of the magnet (see *Multifilamentary Superconducting Composites*). The composite will be twisted along its length to decouple the filaments against eddy currents being set up in the copper matrix when the applied field changes at the surface of the conductor (i.e., the field in the magnet winding is changed). By design of the fabrication process of the conductor, the forces pinning magnetic flux can be varied, so modifying the current field properties of a type-II conductor. Optimizing conductor performance has been a major industrial activity in type-II superconductor manufacture. If the field provided by the magnet is required to vary with time—an extreme case being an ac field—then losses will develop from eddy currents and flux motion hysteresis based on the mechanisms described in this section (see *AC Losses in Superconducting Materials*). The magnet will develop an apparent resistance across its terminals that will place an extra load on the refrigeration system and need special design responses in defining the conductor geometry (see *AC Applications of Superconducting Materials*). (Note that superconducting magnets are resistanceless in the stable dc state but still have inductive and capacitive reactance.) Superconducting magnets can be energized, brought to a stable dc condition and a superconducting link established across their terminals. In this way, the dc field can be maintained without an external source of power and the field-creating current persists in the winding to provide a very stable magnetic field. A magnet operating in this way is said to be in "persistent mode."

3. Industrial Exploitation of Superconducting Magnets

Initially, uses of superconducting magnets were confined to laboratory physics experiments, starting in the 1950s when type-II materials became available. The use of superconductivity was promoted by the availability of stable high fields. The practicality of achieving high fields (2–3 T) over several cubic meters of space, gave rise to a number of programs to provide magnets for particle physics experiments. These magnets were usually government funded and built in the laboratories where the experiments were designed. Manufacturing superconducting magnets on an industrial basis originated with the supply of instrument systems to laboratories using small magnets. A very important use of superconducting magnets has been in providing for nuclear magnetic resonance (NMR) experiments, initially for spectroscopy and, since 1980, for whole-body imaging.

Low-temperature superconducting magnets range in cost from small laboratory magnets in the range US$10 000–20 000 up to several millions of US dollars for large factory made systems. Many of the obvious uses of superconducting magnets have not been accepted partly on cost grounds, but questions of reliability of the superconducting winding and refrigeration system often dominate economic assessments of deploying a superconducting magnet. Examples of magnet applications are compact synchrotrons for x-ray lithography of chips, levitated high-speed transport, energy generation by fusion and by magnetohydrodynamics, medical imaging scanners and NMR spectrometers deployed to study molecules important in biochemistry (see *Superconducting Materials: Large-Scale Applications*). The world market for helium superconducting magnets is now some US$500 million per year.

The future growth prospects for superconducting magnet markets will be hugely affected by progress in the development of high-temperature ceramic superconductors (HTSs). The deployment of superconducting magnets could be much more widespread if ceramic conductors become available for operation between 30 K and 80 K. Although ceramics bring problems of brittle mechanical behavior, they may well improve the reliability of superconducting magnets by improving the stability of the superconducting state. The central reason for this assertion is that operation of a magnet above approximately 30 K means the specific heat can be one or two orders of magnitude greater than at 4 K.

The possible benefits to the reliability of superconducting magnets to be obtained from ceramic superconductors can be considered against the boundary conditions discussed for helium magnets.

3.1 Environmental Sources of Heat

Operation at temperatures (50–80 K) where material specific heats are approaching those of room temperature greatly simplifies cryostat design. Refrigeration plant becomes more efficient and heat leaks into the magnet cause smaller temperature rises. Thus, more complex winding structures become feasible and machines involving rotating cryogenic seals, such as generators, can be designed.

3.2 Mechanical Sources of Heat

Because larger heat pulses are necessary for a given size of temperature excursion, larger amounts of work done by conductor movement can be accommodated. There is also the possibility of advantageously increasing the temperature difference between the operating temperature and the critical temperature of the conductor. It seems likely that movements of the order of 1.0 mm could be accommodated before thermal effects threatened to send a region of conductor normal.

3.3 Conductor Stability

A very small portion of a magnet winding may have been driven into the normal state, and, by virtue of the cooling arrangements and the cross-sectional area of copper surrounding the superconducting material, the region may become superconducting after the disturbance ceases. Whether the hot spot propagates is dependent on the difference between the conductor critical temperature and the operating temperature of the magnet. The opportunity exists by using ceramic conductors to increase this temperature difference and to increase the critical length of conductor a disturbance must cause before the hot zone propagates. Calculations by Wilson indicate that critical propagation dimensions in ceramic conductors would be, at least, an order of magnitude greater than for helium conductors. The energy needed to create a critical hot spot is also increased dramatically from, typically, 10^{-5} J to 1.0 J.

3.4 Electromagnetic Sources of Heat

The cross-sectional diameter of the superconductor, usually 10–50 μm for operation in liquid helium, can be increased considerably by operation at 60 K or 80 K with ceramics. The higher specific heat and the possibility of a larger difference between operating and critical temperature could increase the stable superconductor diameter up to 2–3 mm. However, because the electrical and thermal conductivities of copper both increase with temperature, dynamic stability is not greatly altered by raising the operating temperature of a copper-stabilized conductor. It may be that single ceramic core conductors, amenable to low-cost production, will prove satisfactory for the manufacture of stable magnets.

3.5 Magnet Protection

Because the magnetic field of a superconducting magnet stores considerable energy (stored energies up to 1 MJ are common), some means of safe removal of this energy is necessary when the winding goes normal. The field energy is converted to heat in the magnet winding and an external protection circuit.

It is important to design the superconducting coil and conductor such that once the propagation of the hot zone becomes inevitable, it will move rapidly throughout the whole winding. This avoids the temperature of the initial hot spot rising to the melting point of the conductor, as might occur if all the field energy were dissipated in too small a volume of the winding. It also assists in reducing internal voltages in the winding. The velocity of propagation of the normal zone reduces as the specific heat of the conductor is raised. Special attention to the conductor and winding design and to protection arrangements for ceramic windings is likely to be an important part of developing rugged superconducting magnets using these new materials.

4. Conclusion

Low-temperature superconductors have provided for a wide range of high-field magnets which can be economically deployed when environmental conditions accommodate complex cryogenic engineering. Most large application areas have involved the use of dc fields. Ceramic superconductors hold out the hope of much wider deployment of superconducting magnets. This is largely because more rugged and reliable field windings should be possible and simpler, self-contained refrigeration should make a greater range of magnet geometries realizable. Devices utilizing changing or ac fields should also become economically attractive.

See also: AC Applications of Superconducting Materials; Magnetic Energy Storage; Magnetic Separation; Multifilamentary Superconducting Composites; Nuclear Magnetic Resonance Imaging; Superconducting Materials: Large-Scale Applications.

Bibliography

Foner S, Schwartz B B 1974 *Superconducting Machines and Devices*. Plenum, New York

Lynton E A 1969 *Superconductivity*. Methuen, London

Montgomery D B 1969 *Solenoid Magnet Design*. Wiley, New York

Plonus M A 1978 *Applied Electromagnets*. McGraw-Hill, New York

Saint-James D, Sarma G, Thomas E J 1969 *Type II Superconductivity*. Pergamon, Oxford

Wilson M N 1983 *Superconducting Magnets*. Clarendon, Oxford

I. L. McDougall
[Oxford Instruments, Eynsham, UK]

Superconducting Materials: An Overview

Superconductivity is a truly multidisciplinary field. Inputs from physics, materials science, electrical engineering and chemistry are needed to understand the behavior of superconducting materials. A knowledge of how composition and microstructure affect properties and device performance is essential to optimize figures of merit. The information required comes from empirical correlations, phenomenological behavior and theory. It is necessary to be familiar with concepts which come from Mendeleev's periodic table, the phenomenological theory of Ginzburg and Landau and the microscopic theory of Bardeen, Cooper and Schrieffer (BCS theory) in order to predict how superconducting materials will respond to changes in structure, composition, microstructure and external fields. (It is remarkable that the periodic table which was originally constructed to reflect atomic configurations on an energy scale of volts and subsequently found to be relevant for nuclei on a megavolt scale is also relevant for superconductors on a millivolt scale.)

Superconductivity was a laboratory mystery starting with its discovery by the Dutch low-temperature physicist H. Kamerlingh Onnes (Onnes 1911). Onnes had liquefied helium gas for the first time only a few years earlier and in so doing extended the range of available temperatures down to 1 K. This advance made it possible to study material properties over three decades of temperature rather than just two. The electrical resistance of a sample of mercury being studied to see how pure metals behave suddenly became too small to detect when the mercury was cooled below 4.2 K (see Fig. 1). Onnes and his collaborators at Leiden University soon found that superconductivity is not confined to pure mercury alone. Corresponding sudden decreases of resistance were found in tin and lead and even mercury amalgams at various low temperatures. Since the drop was sudden, the temperature at which it occurred was recognized as a critical temperature T_c. The question immediately arose of how small was the remaining resistance. To answer this, the Leiden researchers induced a supercurrent in a lead coil and found no measurable decay of the current by observing the associated magnetic moment over the 24 h it was kept below T_c (e.g., de Bruyn Ouboter 1987). Further experiments and theory have shown that even if one waited the life of the universe there would be no measurable decay of induced current. Hence, the prefix *super* is justified.

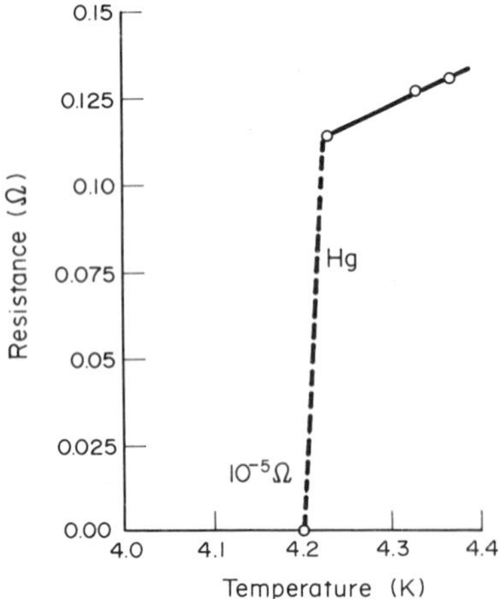

Figure 1
Resistance of a specimen of mercury vs absolute temperature. This plot by Kamerlingh Onnes marked the discovery of superconductivity (H. Kamerlingh Onnes 1911, *Akad. van Wetenschappen (Amsterdam)* **14**(113):818. Further historical references are given by Gorter (1964))

It is now known that superconductors have other spectacular properties which derive from the fact, first noted by Fritz London (London 1935), that superconductivity is a *macroscopic* quantum phenomenon. Humans are not accustomed to encountering quantum phenomena in macroscopic matter which is one reason why superconductors are so fascinating.

The Leiden experiments already provided an important clue to the mechanism responsible for the superconductivity although it may not have been obvious at the time. Highly conducting good metals such as copper do not become superconductors; poorly conducting ones do. It took almost 50 years before this clue was made transparent by the microscopic theory (see *Superconducting Materials: BCS and Phenomenological Theories*). Briefly, BCS theory shows that the transition is caused by an attractive interaction between the itinerant charge carriers (electrons or holes) mediated by the lattice which causes them to form pairs (Cooper pairs) and condense into the superconducting ground state. The attractive interaction responsible for the pairing is caused by the virtual exchange of phonons, the quantized vibrational modes of the lattice. These are the same lattice vibrations which, when thermally excited, scatter the conduction electrons—a process that results in the temperature-dependent electrical resistance. BCS theory provides the understanding for why poor metals are good superconductors: a strong electron–phonon interaction gives high resistance above T_c and strong pairing below. Almost all superconductors that have been studied can be explained by phonon-induced pairing. The BCS expression for T_c can be written in a form which is applicable for most superconductors simply as

$$T_c = \langle\omega\rangle \exp - (\lambda - \mu^*)^{-1} \qquad (1)$$

The prefactor is an average over the lattice vibrational spectrum, which for many purposes can be taken simply as the Debye temperature θ_D, and λ and μ^* are, respectively, averages over the Fermi surface of the attractive interaction and the coulomb (electron–electron) repulsion.

At the present time there are strong indications that the pairing interaction in the remarkable high-T_c copper oxide superconductors discovered in 1986 by Bednorz and Müller is not due to phonons; rather other degrees of freedom are responsible. The mechanisms responsible for the pairing interaction for those materials are under lively theoretical and experimental investigation; it is not clear to what extent, if any, the approximations that lead to the simple expression in Eqn. (1) are valid (see *Superconducting Materials: Basic Theoretical Concepts*).

Because the exponent in Eqn. (1) is difficult to calculate the equation is not very useful for finding new superconductors. Predicting the existence of new superconductors with new compositions of matter has not generally been a successful application of theory.

Discoverers of new superconductors have used the same method as Kamerlingh Onnes, namely cooling a material and observing some electrical or magnetic signal, which can even be due to a trace amount, which is a signature of the superconducting state. They have used intuition, guidance from the presence or absence of superconductivity in materials with related crystal structures, a lot of persistence and, in many cases, good fortune. Once superconductivity in a new class of materials has been discovered, then both theory and experiment become useful for guiding further research.

Present day understanding of superconductivity rests, first, on the occurrence of superconductors in a wide variety of materials many of which are mentioned later in this article, second, on microscopic and phenomenological theory tested by experiment and, third, on the performance of model systems and devices into which the superconducting materials have been incorporated. A useful perspective for the survey is gained by continuing the brief historical sketch of important achievements.

1. The Early Years

Shortly after the discovery of the superconducting transition in mercury, Onnes and his co-workers found that, at temperatures below T_c, a magnetic field of a few hundred gauss applied to the sample caused the superconductivity to disappear. Normal electrical resistance was restored at a well-defined value of this quenching or critical magnetic field H_c; it was also discovered that superconductivity could be quenched by a sufficiently high critical electrical current density J_c.

The possibility of using superconductivity for electric power transmission or in electrical equipment such as generators or motors seemed at first to be ruled out by the low quenching fields and currents found in these early experiments; however, these limitations have to a great extent been removed by further advances (see *Critical Current Density*; *Magnetic Properties of Superconductors*).

After World War One, research on superconductors was expanded from Leiden to laboratories in Germany, the Soviet Union, the UK, Canada and the USA as cryogenic technology gradually spread around the world. By 1928 the known superconductors were all low-melting polyvalent post-transition metals. The search for new superconductors was begun in other laboratories. W. Meissner and his co-workers in Berlin took an important new direction and they discovered many new ones (Shoenberg 1960). They investigated the occurrence of superconductivity in the transition elements of the periodic table, which have atoms with incomplete *d* shells, and in their compounds (see *Superconducting Materials: Elements*). These elements are high-melting refractory metals and are difficult to prepare in pure form. Initially they were known as hard superconductors partly because they were brittle due to the presence of dissolved gases such as oxygen and partly because the superconductivity was harder to destroy by magnetic fields than in the nontransition metal elements. This did not prevent discovery that, in fact, new superconductors are often discovered in impure systems. Meissner and his colleagues found most of the elements of groups IV and V, that is, titanium, zirconium, niobium, vanadium and tantalum, to be superconducting. It is noteworthy that niobium in one form or another is the material of choice for all low-temperature superconducting applications at present. Its alloys and compounds are used in making wires and cables for large-scale applications such as magnets for magnetic resonance imaging (MRI) scans, and niobium tunnel junctions are the basis for electronic applications such as high-speed digital circuits, the voltage standard and the most sensitive detectors of electromagnetic signals. Meissner's group took a bold step when they extended their investigations to transition metal compounds and were rewarded by finding even more superconductors. They found a large number of carbides and nitrides with the rock salt structure including NbC with $T_c > 10$ K (see *Nitride and Carbonitride Superconductors*). (The difficulty in preparing pure phases resulted in some false assignments. For instance, TiC and HfC were incorrectly found to be superconducting probably due to the presence of small amounts of their corresponding nitrides. The ability of trace amounts of a superconducting phase to give large superconducting signals has proved to be an important factor in the discovery of new materials as well as being responsible for many false claims of new superconductors.) Meissner's group found the first copper-containing superconductor in 1929. Copper sulfide was found to have a T_c above 1 K. The best known electrical conductor, copper, itself not a superconductor, was turned into one by heating in sulfur, this being another reason why superconductivity was considered to be such a mystery. CuS, although it has a rather complex crystal structure (covellite), is probably a BCS superconductor described by Eqn. (1).

Working with pure elements, Meissner and Ochsenfeld discovered that superconductors exhibit a property almost as unexpected as zero resistance (see *Superconducting Materials: Basic Theoretical Concepts*). When cooled below T_c in the presence of a magnetic field, the flux that penetrates the superconductor above T_c is completely expelled, a behavior known as the Meissner effect. On the other hand, the Meissner effect, that is, the expulsion of flux, is separate and distinct from zero resistance and indicates reversible thermodynamic behavior. The exclusion of flux when the magnetic field applied to a superconductor is changed is a property predicted for any zero-resistance metal by Maxwell's equations.

In 1935, F. and H. London developed a set of electrodynamic equations which provided a framework for the Meissner effect and showed that magnetic field shielding was associated with a thin penetration layer λ (typically $\sim 10^{-5}$ cm), which carries a persistent supercurrent. The London equations also explained the microwave behavior of superconductors and correctly predicted the superconducting behavior of thin films. For film thickness $d < \lambda$ there is very little exclusion of flux and, hence, very little increase in energy in an applied field. For example, by studying pure tin films of thickness about 15.5 nm, where λ is about 50 nm, H_c was found to increase by a factor of 30.

Ginzburg and Landau (1950) derived their theory by assuming that the superconducting transition could be described by a complex order parameter which plays the role of the superconducting wave function. The square of the order parameter gives the density of superconducting carriers (see *Superconducting Materials: Physical Properties*). The theory can account for spatial variations at boundaries and the presence of internal magnetic fields in terms of two materials parameters, λ and ξ_{GL}. The magnetic penetration depth λ is the characteristic depth to which the magnetic field penetrates a superconductor, as has already been mentioned, and the Ginzburg–Landau coherence length ξ_{GL} is roughly the distance required for the superconducting state to establish itself, that is, for the order parameter to go from 0 to 1. When $\lambda(T)/\xi_{GL}(T) \equiv \kappa \leqslant 1\sqrt{2}$, the superconductor remains in the Meissner state up to the thermodynamic critical field H_c. Such superconductors have positive normal state–superconducting boundary energies and are called type I. For superconductors for which $\kappa > 1\sqrt{2}$, the energy of the boundary becomes negative above a field H_{c1} so that it becomes energetically favorable to form boundaries. Such superconductors are classified as type II (see *Magnetic Properties of Superconductors*). Abrikosov (1957) found that the solution that gives the maximum number of boundaries and, hence, has the lowest free energy is a mixed state in which the field penetrates the superconductor as a lattice of flux quanta whose cores are of dimension ξ_{GL} (see *Flux Line Lattice*). This mixed type II (or Abrikosov) state exists up to critical fields H_{c2} which can be very high, exceeding 10 T for the ductile superconductor alloys with short mean free paths used in large-scale technology and reaching well over 100 T for the high-T_c cuprates.

Before 1960 the implications of Abrikosov's model were not appreciated. It is now known that early experimental results of lead–bismuth alloys (de Haas and Voogd 1930) with superconducting fields as high as 2.5 T were due to these alloys being type II superconductors which could actually have been used to make useful magnets in the 1920s. Unfortunately, it was not realized that they could carry useful quantities of current. Superconductivity was inferred to exist as a "sponge" with filaments of dimension less than λ. This sponge model which had been invoked to explain the combination of extremely small J_c and enhanced H_c found in the refractory metals such as niobium and tantalum (which at the time contained oxygen, nitrogen and other impurities) was incorrectly assumed to be the universal explanation of enhanced H_c. Only after the experimental discovery of high critical currents at high fields in Nb_3Sn (Kunzler et al. 1961) were properties of high-field superconductors modelled properly and intensive studies of the transition metal superconductors begun. It is perhaps not surprising that the sponge model survived so long; there were only a relatively few investigators engaged in low-temperature research and their research efforts were limited. Commercial helium liquefiers became available in the early 1950s and the number of investigators increased dramatically. The research then proceeded somewhat independently along two paths. One focused on quantitative physical experiments on elemental superconductors which could be obtained in pure form and could be used to test and extend phenomenology using the London and the Ginzburg–Landau theory, and the two-fluid models (Gorter and Casimir 1934). The other path was a search for new superconductors and patterns in the occurrence of superconductivity in unexplored regions of the periodic table with the hopes of being able to find higher values of T_c and simultaneously to uncover new clues concerning the origins of superconductivity as well as the relationship of T_c to electronic structure and crystal structure. The as-then-known superconductors tabulated by Shoenberg (1960) consisted of 30 compounds and some of the elements shown in Fig. 2 clustered mainly in the polyvalent metals and in the middle of the *d*-band elements.

This survey will focus primarily on the developments that occurred from the materials exploration approach path. However, it is worthwhile noting two sets of experiments which provided particularly strong guidance to the mechanism.

The first set of experiments led to the discovery that the values of T_c of tin and mercury vary with isotopic mass as $T_c = \text{const}\ M^{-\alpha}$ with α very close to 0.5 (see *Superconducting Materials: Basic Theoretical Concepts*). Since the velocity of sound and θ_D also have the same square root dependence on mass there is a strong suggestion that lattice vibrations are involved in superconductivity. Even earlier, the search for the mechanism of superconductivity had led Leiden investigators to look for such an isotope effect when "uranium" lead (i.e., the lead that came from the decay of uranium) became available (de Bruyn Ouboter 1987). Unfortunately, their gas thermometer had a reproducibility of 0.01 K which was not quite sensitive enough to resolve the mass dependence and the establishment of a key clue to the mechanism was delayed for a generation. The isotope effect, that is, study of the mass dependence of T_c, has become a

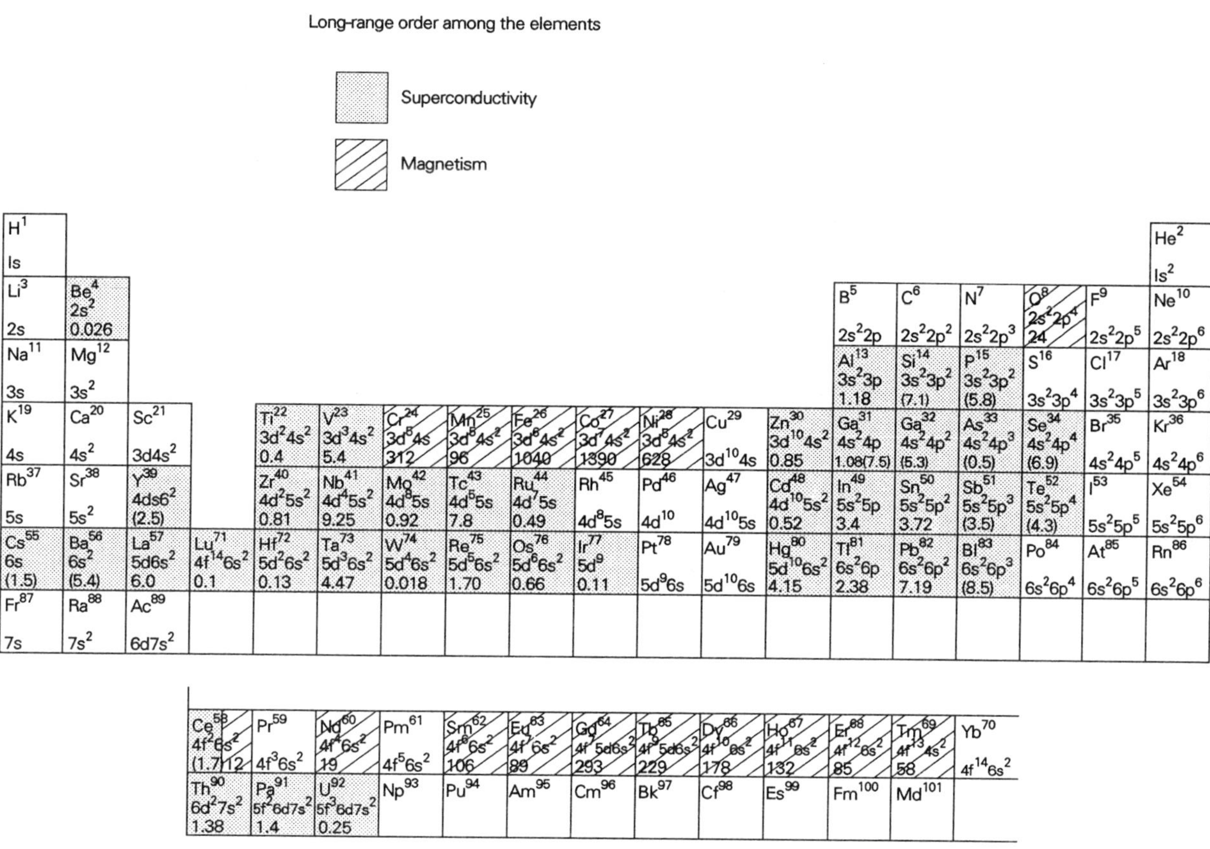

Figure 2
Periodic table of Mendeleev showing magnetic and superconducting elements. Numbers refer to highest ordering temperatures in degrees Kelvin. Those in parentheses indicate highest temperature under high pressure (after White and Geballe 1979)

commonly employed investigative tool for exploring mechanisms in superconductors if isotopes are available and if the superconductors can be prepared with reproducible transition temperatures sharp enough to make small shifts visible. The exponent α is sensitive to the strength of the electron–phonon coupling and other factors as well. Normally, it is positive because for heavier mass there is less phase space available for pairing. However, it is negative in palladium hydride ($T_c(\mathrm{PdH}) = 9.5$ K; $T_c(\mathrm{PdD}) = 11.6$ K) and also in the organic superconductors of the BEDT–TTF family when hydrogen is replaced by deuterium (presumably due to the smaller vibrational amplitude of the deuterium and the anharmonicity of the associated Einstein modes). Values of α can vary widely from negative to greater than $\frac{1}{2}$ because retardation effects, due to phonon velocities being much less than the electron velocity, are mass dependent. The retarded response of the phonons allows the pairing interaction to be localized in space without an accompanying large coulombic repulsion. Retardation offers a rationale for why local effects such as the presence of niobium can enhance superconductivity. For the new high-T_c cuprate superconductors in which it is possible to exchange O^{16} and O^{18}, the values of α are found to vary from around zero for superconductors that have compositions which in any family give the highest values of T_c to values approaching 0.8 when the composition is nonoptimal. Unfortunately the results do not give unambiguous clues as to the mechanisms for superconductivity.

The second set of experiments measured the electronic energy spectra of superconductors. Analysis showed that the electronic contributions to thermal conductivity and heat capacity varied exponentially below T_c, thus defining a gap in the energy spectrum (see *Superconducting Materials: Physical Properties*). Tunnelling experiments are particularly powerful when it is possible to make good superconducting junctions (see *Tunnelling Spectroscopy*; McMillan and

Rowell 1965). Analysis of the voltage dependence of the tunnel conductance using the strong coupling theory permits the extraction of materials parameters that are favorable to the occurrence of superconductivity. Values of λ which enter into the equation for T_c (Eqn. (1)) can be obtained by analysis of the conductance of superconducting tunnel junctions (McMillan and Rowell 1965) and also by heat capacity measurements (McMillan 1968). It is helpful to write λ in terms of normal state parameters:

$$\lambda = N(0)\langle I^2\rangle / M\langle\omega^2\rangle \qquad (2)$$

where $N(0)$ is the density of states at the Fermi surface, $\langle I^2\rangle$ is the Fermi-surface-averaged square of the electron–phonon matrix element, M is the mass and $\langle\omega^2\rangle$ is an averaged phonon frequency. The parameters of Eqns. (1, 2) are useful for understanding the behavior of T_c in any given system as a function of composition. The inverse proportionality to $\langle\omega^2\rangle$ provides the rationale for the empirical observation that T_c maximizes near a phase boundary where phonon softening occurs.

2. Searching for New Superconductors

After World War II, investigators led initially by B. T. Matthias and J. K. Hulm (Hulm and Matthias 1951, Hardy and Hulm 1954) in the USA and others including N. E. Alexseevski in the Soviet Union (see, for example, references in Roberts 1976) made renewed efforts to find new superconductors. They attempted to determine the structural and electronic features favorable for the occurrence of superconductivity. Their timing was good for several reasons including the the availability of much purer materials (the use of metallurgical processes developed during the Manhattan Project made pure rare-earth metals and transition metals available for the first time) and better methods of synthesis (the employment of arc furnaces with water-cooled copper hearths and the emergence of high-vacuum processing for removing oxygen made it possible to prepare many new structures and compositions with high melting points and with little contamination). Particularly fruitful was the study of binary solid solution alloys of the *d*-band transition metals. This work revealed relatively high-T_c binary materials with high critical fields and ultimately led to high-field superconducting magnets, an important new technology which developed rapidly after 1962 based on the alloy niobium–titanium (see *Niobium–Titanium Alloys*).

The search for new and improved superconductors continued on an international scale and rather complete knowledge of elemental superconductors, their solid solutions and their binary compounds accrued. Ternary systems were only modestly explored. More recently, however, extensive studies have been made on the magnetic and superconducting ordering that take place in the Chevrel phases such as $PbMo_6Se_8$ (see *Chevrel Phases*) and in the rare-earth rhodium borides (see, for example, Maple and Fischer 1982). The lack of understanding of the electrical and magnetic behavior of multicomponent, multisited (i.e., three or more sites per unit cell) structures is underscored by the unexpected and startling discovery of high-temperature superconductivity in the low-carrier-density copper-oxide-layered structures by Bednorz and Müller in 1986. Earlier work on low-carrier-density perovskite $SrTiO_3$ (Schooley et al. 1965, Hulm et al. 1970) showed that T_c as a function of carrier density n goes over a maximum at very low levels of doping. This is also the case in copper oxide ceramics, although the levels of doping are almost two orders of magnitude higher.

A further question so far unsettled is whether all pure metals will ultimately become superconducting if cooled down to near absolute zero (see *Superconducting Materials: Elements*). The strongly magnetic metals can probably be excluded from this hypothesis. The question then narrows to whether or not the Fermi sea is unstable at $T=0$ and if so does the instability always lead to superconductivity? Examination of the periodic table shows that all nonmagnetic metals—with the exception of the alkali metals, the alkaline-earth metals except beryllium and the noble metals including palladium, platinum and rhodium—become superconducting (see Fig. 2). (Cesium becomes superconducting under pressure presumably because a substantial *d*-band character is mixed into the conduction band; barium becomes superconducting at high pressure, see Table 1.) By extrapolating the behavior of dilute solutions of metals such as gallium in gold it has been estimated (Hoyt and Mota 1976) that pure gold will become superconducting at a sufficiently low temperature. In terms of Eqn. (1) the occurrence or nonoccurrence of a superconducting ground state depends on whether the phonon attractive interaction is larger or smaller than the coulomb repulsive term that is, on the sign of the $\lambda-\mu^*$ term. It is likely that the question will never be answered by experiment even if $\lambda>\mu^*$, because arbitrarily small values of the difference give exponentially small T_c values.

Many superconductors are solid solutions involving elements, intermetallic phases or even metastable phases and differ only slightly from the parent systems from which they are derived. The sensitivity of the T_c of a given composition to structure, order, coordination and impurities varies widely (Buckel 1991). There is thus an arbitrariness in trying to estimate the total number of known superconductors. The last complete survey of superconducting materials (Roberts 1976) has almost 2000 entries (including negative reports). A much more comprehensive tabulation which reviews the more than 40 000 papers on superconductors which have been published between 1913 and

Table 1
Collapsed phases of nontransition elements and compounds (obtained by low-temperature deposition or high pressures)

	T_c (range) (K)	Pressure range (kbar)	Comments
Cs	1.5	>125	*d*-electron hybridization
Ba	1–5.4	55–190	Ba II, III and IV
Ga	6–8	35	several phases
Si	6.7–7.0	120–130	
Ge	5.3	115	
P	5.8	170	
As	0.5	140	
Sb	2.6	120	can be pressure quenched
Bi	3.9–8.5	25–100	six phases; surface clusters
Se	6.9	130	
Te	2–3.5	50	
InTe	2.2		
GaAs	4.8	260	
InSb	2–4	37–125	two phases

1988 is given by Flükiger and Klose (1991). A database covering the high-T_c superconductors with emphasis on applications is maintained on a yearly basis by J. Talvacchio (Talvacchio 1989–1990).

3. Nontransition Metals

The alkali and alkaline-earth metals are mostly not superconducting down to the temperatures to which they have been tested. Exceptions are beryllium and cesium (Table 1). Beryllium has a high Debye temperature (1260 K) which distinguishes it from other alkaline earths and presumably raises the prefactor in Eqn. (1) sufficiently to overcome the small $\lambda-\mu^*$. Cesium, under pressure, may become transition-metal-like with the hybridization of 5*d* wave functions with the conduction band.

It is possible to increase the low coordination number of polyvalent nontransition metal semiconductors and low-density semimetals or metals by quenching them from a more highly coordinated high-temperature phase and/or high-pressure phase, or by measuring them at high pressures (see *Superconducting Materials Under Pressure*). When this is done the collapsed phase either becomes superconducting or, if it is already superconducting, has its T_c increased. Representative examples are listed in Table 1. Tin is the prototypical example. Grey tin is a 4-coordinated semimetal with the diamond structure. When "quenched" from room temperature it retains its 4+2 (or 6) metallic coordination and becomes superconducting. In a closely related way lighter group IV elements, and covalent semiconductor III–V compounds, can be collapsed by pressure or vapor quenching into more highly coordinated structures which are metallic superconductors. These superconductors can all be expected to have a normal isotope effect. The increase of T_c in going from tin to silicon is expected from consideration of the change in mass alone.

Hydrogen, the lightest element, is expected to have the highest Debye temperature as an atomic solid. Calculations show that it should first transform into a molecular metal and then to a monatomic superconducting metal at presently unattainably high pressures (see *Superconducting Materials Under Pressure*). The monatomic metal is predicted to have a hexagonal structure (with the stacking sequence of samarium) with T_c about 150 K at pressures in excess of 4 Mbar (Barbee et al. 1989). PdH and Th_4H_{15} are superconductors in which hydrogen vibrations play a role which, as already mentioned, can lead to a negative isotopic shift of T_c. In the superconducting hydrides the 1*s* band crosses the Fermi level.

There are few cases of extended ranges of solid solutions among the polyvalent nontransition metal elements in which it is possible to study the variation of T_c over wide changes in carrier concentration while keeping the structure and the mass almost constant (Livingston 1963). Extended solid solutions are seen in the lead–bismuth and lead–indium systems (Evetts and Wade 1970) and in the lead–thallium system (Bon Mardion et al. 1965). In the lead–bismuth system T_c increases monotonically up to 9 K which empirically seems to be the limit for polyvalent nontransition metal elemental and binary phases. Vapor-deposited lead–bismuth films are well suited for making good tunnel junctions; λ increases from 1.55 for pure lead to a record of 2.6 for the most bismuth-rich fcc phase (Dynes and Rowell 1975).

4. Transition d- and f-Band Metals

The transition metal elements with unfilled *d* bands as well as the rare-earth metals with unfilled *f* shells form a large class of superconductors which exhibit a variety of ordering phenomena in a wide range of structures. Known transition temperatures range from $T_c \simeq 0.01$ K for tungsten to $T_c \simeq 23$ K for Nb_3Ge. Solid solutions and simple alloys will be considered in this section.

One of the most interesting patterns in the occurrence of superconductivity is the universal behavior of T_c as a function of the simple average valence electron-to-atom ratio, the useful Matthias rule (Matthias 1955, Hulm and Matthias 1981). T_c is plotted for the 3*d*, 4*d* and 5*d* transition metals and their solid solutions as a function of this ratio in Fig. 3. Roughly symmetrical behavior is found around a minimum near the half-filled *d* band between molybdenum and tungsten. The curves are drawn from data for near-neighbor solid solutions; roughly similar curves are found for somewhat more widely separated neighbours. However, when the charge contrast becomes too great the simple averaging breaks down and intermediate phases are formed. To illustrate how the simple averaging is useful and practical consider the following. The T_c of titanium is raised by the addition of any transition metal to the right of it in the periodic table (Matthias et al. 1963). In contrast, while the T_c of niobium is increased by additions of titanium, zirconium or hafnium, it is lowered by additions of chromium, molybdenum, tantalum, tungsten and so on. The transition temperature of zirconium is raised from below 1 K to 11 K by the addition of 14% of rhodium or by the addition of 88% of niobium; both solid solutions have an electron-to-atom ratio of 4.7 which is the maximum of the less-than-half-filled *d* shell (Fig. 3). By studying isoelectronic alloys and elements it is found that alloying does not affect T_c significantly. The insensitivity of T_c is in accordance with Anderson's theorem concerning "dirty superconductivity," that is, increased scattering caused by nonmagnetic impurities or defects in the lattice will, to first order, affect superconductivity only by smearing over features of the Fermi surface (Anderson 1959).

The Matthias rule can be refined at the price of including a second significant parameter, the electronegativity difference between constituents, along with the average electron-to-atom ratio (e.g., Vonsovsky et al. p. 247). With this second and relatively easily obtained parameter, it is possible to "predict" (after the fact) a minimum in T_c as a function of alloying in the isoelectronic vanadium–niobium system. The Matthias rule depends, at least around the middle of the *d* band filling, on the validity of the rigid band approximation. In this model the energy bands remain fixed as the electron density is changed by alloying. More subtle behavior cannot, of course, be explained so simply. Asymmetry in the behavior of T_c between the more-than and the less-than-half-filled overall *d* band, that is, around six electrons per atom (Fig. 3), can be understood in terms of the reversal of the sign of the spin–orbit coupling parameter.

It must be recognized that T_c is not the only factor determining which superconductors are of technological interest. For magnets the upper critical field H_{c2} and the critical current density J_c, which is dependent on pinning of the flux quanta by metallurgical imperfections, are of equal importance (see *Critical Current Density*; *Flux Pinning and the Summation of Pinning Forces*). Hence, MoRe solid solutions exhibit higher T_c values than NbTi (15 K vs 10 K) and are ductile yet they have never been used for magnet construction outside of an initial demonstration (Kunzler et al. 1961). This is essentially because they have a relatively long electron mean free path leading to a relatively low κ and thus a correspondingly low H_{c2}.

When deviations of T_c from the expected electron-to-atom ratio behavior are found they reveal either new physics or unexpected metallurgy. The decrease in the transition temperature of lanthanum by additions of small amounts of isoelectronic magnetic rare-earth elements, rather than being negligible, amounts to several degrees per percent of lanthanum (Suhl and Matthias 1959). This apparent violation of the Matthias rule as well as the Anderson theorem of dirty superconductivity gave the first direct indication that scattering by magnetic 4*f* ions reduces T_c far beyond what would be expected for ordinary impurity scattering. Systematic studies then showed that the spin–spin exchange interaction between the charge carriers and the rare-earth ions, which breaks time reversal symmetry and thus breaks up the Cooper pairing, is responsible. A larger depression which can be as high as 100 degrees per percent is found for localized magnetic *d*-element impurities where there is stronger hybridization (see *Coherence Length, Proximity Effect and Fluctuations*). The presence of trace amounts of iron in molybdenum was enough to prevent the superconductivity of molybdenum from being discovered until the trace impurities of iron were removed (Geballe et al. 1962). The theory of pair breaking by the magnetic scattering (Abrikosov and Gor'kov 1961) has been extensively tested and is well understood in most situations. The competition between magnetic and superconducting ordering gives rise to a rich variety of phenomena. The nature of magnetic interactions in the heavy fermion systems and high-T_c cuprates is still the subject of intense theoretical and experimental investigation.

A breakdown of the expected dependence of T_c on the electron-to-atom ratio can also signal a lack of

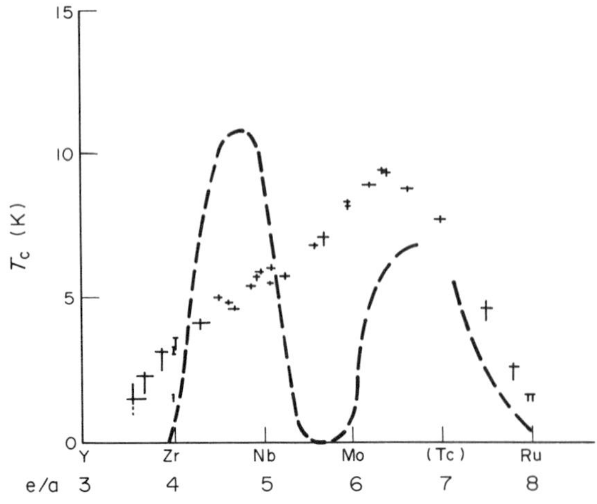

Figure 3
T_c as a function of electron-to-atom ratio: ---, crystalline alloys (after Matthias 1955); +, amorphous alloys (after Collver and Hammond 1973)

understanding of the relevant phase diagram. The flow of information can then be reversed and the observed superconducting behavior used to identify unsuspected metallurgical behavior as for example occurred in the studies of dilute solutions of iron in titanium (Raub et al. 1967). The increase of T_c with nominally small additions of iron is an order of magnitude greater than expected from the Matthias rule and found by alloying titanium with other elements to its right in the periodic table. When first observed this suggested the possibility of an unusual enhancement of the pairing, with the intriguing possibility of its being of magnetic origin. Careful annealing and metallographic studies, however, revealed the presence of a second bcc phase in the hcp parent phase with a volume fraction too small to be detected by x-ray diffraction. The bcc titanium–iron solid solution, enriched in iron by a factor of about ten, was distributed along grain boundaries and was thus connected in closed loops which gave superconducting shielding signals of the magnitude expected for the entire volume, although no Meissner expulsion of flux was observed. (The lack of an appropriate Meissner effect is frequently the case in bulk superconductors where defects can trap or pin the flux as, for example, occurs with commercial superconducting wire.) Given the 10:1 iron-concentration enhancement, the observed increase in T_c was thus in accord with the electron-to-atom ratio. The enhancement of superconducting signals at grain boundaries can be an advantage when searching for new superconductors. Unfortunately, there are cases in the literature where the observed superconductivity is incorrectly attributed to the majority phase and false conclusions are drawn. More than one kind of measurement is needed to avoid making false interpretations (see *Measurements in Superconducting Materials*; Geballe 1983). A particularly simple and frequently effective test is simply to break up the connectivity of the grain boundaries by powdering the sample.

For amorphous alloys where long-range periodic order is not present and crystal momentum is not a good quantum number, the dependence of T_c on the electron-to-atom ratio no longer follows the solid "universal" curve shown in Fig. 3 (Collver and Hammond 1973). A single maximum near the half-filled d band is found. Thus, the amorphous phases behave as expected for simple band filling. Tunnelling spectroscopy shows that the amorphous d band alloys are weakly coupled superconductors ($\lambda < 1$). Even the transverse and the longitudinal acoustic branches are no longer distinct (Kimhi and Geballe 1980) presumably because in addition to topological disorder there is a distribution of interatomic force constants. In contrast, the nontransition amorphous alloys, which are more free-electron-like, show well-defined transverse and acoustic branches and are strongly coupled as has already been discussed (see *Amorphous Superconductors*).

5. Intermediate Phases

Superconductors are found in a large number of phases that are intermediate between the end members (Matthias et al. 1963, Savitskii et al. 1973). These phases, which can be complex with many sites per unit cell, either form at a given composition as "line compounds" or over ranges of compositions as solid solutions. Models used to discuss the stability of the phases include electronic terms derived from the one-electron energy bands, entropy and geometrical considerations such as packing efficiency and the fit of the metallic radii. When the filling of the conduction band is the dominant factor in determining the structure, the compound is referred to as an electron compound. It is not surprising that the T_c values of such compounds are generally distributed in the temperature decade between 1 K and 10 K like elements and alloys from which they form. Likewise, trends can be well predicted in terms of the electron-to-atom ratio. Superconductivity occurs in most cases when there is no magnetic scattering to break the Cooper pairs. For example, bismuth forms dozens of superconducting compounds which have metallic coordination with values of $T_c < 5$ K, such as the Laves phase (see Sect. 5.3) compound Bi_2K or the NiAs compound BiPd. They have properties roughly similar to what is expected by simply averaging over the periodic table in a further generalization of the Matthias rule to accommodate the filled d band.

The sequence in which binary phases form as the d band is filled includes A_3B (β-tungsten), A_2B (Laves phases, AB (NaCl or CsCl), AB_3 and AB_5 all of which tend to be ordered in most cases with fixed composition. In addition there are many more intermediate pseudobinary phases with solid solutions which encompass a wide range of compositions. These can have complex unit cells such as the tetragonal sigma phase with 30 sites per unit cell or the cubic alpha manganese phase with 58 sites per unit cell. Almost all of them are well-behaved superconductors in that T_c behavior can be understood roughly in terms of the Matthias–Miedema rules and the parameters of Eqns. (1,2) (Vonsovsky et al. 1982). Complex structures are found particularly between A atoms with less-than-half-filled d shells and B atoms which are nontransition metals, or transition metals which have greater-than-half-filled d shells. Differences in electronegativity result in charge transfer and a gain in the configurational energy (Hund's rule). The great stability of the B1–NaCl structures which have the highest known melting points (e.g., HfN:$T_c = 8.8$ K, $T_{melting} = 3600$ K) is due in part to the part ionic–covalent nature of the bonding.

A second "rule" deduced by Matthias from studies of binary and pseudobinary systems is that higher symmetry (e.g., cubic) phases have higher T_c values than those with lower symmetry. This deduction can be attributed to the fact that transitions to lower-symmetry phases frequently remove portions of the Fermi surface and consequently lower the density of states and, hence, reduce λ (Eqn. (2)). However, this rule is of limited value because it is not appropriate for ternary compounds where the superconductivity can be dominated by soft phonons and other not-as-yet clearly identified mechanisms. The inapplicability of this rule to the high-T_c cuprates was already evident in earlier studies of the alkali metal tungsten oxides which are known as bronzes because of their colour. In these systems the cubic phases are not superconducting; however, when the symmetry is reduced to tetragonal or hexagonal phases they become superconducting. The maximum T_c values (again below 10 K) occur near the phase boundaries as expected when phonon softening is responsible. In what follows some of the more important intermediate phase structural types are discussed in somewhat more detail. (In discussing crystal structures, it is common to use the crystal types of the Strukturbericht system, as will be used here.) References to the original work up to 1976 can be found (Roberts 1976) unless otherwise noted.

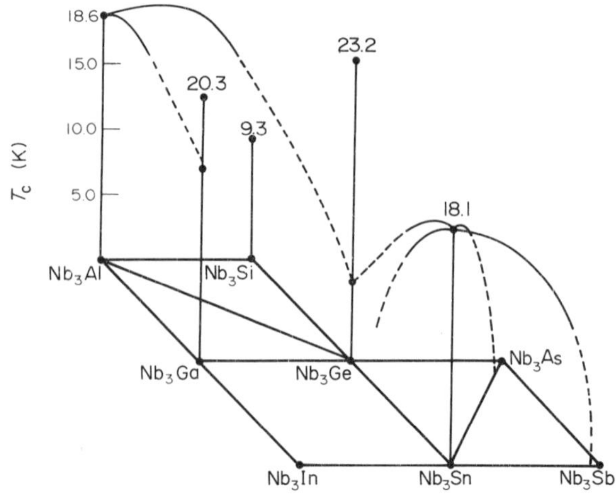

Figure 4
Niobium-based, high-T_c binary A15 compounds and the T_c values of some of their pseudobinary solid solutions

5.1 A15 Structure

The most important A15 family is a cubic structure with the stoichiometric composition A_3B and it contains all the highest superconductors known prior to 1986 (see *A15 Compound Superconductors*). It has special structural features and unusual electronic and lattice properties in the normal state. Of particular note are the low-temperature martensitic transitions which occur in Nb_3Sn and V_3Si at temperatures not far above T_c. The relationship between these properties and the enhanced superconductivity has been a challenging problem. The most prominent structural feature is the nonintersecting orthogonal chains of A atoms which run along the cube directions. In these directions the atoms are closer together than they are in the respective elements. T_c is very sensitive to order. For example, Nb_3Au in the ordered state has a T_c of 11 K, whereas the same composition in the completely disordered (bcc) structure has a T_c of 1.2 K. The enhanced T_c of compounds with the A15 structure is unusually susceptible to degradation by antisite disorder, or to an excess of niobium which substitutes on the B site. All the A15 compounds that have T_c values above 18 K contain niobium as the A element and nontransition three- and four-valent elements as the B atom; as shown in Fig. 4, Nb_3Sn ($T_c = 18$ K) forms from the melt at the stoichiometric composition. The higher-T_c compounds with aluminum, gallium and germanium form from the melt as niobium-rich phases with low values of T_c. Special quench and anneal procedures must be employed to increase the concentration of the B atom and obtain the increased T_c. For example, the A15 phase of NbGe is formed from the melt with a niobium-rich composition with the excess niobium on the germanium sites and $T_c \simeq 6$ K. Tunnelling studies with Nb_3Al (Kwo and Geballe 1982) show that λ increases and the vibrational spectrum softens as the ordered A15 with near-stoichiometric composition is approached. Once again the highest T_c values are found at or just beyond the limit of stability. Sputtered niobium and germanium films which essentially are quenched from the vapor phase can be nearly stoichiometric with T_c onset near 23 K, and for the 15 years prior to the cuprates such films had the highest known T_c. The A15 structure was originally thought to be a polymorph of elemental tungsten and is still frequently referred to as the β-tungsten structure. A small amount of oxygen (not detected when the structure was first characterized) is necessary to stabilize tungsten suboxide in the A15 structure in which it has a T_c in the neighborhood of 3 K. A15 structures with transition metals on the B sites (as well as on the A sites) such as Nb_3Os do not exhibit high T_c values or other unconventional properties. They can occur with compositions well removed from the 3:1 ratio, such as Nb_2Ir ($T_c = 2$ K) which has up to 25% of the iridium on the niobium sites.

5.2 NaCl (B1) Structure

Transition metals of groups IV, V and VI form superconductors wth T_c values greater that or equal to 10 K with carbides and nitrides in the B1 (rock salt) structure. These compounds are frequently but inappropriately referred to as interstitial because the nonmetal ions are roughly the same size as the metallic ones. A substantial number of vacancies may occur on either site or on both sites; for example, TiO ($T_c = 1$ K) with a stoichiometric 1:1 ratio has 15%

vacancies on each site. As a result the compounds are poor electrical conductors in the normal state and have useful technological properties such as high upper critical fields and critical currents and are resistant to the destruction of superconducting properties by neutron radiation. With one exception all the highest-T_c compounds in this family are based on niobium alloys. NbN ($T_c = 15$ K) has 9% vacancies on the nitrogen sites (see *Nitride and Carbonitride Superconductors*). The Fermi level is in a strongly hybridized *d* band with a relatively low density of states. The exception is metastable low-temperature hydrogen-implanted palladium–silver alloys where T_c values up to 17 K have been reached (Stritzker 1983). In these systems there is a substantial coupling to the low-energy optic H (or D) phonons, and the hydrogen-derived *s* band crosses the Fermi level as has already been noted. Over some regions of composition there is a negative isotope effect which can be explained by the smaller dispersion of the deuterium vibrations.

5.3 Laves Phase (C14, C15 and C16) Structures

The Laves phases with the formula AB_2 form in three structures types, cubic, hexagonal and tetragonal. The B atoms form in a diamondlike packing and form covalent–metallic bonds, a feature which discourages deviations from stoichiometry (Savitskii et al. 1973). Magnetic rare earths on the A site have an unusually small pair-breaking effect (i.e., small *s–f* overlap) for a binary compound; as a consequence $CeCo_2$ is superconducting ($T_c = 1.5$ K). While $GdRu_2$ is magnetic, the substitution of cerium in $CeCo_2$ by as much as 8% of gadolinium has almost no effect on T_c. The competition between superconducting ordering in $CeRu_2$ ($T_c = 5$ K) and magnetic ordering in other rare-earth (RE)Ru_2 compounds has been studied by numerous investigators and it has been established that there is a region where ferromagnetism and superconductivity overlap, although the range of the ferromagnetic correlation is only about 8 nm. Superconductivity and magnetism can coexist more readily in ternary structures when there can be less exchange interaction between moments on one set of sites and conduction states on the others.

5.4 α Manganese (A12), σ (D8b) and Other Binary Phases with Large Unit Cells

The α and σ phases have wide ranges of composition and contain large numbers of superconductors. While there are five nonequivalent positions in the σ phase they are only occupied by transition metals. The α manganese has 58 atoms distributed among four nonequivalent sites. It also forms superconductors only with transition metals with values of T_c rarely above 10 K; $MoTc_3$ ($T_c = 15$ K) is an exception. A detailed review of the characteristics of these compounds is given in the book by Savitskii et al. (1973).

6. Unusual Superconductors

Unusual superconductors can be usefully approached from two different limits. The first is in the *local* limit where single-center (e.g., Ce $4f^1$) or two-center (e.g., CuO) configurations are important. The challenge is to understand how on-site correlation enters into the band structure. The retarded nature of the superconducting interaction gives some credence for the importance of the local electronic configuration in the niobium-based superconductors already discussed, but new models are needed to explain high-T_c and heavy fermion superconductivity. The second *extended* or nonlocal limit reflects structural features. Those structures that have three or more different sites and have three or more components are relatively the least explored and have the most potential for unusual and or enhanced T_c for two reasons. First, the integrity and the stability can be maintained by a strongly bonded framework with filled bands below the Fermi level while strong conduction band correlations occur on other sites. Second, there can be much more flexibility in manipulating the crystal chemistry, for example, in introducing charge carriers by doping on sites that are not hybridized into the conduction band states. This occurs in the high-T_c superconductors. Here a natural type of modulation doping takes place which in effect is comparable to the artificial modulation doping which has been so successfully used in the III–V semiconductors.

6.1 Heavy Fermions

Cerium and uranium with extended 4*f* and 5*f* electrons are the key constituents of the remarkable class of heavy fermion superconductors discovered by Steglich and Aarts (1979) (see *Heavy Fermion Metals*). The electronic (Sommerfeld) heat capacity coefficients in these superconductors are two to three orders of magnitude larger than in conventional metals at low temperatures. The large effective masses of the charge carriers at low temperatures result from the strong interaction between the *f* electrons and the conduction band carriers. At temperatures below 1 K in $CeCu_2Si_2$, UPt_3, UBe_{13} and URu_2Si_2 the heavy electrons condense into paired superconducting states. In at least two of them (UPt_3 and URu_2Si_2) the ground state also involves weak antiferromagnetism. The field has been reviewed by Ott (1989). There are other heavy fermion compounds such as $CeCu_6$ and $CeAl_3$ where the ground state is metallic but does not order. The nonmonotonic changes of T_c with thorium doping on the uranium sites in UBe_{13}, along with the nonexponential behavior of properties such as heat capacity, are evidence for an unconventional pairing involving higher angular momentum *p*- or *d*-state pairs. This can give rise to complex order parameters and anisotropic ordered phases as is found in He^3. There are other uranium compounds U_6X (X = Mn, Fe, Co, Ni) with somewhat enhanced heat capacities.

Other uranium compounds U_3Ir, U_3Si, U_3Os and U_2Ti are conventional superconductors with $T_c \geqslant 1$ K. Uranium metallic glasses $U_{1-x}Fe_x$ and $U_{1-x}Co_x$ can be prepared over a wide uranium-rich concentration range. They have T_c values and estimated densities of states similar to that of U_6X and show some evidence for spin fluctuations (paramagnons).

6.2 Metal–Nonmetal Compounds

In metal–nonmetal compounds the lighter nonmetallic elements favor higher values of T_c not only because they have higher vibrational energies (see discussion of hydrides in the preceding section) but also because they form stronger bonds. A class of interesting compounds which involve light elements are the boron- and beryllium-rich phases such as YB_6 ($T_c = 7$ K). However, these compounds do not have strong metal–nonmetal bonds, they are binaries and the T_c values are below 10 K. The three-dimensional boron framework requires two electrons per six borons for stabilization. YB_6 therefore has one itinerant electron per unit cell. There are closely related binary dodecaborides such as ZrB_{12}($T_c = 6$ K). The beryllium-rich compounds can be doped with transition metals. $Be_{22}Re$ ($T_c = 9$ K), or nontransition metals, $Be_{13}Ga$ ($T_c = 6$ K). They have high Debye temperatures and values of T_c much like those of the condensed amorphous phase of pure beryllium ($T_c = 10$ K).

The pnictides and halides do not form superconductors with enhanced T_c values—in fact outside of the B1 NbN class of superconductors there are relatively few others. The polymer $(SN)_x$($T_c = 1$ K) is of interest because of its quasi-one-dimensionality and its lack of a transition metal as a component. The Peierls transition which might be expected to dominate for a one-dimensional system with a half-filled band (SN has 11 electrons) is suppressed because of interchain interactions (Greene et al. 1975). The halogens form mainly ionic bonds which are not conducive for superconductivity. In fact very few halide compounds are metallic, almost all are Mott insulators. Ag_2F, however, is a layered metal which becomes superconducting well below 1 K.

There exists an important series of ternary chalcogens and borides in which there is a strong interplay of magnetism originating on one set of sites and superconductivity based on the others. The most widely studied are the Chevrel phases (e.g., $REMo_6S_8$ and $REMo_6Se_8$) (see *Chevrel Phases*) and the rhodium rare-earth borides ($RERh_4B_4$) (see, for example, Maple and Fischer 1982), where it is possible for magnetic and superconducting states to coexist. The series $Eu_xSn_{1-x}Mo_6S_8$ exhibits the novel effect predicted by Jaccarino and Peter in 1962 of having the superconductivity turned on by an increase in field (Meul et al. 1984). Presumably this happens when the polarization of the conduction electrons caused by negative exchange interaction with the europium ions is cancelled by the external field. In $ErRh_4B_4$ the novel effect of turning on superconductivity by an increase in temperature (sometimes termed reentrant) occurs because the ferromagnetic Curie temperature is below the superconductivity T_c. In the competition between ferromagnetism and superconductivity the ferromagnetism is stronger and thus the ground state is ferromagnetic.

The Chevrel phases based on $PbMo_6S_8$ ($T_c \simeq 12$ K) have the highest H_{c2} values of any three-dimensional superconductor, reaching values greater than 60 T, and have potential for making high-field magnets.

6.3 Carbon-Containing Compounds

There are five classes or carbon compounds that become superconducting. The B1 carbides known since the early work of Meissner have already been mentioned. A second class, sesquicarbides, are binary compounds with enhanced T_c values. The carbon–carbon bonds in these compounds are shorter than in the conventional carbon double bond. T_c for Y_2C_3 is 11 K. It can be raised to 17 K by thorium substitution on the yttrium sites using high-pressure techniques (Krupka et al. 1969).

A third class of carbon compounds, also binary, is formed by intercalating graphite with alkali metals. The first-stage compounds (so called because there is one layer of alkali metal for each graphite layer) RbC_8, KC_8 and CsC_8 all have T_c values below 1 K, although some more complex graphite intercalcation compounds have values of T_c above 1 K. It is also possible to intercalate with anions and introduce holes into the conduction band of the carbon; while compounds such as C_8Br are good metals they do not become superconducting above 0.3 K and thus are an exception to the empiricism that hole conduction generally leads to a higher T_c than electron conduction.

Carbon forms stable molecules of 60 atoms, C_{60}, which have come to be known as buckyballs or fullerenes because of their isomorphism with the geodesic design of Buckminster Fuller. A recently discovered unusual class of superconductors is formed by intercalating C_{60} polyhedra with the alkali metals to form air-sensitive metals. Onset T_c values of approximately 30 K for Rb_3C_{60} and 20 K for K_3C_{60} have been found (see *Organic Superconductors: Doped C_{60} (Fullerene)*). At the time of this writing higher onsets have been reported for cesium- and thallium-containing compounds. The T_c is unusually sensitive to the lattice constant, being rapidly suppressed by the closer approach of the C_{60} polyhedra to each other. It seems reasonable to assume that the strong pairing interaction is in the bands formed from the carbon π orbitals on each polyhedron. Important questions under active investigation are: how closely is the superconductivity related to that of the organic superconductors and is the pairing mediated primarily by electron–electron correlations or by an unusual phonon spectrum?

6.4 Organic Superconductors

Almost all organic solids are insulators; however, in the 1960s organic charge transfer salts were discovered by workers at Du Pont which are metallic at room temperature and highly anisotropic. The structure consists of stacks of planar donor aromatic ions separated by anions. The high conductivity is along the stack direction due to charge transfer into bands formed by the overlap of the pi bonds on adjacent rings. The suggestion that this high conductivity was due to superconductivity or superconducting fluctuations, later shown to be incorrect, together with a theoretical model of superconductivity in one-dimensional organic polymers advanced earlier (Little 1964) stimulated a great deal of research. The instability of (quasi) one-dimensional Fermi *sp* surfaces leads to various kinds of instabilities (charge and spin density waves, metal–insulator transitions) which compete with superconductivity. Jérome et al. (1980) discovered superconductivity in the charge transfer salt $(TMTSF)_2PF_6$ by suppressing competing transitions sufficiently with the application of 12 kbar pressure. The planar donor stacks based upon tetrathio (or tetraseleno) "$>$S=S$<$" cores are shown in Fig. 5. They transfer charge to acceptor anions which can be isolated halides, tetrahedral (ReO_4^-), octahedral (PF_6^-), planar (NO_3^-) or linear (I_3^-) anions. Of special interest are polymeric anions such as $Cu[N(CN)_2]Br^-$ (Saito and Kagoshima 1990). The metallic salt $\kappa(ET)_2$-$Cu[(N(CN)_2)]Br$ (where, as shown in Fig. 5, BEDT–TTF is abbreviated as ET) has a $T_c \simeq 12$ K. The chlorine analog is not superconducting at atmospheric pressure but becomes so, at a record high $T_c \simeq 13$ K, at modest pressures (Williams et al. 1991). These structures have interstack bonds and, hence, are more nearly two dimensional than one dimensional. During the 1980s a total of more than 30 organic superconductors has been found (see *Organic Superconductors: Anisotropic Molecular Metals*). If chemical methods can be found that suppress the competing instabilities which exist in other organic charge transfer compounds it is possible that much higher T_c values will be discovered in this potentially very large class of superconductors.

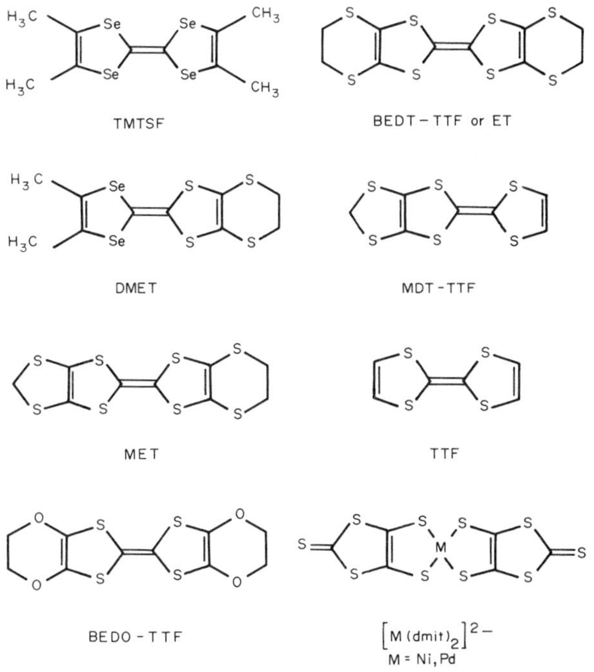

TMTSF; BEDT–TTF or ET; DMET; MDT-TTF; MET; TTF; BEDO-TTF; $[M(dmit)_2]^{2-}$, M = Ni, Pd

Figure 5
Molecular cations of organic superconductor charge transfer salts

6.5 Multicomponent Oxides and Chalcogenides

The family of perovskite (related) copper oxides in which Bednorz and Müller (1986) discovered high-T_c superconductivity includes all of the currently known superconductors above 50 K. Extensive experiments demonstrate that the superconducting strong pairing occurs in properly doped layers of CuO_2. The fact that $(LaSr)_2CuO_4$ and some of its low-temperature metallic properties were investigated well before the discovery of superconductivity (Goodenough et al. 1973) illustrates how, even when the composition is proper, the path to discovery can be thwarted by incorrect processing. For the minority superconducting phase that Bednorz and Müller originally detected to appear at all, the doping, the reaction temperature and the oxygen pressure, all independent synthesis parameters, had to be within restricted ranges. The same goes for the discovery of T_c in YBaCuO above liquid nitrogen temperatures (Wu et al. 1987).

It is impossible in any survey to do justice to the intense activity in synthesis, property measurement, theoretical modelling and device applications which have been underway in exploring the new copper oxide superconductors since the mid-1980s. Much is covered in other entries in this treatise. There are many superconductors with $T_c > 40$ K which presently range up to 125 K all belonging to layered perovskites (see *Oxide Superconductors: Solid-State Chemistry*). Representative stoichiometric formulae are given in Table 2, although it must be understood that the superconductors are in general not stoichiometric compounds. Even in single crystals there are cation substitutions, oxygen vacancies, or interstitials stacking faults, twins and other defects. The outstanding structural feature is the layering which gives rise to strong anisotropy and quasi-two-dimensional behavior. The common feature of all the high-T_c superconductors is the presence of CuO_2 planes (see Fig. 6). These occur in sequences of either 1, 2, 3 or more negatively charged CuO_2 planes separated within each sequence by 0, 1, 2 or more positively charged planes of calcium, yttrium or other rare earths. The copper ions are fourfold, fivefold or sixfold coordinated by oxygen. These sequences in turn are separated by intergrowth planes which act as doping agents which can oxidize or reduce the CuO_2 planes by arbitrary concentrations of electrons or holes. These intergrowth planes can be La(Sr or Ba)O as in the original $(LaBa)_2CuO_4$; or BaO and CuO chain planes in the $T_c = 90$ K $YBa_2Cu_3O_7$; or double layers of thallium or bismuth oxides in the thallium and bismuth compounds which have T_c values greater than 100 K; or even more complex intergrowths. In some cases the intergrowth planes have states that cross the Fermi level and thus participate in the conductivity and superconductivity.

That the electrical conductivity and the superconductivity are largely confined to the CuO_2 planes is made evident by the very much greater in-plane values for the normal state conductivity together with the effects of chemical doping (see *Oxide Superconductors: Physical Properties*). The in-plane conductivity can be as much as five orders of magnitude greater than in the perpendicular direction. With few exceptions the substitution by magnetic rare earths for yttrium or lanthanum on the sites directly adjacent to the CuO_2 planes has absolutely no effect on T_c. On the other hand, doping on the CuO_2 plane by either magnetic (iron, cobalt, nickel) or nonmagnetic (gallium, aluminum) ions causes a marked depression of T_c. The undoped parent compounds are antiferromagnetic insulators with Néel temperatures of several hundred degrees due to ordering of the spin $\frac{1}{2}$ (d^9) copper ions. For the parent compounds in $YBa_2Cu_3O_6$, the moment per Cu is 0.66 Bohr magnetons, close to what is expected for a square planar $S=\frac{1}{2}$ Heisenberg model. The energy gap is due to coulomb repulsion in the excited states which involve either ligand states (charge transfer) or d states (Mott–Hubbard). With only a few percent of doping the Néel state is destroyed and a metallic state appears. Superconductivity in all the high-T_c (>40 K) phases is achieved by hole doping into the less-than-half-filled (lower Hubbard) band. $(NdCe)_2CuO_4$ and

Table 2
Representative high-T_c copper oxide superconductors

Nominal formula	T_c (approximate) (K)
Hole conductors	
$La_{1.85}Sr_{0.15}CuO_{4+y}$	37
$YBa_2Cu_3O_{7-y}$	90
$Bi_2Sr_2CuO_6$	10
$Bi_2Sr_2CaCu_2O_8$	90
$Bi_2Sr_2Ca_2Cu_3O_{10}$	110
$Tl_2Ba_6Ca_2Cu_3O_{10}$	125
Electron conductors	
$Nd_{1.85}Ce_{0.15}CuO_4$	24

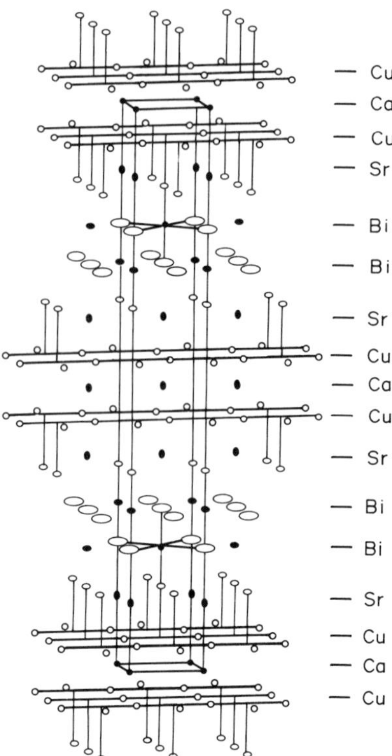

Figure 6
Ideal crystal structure of $Bi_2Sr_2CaCu_2O_8$ superconductor. Cu–O bonds are shown. An ideal subcell is indicated

related compounds with the T′ structure (which is closely related to the structure of La_2CuO_4 but has only four oxygens as near neighbors to the copper) can be electron doped and become superconducting (T_c < 30 K).

The question of whether a single CuO_2 layer, properly doped, can become superconducting has been addressed by studying trilayers and superlattices in which a thin superconducting layer is epitaxially sandwiched between insulating or lower T_c sheets. Most of the superlattice studies have been done using interleaving unit cells of $PrBa_2Cu_3O_7$ (which is an insulator) with the corresponding superconducting yttrium compound. Perhaps the most conclusive work has been done in the bismuth cuprates where the epitaxial match is optimal and charge transfer across the superlattice interfaces is minimal. Here, the T_c (~85–75 K) of $Bi_2Sr_2CaCu_2O_8$ (2212) remains almost constant even when only one-half of a unit cell of the 2212 is sandwiched between five unit cells of the low-T_c(10 K)$Bi_2Sr_2CuO_6$ (2201) phase.

The discovery of superconductivity in tetragonal $BaBi_{0.25}Pb_{0.75}O_3$ (T_c = 13 K) (Sleight et al. 1975) was the first evidence that enhanced superconductivity could be found in oxides of the perovskite family.

Since then superconductivity with $T_c > 30$ K has been found by doping with potassium or rubidium on the barium site which is not hybridized into the conduction band. Superconductivity is also found in oxides with the cubic spinel structures although not nearly so frequently as in the perovskites. Again the oxides with their stronger bonds have higher T_c values than the chalcogens. The T_c of $LiTi_2O_4$ (13.7 K) is higher than that of $CuRh_2S_4$ or $CuRhSe_4$ which have $T_c \simeq 5$ K.

Sulfides, selenides and tellurides have more extended wave functions than the oxides and, hence, tend to be more metallic and less magnetic than their more ionic oxide counterparts. The electronic and crystal structures of these phases are related. When the electron configuration of the M in the MX_2 compound is d^0 (as in TiS_2) or d^3 (as in ReS_2 or $LiMoS_2$) the coordination of M is octahedral. When the configuration is d^1 or d^2, the coordination of M is trigonal prismatic. In the case of $TaX_2(d^1)$ the trigonal prismatic coordination is the preferred coordination in the room-temperature-stable polymorphy, 2H-TaX_2. However, at high temperatures the octahedrally coordinated polymorphs 1T-TaX_2 are stable. 1T-TaX_2 can be retained at room temperature by quenching from above 800 °C. The group V compounds are metallic at high temperatures and have charge-density waves which partially gap the Fermi surface at lower temperatures. These instabilities compete with superconductivity, which has T_c values up to 7.2 K (2H-$NbSe_2$). In the IT phase the instabilities give rise to charge-density waves which completely wipe out the Fermi surface. The 2H phase has interesting sequences of incommensurate and commensurate charge-density waves (e.g., White and Geballe 1979). The weak van der Waals forces between the dichalcogenide layers make it possible to intercalate a wide variety of organic and inorganic molecules. Intercalation destroys the charge density waves in 2H-TaS_2. The resulting increase in T_c from below 1 K to 3 K is attributable to the restoration of the Fermi surface and the resulting increase in the density of states. Figure 7 illustrates changes in T_c for a series of aliphatic amine intercalates as a function of the length of the amine. Initially the amine lies parallel to the layers (constant interlayer spacing) and the reduced charge transfer per unit area results in a monotonically decreasing T_c. For the longer amines (C > 8) the amines are perpendicular to the layers, the layer spacing increases while T_c remains constant indicating that the bond density between the amine and the dichalcogenide, which is also constant, determines the T_c. Superconductivity appears to persist in the very two-dimensional limit, as in the high-T_c cuprates. $LiNbO_2$ has the layered 2H structure and can be made superconducting by removing some lithium; T_c is reported to reach 5.5 K for a lithium–niobium ratio of 0.45 (Geselbracht et al. 1990).

The triselenides $NbSe_3$ and $TaSe_3$ are more one dimensional than the dichalcogenides and in these structures charge-density waves wipe out part of the Fermi surface. The charge-density waves are weakly pinned by random defects. The compounds are particularly interesting due to the fact that the pinning can be overcome by moderately weak electric fields. Superconductivity below 4 K can easily be induced by modest pressures which suppress the charge-density waves; the superconductivity observed at ambient pressure is probably not a bulk effect.

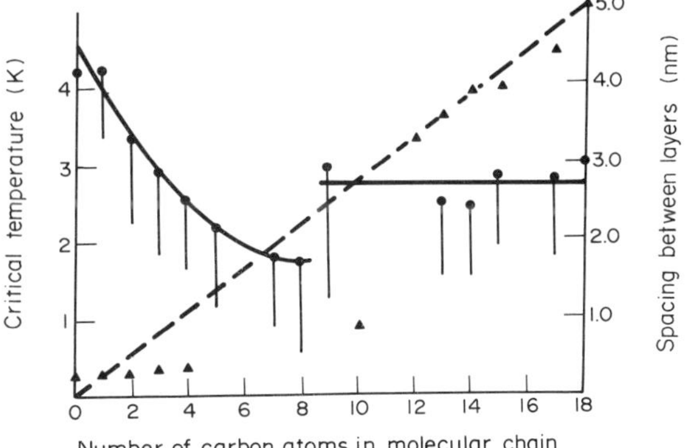

Figure 7
The superconducting critical temperature (●) and the dilation (▲) in interlayer spacing of TaS_2 intercalation compounds with normal aliphatic primary amines up to octadecylamine (after Gamble et al. 1971)

7. Concluding Remarks

Superconductivity is a widely occurring phenomenon and includes elements all the way from beryllium to americium. In view of the wide variety of superconductors which have already been discovered and the increasing amount of research which is now being carried out, there is reason to be optimistic that new discoveries are still to be expected and that this survey will soon be outdated. It seems reasonable to assume that the number of compounds which have yet to be synthesized will substantially exceed those which already have been made. New methods which avoid kinetic bottlenecks and allow equilibrium situations to be achieved on the one hand and permit the quenching in of nonequilibrium phases on the other make it possible to obtain new natural and artificial structures. Vapor-phase deposition using ever-improving vacuum technology and understanding of epitaxial growth has great potential. Solid-state chemistry can also be employed in imaginative ways such as in the synthesis of new alkali metal nitrides (Di Salvo 1990). An almost unlimited number of conceivable new phases are possible; more understanding is needed to define the most promising. At the present time it can only be suggested that systems with potential for high T_c are structures where strongly interacting states on one set of sites can be screened by controlled amounts of doping on another set of sites which in turn are strongly bonded with sufficient elastic energy to maintain the structure.

Bibliography

Abrikosov A A 1957 On the magnetic properties of superconductors of the second group. *Sov. Phys. JETP (Engl. Transl.)* 5: 1174–82

Abrikosov A A, Gor'kov L P 1961 Contribution to the theory of superconducting alloys with paramagnetic impurities. *Sov. Phys. JETP (Engl. Transl.)* 12: 1243–53

Anderson P W 1959 Theory of dirty superconductors. *J. Phys. Chem. Solids* 11: 26–30

Barbee T W III, Garcia A, Cohen M L, Martins J L 1989 Theory of high pressure phases of hydrogen. *Phys. Rev. Lett.* 62: 1150–3

Bardeen J, Cooper L N, Schrieffer J R 1957 Theory of superconductivity. *Phys. Rev.* 108: 1175–204

Bednorz J G, Müller K A 1986 Possible high-T_c superconductivity in the Ba–La–Cu–O system. *Z. Phys. B* 64: 189–93

Bon Mardion G, Goodman B B, Lacaze A 1965 Les supraconducteurs de la deuxième espèce du système plomb–thallium. *J. Phys. Chem. Solids* 26: 1143–55

Buckel W 1991 *Superconductivity: Fundamentals and Applications*. VCH, Weinheim, Germany

Collver M M, Hammond R H 1973 Superconductivity in amorphous transition metal alloy films. *Phys. Rev. Lett.* 30: 92–5

de Bruyn Ouboter R 1987 Superconductivity: discourses during the early years of low temperature research at Leiden. *IEEE Trans. Magn.* 23: 355–70

de Haas W J, Voogd J 1930 The influence of magnetic fields on supraconductors. Communications from the Physical Laboratory of the University of Leiden, No. 124

Di Salvo F J 1990 Solid state chemistry—a rediscovered chemical frontier. *Science* 247: 649–55

Dynes R C, Rowell J 1975 Influence of electrons-per-atom ratio and phonon frequencies on the superconducting transition temperature of lead alloys. *Phys. Rev. B* 11: 1884–94

Evetts J E, Wade J M A 1970 Superconducting properties and the phase diagrams of the Pb–Bi and Pb–In alloy systems. *J. Phys. Chem. Solids* 31: 973–82

Flükiger R, Klose W (eds.) 1991 *Landolt–Bornstein new series*, Group III, Vol. 21. Springer, Berlin

Gamble F R, Osiecki J H, Cais M, Pisharody R, Di Salvo F, Geballe T H 1971 Intercalation complexes of Lewis bases and layered sulfides: a large class of new superconductors. *Science* 174: 493–500

Geballe T H 1983 Empirical approach to superconductivity. In: Deaver B, Ruvalds J (eds.) 1983 *Advances in Superconductivity*. Plenum, New York

Geballe T H, Matthias B T, Corenzwit E, Hull G W Jr 1962 Superconductivity in molybdenum. *Phys. Rev. Lett.* 8: 313

Geselbracht M J, Richardson T J, Stacy A M 1990 Superconductivity in the layered compound Li_xNbO_2. *Nature* 345: 324–6

Ginzburg V L, Landau L 1950 On the theory of superconductivity. *Zh. Eksp. Teor. Fiz.* 20: 1064 English translation in: Ter Haar D (eds.) 1965 *Collected Papers of L. D. Landau*. Gordon and Breach, New York, pp. 546–68

Goodenough J B, Demazeau G, Pouchard M, Hagenmuller P 1973 Sur une nouvelle phase oxygenée du cuivre + III: $SrLaCuO_4$. *J. Solid State Chem.* 8: 325–30

Gorter C J 1964 Superconductivity until 1940 in Leiden and as seen from there. *Rev. Mod. Phys.* 36: 1–7

Gorter C J, Casimir H B G 1934 On supraconductivity I. *Physica* 1: 306–20

Greene R L, Street G B, Suter L J 1975 Superconductivity in polysulfur nitride $(SN)_x$. *Phys. Rev. Lett.* 34: 577–9

Hardy G F, Hulm J K 1954 The superconductivity of some transition metal compounds. *Phys. Rev.* 93: 1004–16

Hoyt R F, Mota A C 1976 Superconductivity in *α*-phase alloys of Cu, Ag and Au. *Solid State Commun.* 16: 139–42

Hulm J K, Ashkin M, Deis D W, Jones C K 1970 Superconductivity in semiconductors and semi-metals. *Prog. Low Temp. Phys.* 4: 205–42

Hulm J K, Matthias B T 1951 New superconductor borides and nitrides. *Phys. Rev.* 82: 273–4

Hulm J K, Matthias B T 1981 Overview of superconducting materials development. In: Foner S, Schwartz B B (eds.) 1981 *Superconductor Materials Science: Metallurgy, Fabrication, and Applications*. Plenum, New York

Jérome D, Mazaud A, Ribault M, Bechgaard K 1980 Superconductivity in a synthetic organic conductor $(TMTSF)_2PF_6$. *J. Phys. Lett.* 41: L95-8

Kimhi D B, Geballe T H 1980 Superconducting tunneling in the amorphous transition metals Mo and Nb. *Phys. Rev. Lett.* 45: 1039–42

Krupka M C, Ziorgi A L, Krikorian A L, Sklarz E G 1969 High-pressure synthesis and superconducting properties of yttrium sesquicarbide. *J. Less-Common Met.* 17: 91–8

Kunzler J E, Buehler E, Hsu F S L, Wernick J H 1961 Superconductivity in Nb_3Sn at high current density in a magnetic field of 88 kgauss. *Phys. Rev. Lett.* 6: 89–91

Kwo I R, Geballe T H 1982 Tunneling into the A15 compounds. *Physica B* 109–10: 1665–70

Little W A 1964 Possibility of synthesizing an organic superconductor. *Phys. Rev.* 134: 1416–24

Livingston J D 1963 Magnetic properties of superconducting lead-base alloys. *Phys. Rev.* 129: 1943–9

London F, London H 1935 Supraleltung und Diamagnetismus. *Physica* 2: 341–54

McMillan W L 1868 Transition temperature of strong-coupled superconductors. *Phys. Rev.* 167: 331–4

McMillan W L, Rowell J M 1965 Lead phonon spectrum calculation from superconducting density of states. *Phys. Rev. Lett.* 14: 108–12

Maple M B, Fischer Ø (eds.) 1982 *Superconductivity in Ternary Compounds I and II*. Springer, Berlin

Matthias B T 1955 Empirical relation between superconductivity and the number of valence electrons per atom. *Phys. Rev.* 97: 74–6

Matthias B T, Geballe T H, Compton V B 1963 Superconductivity. *Rev. Mod. Phys.* 35: 1–22

Meul H W, Rossel C, Decroux M, Fischer O, Remenyi G, Briggs A 1984 Observation of magnetic-field-induced superconductivity. *Phys. Rev. Lett.* 53: 497–500

Onnes H Kamerlingh 1911 On the sudden change in the rate at which the resistance of mercury disappears. Communications from the Physical Laboratory of the University of Leiden, No. 124c (25 Nov 1911)

Ott H R 1989 Unconventional superconductivity in heavy electron systems. *Physica C* 162: 1669–72

Raub E, Raub Ch J, Roschel E, Compton V B, Geballe T H,

Matthias B T 1967 The α Te–Fi solid solution and its superconducting properties. *J. Less-Common Met.* 12: 36–40
Roberts B W 1976 Survey of superconductive materials and critical evaluation of selected properties. *J. Phys. Chem. Ref. Data* 5(3): 581–821
Saito G, Kagoshima S (eds.) 1990 *The Physics and Chemistry of Organic Superconductors*. Springer, Berlin
Savitskii E M, Baron V V, Efimov Yn V, Bychkova M I, Myzendova L F 1973 *Superconducting Materials*. Plenum, New York
Schooley J F, Hosler W R, Ambler E, Becker J H, Cohen M L, Koonce C S 1965 The dependence of the superconduction transition temperature on carrier concentration in semiconducting $SrTiO_3$. *Phys. Rev. Lett.* 14: 305–7.
Shoenberg D 1960 *Superconductivity*, Cambridge Monographs on Physics. Cambridge University Press, Cambridge
Sleight A W, Gillson J L, Bierstedt P E 1975 High temperature superconductivity in the $BaPb_{1-x}Bi_xO_3$ system. *Solid State Commun.* 17: 27–8
Steglich F, Aarts J 1979 Superconducting in the presence of strong Pauli paramagnetism: $CeCu_2Si_2$. *Phys. Rev. Lett.* 43: 1892–6.
Stritzker B 1983 Ion implantation—a promising technique for the production of new superconducting materials. In: Deaver B, Ruvalds J (eds.) 1983 *Advances In Superconductivity*, 13. Plenum, New York
Talvacchio J 1989–1990 Bibliography of high-T_c superconductivity. *J. Supercond.* 2: 1–210; 3: 1–153
Vonsovsky S V, Izyumov Yu A, Kurmaev E Z 1982 *Superconductivity of Transition Metals, Their Alloys and Compounds*. Springer, Berlin
Williams J M, Schultz A J, Geiser U, Carlson K D, Kini A M, Wang, H H, Kwok W, Whangbo M H, Schirber J E 1991 Organic superconductors—new benchmarks. *Science* 251: 1501–8
Wu M K, Ashburn J R, Torng C J, Hor P H, Meng R L, Gao L, Huang Z J, Wang Y Q, Chu C W 1987 Superconductivity at 93 K in a new mixed-phase YBaCuO compound system at ambient pressure. *Phys. Rev. Lett.* 58: 908–10

T. H. Geballe and J. K. Hulm
[Stanford University, Stanford, California, USA]

Superconducting Materials: Basic Theoretical Concepts

Superconductivity is a common phenomenon in metals at low temperatures. To date all known examples are electron paired states, as revealed by flux quantization experiments. The Bardeen–Cooper–Schrieffer (BCS) theory is a spectacularly successful theory of electron pairing (see *Superconducting Materials: BCS and Phenomenological Theories*). It is based on the observation that the electron gas in a metal is unstable at low temperature with respect to even very weak attractive forces between electrons. Attractive forces are often provided by low-frequency (phonon) distortions of the background ionic lattice, which accounts for superconductivity in most metals with low transition temperatures (low T_c).

In fact, the pairing phenomenon occurs even in situations very different from those originally envisioned by BCS. Extensions of the original theory exist which successfully describe situations where, for example, the pairing force is very strong ("local pairing") or where the superconducting electrons are confined to layers or films. Pairing mechanisms other than the phonon mechanism have been proposed in some materials; for instance, magnetic fluctuations may give effective attractions in some magnetic metals.

High-temperature cuprate superconductors have posed the greatest challenge to existing theory. The task in cuprates is to account not only for superconductivity but also for a striking set of "normal state" properties observed above the superconducting T_c (see *Oxide Superconductors: Physical Properties*). These unusual metallic properties are consequences of strong electron repulsive forces (or correlations) and of electronic two dimensionality. Several mechanisms for a pairing-type instability from this unusual metallic state are being investigated, but an unambiguous experimental determination of the mechanism or mechanisms operating in these systems does not yet exist.

Some of the more relevant post-BCS developments will be reviewed in Sects. 1–3 before a more detailed discussion of the current status of high-T_c theory in Sect. 4.

1. Pairing Theory: Weak vs Strong Coupling

In BCS theory the interactions responsible for superconductivity are assumed to be weak compared with the Fermi energy. In this situation (weak coupling) electron pairing occurs at a definite temperature, the transition temperature T_c. At the same time, long-range phase coherence between pairs is established. Phase coherence is the essential ingredient for superconductivity, since it is this feature which provides rigidity (Anderson 1984) against an external magnetic field (Meissner effect). There is an energy cost or gap Δ required to break a pair. In weak coupling theory $2\Delta = 3.53\, k_B T_c$ and T_c is given by

$$k_B T_c = 1.14\, \hbar\omega_D \exp(-1/\lambda^*) \qquad (1)$$

where $\hbar\omega_D$ is the Debye energy, k_B is Boltzmann's constant and λ^* is a dimensionless electron–phonon coupling strength. In the weak coupling limit λ^* is much less than one and T_c is low. One obvious way to achieve higher T_c is to find materials with larger values of λ^*.

As λ^* increases, the BCS picture is gradually modified (Nozieres and Schmitt-Rink 1985). In a hypothetical limit where the pairing energy is much larger than the Fermi energy ($\lambda^* \gg 1$), electron pairs are tightly bound and exist as preformed bosons above T_c. (Electrons have Fermi statistics meaning that the Pauli exclusion principle applies; electrons of like spin cannot occupy the same momentum state. Tightly bound pairs of fermions form bosons to which the exclusion principle does not apply.) Bosons condense into the lowest energy state at a temperature of the order of the Bose–Einstein temperature, at which point phase coherence is established between the bosons. T_c is approximately (in three dimensions)

$$T_{BE} = \frac{2\pi\hbar^2}{M}\left(\frac{n}{2.612}\right)^{2/3} \quad (2)$$

where M is the boson mass, and n is the pair density.

Equation (2) does not promise an unlimited increase of T_c. The reason for this is that tightly bound pairs become heavier as their binding energy increases, leading eventually to a decrease of T_c according to Eqn. (2). The heavy pairs are often called bipolarons. Above T_c there is no phase coherence between bipolarons, which form a nondegenerate Boltzmann gas.

This hypothetical strong coupling state is markedly different from BCS. The energy required to break a pair is large: thus $2\Delta \gg 3.53\, kT_c$ is a signature of strong coupling. The excitation spectrum consists of collective density oscillation of the bosons with a finite excitation energy cost: $\omega_p = (16\pi n_0 e^2/M)^{1/2}$ (the plasma frequency). In contrast single fermion excitations from the BCS state exist well below the plasma frequency. Despite these differences there is continuity between weak and strong coupling limits of the pairing theory. This is a significant point. It means, for instance, that a tunnelling supercurrent between strong and weak coupling superconductors is possible.

"Strong coupling corrections" are commonly observed (e.g., in the A15s, Chevrel phases) and these phonon superconductors are intermediate between weak and strong coupling limits. However, it is controversial as to whether the true strong coupling limit with mobile bosons has really been seen in any material. It may be that in some cases bipolarons do form, but they are heavy objects easily trapped by imperfections in the crystal lattice.

2. *Superconductivity of Correlated Metals*

The discussion of Sect. 1 assumed the existence of a short-range or local attractive interaction. This seems surprising, but in phonon superconductors the coulomb interaction is ineffective at destroying pairing. The lattice distortions responsible for the attractive force persist over long times; they do not have to directly compete against the instantaneous coulomb interaction. Thus there is a pairing amplitude for electrons of opposite spin on the same site, as well as an off-site amplitude.

In systems where the local repulsions are very strong an on-site pair amplitude is unfavorable. This does not preclude superconductivity, but favors relative orbital angular momentum of the electrons in a pair since this enforces a *node* in the pair wave function where the particles coincide. Liquid ^{3}He was the first example of non-s-wave pairing in a correlated Fermi system (Leggett 1975). In liquid ^{3}He BCS pairing occurs in the triplet or parallel nuclear spin channel in the millikelvin temperature range. Antisymmetry of the Cooper pair wave function requires that the orbital angular momentum quantum number is odd, so that s wave is excluded. (^{3}He is also an indisputable example of pairing via a nonphonon mechanism; the p-wave pairing is mediated by the exchange of spin fluctuations.)

Profound modifications of the BCS theory appear when dealing with non-s-wave superconductors. First, the gap function may now vanish along lines or points of the Fermi surface whereas the Fermi surface is fully gapped in the s-wave case. For instance, a representative d-wave state has an excitation spectrum $E(\boldsymbol{k})^2 = \Delta^2 \sin^2(k_x)\sin^2(k_y)$ where $\boldsymbol{k}$ is a momentum vector on the Fermi surface and thus has lines of nodes along $k_x = 0$ or $k_y = 0$. This modifies the temperature dependence of the Meissner screening constant (penetration depth) and thermodynamics below T_c with respect to the s-wave case. It also tends to make the superconductivity more sensitive to the presence of impurities. Second, there may be degenerate superconducting phases, which means that extra internal degrees of freedom are needed to specify the state of the system, in addition to the pair density and phase. For instance, a pure d-wave superconductor has five possible projections of orbital momentum along a given axis. In the magnetic material UPt_3, which is a d-wave superconductor, this degeneracy is lifted by the crystal lattice structure and other symmetry breaking interactions, and particular linear combinations are favorable. These different superconducting states have been resolved experimentally (see *Heavy Fermion Metals*).

3. *Superconductivity in Two Dimensions*

Superconductivity in the classical sense does not exist in a spatially two-dimensional system with a phase symmetry such as the phase of Cooper pairs in a superconducting film. The reason for this is that thermal fluctuations of the phase lead to a local fluctuation which is bigger than 2π. Since states of a given global phase are degenerate, long-wavelength fluctuations of the phase cost very little energy (see *Coherence Length, Proximity Effect and Fluctuations*). At temperature T the amplitude of the fluctuation is $\langle\varphi_k^2\rangle \simeq T/|\boldsymbol{k}|^2$ where $\boldsymbol{k}$ is the wave vector of the phase distortion. To find the fluctuation at a particular point $\boldsymbol{r}$ the contributions of all of these modes must be added. In two dimensions this is $\langle\varphi(\boldsymbol{r})^2\rangle \simeq T\int dk/k$ which is very large. Thus long-range phase coherence, the usual criterion for superconductivity, is absent.

Kosterlitz, Thouless (KT) and Berezinskii showed, however, that a new type of superconducting transition occurs at finite temperature in two-dimensional systems (Kosterlitz and Thouless 1973, Halperin and Nelson 1979). The transition in this case is from a state with ohmic dissipation to a state with a nonlinear IV characteristic, $V \simeq I^{a(T)}$ with $a(T) \geqslant 3$. The KT transition is a defect mediated transition; namely, at a critical temperature T_v it is entropically favorable for free flux vortices and antivortices to appear in the system. Below T_v the vortex–antivortex pairs are bound. The absence of free vortices is the condition required for dissipationless flow of current.

The vortex unbinding transition shows up as a suppression of the condensate *fluctuation* (which has both amplitude and phase) even though it does not have a finite average belowT_v. The range over which phase coherence is maintained is described by the spatial correlation of the gap parameter:

$$\langle\Delta(r)\Delta(0)\rangle \simeq \begin{cases} \Delta^2 & T<T_c \text{ three dimensional} \\ \Delta^2 e^{-r/\xi(T)} & T>T_c \text{ three dimensional} \\ \\ \Delta^2 r^{-p(T)} & T<T_c \text{ two dimensional} \\ \Delta^2 e^{-r/\xi^{KT}(T)} & T>T_c \text{ two dimensional} \end{cases}$$

Note that phase coherence is maintained over only finite distances above the transitions.

The predictions of KT theory for the quantities $\xi^{KT}(T)$ (which translates into the temperature dependence of the resistance above T_v) and $a(T)$ have been confirmed in work on superconducting thin films. In a layered system, where layers of two-dimensional superconducting material are coupled together, long-range phase coherence does occur even for weak coupling. Nevertheless, even in this case significant features of the vortex unbinding transition can be observed (see *Flux Line Lattice*).

4. Microscopic Theory of Cuprate Superconductors

Despite its historical role, the electron–phonon interaction has not played a predominant role in the microscopic theory of cuprate superconductors. The electron–electron interaction in cuprates is the source of observed magnetic effects and unusual normal state features. Even before a mechanism can be considered it is clear that the electronic structure itself must be untangled.

The same electronic structure is common to all cuprate superconductors (Pickett 1989). The important electrons for superconductivity move in CuO_2 sheets present in various structures. The copper ions lie on the sites of a square lattice and the oxygens lie on the bonds. Band structure calculations show the existence of a broad band of hybridized copper $3d_{x^2-y^2}$ and oxygen $2p_{x,y}$ orbitals crossing the Fermi level. In all cases the relevant electrons move in this band (see *Oxide Superconductors: Physical Properties*).

In stoichiometric La_2CuO_4, electron motion is restricted by extreme correlation effects. From valence counting (La^{3+}, O^{2-}, Cu^{2+}) there is on average one electron per copper site, the copper $3d^9$ state. These electrons are not mobile, however, due to the large coulomb energy cost U for placing an additional electron on a copper site ($3d^{10}$). Thus La_2CuO_4 is an insulator, termed a Mott insulator (Mott 1974). However, the electron spin in copper $3d^9$ is left relatively free to fluctuate and spin fluctuations form the low-energy electronic excitations of the system. The preferred arrangement of spins is antiferromagnetic; the microscopic model describing the degrees of freedom of the spins in a Mott insulator is the spin $\frac{1}{2}$ Heisenberg model:

$$H \simeq J \sum_{\langle ij\rangle} S_i . S_j \tag{3}$$

where $\boldsymbol{S}_i = \{S_i^x, S_i^y, S_i^z\}$ is the electron spin localized about copper site i and the sum is taken over copper near neighbors. The magnetic interaction between cuprate layers is very weak (typically $J\perp \simeq 10^{-5}$ J) but is important in determining the antiferromagnetic ordering temperature $T_N \simeq 250$ K. The ground state and low-lying excitations of the model in Eqn. (3) are well understood. Above T_N the system behaves as a two-dimensional antiferromagnet. The spins are correlated over very large distances at temperatures $T \ll J$: $\xi \simeq a_0 \exp(J/T)$. These antiferromagnetic correlations appear clearly in inelastic neutron scattering experiments.

The Mott insulator can be made into a metallic conductor by introducing a sufficient density of dopants (i.e., either by introducing extra holes or electrons). For instance, substituting $Sr^{2+}$for La^{3+} has the effect of removing electrons for the cuprate layers and vacancy sites Cu^{3+} are mobile. The electronic structure in the doped case is a complex question, the principal issue being whether one or more electronic bands are relevant at the low-energy scales (< 0.1 eV) important for superconductivity. Numerical studies of clusters of four CuO_2 units (Hybertsen et al. 1990) have been used to map the problem onto general many-body models which retain explicitly the on-site and off-site coulomb repulsions, and the hybridization between $Cud_{x^2-y^2}$ and $Op\sigma_{x,y}$ orbitals and the $Op_x - Op_y$ overlaps. With these realistic parameters, the applicability of one-band models to the low-energy part of the spectrum can be investigated.

The simplest satisfactory model resulting from this procedure consists of the magnetic exchange term in

Eqn. (3) supplemented by a hopping or kinetic energy part for vacancies:

$$H = -t \sum_{\langle ij \rangle} c^{\dagger}_{i\sigma} c_{j\sigma} + J \sum_{\langle ij \rangle} S_i . S_j \qquad (4)$$

Here $c^{\dagger}_{i\sigma}$ places an electron of spin $\sigma = \pm\frac{1}{2}$ in an orbital localized about a copper site; t is the hopping matrix element for a hop to a neighboring site j on the square lattice. The local coulomb interaction U is still effective in preventing double occupancy. Experimentally the hopping energy $t \simeq 0.1$–0.3 eV is larger than the magnetic exchange $J \simeq 0.1$ eV.

In comparison with the very complete understanding of Eqn. (3) the two-dimensional t–J model is not well understood. Nevertheless, circumstantial experimental support for the model exists. For instance, neutron scattering reveals the persistence of magnetic moments in the superconducting materials as required by Eqn. (4). Moreover, a small concentration of mobile carriers $\delta \ll 1$ is sufficient to destroy long-range magnetic order (Ginsberg 1989). (The spin correlation length ξ is short.) The competition between kinetic energy of carriers and magnetic order is correctly described by the model in Eqn. (4) and derives from the fact that vacancy motion scrambles the spin background. Similarly, nuclear magnetic resonance experiments which are a local probe of the electron spins have been successfully interpreted using this model. The magnetic relaxation observed on different sites of the structure is compatible with the existence of a single spin fluid of copper moments, as implied by Eqn. (4).

The phase diagram of Eqn. (4) is defined by the different regimes of temperature, doping level and the ratio t/J. The present consensus of opinion is summarized in Fig. 1. A variety of ordered states appear to be stable in the low-doping region; for instance magnetic polarons where vacancies distort the antiferromagnetic background locally but leave it (more or less) intact on longer scales or states with interesting types of short-range order such as "flux" states. The symmetry of these states is different from the "Fermi liquid state" (i.e., conventional metal state) which has a closed Fermi surface. In real systems the low-doping region is very sensitive to the effects of disorder. At higher concentrations of charge carriers, the most stable known states of the model have the same symmetry as the Fermi liquid state.

Experimentally, the Fermi surface can be mapped by angle-resolved photoemission experiments (ARPES). These give a measure of the density of electron states $\rho(k, \omega)$ in momentum space and in the occupied energy region defined as $\omega < 0$. In a Fermi liquid a sharply defined electron excitation exists close to the Fermi surface. In ARPES this should appear as a sharp peak which disperses with angle or wave number. A sharp structure *is* observed experimentally at doping levels at which the system superconducts

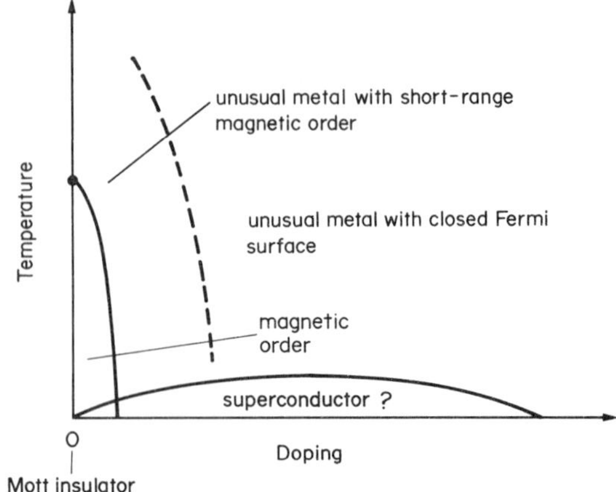

Figure 1
Quasi-two-dimensional t–J model phase diagram

and the location of the surface in momentum space corresponds well with expected geometry of a Fermi liquid state. However, paradoxically, the normal state transport properties, linear resistivity, temperature-dependent Hall voltage, thermopower and tunnelling, Copper site nuclear magnetic relaxation, electronic Raman background and so on are not those of a Fermi liquid (Ginsberg 1989, Anderson 1990).

In ordinary metals electron–electron scattering is severely restricted due to the Pauli exclusion principle: the scattering rate of an electron excited with energy ω above the Fermi surface vanishes as ω^2. Thus at low energies (or temperatures) the excitation is very sharp. There is evidence that this rule is violated in cuprate superconductors. Phenomenologically, a "marginal Fermi liquid" hypothesis may account for some of the data better, where it is assumed that the scattering rate is proportional to ω, rather than ω^2. The single particle excitations are not as well defined as in a Fermi liquid, but remain sharp enough to account for the ARPES data, as well as giving a qualitative account of the normal state features.

Can the microscopic model in Eqn. (4) account for the lack of Fermi liquid properties? Detailed analysis of one-dimensional equivalents of Eqn. (4) show non-Fermi-liquid behavior. The reason for this is that in one dimension the underlying excitations are *independent* charge and spin excitations: when an electron is removed as in a photoemission experiment, the leftover charge moves away with velocity $v_c \simeq t$ and the leftover spin moves away with velocity $v_s \simeq J$. In a Fermi liquid, the spin and charge move together with the Fermi velocity v_F. Analogous behavior in two dimensions, though unproven, could account for some of the observations and the marginal Fermi liquid hypothesis. The coupling of spin and charge degrees of freedom in two dimensions continues to be the subject of intense investigation.

Phenomenologically, the superconducting state of the cuprates is not unlike that of low-T_c materials. For instance, a supercurrent can flow between low- and high-T_c material. The issue of the symmetry of the gap parameter (Annett 1990) is not definitively settled although the evidence favors a fully gapped state. There is also evidence for a gaplike precursor in the electronic structure above T_c in, for example, oxygen-deficient Y–Ba–Cu–O; this is not unlike precursor effects seen in strong coupling BCS materials such as the A15 compounds (see *A15 Compound Superconductors*) and some of the considerations of Sect. 1 may apply here. KT behavior (Sect. 3) is observed in the highly two-dimensional material Bi–Sr–Ca–Cu–O.

There are several competing proposals for superconductivity within the model given in Eqn. (4), although the issue is somewhat confused by a tendency towards phase separation within this model. The great difference in normal state features makes it unclear whether the probable magnetic mechanism in U Pt_3 (the heavy electron superconductor mentioned in Sect. 2) operates here. Bipolaron formation (as outlined in Sect. 1) has been suggested within the t–J model, arising from magnetic instead of lattice distortions. An even more exotic mechanism, called the anyon or semion mechanism, has been studied (Wilcek 1991) which is based on electronic two dimensionality. Two-dimensional systems can support a special type of excitation whose statistical behavior is intermediate between fermion and boson. For example, when a semion pair is interchanged, the wave function acquires a phase $\varphi = \pi/2$: $\psi(x,y) = \exp(i\varphi)\psi(y,x)$. (For fermions $\varphi = \pi$, for bosons $\varphi = 0$.) States supporting excitations with $\varphi = \pi/2$ have been proposed for the t–J model. Numerical studies indicate flux quantization in units of $h/2e$ for semion systems on small lattices, the signature of pair superconductivity. The phase factor associated with an isolated semion is unfavorable energetically and pairing occurs in order to screen out the unwanted phase. At present, however, the evidence for anyons in cuprates is mostly negative.

It is important to remember that the t–J model is at best an approximation valid for the lowest lying electronic excitations of the system. Thus, while it undoubtedly captures many features of the normal state, it is not known whether it can account for the observed superconductivity as well. Residual interactions which have been omitted from the model may be responsible for superconductivity but these have not been conclusively identified. In addition to the coupling described by Eqn. (4), one must include the interlayer hopping (Wheatley et al. 1988), the coupling to structural lattice distortions, the long-range part of the coulomb interaction or even a weak admixture of higher lying electronic states such as the Cud_z^2 orbital. Any or a combination of these may be capable of inducing superconductivity as an instability of the strongly correlated normal state but it should be remembered that a sharply defined electron excitation is required for BCS theory to be straightforwardly applied.

5. Conclusion

Superconductivity is a field which has stimulated rapid parallel developments between materials science and chemistry as well as mathematical physics in recent years. While fermion pairing remains the cornerstone of theory, unexpected connections between superconductivity and the fractional quantum Hall phenomenon of semiconductor physics, low-dimensional magnetism and gauge theories of high-energy physics have become important. Progress has been made in the problem of strongly interacting electrons, which so far is a characteristic of all superconductors with high T_c.

Bibliography

Anderson P W 1984 *Basic Notions of Condensed Matter Physics*. Benjamin-Cummins, Reading, MA

Anderson P W 1990 Common trends in particle and condensed matter physics. *Phys. Rep.* 184: 2–4

Annett J F 1990 Symmetry of the order parameter for high-temperature superconductivity. *Adv. Phys.* 39:83–126

Bulaevskii L N, Ginzburg V L, Sobyanin A A 1989 Macroscopic theory of superconductors with short coherence length. *Sov. Phys. JETP* 68:1499–510

Ginsberg D M (ed.) 1989 *Physical Properties of High Temperature Superconductors,* Vols. 1–3. World Scientific, Singapore

Halperin B I, Nelson D 1979 Resistive transition in superconducting films. *J. Low Temp. Phys.* 36:599–616

Hybertsen M, Stechel E B, Schluter M, Jennison D R 1990 Renormalization from density-functional theory to strong-coupling models for electronic states in Cu–O materials. *Phys. Rev. B* 41:1068–72

Kosterlitz J M, Thouless D J 1973 Ordering, metastability and phase transitions in two dimensional films. *J. Phys. C* 6:1181–203

Leggett A J 1975 A theoretical description of the new phases of liquid ^{3}He. *Rev. Mod. Phys.* 47:331–414

Mott N 1974 *Metal Insulator Transitions*. Taylor Francis, New York

Nozieres P, Schmitt-Rink S 1985 Bose condensation in an attractive fermion gas: from weak to strong coupling. *J. Low Temp. Phys.* 59:195–211

Pickett W 1989 Electronic structure of high T_c oxide superconductors. *Rev. Mod. Phys.* 61:433–512

Wheatley J, Hsu T, Anderson P W 1988 Interlayer effects in high-T_c superconductors. *Nature* 333:121

Wilcek F 1991 Anyon superconductivity. *Sci. Am.* May: 58–65

J. M. Wheatley
[University of Cambridge, Cambridge, UK]

Superconducting Materials: BCS and Phenomenological Theories

At very low temperatures many metals and alloys undergo a second-order phase transition to a superconducting state in which there is no resistance to flow of electricity. Superconductors exhibit other remarkable phenomena such as exclusion of magnetic flux from the interior (the Meissner effect) and flux quantization. Phenomenological theories were developed to describe semiquantitatively different aspects of superconductivity, including the Gorter–Casimir two-fluid model for thermal properties and the London theory for electromagnetic properties. In spite of many attempts, it was nearly 50 years from the discovery of superconductivity in 1911 before a microscopic theory based on pairing of electrons was derived from quantum mechanics. The theory has been able to account quantitatively for most properties of superconductors as well as to predict new unexpected phenomena. Transition temperatures can be derived essentially from first principles for the simpler metals, but the theory has been less successful in predicting new compounds with high transition temperatures.

1. Outline of Phenomenological Theories

Superconductivity was discovered by Kamerlingh Onnes, only three years after he had liquefied helium. He found that the resistance of a rod of frozen mercury drops suddenly to zero when cooled to near the boiling point of helium (4.2 K). It was soon found that superconductivity is a common phenomena, exhibited by many metals, alloys, and intermetallic compounds when cooled below a critical temperature T_c. To show that the resistance really vanishes, Onnes showed that a persistent current flows indefinitely in a ring or solenoid with no voltage applied. In steady state, the electric field in a superconductor vanishes ($E = 0$).

It was not until 1933 that Meissner discovered a property that is even more basic; a superconductor excludes a magnetic field ($B = 0$). If a sphere or other simply connected body is placed in a magnetic field and cooled below T_c, the flux in the interior is expelled. If the field were applied when the sphere is in the superconducting state, currents would be set up near the surface to keep the flux inside from changing. They would flow indefinitely, so that B would remain zero in the interior. What Meissner showed is that the state with the flux excluded is the unique thermodynamically stable state of a superconductor. Thus the currents flowing near the surface that counteract the external field are stable, not metastable.

Early attempts to account for the vanishing resistance were aimed at trying to understand why the electrons making up the supercurrent flow are not scattered as they are in the normal state. After the discovery of the Meissner effect, the focus shifted to the nature of the phase transition in the electronic structure that could lead to such remarkable properties.

Superconductivity can be understood only in terms of quantum theory; superconductors are quantum systems on a macroscopic scale. The current flowing around a superconducting ring is quantized such that the total magnetic flux threading the ring is an integral multiple of a flux unit, $\Phi_0 = h/2e = 2.07 \times 10^{-15}$ Wb.

There is a critical magnetic field as well as a critical temperature. Because flux is excluded from the interior, there is an increase in magnetic energy of $\frac{1}{2}\mu_0 H^2$ per unit volume, where H is the applied field. When this magnetic energy more than makes up for the free energy differences $G_s(0) - G_n$ between superconducting and normal phases in the absence of a field, the system reverts to normal. Thus the critical magnetic field H_c is given by

$$\tfrac{1}{2}\mu_0 H_c^2 = G_n - G_s(0) \tag{1}$$

The free energy difference and critical fields decrease with increasing temperature and vanish at the critical temperature T_c. There is an approximate law of corresponding states:

$$H_c = H_0(1 - (T/T_c)^2) \tag{2}$$

where H_0 is the critical field at $T = 0$ K. Equation (1) gives a relation between $H_c(T)$ and the thermodynamic quantities, such as heat capacity, that can be derived from the free energy. Experiments in the early 1930s by Keesom and co-workers at Leiden verified these relations experimentally.

In 1934, Gorter and Casimir showed that Eqn. (2) and other properties could be accounted for by a phenomenological two-fluid model in which there is an electron charge density $\rho_s (\rho_s = n_s e)$ associated with the superfluid condensate and a density ρ_n associated with electrons excited out of the condensate and subject to the usual scattering processes, with the total density $\rho = \rho_s + \rho_n$. The total current is then $\boldsymbol{J} = \rho_s \boldsymbol{v}_s + \rho_n \boldsymbol{v}_n$, where $\boldsymbol{v}_s$ and $\boldsymbol{v}_n$ are the velocities of the two components. In the absence of an electric field, $\boldsymbol{v}_n = 0$ and $\boldsymbol{J}_s = \rho_s \boldsymbol{v}_s$. There is no entropy associated with ρ_s, so a supercurrent carries no entropy.

A little later, in 1935, F. London and H. London proposed their famous phenomenological equations to describe the electrodynamics of superconductors. To account for infinite conductivity and the Meissner effect, they added to Maxwell's equations an equation relating the current to magnetic field:

$$\boldsymbol{J}_s(r) = -\boldsymbol{A}(r)/\Lambda \tag{3}$$

where $\Lambda = m/n_s e^2 \mu_0$ and $\boldsymbol{A}$ is the vector potential expressed in a gauge such that $\nabla . \boldsymbol{A} = 0$. The time derivative gives the usual equation for acceleration of the electron in an electric field in the absence of dissipation. Equation (3) leads in addition to the Meissner effect.

F. London suggested that superconductivity is a quantum phenomenon associated with the rigidity of the wave function describing the response of the system to changes of the momentum $\boldsymbol{p}$ of electrons in a magnetic field. The current is given by a sum over all electrons of $e\langle \boldsymbol{p}-e\boldsymbol{A}\rangle/m$ averaged over the wave function of the superconducting condensate. Equation (3) follows if $\langle \boldsymbol{p}\rangle = 0$ even in the presence of a magnetic field (described in the London gauge). In a normal metal $\langle \boldsymbol{p}\rangle$ changes in such a way that $\langle \boldsymbol{p}-e\boldsymbol{A}\rangle$ is very small, giving just the small Landau diamagnetism. Particularly in his book, published in 1950, London emphasized that the rigidity of the wave functions in a magnetic field could arise from a long-range order in the momentum distribution of the electrons.

According to the London theory, the shielding currents near the surface of a superconductor in a magnetic field flow in a layer of average thickness determined by a penetration depth $\lambda = \lambda_L = \Lambda^{1/2}$, determined by n_s, the density of superconducting electrons. In the late 1940s, Pippard, working at Cambridge, found that the penetration depth increases markedly with added impurities. Impurities decrease the scattering mean free path of electrons in the normal state but have little effect on their density. For this and other reasons, he proposed a nonlocal form of the London relation in Eqn. (3) in which $\boldsymbol{J}_s(\boldsymbol{r})$ is given by an integral of $\boldsymbol{A}(\boldsymbol{r})$ over a region surrounding the point r of size determined by a coherence distance ξ. If l is the electron mean free path,

$$\xi^{-1} = \xi_0^{-1} + l^{-1} \tag{4}$$

where ξ_0 is the coherence distance in the pure metal, typically of order 10^{-7} m.

About the same time (1950), Ginzburg and Landau proposed a different sort of modification of the London equations. They were interested in trying to understand the energy of the boundary between superconducting and normal domains in the intermediate state. The intermediate state consists of a series of normal and superconducting domains in the form of slabs parallel to the applied field. The field $\boldsymbol{B}$ is equal to the critical field in the normal domains and vanishes in the interior of the superconducting domains. The change from $n_s = 0$ in the normal domains to the equilibrium value in the superconducting domains occurs gradually across the boundary over a distance determined by the two lengths in the problem, ξ and λ.

Thus they needed to have a theory that describes changes of $n_s(\boldsymbol{r})$ in space. Further, in accord with Landau's theory of second-order phase transitions, they described the ordered low-temperature phase by an order parameter that vanishes as $T \rightarrow T_c$. Ginzburg and Landau showed remarkable insight in proposing on phenomenological grounds that the order parameter be a complex function $\Psi(\boldsymbol{r}) = |\Psi(\boldsymbol{r})|\exp[i\varphi(\boldsymbol{r})]$ with amplitude and phase. The order parameter describes both the density n_s and the velocity $\boldsymbol{v}_s$ of superconducting condensate:

$$\left.\begin{aligned} n_s^* &= |\Psi(\boldsymbol{r})|^2 \\ m^*\boldsymbol{v}_s &= \hbar\nabla\varphi(\boldsymbol{r}) - e^*\boldsymbol{A}(\boldsymbol{r}) \end{aligned}\right\} \tag{5}$$

Because of pairing, it is now known that $m^* = 2m$, $e^* = 2e$ and $n_s^* = n_s/2$.

Since $\Psi(\boldsymbol{r})$ must be single valued, the line integral of $\nabla\varphi$ around a closed loop must be a multiple of 2π. If the loop is in the interior of a superconductor where $\boldsymbol{v}_s = 0$ (such as around the interior of a torus containing a persistent current), the enclosed flux $\Phi = \oint \boldsymbol{A}.d\boldsymbol{l} = (h/e^*)\oint \nabla\varphi d\boldsymbol{l}$ must be a multiple of a flux unit $\Phi_0 = h/e^*$.

By expanding the free energy in powers of $\Psi(\boldsymbol{r})$, they derived a nonlinear Schrödinger-like equation to determine $\Psi(\boldsymbol{r})$. This equation was coupled with Maxwell's equations to determine $\Psi(\boldsymbol{r})$ and thus the supercurrent flow subject to appropriate boundary conditions. The expansion is presumed to be valid just below T_c where $\Psi(\boldsymbol{r})$ is small. They used the equations to determine the boundary energy as well as other properties of superconductors.

To have a Meissner effect, the boundary energy must be larger than a critical value. Otherwise the superconductor could break up into a structure with very thin normal regions that would allow the flux to penetrate a distance λ on either side. The decrease in magnetic energy from the flux penetration would more than compensate for the higher free energy of the normal regions. Ginzburg and Landau found that the criterion could be described by a parameter $\kappa = \lambda/\xi$. If $\kappa < 1/\sqrt{2}$, the superconductor is stable against normal domain formation for all fields less than H_c and there is a perfect Meissner effect. If $\kappa > 1/\sqrt{2}$, the flux begins to penetrate when the field is greater than a lower critical field $H_{c1} < H_c$, but the metal remains superconducting up to a higher critical field, $H_{c2} > H_c$. Since magnetic flux can penetrate for $H_{c1} < H < H_{c2}$, the increase in magnetic energy is less than that for a complete Meissner effect, so that the transition to the normal state does not occur until the higher critical field is reached. The upper critical field H_{c2} may be as large as 20 T or greater for some intermetallic compounds. High-field superconducting magnet wire is made with such materials.

Superconductors with $\kappa < 1/\sqrt{2}$ that exhibit a perfect Meissner effect are called type I, those with $\kappa > 1/\sqrt{2}$ are called type II (see *Magnetic Properties of Superconductors*). Type-II superconductors were first investigated by Shubnikov at Kharkov in the Soviet Union in 1937 and 1938. His work stopped when he fell victim to a purge.

It was first thought that flux penetrates type-II superconductors through thin layers of normal domains as it does in the intermediate state. Then

Abrikosov derived another solution, published in 1957, of the Ginzburg–Landau equations that he interpreted as a periodic array of quantized vortex lines. The order parameter $\Psi(\boldsymbol{r})$ goes to zero on the axis of the lines. Supercurrent circulating around the axis gives rise to one unit of flux Φ_0. With this theory, Abrikosov was able to account in a quantitative way for the magnetization curves observed by Shubnikov.

In addition to the Ginzburg–Landau theory, the year 1950 was notable for the experimental discovery that T_c depends on isotopic mass (indicating that the motion of the ions in the metal must be involved) and H. Fröhlich's independent suggestion that interactions between electrons and lattice vibrations (or their quanta, phonons) are responsible for superconductivity. Results of the experiments could be accounted for if T_c varies inversely with the square root of the isotopic mass.

Onnes just missed discovering the isotope effect 30 years earlier. He attempted to detect a difference in critical temperature between ordinary lead and lead that came from radioactive decay of uranium. However, his measurements were not sufficiently precise to find the small difference in T_c arising from the slight difference in mass of the lead nuclei from the two sources.

Attempts by Fröhlich and by Bardeen in 1950 to explain superconductivity in terms of changes in the energy of individual electrons from interaction with the phonons failed. Pines and Bardeen showed that the electron–phonon interaction leads to an effective attractive interaction for electrons which have energies within a phonon energy of the Fermi energy E_F. It was thought that it would be necessary to take this attractive interaction into account in a successful theory.

2. *Pairing Theory of Superconductivity*

In 1957, Bardeen, Cooper and Schrieffer gave a theory of superconductivity based on pairing of electrons in the superconducting ground state. Pairing occurs in such a way as to take advantage of the attractive phonon-induced interaction between electrons. An energy gap for quasiparticle excitations appears at the Fermi surface, in confirmation of experiments in the early 1950s of heat capacity and of thermal conductivity. The theory, known as the BCS theory from the initials of the authors, has been successful in accounting in quantitative detail for a wide variety of properties of superconductors, including the isotope effect, the Meissner effect, persistent currents, a second-order phase transition, an expression for the supercurrent in the form suggested by Pippard, penetration depths and their temperature dependence, and other electromagnetic and thermal properties.

Normal metals are well described by a Fermi sea of electrons and quasiparticle excitations of electrons from the sea. Each quasiparticle state is defined by a wave vector $\boldsymbol{k}$ and spin σ. At $T = 0$ K, all states with energies less than the Fermi energy E_F are occupied; those above are unoccupied. Low-lying excited configurations can be described by the wave vectors of the occupied states above E_F and unoccupied states or holes below E_F. If the energy ξ_k is measured from the Fermi energy, the total excitation energy is given by

$$W_{exc} = \sum_{\substack{occ \\ k > k_F}} \xi_k + \sum_{\substack{occ \\ k < k_F}} |\xi_k| \tag{6}$$

Self-energies from effects of coulomb and phonon interactions are included in the quasiparticle energies ξ_k.

According to BCS, the ground-state wave function of the superconducting condensate is a linear combination of a selected set of wave functions Ψ_i of low-lying excited configurations of the normal metal:

$$\Psi_s = \sum a_i \Psi_i \tag{7}$$

The configurations Ψ_i which enter the sum are restricted to those in which the normal quasiparticle states are occupied in pairs of opposite spin and momentum $(\boldsymbol{k}\uparrow, -\boldsymbol{k}\downarrow)$, such that if in a given configuration one of the two states is occupied, the other is also. An interaction between electrons gives matrix elements between two configurations in which a pair $\boldsymbol{k} \equiv (\boldsymbol{k}\uparrow, -\boldsymbol{k}\downarrow)$ is occupied and the pair $\boldsymbol{k}'$ unoccupied in one configuration and $\boldsymbol{k}'$ occupied and $\boldsymbol{k}$ unoccupied in the other. With paired configurations, the phases of the wave functions can be chosen so that all matrix elements between such configurations are negative for an attractive interaction V. If the coefficients a_i have the same sign or phase, the matrix elements V_{ij} add coherently to give a negative contribution to the energy:

$$\int \Psi_s^* V \Psi_s d\tau = \sum a_i^* a_j V_{ij} \tag{8}$$

The wave function in Eqn. (7) was written by BCS in terms of creation operators, $c_{k\sigma}^*$ for putting a particle in the state $\boldsymbol{k}$, σ:

$$\Psi_s = \prod_k (u_k + v_k c_{k\uparrow}^* c_{-k\downarrow}^*) \Psi_0 = \sum_N a_N \psi_N \tag{9}$$

Here v_k^2 is the probability that the pair $(\boldsymbol{k}\uparrow, -\boldsymbol{k}\downarrow)$ is occupied and $u_k^2 = 1 - v_k^2$ is the probability that it is unoccupied. The wave function is not an eigenstate of the number of pairs N, but the coefficients u_k and v_k can be chosen so that the distribution is sharply peaked about the number in the normal ground state below E_F.

For a simple model in which it is assumed that $V_{ij} = -V$ for $|\xi_k| < \hbar\omega_{ph}$ and zero otherwise, where $\hbar\omega_{ph}$ is an average phonon energy, the following results are obtained by a variational method:

$$u_k^2 = 1 - v_k^2 = \frac{1}{2}\left(1 + \frac{\xi_k}{E_k}\right) \tag{10a}$$

$$E_k = \left(\xi_k^2 + \Delta^2\right)^{1/2} \text{ for } |\xi_k| < \hbar\omega_{ph} \tag{10b}$$

$$\Delta = 2\hbar\omega_{ph} \exp[-1/N(0)V] \tag{10c}$$

$$G_s - G_N = -\tfrac{1}{2}N(0)\Delta^2 \tag{10d}$$

where $N(0)$ is the normal state density of states at the Fermi level.

A superconductor has a set of quasiparticle excitations in one-to-one correspondence with those of the normal metal. The energy of such an excitation is E_k, corresponding to that of a particle above the Fermi sea or a hole below. Since excitations are formed in pairs from the ground state, the minimum energy required for creating a pair of excitations at the Fermi surface is the energy gap 2Δ.

The probability that a state k is occupied at finite temperatures is given by the Fermi function

$$f_k = \frac{1}{1 + \exp(\beta E_k)} \tag{11}$$

where $\beta^{-1} = k_B T$. The gap parameter Δ decreases with increasing temperature and goes to zero at the critical temperature T_c. The value of $\Delta(T)$ can be found from the equation

$$1 = N(0)V \int_0^{\hbar\omega_{ph}} \frac{(1 - 2f_k)d\xi_k}{E_k} \tag{12}$$

This equation is commonly referred to as the BCS gap equation. Near T_c, $\Delta(T)/\Delta(0) \simeq 1.74[1 - (T/T_c)^2]^{1/2}$. The gap goes to zero as $T \to T_c$ and there are no solutions for $T > T_c$. The value of T_c is given by

$$k_B T_c = 1.13\hbar\omega_{ph} \exp[-1/N(0)V] \tag{13}$$

The energy gap at $T = 0$ is proportional to T_c:

$$E_g = 2\Delta(0) = 3.52\, k_B T_c \tag{14}$$

In this weak coupling approximation there is only one parameter, $\Delta(0)$ or $k_B T_c$, rather than two, $\hbar\omega_{ph}$ and $N(0)V$. Weak coupling means that $k_B T_c$ is very small compared with the phonon energies so that the phonon spectrum is not important. In this limit there is a law of corresponding states since everything scales with T_c. Strong coupling and effects of coulomb interactions will be discussed later.

With a ground state and set of quasiparticle excitations it is possible to calculate transition probabilities and various response functions without much more difficulty than for normal metals. The main complications arise from coherence factors that are a result of the pairing. In normal metals scattering from $\boldsymbol{k}\sigma$ to $\boldsymbol{k}'\sigma'$ is independent in first approximation from all other scattering. In a superconductor it is possible to go from $\boldsymbol{k}\uparrow$ to $\boldsymbol{k}'\uparrow$ either directly or by scattering from $-\boldsymbol{k}'\downarrow$ to $-\boldsymbol{k}\uparrow$ and these two contributions must be added coherently. Depending on the nature of the scattering, they add constructively or destructively.

The coherence factors can be calculated most easily by a reformulation of the BCS theory given independently by Valatin and by Bogoliubov. They define quasiparticle operators for a superconductor as linear combinations of creation and destruction operators for normal quasiparticles:

$$\gamma_{k\uparrow}^* = u_k c_{k\uparrow}^* - v_k c_{-k\downarrow} \tag{15a}$$

$$\gamma_{-k\downarrow}^* = u_k c_{-k\downarrow}^* + v_k c_{k\uparrow} \tag{15b}$$

These operators do not conserve particle number and are designed to operate on wave functions of the type of Eqn. (9) that are not eigenstates of the number of pairs N. The ground state Ψ_s is the vacuum state for quasiparticle excitations:

$$\gamma_{k\uparrow}\Psi_s = \gamma_{-k\downarrow}\Psi_s = 0 \tag{16}$$

When there is current flow, the paired states ($\boldsymbol{k}_1\uparrow$, $\boldsymbol{k}_2\downarrow$) all have precisely the same momentum $\boldsymbol{p}_s = \hbar(\boldsymbol{k}_1 + \boldsymbol{k}_2)$ so that any pair can be scattered into another pair. There is thus a long-range order of the momentum distribution as envisaged by London. Individual electrons can be scattered, but such scattering does not change the common momentum of the paired states.

As in the Ginzburg–Landau theory, the canonical momentum $\boldsymbol{p}_s$ can be expressed as the gradient of a phase φ (see *Josephson Effects in Weak Links*). There is an arbitrary phase factor in the ground-state wave function since v_k can be replaced by $v_k \exp(i\varphi)$ without changing the energy. The degenerate wave function is an eigenstate of phase

$$\Psi_s(\varphi) = \prod_k \left(u_k + v_k \exp(i\varphi) c_{k\uparrow}^* c_{-k\downarrow}^*\right)\Psi_0$$

$$= \sum_N a_N \exp(iN\varphi)\Psi_N \tag{17}$$

The phase and the number of pairs N are conjugate variables, with $N_{op} = -i\partial/\partial\varphi$. If the number of pairs is fixed, the phase is undetermined. The phase corresponds to that of the Ginzburg–Landau theory, with $\boldsymbol{p}_s = \hbar\nabla\varphi$ the momentum of a pair.

It was trying to understand whether phase differences between two superconductors connected by a weak link could be observed that led Josephson to predict that a supercurrent should flow through a tunnel junction separating two superconductors differing in phase (see *Josephson Effects in Weak Links*).

The importance of phase as a variable describing the superconducting state had been stressed by Anderson. The Josephson effect was first observed by Anderson and Rowell. It has led to many practical applications.

3. Gor'kov–Eliashberg Formulation

Not long after the first publications of the BCS theory, Gor'kov used the method of thermal Green's functions to treat problems in which there is a space variation of the order parameter $\Delta(\boldsymbol{r})$ as well as problems in which there is strong coupling so that it is necessary to include the energy dependence of Δ. One of his first applications was to show that the theory reduces to the Ginzburg–Landau theory near T_c, with $\Delta(\boldsymbol{r})$ playing the role of the order parameter $\Psi(\boldsymbol{r})$. An important difference is that the charge of a pair, $2e$, replaces the electronic charge e.

Eliashberg used the Gor'kov formulation to derive a gap equation sufficiently general to include strong coupling. The electron–phonon interaction enters the integral equation in the form $\alpha^2(\omega)F(\omega)$, where $F(\omega)d\omega$ gives the number of phonons with energies in the interval $\hbar d\omega$ and $\alpha(\omega)$ is the electron–phonon coupling constant averaged over all directions in space. There are two coupled integral equations for $\Delta(\omega)$ and for the mass renormalization factor $Z_s(\omega)$ from the electron–phonon interaction, both of which are in general complex.

For weak coupling, Z_s is the same as Z_n and is the factor by which the density of states in energy is enhanced by the electron–phonon interaction. Expressed in terms of an interaction parameter λ, $Z_s = Z_n = 1 + \lambda$ where

$$\lambda = 2\int_0^\infty \alpha^2(\omega)F(\omega)d\omega/\omega \tag{18}$$

The parameter λ corresponds to $N(0)V_{ph}$ of the BCS theory, with the phonon energy $\hbar\omega_{ph}$, defined by

$$\hbar\omega_{ph} = 2\int_0^\infty \alpha^2(\omega)F(\omega)d\omega/\lambda \tag{19}$$

The repulsive Coulomb interaction between electrons can be included by renormalizing the screened interaction V_c to the same energy range as the phonons. The interaction constant $\mu = N(0)\langle V_c\rangle$ for scattering between states extending over an energy of the order of E_F from the Fermi surface is replaced by a renormalized constant

$$\mu^* = \mu/[1 + \mu\ln(E_F/\hbar\omega_{ph})] \tag{20}$$

for scattering within an energy $\hbar\omega_{ph}$ of E_F. Virtual scattering to states outside of this range is included in the renormalized interaction. The expression for T_c for weak coupling is then

$$k_B T_c = 1.13\,\hbar\omega_{ph}\exp[-1/(\lambda - \mu^*)] \tag{21}$$

Normally, μ^* varies only slowly with electron density and is typically in the range 0.10–0.15. It may, however, change the exponent α in the isotope effect $T_c \sim M^{-\alpha}$ from the value $\frac{1}{2}$ because μ^* can vary with isotopic mass through its dependence on $\hbar\omega_{ph}$.

McMillan used the Eliashberg equations to calculate T_c for a series of transition metals and found that the results could be fitted approximately by the semiempirical expression

$$k_B T_c = \left(\frac{\Theta_D}{1.45}\right)\exp\left[-\frac{1.04(1+\lambda)}{[\lambda - \mu^*(1+0.62\lambda)]}\right] \tag{22}$$

where Θ_D is the Debye temperature. Values of λ range from about 0.25 for metals with low T_c to above 1.25 for strongly coupled superconductors with high T_c.

Metals and compounds with high T_c are those with a reasonably large λ and high phonon energies as defined by Eqn. (19). It is difficult to find materials that satisfy both criteria because a large λ is usually associated with a soft lattice and consequently low phonon frequencies. Examples of soft metals are mercury ($T_c \simeq 4.2$ K) and lead ($T_c \simeq 7.2$ K). A high density of states $N(0)$ such as is found with transition metals is helpful, but within a range of alloys it is often found that λ increases inversely with the density of states. The matrix elements of the electron–phonon coupling go in the opposite direction and more than compensate for the decrease in $N(0)$. Materials with high T_c have been discovered by empirical methods rather than by suggestions from theory.

The Gor'kov–Eliashberg theory has been used to calculate T_c essentially from first principles for simple metals and alloys such as aluminum (weak coupling), tin (moderate coupling) and lead (strong coupling). The phonon spectra are derived from neutron scattering data and a pseudopotential method is used to calculate $\alpha^2(\omega)$. The good agreement with experimental values leaves little doubt about the basic validity of the theory.

Of even greater interest is the application of equations to derive $\alpha^2(\omega)F(\omega)$ empirically from the current–voltage characteristics of a tunnel junction between a normal metal (N) and the superconductor (S) separated by a thin insulating layer (see *Tunnelling Spectroscopy*). At low temperatures $(dI/dV)_{NS}$ is proportional to the density of states in the superconductor. The ratio to the conductance measured when both sides are normal is for a voltage V corresponding to $eV = E = \hbar\omega$

$$\left(\frac{dI}{dV}\right)_{NS}\Big/\left(\frac{dI}{dV}\right)_{NN} = \frac{N_S(E)}{N(0)} = \frac{E}{(E^2 - \Delta^2)^{1/2}} \tag{23}$$

where $N(0)$ is the uniform density of states in the normal metal and the expression on the right is for weak coupling. The current vanishes when $E < \Delta$.

For strong coupling, when $\Delta(E)$ is complex, Schrieffer and co-workers showed that

$$\frac{N_S(E)}{N(0)} = \mathrm{Re}\left\{\frac{E}{[E^2 - \Delta(E)^2)]^{1/2}}\right\} \qquad (24)$$

If the conductance is measured over a broad energy range, Kramers–Krönig relations can be used to determine both the real and imaginary parts of $\Delta(E)$ and thus the complete Green's functions required to calculate equilibrium and transport properties. This procedure has given excellent agreement between theory and experiment, even when there is strong coupling.

McMillan devised a computer program to determine $\alpha^2(\omega)F(\omega)$ directly from tunnelling data with use of the Eliashberg integral equations. This method has been used for a large number of superconducting elements and compounds to obtain detailed information about $\alpha^2(\omega)F(\omega)$ and its variation with energy. It is necessary to estimate μ^* from the electron density; the only other variable in the equations is then $\alpha^2(\omega)F(\omega)$. Generally, the peaks of $\alpha^2(\omega)F(\omega)$ coincide with the peaks of the phonon density of states $F(\omega)$ measured by neutron scattering.

4. *Basic Features and Extensions of the Pairing Theory*

At low temperatures, the normal state is described as a weakly interacting gas of quasiparticles and phonons. The quasiparticles correspond to electrons above or holes below the Fermi surface.

The fully renormalized quasiparticles are in one-to-one correspondence with those of a gas of noninteracting particles. Because of interactions, the quasiparticle excitations are not exact eigenstates of the normal metal but have a finite lifetime (see *Nonequilibrium Superconductivity*). According to the Landau Fermi liquid theory, the lifetime varies at low temperature as ω^{-2}, where ω is the quasiparticle energy relative to that of the Fermi surface. Thus the approximation treating the quasiparticles as independent becomes better and better as the temperature is decreased.

The wave functions of the superconducting state are described by linear combinations of low-lying normal-state configurations. If the energies of the normal-state configurations are derived empirically, the free energy difference between normal and superconducting states may be derived with far greater precision than would be possible by a first-principles calculation of either phase by itself.

What is needed is the matrix element for scattering of a pair of quasiparticles in state $(\boldsymbol{k}_1, \boldsymbol{k}_2)$ to $(\boldsymbol{k}_1', \boldsymbol{k}_2')$ with conservation of wave vector:

$$\boldsymbol{k}_1 + \boldsymbol{k}_2 \rightarrow \boldsymbol{k}_1' + \boldsymbol{k}_2' \qquad (25)$$

When the interaction is mediated by the electron–phonon interaction it may be regarded as occurring in two steps:

$$\boldsymbol{k}_1 \rightarrow \boldsymbol{k}_1' + \boldsymbol{q}_{\mathrm{ph}} \qquad (26a)$$

$$\boldsymbol{k}_2 + \boldsymbol{q}_{\mathrm{ph}} \rightarrow \boldsymbol{k}_2' \qquad (26b)$$

where $\boldsymbol{q}_{\mathrm{ph}}$ is the wave vector of the phonon. An alternative process is the same as in Eqn. (26) with $\boldsymbol{k}_1$ and $\boldsymbol{k}_2$ interchanged. Energy is conserved only in the final step. The superconducting phase is formed from a subset of low-lying normal configurations between which the matrix elements are predominantly negative, corresponding to an attractive interaction.

The quasiparticles in the subset are associated in pairs each of which has exactly the same momentum as every other pair. When there is not current flow the paired states have opposite spin and momentum.

One of the earliest extensions of the original BCS theory, by Mattis and Bardeen, was to include effects of a scattering mean free path l. In this case, the momentum $\boldsymbol{k}$ is not a good quantum number to describe the quasiparticles. A quasiparticle state $\psi\uparrow$ is paired with the time-reversed state $\psi^*\downarrow$. The main effect is to replace the inverse of the Pippard distance, ξ_0^{-1}, with $\xi^{-1} + l^{-1}$, as in Eqn. (4).

Not long after the microscopic theory was published suggestions were made to get an attractive interaction between quasiparticles other than that mediated by phonons. These are not necessarily alternatives to the phonon mechanism but may supplement it so as to enhance the transition temperature. Although none has been firmly established, it is likely that something besides phonons is required to account for the high transition temperature of compounds with copper oxide layers, first discovered by Bednorz and Müller (see *Superconducting Materials: Basic Theoretical Concepts*).

Other low-lying excitations that have been suggested are, like phonons, defined by a wave vector. They include excitons (creation of electron–hole pairs), acoustic plasmons (in a ground state with equal numbers of conduction electrons and holes in overlapping bands) and spin and charge-density wave excitations (where there is nesting of opposite faces of the Fermi surface).

Little and, independently, Ginzburg suggested an exciton mechanism in which there is physical separation of the excited electron–hole pair and the carriers that give the current flow. Little suggested conduction in linear chains surrounded by polarizable side chains. Ginzburg suggested layer structures with conductive layers alternating with polarizable layers in which the electron–hole excitations take place.

Both may have found realizations in two classes of compounds with abnormally high superconducting transition temperatures, although it is still far from certain that it is the exciton mechanism that is responsible. Ginzburg's suggested geometry is similar to that of the high-T_c copper oxide layer structures. Little's is similar to that of organic linear-chain compounds that have been found by Williams and co-workers to have transition temperatures as high as 13 K. However it has been found that in both systems the normal metal is not described by a simple Fermi liquid. More radical departures from the usual BCS theory than getting enhanced attractive interactions may be required. In both systems, experiments indicate that in the normal state the reciprocal of the quasiparticle lifetime varies as ω rather than as ω^2. The energy gap in planar directions in the high-T_c oxides is about $8k_BT_c$, much larger than the BCS value of $3.52\,k_BT_c$ for weak coupling. Some extensions and changes of the BCS theory are required to account for the high-T_c oxides and organic linear-chain superconductors. It is uncertain whether or not these changes are minor or require a major restructuring of the theory. The theories that have been proposed are discussed in the article *Superconducting Materials: Basic Theoretical Concepts*.

5. Applications to Other Areas of Physics

The concepts and mathematical methods of the theory of superconductivity have been applied successfully to the superfluid phase of liquid helium of isotopic mass three, to the structure of nuclei, to pulsars and neutron stars and to the elementary particles of high-energy physics. Liquid ^{3}He, like electrons in metals, obeys Fermi–Dirac statistics. The liquid undergoes a pairing transition to a superfluid phase below 3 mK. The pairing is of parallel spins in a triplet state ($S=1$) rather than the singlet spin pairing ($S=0$) that occurs with electrons in superconducting metals. Extensions of superconductivity theory have been applied with great success to account for the pairing energy of neutrons and protons in nuclei. The pairing energy gap decreases with increasing atomic number. The biggest superfluid objects in the universe are pulsars and neutron stars in which there is evidence that both neutrons and protons undergo pairing transitions.

To define the state of a superconductor it is necessary to specify the superfluid velocity $\boldsymbol{v}_s$ in addition to the other macroscopic variables, such as temperature and pressure. This is associated with what is called broken-gauge symmetry. Analogous concepts are now being used widely to better understand the structure of the particles (elementary excitations) of high-energy physics.

See also: High-Frequency Impedance of Superconducting Materials; Nonequilibrium Superconductivity; Optical Properties of Superconductors; Superconducting Materials: An Overview; Superconducting Materials: Basic Theoretical Concepts

Bibliography

Bardeen J 1973 Electron–phonon interactions and superconductivity. In: Haken H, Wagner M (eds.) 1973 *Cooperative Phenomena*. Springer, Berlin

Bardeen J, Schrieffer J R 1961 Recent developments in superconductivity. In: Gorter C J (ed.) 1961 *Progress in Low Temperature Physics*, Vol. 3. North-Holland, Amsterdam

Chilton F (ed.) 1971 *Superconductivity: Proc. Int. Conf. Science of Superconductivity*. North-Holland, Amsterdam

de Gennes P G 1966 *Superconductivity in Metals and Alloys*. Benjamin, New York

London F 1950 *Superfluids*, Vol. 1. Wiley, London

Parks R D (ed.) 1969 *Superconductivity*, Vols. 1, 2. Dekker, New York

Rickayzen G 1965 *Theory of Superconductivity*. Wiley, London

Shoenberg D 1952 *Superconductivity*. Cambridge University Press, Cambridge

Schrieffer J R 1964 *Theory of Superconductivity*. Benjamin, New York

Tinkham M 1975 *Introduction to Superconductivity*. McGraw-Hill, New York

J. Bardeen[†]
[University of Illinois, Urbana, Illinois, USA]

Superconducting Materials: Elements

Forty-three of the 88 naturally occurring elements are superconductors under suitable experimental conditions and two of the 18 artificial, or man-made, elements, namely technetium and americium, are also superconductors. Measured superconducting transition temperatures T_c range from a low of 325 μK to a high of 9.2 K (see Fig. 1). Matthias (1969) emphasized the use of the periodic table as a key to the prediction of new superconducting elements and alloys. By noting the correlation between T_c and such normal state properties as crystal structure, melting points and the number of electrons per atom e/a, he was able to formulate a set of rules for the occurrence of superconductivity. The use of the "Matthias rules" has led to the discovery of new superconducting elements and many new and useful superconducting alloy systems. Recent advances in theory, coupled to increased computational capabilities, have produced first-principle calculations of T_c for the elements. Thus there are both theoretical and empirical predictions of new superconducting elements with T_c values ranging from below 10^{-4} K to room temperature and above. Recently perfected techniques combining ultralow temperatures (of the order of microkelvin) and ultrahigh pressures (of the order of 10^2 GPa) have made meaningful tests of these predictions possible.

When discussing superconductivity among the elements it has become customary to group the elements into three categories: (a) the simple elements, (b) the transition metals, and (c) the rare-earth and actinide metals.

Figure 1
Superconductivity in the elements of the periodic table: elements with a second, stable, superconducting phase with lower T_c values are denoted by an asterisk

1. Simple Metals (s–p Elements)

The simple elements are those whose outer electrons occupy *s* and *p* sublevels with inner sublevels being filled. This category contains metals, metalloids and nonmetals. Whereas standard chemistry textbooks place copper, silver and gold among the transition elements, this definition requires them to be considered as simple metals. There are ten simple metals that become superconducting when cooled from room temperature at an ambient pressure of 1 atm (Fig.1). T_c values range from 0.026 K for beryllium to 7.2 K for lead. This category of superconductor has e/a values of 2–4. However, by subjecting other simple elements to high pressures in order to produce new metallic phases and cooling these phases under high pressures, 12 additional *s–p* elements become superconductors. In this case T_c is a function of pressure and varies from 0.02 K for arsenic at 10 GPa to 8.5 K for bismuth at 9 GPa. In general *s–p* metals are type I superconductors and suitably prepared samples exhibit ideal superconducting behavior (see *Superconducting Materials: Physical Properties*).

Empirically, T_c for these simple metals is a monotonically increasing function of e/a and is proportional to the density of electron states at the Fermi level $N(0)$. Theory relates T_c with two normal state parameters: the Debye temperature Θ_D and $N(0)$. The former represents a measure of the phonon spectrum of the element whereas the latter is associated with the electronic structure (see *Superconducting Materials: BCS and Phenomenological Theories*). Band theory calculations of $N(0)$ provide the basis for first-principle calculations of T_c. Both theory and the Matthias rules predict that all simple metals should be superconducting with T_c being very low for e/a values of 1 and 2. For example, magnesium, lithium, barium, calcium, gold and silver are predicted to be superconductors with values of T_c of the order of 10^{-3} K or 10^{-2} K. The use of ultralow-temperature techniques to cool these metals to temperatures as low as 10^{-4} K has shown these estimates for T_c to be too high. For example, data on superconducting dilute binary alloys obtained in the 10^{-3} K temperature range indicate that pure silver, gold and copper would have to be cooled to 10^{-9} K in order to become superconductors (Mota et al. 1989). Present day techniques do not permit lattice temperatures to be produced much smaller than 10^{-5} K.

The use of high pressure has proved to be the best way to produce new superconducting *s–p* elements as evidenced by the successes to date (see *Superconducting Materials Under Pressure*). Bundy and Dunn (1980) report on the appearance of superconductivity in the ultrahigh-pressure metallic phases of sulfur and of xenon, the first rare gas to become a superconductor, and this has added credence to the prediction that the high-pressure metallic phase of hydrogen will be superconducting; theoretical estimates for its T_c go as high as room temperature. Unfortunately, all of the superconducting high-pressure phases revert back to their nonsuperconducting phases upon release of the high pressure. For technological purposes a metastable high-pressure superconducting phase is required (i.e., one which will remain superconducting at 1 atm of pressure).

The possibility that such metastable phases may exist at low temperatures is suggested by results obtained when thin films are condensed at low temperature. Superconducting phases of *s–p* elements which are not usually superconductors can be produced by low-temperature deposition (see *Thin Films, Multilayers and Devices, Superconducting*). Bismuth films condensed at low temperatures are superconducting with T_c values as high as 6 K. Thin films of magnesium, calcium, barium and strontium deposited at low temperatures are reported to be superconductors with T_c values in the 3–4 K range (Roberts 1976). In general, if a superconducting phase can be produced by low-temperature deposition, it can also be produced via high-pressure techniques (Brandt and Ginsberg 1971). Since this is true only in the case of barium, the reported superconductivity of the other three elements is suspect and therefore a question mark is associated with their T_c values as given in Fig. 1.

Low-temperature deposition of films of known *s–p* superconductors have produced enhanced T_c values. The most striking case is beryllium where the T_c of such films can be as high as 9 K, a factor of 300 larger than the bulk T_c value. Theory accounts for this in terms of a "softening" of the phonon spectrum. Films with enhanced T_c values are metastable and revert back to the low-T_c phase upon warming above a transformation temperature of around 100 K.

2. Transition Metals (s–d Elements)

Focusing attention on the transition metals with incomplete *d* sublevels, it can be seen from Fig. 1 that there are 14 metals which superconduct at atmospheric pressure. T_c values range from 325 μK for rhodium to 9.2 K for niobium. Transition metals have high melting points when compared with those of *s–p* metals and sample purity has been a major problem. In the case of molybdenum, iron impurities (~200 ppm) produced the erroneous result that pure molybdenum was nonsuperconducting to temperatures as low as 0.003 K. Purity problems continue to contribute to uncertainties in the value of T_c for titanium and hafnium. Impurities can also lead to high values for T_c. The artificially produced element technetium was originally reported to have the highest T_c of any element (11.2 K). However, subsequent work has established that the T_c for technetium is 7.8 K, which is the second highest T_c of any element at 1 atm of pressure. High-pressure studies have produced three additional *s–d* superconductors: scandium, yttrium and lutetium. As in the case of the simple metals, T_c is a function of pressure and ranges from 0.35 K for scandium at 21 GPa to 2.8 K for yttrium at 15 GPa (Wittig et al. 1979).

A plot of T_c vs e/a (see Fig. 2) reveals a behavior totally different from that observed for the simple metals. These data gave rise to two beliefs that have dominated much of the research, both theoretical and experimental, since the early 1970s. These beliefs are (a) if it is required to synthesize an alloy or intermetallic compound with a high T_c value, then the constituents should be such that the average e/a value should be either 4.7 or 6.4; and (b) the deep minimum at $e/a = 5$, corresponding to a half-filled *d* shell, indicates that the electron–phonon interaction of the Bardeen–Cooper–Schrieffer (BCS) theory may not be the dominant cause of superconductivity in the *s–d* metals. This latter belief was enhanced by the discovery of the absence of any isotope effect in ruthenium and zirconium and nonideal values of the exponent of the isotope effect in osmium and ruthenium (see *Superconducting Materials: An Overview*). Refinements in theory now account for this behavior without recourse to a different pairing interaction. The recent discovery of superconducting ceramics with values of T_c in the 90–110 K range has focused attention away from the Matthias rules. However, this has once again served to emphasize the apparent importance of transition elements in high-T_c compounds as well as renewing theoretical efforts to

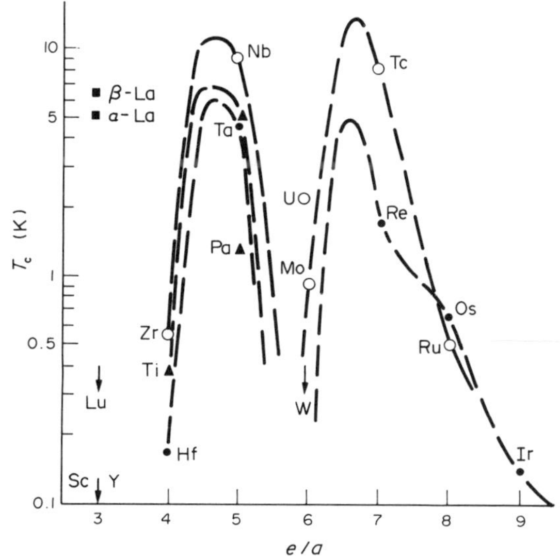

Figure 2
Superconducting transition temperature vs number of electrons per atom for the *s–d* and *s–f* superconductors: ▲ 3*d*; ○ 4*d*; ● 5*d* (after Matthias 1969. © Gordon and Breach, New York. Reproduced with permission)

identify a nonphononic interaction which can lead to electron pairing in these high-T_c ceramics.

Thin films of iron and strontium condensed at helium temperatures have been reported superconducting at 4.2 K and 3.6 K, respectively. In addition, ion-beam-sputtered films of chromium and films of palladium irradiated with He^+ ions have been reported superconducting with T_c values of 1.5 K and 3.2 K, respectively. The lack of additional proof for the superconductivity of these elements is the reason for the question marks in Fig. 1.

Transition metals are, in general, type I superconductors with only niobium and vanadium being borderline type II superconductors. The vast majority of nonmagnetic *s–d* transition metals are superconductors (at pressures of 1 atm) and their variation in T_c can be adequately described quasi-phenomenologically (Rowell 1976). Band theory permits the calculation of T_c from first principles. These calculations and empirical predictions based on studies of alloys indicate that yttrium, scandium, lutetium and platinum should be superconductors with T_c values in the 10^{-2}–10^{-3} K range (at 1 atm pressure). As noted in Fig. 1, yttrium, scandium and lutetium do have high-pressure superconducting phases. The nearly ferromagnetic palladium is an enigma. Advances in ultralow-temperature techniques have resulted in the discovery that rhodium is a type I superconductor with the lowest T_c ever reported—325 μK (Buchal et al. 1983). This noteworthy experimental accomplishment attests to the validity of the scheme of Matthias whereby the values of T_c of binary alloy systems are extrapolated to predict the T_c values of the nonsuperconducting constituents. It was this type of measurement that allowed Mota et al. (1989) to derive T_c values of 10^{-9} K for silver and copper.

The BCS microscopic theory based on the electron–phonon mediated pairing adequately accounts for the superconducting properties of the *s–p* metals. Subsequent refinements of the theory have been able to account for the behavior of the *s–d* transition metals as well. A concept gaining in importance is that of *p*-wave pairing (see *Superconducting Materials: Basic Theoretical Concepts*).

3. Lanthanide and Actinide Metals (s–f Elements)

The lanthanide and actinide metals consist of 28 metals, 11 of which are no longer found in nature but may be artificially produced. There are four superconductors at pressures of 1 atm: lanthanum, thorium, protactinium and americium. There are also two which possess high-pressure phases which are superconducting: cerium and uranium. T_c is a function of pressure and the T_c values are 1.9 K at 4.0 GPa for cerium and 2.4 K at 1 GPa for uranium. In the case of uranium, it was originally believed that it was superconducting at a pressure of 1 atm, but recent results indicate that it is the high-pressure phase which is superconducting. Americium, the most recently discovered *s–f* superconductor, and protactinium are highly radioactive. Actinium has been predicted to be a high-T_c superconductor (10 K); however, the high radioactivity of this element has prevented verification of this (Smith 1980).

Superconductivity among the *f*-band metals is of interest to the solid-state chemist who uses it as a tool to study the itinerant vs the localized nature of the *f* electrons. If itinerant, the *f* electrons lead to superconductivity; if localized, they lead to ferromagnetism (see *Rare-Earth Metals and Alloys*). This subject is somewhat esoteric and the role of *f* electrons in lanthanum has been the subject of much controversy (Wittig et al. 1979). Several unique features are apparent in the superconducting properties of the *f*-band metals. For example, lanthanum exists in two stable phases that are superconducting. The β phase has a T_c that is especially sensitive to pressure. It increases from 5 K at 1 atm to 13 K at 175 atm, a record high T_c for the elements. Another unique feature is that uranium shows a positive isotope effect.

In spite of these peculiar aspects, it is felt that *f*-band superconductivity can be understood in a manner similar to that of *d*-band superconductivity. No exotic mechanism need be evoked to account for the superconductivity of the *f*-band metals (Smith 1980).

See also: Superconducting Materials: An Overview; Superconducting Materials Under Pressure

Bibliography

Brandt N B, Ginsburg N I 1971 Superconductivity at high pressure. *Sci. Am.* 224(4): 83–94

Buchal C, Pobell F, Mueller R M, Kubota M, Owens-Bradley J R 1983 Superconductivity of rhodium at ultra-low temperature. *Phys. Rev. Lett.* 50: 64–7

Bundy F P, Dunn K J 1980 Electrical behavior of Te, Se and S at very high pressures and low temperatures: superconducting transitions. *Phys. Rev. B* 22: 3157–64

Matthias B T 1969 Systematics of superconductivity. In: Wallace P R (ed.) 1969 *Superconductivity*, Vol. 1. Gordon and Breach, New York, pp 227–93

Mota A C, Visini P, Pollini A 1989 Magnetic properties of proximity-induced superconducting copper and silver. *J. Low Temp. Phys.* 76(5/6): 465–512

Roberts B W 1976 Survey of superconducting materials and critical evaluation of selected properties. *J. Phys. Chem. Ref. Data* 5: 581–821

Rowell J M 1976 A simple T_c equation. *Solid State Commun.* 19: 1131–2

Smith J L 1980 Superconductivity in the actinides. In: Suhl H, Maple M B (eds.) 1980 *Superconductivity in the d- and f-Band Metals*. Academic Press, New York, pp. 37–48

Wittig J 1978 New hypothesis concerning the commencement of the rare earth series at high pressures. In: Chu C W, Woolam J A (eds.) 1978 *High-Pressure and Low-Temperature Physics*. Plenum, New York, pp. 467–81

Wittig J, Probst C, Schmidt F A, Gschneider K A Jr 1979 Superconductivity in a new high pressure phase of scandium. *Phys. Rev. Lett.* 42: 469–72

R. A. Hein
[University of Washington, Seattle, Washington, USA]

Superconducting Materials: Large-Scale Applications

The replacement of normal conducting metals by superconductors with high critical current densities and high critical magnetic flux densities can result in considerable improvements in large electrical devices, especially with respect to efficiency, weight and size. In addition, the use of superconductors opens up new solutions for energy problems which cannot be realized on the basis of existing technologies. Examples are magnetic confinement for thermonuclear fusion, magnetohydrodynamic (MHD) energy conversion and inductive energy storage (see *Magnetic Energy Storage*).

At present, superconducting NbTi and Nb_3Sn multifilament conductors, commercially available in a large variety of configurations, are used. They exhibit critical temperatures of approximately 10 K and 18 K, respectively. In consequence, they have to be operated in liquid helium (boiling point 4.2 K). The feasibility, technical performance and superiority of plants and systems equipped with these superconductors has already been demonstrated by a number of large-application-oriented demonstration projects. These included high-power cables, motors for ship propulsion, turbogenerators, magnets for ore separation, superconducting magnetic energy storage (SMES), magnetically levitated high-speed trains (MAGLEV), magnets for fusion test facilities and MHD generator prototypes. However, large-scale application of these superconductors has been restricted to magnets for magnetic resonance imaging (MRI) in medicine and for high-energy physics. Superconducting turbogenerators, magnets for separation, SMES and MAGLEV may be additional applications within a reasonable space of time.

The main reasons for the limited usage of superconducting equipment has apparently been the nonacceptable high hysteretic losses which are generated in these superconductors when ac magnetic fields are applied and the necessity of cooling with liquid helium which is rather complicated, inefficient and expensive. Recently, NbTi multifilament conductors with filament diameters in the submicrometer range (0.1–0.5 μm) embedded in highly resistive CuNi matrices have been developed which in principle allow operation at power frequencies with tolerable losses (Cave et al. 1989) (see *AC Applications of Superconducting Materials*). The existence of these superconductors will open the door for ac applications of metallic superconductors at 4.2 K. In addition the appearance of the new high-T_c superconductors with their ability to operate with liquid nitrogen cooling has considerably increased the prospects of a widespread application of superconductivity in magnet and power engineering. Cooling with liquid nitrogen is more effective, reliable and less expensive than with liquid helium. However, before these favorable prospects can be realized a series of problems have to be solved, such as high current densities in notable magnetic fields ($\geqslant 10^5$ A cm^{-2} at 5 T and 77 K) and workable conductors. The progress achieved with respect to this during the past few years has been considerable (see *Critical Current Density*).

The following evaluations of a change from 4 K to 77 K superconductivity for the various fields of application were made on the basis of a critical current density of 10^5 A cm^{-2} at 77 K, independent of the magnetic field strength.

1. Magnets for Magnetic Resonance Imaging and Spectroscopy

The magnetic resonance (MR) imaging and spectroscopy of living organisms are rapidly evolving new medical diagnostic methods. They are based on high-frequency signals emitted by protons or other nuclei (e.g., phosphorus and sodium) in a living body when aligned in a highly homogeneous magnetic field and excited by high-frequency electromagnetic radiation. The received high-frequency signals are converted by a fast processor into cross-sectional images of the whole body or can be used for located chemical shift spectroscopy of compounds, which is important for studies of metabolism (see *Nuclear Magnetic Resonance Imaging*).

Magnetic flux densities from 0.15 T to approximately 1.5 T are used for whole-body imaging; for chemical shift spectroscopy field strengths of at least 1–2 T are necessary. Conventional electromagnets or permanent magnets are sufficient for the low-field region. For flux densities $\geqslant 0.5$ T only superconducting magnets are suitable. The image quality is increased considerably when higher flux densities are applied. Therefore approximately 85% of existing MR installations (more than 4000 all over the world) are equipped with superconducting magnets operating at flux densities between 0.5 T and 2 T (see *Superconducting Magnets*). For research purposes magnets with flux densities of up to 4 T have been built (Vetter et al. 1988). In general, the bore of these magnets is about 1 m and they are thus capable of housing the patient and the other equipment necessary for imaging or spectroscopy. They are usually 2–3 m long with an outer diameter of around 2–2.5 m (Reichert et al. 1988).

The high spatial homogeneity of these magnets (better than ± 10 ppm in a sphere of 0.5 m diameter) is achieved by using multicoil assemblies and additional shimming. The high temporal stability of the magnets ($\Delta B/B < 10^{-7}\,h^{-1}$) is accomplished by persistent mode operation. All magnets built so far have been constructed with NbTi–Cu multifilament conductors (see *Multifilamentary Superconducting Composites*). The present market for MR magnets is around 800 units per year.

Considering a commercial 1 T magnet and assuming an electromagnetic performance and price equivalent to that of NbTi superconductors the use of a 77 K superconductor will reduce the costs of the magnet by approximately 15% and additionally lead to a saving of around US$100 000–200 000 in cooling costs over a period of five years. Since the magnet costs amount to about one-third of the total plant costs, a reduction of about 5% is obtained for the plant costs. In addition, the geographically unlimited availability of liquid nitrogen will open new markets in regions which suffer from a lack of helium supply.

2. *Superconducting Magnets for High-Energy Physics*

Before MRI was established, high-energy physics had been the most important area for large-scale applications of superconductors. A series of large superconducting dc particle detector magnets and beam guiding magnets (dipoles and quadrupoles) have been built and successfully operated for several years in various laboratories. They yield remarkable savings in electrical energy consumption, resulting in reduced operation costs. Because of the higher magnetic flux densities achievable in superconducting magnets, size reduction of the magnets and, therefore, buildings is possible.

The same applies to the 1000 GeV ring-shaped proton accelerator (TEVATRON) at the Fermi Laboratory in the USA, which is equipped with approximately 1000 superconducting dipole and quadrupole magnets along a periphery of 6 km and has been in operation since 1984. A similar accelerator (HERA) is being commissioned at DESY, Hamburg (Wolf 1992). Giant supermachines of this type which will consist of several thousand superconducting dipole and quadrupole magnets with a total length of up to 80 km and energies up to 20 TeV are in the state of planning and design in the USA (the Superconducting Super Collider (SSC)) and at CERN (the Large Hadron Collider (LHC)). The conductors provided for these accelerators are cables of transposed NbTi multifilament conductors for the SSC and most probably, Nb_3Sn multifilament conductors in the case of the LHC. Concerning especially the SSC, enormous efforts have been made with respect to improvement of the filament current density ($3.5 \times 10^5\,A\,cm^{-2}$ at 5 T and 4.2 K) and to small hysteresis effects of the windings in order to achieve high magnetic field quality (Peoples 1989).

The replacement of the metallic superconductors by high-T_c superconductors would have, in a first approximation, no influence on the installation costs (the saving in cryogenic equipment is compensated by the need for additional vacuum pumping systems for the beam tube because of lack of the cryopumping effect of 4.2 K cold beam tube walls) and cause 25% lower operation costs by a corresponding reduction of the total power consumption. Based on experience, technical problems have to be expected with regard to the realization of the specified high field accuracy of the dipole magnets.

3. *Magnetic Separation and Filtration*

In a magnetic separation process, ferromagnetic or paramagnetic particles are separated by magnetic forces from a mass stream of nonmagnetic material. The magnetic force is proportional to the magnetizing field strength H, to the particle susceptibility χ and size V_p and to a magnetic field gradient ∇H applied to the particle:

$$F_p \simeq \chi V_p \mu_0 H \nabla H \qquad (1)$$

where μ_0 is the magnetic field constant.

Magnetic separation with conventional magnets is already used widely, especially in the enrichment of ferromagnetic iron ores. However, for weakly ferromagnetic or paramagnetic ores with small particle sizes, expensive chemical flotation has to be utilized. The enhancement of the magnetizing field by the use

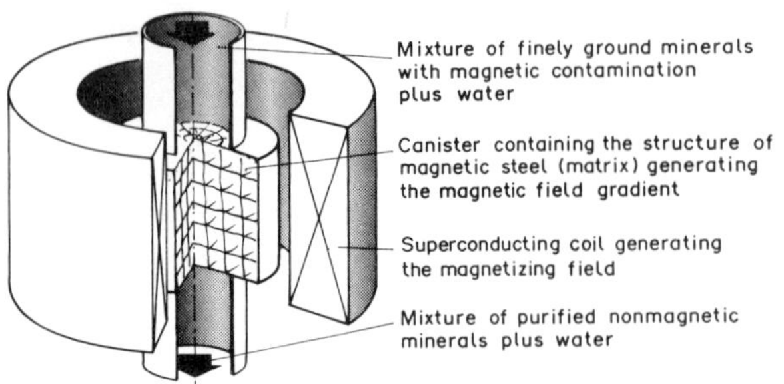

Figure 1
Schematic of a high-gradient separator (matrix separator) for purification of minerals

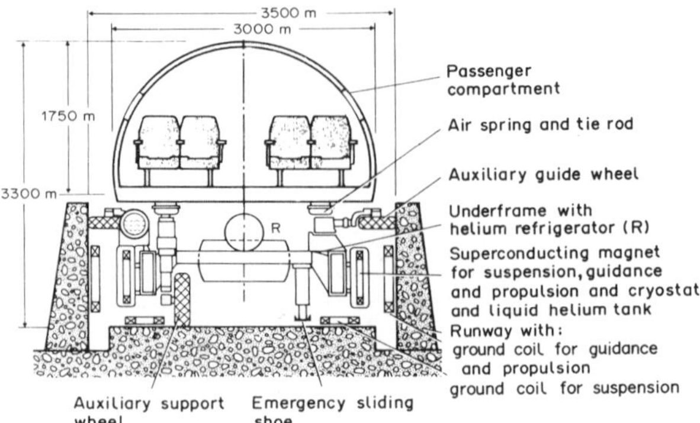

Figure 2
Cross section of an advanced levitated train test vehicle (MLU001, Railway Technical Research Institute, Japan)

of superconducting magnets in separators is a means of replacing chemical flotation methods. Other areas of potential applications of high-field and high-gradient separation under discussion are the removal of weakly magnetic impurities from minerals, purification of water and desulfurization of coal, as well as chemical and biological applications (see *Magnetic Separation*).

Several pilot plants with superconducting magnets have been successfuly operated (Watson 1988). A favored separator type which is especially applied to Kaolin purification is the high-gradient magnetic or matrix separator (see Fig. 1). The magnetizing field is generated by a superconducting solenoid and the gradient by a stack of fine magnetic steel meshes (matrix). Magnetic particles are captured and deposited at the surface of the steel structure, whereas nonmagnetic material passes without hindrance. When the matrix is saturated by magnetic material, the magnetic field is switched off or the matrix (contained in a canister) is transferred into a field-free space and cleaned by water.

As can be demonstrated by pilot plants, the higher capital costs of the superconducting magnets are offset by their considerably lower power consumption. By the application of high-T_c superconductors and operation at 77 K, a reduction in magnet costs (compared with 4 K) of about 20% can be expected and a further decrease in magnet power consumption by a factor of three. Together with a higher reliability and availability of operation, these factors will lead to enhanced usage of superconducting plants.

4. Electrodynamic Levitation System for High-Speed Trains

Magnetic suspension and propulsion are basic requirements for a tracked high-speed ground transportation system with speeds of around 500 km h^{-1} and low environmental impact. In the electrodynamic system, persistent current carrying superconducting dc coils are installed in tandem arrangement at both sides of the chassis of a passenger vehicle. These superconducting coils interact with stationary suspension, guidance and propulsion coils—loops or sheets of copper or aluminum arranged along the track at ambient temperature.

Figure 2 shows schematically the principle of an advanced version of this system. Suspension and guidance of the vehicle are accomplished by repulsive forces between the superconducting coils on the chassis and the coils on the track, arising from currents induced in the latter by the magnetic flux of the superconducting coils when they are moved with the vehicle. Propulsion is effected by sectionally charging the stationary propulsion ground coils (acting also as guidance coils) by a three-phase current of variable frequency which causes a travelling electromagnetic wave of variable speed along the track. By interaction of this wave with the dc magnetic flux of the superconducting coils in the vehicle, a propulsion force is generated synchronously moving with the vehicle (ironless linear synchronous motor with active track). The main advantages of this system are

(a) wide air gaps ($\geqslant$ 100 mm) for suspension, guidance and propulsion because of the high flux densities available with superconducting magnets;

(b) suspension and guidance are statically stable due to the repulsive forces—no gap control is required;

(c) the problems generally associated with the transmission of high electrical powers at high speeds by sliding contacts are avoided by the linear synchronous motor with active track; and

(d) no magnet power supplies on the vehicle because of persistent mode operation of the superconducting coils.

The most advanced project is that being undertaken in Japan by the Railway Technical Research Institute

(RTRI) in Tokyo and in Miyazaki on Kyushu island (until 1984 the Japanese National Railway (JNR) was responsible). In 1979 on a 7 km long test track at Miyazaki a speed of 517 km h^{-1} was attained with a 10 t test vehicle. In 1987 one 10 t section of the three-sectional test vehicle MLU 001 (MLU from magnetic levitation in a U-shaped guideway) attained 401 km h^{-1} on the same test track with passengers on board. In 1988 tests were commenced with a 17 t, 22 m long vehicle, MLU 002, which is capable of carrying 44 passengers. It is the precursor of a final test vehicle for public transportation and has achieved a speed of more than 400 km h^{-1}. The tests will be continued by the Japanese Railway Technical Research Institute on a new 43 km long test track in the Yamanashi Prefecture west of Tokyo, aiming at a speed of 500 km h^{-1} (ISTEC 1990).

Only NbTi multifilament conductors cooled at 4.2 K were considered for the superconducting magnets. Since the magnet costs, inclusive of cryogenic equipment (refrigerator), only present a minor portion of the total system costs ($\sim$ 1%), the change from 4 K to 77 K cooling is of negligible importance.

5. *Superconducting Magnet Energy Storage*

Superconducting magnet energy storage (SMES), the idea of using very large superconducting coils as a means of inductive energy storage in power grids for load balancing, is based on the fact that dc currents flow infinitely in a superconducting coil short circuited at its ends. The interconnection between the dc storage coil and the ac grid for charging and discharging is accomplished by a thyristor converter, the power of which determines the power of the plant. Studies carried out on 1000 MW systems with a storage capacity of 4000–5000 MWh predict a storage efficiency as diurnal storage of 93–95%, which is much higher than that of pumped hydrostorage ($\sim$75%). A 5000 MWh storage design (Loyd et al. 1987) provides a cylindrical four-layer coil, 1000 m in diameter and 19 m high, mounted in an annular open-top tunnel mined into the rock just below ground. The rock serves as a cheap support for the considerable radial magnetic forces. A 200 kA conductor, composed of NbTi–Cu multifilamentary conductors, a high-purity aluminum stabilizer and high-strength aluminum alloy and superfluid helium at 1.8 K as coolant is used in this design. Most advanced is a joint SDI/EPRI project in the USA which aims at a 20 MWh engineering test model (ETM) suitable for both a pulsed power source for lasers and a prototype for power levelling (see *Magnetic Energy Storage*).

A change from liquid helium to liquid nitrogen cooling would reduce the capital costs of a large storage system by about 10% which would most probably be compensated by the worsening of the electromagnetic and mechanical properties of the conductor materials (stabilizer, reinforcement) at 77 K. A positive consequence would be the enhancement of the storage efficiency to approximately 90% when operated as a weekly storage.

6. *Magnet Systems for Fusion Reactors and MHD Generators*

6.1 *Fusion Reactors*

As far as magnetic confinement for thermonuclear fusion is concerned, large superconducting magnets generating high magnetic flux densities at low power consumption are the only way to achieve acceptable plant efficiencies. Three types of magnet configurations are under consideration in particular: the tokamak (Fig. 3a), the stellarator (Fig. 3b) and the tandem mirror concept (Fig. 3c). The main superconducting components of the tokamak arrangement are a toroidal dc field coil which encloses the plasma ring and pulsed ac poloidal field coils which act as primary windings of a transformer and induce a current in the plasma ring (secondary transformer winding). The superposition of the magnetic fields of this current and that of the toroidal coil results in a magnetic field configuration with helically circulating field lines which accomplishes the plasma confinement. A second purpose of the plasma current is plasma heating. With the stellarator a similar field configuration is generated only by currents in external coils, but the topology is very complex (Fig. 3b). Flux densities of 5–6 T are required on the plasma axis of tokamak arrangements, corresponding to maximum flux densities of 12 T at the coil windings, achievable only with Nb_3Sn. The dimensions of the toroidal coil of an advanced 1000 MW (electric) tokamak reactor

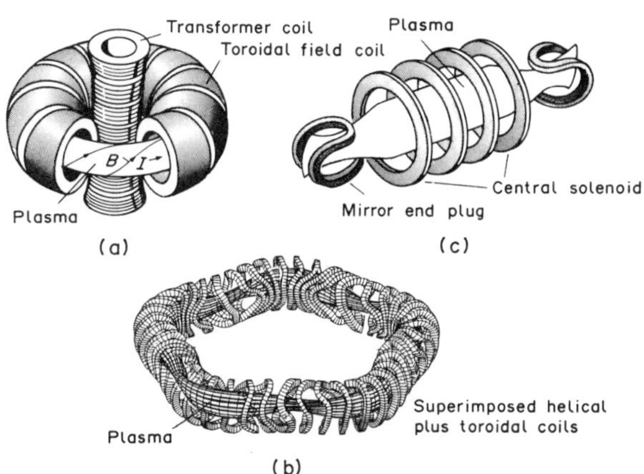

Figure 3
Principal magnet arrangements for plasma confinement in fusion reactors: (a) tokamak, (b) stellarator and (c) tandem mirror

design are minor diameter 7 m and major diameter 10 m. The magnetic field requirements of stellarator systems are less severe and can be met most probably by NbTi.

The tandem mirror system consists of a long central dc solenoid, which radially confines a plasma cylinder, and a set of dc coils at both ends of the solenoid, which act as mirror end plugs, preventing the hot plasma from axial escape. Similar to the toroidal dc coil of the tokamak, the central solenoid of the tandem mirror is assembled from several single-disk coils. Reactor designs with electrical power ratings of 1000 MW contain central solenoids with lengths $\geqslant$ 100 m and inner diameters of 7–9 m. A magnetic flux density of 3–4 T is required at the solenoid axis, corresponding to maximum fields at the windings of 5–7 T. Therefore, NbTi conductors can be used. The flux densities of the end plug coils may partly reach values of 15–20 T, which can be accomplished only by utilizing Nb_3Sn conductors.

In order to demonstrate the technological feasibility of the large magnet systems needed for commercial reactors, large superconducting tokamak test facilities (scale-down factor $\frac{1}{3}$–$\frac{1}{5}$) have been constructed and successfully operated. Examples are the Large Coil Task (LCT) at the US Oak Ridge National Laboratory, a coil arrangement of six D-shaped NbTi magnets, and TORE–SUPRA in France (Turck 1989). The same is true for a large tandem mirror machine, the Mirror Fusion Test Facility B, at the US Livermore National Laboratory (Kozman et al. 1987). At the Institute of Plasma Physics (IPP) at Garching, Germany, a superconducting stellarator (W7X) is in the planning stage. Design studies for the next generation of superconducting tokamaks such as the Next European Torus (NET) and the International Thermonuclear Experimental Reactor (ITER) are under way.

6.2 MHD Plants

MHD generators allow direct conversion of heat enthalpy to dc electrical energy. This is accomplished by passing ionized high-temperature gas at high speed between electrodes arranged along a channel and in a magnetic field perpendicular to the gas stream. MHD generators were intended to be used as a first stage in combination with conventional steam power plants, thus considerably improving the overall efficiency of the total plant. To achieve the efficiencies aimed at (say 50%) enthalpy extraction from the hot gas of at least 20% is required, which demands magnetic flux density values of 6 T or more. Because of these high flux densities and the enormous size of final commercial generators (dipole magnets 25 m long with 5 m open diameter are needed), only superconducting magnets can be used. The properties of NbTi multifilament conductors are sufficient. A series of scaled-down superconducting magnets or pilot generators has been built (Marston et al. 1981). In the meantime, programs for the construction of large base load facilities (500 MW) have been terminated worldwide for technical and economic reasons.

The direct installation costs for the magnet systems of fusion reactors or MHD generators are about 10% of the total plant costs, the portion for the refrigerators being about 1%. The amount of electrical energy consumed by the cooling plants is also in the vicinity of 1% of the total generated electrical energy. Thus no dramatic effects on installation and operation costs can be expected with regard to both areas of application by the replacement of the 4 K by the 77 K superconductors. The technical problems yet to be solved are independent of superconductivity.

7. *Rotating Machinery with Superconducting Windings*

Because of their ac loss behavior, the application of metallic multifilament superconductors was, until recently, restricted to the dc field windings of dc machines and synchronous machines. With respect to the latter, large power station synchronous turbogenerators are the most attractive candidates. Estimates have indicated that superconducting generators are economically competitive at unit ratings $\leqslant$ 800 MW (Intichar and Lambrecht 1989). Therefore large engineering projects are being pushed ahead with high priority particularly in Japan, Russia and Germany.

In principle the machine design concepts of the different projects are the same: only the rotating dc dipole field winding is built up by superconductors (preferably NbTi) which are laid in the slots of a cylindrical stainless steel coil former. The stationary three-phase armature windings consist of water-cooled stranded and transposed copper cable conductors. Because of the high magnetic flux density attainable with the superconducting field winding, no active magnetic iron is used, either in the rotor or in the armature winding (suppression of the stray field is achieved by an external laminated iron shield). Consequently, the armature ampere loading is considerably increased, which together with a higher magnetic flux density in the air gap leads to remarkable savings in weight and volume (by a factor of around two). Due to the low losses of the superconducting field winding (only thermal losses) and decreased frictional losses of the rotor (reduced length), the efficiency is considerably improved, for example from 98.8% for conventional machines to 99.5%.

Further improvements with respect to weight, volume and efficiency can be expected for the future, when the armature windings will be equipped with the new NbTi multifilament superconductors with filament diameters of around 0.1 μm.

Figure 4 shows the conceptual design of a superconducting rotor. The field winding is enclosed in a shrink fitted stainless steel torque tube and is sus-

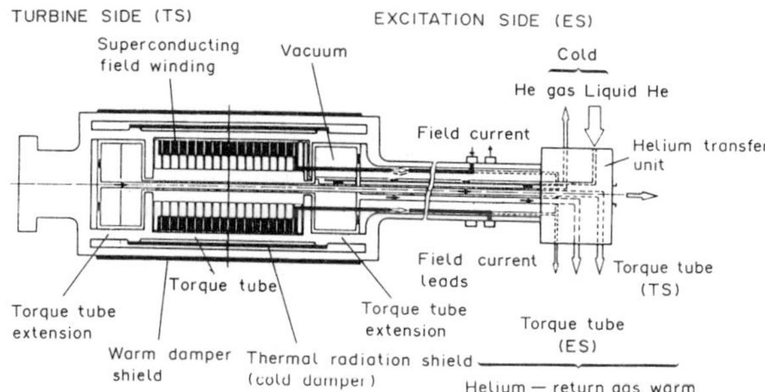

Figure 4
Schematic configuration of the superconducting rotor of a superconducting turbogenerator

pended by the helium-vapor-cooled tube extensions within the evacuated warm stainless steel rotor cylinder. Torque transmission from the shaft (300 K) to the field winding (4 K) with tolerable thermal losses is achieved by this construction. Thermal radiation losses are reduced by a cold thermal radiation shield. Warm and cold damper shields prevent undesired interaction between ac armature fields and the field winding during transient operations (e.g., short circuit). The helium transfer unit delivers liquid helium to the rotating field winding and extracts warm and cold helium gas from the rotor. The cooling system of the winding is based on forced convection under the influence of the enormous centrifugal forces.

Prototype generators with considerable power outputs are under construction. These include a 400 MVA unit at Siemens, Germany (test operation in 1995), two 70 MVA machines (ready for operation in 1994) developed by Japanese industry within the Super GM program and a 300 MW generator built by the All Union Electrotechnical Institute in Leningrad (operation in 1990). With regard to dc machines, homopolar motors and generators for ship propulsion systems have been investigated. The main problem with this type of machine is the necessity for a rotating collection of large currents, which has not yet been satisfactorily solved.

When the same technical concept (i.e., the field winding is superconducting) is assumed for a 77 K generator, capital cost savings of about 10% can be expected when compared with a 4.2 K machine. This will shift the economical break-even point to a power of the order of 400 MVA and therefore considerably increase the market for superconducting generators. In addition, a further reduction of the field winding losses (800 MVA generator: 400 kW (4.2 K) to $\sim$40 kW (77 K)) has been calculated.

8. Superconducting High-Power Cables

In the vicinity of built-up areas, transmission has to be accomplished underground by cables. For powers

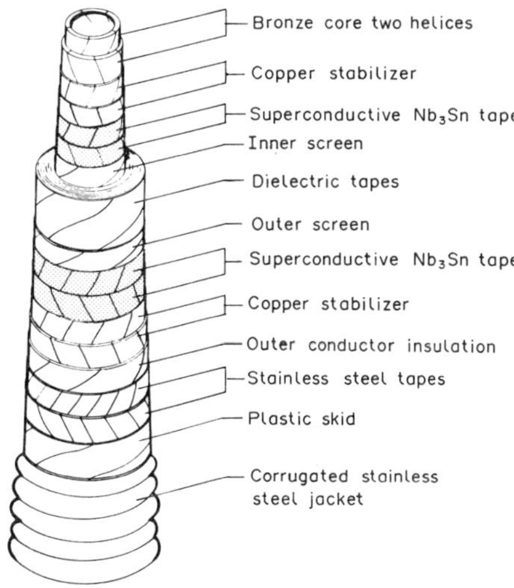

Figure 5
General assembly diagram of flexible superconducting power transmission cable core (Brookhaven design)

of $\geqslant$1000 MW, forced-cooled cables are suitable. Due to low losses and high power densities (low space requirements), superconducting cables cooled by a forced flow of helium have been considered to be the most promising. The main emphasis has been placed on the development of three-phase ac cables.

The ac cable designs proposed and realized have many common features (see Fig. 5). The superconducting cable cores are of the flexible, coaxial and hollow type and are built up with niobium or Nb_3Sn conductors. Because of low magnetic field values ($\sim$0.1 T) in the cable the ac losses are at an acceptable level. The electrical insulation consists of wound plastic tapes with low dielectric losses at low temperatures, such as polyethylene or polypropylene. The cores are contained in a flexible or rigid stainless steel tube through which helium is forced. This tube is surrounded by a flexible or rigid cryogenic insulation consisting of vacuum, radiation shields and an outer ambient temperature tube.

Because of the high current densities of superconductors, superconducting cables are designed for high currents and comparatively low voltages (420 kV maximum). The feasibility of such cables has, in principle, been proved by a series of prototypes. The most advanced project was that of the Brookhaven National Laboratory, USA (Forsyth and Thomas 1986), in the form of a 138 kV, 1000 MVA Nb_3Sn test cable of 115 m length where two flexible cable cores were contained in a rigid cryogenic envelope (concluded in 1986). Studies have indicated, however, that the economic break-even point of 4 K superconducting cables will be in the vicinity of 4000 MW per circuit, when compared with new advanced conventional systems (water-cooled or SF_6 gas-insulated systems). Such power levels will not be needed in the foreseeable future.

Changing from a 4 K to a 77 K superconductor would lead to simplification of the cooling system resulting in a reduction of investment costs of about 25%. In addition, the transmission losses are considerably decreased. When capitalized loss costs are taken into account, the economic break-even point of superconducting cables is shifted to the vicinity of 1000 MW per circuit. A few conventional cables with transmission capacities of this size are already in operation.

9. *Superconducting Transformers*

The availability of the new low-ac-loss NbTi submicrometer multifilament conductors with high-resistivity matrix has considerably increased the interest in superconducting transformers. Figure 6 shows the basic structure of a single-phase 4 K superconducting transformer. The iron core is at room temperature for more effective removal of core losses and is cooled with oil. The high- and low-voltage windings are housed in a cryovessel which partially encloses the iron core and is made of glass-fiber-reinforced plastic to avoid eddy current losses. The windings are cooled with liquid helium. Since, at least partially, the coolant has also to act as a dielectric medium, preferably subcooled liquid helium should be used in order to avoid gas bubbles and unwanted partial discharges. Further important components are the low- and high-voltage bushings since they contribute significantly to the losses of a superconducting transformer.

Superconducting transformers possess the following main advantages.

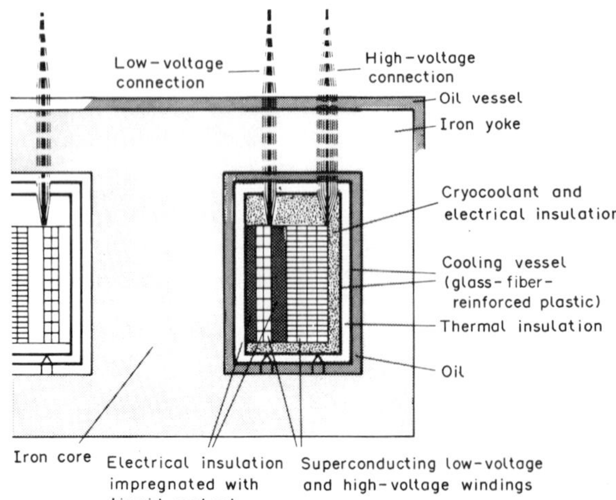

Figure 6
Schematic cross section of a superconducting single-phase transformer

(a) There are lower 50/60 Hz winding losses.

(b) Low winding losses permit a larger number of ampere turns, which enables a reduction of the magnetic flux. This leads to a considerable reduction in iron core losses and weight.

(c) Fault current limiting behavior occurs by transition of the superconducting windings into the resistive normal conducting state.

(d) When fault current limitation occurs, simplification of the mechanical winding design and higher overall current density are the consequences, which result in a drastic reduction of the coil weights.

In recent years a number of relatively small single-phase model transformers with power of the order of 100 kVA have been constructed with the aid of submicrometer NbTi filament wires. These successfully demonstrated the advantages of superconducting technology, particularly the current limiting effect when a short circuit occurs. Interesting examples in this context are a single-phase, 60 Hz/72 kVA transformer built by the Kyushu University in Japan and a 50 Hz/220 kVA transformer built by the Alsthom Company and the Laboratories de Marcoussis in France (see *AC Applications of Superconducting Materials*).

The effects of a 77 K superconductor were studied on the basis of a single-phase 850 MVA generator transformer. When the same loss mechanism as for metallic superconductors is assumed and the higher efficiency of liquid nitrogen cooling is considered, conductors with superconducting filaments or layers with a thickness of about 10 μm can be used. The total mass (without cooling oil) was estimated to be about 100 t compared with the 520 t of a conventional machine; total losses of 240 kW were calculated in comparison with the 1820 kW of a conventional transformer (Bogner 1989). If it is assumed that the material savings in the superconducting version are balanced by a more complicated technology, then the advantage of the 77 K superconducting transformer results solely from the capitalization of the far lower losses.

10. *Fault Current Limiters*

Increases in the installed electrical power and higher power densities in the networks result in greater values of fault currents in the event of short circuits. At higher voltage levels (245 kV, 420 kV), the conventional measures available for limiting the fault current are restricted. A superconducting current limiter which utilizes the controlled rapid change from the superconducting state to the normal conducting high-resistive condition to achieve effective current limitation in the event of a short circuit would therefore be particularly attractive for relatively high voltage

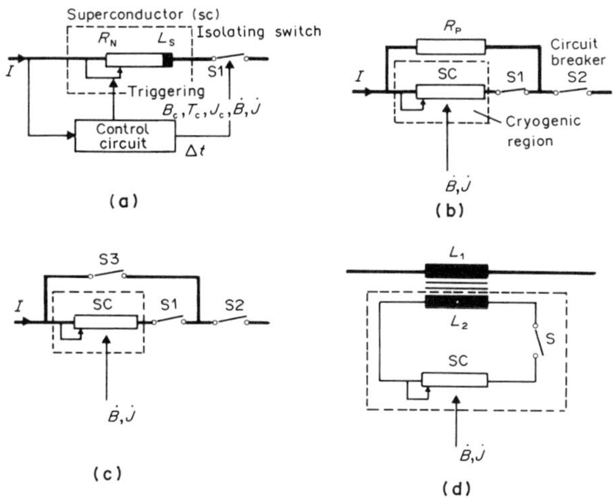

Figure 7
Various switching schemes of superconducting fault current limiters

levels. By using such current limiters, most of the negative effects (e.g., high dynamic loads) which fault currents exert on the network and its components could be eliminated.

Figure 7 shows four different current limiter circuits based on the rapid ($\leqslant$ 1 ms) transition of a superconducting wire or tape of definite length from the superconducting to the normal conducting state. In order to avoid excessively high overvoltage during rapid switching and also to keep the magnetic field at the superconductor small (high critical current density, small ac losses), the superconductors are arranged in low-inductance bifilar configurations.

Although the arrangement in Fig. 7a is the simplest one, the arrangement in Fig. 7b has been suggested most frequently. The advantage of the inductive solution in Fig. 7d is that it operates without lossy current leads. The arrangement in Fig. 7c appears to be the most complicated. In the circuit version (Fig. 7b) the resistance of the superconductor in the normal state is much greater than R_p. Therefore, after the transition of the superconductor into the normal state the current is commutated into R_p, which limits the current. The superconductor is released from the residual current by opening the isolating switch S_1 and the limited current is interrupted by the circuit breaker S_2 (50 ms after occurrence of short circuit).

Superconducting current limiters have particularly been dealt with in "feasibility studies" and their behavior simulated with computers. An example is a study commissioned by the Electrical Power Research Institute (EPRI), USA, in 1976, where the task was to limit the fault current to 20 kA in a 145 kV network with a nominal current I_N of 2 kA. Thin layers of NbN deposited on plastic tapes were used as conductors in this study. More recently, a single-phase 25 kV, 5000 kVA current limiter was realized by the Laboratories de Marcoussis on the basis of the new submicrometer NbTi multifilament conductors (see *AC Applications of Superconducting Materials*). Current limitation occurred at a peak value after short circuit of $1.4I_N$. The behavior of the device was satisfying with the exception of the reclosing time (time between occurrence of normal state and full recovery of the superconductivity) which was too long with respect to network requirements. Detailed investigations of this problem will have to be carried out in the future.

To study the consequences of a change from 4 K to 77 K superconductors, two single-phase limiters (500 MVA, 110 kV, 2.5 kA) were considered, one consisting of NbN and one of the new ceramic superconductors. The obvious advantage of the 77 K version lies in its lower operating losses which amount to only about 30% of those of the 4 K model. In both cases over 90% of the losses are due to the current leads. The associated reduction of refrigeration system costs amounts to around 60%.

Bibliography

Bogner G 1989 Kabel, Schalter und Transformatoren. *VDI Ber. (Ver. Dtsch. Ing.)* 733: 153–69

Cave J R, Fevrier A, Verhage T 1989 Reduction of ac losses in ultra-fine multifilamentary NbTi wires. *IEEE Trans. Magn.* 25: 1945–8

Forsyth E B, Thomas R A 1986 Performance summary of the Brookhaven superconducting power transmission system. *Cryogenics* 26: 599–614

International Superconductivity Technology Center (ISTEC) Japan 1990 Feasibility study in application fields on superconductivity. *Executive Summary Report* March: 4–5

Intichar L, Lambrecht D 1989 Entwicklung supraleitender Kraftwerksgeneratoren. *VDI Ber. (Ver. Dtsch. Ing.)* 733: 121–35

Kozman T, Shimer D, Van Scant J, Zbasnik J 1987 Construction and testing of the Mirror Fusion Test Facility magnets. *IEEE Trans. Magn.* 23: 1448–63

Loyd R J, Schoenung S M, Nakamura T, Hassenzahl W V, Rogers J D, Purcell J R, Lienrance D W, Hilal M A 1987 Design advances in superconducting magnetic energy storage for electric utility load leveling. *IEEE Trans. Magn.* 23: 1323–30

Marston P G, Thome R J, Dawson AM, Bobrov E S, Hatch A M 1981 ETF magnet design alternatives for the national MHD-program. *IEEE Trans. Magn.* 17: 352–5

Peoples J 1989 Status of the SSC superconducting magnet program. *IEEE Trans. Magn.* 25: 1444–54

Reichert T, Bittner G, Heinzelmann K G, Söldner L, Vetter J 1988 Siemens Helicon: a novel 2-tesla superconducting whole-body MR magnet. *IEEE Trans. Magn.* 24: 1272–5

Turck B 1989 TORE SUPRA: a tokamak with superconducting toroidal coils status report after the first plasmas. *IEEE Trans. Magn.* 25: 1473–80

Vetter J, Ries G, Reichert T 1988 A 4-tesla superconducting whole-body magnet fpr MR imaging and spectroscopy. *IEEE Trans. Magn.* 24: 1285–7

Watson J H P 1988 Status of superconducting magnetic separation. *Proc. 12th Int. Cryogenic Engineering Conf.* Butterworth, Guildford, UK, pp. 339–44

Wolff S 1992 Review of accelerator magnet design in the world. *IEEE Trans. Magn.* 28: 96–103

G. Bogner
[Siemens, Erlangen, Germany]

Superconducting Materials: Physical Properties

To what extent are the high-temperature superconductors similar to the classic superconductors that were explored from 1911 to 1986? This is currently a controversial issue of critical importance for the development of a sound theoretical understanding of superconducting phenomena (see *Superconducting Materials: Basic Theoretical Concepts*; *Superconducting Materials: BCS and Phenomenological Theories*). It is clear that the new oxide superconductors have unprecedented high values of the superconducting transition temperature T_c and that they have very small values of the superconducting coherence length ξ. In addition, they are much more anisotropic than any of the classic superconductors, except for $NbSe_2$ and related materials. On the other hand, many of the classic properties, such as flux quantization and electron pairing, have also been observed in the new materials, and many other properties, such as electron tunnelling, the electromagnetic penetration depth and thermal conductivity, are at least qualitatively similar to the corresponding properties of the classic superconductors. The binding energy Δ per electron that gives the Cooper pair stability must, however, be enhanced by an electron–electron interaction, since the electron–phonon interaction appears, by itself, to be too weak. With this increased interaction, it may reasonably be hoped that an extended BCS theory will apply to the high-T_c superconductors, although many quantitative differences have been observed and there are indications of low-lying energy states within the superconducting gap. Hence, it is appropriate to review, in this article, the properties of the classic, BCS superconductors. A survey of the particular characteristics of high-T_c superconductors is given in the article *Oxide Superconductors: Physical Properties*.

The thermodynamic properties of superconductors are dominated by the temperature dependence of the energy gap parameter Δ because the thermal excitation of quasiparticles requires a minimum energy Δ per quasiparticle. Since Δ is almost independent of the temperature T up to about half the transition temperature T_c, the number of quasiparticles (unpaired electrons) in the sample is approximately proportional to $\exp(-\Delta/k_BT)$ in that temperature range, and so is the superconducting electronic specific heat C_{es} of the quasiparticles. As T approaches T_c, Δ begins to decrease and then drops rapidly to zero. This rapid drop causes the number of quasiparticles to increase so quickly that, even though the minimum energy of each quasiparticle is decreasing, the total energy of all the quasiparticles rises just below T_c. This rise in energy is associated with a peak in C_{es} at T_c. As the temperature rises above T_c, the specific heat falls to the normal state value C_{en}, which is proportional to T for further increases in the temperature. These observations are summarized in Fig. 1. The fractional drop in the specific heat at T_c is predicted by the Bardeen–Cooper–Schrieffer (BCS) theory of superconductivity to obey the relation

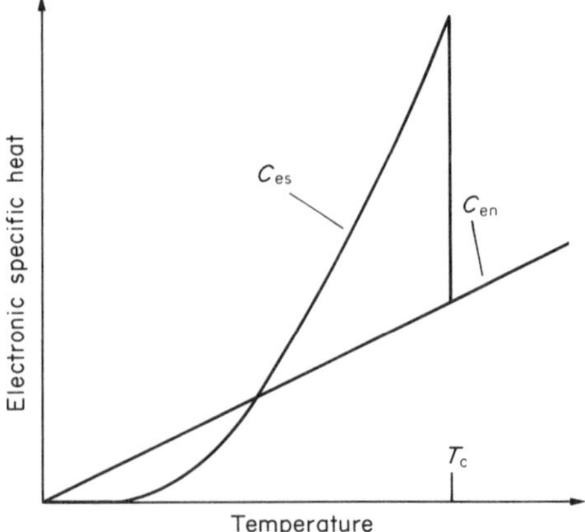

Figure 1
Dependence of the electronic specific heat C_e on temperature for the superconducting state (C_{es}) and for the normal state (C_{en})

$$[(C_{es}-C_{en})/C_{en}]_{T_c}=1.43 \tag{1}$$

Observed values of this fractional drop range from about 1.3 to 2.7. The largest values are expected and observed for bulk superconductors such as mercury and lead, in which Δ at absolute zero is larger than the BCS-predicted value of $1.76\,kT_c$ because of strong electron–phonon coupling (see *Superconducting Materials: BCS and Phenomenological Theories*). Values slightly less than 1.43 may be expected to result from anisotropy of Δ; that is, a dependence of Δ on the direction of motion of a quasiparticle in the crystal lattice. (This anisotropy is usually no more than a few percent (see *Anisotropy in Superconducting Materials*).)

In 1933, Meissner and Ochsenfeld discovered that a superconductor expels magnetic flux as its temperature is lowered. Actually, it is now known that the flux expulsion is complete only in some superconductors, called type I, and that it can be inhibited by voids or irregularities in the superconductor. However, the discovery of the flux expulsion, which has come to be called the Meissner effect, opened the way for a thermodynamic analysis of type-I superconductors, which include all of the superconducting elements in unstrained, bulk form except niobium and possibly vanadium.

The magnetic flux is completely expelled from a type-I superconductor when the temperature is lowered enough so that the expulsion lowers the Gibbs free energy G. (All thermodynamic quantities here will be expressed per volume. The subscripts n and s will refer to the normal and superconducting states, respectively.) As the temperature is lowered, the difference $(G_n - G_s)$ increases (Reif 1965). Correspondingly, there is an increase in the critical magnetic field H_c which is required to drive the material into the normal state (see *Magnetic Properties of Superconductors*). G_n is independent of the applied field H for practical purposes, but G_s rises with increasing H because the expulsion of flux by the superconductor crowds the magnetic field lines together outside of the sample, storing increased energy. These quantities are related as follows by the laws of thermodynamics:

$$G_n - G_s = \mu_0(H_c^2 - H^2)/2 \tag{2}$$

where μ_0 is the permeability of free space. The entropy $S = -(\partial G/\partial T)_H$, so the entropy difference between the two states is related to H_c:

$$S_n - S_s = -\mu_0 H_c(dH_c/dT) \tag{3}$$

Since $C = T(dS/dT)$,

$$C_n - C_s = -T\mu_0[H_c(d^2H_c/dT^2) + (dH_c/dT)^2] \tag{4}$$

If there is no applied field, the transition takes place at T_c, where $H_c = 0$. Therefore,

$$(C_n - C_s)_{T_c} = -T\mu_0(dH_c/dT)^2_{T_c} \tag{5}$$

Equation (3) shows that S is continuous at the superconducting transition if $H = 0$, but is discontinuous if $H > 0$. Together with the discontinuity in C (for $T > 0$), indicated in Eqn. (4), this observation means that the superconducting transition is second order in the absence of an applied field and first order in the presence of an applied field. As $T \to 0$, Eqn. (4) shows that C_s must approach C_n, which vanishes in that limit. The BCS theory predicts that over a wide range of temperatures, not too near T_c or to zero,

$$C_s \cong 9.17\, \gamma T_c \exp(-1.5\, T_c/T) \tag{6}$$

where γ is the normal-state specific-heat coefficient ($C_n = \gamma T$).

Once the theory is used to calculate G_n and G_s for $H = 0$, the temperature dependence of H_c can be calculated from Eqn. (2). The result indicates that H_c should have a nearly quadratic dependence on T. It is possible to define a function D of the reduced temperature T/T_c to be the deviation of $H_c(T)/H_c(0)$ from this quadratic dependence:

$$D(T/T_c) = [H_c(T)/H_c(0)] - [1 - (T/T_c)^2] \tag{7}$$

Figure 2 shows the BCS-predicted and the observed dependences of D on T/T_c. As for specific heat data, the superconductors in which electron–phonon coupling is strong differ from the theory. These differences have been accounted for by theories which build on the BCS theory, but which include the effects of strong coupling (see *Superconducting Materials: BCS and Phenomenological Theories*).

The electromagnetic properties of superconductors provide the basis for most applications of these materials. The properties depend in characteristic ways on the angular frequency ω. It is convenient to consider separately the low-frequency properties and the high-frequency properties. Low frequencies here mean those for which the photon energy $\hbar\omega$ is much less than 2Δ, the minimum energy required to break a Cooper pair and create two quasiparticles. At dc or a low frequency, the Meissner effect provides the starting point for understanding superconductivity in a type-I superconductor. Deep in the interior of a superconducting sample, where the magnetic field $\boldsymbol{H} = 0$, the current density $\boldsymbol{J}$ must vanish, since one of Maxwell's equations states that if $\varepsilon\partial\boldsymbol{E}/\partial t \ll \boldsymbol{J}$ (where ε is the permittivity and $\boldsymbol{E}$ is the electric field) then

$$\nabla \times \boldsymbol{H} = \boldsymbol{J} \tag{8}$$

However, there must be shielding currents flowing near the surface of the sample, generating a magnetic field which cancels the applied field within the sample. The shielding-current density must be finite, so it must be flowing in a region of space with a nonzero depth,

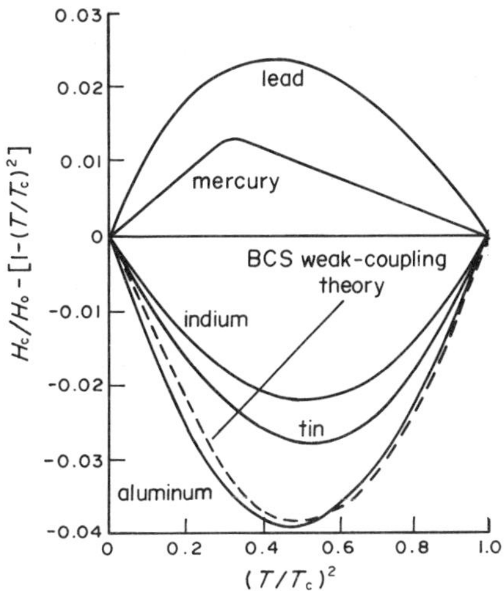

Figure 2
Dependence of the deviation function D on the reduced temperature T/T_c for various superconductors: theoretical values are also shown (after Scalapino 1969. © Marcel Dekker, New York. Reproduced with permission)

and the applied field must penetrate into this depth. London and London (1935a,b) suggested that the Meissner effect can be understood if, in addition to Maxwell's equations, the superconductor obeys the equation

$$\nabla \times \boldsymbol{J} = -\boldsymbol{H}/\Lambda \quad (9)$$

where Λ is related to the electron mass m and charge e and the number n_s of superconducting electrons per volume:

$$\Lambda = m/n_s e^2 \mu_0 \quad (10)$$

The relation of Eqn. (9) to field penetration into the surface of the sample is easy to see. Taking the curl of Eqn. (8) and using Eqn. (9), then

$$\nabla \times (\nabla \times \boldsymbol{H}) = \nabla \times \boldsymbol{J} = -\boldsymbol{H}/\Lambda \quad (11)$$

But

$$\nabla \times (\nabla \times \boldsymbol{H}) = \nabla(\nabla.\boldsymbol{H}) - \nabla^2 \boldsymbol{H} \quad (12)$$

and $\nabla.\boldsymbol{H} = 0$. Therefore,

$$\nabla^2 \boldsymbol{H} = \boldsymbol{H}/\lambda_L^2 \quad (13)$$

where the constant

$$\lambda_L = \Lambda^{1/2} \quad (14)$$

is called the London penetration depth. In the case of a field $\boldsymbol{H}_0$ applied parallel to the surface of the sample, the solution of Eqn. (13) is

$$\boldsymbol{H} = \boldsymbol{H}_0 \exp(-x/\lambda_L) \quad (15)$$

where x is the distance into the sample.

The actual penetration depth λ can be measured by a variety of means, such as determinations of the magnetic susceptibility of small samples (see *Measurements in Superconducting Materials*). It is found that λ tends to infinity as T approaches T_c. This is easy to understand in terms of Eqn. (10). As T_c is approached, more electrons are excited, becoming quasiparticles, which act as normal (nonsuperconducting) current carriers. Consequently, n_s goes towards zero, and Λ increases without limit. However, two experimental results indicate that this picture is not complete, or needs to be modified. In the first place, the measured values (Tinkham 1975) of λ are considerably larger than λ_L. Second, it is observed that λ increases if the electron mean free path is decreased by impurities or strains, and this contradicts Eqn. (10). Furthermore, in a region of the sample in which H is varying, it is strange that Eqn. (9) indicates that the behavior of $\boldsymbol{J}$ at a given location should be determined by $\boldsymbol{H}$ only at that location, rather than in a region with size roughly equal to the electron mean free path l. For these reasons and others, Pippard (1953) proposed a nonlocal relation between $\boldsymbol{J}$ and the vector potential $\boldsymbol{A}$, analogous to a similar expression which had been introduced for normal metals with long electron mean free paths. Pippard's relation is

$$\boldsymbol{J}(\boldsymbol{r}) = (-3/4\pi\mu_0\xi_0\Lambda)\int\{\boldsymbol{R}[\boldsymbol{R}.\boldsymbol{A}(\boldsymbol{r}')]/R^4\} \times \exp(-R/\xi_0)\exp(-R/\alpha l)d^3\boldsymbol{r}' \quad (16)$$

where ξ_0 is the ideal Pippard coherence length, $\boldsymbol{R} = \boldsymbol{r} - \boldsymbol{r}'$ and α is a number of order unity.

The BCS theory yields a relation for $\boldsymbol{J}$ which is very close to that given by Eqn. (16), so this equation is commonly used in treating superconducting currents. It correctly predicts the dependence of the penetration depth on temperature and on electron mean free path. The BCS theory indicates that the constants ξ_0 and Λ in Eqn. (16) depend on temperature; ξ_0 decreases by 25% and Λ increases without limit as T approaches T_c. At $T = 0$,

$$\xi_0 = \hbar v_F/\pi\Delta \quad (17)$$

where v_F is the Fermi velocity. Equation (16) indicates that an electromagnetic field at a point in a superconducting sample disturbs the current density in a region with size of the order of

$$\xi = [\xi_0^{-1} + (\alpha l)^{-1}]^{-1} \quad (18)$$

This quantity ξ is therefore called the electromagnetic coherence length. As l is limited by impurities or strains, ξ decreases and so does $\boldsymbol{J}$ as given by Eqn. (16). The shielding of an applied magnetic field is correspondingly decreased and therefore λ increases.

Electromagnetic energy can be absorbed even at photon energies $\hbar\omega$ which are less than 2Δ if $T > 0$, since the thermally excited quasiparticles are present. This energy absorption is small for such low frequencies if $T < T_c/2$. As T_c is approached, the number of quasiparticles increases. Furthermore, 2Δ decreases, eventually becoming less than $\hbar\omega$, so the breaking of Cooper pairs begins to occur, absorbing more electromagnetic energy. These effects can be calculated according to the BCS theory (see *High-Frequency Impedance of Superconducting Materials*; Mattis and Bardeen 1958, Abrikosov et al. 1959). An equation similar to Eqn. (16) gives the current density, but with the factor $\exp(-R/\xi_0)$ replaced by a function of ω as well as R and T. This function is a complicated one, and use of the theory is made difficult by the necessity of obtaining a self-consistent solution to Maxwell's equations, taking into account the fields generated by the induced currents and the impedance mismatch at the surface of the sample (Ginsberg and Hebel 1969). The theory is in good agreement with experiment. The electromagnetic absorption decreases as T is lowered, as shown in Fig. 3. There are two simple reasons for this. One is the decreased number of thermally excited quasiparticles. The other is the decreased size of λ, lowering the size of the energy-absorbing part of the sample. As shown in Fig. 3, the energy absorption increases sharply as the photon energy $\hbar\omega = h\nu$ is increased above 2Δ, which is approximately $3.5\,k_B T_c$ for $T < T_c/2$. The corresponding frequencies are very large; for example, for tin, $T_c = 3.72$ K and $3.5\,k_B T_c/h = 2.7 \times 10^{11}$ Hz.

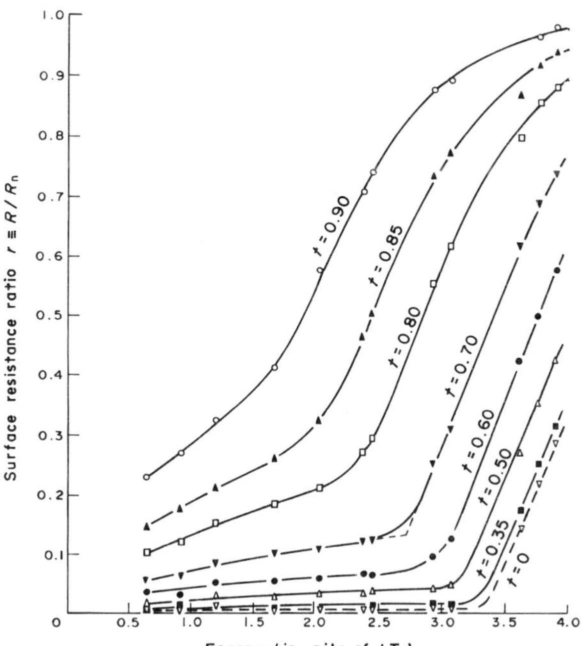

Figure 3
Experimental curves of the surface resistance ratio r of superconducting aluminum as a function of reduced frequency for various values of reduced temperature t (after Biondi and Garfunkel 1959)

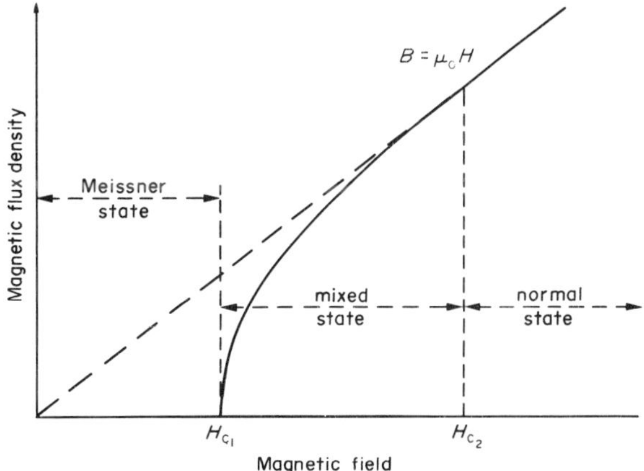

Figure 4
Dependence of magnetic flux density B on the magnetic field H for a type-II superconductor

The transformation of a type-I superconducting sample into the normal state occurs at a single value of a uniform applied field H if the sample is a sheet or cylinder which is very long in the direction of the field compared with its other dimensions. Otherwise, the transformation occurs in various parts of the sample at different values of H because the magnetic field generated by the superconducting shielding currents is nonuniform. This inhomogeneous state is called the intermediate state. The normal and superconducting regions are of macroscopic size (see *Magnetic Properties of Superconductors*).

For a type-II superconductor, the sample becomes normal in a very different way as H is increased. For small values of H, the Meissner effect is observed, and the magnetic flux is excluded from the sample by the shielding currents. However, as H is increased past a value called the lower critical field H_{c1}, the magnetic flux begins to enter the sample gradually in the form of isolated, microscopic, threadlike regions, called fluxoids, flux vortices or flux lines, and the so-called mixed state is formed. Each fluxoid contains one quantum of flux Φ_0, given by

$$\Phi_0 = h/2e = 2.07 \times 10^{-15}\ \mathrm{T\ m^2} \tag{19}$$

The fluxoid is surrounded by circulating currents which shield the rest of the material from the flux. With increasing applied field, more and more of these fluxoids are formed, so they crowd closer together. As they begin to overlap appreciably, a field called the upper critical field H_{c2} is reached, and the sample makes a second-order transition into the normal state, except for a thin region near the surface where it is parallel to $\boldsymbol{H}$. The so-called surface sheath goes normal at a still higher field, H_{c3}. The dependence of the magnetic flux density B on magnetic field H for a type-II superconductor is shown in Fig. 4: $\boldsymbol{B} = \mu_0(\boldsymbol{H} + \boldsymbol{M})$, where flux $\boldsymbol{M}$ is the magnetization. In the Meissner state $B = 0$, so $\boldsymbol{M} = -\boldsymbol{H}$. The dependence of M on H was predicted by Abrikosov (1957), whose theoretical work was founded on a phenomenological theory of Ginzburg and Landau (1950), formulated before the BCS theory, and shown later by Gor'kov (1959) to be derivable from it (see *Flux Line Lattice*; *Superconducting Materials: BCS and Phenomenological Theories*). Experiments have verified the type-II dependence of M on H. The surface sheath was predicted by Saint-James and de Gennes (1964), also on the basis of the Ginzburg–Landau (GL) theory.

The GL theory introduces a superconducting order parameter $\Psi = |\Psi| \exp(\mathrm{i}\varphi)$ which can depend on position $\boldsymbol{r}$ as well as time. This function is a complex effective superfluid wavefunction whose amplitude goes to zero when the material becomes normal. The theory assumes that if Ψ is small and varies slowly in space the free energy density of the material can be expanded in terms of $|\Psi|^2$ and $|\nabla\Psi|^2$. The latter term is introduced to take account of the observed property that a material containing a superconducting and a normal region has a wall of finite width between them, in which the transformation between the two states occurs gradually. This term is sometimes referred to as the "strain" term since it takes account of the rigidity or stiffness of the superfluid wavefunction.

The GL expression for the local Helmholtz free energy density referred to its normal conducting state in zero field is

$$f_s - f_n = \alpha|\Psi|^2 + \frac{\beta}{2}|\Psi|^4 + \frac{1}{2m^*}\left|\left(\frac{\hbar\nabla}{i} - e^*\boldsymbol{A}\right)\Psi\right|^2 + \tfrac{1}{2}\mu_0 h^2 \quad (20)$$

where h is the local magnetic field and $e^* = 2e$ is the charge of a pair of electrons. The form of Eqn. (20) leaves the interpretation of $|\Psi|^2$ undetermined since $|\Psi|^2$ may be scaled if α, β and m^* are simultaneously scaled. It is usual to set $m^* = 2m$, the mass of two free electrons. With this convention $|\Psi|^2 = n_s^*$ can, in the free electron approximation, be interpreted as the number density of Cooper pairs in the superconducting condensate. In this limit $n_s^* e^{*2}/m^* = n_s e^2/m$ so the London penetration depth λ_L is unchanged by pairing. In real materials the situation is more complicated and $|\Psi|^2 = n_s^* = \frac{1}{2}n_s$ does not correspond to any simple electron density for the material (Tinkham 1975).

If the free energy density is integrated over all space and then minimized with respect to local changes in $\boldsymbol{A}$ and Ψ, two coupled differential equations are obtained. These govern the equilibrium variation of $\boldsymbol{A}$ and Ψ with position, given α and β and particular boundary conditions, and are known, respectively, as the first and second Ginzburg–Landau equations

$$\alpha\Psi + \beta|\Psi|^2\Psi + \frac{1}{4m}\left(\frac{\hbar\nabla}{i} - 2e\boldsymbol{A}\right)^2\Psi = 0 \quad (21)$$

$$\boldsymbol{J} = \nabla \times \boldsymbol{H} = \frac{e\hbar}{2mi}(\Psi^*\nabla\Psi - \Psi\nabla\Psi^*) - \frac{2e^2|\Psi|^2\boldsymbol{A}}{m} \quad (22)$$

The boundary condition at an insulating surface is that the value of $(i\hbar\nabla + 2e\boldsymbol{A})\Psi$ in the direction normal to the surface is zero. At a superconductor–normal metal interface with no current across the interface its value in the same direction is no longer zero and can be expressed as $(i/b)\Psi$ where b is a real constant (see also *Coherence Length, Proximity Effect and Fluctuations*).

The GL theory indicates that a superconductor is type II if the ratio of the penetration depth λ to a coherence length ξ_{GL} is large enough. The Ginzburg–Landau coherence length ξ_{GL} is the distance to which a disturbance in $|\Psi|^2$ extends in the material (see *Coherence Length, Proximity Effect and Fluctuations*). It should not be confused with the electromagnetic coherence length ξ which has already been discussed although it is also equal to ξ_0 given by Eqn. (17), if $l \gg \xi_0$ and $T = 0$. Unlike ξ, ξ_{GL} becomes very large as T approaches T_c:

$$\xi_{GL}(T) \cong \xi_{GL}(0)[1 - (T/T_c)^4]^{-1/2} \quad (23)$$

The dependence of ξ_{GL} on l is different from that of ξ:

$$\xi_{GL}^2 = 0.55\, \xi_0^2\, \chi(\alpha)(1 - T/T_c)^{-1} \quad (24)$$

where $\alpha = 0.882\, \xi_0/l$ is the impurity parameter and $\chi(\alpha)$ is the Gor'kov impurity function (Parks 1969). For values of $l \ll \xi_0$,

$$\xi_{GL}(0) \cong (\xi_0 l)^{1/2} \quad (25)$$

The theory introduces a parameter κ, known as the Ginzburg–Landau parameter, defined by

$$\kappa = \lambda/\xi_{GL} \quad (26)$$

If $\kappa > 1/\sqrt{2}$, the material is type II. The reason why large values of κ lead to type-II behavior is as follows. The material in each fluxoid is in or near to the normal state, and the fluxoid must have a thickness of about ξ_{GL}. This costs an amount of free energy which is proportional to ξ_{GL} since $G_n > G_s$. However, some free energy is returned because the magnetic flux in the fluxoid can spread out by penetrating into the surrounding superconducting material, and this gives back an amount of free energy which is proportional to λ. If λ/ξ_{GL} is large enough, the total effect is favourable and fluxoids are formed, so the material is a type-II superconductor. H_{c2} can be simply expressed in terms of κ:

$$H_{c2} = \kappa H_c \sqrt{2} \quad (27)$$

where H_c is defined now in terms of G_n and G_s by Eqn. (2) with H set equal to 0. H_{c3} is given by

$$H_{c3} = 1.695\, H_{c2} \quad (28)$$

near T_c according to theory and in accordance with experimental observations.

For calculating the detailed spatial variation of the order parameter and magnetic field the GL theory is commonly expressed in reduced units and in terms of a reduced-order parameter $\psi(\boldsymbol{r}) = \psi_0 \exp(i\varphi)$, where $|\psi|^2 = |\Psi|^2/|\Psi_\infty|^2$ and $|\Psi_\infty|^2 = -\alpha/\beta$ corresponds to the equilibrium value of Ψ deep in the interior of the superconductor. The reduced-order parameter may also be expressed as $\psi(\boldsymbol{r}) = \Delta(\boldsymbol{r})/\Delta_{BCS}$ where $\Delta(\boldsymbol{r})$ is the gap parameter or pair potential of the BCS theory and Δ_{BCS} is the maximum value $|\Delta|$ can attain in equilibrium at that temperature. In reduced units, $\psi \lesssim 1$ and the GL equations can be expressed very simply in terms of the single parameter κ (see *Flux Line Lattice*). However, for a quantitative treatment of fluxoid pinning in inhomogeneous materials that correctly links pinning forces to the variation of particular material parameters, the full two-parameter expression for the free energy given in Eqn. (20) is required (see *Flux Pinning and the Summation of Pinning Forces*). The two GL parameters α and β can be expressed directly in terms of H_{c2} and κ (when $m^* = 2m$),

$$\alpha = -\left(\frac{e\hbar}{2m}\right)\mu_0 H_{c2} \tag{29}$$

$$\beta = 2\left(\frac{e\hbar}{2m}\right)^2 \mu_0 \kappa^2 \tag{30}$$

The phenomenological order parameter of the GL theory is linked to the microscopic theory for $T \simeq T_c$ by the expressions

$$C|\Delta|^2 = (\hbar^2/4m)|\Psi|^2 = \mu_0 H_c^2 \xi_{GL}^2 |\psi|^2 \tag{31}$$

where $C = \xi_{GL}{}^2 N(0)(1 - T/T_c) = 0.55\,\xi_0{}^2 N(0)\chi(\alpha)$ and $N(0)$ is the density of states at the Fermi level in the normal state (de Gennes 1966).

When impurities are introduced into a superconducting material, the electron mean free path is decreased. This decreases ξ_{GL}, according to Eqn. (24). It also increases λ, as already discussed. Therefore, κ is increased. In this way, any type-I superconductor can be made type II by a sufficient amount of impurities. The values of $\mu_0 H_c$ in type-I superconductors are all less than 0.1 T. For type-II superconductors, values of $\mu_0 H_{c2}$ range up to about 100 T. All of the materials from which superconducting magnets are commonly made are therefore type II.

The generation of electrical currents has already been discussed. Now two other transport processes are briefly described—ultrasonic attenuation and thermal conductivity, both in the absence of an applied field. The following remarks apply to both type-I and type-II superconductors.

The attenuation of transverse ultrasonic waves is somewhat complicated, because transverse electric currents are generated and they are screened. Therefore, this discussion is restricted to longitudinal waves. For typical frequencies v, which are less than 10^9 Hz, $hv = \hbar\omega$ is much less than Δ and $k_B T$. For these frequencies, the ratio of the ultrasonic attenuation in the superconducting state to that in the normal state is given by the theory as

$$\alpha_s/\alpha_n = 2/[1 + \exp(\Delta/k_B T)] \tag{32}$$

This ratio falls steeply as T is lowered from T_c, and approaches an exponential dependence on T^{-1} for $T \ll T_c$. These features are easy to understand when it is remembered that the ultrasonic attenuation is a manifestation of energy absorption by thermally excited quasiparticles.

In actual materials, the ultrasonic attenuation has additional terms. An example is a term arising from the damping of the ultrasonically induced motion of dislocations. These other sources of attenuation are sometimes nonlinear in the amplitude of the waves and depend on the temperature (Ginsberg and Hebel 1969, Tinkham 1975).

The thermal conductivity K_T of superconductors is a sum of two terms:

$$K_T = K_e + K_p \tag{33}$$

where the term K_e is the part of K_T contributed by the electrons (actually, the quasiparticles) and the term K_p is contributed by heat carried by the phonons (quantized lattice vibrations). Each of these types of heat carriers scatters from the other and from lattice imperfections, such as dislocations, grain boundaries and the sample boundaries. As T is lowered, there are fewer thermally excited quasiparticles to carry heat, and K_e must be proportional to $\exp(-\Delta/k_B T)$ if $T \ll T_c$, in agreement with Eqn. (32). When T is near T_c, the term K_e in Eqn. (33) falls gradually with decreasing temperature if the quasiparticles are scattered mainly by static lattice imperfections. If they are scattered mainly by the phonons, then K_e falls sharply. (This occurs for metals which have low Debye temperatures; these are the same superconductors in which the electron–phonon coupling is strong.) As the number of quasiparticles decreases with decreasing temperature, the phonons travel further between collisions with quasiparticles, and K_p can rise for a while as T falls, even though there are fewer phonons around to carry heat. At very low temperature, K_T always decreases with decreasing T; but near T_c, it can have a variety of temperature dependences. For the pure superconducting elements, the K_e term in Eqn. (33) is dominant, and the total thermal conductivity K_T falls as T is lowered near T_c because of the decreasing magnitude of K_e. However, for alloys, in which the electron mean free path is severely limited, the K_p term in Eqn. (33) may be dominant and K_T can therefore increase with decreasing T near T_c.

See also: Amorphous Superconductors; Oxide Superconductors: Physical Properties; Superconducting Materials: An Overview; Superconducting Materials: Basic Theoretical Concepts; Superconducting Materials: BCS and Phenomenological Theories

Bibliography

Abrikosov A A 1957 Magnetic properties of superconductors of the second group. *Sov. Phys. JETP (Engl. Transl.)* 5: 1174–82

Abrikosov A A, Gor'kov L P, Khalatnikov I M 1959 A superconductor in a high frequency field. *Sov. Phys. JETP (Engl. Transl.)* 8: 182–9

Biondi M A, Garfunkel M P 1959 Millimeter wave absorption in superconducting aluminum, I. Temperature dependence of the energy gap. *Phys. Rev.* 116: 853–67

de Gennes P G 1966 *Superconductivity in Metals and Alloys.* Benjamin, New York

Ginsberg D M, Hebel L C 1969 Nonequilibrium properties: comparison of experimental results with predictions of the BCS theory. In: Parks 1969, Vol. 1, Chap. 4

Ginzburg V L, Landau L D 1950 K teorii cverkhrovo dimosti. *Zh. Eksp. Teor. Fiz.* 20: 1064–82

Gor'kov L P 1959 Microscopic derivation of the Ginzburg–Landau equations in the theory of superconductivity. *Sov. Phys. JETP (Engl. Transl.)* 9: 1364–6

Gor'kov L P 1960 Theory of superconducting alloys in a strong magnetic field near the critical temperature. *Sov. Phys. JETP (Engl. Transl.)* 10: 998–1004

London H, London F 1935a Supraleitung und Diamagnetismus. *Physica* 2: 341–54

London H, London F 1935b The electromagnetic equations of the supraconductor. *Proc. R. Soc. London, Ser. A* 149: 71–88

Mattis D C, Bardeen J 1958 Theory of anomalous skin effect in normal and superconducting metals. *Phys. Rev.* 111: 412–17

Meissner W, Ochsenfeld R 1933 Ein neuer Effekt bei Eintritt der Supraleitfähigkeit. *Naturwissenschaften* 21: 787–8

Parks R D (ed.) 1969 *Superconductivity*, Vols. 1, 2. Dekker, New York

Pippard A B 1953 An experimental and theoretical study of the relation between magnetic field and current in a superconductor. *Proc. R. Soc. London, Ser. A* 216: 547–68

Reif F 1965 *Fundamentals of Statistical and Thermal Physics*. McGraw-Hill, New York

Saint-James D, de Gennes P G 1964 Onset of superconductivity in decreasing fields. *Phys. Lett.* 7: 306–8

Scalapino D J 1969 The electron–phonon interaction and strong-coupling superconductors. In: Parks 1969, Vol. 1, Chap. 10, pp. 449–560

Tinkham M 1975 *Introduction to Superconductivity*. McGraw-Hill, New York

D. M. Ginsberg
[University of Illinois, Urbana, Illinois, USA]

Superconducting Materials: Specification

The specification of superconductors is needed both in the purchase of practical superconductors and in the characterization of superconductor materials for research. Parameters such as the critical temperature T_c and the upper critical field H_{c2} are intrinsic to the material and are usually adequately specified just by specifying the material (i.e., by the ratio of elements in the superconducting alloy or compound). Extrinsic parameters that vary with processing technique and conductor design, on the other hand, are much more variable and need to be specified and measured on a conductor-to-conductor basis, especially for the procurement of superconductors for practical applications (see *Multifilamentary Superconducting Composites*). Such parameters include the physical shape of the conductor, critical current, matrix resistivity and mechanical properties. In this article, the emphasis will be on practical superconductor materials and, primarily, specification of their extrinsic parameters.

Furthermore, the article concentrates on the specification of single-strand multifilamentary conductors and does not include additional parameters needed for cable specification. The single-strand multifilamentary parameters, however, form the basis for most conductor specification and are a key element for the more complex cable conductors.

The extrinsic parameters of single-strand conductors are described here under three general headings: physical, electrical and mechanical specifications. An example of a set of typical superconductor specifications for a Nb_3Sn superconductor is given in Table 1.

Table 1
Example of superconductor wire specification

Specification	Typical value	Test specimen or sample
Physical		
superconductor material	(Nb–1.5 at.%Ti)$_3$Sn	
stabilizer material	oxygen-free copper	
wire diameter	0.808 ± 0.005 mm	each 150 m, middle and end of each billet
copper-to-noncopper ratio	(1.5 ± 0.1):1	both ends of each billet
filament size and spacing	6 μm >1.0 μm	both ends of each billet
twist pitch	1.0 ± 0.1 clockwise twists cm^{-1}	beginning and end of each continuous wire
piece length	1.5 km	
Electrical		
critical current (min at 12 T)	164 A at 10^{-14} Ω m and 4.22 K for magnetic field ⊥ wire axis	beginning and end of each continuous wire length
resistance max R_{295K} max R_{19K}	 565 μΩ cm^{-1} 6.8 μΩ cm^{-1}	beginning and end of each continuous wire length
Mechanical		
prestrain	0.15–0.25%	both ends of each billet
irreversible strain	>0.7%	both ends of each billet

1. Physical Specifications

1.1 Superconductor Material

Superconductor material should be specified in order to determine T_c and H_{c2}. Also, the mechanical properties inherent to the material will need to be considered in application of the material. For example, NbTi superconductors are easy to handle mechanically, because they are ductile. However, they are commonly used in magnet applications up to a magnetic field of only about 8 T. Nb_3Sn superconductors, on the other hand, are commonly used at higher fields up to about 14 T, but, being brittle, they require special handling and special mechanical design for applications.

Furthermore, the specific superconductor composition should be prescribed because some compositions are better for obtaining high current density J_c at high magnetic field, while other compositions are better for low fields. For example, in NbTi superconductors, the nominal composition of niobium and titanium should be prescribed along with alloy homogeneity. For Nb_3Sn, the superconductors either can be binary (just niobium and tin) or third (and sometimes fourth) elements, such as titanium or tantalum, can be added in order to increase J_c at very high magnetic fields.

1.2 Stabilizer Material

The type of material used as a stabilization matrix (see Fig. 1) should be specified in order to ensure adequate electrical and thermal conductivity to protect the conductor against thermal runaway. For example, oxygen-free copper is a common stabilizer material. Specification of the type of stabilizer material ensures that the conductor will be adequate to meet the residual resistivity ratio requirements discussed in Sect. 2.2.

1.3 Conductor Size

The cross-sectional dimensions of the conductor should be specified so that they are consistent with the current capability of the power supply and the desired magnet inductance. The tolerance for variation in these dimensions can be critical for some magnet applications. The cross-sectional dimensions are also needed to calculate the critical current density from the critical current, described in Sect. 2.1.

1.4 Copper-to-Noncopper Ratio

The copper-to-noncopper ratio specifies the volume of stabilizer material relative to the superconductor material. It must not be too low, or the conductor will be unstable and susceptible to thermal runaway. If the ratio is too high, on the other hand, the overall critical current density suffers. Values from 1:1 to 5:1 are typical, with the higher ratios needed to ensure stability in large magnets where the stored energy can be very great.

The copper-to-noncopper ratio is usually measured by weighing a sample of the conductor, chemically etching away the copper with a 50% nitric acid solution, drying the remaining superconducting filaments, and reweighing. The copper mass is obtained from the difference. The copper-to-noncopper ratio is then determined by taking the ratio of the weight of copper to the remains and dividing by the ratio of specific gravities for copper and the noncopper material.

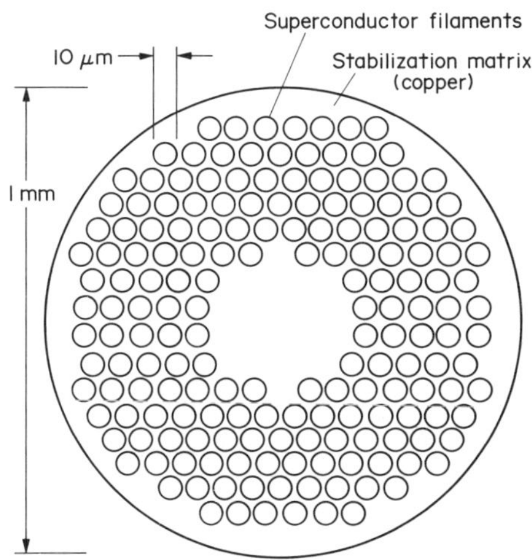

Figure 1
Cross-sectional view of a typical NbTi, copper-stabilized superconducting composite (after Ekin 1983b)

1.5 Filament Size and Spacing

The diameter of the superconducting filaments should be specified in order to ensure that hysteretic ac losses in the conductor filaments are reduced to an acceptable level. The smaller the filament size, the lower the losses. Too small a diameter, however, results in filament cross-sectional inhomogeneities and a lowering of the critical current density. A filament diameter of about 5 μm is a typical compromise.

The spacing between filaments must also be kept above a minimum value, typically 1 μm or more, edge to edge, in order to electrically decouple neighboring filaments. If the filaments couple, low-field ac losses can reach unacceptable levels.

Filament size and spacing are usually determined from a photomicrograph of a polished cross section of the conductor. Several filament diameters and spacings are measured and the results averaged.

Instead of specifying filament diameter and spacing, the maximum acceptable level of ac loss can be stipulated, along with the method by which it should be measured. This is usually preferable because it ensures a minimum performance level. Two methods are in common use: magnetization and calorimetry. The calorimetric method (using a boil-off technique, for example) measures a total ac loss and is therefore usually favored, but it is more difficult than magnetization measurements.

1.6 Twist Pitch

The filaments in a practical superconductor are twisted to reduce ac loss heating that would otherwise occur when the conductor is subjected to magnetic fields that change with time (see *AC Applications of Superconducting Materials*; *AC Losses in Superconducting Materials*). The rates of change can be particularly large in pulsed magnet applications, but can be significant in dc magnets as well, because of the need to charge and discharge the magnet. The number of twists per unit length, or the twist pitch (the length of one twist), is specified in order to ensure that ac losses are reduced to an acceptable level. A twist pitch of about one centimeter is typical for a small (1 mm diameter) conductor.

1.7 Piece Length

The minimum acceptable length of continuous superconductor is usually specified in order to avoid an excessive number of joints between lengths. Sometimes the minimum length is determined by requirements on the joint location as well. For example, in superconducting electromagnets, the piece length needs to be long enough that the joints between conductors can be made in a low-magnetic-field region.

2. *Electrical Specification*

2.1 Critical Current

The critical current I_c is perhaps the most important parameter from a practical standpoint. Because the rise in the voltage–current (V–I) characteristic of a superconductor is not infinitely sharp (see Fig. 2), the critical current is arbitrary, unless a criterion is specified to determine the critical current (see *Critical Current Density*). This is usually done by using an electric field criterion or resistivity criterion. For example, the critical current I_c can be defined as the current at which the electric field E along the conductor reaches a specific value, such as 1 μV cm^{-1}. This is shown by the intersection of the V–I curve and the horizontal dashed line in Fig. 2. Alternatively, I_c can be defined as the current at which the resistivity of the conductor rises to a specific value, such as 10 pΩ cm, shown by the dashed line passing through the origin in Fig. 2. When using the resistivity criterion, the cross-sectional area to be used in the determination should be specified. Finally, when the rise in the V–I characteristic is very gradual, much of the arbitrariness and variability is eliminated by using the offset criterion. As shown in Fig. 2, it is determined by taking the tangent to the V–I curve at a specified electric field.

The critical current density J_c is defined as the critical current divided by the cross-sectional area A of the superconductor,

$$J_c \equiv I_c/A$$

Usually the area A is taken as just the cross-sectional area of the conductor minus the cross-sectional area of the copper stabilizer; this is commonly referred to as the noncopper area.

Sometimes the shape of the voltage–current curve is also specified to indicate the quality of the superconductor. A sharp transition is desirable, indicating the rapid disappearance of resistivity below the nominal value of I_c. A common method for representing the sharpness of the resistive transition is through the use of the n value defined by a simple power law

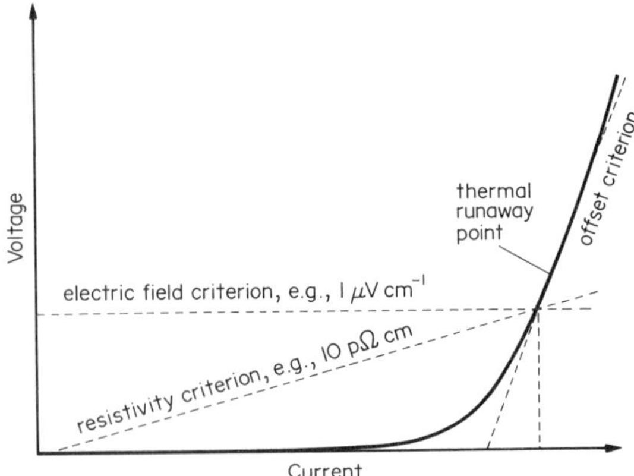

Figure 2
The voltage–current characteristic of a practical superconductor, showing electric-field and resistivity criteria for determining critical current

$$V \propto I^n$$

or equivalently

$$E \propto J^n$$

Here E is the electric field along the conductor and J is the current density. Values of the power n are easily determined by plotting the V–I or E–J curves on a log–log plot and determining the slope of the curve at a designated electric field (the curve is typically linear over a considerable range of electric field). Values of n generally lie between 20 and 100 for high-quality superconductors (see *Resistive Transition and Flux Flow in Superconducting Materials*). Lower n values usually indicate inhomogeneities in the local critical current density. Other methods of fitting the shape of the V–I curve are also used, usually involving more fitting parameters. However, this empirical formula is simple, involves only a single fitting parameter (other than a proportionality constant) and, consequently, is easy to use.

The critical current is also strongly dependent on the magnetic field magnitude (see Fig. 3) and direction. Typically, I_c values are prescribed at one or more magnetic fields, usually for the magnetic field applied perpendicular to the wire axis. This perpendicular orientation for the magnetic field usually corresponds to the worst case (lowest I_c), but is the most common orientation in superconductor magnet applications.

The critical current is also strongly dependent on temperature (see Fig. 3). Usually the temperature at which liquid helium boils at atmospheric pressure is used, that is, 4.2 K. If the critical current measurement cannot be carried out at precisely 4.2 K, small temperature corrections can be made to the measured I_c values using the "linear temperature" formula

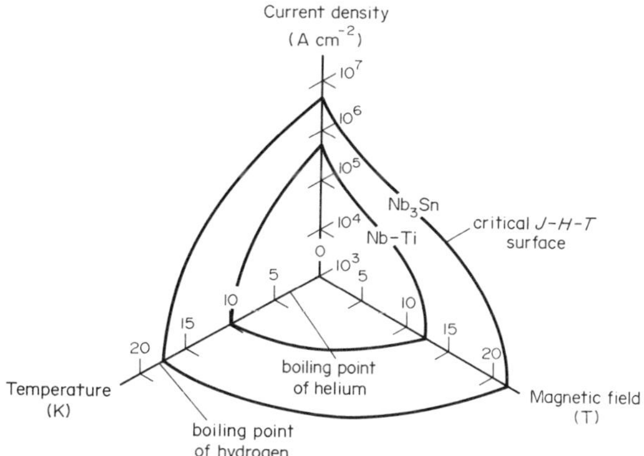

Figure 3
Interrelation of critical current density, magnetic field and temperature for Nb_3Sn and NbTi (courtesy of W. A. Fietz). In the region between the origin and the critical *J–B–T* surface the material is superconducting; beyond this critical surface the material is in the normal state

$$I_c/I_m = (T_c - T)/(T_c - T_m)$$

where T_c is the transition temperature at the specified magnetic field, I_m is the current measured at temperature T_m and I_c is the critical current at the specified temperature T.

Usually, the critical current is measured on samples taken from each continuous length of conductor. For testing, these samples are wound typically onto a cylindrical sample holder, such as shown in Fig. 4, and a four-terminal transport I_c measurement is carried out in accordance with a specific measurement procedure such as American Society for Testing and Materials (ASTM) Standard Test Method B714–82. A minimum length of sample for the critical current measurement should be prescribed, since inhomogeneities along the conductor have a greater effect on I_c over longer conductor lengths. A 25 cm length of sample between the voltage taps in Fig. 4 would be a typical test length. Long test lengths also allow a more sensitive measurement of low electric fields along the conductor, which permits a sensitive critical current criterion to be used.

2.2 *Resistance at Room Temperature and* T_c

The conductor resistance at room temperature (R_{295K}) is determined mainly by the amount of copper matrix. Thus, it can be used to provide an independent measure of the volume of copper to superconductor. The conductor resistance just above the superconducting transition temperature R_{T_c} is mainly determined by the purity of the matrix material. Thus, the resistance just above T_c is important because it gives a measure of the ability of the matrix to stabilize and protect the conductor from thermal runaway, should a disturbance raise a portion of the superconductor above T_c. Typically the ratio, R_{295K}/R_{T_c}, called the residual resistance ratio (RRR), is specified. The higher the RRR value, the greater the purity of the copper matrix material and the better the stabilization of the conductor. Values of RRR of about 70 or more are typically achieved using oxygen-free copper.

Room-temperature resistance is usually measured using a four-terminal resistance measurement technique, similar to the critical current measurement, except that the *V–I* curve is linear and the measuring current is kept small to avoid sample heating. Temperature control for measuring R_{T_c} just above T_c can be achieved fairly easily by lowering the sample into a liquid helium bath and then raising it above the liquid helium a couple of centimeters and recording the resistance as the temperature rises through T_c. The superconductor is thus used as a fixed-point thermometer: as the temperature of the conductor increases, the resistance rises rapidly at T_c, then becomes constant just above T_c. This region of relatively constant resistance on warming is taken to be R_{T_c}. The RRR is then just the quotient R_{295K}/R_{T_c} of these two measured resistances.

3. *Mechanical Parameters*

For compound superconductors, such as Nb_3Sn, V_3Ga or Nb_3Al, several mechanical parameters need to be specified because of the sensitivity of the critical current of these materials to strain.

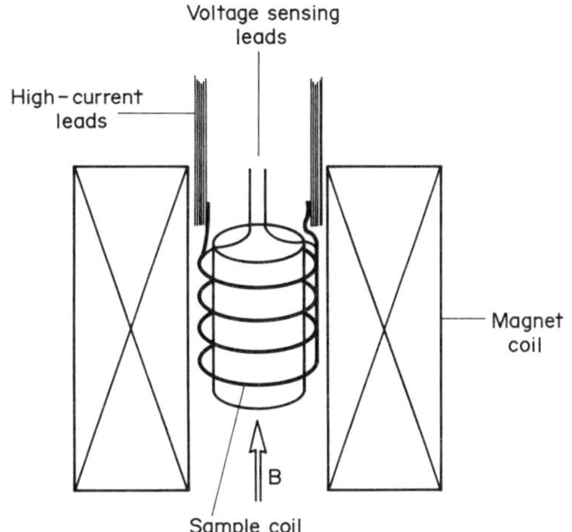

Figure 4
A typical cylindrical sample holder for measuring the critical current of a spirally wrapped superconducting wire sample (adapted from Goodrich and Fickett 1982). The sample holder assembly is placed inside the bore of a solenoid magnet that provides a background field along the axis of the holder (perpendicular to the sample). In this way, a long superconducting sample can be tested in the confined space of a high-field background magnet

3.1 Prestrain

The different materials that make up a composite superconductor (stabilizer matrix, tin–bronze matrix for Nb_3Sn conductors, superconducting material and diffusion barriers) all thermally contract by different amounts as the conductor is cooled from its fabrication temperature (typically 700 °C). This places the superconducting material under an initial prestrain ε_m, usually compressive, which arises from differential thermal contraction of the various materials making up the conductor.

The value of the prestrain ε_m is also needed to compare critical currents in different samples equally. Either compressive or tensile strain ε degrades the superconductor; thus, because of the inital compressive prestrain, the critical current will start at a degraded value (see Fig. 5), then pass through a maximum at zero "intrinsic" strain where the applied tensile strain removes all the precompression and the superconducting material itself is in a relatively strain-free state, and then again decrease when the intrinsic strain within the superconducting material becomes tensile. This results in the peaked curves shown in Fig. 5. The value of the prestrain can be determined by applying tension to a conductor and measuring the strain where the critical current passes through a maximum. Critical currents should be compared at the peak, or at comparable intrinsic strain relative to the peak, for the purpose of selecting conductors.

Because of the effect of strain on the critical current, the compressive prestrain ε_m from differential thermal contraction should be specified in order to set strain limits in designing magnets and specifying reinforcing structures for magnets. Ideally, the superconductor should operate just on the compressive side of maximum (about 0.1–0.2% intrinsic compressive strain) when the conductor is under full operating load. In such a case J_c is near maximum, but there is still a sufficient margin for additional strain.

3.2 Irreversible Strain

Irreversible strain specifies the strain beyond which the conductor is permanently damaged and J_c is irreversibly degraded. The importance of this parameter is that it sets a strain-handling limit in bending and spooling a conductor during magnet fabrication. It also defines the ultimate practical strain the conductor can tolerate in the mechanical design of the conductor support structure.

The irreversible strain is determined by measuring the critical current of a conductor as a function of tensile strain and periodically unloading the conductor to determine if the J_c vs ε curve retraces itself, that is, to see whether it is reversible. For example, in Fig. 6, the critical current on unloading from strain point A was measured to be at the + marked A′, right on the original curve. The same was true after unloading from strain point B; that is, the critical current was reversible. However, after unloading from strain point C, the critical current was irreversible as shown by point C′, which falls below the original I_c vs strain curve. Thus, the reversibility of the I_c vs ε curve is lost and permanent damage occurs to the superconducting material when the conductor is strained beyond point B, which is defined as the irreversible strain limit.

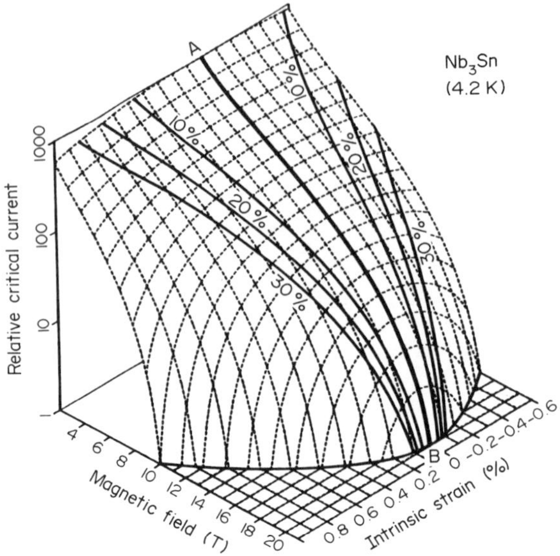

Figure 5
J–B–ε critical surface for Nb_3Sn showing the reversible degradation of critical current at either tensile or compressive strain as a function of magnetic field (after Ekin 1983a). Line A–B represents the maximum (strain-free) value of the critical current as a function of magnetic field. Corresponding pairs of curves on either side of line A–B show the strain window for mechanical design which will result in a critical current within the indicated percentage of maximum

4. Future Specification Needs

Future needs for specification parameters will change with both applications and materials. For example, in new applications, superconductor magnets that use very rapidly changing ac currents will impose severe ac-loss requirements on the conductor. This problem is overcome by using very-small-diameter filaments and a short twist pitch (see *AC Applications of Superconducting Materials*). However, the very small filament diameter may well require a filament uniformity specification to avoid a lowering of the critical current from localized necking of the filament. Alternatively, the critical current may have to be specified at a very low criterion level to discriminate against the low voltages associated with such filament nonuniformity.

In the materials area, the advent of high-T_c super-

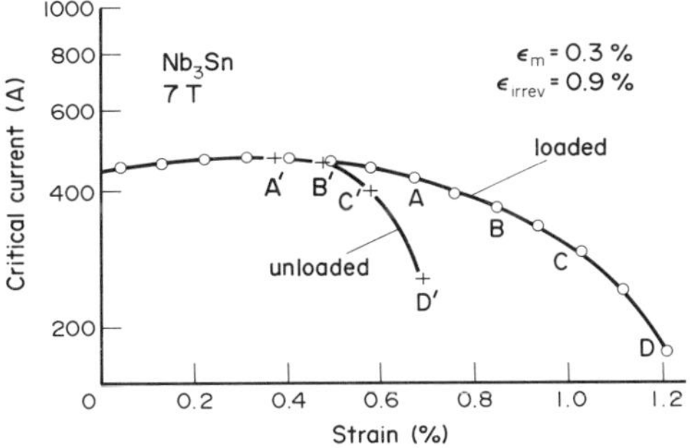

Figure 6
Illustration of the irreversible strain limit (after Ekin 1981). Corresponding loaded and unloaded data points are labelled by letters (A, B, C, etc.). The irreversible strain limit ε_{irrev} occurs just beyond point B, after the conductor has been loaded to about 0.9%

conductors brings with it a new set of specifications. Mechanical parameters will become particularly important in these brittle ceramic materials. Also, the very low critical current densities of some bulk high-T_c conductors and the gradual rise of the V–I characteristic (see Fig. 2) make the specification of a critical current somewhat arbitrary with conventional criteria. More universal critical current criteria such as the offset criterion are being used to meet this need.

See also: Measurements in Superconducting Materials; Multifilamentary Superconducting Composites; Resistive Transition and Flux Flow in Superconducting Materials; Superconducting Magnets

Bibliography

Annual Book of ASTM Standards 1982a Standard definitions of terms relating to superconductors, ASTM B713–82. American Society for Testing and Materials, Philadelphia, PA

Annual Book of ASTM Standards 1982b Standard test method for dc critical current of composite superconductors, ASTM B714–82. American Society for Testing and Materials, Philadelphia, PA

Clark A F, Ekin J W 1977 Defining critical current. *IEEE Trans. Magn.* 13: 38–40

Ekin J W 1981 Mechanical properties and strain effects in superconductors. In: Foner S, Schwartz B (eds.) 1981 *Superconducting Materials Science: Metallurgy, Fabrication and Applications.* Plenum, New York, pp. 455–510

Ekin J W 1983a Four dimensional *J–B–T–ε* critical surface for superconductors. *J. Appl. Phys.* 54: 303–6

Ekin J W 1983b Superconductors. In: Reed R P, Clark A F (ed.) 1983 *Materials at Low Temperatures.* American Society for Metals, Metals Park, OH, pp. 465–513

Ekin J W 1987 Irregularity in Nb–Ti filament area and electric field vs current characteristics. *Cryogenics* 27: 603–7

Ekin J W 1989 Offset criterion for determining superconductor critical current. *Appl. Phys. Lett.* 55: 905–7

Goodrich L F, Bray S L 1990 High-T_c superconductors and critical current measurement. *Cryogenics* 30: 667–7

Goodrich L F, Fickett F R 1982 Critical current measurements: a compendium of experimental results. *Cryogenics* 22: 225–41

Larbalestier D C, West A W, Starch W, Warnes W, Lee P, McDonald W K, Olarey P, Hemachalam K, Zeitlin B, Scanlan R 1985 High critical current densities in industrial scale composites made from high homogeneity Nb 46.5 Ti. *IEEE Trans. Magn.* 21: 269–72

Plummer C J G, Evetts J E 1987 Dependence of the resistive transition on composite inhomogeneity in multifilamentary wires. *IEEE Trans. Magn.* 23: 1179–82.

J. W. Ekin
[National Institute of Standards and Technology, Boulder, Colorado, USA]

Superconducting Materials Under Pressure

In general, pressure will induce superconductivity in any material. Pressure increases the overlap of electronic wave functions and thus causes a metallic transition while magnetism is quenched because unpaired spins take more space then paired ones. Since all nonmagnetic metals are believed to be superconductors, all materials are expected to become superconductors at sufficiently high pressure.

At intermediate pressures, this very general rule is not followed in all substances. In particular, phase changes may induce much more complicated behavior. Also, superconductivity is a rather subtle phenomenon, involving several material parameters, each differently affected by pressure. For those materials that are adequately described by the Bardeen–Cooper–Schrieffer (BCS) theory, these parameters are mainly phonon frequency, electron–phonon coupling strength and the density of electronic states at the Fermi level. In the recently discovered cuprate perovskites, the main effect of pressure seems to be a modification of the number of charge carriers in the copper–oxygen planes.

Of course pressure not only affects the critical temperature T_c but also other superconductor characteristics such as the critical field H_c, the coherence length ξ and the London penetration depth λ_L.

1. *Phase Changes*

Pressure may induce structural, electronic or magnetic phase changes. Structural phase changes will in general be accompanied by an increase in the coordination number of the atoms which makes an increase in bond length possible. Phase changes can also be induced by a changing electronic structure. A clear example is bismuth (see Fig. 1). The number of carriers in the nonsuperconducting ambient pressure

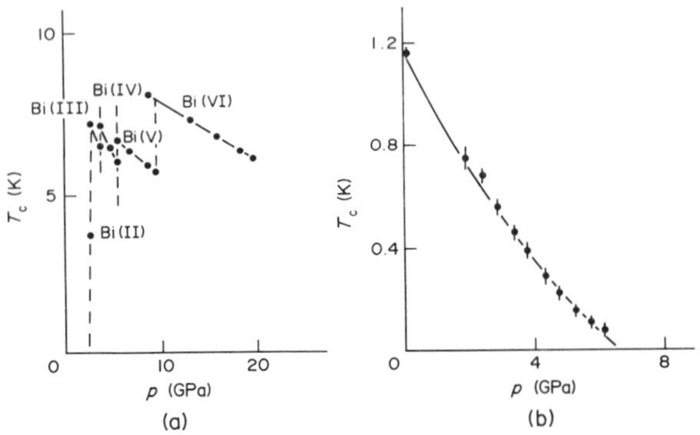

Figure 1
Critical temperature T_c of (a) bismuth and (b) aluminum as a function of pressure

phase Bi(I) decreases to approximately 0 at 2 GPa (1 GPa = 10 kbar), where a phase transition to Bi(II) occurs. This phase and all higher pressure phases shown are superconducting.

Phase changes involving magnetic structure are important because of the mutual exclusion of macroscopic magnetization and superconductivity. Pressure suppresses the magnetism in the itinerant ferromagnet Y_9Co_7 and increases the superconducting T_c (see below).

2. Relation Between T_c, H_c, ξ, λ and Other Parameters

For a given phase, several useful relations can be derived. Within the BCS framework (see *Superconducting Materials: BCS and Phenomenological Theories; Superconducting Materials: Physical Properties*), for an isotropic superconductor where pairing is due to an attractive interaction V for those electrons that are less than $\hbar\omega_{ph}$ from the Fermi level E_F, and where the relevant electron density is thus the density of states at the Fermi level $N(0)$, the critical temperature is given by

$$k_B T_c = 1.14\, \hbar\omega_{ph} \exp[-1/N(0)V] \tag{1}$$

If the interaction V is due to phonons, $\hbar\omega_{ph}$ may be replaced by $k_B\Theta_D$, where Θ_D is the Debye temperature. For our purpose also a slightly more elaborate formula due to McMillan is of importance:

$$T_c = \frac{\Theta_D}{1.45} \exp\left[-\frac{1.04(1+\lambda)}{\lambda - \mu^*(1+0.62\,\lambda)}\right] \tag{2}$$

The pressure dependence of the Coulomb repulsion term μ^* is very small and will be neglected here. The effective electron–phonon coupling parameter λ is given by $\lambda = N(0)\langle I^2\rangle / M\langle\omega^2\rangle$, where $\langle I^2\rangle$ is the Fermi surface averaged square of the electron–phonon matrix element, M is the mass and $\langle\omega^2\rangle$ is an averaged phonon frequency. The most important effect of pressure is the increase of the phonon energies. This results in a decrease of λ and hence of T_c. Indeed, experimentally in the simple systems pressure seems to destroy superconductivity. The M in the expression for λ is the mass of the ion and explains the isotope effect: lighter isotopes generally have lower T_c.

It is illustrative to calculate (in the framework of the BCS theory) the pressure dependence of a number of important parameters in the very simple isotropic case with a spherical Fermi surface and a fixed total number of charge carriers.

For small λ, Eqn. (2) can be written as $T_c = (\Theta_D/1.45)\exp(-1/\lambda)$, hence:

$$\frac{1}{T_c}\frac{dT_c}{dp} = \frac{\gamma}{B} + \frac{1}{\lambda^2}\frac{d\lambda}{dp}$$
$$\simeq \frac{\gamma}{B}\left(1 - \frac{2}{\lambda}\right) \tag{3}$$

where p is the pressure, with the Grüneisen parameter $\gamma = -(V/\Theta_D)(d\Theta_D/dV)$ and the bulk modulus $B = -V(dp/dV)$. For the second line of Eqn. (3) the approximation $\lambda = C/M\Theta_D^2$ was used, where C is a constant, with the further crude empirical approximation that the pressure dependences of $N(0)$ and $\langle I^2\rangle$ cancel.

The pressure dependence of the coherence length at zero temperature $\xi(0) = 0.18\,\hbar v_F / k_B T_c$ is given by

$$\frac{1}{\xi(0)}\frac{d\xi(0)}{dp} = -\frac{1}{3B} - \frac{1}{T_c}\frac{dT_c}{dp} \tag{4}$$

The pressure dependence of the London penetration depth at zero temperature $\lambda_L = [m/(n_s e^2 \mu_0)]^{1/2}$ is given by

$$\frac{1}{\lambda_L}\frac{d\lambda_L}{dp} = -\frac{1}{2B} \tag{5}$$

The pressure dependence of the Ginzburg–Landau parameter $\kappa = \lambda/\xi$ is

$$\frac{1}{\kappa}\frac{d\kappa}{dp} = -\frac{1}{6B} + \frac{1}{T_c}\frac{dT_c}{dp} \tag{6}$$

The pressure dependence of the upper critical field $B_{c2} = \Phi_0/(2\pi\xi^2)$ is given by

$$\frac{1}{B_{c2}}\frac{dB_{c2}}{dp} = \frac{2}{3B} + \frac{1}{T_c}\frac{dT_c}{dp} \tag{7}$$

The pressure dependence of the scattering length for impurity scattering l is:

$$\frac{1}{l}\frac{dl}{dp} = -\frac{1}{3B} \tag{8}$$

The results of these equations after substituting data for some metals are given in Table 1. In spite of the

Table 1
Superconducting parameters for a few metals: the calculated values were obtained from Eqns. (1)–(7)

Metal	Aluminum	Lead	Tantalum	Niobium	Iridium
T_c (K)	1.14	7.19	4.48	9.5	0.14
B_c (T)	0.0105	0.0803	0.0831	0.198	0.0019
Θ_D	423	102	247	241	425
γ	2.14	2.74	1.69	1.58	2.39
B (GPa)	74	44	204	174	362
λ_L (nm)	50	39		44	
ξ (nm)	1600	87		38	
λ	0.17	0.38	0.25	0.31	0.12
dT_c/dp (K GPa^{-1})					
experimental	−0.29	−0.38	−0.26	−0.02	−0.005
calculated	−0.36	−1.9	−0.26	−0.47	−0.01
dB_c/dp (T GPa^{-1})					
experimental	−0.002	−0.0073		−0.011	
calculated	−0.003	−0.020	−0.005	−0.009	−0.0002
$d(\ln \lambda_L)/dp$ (GPa^{-1})[a]	−0.007	−0.01	−0.002	−0.003	−0.001
$d(\ln \xi)/dp$ (GPa^{-1})[a]	0.3	0.3	0.06	0.05	0.1
$d(\ln \kappa)/dp$ (GPa^{-1})[a]	−0.3	−0.3	−0.06	−0.05	−0.1

a all calculated

crudeness of the approximations made, these results are in reasonable agreement with experimental results for the nontransition metals.

3. Elemental Superconductors

Somewhat surprisingly, the superconducting properties of the elements have not yet thoroughly been investigated as a function of pressure. In Table 2 all known values for dT_c/dp are given for all elements that are superconductors at zero pressure or become superconducting at high pressure. The highest value of T_c obtained reliably is 13 K for lanthanum at 17 GPa and also for phosphorus at 30 GPa. Values of approximately 30 K have been reported for sulfur. However, the phase diagram of sulfur is very complicated and measurements are not reproducible. The largest positive value for dT_c/dp is found for the α phase of uranium.

Structural changes are generally accompanied by a change in T_c and dT_c/dp and are sometimes responsible for the occurrence of pressure-induced superconductivity. In those materials that are (very) poor conductors at room temperature, superconductivity is often made possible by increased conductivity due to pressure-induced band overlap.

4. Intermetallic Alloys

In some alloys of metals, in particular the A15 compounds (see *A15 Compound Superconductors*), both a higher T_c and a higher upper critical field H_{c2} than in any element (at ambient pressure) are realized. Unfortunately, for the A15 compounds with a high T_c, the dT_c/dp is negative and thus T_c cannot be increased with pressure; however, for low-T_c compounds dT_c/dp may be large (see Table 3). The high T_c values are probably related to phonon softening (which increases λ) due to lattice instability. In this situation increasing pressure may increase T_c if it brings the system closer to instability, but increasing pressure to beyond instability will decrease T_c again.

A quite different class of materials is that of superconductors with magnetic ions. The T_c in Y_9Co_7 increases with pressure due to the suppression of its itinerant ferromagnetism. Interestingly, long-range ferromagnetic and superconducting order parameters seem to coexist, although they vary spatially.

In Table 3 some values are quoted for the so called heavy fermion systems (see *Heavy Fermion Metals*), characterized by a very large value for the linear coefficient of the specific heat: $\gamma \simeq$ 100–2000 mJ K^{-2} mol^{-1} and a very large specific heat anomaly at T_c.

5. Superconductivity in Metallic Semiconductors

A number of semiconductors at ambient pressure have already been found to become metallic and superconducting under pressure. Examples are the high-pressure phases of GaAs: GaAs(II) (17–24 GPa) and GaAs(III) (24–60 GPa). At 20 GPa, GaAs has a $T_c \simeq 4.5$ K and a $dT_c/dp = -0.05$ K GPa^{-1}.

Table 2
Critical temperature and pressure derivative for the elements: P denotes the pressure range of validity

Element	P (GPa)	T_c (K)	dT_c/dp (K GPa^{-1})	Element	P (GPa)	T_c (K)	dT_c/dp (K GPa^{-1})
Al	0	1.14	−0.29	Os	0	0.655	−0.02
Am	0	0.6		P	8–30	5.5	+0.03
As	13–20	0.1	+0.1		10–30	4	+0.3
	24	2.7		Pa	0	0.4	
Ba (III), hcp	6.5–10.5	1.2	+0.3	Pb	0	7.193	−0.38
Ba (III)	10.5–12.5	2.4	+1.3	Re	0	1.7	−0.023
Ba (IV)	12.5–47	5.0	−0.03	Ru	0	0.51	+0.003
Ba (V)	27–52	7.2	−0.17	S	?	15–30	
Be	0	0.026		Sb	5–9	0.1	+0.12
Bi (III)	2.7–4	7	−0.45		9–23	3.8	−0.13
Bi (IV)	4–6.3	7	−0.5	Sc	18–22	0.05–0.35	+0.1
Bi (V)	6.3–10	6.8	−0.3	Se	17	4.9	
Bi (VI)	9–40	8.5	−0.2		33	5.5	
Cd	0	0.56	−0.18	Si	16–18	7	+0.8
Ce	2–4	0.02	+0.02		18–37	8	−0.3
	4–12	1.8	−0.06		37–40	3	+0.4
Cs	11–12	0.05	+0.3		40	3.5	0.0
	12–14	1.5	−0.1	Sn	0	3.722	−0.50
Ga (I)	0–3.5	1.091	−0.18	Sr	21	0	+0.25
Ga (II)	3.5	6.46		Ta	0	4.483	−0.026
Ge	11.5	5.35	−0.2	Tc	0	7.77	−0.13
Hf	0	0.12		Te	3–6	2.5	+1.0
Hg α	0	4.153	−0.37		6–8	4.2	−0.02
Hg β	0	3.95	−0.48		8–15	4.2	
In	0	3.40	−0.43	Th	0	1.368	−0.15
Ir	0	0.14	−0.005	Ti	0	0.26	+0.006
α-La	0–2.3	4.9	+1.4	Tl	0	2.39	+0.23
β-La	2.3–5	6.00	+1.1	α-U	0–0.5	0.68	+4.0
	5–15	10	+0.3		0.5–1.4	2.2	0.0
	15–27	13	−0.3		1.4	2.2	−1.2
Li	22–23	7		V	0	5.38	+0.063
Lu	0	0.1		W	0	0.012	
	4.5–22	0.02	+0.15	Y	11–16	1.3	+0.37
	22	2.8			15–24	2	+0.6
	28	1.5			24–32	7	+0.4
Mo	0	0.92	−0.014	Zn	0	0.875	−0.16
Nb	0	9.50	−0.02	Zr	0	0.65	+0.15

Table 3
Pressure dependence of T_c and B_{c2} for some selected compounds: the value of B_{c2} was in some cases obtained by parameter-dependent extrapolation

Compound	T_c (K)	dT_c/dp (K GPa^{-1})	B_{c2} (T)	dB_{c2}/dp (T GPa^{-1})
Nb_3Ge	23	−0.25	39	
Nb_3Sn	18	−0.22	23	−1.7
V_3Si	17	+0.31	20	+2.7
V_3Ge	6	+0.81		
Y_9Co_7	2	+0.7	0.7	+0.3
UPt_3($\gamma = 422$)	0.5	−0.13	1	−0.5
U_6Mn ($\gamma = 99$)	2.2	−0.22		
U_6Fe ($\gamma = 155$)	3.8	0		
U_6Co ($\gamma = 126$)	2.3	+0.22		

6. *Superconductivity in Intercalation Compounds*

Most intercalates studied under pressure consist of alkali metals (sometimes mixed with mercury) and graphite, the superconductivity taking place mainly in the metal layer. Some intercalates only become superconducting under pressure. In NaC_2, NaC_3 and LiC_2, pressurization up to approximately 3 GPa and quenching to ambient pressure at low temperature leads to values of T_c of 5 K, 3.8 K and 2 K, respectively. The dT_c/dp seems to be negative in these compounds. Examples are given in Table 4.

7. *Ternary Compounds*

Very large values for dT_c/dp, an order of magnitude larger than in most elements or binary compounds, both positive and negative, are found in the ternary molybdenum chalcogenides MMo_6X_8 (X = S,Se), where the Mo_6X_8 clusters play an essential role for superconductivity (see Table 5). This is not too surprising since semiconductors, metals, ferromagnets and antiferromagnets are found in this class of materials with only minor variations in composition or structure. In the vicinity of a structural phase transition the electron–phonon coupling can be large. In nearly all materials, T_c has a clear maximum as a function of pressure.

This phenomenon is even clearer in the $A_2Mo_6X_6$ (A = Na, K, Rb; X = Se, Te) compounds, which are not superconducting at ambient pressure, but have a maximum T_c of 2–3 K at 0.5–1.5 GPa. These compounds are insulators due to Peierls distortion of the quasi one-dimensional chains; pressure reduces the anisotropy and makes them metallic again. Under pressure a marked downward curvature in their resistivity is observed which is interpreted as the onset of superconductivity.

Pressure is also needed to make $Eu_{0.9}Ho_{0.1}Mo_6S_8$ a bulk superconductor. T_c shows a maximum at roughly 1 GPa and magnetic field measurements show a reentrant behavior at lower pressures which decreases with increasing pressure.

Due to the presence of several structural phases, $Cu_{1.5}Mo_3S_4$ shows a complex pressure dependence with a maximum dT_c/dp of 9 K GPa^{-1} (see Fig. 2).

The ternary rare-earth transition metal borides have a much smaller dT_c/dp with some slight, but significant, nonlinearities that are not understood occurring in $LuRh_4B_4$ and YRh_4B_4. Variation of T_c seems to be due mainly to the magnetic properties of the rare-earth ions.

Quite suprisingly $X_2Fe_3Si_5$ (X = Lu,Sc,Y) are superconducting despite the presence of iron. Especially when X is yttrium the dT_c/dp is initially very large: $dT_c/dp = +3.7$ KGPa^{-1}, which is probably due to a complicated, nested Fermi surface which rapidly changes with pressure, thus changing $N(0)$.

8. *Organic Superconductors*

Following an idea by Little (1964), who suggested that electrons moving along an unsaturated carbon chain with side molecules (which greatly enhance the polarizability) would feel a large effective electron–electron interaction, a number of organic materials such as TMSF, BEDT–TTF, DMET, MDT–TTF and $M(dmit)_2$ have been synthesized and studied (see *Organic Superconductors: Anisotropic Molecular Metals*; Jerome 1982). They form one-dimensional chains or stacks, leading to quasi one-dimensional conductivity. In a pure one-dimensional conductor macroscopic superconductivity is destroyed at finite temperature due to thermodynamic fluctuations. Hence, coupling between the chains or stacks is essential. The observed maximum T_c of approximately 13 K is the intrachain/stack ordering temperature, while the paraconductivity may extend up to approximately 40 K.

A Peierls distortion can destroy conductivity in a one-dimensional system. In some of the systems mentioned previously pressure is needed to destroy this distortion and induce or enhance superconductivity. Very complicated behavior as a function of pressure and temperature is observed due to competition between several possible ground states: Peierls distortion, spin density waves, charge density waves, Mott–Hubbard charge localization, spin–Peierls distortion, superconductivity and normal metal states. Experiments at high pressure must be carried out very carefully due to the extreme sensitivity of the behavior of these materials to shear stresses.

9. *Metallic Hydrogen*

Theoretical predictions that solid molecular hydrogen will dissociate at high enough pressure to form metallic atomic hydrogen with a $T_c \simeq$ 150–300 K have led to keen interest from experimentalists. Some evidence has been reported for metallization around 150 GPa; however, the sample in this case was almost certainly still in the molecular phase.

The theoretical dissociation pressure seems to converge towards 200–400 GPa, which should be within experimental reach, the highest reported pressure on hydrogen being close to 300 GPa. However, no resistivity measurements have been done above 100 GPa, while a magnetic determination of T_c would probably be even more difficult on these very small samples. Another difficulty is that, while the equation of state seems to be reasonably well established at these

Table 4
Superconductivity in two intercalates: the ⊥ sign corresponds to a magnetic field perpendicular to the graphite layers; // means parallel

Compound	T_c (K)	dT_c/dp (K GPa^{-1})	$B_{c2//}$ (T)	$B_{c2\perp}$ (mT)	$dB_{c2//}/dp$ (T GPa^{-1})	$dB_{c2\perp}/dp$ (mT GPa^{-1})
$KHgC_8$	1.88	−0.63	4.8	7.8	−5.5	−2
$RbHgC_8$	1.41	−0.53	11.0	4.8	−14.0	−1

Table 5
Critical temperatures and pressure dependence of some selected ternary compounds

Compound	T_c (K)	dT_c/dp ($K\,GPa^{-1}$)
$Sn(II)Mo_6S_8$	13	−0.5
$ScMo_6S_8$	2.6	+1.0
YMo_6S_8	2.6	−0.5
$ErRh_4B_4$	8.5	+0.05
$LuRh_4B_4$	11.2	0.0
$Lu_2Fe_3Si_5$	6.0	−0.7

pressures, pressure calibration still poses some problems, which may imply that the experimental pressure quoted above must be corrected downwards.

Because of these problems, which are all related to the very high pressures needed, the idea of reducing this pressure by adding metallic impurities was welcomed. In one interpretation, this has stimulated work on metal hydrides, which will be discussed in Sect. 10.

10. Superconducting Metal Hydrides

The possibility of superconducting metallic hydrogen suggests that interstitial hydrogen might enhance the T_c of other metals. In palladium ($T_c < 0.001$ K) and thorium ($T_c = 1.3$ K) superconductivity is indeed induced or greatly enhanced by the presence of hydrogen: for PdH $T_c = 8.8$ K and for Th_4H_{15} $T_c = 9.2$ K. However, this is not the general rule: in superconductors which already have a high T_c such as Nb_3Sn or niobium, T_c decreases with increasing hydrogen concentration.

The role of pressure is illustrated in palladium hydrides, where hydrogen occupies octahedral intersitial sites. With a pressure of roughly 4 GPa, generated with a cryogenic diamond anvil cell, stoichiometric PdH and PdD have been prepared in equilibrium conditions. At low temperatures the stoichiometry can be maintained during pressure release. Apart from their high T_c the palladium hydrides display another interesting effect: the dependence of T_c on the atomic mass of the hydrogen isotope is opposite to that predicted by a simple BCS theory: at zero pressure T_c for PdH ($T_c = 8.8$ K) is less than that for PdD ($T_c = 11$ K), while $dT_c/dp \simeq -1.8\ K\,GPa^{-1}$ for both isotopes. This apparent conflict with BCS theory has been satisfactorily resolved by including the anharmonicity of the palladium–hydrogen potential and the large zero-point motion of the hydrogen isotope in the BCS formula (Eqn. (2)).

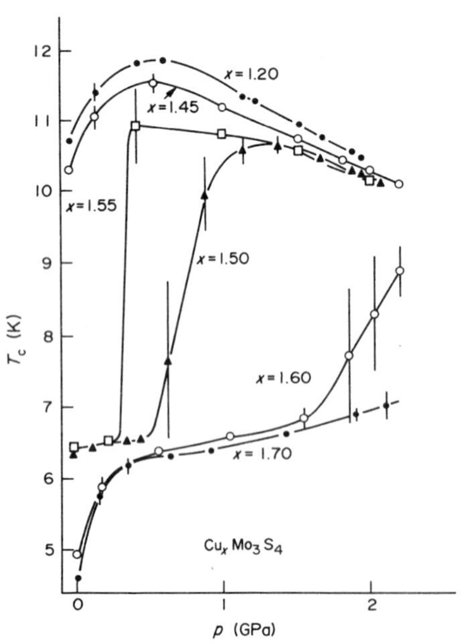

Figure 2
Critical temperature T_c of six $Cu_xMo_3S_4$ compounds as a function of pressure (after Vining and Shelton 1984. © Elsevier, New York. Reproduced with permission)

11. Cuprate Perovskites

Of all materials, the highest values of T_c are found in the class of cuprate perovskites. It seems unlikely that they can be described by BCS theory with phonon mediated electron–electron coupling. Much current effort is pointed towards the determination of the true mechanism for superconductivity, and pressure experiments play a significant role in the sifting of theories. The perovskites are layered materials; the

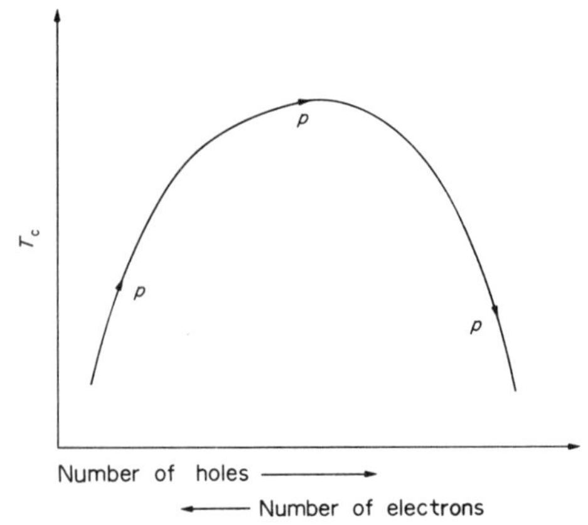

Figure 3
Critical temperature T_c as a function of the carrier concentration: dT_c/dp is proportional to the slope of this curve for compounds with apical oxygen

layer(s) important for superconductivity have CuO_2 stoichiometry and consist of a rectangular array of copper and oxygen atoms (see *Oxide Superconductors: Solid-State Chemistry*). Outside the layer and above or below each copper atom there may or may not be an additional oxygen atom, called apical oxygen because it is at the apex of a small pyramid which it defines together with the four oxygen atoms surrounding the nearest copper atom in the layer. Compounds with one, two, three or four of these layers per unit cell have a maximum T_c of roughly 90 K, 110 K, 130 K or 120 K, respectively.

Soon after the discovery of superconductivity with $T_c \simeq 30$ K in $La_{2-x}Ba_xCuO_4$ by Bednorz and Müller, its large pressure derivative $dT_c/dp \simeq 5$ K GPa^{-1} led to the highly successful idea of replacing lanthanum (ionic radius 0.104 nm) by the smaller yttrium (ionic radius 0.088 nm), thereby creating a chemical pressure and thus a higher T_c at zero pressure. The resulting compound $YBa_2Cu_3O_7$ has a $T_c = 91$ K and a very small dT_c/dp. As can be seen from the stoichiometry, the replacement of lanthanum by yttrium also led to a double CuO_2 layer, probably thereby yielding extra enhancement of T_c. Later compounds with three adjacent layers and $T_c \gtrsim 120$ K were discovered. All compounds mentioned above display a holelike Hall effect and have apical oxygen atoms at the outside of the aggregate of the $n CuO_2$ layers, that is if $n = 1$ there are apical oxygens both below and above this layer. Also, cuprate perovskites with an electronlike Hall effect were discovered. These, up to now, all have a single CuO_2 layer ($n = 1$) and no apical oxygen.

The number of carriers in the planes can be changed chemically by doping. From these substitution compounds it has become clear that T_c as a function of number of carriers is a bell-shaped curve, as shown in Fig. 3.

The effect of pressure is illustrated in Fig. 4, where $d(\ln T_c)/d(\ln V) = -(B/T_c)\, dT_c/dp$ is plotted vs T_c. Here, V is the molar volume and B is the bulk modulus. Although a large scatter is present, it is clear that compounds with a high T_c have a small $d(\ln T_c)/d(\ln V)$. If T_c is low and if an apical oxygen is present $d(\ln T_c)/d(\ln V)$ is negative. If T_c is low and no apical oxygen is present, the $d(\ln T_c)/d(\ln V)$ is positive.

There are some experiments (Van Eenige et al. 1990) over a larger pressure range, for example on $YBa_2Cu_4O_8$ where T_c as a function of pressure is first found to increase from 80 K at zero pressure to 108 K at approximately 10 GPa and then to decrease again to 100 K at approximately 20 GPa, thus following a bell-shaped curve in agreement with the idea mentioned.

One of the remarkable facts of high-T_c superconductors is the large dT_c/dp of $YBa_2Cu_4O_8$, which is 5 K GPa^{-1} at zero pressure, compared with the almost pressure-insensitive $CaLaBaCu_3O_7$ where T_c only increases from 79 K at zero pressure to 86 K at 46 GPa (see Fig. 5).

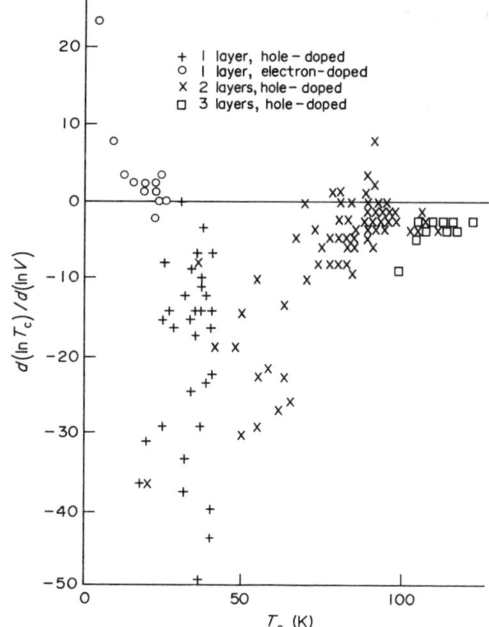

Figure 4
Relative change in T_c per relative change in molar volume as a function of T_c for all cuprate perovskites

For hole-doped materials it has been found that pressure increases the number of holes in the layers. The pressure dependence can then be explained using Fig. 3: dT_c/dp is determined by the starting point at zero pressure on this curve. Hence it is clear that $dT_c/dp \simeq 0$ close to the maximum of the curve and is large and either positive or negative far away from the maximal T_c. This is indeed the behavior observed experimentally. This curve can also explain experiments over a large pressure range: if a compound with a small number of holes (low T_c, left part of the curve) is pressurized, T_c will first increase, go through a maximum and then decrease, as observed experimentally. Apparently the number of holes in $CaLaBaCu_3O_7$ is much less sensitive to pressure than for $YBa_2Cu_4O_8$.

It is thus clear that a large $|dT_c/dp|$ can be taken as an indication that the optimal T_c as function of composition has not yet been reached for the material studied. Experiments at high pressure are thus very helpful in determining the potential of a group of materials. Furthermore, experiments at high pressure yield extra parameters (such as dT_c/dp and dH_{c2}/dp) which can be used to test theoretical models.

See also: Superconducting Materials: An Overview; Superconducting Materials: BCS and Phenomenological Theories; Superconducting Materials: Elements; Superconducting Materials: Physical Properties

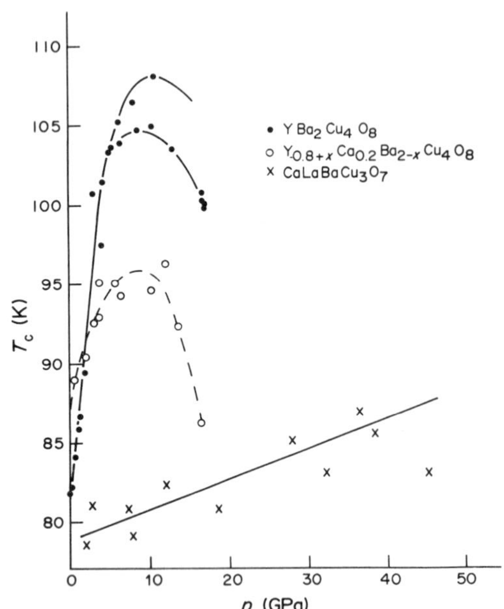

Figure 5
Critical temperature T_c of $CaLaBaCu_3O_7$, $Y_{0.8+x}Ca_{0.2}Ba_{2-x}Cu_4O_8$ and two samples of $YBa_2Cu_4O_8$ as a function of pressure (after Van Eenige et al. 1990).

Bibliography

Ashcroft N W 1981 Ordered ground states of metallic hydrogen and deuterium. In: Schilling J S, Shelton R N (eds.) 1981 *Physics of Solids Under High Pressure*. North-Holland, Amsterdam, pp. 155–60

Benneman K H, Garland J W 1972 Theory for superconductivity in *d*-band metals. In: Douglas D H (ed.) 1972 *Superconductivity in d- and f-band Metals*. American Institute of Physics, New York, pp. 103–37

Hemmes H K, Driessen A, Griessen R, Gupta M 1989 Isotope effects and pressure dependence of the T_c of superconducting stoichiometric PdH and PdD synthesized and measured in a diamond anvil cell. *Phys. Rev. B* 39: 4110–18

Jerome D 1982 Organic superconductivity. *Physica B* 109, 110: 1447–60

Little W A 1964 Possibility of synthesizing an organic superconductor. *Phys. Rev. A* 134: 1416–24

Smith T F 1972 Determination of pressure dependence of T_c for *d*- and *f*-band superconductors. In: Douglas D H (ed.) 1972 *Superconductivity in d- and f-band Metals*. American Institute of Physics, New York, pp. 293–319

Van Eenige E N, Griessen R, Wijngaarden R J, Karpinski J, Kaldis E, Rusiecki S, Jilek E 1990 Superconductivity at 108 K in $YBa_2Cu_4O_8$ at pressures up to 12 GPa. *Physica C* 168: 482–8

Vining C B, Shelton R N 1984 Superconductivity in ternary compounds at high pressure. In: Homan C, MacCrone R K, Whalley E (eds.) 1984 *High Pressure in Science and Technology*. Materials Research Society Symp. Proc., Vol. 22. MRS, Pittsburgh, PA, Pt. 1, pp. 29–38

Wijngaarden R J, Griessen R 1989 High pressure studies. In: Narlikar A (ed.) 1989 *Studies of High Temperature Superconductors*, Vol. 2, Nova, Commack, NY, pp. 29–77

Wijngaarden R J, Scholtz J J, Van Eenige E N, Griessen R 1991 Can high pressure experiments on high T_c superconductors put theory under pressure? *High Press. Res.* 7: 33–7

Wijngaarden R J, Scholtz J J, Van Eenige E N, Griessen R 1992 High-T_c superconductors under very high pressure. In: Hocheimer H D, Etters R D (eds.) 1992 *Frontiers of High Pressure Research*, NATO ASI Series, Vol. 286. Plenum, New York, pp. 399–417

Wijngaarden R J, Van Eenige E N, Scholtz J J, Griessen R 1990 High temperature superconductors under pressure: experimental review and test on the RVB model. *High Press. Res.* 3: 105–7

Wittig J 1984 Superconductivity in elements and binary systems at high pressure. In: Homan C, MacCrone R K, Whalley E (eds.) 1984 *High Pressure in Science and Technology*. Materials Research Society Symp. Proc. Vol. 22. MRS, Pittsburgh, PA, Pt. 1, pp. 17–28

R. J. Wijngaarden and R. Griessen
[Vrije Universiteit Amsterdam, Amsterdam, The Netherlands]

T

Temperature Compensation Materials

Temperature compensation materials are magnetic materials whose permeability falls rapidly with increasing temperature so as to compensate for the temperature effects of a permanent magnet in a magnetic circuit. The compensation material is used as a shunt to divert a portion of the magnetic flux from the air gap. At low temperature, the shunt diverts a greater portion of the flux on account of a relatively high permeability. As the temperature rises, the falling permeability of the shunt diverts less of the flux from a weaker permanent magnet, thus maintaining a constant flux across the air gap.

As most devices operate near room temperature, common temperature compensation materials are iron–nickel alloys in the 30–32% nickel range, where the Curie temperatures are in the vicinity of 40–200 °C. Figure 1 shows the permeability at a magnetic field strength of 37 A cm^{-1} as a function of temperature for two types of commercial temperature compensation alloys. Since the Curie temperature falls steeply with composition, the control of nickel content is very critical for these alloys. A change of 0.25% nickel can result in a change of 10 °C in Curie temperature.

In addition, the flux density increases with magnetic field strength. Therefore, the permeability is dependent on the field strength and can increase or go through a maximum with temperature at low fields, as shown in Fig. 2.

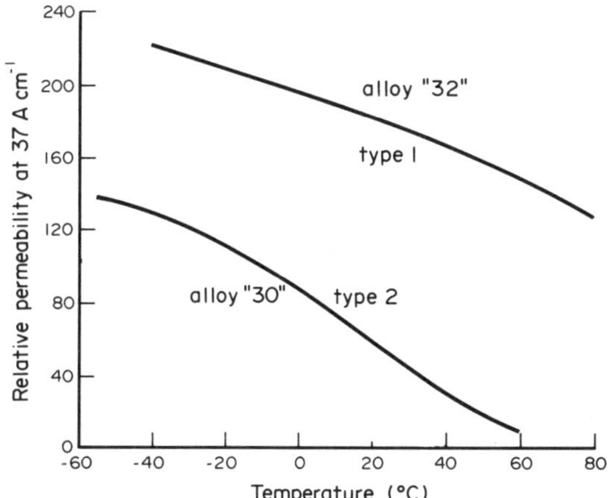

Figure 1
Permeability at a magnetic field strength of 37 A cm^{-1} as a function of temperature for two types of commercial nickel–iron temperature compensation alloys. Type 1 contains ~32% nickel and type 2 contains ~30% nickel

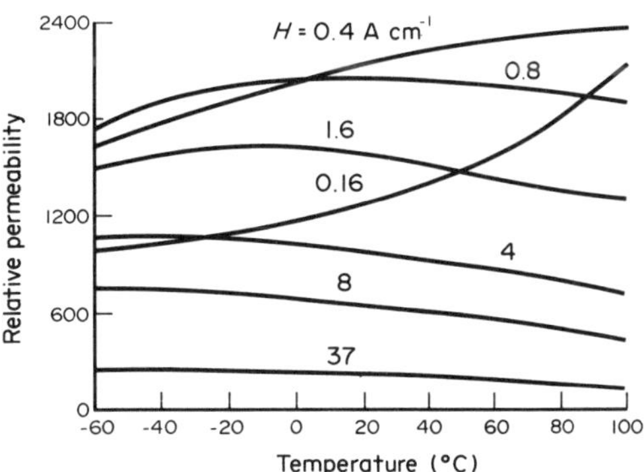

Figure 2
Permeability–temperature curves at various values of magnetic field strength H for alloy "32" (type 1)

Other materials having similar properties are Fe–Ni–Mo, Fe–Ni–Cr and nickel–copper alloys, as well as soft ferrites. Chromium and molybdenum additions tend to reduce the sensitivity of the alloy to small variations in nickel content.

Temperature compensation magnetic circuits are often employed in measuring instruments, relays, voltage regulators, tachometers and speedometers.

See also: Magnetic Materials: An Overview

Bibliography

Chin G Y, Wernick J H 1980 Soft magnetic metallic materials. In: Wohlfarth E P (ed.) 1980 *Ferromagnetic Materials*, Vol. 2. North-Holland, Amsterdam, pp. 55–188

Heck C 1974 *Magnetic Materials and Their Applications*. Crane, Russak & Co., New York, Chap. 18

G. Y. Chin†
[AT&T Bell Laboratories, Murray Hill, New Jersey, USA]

J. H. Wernick
[Bellcore, Red Bank, New Jersey, USA]

Thin Films and Multilayers, Magnetic

Modern magnetic thin-film and multilayer research is focused on the unique properties of surfaces and interfaces. This has created a renaissance of interest in a field with a long-standing tradition. Materials of prime interest include magnetic surfaces, overlayers and sandwiches, and repetitive multilayers that form superlattices and compositionally modulated films.

Motivations for the renewed interest are attributed to the discovery of new materials, new properties and new techniques. These developments have been made possible by a number advances: advances in vacuum technology have ushered in refined film-growth methods based on sputtering and on ultrahigh-vacuum evaporation, also known as molecular-beam epitaxy. This, in turn, has given rise to sophisticated characterization techniques, such as the plethora of electron spectroscopies and diffraction approaches that have bootstrapped the entire thin-film field, and specialized spin-polarized electron adaptations that are sensitive to surface magnetic phenomena and that can be spatially resolved on the submicrometer-to-nanometer scale. Materials scientists are intensely interested in new and altered phases that can be epitaxially stabilized. Magnetic-information technologists are eager to adapt artificially layered magnetic materials to the marketplace. Computationalists now have high-performance machines to aid their ability to make realistic predictions that synergistically focus experimental efforts.

Comprehensive introductions to traditional thin-film magnetism are given by Prutton (1964) and Soohoo (1965). Recent topical overviews are given by Bader (1990), Dürr et al. (1990), Falicov et al. (1990), Fert (1990) and Prinz (1990). The field is rapidly evolving so progress is conveyed herein by highlighting prototypical examples, rather than by attempting all-inclusive tabulations.

1. Surface Anisotropy, Epitaxy and Domain Imaging

The surface magnetic anisotropy has been predicted and measured to be qualitatively different from that of the bulk (see *Anisotropy in Magnetic Materials*). Perpendicular (or vertical) easy axes of magnetization have been reported for a number of monolayer-range films of iron, for example, grown epitaxially on various substrates. As the thickness of the iron overlayer increases a critical thickness d_c is reached where the easy axis reorients in plane due to the demagnetizing field $M_s \simeq 2\ \mathrm{MA\ m^{-1}}$ (20 kOe). To provide a realistic range of the parameters of interest, it is estimated from the expression $2H_k^s \sim M_s d_c$ that for a d_c value of 5 monolayers (ML) the surface anisotropy field H_k^s is $\sim 5\ \mathrm{MA\ m^{-1}}$ where $2H_k^s$ takes into account the two interfaces (iron–vacuum and iron–substrate). Present activities concern the origin of this strong surface-anisotropy field. Band structure calculations (Gay and Richter 1986) relate it to the spin–orbit interaction at the surface. Epitaxial strain is also considered a stabilizing factor. These considerations need further exploration using highly accurate numerical techniques and microscopy correlations to detect strain-relieving misfit dislocations.

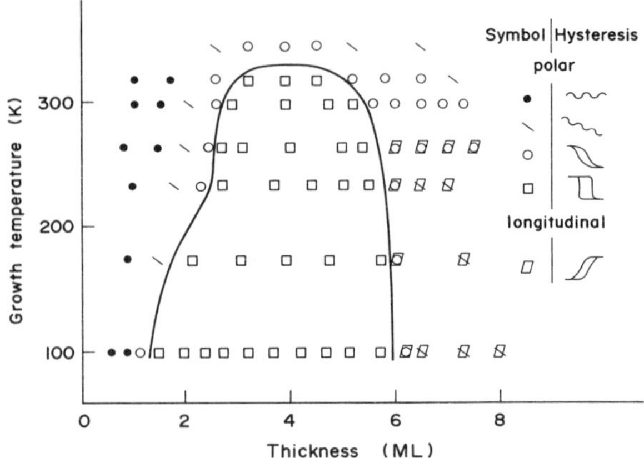

Figure 1
Growth conditions to stabilize vertical easy axes of magnetization in ultrathin fcc Fe–Cu(100) films, based on surface magnetooptic Kerr effect measurements by Liu et al. 1988

Figure 1 shows an anisotropy "phase diagram" constructed from magnetooptic hysteresis loops for iron grown on Cu(100). The region of vertical easy axes is delineated in the figure; the d_c value is 6 ML and, for the thinnest films and the highest growth temperatures, the loss of vertical easy axes is attributed to interfacial mixing effects. Iron grows pseudomorphically on Cu(100) in the fcc or γ phase, which is a bulk iron phase that is thermodynamically stable at high temperatures. Figure 2 summarizes results for four epitaxial systems; approximate d_c values are indicated in the left panels and the high coercivities H_c that peak in the ML range are shown in the right panels. The iron initially grows pseudomorphically on each substrate. This leads to the bcc or α phase of iron on Au(100), a new body-centered-tetragonal (bct) iron phase on Pd(100), and trigonally distorted Fe(111) layers on Ru(0001).

The systems considered in Fig. 2 display hysteresis loops with high remanence when vertically magnetized (cf. Fig. 3 for Fe–Pd(100)). However, the vertical magnetization can decay into an equilibrium domain structure in some systems. This case has been treated theoretically and was recently observed in imaging experiments on Co–Au(111) films. The image in Fig. 4 was obtained using a scanning electron microscope with polarization analysis (SEMPA) (see *Magnetization Imaging*). The experiment involves a fine electron beam that impinges on the sample to emit low-energy secondary electrons whose polarization is measured with a spin detector based on Mott-scattering principles. The Mott detector measures the backscattering asymmetry of the emitted electrons after they are accelerated into a heavy (high-Z) target. A left (right) scattering asymmetry results from the spin–orbit in-

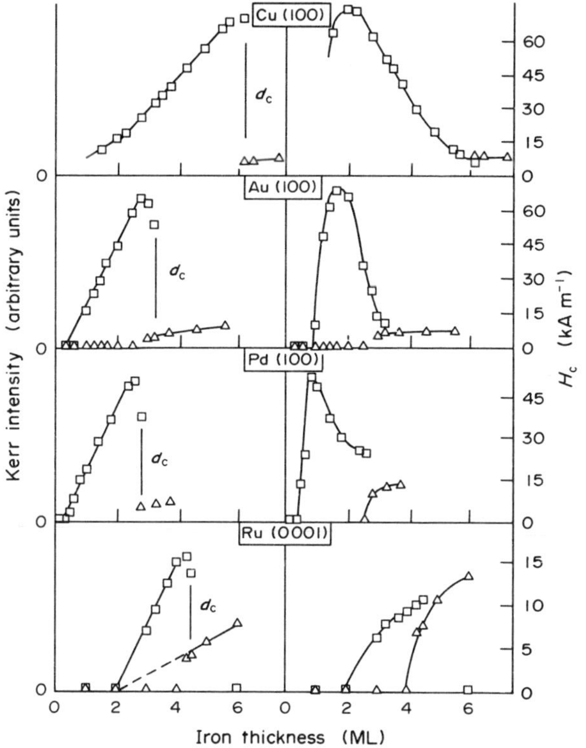

Figure 2
Critical thicknesses for vertical easy axes (left) and coercivities (right) for epitaxial iron films grown on the indicated substrates. Squares refer to polar and circles to longitudinal Kerr hysteresis-loop heights and half-widths (see Bader 1990)

teraction due to the spin-up (down) polarization asymmetry. The imaging is surface sensitive on a nanometer scale because of the limited escape depth of low-energy (~100 eV) electrons. SEMPA is a powerful new probe with an ultimate spatial resolution (~0.01 μm) that permits imaging even across domain walls.

2. *Magnetic Ordering Temperatures and Magnetic Moments*

Epitaxial magnetic overlayers serve as ideal realizations of two-dimensional Heisenberg spin systems. For an isotropic two-dimensional Heisenberg model there is no ordering at finite temperature. This is known as the Mermin–Wagner theorem (Mermin and Wagner 1966). The systems considered in Sect. 1 have strong anisotropies that trigger their finite Curie temperatures T_C. It is known theoretically that even an arbitrarily small anisotropy endows such systems with two-dimensional Ising-like transitions. The surface anisotropy is uniaxial and induces vertical easy axes, while systems possessing in-plane easy axes might also be described by *xy* models. Figure 5 shows the variation in T_C vs iron thickness for Fe–Pd(100). The thinnest films have vertical easy axes, while the thicker films are magnetized in plane. A novelty of the system is that band structure calculations predict that the iron induces magnetic moments in the interfacial palladium layers. This is analogous to the well-known giant-moment effect for dilute iron alloys in a palladium host. The measured T_C values in Fig. 5 are substantial. They extrapolate to a bulk value denoted $T_{C\mathrm{bulk}*}$ that differs from that of ordinary iron because of the new epitaxial structure. The dashed curves in Fig. 5 are for Monte-Carlo simulations of two-dimensional Ising models (using the related bcc structure) as a function of the coupling constant between the iron spins and interfacial palladium spins (see also the seminal treatment by Binder and Hohenberg 1976). The comparison in Fig. 5 indicates that coupled palladium moments are needed to obtain agreement with experiment. Thus, induced moments in ordinarily paramagnetic substrates have profound influence on the magnetic properties. Another aspect of Fig. 5 is that the submonolayer trend in T_C values is indicative of an interconnected surface morphology with random site vacancies, rather than the more familiar morphology of disconnected two-dimensional islands. The influence of epitaxial strain on the T_C variation is also an active area of concern.

Early researchers identified magnetic "dead" layers with no moments at the surface, but these were later attributed to contamination problems. Band structure computationalists now tend to predict enhanced surface moments relative to bulk values. For 1-ML Fe–Pd(100), the interfacial moments calculated for palla-

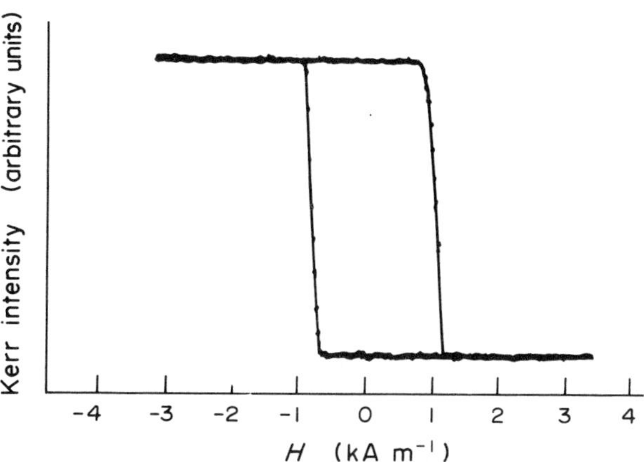

Figure 3
Hysteresis loop for 1.2 ML of Fe–Pd(100) with vertical easy axes (see Bader 1990)

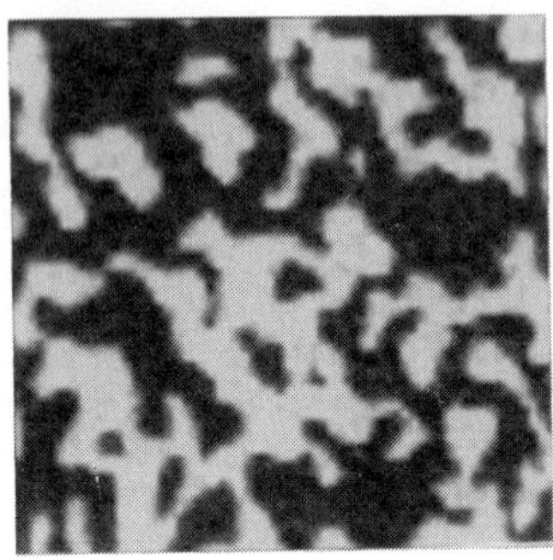

Figure 4
Magnetic domain structure of a 3-ML Co–Au(111) film showing vertically oriented domains, where white are "up" and black are "down" domains (after Allenspach et al. 1991)

dium (Blügel et al. 1989) are around $0.3\,\mu_B$, the next palladium layer down has around $0.15\,\mu_B$, and the iron moments are enhanced to around $3.1\,\mu_B$ vs $2.2\,\mu_B$ for bulk iron. Enhanced moments at the surface of itinerant magnets are attributed to reduced coordination; the surface is envisioned to represent a case intermediate between that of full moment formation for a free atom and reduced moment formation of a bulk atom due to *d*-band hybridization. The challenge is to provide direct experimental confirmation of the theoretical predictions of enhanced moments. This has yet to be satisfactorily achieved because it requires quantitative measurements, unlike the domain-imaging challenge that requires qualitative contrast or the anisotropy question that requires a qualitative distinction between in-plane and vertical easy axis. The theoretical techniques for predicting moments at surfaces and in epitaxial phases involve two different band-structure approaches to self-consistently evaluate the total energy. The approach used in the Fe–Pd(100) case is based on "slab" codes, which can take into account hybridization with the substrate and differences in interlayer separations due to epitaxial strain and/or surface relaxation. The other approach uses bulk codes and considers artificial three-dimensional structures with unusual lattice spacings. A rich example of the latter is shown in Fig. 6 for the γ-Fe structure. The calculations identify nonmagnetic (NM), high-spin (HS), low-spin (LS) and antiferromagnetic (AF) phases as a function of Wigner–Seitz (WS) radius. Experimental points are indicated at the copper lattice spacing for (a) AF γ-Fe coherent precipitates (circle) in a copper-rich alloy, (b) an early magnetometry measurement (triangle) on epitaxial Fe–Cu(111), and (c) a very recent estimate (hatched) for Fe–Cu(100) based on an analysis of the spin polarization of the secondary-electron background measured in photoemission experiments using synchrotron radiation obtained from an undulator source. An alternative to testing moment predictions has become to test band structure predictions of hyperfine fields at surfaces. This quantity is accessible via Mössbauer spectroscopy (Gradmann et al. 1989), especially if conversion-electron detection is utilized to enhance surface sensitivity.

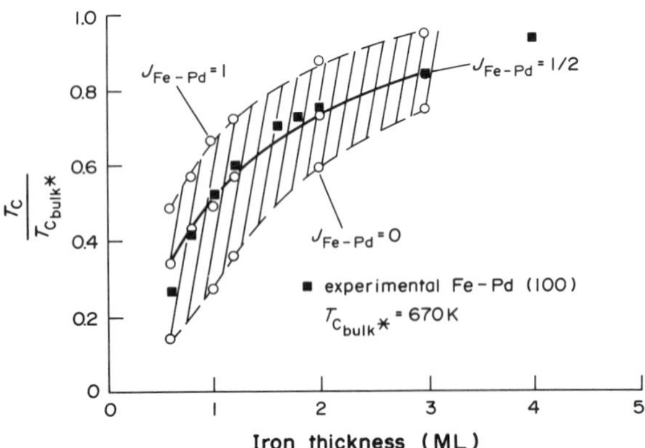

Figure 5
T_C trends in Fe–Pd(100) vs iron thickness: squares are experimental (Liu and Bader 1990), circles and curves are Monte-Carlo calculations with different couplings $J_{Fe\text{-}Pd}$ between the iron and induced palladium spins (courtesy of K. J. Strandburg)

A celebrated moment-enhancement case (see Falicov et al. 1990) involves the surface of Cr(100), for which a ferromagnetic value of up to around $3.6\,\mu_B$ is calculated vs the AF bulk value of about $0.6\,\mu_B$. The experimental difficulty is that the giant ferromagnetic moments at the surface are expected to decay to the bulk value over the distance of a number of layers that are AF coupled to each other. Given that even carefully prepared samples have atomic steps about every 10–20 nm, extreme spatial resolution is needed to prevent averaging over adjacent terraces of opposite spin orientation. Recently, a spin-polarized scanning tunnelling microscope (STM) was used to verify that adjacent Cr(100) terraces are oppositely spin aligned. STM is noted for the possibility of providing atomic resolution. The STM was operated in a constant-current mode in which a piezodriven servosystem adjusts the height of the tip as it is scanned across the surface. Using a magnetized CrO_2 tip yielded effective step heights that alternate between $h_1 = h + \delta$ and $h_2 = h - \delta$, where $h = 0.14$ nm is the actual step height and $\delta = 0.02$ nm. The values alternate because the spin-dependent contribution to the tunnel current increases or decreases as the relative spin-orientation of the tip and terrace change from being parallel to antiparallel aligned (see Fig. 7). A determination of the absolute moment value is still elusive and would require a quantitative analysis of the tunnelling processes.

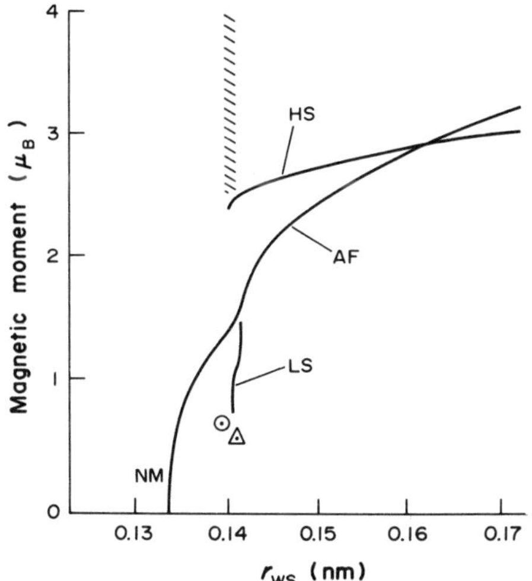

Figure 6
The magnetic moment of γ-Fe vs Wigner–Seitz radius (courtesy of P. D. Johnson): the calculated curves are after Morruzi et al. (1989)

3. Spin-Wave and Critical Behavior

The Bloch $T^{3/2}$ law describes how three-dimensional ferromagnetic spin waves govern the approach of the bulk magnetization M to its low-T limiting form. In two dimensions this becomes a linear approach when anisotropy is appropriately incorporated into the derivation. For a semi-infinite solid the surface magnetization still obeys a $T^{3/2}$ law, but the prefactor, or stiffness constant, can be strikingly altered (e.g., by factors of three) since it becomes sensitive to the ratio of the surface-to-bulk magnetic coupling constants (Mathon and Phan 1989). Bulk AF materials are generally expected to exhibit a T^3 law. Anisotropically coupled AF materials should obey a T^2 law when they simulate weakly coupled layers (as in undoped La_2CuO_4, a naturally layered material which is the precursor to the high-temperature superconducting system (see *Oxide Superconductors: Physical Properties*)). The T^2 law is predicted to cross over to a quasi-two-dimensional $T \ln T$ behavior as thermal fluctations overcome the weak interlayer coupling (Singh et al. 1990). The verification and elucidation of these expectations represents another active area of research.

At high T the epitaxial monolayer-range systems act as anisotropic two-dimensional Heisenberg systems and belong to the two-dimensional Ising-type universality class. Recent experimental determinations of the effective critical exponents yield two-dimensional Ising values (Liu and Bader 1990), but unusual and perplexing values have also been reported (Dürr et al. 1990) and are the subject of further studies. The two-dimensional Ising magnetization exponent is $\beta_c = 1/8$, where $M \rightarrow (1 - T/T_C)^{\beta_c}$ as $T \rightarrow T_C$. While monolayer-range films are ideal for testing critical phenomena predictions, thicker films should cross over to two-dimensional behavior near T_C since the correlation length diverges as T_C is approached. Critical-phenomena measurements are facilitated in two-dimensional systems because the region that exhibits critical fluctuations seems to be particularly broad (> 10% of T_C), and the $\beta_c = 1/8$ exponent is so steep that M remains large over much of this range.

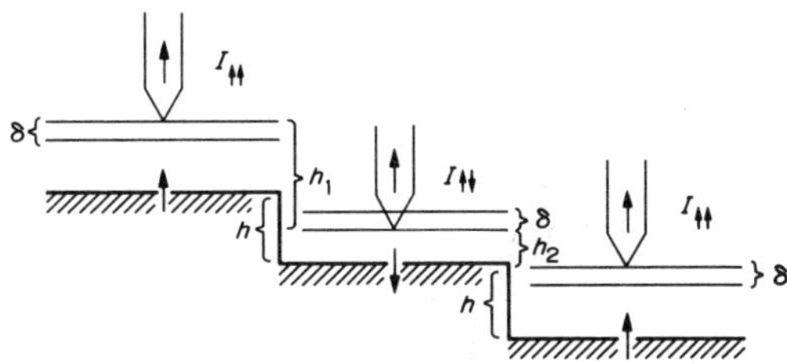

Figure 7
Schematic of a spin-polarized STM scan over alternately magnetized terraces of Cr(100) surface separated by monatomic steps (after Wiesendanger et al. 1990)

4. Magnetooptics and Magnetotransport in Multilayers

Superlattices and multilayers are the materials of choice for applications, as well as for many basic studies. This is because they do not require preservation in stringent vacuum conditions and they are amenable to conventional characterization techniques, such as x-ray and neutron diffraction, magnetometry, and so on. Although these systems possess some of their own unique characteristics (such as new spin-wave branches observed in Brillouin light scattering as discussed by Falicov et al. (1990)), they also preserve and amplify interesting surface and interfacial properties. Two types of superlattices will be considered: candidate materials for magnetooptic media, and a new class of materials that exhibits oscillatory ferromagnetic–AF couplings as a function of the thickness of the component that separates the magnetic layers. The former system is typified by cobalt–platinum and the latter by iron–chromium, cobalt–rutherium or cobalt–copper, which also exhibit giant, negative magnetoresistive anomalies when AF coupled.

Interest in magnetooptic (MO) films stems from their recent commercial viability as erasable optical media (see *Magnetooptic Storage Media*). MO media complement conventional magnetic-storage media due to their high bit-packing densities and their removability and portability. The current generation utilizes amorphous rare-earth–transition metal (RETM) alloys (e.g., a-TbFeCo). Undercoatings and overcoatings are used to tune optical and thermal properties (e.g., to confine heat dissipation in magnetothermal writing), as well as corrosion resistance and tribological properties. Amorphous active layers are useful to reduce problematic media noise, which is adversely affected by grain boundaries. The RE component provides a single-atom anisotropy that yields vertical easy axes in the media, but does not contribute significantly to the MO signal at optical wavelengths because the localized $4f$ levels are deeply bound (~ 7 eV). A vertical medium is desirable because the "read" operation relies on the polar Kerr signal which is much stronger than the other (longitudinal or transverse) Kerr signals.

Epitaxial TM superlattices may become attractive alternatives to the a-RETM media. Vertical media can be stabilized via the surface anisotropy, eliminating the need for the RE component. Recent interest involves cobalt–platinum superlattices (see Fig. 8). The signal at short wavelength (4 eV) is comparable with that of the bulk cobalt film, which is unusable because its vertical axes are hard. The rotation φ_K of around 0.4° is less than that of a-TbFeCo ($\sim 1°$), but the reflectivity R is higher, yielding a comparable figure of merit ($\varphi_K{}^2 R$). The important aspect to emphasize is that the platinum is strongly magneto-optically active due to its induced moments at the interface, which are similar in magnitude to those of palladium. Platinum outperforms palladium, however, because the MO signal arises from the spin–orbit interaction, which is enhanced in platinum because of its high Z value. The particular enhancement at 4 eV photon energy in Fig. 8 is attributed to the total spin polarization of the bottom of the platinum d band. Thus, surface and interfacial properties control many

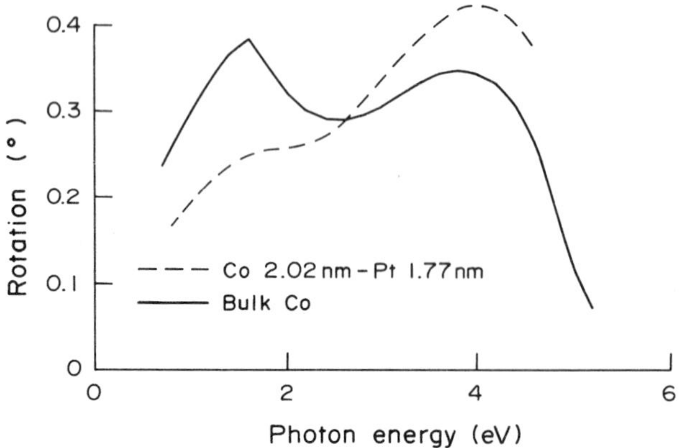

Figure 8
Polar Kerr rotation of a cobalt–platinum superlattice and a cobalt film (after Zeper et al. 1989)

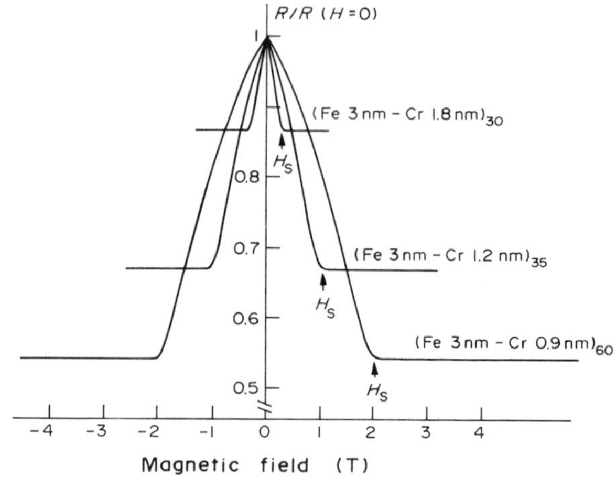

Figure 9
Magnetoresistance of iron–chromium superlattices at 4.2 K (see Fert 1990)

of the desirable features of this emerging prototype.

The new class of magnetic materials with AF interlayer magnetic coupling have intriguing magnetoresistive properties (see *Magnetotransport Phenomena*), as shown in Fig. 9 for iron–cobalt superlattices. Although the properties of these materials are not well understood, the materials systematics that are unfolding are serving to guide the formulation of theoretical models. Each iron layer is ferromagnetic, but adjacent iron layers can AF couple if the thickness of the chromium spacer layer is set to around 1.2 nm, or a multiple thereof. The effect does not require the existence of chromium magnetic moments, because cobalt–ruthenium and cobalt–copper behave similarly. The spacer layer may, however, be affected by the presence of intrinsic or dissolved magnetic atoms. Also, a superlattice configuration is not prerequisite for the effect; an Fe–Cr–Fe sandwich, for example, will suffice, although the magnitude of the transport anomaly tends to be larger in superlattices due to the preponderance of interfaces. Intrinsic AF coupling also is not prerequisite for the transport anomaly; ferromagnetic Co–Au–Co sandwiches with independent H_c values for the two cobalt layers exhibit analogous transport anomalies in the magnetic field region between the two H_c values, where the layers are coupled antiparallel. The transport anomaly is related to strong spin-dependent scattering processes at the interfaces. If alternate magnetic layers have opposite spin orientation, both spin-up and spin-down carriers will encounter interfaces at which they scatter strongly. However, if a magnetic field aligns the spin orientation of all the magnetic layers in the same direction, one carrier type will experience a weaker scattering cross section and will "shunt" the current (the phenomenon is sometimes referred to as a "spin valve" effect). Thus, the ferromagnetically aligned films have reduced resistance with respect to the AF-aligned films, as is shown in Fig. 9. Research in this area has focused on a number of objectives:

(a) to develop microscopic models of the transport phenomena;

(b) to explore the RKKY coupling mechanism, and its modifications due to interfacial roughness and Fermi-surface effects (see *Magnetic Materials: Basic Concepts*);

(c) to search for additional members of this class of materials in order to gain insights from materials systematics; and

(d) to explore film processing parameters that result in optimally enhanced magnetoresistive anomalies.

The practical motivation is that these materials can be used as miniature magnetoresistive sensors, such as in magnetic recording pick-up heads.

The vigorous activity in the field of magnetic thin films and multilayers assures that the materials challenges of the twenty-first century will be met, in part, by a new generation of artificially layered magnetic materials based on new, epitaxially stabilized structures.

Acknowledgement

This work has been supported by the US Department of Energy, BES-Materials Sciences, under contract number W-31-109-ENG-38.

See also: Magnetic Materials: Basic Concepts; Magneto-optic Storage Media; Magnetotransport Phenomena

Bibliography

Allenspach R, Stampanoni M, Bischof A 1991 Magnetic domains in thin epitaxial Co/Au(111) films. *Phys. Rev. Lett.* 65: 3344–7

Bader S D 1990 Thin film magnetism. *Proc. IEEE* 78: 909–22

Binder K, Hohenberg P C 1976 Monte Carlo and series expansion investigations of magnetic surfaces. *IEEE Trans. Magn.* 12: 66–74

Blügel S, Drittler B, Zeller R, Dederichs P H 1989 Magnetic properties of 3*d* transition metal monolayers on metal substrates. *Appl. Phys. A* 49: 547–62

Dürr W, Kerkmann D, Pescia D 1990 Experimental advances in two-dimensional magnetism. *Int. J. Mod. Phys. B* 4: 401–35

Falicov L M, Pierce D T, Bader S D, Gronsky R, Hathaway K B, Hopster H J, Lambeth D N, Parkin S S P, Prinz G, Salamon M, Schuller I K, Victora R H 1990 Surface, interface, and thin-film magnetism. *J. Mater. Res.* 5: 1299–340

Fert A 1990 Magnetic and transport properties of metallic multilayers. *Mater. Sci. Forum* 59–60: 439–80

Gay J G, Richter R 1986 Spin anisotropy of ferromagnetic films. *Phys. Rev. Lett.* 56: 2728–31

Gradmann U, Przybylski M, Elmers H J, Liu G 1989 Ferromagnetism in the thermodynamically stable monolayer Fe(110) on W(110), coated by Ag. *Appl. Phys. A* 49: 563–71

Li C, Freeman A J, Jansen H J F, Fu C L 1990 Magnetic anisotropy in low-dimensional ferromagnetic systems: Fe monolayers on Ag(001), Au(001) and Pd(001) substrates. *Phys. Rev. B* 42: 5433–42

Liu C, Bader S D 1990 Two-dimensional magnetic phase transition of ultrathin iron films on Pd(100). *J. Appl. Phys.* 67: 5758–60

Liu C, Moog E R, Bader S D 1988 Polar Kerr-effect observation of perpendicular surface anisotropy for ultrathin fcc Fe grown on Cu(100). *Phys. Rev. Lett.* 60: 2422–5

Mathon J, Phan M S 1989 Temperature dependence of the magnetisation in a magnetically inhomogeneous surface. *J. Phys. Condens. Matter* 1: 7983–7

Mermin N D, Wagner H 1966 Absence of ferromagnetism or antiferromagnetism in one- and two-dimensional isotropic Heisenberg models. *Phys. Rev. Lett.* 17: 1133–6; 1307

Moruzzi V L, Marcus P M, Kübler J 1989 Magnetovolume instabilities and ferromagnetism versus antiferromagnetism in bulk fcc iron and manganese. *Phys. Rev. B* 39: 6957–61

Parkin S S P, More N, Roche K P 1990 Oscillations in exchange coupling and magnetoresistance in metallic superlattice structures: Co/Ru, Co/Cr, and Fe/Cr. *Phys. Rev. Lett.* 64: 2304–7

Prinz G A 1990 Hybrid ferromagnetic–semiconductor structures. *Science* 250: 1092–7

Prutton M 1964 *Thin Ferromagnetic Films.* Butterworth, London

Singh A, Tesanovic Z, Tang H, Xiao G, Chien C L, Walker J C 1990 Magnetic dynamics in copper-oxide-based antiferromagnets: the role of interlayer coupling. *Phys. Rev. Lett.* 64: 2571–4

Soohoo R F 1965 *Magnetic Thin Films.* Harper and Row, New York

Wiesendanger R, Güntherodt H-J, Güntherodt G, Gambino R J, Ruf R 1990 Observation of vacuum tunneling of spin-polarized electrons with the scanning tunneling microscope. *Phys. Rev. Lett.* 65: 247–50

Zeper W B, Greidanus F J A M, Carcia P F, Fincher C R 1989 Perpendicular magnetic anisotropy and magneto-optical Kerr effect of vapor deposited Co/Pt-layered structures. *J. Appl. Phys.* 65: 4971–5

S. D. Bader
[Argonne National Laboratory,
Argonne, Illinois, USA]

Thin Films, Multilayers and Devices, Superconducting

Thin films are an essential part of all electronic technologies. This is particularly true for superconducting instrumentation, sensors and electronics, in which the active devices themselves are made from deposited films, often a sequence of films made as a multilayer structure. This is in contrast to some semiconductor technologies, in which devices are often created by doping of bulk single crystals, say silicon or GaAs, and are then linked together with interconnects and to passive components (capacitors, resistors) that are made from deposited films. In the most highly developed superconducting technology, namely Josephson junction circuits operating at 4.2 K, the entire circuit of junctions, resistors, capacitors, insulators and interconnects is made from films that are deposited at room temperature (a significant difference from semiconductor processing, which is carried out at high temperatures). Thus, the control of thin-film processes is especially critical to the success of superconducting electronics technologies. The important film synthesis methods, the properties of superconducting films and the devices made from them are the subjects of this article, which therefore forms a link between articles that are concerned predominantly with different materials or physical phenomena and articles that deal with low-current applications.

The very significant differences between low-temperature superconducting (LTS) materials and high-temperature superconducting (HTS) materials have already resulted in some divergence in the approaches being used to make, understand and use LTS and HTS films. In each section of this article LTS films will be described first, then the differences in the HTS cases will be identified. It is important to note that LTS film studies and technology date from the late 1950s and are, hence, well established, whereas HTS films are very recent and the field is subject to rapid change.

1. Thin Films

1.1 Thin-Film Synthesis

The various deposition methods that are available to make thin films of all kinds can be readily used for superconductors. These methods are thermal and electron-beam evaporation and sputtering. Many elemental superconductors that have been used for both science and applications studies (e.g., lead, tin, indium, aluminum, zinc) can be made as films by the simplest methods, which is thermal evaporation in a moderate vacuum (10^{-4} Pa or 10^{-6} torr) onto a substrate at room temperature. The nature of the substrate is not critical, glass is quite adequate. Alloys of these elements can be made by evaporation from two sources or sometimes (e.g., PbBi) from a single source.

Films of greater technological interest (e.g, niobium, tantalum and their alloys with other transition metals) are not easy to evaporate thermally, but electron-beam evaporation and sputtering are straightforward for such materials. Magnetron sputtering sources have become the method of choice for the niobium films used in Josephson circuits (see *Superconducting Device Fabrication*). Alloys can be made from a sputtering target of the appropriate composition. The superconducting properties of transition metals are more sensitive to impurities, such as oxygen, than the *s–p* metals and a better vacuum system must be used, say $\leqslant 10^{-6}$ Pa (10^{-8} torr). The transition metals make excellent getters, so placing a liquid-nitrogen-cooled surface near the substrate very effectively lowers the system pressure. The substrate temperature can be 300 K for niobium film deposition, but the resistance ratio of clean films made at this temperature will be limited by the grain size. Higher substrate temperatures are needed for some transition metals (e.g., tantalum) and to increase the grain size of niobium. In particular, high-quality epitaxial films with long electronic mean free paths, such as those used in SIS particle detectors, and epitaxial superlattices with mean free paths extending over many layers (see later) are grown at temperatures as high as 800 °C.

To form films of compounds based on niobium which have higher critical temperatures, such as NbN, NbCN (see *Nitride and Carbonitride Superconductors*) and the family of A15 compounds Nb_3Sn, Nb_3Ge, Nb_3Al, (see *A15 Compound Superconductors*) somewhat more complicated methods are used. The nitrides and carbonitrides, which are important materials for Josephson circuits operating at 8–10 K, are made by sputtering niobium targets in argon plus nitrogen or nitrogen plus methane onto substrates at 100–1000 °C (Braginski and Talvacchio 1990). The A15 compounds have generally been made by coevaporation from electron-beam sources, say niobium and tin, but cosputtering from multiple targets or sputtering from a single mixed target can also be used. The geometry of the evaporation sources (see Fig. 1) creates a spread of compositions across an array of substrates (Hammond 1975). This allows measurements to explore the effects of stoichiometry on the physical properties of the films. To form the crystalline structure of the compound, a heated substrate has to be used. Typical optimum temperatures are 880 °C for Nb_3Sn, 900 °C for Nb_3Ge. These high deposition temperatures have limited the interest in A15 compounds for electronics applications, although values over 700 °C are now accepted as necessary for HTS films.

To prepare films of the HTS materials, many of the techniques described in this section are being used (see *Oxide Superconductors: Thin-Film Deposition*). The presence of three elements plus oxygen in the HTS materials leads to increased complexity of the deposition systems, particularly those for cosputtering, sequential sputtering (a fraction of a monolayer of each material is deposited sequentially and repeated many times), coevaporation (MBE) and sequential evaporation. Two comparatively simple methods have quickly become popular and reproducibly yield very-high-

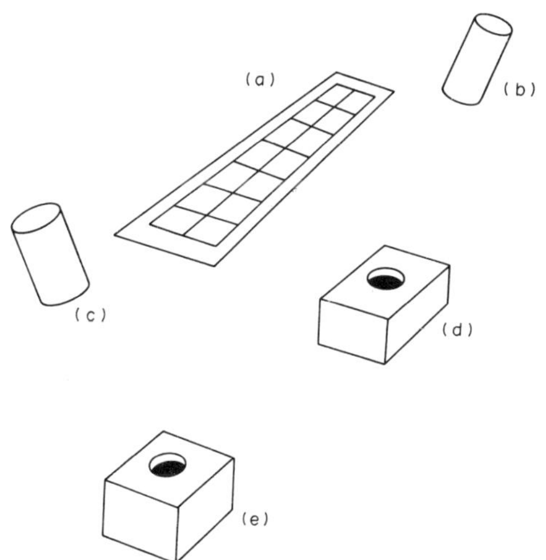

Figure 1
Arrangement of sources and substrates to achieve a spread of compositions: (a) array of substrates heated to 800 °C; (b) niobium deposition-rate monitor; (c) tin deposition-rate monitor; (d) niobium electron-beam source; and (e) tin electron-beam source

quality films. These are off-axis sputtering (Eom et al. 1989) and laser ablation (Inam et al. 1988). As in the case of A15 compounds, HTS films are deposited at high temperatures (600–750 °C) with the best films being made at ⩾ 700 °C. Unlike the transition metals, HTS films are deposited in high oxygen pressures (sometimes 250 Pa) and, hence, the base pressure of the vacuum system is less relevant. One film process that is widespread for semiconductors, chemical vapor deposition (Li et al. 1991), is being tried for HTSs with some success. It has the potential to be a commercial process for films of large area. At present typical film substrate sizes are 4 in or 5 in diameter wafers for niobium circuits, with plans in one laboratory to move to 8 in and 2 in diameter for HTS devices.

A very significant difference between LTS and HTS films is in the choice of substrates. Epitaxy between the substrate and film is essential for HTS materials, hence, a very much more limited choice of substrates is available. Because of the high deposition temperatures, substrates for both HTS and A15 films must not react chemically with the film at temperatures up to around 800 °C. For most LTS films this is not a consideration, but for those applications requiring epitaxial films with long electronic mean free paths, the substrate again must be chosen to have a good lattice match and to be nonreactive.

1.2 Film Characterization

Given the possibility that impurities introduced into films during deposition will change their properties from those exhibited by bulk samples of the same material, characterization is an important part of thin-film research. Important measurements include: the resistivity of the film at 300 K, which will be increased by impurities and disorder; the resistivity ratio between 300 K and (say) 10 K, which is strongly reduced by impurities and electron scattering from the film surfaces; the transition temperature T_c, which is generally decreased by impurities but can sometimes be increased by disorder; the critical current density J_c, which can be most easily measured by making a constriction in the film; the stoichiometry, which can be determined by Rutherford backscattering, inductively coupled plasma spectroscopy or energy/wavelength dispersive x-ray analysis; and the crystalline perfection, which is studied by x-ray scattering or channelling techniques.

Of these characterization techniques, one of the simplest and most informative is the measurement of resistivity and resistivity ratio (ρ_{300K}/ρ_{10K}). For example, $\rho_{300K} \simeq 15\ \mu\Omega$ cm for niobium implies an electronic mean free path of 3 nm. As phonon scattering is reduced with decreasing temperature, scattering from grain boundaries and lattice defects (or the film surface) becomes dominant, for films deposited in a poor vacuum system scattering by included impurities becomes important. Thus, a resistivity ratio of 7 implies a mean free path at 10 K of 21 nm, whereas 50 implies 150 nm. The ratios of 7 and 50 are typical of niobium films made on room-temperature and heated substrates. If the film thicknesses are larger than the mean free path, the resistivity ratio is an immediate indicator of film quality. In some epitaxial niobium films, resistivity ratios as large as 300–400 in 2 μm thick films have been reported. For many materials (e.g., silver, niobium, molybdenum, tin, gallium) precision deposition techniques enable films to be deposited with mean free paths that are so long at low temperatures that the resistance ratio is generally limited by surface scattering, unless the films become very thick.

These measurements are commonly used for both LTS and HTS superconducting films. In the HTS case, an important additional parameter of interest is the surface impedance R_s at frequencies in the range from 1 GHz to 100 GHz. The importance of microwave applications of HTS films has created the need to measure R_s from 4.2 K to at least 77 K (see *Microwave Applications of Superconducting Materials*). As R_s for good HTS films is at least an order of magnitude below that of copper, a copper cavity cannot be used for the measurements at 77 K. A niobium cavity can be used at 4.2 K. No standard measurement method exists, but a technique which determines the Q of a resonator made of two HTS films facing each other is becoming popular (Taber 1990).

1.3 Physical Properties of Films

Many of the properties of superconducting films are discussed in other articles, in that they are identical to those of the bulk material. However, thin films are often used in preference to single crystals or polycrystalline samples, for three reasons. First, they allow certain experiments to be carried out more easily. Examples are high-frequency properties (see *Optical Properties of Superconductors*), where a film can be prepared with a smooth clean surface more easily than the bulk; nonequilibrium properties (see *Nonequilibrium Superconductivity*), where the sample volume can be made small; Josephson effects in weak links (see *Josephson Effects in Weak Links*), which are most easily defined in films; tunnelling spectroscopy (see *Tunnelling Spectroscopy*) for which tunnel junctions are prepared by oxidizing clean film surfaces; and proximity effect studies (see *Coherence Length, Proximity Effect and Fluctuations*), in that films can easily be made with thickness comparable to a coherence length (at least in LTS materials).

The second reason for using films for scientific studies is that they can be made to exhibit properties that are different from those of the same bulk material. These differences can arise because of dimensional effects, as films can be grown very thin in the dimension normal to the substrate and can be patterned to be small in the plane. When the thickness of the film is less than the coherence length, it becomes two dimensional, in that its properties cannot vary in the direction normal to its surfaces. These dimensional (including dimensional crossover effects) effects will be discussed in Sect. 2.

Another difference between film and bulk materials can be created because the structure and electronic properties of the film are changed by the deposition process. This is generally a result of crystalline disorder (see *Radiation Effects on Superconducting Materials*). If the film is deposited on a substrate that is too cold to allow ordering to occur, a very-small-grained or even amorphous film can result (see *Amorphous Superconductors*). An early example was the nonsuperconducting semimetal bismuth, which became superconducting up to 6 K when deposited on a substrate held at 4.2 K. The T_c of gallium is similarly raised from 1.1 K to 8.5 K by such "quench condensation." Such films were the first amorphous metals (Brandt and Ginzburg 1960, Buckel Gey 1963). Aluminum films can have a T_c up to 6 K compared to the bulk 1.2 K, the T_c depending largely on the conditions of deposition and the strain induced by the substrate. Such changes in T_c are due to modification of both the phonon spectrum and the density of electronic states of the material (see *Superconducting Materials: BCS and Phenomenological Theories*; *Tunnelling Spectroscopy*). In crystalline transition metals, with their well known dependence of T_c on electron-to-atom ratio (see *Superconducting Materials: An Overview*), a remarkably different dependence is observed when films are deposited at low temperatures to make them amorphous. Such quench-condensed films have high resistivities and small resistance ratios (close to 1).

A third reason for using films is for superconducting applications in devices and electronics, as described later in this article and in other more specific contributions on particular application areas.

The same reasons for interest in the physical properties of LTS films apply to HTS materials. However, the radically different properties of the HTS oxides do create some different results. For example, the oxides have anisotropic and very short coherence lengths (see *Anisotropy in Superconducting Materials*). Most HTS films are grown with the *c*-axis normal to the substrate, with the copper oxide sheets in the plane of the substrate. The *c*-axis coherence length is only a few tenths of nanometers. Thus, the surfaces of HTS films, unlike their LTS counterparts, are most unlikely to have properties representative of the bulk material (even those properties along the *c* axis). This is because any change in stoichiometry, for example loss of oxygen, or disorder, including that due to surface reconstruction of the lattice, can affect a surface layer whose thickness is comparable to, or greater than, a coherence length. Thus, any experiments that are surface sensitive, such as tunnelling, proximity effects and photoemission studies, are unlikely to measure bulk properties of HTS materials in measurents on *c*-axis films. Recently, films with other orientations have been deposited (see *Oxide Superconductors: Thin-Film Deposition*); measurements on these films are likely to reveal more of the bulk properties of HTS materials. In LTS materials the coherence length is generally much longer than the thickness of any surface layer, although niobium and A15 films can exhibit problems of surface degradation unless care is taken.

Effects of dimensionality are also different in HTS materials, in that even in the bulk they are strongly anisotropic and can be considered pseudo two dimensional. It is also not possible to produce films thinner than the coherence length along the *c* axis. The coherence length is only 4 nm in the *a–b* plane of YBCO, so even *a*-axis films are difficult to make thin compared to this length. The effect on physical properties of making the films thin is most easily studied in superlattices (see Sect. 2) but remarkably thin HTS films have been made that remain superconducting to quite high temperatures.

The properties of both films and bulk HTS materials can differ significantly from optimum values, particularly due to loss of oxygen, to crystalline disorder and to doping effects. The trend of these effects is to decrease T_c and increase resistivity until the material becomes insulating at low temperatures (Greene and Bagley 1990).

There is one parameter that is very different in films compared to bulk single crystals, especially in the case of YBCO. This is the value of critical current J_c. All good HTS films with T_c close to 90 K, independent of their method of preparation, have J_c values at all temperatures $< T_c$ much higher than those for single crystals. It appears that a pinning center is rather readily produced in thin films that is difficult to reproduce in crystals and that this defect is common to all "high quality" YBCO films to date.

2. Superlattices

As an object of scientific study, multilayers or superlattices are an interesting class of "engineered" materials in the fields of semiconductors, magnetism (see *Thin Films and Multilayers, Magnetic*), optics, x-ray optics and superconductivity (Shinjo and Takeda 1987). They are made by sequentially depositing two or more materials many times (e.g., ABABAB...) The individual film thicknesses of A and B approach one atomic layer in some cases. This repetition of the film sequence distinguishes them from the multilayer device structures to be discussed later.

In semiconductors the long electron wavelength (~10 nm) implies that the bulk electronic band structure can be radically changed in the direction normal to the layers of a superlattice, once the individual layers have thickness comparable to this wavelength. This regime is relatively straightforward to reach with current growth capabilities such as MBE. Thus, semiconductor superlattices have become a field of intense activity, for both their research interest and device potential. In metals, the short electron wavelength (a few tenths of nanometers) implies that such band structure changes cannot be observed. When metals become superconducting, however, the electron wavelength is replaced by the coherence length as the important length parameter. In LTS materials, it is easy to make superlattices with layer thickness much less than coherence lengths (typically tens of nanometers). Hence, there is scientific interest in superconducting superlattices, in which the components A and B can be superconductor (A) with another superconductor, metal, semiconductor or insulator (B).

It is important to note that in HTS superlattices, at least those with *c* axis normal to the substrate, it is again not possible to make the layer thickness thinner than the coherence length, which is only a few tenths of nanometers. If the HTS materials were strictly two dimensional, layering them into a superlattice should have no effect on T_c unless structural effects occurred (see later).

Unlike semiconductor superlattices, at present there are no important devices made from superconducting superlattices (in contrast to multilayer device structures), although the Nb/Al–Al oxide–Nb trilayer process that is now the metallization of choice for Josephson circuits was a direct result of studies of tunnelling in niobium–aluminum superlattices (Geerk et al. 1982).

2.1 Superlattice Synthesis

The vacuum equipment that is commercially available makes it relatively easy to prepare superconducting superlattices. A substrate is exposed to alternating sources of materials, for example by moving shutters in front of electron-beam sources, or by rotating or oscillating the substrate in front of two sputtering sources. Magnetron sputtering sources are particularly suitable because of the stability of their deposition rate. It is often important to heat the substrate, as the resulting films (e.g., niobium, tantalum) will then have larger grain size, longer electron mean free path and longer coherence lengths. Superlattices of HTS materials have been made by the same methods. Obviously a heated substrate (~700 °C) is always necessary. In addition, laser deposition lends itself ideally to the synthesis of HTS superlattices. Two targets are moved alternately into the focus of the laser beam and the layer thickness is controlled by the number of laser shots allowed to impinge on each target (Lowndes et al. 1990).

2.2 Superlattice Characterization

As a superlattice has an additional periodicity of the structure in the growth direction, which is superimposed on the crystal lattice spacing, an important measurement is the standard θ–2θ x-ray scan. An example is shown in Fig. 2, where the subsidiary maxima are due to x-ray reflections from the superlattice (Somekh et al. 1989). The position of these maxima can be used to determine the superlattice wavelength (the sum of the A and B layer thicknesses) and their intensities are related to the interfacial roughness between layers.

Another important measurement is simply the net resistivity of the superlattice (in the case of two metals). If the electron mean free path is long, that is, the electrons move freely through the interfaces between A and B, then the resistivity will have a value between the values of A and B in their bulk form. However, if the electrons are scattered at the interfaces, or if the films themselves become structurally disordered due to the layering process or to strain, then the resistivity will be higher than the values of A or B. Even in superlattices of superconductors with other materials (insulators, semiconductors), or in HTS superlattices, the resistivity and resistance ratio are very important parameters to measure.

Transmission electron microscopy can yield elegant pictures of superlattices and can be used to give the orientation of the individual layers or grains within the layers. In the HTS materials, the layered nature of the oxide structure gives even more striking results. One example (see *Oxide Superconductors: Electron Microscopy* Fig. 11) shows that within the one layer there is a very abrupt change from YBCO to PrBCO, presumably the result of ledge growth of the layers of unit cell dimensions, that is, the layer was growing as one unit cell of YBCO say when the source was changed to PrBCO, which continued the growth to complete the layer. This ledge growth is also clearly evident in scanning tunnelling microscope profiles of

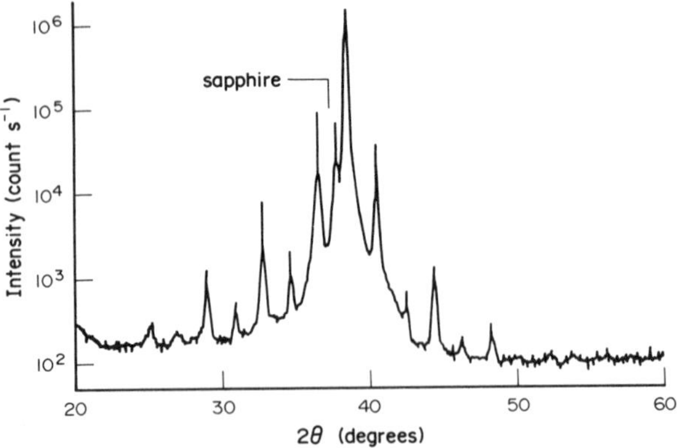

Figure 2
CuK alpha x-ray trace of niobium–tantalum epitaxial multilayer (4.9 nm wavelength) grown on A-plane sapphire to give (110) niobium–tantalum epitaxy; from the superlattice intensities the interface roughness can be estimated at 0.22 nm (after Somekh et al. 1989)

the surface of YBCO films (not superlattices), in which the layers of one unit cell in height are shown around screw growth dislocations (see *Oxide Superconductors: Thin-Film Deposition* Fig. 2).

2.3 Physical Properties of Superlattices

The superconducting property of LTS superlattices that has received by far the greatest attention is the upper critical field H_{c2}, measured with the field applied parallel to the layers. This is because the dimensionality of the superconducting state is readily determined from the form of the dependence of H_{c2} on temperature T.

As an example, consider the specific case of a superlattice of niobium–germanium (germanium films are insulators at low temperatures) (Ruggiero et al. 1982). If the niobium layers are thick ($t \gg \xi$), then the superlattice behaves like bulk niobium, that is, it is three dimensional. If the niobium layers are thin ($t \ll \xi$), and the germanium is thick, then the niobium layers are isolated from each other and are two dimensional. However, if the niobium layers are again, thin, and the germanium layers are thin enough to allow appreciable coupling by tunnelling between the niobium layers, then bulk three-dimensional behavior is again recovered, as the superlattice appears again as uniform material over the length scale of ξ. Interesting crossover effects have been observed as a function of temperature, in that the coherence length is much longer near T_c than it is at lower temperatures. Hence, it is possible to find two regimes, one at higher temperatures where $\xi > t$ and three-dimensional behavior is observed and another at low temperatures with $\xi < t$ that exhibits two-dimensional behavior. As these two regimes have different temperature dependences of H_{c2}, namely $H_{c2} \simeq T_c - T$ for three dimensions and $H_{c2} \simeq (T_c - T)^{1/2}$ for two dimensions, the crossover between them is observed in the plot of H_{c2} vs T (see Fig. 3). These regimes can be thought of as the situations where the cores of the flux vortices either extend across multiple layers as Abrikosov vortices (three dimensional) or sit between the niobium layers in the germanium layers as Josephson vortices (two dimensional) (see *Flux Line Lattice*).

A second type of superlattice utilizes superconducting and normal metals (e.g., SNSNSN...). The coupling between the S layers is by the proximity effect induced in the N layer (see *Coherence Length, Proximity Effect and Fluctuations*). A widely studied example is niobium–copper superlattices, in which one unusual effect occurs for small periods ($\lesssim 2$ nm). Both the niobium and copper layers appear to become disordered, with a strong increase in resistivity, a decrease in resistance ratio and decrease in the T_c of the niobium. Dimensional crossover effects can also be observed in SN superlattices as the N layer becomes longer than its effective coherence length, which in copper in niobium–copper superlattices was demonstrated to be about 20 nm (Benerjee et al. 1983).

More recently, a further novel transition was predicted (Takahashi and Tachiki 1986) and observed (Karkut et al. 1988) in superlattices of two supercon-

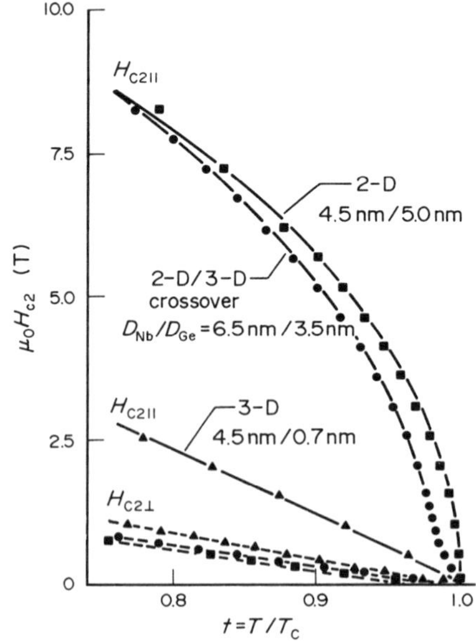

Figure 3
Upper critical fields near T_c for Nb–Ge multilayers. Systematically varying the germanium thickness between the two-dimensional niobium layers effects a progression from anisotropic three-dimensional behavior $[H_{c2\parallel}(T) \sim (T_c - T)]$ to "crossover" behavior and finally to decoupled two-dimensional behavior $[H_{c2\parallel}(T) \sim (T_c - T)^{1/2}]$ (after Ruggiero et al. 1980)

ductors with very similar T_c but with very different mean free paths, for example, niobium and a niobium–titanium alloy. In this case two crossover transitions were observed. Near T_c the vortices are located in the niobium–titanium but are large enough to extend across the niobium (three dimensional, $H_c \simeq (T_c - T)$). At a lower temperature ξ_{Nb} becomes less that t_{Nb} and the vortices relocate into the niobium layers and are confined within them (two dimensional, $H_{c2} \simeq (T_c - T^{1/2})$) (see Fig. 4). At even lower temperatures, ξ_{NbTi} becomes $< t_{NbTi}$ and the vortices become two dimensional within the niobium–titanium layers, which have a very high H_{c2}.

Work on HTS superlattices is naturally at an earlier stage, although the layered nature of the oxide materials has allowed superlattices of very impressive perfection to be grown. Perhaps the most interesting question to be answered to date is the value of T_c as the individual HTS layers are made thinner and as the coupling across an intermediate layer is made weaker. The intermediate layer is, for example, PrBCO which is a lattice-matched oxide that is a Mott insulator that exhibits high electrical resistivity at low temperatures. The results of Fig. 5 show that even layers of YBCO that are one unit cell thick have a T_c of 20 K when widely separated (Lowndes et al. 1990). The T_c increases as the coupling increases, reaching 56 K when the PrBCO is one unit cell thick.

It is evident that HTS materials with a pronounced perovskite layer structure behave in many respects as intrinsic atomic-scale multilayers, the alternating layers being the copper oxide planes (S) and the redox spacer layers (N or I) (see *Oxide Superconductors: Solid-State Chemistry*). In the more complex HTS materials the doping levels on the two types of layer may be varied independently, their character changing from ... SNSNSN ... to ... SISISIS ... depending on the preparation conditions. The interpretation and modelling of HTS materials as intrinsic multilayer structures is currently an active area of considerable importance both for their basic physical properties (see *Anisotropy in Superconducting Materials*) and for their behavior in magnetic fields (see *Flux Line Lattice*).

3. Superconducting Device Structures

Superconducting devices are rather simple structures compared to those that are commonplace in semiconductor technology. This perhaps reflects the orders of magnitude difference in commitments of time, personnel and funds to the two fields since the mid-1940s, which are themselves measures of the relative commercial importance of the two fields to date. Some superconducting devices can be made in a single film of material, the most common uses two films and the most complex requires only three. Despite this, they perform logic functions faster than the most advanced semiconductor devices and at a level of power dissipation which is orders of magnitude smaller (see *Superconducting Devices and Electronics*). Whether they will ever become more complex, or indeed need to do so, is an open question.

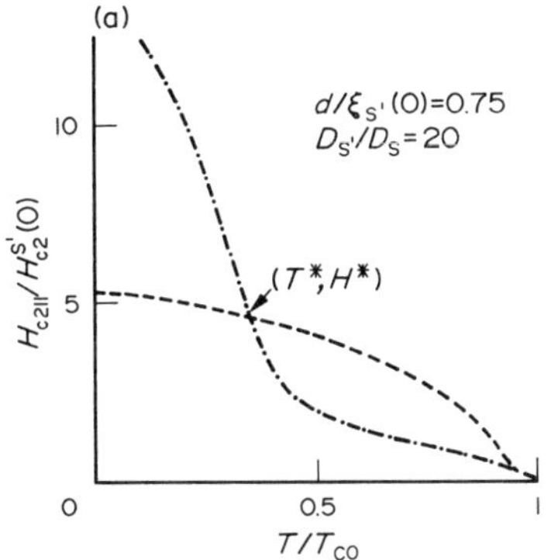

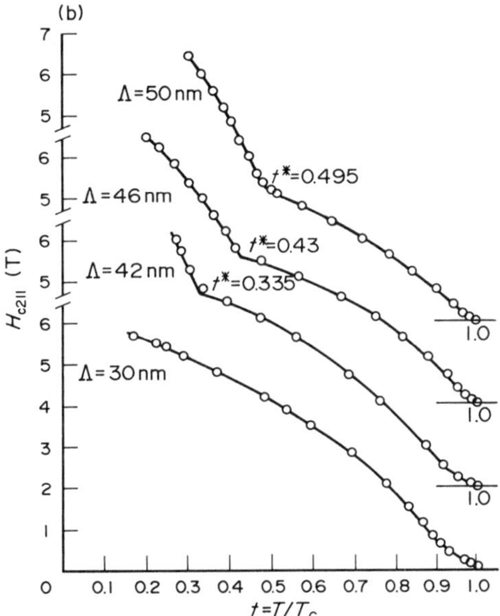

Figure 4
(a) Theoretical temperature dependence of the parallel critical fields for a multilayer SS′SS′SS′ .... The dashed and dash-dotted curves indicate the upper critical fields when the superconducting order parameter nucleates in the S and S′ layers respectively. The observable $H_{c2\parallel}$ is the higher one of the two critical fields at each temperature. The point (T^*, H^*) is a multicritical crossover point, d is the layer thickness and D is the diffusivity (after Takahashi and Tachiki 1986). (b) Resistive measurements of $H_{c2\parallel}$ vs $t(= T/T_c)$ for Nb/NbTi superlattices with layers of equal thickness showing behavior at t^* for samples with different wavelength Λ (after Karkut et al. 1988)

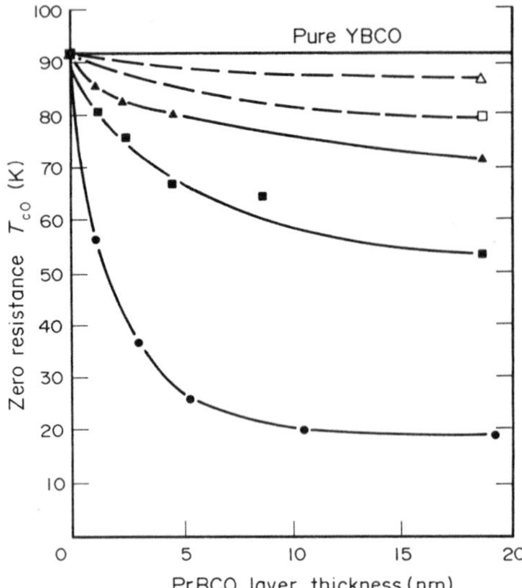

Figure 5
Critical temperature of $(YBCO)_M/(PrBCO)_N$ superconducting multilayers ($m = 1$ to 8 unit cells thick) vs the separating PrBCO layer thickness: △, eight-cell YBCO; □, four-cell YBCO; ▲, three-cell YBCO; ■, two-cell YBCO; ●, single-cell YBCO (after Lowndes et al. 1990)

3.1 Tunnel Junctions and Weak Link Structures

Superconductors are affected strongly by magnetic fields and only weakly by electric fields. This magnetic field sensitivity is the basis for many electronic devices. In a superconducting film, or film adjacent to a superconductor, a local region is produced where the superconducting state can be controlled by external magnetic fields. The cryotron (Buck 1956), which was the earliest superconducting electron device, used this principle. This device, which was invented in 1954 and therefore predates the Josephson effect, used two films (see Fig. 6) separated by an insulator. A current through the control film created a magnetic field which switched the second film from its superconducting to its normal state. Recovery from the normal state was not fast (by modern standards). With the discovery of the Josephson effect (see Josephson 1962; *Josephson Effects in Weak Links*) a much faster device became possible. The Josephson junction (see Fig. 7) consists of two superconducting films separated by a thin insulator (< 1.5 nm thick). This insulator allows the tunnelling of pairs of electrons (the Josephson current) and single electrons (the quasiparticle current). These result in the tunnelling currents shown in Fig. 8 as a function of the voltage across the insulator. For an ideal junction there would be a discontinuous jump at $2\Delta/e$ and an exponentially small subgap tunnelling current proportional to $\exp(-\Delta/k_BT)$ (Van Duzer and Turner 1981). In real systems there is a smearing due to lifetime effects and nonuniformity in the films. In addition there is invariably a subgap leakage current. The sources of subgap leakage vary from device to device and depend very much on the fabrication of the device. Leakage can for instance arise from electron states within the tunnel barrier, or pinholes in the barrier itself. At low temperatures for very-high-quality low-leakage barriers, small subgap currents due to two-particle or many-particle tunnelling can be observed (Kirk et al. 1991). If a small magnetic field is applied to the junction parallel to the films, the Josephson current is reduced to zero when one quantum of magnetic flux is in the area given by the width of the junction multiplied by the sum of the London penetration depths of the two films (Rowell 1963). In practical devices, this is about 0.1–1 mT. It is important to realize that, in contrast to the cryotron, it is only the Josephson current that is switched off by the magnetic field. The two films comprising the junction remain entirely superconducting at these low magnetic fields. In addition to control by magnetic field, the junction can also be switched by increasing the current until the maximum Josephson current is exceeded.

The Josephson tunnel junction has become the device of choice for LTS electronic, instrument and sensor applications (Ruggiero and Rudman 1990). It is fast and consumes little power. Its modern form, made from Nb/Al–Al oxide–Nb, is described in detail in *Superconducting Device Fabrication*, and its applications in *Biomagnetic Measurements*; *Millimeter-*

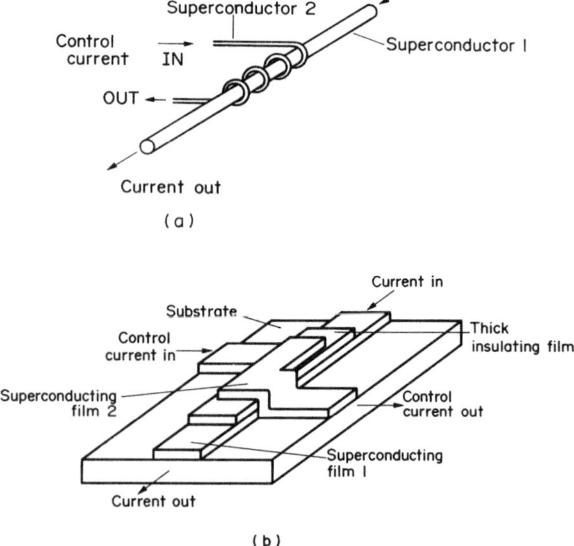

Figure 6
Cryotron types: (a) wire-wound type, (b) thin-film type

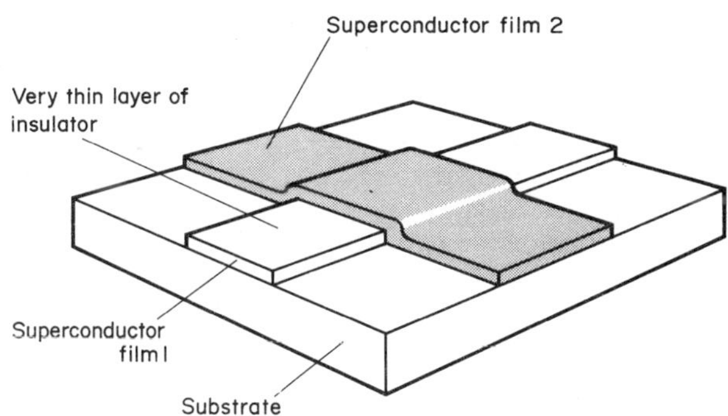

Figure 7
Josephson junction (SIS tunnel junction)

Wave Detection; *SQUID Design, Fabrication and Application; Superconducting Devices and Electronics*; and *Voltage-Standard Devices*.

Soon after the theoretical discovery and experimental verification of the Josephson tunnel junction, it was realized that Josephson effects could be observed in many even simpler structures, which are now collectively known as weak links (Anderson and Dayem 1964; *Josephson Effects in Weak Links*). No tunnelling occurs in any of these structures, they rely instead on creating a region of weak coupling between two superconductors by various means other than an oxide layer. Some examples are shown in Fig. 9. The simplest is a contact between a pointed wire of one superconductor and a second bulk superconductor ("point contact"). The thin-film version is a small constriction defined in a film (a "weak link", "bridge" or "microbridge"). A "variable thickness bridge" has the constriction both thinner and narrower than the surroundings (banks). The coupling region can also be made from a normal metal, which is made weakly superconducting by the proximity effect (SNS junction).

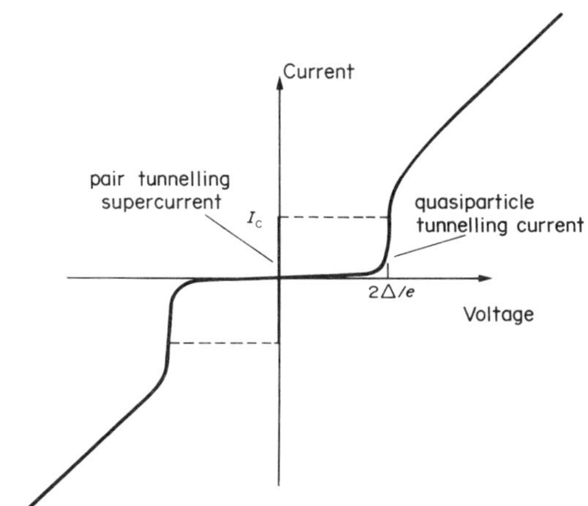

Figure 8
I–V characteristic of Josephson junction for the case where the two superconducting films have the same energy-gap parameter Δ, the characteristic shows some subgap leakage current

In all these Josephson weak link devices, the length of the coupling region must be comparable to the superconducting coherence length in that region. In LTS materials and normal metals, standard film growth and lithographic techniques can produce features of the sizes required (generally $\lesssim 1\ \mu m$). In HTS materials, this is obviously not possible.

Unlike the semiconducting transistor, superconducting devices do not have gain. At a fixed current level, they simply switch between the zero voltage and finite voltage states with application of magnetic field. A few three-terminal devices have been investigated that are more transistorlike in design, but not yet in operation (Kleinsasser and Gallagher 1990). One example is the superconducting FET (see Fig. 10), in which two closely spaced superconducting contacts induce a weak coupling region on the surface of a semiconductor by the proximity effect. This superconducting channel is then switched on and off by application of an electric field from the gate. To date, large gate voltages are required to achieve switching and the devices are not practical.

As of mid-1991, the device scene in HTS materials is one of great activity and some uncertainty. All the devices made in LTS materials are being investigated, with varied levels of success. It is clear that the short coherence lengths of the HTS materials creates a real challenge in device design and processing. It does seem likely that junctions of the weak link type will emerge as more important than tunnel junctions, at least in those HTS materials with $T_c > 77\,K$. The reasons for this will be given later.

One HTS device has been created which has no LTS counterpart. A control line modulates the flux density in a weakened region of a second film (channel). It can be called a flux-flow cryotron (Martens et al. 1991) in that the channel is switched from zero resistance into its flux flow state (see *Resistive Transition and Flux Flow in Superconducting Materials*), rather than into its normal state. Its value in interface circuits has already been demonstrated, in that its output voltage is quite large (tens to hundreds of millivolts).

3.2 Device Processing: LTSs

The LTS weak link devices are relatively easy to make with current lithographic techniques. Optical, electron-beam and ion-beam patterning have all been used to create the weak regions of dimensions 1 μm or less in a variety of materials. The details will not be discussed here.

Tunnel junctions exhibiting quasiparticle tunnel-

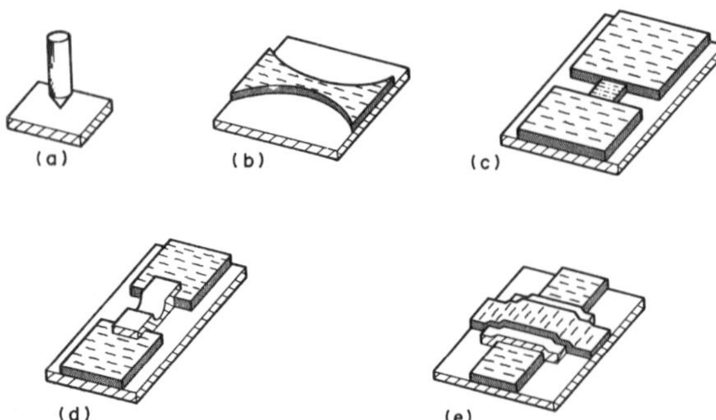

Figure 9
Josephson weak link structures: (a) point contact, (b) constriction, (c) variable thickness bridge, (d) SNS planar and (e) SNS sandwich

ling were first made in the late 1950s by the very simple process of evaporating a strip of aluminum film using a shadow mask, exposing it to the air for a short time (~1 min), then evaporating a crossing strip of lead to produce an Al–Al oxide–Pb tunnel junction (Giaever 1960). The brief exposure to air grows the required thickness of aluminum oxide (~1.5 nm) which completely covers the aluminum film even over relatively large areas of a few square millimeters. It is important that there are no defects or "pinholes" in the oxide, otherwise a metallic short circuit occurs. This basic oxidation process has been used for the vast majority of tunnel junctions made on many different materials, and is used in the current commercial process (see *Superconducting Device Fabrication*).

Other materials that have been used to make oxide tunnel junctions include lead, tin, niobium, tantalum and the alloys and A15 compounds of niobium. In some cases the oxidation is not so easy as it is for aluminum and heating in oxygen is required. Because of its relatively high T_c (7.2 K), lead and its alloys received many years of study as the metallizations for Josephson circuits. The problems that were encountered illustrate the difference between the research mode of making a few successful junctions for scientific study, and a manufacturing mode of reproducibly making circuits that require many thousands of junctions. The variation of critical current density across these junctions must be small and they must survive both shelf storage and many thermal shocks as they are cycled from 300 K to 4.2 K and back to 300 K. Because of the failure of lead, no satisfactory metallization for Josephson circuits existed for 20 years (1963–1983). Niobium was an obvious choice ($T_c = 9.2$ K), but oxidation of niobium itself produces a mixture of oxides with insulating, semiconducting and even metallic properties. There is probably also degradation of the niobium film surface. This problem was finally solved by immediately depositing a thin layer (< 10 nm) of aluminum on top of the niobium film in the same vacuum system, then oxidizing the aluminum to form an aluminum oxide, rather than niobium oxide, tunnel barrier (Geerk et al. 1982).

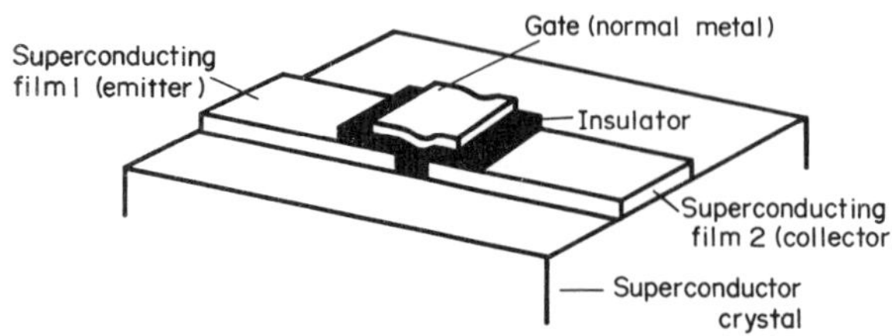

Figure 10
Superconducting three-terminal device (FET type)

Such "artificial" barriers have been used for a number of other materials that do not readily oxidize themselves. The best known example is MgO, which is deposited as a thin layer on top of NbN to form NbN–MgO–NbN tunnel junctions (Shoji et al. 1985). The different types of "sandwich" tunnel junctions are shown in Fig. 11.

The sizes of sandwich junctions can be as small as 1μm × 1μm using standard optical lithography and submicrometer sizes using deep-UV sources. Another way to make very small junctions is to oxidize the edge of a film that has been cut by ion milling. Such an "edge junction" is shown in Fig. 11. The small capacitance of such structures makes them suitable for high-frequency SIS detectors used in radio astronomy studies.

3.3 Device Processing: HTSs

The new oxide materials with $T_c > 77$ K are demanding the invention of a new device technology. They readily form tunnel junctions with counterelectrodes of low-T_c materials (Geerk et al. 1988), either because their surfaces lose oxygen and become insulating, or because the counterelectrode itself (e.g., lead) removes the oxygen to form the insulator. However, such junctions have no technological value. Obviously there is no Josephson current until the temperature is below T_c of the counter electrode and, at least to date (1991), none of them exhibits the *I–V* characteristic of Fig. 8 with well-defined Josephson current and energy gap. This comment applies to all the anisotropic oxides. Whether it is an intrinsic property remains to be seen.

The isotropic oxides (e.g., BaPbBi oxide, BaKBi oxide) are strikingly different. A "classical" *I–V* characteristic of the type shown in Fig. 8 has been observed by simply touching a BKBO single crystal to a BKBO film and work is underway to produce similar results in a thin-film sandwich structure. At present, the maximum T_c in an isotropic oxide superconductor is about 30 K. However, this would allow a

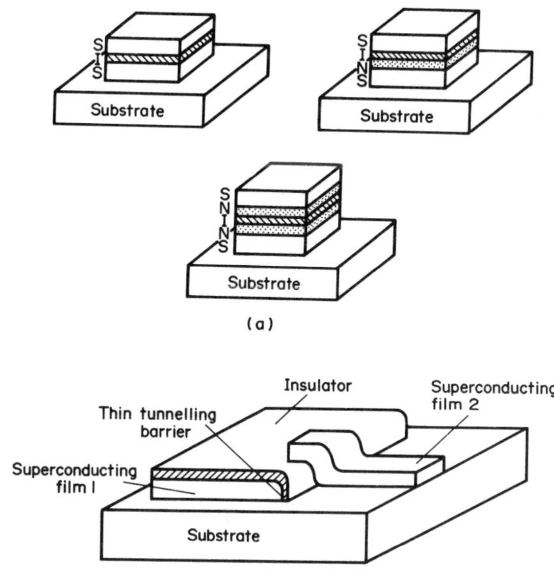

Figure 11
(a) Three types of multilayer sandwich structure Josephson junction, SIS, SNIS and SNINS; (b) edge-type Josephson junction

Josephson technology similar to that developed for niobium to be used at 12–15 K, well within the range of closed cycle refrigerators of reasonable cost. Thus, these isotropic oxides might become important metallizations for electronics applications.

In the absence of any useful tunnel junctions in HTS materials, the weak link devices are of more importance than they were in the LTS field (although it is true that most of the SQUID-based LTS magnetometers that have been sold have in fact used such weak links rather than tunnel junctions). A great variety of device structures are being investigated at present (mid-1991). Many of them make SQUIDs operating at 77 K. Whether any of them will become the basis of a genuine circuit technology remains to be seen. They include SNS devices where the N layer can be silver (or silver–gold alloy). In one geometry the silver covers a break in the HTS film induced by a step in the substrate. Other N layers are semiconducting and "metallic" oxides, in various geometries (Simon 1991). The possible presence of a proximity effect in such oxides is an interesting issue.

To define a weak link constriction, or region of reduced J_c and T_c, of a size comparable to the very short coherence lengths, is difficult in HTS materials. However, electron-beam writing and focused ion-beam milling have been used with some success, and the use of patterning by a scanning tunnelling microscope has been proposed. In one structure nature itself seems to provide a structure of the correct dimensions, namely a boundary between two grains in the film. The cleanest examples of such devices are grain boundary junctions grown on substrates that were cut and deliberately misaligned in the plane by various tilt angles (Gross et al. 1990a). The film, growing epitaxially, replicates the orientation of the substrate. On yttria-stabilized zirconia substrates, the critical current drops exponentially with angle, by four orders of magnitude as the angle is increased from 0 to 40° (see Fig. 12) (Ivanov et al. 1991). SQUIDs of YBCO showing modulation with magnetic field up to 87 K have been produced with such grain boundary junctions (Gross et al. 1990b). (See Fig. 13, which shows the structure, Fig. 14 demonstrating the modulation and also the article *SQUID Design, Fabrication and Application*.)

A step geometry has also been used to form weak link HTS junctions (see Fig. 15). If the angle of the step is steep ($\theta > 45°$) single or multiple grain boundaries form at the discontinuity in the substrate surface (Fig. 15a,b). For shallow junctions ($\theta < 35°$) the epitaxy is preserved at the step and an SNS junction can be formed by depositing a heteroepitaxial layer of normal PrBCO (Fig. 15c) (Gao et al. 1991). In this junction films are grown with the c axis normal to the substrate so that the tunnelling current is in the ab plane, the direction with longest coherence length.

An alternative process called biepitaxy creates a grain boundary in the film by control of the in-plane epitaxy using seed and buffer layers (Char et al. 1991a), rather than by changing the substrate orientation (see *Oxide Superconductors: Thin-Film Deposition*). This process could allow complex circuits of grain boundary junctions to be demonstrated, as the junction can be located at any point on a single-crystal

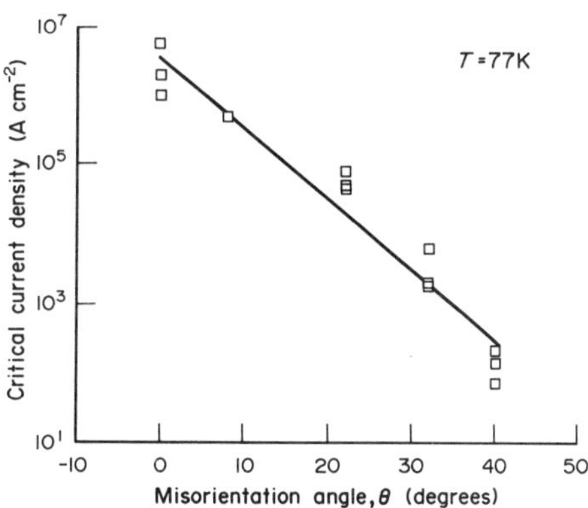

Figure 12
Variation of critical current density with misorientation angle for grain boundaries in YBCO films on yttria-stabilized zirconia (YSZ) substrates

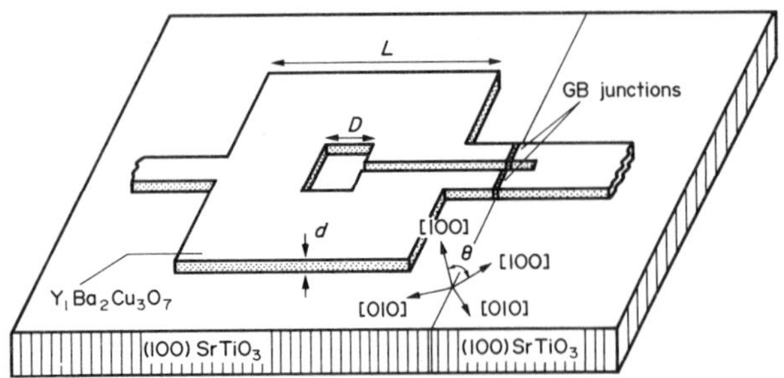

Figure 13
Bicrystal SQUID structure with grain boundary junctions

substrate. The first integrated SQUID magnetometer fabricated on one chip used such biepitaxial grain boundary junctions (Char et al. 1991b).

3.4 Properties of Devices

The ideal BCS current–voltage characteristic of LTS tunnel junctions can be readily understood by the use of semiconductorlike diagrams in which the filled and empty quasiparticle states are shown at an energy 2Δ apart and have $(E^2 - \Delta^2)^{1/2}$ energy dependencies of the density of states (see Fig. 16) (see *Superconducting Materials: BCS and Phenomenological Theories; Tunnelling Spectroscopy*). The superconducting pairs are shown at the Fermi level, the midpoint of the gap. The effect of an applied voltage V is simply to slide the Fermi levels and thus the whole gap structures relative to each other by a distance eV. The effect of temperature is included by a thermal distribution over only the quasiparticle states.

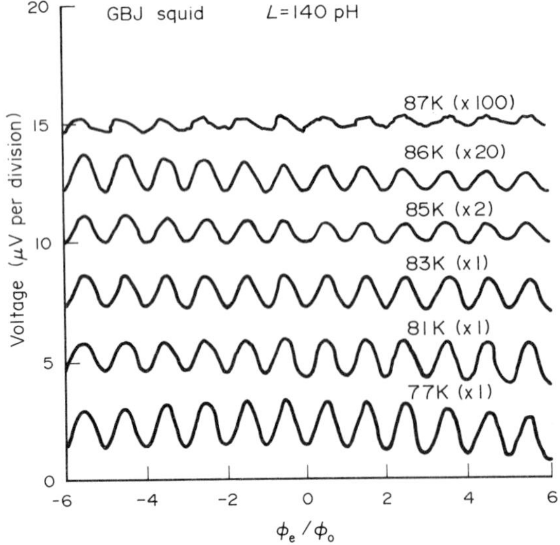

Figure 14
Field modulation of bicrystal SQUID illustrated in Fig. 13 for temperatures between 77 K and 87 K

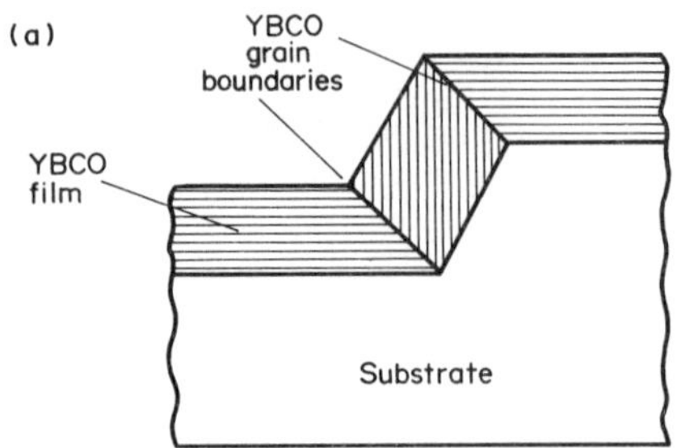

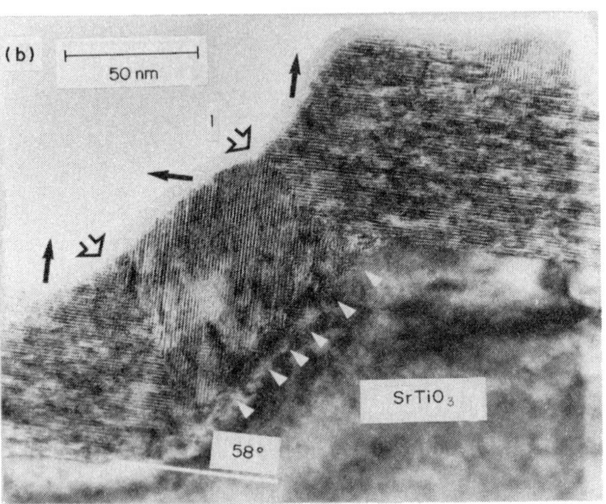

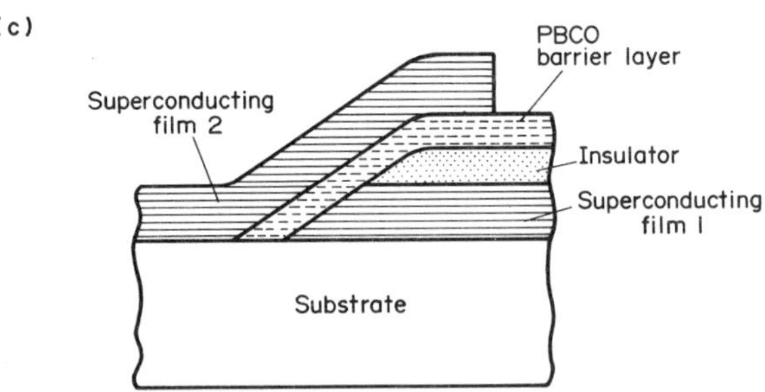

Figure 15
(a) Schematic of steep step junction with two grain boundaries, (b) electron micrograph of 58° step junction in a YBCO film on $SrTiO_3$ (after Jia et al. 1991) and (c) schematic of a shallow step junction with a normal PrBCO layer

The most interesting case is that of two superconductors with different energy gaps, say tin and lead to be specific. The T_c values are 3.7 K and 7.2 K, and the energy-gap parameters Δ = 0.6 meV and 1.4 meV, respectively or 2Δ = 1.2 meV and 2.8 meV. The density of states diagrams are then as shown in Fig. 16a at zero voltage and a temperature significantly below the

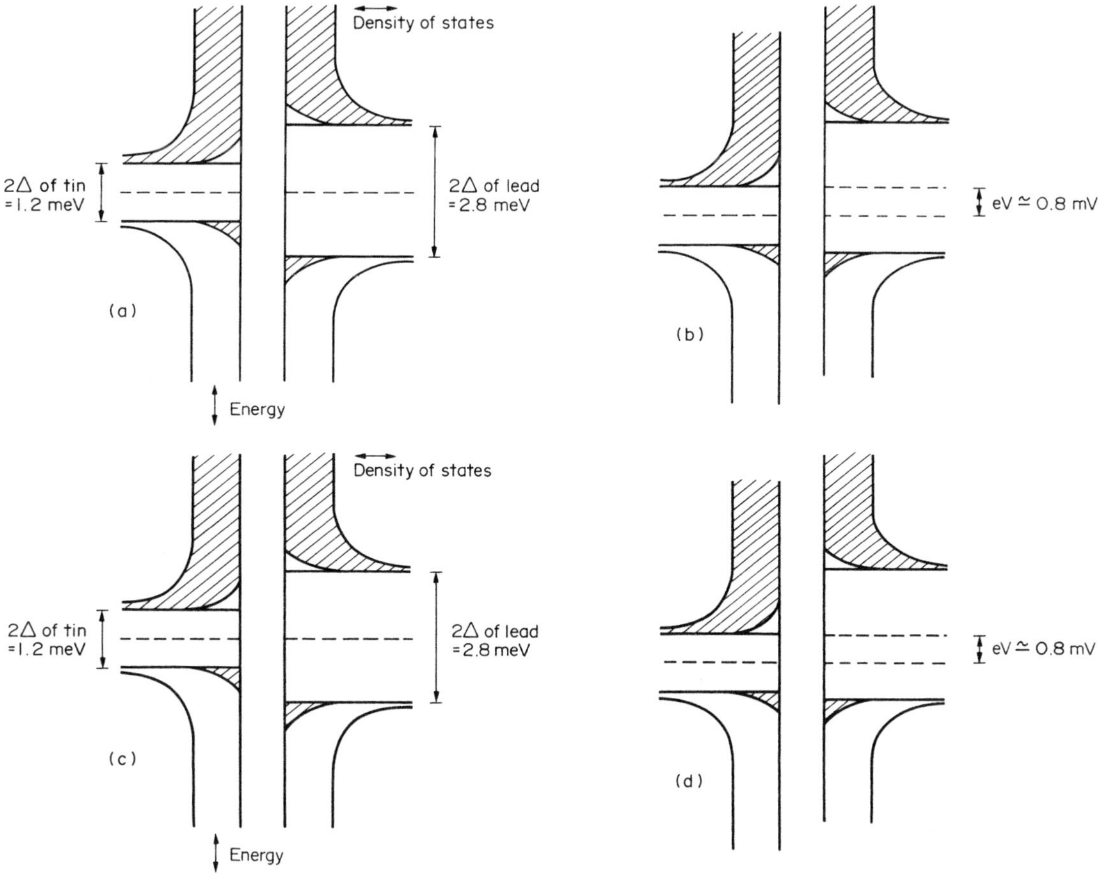

Figure 16
The density of filled and empty quasiparticle states for lead and tin superconducting layers on each side of a tunnel barrier. An applied voltage V leads to a relative displacement of the density of state diagrams, the voltage corresponding to the gap 2Δ is $2\Delta/e$: (a) zero applied voltage; (b) $eV=\Delta_{Pb}-\Delta_{Sn}\simeq 0.8$ mV; (c) $\Delta_{Pb}-\Delta_{Sn}<eV<\Delta_{Pb}+\Delta_{Sn}$, $eV\simeq 1.4$ mV; and (d) $eV=\Delta_{Pb}+\Delta_{Sn}\simeq 2.0$ mV

T_c of tin. As the voltage is increased, a maximum in current occurs at $eV=\Delta_{Pb}-\Delta_{Sn}$ (Fig. 16b), but the current then decreases for larger voltages. This is because the current is the integral over the filled and empty states that are aligned with each other, which is smaller in Fig. 16c than in Fig. 16b. A sudden increase in current occurs for $eV=\Delta_{Pb}+\Delta_{Sn}$ in Fig. 16d. In theory, this should be a discontinuous jump in current at any temperature, but in practice it has a slight width due to lifetime effects and nonuniformities in the films.

At high voltages, the current–voltage characteristic approaches that of the normal state. The quasiparticle branch of the characteristic is shown in Fig. 17; the complete characteristic, including the Josephson current of pairs flowing at $V=0$ exhibits the main features of Fig. 8. Another interesting case is the S–I–N junction, where N is a normal metal, for which the density of states diagram and I–V characteristic are shown in Fig. 18. The reader should be able to recreate the I–V characteristics by following the process of sliding the density of states as described.

In weak link devices, as there is no tunnelling, there is no signature of the energy gap in the I–V characteristic. In many such devices, the I–V characteristic resembles that for a tunnel junction with a low-resistance shunt across it, the resistively and capacitatively shunted junction (RCSJ) (see *Josephson Effects in Weak Links*). The I–V characteristic then appears as in Fig. 19.

An important "figure of merit" for both LTS and HTS devices of both tunnelling and weak link types is the I_cR_N product, which is the voltage defined in Fig. 19. It is related to the signal voltage that is available in a SQUID or circuit. Obviously it should be large (millivolts). It can be $\leqslant$ 3 mV in niobium tunnel

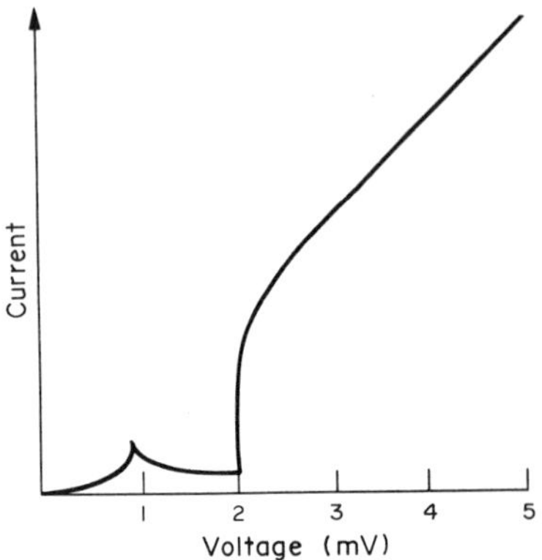

Figure 17
The current–voltage characteristic for the tunnelling of quasiparticles in an Sn–I–Pb junction

junctions, but in many HTS devices, especially the SNS type, it is very small (e.g., tens of microvolts). This makes such structures less interesting for practical applications.

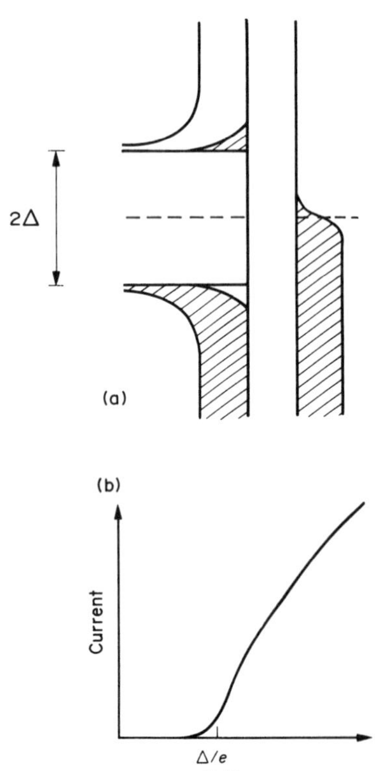

Figure 18
SIN tunnel junction: (a) density of filled and empty states at zero applied voltage, and (b) current–voltage characteristic at finite temperature

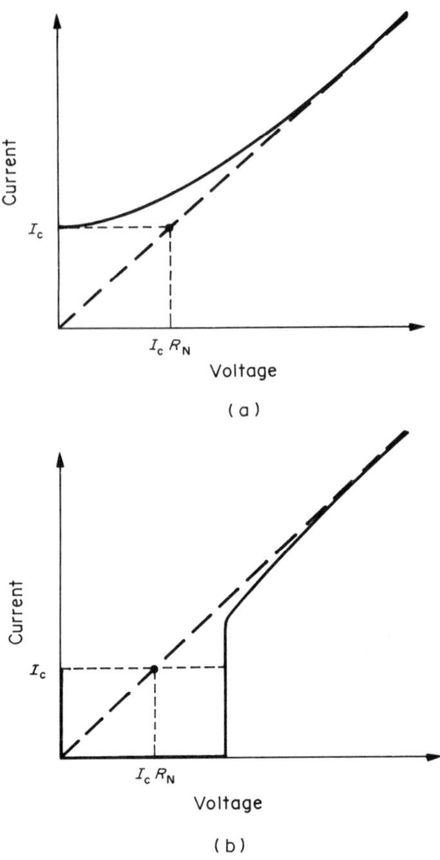

Figure 19
(a) The current–voltage characteristic for a weak link with a resistively shunted junction RSJ type transition, the I_cR_N product for the weak link is indicated; and (b) I_cR_N product for Josephson tunnel junction

A second figure of merit, used for tunnel junctions, is a measure of the subgap leakage current that flows for voltages less than the energy gap compared with the current that would flow at the same voltage in the normal state. This can be expressed as the ratio R_S/R_N calculated at some defined voltage (e.g., 2 mV for niobium devices), or as a voltage V_m defined as R_S times the critical current I_c. In an ideal junction, this current is only due to quasiparticles that are thermally excited above the gap. Hence, at temperatures small compared with T_c, it should be very small. For example, for Pb–I–Pb junctions at 1 K values for R_S/R_N of 10^5 have been observed. For Nb–I–Nb at 4.2 K most of the subgap current is thermally excited and the ratio is about 20–25. Values lower than this indicate a poor quality junction with some non-tunnelling leakage paths for the excess current. The most obvious possibility is a small pinhole in the tunnel barrier.

4. Concluding Remarks

The discovery of the HTS oxides raises interesting questions of how LTS and HTS junction technologies will be used, assuming that the latter can be developed. There are strong signs that this will happen. In some areas of application, it seems likely that HTS and LTS technologies will compete with each other, as they do with silicon and GaAs in some cases.

In one example, the most sensitive SQUIDs for measuring magnetic signals from the brain might well remain at 4.2 K, whereas for other measurements, where portability of the equipment is needed (e.g., nondestructive evaluation of pipes) 77 K operation will be favored (see *SQUID Design, Fabrication and Application*).

The majority of the digital circuits that have been built with niobium tunnel junctions employ "voltage state logic," in which the two states of the junction are at zero voltage and about 2.9 mV. Power dissipation is the product of the current and this voltage for the time that the junction remains switched. If HTS junctions of the same type could be made, the advantages of the low power dissipation of superconducting technology would be partly lost. The voltage (gap) would be ten times larger, and the current also ten times larger (to scale with the operating temperature); hence, the power would be 100 times larger.

However, there is a class of superconducting circuits, known as QFP (quantum flux parametron) or SFQ (single flux quantum) circuits, that avoid this problem and, hence, regain the advantage of the very low power dissipation of Josephson technology, even at temperatures up to 77 K (see *Superconducting Devices and Electronics*). The basic circuit element is the SQUID and power is dissipated only when a quantum of flux (which is the bit of information in QFP) moves in or out of the SQUID loop. The momentary voltage pulse is the bit of information in SFQ circuits. Such circuits do not need Josephson junctions of the type shown in Fig. 8; in fact they can use weak link devices. Thus, they appear to be the most likely candidates for initial demonstrations of HTS high-speed electronics.

Another divergence between LTS and HTS that is already becoming apparent is the extensive use of fully epitaxial multilayer devices in HTS. In this materials aspect, HTS begins to resemble III–V semiconductor technology much more than it does conventional superconductor (niobium) technology. As an example, an integrated HTS SQUID magnetometer has been fabricated which utilized 15 epitaxial oxide layers of superconductors and insulators, plus a final silver contact layer (see *Oxide Superconductors: Thin-Film Deposition* Fig. 3; Char et al. 1991a). Indeed, the field of superconducting thin films, multilayers and devices is changing and progressing rapidly.

See also: Oxide Superconductors: Thin-Film Deposition; Superconducting Device Fabrication; Superconducting Devices and Electronics; Superconducting Materials: An Overview; Thin Films and Multilayers, Magnetic

Bibliography

Anderson P W, Dayem A H 1964 *Phys. Rev. Lett.* 13: 195

Banerjee I, Yang Q S, Falco C M, Schuller I K 1983 Anisotropic critical fields in superconducting superlattices. *Phys. Rev. B* 28: 5037–40

Braginski A I, Talvacchio J 1990 In: Ruggiero S T, Rudman D A (eds.) 1990 *Superconducting Devices.* Academic Press, San Diego, CA, Chap. 8, p. 273

Brandt N B, Ginzburg N I 1960 *Zh. Eksp. Teor. Fiz.* 39: 1554

Buck D A 1956 *Proc. IRE* 44: 482

Buckel W, Gey W 1963 *Z. Phys.* 176: 336

Char K, Colclough M S, Garrison S M, Newman N, Zaharchuk G 1991a Biepitaxial grain-boundary junctions in $YBa_2Cu_3O_7$. *Appl. Phys. Lett.* 59: 733–5

Char K, Colclough M S, Lee L P, Zaharchuk G 1991b Extension of the biepitaxial Josephson junction process to various substrates. *Appl. Phys. Lett.* 59: 2177–9

Eom C B, Sun J Z, Yamamoto K, Marshall A F, Luther K E, Geballe T H, Cadermann S S 1989 *In situ* grown $YBa_2Cu_3O_{7-\delta}$ thin films from single-target magnetron sputtering. *Appl. Phys. Lett.* 55(6): 595–7

Gao J, Aarnink W A M, Gerritsma G J, Veldhuis D, Rogalla H 1991 Preparation and properties of all high T_c SNS-type edge dc SQUIDS. *IEEE Trans. Magn.* 27: 3062–5

Geerk J, Gurvitch M, McWhan D B, Rowell J M 1982 *Physica* 109: 1775

Geerk J, Xi X X, Linker G 1988 Electron tunnelling into thin films of $Y_1Ba_2Cu_3O_7$. *Z. Phys. B* 73: 329–36

Giaever I 1960 *Phys. Rev. Lett.* 5: 147

Greene L H, Bagley B G 1990 In: Ginsberg D M (ed.) 1990 *Physical Properties of High Temperature Superconductors II.* World Scientific, Singapore, p. 509

Gross R, Chaudhari P, Dimos D, Gupta A, Koren G 1990a Thermally activated phase slippage in higher T_c grain boundary Josephson junctions. *Phys. Rev. Lett.* 64: 228–31

Gross R, Chaudhari P, Kawasaki M, Ketchen M B, Gupta A 1990 Low noise $YBa_2Cu_3O_{7-\delta}$ grain boundary junction dc SQUIDS. *Appl. Phys. Lett.* 57: 727–9

Hammond R H 1975 IEEE *Trans. Magn.* 11: 201

Inam A, Hegde M S, Wu X D, Venkatesan T, England P, Miceli P F, Chase E W, Chang C C, Tarascon J M, Wachtman J B 1988 As-deposited High-T_c and J_c superconducting thin films made at low temperatures. *Appl. Phys. Lett.* 53: 908–9

Ivanov Z G, Nilsson P A, Winkler D, Brorsson G, Claeson T 1991 *Proc. SQUID 1991* (in press)

Jia C L, Kabius B, Urban K, Herrman K, Cui G J, Schubert J, Zander W, Braginski A I, Heiden C 1991 Microstructure of epitaxial $YBa_2Cu_3O_7$ films on step edge $SrTiO_3$ substrates. *Physica C* 175: 545–54

Josephson B D 1962 *Phys. Lett.* 1: 251

Karkut M G, Matijaservic V, Antognazza L, Triscone J-M, Missert N, Beasley M R, Fischer Ø 1988 Anomalous upper critical fields of superconducting multilayers—verification of the Takahashi–Tachiki effect. *Phys. Rev. Lett.* 60: 1751–4

Kirk E C G, Blamire M G, Somekh R E, Evetts J E, Van

Vechten D, Lovellette M N 1991 Ultra-high quality Nb/AlO_x/Nb tunnel junctions with epitaxial base layers. *IEEE Trans. Magn.* 27: 3137–40

Kleinsasser A W, Gallagher W J 1990 In: Ruggiero S T, Rudman D A (eds.) 1990 *Superconducting Devices*. Academic Press, San Diego, CA, Chap. 9, p. 325

Li Y Q, Zhao J, Chern C S, Huang W, Kulesha G A, Lu P, Gallois B, Norris P, Kear B, Cosandey F 1991 High critical current densities in $YBa_2Cu_3O_{7-x}$ thin films formed by metallorganic chemical vapor deposition at 730 °C. *Appl. Phys. Lett.* 58: 648–50

Lowndes D H, Norton D P, Budai J D 1990 Superconductivity in nonsymmetric epitaxial $YBa_2Cu_3O_{7-x}$/$PrBa_3Cu_3O_{7-x}$ superlattices—the superconducting behavior of Cu–O bilayers. *Phys. Rev. Lett.* 65: 1160–3

Martens J S, Ginley D S, Beyer J B, Nordman J E, Hohenwarter G K G 1991 A Josephson junction to FET high-speed line driver made of TlCaBaCuO. *IEEE Trans. Magn.* 27: 3284–8

Rowell J M 1963 *Phys. Rev. Lett.* 11: 200

Ruggiero S T, Barbee T W, Beasley M R 1980 *Phys. Rev. Lett.* 45: 1299

Ruggiero S T, Barbee T W, Beasley M R 1982 *Phys. Rev. B* 26: 4894

Ruggiero S T, Rudman D A (eds.) 1990 *Superconducting Devices*. Academic Press, San Diego, CA

Shinjo T, Takeda T 1987 *Metallic Superlattices*. Elsevier, Amsterdam

Shoji A, Aoyagi M, Kosaka S, Shinokik F, Hayakawa H 1985 Niobium nitride Josephson tunnel junctions with magnesium oxide barriers. *Appl. Phys. Lett.* 46: 1098–100

Simon R 1991 High-T_c thin films and electronic devices. *Phys. Today* 44(6): 64–70

Somekh R E, Shih W C, Dyrbye K, Huang K H, Baxter C S 1989 *The Sputter Deposition of Metal Multilayers*, X-Ray Instrumentation, SPIE Vol. 1140. Society of Photo-Optical Instrumentation Engineers, Bellingham, WA, pp. 453–63

Taber R C 1990 A parallel plate resonator technique for microwave loss measurements on superconductors. *Rev. Sci. Instrum.* 61: 2200–6

Takahashi S, Tachiki M 1986 Theory of the upper critical field of superconducting superlattices. *Phys. Rev. B* 33: 4620–31

Van Duzer T, Turner C W 1981 *Principles of Superconductive Devices and Circuits*. Elsevier, Amsterdam

J. M. Rowell
[Conductus Inc., Sunnyvale, California, USA]

Transport Currents and Force-Free Current Flow

When electric currents are applied to type II superconductors, the spatial distribution of the current density inside the material is much more complex than in normal metals. To understand how this complexity arises, how transport currents are distributed and how force-free current flow can arise, one must begin with an explanation of how magnetic fields penetrate into type II superconductors.

When a sufficiently large magnetic field is applied to a type II superconductor, magnetic flux does not enter the specimen uniformly, but in quantized bundles called vortices (other names being fluxons, flux lines or fluxoids), each carrying magnetic flux $\Phi_0 = h/2e = 2.07 \times 10^{-15}\,\mathrm{T\,m^2}$, where h is Planck's constant and e is the elementary charge (see *Magnetic Properties of Superconductors*). The local magnetic induction $\boldsymbol{b}$ inside the superconductor can be regarded as a vector superposition of individual tubelike magnetic flux contributions carried by the vortices (Clem 1981a). Each of these flux distributions is centered on a vortex axis, but decays exponentially outside a cross-sectional area of order $\lambda^2(B, T)$. Here $\lambda(B, T) \simeq \lambda(T)[1 - B/B_{c2}(T)]^{-\frac{1}{2}}$ is an effective penetration depth depending on the weak-field London penetration depth $\lambda(T)$, the magnitude B of the average magnetic induction $\boldsymbol{B} = \langle \boldsymbol{b} \rangle$ and the upper critical field $B_{c2}(T)$ (Clem 1975, Brandt 1977, Kogan 1978). In very low magnetic fields, the effective penetration depth increases with temperature T and diverges at the superconducting critical temperature T_c; at higher fields it diverges at a temperature $T_{c2}(B)$, the temperature at which $B_{c2}(T) = B$.

In a homogeneous superconductor not subject to an applied electric current, the equilibrium distribution of vortices in the superconductor is determined by a balance between the applied magnetic field, which acts to drive vortices into the material, and the repulsive interaction between vortices, which acts to keep the vortices apart. The combination of these two effects leads to a nearly uniform vortex distribution, in which the vortices are locally parallel to each other and nearly equally spaced. The local vortex density n (number of vortices per unit area) is related to the magnitude of the average magnetic induction via $B = n\Phi_0$. For most type II superconductors of interest, such as NbTi, Nb_3Sn and the high-temperature superconductors, there is an experimentally wide range of magnetic inductions for which the bulk of the material remains superconducting and the spacing a_0 between vortices obeys $a_0 \ll \lambda$. (Consideration here is only given to samples whose linear dimensions are much larger than the effective penetration depth $\lambda(B, T)$, which typically is of the order of 0.1 µm.) Under this high-vortex-density condition, the magnetic flux contributions of the individual vortices strongly overlap (Schelten et al. 1972). The resulting local magnetic induction $\boldsymbol{b}$, which points along the vortex axes, is then nearly uniform, except for local maxima on the vortex axes and shallow minima between vortices (see *Flux Line Lattice*).

1. Lorentz Force and Helmholtz Free Energy

If the vortex array is not in thermodynamic equilibrium, the Lorentz forces acting on the array to restore equilibrium must be calculated from the

Helmholtz free energy density $F(\boldsymbol{B}, T)$. Although this quantity is very difficult to calculate from theory in general, its approximate behavior as a function of $\boldsymbol{B}$ and T is known. For very low flux density, such that the intervortex spacing obeys $a_0 \gg \lambda$, we have $F(\boldsymbol{B}, T) \simeq B\varepsilon_1(T)/\Phi_0$, where $\varepsilon_1(T)$ is the energy cost per unit length of creating a vortex inside the superconductor at temperature T. (In anisotropic materials, $\varepsilon_1(T)$ depends on the direction of the vortex relative to the crystalline axes (see *Anisotropy in Superconducting Materials*), but this article will focus on the behavior in isotropic materials.) For extreme type II superconductors with large Ginzburg–Landau parameter ($\kappa \gg 1$) and for high magnetic flux densities such that $a_0 \ll \lambda$, $F(\boldsymbol{B}, T) \simeq \mu_0 H_c^2/2 + B^2/2\mu_0$, where $H_c(T)$ is the bulk thermodynamic critical field. In general, the thermodynamic field $\boldsymbol{H}$ is defined via $\boldsymbol{H} = \nabla_B F(\boldsymbol{B}, T)$ (Campbell and Evetts 1972). For isotropic superconductors, F is independent of the orientation of $\boldsymbol{B}$ relative to the crystalline axes, so that $\boldsymbol{H}$ is parallel to $\boldsymbol{B}$. Then $H(B, T) \simeq H_{c1}(T) = \varepsilon_1(T)/\Phi_0$ for low magnetic inductions obeying $B \ll \mu_0 H_{c1}$ ($a_0 \gg \lambda$), where H_{c1} is the lower critical field. Also $H(B, T) \simeq B/\mu_0$ for large magnetic inductions obeying $B \gg \mu_0 H_{c1}$ ($a_0 \ll \lambda$).

The Lorentz force density, which acts to drive the vortices towards local equilibrium, is $\boldsymbol{F}_L = \boldsymbol{J} \times \boldsymbol{B}$, where the current density $\boldsymbol{J}$ is defined from $\boldsymbol{H}$ (not $\boldsymbol{B}$) via Ampère's law, $\boldsymbol{J} = \nabla \times \boldsymbol{H}$ (Friedel et al. 1963, Anderson and Kim 1964). The current density $\boldsymbol{J}$ can be resolved into a parallel component $\boldsymbol{J}_\parallel$ (the projection of $\boldsymbol{J}$ along $\boldsymbol{B}$) and a perpendicular component $\boldsymbol{J}_\perp = \boldsymbol{J} - \boldsymbol{J}_\parallel$. Note that only the perpendicular component $\boldsymbol{J}_\perp$ contributes to the Lorentz force $\boldsymbol{F}_L$.

2. *Flux Flow and Viscous Forces in Homogeneous Materials*

The vortex array responds by moving collectively in the direction of the Lorentz force. (The Hall effect, which is important only for very pure superconductors, will be ignored here). The local vortex velocity $\boldsymbol{v}$ can be determined from a balance between the driving force $\boldsymbol{F}_L$ and the viscous drag force $-n\eta\boldsymbol{v}$ per unit volume, where η is the viscous drag coefficient per unit length of vortex (Kim et al. 1964), such that $\boldsymbol{v} = (\boldsymbol{J}_\perp \times \boldsymbol{B})/n\eta$. This collective vortex motion generates an electric field $\boldsymbol{E} = \boldsymbol{B} \times \boldsymbol{v}$ (Josephson 1965), which has only a component $\boldsymbol{E}_\perp$ perpendicular to $\boldsymbol{B}$. When the vortices move collectively, all in the same direction, there can be no component of the electric field parallel to $\boldsymbol{B}$ (i.e., $\boldsymbol{E}_\parallel = 0$). From the force balance equation, $\boldsymbol{E}_\perp = \rho_f \boldsymbol{J}_\perp$, where $\rho_f = B\Phi_0/\eta$ is called the flux flow resistivity (Kim et al. 1965). Combination of this equation with Ampère's law and Faraday's law yields a magnetic flux diffusion equation similar to that governing the penetration of magnetic flux and the decay of eddy currents in an ordinary metal. In any case, when a homogeneous superconductor is subjected to an applied magnetic field, but in the absence of an externally applied current, the vortex distribution relaxes with time to a state in which $\boldsymbol{J} = 0$ and $\boldsymbol{F}_L = 0$.

The behavior of a homogeneous current-carrying type II superconductor subjected to a perpendicular applied magnetic field is somewhat more complicated but still can be understood easily in terms of these ideas. (Simple examples are (a) a current-carrying infinite superconducting slab in a magnetic field applied perpendicular to the current direction but parallel to the slab surfaces and (b) a current-carrying cylindrical superconducting wire in a perpendicular applied magnetic field.) Although a superconductor under such conditions cannot achieve a state of thermodynamic equilibrium, it can achieve a quasi-equilibrium state described in terms of steady-state flux flow (see *Resistive Transition and Flux Flow in Superconducting Materials*). Inside the superconductor, the vortex array generates a nearly uniform magnetic flux density $\boldsymbol{B}$. The current density $\boldsymbol{J}$ distributes itself almost uniformly throughout the cross section of the superconductor and acts upon the vortex array. The resulting Lorentz force density $\boldsymbol{F}_L$, which is perpendicular to $\boldsymbol{J}$ and $\boldsymbol{B}$, causes a nearly uniform flow of vortices from one side of the superconductor to the other. The Lorentz force, which is best thought of as being initially applied to the vortices, is in turn transmitted via the viscous drag force to the crystalline lattice. This gives rise to the usual mechanical Lorentz force expected for a conductor carrying a current in a perpendicular magnetic field. New vortices are generated on the side of the superconductor where the self-field of the current adds to the applied magnetic field. These vortices move through the material under the influence of the Lorentz force and annihilate on the other side, where the self-field of the current partially cancels the applied magnetic field.

3. *Transport Currents in Parallel Applied Fields*

The behavior of a homogeneous current-carrying type II superconductor subjected to a *parallel* applied magnetic field (Campbell and Evetts 1972) is not easily understood in the aforementioned terms and remains an incompletely solved puzzle in the field of superconductivity. What is surprising is that, for the same experimental constraints (i.e., given values of the applied field $\boldsymbol{H}_a$ and current I), the basic equations given possess an infinite number of solutions (Campbell and Evetts 1972, Walmsley 1972, Timms and Walmsley 1975). All these solutions correspond to force-free current flow: as $\boldsymbol{B}$ varies, so does $\boldsymbol{J}$, but in a way that $\boldsymbol{J}$ is always parallel to $\boldsymbol{B}$ (i.e., $\boldsymbol{J} = \boldsymbol{J}_\parallel$). Because $\boldsymbol{J}_\perp = 0$, the Lorentz force vanishes ($\boldsymbol{F}_L = \boldsymbol{J} \times \boldsymbol{B} = 0$); hence, such magnetic flux distributions are said to be force free.

To demonstrate such a solution, consider the behavior of an infinite superconducting slab of thickness w, centered on the plane $x = 0$, subjected to a parallel applied magnetic field $\boldsymbol{H}_a = H_a\hat{z}$ and an applied current of average density $\boldsymbol{J}_a = J_a\hat{z}$. The self-field of this applied current is $\boldsymbol{H}_I(\pm w/2) = \pm \hat{y}J_a w/2$ at the two surfaces ($x = \pm w/2$).

Consider an arbitrary magnetic flux density distribution $\boldsymbol{B}(x)$ that depends only on the x coordinate. Let $\boldsymbol{B}(x) = B(x)\hat{\alpha}(x)$, where $B(x) = |\boldsymbol{B}(x)|$ and $\hat{\alpha}(x) = \hat{y}\sin\alpha(x) + \hat{z}\cos\alpha(x)$ is the unit vector along $\boldsymbol{B}(x)$. Let $\hat{\beta}(x) = \hat{\alpha}(x) \times \hat{x} = \hat{y}\cos\alpha(x) - \hat{z}\sin\alpha(x)$. Then the corresponding current density $\boldsymbol{J}$ inside the superconductor, obtained from Faraday's law, can be resolved into components along $\hat{\alpha}(x)$ and $\hat{\beta}(x)$ as follows: $\boldsymbol{J} = \nabla \times \boldsymbol{H} = J_{\parallel}(x)\hat{\alpha}(x) + J_{\perp}(x)\hat{\beta}(x)$. For simplicity, attention will be restricted to high-κ type II superconductors in which the magnetic flux density obeys $B \gg \mu_0 H_{c1}$, such that $\boldsymbol{B} = \mu_0\boldsymbol{H}$ to good approximation. Then,

$$J_{\parallel}(x) = \mu_0^{-1}B(x)\frac{d\alpha(x)}{dx} \tag{1}$$

$$J_{\perp}(x) = -\mu_0^{-1}\frac{dB(x)}{dx} \tag{2}$$

To satisfy the boundary conditions that $\boldsymbol{B}(\pm w/2) = \mu_0[\boldsymbol{H}_a + \boldsymbol{H}_I(\pm w/2)]$, it is necessary to have $B(\pm w/2) = \mu_0[H_a^2 + (J_a w/2)^2]^{1/2}$ and $\alpha(\pm w/2) = \pm \tan(J_a w/2H_a)$.

The requirement of force-free equilibrium ($\boldsymbol{F}_L = 0$) places no restriction upon $J_{\parallel}(x)$ but dictates only that $J_{\perp}(x) = 0$. From Eqn. (2) it is thus found that the magnitude of $\boldsymbol{B}(x)$ in the force-free case is constant, with the value $B_0 \equiv \mu_0[H_a^2 + (J_a w/2)^2]^{1/2}$, but that its angle $\alpha(x)$ can be any function of x that satisfies the boundary conditions at $\pm w/2$. The component of $\boldsymbol{J}$ along $\boldsymbol{B}$ (i.e., $J_{\parallel}(x) = \mu_0^{-1}B_0\, d\alpha(x)/dx$ from Eqn. (1)) can thus be arbitrarily large. A sketch of a force-free solution is shown in Fig. 1.

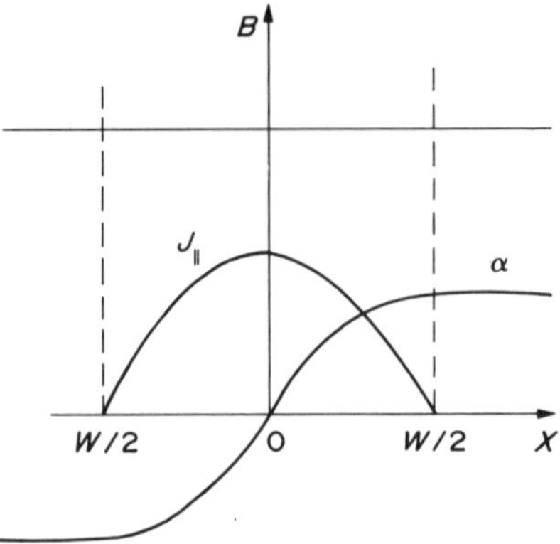

Figure 1
Sketch of a force-free magnetic field and current distribution in a type II superconducting slab of thickness w. Here, B and α are the magnitude and angle of the magnetic induction $\boldsymbol{B}$ and $J_{\parallel}$ is the component of the current density $\boldsymbol{J}$ that is parallel to $\boldsymbol{B}$ (Eqn. (1)). The component of $\boldsymbol{J}$ that is perpendicular to $\boldsymbol{B}$ (Eqn. (2)) vanishes ($J_{\perp} = 0$), so that the Lorentz force density is zero ($\boldsymbol{F}_L = \boldsymbol{J} \times \boldsymbol{B} = 0$)

The nonuniqueness of the force-free solutions is not restricted to the idealized geometry discussed previously. A similar arbitrariness of force-free solutions has been found in cylindrical geometry for the closely related case of a current-carrying wire subjected to a parallel magnetic field (Clem 1976). Both the nonuniqueness of the force-free solutions and the resulting apparent freedom for $J_{\parallel}$ to be arbitrarily large seem unphysical. Theoretical studies of the local stability of force-free solutions have led to a partial resolution of these problems. Perhaps the simplest of these is a calculation of the stability of a straight vortex at the center of a superconducting current-carrying cylinder subjected to a parallel applied magnetic field (Clem 1977). Although the vortex experiences no Lorentz force so long as it remains straight, the calculation showed that tiny helical perturbations along the length of the vortex become unstable to exponential growth if the applied current rises above a certain threshold value. Other calculations have shown that arrays of vortices subjected to parallel currents in force-free configurations become unstable to various kinds of instabilities, which break the symmetry of the array (Brandt 1980, 1981, Clem and Yeh 1980). The growth of these instabilities involves the relative motion of segments of different vortices. This apparently leads inevitably to flux line cutting, the intersection and cross-joining of adjacent vortices (Campbell and Evetts 1972, Walmsley 1972).

Figure 2 illustrates the process of flux line cutting. Because no microscopes or other probes are available that can visualize vortices with the detail shown in Fig. 2, it has been difficult to design totally convincing experimental proofs that this process actually occurs. The most persuasive demonstrations have been made

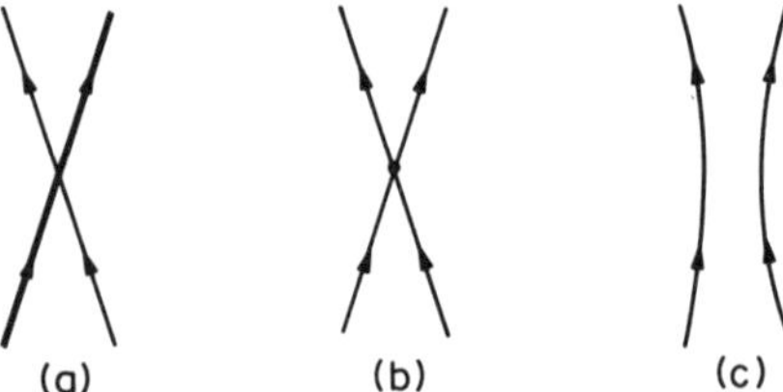

Figure 2
Sketch of the process of flux line cutting: (a) a current density in the vertical direction (not shown) drives together two nonparallel vortices (the bold line indicates the vortex in the foreground), (b) the vortex axes intersect and cross-join, and (c) the resulting vortices move apart and straighten because of repulsive interactions and line tension

by Cave et al. (1978), Fillion et al. (1979) and Blamire and Evetts (1985, 1986).

4. Inhomogeneous Materials: Critical State

In the previous discussion, only homogeneous type II superconductors were considered. Actual superconducting materials, on the other hand, contain numerous imperfections, such as precipitates, dislocations, twin boundaries and grain boundaries, which serve as local pinning sites for the vortices (see *Flux Pinning and the Summation of Pinning Forces*). The magnetic flux distributions in such superconductors thus cannot be truly force free. To describe the evolution of the magnetic flux distribution inside the superconductor, the following critical state description is usually used (see *Critical State Model*; Bean 1962, 1964, Kim et al. 1962, 1963, Kim and Stephen 1969, Campbell and Evetts 1972). As a magnetic field is applied to an inhomogeneous superconductor, vortices are nucleated at the surface and push their way towards the center of the sample under the influence of the Lorentz force $\boldsymbol{F}_L$. This force, however, is opposed by pinning forces and the vortices can move only when the magnitude of $\boldsymbol{F}_L$ locally exceeds the pinning force density F_p. This quantity is usually related, via $F_p = J_c B$, to the critical depinning current density J_c measured in experiments in which the current density $\boldsymbol{J}$ is perpendicular to the magnetic flux density $\boldsymbol{B}$. In general, $J_c(B, T)$ is a function of B and the absolute temperature T (see *Critical Current Density*). It is normally largest when B and T are zero, decreases monotonically as B and T increase and vanishes when B is equal to the upper critical field $B_{c2}(T)$ or T is equal to the superconducting transition temperature T_c. For simple calculations at a given temperature, the B dependence of J_c can be neglected, as in the original Bean model (Bean 1962, 1964), but to obtain excellent agreement between theory and experiment the full B dependence of J_c must be accounted for in the calculations.

For example, Fig. 3 is a sketch of the magnetic induction and the current density distribution in the critical state of an inhomogeneous type II superconducting slab under the conditions of a slowly increasing applied magnetic field $\boldsymbol{H}_a = H_a\hat{z}$. (For simplicity, attention will continue to be restricted to high-κ type II superconductors in which the magnetic flux density obeys $B \gg \mu_0 H_{c1}$, such that $\boldsymbol{B} = \mu_0 \boldsymbol{H}$ to good approximation.) The slopes of B vs x are determined from the force balance equation $F_L = F_p$, which is equivalent to

$$J_\perp(x) = \pm J_c \qquad (3)$$

When combined with Eqn. (2), this equation states that $B(x)$ consists of regions for which, inside the superconductor, the magnitude of $dB(x)/dx$ has a critical value, proportional to J_c. The induced current $\boldsymbol{J}$ has no component parallel to $\boldsymbol{B}$ ($J_\parallel = 0$), so that

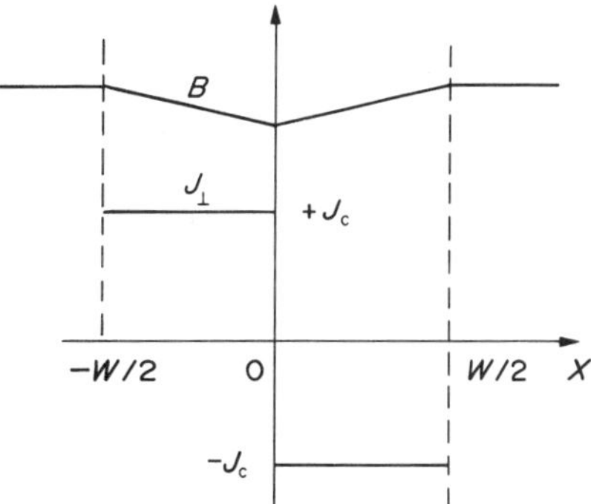

Figure 3
Sketch of the critical state magnetic field $\boldsymbol{B} = B\hat{z}$ and current density distribution $\boldsymbol{J} = J_\perp \hat{y}$ in an inhomogeneous type II superconducting slab of thickness w subjected to a slowly increasing applied magnetic field $\boldsymbol{H}_a = H_a\hat{z}$. The depinning critical current density is J_c. In this case, the component of $\boldsymbol{J}$ that is parallel to $\boldsymbol{B}$ (Eqn. (1)) vanishes ($J_\parallel = 0$)

(Eqn.(1)) $\hat{\alpha} = \hat{z}$ and $\hat{\beta} = \hat{\alpha} \times \hat{x} = \hat{y}$. The perpendicular component (Eqn.(2)) is $J_\perp(x) = -J_c$ for $x > 0$ and $J_\perp(x) = +J_c$ for $x < 0$. For both the regions $x > 0$ and $x < 0$, the Lorentz forces on the vortices are directed towards the center and the pinning forces, which oppose the inward motion of the vortices, are directed outwards.

Another example of a sample in the critical state with $\boldsymbol{J}$ perpendicular to $\boldsymbol{B}$ is sketched in Fig. 4. Here the magnitude of the applied transport current density $\boldsymbol{J}_a = J_a\hat{y}$ is at the critical value J_c at the onset of flux flow. Here, $B(x)$ is obtained from Eqns. (2) and (3), subject to the boundary condition $B(\pm w/2) = \mu_0(H_a \mp J_c w/2)$.

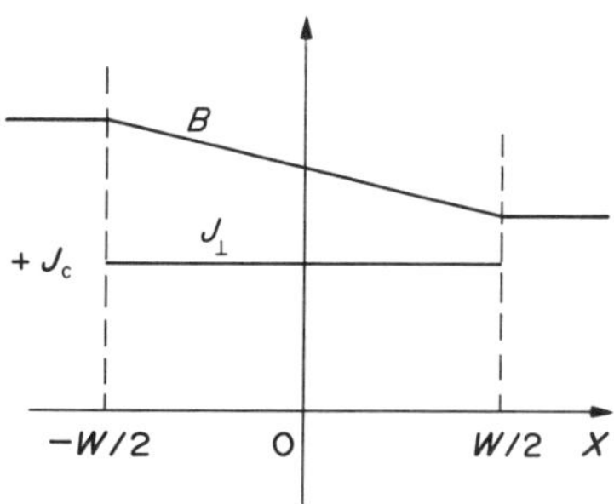

Figure 4
Sketch of the critical state magnetic field $\boldsymbol{B} = B\hat{z}$ and current density distribution $\boldsymbol{J} = J_\perp \hat{y}$ in an inhomogeneous type II superconducting slab of thickness w in an applied magnetic field $\boldsymbol{H}_a = H_a\hat{z}$. The magnitude of the transport current density $J_a = J_\perp$ is just equal to the depinning critical current density J_c and $B(\pm w/2) = \mu_0(H_a \mp J_c w/2)$. The component of $\boldsymbol{J}$ that is parallel to $\boldsymbol{B}$ (Eqn. (1)) vanishes ($J_\parallel = 0$)

As mentioned earlier, the current density $\boldsymbol{J}$ in general has components both perpendicular and parallel to $\boldsymbol{B}$. Since only the perpendicular component $\boldsymbol{J}_\perp$ leads to the Lorentz force, the critical depinning current density will henceforth be denoted by the symbol $J_{c\perp} = F_p/B$, rather than J_c. When $|\boldsymbol{J}_\perp| < J_{c\perp}$, the vortices may be in local thermodynamic equilibrium on a small length scale, but associated with the pinning centers are energy barriers that cause the vortex distribution to be locally stuck in metastable states. When $|\boldsymbol{J}_\perp| = J_{c\perp}$, the vortex array is in a critical state, in which the magnitude of the Lorentz force density is equal to the pinning force density. When $|\boldsymbol{J}_\perp| > J_{c\perp}$, the vortex distribution is unstable against collective vortex motion in the direction of the Lorentz force. Such motion gives rise to an electric field that has only a component $\boldsymbol{E}_\perp$ parallel to $\boldsymbol{J}_\perp$ (Josephson 1965).

5. Critical State Model with Parallel Field Component

A critical state theory accounting in an analogous way for $\boldsymbol{J}_\parallel$, the parallel component of $\boldsymbol{J}$, as well as for the flux line cutting effects that occur above a critical value of $J_\parallel$, has been developed by a number of authors (Cave et al. 1978, Boyer et al. 1980, Clem 1982, Perez-Gonzalez and Clem 1985a,b). This theory contains both the usual depinning critical current density $J_{c\perp}$ and a critical parallel critical current density $J_{c\parallel}$. When $|\boldsymbol{J}_\parallel| < J_{c\parallel}$, there exist energy barriers against flux line cutting that cause the vortex distribution to be locally stuck in metastable states. When $|\boldsymbol{J}_\parallel| = J_{c\parallel}$, the vortex array is in a critical state on the threshold of the flux line cutting instability. When $|\boldsymbol{J}_\parallel| > J_{c\parallel}$, the vortex distribution undergoes flux line cutting and complicated motions of countermoving vortex segments, all of which generates a macroscopic electric field component $\boldsymbol{E}_\parallel$ parallel to $\boldsymbol{J}_\parallel$ (Clem 1981b, 1982).

Despite the simplicity of these ideas, calculations of the time evolution of the magnetic flux distribution in a superconducting slab subjected to parallel time-varying magnetic fields and parallel applied or induced electric currents are surprisingly complex. Such calculations make use of (a) Eqns. (1) and (2) for the parallel and perpendicular current components, (b) Faraday's law, which takes the form (Perez-Gonzalez and Clem 1985a,b)

$$\frac{\partial E_\parallel}{\partial x} = \frac{B\partial\alpha}{\partial t} + E_\perp\frac{\partial\alpha}{\partial x} \tag{4}$$

$$\frac{\partial E_\perp}{\partial x} = -\frac{\partial B}{\partial t} - E_\parallel\frac{\partial\alpha}{\partial x} \tag{5}$$

and (c) equations that express the dependence of $E_\perp$ on $J_\perp$ and $E_\parallel$ on $J_\parallel$ above the corresponding threshold values $J_{c\perp}$ and $J_{c\parallel}$.

The simplest solutions of Eqns. (4) and (5) occur just above the critical current of the slab, when a tiny steady-state electric field is produced throughout the sample cross section (Clem and Perez-Gonzalez 1986); in this case unique solutions for the profiles of $B(x)$ and $\alpha(x)$ can be obtained under special conditions by simultaneously solving the two critical state conditions

$$J_\parallel(x) = \pm J_{c\parallel} \tag{6}$$

$$J_\perp(x) = \pm J_{c\perp} \tag{7}$$

This extended critical state theory with two critical current densities $J_{c\perp}$ and $J_{c\parallel}$ appears to provide a correct explanation of the main features found in the hysteretic behavior of type II superconductors subjected to time-varying magnetic fields and applied or induced electrical current densities with large components parallel to the vortex directions (Cave and LeBlanc 1982, Perez-Gonzalez and Clem 1985b). Some experimental results, however, are not explained by this theory. For example, in the case of cylindrical current-carrying superconductors subjected to a parallel magnetic field, puzzling results include the observation of (a) oscillatory time dependence of the longitudinal electric field (Timms and Walmsley 1976), (b) helical zones of different behavior on the cylinder surface (Irie et al. 1975) and (c) regions of reversed longitudinal electric field (Cave and Evetts 1978).

6. Anisotropic Superconductors

The previous theoretical approaches for understanding force-free currents have focused on the behavior of magnetically isotropic superconductors. It evidently will be necessary, however, to extend these ideas to account for the strong magnetic anisotropy found in the high-temperature oxide superconductors (see *Anisotropy in Superconducting Materials*). Recent experiments on the resistive transition of high-temperature superconductors subjected to magnetic fields both parallel and perpendicular to the current direction (Iye et al. 1989, Woo et al. 1989) have yielded results that are in conflict with the isotropic theory but which can be understood qualitatively if strong anisotropy is taken into account (Kes et al. 1990).

See also: Critical Current Density; Critical State Model; Flux Line Lattice; Flux Pinning and the Summation of Pinning Forces; Magnetic Properties of Superconductors; Resistive Transition and Flux Flow in Superconducting Materials

Bibliography

Anderson P W, Kim Y B 1964 Hard superconductivity: theory of the motion of Abrikosov flux lines. *Rev. Mod. Phys.* 36: 39–43

Bean C P 1962 Magnetization of hard superconductors. *Phys. Rev. Lett.* 8: 250–3

Bean C P 1964 Magnetization of high-field superconductors. *Rev. Mod. Phys.* 36: 31–9

Blamire M G, Evetts J E 1985 Critical currents and flux cutting in thin superconducting films. *Appl. Phys. Lett.* 46: 1181–3

Blamire M G, Evetts J E 1986 Critical cutting force between flux vortices in a type-II superconductor. *Phys. Rev. B* 33: 5131–3

Boyer R, Fillion G, LeBlanc M A R 1980 Hysteresis losses and magnetic phenomena in rotating disks of type-II superconductors. *J. Appl. Phys.* 51: 1692–701

Brandt E H 1977 Elastic energy of the vortex state in type II superconductors: I, high inductions. *J. Low Temp. Phys.* 26: 709–33

Brandt E H 1980 Helical vortex instability limits in longitudinal current in type II superconductors. *Phys. Lett. A* 79: 207–9

Brandt E H 1981 Longitudinal critical currents in type II superconductors: I, helical vortex instability in the bulk; II, helical vortex instability near the surface. *J. Low Temp. Phys.* 44: 33–57; 59–72

Campbell A M, Evetts J E 1972 Flux vortices and transport currents in type-II superconductors. *Adv. Phys.* 21: 199–428

Cave J R, Evetts J E 1978 Static electric potential structures on the surface of a type II superconductor in the flux flow state. *Philos. Mag. B* 37: 111–8; errata, ibid. 663

Cave J R, Evetts J E, Campbell A M 1978 AC flux penetration measurements on force free configurations in reversible type II superconductors. *J. Phys. (Paris)* 39 (C6): 614–16

Cave J R, LeBlanc M A R 1982 Hysteresis losses and magnetic phenomena in oscillating disks of type II superconductors. *J. Appl. Phys.* 53: 1631

Clem J R 1975 Phenomenological theory of the local magnetic field in type-II superconductors. In: Krusius M, Vuorio M (eds.) 1975 *Low Temperature Physics-LT14*, Vol. 2. North-Holland, Amsterdam, pp. 285–8

Clem J R 1976 On the establishment of force-free configurations in type-II superconductors. *Phys. Lett. A* 59: 401–2

Clem J R 1977 Spiral-vortex expansion instability in type-II superconductors. *Phys. Rev. Lett.* 38: 1425–8

Clem J R 1981a Flux-flow noise in superconductors. *Phys. Rep.* 75: 1–55

Clem J R 1981b Steady-state flux-line cutting in type-II superconductors: a double-cutting model. *Physica B* 107: 453–4

Clem J R 1982 Flux-line-cutting losses in type-II superconductors. *Phys. Rev. B* 26: 2463–73

Clem J R, Perez–Gonzalez A 1986 Internal-magnetic-field distribution at the critical current of a type-II superconductor subjected to a parallel magnetic field. *Phys Rev. B* 33: 1601–10

Clem J R, Yeh S. 1980 Flux-line-cutting threshold in type-II superconductors. *J. Low Temp. Phys.* 39: 173–89

Fillion G, Gauthier R, LeBlanc M A R 1979 Direct evidence of flux cutting in type-II superconductors. *Phys. Rev. Lett.* 43: 86–9

Friedel J, de Gennes P G, Matricon J 1963 Nature of the driving force in flux creep phenomena. *Appl. Phys. Lett.* 2: 119–22

Irie F, Izaki T, Yamafuji K 1975 Flux-flow-like state of a Pb–T1 rod in a longitudinal field. *IEEE Trans. Magn.* 11: 332–5

Iye Y, Nakamura S, Tamegai T 1989 Absence of current direction dependence of the resistive state of high temperature superconductors in magnetic fields. *Physica C* 159: 433–8

Josephson B D 1965 Potential differences in the mixed state of type II superconductors. *Phys. Lett.* 16: 242–3

Kes P H, Aarts J, Vinokur V M, van der Beek C J 1990 Dissipation in highly anisotropic superconductors. *Phys. Rev. Lett.* 64: 1063–6

Kim Y B, Hempstead C F, Strnad A R 1962 Critical persistent currents in hard superconductors. *Phys. Rev. Lett.* 9: 306–9

Kim Y B, Hempstead C F, Strnad A R 1963 Magnetization and critical supercurrents. *Phys. Rev.* 129: 528–35

Kim Y B, Hempstead C F, Strnad A R 1964 Resistive states of hard superconductors. *Rev. Mod. Phys.* 36: 43–5

Kim Y B, Hempstead C F, Strnad A R 1965 Flux-flow resistance in type-II superconductors. *Phys. Rev. A* 139: 1163–72

Kim Y B, Stephen M J 1969 Flux flow and irreversible effects. In: Parks R D (ed.) 1969 *Superconductivity*, Vol. 2. Dekker, New York, pp. 1107–65

Kogan V G 1978 On the thermodynamics of the inhomogeneous mixed phase of an ideal type II superconductor. *J. Low Temp. Phys.* 32: 419–38

Perez–Gonzalez A, Clem J R 1985a Response of type-II superconductors subjected to parallel rotating magnetic fields. *Phys. Rev. B* 31: 7048–58

Perez–Gonzalez A, Clem J R 1985b Magnetic response of type-II superconductors subjected to large-amplitude parallel magnetic fields varying in both magnitude and direction. *J. Appl. Phys.* 58: 4326–35

Schelten J, Ullmaier H, Lippmann G 1972 Local magnetic field distributions in superconducting niobium at 4.2 K by neutron diffraction. *Z. Phys.* 253: 219–31

Timms W E, Walmsley D G 1975 Spiral fluxlines in irreversible superconducting cylinders. *J. Phys. F* 5: 287–306

Timms W E, Walmsley D G 1976 Magnetic and resistive behaviour of a superconducting Nb–54 at% Ta cylinder. *J. Phys. F* 6: 2107–18

Walmsley D G 1972 Force free magnetic fields in a type II superconducting cylinder. *J. Phys. F* 2: 510–28

Woo K C, Gray K E, Kampwirth R T, Kang J H, Stein S J, East R, McKay D M 1989 Lorentz-force independence of resistance tails for high-temperature superconductors in magnetic fields near T_c. *Phys. Rev. Lett.* 63: 1887–9

J. R. Clem
[Iowa State University, Ames, Iowa, USA]

Tunnelling Spectroscopy

The tunnelling effect is basic to quantum mechanics, in which the probability $\Psi^*\Psi$ of finding an electron beyond a barrier of width t and height φ_B is nonzero of order $\exp[-2(2m\varphi_B/\hbar^2)^{1/2}t]$. Here, m is the electron mass and $2\pi\hbar$ is Planck's constant, 4.14×10^{-12} meV s. Hence, a voltage-biased capacitor-like structure with a sufficiently thin ($t = 1$–5 nm) uniform insulating barrier (insulator) between two normal metallic electrodes passes an electron current, which is approximately ohmic, $I = G_nV$, for $V \ll \varphi_B/e$. The tunnelling current density, a sensitive function of φ_B and t, can be large (e.g., $\sim 10^7\,\mathrm{A\,m^{-2}}$). Tunnelling spectroscopy, most effective with superconductors, employs the tunnelling effect in normal–insulator–superconductor (NIS), superconductor–insulator–superconductor (SIS) and other structures to determine energy gaps, densities of states and the origin of superconductivity through the effective phonon spectrum $\alpha^2F(\omega)$ and its possible generalization in unconventional superconductors (see *Superconducting Materials: BCS and Phenomenological Theories*). Spectroscopy is possible since the bias V appears only across the barrier, and the most energetic electron elastically tunnelling from a normal metal appears at energy eV in the opposite electrode. Hence, the conductance dI/dV gives a spectroscopy of the density of excitations or final states $E = eV$ in the electrode with resolution limited only by the thermal energy k_BT.

1. Superconductive Tunnelling

1.1 Spectra of NIS Junctions

Giaever (1960) showed that the normalized conductance $G_s/G_n = N_T(E = eV) = [(dI/dV)_{ns}/(dI/dV)_{nn}]$ of an NIS tunnel junction measures the excitation spectrum $N(E)$ or density of states of the superconductor S, supporting the BCS prediction $N_{BCS}(E) = |E|/(E^2 - \Delta_0^2)^{1/2}$, for $|E| > \Delta_0$ and $N_{BCS}(E) = 0$ for $|E| < \Delta_0$ (see Fig. 1). Thus, the normalized conductance of an NIS tunnel junction can provide a precise measure of the energy gap Δ_0 of a BCS superconductor. Quantitative work at higher T entails deconvolution of the thermal broadening:

$$N_T(eV) = \int_{-\infty}^{\infty} N_{BCS}(E)[-df(E - eV)/d(eV)]dE \quad (1)$$

where $f(E) = [1 + \exp(E/k_BT)]^{-1}$.

(a) *Strong coupling and Eliashberg inversion*. The deviations from $N_{BCS}(eV)$ evident in Fig. 1 near $eV = 4\Delta$ and 8Δ, in fact appearing at $eV = \hbar\omega_{ph} + \Delta_0$, with ω_{ph} a peak in the phonon spectrum $F(\omega)$ of lead and Δ_0 the gap of lead, indicate strong coupling superconductivity. Quantitative determination from precise tunnelling data of the effective phonon spectrum $\alpha^2F(\omega)$ and the Coulomb pseudopotential μ^*, making essential use of Eliashberg's strong coupling theory of superconductivity, is due to McMillan and Rowell (1969). In Eliashberg's theory, Δ_0 is generalized to complex $\Delta(E)$, and $N(E) = \mathrm{Re}[E/(E^2 - \Delta^2)^{1/2}]$. This quantitatively treats phonon anomalies as seen in Fig. 1 which arise from structure in $\Delta(E)$ near critical points $E = \hbar\omega_{ph}$ of the phonon spectrum $F(\omega)$ of the superconductor (see *Superconducting Materials: BCS and Phenomenological Theories*). A review of extensions to the theory is given by Wolf (1985).

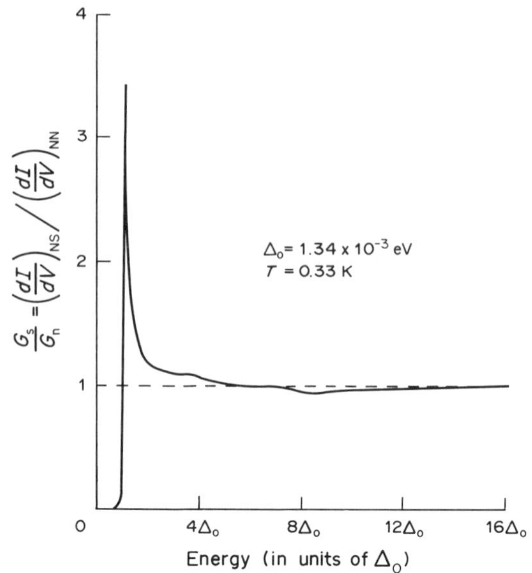

Figure 1
Normalized conductance of Mg–MgO–Pb (NIS) sandwich, measured at 0.33 K, vs eV/Δ_0. Note the sharpness of the gap, $\Delta_0 = 1.34$ meV for lead, and phonon-related strong coupling anomalies near $4\Delta_0$ and $8\Delta_0$, from which precise determination of $\alpha^2F(\omega)$ and μ^* is possible (after Giaever et al. 1962)

(b) *Anisotropic gaps, pairbreaking and gapless superconductivity*. Quantitative study of G_s/G_n via Eqn. (1) can identify other forms of superconductivity and junction structures. For example, if the gap is $\Delta(\boldsymbol{k})$, a function of the wave vector in k space and possibly zero in some directions of $\boldsymbol{k}$ (e.g., with p- or d-type symmetry), $N_T(E) = \int N_{BCS}[\Delta(\boldsymbol{k})]d^3\boldsymbol{k}$. Pairbreaking at a rate Γ (e.g., by magnetic impurities or by strong coupling) can change $N_T(E)$ to characteristic forms, one being obtained by replacing E by $E + i\Gamma$ (see *Nonequilibrium Superconductivity*). A thin normal metal "proximity" layer between the S electrode and the barrier also produces characteristic effects (see Fig. 2).

1.2 Josephson Tunnelling in SIS Junctions

In S_1IS_2 junctions between two superconductors, a dc Josephson supercurrent, $I_J = I_c \sin\varphi$, with $\varphi = \theta_2 - \theta_1$ the difference in the superconducting phases on the two sides of the junction, occurs at $V = 0$ (see *Josephson Effects in Weak Links*). This effect is expected at sufficiently small junction resistance (high I_c) that the Josephson coupling energy $hI_c/2e$ exceeds both k_BT and the ambient noise power in the experiment—to provide phase locking. The rapid Fraunhofer-like $I_c(B)$ is a useful verification of true Josephson SIS behavior. This is $I_c(B) = I_c(0)\,\mathrm{abs}[\sin(\pi\Phi_B/\Phi_0)/(\pi\Phi_B/\Phi_0)]$, with Φ_B the magnetic flux threading the junction, and

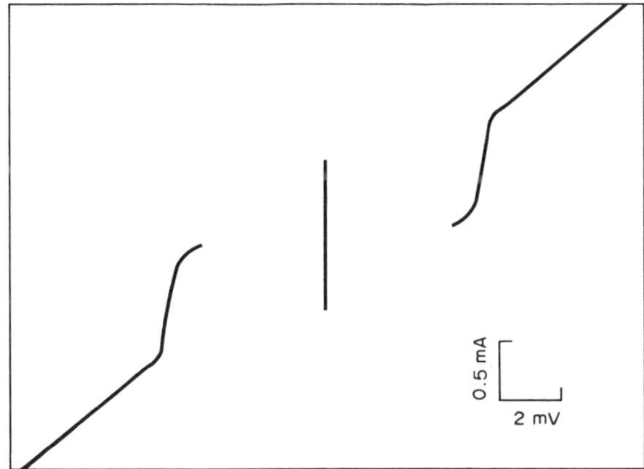

Figure 2
Josephson supercurrent and sum-gap quasiparticle branches of a 10 μm × 10μm NbN–MgO–NbN computer junction (Shoji et al. 1982) with amorphous RF sputtered MgO barrier: the proximity layer produces a "ski jump" anomaly above the sum-gap edge (oscilloscope trace)

$\Phi_0 = h/2e = 2.07 \times 10^{-15}$ Wb, the magnetic flux quantum. As a consequence of Josephson's ac effect, $d\varphi/dt = 2eV/\hbar$, Shapiro steps of spacing $V = \hbar\omega/2e$ and characteristic intensity dependence appear in the $I(V)$ of a junction irradiated with microwaves (see *Voltage-Standard Devices*). An excellent review is given by Barone and Paterno (1982). In addition to the superconducting branch of the S_1IS_2 junction, the quasiparticle branch of $I(V)$ at zero temperature has sharp onsets at the sum gap $|eV| = \Delta_1 + \Delta_2$; and at finite temperature shows a cusplike current $I(V) = \text{const} \times \ln[|eV - (|\Delta_1 - \Delta_2|)|] \times [\exp(\Delta_1/k_BT) - \exp(\Delta_2/k_BT)]$. At high bias $(eV \gg \Delta_1 + \Delta_2)$, I approaches G_nV (Fig. 2) (see *Thin Films, Multilayers and Devices, Superconducting*).

2. *Fabrication of Tunnelling Junctions*

2.1 *Oxide and Deposited Insulator Thin-Film Junctions*

The original junctions of Giaever (1960) were formed by self-limiting thermal oxidation of an evaporated metal strip, forming a thin uniform oxide layer, followed by evaporation of a second crossing strip. Conventional thin-film tunnelling spectroscopy of metals such as gold and niobium which do not form suitable uniform oxides has been problematic; uniformity and freedom from pinholes of directly deposited layers has been achieved with care in some cases. These include rf-sputtered MgO (Shoji et al. 1982), electron beam evaporated Al_2O_3 (Moodera et al. 1982), evaporated amorphous silicon (Rudman and Beasley 1980), and by deposition of subsequently oxidized wetting metal proximity layers, especially aluminum (Wolf et al. 1980, Gurvitch et al. 1983). The last method has become the basis for Josephson computer junction manufacture using refractory superconductors (see *Superconducting Device Fabrication*). Current reviews of these topics are provided by Donaldson (1989) and Hayakawa (1989).

2.2 *Mechanically Controlled Vacuum Barriers*

Variable piezoelectrically controlled "vacuum" or liquid helium gaps, continuously x, y and z adjustable up to about 1 μm and stable to better than 10^{-11} m are now available using techniques developed in scanning tunnelling microscopy (e.g., Kuk and Silverman 1989). A review of low-temperature STM gap spectroscopy of high-temperature superconductors (HTSs) has been given by Kirtley (1990). Large area mechanically z-adjusted tunnel electrodes permit "vacuum tunnelling." In the "break-junction" method a crystalline sample is mechanically stressed to fracture under liquid helium (or in vacuum), and mechanically readjusted to bring the resulting clean surfaces to tunnelling distance.

2.3 *Point Contact Junctions*

Mechanically adjusted point contacts, with the two metals mechanically touching through an oxide or other insulating layer, are sometimes useful. For a recent example see the study by Huang et al. (1990) of $Ba_{1-x}K_xBiO_3$.

2.4 *Natural Barriers*

Charge readjustment occurs within a screening length $\lambda = (2\varepsilon_0\varphi_B/N_ie^2)^{1/2}$ of the surface of a semiconductor of shallow impurity density N_i; the resulting natural Schottky barrier of height φ_B is inherently uniform and has permitted tunnelling spectroscopy with metals deposited directly on doped-silicon surfaces, for example. Gurvitch et al. (1989) have described etching processes for YBCO which form a tunnelling barrier between the material and lead-evaporated films.

2.5 *Proximity Junctions NIN′S*

The tunnelling density of states of N′S bilayers in the clean limit and N′ of thickness 2–10 nm is known in terms of the gaps $\Delta(E)_{s,n}$ (Arnold 1978, Wolf and Arnold 1982). This permitted, for example, determination of the Eliashberg function $\alpha^2F(\omega)$ for the normal metal magnesium, inducing superconductivity by proximity to niobium in an Ag–MgO–MgNb junction (see *Coherence Length, Proximity Effect and Fluctuations*; Burnell and Wolf 1982).

3. *Other Forms of Spectroscopy*

3.1 *Inelastic Thresholds*

Small increases in $G(V)$ may occur at thresholds $V = e^{-1}\hbar\omega_x$, representing the onset of a second process in which the tunnelling electron excites a mode $\hbar\omega_x$ and appears at the Fermi energy, rather than at eV. Hence, $d^2I/dV^2 = -(dI/dV)^3 d^2V/dI^2$ can provide an inelastic electron tunnelling spectroscopy (IETS), for example of vibrational modes of organic molecules trapped in the oxide of conventional tunnel junctions (Hansma 1982).

3.2 Final State Density and Normal Systems

In planar NIN junctions where N is a noninteracting one-electron metal, there is no useful spectroscopy of band density of states, although weak evidence of band edges is reported.

(*a*) *Two-dimensional final states.* $G(V)$ is proportional to the density of states multiplied by the voltage-dependent tunnel probability if these are two-dimensional as in size quantization in a thin film or at the surface of a semiconductor.

(*b*) *Many-body density of states.* If one metal N exhibits strong electron correlation effects, then $G(V)$ may represent a many-body density of states (again multiplied by the voltage-dependent tunnelling probability factor), as, for example, near the metal–insulator transition in niobium–silicon (Hertel et al. 1983).

(*c*) *Scanning tunnelling microscope.* In scanning tunnelling spectroscopy, in which tunnelling occurs not from a planar electrode but from a tip which may be approximated by a single atom, a local density of states is obtained in the form $(dI/dV)[I/V]^{-1}$ which is revealing of the electronic structure of reconstructed silicon surfaces (e.g., Stroscio et al. 1986).

4. Tunnelling Spectroscopy of High-T_c Cuprates and Bismuthates

Tunnelling spectra of cuprates are complicated by the anisotropy and small coherence lengths. Giving preference to conventionally obtained spectra (e.g., Gurvitch et al. 1989) over STM results (reviewed by Kirtley 1990) it appears that the cuprates are gapless superconductors. Whether this arises by unconventional superconductivity in the CuO_2 planes (e.g., Lenck et al. 1990) or possibly from the layering of normal and superconducting planes (Tachiki et al. 1990) is unclear. Evidence of high-temperature pair breaking in cuprates from tunnelling is given by Wolf et al. (1991). The superconductivity of the bismuthates appears with a full conventional gap in the recent data of Huang et al. (1990).

See also: Josephson Effects in Weak Links; Nonequilibrium Superconductivity; Superconducting Materials: BCS and Phenomenological Theories; Thin Films, Multilayers and Devices, Superconducting

Bibliography

Arnold G B 1978 *Phys. Rev. B* 18: 1076

Barone A, Paterno G 1982 *Physics and Applications of the Josephson Effect.* Wiley, New York

Burnell D M, Wolf E L 1982 Tunneling determination of electron–phonon interaction in Mg. *Phys. Lett. A* 90: 471

Donaldson G B 1989 Fabrication of tunnel junction structures. In: Weinstock H, Nisenoff M (eds.) 1989 *Superconducting Electronics.* Springer, Berlin

Giaever I 1960 *Phys. Rev. Lett.* 5: 147; 464

Giaever I, Hart H R, Megerle K 1962 *Phys. Rev.* 126: 941

Gurvitch M, Valles J M Jr, Cucolo A M, Dynes R C, Garno J P, Schneemeyer L F, Waszczak J V 1989 Reproducible tunnelling data on chemically etched single crystals of $YBa_2Cu_3O_7$. *Phys. Rev. Lett.* 63: 1008–11

Gurvitch M, Washington M A, Huggins H A 1983 High-quality refractory Josephson tunnel junctions utilizing thin aluminum layers *Appl. Phys. Lett.* 42: 472–4

Hansma P K 1982 *Tunneling Spectroscopy: Capabilities, Applications and New Techniques.* Plenum, New York

Hayakawa H 1989 Josephson LSI technology and circuits. In: Weinstock H, Nisenoff M (eds.) 1989 *Superconducting Electronics.* Springer, Berlin

Hertel G, Bishop D J, Spencer E G, Rowell J M, Dynes R C 1983 Tunneling and transport measurements at the metal–insulator transition of amorphous Nb–Si. *Phys. Rev. Lett.* 50: 743–6

Huang Q, Zasadzinski J F, Gray K E, Richards D R, Hinks D G 1990 Ideal tunneling characteristics in $Ba_{1-x}K_xBiO_3$ point-contact junctions with Au and Nb tips. *Appl. Phys. Lett.* 57: 2356–8

Kirtley J R 1990 *Int. J. Mod. Phys. B* 4: 201

Kuk Y, Silverman P J 1989 Scanning tunneling microscope intrumentation. *Rev. Sci. Instrum.* 60: 165–80

Lenck S, Wermbter S, Tewordt L 1990 Effect of phonons and antiferromagnetic spin fluctuations on the pairing and density of states in high-T_c superconductors. *J. Low Temp. Phys.* 80: 269–83

McMillan W L, Rowell J M 1969 Tunnelling and strong coupling superconductivity. In: Parks R D (ed.) 1969 *Superconductivity*, Vol. 1. Dekker, New York

Moodera J S, Meservey R, Tedrow P M 1982 *Appl. Phys. Lett.* 41: 488

Rudman D, Beasley M R 1980 *Appl. Phys. Lett.* 36: 1010

Shoji A, Shinoki F, Kosaka S, Aoyagi J, Hayakawa H 1982 *Appl. Phys. Lett.* 41: 1097

Stroscio J A, Feenstra R M, Fein A P 1986 Electronic structure of the Si(111) 2 × 1 surface by scanning tunneling microscopy. *Phys. Rev. Lett.* 57: 2579–82

Tachiki M, Takahashi S, Steglich F, Adrian H 1990 Tunnelling conductance in layered oxide superconductors: *Z. Phys. B* 80: 161–6

Wolf E L 1985 *Principles of Electron Tunneling Spectroscopy.* Oxford University Press, New York

Wolf E L, Arnold G B 1982 Proximity electron tunnelling spectroscopy. *Phys. Rep.* 91: 31–102

Wolf E L, Tao H J, Susla B 1991 Tunnelling evidence of strong Cooper-pair-breaking near T_c in cuprate superconductors. *Solid State Commun.* 77: 519–23

Wolf E L, Zasadzinski J, Osmun J W, Arnold G B 1980 *J. Low Temp. Phys.* 40: 19

E. L. Wolf
[Polytechnic University, Brooklyn, New York, USA]

V

Voltage-Standard Devices

The International System of Units, abbreviated SI, has been developed to meet the need for a uniform and consistent set of units. One of the most important and widely used of these units is the volt. The SI definition of the volt is that electromotive force between two points on a conductor carrying a constant current of 1 A when the power dissipated between the two points is 1 W. The base SI electrical unit, the ampere, is defined in terms of the force between two current-carrying wires 1 m apart. The ohm is obtained from the farad which is realized using a calculable capacitor based on electrostatic theory. The realization of the SI volt, therefore, depends on experiments that relate the primary electrical units to mechanical units of length, force and power. These experiments are exceedingly difficult and time consuming.

Precise modern instrumentation requires voltage measurements with a reproducibility exceeding the accuracy with which the SI volt can be realized. To meet this need, realizations of the volt that are stable and reproducible at a level near 0.01 ppm have been developed even though their value relative to the SI volt is uncertain within about 0.4 ppm.

Before 1972, realizations of the volt were made by assigning values to carefully stabilized banks of Weston cells. These Weston cells served as a kind of flywheel to maintain the unit of voltage between the comparison experiments with the SI definition of the volt. Drifts and problems with the transportability of Weston cells limited the uniformity of voltage standards around the world to about 1 ppm. The voltage standard clearly needed an intrinsic definition much like the definition of the meter in terms of the speed of light. Fortunately, such an intrinsic standard became feasible with the discovery of the Josephson effect in 1962.

1. Josephson Voltage Standard

Brian Josephson derived an equation for the current that would flow through a tunnel junction formed by a thin insulating barrier separating two superconductors:

$$I = I_c \sin \frac{4\pi e}{h} \int V\, dt \qquad (1)$$

where I is the junction current, I_c is the critical current (a constant of the junction), V is the junction voltage and e/h is the ratio of the electron charge to Planck's constant. When a dc voltage is applied across the junction, Eqn. (1) shows that the current will oscillate at a frequency $f = 2eV/h$, where $2e/h \simeq 484\ \mathrm{GHz\,mV^{-1}}$. The very high frequency and low level of this oscillation make it difficult to observe directly. However, if an ac current at frequency f is applied to the junction, the junction oscillation tends to phase lock to the applied frequency. During this phase lock, the average junction voltage must equal $hf/2e$. This effect, known as the ac Josephson effect, is observed as a constant-voltage step at $V = hf/2e$ in the current–voltage (IV) curve of the junction. It is also possible for the junction to phase lock to harmonics of f. This results in a series of steps at $V = nhf/2e$, where n is an integer as shown in Fig. 1a.

The ac Josephson effect was initially used to improve the measurement of the constant $2e/h$. These

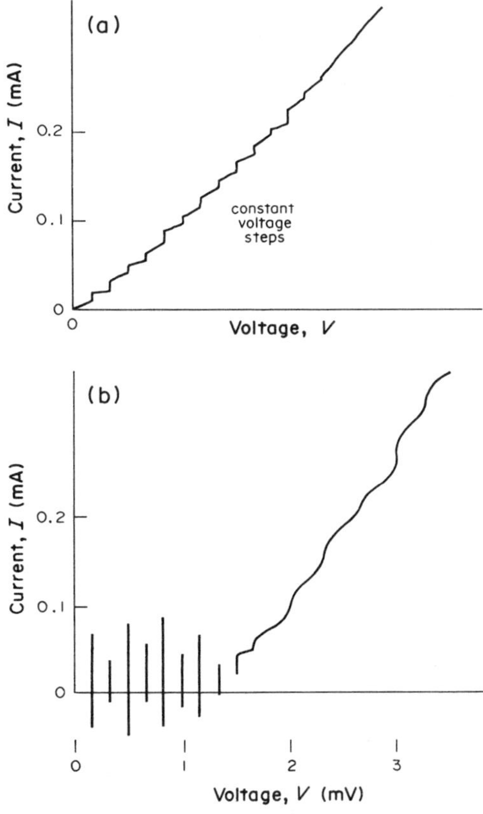

Figure 1
IV curves of a Josephson junction showing the quantum voltage steps induced by microwave radiation for (a) a low-capacitance junction and (b) a high-capacitance junction

measurements were limited by the accuracy of the voltage standard. Thus, in the early 1970s many national standards laboratories assigned a value to the Josephson constant $K_J = 2e/h$ and began using the ac Josephson effect as the practical standard of voltage. Although the Josephson standard does not realize the SI definition of the volt, it provides a very stable reference voltage that can be reproduced anywhere without the need to transfer artifact standards such as Weston cells. By international agreement, K_J was assigned the value 483 597.9 GHz V^{-1}, effective 1 January 1990.

The accuracy of the Josephson voltage–frequency relation $V = nf/K_J$ and its independence from experimental conditions, such as bias current and junction materials, have been subjected to many tests. No significant deviations from this relation have ever been found. In the most precise of these experiments, the voltage difference between two Josephson devices operating at the same frequency is less than 3 parts in 10^{19}.

Although the ac Josephson effect provides a much better voltage reference than standard cells, the first single-junction Josephson standards were difficult to use because they generated very small voltages (1–10 mV). Early attempts to add the voltages across many junctions were limited by the independent bias-current adjustment needed for each junction. The multiple-bias problem has since been avoided by using constant-voltage steps that cross the zero-current axis of the junction. These zero-crossing steps occur when junctions with a large capacitance C are driven by microwave current at a frequency well above the junction's plasma frequency $f_p = (I_c e/\pi hC)^{1/2}$. The plasma frequency is a natural junction resonance defined by the capacitance and inductance of the junction.

An *IV* curve with zero-crossing steps is shown in Fig. 1b. The lack of stable regions between the first few steps confirms that, for small dc bias currents, the junction voltage must be quantized. With a common bias current at or near zero, the voltage across a large array of these junctions must also be quantized. With an array of 10 000 or more junctions, quantized voltages in the range -10–$+10$ V are possible.

2. *Junction Properties*

The important parameters for the Josephson junction shown in Fig. 2 are its length l, width W, critical current density J_c and the rf drive frequency f. The practical realization of an array voltage standard requires a thorough understanding of how these parameters affect the stability of the quantized voltage levels shown in Fig. 1b. Stable operation requires that four conditions be satisfied: (a) l must be small enough that the flux induced through the junction area by the ac magnetic field is much less than $h/2e$, (b) both W and l must be small enough that the lowest resonant cavity mode of the junction is greater than f, (c) the junction plasma frequency f_p which is proportional to $J_c^{1/2}$ must be about $\frac{1}{3}f$, and (d) the junction's critical current $I_0 = WlJ_c$ should be as large as possible to prevent quantum step transitions due to noise. If any of these conditions is violated the junction voltage is likely to switch randomly among several steps and make measurements impossible. Table 1 lists a suitable set of design parameters.

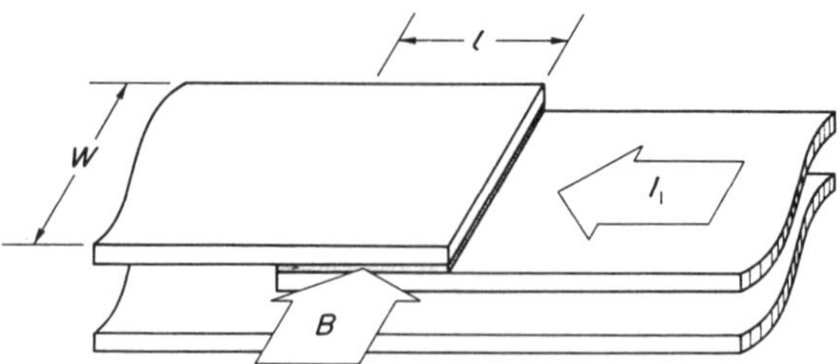

Figure 2
A schematic drawing of a Josephson junction

3. *Array Design*

A Josephson series-array voltage standard requires a circuit design that can deliver nearly uniform microwave power to many thousands of junctions, all of which are connected in series. A typical integrated circuit layout for an 18 992-junction array is shown in Fig. 3. The rf drive is collected from a waveguide by a finline antenna, split 16 ways and injected into 16 linear series-connected arrays of 1187 junctions each. The arrays are separated from a superconducting ground plane by SiO dielectric and operate as microstriplines. Each stripline is terminated in a matched load. Addition of the dc voltages across the 16 arrays produces the reference-voltage output. Capacitors prevent the dc voltage from shorting through the rf distribution network. The rf power is applied by inserting the finline end of the chip into a slot parallel to the E field in a WR-12 waveguide. The dc output appears across superconducting pads on the edge of

Table 1
10 V array design parameters

Junction materials	$Nb/Nb_2O_5/PbInAu$
Critical current density (A cm^{-2})	70
Junction length (μm)	12
Junction width (μm)	24
Critical current (μA)	200
Plasma frequency (GHz)	22
Lowest resonant mode (GHz)	175
rf drive frequency (GHz)	70–100
rf drive power at finline input (mW)	5

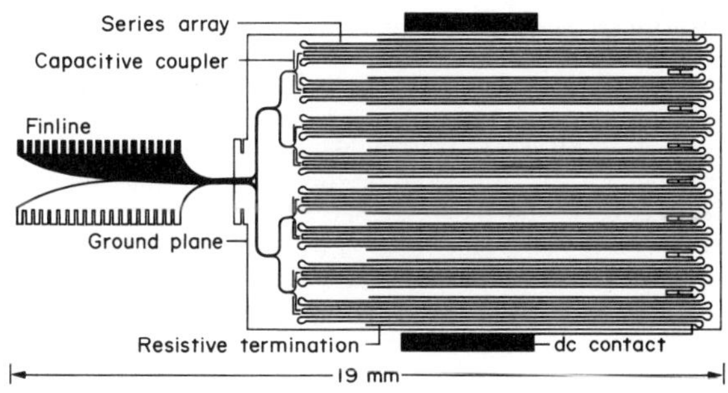

Figure 3
The integrated circuit layout for a large array of Josephson junctions designed for use as a 10 V voltage standard

the chip. The number of junctions in each stripline section is limited by the attenuation in the stripline, which is estimated to be 0.004 dB per junction. After passing through about 1200 junctions the microwave power is significantly below the optimum value for step generation. At 72 GHz, each junction generates about eight quantum voltage levels separated by $f/K_J \simeq 150\,\mu V$. The array shown in Fig. 3 can generate about 150 000 levels spanning the range from −10–+10 V.

Voltage standard chips are fabricated on silicon or glass substrates. The integrated circuit has seven levels: (a) a 300 nm niobium ground plane that shields the circuit from outside interference and defines the impedance of the microstriplines, (b) a 1000 nm layer of SiO that forms the microstripline dielectric, (c) a 200 nm niobium film that forms the lower electrode of the Josephson junctions, (d) a 3 nm metal oxide layer that forms the Josephson tunnelling barrier, (e) windows in a 300 nm SiO film that define the junction areas, (f) a 400 nm film of niobium or lead alloy that forms the junction top electrodes, and (g) a 100 nm resistive film that forms the stripline terminations (see *Superconducting Device Fabrication*).

4. *Voltage-Standard Systems*

A block diagram of a typical voltage-standard system is shown in Fig. 4. The chip is mounted in a magnetically shielded cryoprobe and immersed in liquid helium. A Gunn diode, which operates in the range 70–100 GHz, provides the microwave power to the chip. A microwave counter measures and phase locks the frequency of the Gunn diode to an internal or external frequency reference. A dc current source and a low-frequency triangle wave generator provide current bias to the chip. The oscilloscope displays the triangle wave current as a function of the array voltage. When the array is locked to a constant-voltage step, the oscilloscope displays a vertical line.

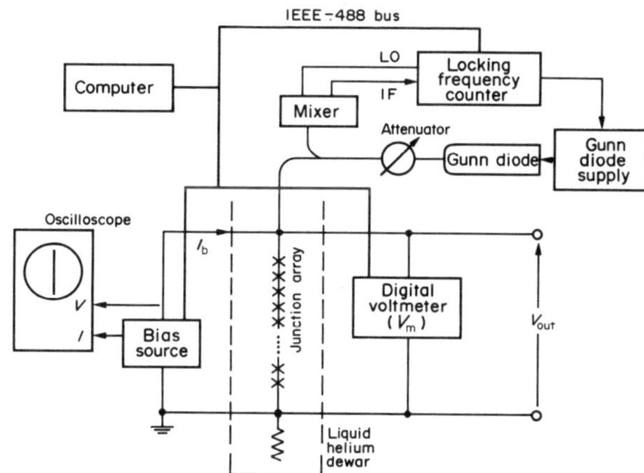

Figure 4
A block diagram of a Josephson voltage-standard system

If even one junction in the array is not locked to a step, the oscilloscope display has a pronounced slope. The voltmeter and frequency counter are connected through the IEEE-488 bus to a computer that can perform calibration, record keeping and diagnostic functions.

Systems like that shown in Fig. 4 are used to calibrate secondary standards, such as Weston cells, Zener references and precise digital voltmeters. These calibrations are greatly simplified by the fact that the Josephson array voltage $V = nf/K_J$ can be set to any arbitrary value in the range −10–+10 V. In a typical calibration, a sensitive meter is used to measure the difference between the array and unknown reference voltages. The parameters n and f are then adjusted to null the difference voltage. The ability to adjust the array voltage eliminates the need for voltage dividers. The typical uncertainty in these calibrations is 0.01 ppm; with great care, 0.001 ppm can be achieved. Greater accuracies are currently irrelevant since there are no secondary standards with even short-term stability better than 0.001 ppm.

The accuracy and stability of voltage standards based on the Josephson effect far surpass those of all other voltage-standard devices. The voltage generated is independent of environmental and material characteristics because it is based on a fundamental constant, the ratio of the elementary charge to Planck's constant. International agreement on the value of this constant ensures uniformity and reproducibility throughout the world.

Acknowledgement

This article was written by an employee of the US Government and is therefore not subject to copyright.

See also: Josephson Effects in Weak Links; Superconducting Device Fabrication; Thin Films, Multilayers and Devices, Superconducting

Bibliography

Field B F, Finnegan T F, Toots J 1973 Volt maintenance at NBS via $2e/h$; a new definition of the NBS volt. *Metrologia* 9: 155–66

Goldman D F, Bell R J 1986 *The International of Units (SI)*, National Bureau of Standards Special Publication 330. US Government Printing Office, Washington, DC

Petley B W 1981 Quantum metrology and electrical standards. In: Cutler P H, Lucas A A (eds.) 1981 *Quantum Metrology and Fundamental Physical Constants*. Plenum, New York, pp. 293–311

Solymar L 1972 *Superconductive Tunnelling and Applications*. Wiley, New York

Taylor B F, Witt T J 1989 New international electrical reference standards based on the Josephson and quantum Hall effects. *Metrologia* 26: 47–62

Van Duzer T, Turner C W 1981 *Principles of Superconductive Devices and Circuits*. Elsevier, Amsterdam

C. A. Hamilton

[National Institute of Standards and Technology, Boulder, Colorado, USA]

LIST OF CONTRIBUTORS

Contributors are listed in alphabetical order together with their addresses. Titles of articles that they have authored follow in alphabetical order. Where articles are coauthored, this has been indicated by an asterisk preceding the title.

Bader, S. D.
Materials Science Division
Building 223
Argonne National Laboratory
9700 South Cass Avenue
Argonne, IL 60439
USA
Thin Films and Multilayers, Magnetic

Barber, Z. H.
Department of Materials Science and Metallurgy
University of Cambridge
Pembroke Street
Cambridge
CB2 3QZ
UK
**Nitride and Carbonitride Superconductors*
**Oxide Superconductors: Thin-Film Deposition*

Bardeen, J.
[deceased; late of University of Illinois at Urbana, USA]
Superconducting Materials: BCS and Phenomenological Theories

Bate, G.
School of Engineering
Santa Clara University
Santa Clara, CA 95053
USA
Magnetic Storage Media

Benford, J. G.
Allegheny Ludlum Corporation
Technical Center
Albama & Pacific Avenues
Brackenridge, PA 15014-1597
USA
**Silicon Steels*

Bogner, G.
Siemens AG
Forschung für Materialwissenschaften und Elektronik
Paul-Gossen-Strasse 100
Postfach 3220
D-8520 Erlangen
Germany
Superconducting Materials: Large-Scale Applications

Brandt, E. H.
Max-Planck-Institut für Metallforschung
Heisenbergstrasse 1
Postfach 800668
D-7000 Stuttgart 80
Germany
**Flux Line Lattice*

Brown, B. S.
Argonne National Laboratory
9700 South Cass Avenue
Argonne, IL 60439
USA
**Radiation Effects on Superconducting Materials*

Buschow, K. H. J.
Philips Research Laboratories
PO Box 80.000
5600 JA Eindhoven
The Netherlands
Rare-Earth Intermetallics

Campbell, A. M.
Department of Engineering
University of Cambridge
Trumpington Street
Cambridge
CB2 1PZ
UK
AC Losses in Superconducting Materials
**Critical State Model*
Measurements in Superconducting Materials

Cave, J. R.
VPR
Hydro-Québec
Varennes
Québec
Canada
**AC Applications of Superconducting Materials*

Chen, C. W.
L-350
Lawrence Livermore National Laboratory
University of California
PO Box 808
Livermore, CA 94550
USA
Radiation Effects on Magnetic Materials

Chin, G. Y.
[deceased; late of AT&T Bell Laboratories, USA]
**Bubble Domain Materials*
**Cobalt–Platinum Alloys*
**Elongated-Single-Domain Magnets*
**Iron–Aluminum and Iron–Aluminum–Silicon Alloys*
**Iron–Chromium–Cobalt Alloys*
**Iron–Cobalt Alloys*
**Magnetic Alloys: Intermediate Coercivity Materials*
Magnetic Materials: An Overview
**Manganese–Aluminum–Carbon Alloys*
**Steels for Magnets*
**Temperature Compensation Materials*

Clark, A. E.
Department of the Navy
White Oak Laboratory
US Naval Surface Warfare Center
10901 New Hampshire Avenue
Silver Spring, MD 20910-5000
USA
Magnetostrictive Materials

Clarke, J.
Department of Physics
University of California
Berkeley, CA 94720
USA
**SQUID Design, Fabrication and Application*

Clem, J. R.
Ames Laboratory
Department of Physics
Iowa State University
Ames, IA 50011-3020
USA
**Anisotropy in Superconducting Materials*
Transport Currents and Force-Free Current Flow

Coles, B. R.
Department of Physics
Imperial College of Science, Technology and Medicine
Prince Consort Road
London
SW7 2BZ
UK
Paramagnetic Materials

Cooper, J. R.
Interdisciplinary Research Centre in Superconductivity
University of Cambridge
West Cambridge Site
Madingley Road
Cambridge
CB3 0HE
UK
**Organic Superconductors: Anisotropic Molecular Metals*
**Organic Superconductors: Doped C_{60} (Fullerene)*

Datta, T.
Department of Physics and Astronomy
University of South Carolina
Columbia, SC 29208
USA
Oxide Superconductors: Physical Properties

De Shazer, L. G.
Mail Code 111-1151
McDonnell Douglas
PO Box 516
St Louis, MO 63166
USA
Magnetooptics

Decroux, M.
Département de Physique de la Matière Condensée
Université de Genève
24 quai Ernest-Ansermet
CH-1211 Genève 4
Switzerland
**Chevrel Phases*

Della Torre, E.
Department of Electrical Engineering and Computer Science
George Washington University
Washington, DC 20052
USA
Micromagnetic Calculations

Deutscher, G.
School of Physics and Astronomy
Tel Aviv University
Ramat Aviv
Tel Aviv 69978
Israel
Coherence Length, Proximity Effect and Fluctuations

Dillon, J. F. Jr
AT&T Bell Laboratories
600 Mountain Avenue
Murray Hill, NJ 07974
USA
Magnetooptic Materials

Donaldson, G. B.
Department of Physics and Applied Physics
University of Strathclyde
John Anderson Building
107 Rottenrow
Glasgow
G4 0NG
UK
**SQUID Design, Fabrication and Application*

Ekin, J. W.
Electromagnetic Technology Division

National Institute of Standards and Technology
Mail Code 814.05
325 Broadway
Boulder, CO 80303-3328
USA
Superconducting Materials: Specification

Epstein, A. J.
Department of Physics
The Ohio State University
Columbus, OH 43210-1106
USA
**Organic and Molecular Ferromagnets*

Evetts, J. E.
Department of Materials Science and Metallurgy
University of Cambridge
Pembroke Street
Cambridge
CB2 3QZ
UK
**Critical Current Density*
**Critical State Model*
**Flux Line Lattice*
**Nitride and Carbonitride Superconductors*
**Oxide Superconductors: Thin-Film Deposition*
Resistive Transition and Flux Flow in Superconducting Materials

Feldmen, M. J.
Department of Electrical Engineering
Hopeman Hall
University of Rochester
Rochester, NY 14627
USA
Millimeter-Wave Detection

Fert, A.
Laboratoire de Physique des Solides
Université Paris-Sud
Bâtiment 510
Centre Universitaire
F-91405 Orsay Cédex
France
**Magnetotransport Phenomena*

Février, A.
Alsthom-Alcatel Recherche
Division Energétique et Matériaux
Route de Nozay
F-91460 Marcoussis Cédex
France
**AC Applications of Superconducting Materials*

Finnemore, D. K.
Ames Laboratory
Institute for Physical Research and Technology
Iowa State University
Ames, IA 50011
USA
Granular Superconductors

Flükiger, R.
Département de Physique de la Matière Condensée
Université de Genève
24 quai Ernest-Ansermet
CH-1211 Genève 4
Switzerland
A15 Compound Superconductors

Friend, R. H.
Department of Physics
Cavendish Laboratory
University of Cambridge
Madingley Road
Cambridge
CB3 0HE
UK
**Organic Superconductors: Anisotropic Molecular Metals*
**Organic Superconductors: Doped* C_{60} *(Fullerene)*

Gardner, C. G.
Department of Electrical Engineering
University of Houston
Houston, TX 77004
USA
Barkhausen Effect

Geballe, T. H.
Department of Applied Physics
Stanford University
Stanford, CA 94305-4090
USA
**Superconducting Materials: An Overview*

Gibbs, M. R. J.
School of Physics
University of Bath
Claverton Down
Bath
BA2 7AY
UK
**Metallic Glasses*
Sensors and Devices, Magnetic

Ginsberg, D. M.
Department of Physics
University of Illinois at Urbana
1110 West Green Street
Urbana, IL 61801
USA
Superconducting Materials: Physical Properties

Goldfarb, R. B.
National Institute of Standards and Technology

Mail Code 814.05
325 Broadway
Boulder, CO 80303-3328
USA
Demagnetizing Factors
Magnetic Units and Material Specification

Goldman, A.
Director of Research
Magnetics Division
Spang and Company
Butler, PA 16001
USA
Powder Cores, Magnetic

Graham, C. D. Jr
Department of Materials Science and Engineering
University of Pennsylvania
3231 Walnut Street
Philadelphia, PA 19104-6272
USA
Measurements in Magnetic Materials

Gray, K. E.
Materials Science Division
Argonne National Laboratory
9700 South Cass Avenue
Argonne, IL 60439
USA
Nonequilibrium Superconductivity

Gregory, E.
IGC Advanced Superconductors Inc.
1875 Thomaston Avenue
Waterbury, CT 06704
USA
Multifilamentary Superconducting Composites

Griessen, R.
Faculteit der Natuurkunde en Sterrenkunde
Vrije Universiteit Amsterdam
De Boelelaan 1081
1081 HV Amsterdam
The Netherlands
Flux Creep Phenomena
**Superconducting Materials Under Pressure*

Gronsky, R.
National Center for Electron Microscopy
Lawrence Berkeley Laboratory
University of California
Berkeley, CA 94720
USA
**Oxide Superconductors: Electron Microscopy*

Grundy, P. J.
Department of Pure and Applied Physics
University of Salford
Salford
M5 4WT
UK
Domains and Domain Walls

Halbritter, J.
Kernforschungszentrum Karlsruhe GmbH
Institut für Materialforschung I
Postfach 3640
D-7500 Karlsruhe
Germany
Superconducting Cavities

Hall, L. D.
University of Cambridge School of Clinical Medicine
Herchel Smith Laboratory for Medicinal Chemistry
University Forvie Site
Robinson Way
Cambridge
CB2 2PZ
UK
Nuclear Magnetic Resonance Imaging

Hamilton, C. A.
Cryoelectronic Metrology Group
National Institute of Standards and Technology
325 Broadway
Boulder, CO 80303
USA
Voltage-Standard Devices

Hashimoto, T.
Department of Applied Physics
Tokyo Institute of Technology
Oh-okayama
Meguro
Tokyo 152
Japan
Magnetocaloric Effect

Hasuo, S.
Exploratory Devices Laboratory
Fujitsu Laboratories Ltd
10-1 Morinosato-Wakamiya
Atsugi 243-01
Japan
**Superconducting Device Fabrication*

Hein, R. A.
Physics Department FM-15
University of Washington
Seattle, WA 98195
USA
Superconducting Materials: Elements

Herzer, G.
Development Department

Rapid Solidification Technology
Vacuumschmelze GmbH
Postfach 2253
D-6450 Hanau 1
Germany
**Nanocrystalline Soft Magnetic Materials*

Huebener, R. P.
Lehrstuhl für Experimentalphysik II
Physikalisches Institut
Universität Tübingen
Morgenstelle 14
D-7400 Tübingen
Germany
Magnetic Properties of Superconductors

Hulm, J. K.
Department of Applied Physics
Stanford University
Stanford, CA 94305-4090
USA
**Superconducting Materials: An Overview*

Imamura, T.
Fujitsu Laboratories Ltd
10-1 Morinosato-Wakamiya
Atsugi 243-01
Japan
**Superconducting Device Fabrication*

Jagielinski, T. M.
Eastman Kodak Company
San Diego Research Laboratories
3985 Sorrento Valley Boulevard
San Diego, CA 92121-1402
USA
Recording Head Technology

Jakubovics, J. P.
Department of Metallurgy and Science of Materials
University of Oxford
Parks Road
Oxford
OX1 3PH
UK
Magnetization Imaging

Jin, S.
AT&T Bell Laboratories
600 Mountain Avenue
Murray Hill, NJ 07974
USA
Oxide Superconductors: Ceramic Processing

Kes, P. H.
Kamerlingh Onnes Laboratorium
Rijksuniversiteit Leiden
Nieuwsteeg 18
Postbus 9506
2300 RA Leiden
The Netherlands
**Critical Current Density*
Flux Pinning and the Summation of Pinning Forces

Kim, Y. S.
Bell Laboratories
Allentown, PA 18103
USA
**Ferrites, Soft*

Kogan, V. G.
Ames Laboratory
Department of Physics
Iowa State University
Ames, IA 50011-3020
USA
**Anisotropy in Superconducting Materials*

Kools, F.
Philips Composants
Division Ferrites
41 Rue Pierre Brossolette
F-27016 Evreux
France
Ferrites, Hard

Kouvel, J. S.
Department of Physics (M/C 273)
College of Liberal Arts and Sciences
University of Illinois
PO Box 4348
Chicago, IL 60680
USA
Magnetic Materials: Basic Concepts

Kovac, Z.
IBM
Thomas J. Watson Research Center
PO Box 218
Yorktown Heights, NY 10598
USA
**Inks, Magnetic*

Kronmüller, H.
Max-Planck-Institut für Metallforschung
Institut für Physik
Heisenbergstrasse 1
Postfach 800665
D-7000 Stuttgart 80
Germany
Coecivity and Domain Wall Pinning

Kryder, M. H.
Director
Magnetics Technology Center
Carnegie Mellon University
Pittsburgh, PA 15213
USA
Magnetooptic Storage Media

Lancaster, M. J.
School of Electronic and Electrical Engineering
University of Birmingham
Edgbaston
Birmingham
B15 2TT
UK
**Microwave Applications of Superconducting Materials*

Larbalestier, D. C.
Applied Superconductivity Center
Engineering Research Building
College of Engineering
University of Wisconsin–Madison
1500 Johnson Drive
Madison, WI 53706-1687
USA
**Niobium–Titanium Alloys*

Lee, P. J.
Applied Superconductivity Center
Engineering Research Building
College of Engineering
University of Wisconsin–Madison
1500 Johnson Drive
Madison, WI 53706-1687
USA
**Niobium–Titanium Alloys*

Likharev, K. K.
Department of Physics
State University of New York
Stony Brook, NY 11794-3800
USA
Superconducting Devices and Electronics

Livingston, J. D.
Department of Materials Science and Engineering
77 Massachusetts Avenue
Massachusetts Institute of Technology
Cambridge, MA 02139
USA
Neodymium–Iron–Boron Alloys

Lottis, D. K.
Laboratoire de Physique des Solides
Université Paris-Sud
Bâtiment 510
Centre Universitaire
F-91405 Orsay Cédex
France
**Magnetotransport Phenomena*

Luborsky, F. E.
Corporate Research and Development Center
General Electric Company
PO Box 8
Schenectady, NY 12345
USA
**Metallic Glasses*

MacCrone, R. K.
Materials Engineering Department
Rensselaer Polytechnic Institute
Troy, NY 12180-3590
USA
Oxide Glasses as Magnetic Materials

McCurrie, R. A.
Department of Electrical Engineering
University of Bradford
Bradford
BD7 1DP
UK
Alnicos

McDougall, I. L.
Oxford Instruments Ltd
Scientific Research Division
Station Road
Eynsham
Witney
Oxfordshire
OX8 1TL
UK
Superconducting Magnets

McEwen, K. A.
Department of Physics
Birkbeck College
University of London
Malet Street
London
WC1E 7HX
UK
Rare-Earth Metals and Alloys

MacFarlane, J. C.
Division of Applied Physics
CSIRO
Bradfield Road
West Lindfield
PO Box 218
Lindfield, NSW
Australia
**SQUID Design, Fabrication and Application*

Major, R. V.
Director, Research and Development
Telcon Ltd
Napier Way
Crawley
West Sussex
RH10 2RB
UK
Shields, Magnetic

Miller, J. S.
E I du Pont de Nemours Inc
Central Research and Development
Experimental Station E328/266
PO Box 80328
Wilmington, DE 19880-0328
USA
**Organic and Molecular Ferromagnets*

Mitchell, T. E.
Center for Materials Science
Mail Stop K765
Los Alamos National Laboratory
Los Alamos, NM 87545
USA
**Oxide Superconductors: Electron Microscopy*

Overshott, K. J.
Principal
Gwent College of Higher Education
College Crescent
Caerleon
Newport
Gwent
NP6 1XJ
UK
Power Loss in Magnetic Materials

Pendleton, W. W.
1542 Clinton Street
Muskegon, MI 49442
USA
Magnet Wire

Pfeifer, F.
Vacuumschmelze GmbH
Postfach 109
D-6450 Hanau
Germany
Nickel–Iron Alloys

Pfotenhauer, J. M.
Applied Superconductivity Center
Engineering Research Building
College of Engineering
University of Wisconsin–Madison
1500 Johnson Drive
Madison, WI 53706-1687
USA
Magnetic Energy Storage

Pinnel, M. R.
AT&T Bell Laboratories
6200 East Broad Street
Columbus, OH 43213
USA
Reed Contact Materials

Pizzella, V.
Consiglio Nazionale Delle Ricerche
Istituto di Elettronica dello Stato Solido
Via Cineto Romano 42
I-00156 Roma
Italy
**Biomagnetic Measurements*

Pointon, A. J.
Department of Applied Physics and Physical Electronics
Portsmouth Polytechnic
Park Building
King Henry I Street
Portsmouth
PO1 2DZ
UK
Microwave Materials and Devices, Magnetic

Porch, A.
School of Electronic and Electrical Engineering
University of Birmingham
Edgbaston
Birmingham
B15 2TT
UK
**High-Frequency Impedance of Superconducting Materials*
**Microwave Applications of Superconducting Materials*

Rao, B. U. M.
Department of Physics
Oakland University
Rochester, MI 48309-4401
USA
**High-Frequency Properties of Magnetic Materials*

Rastogi, P. K.
Inland Steel Research Laboratories
3001 East Columbus Drive
East Chicago, IN 46312
USA
Iron and Low-Carbon Steels for Magnets

Romani, G. L.
Instituto di Fisica Medica
Università G. D'Annunzio
Via dei Vestini
I-66100 Chieti
Italy
**Biomagnetic Measurements*

Rosensweig, R. E.
Exxon Research and Engineering Company
Clinton Township
Route 22 East
Annandale, NJ 08801
USA
Fluids, Magnetic

Rowell, J. M.
Chief Technical Officer
Conductus Inc.
969 West Maude Avenue
Sunnyvale, CA 94086
USA
Thin Films, Multilayers and Devices, Superconducting

Sambucetti, C. J.
IBM Deutschland GmbH
Postfach 266
D-7032 Sindelfingen
Germany
**Inks, Magnetic*

Savage, H. T.
Instituto de Magnetismo Aplicado
Laboratorio "Salvador Velayo"
Apartado de Correos 155
E-28230 Las Rozas
Madrid
Spain
Magnetoelastic Phenomena

Schenk, A.
Siemens AG
FTE 2 FT 1
Forschungszentrum
D-8520 Erlangen
Germany
Electromagnetic Forming

Seeber, B.
Groupe de Physique Appliquée
Université de Genève
20, Rue de l'Ecole-de-Médecine
CH-1211 Genève 4
Switzerland
**Chevrel Phases*

Sleight, A. W.
Department of Chemistry
Oregan State University
Corvallis, OR 97331-4003
USA
Oxide Superconductors: Solid-State Chemistry

Somekh, R. E.
Department of Materials Science and Metallurgy
University of Cambridge
Pembroke Street
Cambridge
CB2 3QZ
UK
**Nitride and Carbonitride Superconductors*
**Oxide Superconductors: Thin-Film Deposition*

Srinivasan, G.
Department of Physics
Oakland University
Rochester, MI 48309-4401
USA
**High-Frequency Properties of Magnetic Materials*

Sundahl, R. C. Jr
Signal Research Laboratory
50 Uop Plaza
Des Plaines, IL 60016
USA
**Ferrites, Soft*

Swartzendruber, L. J.
B150
Materials Building
National Institute of Standards and Technology
Gaithersburg, MD 20899
USA
Nondestructive Evaluation Using Magnetism

Szymczak, H.
Instytut Fizyki
Polskiej Akademii Nauk
Al. Lotników 32/46
02-668 Warszawa
Poland
Anisotropy in Magnetic Materials

Tawara, Y.
Shin-Etsu Chemical Company Ltd
Corporate Research Center
R & D Building A-12F
Kanagawa Science Park
Kawasaki City 213
Japan
Cobalt–Rare-Earth Compounds

Thompson, R. B.
Ames Laboratory
Energy and Mineral Resources Research Institute
Iowa State University

Ames, IA 50011
USA
Electromagnetic–Acoustic Transducers

Tilley, D. R.
Department of Physics
University of Essex
Wivenhoe Park
Colchester
CO4 3SQ
UK
Optical Properties of Superconductors

Tinkham, M.
Physics Department
Lyman Laboratory
Harvard University
Cambridge, MA 02138
USA
Josephson Effects in Weak Links

Trowbridge, C. W.
Vector Fields Ltd
24 Bankside
Kidlington
Oxford
OX5 1JE
UK
Design with Magnetic Materials: Computer Simulation

Tsuei, C. C.
IBM
Thomas J. Watson Research Center
PO Box 218
Yorktown Heights, NY 10598
USA
Amorphous Superconductors

Van Tendeloo, G.
Universiteit Antwerpen (RUCA)
Groenborgerlann 171
B-2020 Antwerp
Belgium
**Oxide Superconductors: Electron Microscopy*

Waldram, J. R.
Department of Physics
Cavendish Laboratory
University of Cambridge
Madingley Road
Cambridge
CB3 0HE
UK
**High-Frequency Impedance of Superconducting Materials*

Warlimont, H.
Am Weinberg 2
D-W 6463 Freigericht
Germany
**Nanocrystalline Soft Magnetic Materials*

Washko, S. D.
Allegheny Ludlum Corporation
Leechburg Works
PO Box 565
Leechburg, PA 15656
USA
**Silicon Steels*

Watson, J. H. P.
Institute of Cryogenics
University of Southampton
University Road
Highfield
Southampton
SO9 5NH
UK
Magnetic Separation

Watson, J. K.
Department of Electrical Engineering
University of Florida
216 Larsen Hall
Gainesville, FL 32611
USA
Design with Magnetic Materials

Weber, H. W.
Atominstitut der Österreichischen Universitäten
Technische Universität Wien
Schuttelstrasse 115
A-1020, Vienna
Austria
**Radiation Effects on Superconducting Materials*

Wernick, J. H.
Bellcore
331 Nemans Spring Road
NVC3A-377
Red Bank, NJ 07701-7020
USA
**Bubble Domain Materials*
**Cobalt–Platinum Alloys*
**Elongated-Single-Domain Magnets*
**Iron–Aluminum and Iron–Aluminum–Silicon Alloys*
**Iron–Chromium–Cobalt Alloys*
**Iron–Cobalt Alloys*
**Magnetic Alloys: Intermediate Coercivity Materials*
**Manganese–Aluminum–Carbon Alloys*
**Steels for Magnets*
**Temperature Compensation Materials*

Wheatley, J. M.
Interdisciplinary Research Centre in Superconductivity

University of Cambridge
West Cambridge Site
Madingley Road
Cambridge
CB3 0HE
UK
Superconducting Materials: Basic Theoretical Concepts

Wijngaarden, R. J.
Faculteit der Natuurkunde en Sterrenkunde
Vrije Universiteit Amsterdam
De Boelelaan 1081
1081 HV Amsterdam
The Netherlands
**Superconducting Materials Under Pressure*

Wohlleben, D. K.
Universität zu Köln
II. Physikalisches Institut
Zülpicher Strasse 77
D-5000 Köln 41
Germany
Heavy Fermion Metals

Wolf, E. L.
Physics Department
Six Metrotech Center
Polytechnic University
333 Jay Street
Brooklyn, NY 11201
USA
Tunnelling Spectroscopy

SUBJECT INDEX

The Subject Index has been compiled to assist the reader in locating all references to a particular topic in the Encyclopedia. Entries may have up to three levels of heading. Where there is a substantive discussion of the topic, the page numbers appear in ***bold italic*** type. As a further aid to the reader, cross-references have also been given to terms of related interest. These can be found at the bottom of the entry for the first-level term to which they apply. Every effort has been made to make the index as comprehensive as possible and to standardize the terms used.